French Dictionary of Environmental Technology
Dictionnaire anglais du génie de l'environnement

Routledge
French Dictionary of Environmental Technology
Dictionnaire anglais du génie de l'environnement

French–English/English–French

Français–Anglais/Anglais–Français

London and New York

First published 1997
by Routledge
11 New Fetter Lane, London EC4P 4EE

Simultaneously published in the USA and Canada by Routledge
29 West 35th Street, New York, NY 10001

Typeset in Monotype Times, Helvetica Neue and Bauer Bodoni by Routledge
Printed in England by TJ International Ltd, Cornwall

Printed on acid-free paper

British Library Cataloguing-in-Publication Data
A catalogue record for this book is available from the British Library

Library of Congress Cataloguing-in-Publication Data
Applied for

ISBN: 0–415–13918–X

French Dictionary of Environmental Technology
Dictionnaire anglais du génie de l'environnement

Project Manager/Gestion de projet
Neil Clements

Managing Editor/Coordination éditoriale
Sinda López

Editorial/Rédaction
Justine Bird Lisa Carden Cécile Guinard Janice McNeillie Robert Timms

Programme Manager/Direction de collection
Elizabeth White

Marketing
Vanessa Markey Rachel Miller

Systems/Informatique
Omar Raman Simon Thompson

Production
Michelle Draycott Nigel Marsh Joanne Tinson

Terminologist/Terminologue
Freddie Plassard

Specialist Consultants/Spécialistes consultés
Dr James P. Lassoie

Pierre Leizour

Hélène Mateescu
AFNOR

Lexicographers & Proofreaders/Lexicographie & Correction des épreuves
Anne Convery Hazel Curties Stephen Curtis Gemma Marren Jeremy Munday Maggie Seaton

Keyboarders/Clavistes
Emmanuelle Bels Sam Edenborough Beatriz Fernández Ilona Lehmann
Rosa Gálvez López David Morgan Fabienne Rangeard Jack Simms Debbie Thomas

Contents/Table de matières

Foreword/Préface

We live in a modern world where the interdependencies within and among sovereign nations grow daily with respect to both their complexity and importance. Without a doubt, the significance of international trade to social development has promoted the wide recognition that monetary dynamics commonly transcend governmental boundaries. We live in a world economy which will continue to shape the nature of international and domestic relations.

Similarly, the development of modern science has led to an understanding of and appreciation for the interconnectedness of the earth's biological and physical systems. Today, widespread concern over such seemingly intractable problems as human-induced global climate change, declining biodiversity and accelerated environmental degradation has forced societies to look for solutions beyond the confines of traditional jurisdictional boundaries. Furthermore, solutions to these environmental problems will continue to elude those who search for salvation only within specific disciplines, as their complexities demand applied interdisciplinary methodologies. We live in a world ecology which recognizes no political demarcations in either time or space.

The seriousness of these ecological and environmental problems is further enhanced by the growing understanding that they are inextricably linked to economic viability and social stability. The plight of many Third-World Countries trying to develop their fiscal and human capital in the face of declining environmental resources serves to reinforce the importance of such a linkage. The environmental, social, and economic problems currently facing Eastern Europe is yet another example. Economic development is not possible without a healthy environment, and maintaining the integrity of ecological systems in turn depends on a healthy economy.

The history of humankind is characterized by the exploitation of the world's natural resources and marked by environmental insults often

Dans le monde moderne où nous vivons, les interdépendances entre états souverains ainsi qu'à l'intérieur de ces états se font chaque jour plus nombreuses et plus complexes. Il ne fait aucun doute qu'en influant sur le développement des sociétés, le commerce international a largement contribué à faire admettre que la dynamique monétaire transcende généralement les frontières nationales et que la mondialisation de l'économie va continuer à façonner la nature des relations tant intérieures qu'internationales.

Parallèlement, les développements de la science moderne nous ont amenés à comprendre et à apprécier à juste titre combien les systèmes biologiques et physiques étaient étroitement imbriqués sur la planète. L'inquiétude générale que suscitent ces problèmes que l'homme a engendrés et qu'il semble avoir tant de mal à résoudre: les modifications climatiques à l'échelle du globe, le déclin de la diversité biologique, l'accélération de la dégradation de l'environnement, force aujourd'hui les sociétés à rechercher des solutions qui vont au-delà des frontières juridictionnelles traditionnelles, et à éviter ceux qui ne cherchent la solution aux problèmes environnementaux qu'au travers de disciplines particulières car la complexité de ces derniers demande une méthodologie interdisciplinaire. La mondialisation de l'écologie ne connaît aucune frontière politique, ni dans l'espace, ni dans le temps.

La conscience grandissante que les problèmes écologiques et environnementaux sont indissolublement liés à la viabilité de l'économie et à la stabilité de la société accentue encore leur gravité. Les difficultés de nombreux pays du Tiers Monde à développer leur capital économique et humain face à des ressources environnementales qui s'épuisent viennent encore accentuer l'importance de ce lien de même que l'exemple des soucis environnementaux, sociaux et économiques qui frappent à l'heure actuelle l'Europe de l'est. Aucun développement économique n'est possible sans un environnement sain et l'intégrité des systèmes

leading to negative effects on human health and welfare. However, until the Industrial Revolution, this impact was greatly limited by the mere size of the human population and its ability to cause widespread, long-lasting and potentially dangerous changes to the environment. Society is now accepting the accelerated environmental degradation associated with its development activities over the past few decades as being an indisputable threat to social stability. The severity and complexity of environmental problems associated with air, water, soil and noise pollution; waste production; and energy use reflect not only our enhanced ability to cause these problems, but also to develop better means to measure and characterize their ecological, social and medical impacts. Science and engineering communities serve a uniquely pivotal role here having developed both the products and processes that can lead to environmental degradation as well as the technology needed to identify and solve specific environmental problems.

It is often easy to blame the science and engineering communities for society's ecological and environmental transgressions. Modern technology has undoubtedly given us the ability to disrupt and perhaps destroy the environmental functioning of our world. However, 'invention' is not the only root of our environmental problems as other segments of society must also share this responsibility because of their unrealistic expectations, misapplication and mismanagement of technology. Hence, despite the various management tools currently available for air, water, soil and noise quality control; waste treatment; and the production of renewable energy, severe problems persist and often worsen as we continue to develop as a society. Fortunately, there is a growing appreciation that long-lasting solutions which are economically, socially, and ecologically sound can only come from new, innovative interdisciplinary approaches. Such thinking is at the heart of society's current emphasis on sustainable development.

If interdisciplinary approaches are truly the key to environmental problem-solving and to attaining sustainability, it is interesting that we remain reticent to embrace and incorporate them into our scientific, engineering, and decision-making practices. Why? The rationale for such actions is logical, widely expounded by university presidents, business leaders, foundation officers and governmental officials, and usually understood and accepted by the general public. Unfortunately, it is much easier to talk about interdisciplinary scholarship and its

écologiques conditionne la santé de l'économie.

L'histoire de l'humanité se caractérise par l'exploitation des ressources naturelles de la planète; elle est marquée d'insultes à l'environnement et des effets négatifs que cela a souvent entraînés sur la santé et le bien-être de l'homme. Jusqu'à la Révolution industrielle, cet impact a été grandement limité du simple fait que le monde était peu peuplé: la population était incapable d'infliger des modifications larges, durables et potentiellement dangereuses à l'environnement. Aujourd'hui la société admet que la dégradation accélérée du milieu environnant que provoquent les activités qu'elle a développées ces dernières décennies constitue une menace irréfutable pour la stabilité sociale. La gravité et la complexité des problèmes environnementaux liés à la pollution de l'air, de l'eau et des sols, au bruit, à la production de déchets et à l'utilisation de l'énergie reflètent à l'évidence notre capacité accrue à engendrer ces problèmes mais témoignent également de la faculté que nous avons d'améliorer les moyens de mesure et de caractérisation des impacts de ces derniers sur l'écologie, la société et la santé. Ayant imaginé à la fois les produits et les processus qui dégradent l'environnement et les moyens nécessaires pour répertorier et résoudre les problèmes environnementaux correspondants, les communautés scientifiques et techniques ont ici un rôle de pivot unique à jouer.

Il serait trop facile de blâmer la science et la technique pour les transgressions écologiques et environnementales de la société. Il ne fait aucun doute que la technologie moderne nous a donné le pouvoir de perturber et peut-être même de détruire l'équilibre environnemental de notre planète. L' 'invention' n'est cependant pas seule à la racine de nos maux en matière d'environnement; d'autres secteurs de la société portent également leur part de responsabilité par le manque de réalisme de leurs attentes et leurs erreurs de mise en oeuvre et de gestion des technologies. En effet, malgré le nombre d'outils actuellement disponibles pour maîtriser la qualité de l'air, de l'eau, des sols, le niveau de pollution acoustique, pour traiter les déchets et pour produire des énergies renouvelables, de graves problèmes subsistent qui souvent s'aggravent du fait du développement continu de notre société. Bien heureusement, de plus en plus de gens commencent à se rendre compte que les solutions à long terme, valables sur les plans économique, social et écologique, ne peuvent venir que d'approches interdisciplinaires, neuves et innovantes. Cette notion est au cœur de ce que la société

application to solving complex environmental problems than it is do it.

One barrier to overcome as we move from the 'talking' to the 'doing' stage is our inability to communicate effectively and accurately. We may live in a world economy and be affected by a world ecology, but we certainly cannot discuss these issues using a world language. Extensive and complex knowledge systems exist throughout the world that remain obscure because of language. The peoples of the world, often facing similar environmental hardships, find cooperation and coordination difficult because of language. And lastly, the world's scientific and engineering communities find themselves hampered in their efforts to find solutions to universal environmental problems because of language. Despite the frustrations and inefficiencies associated with working across the world's many languages, greater problems arise because of the unique languages of the disciplines. The inherent difficulties in communicating clearly among divergent fields of study is a major barrier to interdisciplinary studies, as each discipline carries its own theoretical framework, definitions, underlying assumptions, and relevancy biases into discussions of environmental problems and their possible solutions. Realistic and permanent solutions to interdisciplinary problems will be slow to develop from those unable to fully understand one another.

Routledge has recognized the importance of effective communication by their creation of specialized bilingual dictionaries for translators, researchers and professionals. The production of their *French Dictionary of Environmental Technology* is another important contribution in this series. It contains approximately 20,000 entries in each language, covering such subject areas as the assessment, control and treatment of air, water, soil and noise pollution; the treatment and management of solid waste and wastewater; and the production and management of potable water and renewable energy. Not only are important scientific and engineering subjects covered, but the dictionary also includes topics directly related to the marketing, policy, legal and safety aspects of environmental management. Hence, this reference tool should be useful to a wide group of environmental professionals.

This dictionary will certainly enhance communication among those committed to finding real-world solutions to the many environmental problems confronting modern society. It will help to concentrate and coordinate the experiences and capabilities of the major scientific and engineering

désigne maintenant sous le terme de développement durable.

Si l'approche interdisciplinaire est véritablement la clé des problèmes environnementaux et du développement durable, il est intéressant de noter combien nous demeurons réticents quand il s'agit de l'envisager et de la mettre en pratique dans notre démarche scientifique, technique et décisionnelle. La logique de cette approche est pourtant largement exposée par les présidents d'universités, les grands chefs d'entreprise, les dirigeants de fondations et les responsables gouvernementaux; elle est de même généralement comprise et acceptée par le grand public. Alors pourquoi? Parce qu'il est malheureusement moins facile d'agir que de disserter savamment de l'approche interdisciplinaire et de son application à la solution de problèmes environnementaux complexes.

Il est un obstacle qu'il faut en effet surmonter pour pouvoir passer de la parole aux actes et c'est notre incapacité à communiquer de façon efficace et précise. Nous vivons peut-être dans un monde où l'économie est 'internationale', où l'écologie est 'internationale', mais il est certain que le langage que nous adoptons pour parler de ces questions n'est pas, lui, 'international'. De larges et complexes pans de la connaissance qui existent de par le monde nous demeurent étrangers à cause de la langue. Confrontés à des difficultés environnementales similaires, les peuples du monde trouvent souvent difficile de coopérer et de coordonner leurs actions à cause de la barrière de la langue. Et c'est toujours la langue qui bloque la communauté mondiale des hommes de sciences et de techniques dans leurs efforts pour apporter des solutions à des problèmes environnementaux universaux.

Aux frustrations et aux inefficacités associées au fait de travailler dans la multitude des langues du monde, viennent s'ajouter les problèmes liés aux langages de spécialités. Les difficultés inhérentes à une communication claire entre domaines de recherche divergents constituent un frein majeur aux études interdisciplinaires, chaque discipline ayant son cadre théorique propre, ses définitions particulières, ses hypothèses fondamentales et apportant sa propre manière de voir dans les discussions des problèmes environnementaux et de leurs possibles solutions. Cette incapacité à comprendre parfaitement l'autre ralentit l'élaboration de solutions réalistes et permanentes à ces questions interdisciplinaires.

Routledge a bien senti l'importance de la communication quand il a décidé de mettre au point des dictionnaires bilingues spécialisés à l'usage des

communities found in French- and English-speaking countries. In addition, the diversity of subjects covered will also facilitate all-important interdisciplinary communication, including the application of technology to decision-making and management processes. This dictionary may not provide the precise path to a sustainable future for humankind, but as a widely useful and comprehensive reference, it should most definitely be of invaluable assistance along the way.

James P. Lassoie, PhD
Dept of Natural Resources and past Director of the Center for the Environment
Cornell University
New York

traducteurs, des chercheurs et des professionnels. La publication du *Dictionnaire anglais du génie de l'environnement* ajoute une pierre à l'édifice. Ce dictionnaire renferme environ 20 000 entrées dans chaque langue et couvre des domaines tels que l'évaluation, la maîtrise et le traitement de la pollution de l'air, de l'eau, des sols, et de la pollution acoustique; le traitement et la gestion des déchets solides et des eaux usées, ainsi que la production et la gestion de l'eau potable et des énergies renouvelables. Il ne traite pas uniquement de sujets scientifiques et techniques majeurs mais également de sujets directement liés aux aspects marketing, politiques, juridiques et à la sécurité du management de l'environnement. Il devrait donc être utile à un large éventail de professionnels de l'environnement.

Ce dictionnaire améliorera certainement la communication entre ceux qui se sont engagés à trouver des solutions concrètes aux nombreux problèmes auxquels doit faire face la société moderne. Il aidera à concentrer et à coordonner les expériences et les capacités des communautés scientifiques et techniques majeures des pays anglophones et francophones. Il facilitera également, grâce à la diversité des sujets traités, cette communication interdisciplinaire si importante en intégrant la technologie aux processus de décision et de gestion. Peut-être ce dictionnaire ne représente-t-il pas la panacée à ce futur durable dont rêve l'humanité mais par les références des plus utiles et complètes qu'il renferme, il offre une aide inestimable sur le chemin qui y conduit.

Traduit de l'anglais par Hélène Mateescu

Introduction/Introduction

Following on from the programme of general technical and business dictionaries launched in October 1994, the *French Dictionary of Environmental Technology* is the latest in a series of specialist dictionaries to be published by Routledge. Given the growing importance of enviromental technology in everyday life and the fact that English and French as a language combination has become increasingly important over recent years, this dictionary will be a valuable resource for specialist translators, terminologists and experts working in all fields of environmental technology. Approximately 20,000 entries in each language cover terminology used in areas such as Air Pollution & Air Quality Control, Soil Contamination & Remediation and Renewable Energy Sources.

It would not have been possible to compile this dictionary within a realistic timescale, and to the standard achieved, without the use of a custom-designed database. The database used is a relational one: term records for each language are held in separate files, with further files consisting only of link records. Links between terms within a single-language file represent cross-references, themselves of a wide variety of types: spelling variants, geographical variants and abbreviations.

The content of the database for this dictionary was compiled and vetted by a team of English and French terminologists with current practical experience of a narrowly-defined subject area within environmental technology in order to ensure its currency, the accuracy of translations and the adequacy of coverage. Finally, each language file was reviewed by editors to ensure coverage of North American terms and spelling variants, which are clearly labelled and distinguished.

The compilation and editing of the database of terms was, however, only the first stage in the making of the dictionary. Within the database, the distinction between source and target languages is not meaningful, but for this printed dictionary it was necessary to format the data to produce separate French–English and English–French sec-

Le *Dictionnaire anglais du génie de l'environnement* est le dernier d'une série de dictionnaires spécialisés bilingues que Routledge publie dans le cadre d'un programme lancé en octobre 1994, et dont les premières publications sont les dictionnaires de technique générale et les dictionnaires des affaires, du commerce et de la finance. Non seulement l'environnement joue un rôle croissant dans notre vie quotidienne, mais la combinaison de langues anglais et français, de plus en plus utilisée aujourd'hui, fait de ce dictionnaire un outil précieux pour les traducteurs techniques, terminologues et spécialistes travaillant dans le domaine du génie de l'environnement. Les 20 000 entrées dans chaque langue englobe une terminologie relative à des disciplines telles que la pollution atmosphérique & le contrôle de la qualité de l'air, la contamination des sols & les remèdes et les sources d'énergies renouvelables.

La sortie de ce dictionnaire dans les délais et conformément aux critères prévus imposait la compilation d'une base de données spécifique qui est de type relationnel: dans chaque langue, les termes sont stockés dans des fichiers séparés, tandis que d'autres fichiers ne se composent que de liens. Les liens entre les différents termes d'un même fichier de langue représentent des renvois, eux-mêmes de différents types: variantes orthographiques, variantes géographiques et abréviations.

La base de données a été compilée et approuvée par une équipe de terminologues anglais et français, ayant une connaissance pratique approfondie des domaines s'appliquant au génie de l'environnement afin de garantir l'actualité et la précision des explications, ainsi que l'étendue des sujets traités. Enfin, chaque fichier de langue a été contrôlé par des lexicographes des pays concernés, chargés de veiller au respect des variantes géographiques et orthographiques. Dans chaque langue, les formes et l'orthographe (anglaises, américaines ou canadiennes) sont clairement identifiées.

La création et la vérification de la base de données terminologique n'ont cependant représenté que la

tions. The data was processed by a further software module to produce two alphabetic sequences, of French headwords with English translations and vice-versa, each displaying the nesting of compounds, ordering of translations, style for cross-references of different types, and other features according to the algorithm.

At this stage the formatted text was edited by a team of experienced English and French lexicographers whose task it was to eliminate duplication or inconsistency; edit the contextual information; and to remove terms that were too general for inclusion in a specialist dictionary.

The editorial team

première étape de la constitution du dictionnaire. La distinction entre les langues cible et source n'étant pas significative au niveau de la base de données, nous avons dû, pour la version imprimée, mettre les données en forme pour obtenir des parties français-anglais et anglais-français séparées. Les données ont alors été traitées par un logiciel permettant de produire deux séquences alphabétiques, des mots-clés français avec leurs traductions anglaises et vice-versa avec, dans chaque cas, indication de l'imbrication des composants, de l'ordre des traductions, du style des renvois de différents types, et d'autres fonctions, selon l'algorithme.

A ce stade, le texte mis en forme a été vérifié par une équipe de lexicographes français et anglais expérimentés, qui avaient pour mission d'éliminer les doublons et les incohérences, de vérifier les informations et les explications contextuelles, et de supprimer les termes soit trop généraux, soit trop spécialisés pour être inclus dans un dictionnaire technique.

Les éditeurs

Features of the dictionary/
Caractéristiques du dictionnaire

The following text extracts illustrate the principal features of the dictionary. For a more detailed explanation of each of these features and a full guide to using the dictionary, see pages xix–xx.

Les éclatés ci-contre mentionnent les principales caractéristiques de ce dictionnaire. Pour une explication détaillée de chacune de ces caractéristiques et un guide complet de l'utilisation du dictionnaire, reportez-vous aux pages xxi–xxiii.

approvisionner *vt* DIST EAU, ÉN RENOUV, POLL ATM, POLL EAU supply

aptitude: ~ **à l'emploi** *f* GÉNÉRAL usability; ~ **à la mise en décharge** *f* POLL EAU landfilling ability; ~ **à l'usage** *f* GÉNÉRAL usability

Cross-references from abbreviations to their full forms are shown for both French and English — **AQ** *abrév* (*assurance-qualité*) GESTION QA (*quality assurance*) — *Abréviation du terme donnée à l'entrée pour la forme intégrale*

aquaculture *f* DIST EAU aquaculture

aqueduc *m* DIST EAU aqueduct

aqueux *adj* DIST EAU aqueous

aquiclude *f* DIST EAU aquiclude

aquiculture *f* DIST EAU aquaculture

aquifère[1] *adj* DIST EAU aquiferous

Superscript numbers denote homographs — **aquifère**[2] *m* DIST EAU aquifer; ~ **captif** DIST EAU confined aquifer — *Les numéros en exposant indiquent des homographes*

aquifuge *m* DIST EAU aquifuge

Ar *abrév* (*argon*) DÉCHETS, DIST EAU, POLL ATM Ar (*argon*)

A colon introduces a compound where the headword has no technical value — **arbre**: ~ **d'admission** *m* DIST EAU, SOLS entrance shaft; ~ **barillet** *m* DIST EAU, POLL EAU *pompe HP* pump-plunger shaft; ~ **de commande** *m* DÉCHETS, ÉN RENOUV control shaft; ~ **de pompe** *m* DIST EAU, POLL EAU *mines* pump shaft; ~ **de rotor** *m* ÉN RENOUV rotor hub; ~ **de rotor équipé d'un mécanisme de réglage des pales** *m* ÉN RENOUV rotor hub with adjustable blade mechanism — *Deux-points indique un mot dépourvu de valeur technique et introduit un mot composé*

architecte: ~**-paysagiste** *m* GESTION landscape architect

architecture: ~ **solaire** *f* ÉN RENOUV solar architecture

argile: ~ **marneuse** *f* DIST EAU marly clay

argon *m* (*Ar*) DÉCHETS, DIST EAU, POLL ATM argon (*Ar*)

armature: ~ **de pompe** *f* POLL EAU pump gear

armer *vt* DIST EAU, POLL EAU reinforce

aromatique *adj* DIST EAU, POLL ATM aromatic

arrachage *m* POLL EAU *industrie papetière* flaking

Genders are indicated at French nouns — **arrêt**: ~ **d'implantation** *m* DIST EAU settlement stoppage; ~ **par épuisement** *m* DÉCHETS, POLL ATM *nucléaire* burnout; ~ **de pourrissage** *m* DIST EAU rotting standstill; ~ **du tassement** *m* SOLS settlement stoppage — *Genre de chaque nom indiqué pour le nom français*

arrosage *m* DIST EAU, POLL EAU watering, spraying

arroseur *m* POLL EAU sparger; ~ **rotatif** DIST EAU rotating sprayer

arsenic *m* (*As*) SOLS arsenic (*As*)

British English and American English spelling variants are given in full and labelled accordingly — **article**: ~ **en cellulose moulée** *m* DÉCHETS molded pulp article (*AmE*), moulded pulp article (*BrE*) — *Informations complètes sur les variantes américaines et anglaises suivies de codes géographiques*

artificiel *adj* DIST EAU artificial, SOLS man-made

As *abrév* (*arsenic*) SOLS As (*arsenic*)

asbestose *m* POLL ATM asbestosis

ascension: ~ **du panache** *f* POLL ATM plume rise

aseptique *adj* DÉCHETS aseptic

aspect: ~ **des déchets** *m* DIST EAU appearance of waste

asperseur *m* POLL EAU *agriculture* sprinkler; ~ **longue portée** POLL EAU long-throw sprinkler

Compound terms are nested alphabetically at the first element — **aspersion** *f* POLL EAU sprinkling, spraying; ~ **à l'air** SOLS air sparging; ~ **d'eaux usées** POLL EAU wastewater sprinkling; ~ **de liquide de lavage** POLL ATM scrubbing liquid jet — *Termes composés imbriqués en ordre alphabétique sous le premier élément*

asphaltisation *f* DÉCHETS, DIST EAU asphalt sealing

aspirateur: ~ **central** *m* SÉCURITÉ central vacuum-cleaning system; ~ **et filtre pour poussières et copeaux** *m* SÉCURITÉ suction and filter installation for dust and chippings; ~ **à goulet** *m* POLL MAR gulley sucker

Subject area labels given in alphabetical order show appropriate translation — **aspiration** *f* DIST EAU, POLL ATM, POLL EAU suction, SOLS *de courant* drawing; ~ **d'air pur** POLL ATM clean-air suction; ~ **des gaz** DÉCHETS gas suction; ~ **isocinétique** ÉCHANT isokinetic suction — *Domaines en ordre alphabétique indiquent la traduction appropriée à chaque domaine*

aspirer *vt* DIST EAU, POLL ATM, POLL EAU suck in

emplaced: ~ **waste** *n* WASTE déchets mis en place *m pl*
empties *n pl* WASTE emballages vides *m pl*, vidanges *f pl*
empty: ~ **trip** *n* WASTE *transport* voyage à vide *m*
emptying *n* WASTE *of container* dépotage *m*, vidage *m*, WATER POLL *of wastewater canals* vidange *f*

Abréviation du terme donnée à l'entrée pour la forme intégrale — *Cross-references from abbreviations to their full forms are shown for both French and English*

EMS *abbr* (*environmental management system*) MANAGE SME (*système de management environnemental*)
Emscher: ~**-type settling tank** *n* WATER POLL fosse Emscher *f*
emulsified[1] *adj* WASTE, WATER POLL, WATER SUPP émulsifié, émulsionné
emulsified[2]: ~ **oil** *n* WATER POLL huile émulsionnée *f*

Genre de chaque nom indiqué pour la traduction française — *Genders are indicated at French translations*

emulsifier *n* WASTE, WATER POLL, WATER SUPP émulsifiant *m*, émulsionnant *m*
emulsify *vt* WASTE, WATER POLL, WATER SUPP émulsifier, émulsionner
emulsifying: ~ **agent** *n* SOIL agent émulsifiant *m*, WASTE séparateur de fractionnement *m*; ~ **liquid** *n* SAFETY liquide émulsifiant *m*

Les numéros en exposant indiquent des homographes — *Superscript numbers denote homographs*

emulsion[1]: ~**-coated** *adj* WASTE, WATER SUPP couché par émulsion, revêtu par émulsion
emulsion[2] *n* WASTE, WATER POLL, WATER SUPP émulsion *f*; ~ **adhesive** WASTE colle en émulsion *f*; ~ **breaker** MARINE, WATER POLL désémulsifiant *m*, WATER SUPP briseur d'émulsion *m*, casseur d'émulsion *m*, désémulsifiant *m*; ~ **composition** WATER POLL composition d'une émulsion *f*; ~ **separation** WATER POLL, WATER SUPP séparation d'une émulsion *f*
enamel: ~ **board** *n* WASTE carton glacé *m*
enameling *AmE see enamelling BrE*

Informations complètes sur les variantes américaines et anglaises suivies de codes géographiques — *British English and American English spelling variants are given in full and labelled accordingly*

enamelling *n BrE* WASTE, WATER SUPP émaillage *m*, laquage *m*, vernissage *m*
encapsulate *vt* NOISE enrober, SOIL confiner, WASTE confiner, enrober, sceller, WATER SUPP *nuclear waste* enrober, sceller
encapsulated *adj* NOISE enrobé, SOIL confiné, WASTE scellé, confiné, WATER SUPP *nuclear waste* enrobé, scellé

Domaines en ordre alphabétique indiquent la traduction appropriée à chaque domaine — *Subject area labels given in alphabetical order show appropriate translation*

encapsulation *n* NOISE enceinte acoustique *f*, WASTE, WATER SUPP enrobage *m*, scellement *m*, confinement *m*; ~ **measure** SOIL mesure de confinement *f*; ~ **using inert materials** SOIL encapsulage avec des matériaux inertes *m*
enclosed: ~ **emptying** *n* WASTE vidage intégré *m*; ~ **sea** *n* AIR, MARINE mer fermée *f*
enclosure *n* NOISE *to prevent sound propagation* enceinte *f*
encoding *n* SOIL, WASTE codage *m*, encodage *m*

Deux-points indique un mot dépourvu de valeur technique et introduit un mot composé — *A colon introduces a compound where the headword has no technical value*

Termes composés imbriqués en ordre alphabétique sous le premier élément — *Compound terms are nested alphabetically at the first element*

end: ~**-frame member** *n* SOIL barre pour cadre d'extrémité *f*; ~**-gas purity** *n* AIR pureté du gaz final *f*; ~**-of-life vehicle** *n* WASTE véhicule en fin de vie *m*; ~**-of-line packaging** *n* WASTE emballage en fin de chaîne de production *m*; ~**-of-pipe technology** *n* AIR, WATER POLL *waste water* technologie en bout de chaîne *f*, technologie en fin de canalisation *f*; ~**-of-the-pipe measure** *n* MANAGE mesure en aval *f*, mesure en fin de cycle *f*; ~ **panel** *n* WASTE panneau d'extrémité *m*; ~ **product** *n* AIR produit final *m*; ~ **product generation** *n* AIR obtention de produits finals *f*
endanger *vt* SAFETY mettre en danger
endangered[1] *adj* POLICY menacé SAFETY mis en danger
endangered[2]: ~ **species** *n pl* POLICY espèce menacée *f*
endogenous: ~ **catabolism** *n* WATER POLL catabolisme endogène *m*; ~ **respiration** *n* WATER SUPP respiration endogène *f*
endothermic[1] *adj* AIR, WATER SUPP endothermique
endothermic[2]: ~ **reaction** *n* AIR, WATER SUPP *solar power* réaction endothermique *f*

Using the dictionary

Range of coverage

This is a specialist French–English/English–French dictionary that covers the whole range of environmental technology and the scientific knowledge that underlies it. It contains a broad base of terminology drawn from areas such as Waste Treatment & Management, Water Supply & Drinking Water and Air Polution & Air Quality Control.

Selection of terms

We have aimed to include the essential vocabulary of each subject area, and the material has been checked by leading subject experts to ensure that both the English and the French terms are accurate and current, that the translations are valid equivalents, and that there are no gaps in coverage.

Placement of terms

All terms are ordered alphabetically. This is also the policy for hyphenated compounds.

Stoplists

All terms are entered at their first element except where that element is an article, preposition, conjunction, pronoun or other delexicalized word. In such cases, the term is entered at its next valid element.

Open compounds consisting of more than one element are listed at their first element. In these nested listings, the simple form is replaced by a swung-dash (~). For example:

marine: ~ **biota** *n* MARINE faune et flore marines *f pl*; ~ **climate** *n* MARINE climat marin *m*; ~ **dredge** *n* MARINE drague de mer *f*

When this first element is itself a headword with a technical sense, compounds follow the simple form in alphabetical order. For example:

catalyst *n* AIR, ANALYSIS, SOIL catalyseur *m*; ~ **bed** AIR lit catalytique *m*; ~ **carrier** AIR support de catalyse *m*; ~ **charge** AIR charge catalytique *f*

If the first element is not itself translated, a colon precedes the compounds. For example:

dual: ~-**bed bulk material catalyst** *n* AIR catalyseur massique sur lits superposés *m*

Ordering of terms

All terms, including hyphenated compounds, are nested according to part of speech. For example:

double[1]: ~-**cycle** *adj* AIR à double cycle
double[2]: ~-**beam photometer** *n* ANALYSIS photomètre à double faisceau *m*; ~-**belt press** *n* WASTE, WATER SUPP presse à double bande *f*

Within nests, articles, pronouns and prepositions are ignored in determining the sequence of nested open compounds. For example:

four: ~ **à bande** *m* SOLS band oven; ~ **chauffé** *m* ÉCHANT heated kiln; ~ **à colonne chauffante** *m* ÉCHANT heated-column oven; ~ **de contact** *m* POLL ATM contact reactor

Abbreviations and acronyms written in upper case appear after entry terms of the same form written in lower case. For example:

cap *vt* WASTE capsuler
CAP *abbr* (*controlled-atmosphere packaging*) WASTE conditionnement sous atmosphère contrôlée *m*

Terms containing figures and symbols are alphabetized according to the usual expansion when written out in full. For example:

nickel *n* (*Ni*) AIR, SOIL nickel *m* (*Ni*)
90%: ~ **threshold value range** *n* ANALYSIS seuil de 90% *m*
NIPTS *abbr* (*noise-induced permanent threshold shift*) NOISE dépassement permanent de seuil induit par le bruit *m*

Homographs

Every term is accompanied by a label indicating its part of speech. For a complete list of these labels and their expansions, please see page xxv.

When terms beginning with the same element fall

into two or more part-of-speech categories, the different categories are distinguished by a raised number immediately following the headword. The sequence is adjective, adverb, noun, verb, followed by less frequent parts of speech. For example:

adhesive[1] *adj* WATER POLL adhésif
adhesive[2] *n* AIR, WASTE adhésif *m*; ~ **agent** *n* AIR agent adhésif *m*; ~ **film** *n* WASTE film adhésif *m*, pellicule adhésive *f*; ~ **force** *n* AIR force d'adhésion *f*, ANALYSIS *chromatography* adhérence *f*, pouvoir adhérent *m*; ~ **tape** *n* WASTE ruban adhésif *m*

Ordering of translations

Every term is accompanied by one or more labels indicating the technological area in which it is used. For a complete list of these labels and their expansions, please see pages xxv–xxvi.

Where the same term is used in more than one technological area, multiple labels are given as appropriate. These labels appear in alphabetical order. Where a term has the same translation in more than one technological area, this translation is given after the sequence of labels. For example:

écrémage *m* DIST EAU, POLL EAU, POLL MAR skimming

Where a term has different translations according to the technological area in which it is used, the appropriate translation is given after each label or set of labels. For example:

sonder *vt* ÉCHANT sound, sense, ÉN RENOUV drill

Supplementary information

The gender is given for every French noun term in the French–English section as well as for every French translation that is a noun in the English–French section. In the case of compound terms this is the gender of the noun head. For example:

droplet *n* AIR gouttelette *f*; ~ **diameter** AIR diamètre de gouttelette *m*; ~ **dispersion** AIR dispersion de gouttelettes *f*

In many cases additional data is given about a term in order to show how it is used. Such contextual information can be:

(a) the typical subject or object of a verb, for example:

amorcer *vt* DIST EAU, POLL EAU *pompe* prime

(b) typical nouns used with an adjective, for example:

sour *adj* AIR *gas* corrosif

(c) words indicating the reference of a noun, for example:

sone *n* ANALYSIS *phonometric unit* sone *m*

(d) a narrower subject area than indicated by the label, for example:

asperseur *m* POLL EAU *agriculture* sprinkler

When various different translations apply in the same subject area, contextual information is also used to show which translation is appropriate in different circumstances. For example:

demineralization *n* WATER POLL, WATER SUPP *removal of minerals* déminéralisation *f*, *removal of salt* dessalage *m*, dessalement *m*

Cross-references

All terms not universally applicable are accompanied by a geographical label. In the English–French section, British and American terms are covered, and these are differentiated by regional labels. In the case of lexical variants, full information including translations and cross-references to other regional forms is given at each entry. For example:

muffler *n AmE* (*cf silencer BrE*) NOISE silencieux *m*

silencer *n BrE* (*cf muffler AmE*) NOISE silencieux *m*

Geographical variants, both spelling and lexical, are given in full when they are translations. For example:

coloration *f* ÉCHANT coloring (*AmE*), colouring (*BrE*)

Both abbreviations and their full forms are entered in the main body of the dictionary in alphabetical sequence. Full information including translations and cross-references to the full form or abbreviation as appropriate is given at each entry.

Where an abbreviation in the source language translates to an abbreviation in the target language, the full form in each case is given after the abbreviation in brackets and italics. For example:

DCO *abbr* (*demande chimique en oxygène*) DÉCHETS, DIST EAU, POLL ATM, POLL EAU COD (*chemical oxygen demand*)

If a source language abbreviation does not translate to a target language abbreviation, the translation of the abbreviation is in roman. For example:

SCC *abbr* (*secondary combustion chamber*) WATER SUPP chambre de combustion secondaire *f*

Comment utiliser ce dictionnaire

Domaines traités

Ce dictionnaire est un ouvrage spécialisé français–anglais/anglais–français qui couvre l'ensemble du génie de l'environnement et des connaissances scientifiques sur lesquelles elles reposent. Il comporte une importante base terminologique extraite de disciplines comme le traitement & la gestion des déchets, la distribution de l'eau & eau potable et la pollution atmosphérique & le contrôle de la qualité de l'air.

Sélection des termes

Les termes qu'il contient se rapportent au vocabulaire de base de chaque discipline, et ont été minutieusement vérifiés par des spécialistes de haut niveau, chargés de veiller à l'exactitude et à l'actualité des termes anglais et français, à la parfaite correspondance des termes, et à la cohérence de l'ensemble.

Classement des termes

Tous les termes sont classés par ordre alphabétique. Les termes composés apparaissent sous leur premier élément, quelle que soit la structure sémantique de l'ensemble.

Termes ignorés

Tous les termes apparaissent sous leur premier élément, sauf dans le cas où cet élément est un article, une préposition, une conjonction, un pronom ou tout autre terme depourvu de valeur lexicale. Lorsque de tels cas se présentent, le terme apparaît sous l'élément significatif le plus proche.

Les locutions ou les syntagmes figurent sous le premier élément. Dans les listes imbriquées, la forme simple est remplacée par un 'tilde' (~). Exemple:

marine: ~ **biota** *n* MARINE faune et flore marines *f pl*; ~ **climate** *n* MARINE climat marin *m*; ~ **dredge** *n* MARINE drague de mer *f*

Lorsque ce premier élément est un mot-vedette ayant un sens technique, les mots composés suivent la forme simple et sont classés par ordre alphabétique. Exemple:

catalyst *n* AIR, ANALYSIS, SOIL catalyseur *m*; ~ **bed** AIR lit catalytique *m*; ~ **carrier** AIR support de catalyse *m*; ~ **charge** AIR charge catalytique *f*

Si le premier élément n'est pas traduit, ses composants sont précédés de deux-points. Exemple:

dual: **~-bed bulk material catalyst** *n* AIR catalyseur massique sur lits superposés *m*

Ordre des termes

Les locutions et termes comportant des traits-d'union sont classés selon leur catégorie grammaticale. Exemple:

double[1]: **~-cycle** *adj* AIR à double cycle
double[2]: **~-beam photometer** *n* ANALYSIS photomètre à double faisceau *m*; **~-belt press** *n* WASTE, WATER SUPP presse à double bande *f*

Les articles, prépositions et pronoms ne sont pas pris en compte lors du tri. Exemple:

four: ~ **à bande** *m* SOLS band oven; ~ **chauffé** *m* ÉCHANT heated kiln; ~ **à colonne chauffante** *m* ÉCHANT heated-column oven; ~ **de contact** *m* POLL ATM contact reactor

Les abréviations et acronymes en majuscules figurent après les termes de même forme en minuscules. Exemple:

cap *vt* WASTE capsuler
CAP *abbr* (*controlled-atmosphere packaging*) WASTE conditionnement sous atmosphère contrôlée *m*

Les termes contenant des chiffres ou des symboles sont classés comme s'ils étaient écrits en lettres. Exemple:

nickel *n* (*Ni*) AIR, SOIL nickel *m* (*Ni*)
90%: ~ **threshold value range** *n* ANALYSIS seuil de 90% *m*
NIPTS *abbr* (*noise-induced permanent threshold shift*) NOISE dépassement permanent de seuil induit par le bruit *m*

use *n* GEN utilisation *m*, *working* fonctionnement *m*

Homographes

Chaque terme est accompagné d'une abréviation indiquant sa fonction grammaticale. La liste complète de ces abréviations figure page xxv.

Un terme ayant plusieurs fonctions grammaticales figure sous une entrée séparée pour chaque fonction. Ces entrées se distinguent par un chiffre en exposant qui suit immédiatement le mot-vedette. La séquence est la suivante: adjectif, adverbe, nom, verbe, suivie des fonctions moins fréquentes. Exemple:

adhesive^1 *adj* WATER POLL adhésif
adhesive^2 *n* AIR, WASTE adhésif *m*; ~ **agent** *n* AIR agent adhésif *m*; ~ **film** *n* WASTE film adhésif *m*, pellicule adhésive *f*; ~ **force** *n* AIR force d'adhésion *f*, ANALYSIS *chromatography* adhérence *f*, pouvoir adhérent *m*; ~ **tape** *n* WASTE ruban adhésif *m*

Ordre des traductions

Chaque terme est accompagné d'une ou plusieurs abréviations indiquant le domaine auquel il se rapporte. La liste complète de ces abréviations figure pages xxv–xxvi.

Lorsqu'un terme est utilisé dans plusieurs disciplines, il est suivi des abréviations appropriées, en ordre alphabétique. Lorsqu'un terme est traduit de la même manière dans différentes disciplines, la traduction figure après la liste des abréviations. Exemple:

écrémage *m* DIST EAU, POLL EAU, POLL MAR skimming

Lorsqu'un terme peut se traduire de différentes façons, selon les disciplines, sa traduction figure après chaque abréviation ou série d'abréviations. Exemple:

sonder *vt* ÉCHANT sound, sense, ÉN RENOUV drill

Informations complémentaires

Le genre de chaque nom est indiqué dans la partie français–anglais, ainsi que pour chaque traduction française s'il s'agit d'un nom dans la partie anglais–français. Pour les expressions ou termes composés, le genre est celui de son mot-vedette, et non celui de l'élément final. Exemple:

droplet *n* AIR gouttelette *f*; ~ **diameter** AIR diamètre de gouttelette *m*; ~ **dispersion** AIR dispersion de gouttelettes *f*

Dans de nombreux cas, des informations complémentaires expliquent l'utilisation du terme concerné. Exemple:

(a) le sujet ou l'objet typique d'un verbe:

amorcer *vt* DIST EAU, POLL EAU *pompe* prime

(b) les noms typiques utilisés avec un adjectif:

sour *adj* AIR *gas* corrosif

(c) les mots indiquant la référence d'un nom:

sone *n* ANALYSIS *phonometric unit* sone *m*

(d) un domaine plus limité que celui indiqué par l'abréviation:

asperseur *m* POLL EAU *agriculture* sprinkler

Lorsqu'un terme peut donner lieu à différentes traductions, des informations contextuelles indiquent également la traduction appropriée. Exemple:

demineralization *n* WATER POLL, WATER SUPP *removal of minerals* déminéralisation *f*, *removal of salt* dessalage *m*, dessalement *m*

Renvois

Tous les termes non universels sont accompagnés d'une mention géographique. Dans la partie anglais–français, les variantes linguistiques (entre l'anglais britannique et l'anglais américain) sont également accompagnées de la mention géographique qui convient.

En cas de variantes lexicales, les informations les concernant et notamment les traductions et renvois à d'autres formes régionales sont fournies systématiquement à chaque entrée. Exemple:

muffler *n AmE* (*cf silencer BrE*) NOISE silencieux *m*

silencer *n BrE* (*cf muffler AmE*) NOISE silencieux *m*

Les variantes géographiques, tant orthographiques que lexicales, sont indiquées intégralement lorsqu'il s'agit de traductions. Exemple:

coloration *f* ÉCHANT coloring (*AmE*), colouring (*BrE*)

Les abréviations et leurs formes intégrales figurent dans le corps principal du dictionnaire, classées par ordre alphabétique. Les informations complètes et notamment leurs traductions et les renvois à leur abréviation ou à leur forme intégrale, le cas échéant sont indiquées pour chaque entrée.

Lorsque qu'une abréviation dans la langue source est traduite par une abréviation dans la langue cible, la forme développée est indiquée dans chaque cas entre parenthèses en italique après l'abréviation. Exemple:

DCO *abbr* (*demande chimique en oxygène*) DÉCHETS, DIST EAU, POLL ATM, POLL EAU COD (*chemical oxygen demand*)

Si une abréviation dans la langue source n'est pas traduite par une abréviation dans la langue cible, la traduction de l'abréviation est en caractères romains. Exemple:

SCC *abbr* (*secondary combustion chamber*) WATER SUPP chambre de combustion secondaire *f*

Abbreviations used in this dictionary/Abréviations utilisées dans ce dictionnaire

Parts of speech/Catégories grammaticales

abbr	abbreviation	abréviation
abrév	abréviation	abbreviation
adj	adjective	adjectif
adv	adverb	adverbe
f	feminine	féminin
f pl	feminine plural	féminin pluriel
loc	locution	phrase
m	masculine	masculin
m pl	masculine plural	masculin pluriel
n	noun	nom
n pl	noun plural	nom pluriel
phr	phrase	locution
pref	prefix	préfixe
prep	preposition	préposition
vi	intransitive verb	verbe intransitif
v réfl	verbe réfléchi	reflexive verb
vt	transitive verb	verbe transitif

Geographic codes/Codes géographiques

AmE	American English	anglais américain
BrE	British English	anglais britannique

Subject area labels/Domaines

AIR	Air Pollution & Air Quality Control	Pollution atmosphérique & contrôle de la qualité de l'air
ANALYSIS	Sampling, Measurement & Analysis	Échantillonnage, analyse & mesure
GEN	General Environmental Technology	Général
MANAGE	Management & Marketing	Gestion & marketing
MARINE	Marine Pollution	Pollution marine
NOISE	Noise Pollution	Pollution sonore
POLICY	Environmental Policy	Politique de l'environnement
RENEW ENERG	Renewable Energy Sources	Sources d'énergie renouvelables
SAFETY	Safety Engineering	Mesures de sécurité & équipements associés
SOIL	Soil Contamination & Remediation	Contamination des sols & remèdes
WASTE	Waste Treatment & Management	Traitement & gestion des déchets

WATER POLL	Water Pollution & Wastewater Treatment	Pollution des eaux & traitement des eaux usées
WATER SUPP	Water Supply & Drinking Water	Distribution de l'eau & eau potable
DÉCHETS	Traitement & gestion des déchets	Waste Treatment & Management
DIST EAU	Distribution de l'eau & eau potable	Water Supply & Drinking Water
ÉCHANT	Échantillonnage, analyse & mesure	Sampling, Measurement & Analysis
ÉN RENOUV	Sources d'énergie renouvelables	Renewable Energy Sources
GÉNÉRAL	Général	General Environmental Technology
GESTION	Gestion & marketing	Management & Marketing
POL ENV	Politique de l'environnement	Environmental Policy
POLL ATM	Pollution atmosphérique & contrôle de la qualité de l'air	Air Pollution & Air Quality Control
POLL EAU	Pollution des eaux & traitement des eaux usées	Water Pollution & Wastewater Treatment
POLL MAR	Pollution marine	Marine Pollution
POLL SON	Pollution sonore	Noise Pollution
SÉCURITÉ	Mesures de sécurité & équipements associés	Safety Engineering
SOLS	Contamination des sols & remèdes	Soil Contamination & Remediation

Registered trademarks

Note that neither the presence nor the absence of the symbol ® should be regarded as affecting the legal status of any trademark.

Marques déposées

Veuillez noter que l'absence ou la présence de la mention ® est sans effet sur le statut légal d'une marque déposée, quelle qu'elle soit.

FRANÇAIS–ANGLAIS
FRENCH–ENGLISH

A

abaisser *vt* DIST EAU *charge* lower
abattement *m* POL ENV tax relief; **~ de la couleur** POLL EAU color reduction (*AmE*), colour reduction (*BrE*), color removal (*AmE*), colour removal (*BrE*)
abilité *f* GÉNÉRAL usability
abondant: peu ~ *adj* ÉN RENOUV, POL ENV, POLL ATM, SOLS scarce
abrasif *adj* DIST EAU, POLL ATM abrasive
abrasion *f* DIST EAU, POLL ATM abrasion
abreuver *vt* DIST EAU *pompe* prime
abroger *vt* POL ENV repeal
ABS *abrév* (*acrylonitrile-butadiène-styrène*) DÉCHETS ABS (*acrylonitrile-butadiene-styrene*)
absence: **~ d'effet** *f* SÉCURITÉ no-effect level; **~ de mise en œuvre** *f* POL ENV lack of implementation; **~ d'oxygène** *f* POLL EAU lack of oxygen; **~ totale de perturbation observée** *f* SÉCURITÉ no observed adverse-effect level (*NOAEL*); **~ totale d'émissions de CO_2** *f* POLL ATM zero CO_2 emissions
absorbable *adj* GÉNÉRAL absorbable
absorbant[1] *adj* GÉNÉRAL absorbent; **très ~** POLL ATM, POLL SON highly-absorbent
absorbant[2] *m* GÉNÉRAL absorbent; **~ basique** POLL ATM basic absorbent; **~ désorbé** POLL ATM desorbed absorbent; **~ gazeux** POLL ATM gas absorbent; **~ zéolitique** POLL ATM zeolitic absorbent
absorbat *m* POLL ATM absorbate
absorbé *adj* GÉNÉRAL absorbed
absorber *vt* GÉNÉRAL absorb, sorb; **~ chimiquement** POLL ATM chemisorb
absorbeur *m* GÉNÉRAL absorber; **~ acoustique** POLL MAR, POLL SON sound absorber; **~ acoustique aérien** POLL SON overhead sound absorber; **~ acoustique pneumatique** POLL MAR, POLL SON pneumatic sound absorber; **~ acoustique surélevé** POLL SON overhead sound absorber; **~ d'aigus** POLL SON treble absorber; **~ basse fréquence** POLL SON low-frequency absorber; **~ à cavités** POLL SON cavity absorber; **~ de charbon actif** POLL ATM, POLL EAU activated-carbon canister; **~ compact** POLL SON compact absorber; **~ en fibre minérale** POLL SON mineral fiber absorber (*AmE*), mineral fibre absorber (*BrE*); **~ en fibre de verre** POLL SON glass-fiber absorber (*AmE*), glass-fibre absorber (*BrE*); **~ de film ruisselant** POLL ATM falling-film absorber; **~ de film tombant** POLL ATM falling-film absorber; **~ d'infrarouge non dispersif** ÉCHANT nondispersive infrared absorber; **~ intégré** POLL SON structure-borne absorber; **~ à lit fluidisé** POLL ATM fluidized-bed absorber; **~ de particules** POLL ATM spray absorber; **~ à plateau souple** POLL SON flexible-plate absorber; **~ à plateaux** POLL ATM tray absorber; **~ à plateaux rigides** POLL SON rigid-plate absorber; **~ à pulvérisation par voie sèche** POLL ATM spray-dry absorber (*SDA*); **~ de résonance** POLL SON resonance absorber; **~ de séchage par pulvérisation** POLL ATM spray-dry absorber (*SDA*); **~ stratifié** POLL SON sheet absorber
absorptance *f* GÉNÉRAL absorbency
absorptiomètre: **~ à infrarouge** *m* ÉCHANT infrared absorption measuring device
absorption *f* GÉNÉRAL absorption; **~ gazeuse** POLL ATM gas absorption; **~ infrarouge non dispersive** ÉCHANT nondispersive infrared absorption (*NDIRA*); **~ de lumière** ÉCHANT optical absorption; **~ optique** ÉCHANT optical absorption; **~ optique des infrarouges** ÉCHANT infrared optical absorption; **~ optique des ultraviolets** ÉCHANT ultraviolet optical absorption; **~ optique visible** ÉCHANT visible optical absorption; **~ de rayonnement** ÉCHANT, POLL ATM radiation absorption; **~ de résonance** POLL SON resonance absorption; **~ à sec** POLL ATM dry absorption; **~ du son** POLL SON sound absorption; **~ sur carbone actif** SOLS activated-carbon absorption; **~ ultraviolette** DIST EAU, ÉCHANT ultraviolet absorption
absorptivité *f* POLL ATM absorptivity
acaricide *m* POLL ATM acaricide
acarien *m* POLL ATM acaride
accélérateur: **~ d'électrons** *m* POLL ATM electron accelerator
accélération *f* POLL ATM acceleration; **~ catalytique** POLL ATM catalytic acceleration; **~ centrifuge** POLL ATM centrifugal acceleration; **~ de la vitesse du gaz** POLL ATM gas acceleration
accéléré *adj* POLL ATM accelerated
accélérer *vt* POLL ATM *particule* accelerate
accéléromètre *m* ÉCHANT accelerometer
acceptable: **~ pour l'environnement** *adj* DÉCHETS, DIST EAU *risque* environmentally-acceptable
accessoire: **~ d'échantillonnage** *m* ÉCHANT sampling aid
accident: **~ maximal de référence** *m* SÉCURITÉ maximum credible accident; **~ mortel** *m* SÉCURITÉ fatal accident; **~ du travail** *m* SÉCURITÉ industrial accident
acclimatation *f* POLL ATM, POLL EAU acclimation, acclimatization
acclimater *vt* POLL ATM, POLL EAU acclimate, acclimatize
accord: **~ bilatéral en matière d'environnement** *m* POL ENV bilateral environmental agreement; **~ librement souscrit** *m* POL ENV voluntary agreement; **~ relatif à l'environnement** *m* POL ENV environmental agreement
accoutumance *f* POLL ATM, POLL EAU habituation
accrétion *f* DIST EAU accretion
accroissement: **~ démographique** *m* POL ENV population growth
accumulateur: **~ de chaleur** *m* POLL ATM heat accumulator; **~ au plomb** *m* POLL ATM lead-acid battery; **~ usagé** *m* DÉCHETS used storage battery
accumulation *f* GÉNÉRAL accumulation; **~ de boues** POLL EAU sludge accumulation; **~ de contaminants** POLL ATM, SOLS accumulation of contaminants; **~ de métaux lourds** POLL ATM, SOLS heavy metals accumulation; **~ de polluants** POLL ATM, SOLS accumulation of pollutants; **~ de polluants absorbant la lumière** ÉCHANT optically-absorptive pollutant accumulation; **~ de solvants** POLL ATM solvent accumulation; **~ de substances nocives** POLL ATM, SOLS accumulation of harmful substances
accumuler *vt* GÉNÉRAL accumulate; **~ par voie biologique** POLL ATM bioaccumulate
acétate: **~ de cellulose** *m* DIST EAU, POLL EAU cellulose acetate
acétone *f* DÉCHETS, POLL ATM acetone
acide[1] *adj* GÉNÉRAL acid, acidic
acide[2] *m* GÉNÉRAL acid; **~ éthylènediamine-tétracétique** ÉCHANT ethylenediamine tetracetic acid (*EDTA*); **~ gras** POLL EAU fatty acid; **~ nitrique** (HNO_3) DIST EAU, POLL ATM nitric acid (HNO_3); **~ perdu** DÉCHETS waste acid; **~ propionique** POLL EAU propionic acid; **~ récupéré** DÉCHETS waste acid; **~ résiduaire** DÉCHETS waste acid; **~ sulfhydrique** DÉCHETS, DIST EAU hydrogen sulphide (*BrE*), hydrogen sulfide (*AmE*); **~ sulfureux** (H_2SO_3) POLL

ATM sulfurous acid (*AmE*) (H_2SO_3), sulphurous acid (*BrE*) (H_2SO_3); ~ **sulfurique** (H_2SO_4) POLL ATM sulfuric acid (*AmE*) (H_2SO_4), sulphuric acid (*BrE*) (H_2SO_4)
acidifiant[1] *adj* GÉNÉRAL acidifying
acidifiant[2] *m* DIST EAU, POLL ATM, SOLS acid former
acidification *f* GÉNÉRAL acidification; ~ **anthropique** SOLS anthropogenic acidification; ~ **artificielle** SOLS artificial acidification; ~ **en cascade** SOLS cascade acidification; ~ **naturelle** DIST EAU, POLL ATM, SOLS natural acidification
acidifier *vt* GÉNÉRAL acidify
acidité[1]: **d'~ moyenne** *adj* SOLS moderately acidic
acidité[2] *f* GÉNÉRAL acidity; ~ **atmosphérique** POLL ATM atmospheric acidity; ~ **potentielle** SOLS potential acidity
acier: ~ **de cerclage** *m* DÉCHETS strapping steel
acoustique[1] *adj* POLL SON acoustic
acoustique[2]: ~ **des locaux** *f* POLL SON room acoustics
acquisition: ~ **des données** *f* ÉCHANT data acquisition
acquitter *vt* SÉCURITÉ *alarme* acknowledge
acrylonitrile: **~-butadiène-styrène** *f* (*ABS*) DÉCHETS acrylonitrile-butadiene-styrene (*ABS*)
actif *adj* POLL ATM, SOLS active
actinomètre *m* ÉCHANT actinometer
action: ~ **d'amortissement** *f* ÉCHANT, POLL ATM buffering action; ~ **de filtration** *f* POLL ATM filtering operation; ~ **phytosanitaire** *f* GESTION pest control
activation *f* POLL ATM, POLL EAU, SOLS activation
activé *adj* POLL ATM, POLL EAU, SOLS activated
activité: ~ **achevée** *f* POL ENV completed activity; ~ **bactérienne** *f* POLL ATM, POLL EAU, SOLS bacteria activity; ~ **catalytique** *f* POLL ATM catalytic activity; ~ **datée** *f* POL ENV scheduled activity; ~ **environnementale** *f* GESTION, POL ENV environmental activity; ~ **fictive** *f* POL ENV dummy activity; ~ **ionique** *f* ÉCHANT ion activity; ~ **liée à l'environnement** *f* GESTION, POL ENV environmental activity; ~ **métabolique** *f* POLL EAU metabolic activity; ~ **microbienne** *f* SOLS microbial activity; ~ **par force centrifuge** *f* POLL ATM centrifugal force activity; ~ **programmée** *f* POL ENV scheduled activity; ~ **résiduelle** *f* POLL ATM residual activity
activités: ~ **de recyclage** *f pl* POL ENV recycling economy
actualisation *f* POL ENV modernization
actualiser *vt* POL ENV modernize
acyclique *adj* POLL ATM acyclic
adaptation: ~ **sensorielle** *f* ÉCHANT sensory adaptation
additif *m* DÉCHETS, DIST EAU, POLL ATM, SOLS additive; ~ **antidétonant** POLL ATM anti-knock additive; ~ **au plomb** DIST EAU lead additive
adduction: ~ **d'eau** *f* DIST EAU water delivery
adhérence *f* DÉCHETS adherence, ÉCHANT adhesive force, POLL ATM adherence
adhérer *vi* DÉCHETS, POLL ATM *particules* adhere; ~ **par aspiration** POLL ATM adhere by suction
adhésif[1] *adj* DÉCHETS, POLL ATM *poussière* sticky, POLL EAU adhesive
adhésif[2] *m* DÉCHETS, POLL ATM adhesive, bonding agent; ~ **à base d'acétate** DÉCHETS acetate adhesive; ~ **de contact** DÉCHETS contact adhesive; ~ **mixte** DÉCHETS mixed adhesive; ~ **pouvant être refondu** DÉCHETS repulpable adhesive; ~ **thermocollant** DÉCHETS heat-sealing adhesive; ~ **universel** DÉCHETS all-purpose adhesive
adhésion *f* DÉCHETS, POLL ATM adhesion
adhésivité *f* POLL ATM stickiness, POLL EAU adhesiveness
adiabatique *adj* POLL ATM adiabatic
adjuvant: ~ **d'échantillonnage** *m* ÉCHANT sampling aid; ~ **de filtration** *m* DIST EAU, POLL ATM filter aid; ~ **de floculation** *m* DIST EAU flocculation aid; ~ **de sablage** *m* DIST EAU grit additive
admissible *adj* DÉCHETS tolerable
admission *f* DIST EAU, POLL ATM inlet, POLL EAU *d'eau* inflow; ~ **d'adsorbants** POLL ATM adsorbent inlet; ~ **d'air** POLL ATM air intake; ~ **d'air purgé** POLL ATM purge-air inlet; ~ **d'air de refroidissement** POLL ATM cooling-air inlet; ~ **de fumées** POLL ATM fume inlet; ~ **de gaz** POLL ATM gas inlet; ~ **de gaz pollué** POLL ATM dirt-gas inlet; ~ **latérale** POLL ATM transverse entry; ~ **de liquide** POLL ATM liquid inlet; ~ **de réfrigérant** POLL ATM coolant inlet; ~ **du rotor** ÉN RENOUV input of rotor; ~ **tangentielle** POLL ATM transverse entry
adopter *vt* POL ENV *loi* pass
adouci *adj* DIST EAU softened
adoucir *vt* DIST EAU soften
adoucissement *m* DIST EAU softening; ~ **de l'eau potable** DIST EAU drinking-water softening; ~ **mécanique** DIST EAU mechanical softening; ~ **par échange ionique** DIST EAU ion-exchange softening
adoucisseur *m* DIST EAU softener; ~ **par échange ionique** DIST EAU ion-exchange softener
adsorbant[1] *adj* GÉNÉRAL adsorbent, adsorptive
adsorbant[2] *m* GÉNÉRAL adsorbent; ~ **amorphe** POLL ATM amorphous adsorbent; ~ **au carbone** POLL ATM adsorbent containing carbon; ~ **chargé** POLL ATM loaded adsorbent; ~ **poreux** POLL ATM porous adsorbent; ~ **régénéré** POLL ATM regenerated adsorbent
adsorbat *m* GÉNÉRAL adsorbate
adsorbé *adj* GÉNÉRAL adsorbed
adsorber *vt* GÉNÉRAL adsorb, sorb; ~ **physiquement** POLL ATM physically adsorb
adsorbeur *m* GÉNÉRAL adsorber; ~ **à anneau** POLL ATM ring adsorber; ~ **à charbon actif** POLL ATM activated-carbon adsorber; ~ **au coke actif** DÉCHETS, POLL ATM activated-coke adsorber; ~ **à lit fixe** POLL ATM, POLL EAU fixed-bed adsorber, stationary-bed adsorber; ~ **à lit fixe discontinu** POLL ATM, POLL EAU discontinuous fixed-bed adsorber; ~ **à lit fixe horizontal** POLL ATM, POLL EAU horizontal fixed-bed adsorber; ~ **à lit fixe vertical** POLL ATM, POLL EAU vertical fixed-bed adsorber; ~ **à lit fluidisé** POLL ATM, POLL EAU fluidized-bed adsorber, moving-bed adsorber; ~ **à lit fluidisé à étage unique** POLL ATM single-stage fluidized-bed adsorber; ~ **à lit fluidisé multi-étage** POLL ATM multi-stage fluidized-bed adsorber; ~ **à lit mobile** POLL ATM, POLL EAU moving-bed adsorber; ~ **à lit mobile horizontal** POLL ATM horizontal moving-bed adsorber; ~ **à lit mobile vertical** POLL ATM, POLL EAU vertical moving-bed adsorber; ~ **de réservoir** POLL ATM tank adsorber; ~ **rotatif** POLL ATM rotating adsorber; ~ **secondaire** POLL ATM secondary adsorber
adsorption *f* GÉNÉRAL adsorption; ~ **chimique** POLL ATM chemical adsorption; ~ **de gaz nocif** POLL ATM harmful gas adsorption; ~ **à l'indicateur fluorescent** ÉCHANT fluorescent indicator adsorption (*FIA*); ~ **initiale à dosage optimal de polymères** POLL EAU initial adsorption at optimum polymer dosage; ~ **par voie sèche** POLL ATM dry adsorption; ~ **physique** POLL ATM physical adsorption; ~ **secondaire** POLL ATM, POLL EAU secondary adsorption; ~ **sur charbon actif** POLL ATM activated-carbon adsorption
aérage *m* DÉCHETS, DIST EAU, POLL ATM, SÉCURITÉ ventilation
aérateur *m* DIST EAU aerator; ~ **d'essai** POLL EAU testing aerator; ~ **spiral** POLL EAU spiral aerator
aération *f* DÉCHETS ventilation, aeration, DIST EAU, POLL ATM vent, ventilation, aeration, SÉCURITÉ ventilation, aeration, SOLS aeration; ~ **avec injection d'air** POLL EAU jet aeration; ~ **avec réduction progressive** POLL EAU tapered aeration; ~ **du compost** DÉCHETS, SOLS compost aeration; ~ **des cours d'eau** POLL EAU stream aeration; ~ **décroissante** POLL EAU tapered aeration; ~ **étagée** DIST EAU, POLL EAU stepped aeration; ~ **forcée** SOLS forced aeration; ~ **à jet** POLL EAU jet aeration; ~ **par air surpressé** POLL EAU diffused-air aeration; ~ **par barbotage** DIST EAU bubbling aeration; ~ **par bulles moyennes** POLL EAU medium-size bubble aeration; ~ **par diffusion** POLL EAU diffused-air aeration; ~ **par grosses bulles** POLL EAU large bubble

aeration; ~ **par pulvérisation** DIST EAU, POLL ATM spray aeration; ~ **par roue à ailettes** POLL EAU paddle-wheel aeration; ~ **prolongée** DIST EAU, POLL ATM extended aeration; ~ **de sécurité** SÉCURITÉ protective aeration; ~ **sur une période prolongée** POLL EAU extended-period aeration; ~ **de surface** POLL EAU surface aeration

aéré: ~ **thermiquement** *adj* ÉCHANT thermally-aerated

aérer *vt* DÉCHETS ventilate, DIST EAU ventilate, air, ÉCHANT aerate, POLL ATM aerate, ventilate, POLL EAU aerate, SÉCURITÉ aerate, ventilate, SOLS aerate

aérien *adj* POLL ATM airborne

aérobie *adj* DIST EAU aerobic

aéroglissière *f* DÉCHETS air slide

aéromètre *m* ÉCHANT aerometer

aéroréfrigérant *m* ÉN RENOUV cooling tower

aérosol *m* DÉCHETS, DIST EAU, POLL ATM aerosol; ~ **acide** POLL ATM acid aerosol; ~ **de condensation** POLL ATM condensation aerosol; ~ **de dispersion** POLL ATM dispersion aerosol; ~ **salin** POLL ATM salt aerosol

affaibli *adj* DIST EAU cushioned, dampened, impaired, ÉCHANT attenuated, weakened

affaiblir *vt* DIST EAU cushion, dampen, impair, ÉCHANT alternate, weaken

affaiblissement *m* DIST EAU impairment, ÉCHANT attenuation, weakening; ~ **d'insertion** POLL SON insertion loss; ~ **de la propagation** POLL SON propagation deadening

affaire: ~ **viable** *f* GESTION sustainable business

affaissement *m* POLL ATM collapse

affermage *m* GESTION leasing

affichage: ~ **complémentaire** *m* ÉCHANT additional display; ~ **conversationnel** *m* ÉCHANT dialog display (*AmE*), dialogue display (*BrE*); ~ **à cristaux liquides** *m* ÉCHANT liquid crystal display (*LCD*); ~ **à diode électroluminescente** *m* ÉCHANT LED display; ~ **de lecture** *m* ÉCHANT read-out display; ~ **principal** *m* ÉCHANT main display; ~ **simultané** *m* ÉCHANT simultaneous read-out

affinage *m* DIST EAU, POLL ATM, POLL MAR refining

affleurement: ~ **de déchets** *m* DÉCHETS disinterment of waste

affluent *m* DIST EAU influent, POLL EAU *d'un fleuve* influent, tributary

affluer *vi* DIST EAU *dans la mer* run, POLL ATM stream in, POLL EAU run

affouillement *m* DIST EAU underwashing, ÉCHANT, SOLS washout

affouiller *vt* DIST EAU scour, undermine

AFNOR *abrév* (*Association Française de Normalisation*) ÉCHANT, POL ENV AFNOR (*French Standards Association*)

âge: ~ **des boues** *m* POLL EAU sludge age; ~ **de décharge** *m* DÉCHETS landfill age

agence: ~ **de bassin** *f* DIST EAU river authority, POL ENV *gestion de l'eau* river-basin authority; ~ **de l'eau** *f* DIST EAU, POL ENV water authority; ~ **de protection de l'environnement** *f* (*EPA*) GESTION, POL ENV environmental protection agency (*EPA*)

agent: ~ **adhésif** *m* POLL ATM adhesive agent; ~ **anticorrosion** *m* DIST EAU corrosion inhibitor; ~ **antidétonant** *m* POLL ATM antiknock agent; ~ **antimicrobien** *m* DIST EAU biological agent; ~ **antimousse** *m* DIST EAU antifoaming agent; ~ **antipoussière** *m* POLL ATM antidust agent, dust-binding agent; ~ **biologique** *m* DIST EAU biological agent; ~ **complexant** *m* POLL EAU complexing agent; ~ **défoliant** *m* SOLS defoliating agent; ~ **déshydratant** *m* POLL ATM drying medium; ~ **dispersant** *m* ÉCHANT, POLL ATM, POLL EAU, POLL MAR, SOLS dispersant, dispersing agent; ~ **émulsifiant** *m* SOLS emulsifying agent; ~ **épaississant** *m* ÉN RENOUV bodying agent; ~ **d'épuration** *m* POLL ATM, SOLS scavenger; ~ **d'extraction** *m* SOLS extraction agent; ~ **filtrant** *m* POLL ATM, POLL EAU filter medium, filtering agent; ~ **de flottation** *m* POLL EAU, SOLS flotation agent; ~ **gonflant** *m* DÉCHETS blowing agent; ~ **de gonflement** *m* SOLS bulking agent; ~ **d'imprégnation** *m* DÉCHETS impregnating agent; ~ **de lavage** *m* POLL ATM washing agent; ~ **de lavage chargé** *m* POLL ATM loaded washing agent; ~ **de lavant recirculant** *m* POLL ATM recirculating washing agent; ~ **de liaison** *m* DIST EAU binding agent; ~ **méthanogène** *m* POLL EAU methane former; ~ **du milieu** *m* POLL ATM environmental medium; ~ **moteur** *m* POLL ATM propellant; ~ **mouillant** *m* POLL ATM, POLL MAR wetting agent; ~ **neutralisant** *m* DÉCHETS, POLL ATM, SÉCURITÉ neutralizing agent; ~ **neutralisateur** *m* DÉCHETS, POLL ATM, SÉCURITÉ neutralizing agent; ~ **nocif** *m* POLL EAU harmful agent; ~ **orange** *m* POLL ATM agent orange; ~ **oxydant** *m* POLL ATM, POLL EAU, SOLS oxidizing agent; ~ **pathogène** *m* DIST EAU pathogen, POLL EAU pathogen, clarifier; ~ **physique** *m* DIST EAU physical agent; ~ **polluant** *m* SÉCURITÉ polluting agent; ~ **de précipitation** *m* DÉCHETS, DIST EAU, POLL ATM coagulant; ~ **protecteur** *m* DIST EAU protective agent; ~ **réducteur** *m* ÉCHANT, POLL ATM, POLL EAU reducing agent; ~ **repousseur** *m* DIST EAU, POLL MAR herding agent; ~ **retardateur** *m* DIST EAU retarding agent; ~ **solidifiant** *m* DIST EAU, SOLS solidifying agent; ~ **de surfactage** *m* POLL MAR sinking agent; ~ **tensioactif** *m* POLL ATM, POLL MAR, SOLS surface-active agent, DIST EAU, POLL EAU, SOLS surfactant, POLL MAR sinking agent, surface-active agent; ~ **toxique** *m* DÉCHETS, ÉCHANT, POLL ATM, SÉCURITÉ, SOLS toxic agent; ~ **de trempe** *m* POLL ATM quencher; ~ **de trempe en suspension** *m* POLL ATM quencher suspension

agglomérant *m* DIST EAU binding agent

agglomérat *m* DÉCHETS, DIST EAU, POLL ATM, SOLS cluster; ~ **goutte-particule** POLL ATM particle-drop agglomerate

agglomérateur *m* DÉCHETS, DIST EAU, POLL ATM, SOLS agglomerator

agglomération *f* DÉCHETS, DIST EAU, POLL ATM, SOLS agglomeration; ~ **urbaine** GESTION, POL ENV urbanized area

aggloméré *m* DÉCHETS, DIST EAU, POLL ATM, SOLS agglomerate; ~ **de goudron** POLL MAR tar ball

agglomérer *vt* DÉCHETS, DIST EAU, POLL ATM, SOLS agglomerate

agglutinant *m* DÉCHETS, DIST EAU, POLL ATM binding agent

agglutination *f* DÉCHETS, DIST EAU, POLL ATM agglutination

agitateur *m* DÉCHETS agitator, POLL ATM agitating arm, SOLS agitator, stirrer; ~ **à immersion** ÉCHANT immersion stirrer; ~ **magnétique** POLL EAU *brassage des boues* magnetic stirrer

agiter *vt* ÉCHANT *flacon d'échantillonnage* shake

agrafage *m* DÉCHETS *de balles* clipping, stapling; ~ **latéral** DÉCHETS lateral stapling; ~ **sur coin** DÉCHETS corner stapling

agrafe *f* DÉCHETS staple

agrafer *vt* DÉCHETS *balles* clip, staple

agrafeuse: ~ **sur coin** *f* DÉCHETS corner stapling machine

agrégat *m* DIST EAU, POLL ATM, SOLS aggregate

agression: ~ **acide** *f* SOLS acid stress

agressivité: ~ **chimique** *f* DÉCHETS, DIST EAU chemical aggressivity

agri: ~-**énergie** *f* ÉN RENOUV energy farming

agriculture: ~ **biologique** *f* POL ENV, SOLS organic farming; ~ **industrielle** *f* POL ENV, SOLS factory farming; ~ **intensive** *f* POL ENV, SOLS intensive farming; ~ **sur brûlis** *f* POL ENV, POLL ATM slash-and-burn agriculture

agrippe: ~-**col** *m* DÉCHETS clip-on carrier

agro: ~-**écologie** *f* GESTION, POL ENV agro-ecology

ailette *f* ÉN RENOUV *de turbine* blade, POLL ATM *de ventilateur* vane, POLL EAU paddle

aimant *m* DÉCHETS, DIST EAU, ÉCHANT magnet; ~ **aérien** DÉCHETS overhead magnet~ **permanent** DÉCHETS, DIST EAU, ÉCHANT permanent magnet

aimantation[1]: ~ **rémanente** *f* POLL ATM retentivity

aimantation[2]: **à ~ permanente** *loc* DIST EAU permanent magnetic

air: **~ admis** *m* POLL ATM intake air; **~ ambiant** *m* POLL ATM ambient-air, air ambient; **~ d'appoint** *m* POLL ATM auxiliary air; **~ chargé de solvants** *m* POLL ATM, SÉCURITÉ solvent-laden air; **~ chargé en substances odoriférantes** *m* POLL ATM odor-substance-laden air (*AmE*), odour-substance-laden air (*BrE*); **~ de combustion** *m* POLL ATM combustion air; **~ de combustion préchauffé** *m* POLL ATM, SOLS preheated combustion air; **~ de combustion primaire** *m* DIST EAU, POLL ATM primary combustion air; **~ comprimé** *m* POLL ATM compressed air; **~ de désorption** *m* POLL ATM desorption air; **~ en excès** *m* POLL ATM excess air; **~ extérieur** *m* POLL ATM outer air; **~ impropre à la respiration** *m* POLL ATM, SÉCURITÉ air unfit for respiration; **~ irrespirable** *m* POLL ATM, SÉCURITÉ air unfit for respiration; **~ des lieux de travail** *m* POLL ATM, SÉCURITÉ workplace atmosphere; **~ oxydant** *m* POLL ATM oxidation air; **~ poussiéreux** *m* POLL ATM, SÉCURITÉ dust-laden air; **~ primaire** *m* POLL ATM primary air; **~ pur** *m* DÉCHETS, DIST EAU, POLL ATM clean air; **~ de refroidissement** *m* POLL ATM cooling air; **~ résiduel épuré** *m* POLL ATM cleaned waste air; **~ respirable** *m* POLL ATM, SÉCURITÉ breathable air; **~ sec** *m* POLL ATM dry air; **~ secondaire** *m* POLL ATM, POLL EAU secondary air

aire: **~ de décharge** *f* DÉCHETS landfill surface; **~ de loisirs** *f* POL ENV recreation area; **~ de séchage des boues** *f* POLL EAU sludge-drying area; **~ de stockage** *f* DÉCHETS storage area

ajoutoir *m* POLL ATM nozzle pipe

ajutage *m* DIST EAU jet, nozzle, POLL EAU jet

Al *abrév* (*aluminium*) DIST EAU, SOLS Al (*aluminium BrE, aluminum AmE*)

alarme: **~ antidéflagrante** *f* ÉCHANT explosion warning device

albédo *m* POLL ATM albedo

albumine *f* POLL ATM albumen

alcalin *adj* DIST EAU, POLL ATM, POLL EAU, SOLS alkaline

alcalinité *f* DIST EAU, POLL ATM, POLL EAU, SOLS alkalinity; **~ initiale** POLL ATM pre-acidification alkalinity

alcane *m* POLL ATM alkane

alcène *m* POLL ATM alkene

alcool *m* POLL ATM, POLL EAU alcohol; **~ gras** POLL EAU fatty alcohol

alerte: **~ au brouillard photochimique** *f* POLL ATM smog alarm, smog warning; **~ aux fuites** *f* SÉCURITÉ leakage warning; **~ à l'ozone** *f* POLL ATM ozone warning; **~ au smog** *f* POLL ATM smog alarm, smog warning

algue *f* DIST EAU alga, POLL MAR alga, seaweed

algues *f pl* DIST EAU algae, POLL MAR algae, seaweed; **~ marines** POLL MAR seaweed; **~ rouges** DIST EAU red algae

alicyclique *adj* POLL ATM alicyclic

alimentation *f* DIST EAU supply, feed, ÉN RENOUV, POLL ATM supply, POLL EAU supply, feeding; **~ basse pression** DIST EAU low-pressure supply; **~ en eau** DIST EAU water supply; **~ en eau à distance** DIST EAU distant water supply; **~ en eau potable** DIST EAU drinking-water supply; **~ étagée en combustible** POLL ATM staged fuel feeding; **~ manuelle** DÉCHETS manual feed; **~ en oxygène** POLL EAU oxygen supply; **~ en précipitats** POLL EAU precipitate feeding

alimenté: **~ à l'énergie solaire** *adj* ÉN RENOUV solar-powered; **~ par des combustibles fossiles** *adj* ÉN RENOUV fossil-fuel driven, fossil-fuel fired; **~ par l'énergie photovoltaïque** *adj* ÉN RENOUV photovoltaic-powered; **~ par l'énergie solaire** *adj* ÉN RENOUV charged with solar power, solar-driven; **~ par gravité** *adj* ÉN RENOUV gravity-fed; **~ par des piles** *adj* GÉNÉRAL battery-powered

alimenter *vt* DIST EAU, ÉN RENOUV, POLL ATM, POLL EAU feed, supply, SOLS feed, feed in; **~ en eau** DIST EAU water; **~ tangentiellement** POLL ATM feed tangentially

aliments: **~ protéiques** *m pl* ÉN RENOUV protein feed

aliphatique *adj* DÉCHETS, POLL ATM aliphatic

alkylène *m* POLL ATM , alkylene

allège *f* POLL MAR lighter

allègement *m* POLL MAR lightening

alléger *vt* POLL MAR lighten

alliage: **~ au platine** *m* POLL ATM platinum alloy

allochtone *adj* DIST EAU allochthonous

allumer *vt* DÉCHETS, POLL ATM, POLL EAU ignite

allumoir *m* POLL MAR lighter

alluvionnement *m* DIST EAU aggradation

alluvions *f pl* DIST EAU alluvial deposit; **~ fluviales** DIST EAU fluvial alluvium

altération *f* DIST EAU, POLL MAR, SOLS *par les agents atmosphériques* weathering, POLL SON impairment

altéré *adj* POLL SON impaired, damaged

altérer *vt* POLL SON impair, damage

alternateur *m* ÉN RENOUV generator; **~ asynchrone** ÉN RENOUV induction generator; **~ haute tension** ÉN RENOUV high-voltage generator; **~ synchrone** ÉN RENOUV synchronous generator

alterner *vt* POLL ATM alternate

alumine: **~ activée** *f* POLL ATM activated alumina

aluminium *m* (*Al*) DIST EAU, SOLS aluminium (*BrE*) (*Al*), aluminum (*AmE*) (*Al*); **~ de première fusion** DIST EAU primary aluminium (*BrE*), primary aluminum (*AmE*); **~ de récupération** DIST EAU secondary aluminium (*BrE*), secondary aluminum (*AmE*); **~ usagé** DIST EAU used aluminium (*BrE*), used aluminum (*AmE*)

alvéole: **~ de décharge** *m* DÉCHETS landfill cell

âme *f* ÉN RENOUV *de bobine* core

amélioration *f* POLL ATM upgrading, SOLS amelioration

aménagement *m* ÉN RENOUV *de centrale solaire* installation, POL ENV planning; **~ des cours d'eau** DIST EAU river works; **~ d'une décharge** DÉCHETS *installation* landfill building, landfill installation; **~ de décharge** DÉCHETS *projet* waste landfill planning; **~ de l'environnement** POL ENV environmental planning; **~ intérieur d'une décharge** DÉCHETS landfill structure; **~ du paysage** ÉN RENOUV landscaping; **~ d'une pente** DIST EAU slope recess; **~ rural** POL ENV rural development; **~ d'un site** DIST EAU site operation; **~ d'une station d'épuration d'eaux usées** POLL EAU wastewater treatment plant installation; **~ du territoire** POL ENV town and country planning

aménager *vt* ÉN RENOUV *ressource* harness, POL ENV *site* develop

amende *f* POL ENV fine

amendement *m* SOLS soil conditioning, soil improvement; **~ aux électrons** SOLS electron amelioration

amenée *f* POLL EAU *température d'admission* inlet

amenuisement: **~ de la couche d'ozone** *m* POLL ATM ozone shrinkage

ameublir *vt* DIST EAU, SOLS loosen

amiante *f* POLL ATM, SÉCURITÉ asbestos

Amis: **~ de la Terre** *m pl* POL ENV Friends Of The Earth

ammoniac *m* DÉCHETS, DIST EAU, POLL ATM, SOLS ammonia; **~ total** POLL EAU total ammonia

ammoniaque *m* DÉCHETS, DIST EAU, POLL ATM, SOLS ammonia; **~ totale** POLL EAU total ammonia

ammonisation *f* DÉCHETS, DIST EAU, POLL ATM, SOLS ammoniation

amont[1]: **d'~** *adj* DIST EAU, POLL ATM upstream

amont[2]: **en ~** *adv* DIST EAU, POLL ATM upstream

amorçage *m* DIST EAU, POLL EAU *d'une pompe* priming; **~ d'arc** POLL ATM arc-over

amorcer *vt* DIST EAU, POLL EAU *pompe* prime

amorceur *m* DIST EAU priming cock

amorphe *adj* POLL ATM amorphous

amortir *vt* ÉCHANT, POLL ATM buffer, POLL SON *vibrations*

attenuate, cushion, damp, *ventilateur* silence, *son* attenuate, silence, deaden
amortissement *m* ÉCHANT, POLL ATM buffering, POLL SON *vibrations* attenuation, cushioning, damping, *ventilateur* silencing, *son* attenuation, silencing, deadening; **~ par friction d'air** ÉCHANT air friction damping; **~ du son** POLL SON sound damping, sound deadening; **~ sonore** POLL SON sound damping, sound deadening; **~ des vibrations** POLL SON vibration damping
amortisseur *m* ÉCHANT, POLL ATM, POLL SON damper; **~ absorbant le son** POLL SON sound-absorbent damper; **~ acoustique** POLL SON sound damper; **~ de bruit** POLL SON sound deadener, sound-absorbent damper
ampèremètre *m* ÉCHANT ammeter
ampèremétrie *f* ÉCHANT amperometry
ampèremétrique *adj* ÉCHANT amperometric
amplificateur *m* ÉCHANT amplifier; **~ de mesure** ÉCHANT measuring amplifier
amplitude: **~ de la marée** *f* ÉN RENOUV tidal range
ampoule *f* DÉCHETS ampoule; **~ graduée** ÉCHANT volumetric flask
amylacé *adj* ÉN RENOUV starchy
anaérobie *adj* DIST EAU anaerobic, oxygen-free
analysable *adj* ÉCHANT analysable (*BrE*), analyzable (*AmE*)
analyse *f* ÉCHANT analysis; **~ de l'air ambiant** ÉCHANT ambient-air analysis; **~ arbitraire** ÉCHANT arbitrary analysis; **~ de bout en bout** GESTION, POL ENV cradle-to-grave analysis; **~ de carotte** ÉCHANT *géologie* drilling-core analysis; **~ du champ périphérique** ÉCHANT surrounding-field analysis; **~ colorimétrique** ÉCHANT color analysis (*AmE*), colour analysis (*BrE*); **~ des composés organo-halogénés adsorbables** ÉCHANT AOX analysis; **~ conductimétrique** ÉCHANT conductometric analysis; **~ continu** ÉCHANT, POLL EAU on-line analysis; **~ de la couleur** ÉCHANT color analysis (*AmE*), colour analysis (*BrE*); **~ coût-avantage** GESTION cost-benefit analysis; **~ des coûts** GESTION cost analysis; **~ du cycle de vie** GESTION, POL ENV cradle-to-grave analysis, life-cycle analysis (*LCA*); **~ des déchets** ÉCHANT analysis of residue; **~ de diffusion** ÉCHANT *chimie* diffusion analysis; **~ de dilution** ÉCHANT *olfactométrie* dilution analysis; **~ des données** ÉCHANT data analysis; **~ de l'eau** ÉCHANT water analysis; **~ de l'eau potable** ÉCHANT drinking water analysis; **~ d'échantillon** ÉCHANT sample analysis; **~ élémentaire** ÉCHANT elemental analysis (*EA*); **~ d'émission de rayons X induite par des protons** ÉCHANT PIXE analysis, proton-induced X-ray emission analysis; **~ des émissions** ÉCHANT emissions analysis; **~ d'émissions acoustiques** ÉCHANT acoustic emission analysis, AE analysis; **~ de l'environnement** ÉCHANT, GESTION environmental analysis; **~ des fluorures dans les eaux usées** ÉCHANT fluoride analysis in wastewater; **~ de gaz de décharge** ÉCHANT landfill gas analysis; **~ de gaz à long terme** ÉCHANT long-term gas analysis; **~ de gaz volumétrique** ÉCHANT volumetric gas analysis; **~ granulométrique** ÉCHANT grain-size analysis, screen analysis; **~ gravimétrique** ÉCHANT gravimetric analysis; **~ gravimétrique des gaz** ÉCHANT gravimetric gas analysis; **~ hors site** ÉCHANT off-site analysis; **~ hydrogéologique** DIST EAU, ÉCHANT, SOLS hydrogeological analysis; **~ d'image** ÉCHANT image analysis; **~ immédiate** ÉCHANT proximate analysis; **~ d'impact** GESTION impact analysis; **~ in vitro** DIST EAU in vitro analysis; **~ in vivo** DIST EAU in vivo analysis; **~ instrumentale** ÉCHANT instrumental analysis; **~ intégrale** ÉCHANT full analysis; **~ d'une ligne de produits** GESTION product-line analysis; **~ de masse avec microsonde laser** ÉCHANT laser microprobe mass analysis; **~ des métaux lourds** ÉCHANT heavy-metal analysis; **~ du milieu atmosphérique** ÉCHANT ambient-air analysis; **~ multicritère** DIST EAU multicriteria analysis (*MCA*); **~ néphélométrique** ÉCHANT nephelometric analysis; **~ d'odeur** ÉCHANT odor analysis (*AmE*), odour analysis (*BrE*); **~ de l'odeur des eaux usées** DÉCHETS, ÉCHANT odor analysis in wastewater (*AmE*), odour analysis in wastewater (*BrE*); **~ olfactive** ÉCHANT odor analysis (*AmE*), odour analysis (*BrE*); **~ olfactive des eaux usées** DÉCHETS, ÉCHANT odor analysis in wastewater (*AmE*), odour analysis in wastewater (*BrE*); **~ olfactive qualitative** ÉCHANT qualitative odor analysis (*AmE*), qualitative odour analysis (*BrE*); **~ olfactive quantitative** ÉCHANT quantitative odor analysis (*AmE*), quantitative odour analysis (*BrE*); **~ des ordures ménagères** ÉCHANT domestic waste analysis; **~ par activation** ÉCHANT activation analysis; **~ par dilution** ÉCHANT dilution analysis; **~ par élément** ÉCHANT element-specific analysis; **~ par fluorescence** ÉCHANT fluorescence analysis; **~ par goutte** ÉCHANT spot analysis; **~ par séparation électrochimique** ÉCHANT electrochemical stripping analysis (*ESA*); **~ par sorption** ÉCHANT sorption analysis; **~ par tamisage** ÉCHANT sieve analysis; **~ par triage** ÉCHANT sorting analysis; **~ de phase** ÉCHANT phase analysis; **~ des points faibles** ÉCHANT analysis of weakest points; **~ préliminaire** ÉCHANT preliminary analysis; **~ de pression de vapeur** ÉCHANT vapor pressure analysis (*AmE*), vapour pressure analysis (*BrE*); **~ principale** ÉCHANT main analysis; **~ de profil environnemental** POL ENV environmental profile analysis; **~ qualitative** ÉCHANT qualitative analysis; **~ qualitative des gaz** ÉCHANT qualitative gas analysis; **~ qualitative des odeurs** ÉCHANT qualitative odor analysis (*AmE*), qualitative odour analysis (*BrE*); **~ quantitative** ÉCHANT quantitative analysis; **~ quantitative des gaz** ÉCHANT quantitative gas analysis; **~ quantitative des odeurs** ÉCHANT quantitative odor analysis (*AmE*), quantitative odour analysis (*BrE*); **~ de rendement** ÉCHANT efficiency analysis; **~ de risque** DIST EAU, POL ENV risk analysis; **~ de routine** ÉCHANT routine analysis; **~ sensorielle** ÉCHANT sensory analysis; **~ du sol** ÉCHANT, SOLS soil analysis; **~ spécifique** ÉCHANT element-specific analysis; **~ spectrale** ÉCHANT spectral analysis; **~ d'une substance isolée** ÉCHANT individual substance analysis; **~ de substances dangereuses** ÉCHANT harmful substances analysis; **~ sur échantillon** ÉCHANT sample analysis; **~ sur site** ÉCHANT, SOLS on-site analysis; **~ sur le terrain** ÉCHANT, SOLS field analysis; **~ sur la totalité du cycle** GESTION, POL ENV cradle-to-grave analysis, life-cycle analysis (*LCA*); **~ de sûreté** SÉCURITÉ safety analysis; **~ au tamis** ÉCHANT sieve analysis; **~ thermique** ÉCHANT thermal analysis; **~ thermique différentielle** (*ATD*) ÉCHANT differential thermal analysis (*DTA*); **~ totale** ÉCHANT total analysis; **~ en trace** ÉCHANT *des éléments* trace analysis; **~ turbidimétrique** ÉCHANT turbidimetric analysis; **~ de la valeur d'usage** GESTION use-value analysis; **~ visuelle** ÉCHANT visual analysis; **~ de volatilité** ÉCHANT volatility analysis; **~ volumétrique** ÉCHANT volumetric analysis, headspace analysis; **~ volumétrique par titrage** ÉCHANT volumetric analysis by titration; **~ volumétrique statique** ÉCHANT static headspace analysis (*SHS*)
analyser *vt* ÉCHANT analyse (*BrE*), analyze (*AmE*); **~ par oxydimétrie** ÉCHANT analyse oxidimetrically (*BrE*), analyze oxidimetrically (*AmE*)
analyseur *m* ÉCHANT analyser (*BrE*), analyzer (*AmE*); **~ à balayage** ÉCHANT scanner; **~ de carbone organique total** ÉCHANT total organic carbon analyser (*BrE*) (*TOC analyser*), total organic carbon analyzer (*AmE*) (*TOC analyzer*); **~ de composés organiques volatils à infrarouge** ÉCHANT infrared volatile organic compound analyser (*BrE*) (*IR-VOC analyser*), infrared volatile organic compound analyzer (*AmE*) (*IR-VOC analyzer*); **~ en continu** ÉCHANT, POLL EAU on-line analyser (*BrE*), on-line analyzer (*AmE*); **~ coulométrique** ÉCHANT coulometric analyser (*BrE*), coulometric analyzer (*AmE*); **~ des eaux usées** ÉCHANT

wastewater analyser (*BrE*), wastewater analyzer (*AmE*); **~ de fluorescence X portable** ÉCHANT portable X-ray fluorescence analyser (*BrE*), portable X-ray fluorescence analyzer (*AmE*); **~ de fréquence** ÉCHANT frequency analyser (*BrE*), frequency analyzer (*AmE*); **~ de fumées** ÉCHANT flue-gas tester; **~ de fumées type Orsat** ÉCHANT Orsat gas analyser (*BrE*), Orsat gas analyzer (*AmE*); **~ de gaz** ÉCHANT gas analyser (*BrE*), gas analyzer (*AmE*); **~ de gaz à l'état de trace** ÉCHANT gas trace analyser (*BrE*), gas trace analyzer (*AmE*); **~ de gaz à infrarouge** ÉCHANT infrared gas analyser (*BrE*), infrared gas analyzer (*AmE*); **~ de gaz spécifique d'un composant** ÉCHANT component-specific gas analyser (*BrE*), component-specific gas analyzer (*AmE*); **~ granulométrique** ÉCHANT particle size analyser (*BrE*), particle size analyzer (*AmE*); **~ informatique** ÉCHANT computerized analyser (*BrE*), computerized analyzer (*AmE*); **~ d'infrarouge non dispersif** ÉCHANT nondispersive infrared analyser (*BrE*), nondispersive infrared analyzer (*AmE*); **~ d'intensité acoustique** ÉCHANT sound intensity analyser (*BrE*), sound intensity analyzer (*AmE*); **~ magnétique d'oxygène** ÉCHANT magnetic oxygen analyser (*BrE*), magnetic oxygen analyzer (*AmE*); **~ de masse quadripolaire** ÉCHANT quadrupole-mass analyser (*BrE*), quadrupole-mass analyzer (*AmE*); **~ de niveau sonore** ÉCHANT, POLL SON noise level analyser (*BrE*), noise level analyzer (*AmE*); **~ d'ozone** ÉCHANT ozone analyser (*BrE*), ozone analyzer (*AmE*); **~ portable portatif** ÉCHANT portable hand-held analyser (*BrE*), portable hand-held analyzer (*AmE*); **~ de la qualité atmosphérique** ÉCHANT ambient-air quality analyser (*BrE*), ambient-air quality analyzer (*AmE*); **~ séquentiel** ÉCHANT sequential analyser (*BrE*), sequential analyzer (*AmE*); **~ en temps réel** ÉCHANT real-time analyser (*BrE*), real-time analyzer (*AmE*)

analytique *adj* ÉCHANT analytical

analytiquement: **~ pur** *adj* ÉCHANT analytically-pure

anémomètre *m* ÉCHANT anemometer, wind sensor; **~ à coupelles** ÉCHANT cup anemometer; **~ à hélice** ÉCHANT *anémomètre avec girouette* impeller anemometer; **~ à résistance électrique** ÉCHANT hot-wire anemometer

ange: **~ bleu** *m* POL ENV *éco-étiquetage en Allemagne* blue angel

anhydre *adj* DIST EAU anhydrous, water-free

anhydride: **~ carbonique** *m* (CO_2) DIST EAU carbonic anhydride (CO_2); **~ hypo-azoteux** *m* (N_2O) POLL EAU nitrous oxide (N_2O); **~ nitrique** *m* (N_2O_5) POLL EAU nitrogen pentoxide (N_2O_5), dinitrogen pentoxide (N_2O_5); **~ sulfureux** *m* (SO_2) DIST EAU sulfur anhydride (*AmE*) (SO_2), sulphur anhydride (*BrE*) (SO_2); **~ sulfurique** *m* (SO_3) DIST EAU sulfuric anhydride (*AmE*) (SO_3), sulphuric anhydride (*BrE*) (SO_3)

anhydrite *m* POLL ATM anhydrite; **~ d'azote** (N_2O_5) POLL EAU nitrogen pentoxide (N_2O_5), dinitrogen pentoxide (N_2O_5)

anneau: **~ en béton** *m* DÉCHETS *pour tubage* concrete ring of tubings; **~ correspondant** *m* POLL SON attached shell; **~ d'écoulement** *m* DIST EAU volute insert; **~ de Pall** *m* POLL ATM *garnissage de colonne* pall ring; **~ Raschig** *m* POLL ATM *garnissage de colonne* Raschig ring; **~ support** *m* POLL ATM support ring

annuler *vt* POL ENV repeal

anode *f* POLL ATM, SOLS anode

anodin *adj* POLL ATM innocuous

anthracite *f* POLL ATM glance coal, hard coal

anthropique *adj* POLL ATM, SOLS anthropogenic, anthropogenically-caused, human-induced

anthroposphère *f* SOLS anthroposphere

antibruit *adj* POLL SON antinoise

anti-colmatage *adj* DIST EAU, POLL EAU nonclogging

anticorrosion *adj* ÉN RENOUV corrosion-preventive, POLL ATM corrosion-resistant

antidéflagrant *adj* DÉCHETS explosion-proof

antigélif *m* DÉCHETS *béton* frost-preventive agent

antimoine *m* (*Sb*) DIST EAU antimony (*Sb*)

antimousse *m* DIST EAU antifoaming agent

antirouille *adj* DIST EAU rust-preventive, rust-resistant

anti-vieillissement *adj* DÉCHETS ageing-resistant

appareil *m* GÉNÉRAL apparatus, device; **~ d'absorption nu** POLL ATM absorption apparatus without internal fittings; **~ d'absorption sans aménagement interne** POLL ATM absorption apparatus without internal fittings; **~ d'assainissement de l'air** POLL ATM air purifier; **~ déprimogène** DIST EAU suction pump, ÉCHANT pressure-difference device; **~ d'élutriation** DÉCHETS, ÉCHANT, POLL ATM, POLL EAU elutriator; **~ d'exhaure** DIST EAU *mines* pumping plant; **~ de lavage par voie humide** POLL ATM wet-scrubbing device; **~ de mesure électrique** SÉCURITÉ electric measuring apparatus; **~ de mesure de l'exposition au bruit** ÉCHANT noise exposure meter; **~ de mesure des fumées d'Orsat** ÉCHANT Orsat apparatus; **~ de mesure industriel** ÉCHANT commercially-available measuring device; **~ de mesure de la teneur en hydrocarbures** POLL EAU, SOLS oil-content meter; **~ de mesure des traces** ÉCHANT trace-measuring device; **~ de précombustion** SOLS precombustor; **~ de protection respiratoire** SÉCURITÉ respiratory-protection apparatus; **~ à pulvérisation d'eau** DIST EAU pulverizer; **~ de récupération** DIST EAU recovery device; **~ de remplissage** SOLS filler; **~ respiratoire** SÉCURITÉ air breathing apparatus; **~ respiratoire fourni** SÉCURITÉ supplied-air breathing apparatus; **~ respiratoire industriel à air comprimé** SÉCURITÉ compressed-air industrial breathing apparatus; **~ respiratoire à oxygène** SÉCURITÉ oxygen breathing apparatus; **~ de tri** DÉCHETS checking apparatus

apparition: **~ d'odeurs** *f* POLL ATM, POLL EAU odor formation (*AmE*), odour formation (*BrE*)

appauvrissement: **~ de la couche d'ozone** *m* POLL ATM ozone-layer depletion; **~ de la couche d'ozone stratosphérique** *m* POLL ATM stratospheric ozone-layer depletion; **~ en oxygène** *m* ÉCHANT, POLL ATM, POLL EAU oxygen depletion; **~ en ozone** *m* POLL ATM ozone depletion; **~ du sol** *m* SOLS soil exhaustion

applicabilité *f* DIST EAU, ÉN RENOUV, POLL ATM applicability

application *f* GÉNÉRAL application, implementation; **~ agricole** POLL EAU land application; **~ de chauffage** ÉN RENOUV heating purpose; **~ à l'échelle industrielle** ÉN RENOUV, POLL ATM commercial-scale application; **~ énergétique** DÉCHETS energy use, ÉN RENOUV energy purpose, energy use, POLL ATM energy use; **~ de l'énergie solaire** ÉN RENOUV solar energy use; **~ de l'énergie solaire à grande échelle** ÉN RENOUV large area use of solar energy; **~ de ferraille** DÉCHETS scrap application; **~ à grande échelle** ÉN RENOUV large area use; **~ de papier recyclé** DIST EAU recycled paper application; **~ photoélectrique décentralisée** ÉN RENOUV decentralized photoelectric use; **~ de plastique** DIST EAU plastics application; **~ sur place** SOLS on-site application; **~ en surface** POLL EAU surface application; **~ technologique** ÉN RENOUV, POLL ATM, POLL MAR, POLL SON, SÉCURITÉ, SOLS technology application; **~ de verre recyclé** DIST EAU recycled-glass application

appliquer *vt* GÉNÉRAL apply, implement

apport *m* POLL ATM, POLL EAU intake; **~ d'air** POLL ATM air supply; **~ d'eau** DIST EAU water inflow, influent, POLL EAU influent; **~ d'eau douce** DÉCHETS freshwater introduction; **~ extérieur** DIST EAU, POLL ATM external input; **~ d'oxygène** DIST EAU, ÉN RENOUV oxygen introduction, POLL EAU oxygen uptake

apposition: **~ d'un code à barres** *f* DÉCHETS, DIST EAU *identification des produits* bar coding

approche: **~ sectorielle** *f* POL ENV sectoral approach

approvisionnement *m* DIST EAU, ÉN RENOUV, POLL ATM, POLL EAU delivery, supply; **~ collectif en eau potable** DIST EAU public drinking-water supply; **~ en eau dans le secteur**

résidentiel DIST EAU domestic water supply; **~ en eau douce** POLL ATM freshwater supply; **~ en énergie** POLL ATM energy supply; **~ d'énergie classique** ÉN RENOUV conventional energy supply; **~ mondial en énergie** ÉN RENOUV global energy supply; **~ en oxygène** DIST EAU, POLL ATM oxygen supply; **~ en oxygène stœchiométrique** POLL ATM stoichiometric oxygen supply
approvisionner *vt* DIST EAU, ÉN RENOUV, POLL ATM, POLL EAU supply
aptitude: **~ à l'emploi** *f* GÉNÉRAL usability; **~ à la mise en décharge** *f* POLL EAU landfilling ability; **~ à l'usage** *f* GÉNÉRAL usability
AQ *abrév* (*assurance-qualité*) GESTION QA (*quality assurance*)
aquaculture *f* DIST EAU aquaculture
aqueduc *m* DIST EAU aqueduct
aqueux *adj* DIST EAU aqueous
aquiclude *f* DIST EAU aquiclude
aquiculture *f* DIST EAU aquaculture
aquifère[1] *adj* DIST EAU aquiferous
aquifère[2] *m* DIST EAU aquifer; **~ captif** DIST EAU confined aquifer
aquifuge *m* DIST EAU aquifuge
Ar *abrév* (*argon*) DÉCHETS, DIST EAU, POLL ATM Ar (*argon*)
arbre: **~ d'admission** *m* DIST EAU, SOLS entrance shaft; **~ barillet** *m* DIST EAU, POLL EAU *pompe HP* pump-plunger shaft; **~ de commande** *m* DÉCHETS, ÉN RENOUV control shaft; **~ de pompe** *m* DIST EAU, POLL EAU *mines* pump shaft; **~ de rotor** *m* ÉN RENOUV rotor hub; **~ de rotor équipé d'un mécanisme de réglage des pales** *m* ÉN RENOUV rotor hub with adjustable blade mechanism
architecte: **~-paysagiste** *m* GESTION landscape architect
architecture: **~ solaire** *f* ÉN RENOUV solar architecture
argile: **~ marneuse** *f* DIST EAU marly clay
argon *m* (*Ar*) DÉCHETS, DIST EAU, POLL ATM argon (*Ar*)
armature: **~ de pompe** *f* POLL EAU pump gear
armer *vt* DIST EAU, POLL EAU reinforce
aromatique *adj* DIST EAU, POLL ATM aromatic
arrachage *m* POLL EAU *industrie papetière* flaking
arrêt: **~ d'implantation** *m* DIST EAU settlement stoppage; **~ par épuisement** *m* DÉCHETS, POLL ATM *nucléaire* burnout; **~ de pourrissage** *m* DIST EAU rotting standstill; **~ du tassement** *m* SOLS settlement stoppage
arrosage *m* DIST EAU, POLL EAU watering, spraying
arroseur *m* POLL EAU sparger; **~ rotatif** DIST EAU rotating sprayer
arsenic *m* (*As*) SOLS arsenic (*As*)
article: **~ en cellulose moulée** *m* DÉCHETS molded pulp article (*AmE*), moulded pulp article (*BrE*)
artificiel *adj* DIST EAU artificial, SOLS man-made
As *abrév* (*arsenic*) SOLS As (*arsenic*)
asbestose *m* POLL ATM asbestosis
ascension: **~ du panache** *f* POLL ATM plume rise
aseptique *adj* DÉCHETS aseptic
aspect: **~ des déchets** *m* DIST EAU appearance of waste
asperseur *m* POLL EAU *agriculture* sprinkler; **~ longue portée** POLL EAU long-throw sprinkler
aspersion *f* POLL EAU sprinkling, spraying; **~ à l'air** SOLS air sparging; **~ d'eaux usées** POLL EAU wastewater sprinkling; **~ de liquide de lavage** POLL ATM scrubbing liquid jet
asphaltisation *f* DÉCHETS, DIST EAU asphalt sealing
aspirateur: **~ central** *m* SÉCURITÉ central vacuum-cleaning system; **~ et filtre pour poussières et copeaux** *m* SÉCURITÉ suction and filter installation for dust and chippings; **~ à goulet** *m* POLL MAR gulley sucker
aspiration *f* DIST EAU, POLL ATM, POLL EAU suction, SOLS *de courant* drawing; **~ d'air pur** POLL ATM clean-air suction; **~ des gaz** DÉCHETS gas suction; **~ isocinétique** ÉCHANT isokinetic suction
aspirer *vt* DIST EAU, POLL ATM, POLL EAU suck in
assainir *vt* POLL EAU cleanse, SOLS remediate, rehabilitate, *vieille décharge* process, *sol* clean up
assainissement *m* POLL EAU cleansing, SOLS remediation, rehabilitation, *d'une vieille décharge* processing, *du sol* clean-up; **~ biologique** SOLS bio-remediation; **~ biologique sur place** SOLS in-situ bio-remediation; **~ électrocinétique** SOLS electrokinetic remediation; **~ final** SOLS final remediation; **~ du milieu** POL ENV environmental sanitation; **~ partiel du sol** SOLS partial soil remediation
asséché *adj* POLL EAU *produit* dewatered
assèchement *m* GÉNÉRAL dewatering, drainage, *avec une pompe* pumping, pumping out
assécher[1] *vt* GÉNÉRAL dewater, drain, *avec une pompe* pump, pump out
assécher[2]: **s'~** *v réfl* DIST EAU run dry
asservissement *m* DIST EAU *d'une pompe* controls, POLL EAU *d'un mécanisme* interlocking
assiette: **~ des redevances** *f* POL ENV basis for charges, charges quota
Association: **~ Française de Normalisation** *f* (*AFNOR*) ÉCHANT, POL ENV French Standards Association (*AFNOR*)
associer: **s'~** *v réfl* DIST EAU, POLL ATM coalesce
assourdissement *m* POLL SON, SÉCURITÉ noise abatement
assurance: **~ contre les risques professionnels** *f* SÉCURITÉ industrial insurance; **~ professionnelle** *f* SÉCURITÉ industrial insurance; **~-qualité** *f* (*AQ*) GESTION quality assurance (*QA*)
ATD *abrév* (*analyse thermique différentielle*) ÉCHANT DTA (*differential thermal analysis*)
atlas: **~ de l'environnement** *m* POL ENV environment atlas
atmosphère *f* POLL ATM atmosphere; **~ ambiante** POLL ATM ambient atmosphere; **~ contrôlée** DÉCHETS, DIST EAU controlled atmosphere (*CA*); **~ présentant des risques d'explosion** SÉCURITÉ explosive atmosphere; **~ terrestre** POLL ATM earth's atmosphere
atmosphérique *adj* POLL ATM airborne, atmospheric
atomisation *f* GÉNÉRAL *d'un liquide* atomization, spraying, *d'un solide* atomization, pulverizing
atomiser *vt* GÉNÉRAL *liquide* atomize, spray, *solide* atomize, pulverize
atomiseur *m* GÉNÉRAL atomizer, spray; **~ à buse** POLL ATM nozzle atomizer
atteinte[1]: **~ à l'environnement** *f* POL ENV, POLL ATM, SOLS environmental damage; **~ portée à la végétation** *f* ÉN RENOUV damage to vegetation; **~ à la santé** *f* SÉCURITÉ damage to health
atteinte[2]: **porter ~ à** *vt* GÉNÉRAL damage
attenant *adj* SOLS close-coupled
atténuateur *m* POLL SON attenuator; **~ d'absorption** POLL SON absorptive attenuator; **~ réactif** POLL SON reactive attenuator
atténuation *f* ÉCHANT attenuation, weakening, POL ENV *d'une décision* mitigation, POLL SON attenuation; **~ du son** POLL SON sound attenuation; **~ des vibrations** POLL SON vibration attenuation
atténuer *vt* ÉCHANT attenuate, weaken, POL ENV *décision* mitigate, POLL SON attenuate
attraction: **~ électrostatique** *f* POLL ATM electrostatic attraction; **~ universelle** *f* DIST EAU, ÉN RENOUV, POLL EAU gravitational force
attribuer *vt* ÉCHANT apportion
attribution: **~ d'une classe** *f* DÉCHETS grading
attrition *f* POLL ATM attrition
aube *f* POLL EAU paddle blade; **~ directrice** DIST EAU *aux turbines*, ÉN RENOUV, POLL ATM guide vane; **~ de turbine** ÉN RENOUV turbine blade
audibilité *f* POLL SON audibility
audible *adj* POLL SON audible
audiomètre *m* ÉCHANT, SÉCURITÉ audiometer
audiométrie: **~ de seuil** *f* SÉCURITÉ threshold audiometry

audit: **~ de l'environnement** *m* DIST EAU, GESTION, POL ENV eco-audit, environmental audit; **~ pilote** *m* POL ENV pilot audit; **~ du système de management environnemental** *m* GESTION environmental management system audit
audité *m* GESTION auditee
auditeur *m* GESTION auditor
audition *f* POL ENV public hearing
augmentation: **~ du pH** *f* SOLS pH-value increase; **~ de pression** *f* POLL ATM pressure increase; **~ du prix de l'énergie** *f* ÉN RENOUV rising energy prices; **~ de la valeur du pH** *f* POLL EAU rising of pH-value
auto-échantillonnage *m* ÉCHANT autosampling
auto-échantillonneur *m* ÉCHANT autosampler
autoécologie *f* GESTION, POL ENV autoecology
auto-épuration *f* DIST EAU, POLL EAU, SOLS self-purification; **~ naturelle** DIST EAU, POLL EAU natural self-purification, self-purifying capacity
automobile: **~ à faible consommation** *f* POLL ATM low-energy automobile
autonettoyage *m* POLL EAU self-cleaning
autonettoyant *adj* POLL EAU self-cleaning
autonome *adj* ÉCHANT off-line
autorisation *f* POL ENV *action* permission, authorization, *document* licence (*BrE*), license (*AmE*), permit; **~ d'appel** DIST EAU appeal authorization; **~ partielle** POL ENV partial authorization
autoriser *vt* POL ENV authorize, permit
autorité: **~ compétente** *f* POL ENV competent authority, responsible authority; **~ de tutelle** *f* POL ENV supervisory authority
autosurveillance *f* POL ENV self-monitoring
autotherme *adj* POLL ATM autothermic
aval[1]: **d'~** *adj* DIST EAU, POLL ATM downstream
aval[2]: **en ~** *adv* DIST EAU, POLL ATM downstream
avance *f* DIST EAU *de la marée* priming
aven *m* POLL EAU sinkhole, swallow hole
avion: **~ d'observation** *m* POLL MAR spotter plane
avis: **~ d'expert en matière d'environnement** *m* POL ENV environmental expert opinion
axe: **~ focal** *m* ÉN RENOUV *énergie solaire* focal line; **~ de mesure** *m* ÉCHANT measuring shaft; **~ du panache** *m* POLL ATM plume axis
axé: **~ sur les effets** *adj* SOLS effect-oriented; **~ sur l'élimination** *adj* DÉCHETS disposal-oriented; **~ sur l'environnement** *adj* GESTION, POL ENV ecology-oriented, environmentally-oriented; **~ sur la protection de l'environnement** *adj* GESTION, POL ENV environmental-protection-oriented; **~ sur le recyclage** *adj* DÉCHETS, DIST EAU recycling-oriented; **~ sur les sources** *adj* SOLS source-oriented
axial *adj* POLL ATM axial
azote *m* (*N*) POLL ATM, POLL EAU nitrogen (*N*); **~ Kjeldahl** DIST EAU Kjeldahl nitrogen
azotomètre *m* ÉCHANT azotometer, nitrometer

B

bac *m* DIST EAU pit; **~ à boues** POLL EAU silt trap; **~ catalytique** POLL ATM catalyst container; **~ de collecte** ÉCHANT collecting receptacle; **~ de compostage** DIST EAU compost container; **~ de décantation** DIST EAU settling tank; **~ à déchets** DÉCHETS waste container; **~ de dissolution** POLL EAU solution tank; **~ d'échantillonnage** ÉCHANT sample container; **~ à écumes** POLL EAU skimming tank; **~ d'évaporation** DIST EAU evaporation pan; **~ de filtration** POLL EAU filtration vat; **~ à humus** DIST EAU, POLL EAU humus tank; **~ jaugeur** ÉCHANT gaging tank (*AmE*), gauging tank (*BrE*), measuring tank; **~ de manutention** DÉCHETS tote box; **~ récepteur** DÉCHETS catch tank; **~ de titrage** ÉCHANT titration vessel

bâche *f* DIST EAU cistern, tank; **~ alimentaire** DÉCHETS, POLL ATM feed tank; **~ d'eau brute** DIST EAU, POLL EAU untreated-water tank; **~ d'eau non-traitée** DIST EAU, POLL EAU untreated-water tank

bactéries *f pl* POLL ATM, POLL EAU, SOLS bacteria; **~ ammonifiantes** POLL EAU ammonifying bacteria; **~ anaérobies** POLL EAU anaerobic bacteria; **~ anaérobies facultatives** POLL EAU facultative anaerobic bacteria; **~ autotrophes** POLL EAU autotrophic bacteria; **~ coliformes** ÉCHANT coliform bacteria; **~ dénitrifiantes** POLL EAU denitrifying bacteria; **~ hétérotrophes** POLL EAU heterotrophic bacteria; **~ mésophiles** POLL EAU mesophilic bacteria; **~ méthanogènes** POLL EAU methanogenic bacteria; **~ mortes** POLL EAU dead bacteria; **~ nitrifiantes** POLL EAU nitrifying bacteria; **~ nitriques** POLL EAU nitric bacteria; **~ pathogènes** DIST EAU, POLL EAU pathogenic bacteria; **~ pouvant se développer en milieu anaérobie** POLL EAU facultative anaerobic bacteria; **~ revivifiables** POLL EAU viable bacteria; **~ en suspension** POLL EAU suspended bacteria; **~ thermophiles** POLL EAU thermophilic bacteria; **~ viables** DIST EAU viable bacteria; **~ vivantes** POLL EAU live bacteria

baffle: **~ à chevrons** *m* POLL ATM chevron baffle

bagasse *f* ÉN RENOUV bagasse

bague: **~ de tuyauterie** *f* DIST EAU ring of tubings

baignade *f* POL ENV bathing place

bain: **~ fluidisé** *m* DIST EAU fluidized bed

baisse *f* DIST EAU fall; **~ au-dessous du point de rosée** POLL ATM dropping below dew point; **~ de niveau** DIST EAU, POLL MAR level reduction; **~ du niveau d'eau** DIST EAU, POLL MAR water level recession; **~ de pH** ÉCHANT, POLL ATM, POLL EAU pH-depression; **~ de pression** POLL ATM pressure drop; **~ d'utilisation** DIST EAU, POL ENV use reduction

baisser *vt* POLL ATM *température* lower

bajoyer *m* DIST EAU lock side wall, river wall

balai: **~ d'alimentation** *m* DÉCHETS feeder broom; **~ circulaire** *m* DÉCHETS, DIST EAU circular broom; **~ cylindrique** *m* DÉCHETS, DIST EAU cylinder broom; **~ mécanique** *m* DIST EAU mechanical sweeper

balance *f* ÉCHANT scales; **~ de Mohr** ÉCHANT Mohr balance

balayage *m* DÉCHETS *des rues* cleaning, scavenging, sweeping, POLL ATM *de gaz* scavenging; **~ des données mesurées** ÉCHANT measuring-data scanning

balayer *vt* DÉCHETS *des rues* clean, scavenge, sweep, POLL ATM *gaz* scavenge

balayeuse *f* DÉCHETS *engin de nettoiement de la voierie* street cleaner, street sweeper, DIST EAU road-sweeping lorry (*BrE*), road-sweeping truck (*AmE*); **~ aspiratrice** DÉCHETS suction blower; **~ mécanique** DÉCHETS mechanical sweeper; **~ à vide** DÉCHETS *nettoyage des rues* vacuum street-sweeper

balistique *adj* DÉCHETS ballistic

ballast: **~ calcaire** *m* POLL ATM limestone ballast

balle: **~ de déchets** *f* DÉCHETS waste bale; **~ de ferraille** *f* DÉCHETS scrap bale; **~ de matériaux de récupération** *f* DIST EAU secondary resources bale; **~ de paille** *f* ÉN RENOUV straw bale

ballon: **~ d'air surpressé** *m* DIST EAU diffused-air tank; **~ captif** *m* POLL ATM captive balloon; **~ rond** *m* ÉCHANT round-bottomed flask; **~-sonde** *m* ÉCHANT balloon sonde

banc *m* DIST EAU *de vase* bank, *assemblage* plant; **~ de compactage des floculants** DIST EAU compact flocculation plant; **~ de décarbonisation** DIST EAU decarbonization plant; **~ électrolytique** ÉN RENOUV electrolysis installation; **~ d'essai de compression** DÉCHETS, ÉCHANT compression test machine; **~ d'étalonnage** ÉCHANT calibration unit; **~ de filtration** DIST EAU filtering unit, POLL ATM filtration plant, SÉCURITÉ filter plant; **~ de filtration expérimental** DIST EAU test filtration plant; **~ de flottation** POLL EAU flotation plant; **~ d'osmose inverse** DIST EAU reverse osmosis plant; **~ de plastification** DIST EAU plasticizing unit; **~ rocheux** DIST EAU rock bed

bande: **~ d'absorption** *f* ÉCHANT absorption band; **~ d'amenée** *f* DÉCHETS, ÉN RENOUV conveyor belt; **~ cyclable** *f* POL ENV bicycle path; **~ d'étiquetage** *f* DÉCHETS label film; **~ de papier de mesure du pH** *f* ÉCHANT pH-paper strip; **~ de protection de surface** *f* DÉCHETS surface protection tape; **~ de serrage par pince** *f* DÉCHETS clip band; **~ de tamisage** *f* POLL MAR sifting belt; **~ transporteuse** *f* DÉCHETS, ÉN RENOUV band conveyor, conveyor belt; **~ transporteuse aérienne** *f* DÉCHETS aerial conveyor; **~ transporteuse ascendante** *f* DÉCHETS upbrow conveyor belt; **~ transporteuse pour le stockage en balles** *f* DÉCHETS bale storage belt; **~ trieuse** *f* DÉCHETS sorting belt

banderolage *m* DÉCHETS band sealing; **~ de couleur** DÉCHETS colored strapping (*AmE*), coloured strapping (*BrE*)

banderoleuse *f* DÉCHETS band sealer

bang: **~ supersonique** *m* POLL SON supersonic boom

banlieue *f* POL ENV suburbs, urban fringe

barbotage *m* ÉCHANT, POLL EAU bubbling

barboteur *m* ÉCHANT bubbler, immersion stirrer; **~ à gaz avec tête frittée** ÉCHANT gas washing bottle with sintered head

barème: **~ fiscal applicable à l'environnement** *m* POL ENV eco-tax system; **~ d'imposition** *m* POL ENV tax tariff; **~ des redevances** *m* POL ENV charges tariff

barge: **~ d'écrémage** *f* POLL MAR skimming barge

baril *m* POLL MAR drum

barillet *m* DIST EAU pump barrel, pump cylinder, POLL ATM collecting main, POLL EAU pump barrel, pump cylinder

baromètre *m* ÉCHANT barometer

barquette: **~ en carton** *f* DÉCHETS cardboard tray

barrage *m* ÉN RENOUV barrage, dam, lock weir; **~-barrière** POLL MAR fence boom; **~ d'écrémage** POLL MAR skimming barrier; **~ flottant** POLL MAR oil boom; **~ flottant autoréglable** POLL MAR self-adjusting floating weir; **~ hydroélectrique** ÉN RENOUV hydroelectric dam; **~ en maçonnerie** ÉN RENOUV masonry dam; **~ mobile à vannes** ÉN RENOUV sluiceway; **~ récupérateur** POLL MAR skimming barrier; **~ réservoir** DIST EAU storage reservoir;

~ **de retenue** DIST EAU dam; ~ **rideau** POLL MAR curtain boom

barre: ~ **pour cadre d'extrémité** *f* SOLS end-frame member

barrière *f* DÉCHETS barrier, fence; ~ **argileuse** DÉCHETS clay barrier; ~ **flottante** POLL MAR oil boom; ~ **intégrée** DÉCHETS structurally-engineered barrier

bas: ~**-fond** *m* POLL EAU shallow water

base[1] *f* DIST EAU sill, SOLS base; ~ **de données** ÉCHANT database; ~ **de données écologiques** GESTION ecological database; ~ **fiscale** POL ENV tax base; ~ **d'imposition** POL ENV tax base; ~ **support** SOLS carrier material

base[2]: **de** ~ *loc* GÉNÉRAL basic

basique *adj* POLL ATM alkaline

basse[1]: ~ **mer** *f* DIST EAU low water; ~ **pression** *f* POLL ATM, POLL EAU, POLL MAR low pressure; ~ **température** *f* DIST EAU, ÉN RENOUV low temperature

basse[2]: **à** ~ **pression** *loc* POLL ATM, POLL EAU, POLL MAR, SOLS low-pressure; **à** ~ **température** *loc* SOLS low-temperature; **à** ~ **tension** *loc* POLL ATM low-voltage

basses: ~ **eaux** *f pl* DIST EAU low water

bassin *m* DIST EAU, ÉCHANT, ÉN RENOUV, POLL EAU basin, pond, reservoir, tank; ~ **d'aération** POLL EAU aeration basin, aeration tank; ~ **atmosphérique** POLL ATM air shed; ~ **de boues** POLL EAU sludge pond; ~ **de compensation** DIST EAU equalizing tank; ~ **de contact** POLL EAU mixing tank, sludge contact clarifier; ~ **de curage** DIST EAU, POLL EAU settling basin; ~ **de décantation** DIST EAU sedimentation pond, settling basin, POLL ATM *épuration biologique* settling tank, POLL EAU settling tank, sludge drying bed; ~ **de décantation en entonnoir** POLL EAU funnel-shaped settling chamber; ~ **de décantation finale** POLL EAU final sedimentation tank, final settling tank, secondary settler; ~ **de décantation intermédiaire** POLL EAU intermediate settling tank; ~ **de décantation préliminaire** POLL EAU preliminary settling tank; ~ **de dessablement** DÉCHETS detritor, grit chamber, POLL EAU settling basin, detritor, grit chamber; ~ **d'épuration** POLL ATM clarifying tank; ~ **fermé** DIST EAU blind drainage area; ~ **filtrant** DIST EAU, POLL EAU filter bed; ~ **de filtration** DIST EAU, POLL EAU filter bed; ~ **de finition** POLL EAU secondary polishing pond; ~ **de floculation** DIST EAU, POLL EAU flocculation tank, flocculator, flocculator basin; ~ **à flot** ÉN RENOUV tidal basin; ~ **fluvial** DIST EAU river basin; ~ **à fond pyramidal** POLL EAU hopper-bottom tank; ~ **hydrographique** DIST EAU drainage area, drainage basin, river basin, watershed; ~ **hydrologique** DIST EAU drainage area, drainage basin; ~ **d'infiltration** POLL EAU infiltration basin; ~ **jaugé** SOLS calibrated watershed; ~ **de marée** POLL MAR tidal basin; ~ **de maturation** POLL EAU maturation pond; ~ **océanique** POLL MAR ocean basin; ~ **d'oxydation** DÉCHETS oxidation pond, POLL ATM oxidation tank, POLL EAU oxidation pond; ~ **d'oxydation non-aéré** POLL EAU nonaerated oxidation pond; ~ **à poissons alimenté avec des eaux recyclées** POLL EAU wastewater fish pond; ~ **de prélavage** DIST EAU precleaning basin; ~ **de réalimentation** DIST EAU recharge basin; ~ **rectangulaire** POLL EAU rectangular basin, rectangular clarifier; ~ **de réfrigération** POLL EAU cooling pond; ~ **de refroidissement** DIST EAU, POLL EAU cooling pond; ~ **de refus** POLL EAU *industrie minière* tailings pond; ~ **de régénération** POLL EAU regeneration basin; ~ **de rétention** POLL EAU retention tank; ~ **de rétention des eaux pluviales** POLL EAU storm-water retention tank; ~ **de retenue** DIST EAU impounding reservoir, storage reservoir, reservoir, ÉN RENOUV, POLL EAU reservoir; ~ **à schistes** POLL EAU *industrie minière* tailings pond; ~ **secondaire** POLL EAU secondary pond; ~ **de sédimentation** POLL EAU sedimentation tank; ~ **solaire** ÉN RENOUV solar pond; ~ **solaire à l'échelle industrielle** ÉN RENOUV commercial-scale solar pond; ~ **de stabilisation** DIST EAU equalizing tank, POLL EAU stabilization pond; ~ **de stockage** ÉN RENOUV tidal basin; ~ **tertiaire** POLL EAU tertiary pond; ~ **de tranquillisation** DIST EAU settling basin, stilling chamber; ~ **urbain** DIST EAU urban catchment; ~ **versant** DIST EAU drainage area, drainage basin, river basin; ~ **versant expérimental** DIST EAU experimental basin

bâtardeau *m* ÉN RENOUV cofferdam

bâtiment *m* POLL MAR *de pêche* vessel; ~ **de dépollution** DIST EAU, POLL EAU, POLL MAR oil pollution fighter; ~ **des pompes** POLL EAU pump house

battage: ~ **de pilots** *m* DÉCHETS, DIST EAU pile-driving

batterie *f* DÉCHETS battery, DIST EAU, POLL EAU *de cribles* set; ~ **à accumulation rechargeable** ÉN RENOUV rechargable storage battery; ~ **de filtres** POLL ATM filter unit, filtering unit, SÉCURITÉ filter plant; ~ **de filtres pour fibres et poussières** SÉCURITÉ filter plant for dusts and fibres; ~ **d'hydrocyclones** POLL ATM hydrocyclone battery; ~ **de pompes** DIST EAU pump set; ~ **recyclable** DÉCHETS recyclable battery; ~ **usagée** DÉCHETS used battery

battre *vt* DIST EAU beat; ~ **les pieux** DÉCHETS, DIST EAU pile-drive

bauxite *f* DIST EAU, POLL ATM bauxite

Be *abrév* (*béryllium*) POLL ATM Be (*beryllium*)

bec: ~ **Bunsen** *m* ÉCHANT Bunsen burner; ~ **de déversement** *m* POLL ATM flow line

bélier *m* DÉCHETS, DIST EAU pile-driver

benne *f* DÉCHETS skip, pick-up; ~ **étanche à ordures** DÉCHETS packer body; ~ **à ordures** DÉCHETS refuse collection vehicle, dustbin lorry (*BrE*), dustcart, garbage truck (*AmE*); ~ **de ramassage** DÉCHETS scoop; ~ **tasseuse** DIST EAU packer body

benthos *m* DIST EAU, POLL EAU, POLL MAR benthos

benzène *m* POLL ATM benzene

benzofurane *m* SOLS benzofuran

berge *f* DIST EAU *d'une rivière, d'un fleuve* bank

berme: ~ **de séparation** *f* DIST EAU segregation berm

béryllium *m* (*Be*) POLL ATM beryllium (*Be*)

besoin: ~ **d'air effectif** *m* POLL ATM actual air requirement; ~ **d'assainissement** *m* SOLS need for remediation, remediation need

besoins: ~ **de surface** *m pl* ÉN RENOUV surface requirements

béton: ~ **polymérisé** *m* DÉCHETS polymer concrete

bidon *m* DÉCHETS canister, ÉCHANT *d'essence, d'huile* can, POLL MAR drum

bidonville *m* GESTION, POL ENV slum, shantytown

bief *m* DIST EAU *d'un canal* reach, *d'un fleuve* pond

biens: ~ **de consommation** *m pl* DÉCHETS, ÉN RENOUV consumer goods; ~ **irremplaçables** *m pl* DÉCHETS irreplaceables

bilan: ~ **du cycle de vie** *m* GESTION life-cycle inventory; ~ **de dioxyde de carbone** *m* POLL ATM carbon dioxide budget; ~ **d'eau** *m* DIST EAU water balance; ~ **d'eau du sol** *m* DIST EAU soil-water balance; ~ **écologique** *m* DIST EAU, GESTION, POL ENV eco-balance; ~ **des émissions mondiales de soufre** *m* POLL ATM global sulfur budget (*AmE*), global sulphur budget (*BrE*); ~ **énergétique** *m* DÉCHETS, DIST EAU, ÉN RENOUV, POLL ATM energy balance, energy budget; ~ **énergétique naturel** *m* ÉN RENOUV natural energy balance; ~ **d'énergie primaire** *m* ÉN RENOUV primary energy balance; ~ **hydrique** *m* POL ENV water balance; ~ **hydrobiologique** *m* POLL EAU hydrobiological equilibrium; ~ **hydrologique** *m* POL ENV hydrological balance, water balance; ~ **ionique** *m* DIST EAU, POLL ATM ion budget; ~ **matières** *m* DIST EAU material, material-flow analysis, materials balance, GESTION, POL ENV materials balance; ~ **d'oxygène** *m* POLL ATM oxygen budget; ~ **d'ozone** *m* POLL ATM ozone budget; ~ **de soufre** *m* POLL ATM sulfur budget (*AmE*), sulphur budget (*BrE*); ~ **thermique** *m* POLL ATM, POLL EAU heat balance, thermal balance

bilatéral *adj* POL ENV bilateral

bilharziose *f* POLL EAU bilharzia, schistosomiasis

bille: ~ **de céramique** *f* SOLS ceramic pellet
binaire *adj* ÉCHANT binary
bioaccumulatif *adj* POLL ATM, SOLS bioaccumulative
bioaccumulation *f* POLL ATM, SOLS bioaccumulation
bioaugmentation *f* SOLS bioaugmentation
biocapteur *m* ÉCHANT biosensor
biocarburant *m* DÉCHETS waste fuel
biocénose *f* POLL ATM biocenosis (*AmE*), biocoenosis (*BrE*)
biocide *m* SOLS biocide
biocombustible *m* ÉN RENOUV fuel from biomass
biodégradabilité *f* DÉCHETS, DIST EAU, POLL ATM, POLL EAU biodegradability
biodégradable *adj* DÉCHETS, DIST EAU, POLL ATM, POLL EAU biodegradable
biodégradation *f* DÉCHETS, DIST EAU, POLL ATM, POLL EAU biodegradation
bioessai *m* ÉCHANT, POL ENV bioassay
biofilm *m* SOLS biofilm
biofiltration *f* DÉCHETS, DIST EAU, POLL ATM biological filtration, biofiltration
biofiltre *m* DÉCHETS, DIST EAU, POLL ATM biofilter; ~ **à étages** POLL ATM tiered biofilter; ~ **à fond plat** POLL ATM flat-bed biofilter
biogaz *m* DÉCHETS biogas, DIST EAU biogas, digester gas, POLL EAU fermentation gas, sludge gas; ~ **à faible pouvoir calorifique** ÉN RENOUV low-thermal-value biogas (*LBG*); ~ **à pouvoir calorifique moyen** ÉN RENOUV mid-thermal-value biogas (*MBG*)
biogéocénose *f* POLL ATM biogeocenosis (*AmE*), biogeocoenosis (*BrE*)
bioindicateur *m* DIST EAU, SOLS bioindicator
biolaveur *m* POLL ATM biowasher
biologique *adj* GÉNÉRAL biological
biomasse *f* DIST EAU, POLL EAU, SOLS biomass; ~ **excédentaire** POLL EAU excess biomass; ~ **forestière** SOLS forest biomass; ~ **organique** ÉN RENOUV organic biomass; ~ **de séparation** POLL EAU sloughed biomass; ~ **végétale** SOLS crop biomass
biosphère *f* POL ENV biosphere
biostabilisateur *m* DÉCHETS, DIST EAU, POLL ATM biostabilizer
biosurveillance *f* GESTION, POL ENV biomonitoring
biote *m* POL ENV *biomasse* biota
biotechnologie *f* ÉN RENOUV, POL ENV, POLL ATM, POLL EAU biotechnology
biotechnologique *adj* ÉN RENOUV, POL ENV, POLL ATM, POLL EAU biotechnological
biotique *adj* ÉN RENOUV, DIST EAU biotic
biotope *m* POL ENV biotope
biphényle: ~ **polychloré** *m* (*BPC*) DÉCHETS, POLL ATM, SOLS polychlorinated biphenyl (*PCB*)
bisulfite *m* DIST EAU sulfite (*AmE*), sulphite (*BrE*)
bitumage *f* DÉCHETS *de déchets radioactifs*, DIST EAU asphalt sealing
blindage *m* DÉCHETS barrier packaging; ~ **acoustique** POLL SON acoustic shielding, noise shield; ~ **interne** POLL SON indoor shielding; ~ **de protection** DIST EAU, DÉCHETS ~ **de protection contre les rayonnements** SÉCURITÉ radiation shielding
blister: ~ **double coque** *m* DÉCHETS clamshell blister
blocage *m* POLL ATM blockage
blocaille *f* DÉCHETS building rubble
bobine *f* DÉCHETS *papier* coil; ~ **de câble** DÉCHETS *emballage* cable reel
bois *m* ÉN RENOUV wood; ~ **de chauffe** ÉN RENOUV firewood; ~ **à pâte** DÉCHETS pulp wood; ~ **récoltable** ÉN RENOUV harvestable wood
boisement *m* ÉN RENOUV woodland, POL ENV, SOLS afforestation
boîte *f* DÉCHETS can (*AmE*), metal can, tin (*BrE*); ~ **en carton caisse** DÉCHETS container board box; ~ **en carton ondulé** DÉCHETS corrugated board box; ~ **de conserve** DÉCHETS can (*AmE*), metal can, tin (*BrE*); ~ **de conserve en fer blanc** DÉCHETS tin-plated can; ~ **en fer blanc** DÉCHETS can (*AmE*), tin (*BrE*), tin-plated can; ~ **à fond plié** DÉCHETS folded-bottom box; ~ **métallique** DÉCHETS metal box; ~ **à ordures** DÉCHETS garbage can (*AmE*), trash can (*AmE*), dustbin (*BrE*), refuse can, rubbish bin (*BrE*); ~ **à provisions** DÉCHETS carrier box; ~ **réutilisable** DIST EAU reusable box
boîtier: ~ **d'aérosol** *m* DÉCHETS aerosol container; ~ **de compteur** *m* ÉCHANT meter housing; ~ **insonorisant** *m* POLL SON sound-absorbing casing; ~ **protecteur** *m* POLL ATM protective housing
bombe: ~ **aérosol** *f* DÉCHETS aerosol dispenser; ~ **calorimétrique** *f* ÉCHANT bomb calorimeter
bonde *f* DIST EAU plug
bonne: ~ **gestion de produit** *f* DIST EAU product stewardship
bonnes: ~ **pratiques de laboratoire** *f pl* ÉCHANT good laboratory practices (*GLP*)
bord *m* DIST EAU *d'un fleuve* bank; ~ **bridé** DÉCHETS flanged edge; ~ **gommé** DÉCHETS gummed edge; ~ **tombé** DÉCHETS flanged edge
bordereau: ~ **de suivi des déchets industriels** *m* POL ENV industrial waste follow-up form
borne *f* ÉCHANT *mesure* meter terminal
botte *f* DÉCHETS *emballage* bundle
bouchage *m* DIST EAU, POLL EAU *engorgement* clogging, fouling, *étanchement* sealing, plugging; ~ **de dépoussiéreur** POLL ATM scrubber plugging
bouche: ~ **d'aération** *f* DIST EAU, POLL ATM vent; ~**-bouteille** *f* DÉCHETS corking machine; ~ **d'eau** *f* DIST EAU water hydrant; ~ **d'égout** *f* DIST EAU gully hole, POLL EAU drop inlet; ~ **d'incendie** *f* DIST EAU fire hydrant; ~ **murale** *f* DIST EAU wall hydrant
bouché *adj* DIST EAU, POLL EAU *engorgé* clogged, fouled, *étanché* sealed, plugged
boucher *vt* DIST EAU, POLL EAU engorger, clog, foul, *étancher* seal, plug
bouchon *m* DÉCHETS corking plug; ~ **de bouteille** DÉCHETS bottle stopper; ~ **émeri** ÉCHANT ground-glass stopper; ~ **de sécurité** DÉCHETS guarantee cap, SÉCURITÉ child-proof closure (*CPC*); ~ **de verre dépoli** ÉCHANT ground-glass stopper
bouchons: ~ **antibruit** *m pl* POLL SON earplugs; ~ **de protection auditive** *m pl* POLL SON hearing-protection plugs
bouclage *m* DÉCHETS wraparound
boucle *f* ÉN RENOUV loop; ~ **de commande** ÉN RENOUV controlling loop; ~ **de contrôle** DIST EAU, ÉN RENOUV controlling loop; ~ **de lavage principale** POLL ATM main-wash circulation loop; ~ **production-consommation** DIST EAU production-consumption loop; ~ **de la qualité** GESTION quality loop
bouclier: ~ **d'ozone** *m* POLL ATM ozone shield; ~ **de protection contre les rayonnements** *m* SÉCURITÉ radiation shield
boudineuse: ~ **à deux vis** *f* DÉCHETS, DIST EAU *plastique* twin-screw extruder; ~ **à vis jumelée** *f* DÉCHETS, DIST EAU *plastique* twin-screw extruder
boue *f* DIST EAU, POLL EAU sludge; ~ **humique** POLL EAU humus sludge
boues: ~ **activées** *f pl* DIST EAU, POLL ATM, POLL EAU activated sludge; ~ **alluviales** *f pl* POLL EAU alluvial silt; ~ **brutes** *f pl* POLL EAU raw sludge, undigested sludge; ~ **brutes hydrolysées** *f pl* POLL EAU hydrolysed raw sludge (*BrE*), hydrolyzed raw sludge (*AmE*); ~ **cyanurées** *f* POLL EAU cyanide-containing sludge; ~ **décantables** *f pl* POLL EAU settleable sludge; ~ **décantées** *f pl* DIST EAU sediment, POLL EAU sedimented sludge, sediment, settled sludge, POLL MAR sediment; ~ **de décapage** *f pl* DIST EAU, POLL EAU paint sludge; ~ **de déchets vernis** *f pl* DÉCHETS varnish waste

sludge; ~ **de désencrage** *f pl* POLL EAU de-inking sludge; ~ **déshydratées** *f pl* POLL EAU dewatered sludge, dried sludge; ~ **déshydratées mécaniquement** *f pl* POLL EAU mechanically-dewatered sludge; ~ **digérées** *f* POLL EAU digested sludge; ~ **d'eaux résiduaires** *f pl* POLL EAU wastewater sludge; ~ **des eaux usées urbains** *f pl* POLL EAU municipal wastewater sludge; ~ **d'effluents** *f pl* POLL EAU effluent sludge; ~ **d'égout** *f pl* POLL EAU sewage sludge; ~ **d'ensemencement** *f pl* POLL EAU seeding sludge; ~ **épaissies** *f pl* POLL EAU thickened sludge; ~ **épaissies mécaniquement** *f pl* POLL EAU mechanically-thickened sludge; ~ **d'épuration** *f pl* DIST EAU sewage sludge; ~ **en excès** *f pl* POLL EAU excess sludge; ~ **fécales solides** *f pl* POLL EAU faecal solids sludge (*BrE*), fecal solids sludge (*AmE*); ~ **de filtration** *f pl* POLL EAU filtration sludge; ~ **galvaniques** *f pl* POLL EAU galvanic sludge; ~ **grasses** *f pl* POLL EAU ooze mud; ~ **d'hydroxyde** *f pl* POLL EAU hydroxide sludge; ~ **liquides** *f pl* POLL EAU liquid sludge; ~ **de matières fécales fortement polluées** *f pl* POLL EAU highly-polluted faecal solids sludge (*BrE*), highly-polluted fecal solids sludge (*AmE*); ~ **mélangées** *f pl* POLL EAU mixed sludge; ~ **métalliques** *f pl* POLL EAU metallic sludge; ~ **non épaissies** *f pl* POLL EAU unthickened sludge; ~ **non-traitées** *f pl* POLL EAU untreated sludge; ~ **partiellement stabilisées** *f pl* POLL EAU partially-stabilized sludge; ~ **de peinture** *f pl* DIST EAU paint sludge, DÉCHETS waste paint sludge; ~ **phosphatées** *f pl* POLL EAU phosphate sludge; ~ **de pigments** *f pl* POLL EAU pigment sludge; ~ **de précipitation** *f pl* POLL EAU precipitation sludge; ~ **de première décantation** *f pl* POLL EAU preliminary-settled sludge; ~ **primaires** *f pl* POLL EAU primary sludge; ~ **primaires mélangées** *f pl* POLL EAU mixed primary sludge; ~ **putréfiées** *f* POLL EAU digested sludge; ~ **putrescibles** *f pl* POLL EAU putrescible sludge; ~ **de recyclage** *f pl* POLL EAU return sludge; ~ **résiduaires industrielles** *f pl* POLL EAU industrial wastewater sludge; ~ **de retour** *f pl* POLL EAU return sludge; ~ **rouges** *f pl* POLL EAU red mud; ~ **séchées** *f pl* POLL EAU dried sludge; ~ **sèches** *f pl* POLL EAU dry sludge; ~ **secondaires** *f pl* POLL EAU secondary sludge; ~ **stabilisées** *f pl* POLL EAU stabilized sludge; ~ **traitées** *f pl* POLL EAU treated sludge

bouffée *f* POLL ATM puff

bouillir *vt* ÉCHANT boil

bouilloire *f* POLL ATM boiling vessel

bouletage *m* DÉCHETS, DIST EAU, POLL EAU, SOLS pelletization, pelletizing

bouleter *vt* DÉCHETS, DIST EAU, POLL EAU, SOLS pelletize

boulette *f* DÉCHETS, DIST EAU, POLL EAU, SOLS pellet

bourre *f* DÉCHETS wadding

bourse: ~ **des déchets** *f* DÉCHETS waste exchange market

bouteille: ~ **consignée** *f* DÉCHETS returnable bottle, deposit bottle; ~ **de gaz de référence** *f* ÉCHANT calibration gas bottle; ~ **jetable** *f* DÉCHETS disposable bottle; ~ **à jeter** *f* DÉCHETS nonreturnable bottle; ~ **non consignée** *f* DÉCHETS one-way bottle, nonreturnable bottle; ~ **d'oxygène** *f* ÉCHANT oxygen flask; ~ **perdue** *f* DÉCHETS one-way bottle, waste bottle; ~ **en plastique** *f* DÉCHETS plastic bottle, plastics bottle, DIST EAU, ÉCHANT plastic bottle; ~ **en polyéthylène téréphtalate** *f* DÉCHETS PETP bottle, polyethylene terephthalate bottle; ~ **en PVC** *f* DÉCHETS PVC bottle; ~ **recyclée** *f* DÉCHETS recycled bottle; ~ **soufflée** *f* DÉCHETS blown bottle; ~ **en verre perdu** *f* DÉCHETS nonreturnable bottle

bouteur *m* DÉCHETS bulldozer

BPC *abrév* (*biphényle polychloré*) DÉCHETS, POLL ATM, SOLS PCB (*polychlorinated biphenyl*)

branche: ~ **basse pression** *f* POLL ATM low-pressure side; ~ **chaude** *f* POLL ATM hot tubing (*ht*); ~ **froide** *f* ÉN RENOUV cooling-water loop; ~ **haute pression** *f* POLL ATM high-pressure side

branchement *m* DIST EAU branching

bras: ~ **mort** *m* DIST EAU backwater, dead channel, oxbow; ~ **de raclage** *m* DIST EAU rabble arm, skimmer arm; ~ **racleur oscillant** *m* POLL EAU swinging scraper; ~ **tournant** *m* POLL MAR rotatable arm

brassage *m* DÉCHETS agitation, POLL EAU mixing; ~ **à l'air** DIST EAU pneumatic flotation

brasser *vt* DÉCHETS agitate, POLL EAU mix

brasseur *m* DÉCHETS agitator

bride *f* POLL ATM, POLL MAR flange; ~ **de liaison** POLL ATM brace

brides: **à** ~ *loc* POLL ATM, POLL MAR flanged

brique: ~ **insonorisante** *f* POLL SON sound-absorbing brick; ~ **isolante** *f* POLL SON insulating brick; ~ **pleine** *f* POLL SON solid brick; ~ **vitrifiée** *f* DIST EAU clinker

briquette *f* DÉCHETS briquette; ~ **de charbon** ÉN RENOUV coal brick

brisant *m* POLL MAR *rocher* reef

briseur: ~ **d'émulsion** *m* DIST EAU emulsion breaker

British: ~ **Standards Institution** *m* (*BSI*) ÉCHANT British Standards Institution (*BSI*)

broche *f* DÉCHETS staple

brocheuse *f* DÉCHETS stapling machine

bromure *m* DIST EAU bromide

brosse: ~ **cylindrique** *f* DÉCHETS cylinder broom

brouillard *m* POLL ATM fog; ~ **acide** POLL ATM acid fog; ~ **de fines gouttelettes** POLL ATM spray; ~ **industriel** POLL ATM smog; ~ **d'ozone** POLL ATM ozone smog; ~ **photochimique** POLL ATM photochemical smog, smog; ~ **photochimique acide** POLL ATM acid smog

broyage *m* DÉCHETS, DIST EAU *concassage* comminution, crushing, grinding, *déchiquetage* shredding, ÉCHANT grinding, SOLS crushing, grinding; ~ **par voie humide** DÉCHETS, POLL ATM wet grinding; ~ **préliminaire** DIST EAU prior crushing; ~ **sec** DÉCHETS dry grinding; ~ **secondaire** DIST EAU regrinding; ~ **semi-fin** DÉCHETS semi-finish grinding

broyé *adj* DÉCHETS, DIST EAU *concassé* comminuted, crushed, ground, ÉCHANT ground, SOLS crushed, ground

broyer *vt* DÉCHETS, DIST EAU *concasser* comminute, crush, grind, ÉCHANT grind, SOLS crush, grind

broyeur *m* DÉCHETS, DIST EAU crusher, comminutor; ~**-concasseur à marteaux** DIST EAU hammermill crusher; ~ **de déchets** DÉCHETS waste crusher; ~ **final** DÉCHETS final crusher; ~ **grossier** DÉCHETS *charbon* coarse crusher; ~ **à marteaux** DIST EAU hammermill; ~ **à ordures** DÉCHETS refuse crusher; ~ **primaire** DIST EAU primary crusher

bruit *m* POLL SON, SÉCURITÉ noise; ~ **aérien** POLL SON, SÉCURITÉ airborne noise; ~ **aléatoire** POLL SON random noise; ~ **ambiant** POLL SON ambient noise, environmental noise, SÉCURITÉ environmental noise; ~ **blanc** POLL SON white noise; ~ **de la circulation routière** POLL SON traffic noise; ~ **de construction** POLL SON construction noise; ~ **à l'échappement** POLL SON exhaust noise; ~ **et vibrations** *m pl* POLL MAR, POLL SON, SÉCURITÉ noise and vibration; ~ **extérieur** POLL SON extraneous noise; ~ **de fond** POLL SON background noise; ~ **impulsionnel** POLL SON impulse noise; ~ **industriel** POLL SON industrial noise; ~ **inhérent au site** POLL MAR, POLL SON site-related noise; ~ **lié à l'objet** POLL MAR, POLL SON object-related noise; ~ **lié au trafic ferroviaire** POLL SON noise caused by rail traffic; ~ **localisé** POLL SON area-related noise; ~ **masquant** POLL SON masking noise; ~ **de la navigation aérienne** POLL SON noise caused by air traffic; ~ **perturbant** POLL SON disturbing noise; ~ **porté dans les solides** POLL SON solid-borne noise; ~ **recouvrant** POLL SON masking noise; ~ **résiduel** POLL SON residual noise; ~ **du site** POLL SON site-related noise; ~ **de structure** POLL SON structure-borne noise; ~ **sur le lieu de travail** GESTION, POLL SON noise at the workplace; ~ **total** POLL SON total noise; ~ **de ventilateur** POLL SON fan noise

brûlant: ~ **du biocombustible** *adj* DÉCHETS waste-fired
brûler *vt* DÉCHETS, POLL EAU burn, flare; ~ **totalement** SOLS burn out
brûleur *m* DÉCHETS, DIST EAU, POLL ATM, POLL EAU burner; ~ **d'appoint** POLL ATM auxiliary burner; ~ **automatique** DÉCHETS automatic burner; ~ **auxiliaire** DIST EAU auxiliary burner; ~ **de démarrage** POLL ATM start burner; ~ **à faibles émissions de** NO_x POLL ATM low NO_x burner (*LNB*), low NO_x combustor, low nitrogen burner (*LNB*); ~ **à paille** ÉN RENOUV straw burner; ~ **de postcombustion** DÉCHETS, POLL ATM afterburner; ~ **submersible** DIST EAU, POLL EAU submersible burner; ~ **à Venturi** POLL ATM Venturi burner
brume *f* DÉCHETS, POLL ATM haze, mist
brumée *f* POLL ATM smog
brut *adj* DIST EAU *eau* raw, untreated, POLL EAU untreated
bruyant: **peu** ~ *adj* POLL SON low-noise
BSI *abrév* (*British Standards Institution*) ÉCHANT BSI (*British Standards Institution*)
bulle *f* POLL EAU bubble; ~ **d'air** DÉCHETS entrapped air; ~ **de gaz** POLL EAU gas bubble
bureau: ~ **mobile** *m* POL ENV mobile office
burette *f* ÉCHANT burette; ~ **à gaz** ÉCHANT gas burette; ~ **graduée** ÉCHANT measuring jar; ~ **d'Orsat** ÉCHANT Orsat burette
buse: ~ **d'alimentation** *f* DIST EAU, POLL EAU discharge nozzle; ~ **d'aspiration** *f* DIST EAU suction nozzle; ~ **d'atomiseur** *f* POLL ATM atomizer nozzle; ~ **d'évacuation** *f* DIST EAU, POLL EAU discharge nozzle; ~ **d'impacteur** *f* POLL ATM impactor nozzle; ~ **d'injection** *f* POLL ATM injection nozzle; ~ **de pulvérisation haute pression** *f* POLL ATM high-pressure spray nozzle; ~ **de vaporisation** *f* POLL ATM spray nozzle, POLL MAR spray aperture
butadiène *m* POLL ATM butadiene
butane *m* POLL ATM butane
butée *f* POLL MAR thrust
buvable *adj* DIST EAU potable
bypass *m* ÉCHANT bypass

C

C *abrév* (*carbone*) ÉCHANT, POLL ATM C (*carbon*)
Ca *abrév* (*calcium*) DIST EAU, POLL ATM, SOLS Ca (*calcium*)
CA *abrév* (*courant alternatif*) ÉN RENOUV AC (*alternating current*)
cabine: **~ acoustique** *f* POLL SON acoustic booth; **~ avec aération de protection** *f* SÉCURITÉ protective-aerated cab; **~ avec filtre à charbon activé intégré** *f* SÉCURITÉ cab with integrated activated-carbon filter unit; **~ de conduite** *f* SÉCURITÉ cab; **~ insonorisée** *f* POLL MAR, POLL SON, SÉCURITÉ noise protection booth; **~ de protection contre le bruit** *f* POLL MAR, POLL SON, SÉCURITÉ noise protection booth
câble: **~ de dérivation** *m* SOLS stub cable; **~ de sonde** *m* ÉCHANT measuring-head cable
cache: **~-poussières** *m* POLL ATM dust shield
cachet: **~ en plomb** *m* DÉCHETS lead seal
cadence: **~ de traitement** *f* POL ENV processing speed
cadmium *m* (*Cd*) SOLS cadmium (*Cd*)
cadre: **~ d'astreinte** *m* SÉCURITÉ duty engineer; **~ de pulvérisation** *m* POLL ATM spray frame; **~ suspendu** *m* POLL ATM suspended frame
cage: **~ de cabine** *f* SÉCURITÉ cab enclosure; **~ d'écureuil** *f* DÉCHETS wire cage; **~ d'écureuil support** *f* POLL ATM support wire cage; **~ filtrante** *f* POLL ATM filter cage; **~ de rétention** *f* POLL ATM retention cage; **~ tubulaire** *f* POLL ATM tube cage
cahier: **~ des charges** *m* GESTION job instructions
caillement *m* GÉNÉRAL coagulation
cailler *vt* GÉNÉRAL coagulate
caisse: **~ aspirante** *f* POLL ATM suction box; **~ à bouteilles** *f* DÉCHETS bottle crate; **~ à claire-voies** *f* DÉCHETS crate; **~ compartimentée** *f* DÉCHETS compartment case; **~ pliable** *f* DÉCHETS collapsible case
caisson *m* POLL EAU caisson, POLL SON coffer; **~ en acier** POLL ATM steel housing; **~ d'alternateur** POLL SON generator house; **~ de plongée** POLL EAU caisson
calage: **~ de modèle** *m* ÉCHANT model calibration
calaminage *m* POLL ATM scaling
calamine *f* DIST EAU scale, boiler scale, POLL EAU, SOLS boiler scale
calcaire *m* POLL EAU, SOLS lime scale; **~ pulvérulent** POLL ATM limestone meal
calcin *m* DÉCHETS *recyclage du verre* cullet, broken glass, waste glass cullet, refuse glass, *de chaudière* boiler scale, DIST EAU *recyclage du verre* cullet, broken glass, refuse glass, *de chaudière* boiler scale, POLL EAU, SOLS boiler scale; **~ recyclé** DIST EAU recycled glass cullet
calcination *f* DÉCHETS, POLL ATM calcination
calciner *vt* DÉCHETS, POLL ATM calcinate
calcium *m* (*Ca*) DIST EAU, POLL ATM, SOLS calcium (*Ca*)
calcul *m* ÉCHANT determination, calculation, computation; **~ de concentration** ÉCHANT determination of content; **~ des coûts** GESTION cost calculation; **~ de la DBO** ÉCHANT BOD determination; **~ de la DBO des eaux usées** ÉCHANT BOD determination in wastewater; **~ de la DBO_5** ÉCHANT BOD_5 determination; **~ de la DCO** ÉCHANT COD determination; **~ électrométrique du pH** ÉCHANT electrometric pH-value determination; **~ gravimétrique** ÉCHANT gravimetric determination; **~ de masse volumique** ÉCHANT density determination; **~ de modèle** POLL ATM model calculation; **~ du pH des eaux usées** ÉCHANT determination of wastewater pH-value; **~ de réseau** DIST EAU network calculation; **~ de teneur** ÉCHANT determination of content; **~ de la teneur en sodium des eaux usées** ÉCHANT determination of natrium content in wastewater
calculable *adj* ÉCHANT calculable, computable
caléfaction *f* DÉCHETS *nucléaire*, POLL ATM burnout
caler *vt* POLL MAR *navire* draw
calibrage *m* DÉCHETS grading, sizing, ÉCHANT grading, calibration; **~ d'échantillon** ÉCHANT sample calibration; **~ de modèle** DIST EAU model calibration
calibré *adj* DÉCHETS graded, sized, ÉCHANT graded, calibrated
calibre *m* DÉCHETS *charbon*, DIST EAU particle size; **~ pour fils métalliques** ÉCHANT gage wire (*AmE*), gauge wire (*BrE*); **~ de grain** SOLS grain size
calibrer *vt* DÉCHETS grade, size, ÉCHANT grade, calibrate
caloporteur *m* ÉN RENOUV heat carrier medium, heat transfer medium, working medium, POLL EAU heat transfer medium; **~ liquide** ÉN RENOUV liquid working medium
calorifuge *adj* DÉCHETS, DIST EAU, SÉCURITÉ heat-proof, heat-insulating, *matériau* heat-resistant
calorifugeage *m* DÉCHETS, SÉCURITÉ heat insulation
calorimètre *m* ÉCHANT calorimeter; **~ d'incinération** ÉCHANT incineration calorimeter; **~ de titrage** ÉCHANT titration calorimeter
calorimétrie *f* ÉCHANT calorimetry; **~ différentielle** ÉCHANT differential scanning calorimetry (*DSC*)
calotte: **~ à barbotage** *f* POLL ATM bubble cap; **~ de brume** *f* POLL ATM haze canopy; **~ de gaz** *f* POLL EAU gas hood
cambouis *m* POLL ATM, POLL EAU dirty oil, sludge
camion: **~ benne** *m* DÉCHETS dumper, skip lorry (*BrE*); **~ citerne** *m* POLL MAR road tanker; **~ de collecte des ordures** *m* DÉCHETS refuse collector, refuse lorry (*BrE*), refuse truck (*AmE*); **~ tasseur** *m* DÉCHETS SOLS, compactor vehicle; **~-vidange** *m* DÉCHETS eductor-basin cleaner; **~ à vide** *m* POLL MAR vacuum truck
campagne *f* ÉCHANT collection period, POL ENV campaign; **~ d'affichage** POL ENV poster campaign; **~ d'échantillonnage** ÉCHANT sampling plan; **~ d'information** POL ENV information campaign; **~ de mesure** ÉCHANT measurement plan, measurement program (*AmE*), measurement programme (*BrE*), measuring period; **~ de sensibilisation** POL ENV public awareness campaign
canal *m* DIST EAU, POLL EAU canal, channel, culvert; **~ alluvial** DIST EAU alluvial bed; **~ d'amenée** DIST EAU intake main, POLL EAU delivery channel, intake main; **~ d'assèchement** DIST EAU drainage channel; **~ en bois** DIST EAU *support par chevalets* flume; **~ collecteur** DIST EAU *d'une pompe centrifuge* volute chamber; **~ de décharge** DIST EAU emergency spillway, intake main, POLL EAU discharge gutter, intake main; **~ de dérivation** DIST EAU flume; **~ de déversement** POLL EAU *égout* discharge canal; **~ distributeur** DIST EAU delivery channel, delivery race; **~ d'épuisement** DIST EAU drainage channel; **~ évacuateur de crues** DIST EAU, POLL EAU spillway; **~ d'évacuation** DIST EAU disposal channel, drainage chanel; **~ fermé** DIST EAU closed canal; **~ d'irrigation** DIST EAU catch feeder, irrigation channel; **~ jaugeur à ressaut** DIST EAU standing-wave flume; **~ de mesure** POLL EAU flow-measuring flume; **~ en planches** DIST EAU flume; **~ de prise** DIST EAU flume, penstock; **~ de prise d'eau** DIST EAU intake; **~ Venturi** DIST EAU Venturi flume
canalisation *f* DIST EAU *action* canalization, *tuyau* conduit, pipeline, GESTION canalization, river training, POLL EAU *action* canalization, *tuyau* pipeline; **~ d'alimentation de**

gaz SOLS gas soil pipe; ~ **d'amenée** DIST EAU delivery channel; ~ **de drainage** POLL EAU catchwater drain; ~ **du réseau public** POLL EAU public canalization; ~ **sur passerelle** POLL EAU pipe bridge; ~ **urbaine** POLL EAU urban canalization
canaliser *vt* DIST EAU, GESTION, POLL EAU canalize
cancérigène *adj* SÉCURITÉ carcinogenic
caniveau *m* DIST EAU berm ditch, flume, POLL EAU gully, gutter
canne: ~ **d'échantillonnage** *f* ÉCHANT sampling tube; ~ **de Provence** *f* ÉN RENOUV bamboo reed
canon: ~ **d'eau haute pression** *m* SOLS high-pressure water jet
CaO *abrév* (*oxyde de calcium*) POLL ATM, SOLS CaO (*calcium oxide*)
caoutchouc *m* DÉCHETS rubber; ~ **acrylique** DÉCHETS acrylic rubber; ~ **perdu** DÉCHETS waste rubber; ~ **usagé** DÉCHETS waste rubber
capacité *f* GÉNÉRAL capacity, potential; ~ **d'absorption** DÉCHETS, DIST EAU *puits* absorptive capacity, GÉNÉRAL absorbency, POLL ATM absorption capacity; ~ **d'absorption des protons** POLL ATM proton absorptive capacity; ~ **acoustique** POLL SON acoustic capacity; ~ **d'adsorption** POLL ATM adsorption capacity; ~ **d'agglomération** POLL ATM agglomeration capacity; ~ **d'assimilation** DÉCHETS, DIST EAU assimilative capacity; ~ **d'auto-épuration** POLL EAU purification capacity; ~ **biotique** GESTION, POL ENV biotic potential; ~ **calorifique** POLL ATM thermal capacity, bioaccumulation capacity, heat capacity; ~ **de champ** SOLS field capacity; ~ **de chargement** POLL ATM loading capacity; ~ **de décharge** DÉCHETS landfill capacity; ~ **d'échange cationique** (*CEC*) SOLS cation exchange capacity (*CEC*); ~ **d'échange ionique** POLL EAU ion-exchange capacity; ~ **d'échantillonnage** ÉCHANT sampling capacity; ~ **effective d'une décharge** DÉCHETS field capacity of landfill; ~ **électrique** POLL ATM *électrofiltre* power capacity; ~ **d'épuration** POLL ATM cleaning capacity; ~ **de filtration** POLL ATM filtration capacity; ~ **horaire** DÉCHETS hourly capacity; ~ **d'incinération** DÉCHETS, POLL EAU incineration capacity; ~ **d'infiltration** DIST EAU percolation capacity; ~ **limite** DIST EAU, POLL EAU *charge biotique utile* carrying capacity; ~ **de neutralisation des acides** POLL ATM acid-neutralizing capacity; ~ **nominale** DIST EAU, POLL EAU rated capacity; ~ **de prévision** POLL ATM predictive capability; ~ **de prise en charge** GESTION, POL ENV acceptance capacity; ~ **de protection** POLL SON protection capacity; ~ **de recyclage** DIST EAU recycling capacity; ~ **de refoulement** POLL EAU pump capacity; ~ **de régénération** POLL EAU regeneration capacity, self-restoring capacity; ~ **de résistance** ÉCHANT resistance capacity; ~ **respiratoire** POLL ATM breathing capacity; ~ **de rétention** DÉCHETS holding capacity, retention capacity; ~ **de retraitement** DIST EAU processing capacity; ~ **de ruissellement** DIST EAU run-off capacity; ~ **de séparation** ÉCHANT separating capacity, POLL ATM separation capacity; ~ **de sorption** DIST EAU, POLL ATM sorption capacity; ~ **de stockage** DÉCHETS storage capacity, storage facility, ÉN RENOUV, POLL ATM storage capacity; ~ **thermique** POLL ATM thermal capacity; ~ **de tolérance** GESTION, POL ENV acceptance capacity; ~ **totale** POLL EAU total capacity; ~ **de traitement** POLL ATM, POLL EAU, SOLS processing capacity; ~ **de traitement journalier** DIST EAU, POLL ATM, POLL EAU, SOLS daily processing capacity; ~ **de transfert d'oxygène** POLL EAU *dans l'eau non polluée* oxygen transfer capacity; ~ **de transfert d'oxygène spécifique de l'eau propre** POLL EAU specific oxygen transfer capacity in clean water; ~ **de transfert d'oxygène spécifique des eaux usées** POLL EAU specific oxygen transfer capacity in wastewater; ~ **de transport** DÉCHETS *tube* carrying capacity, DIST EAU, ÉN RENOUV transport capacity; ~ **de tri** DÉCHETS sorting capacity; ~ **d'un véhicule** DÉCHETS vehicle capacity
capillaire *m* ÉCHANT capillary; ~ **à ozone** POLL ATM ozone capillary; ~ **pour prélèvement gazeux** ÉCHANT sample gas capillary
capillarité *f* POLL EAU, SOLS capillarity
capital: ~ **génétique** *m* POL ENV, SÉCURITÉ gene pool
capitonnage *m* DÉCHETS wadding
capnofuge *adj* POLL ATM smoke-preventing
capnographe *m* ÉCHANT capnograph
capot *m* POLL ATM hood; ~ **insonorisant** POLL SON anti-noise hood, sound-absorbing hood; ~ **de protection acoustique** POLL SON noise guard
capsulation *f* DÉCHETS capping
capsule *f* DÉCHETS *de bouteille* bottle cap, cap; ~ **d'aérosol** DÉCHETS aerosol cap; ~**-couronne** DÉCHETS crown closure, crown cork, crown cup; ~ **de gélatine** DÉCHETS gelatine capsule; ~ **à vis préfiletée** DÉCHETS continuous-thread cap, continuous-thread closure
capsuler *vt* DÉCHETS cap
capsuleuse *f* DÉCHETS capper
captage *m* DÉCHETS collection, DIST EAU tapping, catchment, intake, collection, POLL EAU abstraction; ~ **d'eau** DIST EAU water catchment; ~ **d'eaux fluviales** DIST EAU river water tapping; ~ **de gaz** POLL ATM, POLL SON gas capture; ~ **de rivière** DIST EAU river tapping
captation *f* POLL ATM recovery, collection, POLL EAU catchment; ~ **d'énergie solaire** ÉN RENOUV collection of solar heat
capte: ~**-suies** *m* POLL ATM soot catcher
capter *vt* DIST EAU collect, GESTION, POLL ATM, POLL SON capture
capteur *m* ÉCHANT, POLL MAR sensor, sensing device; ~ **de débit** POLL ATM flow sensor; ~ **de débit d'air** POLL ATM airflow sensor; ~ **de déplacement** ÉCHANT displacement transducer; ~ **à électrode** ÉCHANT electrode sensor; ~ **à électrode potentiostatique** ÉCHANT potentiostatic electrode sensor; ~ **d'humidité** ÉCHANT moisture sensor; ~ **de lumière diffuse** ÉCHANT diffused-light sensor; ~ **de lumière parasite** ÉCHANT diffused-light sensor; ~ **modulaire** ÉCHANT sensor module; ~ **d'oxygène** ÉCHANT, POLL ATM oxygen meter, oxygen sensor; ~ **plan** ÉN RENOUV flat-plate collector; ~ **potentiométrique** ÉCHANT potentiometric sensor; ~ **de poussière** DIST EAU dust collector; ~ **de pression** POLL ATM pressure sensor; ~ **de pression de contournement** POLL ATM bypass-pressure sensor; ~ **de proximité** ÉCHANT proximity pick-up, proximity sensor; ~ **solaire** ÉN RENOUV solar collector; ~ **de surface** ÉCHANT surface sensor; ~ **de température** ÉCHANT temperature sensor; ~ **turbidimétrique** ÉCHANT, POLL ATM, POLL EAU turbidimeter; ~ **ultrasonique** ÉCHANT ultrasound sensor
capture *f* DIST EAU *boues activées* substrate utilization; ~ **d'électrons** ÉCHANT electron capture
capturer *vt* POLL ATM capture
capuchon: ~ **à cendres volantes** *m* POLL ATM flue-dust catcher
capuchonnement: ~ **de ventilateur** *m* POLL SON cowling, fan shroud
caractère: ~ **durable** *m* POL ENV sustainability; ~ **entartrant** *m* POLL EAU scale-forming properties; ~ **héréditaire** *m* POL ENV, SÉCURITÉ inherited trait; ~ **inépuisable** *m* ÉN RENOUV inexhaustibility; ~ **marécageux** *m* DÉCHETS marshiness
caractérisation: ~ **du site** *f* DIST EAU site characterization
caractéristiques *f pl* GÉNÉRAL characteristics, properties; ~ **des boues** POLL EAU sludge characteristics; ~ **de décantation** DIST EAU settling characteristics, settling properties; ~ **de la décharge** SOLS site characteristics; ~ **des émissions** ÉCHANT, POLL ATM emissions data; ~ **d'étalonnage** ÉCHANT calibration data; ~ **des matériaux** DÉCHETS properties of materials; ~ **physico-chimiques du milieu** DIST EAU, POLL ATM, POLL EAU physio-chemical environment; ~ **d'une pompe** DIST EAU,

POLL EAU pump characteristics; **~ de séparation** POLL ATM separation properties; **~ du site** ÉCHANT site parameters, ÉN RENOUV site-specific properties, site-specific criteria, POLL ATM site criteria; **~ du sol** SOLS soil characteristics

carbonatation *f* DIST EAU aeration; **~ à bulles fines** POLL EAU fine-bubble aeration

carbonate: **~ de chaux** *m* POLL EAU lime carbonate; **~ de sodium** *m* POLL EAU soda; **~ de soude** *m* DIST EAU soda ash

carbone *m* (*C*) ÉCHANT, POLL ATM carbon (*C*); **~ fixe** ÉN RENOUV fixed carbon; **~ inorganique dissous** DIST EAU, ÉCHANT, POLL ATM dissolved inorganic carbon; **~ inorganique total** ÉCHANT total inorganic carbon (*TIC*); **~ organique assimilable** POLL EAU assimilable organic carbon; **~ organique dissous** (*COD*) DIST EAU, ÉCHANT, POLL ATM dissolved organic carbon (*DOC*); **~ organique total** ÉCHANT, POLL EAU total organic carbon (*TOC*); **~ total** ÉCHANT total carbon (*TC*)

carbonifère *adj* POLL ATM carboniferous

carburant *m* POLL ATM fuel oil; **~ non brûlé** DÉCHETS, POLL ATM unburnt fuel; **~ au plomb** POLL ATM lead-containing fuel; **~ sans plomb** POLL ATM lead-free fuel

carburateur *m* POLL ATM carburetor (*AmE*), carburettor (*BrE*)

carbure *m* POLL ATM, POLL EAU carbide

carburol *m* ÉN RENOUV gasohol

carcasse *f* DÉCHETS waste body; **~ de pneus usés** DIST EAU used-tire carcass (*AmE*), used-tyre carcass (*BrE*); **~ de voiture** DÉCHETS scrap motorcar

carcinogène *adj* SÉCURITÉ carcinogenic

carence: **~ en oxygène** *f* POLL EAU oxygen deficiency

carreau: **~ insonorisant** *m* POLL SON soundproof tile, acoustical tile

carrousel: **~ d'extraction** *m* DÉCHETS carousel extractor

carte: **~ acoustique** *f* POLL SON noise map; **~ marine** *f* POLL MAR nautical chart, sea chart; **~ météorologique** *f* POLL ATM, POLL MAR weather map; **~ des sols** *f* GESTION, SOLS soil map; **~ topographique** *f* GESTION, POL ENV, SOLS survey map; **~ des vents** *f* GESTION, POL ENV, POLL ATM, POLL MAR wind chart

carter *m* POLL ATM crankcase; **~ de colonne d'absorption** POLL ATM absorber sump; **~ compresseur** POLL ATM compressor casing; **~ à huile** POLL EAU, SOLS oil sump; **~ de pompe** POLL EAU pump casing

cartographie: **~ acoustique** *f* POLL SON noise area map; **~ détaillée** *f* ÉCHANT detailed mapping

carton *m* DÉCHETS carton, DIST EAU paper board; **~ deux couches** DÉCHETS duplex board; **~ étanche** DÉCHETS liquid-proof carton; **~ d'expédition cylindrique** DÉCHETS mailing tube; **~ de fabrication mécanique en continu** DÉCHETS machine-made board; **~ fort** DÉCHETS hardboard; **~ glacé** DÉCHETS enamel board, glazed board; **~ gris** DÉCHETS, ÉN RENOUV, POLL SON gray board (*AmE*), grey board (*BrE*), chipboard; **~ d'insonorisation** POLL SON acoustic board; **~ isolant** DÉCHETS, POLL SON insulating board; **~ multicouche** DÉCHETS multiple board; **~ ondulé** DÉCHETS corrugated board, corrugated fiberboard (*AmE*), corrugated fibreboard (*BrE*); **~ ondulé simple rannelura** DIST EAU single-wall corrugated fiberboard (*AmE*), single-wall corrugated fibreboard (*BrE*); **~ ondulé très résistant** DÉCHETS heavy-duty corrugated fiberboard (*AmE*), heavy-duty corrugated fibreboard (*BrE*); **~ paille** DÉCHETS strawboard; **~ pour pancartes** DÉCHETS gray board (*AmE*), grey board (*BrE*); **~ pliant** DÉCHETS folding carton; **~ à rabats** DÉCHETS folding carton; **~ renforcé isolant** POLL SON structural insulation board; **~ de revêtement expérimental** DÉCHETS test liner board; **~ rigide** DIST EAU solid fiberboard (*AmE*), solid fibreboard (*BrE*)

cartonneuse *f* DÉCHETS cardboard machine

cartons: **~ d'emballage de produits de consommation lavés et compactés** *m pl* DÉCHETS washed and squashed consumer waste cartons

casier *m* DÉCHETS crate; **~ de décharge** DÉCHETS landfill cell

casque *m* POLL SON ear defenders, ear protectors, headphones, SÉCURITÉ helmet; **~ antibruit** POLL SON, SÉCURITÉ noise exclusion helmet; **~ de protection** POLL SON, SÉCURITÉ safety helmet

casse: **~-vide** *m* DIST EAU vacuum breaker

casseur: **~ d'émulsion** *m* DIST EAU emulsion breaker

catabolisme: **~ endogène** *m* POLL EAU endogenous catabolism

catalyse *f* POLL ATM, SOLS catalysis; **~ des boues d'eaux résiduaires** POLL EAU wastewater sludge catalysis; **~ de contact** POLL ATM contact catalysis; **~ négative** POLL ATM negative catalysis; **~ de surface** POLL ATM surface catalysis

catalyser *vt* POLL ATM, SOLS catalyse (*BrE*), catalyze (*AmE*)

catalyseur *m* POLL ATM, SOLS catalyst; **~ à base de métaux nobles** DIST EAU, POLL ATM noble metal catalyst; **~ biologique** POLL ATM biological catalyst; **~ céramique** POLL ATM ceramic catalyst; **~ en circuit ouvert** POLL ATM open-loop catalyst; **~ à collecteur d'échappement** POLL ATM exhaust manifold catalyst; **~ dénitrifiant** POLL ATM denox catalyst; **~ fixe** POLL ATM fixed catalyst; **~ fluide** POLL ATM fluid catalyst; **~ géré par ordinateur** POLL ATM computer-controlled catalyst; **~ en grains** POLL ATM granular catalyst, pellet catalyst; **~ hétérogène** POLL ATM heterogeneous catalyst; **~ homogène** POLL ATM homogeneous catalyst; **~ d'hydrocarbures** POLL ATM hydrocarbon catalyst; **~ à lit double** POLL ATM dual-bed catalyst; **~ massique sur lits superposés** POLL ATM dual-bed bulk material catalyst; **~ métallique basique** POLL ATM base-metal catalyst; **~ monolithique à lit double** POLL ATM dual-bed monolithic catalyst; **~ d'oxydation** POLL ATM oxidizing catalyst; **~ d'oxydation classique** POLL ATM conventional oxidation catalyst (*COC*); **~ Pd(II)/Cu(II)** POLL ATM Pd(II)/Cu(II) catalyst; **~ poreux** POLL ATM honeycomb catalyst; **~ de poussière** POLL ATM dust catalyst; **~ primaire** POLL ATM primary catalyst; **~ principal** POLL ATM main catalyst; **~ de réduction** POLL ATM reduction catalyst; **~ secondaire** POLL ATM secondary catalyst; **~ souterrain** POLL ATM underfloor catalyst; **~ stéréospécifique** POLL ATM stereospecific catalyst; **~ supporté** POLL ATM plate catalyst; **~ sur lit fixe** POLL ATM fixed-bed catalyst; **~ sur support** POLL ATM plate catalyst; **~ à trois voies** POLL ATM three-way catalyst (*TWC*); **~ usagé** POLL ATM aged catalyst; **~ en vrac** POLL ATM bulk material catalyst; **~ à zéolite** POLL ATM zeolite catalyst

catalytique *adj* POLL ATM, SOLS catalytic; **~ par voie biologique** POLL ATM biologically catalytic; **~ sélectif** POLL ATM selective catalytic

catastrophe: **~ écologique** *f* GESTION, POL ENV, POLL ATM, SOLS ecological collapse, environmental disaster; **~ naturelle** *f* POL ENV natural disaster

catégorie: **~ de biens** *f* DÉCHETS commodity group; **~ de pollueurs** *f* POL ENV polluter group

cathode *f* SOLS cathode

cation *m* DIST EAU, POLL ATM, POLL EAU, SOLS cation; **~ basique** POLL ATM base cation; **~ échangeable** POLL ATM exchangeable cation

causticité *f* POLL EAU, SOLS causticity

caustique *adj* POLL EAU caustic, lime-aggressive, SOLS caustic

cavitation *f* DIST EAU *d'une pompe* cavitation; **~ d'une pompe** POLL EAU pump cavitation

cavité *f* DIST EAU void

CC *abrév* (*courant continu*) ÉN RENOUV DC (*direct current*)

Cd *abrév* (*cadmium*) SOLS Cd (*cadmium*)

CEC *abrév* (*capacité d'échange cationique*) SOLS CEC (*cation exchange capacity*)

CE_{50} *abrév* SÉCURITÉ *concentration moyenne réelle* CE_{50}

ceinture: ~ **de van Allen** *f* ÉN RENOUV earth's sunbelt; ~ **verte** *f* POL ENV green belt
cellophane *f* DÉCHETS film wrap, cellophane
cellophaneuse *f* DÉCHETS film-wrapping machine
cellule *f* ÉCHANT, ÉN RENOUV cell; ~ **au diséléniure cuivre-indium** ÉN RENOUV *énergie solaire* copper-indium diselenide cell; ~ **d'échantillonnage** ÉCHANT sampling cell; ~ **d'électrode** ÉCHANT electrode cell; ~ **d'électrolyse** ÉCHANT electrolysis cell; ~ **de mesure** ÉCHANT measuring cell; ~ **photovoltaïque** ÉN RENOUV photovoltaic cell, solar cell; ~ **photovoltaïque en couche mince** ÉN RENOUV thin-film photovoltaic cell; ~ **au silicium monocristallin** ÉN RENOUV single-crystalline silicon cell; ~ **au silicium polycristallin** ÉN RENOUV *énergie solaire* polycrystalline silicon cell; ~ **solaire** ÉN RENOUV solar cell
Celluloïd® *m* DÉCHETS, DIST EAU Celluloid®
cellulose *f* DÉCHETS, DIST EAU, ÉN RENOUV, POLL EAU cellulose; ~ **de bois** DÉCHETS mechanical wood pulp; ~ **au sulfate** DIST EAU sulfate cellulose (*AmE*), sulphate cellulose (*BrE*)
CEN *abrév* (*Comité Européen de Normalisation*) ÉCHANT, POL ENV CEN (*European Committee for Standardization*)
cendre *f* DÉCHETS, DIST EAU, POLL ATM, POLL EAU ash; ~ **de chaudière** DÉCHETS boiler ash
cendres: ~ **d'incinération** *f pl* DÉCHETS incineration ash; ~ **de soude** *f pl* DIST EAU soda ash; ~ **volantes** *f pl* DÉCHETS, POLL ATM fly ash
centrale *f* DIST EAU, ÉN RENOUV plant, powerhouse, POLL ATM, POLL MAR, POLL SON, SÉCURITÉ, SOLS plant; ~ **à cycle ouvert** ÉN RENOUV open-cycle plant; ~ **électrique** ÉN RENOUV power station, POLL ATM power plant; ~ **éolienne** ÉN RENOUV wind power plant; ~ **ETM flottante** ÉN RENOUV floating OTEC plant; ~ **ETM terrestre** ÉN RENOUV land-based OTEC plant; ~ **exploitant les courants marins** ÉN RENOUV ocean current power plant; ~ **force-chaleur** DÉCHETS, POLL ATM combined heat and power station (*CHP station*); ~ **géothermique** ÉN RENOUV geothermal power plant; ~ **héliothermique** ÉN RENOUV solar-heat power plant; ~ **hydroélectrique** ÉN RENOUV hydroelectric plant, water power plant; ~ **hydroélectrique avec retenue** ÉN RENOUV storage hydroelectric plant; ~ **marémotrice** ÉN RENOUV, POL ENV tidal power plant; ~ **nouvelle** DIST EAU *construite après 1975* new plant; ~ **photovoltaïque** ÉN RENOUV photovoltaic power plant; ~ **de pointe** ÉN RENOUV peak-load power plant; ~ **prototype de production d'électricité** ÉN RENOUV demonstration power plant; ~ **solaire** ÉN RENOUV solar farm, solar-heat power plant, solar power plant; ~ **solaire basse température** ÉN RENOUV low-temperature solar collector plant; ~ **sous le vent** ÉN RENOUV downwind power plant; ~ **thermique** POLL ATM thermal combustion plant, thermal power plant; ~ **thermique au charbon** POLL ATM coal power station; ~ **thermique au fioul** ÉN RENOUV, POL ENV oil-fired power plant; ~ **thermique au lignite** POLL ATM brown coal power station, lignite power station; ~ **thermique à vapeur** ÉN RENOUV, POL ENV steam power plant; ~ **utilisant l'énergie des vagues** ÉN RENOUV wave power plant; ~ **au vent** ÉN RENOUV upwind power plant
centre: ~ **de collecte** *m* DIST EAU public collection station; ~ **de contrôle** *m* DÉCHETS control center (*AmE*), control centre (*BrE*); ~ **d'enfouissement des déchets** *m* DÉCHETS waste disposal plant; ~ **de recyclage** *m* DIST EAU recycling center (*AmE*), recycling centre (*BrE*); ~ **de retraitement des déchets solides** *m* DIST EAU processing center for recyclable solid waste materials (*AmE*), processing centre for recyclable solid waste materials (*BrE*); ~ **urbain** *m* POL ENV urban core
centrifugation *f* DÉCHETS, POLL ATM, POLL EAU, SOLS centrifugation, centrifuging; ~ **des boues** POLL EAU sludge centrifugation
centrifuge *adj* DÉCHETS, POLL ATM, POLL EAU, SOLS centrifugal
centrifuger *vt* DÉCHETS, POLL ATM, POLL EAU, SOLS centrifuge
centrifugeuse *f* DÉCHETS, POLL ATM centrifuge, POLL EAU centrifuge, spin dryer, SOLS centrifuge; ~ **à boues** POLL EAU sludge centrifuge; ~ **d'extraction** POLL EAU extraction centrifuge; ~ **portative** ÉCHANT hand-held centrifuge
céramisation *f* SOLS ceramization
cerclage *m* DÉCHETS hooping
cercleuse *f* DÉCHETS strapping machine
céréale: ~ **d'hiver** *f* ÉN RENOUV winter grain
céréales *f pl* ÉN RENOUV grain
certificat: ~ **de conformité** *m* GESTION certificate of compliance; ~ **d'inspection** *m* POL ENV certificate of inspection; ~ **d'origine** *m* GESTION certificate of origin; ~ **de solvabilité** *m* GESTION soundness certificate
certification: ~ **de conformité** *f* GESTION, POL ENV certification of conformity
cessation: ~ **d'implantation** *f* DIST EAU settlement stoppage
CFC[1] *abrév* (*chlorofluorocarbone*) DÉCHETS, DIST EAU, POLL ATM CFC (*chlorofluorocarbon*)
CFC[2]: ~ **partiellement halogéné** *m* POLL ATM partially-halogenated CFC; ~ **totalement halogéné** *m* POLL ATM fully-halogenated CFC
CGS *abrév* (*chromatographie gaz-solide*) ÉCHANT GSC (*gas-solid chromatography*)
CG-SM *abrév* (*chromatographie gazeuse par capillarité couplée à un spectromètre de masse*) ÉCHANT GC-MS (*capillary gas chromatography mass spectrometer*)
chaîne: ~ **alimentaire** *f* POL ENV food chain, metabolic chain; ~ **alimentaire marine** *f* POLL MAR marine food chain; ~ **d'attelage** *f* DÉCHETS drag chain; ~ **à drague** *f* DIST EAU drag chain; ~ **d'emballage** *f* DÉCHETS packaging line; ~ **de mise en bouteille** *f* DÉCHETS bottling line; ~ **de montage** *f* DÉCHETS, DIST EAU assembling unit; ~ **à palettes** *f* DÉCHETS flat-top chain conveyor; ~ **de presse en continu** *f* DÉCHETS continuous press line; ~ **de production** *f* DIST EAU production line; ~ **de puits de gaz** *f* DÉCHETS gas well chain; ~ **de remplissage des boîtes** *f* DÉCHETS *emballage* can-filling line
chaleur *f* GÉNÉRAL heat; ~ **dissipée** DÉCHETS, DIST EAU, POLL ATM waste heat; ~ **haute température** ÉN RENOUV high-temperature heat; ~ **humide** POLL ATM damp heat; ~ **industrielle** ÉN RENOUV process heat; ~ **latente** POLL ATM latent heat; ~ **mesurée en kilowatt-heure** (*kWt*) ÉN RENOUV kilowatt-hour heat (*kWt*); ~ **perdue** DÉCHETS, DIST EAU waste heat; ~ **de réaction** ÉCHANT heat of reaction; ~ **récupérée** DIST EAU, POLL ATM recovered heat; ~ **recyclée** DIST EAU, POLL EAU recirculated heating; ~ **résiduelle** DIST EAU waste heat; ~ **sèche** POLL ATM dry heat; ~ **stockée** ÉN RENOUV stored heat; ~ **à température moyenne** ÉN RENOUV mid-temperature heat
chalut *m* POLL MAR trawl net
chalutage *m* POLL MAR trawling
chambre *f* DÉCHETS, DIST EAU, POLL ATM chamber; ~ **de chargement** DÉCHETS charging chamber; ~ **à chicanes** POLL ATM baffle chamber; ~ **climatique** DÉCHETS, POLL ATM climatic chamber; ~ **de combustion** DÉCHETS, POLL ATM, SOLS combustion chamber; ~ **de combustion cyclone** DÉCHETS cyclone burning chamber; ~ **de combustion pour gaz inerte** POLL ATM inert-gas combustion chamber; ~ **de combustion primaire** DIST EAU primary combustion chamber; ~ **de combustion principale** POLL ATM main combustion chamber; ~ **de combustion à revêtement céramique** POLL ATM ceramic-lined combustion chamber; ~ **de combustion secondaire** DIST EAU secondary combustion chamber (*SCC*); ~ **de décantation** DIST EAU, POLL ATM *dépoussiérage mécanique*, POLL EAU settling chamber; ~ **de dégazage** DÉCHETS, DIST EAU degassing chamber; ~ **de désaturation** POLL ATM demister; ~ **de désorption thermique** ÉCHANT thermal desorption chamber; ~ **de digestion** DIST EAU, POLL EAU digestion tank; ~ **de digestion des boues** POLL EAU sludge-digestion tank; ~ **d'eau** DIST

EAU flume; ~ **de flottation** POLL MAR flotation chamber; ~ **d'ionisation** ÉCHANT ionization chamber; ~ **de mélange d'air** POLL SON plenum chamber; ~ **de mélange des gaz** ÉCHANT gas-mixing chamber; ~ **de mesure** ÉCHANT measuring chamber; ~ **de postcombustion** DÉCHETS, POLL ATM afterburner, afterburner chamber, afterburner housing, afterburning chamber; ~ **de pourrissage** DIST EAU rotting chamber, rotting hall; ~ **de précombustion** POLL ATM precombustion chamber; ~ **de pression** POLL ATM pressure chamber; ~ **de réaction** ÉCHANT, POLL ATM reaction chamber; ~ **réceptrice** POLL EAU receiving chamber; ~ **de référence** ÉCHANT reference chamber; ~ **de reprise de gaz** POLL ATM exhaust housing; ~ **de réverbération** POLL SON reverberation chamber; ~ **à sable** DÉCHETS, POLL EAU grit chamber; ~ **de sédimentation** POLL ATM *dépoussiérage mécanique* settling chamber; ~ **de tranquillisation** DIST EAU settling chamber, stilling chamber, POLL EAU settling chamber

champ: ~ **de capteurs** *m* ÉN RENOUV *énergie solaire* collector field; ~ **de charge** *m* POLL ATM *dépoussiérage par électro-filtre* field strength; ~ **de courant continu** *m* POLL ATM direct current voltage field; ~ **de densité** *m* POLL ATM flow field; ~ **diffus** *m* POLL SON diffuse field; ~ **électrique** *m* ÉN RENOUV, POLL ATM electric field, electrical field; ~ **d'épandage** *m* POLL EAU wastewater farm; ~ **de filtration** *m* DIST EAU filter field; ~ **en friche** *m* ÉN RENOUV, SOLS fallow; ~ **d'héliostats** *m* ÉN RENOUV *énergie solaire* collector field; ~ **infini** *m* POLL SON *acoustique* free field; ~ **irrigué** *m* DÉCHETS irrigation field; ~ **libre** *m* POLL SON free field; ~ **magnétique** *m* ÉCHANT magnetic field; ~ **sonore** *m* POLL SON sound field

champignon *m* POLL EAU fungus; ~ **des eaux résiduaires** POLL EAU sewage fungus; ~ **des eaux usées** POLL EAU wastewater fungus

changement: ~ **climatique** *m* POLL ATM climate changes; ~ **climatique d'origine humaine** *m* POLL ATM anthropogenic climate change; ~ **de couleur** *m* ÉCHANT color change (*AmE*), colour change (*BrE*); ~ **de direction** *m* POLL ATM direction change; ~ **microclimatique** *m* POLL ATM microclimatic change

changeur: ~ **de grille de rinçage** *m* DÉCHETS backflush screen changer

chape: ~ **insonorisante** *f* POLL SON sound-absorbing apron

chapeau *m* POLL EAU *boue en cours de décantation* scum, surface film

charbon: ~ **actif** *m* DIST EAU, POLL ATM, POLL EAU, SOLS activated carbon; ~ **actif en bâtonnets** *m* POLL ATM chunk activated carbon; ~ **actif en grains** *m* POLL ATM, POLL EAU granulated activated carbon (*GAC*); ~ **actif poreux** *m* POLL ATM porous activated carbon; ~ **actif en poudre** *m* DIST EAU, POLL EAU powdered activated carbon (*PAC*); ~ **actif régénéré** *m* POLL ATM regenerated activated carbon; ~ **actif saturé** *m* POLL EAU saturated activated carbon

charge *f* DÉCHETS, DIST EAU, ÉCHANT load, POLL ATM charge, load, burden, charging material, POLL EAU load, SOLS charge; ~ **annuelle** DIST EAU annual load; ~ **biotique utile** DIST EAU, POLL EAU carrying capacity; ~ **catalytique** POLL ATM catalyst charge; ~ **de compression** DÉCHETS, ÉCHANT compression load; ~ **critique** SOLS critical load; ~ **dangereuse** DÉCHETS, DIST EAU dangerous load; ~ **de désorbeur** POLL ATM desorber feed; ~ **de diffusion** POLL ATM diffusion charging; ~ **de dolomie** POLL ATM dolomite meal; ~ **d'eau** DIST EAU water pressure; ~ **en eaux usées** POLL EAU wastewater load; ~ **d'équilibre** POLL ATM equilibrium load; ~ **excessive** SOLS excess load; ~ **de gaz pollué** POLL ATM dirty-gas load, dirty-gas loading, waste gas loading; ~ **hydraulique** POLL EAU hydraulic loading; ~ **d'incinération** *m* DIST EAU incineration stock; ~ **de matériau absorbant** POLL SON absorbent-material filler; ~ **de matières solides** POLL ATM solid-matter loading; ~ **de nitrates** POLL EAU nitrate load; ~ **nominale** DIST EAU, POLL EAU rated load; ~ **de particule** POLL ATM particle charging; ~ **de pointe** ÉN RENOUV peak load; ~ **polluante** DIST EAU, POL ENV pollution burden, pollutant load; ~ **de pollution** POLL ATM pollutant load, dust burden, pollution burden; ~ **de poussière** POLL ATM dust burden; ~ **de poussière à l'admission** POLL ATM inlet dust burden; ~ **de poussière en résidus** POLL ATM residue-dust loading; ~ **pulvimétrique** POLL ATM dust burden; ~ **pulvimétrique d'un gaz pollué** POLL ATM dirty-gas dust loading; ~ **de remplissage** DÉCHETS fill mass; ~ **résiduelle** POLL ATM residue load; ~ **de rupture** DÉCHETS breaking load, POLL ATM tensile strength; ~ **de saturation** DIST EAU saturation concentration, POLL ATM saturation load; ~ **secondaire** POLL ATM secondary loading; ~ **de solvants** POLL ATM solvent loading; ~ **de surface** POLL ATM surface load; ~ **thermique** DIST EAU, ÉN RENOUV heat content, thermal load; ~ **utile** DIST EAU, POLL EAU service load; ~ **volumique** POLL ATM volume charge

chargé *adj* POLL ATM charged, laden, loaded; ~ **de gouttes** POLL ATM drop-laden; ~ **de particules** POLL ATM particulate-laden; ~ **de poussières** POLL ATM dust-laden; ~ **de solvants** POLL ATM solvent-laden

chargement *m* DÉCHETS, POLL ATM, POLL EAU, SOLS loading; ~ **des boues sur bateau-citerne** POLL EAU sludge barging; ~ **de choc** POLL EAU shock loading

charger *vt* POLL ATM load

chargeur *m* POLL ATM stoker; ~**-doseur** DÉCHETS dosing feeder; ~ **frontal** POLL MAR front-end loader; ~ **à roues** SOLS wheel loader

chariot: ~ **élévateur automoteur** *m* DÉCHETS fork-lift truck; ~ **élévateur à fourche** *m* DÉCHETS fork-lift truck

charriage: ~ **de sédiment** *m* DIST EAU bed load

charrier *vt* ÉCHANT wash away

chasse *f* DIST EAU scouring, POLL MAR flushing; ~ **basse pression** POLL EAU low-pressure flushing; ~ **du caniveau** POLL EAU flume flushing; ~ **d'eau** DIST EAU water flush; ~ **haute pression** POLL MAR high-pressure flushing

châssis: ~ **de conteneur** *m* DIST EAU container slide

chaudière *f* GÉNÉRAL boiler; ~ **alimentée par des rejets thermiques** DIST EAU, DÉCHETS, POLL ATM, POLL EAU waste heat boiler (*WHB*), waste heat recovery boiler; ~ **à chambre de combustion** DÉCHETS furnace boiler; ~ **à copeau de bois** ÉN RENOUV wood chip boiler; ~ **à paille** ÉN RENOUV straw-fired boiler; ~ **de récupération** DÉCHETS, DIST EAU, POLL ATM, POLL EAU recovery boiler, waste heat boiler (*WHB*); ~ **de récupération de chaleur** DÉCHETS, DIST EAU, POLL ATM, POLL EAU waste heat recovery boiler, waste heat boiler (*WHB*)

chauffage *m* GÉNÉRAL heating, heating up; ~ **au bois** ÉN RENOUV wood firing; ~ **direct** SOLS direct heating; ~ **direct par rayonnement** SOLS direct heating by radiation; ~ **à haute fréquence** SOLS radio-frequency heating; ~ **indirect** SOLS indirect heating; ~ **par induction** POLL ATM induction heating; ~ **par rayonnement d'air chaud** SÉCURITÉ hot-air radiation heating system; ~ **progressif** DIST EAU, POLL ATM incremental heating; ~ **solaire** ÉN RENOUV solar heating; ~ **urbain alimenté par l'énergie solaire** ÉN RENOUV solar district heating; ~ **urbain d'origine solaire** ÉN RENOUV solar district heating supply

chauffe: ~**-eau** *m* POLL ATM warm-water heater

chauffé *adj* GÉNÉRAL heated; ~ **par les déchets** DÉCHETS waste-fired; ~ **par l'énergie solaire** ÉN RENOUV solar-heated

chauffer *vt* GÉNÉRAL heat, heat up

chaufferie *f* ÉN RENOUV heating installation

chauffeur *m* POLL ATM stoker

chausse: ~ **de palette** *f* DÉCHETS pallet hood

chaux *f* DIST EAU, POLL ATM, POLL EAU, SOLS lime; ~ **éteinte** DIST EAU, POLL ATM, POLL EAU, POLL MAR, SOLS slaked lime;

~ vive DIST EAU, POLL ATM, POLL EAU, POLL MAR, SOLS quicklime
chélate *m* SOLS chelate
cheminée *f* ÉN RENOUV, POLL ATM chimney, stack; **~ de dérivation** POLL ATM bypass chimney, bypass stack; **~ d'évacuation des fumées** POLL ATM smokestack
chemisage *m* POLL ATM, POLL EAU, POLL SON lining
chemise *f* DIST EAU *d'une pompe* jacket, liner, lining
chenal: **~ de déversement** *m* DIST EAU drainage channel; **~ d'oxydation** *m* POLL EAU oxidation ditch
chenille *f* DÉCHETS, DIST EAU, SOLS caterpillar; **~ de recyclage** DIST EAU recycling caterpillar
chicane *f* POLL ATM, POLL EAU baffle; **~ transversale** POLL ATM segment baffle
chimie *f* GÉNÉRAL chemistry; **~ atmosphérique** POLL ATM atmospheric chemistry
chimiolyse *f* DÉCHETS chemolysis
chimique *adj* GÉNÉRAL chemical
chimiquement: **~ agressif** *adj* POLL ATM chemically aggressive; **~ résistant** *adj* DÉCHETS, POLL ATM, SÉCURITÉ chemically resistant
chimisorption *f* POLL ATM chemisorption
chloration *f* DÉCHETS, POLL ATM, SOLS chlorination; **~ au point critique** POLL EAU breakpoint chlorination
chlore *m* (*Cl*) DÉCHETS, POLL ATM, SOLS chlorine (Cl)
chloré *adj* DÉCHETS, POLL ATM, SOLS chlorinated
chlorer *vt* DÉCHETS, POLL ATM, SOLS chlorinate
chlorobenzène *m* POLL ATM chlorobenzene
chlorofluorocarbone *m* (*CFC*) DÉCHETS, DIST EAU, POLL ATM chlorofluorocarbon (*CFC*)
chlorophénol *m* POLL ATM chlorophenol
chlorophylle *f* ÉN RENOUV chlorophyll
chlorure *m* POLL ATM, SOLS chloride; **~ de méthylène** POLL ATM methylene chloride; **~ organique extractible** POLL ATM extractable organic chloride (*EOCL*); **~ de sodium** POLL EAU, POLL MAR sodium chloride; **~ de vinyle** POLL ATM vinyl chloride (*VC*)
choc: **~ acide printanier** *m* POLL ATM spring acid shock
choix: **~ des priorités** *m* POL ENV priority determination
chromatogramme *m* ÉCHANT chromatogram; **~ différentiel en phase gazeuse** ÉCHANT differential gas chromatogram; **~ intégral d'un gaz** ÉCHANT integral gas chromatogram; **~ en temps réel** ÉCHANT real-time chromatogram
chromatographe *m* ÉCHANT chromatograph; **~ en phase gazeuse** (*CPG*) ÉCHANT gas chromatograph (*GC*); **~ en phase gazeuse portable** (*CPG portable*) ÉCHANT portable gas chromatograph (*portable GC*)
chromatographie: **~ d'adsorption** *f* ÉCHANT adsorption chromatography; **~ capillaire électrocinétique** *f* ÉCHANT electrokinetic capillary chromatography (*EKC*); **~ en couche mince** *f* ÉCHANT thin-layer chromatography (*TLC*); **~ en couche mince sous haute pression** *f* ÉCHANT high-pressure thin-layer chromatography; **~ d'échange anionique** *f* ÉCHANT anion-exchange chromatography; **~ d'échange d'ions** *f* ÉCHANT ion-exchange chromatography (*IC*); **~ gazeuse alimentée par des piles** *f* ÉCHANT battery-powered gas chromatograph; **~ gazeuse par capillarité couplée à un spectromètre de masse** *f* (*CG-SM*) ÉCHANT capillary gas chromatography mass spectrometer (*GC-MS*); **~ gaz-liquide** *f* ÉCHANT gas-liquid partition chromatography; **~ gaz-solide** *f* (*CGS*) ÉCHANT gas-solid chromatography (*GSC*); **~ HP en couche mince** *f* ÉCHANT high-pressure thin-layer chromatography; **~ liquide à haute performance** *f* (*CLHP*) ÉCHANT high-performance liquid chromatography (*HPLC*); **~ par élution** *f* ÉCHANT elution chromatography; **~ de partage** *f* ÉCHANT partition chromatography; **~ de partage gaz-gaz** *f* ÉCHANT gas chromatography-gas chromatography (*GC-GC*); **~ de partage liquide-liquide** *f* ÉCHANT liquid chromatography-liquid chromatography (*LC-LC*); **~ de partage en phase gazeuse** *f* ÉCHANT gas partition chromatography; **~ en phase gazeuse** *f* (*CPG*) ÉCHANT gas chromatography (*GC*); **~ en phase gazeuse avec pyrolyse** *f* ÉCHANT pyrolysis gas chromatography (*PY-GC*); **~ en phase gazeuse couplée à des détecteurs d'ionisation de flamme** *f* ÉCHANT gas chromatography coupled with flame ionization detectors; **~ en phase gazeuse couplée à des spectromètres de masse** *f* ÉCHANT gas chromatography coupled with mass spectrometers; **~ en phase gazeuse à double détecteur** *f* ÉCHANT double-detector gas chromatography; **~ en phase gazeuse isotherme** *f* ÉCHANT isothermal gas chromatography; **~ en phase gazeuse par capillarité** *f* ÉCHANT capillary gas chromatography; **~ en phase gazeuse par piégeage** *f* ÉCHANT purge-and-trap gas chromatography; **~ en phase gazeuse-chromatographie sur gel perméable** *f* ÉCHANT gel-permeation chromatography gas chromatography (*GPC-GC*); **~ en phase liquide à haute performance** *f* ÉCHANT high-performance liquid chromatography (*HPLC*); **~ en phase liquide sur gel** *f* ÉCHANT liquid gel chromatography; **~ en phase supercritique** *f* (*CPS*) ÉCHANT supercritical-fluid chromatography (*SFC*); **~ sur colonne** *f* ÉCHANT column chromatography; **~ sur gel perméable** *f* ÉCHANT gel permeation chromatography (*GPC*); **~ sur papier** *f* ÉCHANT paper chromatography; **~ de tête en phase gazeuse** *f* ÉCHANT headspace gas chromatography (*HSGC*); **~ volumétrique en phase gazeuse** *f* ÉCHANT headspace gas chromatography (*HSGC*)
chromatographique *adj* ÉCHANT chromatographic; **~ en phase gazeuse** ÉCHANT gas-chromatographic
chrome *m* (*Cr*) POLL ATM, SOLS chromium (*Cr*)
chromogène *m* ÉCHANT chromogen
chronique *adj* SÉCURITÉ chronic
chronométreur: **~ à cadran** *m* ÉCHANT dial-type timer
chronopotentiométrie *f* ÉCHANT chronopotentiometry
chute *f* DIST EAU fall, lift; **~ d'eau** DIST EAU waterfall; **~ libre** DÉCHETS free fall; **~ de pH** ÉCHANT, POLL ATM, POLL EAU pH-drop; **~ de trémie d'alimentation** DÉCHETS hopper-furnace feed chute
chutes: **~ de bois** *f pl* DÉCHETS wood waste
CI_{50} *abrév* (*concentration inhibitrice 50%*) POLL EAU, SÉCURITÉ 50% inhibiting concentration
ciment *m* SOLS cement
cimentation *f* DIST EAU solidification with cement
cimetière: **~ de déchets radioactifs** *m* DIST EAU radioactive burial ground, radioactive graveyard, repository for radioactive waste
circuit: **~ en boucle fermée** *m* POLL ATM closed-loop circulation; **~ de chauffage** *m* ÉN RENOUV heating system; **~ fermé** *m* DÉCHETS closed system, ÉN RENOUV closed cycle (*CC*); **~ interne de recyclage des eaux** *m* POLL EAU in-house water recirculation system; **~ ouvert** *m* DÉCHETS open system; **~ de recirculation** *m* POLL EAU recirculation system; **~ de réfrigération** *m* DIST EAU, ÉN RENOUV cooling system; **~ de refroidissement** *m* DIST EAU, ÉN RENOUV, POLL ATM cooling system
circulation *f* POLL ATM, POLL EAU circulation; **~ d'agent de lavage** POLL ATM washing agent circulation; **~ d'eau** DIST EAU, POLL EAU water circulation; **~ d'eau de lavage** POLL ATM wash-water circulation; **~ à échelle intermédiaire** POLL ATM mesoscale circulation; **~ de lavage principal** POLL ATM main wash circulation; **~ de métaux lourds** POLL ATM, SOLS heavy metals circulation; **~ sur matériau filtrant** POLL ATM filter material circulation; **~ de vapeur** ÉN RENOUV steam circulation
circuler *vt* POLL EAU flow through
cire: **~ d'imprégnation** *f* DÉCHETS impregnating wax
cisaillage: **~ de ferraille** *m* DÉCHETS scrap shear
cisaille: **~ à guillotine** *f* DÉCHETS, DIST EAU guillotine
cisaillement *m* DÉCHETS, POLL ATM shearing, shearing off
cisailler *vt* DÉCHETS, POLL ATM shear, shear off

cisailles: ~ **pour automobiles** *f pl* DÉCHETS vehicle shredder
ciseau: ~ **universel** *m* DÉCHETS universal shredder
citerne *f* DÉCHETS, POLL EAU cistern, tank; ~ **des huiles usées** DÉCHETS slop tank; ~ **rectangulaire** POLL EAU rectangular tank; ~ **souple flottante** POLL MAR dracone, floating flexible tank; ~ **de stockage** DÉCHETS storage tank
Cl *abrév* (*chlore*) DÉCHETS, POLL ATM, SOLS CL (*chlorine*)
CL *abrév* (*concentration létale*) DIST EAU, ÉCHANT, POLL ATM, SÉCURITÉ LC (*lethal concentration*)
CL_{50} *abrév* DIST EAU, ÉCHANT, POLL ATM, SÉCURITÉ (*concentration létale cinquante, concentration létale à 50%*) LC_{50} (*lethal concentration fifty*), median lethal concentration
CL0 *abrév* (*concentration létale zéro*) SÉCURITÉ LC0 (*lethal concentration zero*)
CL_x *abrév* (*concentration létale x*) SÉCURITÉ LC_x (*lethal concentration x*)
clapet: ~ **à commande hydraulique** *m* POLL ATM hydraulically-controlled flap; ~ **de retenue** *m* DIST EAU check valve
clarificateur: ~ **rectangulaire** *m* POLL EAU rectangular basin, rectangular clarifier
clarification: ~ **de l'eau** *f* DIST EAU, POLL EAU water purification; ~ **des eaux usées** *f* POLL EAU wastewater clarification; ~ **primaire** *f* POLL EAU primary clarification
clarifier *vt* DIST EAU purify, ÉCHANT purge, POLL EAU clarify, purify; ~ **par filtre** DIST EAU, ÉCHANT, POLL EAU filter
classable *adj* ÉCHANT able to be classified
classe: ~ **de décharge** *f* DÉCHETS landfill class; ~ **granulométrique** *f* DIST EAU, POLL ATM, POLL EAU particle size class; ~ **de pollution** *f* POL ENV class of pollution; ~ **de précision** *f* ÉCHANT *étalonnage* class of accuracy; ~ **de propagation** *f* POLL ATM propagation class; ~ **de qualité de l'eau** *f* DIST EAU, POLL EAU water quality class, water quality grade; ~ **de transmission acoustique** *f* POLL SON sound transmission class (*STC*)
classifiable *adj* ÉCHANT classifiable, able to be classified
classificateur *m* DÉCHETS, ÉCHANT, POLL ATM, SOLS classifier; ~ **à courant ascendant** DIST EAU up-current classifier; ~ **hydraulique** SOLS hydraulic classifier; ~ **pneumatique** DÉCHETS pneumatic sorter, air separation plant, POLL ATM air classifier; ~ **à vis hélicoïdale** SOLS spiral classifier
classification *f* DÉCHETS, ÉCHANT, SOLS classification; ~ **pneumatique** DÉCHETS, DIST EAU pneumatic classification, POLL ATM air classification
classifier *vt* DÉCHETS, ÉCHANT, SOLS classify
classique *adj* ÉN RENOUV, POLL ATM, POLL SON, SOLS conventional
clause: ~ **explicative** *f* POL ENV explanatory clause
clé: ~ **de mesure** *f* ÉCHANT measuring switch
CLHP *abrév* (*chromatographie liquide à haute performance, chromatographie en phase liquide haute performance*) ÉCHANT HPLC (*high-performance liquid chromatography*)
client *m* GESTION customer
climat *m* POLL ATM climate; ~ **marin** POLL MAR marine climate
climatique *adj* POLL ATM climatic
cloche *f* POLL EAU gas container; ~ **à gaz** POLL EAU gas hood
cloison *f* POLL SON partition
clôture *f* DÉCHETS *décharge* fence
Co *abrév* (*cobalt*) POLL ATM, SOLS Co (*cobalt*)
CO *abrév* (*monoxyde de carbone*) POLL ATM CO (*carbon monoxide*)
CO_2 *abrév* DIST EAU (*anhydride carbonique*) CO_2 (*carbonic anhydride*), POLL ATM (*dioxyde de carbone*) CO_2 (*carbon dioxide*)
coagulant *m* GÉNÉRAL coagulant
coagulation *f* GÉNÉRAL coagulation; ~ **non spécifique** POLL EAU unspecific coagulation; ~ **par piégeage** POLL EAU sweep coagulation; ~ **spécifique** POLL EAU specific coagulation
coaguler[1] *vt* GÉNÉRAL coagulate, thicken
coaguler[2]: **se** ~ *v réfl* GÉNÉRAL coagulate, thicken
coalescence *f* DIST EAU, POLL ATM coalescence
cobalt *m* (*Co*) POLL ATM, SOLS cobalt (*Co*)
COD *abrév* (*carbone organique dissous*) DIST EAU, ÉCHANT, POLL ATM DOC (*dissolved organic carbon*)
codage *m* DÉCHETS, SOLS encoding; ~ **par contact en ligne** DÉCHETS in-line contact coding; ~ **des valeurs mesurées** ÉCHANT coding of measured values
code: ~ **de l'environnement** *m* POL ENV environmental code; ~ **de sécurité** *m* SÉCURITÉ safety code
coefficient *m* ÉCHANT, POLL ATM, POLL SON coefficient; ~ **d'abondance** ÉCHANT *nucléaire* abundance coefficient; ~ **d'absorption** POLL ATM, POLL SON absorption coefficient; ~ **d'absorption acoustique** POLL SON sound absorption coefficient; ~ **d'absorption linéaire** ÉCHANT linear absorption coefficient; ~ **d'absorption spectrale** ÉCHANT spectral absorption coefficient; ~ **de brume** POLL ATM coefficient of haze; ~ **d'écoulement** DIST EAU run-off coefficient; ~ **d'emprise au sol** POL ENV coverage ratio; ~ **d'extinction** ÉCHANT, POLL ATM extinction coefficient; ~ **d'extinction spectrale** ÉCHANT spectral extinction coefficient; ~ **de perméabilité** DÉCHETS, DIST EAU, POLL ATM, SOLS coefficient of permeability; ~ **de plénitude** ÉN RENOUV *éolienne* solidity; ~ **de ruissellement** POLL EAU run-off coefficient; ~ **de viscosité turbulente** POLL ATM *météorologie* eddy viscosity coefficient
coffre *m* DIST EAU coffer, POLL MAR mooring buoy
coffret: ~ **métallique** *m* DÉCHETS metal box
cogner *vt* POLL ATM knock
cohérence *f* ÉCHANT consistency
cohésion *f* POLL ATM cohesion
co-injection *f* DÉCHETS co-injection
coke: ~ **actif** *m* DÉCHETS, POLL ATM activated coke; ~ **actif chargé** *m* POLL ATM loaded activated coke; ~ **actif désorbé** *m* POLL ATM desorbed activated coke; ~ **actif régénéré** *m* POLL ATM regenerated activated coke; ~ **frais** *m* POLL ATM fresh coke
col: ~ **annulaire** *m* POLL ATM, SOLS *laveur à barbotage* annular gap; ~ **de Venturi** *m* POLL ATM Venturi throat
colibacille *m* DIST EAU coli bacillus, colibacterium
coliforme *m* DIST EAU coliform
collage *m* DÉCHETS, POLL ATM *charbon* caking, gluing; ~ **intégral** DÉCHETS full gluing
collant *adj* POLL ATM sticky
colle *f* DÉCHETS glue; ~ **à base d'acétate** DÉCHETS acetate glue; ~ **de contact** DÉCHETS contact adhesive; ~ **en émulsion** DÉCHETS emulsion adhesive; ~ **permanente résistante à la réfrigération** DÉCHETS chill-permanent adhesive
collecté *adj* DÉCHETS, ÉCHANT, POLL ATM, POLL MAR collected
collecte *f* DÉCHETS, DIST EAU, POLL MAR collection; ~ **des déchets** DÉCHETS waste collection; ~ **des déchets dangereux** DÉCHETS hazardous waste collection; ~ **des déchets volumineux** DÉCHETS bulk collection; ~ **distincte** DIST EAU separate collection; ~ **des eaux usées** POLL EAU wastewater collection; ~ **d'échantillons** ÉCHANT sample collection; ~ **de ferraille** DÉCHETS scrap collection; ~ **des gaz hors décharge** DÉCHETS gas collection outside of landfill; ~ **des huiles usées** DÉCHETS waste oil collection; ~ **d'informations sur l'environnement** POL ENV environmental information collection; ~ **des lixiviats** DÉCHETS leachate collection; ~ **des papiers** DÉCHETS paper collection; ~ **par degré de fractionnement** POLL ATM degree-of-fraction collection; ~ **par type** DÉCHETS separate collection by type; ~ **de particules** DÉCHETS, DIST EAU, POLL ATM particulate collection; ~ **privée** DÉCHETS private collection; ~ **de sacs poubelles** DÉCHETS refuse sack collection; ~ **sélective** DÉCHETS selective collection, separate collection; ~ **sélective des déchets** DÉCHETS selective refuse collection, separate refuse collection; ~ **sur un échantillon de population** SÉCURITÉ subpopulation collection; ~ **totale**

POLL ATM total collection; ~ **verticale des gaz** DÉCHETS, DIST EAU, POLL ATM vertical gas collection; ~ **de vieux papiers** DÉCHETS waste paper collection; ~ **en vrac** DÉCHETS bulk collection

collecter *vt* DÉCHETS, ÉCHANT, POLL ATM, POLL MAR collect; ~ **des échantillons** ÉCHANT collect samples

collecteur *m* DÉCHETS collector, ÉN RENOUV receiver, POLL ATM collecting main, POLL EAU receiving tank; ~ **d'admission** POLL ATM intake manifold; ~ **d'air comprimé** POLL ATM compressed-air receiver; ~ **de cendres volantes** POLL ATM flue-dust catcher; ~ **centrifuge** POLL ATM centrifugal collector; ~ **à ciel ouvert** POLL EAU open drain; ~ **circulaire** DIST EAU circular main; ~ **de déchets** DÉCHETS waste collector; ~ **à déflecteur** POLL SON baffle collector; ~ **à deux augets** DIST EAU double-bucket collector; ~ **de données** ÉCHANT data logger; ~ **d'eau brute** DIST EAU raw-water main; ~ **d'eaux usées** POLL EAU wastewater collector; ~ **d'échappement** POLL ATM exhaust manifold; ~-**fossé** DIST EAU drainage ditch; ~ **de fumées** POLL ATM smoke streamer; ~ **humide** POLL ATM wet collector; ~ **inertiel** POLL ATM inertial collector; ~ **mécanique** POLL ATM mechanical collector; ~ **noyé** POLL EAU underdrain; ~ **plan** ÉN RENOUV flat-plate collector; ~ **de poussière** POLL ATM dust catcher, dust receiver; ~ **de précipitation** POLL ATM precipitation collector; ~ **de précipitations atmosphériques** POLL ATM deposit gage (*AmE*), deposit gauge (*BrE*); ~ **principal** DÉCHETS, DIST EAU, POLL EAU main sewer, main collector; ~ **de sable** POLL EAU grit can; ~ **sans concentration** ÉN RENOUV nonconcentrating collector; ~ **secondaire** POLL EAU branch sewer; ~ **solaire à concentration** ÉN RENOUV concentrating collector; ~ **de surface** ÉN RENOUV surface collector; ~ **tubulaire à bain d'huile** ÉN RENOUV oil-filled receiving tube; ~ **à vide** ÉN RENOUV *solaire* mirror-evacuated tube

collectivité: ~ **locale** *f* GESTION, POL ENV local government

colleur *m* DÉCHETS gluing device

collimateur *m* ÉCHANT collimator

collision *f* DIST EAU collision, POLL ATM impingement, collision, SÉCURITÉ collision

colloïdal *adj* DIST EAU, POLL ATM, POLL EAU colloidal

colloïde *m* DIST EAU, POLL ATM, POLL EAU colloid; ~ **de dispersion** POLL EAU dispersion colloid; ~ **micellaire** POLL EAU association colloid; ~ **moléculaire** POLL EAU molecular colloid

colluvion *f* SOLS colluvium

colluvionnement *m* POLL EAU *des lacs* siltage

colmatage *m* DÉCHETS, DIST EAU binding, clogging, GÉNÉRAL *étanchement* sealing, POLL ATM, POLL EAU *engorgement* clogging; ~ **de dépoussiéreur** POLL ATM scrubber clogging

colmaté *adj* GÉNÉRAL *engorgé* clogged, *étanché* sealed

colmater *vt* DÉCHETS, DIST EAU bind, clog, GÉNÉRAL *étancher* seal, POLL ATM, POLL EAU *engorger* clog

colonne: ~ **d'absorption** *f* POLL ATM absorption column, *industrie pétrolière* absorber; ~ **d'adsorption** *f* ÉCHANT adsorption column; ~ **de barbotage en boucle** *f* POLL EAU loop-type bubble column; ~ **biologique à disques rotatifs** *f* SOLS rotating-disk biological contactor; ~ **à bulles** *f* POLL ATM bubble column; ~ **capillaire** *f* ÉCHANT capillary column; ~ **centrale d'arrivée d'eau** *f* DIST EAU influent pipe; ~ **de charbon actif** *f* ÉCHANT activated-carbon column; ~ **chromatographique** *f* ÉCHANT chromatographic column; ~ **à corps de remplissage** *f* DIST EAU, ÉCHANT, POLL ATM packed column; ~ **de désorption** *f* POLL ATM desorption column; ~ **de distillation** *f* ÉCHANT distillation column, POLL ATM distilling column; ~ **d'eau** *f* DIST EAU, ÉCHANT water column; ~ **d'eau oscillante** *f* ÉN RENOUV *énergie des vagues* oscillating water column (*OWC*); ~ **échangeuse d'ions** *f* POLL EAU ion-exchanger column; ~ **d'extraction** *f* DIST EAU extractor; ~ **filtrante** *f* POLL ATM filter column; ~ **garnie** *f* DIST EAU, ÉCHANT, POLL ATM packed column; ~ **à garnissage** *f* DIST EAU, ÉCHANT packed column, POLL ATM packed column, bag column, packed-bed column; ~ **de liquide** *f* ÉCHANT liquid column; ~ **à plateaux** *f* POLL ATM plate column, plate tower, tray-plate column, tray tower; ~ **à plateaux de barbotage** *f* POLL ATM bubble column, bubble-cap column, bubble-cap tray column, bubble-cap tray tower; ~ **à poche** *f* POLL ATM bag column; ~ **de pulvérisation** *f* POLL ATM spray column; ~ **de réaction** *f* ÉCHANT reaction chamber; ~ **de rectification** *f* SOLS stripping tower; ~ **de vapeur montante** *f* POLL ATM vapor riser (*AmE*), vapour riser (*BrE*)

coloration *f* ÉCHANT coloring (*AmE*), colouring (*BrE*); ~ **des eaux usées** ÉCHANT wastewater coloration ; ~ **de flamme** ÉCHANT flame coloration

colorimètre *m* ÉCHANT colorimeter; ~ **numérique portable** ÉCHANT portable digital colorimeter

colorimétrie *f* ÉCHANT colorimetry

colorimétrique *adj* ÉCHANT colorimetric

combiné *adj* SOLS combined

combiner: **se** ~ *v réfl* DIST EAU, POLL ATM coalesce

comblé *adj* DÉCHETS, DIST EAU, POLL EAU, SOLS landfilled; ~ **par du gravier** DIST EAU rubble-filled

comblement *m* DÉCHETS backfilling

combler *vt* DÉCHETS, DIST EAU, POLL EAU, SOLS landfill

comburant *m* POLL ATM, POLL EAU, SOLS oxidant, oxidizer, oxidizing agent

combustibilité *f* GÉNÉRAL combustibility

combustible[1] *adj* GÉNÉRAL combustible

combustible[2] *m* ÉN RENOUV, POLL ATM, POLL EAU fuel; ~ **antidétonant** POLL ATM antiknock fuel; ~ **d'appoint** POLL ATM, POLL EAU make-up fuel, supplementary fuel; ~ **en excès** POLL ATM excess fuel; ~ **fossile** ÉN RENOUV, POLL ATM fossil fuel; ~ **fossile combustible** ÉN RENOUV, POLL ATM combustible fossil fuel; ~ **obtenu à partir de la biomasse** ÉN RENOUV fuel from biomass; ~ **en phase gazeuse** POLL ATM gaseous fuel; ~ **en phase vapeur** POLL ATM vaporous fuel; ~ **primaire** DIST EAU primary fuel; ~ **solide** POLL ATM solid fuel; ~ **sulfureux** POLL ATM sulfurous combustible (*AmE*), sulphurous combustible (*BrE*)

combustibles: **à** ~ **fossiles** *loc* ÉN RENOUV fossil-fuel-driven, fossil-fuel-fired

combustion *f* ÉN RENOUV, POLL ATM, SÉCURITÉ combustion; ~ **catalytique** POLL ATM catalytic combustion; ~ **de combustible** POLL ATM fuel combustion; ~ **de combustible fossile** POLL ATM fossil-fuel combustion; ~ **excessive** SOLS overburning; ~ **incomplète** POLL ATM incomplete burning, incomplete combustion; ~ **de poussière** POLL ATM dust combustion; ~ **spontanée** SÉCURITÉ spontaneous combustion; ~ **stœchiométrique** POLL ATM stoichiometric combustion; ~ **sur grille** POLL ATM grate firing; ~ **sur lit fluidisé** POLL ATM fluidized-bed combustion (*FBC*); ~ **thermique** POLL ATM, POLL EAU thermal combustion

comité: ~ **de l'environnement** *m* POL ENV environment committee

Comité: ~ **Européen de Normalisation** *m* (*CEN*) ÉCHANT, POL ENV European Committee for Standardization (*CEN*)

commande[1]: ~ **à engrenage** *f* POLL EAU gear drive; ~ **de pompe** *f* DIST EAU pump controls; ~ **de pression** *f* POLL ATM pressure governor; ~ **de registre photoélectrique** *f* DÉCHETS photoelectric register control; ~ **à roues dentées** *f* POLL EAU gear drive

commande[2]: **à** ~ **analogique** *loc* ÉCHANT analog-controlled; **à** ~ **hydraulique** *loc* DIST EAU, ÉN RENOUV, POLL EAU hydraulically-controlled; **à** ~ **informatique** *loc* ÉCHANT software-controlled; **à** ~ **manuelle** *loc* ÉCHANT hand-driven, hand-operated

commander *vt* DÉCHETS, ÉCHANT, POLL ATM, POLL MAR control

commercialisable *adj* GESTION marketable

commercialisation *f* GESTION marketing

commercialisé *adj* ÉCHANT, GESTION, ÉN RENOUV, POLL MAR commercially available
commission: ~ **de garantie** *f* GESTION underwriting fee; ~ **de placement** *f* GESTION underwriting fee
Commission: ~ **du Développement Durable** *f* POL ENV Commission on Sustainable Development (*CSD*)
communauté: ~ **benthique** *f* POLL EAU, POLL MAR benthic community; ~ **vivante** *f* DIST EAU living community
communication: ~ **en matière d'environnement** *f* POL ENV environmental communication
communiqué: ~ **de presse** *m* POL ENV press communication, press release
commutateur: ~ **manométrique** *m* POLL ATM manometric switch
compactage *m* DÉCHETS SOLS compaction, compacting; ~ **automatique** DIST EAU self-compaction
compacté *adj* DÉCHETS, DIST EAU, SOLS compacted
compacter *vt* DÉCHETS, DIST EAU, SOLS compact
compacteur *m* DÉCHETS compacting device, compaction equipment, compactor, DIST EAU packer unit, SOLS compaction equipment, compacting device, compactor, packer; ~ **de déchets** DÉCHETS landfill compactor; ~ **intégré** DIST EAU internal compactor; ~ **de refus de dégrillage** POLL EAU screenings bale press
compagnie: ~ **des eaux** *f* DIST EAU, POL ENV water authority; ~ **d'élimination des déchets** *f* DÉCHETS, GESTION waste disposal company, waste disposal firm; ~ **privée d'élimination des déchets** *f* DIST EAU private disposal company; ~ **de service public** *f* GESTION public utility
comparateur: ~ **à cadran** *m* ÉCHANT dial gage (*AmE*), dial gauge (*BrE*)
compartiment *m* DIST EAU *décharge* refuse cell; ~ **d'assainissement** SOLS treatment cell; ~ **de décharge** DÉCHETS landfill cell; ~ **emboîtable** DÉCHETS compartmented insert; ~ **de filtre** POLL ATM filter compartment; ~ **de flottation** SOLS flotation cell; ~ **des gaz** POLL ATM gas compartment; ~ **des gaz propres** POLL ATM clean-gas compartment; ~ **des gaz à traiter** POLL ATM dirty-gas compartment
compatibilisateur *m* DÉCHETS, DIST EAU compatibilizer
compatibilité *f* DÉCHETS, DIST EAU compatibility, ÉCHANT consistency, POL ENV, POLL ATM compatibility; ~ **avec l'environnement** DIST EAU, POL ENV, POLL ATM environmental compatibility; ~ **avec le milieu** DIST EAU, POL ENV, POLL ATM environmental compatibility
compatible: ~ **avec l'environnement** *adj* GESTION, POL ENV environmentally-compatible
compensation *f* POL ENV, SÉCURITÉ compensation; ~ **de pression** DIST EAU pressure compensation; ~ **à zéro** ÉCHANT null balance
compenser *vt* POL ENV, SÉCURITÉ compensate
compétitivité *f* ÉN RENOUV, GESTION, POL ENV competitiveness
complexe: ~ **métallifère** *m* SOLS metal complex; ~ **polymères-xanthate** *m* POLL EAU xanthate polymer complex
complexométrique *adj* ÉCHANT *volumétrie* complexometric
comportement *m* GÉNÉRAL behavior (*AmE*), behaviour (*BrE*); ~ **du client** GESTION customer behavior (*AmE*), customer behaviour (*BrE*); ~ **du consommateur** GESTION consumer behavior (*AmE*), consumer behaviour (*BrE*); ~ **de dispersion** DIST EAU, SOLS dispersion behavior (*AmE*), dispersion behaviour (*BrE*); ~ **hydrodynamique dans le sous-sol** SOLS hydrodynamic behavior in subsoil (*AmE*), hydrodynamic behaviour in subsoil (*BrE*); ~ **au lessivage** DÉCHETS, DIST EAU, POLL EAU, SOLS leaching behavior (*AmE*), leaching behaviour (*BrE*); ~ **des lixiviats** DÉCHETS leachate behavior (*AmE*), leachate behaviour (*BrE*); ~ **à long terme** DÉCHETS, DIST EAU, POLL ATM long-term behavior (*AmE*), long-term behaviour (*BrE*); ~ **des polluants** POLL ATM, SOLS pollutant behavior (*AmE*), pollutant behaviour (*BrE*); ~ **des produits de lixiviation** DÉCHETS, DIST EAU, POLL EAU, SOLS leachate behavior (*AmE*), leachate behaviour (*BrE*); ~ **de retrait** DIST EAU shrinking behavior (*AmE*), shrinking behaviour (*BrE*); ~ **au rétrécissement** DIST EAU shrinking behavior (*AmE*), shrinking behaviour (*BrE*); ~ **au ruissellement** DÉCHETS leaching behavior (*AmE*), leaching behaviour (*BrE*), DIST EAU, POLL EAU leaching behavior (*AmE*), leaching behaviour (*BrE*); ~ **au stockage** DÉCHETS, DIST EAU storage behavior (*AmE*), storage behaviour (*BrE*); ~ **vibratoire** ÉN RENOUV vibration behavior (*AmE*), vibration behaviour (*BrE*)
composant *m* ÉCHANT, POLL ATM, SOLS component, constituent; ~ **d'air résiduel** POLL ATM waste air constituent; ~ **d'un gaz** POLL ATM gas component; ~ **de mesure** ÉCHANT *chaîne de mesure* measurement component; ~ **semiconducteur** ÉN RENOUV *électronique* semiconductor component
composé: ~ **alicyclique** *m* POLL ATM alicyclic compound; ~ **aliphatique** *m* POLL ATM aliphatic compound; ~ **azoté** *m* POLL ATM, POLL EAU nitrogen compound; ~ **isolant** *m* DÉCHETS, POLL SON insulating compound
composés: ~ **aromatiques** *m pl* POLL ATM aromatic compounds; ~ **haloformes** *m pl* POLL EAU haloform compounds; ~ **inorganiques volatils** *m pl* POLL ATM, SOLS volatile inorganic compounds (*VIC*); ~ **organiques halogénés éliminables** *m pl* DIST EAU purgeable organic halogen compounds (*POX*); ~ **organiques semi-volatils** *m pl* SOLS semivolatile organic compounds (*SVOC*); ~ **organiques totaux** *m pl* POLL ATM total organic compounds; ~ **organiques volatils** *m pl* (*COV*) POLL ATM, SOLS volatile organic compounds (*VOC*); ~ **organo-halogénés adsorbables** *m pl* DIST EAU adsorbable organic halogen compounds (*AOX*)
composite *adj* DÉCHETS, DIST EAU composite
composition: ~ **des déchets** *f* DÉCHETS waste composition; ~ **dégraissante** *f* DÉCHETS, ÉN RENOUV degreasing compound; ~ **des eaux de ruissellement** *f* DÉCHETS leachate composition; ~ **des eaux usées** *f* POLL EAU wastewater composition; ~ **d'échantillon** *f* ÉCHANT sample composition; ~ **des effluents gazeux** *f* POLL ATM waste gas composition; ~ **d'une émulsion** *f* POLL EAU emulsion composition; ~ **de fibre** *f* DÉCHETS, POLL ATM fiber composition (*AmE*), fibre composition (*BrE*); ~ **des gaz** *f* POLL ATM gas composition; ~ **des gaz de décharge** *f* DÉCHETS landfill gas composition; ~ **des lixiviats** *f* DÉCHETS leachate composition; ~ **des ordures** *f* DIST EAU refuse composition; ~ **des ordures ménagères** *f* DÉCHETS domestic waste composition
compost *m* DÉCHETS, DIST EAU, SOLS compost; ~ **brut** DIST EAU raw compost; ~ **frais** DÉCHETS fresh compost; ~ **mûri** DIST EAU readied compost; ~ **d'ordures** DIST EAU refuse compost
compostable *adj* DÉCHETS, DIST EAU, SOLS compostable
compostage *m* DÉCHETS, DIST EAU, SOLS composting; ~ **accéléré** DÉCHETS accelerated composting; ~ **aéré en tas** SOLS aerated pile composting (*APC*); ~ **automatique** DIST EAU self-composting; ~ **des boues** POLL EAU sludge composting; ~ **des boues d'eaux résiduaires** POLL EAU wastewater sludge composting; ~ **en conteneur** ÉN RENOUV container composting; ~ **distinct des boues déshydratées** POLL EAU separate composting of dewatered sludges; ~ **en fût** ÉN RENOUV *déchets nucléaires* container composting; ~ **lent** DIST EAU slow composting; ~ **liquide** POLL EAU liquid composting; ~ **mécanique** DÉCHETS mechanical composting; ~ **des ordures** DIST EAU refuse composting; ~ **sur lit non aéré** SOLS unaerated-bed composting
compresser *vt* DÉCHETS, ÉN RENOUV, SOLS compress
compresseur *m* ÉN RENOUV, POLL ATM compressor; ~ **d'air** POLL ATM air compressor; ~ **à canal latéral** SOLS side-channel compressor; ~ **centrifuge** POLL ATM centrifugal compressor; ~ **de gaz chaud** POLL ATM hot-gas compressor; ~ **à lame incurvée vers l'avant** POLL ATM forward-curved

blade compressor; ~ **à lame rétroincurvée** POLL ATM backward-curved blade compressor; ~ **à piston** DIST EAU, POLL EAU piston compressor; ~ **à piston rotatif** POLL ATM rotary-piston compressor; ~ **radial haute pression** SOLS high-pressure radial flow compressor; ~ **de vapeur** POLL ATM vapor compressor (*AmE*), vapour compressor (*BrE*); ~ **à vis** DIST EAU screw compressor

compression *f* GÉNÉRAL compression, pressing

comprimer *vt* GÉNÉRAL compress, press, press down, press together

comptabilité: ~ **écologique** *f* GESTION ecological account

comptage: ~ **des colonies** *m* POLL EAU colony count

compte: ~-**rendu de l'étude d'impact** *m* POL ENV environmental impact assessment report; ~-**rendu de mesure** *m* ÉCHANT measuring report

compteur: ~ **de contamination** *m* DIST EAU contamination meter; ~ **d'eau** *m* DIST EAU water meter; ~ **à gaz** *m* ÉCHANT gas meter; ~ **Geiger-Müller** *m* ÉCHANT Geiger-Müller counter; ~ **de noyaux de condensation** *m* ÉCHANT, POLL ATM condensation nucleus counter; ~ **de particules** *m* ÉCHANT particle counter, particle counting device; ~ **de particules de dispersion** *m* ÉCHANT dispersion particle counter; ~ **de particules solides en suspension** *m* ÉCHANT suspended solid particle meter; ~ **proportionnel** *m* ÉCHANT proportional counter; ~ **de transmittance** *m* ÉCHANT transmittance meter

concaisseur *m* DÉCHETS fraction separator

concassage *m* DÉCHETS fraction separator, grinding, ÉCHANT, POLL ATM fraction separator

concasser *vt* DÉCHETS, DIST EAU, SOLS crush, break, grind

concasseur *m* DÉCHETS, DIST EAU, SOLS crusher, grinder, breaker; ~ **à cylindre** DÉCHETS roll crusher; ~ **à grille** DÉCHETS grid crusher; ~ **à mâchoires** DÉCHETS impact crusher, jaw breaker; ~ **à percussion** DÉCHETS impact crusher; ~ **primaire** DÉCHETS *charbon* coarse crusher

concentration *f* GÉNÉRAL concentration; ~ **acide** POLL ATM acid loading; ~ **d'acides** POLL ATM acid concentration; ~ **d'affluents** POLL EAU influent concentration; ~ **d'air résiduel** POLL ATM waste air concentration; ~ **atmosphérique** POLL ATM ambient-air concentration, atmospheric concentration; ~ **de boues** POLL EAU mud content; ~ **de CO_2** POLL ATM CO_2 concentration; ~ **colloïdale** DIST EAU colloidal concentration; ~ **dans l'atmosphère** POLL ATM atmospheric load, atmospheric loading; ~ **de déchets** DIST EAU residue concentration; ~ **d'échantillon** ÉCHANT concentrated amount of sample; ~ **à effet le plus bas observé** SÉCURITÉ lowest observed effect concentration (*LOEC*); ~ **à effet le plus faible observé** SÉCURITÉ lowest observed effect concentration (*LOEC*); ~ **électrolytique** POLL EAU electrolytic concentration; ~ **d'émissions maximale admissible** ÉCHANT maximum allowable emissions concentration; ~ **de fumée** POLL ATM smoke concentration; ~ **de gaz pollué** POLL ATM dirty-gas concentration; ~ **de gaz propre** POLL ATM clean-gas concentration; ~ **d'huile** POLL EAU oil concentration; ~ **inhibitrice 50%** (CI_{50}) POLL EAU, SÉCURITÉ 50% inhibiting concentration; ~ **instantanée** DIST EAU instantaneous concentration; ~ **ionique** DIST EAU, POLL EAU ionic concentration; ~ **létale** (*CL*) DIST EAU, ÉCHANT, POLL ATM, SÉCURITÉ lethal concentration (*LC*); ~ **létale cinquante** (CL_{50}) DIST EAU, ÉCHANT, POLL ATM, SÉCURITÉ lethal concentration fifty (LC_{50}); ~ **létale à 50%** (CL_{50}) DIST EAU, ÉCHANT, POLL ATM, SÉCURITÉ median lethal concentration, lethal concentration fifty (LC_{50}); ~ **létale x** (CL_x) SÉCURITÉ lethal concentration x (LC_x); ~ **létale zéro** (CL_0) SÉCURITÉ lethal concentration zero (LC_0); ~ **létale zéro totale** SÉCURITÉ total lethal concentration zero (TCL_0); ~ **limite** POLL ATM, SÉCURITÉ limit concentration; ~ **en masse** POLL ATM mass concentration; ~ **de masse sèche** POLL EAU loading of dry mass; ~ **en matières sèches** POLL EAU dry mass concentration; ~ **maximale** ÉCHANT, POLL ATM peak concentration; ~ **maximale admissible** ÉCHANT, POLL ATM *dans l'air ambiant* maximum allowable concentration (*MAC*); ~ **maximale admissible à l'air libre** ÉCHANT, POLL ATM MAC in the free environment, maximum allowable concentration in the free environment; ~ **maximale admissible dans l'environnement** ÉCHANT, POLL ATM MAC in the free environment, maximum allowable concentration in the free environment; ~ **maximale admissible sur le lieu de travail** SÉCURITÉ maximum allowable concentration at the workplace; ~ **maximale d'émissions** DIST EAU, POLL ATM maximum emission concentration; ~ **maximale d'immissions** DIST EAU maximum immission concentration; ~ **maximale de produits toxiques admissible** SÉCURITÉ maximum allowable toxicant concentration (*MATC*); ~ **de métaux lourds** POLL ATM, SOLS heavy-metal concentration; ~ **au niveau du sol** SOLS ground level concentration (*GLC*); ~ **organique maximale admissible** ÉCHANT maximum allowable organic concentration; ~ **d'oxygène** ÉCHANT, POLL EAU oxygen concentration; ~ **en oxygène** ÉCHANT, POLL EAU oxygen concentration; ~ **d'ozone** POLL ATM ozone concentration; ~ **d'ozone dans l'atmosphère** POLL ATM atmospheric ozone concentration; ~ **d'ozone dans l'environnement** POLL ATM natural ozone content; ~ **d'ozone de référence** POLL ATM background ozone concentration; ~ **d'ozone stratosphérique** POLL ATM stratospheric ozone concentration; ~ **d'ozone troposphérique** POLL ATM tropospheric ozone concentration; ~ **des polluants** ÉCHANT, POLL ATM, POLL EAU pollutant concentration; ~ **de poussière** POLL ATM dust concentration, dust loading; ~ **de poussière dans un gaz pollué** POLL ATM dirty-gas dust content; ~ **de poussière dans un gaz propre** POLL ATM clean-gas dust content; ~ **de poussières en suspension dans l'air** POLL ATM, SÉCURITÉ airborne-dust concentration; ~ **de référence** POLL ATM background concentration; ~ **de résidus de polluants** POLL ATM residue concentration of pollutants; ~ **saline** ÉN RENOUV saline concentration; ~ **sans effet observé** (*CSEO*) SÉCURITÉ no observed effect concentration (*NOEC*); ~ **de solvants** POLL ATM solvent concentration; ~ **sous vide** DIST EAU vacuum concentration; ~ **de substances dangereuses** POLL ATM concentration of hazardous substances; ~ **de substances nocives** POLL EAU harmful substance concentration; ~ **de traces** POLL ATM trace concentration; ~ **des vapeurs du sol** SOLS soil vapor concentration (*AmE*), soil vapour concentration (*BrE*); ~ **à vide** DIST EAU vacuum concentration; ~ **volumique** POLL ATM volume concentration

concentré[1] *adj* GÉNÉRAL concentrated

concentré[2] *m* GÉNÉRAL concentrate; ~ **de déchets dangereux** DÉCHETS hazardous waste concentrate

concentrer *vt* DÉCHETS, DIST EAU, ÉN RENOUV *rayons*, POLL ATM, SOLS concentrate

conception *f* DÉCHETS, DIST EAU, GESTION, POLL ATM design; ~ **à coût objectif** GESTION design to cost; ~ **d'une décharge** DÉCHETS landfill design; ~ **de l'environnement** GESTION environmental idea; ~ **modulaire** DÉCHETS modular design; ~ **de la zone de lavage** POLL ATM scrubbing-area design

concession *f* POL ENV leasing

condensat *m* POLL ATM, POLL EAU, SOLS condensate

condensateur: ~ **à PCB** *m* DIST EAU PCB-capacitor

condensation *f* GÉNÉRAL condensation; ~ **d'échantillonnage** ÉCHANT sampling condensation

condenser *vt* GÉNÉRAL condense

condenseur *m* GÉNÉRAL condenser; ~ **de buées** POLL EAU vent condenser; ~ **à contact direct** POLL ATM direct-contact condenser; ~ **à double tube** POLL ATM double-pipe condenser; ~ **indirect** POLL ATM indirect condenser; ~ **par contact** POLL ATM contact condenser; ~ **à plaques spirales** POLL ATM spiral-plate condenser; ~ **de surface**

POLL ATM surface condenser; ~ **tubulaire** POLL ATM shell-and-tube condenser
conditionnement *m* DÉCHETS, DIST EAU, POLL ATM conditioning; ~ **des boues** POLL EAU sludge conditioning; ~ **des boues liquides** POLL EAU liquid-sludge conditioning; ~ **des cendres volantes** POLL ATM fly-ash conditioning; ~ **de gaz pollué** POLL ATM dirty-gas conditioning; ~ **sous atmosphère contrôlée** DÉCHETS controlled-atmosphere packaging (*CAP*); ~ **sous carte** DÉCHETS carded packaging
conditionner *vt* DÉCHETS, DIST EAU, POLL ATM condition
conditionneur: ~ **de sol** *m* SOLS soil conditioner, soil ameliorant
conditions *f pl* GÉNÉRAL conditions; ~ **applicables aux émissions** POL ENV emissions requirements; ~ **atmosphériques** POLL ATM atmospheric conditions; ~ **d'autorisation** POL ENV conditions for authorization; ~ **climatiques** DÉCHETS, POLL ATM climatic conditions; ~ **climatiques éprouvantes** POLL ATM climatic stress; ~ **de décantation** POLL EAU settling conditions; ~ **d'essai** DÉCHETS, DIST EAU test conditions; ~ **d'hygiène et de sécurité** SÉCURITÉ health and safety requirements; ~ **juridiques** POL ENV legal requirements; ~ **de mesurage** ÉCHANT measurement conditions, measuring conditions; ~ **météorologiques** ÉN RENOUV, POLL ATM metereological conditions, weather conditions; ~ **météorologiques favorisant le brouillard photochimique** POLL ATM smog weather conditions; ~ **de mise en décharge** DÉCHETS landfill requirement; ~ **normales** ÉCHANT standard conditions (*SC*); ~ **normales de température et pression** ÉCHANT standard temperature and pressure (*STP*); ~ **d'octroi de licence** POL ENV conditions for licensing; ~ **préliminaires de mesure** ÉCHANT measurement prerequisites; ~ **de rejet** POLL ATM discharge conditions
conductance: ~ **spécifique** *f* ÉCHANT specific conductance
conductibilité *f* ÉCHANT, POLL ATM conductibility; ~ **thermique** ÉCHANT, POLL ATM heat conductivity
conductimètre *m* ÉCHANT conductometer
conductimétrique *adj* ÉCHANT conductometric
conduction: ~ **thermique** *f* ÉCHANT, POLL ATM heat conduction, thermal conduction
conductivimètre *m* ÉCHANT conductivity measuring instrument, conductivity meter; ~ **industriel** ÉCHANT process conductivity meter
conductivité *f* DIST EAU, ÉCHANT, SOLS conductivity; ~ **superficielle** POLL ATM surface conductivity; ~ **thermique** ÉCHANT, POLL ATM heat conductivity
conduire *vt* GESTION conduct, POLL ATM conduct, lead, POLL SON conduct
conduit *m* DIST EAU pipe, POLL ATM duct; ~ **principal** DIST EAU distributing pipe; ~ **de récupération** POLL ATM channel recuperator
conduite *f* DIST EAU, POLL ATM, POLL EAU conduit, pipe, pipeline; ~ **d'adduction** DIST EAU water main; ~ **alvéolée** POLL ATM honeycomb-shaped channel; ~ **d'amorçage** DIST EAU priming pipe; ~ **d'aspiration** POLL ATM suction pipe; ~ **d'aspiration noyée** DIST EAU underwater suction pipe; ~ **en béton armé** DIST EAU, POLL EAU reinforced-concrete pipe; ~ **en béton précontraint** DIST EAU, POLL EAU prestressed-concrete pipe; ~ **de boues de recyclage** POLL EAU return sludge pipe; ~ **centrale** DÉCHETS central pipe; ~ **de chauffage** DÉCHETS *plastique* heating channel; ~ **de cheminée** POLL ATM flue; ~ **à contre-courant** POLL ATM reverse-flow duct; ~ **de débordement** POLL ATM overflow pipe; ~ **de décharge** POLL EAU *égout* effluent sewer, effluent channel; ~ **de dégazage** *m* DIST EAU, ÉN RENOUV degassing pipe; ~ **de dérivation** POLL ATM bypass duct; ~ **distincte** POLL EAU separate canalization; ~ **de distribution** DÉCHETS distributor worm; ~ **de distribution d'eau** DIST EAU water main; ~ **d'eau** DIST EAU water pipe; ~ **d'eau brute** DIST EAU raw-water main; ~ **d'échantillonnage** ÉCHANT sampling line; ~ **d'égout** DÉCHETS discharge worm; ~ **d'évacuation** POLL EAU outlet channel, outlet pipe, discharge pipe; ~ **d'évents** POLL ATM dust-transporting worm; ~ **forcée** DIST EAU penstock, POLL EAU pressure pipe; ~ **de gaz épuré** POLL ATM clean-gas canal; ~ **de gaz pollué** POLL ATM dirty-gas canal; ~ **en grès cérame vitrifié** DIST EAU vitrified-clay pipe; ~ **mixte** POLL EAU mixed canalization; ~ **principale** DÉCHETS, DIST EAU main pipe; ~ **de puits** DIST EAU well pipe; ~ **de refoulement** DIST EAU delivery main, delivery pipe
cône *m* DIST EAU, ÉCHANT, POLL ATM cone; ~ **d'alluvions** DIST EAU alluvial cone; ~ **d'appel** DIST EAU depression cone; ~ **d'aspersion** POLL ATM spraying cone; ~ **de dépression** DIST EAU depression cone; ~ **d'éboulis** DIST EAU alluvial cone; ~ **d'Imhoff** ÉCHANT Imhoff cone; ~ **Seger** ÉCHANT Seger cone
conférence: ~ **sur les procédures d'autorisation** *f* DIST EAU, POL ENV licensing conference
configuration: ~ **spatiale** *f* DIST EAU spatial pattern
confiné *adj* DÉCHETS encapsulated, DIST EAU confined, SOLS encapsulated
confinement *m* DÉCHETS encapsulation, DIST EAU confinement, encapsulation, GÉNÉRAL *des substances nocives* sealing; ~ **en grains** DÉCHETS grain encapsulation
confiner *vt* DÉCHETS encapsulate, DIST EAU confine, SOLS encapsulate
confluent *m* DIST EAU confluence
congélation: ~ **des boues** *f* POLL EAU *conditionnement* sludge freezing; ~ **du sol** *f* SOLS soil freezing
conglutination *f* GÉNÉRAL conglutination, coagulation
conglutiner *vt* GÉNÉRAL conglutinate, coagulate
conimètre *m* ÉCHANT *mesure de particules* conimeter; ~ **à impact** ÉCHANT impactor, impinger; ~ **à impact en cascade** ÉCHANT cascade impactor
conique *adj* POLL ATM conical
connexion: ~ **en étoile** *f* POLL EAU Wye branch, Wye connection
conseil: ~ **en matière d'environnement** *m* POL ENV environmental advice, environmental counseling (*AmE*), environmental counselling (*BrE*)
Conseil: ~ **des ministres de l'environnement** *m* POL ENV Council of environmental ministers
conseiller: ~ **à l'environnement** *m* POL ENV environmental adviser
conservation *f* GÉNÉRAL conservation; ~ **de l'eau douce** DIST EAU freshwater conservation; ~ **de la faune et de la flore sauvage** POL ENV wildlife conservation; ~ **du sol** SOLS soil conservation
conserver *vt* GÉNÉRAL conserve; ~ **à l'abri de la lumière** ÉCHANT keep in the dark; ~ **au frais** ÉCHANT keep cool
consigne: ~ **de bouteille** *f* DÉCHETS bottle deposit; ~ **d'ouverture** *f* DÉCHETS instructions for opening
console: ~ **de visualisation** *f* ÉCHANT display unit
consolidation *f* POL ENV funding
consommable *abrév* DIST EAU drinkable
consommateur[1]: ~ **d'énergie** *adj* ÉN RENOUV, POLL ATM energy-consumptive; **peu** ~ **d'énergie** *adj* ÉN RENOUV, POLL ATM low-energy
consommateur[2] *m* GESTION consumer
consommation *f* GÉNÉRAL use, consumption, POLL ATM, POLL EAU consumption, *d'eau* intake; ~ **d'air** DIST EAU *brûleur* aeration; ~ **d'air comprimé** POLL ATM compressed-air consumption; ~ **de chaleur** POLL ATM heat use; ~ **de chaleur de récupération** DÉCHETS, DIST EAU, POLL ATM, POLL EAU waste heat use, waste heat utilization; ~ **de chaleur résiduelle** DÉCHETS, DIST EAU, POLL ATM, POLL EAU waste heat use, waste heat utilization; ~ **d'eau** DIST EAU water consumption; ~ **d'eau de boisson des ménages** DIST EAU private drinking water consumption; ~ **d'eau fonction de la population** DIST EAU population-related water consumption; ~ **d'eau potable** DIST EAU drinking

water consumption; ~ **effective d'énergie** ÉN RENOUV, POL ENV effective energy use; ~ **d'énergie** DÉCHETS energy consumption, energy use, DIST EAU energy consumption, energy expenditure, energy use, ÉN RENOUV energy consumption, energy expenditure, energy use, POL ENV energy consumption, energy use, POLL ATM energy consumption, energy expenditure, energy use; ~ **d'énergie finale** ÉN RENOUV final energy consumption; ~ **d'énergie par habitant** ÉN RENOUV per capita energy consumption; ~ **d'énergie primaire** ÉN RENOUV primary energy consumption; ~ **globale d'énergie** ÉN RENOUV global energy consumption; ~ **à grande échelle** ÉN RENOUV large area use; ~ **journalière** DIST EAU daily consumption; ~ **maximale** DIST EAU peak consumption; ~ **mondiale d'énergie** ÉN RENOUV global energy consumption; ~ **d'oxygène** POLL EAU oxygen consumption; ~ **de papier** DIST EAU paper consumption; ~ **de papier recyclé** DIST EAU recycled-paper use; ~ **par habitant** DIST EAU per capita consumption; ~ **passive d'énergie thermique d'origine solaire** ÉN RENOUV passive thermal solar energy use; ~ **de pointe** DIST EAU peak demand; ~ **des ressources** DIST EAU resource consumption; ~ **thermique isolée** ÉN RENOUV isolated thermal use; ~ **verte** GESTION green consumerism

consommer *vt* GÉNÉRAL use, consume; ~ **l'énergie stockée** ÉN RENOUV use stored energy

constance: ~ **de lecture** *f* ÉCHANT reading constancy

constante: ~ **d'Avogadro** *f* POLL ATM Avogadro's constant; ~ **d'équilibre** *f* POLL ATM equilibrium constant; ~ **de Planck** *f* POLL ATM Planck's constant

constituant *m* SOLS constituent

constitution: ~ **d'une base de données sur les émissions** *f* POL ENV emissions banking; ~ **d'un dépôt** *f* POLL ATM deposit build-up

construction: ~ **insonorisée** *f* POLL SON low-noise construction; ~ **en stratifié isolant** *f* POLL SON sandwich construction

contact *m* POLL ATM contact

contaminant *m* GÉNÉRAL contaminant, pollutant; ~ **concentré** SOLS contaminant concentrate; ~ **gazeux** POLL ATM gaseous contaminant; ~ **radioactif** POLL ATM, POLL EAU radioactive contaminant; ~ **du sol** SOLS contaminant bound to soil; ~ **thermoadoucissant** DÉCHETS thermo-softening contaminant

contamination *f* GÉNÉRAL contamination, pollution; ~ **des aliments** POLL ATM, POLL EAU, SÉCURITÉ food contamination; ~ **du catalyseur** POLL ATM catalyst contamination; ~ **des eaux** POLL EAU water pollution; ~ **des eaux souterraines** DÉCHETS, DIST EAU, SOLS groundwater contamination; ~ **des filtrats** POLL EAU filtrate contamination; ~ **radioactive** DIST EAU, POLL EAU radioactive pollution; ~ **du sol** SOLS soil contamination; ~ **du sol imputable aux activités humaines** SOLS anthropogenically-caused soil contamination

contaminé *adj* GÉNÉRAL contaminated, contaminant-laden; ~ **par le bruit** POLL SON noise-contaminated

contaminer *vt* DIST EAU, POLL ATM, POLL EAU contaminate, pollute, SÉCURITÉ *radioactivité, chimie* contaminate, SOLS contaminate, pollute

conteneur *m* DÉCHETS, POLL ATM, SOLS container; ~ **de boisson** DÉCHETS beverage container; ~ **de collecte** DÉCHETS bank; ~ **compartimenté** DÉCHETS multichamber container, multiunit container; ~ **de déchets** DÉCHETS waste container, *agriculture* tailings container; ~ **de déchets amovible** DIST EAU roll-out waste container; ~ **de déchets de grande capacité** DÉCHETS roll-out waste container; ~ **de dépôt** DÉCHETS depot container; ~ **d'échantillonnage** ÉCHANT sample container; ~ **enterré** POL ENV ground container; ~ **fixe** DÉCHETS, DIST EAU stationary container system (*SCS*); ~ **de grande contenance** DÉCHETS large-size container; ~ **de grandes dimensions** DÉCHETS large-size container; ~ **gris** DÉCHETS gray container (*AmE*), grey container (*BrE*); ~ **industriel pour vrac** DÉCHETS industrial bulk-container system; ~ **jetable** DÉCHETS disposable container; ~ **de liquide dense** DÉCHETS heavy-medium container; ~ **en matériau composite** DÉCHETS composite container; ~ **de matériaux de récupération** DÉCHETS secondary-resource container; ~ **multicellulaire** DÉCHETS multi-unit container; ~ **multicouche** DÉCHETS multilayer container; ~ **d'otons** DÉCHETS *agriculture* tailings container; ~ **de parc** DÉCHETS depot container; ~ **perdu** DÉCHETS one-way container, nonreturnable container; ~ **en plastique** DÉCHETS, DIST EAU plastic container; ~ **conteneur à plastiques** DÉCHETS, DIST EAU plastic container; ~ **en polyéthylène** DÉCHETS, DIST EAU polyethylene container; ~ **en polyoléfine** DÉCHETS, DIST EAU polyolefin container; ~ **pour verre usagé** DÉCHETS bottle bank; ~ **de stockage** SOLS depot container; ~ **de stockage temporaire en vrac** DÉCHETS intermediate bulk container; ~ **sur roulettes** DÉCHETS caster-equipped container; ~ **de transport** DÉCHETS transport container; ~ **d'usage courant** DÉCHETS general cargo; ~ **à verre** DÉCHETS glass container; ~ **de verre perdu** DÉCHETS waste glass container; ~ **de verre usagé** DÉCHETS waste glass container; ~ **vert** DIST EAU green container

conteneurisable *adj* DÉCHETS, ÉN RENOUV, SOLS containerizable

conteneurisation *f* DÉCHETS, ÉN RENOUV, SOLS containerization

conteneurisé *adj* DÉCHETS, ÉN RENOUV, SOLS containerized

conteneuriser *vt* DÉCHETS, ÉN RENOUV, SOLS containerize

continu *adj* ÉCHANT *procédé* continuous

contrainte *f* POLL ATM, POLL MAR, SOLS stress; ~ **climatique** POLL ATM climatic stress; ~ **s'exerçant sur l'environnement** POLL ATM, SOLS environmental stress; ~ **tangentielle** SOLS transverse strain

contrat: ~ **d'utilisation** *m* POL ENV user's contract

contre: ~**-courant** *m* POLL ATM cross flow, reverse flow; ~**-fossé** *m* DIST EAU berm ditch; ~**-pression** *f* POLL ATM back pressure

contrevenant *m* POL ENV infringer, tortfeasor

contrevenir: ~ **à** *vt* POL ENV, SÉCURITÉ contravene; ~ **à la réglementation** *vt* POL ENV, SÉCURITÉ contravene regulations

contrôlabilité *f* POLL ATM controllability

contrôle *m* GÉNÉRAL *maîtrise* control, *vérification* check, checking, monitoring; ~ **biologique** ÉCHANT biomonitoring; ~ **du bruit** ÉCHANT, POLL SON noise monitoring; ~ **des éluats** ÉCHANT, POLL EAU eluate control; ~ **de granulométrie par tamisage** ÉCHANT sieve analysis; ~ **à long terme** POLL ATM, POLL SON long-term control; ~ **des particules** ÉCHANT, POLL ATM particulates monitoring; ~ **photographique infrarouge** ÉCHANT photographic infrared control; ~ **de la qualité de l'air ambiant** POLL ATM, SÉCURITÉ quality monitoring of ambient air; ~ **de la qualité de l'eau de consommation** POLL EAU, SÉCURITÉ drinking water quality control; ~ **de sécurité** SÉCURITÉ safety check; ~ **de température** POLL ATM *brasserie* attemperation; ~ **de valeur limite** ÉCHANT limit-value control; ~**-qualité** DIST EAU quality control; ~**-qualité d'analyse** ÉCHANT analytical quality control

contrôler *vt* GÉNÉRAL *maîtriser* control, *vérifier* check, monitor

contrôleur: ~ **de bruit** *m* ÉCHANT, POLL MAR, POLL SON noise monitor; ~ **de niveau** *m* ÉCHANT level recorder

conurbation *f* GESTION, POL ENV conurbation

convection *f* ÉCHANT, ÉN RENOUV, POLL ATM, POLL EAU convection; ~ **naturelle** POLL EAU natural convection; ~ **thermique** POLL ATM heat convection

convention *f* POL ENV convention

conversion *f* GÉNÉRAL conversion; ~ **à basse température**

POLL EAU low-temperature conversion; ~ **de la bioénergie** DÉCHETS, DIST EAU biological-energy conversion; ~ **de dette en investissements écologiques** POL ENV debt for nature swaps; ~ **énergétique** ÉN RENOUV, POLL ATM energy conversion; ~ **de l'énergie éolienne** ÉN RENOUV wind energy conversion (*WEC*); ~ **de l'énergie par voie biologique** DÉCHETS, DIST EAU biological energy conversion; ~ **de l'énergie photovoltaïque** ÉN RENOUV photovoltaic-energy generation; ~ **de l'énergie des vagues** ÉN RENOUV wave energy conversion; ~ **de polluants** POLL ATM pollutant conversion; ~ **sous forme d'énergie consommable** ÉN RENOUV conversion into usable energy form; ~ **thermochimique** ÉN RENOUV thermochemical conversion

convertir *vt* GÉNÉRAL convert; ~ **en chaleur** ÉN RENOUV convert to heat; ~ **en électricité** ÉN RENOUV convert to electricity

convertisseur *m* GÉNÉRAL converter; ~ **catalytique** POLL ATM catalytic converter; ~ **catalytique en circuit fermé** POLL ATM catalytic converter closed loop; ~ **catalytique en circuit ouvert** POLL ATM catalytic converter open loop; ~ **catalytique à commande informatique** POLL ATM computer-controlled catalytic converter; ~ **catalytique à trois voies** POLL ATM three-way catalyst (*TWC*); ~ **CC/CA** ÉN RENOUV DC/AC converter; ~ **de mesure** ÉCHANT measuring transducer

convoyeur *m* DÉCHETS, DIST EAU conveyor; ~ **en auge** DÉCHETS, DIST EAU troughed chain conveyor; ~ **en auge avec chaîne d'alimentation** DÉCHETS, DIST EAU feeding-chain trough conveyor; ~ **à bande** DÉCHETS, DIST EAU belt conveyor, POLL EAU transport conveyor belt; ~ **à chaîne** POLL ATM chain conveyor; ~ **à charnières** DÉCHETS, DIST EAU flat-top chain conveyor; ~ **de décharge** DÉCHETS, DIST EAU discharge conveyor; ~ **à godets** DÉCHETS, DIST EAU bucket conveyor; ~ **à plateaux** DÉCHETS, DIST EAU apron conveyor; ~ **à plateaux pivotants** DÉCHETS, DIST EAU pivoting pan conveyor; ~ **à secousses** DÉCHETS oscillating conveyor; ~ **à vis sans fin** DÉCHETS, DIST EAU screw conveyor

copeau: ~ **de bois** *m* ÉN RENOUV wood chip; **copeaux de rabotage** *m pl* ÉN RENOUV shavings

copolymère *m* DÉCHETS, DIST EAU copolymer; ~ **bloc** DÉCHETS block polymer

copolymérisation *f* DÉCHETS, DIST EAU copolymerization

coque: ~ **de cabine** *f* SÉCURITÉ cab enclosure

corbeille: ~ **à papier** *f* DÉCHETS wastepaper basket

corde: ~ **en chanvre** *f* DÉCHETS hemp rope

cordiérite *f* POLL ATM cordierite

cordon: ~ **littoral** *m* POLL MAR littoral fringe; ~ **oléophile** *m* POLL MAR oleophilic belt

corps: ~ **en acier** *m* POLL ATM steel body; ~ **céramique** *m* POLL ATM ceramic body; ~ **de cyclone** *m* POLL ATM cyclone housing; ~ **de décharge** *m* SOLS landfill body; ~ **d'échappement** *m* POLL ATM exhaust housing; ~ **de filtre** *m* POLL ATM filter housing; ~ **granuleux** *m* POLL ATM granulate body; ~ **métallique** *m* POLL ATM metal housing; ~ **noir** *m* ÉCHANT *physique*, POLL ATM black body; ~ **de pompe** *m* DIST EAU, POLL EAU pump barrel, pump case, pump casing, pump cylinder; ~ **protecteur** *m* POLL ATM protective housing; ~ **à revêtement céramique** *m* POLL ATM ceramic-coated body; ~ **de sonde** *m* SOLS drilling shaft; ~ **support** *m* POLL ATM support body

correcteur: ~ **d'étalonnage** *m* ÉCHANT calibration buffer

corrosif *adj* DÉCHETS, DIST EAU, POLL ATM, SÉCURITÉ corrosive

corrosion *f* DÉCHETS, DIST EAU, POLL ATM, SÉCURITÉ corrosion

corrosivité *f* DÉCHETS, DIST EAU, POLL ATM, SÉCURITÉ corrosiveness

cosmique *adj* POLL ATM cosmic

cosse *f* DÉCHETS eyelet, DIST EAU, POLL EAU socket

côté: ~ **admission** *m* POLL EAU suction side; ~ **haute pression** *m* POLL ATM high-pressure side; ~ **utilisable** *m* DÉCHETS operating face

couchage *m* DÉCHETS *papier* coating, *plastique* film coating; ~ **du papier** DÉCHETS paper liner

couche[1] *f* DIST EAU layer, ÉN RENOUV cover, POLL ATM, POLL SON, SOLS layer; ~ **active de catalyseur** POLL ATM catalytic active layer; ~ **d'air** POLL ATM air layer; ~ **d'air statique** POLL ATM static air layer; ~ **alluvionnaire** DIST EAU aggradational deposit; ~ **d'amortissement** POLL SON damped layer; ~ **annelée** POLL ATM ringed layer; ~ **aquifère** DIST EAU aquifer; ~ **argileuse** DÉCHETS clay layer; ~ **de barrage** DÉCHETS, DIST EAU barrier layer; ~ **comprimée** POLL SON constrained layer; ~ **construite** DÉCHETS constructed layer; ~ **de couverture** DÉCHETS *décharge* covering layer, top layer, POLL SON covering sheet; ~ **de déchets** DÉCHETS waste layer; ~ **drainante** DÉCHETS drainage layer; ~ **élastique** POLL SON resilient layer; ~ **étanche** DÉCHETS tight layer; ~ **d'étanchéité** DIST EAU sealing layer; ~ **de fibre** POLL ATM fiber layer (*AmE*), fibre layer (*BrE*); ~ **filtrante** DIST EAU filter bed, POLL ATM filter layer, POLL EAU filter bed; ~ **filtrante du sol** DIST EAU soil filter-layer; ~ **de filtre à gravier** DÉCHETS gravel filter layer; ~ **de filtre morainique** DÉCHETS, POLL ATM morainic filter layer; ~ **fluidifiée** DIST EAU fluidized bed; ~ **de frottement** POLL ATM *météorologie* friction layer; ~ **de gaz propres** POLL ATM clean-gas lid; ~ **de granulats** POLL ATM granulate layer; ~ **granuleuse** POLL ATM granulated layer, SOLS pellet layer; ~ **imperméable** DIST EAU aquifuge; ~ **intermédiaire** POLL SON intermediate layer; ~ **d'inversion** POLL ATM *météorologie* inversion layer; ~ **d'inversion atmosphérique** POLL ATM *métérologie* atmospheric inversion layer; ~ **isolante** DÉCHETS, POLL SON insulating layer; ~ **d'isolation** DÉCHETS, POLL SON insulating layer; ~ **limite** POLL ATM *météorologie* boundary layer; ~ **limite atmosphérique** POLL ATM atmospheric boundary layer; ~ **limite laminaire** POLL ATM laminar boundary layer; ~ **limite planétaire** POLL ATM *météorologie* planetary boundary layer (*PBL*); ~ **liquide** POLL ATM liquid layer; ~ **de membrane** DIST EAU, POLL EAU membrane layer; ~ **de milieu ambiant** SOLS environmental medium compartment; ~ **mince** DÉCHETS, DIST EAU, ÉCHANT, POLL EAU thin layer; ~ **de minéral argileux** SOLS clay mineral layer; ~ **d'ozone** POLL ATM ozone layer; ~ **de particules de poussière** POLL ATM dust-particulate layer; ~ **perméable** DIST EAU aquifer; ~ **de poussière** POLL ATM dust coating, SÉCURITÉ dust cover; ~ **principale d'étanchéité** DÉCHETS, POLL ATM main sealing layer; ~ **protectrice** DIST EAU protective layer; ~ **résiliente** POLL SON resilient layer; ~ **de revêtement** ÉN RENOUV cover, SOLS covering layer; ~ **de sable** DIST EAU sand layer; ~ **semi-perméable** DIST EAU semi-impermeable layer; ~ **de séparation** DIST EAU separation layer; ~ **de sol** SOLS soil layer; ~ **de surface** POLL MAR surface layer; ~ **temporaire élastique** POLL ATM elastic temporary layer; ~ **turbulente** DIST EAU fluidized solid system

couche[2]: **en** ~ *loc* ÉCHANT layered; **à** ~ **mince** *loc* DÉCHETS, DIST EAU, ÉCHANT, POLL EAU thin-layered; **à** ~ **unique** *loc* POLL SON single-sheet

couché: ~ **par émulsion** *adj* DÉCHETS, DIST EAU emulsion-coated

coude *m* POLL ATM elbow

coulable *adj* DIST EAU, POLL ATM pourable

coulage *m* DIST EAU *d'eau* leakage

coulant *adj* DIST EAU, POLL ATM *qui coule* flowing, *capable d'être versé* pourable

couler[1] *vt* DIST EAU *l'eau d'une bâche* run off

couler[2] *vi* DIST EAU flow, run, ÉN RENOUV, POLL ATM flow, POLL EAU flow, run; ~ **goutte à goutte** DIST EAU trickle

couloir *m* POLL ATM downspout

coulomètre *m* ÉCHANT coulometer

coulométrie *f* ÉCHANT coulometry

coulométrique *adj* ÉCHANT coulometric

coup: ~ **de bélier** *m* DIST EAU water hammer

coupe[1]: **~-feu** *adj* DÉCHETS, SÉCURITÉ *dispositif* fireproof
coupe[2] *f* DÉCHETS cup, ÉN RENOUV *du bois* cutting, harvesting; **~ stratigraphique** SOLS soil column
coupelle *f* DÉCHETS cup, base cup, ÉCHANT test crucible; **~ de mesure** ÉCHANT measuring cup
couplage *m* DIST EAU connection, ÉCHANT, POLL ATM coupling; **~ magnétique** POLL EAU magnetic coupling; **~ au réseau** ÉN RENOUV grid connection
couple: **~ moteur** *m* ÉN RENOUV torque
couplé: **~ au réseau** *adj* ÉN RENOUV grid-coupled
coupleur: **~ acoustique** *m* ÉCHANT, POLL SON acoustic coupler
courant *m* DIST EAU, ÉN RENOUV, POLL ATM, POLL EAU current, flow, stream; **~ alternatif** (*CA*) ÉN RENOUV alternating current (*AC*); **~ continu** (*CC*) ÉN RENOUV direct current (*DC*); **~ de densité** DIST EAU, POLL EAU density current; **~ dérivé** DIST EAU underflow; **~ de fond** DIST EAU underflow; **~ ionique** POLL ATM ion current; **~ marin** ÉN RENOUV ocean current; **~ océanique** ÉN RENOUV ocean current; **~ partiel** POLL ATM partial stream, POLL EAU partial current; **~ photoélectrique** ÉN RENOUV photoelectric current; **~ principal** POLL ATM main stream; **~ profond** DIST EAU underflow
courantomètre *m* DIST EAU current meter
courbe: **~ de bruit** *f* ÉCHANT noise curve, noise criterion curve (*NC-curve*); **~ de débit** *f* POLL EAU flow-duration curve; **~ d'étalonnage** *f* ÉCHANT calibration curve; **~ d'évaluation de bruit** *f* ÉCHANT, POLL SON noise rating curve (*NR-curve*); **~ de filtration** *f* POLL ATM filtration curve; **~ normale temps-température** *f* ÉCHANT standard curve time-temperature; **~ de régression** *f* ÉCHANT regression curve; **~ de rendement de niveau** *f* POLL ATM grade efficiency curve; **~ de sédimentation** *f* POLL EAU sedimentation curve
courbure *f* ÉN RENOUV curl
couronne *f* POL ENV urban fringe, POLL ATM annulus, corona
couronnement *m* DIST EAU crest
courroie: **~ d'alimentation** *f* DIST EAU *d'une presse* feeder belt
cours *m* DIST EAU *d'une rivière* course, flow; **~ d'eau** DIST EAU river, watercourse; **~ inférieur** DIST EAU lower course; **~ supérieur** DIST EAU upper course
course *f* DIST EAU lift; **~ de presse** DIST EAU press stroke
coussin: **~ d'air** *m* POLL SON air cushion
coussinet: **~ amortisseur** *m* POLL SON attenuator pad
coût: **~ d'élimination** *m* GESTION disposal cost; **~ global** *m* GESTION life-cycle cost
coûteux *adj* GESTION, POLL ATM cost-intensive
coûts: **~ de collecte** *m pl* GESTION collection costs; **~ de dépollution** *m pl* GESTION cleaning-up costs; **~ énergétiques** *m pl* ÉN RENOUV energy costs; **~ d'exploitation** *m pl* DIST EAU, POL ENV operating costs; **~ externes** *m pl* POL ENV external costs; **~ de recyclage** *m pl* GESTION recycling costs; **~ de retraitement** *m pl* DIST EAU reprocessing costs; **~ de traitement** *m pl* GESTION process costs; **~ de transaction** *m pl* POL ENV transaction costs
couvercle *m* POLL ATM lid; **~ à charnière** DÉCHETS, DIST EAU hinged lid; **~ à languette de déchirage** DÉCHETS tear-tab lid
couvert: **~ forestier** *m* ÉN RENOUV, SOLS canopy, forest cover; **~ herbacé** *m* ÉN RENOUV, SOLS grass cover; **~ végétal** *m* SOLS vegetation cover
couverture *f* DÉCHETS, DIST EAU, ÉN RENOUV cover, covering; **~ définitive** DÉCHETS *décharge* final cover, final covering; **~ intermédiaire** DÉCHETS *décharge* intermediate cover; **~ journalière** DÉCHETS *décharge* daily cover; **~ poreuse** DIST EAU porous cover; **~ du sol** SOLS soil covering; **~ de surface** SOLS surface cover; **~ végétale** SOLS tree cover
couvrir *vt* ÉN RENOUV cover; **~ la demande d'énergie** ÉN RENOUV cover energy demand
COV *abrév* (*composés organiques volatils*) POLL ATM, SOLS VOC (*volatile organic compounds*)
co-voiturage *m* POL ENV car pool

CPG[1] *abrév* ÉCHANT (*chromatographie en phase gazeuse*) GC (*gas chromatography*), (*chromatographe en phase gazeuse*) GC (*gas chromatograph*)
CPG[2]: **~ portable** *m* (*chromatographe en phase gazeuse portable*) ÉCHANT portable GC (*portable gas chromatograph*)
CPS *abrév* (*chromatographie en phase supercritique*) ÉCHANT SFC (*supercritical-fluid chromatography*)
Cr *abrév* (*chrome*) POLL ATM, SOLS Cr (*chromium*)
craquage: **~ catalytique** *m* POLL ATM catalytic cracking
craqueur: **~ catalytique** *m* POLL ATM catalytic cracker
crasses *f pl* POLL ATM slagged dust
crédit: **~ d'émissions** *m* POL ENV emissions credit; **~-bail** *m* POL ENV leasing
crédits: **~ de restructuration** *m pl* GESTION new money
crépine *f* DIST EAU rose; **~ d'aspiration** DIST EAU rose
crête *f* DIST EAU crest; **~ de houle** ÉN RENOUV wave crest; **~ de vague** ÉN RENOUV wave crest
creux: **~ de vague** *m* ÉN RENOUV wave hollow
criblage *m* DIST EAU, ÉCHANT, POLL ATM, POLL MAR screening, sieving, sifting, grading, SOLS jigging
crible *m* DIST EAU, ÉCHANT, POLL ATM, POLL MAR screen, sieve, sift, SOLS jig; **~ pour compost** DÉCHETS, SOLS compost-screening machine; **~ de lavage** DÉCHETS, SOLS jigger; **~ magnétique** DIST EAU screening magnet; **~ pulsant** DÉCHETS, SOLS jigger; **~ à résonance** DIST EAU resonance screen; **~ à tambour** DIST EAU trommel screen
cribler *vt* DIST EAU, ÉCHANT, POLL ATM, POLL MAR screen, sieve, sift, grade, SOLS jig
crique *f* DIST EAU, POLL EAU inlet
cristal: **~ déshydratable** *m* POLL ATM dewaterable crystal; **~ de gypse** *m* POLL ATM gypsum crystal
cristallin *adj* ÉN RENOUV crystalline
cristallisation *f* POLL ATM crystallization
cristalliser[1] *vt* POLL ATM crystallize
cristalliser[2]: **se ~** *v réfl* POLL ATM crystallize
cristallisoir: **~ d'évaporation** *m* POLL ATM evaporation crystallizer
cristallographie *f* ÉCHANT crystallography
cristaux: **~ de soude** *m pl* DIST EAU soda ash, washing soda
critères *m pl* GÉNÉRAL criteria; **~ de la décharge** SOLS site criteria; **~ d'évaluation** POL ENV evaluation criteria, POLL ATM, POLL SON, SOLS assessment criteria; **~ d'implantation** DIST EAU site criteria; **~ juridiques** POL ENV legal standards; **~ qualitatifs** GESTION quality requirements; **~ de qualité** DIST EAU, POLL MAR, SÉCURITÉ quality requirements; **~ de qualité atmosphérique** POLL ATM air-quality criteria; **~ de qualité écologique** GESTION eco-quality criteria; **~ de qualité de l'environnement** GESTION eco-quality criteria; **~ de qualité requis** POLL SON quality requirements; **~ de sécurité** SÉCURITÉ safety requirements
croissance: **~ désordonnée** *f* POLL EAU uncontrolled growth
croûte: **~ terrestre** *f* ÉN RENOUV earth's crust
crue *f* DIST EAU, POLL EAU flood, inundation; **~ annuelle** DIST EAU annual flood
cryobroyage *m* DÉCHETS cryogrinding, cryogenic crushing, freeze-grinding
CSEO *abrév* (*concentration sans effet observé*) SÉCURITÉ NOEC (*no observed effect concentration*)
Cu *abrév* (*cuivre*) DIST EAU, POLL ATM, SOLS Cu (*copper*)
cuiller *f* POLL MAR shovel; **~ de dosage** ÉCHANT measuring scoop
cuivre *m* (*Cu*) DIST EAU, POLL ATM, SOLS copper (*Cu*)
culasse *f* ÉN RENOUV *solaire* housing
culbuteur *m* DÉCHETS tip
culture *f* ÉCHANT, SOLS cultivation; **~ dérobée** ÉN RENOUV catch crop; **~ maraîchère** SOLS vegetable crop; **~ mélangée** POLL EAU mixed culture; **~ nomade** POL ENV, SOLS shifting cultivation; **~ sèche** POL ENV, SOLS dry farming; **~ vivrière en plein champ** POL ENV, SOLS field food crop
curage *m* DIST EAU flushing, cleaning out, scour, scouring,

POLL EAU cleansing, POLL MAR scraping; ~ **des égouts** DIST EAU sewer cleaning; ~ **de réservoir** DIST EAU reservoir cleaning

curer *vt* DIST EAU flush, clean out, scour, POLL EAU cleanse, scrape, POLL MAR scrape

cureuse *f* POLL EAU, POLL MAR scraper

cuve *f* DÉCHETS tank, DIST EAU cistern, vat, POLL ATM tank, vessel; ~ **d'absorption** POLL ATM absorber tank; ~ **d'alimentation** DÉCHETS, POLL ATM feed tank; ~ **à boues activées** POLL ATM activated-sludge tank; ~ **des boues activées secondaires** POLL EAU secondary activated-sludge tank; ~ **de captage** DÉCHETS catch tank; ~ **à combustible** POLL ATM fuel tank; ~ **à débordement d'épaississeur** POLL ATM thickener overflow tank; ~ **de décantation** POLL EAU settling vessel; ~ **de désactivation** POLL EAU *de matériaux radioactifs* decay tank; ~ **de digestion secondaire** POLL EAU secondary digestion tank; ~ **de distillation discontinue** SOLS batch distillation tank; ~ **à ébullition** POLL ATM boiling vessel; ~ **d'échantillonnage** ÉCHANT sampling vessel; ~ **d'écrémage** POLL MAR skimming tank; ~ **d'extraction** POLL EAU extraction vessel, SOLS extraction tank; ~ **de filtration** POLL EAU filtration vat, filtering tank; ~ **de floculation** DIST EAU, POLL EAU flocculation tank; ~ **de flottation** POLL EAU flotation tank; ~ **de gaz riches** POLL ATM rich-gas tank; ~ **à hydrolyse** POLL ATM hydrolizing tank; ~ **à laitier** SOLS slag tank; ~ **de mélange** DIST EAU mixing tank; ~ **de mélange pour chaux éteinte** POLL ATM slaked-lime mixing tank; ~ **pour nutriments en solution** POLL ATM nutrient-solution tank; ~ **de pourrissage** DIST EAU rotting container; ~ **de pré-aération** POLL EAU pre-aeration tank; ~ **de prélavage** POLL ATM prescrubbing tank; ~ **de pré-stockage centrifuge** POLL ATM centrifuge pre-storage tank; ~ **de réaction** POLL EAU reaction tank; ~ **de refroidissement** POLL ATM cooling tank; ~ **de rinçage des particules** POLL ATM rinsing PMT case; ~ **de séparation** POLL ATM phase-break vessel; ~ **de solvants** POLL ATM solvent tank; ~ **de stockage** DÉCHETS, POLL ATM storage tank; ~ **de stockage à air comprimé** POLL ATM compressed-air storage tank; ~ **de stockage du gaz** POLL EAU gas storage tank; ~ **de stockage des lixiviats** DÉCHETS, POLL EAU landfill leachate tank; ~ **de stockage préliminaire d'épaississeur** POLL ATM thickener pre-storage tank; ~ **de stockage temporaire** SOLS interim tank; ~ **tampon** POLL ATM *gaz* buffer tank; ~ **temporaire** POLL ATM temporary tank; ~ **de trempe** POLL ATM quench tank, quencher tank; ~ **d'usine immergée** POLL EAU submerged-plant tank

cuvées: **par** ~ *adv* POLL EAU *décantation* in batch quantities

cuvette *f* DIST EAU, ÉCHANT, POLL EAU basin; ~ **de mesure** ÉCHANT measuring cuvette

cyanure *m* POLL EAU, SOLS cyanide; ~ **libre** POLL EAU free cyanide

cyanuré *adj* POLL EAU cyanide-containing

cycle[1]: ~ **de l'azote** *m* POLL ATM, POLL EAU nitrogen cycle; ~ **biogéochimique** *m* POLL EAU bio-geochemical cycle; ~ **biologique de l'eau** *m* POLL EAU biological water cycle; ~ **du carbone** *m* POLL ATM carbon cycle; ~ **de l'eau** *m* POLL EAU water cycle; ~ **d'écoulement** *m* DIST EAU run-off cycle; ~ **de l'énergie** *m* ÉN RENOUV energy cycle; ~ **externe** *m* POLL ATM bio-geochemical cycle; ~ **de filtration** *m* POLL ATM filtration cycle; ~ **hydrologique** *m* DIST EAU hydrologic cycle; ~ **de lavage** *m* POLL ATM washing cycle; ~ **nutritif** *m* POL ENV nutrient cycle; ~ **ouvert** *m* ÉN RENOUV *thermodynamique* open cycle (*OC*); ~ **du phosphore** *m* POLL ATM phosphorus cycle; ~ **de Rankine fermé** *m* ÉN RENOUV closed Rankine cycle; ~ **de Rankine ouvert** *m* ÉN RENOUV open Rankine cycle; ~ **du soufre** *m* POLL ATM sulfur cycle (*AmE*), sulphur cycle (*BrE*); ~ **d'utilisation** *m* DÉCHETS utilization cycle; ~ **de vie d'un produit** *m* GESTION product life-cycle

cycle[2]: **à** ~ **unique** *loc* POLL ATM single-cycle

cyclone *m* DIST EAU, POLL ATM, SOLS cyclone; ~ **à admission axiale** POLL ATM axial-entry cyclone; ~ **à admission tangentielle** POLL ATM tangential-entry cyclone; ~ **cylindrique** POLL ATM cylindrical cyclone; ~ **de dépoussiérage préliminaire** POLL ATM preseparation cyclone; ~ **de dévésiculeur** POLL ATM cyclone separator; ~ **à haut rendement** POLL ATM high-efficiency cyclone; ~ **haute performance** POLL ATM high-performance cyclone; ~ **monocorps** POLL ATM single-unit cyclone; ~ **multiétages** ÉCHANT multistage cyclone; ~ **multitubulaire** POLL ATM multitube cyclone; ~ **à parois lisses** POLL ATM interior-polished cyclone

cyclones: ~ **en série** *m pl* POLL ATM multi-unit cyclones

cylindre *m* SOLS *de rouleau lisse* roll body; ~ **creux** POLL ATM hollow cylinder; ~ **hydraulique** DÉCHETS hydraulic cylinder; ~ **de pompe** DIST EAU, POLL EAU pump barrel, pump cylinder

cylindrée *f* DIST EAU piston capacity

cylindrique *adj* POLL ATM cylindrical

cytotoxicité *f* SÉCURITÉ cytotoxicity

D

Dacron® *m* POLL ATM Dacron® (*AmE*), Terylene® (*BrE*)
dalle: **~ insonorisante** *f* POLL SON acoustic slab
danger: **~ potentiel** *m* SÉCURITÉ hazard, potential danger
dangereux *adj* GÉNÉRAL dangerous, harmful, hazardous, unsafe; **~ pour l'environnement** POL ENV, POLL ATM, POLL MAR, POLL SON, SOLS harmful to the environment
date: **~ d'audition** *f* POL ENV hearing date; **~ d'audition publique** *f* POL ENV *mécanisme des organisations internationales* public hearing date; **~ limite d'appel** *f* POL ENV appeal deadline
dB *abrév* (*décibel*) ÉCHANT, POLL SON dB (*decibel*)
DBO *abrév* DÉCHETS, DIST EAU, POLL ATM (*demande biologique en oxygène*) BOD (*biological oxygen demand*), (*demande biochimique en oxygène*) BOD (*biochemical oxygen demand*)
DBO$_5$ *abrév* DÉCHETS, DIST EAU, POLL ATM (*demande biologique d'oxygène en cinq jours*) BOD$_5$ (*biological oxygen demand in five days*), (*demande biochimique d'oxygène en cinq jours*) BOD$_5$ (*biochemical oxygen demand in five days*)
DCO *abrév* (*demande chimique en oxygène*) DÉCHETS, DIST EAU, POLL ATM, POLL EAU COD (*chemical oxygen demand*)
DDT *abrév* (*dichloro-diphényl-trichloroéthane*) SOLS DDT (*dichlorodiphenyltrichloroethane*)
débarquement *m* POLL MAR unloading
débarquer *vt* POLL MAR unload
débit *m* DÉCHETS thruput (*AmE*), throughput, DIST EAU *d'une pompe* discharge, flow rate, rate of flow, *d'un compresseur, d'un ventilateur* delivery, ÉCHANT flow rate, rate of flow, throughput, thruput (*AmE*), GÉNÉRAL capacity, POLL ATM, POLL EAU flow rate, rate of flow, throughput, thruput (*AmE*), SOLS zeta potential (*ZP*); **~ d'air** POLL ATM air flow; **~ d'air résiduel** POLL ATM waste air flow, waste air throughput, waste air thruput (*AmE*); **~ annuel** DIST EAU annual flow; **~ croissant** POLL EAU *filtre* rising flow rate; **~ d'eau pompée** DIST EAU pumped-water flow; **~ d'eaux résiduaires** POLL EAU wastewater sludge delivery; **~ d'effluents gazeux** ÉCHANT exhaust-gas flow; **~ d'étiage** DIST EAU low-water discharge; **~ d'évacuation** POLL EAU discharge capacity; **~ de gaz** ÉCHANT gas flow, POLL ATM gas throughput, gas thruput (*AmE*); **~ des gaz brûlés** POLL ATM flue-gas throughput, flue-gas thruput (*AmE*); **~ des gaz de combustion** POLL ATM flue-gas throughput, flue-gas thruput (*AmE*); **~ de gaz pollué** POLL ATM dirty-gas flow; **~ de gaz variable** POLL ATM variable-gas throughput, variable-gas thruput (*AmE*), variable-gas throughput amount, variable-gas thruput amount (*AmE*); **~ global** ÉCHANT *d'une nappe* total flow; **~ journalier** DIST EAU daily water flow; **~ journalier moyen** DIST EAU mean daily flow; **~-masse** DIST EAU mass-flow rate; **~ matière** DÉCHETS *emballage* material flow; **~ maximal** DIST EAU maximum flow; **~ minimal** DIST EAU minimum flow; **~ nominal** DIST EAU design flow; **~ de pointe** DIST EAU peak water flow, POLL EAU peak flow; **~ de réfrigérant** POLL ATM coolant flow rate; **~ de rejet** POLL EAU discharge capacity; **~ de rejets gazeux** POLL ATM waste gas throughput, waste gas thruput (*AmE*); **~ solide** DIST EAU sediment discharge; **~ sortant** DIST EAU, POLL EAU outflow; **~ tributaire du rejet d'eaux souterraines** DIST EAU rate of flow originating from groundwater discharge; **~ de trop-plein** DIST EAU overflow rate; **~-volume** DIST EAU, ÉCHANT, POLL ATM volume flow rate, volume flow; **~ volumique** DIST EAU, ÉCHANT, POLL ATM volume flow rate, volume flow; **~ volumique de gaz pollué** POLL ATM dirty-gas volume flow
débitmètre *m* DIST EAU flow meter, ÉCHANT flow meter, tamping volume meter, flow rate measuring device, POLL ATM airflow sensor, POLL EAU flow meter; **~ capillaire** ÉCHANT capillary flow meter; **~ à flotteur** ÉCHANT rotameter; **~ de gaz** ÉCHANT gas flow counter, gas flow meter; **~ instantané** DIST EAU instant flow meter; **~ massique** ÉCHANT mass-flow rate meter
déblais *m pl* DÉCHETS construction waste, demolition waste, SOLS excavated material
déboisement *m* POLL ATM deforestation, SOLS clear-cutting, deforestation
débordement *m* DIST EAU, POLL ATM overflow; **~ de tamis** DIST EAU screen overflow
déborder *vi* DIST EAU overflow, run over, ÉCHANT flow over, POLL ATM overflow
débouché: **~ possible** *m* DIST EAU, POLL SON, SÉCURITÉ application possibility
déboucher *vi* DIST EAU *dans la mer*, POLL EAU run
débouchés: **~ possibles** *m pl* GESTION marketability
débourbage *m* DIST EAU , POLL EAU clarification, de-sludging, cleansing
débourber *vt* DIST EAU clarify, clean out, POLL EAU clarify, de-sludge
débourbeur *m* DIST EAU de-sludger
débris *m pl* DÉCHETS debris, rubbish, scrap, DIST EAU residue, POLL ATM debris, rubbish; **~ de fer** DÉCHETS junk iron; **~ organiques** DÉCHETS organic debris; **~ de verre** DIST EAU refuse glass
débroussaillant *m* POLL EAU, SOLS bush killer
décalage *m* ÉCHANT lag time
décalaminage *m* DIST EAU decarbonization, descaling, ÉCHANT cleaning, POLL EAU descaling; **~ à la chaux éteinte** DIST EAU decarbonization with slaked lime; **~ rapide** DIST EAU rapid carbonate removal
décalaminé *adj* DIST EAU decarbonized, descaled, cleaned
décalaminer *vt* DIST EAU clean, descale, POLL EAU descale
décantabilité *f* DIST EAU, POLL EAU settleability
décantable *adj* DIST EAU settleable, POLL ATM sedimentable, POLL EAU settleable
décantage *m* GÉNÉRAL decantation
décantation *f* GÉNÉRAL decantation, POLL EAU clarification; **~ finale** POLL EAU final clarification; **~ gravitaire** POLL ATM gravitational settling, gravity settling; **~ intermédiaire** POLL EAU intermediate settling; **~ préliminaire** DIST EAU preliminary settlement; **~ primaire** DIST EAU settling, presedimentation, primary settlement; **~ retardée** POLL EAU hindered settling; **~ secondaire** DIST EAU secondary settlement; **~ statique** DIST EAU, ÉCHANT, POLL ATM, POLL EAU, POLL MAR sedimentation
décanté *adj* GÉNÉRAL sedimented
décanter[1] *vt* GÉNÉRAL decant, POLL EAU clarify
décanter[2]: **se ~** *v réfl* DIST EAU, POLL EAU settle
décanteur *m* GÉNÉRAL decanter, POLL EAU clarifier; **~ centrifuge** ÉN RENOUV, POLL EAU decanting centrifuge; **~ Dortmund** POLL EAU Dortmund tank; **~ final** POLL EAU final settling tank, filtration sedimentation tank; **~ gravitaire** POLL ATM *dépoussiérage* gravity settler; **~ lamellaire** POLL ATM lamellar precipitator; **~ primaire** POLL EAU primary settling basin, primary settling tank; **~ secondaire** POLL EAU secondary settler, secondary settling tank
décapage *m* DIST EAU scour, scouring

décaper *vt* DIST EAU scour
décarbonaté *adj* DIST EAU decarbonated
décarbonisation *f* DIST EAU carbonate removal, decarbonization; **~ à l'eau de chaux** DIST EAU lime softening; **~ rapide** DIST EAU rapid carbonate removal
décarburation *f* DIST EAU decarbonization
décélération *f* POLL ATM deceleration
décharge *f* DÉCHETS waste dump, waste tip (*BrE*), landfill, dump, *action* dumping, DIST EAU *à ordures* landfill, *action* discharge, POLL SON release, site, SOLS site; **~ pour agrégats miniers** DÉCHETS mineral-aggregate landfill; **~ aménagée** DÉCHETS waste bunker; **~ de balles** DÉCHETS, DIST EAU bale landfill; **~ pour boues sèches** POLL EAU dry-sludge disposal site; **~ brute** DÉCHETS open dump; **~ centralisée** DÉCHETS central landfill; **~ à ciel ouvert** DÉCHETS open dump, open-pit landfill, surface landfill; **~ clandestine** DÉCHETS phantom dump; **~ classique à ciel ouvert et usages multiples** DÉCHETS conventional above-ground multi-component landfill; **~ compactée** DÉCHETS compaction landfill; **~ contaminée** SOLS contaminated site; **~ contrôlée** DÉCHETS sanitary landfill, DIST EAU controlled dumping, controlled tipping; **~ contrôlée pour déchets spéciaux** DÉCHETS controlled high-level landfill; **~ couverte** DÉCHETS closed-pit landfill; **~ de déchets** DÉCHETS waste dump, residue landfill, DIST EAU residue landfill; **~ de déchets dangereux** DÉCHETS hazardous waste landfill; **~ pour déchets déchiquetés** DÉCHETS shredded-refuse landfill; **~ de déchets encombrants** DÉCHETS dump for bulky waste; **~ pour déchiquetés** DÉCHETS shredded-refuse landfill; **~ en décomposition** DIST EAU rot landfill; **~ destinée à être décontaminée** SOLS site targeted for decontamination; **~ disruptive** POLL ATM spark-over; **~ de faible capacité** DIST EAU small landfill; **~ fortement compactée** DÉCHETS tightly-compacted landfill; **~ de haute activité** DÉCHETS high-level landfill; **~ inclinée** DÉCHETS valley landfill, ramp landfill, slope landfill, DIST EAU ramp landfill, slope landfill; **~ inconsidérée** DÉCHETS indiscriminate dumping; **~ industrielle** DÉCHETS industrial landfill; **~ inerte** DÉCHETS inert landfill; **~ légèrement compactée** DÉCHETS lightly-compacted landfill; **~ lumineuse** POLL ATM corona discharge; **~ municipale** DÉCHETS municipal waste landfill; **~ non compactée** DÉCHETS, DIST EAU uncompacted tip; **~ non-contrôlée** DÉCHETS, DIST EAU uncontrolled disposal, uncontrolled dumping, *site* open dump; **~ d'ordures broyées** DÉCHETS crushed-waste landfill; **~ en pente** DIST EAU slope landfill; **~ de pneus** DIST EAU used-tire disposal (*AmE*), used-tyre disposal (*BrE*); **~ point noir** DÉCHETS problem site; **~ réservée à un type de déchets** DÉCHETS mono-dump; **~ sauvage** DÉCHETS, DIST EAU fly tipping (*BrE*), illegal waste disposal, uncontrolled disposal, indiscriminate waste disposal, uncontrolled tipping, uncontrolled dumping, *site* phantom dump; **~ sous surveillance renforcée** DÉCHETS high-security landfill; **~ souterraine de déchets dangereux** DÉCHETS underground landfill for hazardous waste; **~ souterraine de déchets spéciaux** DÉCHETS underground landfill for special waste; **~ spécialisée** DÉCHETS, POLL EAU mono-dump; **~ de surface** DÉCHETS surface dumping; **~ en tranchée** DÉCHETS trench landfill; **~ en vallée** DÉCHETS valley landfill
déchargé *adj* POLL ATM unladen
déchargement *m* DIST EAU, unloading, GÉNÉRAL drainage, POLL EAU discharge
décharger *vt* DIST EAU, DÉCHETS unload
déchargeur: **~ automatique** *m* POLL EAU direct discharger
déchet: **~ monstre** *m* DÉCHETS macrowaste
déchets *m pl* DÉCHETS scrap, garbage (*AmE*), tailings, waste, waste materials, waste products, refuse, DIST EAU scales; **~ d'abattoir** DÉCHETS abattoir waste; **~ administratifs** DÉCHETS office waste; **~ agricoles** DÉCHETS, DIST EAU, ÉN RENOUV crop waste, waste crop; **~ alimentaires** DÉCHETS food waste; **~ anatomiques** DÉCHETS anatomical waste, human waste; **~ animaux** DÉCHETS animal wastes; **~ banals** DÉCHETS nonhazardous waste; **~ biodégradables** DÉCHETS, DIST EAU, POLL EAU biodegradable waste; **~ biologiques** DÉCHETS biological waste; **~ biomédicaux** DÉCHETS biomedical waste; **~ de bois** DÉCHETS wood waste; **~ broyés** DÉCHETS, ÉN RENOUV crushed waste; **~ de bureau** DÉCHETS office waste; **~ de caoutchouc** DÉCHETS waste rubber, DIST EAU rubber waste; **~ de carton** DÉCHETS, DIST EAU paper board waste; **~ cellulosiques** DÉCHETS cellulose waste; **~ chlorés** DÉCHETS chlorinated waste; **~ combustibles** DIST EAU combustible waste; **~ combustibles semi-solides** DIST EAU semi-solid combustible waste; **~ du commerce** DÉCHETS trade waste; **~ des commerces** DÉCHETS commercial waste; **~ composites** DÉCHETS composite waste; **~ compostables** DÉCHETS compostable waste; **~ compressibles** DÉCHETS compressible waste; **~ contaminés** DÉCHETS, DIST EAU, POLL EAU *des hôpitaux* pathological waste; **~ dangereux** DÉCHETS hazardous waste; **~ dangereux corrosifs** DÉCHETS, ÉN RENOUV corrosive hazardous waste; **~ dangereux insolubles dans l'eau** DÉCHETS, POLL EAU water-insoluble hazardous waste; **~ dangereux pompables** DÉCHETS, DIST EAU, POLL EAU pumpable hazardous waste; **~ dangereux réactifs** DÉCHETS, DIST EAU, POLL EAU reactive hazardous waste; **~ dangereux semi-solides** DÉCHETS, DIST EAU, POLL EAU semi-solid hazardous waste; **~ dangereux solubles dans l'eau** DÉCHETS, POLL EAU water-soluble hazardous waste; **~ dangereux très toxiques** DÉCHETS, DIST EAU, POLL EAU highly-toxic hazardous waste; **~ de déchiquetage** DÉCHETS, DIST EAU shredder waste; **~ de démolition** DÉCHETS construction waste, demolition waste, ÉN RENOUV demolition waste; **~ de dioxyde de titane** DÉCHETS titanium dioxide waste; **~ domestiques** DÉCHETS domestic waste; **~ électroniques** DÉCHETS electronic scrap; **~ éliminés** DÉCHETS, DIST EAU, POLL EAU, SOLS disposed-of residues; **~ d'emballage** DÉCHETS, DIST EAU packaging waste; **~ encombrants** DÉCHETS, DIST EAU bulky waste; **~ d'exploitation agricole** DÉCHETS farm waste; **~ explosifs** DÉCHETS explosive waste; **~ de fer** DÉCHETS ferrous scrap; **~ fermentés** DÉCHETS fermented waste; **~ horticoles** DÉCHETS park waste; **~ hospitaliers** DÉCHETS hospital waste; **~ huileux** DIST EAU oil waste; **~ de l'industrie pharmaceutique** DÉCHETS pharmaceutical industry waste; **~ industriels** DÉCHETS, DIST EAU, POLL EAU industrial waste, commercial waste, process waste; **~ industriels assimilables aux ordures ménagères** DÉCHETS domestic-like industrial waste; **~ industriels banals** (*DIB*) DÉCHETS domestic-like industrial waste; **~ industriels spéciaux** (*DIS*) DÉCHETS special industrial refuse; **~ inertes** DÉCHETS inert waste; **~ infectieux** DÉCHETS infectious waste; **~ inflammables** DÉCHETS flammable waste; **~ de jardin** DÉCHETS garden waste; **~ de jardinage** DÉCHETS garden waste; **~ lessivables** DÉCHETS leachable waste; **~ ligneux** DÉCHETS woody waste, wood waste; **~ liquides** DÉCHETS, DIST EAU, POLL EAU liquid waste; **~ de marché** DÉCHETS market waste; **~ mélangés** DÉCHETS mixed waste; **~ mêlés** DÉCHETS mixed waste; **~ ménagers** DÉCHETS domestic waste; **~ du métabolisme** DÉCHETS metabolic waste; **~ métalliques** DÉCHETS, DIST EAU, POLL EAU metal scrap, metal waste; **~ miniers** DÉCHETS mineral waste, colliery waste; **~ mis en balle** DÉCHETS waste bale; **~ mis en place** DÉCHETS emplaced waste; **~ monstres** DÉCHETS, DIST EAU bulky waste; **~ nocifs pour l'eau** POLL EAU waste harmful to water; **~ non biodégradables** DÉCHETS, DIST EAU nonbiodegradable waste; **~ non compostables** DÉCHETS, DIST EAU noncompostable waste; **~ non recyclables** DÉCHETS nonrecoverable waste; **~ non-traités** DÉCHETS unprocessed waste, DIST EAU untreated refuse; **~ non triés** DIST EAU unsorted waste; **~ nucléaires**

DÉCHETS nuclear waste; **~ organiques** DÉCHETS organic waste; **~ d'origine animale** *m* DÉCHETS animal waste; **~ d'origine biologique** DÉCHETS bio-refuse; **~ d'origine chimique** DÉCHETS chemical waste; **~ pathogènes** DÉCHETS, DIST EAU, POLL EAU pathological waste; **~ de peinture** DÉCHETS waste paint, DIST EAU varnish waste; **~ pharmaceutiques** DÉCHETS pharmaceutical waste; **~ de plastique** DÉCHETS plastic waste; **~ de plastique après consommation** DIST EAU post-consumer plastics waste; **~ plastiques** DÉCHETS plastics waste, waste plastics, DIST EAU plastics waste; **~ pollués** DÉCHETS waste-containing pollutants; **~ pompables** DÉCHETS, DIST EAU pumpable waste; **~ prétriés** DÉCHETS, DIST EAU, SOLS presorted waste; **~ putrescibles** DÉCHETS, DIST EAU putrescible waste; **~ radioactifs** DÉCHETS, DIST EAU radioactive waste; **~ radioactifs de haute activité** POLL ATM, POLL EAU high-level radioactive waste; **~ radioactifs à vie longue** DÉCHETS, DIST EAU long-life radioactive waste; **~ de raffinerie** DÉCHETS refinery waste; **~ réutilisables** DÉCHETS, POLL ATM re-usable waste products; **~ riches en éléments nutritifs** DÉCHETS nutrient-rich waste; **~ sauvages** DÉCHETS litter; **~ solides** DÉCHETS solid waste, DIST EAU, POLL EAU solid waste; **~ solides déchiquetés** DÉCHETS shredded solid waste; **~ de solidification** DÉCHETS, DIST EAU solidifying waste; **~ solidifiés** DÉCHETS solidified waste; **~ sous forme gazeuse** DÉCHETS, DIST EAU gaseous waste; **~ spéciaux** DÉCHETS, POLL EAU special waste; **~ spécifique d'une production** DÉCHETS, DIST EAU production-specific waste; **~ sulfureux** DIST EAU sulfurous waste (*AmE*), sulphurous waste (*BrE*); **~ textiles** DÉCHETS textile waste; **~ thermiques** DÉCHETS thermal waste; **~ de tôle** DÉCHETS tin-plate waste; **~ toxiques** DÉCHETS, DIST EAU, SÉCURITÉ toxic waste, poisonous waste (*BrE*); **~ traités** DÉCHETS, DIST EAU processed waste; **~ traités aux chlorofluorocarbones** DÉCHETS CFC-processed waste; **~ urbains** DÉCHETS urban waste, municipal waste, DIST EAU municipal waste; **~ urbains solides** DÉCHETS municipal solid waste (*MSW*), DIST EAU municipal solid waste (*MSW*), urban solid waste; **~ valorisables** DÉCHETS usable waste products; **~ verts** DIST EAU green waste; **~ de la voie publique** DÉCHETS street refuse

déchetterie *f* DÉCHETS waste dump, waste plant, DIST EAU public collection station

déchiquetage *m* DÉCHETS shredding, ÉN RENOUV chipping; **~ immédiat en copeaux** ÉN RENOUV *coupe du bois* direct chipping

déchiqueté *adj* DIST EAU shredded

déchiqueter *vt* DIST EAU shred

déchiqueteur: **~ à deux arbres** *m* DÉCHETS twin-cutting shredder

déchiqueteuse *f* DÉCHETS shredding machine, vehicle shredder, cutting shredder, shredder, ÉN RENOUV cutting shredder

déchirure *f* POLL ATM tearing

déchloration *f* DIST EAU dechlorination

déchromer *vt* DIST EAU deplate

décibel *m* (*dB*) ÉCHANT, POLL SON decibel (*dB*)

décision *f* DIST EAU, POL ENV decision; **~ d'aménagement** DIST EAU, POL ENV planning assessment decision; **~ préliminaire** DIST EAU, POL ENV preliminary decision

déclarant *m* POL ENV notifier

déclaration *f* DIST EAU *produit chimique* registration, GESTION reporting; **~ ministérielle** POL ENV ministerial statement; **~ de revenus** GESTION tax return; **~ d'utilité publique** (*DUP*) POL ENV declaration of public utility

décodage: **~ de la valeur mesurée** *m* ÉCHANT decoding of measured value

décolmatage *m* DIST EAU scowing, POLL ATM cleaning, remaining dust removal

décolmater *vt* POLL ATM clean

décoloration *f* ÉCHANT discoloration, decoloration

décolorer *vt* DIST EAU bleach, ÉCHANT discolor (*AmE*), discolour (*BrE*), POLL EAU bleach

décomposer[1] *vt* DÉCHETS, DIST EAU, POLL ATM break down

décomposer[2]: **se ~** *v réfl* DÉCHETS, DIST EAU, POLL ATM break down, decompose; **se ~ par oxydation** *v réfl* POLL EAU decompose oxidatively

décomposition *f* DIST EAU decomposition, POLL EAU *par pourriture* decay, decomposition, SOLS decomposition; **~ de matière première** DIST EAU raw material decomposition; **~ des matières organiques** DÉCHETS, POLL EAU decay of organic matter, breakdown of organic substances; **~ mesurée par la DBO_5** DIST EAU BOD_5 decomposition; **~ microbienne** POLL ATM microbial decomposition, SOLS microbial degradation; **~ par oxydation** POLL EAU oxidative decomposition; **~ thermique** DÉCHETS thermal decomposition

décontaminabilité *f* SOLS decontaminability

décontamination *f* DIST EAU decontamination, elimination of contaminants, POLL ATM, POLL EAU, SÉCURITÉ, SOLS decontamination; **~ de l'air** POLL ATM air decontamination; **~ des eaux usées** POLL EAU wastewater decontamination; **~ électroacoustique in situ** SOLS in-situ electro-acoustic decontamination; **~ partielle du sol** SOLS partial soil decontamination; **~ du sol** SOLS soil decontamination

décontaminé *adj* GÉNÉRAL decontaminated

décontaminer *vt* GÉNÉRAL decontaminate

découper *vt* DIST EAU shear off

décrassage *m* DIST EAU, POLL ATM, POLL EAU *métallurgie* descaling, deslagging, slag removal

décrassant *m* POLL ATM scavenger

décrasser *vt* POLL ATM *gaz* scavenge, descale, deslag

décroissement: **~ de la radioactivité** *m* POLL ATM radioactivity decay

décrue *f* DIST EAU *d'un fleuve* falling stage

dédommagement *m* POL ENV, SÉCURITÉ compensation

dédommager *vt* POL ENV, SÉCURITÉ compensate

déduction: **~ fiscale** *f* POL ENV tax rebate

déferraillage *m* ÉN RENOUV deferrization

déferreur: **~ magnétique de grande puissance** *m* SOLS high-intensity magnet separator

déferrisation *f* DIST EAU deferrization

déficience: **~ auditive induite par le bruit** *f* POLL SON, SÉCURITÉ noise-induced hearing impairment

déficit: **~ hydrique** *m* DIST EAU water deficiency; **~ en oxygène** *m* ÉCHANT oxygen deficit (*OD*); **~ en ozone** *m* POLL ATM ozone deficit

définition: **~ du problème** *f* ÉCHANT problem definition

déflecteur *m* POLL ATM deflector, POLL SON acoustic baffle

déflexion: **~ statique** *f* POLL SON static deflection

défoliant *m* POLL EAU, SOLS defoliant

défoliation *f* POLL EAU, SOLS defoliation

déforestation *f* POLL ATM, SOLS deforestation

déformation: **~ souterraine** *f* DÉCHETS, DIST EAU underground deformation

défrichement *m* POL ENV land clearing

dégagement *m* POLL ATM burning out, POLL SON release; **~ de chaleur** POLL ATM, POLL EAU released heat; **~ de substances nocives** SOLS escape of harmful substances

dégager *vt* POLL EAU, POLL SON, SOLS release

dégât: **~ des eaux** *m* DIST EAU water damage; **~ par compression** *m* DÉCHETS compression damage

dégazage *m* DÉCHETS, DIST EAU, POLL ATM, POLL EAU degasification, degassing; **~ forcé** DÉCHETS forced degassing; **~ en gabion** ÉN RENOUV degassing with gabion; **~ horizontal** DIST EAU, POLL ATM horizontal degassing; **~ en puits** ÉN RENOUV degassing with gas wells; **~ à la sonde** ÉN RENOUV degassing with sonde; **~ tangentiel** DIST EAU, POLL ATM horizontal degassing; **~ vertical** DÉCHETS, DIST EAU vertical degassing

dégazer *vt* DÉCHETS, DIST EAU, POLL ATM, POLL EAU degas, degasify
dégazeur *m* DÉCHETS, DIST EAU, POLL ATM, POLL EAU degasser
dégorgement *m* DIST EAU scour
dégorger *vt* DIST EAU scour
dégradabilité *f* GÉNÉRAL degradability
dégradable *adj* GÉNÉRAL degradable
dégradation *f* DIST EAU *sous l'effet des intempéries* weathering, POLL ATM degradation, damage, POLL EAU degradation, degradation performance, decomposition, POLL MAR, SOLS impairment, weathering; **~ abiotique** POLL EAU abiotic degradation; **~ biologique** DÉCHETS, DIST EAU biological degradation, biodegradation; **~ du catalyseur** POLL ATM catalyst degradation; **~ du climat** POLL ATM climate deterioration; **~ des contaminants** POLL ATM contaminant degradation; **~ continue** SÉCURITÉ persistent deterioration; **~ écologique** POLL ATM, SOLS ecological damage; **~ de l'environnement** POL ENV defacement of the environment, deterioration of the environment, environmental damage, POLL ATM, SOLS environmental damage; **~ du milieu** POL ENV, POLL ATM environmental damage, SOLS environmental collapse, environmental damage; **~ photolytique** SOLS photolytic degradation; **~ des polluants** POLL ATM contaminant degradation; **~ des ressources en eau douce** POLL EAU deterioration of fresh waters; **~ des ressources naturelles** POL ENV, POLL ATM, POLL EAU, SOLS deterioration of natural resources; **~ du sol** SOLS soil degradation; **~ des sols** DIST EAU land degradation; **~ des sols imputable aux activités humaines** SOLS human-induced soil degradation; **~ de substances dangereuses** POLL EAU decomposition of harmful substances; **~ des terres** SOLS land degradation; **~ de la végétation** ÉN RENOUV damage to vegetation
dégradé *adj* GÉNÉRAL degraded, impaired
dégrader *vt* GÉNÉRAL degrade, impair
dégraissage *m* DIST EAU scouring
dégraissé *adj* DIST EAU grease-free, POLL ATM fat-free
dégraisseur *m* DÉCHETS grease trap, grease separator, degreaser, DIST EAU, POLL EAU *plomberie* grease trap, grease separator
dégravoyage *m* POLL EAU grit removal
degré: **~ d'absorption** *m* POLL ATM degree of absorption; **~ d'acidité** *m* SOLS acidity level; **~ d'adhésivité** *m* DÉCHETS *soudure* tack level; **~ d'appauvrissement de la couche d'ozone** *m* POLL ATM degree of ozone depletion; **~ de contamination** *m* SOLS degree of contamination; **~ de décontamination** *m* SOLS degree of decontamination; **~ de dégradabilité** *m* POLL EAU degree of degradability; **~ de dégradation** *m* POLL EAU degree of degradation; **~ de dépoussiérage** *m* POLL ATM degree of dust removal; **~ de déshydratation** *m* POLL EAU degree of dewatering; **~ de désulfuration** *m* POLL ATM degree of desulfurization (*AmE*), degree of desulphurization (*BrE*); **~ de dissociation** *m* ÉCHANT degree of dissociation; **~ de dureté** *m* DIST EAU, ÉCHANT *Brinell* degree of hardness; **~ d'échange** *m* DIST EAU degree of exchange; **~ d'épuration** *m* POLL EAU degree of purification; **~ d'extraction des oxydes d'azote** *m* POLL ATM degree of nitrogen oxides removal; **~ de fermentation** *m* DIST EAU degree of fermentation; **~ de fractionnement** *m* POLL ATM degree of fraction removal; **~ d'humification** *m* POLL EAU degree of humification; **~ hydrotimétrique de l'eau** *m* DIST EAU, POLL EAU water hardness degree; **~ d'hygrométrie** *m* DIST EAU water content, POLL ATM moisture content; **~ hygrométrique** *m* ÉN RENOUV moisture content; **~ de pollution** *m* POLL EAU degree of pollution, pollution level; **~ de pureté** *m* DIST EAU degree of purity; **~ de séparation** *m* POLL ATM, POLL EAU degree of separation; **~ de stabilisation** *m* POLL EAU degree of stabilization
dégressif *adj* POL ENV regressive
dégrèvement: **~ fiscal** *m* POL ENV tax rebate, tax relief
dégrillage *m* DIST EAU, ÉCHANT, POLL ATM screening, POLL EAU *épuration des eaux* screening, *prétraitement des boues* screen cleaning, POLL MAR screening
dégriller *vt* ÉCHANT, POLL ATM, POLL EAU, POLL MAR screen
dégrilleur *m* POLL EAU bar screen, screen cleaner; **~ rotatif autonettoyant** POLL EAU self-cleaning rotary drum screen
dégrossisseur *m* POLL EAU roughing tank
déjections: **~ animales** *f* ÉN RENOUV animal dung
DEL *abrév* (*diode électroluminescente*) ÉCHANT LED (*light-emitting diode*)
délai: **~ d'acheminement** *m* DÉCHETS, DIST EAU delivery time; **~ d'amortissement** *m* ÉN RENOUV pay-back time; **~ d'amortissement d'un investissement énergétique** *m* ÉN RENOUV energy pay-back time; **~ d'appauvrissement** *m* POLL EAU *combustible nucléaire* depletion time; **~ d'épuisement** *m* POLL EAU *combustible nucléaire* depletion time; **~ de filtration** *m* POLL ATM, POLL EAU filtering time; **~ de réaction** *m* POLL ATM reaction time
délayage *m* DÉCHETS, ÉCHANT, POLL ATM dilution
délayé *adj* DÉCHETS, ÉCHANT, POLL ATM diluted
délayer *vt* DÉCHETS, ÉCHANT, POLL ATM dilute
délestage *m* GÉNÉRAL lightening
délibération: **~ sur l'octroi de licence** *f* DIST EAU, POL ENV licensing conference; **~ sur les procédures d'autorisation** *f* DIST EAU, POL ENV licensing conference
délits: **~ à l'encontre de l'environnement** *m pl* POL ENV environmental criminality; **~ relatifs à l'environnement** *m pl* POL ENV environmental crime
délivrance: **~ d'autorisation** *f* DIST EAU, POL ENV licensing
demande: **~ d'autorisation** *f* DIST EAU authorization application, POL ENV licensing application, permit application; **~ biochimique en oxygène** *f* (*DBO*) DÉCHETS, DIST EAU, POLL ATM biochemical oxygen demand (*BOD*); **~ biochimique d'oxygène en cinq jours** *f* (DBO_5) DÉCHETS, DIST EAU, POLL ATM biochemical oxygen demand in five days (BOD_5); **~ biologique en oxygène** *f* (*DBO*) DÉCHETS, DIST EAU, POLL ATM biological oxygen demand (*BOD*); **~ biologique d'oxygène en cinq jours** *f* (DBO_5) DÉCHETS, DIST EAU, POLL ATM biological oxygen demand in five days (BOD_5); **~ chimique en oxygène** *f* (*DCO*) DÉCHETS, DIST EAU, POLL ATM, POLL EAU chemical oxygen demand (*COD*); **~ de chlore** *f* DIST EAU chlorine demand (*CD*); **~ d'eau** *f* DIST EAU water demand; **~ d'eau fonction de la population** *f* DIST EAU population-related water demand; **~ d'eau spécifique de la population** *f* DIST EAU population-specific water demand; **~ d'énergie** *f* ÉN RENOUV energy demand; **~ d'énergie primaire** *f* ÉN RENOUV primary energy demand; **~ globale d'énergie** *f* ÉN RENOUV global energy demand; **~ globale d'énergie dans l'industrie** *f* ÉN RENOUV commercial global energy demand; **~ locale d'eau** *f* DIST EAU area-related water demand; **~ mondiale d'énergie dans l'industrie** *f* ÉN RENOUV commercial global energy demand; **~ mondiale d'énergie dans l'industrie** *f* ÉN RENOUV commercial global energy demand; **~ d'oxygène théorique** *f* POLL EAU theoretical oxygen demand; **~ de permis** *f* DIST EAU permit application; **~ saisonnalisée** *f* GESTION seasonalized demand
démarche *f* GÉNÉRAL procedure
demi: **~-largeur** *f* ÉCHANT *spectrométrie* half-width value
déminéralisation *f* DIST EAU, POLL EAU demineralization
déminéralisé *adj* DIST EAU, POLL EAU demineralized
déminéraliser *vt* DIST EAU, POLL EAU demineralize
démographique *adj* POL ENV population-related
démontage *m* DÉCHETS stripping
démoulage *m* DÉCHETS stripping
dénickeler *vt* DIST EAU deplate
dénitrification *f* GÉNÉRAL denitrification, denox (NO_x *removal*); **~ biologique** DIST EAU biological denitrification; **~ des déchets** DÉCHETS waste denitrification; **~ hétérotrophe** POLL EAU heterotrophic denitrification; **~ in situ** POLL EAU in situ denitrification; **~ intermittente**

POLL EAU intermittent denitrification; **~ non-contrôlée** POLL EAU uncontrolled denitrification; **~ simultanée** POLL EAU simultaneous denitrification; **~ sur lit fluidisé circulant** POLL ATM denitrification in circulating fluidized bed, denox in circulating fluidized bed

dénitrifier *vt* GÉNÉRAL denitrate, denitrify

dénitritation *f* DIST EAU, SOLS denox

dénitrurer *vt* GÉNÉRAL denitrate, remove nitrogen from

dénoyage *m* GÉNÉRAL drainage, draining, SOLS dewatering; **~ à part** POLL EAU separate dewatering; **~ ultérieur** SOLS subsequent dewatering

dénoyer *vt* DÉCHETS, DIST EAU, POLL EAU *mine* pump out, SOLS dewater

densificateur *m* ÉN RENOUV densifier

densifier *vt* DÉCHETS, SOLS densify

densimètre *m* ÉCHANT densimeter

densité *f* DIST EAU specific gravity; **~ absolue** DIST EAU density; **~ d'adsorbants** POLL ATM adsorbent density; **~ apparente** DÉCHETS bulk density; **~ après damage** ÉCHANT tamped density; **~ de charge volumique** POLL ATM density of volume charge; **~ de construction** POL ENV building density; **~ énergétique** DIST EAU, ÉN RENOUV energy density; **~ d'occupation des sols** POL ENV coverage density; **~ relative** DIST EAU relative density; **~ des stations de mesure** ÉCHANT density of measurement stations; **~ en vrac** DÉCHETS bulk density

dépalettisateur *m* DÉCHETS, ÉN RENOUV depalletizer

dépalettisation *f* DIST EAU, ÉN RENOUV depalletization

dépalettiser *vt* DIST EAU, ÉN RENOUV, POLL EAU depalletize

dépassement: **~ de dose** *m* SÉCURITÉ overdose (*OD*)

dépense: **~ d'énergie** *f* DIST EAU, ÉN RENOUV, POLL ATM energy expenditure

dépenses: **~ après fermeture** *f pl* DIST EAU post-closure expenditure

déperdition: **~ de chaleur du moteur arrêté** *f* POLL ATM hot-soak loss

dépérissement: **~ de la végétation** *m* SOLS vegetation loss

dépeuplement: **~ des campagnes** *m* POL ENV depopulation of countryside

dépeupler *vt* POL ENV depopulate

déphénolage *f* POLL EAU dephenolating, dephenolizing

déphénoler *vt* POLL EAU dephenolate, dephenolize

déphosphatation *f* POLL EAU phosphate elimination, phosphate removal

déphosphoration *f* POLL EAU phosphorus removal; **~ physico-chimique** POLL EAU physio-chemical phosphorus removal

déplacement *m* DIST EAU displacement; **~ permanent de seuil** POLL SON permanent threshold shift (*PTS*); **~ permanent de seuil induit par le bruit** POLL SON noise-induced permanent threshold shift (*NIPTS*); **~ de seuil temporaire** POLL SON temporary threshold shift (*TTS*)

dépolarisant *adj* ÉCHANT depolarizing

dépolarisation *f* ÉCHANT depolarization

dépolariser *vt* ÉCHANT depolarize

dépolariseur *m* ÉCHANT *substance ou moyen* depolarizer

dépollué *adj* GÉNÉRAL decontaminated

dépolluer *vt* DIST EAU, POLL ATM decontaminate, POLL MAR clean up, SÉCURITÉ, SOLS decontaminate

dépollution *f* DIST EAU decontamination, pollution control, POL ENV pollution control, POLL ATM, POLL EAU decontamination, POLL MAR clean-up, SÉCURITÉ, SOLS decontamination

dépolymérisation *f* DIST EAU, ÉN RENOUV depolymerization

dépolymériser *vt* ÉN RENOUV depolymerize

déposé *adj* DIST EAU, POLL ATM, POLL EAU sedimented

déposer[1] *vt* DÉCHETS dump, tip, POLL ATM clarify, sediment, POLL EAU deposit

déposer[2]: **se ~** *v réfl* DIST EAU, POLL ATM, POLL EAU settle, deposit, precipitate, sediment

déposition *f* POLL ATM clarification

dépôt *m* DÉCHETS dumping, DIST EAU deposit, dump, sediment, deposition, POLL ATM deposit, POLL EAU deposit, deposition, dump, sediment, POLL MAR sediment, SOLS deposit, deposition, discharge; **~ acide** POLL ATM acid deposit, acid deposition; **~ acide sec** POLL ATM, SOLS dry acid deposit; **~ adhésif** POLL ATM sticky deposit; **~ alluvial** DÉCHETS, DIST EAU alluvial deposit; **~ alluvionnaire** DÉCHETS, DIST EAU alluvial deposit; **~ d'armes** SOLS armament repository site; **~ de boues d'eaux résiduaires sur des herbages** POLL EAU deposition of wastewater sludge on grassland; **~ de boues d'épuration sur des terres à fourrage** POLL EAU deposition of wastewater sludge on ground used for forage; **~ brut** DIST EAU bulk deposition; **~ de carton** SOLS deposition of cardboard; **~ clandestin** DÉCHETS, DIST EAU phantom dump; **~ commun** DÉCHETS co-deposition; **~ contrôlé** DÉCHETS, DIST EAU controlled dumping; **~ côtier** DIST EAU coastal deposit; **~ de déblais** DÉCHETS tip; **~ de déchets** DÉCHETS waste dump; **~ de déchets radioactifs** DÉCHETS, DIST EAU repository for radioactive waste; **~ d'épaves** DIST EAU used-car dump; **~ global** DIST EAU bulk deposition; **~ humide** DIST EAU, POLL ATM wet deposition; **~ de limon** DIST EAU, SOLS deposition of silt, silt deposition; **~ de métaux lourds** POLL ATM, SOLS heavy metals deposition; **~ de polluants** DIST EAU, POLL ATM pollutant deposition; **~ polluant** POLL EAU, SOLS pollution deposition; **~ de poussière** POLL ATM, SOLS dust deposit; **~ poussiéreux** POLL ATM, SOLS dust deposit; **~ public** DIST EAU public collection station; **~ réglementé** DIST EAU regulated deposition; **~ de sable** DIST EAU sand deposition; **~ sauvage** DÉCHETS illegal dumping, uncontrolled tipping, DIST EAU uncontrolled tipping, POL ENV, POLL EAU illegal dumping, uncontrolled tipping; **~ sec** POLL ATM, SOLS dry deposition; **~ souterrain** DÉCHETS subsurface repository, underground depot, DIST EAU underground depot; **~ temporaire de déchets dangereux** DÉCHETS temporary deposit for hazardous waste; **~ total** DIST EAU, POLL ATM, SOLS total deposition; **~ de vase** DIST EAU silt deposition; **~ en vrac** DIST EAU bulk deposition

dépotage *m* DÉCHETS *vidange d'une citerne* emptying

dépotoir *m* DÉCHETS waste tip (*BrE*), junk pile, DIST EAU junk pile

dépoussiérage *m* POLL ATM de-dusting, dust removal, dust control; **~ de gaz chauds** POLL ATM hot-gas dust removal; **~ haute performance** POLL ATM high-performance dust removal; **~ hydraulique** POLL ATM water scrubbing, *gaz* wet scrubbing; **~ intégral** POLL ATM total removal; **~ par voie humide** POLL ATM wet dust removal; **~ préliminaire** POLL ATM preliminary dust removal

dépoussiéré *adj* POLL ATM dust-free, dust-removed

dépoussiérer *vt* DIST EAU, POLL ATM dust, de-dust

dépoussiéreur *m* ÉCHANT dust collector, POLL ATM de-duster, dust remover, dust separator, dust extractor, *par voie humide* scrubber, SÉCURITÉ dust separator; **~ à chicane** POLL SON baffle collector; **~ à colonne garnie** DÉCHETS packed-column scrubber; **~ à contre-courant** POLL ATM reverse-flow precipitator; **~ à déflecteur** POLL ATM deflector separator; **~ électrique** SÉCURITÉ electrical dust-removal installation; **~ électrostatique** DÉCHETS, DIST EAU, POLL ATM *dépoussiérage électrique des gaz* electrostatic precipitator (*ESP*); **~ électrostatique haute tension** POLL ATM high-voltage electrostatic precipitator; **~ électrostatique par condensation** POLL ATM condensation-force wet ESP, condensation-force wet electrostatic precipitator; **~ électrostatique par voie humide** POLL ATM wet electrostatic precipitator (*WEP*); **~ à filtre par voie humide** POLL ATM wetted-filter-type scrubber; **~ haute performance** POLL ATM high-performance precipitator; **~ hydraulique** POLL ATM *gaz* wet scrubber; **~ hydraulique à étage** POLL ATM two-stage gas scrubber; **~ à inertie** POLL ATM *dépoussiérage mécanique* inertial separator; ~ **isolé** SÉCURITÉ individual dust-removal apparatus; ~ **mixte** POLL ATM wet-dry scrubber; ~ **monobloc** POLL ATM compact gas scrubber; ~ **par condensation** POLL ATM condensation scrubber; ~ **par voie humide** POLL ATM *gaz* wet scrubber; ~ **par voie sèche** POLL ATM dry scrubber, dry scrubbing; ~ **à plaques** POLL ATM *dépoussiéreur électrique* plate-wire ESP, plate-wire electrostatic precipitator; ~ **à pulvérisation** POLL ATM *dépoussiérage hydraulique* spray-tower scrubber, spray-column scrubber; ~ **à pulvérisation par voie sèche** POLL ATM spray-dry scrubber; ~ **secondaire** POLL ATM secondary precipitator; ~ **semi-sec** POLL ATM wet-dry scrubber; ~ **à tubes** POLL ATM *dépoussiéreur électrique* tubular ESP, tubular electrostatic precipitator; ~ **tubulaire** POLL ATM tubular precipitator, *dépoussiéreur électrique* tubular ESP, tubular electrostatic precipitator

dépression *f* SOLS negative pressure

déprimomètre *m* ÉCHANT vacuum gage (*AmE*), vacuum gauge (*BrE*)

dérivation *f* DIST EAU submains, ÉCHANT, POLL ATM bypass; ~ **de secours** POLL EAU emergency bypass

dérive *f* DIST EAU, POLL ATM, POLL MAR drift

dériver *vt* DIST EAU, POLL ATM, POLL MAR drift

dérochage *m* DIST EAU *mécanique* scouring

derrick *m* ÉN RENOUV derrick

désactivation *f* DIST EAU inactivation, ÉN RENOUV, POLL ATM deactivation; ~ **du catalyseur** POLL ATM catalyst deactivation

désaérateur *m* DIST EAU, POLL EAU de-aerator; ~ **sous vide** POLL EAU vacuum de-aerator, de-aerator

désaération *f* DIST EAU, POLL EAU de-aeration

désaérer *vt* DIST EAU, POLL EAU de-aerate

désagrégateur *m* SOLS cutter

désagrégation *f* DIST EAU decomposition, weathering, POLL MAR, SOLS weathering

désamiantage *m* POL ENV asbestos removal

descendre *vt* DIST EAU *charge* lower

descente *f* POLL ATM standpipe, POLL EAU drainpipe, downpipe, downspout, *égout* soil pipe; ~ **primaire** POLL ATM primary downpipe; ~ **secondaire** POLL ATM secondary downpipe; ~ **de vidange** POLL EAU drainpipe spout

déschlammage *m* POLL EAU removal of sludge

désembueur *m* ÉN RENOUV de-mister unit, POLL ATM mist eliminator

désempilement *m* DÉCHETS *de palettes* de-stacking

désempiler *vt* DÉCHETS de-stack

désémulsifiant *m* DIST EAU, POLL EAU, POLL MAR demulsifier, emulsion breaker

désémulsification *f* DIST EAU, POLL EAU, POLL MAR demulsification

désémulsifier *vt* DIST EAU, POLL EAU, POLL MAR demulsify

désencrage *m* POLL EAU de-inking; ~ **par flottation** POLL EAU de-inking by flotation

désencrer *vt* POLL EAU de-ink

désertification *f* SOLS desertification, desert formation

désessenciement *m* DÉCHETS *industrie pétrolière* gasoline stripping

désétamage *m* POLL EAU de-tinning

désétamer *vt* POLL EAU de-tin

désétiquetage *m* DÉCHETS *des boîtes* de-labelling

déshalogénation *f* SOLS dehalogenation

déshalogéner *vt* SOLS dehalogenate

désherbant *m* SOLS weedkiller

déshuilage *m* POLL EAU de-oiling, oil removal

déshuiler *vt* POLL EAU de-oil

déshuileur *m* POLL EAU oil trap, SOLS oil separator

déshumidificateur *m* POLL ATM moisture eliminator, air dehumidifier

déshumidification *f* ÉCHANT dehumidification

déshumidifier *vt* ÉCHANT dehumidify

déshydratabilité *f* POLL ATM, POLL EAU dewaterability

déshydratant *m* ÉCHANT drying agent

déshydratation *f* DIST EAU dewatering, POLL ATM dehydration, drying, POLL EAU dewatering; ~ **de la boue** POLL EAU sludge dehydration; ~ **du gypse** POLL ATM gypsum dewatering; ~ **par congélation** POLL EAU dewatering via freezing; ~ **par filtre sous vide** POLL EAU vacuum filter dewatering; ~ **préliminaire** POLL EAU preliminary dewatering; ~ **séparée** POLL EAU separate dewatering

déshydraté *adj* DIST EAU water-free, POLL EAU *produit* dewatered; ~ **mécaniquement** POLL EAU mechanically-dewatered

déshydrater *vt* DIST EAU dewater, POLL ATM dehydrate, dry, POLL EAU dewater

désignation *f* ÉCHANT designation; ~ **de la couleur** ÉCHANT color designation (*AmE*), colour designation (*BrE*)

désinfectant *m* DIST EAU, POLL EAU, SÉCURITÉ disinfectant

désinfecter *vt* DIST EAU, POLL EAU, SÉCURITÉ disinfect

désinfection *f* DIST EAU, POLL EAU, SÉCURITÉ disinfection; ~ **des déchets** DÉCHETS waste disinfection; ~ **aux rayons ultraviolets** DIST EAU UV disinfection, ultraviolet disinfection; ~ **aux rayons UV** DIST EAU UV disinfection, ultraviolet disinfection

désintégrateur *m* DIST EAU *papier* pulper, POLL ATM disintegrator; ~ **de déchets** DÉCHETS waste disintegrator

désintégration *f* DÉCHETS, DIST EAU, SOLS disintegration; ~ **radioactive** ÉN RENOUV, POLL EAU *nucléaire* decay, radioactive decay

désintégrer[1] *vt* DÉCHETS, DIST EAU, SOLS disintegrate

désintégrer[2]: **se** ~ *v réfl* DÉCHETS, DIST EAU, SOLS disintegrate

désionisation *f* POLL EAU deionization

désionisé *adj* POLL EAU deionized

désioniser *vt* POLL EAU deionize

désodation *f* POLL EAU sodium removal

désodorisation *f* DIST EAU, POLL ATM, POLL EAU deodorization, odor control (*AmE*), odor removal (*AmE*), odour control (*BrE*), odour removal (*BrE*)

désodoriser *vt* DIST EAU, POLL ATM, POLL EAU deodorize

desonox *abrév* (*élimination de* SO_2/NO_x) POLL ATM desonox (SO_2/NO_x *removal*)

désorbant *m* POLL ATM desorbent

désorbé *adj* ÉCHANT, POLL ATM, POLL EAU, SOLS desorbed

désorber *vt* ÉCHANT, POLL ATM, POLL EAU, SOLS desorb; ~ **par voie thermique** ÉCHANT desorb thermally

désorption *f* ÉCHANT, POLL ATM, POLL EAU, SOLS desorption; ~ **par champ électrique** ÉCHANT field desorption (*FD*); ~ **thermique** ÉCHANT thermal desorption; ~ **thermique in situ** SOLS in-situ thermal desorption

désoxygénation *f* DIST EAU, POLL EAU deoxygenation

désoxygéner *vt* DIST EAU, POLL EAU deoxygenate

desquamation *f* POLL EAU sloughing

dessabler *vt* POLL EAU degrit

dessableur *m* DÉCHETS detritor, POLL EAU degritter, detritor

dessalage *m* DIST EAU *de l'eau de mer* saline-water conversion, demineralization, desalinization, desalination, POLL EAU demineralization

dessalement *m* DIST EAU *de l'eau de mer* saline-water conversion, demineralization, desalinization, desalination, POLL EAU demineralization

dessaler *vt* DIST EAU desalinate, desalinize, desalt

dessiccateur *m* ÉCHANT desiccator, drying agent, POLL ATM drying medium

dessin *m* ÉCHANT design

déstabilisation *f* POLL EAU destabilization

déstabilisé *adj* POLL EAU destabilized

déstabiliser *vt* POLL EAU destabilize

destructeur[1]: ~ **d'ozone** *adj* POLL ATM *émissions* ozone-damaging

destructeur[2]: ~ **d'ozone** *m* POLL ATM *émissions* ozone destroyer

destruction *f* POLL ATM damage, destruction, POLL EAU, SOLS destruction; **~ aiguë de l'ozone** POLL ATM peak ozone damage; **~ de la couche d'ozone** POLL ATM ozone layer destruction; **~ d'un digesteur** POLL EAU digester disruption; **~ de l'ozone** POLL ATM ozone damage, ozone destruction; **~ photolytique** SOLS photolytic destruction; **~ solaire à haute température** DIST EAU solar high-temperature destruction; **~ de substances dangereuses** SOLS destruction of harmful substances; **~ thermique** SOLS thermal destruction; **~ thermique à l'infrarouge** SOLS infrared thermal destruction

désulfuration *f* DIST EAU, POLL ATM desulfurization (*AmE*) (*DESOX*), desulphurization (*BrE*) (*DESOX*); **~ du charbon** POLL ATM coal desulfurization (*AmE*), coal desulphurization (*BrE*); **~ d'effluents gazeux non recyclables** POLL ATM non-regenerable flue-gas desulfurization (*AmE*), non-regenerable flue-gas desulphurization (*BrE*); **~ des effluents gazeux régénérables** POLL ATM regenerable flue-gas desulfurization (*AmE*), regenerable flue-gas desulphurization (*BrE*); **~ des gaz** DIST EAU, POLL ATM, SÉCURITÉ gas desulfurization (*AmE*), gas desulphurization (*BrE*); **~ des gaz de combustion** POLL ATM flue-gas desulfurization (*AmE*), flue-gas desulphurization (*BrE*) (*FGD*); **~ des gaz d'échappement** DÉCHETS exhaust waste-gas desulfurization (*AmE*), exhaust waste-gas desulphurization (*BrE*); **~ des huiles lourdes** DÉCHETS, DIST EAU heavy oil desulfurization (*AmE*), heavy oil desulphurization (*BrE*); **~ à l'oxyde de magnésium** POLL ATM desulfurization with magnesium oxide (*AmE*), desulphurization with magnesium oxide (*BrE*); **~ sur lit fluidisé circulant** POLL ATM desulfurization in circulating fluidized bed (*AmE*), desulphurization in circulating fluidized bed (*BrE*)

désulfuré *adj* POLL ATM sulfur-free (*AmE*), sulphur-free (*BrE*)

désulfurer *vt* DIST EAU, POLL ATM, SÉCURITÉ desulfurize (*AmE*), desulphurize (*BrE*)

désulfureur *m* POLL ATM sulfur-oxides scrubber (*AmE*), sulphur-oxides scrubber (*BrE*)

désulfurisation: **~ des gaz** *f* POLL ATM flue-gas desulfurization (*AmE*), flue-gas desulphurization (*BrE*) (*FDG*)

détaillant *m* GESTION retailer

détectabilité *f* ÉCHANT detectability

détectable *adj* ÉCHANT detectable

détecter *vt* ÉCHANT detect; **~ qualitativement** ÉCHANT qualitatively detect

détecteur *m* ÉCHANT detector; **~ d'absorption dans l'ultraviolet lointain** ÉCHANT far-ultraviolet absorbance detector (*FUVAD*); **~ amovible** ÉCHANT interchangeable detector; **~ d'azote et de phosphore** ÉCHANT nitrogen-phosphorus detector (*NPD*); **~ de capture d'électrons** ÉCHANT electron-capture detector; **~ de chromatographie en phase liquide** ÉCHANT liquid chromatography detector; **~ de conductivité thermique** ÉCHANT thermal conductivity detector (*TCD*); **~ de déchets** DÉCHETS waste detector; **~ à diodes en série** ÉCHANT diode array detector; **~ de dispersion de la lumière** ÉCHANT light-scattering detector; **~ d'éclats** DÉCHETS metal detector; **~ électrochimique** ÉCHANT electrochemical detector; **~ d'émissions atomiques** ÉCHANT atomic emission detector; **~ de fuite** ÉCHANT leak detector; **~ de fuite sur bouteille** DÉCHETS bottle leak detector; **~ de gaz** ÉCHANT gas detector, gas indicator; **~ de grande sensibilité** ÉCHANT highly-sensitive detector; **~ d'incendie** POLL MAR, SÉCURITÉ fire monitor; **~ intégral** ÉCHANT integral detector; **~ à ionisation de flamme** ÉCHANT flame ionization detector (*FID*); **~ d'ionisation photoélectrique** ÉCHANT photoionization detector (*PID*), portable PID; **~ d'ionisation photoélectrique à microprocesseur** ÉCHANT microprocessor-controlled photoionization detector; **~ de métaux** DÉCHETS metal detector; **~ de monoxyde de carbone** ÉCHANT carbon monoxide detector; **~ par chromatographie en phase gazeuse** ÉCHANT gas chromatography detector (*GCD*); **~ de piège ionique** ÉCHANT ion-trap detector; **~ de poussières** ÉCHANT dust counter; **~ de rayonnement** ÉCHANT radiation detector; **~ de rayonnement ultraviolet et visible** ÉCHANT UV/VIS detector, ultraviolet and visible detector; **~ de rayonnement UV et visible** ÉCHANT UV/VIS spectrometer, ultraviolet and visible spectrometer; **~ de référence** ÉCHANT reference detector; **~ sélectif de masse** ÉCHANT mass-selective detector; **~ à semiconducteur** ÉCHANT semiconductor detector; **~ de spectrométrie de masse** ÉCHANT mass-spectrometry detector (*MSD*); **~ tandem** ÉCHANT tandem detector; **~ thermo-ionique** ÉCHANT therm-ionic detector; **~ de tourbillon** POLL ATM vortex finder

détection *f* ÉCHANT detection; **~ d'azote et de phosphore par chromatographie en phase gazeuse** ÉCHANT gas chromatograph nitrogen-phosphorus detection (*GC-NPD*); **~ de capture d'électrons par chromatographie en phase gazeuse** ÉCHANT gas chromatography-electron capture detection (*GC-ECD*); **~ colorimétrique** ÉCHANT colorimetric detection; **~ de dureté** ÉCHANT hardness detection; **~ des dysfonctionnements** SÉCURITÉ malfunction detection; **~ électrochimique par chromatographie en phase liquide** ÉCHANT liquid chromatography-electrochemical detection (*LC-EC*); **~ d'émissions atomiques** ÉCHANT atomic emission detection (*AED*); **~ d'émissions de plasma** ÉCHANT plasma emission detection (*PED*); **~ de fluorescence par chromatographie en phase liquide** ÉCHANT liquid chromatography-fluorescence detection (*LC-FL*); **~ fluorescente** ÉCHANT fluorescent detection; **~ de fuite** DÉCHETS, ÉCHANT, POLL ATM, SOLS leak detection; **~ d'ionisation de flamme par chromatographie en phase gazeuse** ÉCHANT gas chromatography-flame ionization detection (*GC-FID*); **~ par chimiluminescence** ÉCHANT chemiluminescence detection (*CLD*); **~ par photométrie de flamme** ÉCHANT flame photometric detection (*FPD*); **~ par photométrie de flamme-chromatographie en phase gazeuse** ÉCHANT gas chromatography-flame photometric detection (*GC-FPD*); **~ de pic** ÉCHANT peak detection; **~ de pointe** ÉCHANT peak detection; **~ qualitative** ÉCHANT qualitative detection; **~ quantitative** ÉCHANT quantitative detection; **~ d'UV par chromatographie en phase liquide** ÉCHANT liquid chromatography-UV detection (*LC-UV detection*)

détente *f* DIST EAU expansion, loosening up, SOLS loosening up

détergent[1]: **~-actif** *adj* SOLS detergent, washing-active

détergent[2] *m* POLL EAU detergent; **~ doux** POLL EAU soft detergent; **~ dur** POLL EAU hard detergent; **~ sans phosphate** DIST EAU, POLL EAU phosphate-free washing-up powder

détérioration *f* POLL ATM deterioration, POLL EAU degradation, SOLS deterioration; **~ du catalyseur** POLL ATM catalyst deterioration; **~ due aux intempéries** DIST EAU, POLL MAR, SOLS weathering; **~ du milieu sonore** POLL SON noise ambient impairment

détérioré *adj* POLL ATM, POLL MAR impaired

détériorer[1] *vt* POLL ATM impair, SOLS deteriorate

détériorer[2]: **se ~** *v réfl* SOLS deteriorate

détermination *f* ÉCHANT determination; **~ après distillation sur échantillon** ÉCHANT determination following sample distillation; **~ biologique de la qualité de l'eau** ÉCHANT biological water quality determination; **~ biomacroscopique de la qualité de l'eau** ÉCHANT macroscopic biological water quality determination; **~ de la capacité de neutralisation des acides des eaux usées** ÉCHANT determination of acid capacity in wastewater; **~ colorimétrique** ÉCHANT colorimetric determination; **~ colorimétrique du pH** ÉCHANT colorimetric pH-value determination; **~ complexométrique** ÉCHANT complexometric determination; ~ **de la concentration d'ammonium dans les eaux usées** ÉCHANT determination of ammonium content in wastewater; ~ **de la concentration de dioxyde d'azote** SÉCURITÉ determination of concentration of nitrogen dioxide; ~ **de la concentration de fer dans les eaux usées** ÉCHANT analysis of iron in wastewater; ~ **de la concentration massique de dioxyde de soufre dans l'air ambiant** SÉCURITÉ determination of mass concentration of sulfur dioxide in ambient air (*AmE*), determination of mass concentration of sulphur dioxide in ambient air (*BrE*); ~ **consécutive à la distillation d'échantillon** ÉCHANT determination following sample distillation; ~ **de la demande en eau** DIST EAU determination of water demand; ~ **de la dureté** ÉCHANT hardness determination; ~ **de la dureté totale des eaux usées** ÉCHANT determination of total hardness in wastewater; ~ **iodométrique** ÉCHANT iodometric determination; ~ **de la masse volumique des eaux usées** ÉCHANT determination of wastewater density; ~ **de la masse volumique des gaz** ÉCHANT gas density determination; ~ **de la masse volumique de gaz par gravimétrie** ÉCHANT gravimetric gas density determination; ~ **des matières décantables et non dissoutes dans les eaux usées** ÉCHANT determination of settleable and undissolved matter in wastewater; ~ **des métaux lourds** ÉCHANT heavy-metal determination; ~ **de l'odeur** ÉCHANT odor determination (*AmE*), odour determination (*BrE*); ~ **des oligo-éléments** ÉCHANT trace element determination; ~ **oxydimétrique** ÉCHANT oxidimetric determination; ~ **par photométrie de flamme** ÉCHANT flame photometric determination; ~ **des particules et des composés de plomb** SÉCURITÉ determination of particulate lead and lead compounds; ~ **du point éclair** ÉCHANT flash point determination; ~ **du point de rosée** ÉCHANT determination of dew point; ~ **de la pollution atmosphérique acide** SÉCURITÉ determination of gaseous acid air pollution; ~ **de pression de vapeur** ÉCHANT vapor pressure determination (*AmE*), vapour pressure determination (*BrE*); ~ **qualitative** ÉCHANT qualitative determination; ~ **quantitative** ÉCHANT quantitative determination; ~ **de la teneur en humidité** DÉCHETS moisture determination, ÉCHANT humidity content determination; ~ **volumétrique** ÉCHANT volumetric determination

détersif *m* POLL EAU detergent

détonation *f* POLL ATM knocking

détournement: ~ **d'écoulement** *m* POLL ATM flow deviation

détoxication *f* DIST EAU, POLL ATM, POLL EAU, SOLS detoxification; ~ **par lots** POLL EAU detoxification under batch conditions

détoxiquer *vt* DIST EAU, POLL ATM, POLL EAU, SOLS detoxify

détrempé *adj* SOLS sodden

détremper *vt* ÉCHANT temper

détritiphage *m* DÉCHETS scavenger, detritus feeder

détritivore *m* DÉCHETS detritus feeder

détritus *m pl* DÉCHETS detritus, litter, refuse, DIST EAU, POLL EAU *prétraitement des boues* detritus; ~ **de criblage** DIST EAU, POLL EAU screenings; ~ **ramassés sur la voie publique** DÉCHETS road sweepings

détrôner *vt* POL ENV supersede

détruire *vt* POLL ATM, POLL EAU, SOLS destroy

dévastation: ~ **du sol** *f* DIST EAU land degradation, land disturbance

développement *m* GESTION *d'un procédé*, POL ENV development; ~ **compatible avec l'environnement** POL ENV environmentally-compatible development; ~ **durable** POL ENV sustainable development

développer *vt* GESTION *procédé*, POL ENV develop

déversement *m* DÉCHETS tip, DIST EAU, POLL EAU, POLL MAR *de déchets dangereux* spillage, spill, discharge, POLL MAR spill; ~ **avec compactage** DÉCHETS tipping with compaction; ~ **de déchets** DÉCHETS waste tip (*BrE*); ~ **délibéré** POLL EAU deliberate dumping; ~ **des eaux fluviales** POLL EAU river discharge; ~ **d'eaux usées** POLL EAU outfall of sewage; ~ **d'hydrocarbures** POLL MAR oil leakage, oil spill, oil spillage; ~ **illégal** DÉCHETS, POL ENV, POLL EAU illegal dumping; ~ **illicite** DÉCHETS, POL ENV, POLL EAU illegal dumping; ~ **industriel** POLL EAU industrial outfall; ~ **en mer** POLL MAR marine dumping

déverser[1] *vt* DÉCHETS dump, tip, POLL EAU, POLL MAR discharge, dump, SOLS discharge

déverser[2]: **se** ~ *v réfl* DIST EAU, POLL ATM discharge, overflow

déversoir *m* ÉN RENOUV weir, POLL EAU discharger, outfall, outlet, weir, POLL MAR spill, weir; ~ **dénoyé** POLL EAU freefall weir; ~ **direct** POLL EAU direct discharger; ~ **d'eaux pluviales** POLL EAU storm-water overflow, storm-water overflow tank; ~ **des eaux résiduaires** POLL EAU wastewater outfall; ~ **flottant autoréglable** POLL MAR self-adjusting floating weir; ~ **indirect** POLL EAU indirect discharger; ~ **de mesure** DIST EAU measuring weir; ~ **à nappe noyée** DIST EAU drowned weir; ~ **noyé** DIST EAU drowned weir, submerged weir; ~ **de sortie** POLL EAU effluent weir; ~ **de trop-plein** POLL ATM overflow weir, POLL EAU discharge hopper, overflow weir

dévésiculeur *m* DÉCHETS mist eliminator, POLL ATM mist eliminator, droplet separator

dévier *vi* POLL ATM deviate

devoir: ~ **de prévention** *m* POL ENV duty of prevention

dézincifier *vt* DIST EAU dezincify

dézingage *m* DIST EAU dezincification

diagnostic *m* ÉCHANT diagnosis

diagramme: ~ **d'écoulement** *m* DIST EAU flow pattern

dialyse *f* ÉCHANT, POLL EAU dialysis

diamagnétique *adj* ÉCHANT diamagnetic

diamètre: ~ **aérodynamique** *m* ÉCHANT, POLL ATM aerodynamic diameter; ~ **de cheminée** *m* POLL ATM stack diameter; ~ **équivalent** *m* POLL ATM *étude des polluants particulaires* equivalent diameter; ~ **de fibre** *m* POLL ATM fiber diameter (*AmE*), fibre diameter (*BrE*); ~ **de gouttelette** *m* POLL ATM droplet diameter; ~ **de particule** *m* POLL EAU particle diameter; ~ **de pore** *m* POLL ATM pore diameter

diaphonie *f* POLL SON crosstalk

dibenzofuranne: ~ **polychloré** *m* (*PCDF*) DÉCHETS, SOLS polychlorinated dibenzofuran (*PCDF*)

dichloro: ~**-diphényl-trichloroéthane** *m* (*DDT*) SOLS dichlorodiphenyltrichloroethane (*DDT*)

dichlorodifluorométhane *m* POLL ATM dichlorodifluoromethane

dichlorométhane *m* POLL ATM dichloromethane

différence: ~ **de pression** *f* ÉCHANT pressure difference; ~ **de pression partielle** *f* ÉCHANT partial pressure difference

diffraction *f* ÉCHANT diffraction; ~ **de la lumière** ÉCHANT diffraction of light, light scattering

diffuser *vt* GÉNÉRAL diffuse, scatter

diffuseur *m* GÉNÉRAL diffuser; ~ **à disques** POLL EAU disc aerator (*BrE*), disk aerator (*AmE*); ~ **à plaques** POLL EAU plate diffuser

diffusion *f* GÉNÉRAL diffusion, scattering; ~ **de Mie** ÉCHANT, POLL ATM Mie scattering; ~ **de Rayleigh** ÉCHANT, POLL ATM Rayleigh scattering; ~ **du rayonnement vers l'avant** ÉCHANT forward scattering of radiation; ~ **souterraine** POLL EAU *système d'aération* subsurface diffusion

digesteur *m* DIST EAU, POLL EAU digester, sludge digester; ~ **chauffé** POLL EAU heated digester; ~ **à deux étages** POLL EAU two-stage digester; ~ **à un étage** POLL EAU single-stage digester

digestion *f* DIST EAU, POLL EAU digestion; ~ **aérobie** DIST EAU aerobic digestion; ~ **anaérobie** POLL EAU anaerobic digestion; ~ **avec apport de compost** DIST EAU, POLL EAU digestion with added composting; ~ **des boues** POLL EAU settling time, sludge digestion; ~ **mésophile** POLL EAU mesophilic digestion; ~ **méthanique** POLL EAU methane digestion; ~ **thermophile** POLL EAU thermophilic digestion; ~ **thermophile-mésophile** POLL EAU thermophilic-mesophilic digestion; ~ **thermophile-mésophile en deux étapes** POLL EAU two-stage thermophilic-mesophilic digestion

digue *f* DIST EAU embankment, ÉN RENOUV dyke

dilacérateur *m* DÉCHETS, DIST EAU shredder, comminutor, disintegrator, ÉN RENOUV disintegrator

dilacération *f* DIST EAU shredding

dilatation: ~ **thermique** *f* POLL ATM thermal heat expansion

dilué *adj* GÉNÉRAL diluted

diluer *vt* GÉNÉRAL dilute

dilution *f* GÉNÉRAL dilution; ~ **des effluents gazeux** POLL ATM waste gas dilution

dimension: ~ **de particules** *f* DÉCHETS *charbon* particle size

dimensionnement: ~ **de filtre** *m* POLL ATM filter dimensioning

diminuer *vt* POLL ATM *température* lower

diminution *f* DIST EAU reduction, POLL ATM thinning; ~ **de la couche d'ozone** POLL ATM ozone loss; ~ **du temps de rétention des effluents gazeux** POLL ATM reduction of waste gas retention time

DIN *abrév* (*Institut allemand de normalisation*) ÉCHANT DIN (*German Institute for Standards*)

diode: ~ **électroluminescente** *f* (*DEL*) ÉCHANT light-emitting diode (*LED*)

dioxine *f* SOLS dioxin

dioxyde: ~ **d'azote** *m* (NO_2) POLL ATM, POLL EAU, SÉCURITÉ nitrogen dioxide (NO_2); ~ **de carbone** *m* (CO_2) POLL ATM carbon dioxide (CO_2); ~ **de carbone libre** *m* DIST EAU free carbon dioxide; ~ **de carbone total** *m* POLL ATM total carbon dioxide; ~ **de soufre** *m* (SO_2) POLL ATM sulfur dioxide (*AmE*) (SO_2), sulphur dioxide (*BrE*) (SO_2)

diplomatie *f* POL ENV diplomacy

direct *adj* GÉNÉRAL direct, on-the-spot

direction *f* DIST EAU *d'un courant* drift, direction, GESTION, POL ENV division, directorate, *d'une entreprise* management, POLL ATM *des gaz* direction, POLL MAR *d'un courant* drift, POLL SON *d'un son* direction; ~ **de l'incidence du son** POLL SON sound incidence direction

directive *f* GESTION, POL ENV directive, guideline; ~ **d'achat écologique** GESTION, POL ENV ecological buying guideline; ~ **applicable à l'environnement** GESTION, POL ENV environmental directive; ~ **sur les principaux risques industriels** GESTION, POL ENV directive on major industrial hazards; ~ **technique** GESTION, POL ENV technical directive

Directive: ~ **du Conseil** *f* GESTION, POL ENV *UE* Council Directive

DIS *abrév* (*déchets industriels spéciaux*) DÉCHETS special industrial refuse

discernable *adj* ÉCHANT distinguishable

discontinu *adj* GÉNÉRAL intermittent, discontinuous

discontinuité: ~ **hydraulique** *f* DÉCHETS, POLL EAU leach

dislocation *f* DIST EAU *cristaux* dislocation, displacement

dispersant *m* GÉNÉRAL dispersant agent, dispersant; ~ **du pétrole** POLL EAU oil dispersant

dispersé *adj* GÉNÉRAL dispersed

disperser *vt* GÉNÉRAL disperse

dispersion *f* GÉNÉRALdispersion, scattering; ~ **des gaz** POLL ATM gas dispersion; ~ **de gouttelettes** POLL ATM droplet dispersion; ~ **des hydrocarbures** POLL EAU spilled oil; ~ **de poussière** POLL ATM dust dispersion; ~ **verticale** POLL ATM vertical dispersion

dispositif *m* GÉNÉRAL device; ~ **antibruit** POLL SON noise guard; ~ **antibuée** POLL ATM demister, demister unit; ~ **antiexplosion** POLL ATM explosion suppression device; ~ **d'auto-prélèvement** ÉCHANT autosampler; ~ **de changement de grille** DIST EAU screen-changing device; ~ **de codage de fond** DÉCHETS bottom coder; ~ **de collecte** DIST EAU pick-up mechanismPOLL MAR collection device; ~ **collecteur** POLL EAU oil separator, oil trap; ~ **de commande** DIST EAU control device; ~ **de coupure** DIST EAU shutoff device; ~ **de dénitration** POLL ATM denoxer; ~ **de désembuage** ÉN RENOUV de-mister unit; ~ **d'élimination des buées** DÉCHETS *industrie papetière* mist eliminator; ~ **d'épuration de l'air** SÉCURITÉ clean-air device; ~ **de frappage** POLL ATM *électrodes des dépoussiéreurs* rapper; ~ **homme mort** SÉCURITÉ *transmission de signal d'alarme* dead man's switch; ~ **d'interclassement** POLL ATM collating device; ~ **de lavage par voie humide** POLL ATM wet-scrubbing device; ~ **de mesure** ÉCHANT measuring device; ~ **de mesure multi-usages** ÉCHANT polymeter; ~ **de prélèvement d'échantillon** ÉCHANT sampling device; ~ **de prémélange** DIST EAU premixer; ~ **de récupération** DIST EAU, POLL ATM, POLL MAR recovery device; ~ **de réfrigération** DIST EAU, ÉN RENOUV, POLL ATM cooling system; ~ **de réglage du couple moteur** POLL ATM torque adjuster; ~ **de renouvellement de tamis à contre-courant** DIST EAU backflush screen changer; ~ **de reprise des boues** POLL EAU return-sludge elevator; ~ **de séparation par voie humide** POLL ATM wet-scrubbing device; ~ **de serrage** DÉCHETS *emballage* clam pack; ~ **de transmission des valeurs mesurées** ÉCHANT device for measured value transmission; ~ **d'urgence** GÉNÉRAL contingency plan

disque: ~ **biologique** *m* POLL EAU biodisc (*BrE*), biodisk (*AmE*), rotating biological contactor (*RBC*); ~ **de Rayleigh** *m* ÉCHANT Rayleigh disk

dissimilation *f* POLL EAU dissimilation

dissipation: ~ **d'énergie** *f* ÉN RENOUV energy loss

dissociable *adj* ÉCHANT, POLL EAU dissociable

dissociation *f* ÉCHANT, POLL EAU dissociation

dissocier *vt* POLL ATM break down

dissolution *f* DIST EAU, ÉCHANT, POLL ATM, POLL EAU dissolution; ~ **chimique** DIST EAU solvolysis

dissoudre *vt* DIST EAU *encre*, ÉCHANT, POLL ATM dissolve, POLL EAU dissolve, solubilize; ~ **l'encre d'impression de fibre** DIST EAU dissolve printed color from fiber (*AmE*), dissolve printed colour from fibre (*BrE*)

dissous *adj* DIST EAU, ÉCHANT, POLL ATM dissolved, POLL EAU solubilized, *dans l'eau* dissolved

distillat *m* DIST EAU distillation, ÉCHANT distillate, distillation, POLL EAU, SOLS distillation

distillation *f* DIST EAU, ÉCHANT, POLL EAU, SOLS distillation; ~ **discontinue** SOLS batch distillation; ~ **à l'évaporateur** SOLS evaporator distillation; ~ **instantanée** DIST EAU flash distillation; ~ **par membrane** POLL EAU membrane distillation; ~ **sèche** SOLS dry distillation; ~ **à la vapeur d'eau** ÉCHANT steam distillation

distiller[1] *vt* DIST EAU, ÉCHANT, POLL EAU, SOLS distil (*BrE*), distill (*AmE*)

distiller[2]: **se** ~ *v réfl* DIST EAU, ÉCHANT, POLL EAU, SOLS distil (*BrE*), distill (*AmE*)

distorsion: ~ **de la valeur mesurée** *f* ÉCHANT measured-value distortion

distributeur *m* GÉNÉRAL distributor; ~ **d'échantillon** ÉCHANT sample dispenser; ~ **d'étiquettes** DÉCHETS label dispenser; ~ **de liquide** POLL ATM liquid distributor; ~ **mécanique** POLL ATM stoker; ~ **rotatif** POLL EAU rotary distributor; ~ **à va-et-vient** DIST EAU *égout* traveling distributor (*AmE*), travelling distributor (*BrE*); ~ **à vis** SOLS feeder worm

distribution *f* GÉNÉRAL distribution; ~ **d'air** POLL ATM air distribution; ~ **de chaleur basse température** ÉN RENOUV low-temperature heat distribution; ~ **d'eau** DIST EAU water distribution; ~ **d'eau en milieu rural** DIST EAU rural water supply; ~ **d'eau potable** DIST EAU drinking-water supply; ~ **de fréquence** ÉCHANT frequency distribution; ~ **de gaz** POLL ATM gas distribution; ~ **géographique** DIST EAU, POLL ATM spatial distribution; ~ **publique d'eau** DIST EAU, POLL EAU public water supply; ~ **spatiale** DIST EAU, POLL ATM

spatial distribution; ~ **verticale** POLL ATM, SOLS *de l'ozone* vertical distribution
diversité *f* GESTION, POL ENV *des ressources génétiques* diversity
diviseur: ~ **de faisceau** *m* ÉCHANT *optique laser* beamsplitter
DJA *abrév* (*dose journalière admissible*) SÉCURITÉ ADI (*acceptable daily intake*)
DL *abrév* (*dose létale*) DIST EAU, ÉCHANT, POLL ATM, SÉCURITÉ LD (*lethal dose*)
DL_{50} *abrév* (*dose semi-létale*) SÉCURITÉ LD_{50} (*lethal dose fifty, median lethal dose*)
DL_x *abrév* (*dose létale x*) SÉCURITÉ LD_x (*lethal dose x*)
DMA *abrév* (*dose maximale admissible*) SÉCURITÉ MPD (*maximum permissible dose*)
document: ~ **confidentiel** *m* POL ENV confidential document
documentation: ~ **d'entreprise** *f* POL ENV firm-specific documents; ~ **à usage interne** *f* POL ENV firm-specific documents
domaine: ~ **d'activité** *m* SOLS operative range
dommage: ~ **biologique** *m* SÉCURITÉ biological damage, biological harm
dommageable *adj* SÉCURITÉ injurious
données: ~ **antérieures** *f pl* DIST EAU historical data; ~ **brutes** *f pl* ÉCHANT raw data; ~ **climatiques** *f pl* GESTION, POL ENV climatic data; ~ **historiques** *f pl* DIST EAU historical data; ~ **manquantes** *f pl* ÉCHANT missing data; ~ **recueillies sur le terrain** *f pl* ÉCHANT field data; ~ **relatives à la qualité de l'air** *f pl* ÉCHANT, POLL ATM air quality data; ~ **saisies** *f pl* ÉCHANT input
dopage *m* ÉN RENOUV, POLL ATM doping
doper *vt* ÉN RENOUV, POLL ATM dope
dosage *m* DÉCHETS, DIST EAU dosage, dosing, ÉCHANT dosage, dosing, *de composés* determination, POLL ATM, POLL EAU dosage, dosing; ~ **d'adsorbants** POLL ATM adsorbent dosage; ~ **d'air** POLL ATM air dosage; ~ **de la concentration d'ammonium dans les eaux usées** ÉCHANT determination of ammonium content in wastewater; ~ **de la concentration de fer dans les eaux usées** ÉCHANT analysis of iron in wastewater; ~ **de la concentration massique de dioxyde de soufre dans l'air ambiant** SÉCURITÉ determination of mass concentration of sulfur dioxide in ambient air (*AmE*), determination of mass concentration of sulphur dioxide in ambient air (*BrE*); ~ **d'échantillon** ÉCHANT sample dosing; ~ **immunologique** ÉCHANT immunoassay; ~ **iodométrique** ÉCHANT iodometric determination; ~ **des matières décantables et non dissoutes dans les eaux usées** ÉCHANT determination of settleable and undissolved matter in wastewater; ~ **des métaux lourds** ÉCHANT heavy-metal determination; ~ **des oligo-éléments** ÉCHANT trace element determination; ~ **optimal de polymères** POLL EAU optimum polymer dosage; ~ **oxydimétrique** ÉCHANT oxidimetric determination; ~ **par photométrie de flamme** ÉCHANT flame photometric determination; ~ **des particules et des composés de plomb** SÉCURITÉ determination of particulate lead and lead compounds; ~ **qualitatif** ÉCHANT qualitative determination; ~ **quantitatif** ÉCHANT quantitative determination; ~ **séquentiel** POLL EAU sequential dosage
dose *f* DÉCHETS, DIST EAU, ÉCHANT, POLL EAU dose; ~ **absorbée** SÉCURITÉ absorbed dose; ~ **collective** ÉCHANT, POLL ATM group collective dose; ~ **collective reçue par un groupe de population** POLL ATM subpopulation collective dose; ~ **cumulée** SÉCURITÉ cumulative dose; ~ **effet** DIST EAU, ÉCHANT dose response; ~ **génétiquement significative** POLL ATM, SÉCURITÉ genetically significant dose; ~ **journalière admissible** (*DJA*) SÉCURITÉ acceptable daily intake (*ADI*); ~ **létale** (*DL*) DIST EAU, ÉCHANT, POLL ATM, SÉCURITÉ lethal dose (*LD*); ~ **létale 50** (*DL_{50}*) SÉCURITÉ lethal dose fifty (*LD_{50}*), median lethal dose (*LD_{50}*); ~ **létale indéterminée** SÉCURITÉ lethal dose x; ~ **létale x** (*DL_x*) SÉCURITÉ lethal dose x (*LD_x*); ~ **létale zéro totale** SÉCURITÉ total lethal dose zero (*TDL_0*); ~ **maximale admissible** (*DMA*) SÉCURITÉ maximum permissible dose (*MPD*); ~ **maximale tolérée** SÉCURITÉ maximum tolerated dose (*MTD*); ~ **moyenne mensuelle** ÉCHANT, POLL ATM, SÉCURITÉ average monthly dose; ~ **de rayonnement** POLL ATM, SÉCURITÉ radiation dose; ~ **de rayonnement absorbée** (*rad*) ÉCHANT radiation absorbed (*rad*); ~ **reçue par une population** POLL ATM, SÉCURITÉ population dose; ~ **semi-létale** (*DL_{50}*) SÉCURITÉ lethal dose fifty (*LD_{50}*), median lethal dose (*LD_{50}*); ~ **sublétale** SÉCURITÉ sub-lethal dose
doser *vt* DÉCHETS, DIST EAU dose, ÉCHANT dose, apportion, POLL EAU dose
doseur *m* DÉCHETS dosing apparatus, ÉCHANT dosage device
doseuse *f* DIST EAU dosing machine
dosifilm *m* ÉCHANT *radioprotection* film badge
dosimètre *m* ÉCHANT dosimeter; ~ **acoustique** ÉCHANT noise dosimeter; ~ **chimique** ÉCHANT chemical dosimeter; ~ **exoélectronique** ÉCHANT exoelectron dosimeter; ~ **à long terme** ÉCHANT long-term dosimeter; ~ **photographique individuel** ÉCHANT *radioprotection* film badge; ~ **à thermoluminescence** ÉCHANT thermoluminescence dosimeter (*TLD*)
dossier: ~ **d'étalonnage** *m* ÉCHANT calibration record; ~ **d'impact sur l'environnement** *m* POL ENV, GESTION environmental impact statement (*EIS*)
double[1]: ~ **filtration à contre-courant** *f* DIST EAU double counter-current filtration; ~ **paroi** *f* POLL SON double wall
double[2]: **à ~ couche** *loc* POLL SON double-sheet; **à ~ cycle** *loc* POLL ATM double-cycle; **à ~ feuille** *loc* POLL SON double-leaf, double-sheet; **à ~ tube** *loc* POLL ATM double-pipe
doublé *adj* POLL ATM, POLL SON lined
doublure *f* DÉCHETS liner bag, lining paper, lamination
douille *f* DIST EAU, POLL EAU socket
drague: ~ **de mer** *f* POLL MAR marine dredge
draguer *vt* POLL EAU, POLL MAR dredge
dragueuse *f* POLL EAU, POLL MAR dredger
drainabilité *f* POLL EAU dewaterability
drainage[1] *m* GÉNÉRAL drainage, dewatering; ~ **des eaux de ruissellement** DÉCHETS leachate drainage; ~ **des eaux de ruissellement sur une couche de gravier** DÉCHETS, POLL EAU leachate drainage with filtering rubble; ~ **de fond** DÉCHETS bottom drainage; ~ **naturel** DIST EAU natural drainage; ~ **profond** DIST EAU underground drainage; ~ **souterrain** POLL EAU underdrainage
drainage[2]: **à ~ naturel** *loc* SOLS self-draining
drainé *adj* SOLS dewatered, drained
drainer *vt* SOLS dewater, drain
droit *m* POL ENV levy; ~ **civil** POL ENV civil law; ~ **de consommation** GESTION, POL ENV excise duty; ~ **de l'eau** DIST EAU, POL ENV, POLL EAU water law; ~ **de l'environnement** POL ENV environmental law; ~ **à un environnement vivable** POL ENV amenity rights; ~ **fiscal** POL ENV tax law; ~ **international** POL ENV international law; ~ **à payer par l'usager** POL ENV user charge; ~ **pénal** POL ENV penal law; ~ **pénal en matière d'environnement** POL ENV environmental criminal law; ~ **de raccordement au tout-à-l'égout** POL ENV sewerage connection duty; ~ **du recyclage** DIST EAU, POL ENV recycling economy law
droits: ~ **d'émissions négociables** *m pl* POL ENV tradeable emissions rights; ~ **d'émissions négociables à l'échelle internationale** *m pl* POL ENV international tradeable emissions rights; ~ **de jouissance de l'environnement** *m pl* POL ENV environmental use rights; ~ **d'utilisation de l'environnement** *m pl* POL ENV environmental use rights; ~ **d'usage de l'environnement négociables** *m pl* POL ENV tradeable environmental use rights
DUP *abrév* (*déclaration d'utilité publique*) POL ENV declaration of public utility

durabilité: **~ à long terme** *f* POLL ATM, POLL MAR, POLL SON long-term durability

durable *adj* GÉNÉRAL durable, POL ENV *développement*, SOLS sustainable

durée: **~ d'adsorption** *f* POLL ATM adsorption duration; **~ d'analyse** *f* ÉCHANT analysis time; **~ de la digestion** *f* DIST EAU, POLL EAU digestion period; **~ d'écoulement** *f* POLL EAU run-off duration; **~ d'emmagasinage** *f* DÉCHETS storage durability; **~ d'exposition** *f* ÉCHANT exposure time; **~ d'observation** *f* ÉCHANT observation time; **~ d'une rafale** *f* ÉN RENOUV wind duration; **~ de ruissellement** *f* POLL EAU run-off duration; **~ de service** *f* ÉN RENOUV operation time; **~ du traitement** *f* SOLS treatment duration, treatment time; **~ de vie** *f* DÉCHETS, ÉCHANT, GESTION, POLL ATM, POLL EAU, SOLS lifespan, lifetime, life; **~ de vie du catalyseur** *f* POLL ATM catalyst lifetime; **~ de vie dans l'atmosphère** *f* POLL ATM *substance* atmospheric lifetime; **~ de vie utile** *f* SÉCURITÉ serviceable life

dureté: **~ alcaline** *f* DIST EAU *temporaire* alkaline temporary hardness; **~ carbonatée** *f* POLL EAU carbonate hardness; **~ de l'eau** *f* DIST EAU, POLL EAU water hardness; **~ non alcaline** *f* DIST EAU *permanente* nonalkaline permanent hardness; **~ permanente** *f* DIST EAU, POLL EAU permanent hardness; **~ temporaire** *f* DIST EAU, POLL EAU temporary hardness; **~ totale** *f* DIST EAU, POLL EAU total hardness

duromère *f* SOLS *plastique* duromer

dynamique[1] *adj* DÉCHETS, DIST EAU, ÉCHANT, POLL MAR dynamic

dynamique[2]: **~ des écosystèmes** *f* POL ENV ecosystem dynamics

dynamomètre *m* ÉCHANT dynamometer

dysfonctionnement *m* ÉCHANT, SÉCURITÉ operating error, malfunction

E

eau *f* GÉNÉRAL water; **~ agressive** DIST EAU aggressive water; **~ alimentaire** DIST EAU pump water; **~ alimentaire de chaudière** POLL ATM boiler feedwater; **~ d'amont** DIST EAU water head; **~ d'appoint** POLL ATM make-up water; **~ artésienne** DIST EAU artesian water; **~ baneuse** POLL EAU muddy water; **~ brute** DIST EAU, POLL EAU raw water, untreated water; **~ calcaire** DIST EAU hard water; **~ de cale** POLL MAR bilge water; **~ calme** DIST EAU quiet water; **~ capillaire** DIST EAU, POLL EAU, SOLS capillary water; **~ capturée** DIST EAU, POLL EAU retained water; **~ chargée** POLL EAU muddy water; **~ de chaudière** POLL ATM boiler water; **~ de circulation** POLL ATM circulating water; **~ claire** DIST EAU clear water; **~ condensée** DÉCHETS, POLL ATM condensed water; **~ connée** POLL EAU connate water; **~ de consommation** DIST EAU, POLL EAU consumption water, tap water; **~ corrosive** DIST EAU active water; **~ courante** DIST EAU running water; **~ crue** DIST EAU raw water; **~ déminéralisée** DIST EAU, POLL EAU demineralized water; **~ désionisée** DIST EAU, POLL EAU deionized water; **~ désoxygénée** DIST EAU, POLL EAU deoxygenated water; **~ distillée** ÉCHANT distilled water; **~ de distribution** DIST EAU, POLL EAU tap water; **~ douce** DÉCHETS freshwater, DIST EAU freshwater, soft water, POLL ATM, POLL EAU, POLL MAR freshwater; **~ dure** DIST EAU hard water; **~ d'essai** ÉCHANT test water; **~ étrangère** POLL EAU external water; **~ de fabrication** DIST EAU, POLL EAU process water; **~ fluviale** DIST EAU, POLL EAU river water; **~ de fuite** DIST EAU leak water; **~ gravitaire** SOLS gravitational water; **~ grise** POLL EAU gray water (*AmE*), grey water (*BrE*); **~ hygroscopique** DIST EAU hygroscopic water; **~ industrielle** DIST EAU, POLL EAU process water; **~ d'infiltration** DÉCHETS, DIST EAU seepage water, percolating water, leakage water; **~ interstitielle** DIST EAU interstitial water; **~ d'irrigation** DIST EAU irrigation water; **~ de jour** DIST EAU, POLL EAU surface water; **~ juvénile** DIST EAU juvenile water; **~ de lavage** DIST EAU, POLL ATM backwash water, wash water; **~ de lessivage** DIST EAU, POLL EAU leachate; **~ libre** DIST EAU unconfined water; **~ limpide** DIST EAU clear water; **~ littorale** POL ENV, POLL MAR inshore water; **~ lourde** POLL EAU heavy water; **~ maigre** POLL EAU, POLL MAR shallow water; **~ de mer** DIST EAU, ÉN RENOUV, POLL MAR ocean water, seawater; **~ de mer chaude** ÉN RENOUV warm seawater; **~ de mer froide** ÉN RENOUV cold seawater; **~ de mer superficielle** ÉN RENOUV surface seawater; **~ de mer de surface** ÉN RENOUV surface seawater; **~ météorique** DIST EAU rainwater; **~ minérale** DIST EAU mineral water; **~ motrice** ÉN RENOUV service water; **~ neuve** DIST EAU new water; **~ noire** POLL EAU black water; **~ non traitée** DIST EAU, POLL EAU untreated water; **~ d'orage** POLL EAU storm water; **~ oxygénée** POLL ATM, POLL EAU hydrogen peroxide; **~ de pénétration dans les frondaisons** POLL ATM throughfall; **~ de pente** DIST EAU slope water; **~ de percolation** DÉCHETS, DIST EAU percolating water; **~ peu profonde** POLL EAU, POLL MAR shallow water; **~ de pluie** DIST EAU, POLL EAU rainwater; **~ pluviale** DIST EAU, POLL EAU rainwater, storm water; **~ pluviale polluée** POLL EAU polluted storm water; **~ polluée** POLL EAU, POLL MAR polluted water; **~ potable** DIST EAU drinking water; **~ précipitée** DIST EAU, POLL EAU rainwater; **~ de procédé** DIST EAU, POLL EAU process water; **~ de production** DIST EAU production water; **~ propre** DIST EAU clean water; **~ de puits** DIST EAU pump water; **~ pure** DIST EAU clean water, pure water; **~ de purge** DÉCHETS drain water; **~ de recirculation** DIST EAU, POLL EAU recirculating water; **~ de refroidissement** POLL ATM cooling water; **~ régénérée** POLL EAU reclaimed water, recovered water; **~ de remous** DIST EAU backwater; **~ de réserve** DIST EAU, ÉN RENOUV stored water; **~ de réservoir** DIST EAU reservoir water; **~ de rétention** DIST EAU, POLL EAU retained water; **~ de rétention capillaire** DIST EAU, SOLS capillary water; **~ de retenue** DIST EAU reservoir water; **~ de retour** DIST EAU, POLL EAU recirculating water; **~ de rinçage** DIST EAU, SOLS backwash water, rising water; **~ de rivière** DIST EAU, POLL EAU river water; **~ de ruissellement** DÉCHETS, DIST EAU trickling water, run off; **~ de ruissellement sur les troncs** POLL ATM stemflow; **~ salée** DIST EAU saline water, salt water; **~ saumâtre** DIST EAU saline water, salt water; **~ de sol** SOLS soil water; **~ stockée** DIST EAU, ÉN RENOUV stored water; **~ suintée** DIST EAU vadose water; **~ supercritique** SOLS supercritical water; **~ de traitement** DIST EAU, POLL EAU process water; **~ trouble** POLL EAU turbid water; **~ turbide** POLL EAU turbid water; **~ ultrapure** DIST EAU, SOLS ultrapure water; **~ à usage industriel** POLL EAU industrial water; **~ usée décantée** POLL EAU settled water; **~ utile** ÉN RENOUV service water; **~ de vidange** DÉCHETS drain water; **~ vive** DIST EAU running water

eaux: **~ d'égout** *f pl* POLL EAU sewage, wastewater; **~ fortement polluées** *f pl* POLL EAU highly-polluted waters; **~ gadoueuses** *f pl* POLL EAU faecal wastewater (*BrE*), fecal wastewater (*AmE*); **~ intérieures** *f pl* DIST EAU inland waters; **~ non-épurées** *f pl* DIST EAU, POLL EAU untreated wastewater; **~ pluviales acides** *f pl* POLL EAU acid storm water; **~ polluées** *f pl* POLL EAU, POLL MAR polluted waters; **~ polluées par les hydrocarbures** *f pl* POLL EAU oil-polluted wastewater; **~ résiduaires** *f pl* DÉCHETS, POLL EAU wastewater; **~ résiduaires acides** *f pl* DÉCHETS, POLL EAU acid wastewater; **~ résiduaires brutes** *f pl* DÉCHETS, POLL EAU raw wastewater; **~ résiduaires décantées** *f pl* DÉCHETS, POLL EAU settled wastewater; **~ résiduaires de distillerie** *f pl* DÉCHETS, POLL EAU distillery wastewater; **~ résiduaires industrielles** *f pl* DÉCHETS, POLL EAU industrial wastewater; **~ résiduaires métallifères** *f pl* DÉCHETS, POLL EAU metalliferous wastewater; **~ souterraines** *f pl* DIST EAU groundwater, subsurface water, subterranean water, underground water; **~ usées** *f pl* POLL EAU sewage, wastewater; **~ usées brutes** *f pl* POLL EAU raw sewage; **~ usées décantées** *f pl* POLL EAU settled water; **~ usées épurées** *f pl* POLL EAU purified wastewater; **~ usées fortement colorées** *f pl* POLL EAU heavily-colored wastewater (*AmE*), heavily-coloured wastewater (*BrE*); **~ usées fortement polluées** *f pl* POLL EAU highly-polluted wastewater; **~ usées fortement teintées** *f pl* POLL EAU heavily-dyed wastewater; **~ usées ménagères** *f pl* POLL EAU domestic wastewater, household wastewater; **~ usées mercurielles** *f pl* POLL EAU mercurial wastewater; **~ usées métallifères** *f pl* DÉCHETS, POLL EAU metalliferous wastewater; **~ usées municipales** *f pl* POLL EAU municipal wastewater; **~ usées municipales fortement contaminées** *f pl* POLL EAU highly-contaminated municipal wastewater; **~ usées nauséabondes** *f pl* DIST EAU, POLL ATM maolodorous wastewater; **~ usées non-traitées** *f pl* POLL EAU untreated wastewater; **~ usées d'origine fécale** *f pl* POLL EAU faecal wastewater (*BrE*), fecal wastewater (*AmE*); **~ usées peu épurées** *f pl* POLL EAU poorly-treated wastewater; **~ usées régénérées** *f pl* POLL

EAU reclaimed wastewater, recovered wastewater; ~ **usées résultant de l'extraction des cendres** *f pl* POLL EAU wastewater in ash removal; ~ **usées sanitaires** *f pl* POLL EAU sanitary wastewater; ~ **usées troubles** *f pl* POLL EAU turbid wastewater; ~ **usées turbides** *f pl* POLL EAU turbid wastewater; ~ **usées d'usine d'incinération** *f pl* POLL EAU wastewater from waste incineration plant; ~ **vadoses** *f pl* DIST EAU vadose water

éboulement *m* SOLS landslip, landslide

éboulis *m* DIST EAU rubble

ébulliométrie *f* ÉCHANT boiling analysis

ébullition *f* ÉCHANT boiling

écaillage *m* POLL EAU *peinture* flaking

écailleux *adj* POLL ATM flaky, scale-like

écart: ~ **de mesure** *m* ÉCHANT measuring-range deviation

échange *m* GÉNÉRAL exchange; ~ **anionique** SOLS anion exchange; ~ **atmosphérique** POLL ATM air exchange; ~ **cationique** DIST EAU, POLL EAU, SOLS cation exchange; ~ **ionique** DIST EAU, ÉCHANT, POLL EAU ion exchange (*IE*); ~ **d'ions** DIST EAU, ÉCHANT, POLL EAU ion exchange (*IE*); ~ **de sol** SOLS soil exchange; ~ **thermique** DÉCHETS, ÉN RENOUV, POLL ATM heat exchange

échangeable *adj* POLL ATM, POLL EAU exchangeable

échanger *vt* POLL ATM, POLL EAU exchange

échangeur *m* POLL ATM, POLL EAU exchanger; ~ **d'air de désorption** POLL ATM desorption air heat exchanger; ~ **cationique** SOLS cation exchanger; ~ **de chaleur** DÉCHETS, ÉN RENOUV, POLL ATM heat exchanger; ~ **de chaleur à ailettes** POLL ATM flanged heat exchanger; ~ **de chaleur compact** POLL ATM compact heat exchanger; ~ **de chaleur fumée-fumée** DÉCHETS flue-gas to flue-gas heat exchanger; ~ **de chaleur pour gaz de combustion** POLL ATM flue-gas heat exchanger; ~ **de chaleur lamellaire** POLL ATM lamellar heat exchanger; ~ **de chaleur par convection** POLL ATM convection heat exchanger; ~ **de chaleur en phase gazeuse** POLL ATM gas heat exchanger; ~ **de chaleur à plaques spirales** POLL ATM spiral-plate heat exchanger; ~ **de chaleur à récupération** POLL ATM recuperative heat exchanger; ~ **de chaleur régénérée** POLL ATM regenerative heat exchanger; ~ **de chaleur à tubes doubles** POLL ATM double-pipe heat exchanger; ~ **de chaleur à tubes jumelés** POLL ATM double-pipe heat exchanger; ~ **de chaleur tubulaire** POLL ATM shell-and-tube heat exchanger; ~ **d'ions** POLL EAU ion exchanger; ~ **d'ions à lit fixe** POLL EAU fixed-bed ion exchanger; ~ **d'ions à lits mélangés** POLL EAU mixed-bed ion exchanger

échantillon *m* ÉCHANT sample; ~ **aléatoire** ÉCHANT random sample; ~ **d'analyse** ÉCHANT analytical sample; ~ **cible** ÉCHANT target sample; ~ **composite** ÉCHANT composite sample; ~ **de débit** ÉCHANT, POLL ATM flow sample; ~ **dilué** ÉCHANT diluted sample, dilution sample; ~ **d'eaux usées** ÉCHANT, POLL EAU wastewater sample; ~ **d'essai** ÉCHANT test probe; ~ **évalué** ÉCHANT assessed sample; ~ **faiblement concentré** ÉCHANT weak sample; ~ **faiblement dosé** ÉCHANT weak sample; ~ **filtré** ÉCHANT filtered sample; ~ **flottant** ÉCHANT floating sample; ~ **fortement concentré** ÉCHANT strong sample; ~ **fortement dosé** ÉCHANT strong sample; ~ **isolé** ÉCHANT individual sample; ~ **de laboratoire** ÉCHANT laboratory sample; ~ **lambda** POLL ATM lambda probe; ~ **localisé** ÉCHANT spot sample; ~ **mesuré** ÉCHANT measuring probe; ~ **moyen** ÉCHANT average sample; ~ **non filtré** ÉCHANT unfiltered sample; ~ **non remanié** ÉCHANT, SOLS undisturbed sample; ~ **non remué** ÉCHANT, SOLS undisturbed sample; ~ **perturbé** ÉCHANT disturbed sample; ~ **pilote** ÉCHANT pilot sample; ~ **ponctuel** ÉCHANT snap sample; ~ **porté à ébullition** ÉCHANT boiled sample; ~ **prélevé** ÉCHANT assay; ~ **préliminaire** ÉCHANT preliminary sample; ~ **pris au hasard** ÉCHANT *industrie minière* grab sample; ~ **de recherche** ÉCHANT exploratory sample; ~ **de référence** ÉCHANT background sample; ~ **sédimentaire** ÉCHANT sediment sample

échantillonnage *m* ÉCHANT sampling; ~ **anisocinétique** ÉCHANT anisokinetic sampling; ~ **automatique** ÉCHANT automatic sampling; ~ **en continu** ÉCHANT continuous sampling; ~ **correspondant** ÉCHANT corresponding sampling; ~ **en coupe** ÉCHANT cross-section sampling; ~ **isocinétique** ÉCHANT isokinetic sampling; ~ **laminaire** ÉCHANT layer sampling; ~ **proportionnel** ÉCHANT proportional sampling; ~ **en temps réel** ÉCHANT real-time sampling; ~ **à volume constant** ÉCHANT, POLL ATM constant volume sampling (*CVS*)

échantillonner *vt* ÉCHANT sample

échantillonneur *m* ÉCHANT sampler; ~ **d'absorption par rayonnement bêta** ÉCHANT beta-ray absorption sampler; ~ **à chocs** ÉCHANT impactor; ~ **à chocs en cascade** ÉCHANT cascade impactor; ~ **à courant direct** ÉCHANT direct air sampler; ~ **de dépôt de sol** ÉCHANT bottom deposit sampler, soil deposit sampler; ~ **d'eau** ÉCHANT water sampler; ~ **à grand volume** ÉCHANT high-volume sampler; ~ **laser** ÉCHANT laser sampler; ~ **mobile** ÉCHANT mobile sampler; ~ **à prélèvement proportionnel au débit** ÉCHANT sampler for flow-proportional samples; ~ **de purge et de piégeage** ÉCHANT purge-and-trap sampler; ~ **de sable** ÉCHANT, POLL EAU sand trap; ~ **sectoriel** ÉCHANT sector sampler; ~ **stationnaire** ÉCHANT stationary sampler; ~ **de tête** ÉCHANT *chromatographie* headspace sampler

échappement *m* DIST EAU leakage; ~ **de gaz** ÉCHANT gas release

échapper[1] *vi* ÉCHANT escape

échapper[2]: **s'~** *v réfl* ÉCHANT escape

échelle[1]: ~ **d'analyse** *f* ÉCHANT analytical scale; ~ **de Bacharach** *f* ÉCHANT *mesure d'opacité* Bacharach scale; ~ **de dosage** *f* POLL ATM dosing scale; ~ **d'étiage** *f* DIST EAU water gage (*AmE*), water gauge (*BrE*); ~ **industrielle** *f* ÉN RENOUV utility scale; ~ **de longueurs d'onde** *f* ÉCHANT wavelength scale; ~ **pilote** *f* GESTION pilot plant scale; ~ **de pression de vapeur** *f* ÉCHANT vapor pressure scale (*AmE*), vapour pressure scale (*BrE*); ~ **pulvimétrique** *f* ÉCHANT dust scale; ~ **de Ringelmann** *f* ÉCHANT, POLL ATM Ringelmann scale; ~ **très sensible** *f* ÉCHANT highly-sensitive scale

échelle[2]: **à grande ~** *loc* ÉN RENOUV, POLL EAU large-scale; **à l'~ industrielle** *loc* ÉN RENOUV, POLL ATM on a commercial scale, on an industrial scale; **à ~ moyenne** *loc* ÉN RENOUV medium-scale; **à ~ échelle** *loc* ÉN RENOUV small-scale

échelon: ~ **administratif inférieur** *m* POL ENV lower-level administration

échelonnement: ~ **des doses** *m* POLL ATM temporal dose distribution

échouage *m* POLL MAR grounding

échouement *m* POLL MAR grounding

échouer *vi* POLL MAR to run aground

éclaircie *f* ÉN RENOUV cutting, thinning

éclairement *m* POLL MAR lightening; ~ **planétaire** POLL ATM global irradiance

éclairer *vt* POLL MAR lighten

écluse *f* DIST EAU lock, sluice; ~ **rotative** DÉCHETS *transport de matériaux pulvérulents* rotary valve; ~ **à roue cellulaire** POLL ATM *réfractaires* cellular wheel sluice

éco- *préf* GÉNÉRAL eco-

éco-audit *m* DIST EAU, GESTION, POL ENV eco-audit

éco-bilan *m* DIST EAU, GESTION, POL ENV eco-balance; **~-pauvre** GESTION, POL ENV lean eco-balancing

éco-climat *m* GESTION, POL ENV eco-climate

éco-détaillant *m* GESTION eco-retailer

éco-étiquetage *m* DÉCHETS, GESTION, POL ENV eco-labeling (*AmE*), eco-labelling (*BrE*)

éco-étiquette *f* DÉCHETS, GESTION, POL ENV eco-label

écologie *f* GÉNÉRAL ecology; ~ **animale** GESTION, POL ENV animal ecology; ~ **des cultures agricoles** GESTION, POL ENV crop ecology; ~ **de l'individu** GESTION, POL ENV autoecology; ~ **marine** GESTION, POL ENV, POLL MAR marine ecology; ~ **végétale** GESTION, POL ENV plant ecology

écologique *adj* GÉNÉRAL ecological, ecologically-relevant, ecology-oriented, environmentally-oriented, *défense de l'environnement* environmentally-conservationist

écologiquement: ~ **durable** *adj* GESTION, POL ENV environmentally-sustainable

écologisme *m* GESTION, POL ENV environmentalism

écologiste *m* GESTION, POL ENV ecologist, environmental activist, environmentalist

économie: ~ **de l'environnement** *f* POL ENV environmental economics; ~ **d'espace** *f* DÉCHETS economy of space; ~ **de marché** *f* GESTION, POL ENV market economy; ~ **de recyclage** *f* DIST EAU, ÉN RENOUV, POL ENV recycling economy; ~ **de recyclage viable** *f* ÉN RENOUV, POL ENV sustainable recycling economy; ~ **viable** *f* POL ENV sustainable economy

économies: ~ **d'énergie** *f pl* DÉCHETS, ÉN RENOUV, GESTION energy conservation, energy saving; ~ **de ressources** *f pl* DIST EAU resource conservation

économique *adj* GESTION, POL ENV cost-effective, economic

économisant: ~ **l'énergie** *adj* DÉCHETS, ÉN RENOUV, GESTION energy-saving

économiser *vt* GÉNÉRAL save, conserve; ~ **l'énergie** DÉCHETS, ÉN RENOUV, GESTION conserve energy, save energy

économiste: ~ **spécialisé dans les questions d'environnement** *m* POL ENV environmental economist

écope *f* DIST EAU, POLL MAR scoop

éco-régulation *f* GESTION eco-controlling

écosphère *f* POLL ATM, POLL MAR, SOLS ecosphere

écosystème *m* GÉNÉRAL ecosystem; ~ **anthropique** POLL ATM, SOLS anthropogenic ecosystem; ~ **dépendant** POLL ATM, POLL MAR, SOLS dependent ecosystem; ~ **marin** POLL MAR marine ecosystem, open-ocean ecosystem; ~ **mondial** POLL ATM, POLL MAR, SOLS global ecosystem; ~ **naturel** POLL ATM, POLL MAR, SOLS natural ecosystem; ~ **planétaire** POLL ATM, POLL MAR, SOLS global ecosystem; ~ **tributaire** POLL ATM, POLL MAR, SOLS dependent ecosystem; ~ **urbain** POLL ATM, SOLS urban ecosystem

écotaxe *f* POL ENV eco-tax

écotoxicité *f* POLL EAU ecotoxicity

écotoxicologie *f* POL ENV, POLL EAU, SÉCURITÉ ecotoxicology

écoulement *m* DIST EAU flow, outflow, discharge, leakage, runoff, *en régime transitoire* transient flow, GÉNÉRAL drainage, POLL ATM flow, POLL EAU discharge, outflow; ~ **annuel** DIST EAU annual runoff; ~ **de boues résiduaires** POLL EAU wastewater sludge delivery; ~ **d'eaux résiduaires** POLL EAU wastewater stream; ~ **d'eaux usées** POLL EAU wastewater flow, wastewater stream; ~ **en film ruisselant** POLL EAU trickling-film flow; ~ **de filtrats** POLL EAU filtrate flow; ~ **gravitaire** DIST EAU gravity flow; ~ **de matières épaisses** POLL EAU thick matter flow; ~ **moléculaire** DIST EAU molecular flow; ~ **non perturbé** POLL ATM streamline flow; ~ **parallèle** POLL ATM co-current flow; ~ **principal** POLL ATM main flow; ~ **radial** POLL ATM radial flow; ~ **en régime laminaire** POLL ATM laminar flow; ~ **sous pression** POLL EAU pressure drainage; ~ **supercortical** POLL ATM stemflow; ~ **sur la couche limite** POLL ATM boundary layer flow; ~ **sur la couche limite laminaire** POLL ATM laminar boundary layer flow; ~ **sur la surface de séparation** POLL ATM boundary layer flow; ~ **tourbillonnaire** POLL ATM vortex flow; ~ **turbulent** DIST EAU, POLL ATM eddy flow, turbulent flow; ~ **variable d'effluents gazeux dans un courant gazeux** POLL ATM variable gas waste-gas flow

écouler: **s'**~ *v réfl* DIST EAU run off, run out, *rivière* flow, POLL EAU *cours d'eau* flow

écouteurs *m pl* POLL SON headphones

écran *m* POLL SON screen; ~ **acoustique** POLL SON acoustic baffle, acoustic barrier, acoustic screen, acoustic shield, noise screen, noise shield, sound barrier, sound-absorbing barrier; ~ **d'affichage** ÉCHANT display unit; ~ **d'affichage à cristaux liquides** ÉCHANT liquid-crystal display screen; ~ **anti-bruit** POLL SON noise screen, noise shield; ~ **anti-émission** DÉCHETS emissions barrier; ~ **à cristaux liquides à matrice active** ÉCHANT active matrix liquid-crystal display (*AMLCD*); ~ **à cristaux liquides à matrice passive** ÉCHANT passive matrix liquid-crystal display (*PMLCD*); ~ **à cristaux liquides à transistor en couches minces** ÉCHANT thin film transistor liquid-crystal display (*TFT-LCD*); ~ **à émission de champ** ÉCHANT field emission display (*FED*); ~ **énergétique** DIST EAU, POLL EAU *nucléaire* energy barrier; ~ **externe** POLL SON outdoor shielding; ~ **hydrostatique** DIST EAU, POLL EAU hydrostatic barrier; ~ **imperméable** SOLS impermeable barrier; ~ **luminescent** ÉCHANT electroluminescent display (*EL*); ~ **à micropointes** ÉCHANT field emission display (*FED*); ~ **d'ozone** POLL ATM ozone shield; ~ **photoélectrique** DÉCHETS photoelectric light barrier; ~ **à plasma** ÉCHANT plasma display panel (*PDP*); ~ **plat** ÉCHANT flat panel display; ~ **de protection** POLL ATM, SÉCURITÉ protective shield; ~ **de protection contre les rayonnements** SÉCURITÉ radiation shield; ~ **renforcé** DIST EAU high barrier; ~ **souterrain** DÉCHETS, SOLS subterraneous curtain, subterraneous curtain plate; ~ **de vapeur d'eau** POLL ATM water vapor barrier (*AmE*), water vapour barrier (*BrE*)

écrasement *m* DÉCHETS, DIST EAU, SOLS crushing

écraser *vt* DÉCHETS, DIST EAU, SOLS crush

écrémage *m* DIST EAU, POLL EAU, POLL MAR skimming

écrémer *vt* DIST EAU, POLL EAU, POLL MAR skim

écrémeur *m* DIST EAU, POLL EAU scum skimmer, skimmer, POLL MAR skimmer; ~ **autopropulsé** POLL MAR self-propelled skimmer; ~ **à corde flottante** POLL MAR rope skimmer

écrémeuse *f* POLL MAR licker; ~ **à aspiration directe** POLL MAR *marée noire* direct suction skimmer; ~ **à disque** POLL MAR disc skimmer (*BrE*), disk skimmer (*AmE*)

écumage *m* DIST EAU, POLL EAU, POLL MAR skimming

écume *f* POLL EAU, POLL MAR foam, froth, scum

écumer *vt* DIST EAU, POLL EAU, POLL MAR skim

écumeuse *f* DIST EAU, POLL EAU, POLL MAR scum skimmer, skimmer

écumoire *f* DIST EAU, POLL EAU, POLL MAR skimmer, scum skimmer

effectuer *vt* GESTION, POLL ATM carry out

effet *m* GÉNÉRAL effect; ~ **aigu** DIST EAU, POLL ATM acute effect; ~ **de biofiltre** DÉCHETS, DIST EAU biofilter effect; ~ **de canalisation** POLL ATM chaneling (*AmE*), channelling (*BrE*); ~ **chronique** SÉCURITÉ chronic effect; ~ **couronne** POLL ATM corona effect; ~ **délétère** POLL ATM, POLL EAU, SOLS deleterious effect; ~ **désinfectant** POLL EAU disinfection effect; ~ **desséchant** ÉCHANT dehumidification effect; ~ **externe** POL ENV external effect, externality; ~ **filtrant** POLL ATM filtration effect; ~ **d'injection** POLL ATM injector action; ~ **Joule** SOLS Joule heating effect; ~ **létal** SÉCURITÉ lethal effect; ~ **à long terme** POL ENV, POLL ATM, POLL MAR, POLL SON long-term effect; ~ **nocif** DIST EAU, POLL ATM, SÉCURITÉ harmful effect; ~ **de nuisance sonore** POLL SON noise nuisance effect; ~ **nuisible** DIST EAU, POLL ATM, SÉCURITÉ harmful effect; ~ **photovoltaïque** ÉN RENOUV photovoltaic effect; ~ **physiologique** SÉCURITÉ physiological effect; ~ **préjudiciable** DIST EAU, POLL ATM, SÉCURITÉ harmful effect; ~ **réciproque** ÉCHANT reciprocal effect; ~ **repousseur** POLL MAR *pollution marine* herder effect; ~ **répulsif** POLL ATM repellency effect; ~ **résiduel de sols contaminés** SOLS residual impact of soils with contaminants; ~ **de séparation** POLL ATM separation effect; ~ **de serre** POLL ATM, POLL EAU, SOLS greenhouse effect, global warming potential; ~ **de serre anthropique** POLL ATM anthropogenic greenhouse effect; ~ **de serre induit par le dioxyde de carbone** POLL ATM carbon dioxide greenhouse effect; ~ **de serre naturel** POLL ATM natural greenhouse effect; ~ **sublétal** SÉCURITÉ sub-lethal effect; ~ **sur l'environnement** POLL ATM, POLL EAU, POLL MAR, POLL SON, SOLS environmental effect; ~ **synergique** DIST EAU, POLL ATM synergetic effect; ~ **tampon** ÉCHANT, POLL ATM buffering action, buffer effect, buffering effect; ~ **toxique** DIST EAU, ÉCHANT, POLL ATM, SÉCURITÉ toxic effect; ~ **toxique cumulatif** DIST EAU, POLL ATM, SÉCURITÉ cumulative toxic effect; ~ **Tyndall** ÉCHANT *optique* Tyndall effect

efficace: ~ **par rapport au coût** *adj* GESTION, POL ENV cost-effective

efficacité *f* GÉNÉRAL efficiency, effectiveness; ~ **de destruction et d'élimination** SOLS destruction and removal efficiency (*DRE*); ~ **de filtration** POLL ATM filter efficacy; ~ **par rapport au coût** GESTION cost effectiveness; ~ **d'un procédé** ÉCHANT, POLL ATM process effectiveness; ~ **de traitement** DIST EAU, SOLS process effectiveness

effilochage: ~ **du panache** *m* POLL ATM plume shred

effluent *m* DÉCHETS, POLL EAU effluent, waste product; ~ **d'aéroréfrigérant** POLL ATM cooling tower exhaust air; ~ **atmosphérique** POLL ATM atmospheric fumes; ~ **de carbonisation à basse température** POLL EAU low-temperature carbonization wastewater; ~ **des centrales nucléaires** POLL ATM effluent from nuclear power stations; ~ **gazeux** DÉCHETS, DIST EAU exhaust gas, gas effluent, POLL ATM exhaust gas, gas effluent, waste gas; ~ **gazeux acide** DÉCHETS, POLL ATM acid waste gas; ~ **gazeux chargé de particules solides** POLL ATM solid-containing waste air; ~ **gazeux chargé de solvants** POLL ATM solvent-containing waste air, solvent-laden waste air; ~ **gazeux chaud** POLL ATM hot waste gas; ~ **gazeux de combustion** POLL ATM combustion waste gas; ~ **gazeux dépoussiéré** POLL ATM dust-free waste air; ~ **gazeux exempt de solvants** POLL ATM solvent-free waste air; ~ **gazeux non traité** POLL ATM raw waste gas; ~ **gazeux odorant** POLL ATM odor-laden waste air (*AmE*), odour-laden waste air (*BrE*); ~ **gazeux poussiéreux** POLL ATM dust-laden flue gas; ~ **gazeux toxique** POLL ATM toxic waste gas; ~ **industriel** DÉCHETS, DIST EAU, POLL EAU industrial waste, industrial effluent; ~ **liquide** DÉCHETS, DIST EAU, POLL EAU liquid waste; ~ **liquide dangereux** DÉCHETS, POLL EAU hazardous liquid waste; ~ **liquide du lavage des gaz** POLL EAU wastewater from flue-gas scrubbing; ~ **non épuré** POLL ATM uncleaned waste air; ~ **radioactif** POLL EAU radioactive wastewater; ~ **thermique** POLL ATM thermal effluent; ~ **traité** DIST EAU, POLL EAU treated effluent; ~ **urbain** DIST EAU, POLL EAU municipal effluent; ~ **urbain peu contaminé** DIST EAU, POLL EAU low-contaminated municipal wastewater

effriter: s'~ *v réfl* SOLS crumble, disaggregate

égalisation *f* POLL EAU equalization

égout *m* DÉCHETS, DIST EAU, POLL EAU sewer; ~ **captant** DÉCHETS, DIST EAU, POLL EAU intercepting sewer; ~ **à ciel ouvert** DÉCHETS, DIST EAU, POLL EAU open sewer; ~ **collecteur** DÉCHETS, DIST EAU, POLL EAU main sewer; ~ **gravitaire** DÉCHETS, DIST EAU, POLL EAU gravity sewer; ~ **latéral** DÉCHETS, DIST EAU, POLL EAU lateral sewer; ~ **pluvial** DÉCHETS, DIST EAU, POLL EAU storm-water main, storm-water sewer; ~ **principal** DÉCHETS, DIST EAU, POLL EAU main sewer

égrenage *m* DÉCHETS fraction separator, grinding, ÉCHANT, POLL ATM fraction separator

égreneur *m* DÉCHETS fraction separator

éjecteur *m* POLL EAU *chlorométrie* injector

élastique *adj* POLL SON, SÉCURITÉ resilient

élastomère *m* DÉCHETS, POLL SON elastomer

électricité *f* DIST EAU, ÉN RENOUV, POLL ATM electricity, electrical power, power; ~ **produite** ÉN RENOUV generated electricity, generated power; ~ **produite à partir des courants marins** ÉN RENOUV ocean current power

électrique *adj* GÉNÉRAL, electric, electrical

électro[1]: ~**-osmotique** *adj* SOLS electro-osmotic

électro[2]: ~**-osmose** *f* SOLS electro-osmosis

électroacoustique *adj* POLL SON electroacoustic

électroanalytique *adj* ÉCHANT electroanalytic

électrochimique *adj* ÉCHANT electrochemical

électrocinétique *adj* ÉCHANT, SOLS electrokinetic

électrode *f* DIST EAU, ÉCHANT electrode; ~ **collectrice** DÉCHETS, ÉCHANT collecting electrode; ~ **combinée de mesure du pH** ÉCHANT combination pH-electrode; ~ **de décharge haute tension** DIST EAU, POLL ATM high-voltage discharge electrode; ~ **de dépôt** DÉCHETS, ÉCHANT collecting electrode; ~ **d'émission** POLL ATM charging-electrode wire; ~ **émissive** POLL ATM charging-electrode wire; ~ **extérieure** ÉCHANT outer electrode; ~ **indicatrice** ÉCHANT *mesure du pH* indicating electrode; ~ **interne** ÉCHANT interior electrode; ~ **de mesure** ÉCHANT measuring electrode; ~ **de mesure du pH** ÉCHANT pH-electrode; ~ **négative** POLL ATM negative electrode; ~ **positive** POLL ATM positive electrode; ~ **de référence** ÉCHANT reference electrode; ~ **sélective d'ions** ÉCHANT ion-selective electrode (*ISE*); ~ **sélective d'ions avec membrane en verre** ÉCHANT ion-selective electrode with glass membrane; ~ **sensible aux gaz** ÉCHANT gas-sensitive electrode; ~ **sensible aux ions** ÉCHANT ion-sensitive electrode

électrodialyse *f* DIST EAU, ÉCHANT, POLL EAU electrodialysis

électrofiltre *m* DÉCHETS, DIST EAU, POLL ATM *dépoussiérage* electrofilter, electrostatic precipitator (*ESP*); ~ **basse tension** POLL ATM low-voltage electrostatic precipitator (*low-voltage ESP*); ~ **à contre-courant** POLL ATM reverse-flow precipitator; ~ **haute tension** POLL ATM *dépoussiérage* high-voltage electrostatic precipitator; ~ **humide** DÉCHETS wet electrofilter; ~ **par voie humide** POLL ATM wet electrostatic filter; ~ **à plaques** POLL ATM plate-wire electrostatic precipitator (*plate-wire ESP*); ~ **tubulaire** POLL ATM *dépoussiérage* tubular electrostatic precipitator (*tubular ESP*)

électroflottation *f* POLL EAU electroflotation

électrolyse *f* DIST EAU, POLL EAU, SOLS electrolysis; ~ **sur lit fixe** POLL EAU fixed-bed electrolysis

électrolyte *m* DIST EAU, ÉCHANT electrolyte

électromagnétique *adj* DÉCHETS, DIST EAU, POLL ATM, POLL EAU electromagnetic

électrométrique *adj* ÉCHANT electrometric

électron *m* ÉCHANT, POLL ATM electron; ~ **accéléré** POLL ATM accelerated electron; ~ **libre** POLL ATM free electron

électronique *adj* DÉCHETS, DIST EAU, ÉCHANT electronic

électrophorèse *f* ÉCHANT, SOLS electrophoresis; ~ **capillaire** ÉCHANT capillary electrophoresis (*CE*); ~ **capillaire couplée à la spectrométrie de masse** ÉCHANT capillary electrophoresis-mass spectrometry (*CE-MS*); ~ **capillaire sur gel** ÉCHANT capillary gel electrophoresis (*CGE*); ~ **sur couches minces** ÉCHANT thin-layer electrophoresis; ~ **de zone capillaire** ÉCHANT capillary zone electrophoresis (*CZE*)

électrophorétique *adj* ÉCHANT, SOLS electrophoretic

électrostatique[1] *adj* DÉCHETS, DIST EAU, POLL ATM, POLL EAU electrostatic

électrostatique[2] *f* DÉCHETS, DIST EAU, POLL ATM, POLL EAU electrostatics

électrovanne *f* POLL ATM solenoid valve

élément *m* DÉCHETS, ÉCHANT, POLL ATM, SOLS component, element; ~ **adsorbé** POLL ATM adsorbed component; ~ **alvéolaire** POLL ATM honeycomb element; ~ **du catalyseur** POLL ATM catalyst element; ~ **à col sonique** POLL ATM ring jet element; ~ **constitutif** POLL ATM constituent; ~ **constitutif du milieu** ÉCHANT medium-specific component; ~ **constitutif de la poussière** POLL ATM dust component; ~ **entrant dans la détermination** ÉCHANT determination component; ~ **à l'état de trace** ÉCHANT, POLL ATM trace element; ~ **extrait** POLL EAU extracted component; ~ **filtrant** DIST EAU filtering unit, POLL ATM, POLL EAU, SÉCURITÉ filter element, filtering unit; ~ **de gaz pollués** POLL ATM waste gas component; ~ **gazeux agressif** POLL ATM aggressive gas component; ~ **limite** POLL ATM *granulométrie* near-size particle; ~ **de mesure** ÉCHANT measuring element; ~ **du milieu** ÉCHANT medium-specific component; ~ **mobile** DIST EAU, POLL ATM mobile component; ~ **nutritif** POLL EAU, SOLS nutrient; ~ **nutritif végétal** POLL EAU, SOLS plant nutrient; ~ **oligodynamique** DIST EAU, POLL EAU micronutrient; ~ **Peltier** ÉCHANT Peltier element; ~ **photovoltaïque** ÉN RENOUV photovoltaic element; ~ **semiconducteur** ÉN RENOUV semiconductor component; ~ **de sertissage** DÉCHETS can sealing compound; ~ **structurel insonorisant** POLL SON noise-absorbing structural element; ~ **support** POLL ATM support element; ~ **de tension** POLL MAR tension member; ~**-trace** POLL ATM trace element

éléments: ~ **constitutifs d'échantillon** *m pl* ÉCHANT sample components; ~ **d'échantillonnage** *m pl* ÉCHANT sample components; ~ **particulaires adhésifs** *m pl* POLL ATM tacky particulate components

élevage: ~ **hors sol** *m* ÉN RENOUV feedlot

élévation: ~ **du niveau d'eau** *f* DIST EAU rising of water level; ~ **de température** *f* POLL ATM temperature rise; ~ **de la température terrestre** *f* POLL ATM raising of the earth's temperature

éliminateur: ~ **de gaz carbonique** *m* DIST EAU decarbonator

élimination *f* GÉNÉRAL removal, disposal; ~ **de l'amiante** DÉCHETS asbestos removal; ~ **biologique** SOLS biological elimination; ~ **des cendres** DIST EAU ash removal; ~ **commune** DÉCHETS *déchets compatibles* co-disposal; ~ **des contaminants** DIST EAU separation; ~ **des déchets** DÉCHETS garbage disposal (*AmE*), refuse disposal, waste disposal, DIST EAU residue disposal, GESTION, POL ENV garbage disposal (*AmE*), refuse disposal, waste disposal, POLL EAU residue disposal; ~ **distincte** DIST EAU separate disposal; ~ **hors site** SOLS off-site disposal; ~ **illégale des déchets** DÉCHETS illegal refuse disposal, illegal waste disposal; ~ **inoffensive** DIST EAU, SÉCURITÉ safe disposal; ~ **de matière organique** DIST EAU, POLL ATM removal of organic matter; ~ **en mer** POLL MAR ocean disposal; ~ **de NO_x** (*denox*) DIST EAU, POLL ATM, SOLS NO_x removal (*denox*); ~ **des ordures** (*cf enlèvement des ordures*) DÉCHETS, GESTION, POL ENV garbage disposal (*AmE*), refuse disposal, waste disposal; ~ **par aspiration** POLL EAU removal by suction; ~ **à part** DIST EAU separate disposal; ~ **des plastiques** DÉCHETS, SOLS disposal of plastics; ~ **de polluants** DIST EAU, POLL ATM removal of pollutants; ~ **de la poussière résiduelle** POLL ATM remaining dust removal; ~ **de SO_2/NO_x** (*desonox*) POLL ATM SO_2/NO_x removal (*desonox*); ~ **solaire des déchets chimiques** DIST EAU solar disposal of chemical wastes; ~ **des solvants résiduels** SOLS residual solvent removal; ~ **en toute sécurité** DIST EAU, SÉCURITÉ safe disposal

éliminer *vt* GÉNÉRAL remove, dispose of; ~ **le chlore** SOLS dechlorinate; ~ **par criblage** DIST EAU sieve out; ~ **par lavage** DIST EAU wash out, GÉNÉRAL wash away; ~ **par tamisage** DIST EAU sieve out

élingue *f* DIST EAU, POLL MAR sling

éluant *m* ÉCHANT, POLL EAU *chromatographie* eluant

éluat *m* ÉCHANT, POLL EAU *chromatographie* eluate

éluer *vt* ÉCHANT, POLL EAU elute

élus: ~ **locaux** *m pl* POL ENV local politicians

élution *f* ÉCHANT, POLL EAU elution

élutriation *f* DÉCHETS, ÉCHANT, POLL ATM, POLL EAU elutriation

émaillage *m* DÉCHETS, DIST EAU enameling (*AmE*), enamelling (*BrE*)

emballage *m* DÉCHETS *matériel* package, packaging, packing, *action* packaging, packing; ~ **alvéolé** DÉCHETS bubble pack; ~ **en carton** DÉCHETS cardboard packaging; ~ **composé** DÉCHETS combined packaging; ~ **composite** DÉCHETS composite packaging; ~ **consigné** DÉCHETS returnable pack, returnable packaging; ~ **contractuel** DÉCHETS contract packaging; ~ **coque** DÉCHETS bubble pack; ~ **économique** DÉCHETS economy-size pack; ~ **efficace** DÉCHETS efficient packaging; ~ **extensible** DÉCHETS stretch wrapping; ~ **à fenêtre** DÉCHETS window packaging; ~ **en fin de chaîne de production** DÉCHETS end-of-line packaging; ~ **formé sous vide** DÉCHETS vacuum-formed package; ~ **fraîcheur** DÉCHETS freshness seal; ~ **individuel** DÉCHETS unit pack; ~ **industriel** DÉCHETS industrial packing, commercial packaging; ~ **industriel usagé** DÉCHETS used commercial packaging; ~ **intérieur** DÉCHETS interior wrapping; ~ **interne** DÉCHETS interior wrapping; ~ **jetable** DÉCHETS throwaway pack; ~ **lithographié** DÉCHETS lithographed package; ~ **en mousse de plastique** DÉCHETS plasti-foam packaging; ~ **non biodégradable** DÉCHETS nonbiodegradable packaging; ~ **en papier** DÉCHETS paper wrapping; ~ **pelable** DÉCHETS peel-off wrapping; ~ **perdu** DÉCHETS nonreturnable packaging, one-way pack, throwaway pack; ~ **en plastique** DÉCHETS plastics packaging; ~ **pliable et réutilisable** DÉCHETS collapsible and reusable packaging system; ~ **en polyéthylène expansé** DÉCHETS expanded polythene packaging; ~ **de protection** DÉCHETS barrier packaging; ~ **de récupération** DÉCHETS secondary packaging; ~ **repris** DÉCHETS returnable packaging; ~ **rétractable pour palette** DÉCHETS pallet shrink wrapping; ~ **réutilisable** DÉCHETS reusable packaging; ~ **secondaire** DÉCHETS secondary packaging; ~ **de sécurité** SÉCURITÉ child-resistant packaging; ~ **souple** DÉCHETS flexible package; ~ **sous film** DÉCHETS blister pack; ~ **sous film thermorétractable sans support** DÉCHETS unsupported shrink wrapping; ~ **sous vide** DÉCHETS vacuum pack; ~ **stratifié** DÉCHETS laminated pack; ~ **sur plateau** DÉCHETS tray packing; ~ **thermoformé** DÉCHETS blister pack; ~ **transparent** DÉCHETS visual pack; ~ **de transport** DÉCHETS transport packaging; ~ **à usage alimentaire** DÉCHETS food packaging; ~ **à usages multiples** DÉCHETS multipack; ~ **vide** DÉCHETS dummy

emballages: ~ **vides** *m pl* DÉCHETS empties

emballé *adj* DÉCHETS packed

emballer *vt* DÉCHETS pack, package

embase *f* DÉCHETS base cup

embouteilleuse *f* DÉCHETS bottling machine

emboutissage *m* DÉCHETS forming, DIST EAU pressing

embranchement *m* DIST EAU branching; ~ **de canal** DIST EAU canal junction

embrochage *m* DIST EAU plugging

embrun *m* POLL ATM spray

émetteur *m* DIST EAU, POLL ATM, POLL SON emitter; ~ **de bruit** POLL SON noise emitter; ~ **de lumière** ÉCHANT light sender; ~ **de polluants** POLL ATM, SOLS pollution emitter; ~ **de pollution** POLL ATM pollution emitter; ~ **de rayonnement** DIST EAU radiation emitter; ~ **de rayons ultraviolets** POLL EAU ultraviolet radiator; ~ **sonore** POLL SON sound emitter

émettre *vt* DIST EAU, POLL ATM emit, POLL EAU release, POLL SON emit, release, SOLS release

émis *adj* DIST EAU, POLL ATM, POLL SON emitted

émissaire *m* DIST EAU draining ditch, POLL EAU outfall, outlet

émission *f* DIST EAU discharge, emission, POLL ATM discharge, emission, release; ~ **de lixiviats sur une pente** DIST EAU slope leachate emission; ~ **de rayons X induite par des protons** ÉCHANT proton-induced x-ray emission (*PIXE*)

émissions *f pl* POLL ATM, POLL EAU, POLL SON emissions; ~ **acoustiques** ÉN RENOUV, POLL ATM, POLL SON acoustical emissions, acoustic emissions (*AE*), noise emissions; ~ **à l'admission** POLL ATM inlet emissions; ~ **atmosphériques** POLL ATM air emissions; ~ **atmosphériques polluantes**

POLL ATM air pollution emissions; ~ **de carburateur** POLL ATM carburetor emissions (*AmE*), carburettor emissions (*BrE*); ~ **de carter** POLL ATM crankcase emissions; ~ **chargées de vapeur** POLL ATM vapor-laden emissions (*AmE*), vapour-laden emissions (*BrE*); ~ **de CO** POLL ATM CO emissions; ~ **collantes** POLL ATM sticky emissions; ~ **de contaminants** POLL ATM contaminant emissions; ~ **de dioxyde de soufre** POLL ATM sulfur dioxide emissions (*AmE*), sulphur dioxide emissions (*BrE*); ~ **à l'échappement** POLL ATM exhaust emissions; ~ **d'évaporation** POLL ATM evaporative emissions; ~ **fugaces** POLL ATM fugitive emissions; ~ **de fumées** POLL ATM smoke emissions; ~ **de fumées sous forme de particules** POLL ATM particulate smoke emissions; ~ **de gaz** ÉN RENOUV, POLL ATM gas emissions; ~ **gazeuses** ÉN RENOUV gas emissions, POLL ATM gas emissions, gaseous emissions, waste gas emissions; ~ **d'hydrocarbures** POLL ATM hydrocarbon emissions; ~ **liées à l'entretien permanent** POLL ATM maintenance emissions; ~ **locales** POLL ATM local emissions; ~ **de métaux lourds** POLL ATM heavy-metal emissions; ~ **nationales** POLL ATM domestic emissions; ~ **nocives** POLL ATM harmful emissions; ~ **d'odeurs** POLL ATM, POLL EAU odor emissions (*AmE*), odour emissions (*BrE*); ~ **olfactives** POLL ATM, POLL EAU odor emissions (*AmE*), odour emissions (*BrE*); ~ **d'origine industrielle** POLL ATM industrial emissions; ~ **d'oxydes d'azote** POLL ATM NO_x emissions; ~ **en phase vapeur** POLL ATM vapor emissions (*AmE*), vapour emissions (*BrE*); ~ **planétaires** POLL ATM global emissions; ~ **polluantes** POLL ATM contaminant emissions, pollutant emissions; ~ **de polluants** POLL ATM contaminant emissions, pollutant emissions; ~ **de poussière** POLL ATM dust emissions; ~ **pulvérulentes** POLL ATM dust emissions; ~ **radioactives** POLL ATM radioactive emissions; ~ **du réservoir de carburant** POLL ATM fuel tank emissions; ~ **résiduaires** DIST EAU, POLL ATM residue emissions; ~ **résiduelles** *f* POLL EAU residual emissions; ~ **de solvants** POLL ATM solvent emissions; ~ **sonores** POLL SON noise emissions; ~ **en sortie de cheminée** POLL ATM chimney emissions; ~ **en sortie de tuyau** POLL ATM tailpipe emissions; ~ **de source étrangère** POLL ATM foreign emissions; ~ **sous forme de particules** POLL ATM particle emissions, particulate emissions; ~ **de substances odoriférantes** POLL ATM odoriferous substance emissions; ~ **thermiques** POLL ATM heat emissions; ~ **de vapeur** POLL ATM vapor emissions (*AmE*), vapour emissions (*BrE*); ~ **volatiles** POLL ATM volatile emissions

émissivité *f* POLL ATM emissivity

émondes: ~ **ligneuses** *f pl* ÉN RENOUV ligneous prunings

empaillassonage *m* DÉCHETS interior wrapping

empêcher *vt* DIST EAU, POLL MAR prevent

empierré *adj* DÉCHETS stone-filled

empiéter *vi* POLL ATM impinge

empilement: ~ **de cellules** *m* POLL ATM cell stack

empiler *vt* POLL ATM stack

empoisonnement *m* DIST EAU, POLL ATM, SÉCURITÉ poisoning; ~ **de l'air** POLL ATM air poisoning; ~ **du catalyseur** POLL ATM catalyst contamination, catalyst poisoning; ~ **chronique** SÉCURITÉ chronic poisoning; ~ **au dioxyde de soufre** POLL ATM sulfur dioxide poisoning (*AmE*), sulphur dioxide poisoning (*BrE*); ~ **de l'eau** POLL EAU water poisoning; ~ **immédiat** SÉCURITÉ direct poisoning; ~ **par le mercure** POLL EAU mercury poisoning; ~ **au plomb** SÉCURITÉ lead poisoning

empoisonner *vt* DIST EAU, POLL ATM, SÉCURITÉ poison

emporter *vt* ÉCHANT wash away, GÉNÉRAL remove, SOLS transport away

empoussiérage *m* POLL ATM dusting

empreinte: ~ **acoustique** *f* POLL SON noise carpet

émulsifiant *m* DÉCHETS, DIST EAU, POLL EAU emulsifier

émulsifié *adj* DÉCHETS, DIST EAU, POLL EAU emulsified

émulsifier *vt* DÉCHETS, DIST EAU, POLL EAU emulsify

émulsion *f* DÉCHETS, DIST EAU, POLL EAU emulsion

émulsionnant *m* POLL EAU emulsifier

émulsionner *vt* POLL EAU emulsify

encaisseuse *f* DÉCHETS boxing machine

encammionner *vt* DÉCHETS load onto lorries (*BrE*), load onto trucks (*AmE*)

encapsulage: ~ **avec des matériaux inertes** *m* SOLS encapsulation using inert materials

encapsuleuse *f* DÉCHETS bottle sealing machine

encart: ~ **écologique** *m* GESTION eco-window

encartonnage: ~ **par pellicules** *m* DÉCHETS film cartoning

encartonneuse *f* DÉCHETS cartoner, cartoning machine; ~ **verticale** DÉCHETS vertical cartoner, vertical cartoning machine

enceinte *f* POLL SON enclosure; ~ **acoustique** POLL SON acoustic enclosure, encapsulation; ~ **intégrale** POLL SON complete enclosure; ~ **de matériel** POLL SON machine enclosure; ~ **partielle** POLL SON partial enclosure; ~ **à repoussièrement contrôlé** POLL ATM controlled dust content chamber; ~ **de site** DIST EAU site closure

encercler *vt* POLL MAR *pollution* corral

enchevêtrement *m* POLL EAU interlocking

encodage *m* DÉCHETS, SOLS encoding; ~ **des valeurs mesurées** ÉCHANT coding of measured value

encolleuse: ~ **sur marge** *f* DÉCHETS margin gluer

encombrant *adj* DÉCHETS bulky

encrassement *m* DÉCHETS, DIST EAU fouling, ÉN RENOUV *de moteur* coking, POLL ATM clogging, POLL EAU fouling

encrassé *adj* POLL ATM clogged

encrasser *vt* DÉCHETS, DIST EAU foul, ÉN RENOUV *moteur* coke, POLL ATM clog, POLL EAU foul

endigage *m* DÉCHETS embankment

endommagé *adj* GÉNÉRAL impaired, damaged, harmed

endommager *vt* GÉNÉRAL harm, damage, impair

endothermique *adj* DIST EAU, POLL ATM endothermic

enduit: ~ **insonorisant** *m* POLL SON acoustic plaster

énergétique *adj* DIST EAU, ÉN RENOUV, POLL ATM, SOLS energetic, energy-related

énergie *f* DIST EAU, ÉN RENOUV, POLL ATM, SOLS energy; ~ **acoustique** POLL SON sound power; ~ **d'activation** POLL ATM activation energy; ~ **cinétique** DIST EAU velocity head, ÉN RENOUV, POLL ATM kinetic energy, POLL EAU velocity head; ~ **confinée** ÉN RENOUV *en milieu clos* contained energy; ~ **contenue** ÉN RENOUV contained energy; ~ **des courants marins** ÉN RENOUV ocean current energy; ~ **électrique** ÉN RENOUV electric energy, electrical energy, electrical power; ~ **éolienne** ÉN RENOUV wind energy, wind power; ~ **finale** ÉN RENOUV final energy; ~ **de fission** ÉN RENOUV fission power; ~ **géothermique** DIST EAU, ÉN RENOUV, POLL ATM geothermal energy; ~ **géothermique stockée** ÉN RENOUV, POLL ATM stored geothermal energy; ~ **hélio-électrique** ÉN RENOUV solar power; ~ **hydro-électrique** ÉN RENOUV water power; ~ **irradiée** ÉN RENOUV irradiated energy; ~ **marémotrice** ÉN RENOUV tidal energy, tidal power; ~ **mécanique** ÉN RENOUV mechanical energy; ~ **océanique** ÉN RENOUV ocean energy; ~ **de pression** POLL ATM pressure energy; ~ **rayonnée** ÉN RENOUV radiation energy; ~ **renouvelable** ÉN RENOUV renewable energy; ~ **solaire** ÉN RENOUV solar energy, solar power; ~ **solaire concentrée** ÉN RENOUV concentrated solar energy; ~ **solaire stockée** ÉN RENOUV stored solar energy; ~ **sonore** POLL SON sound energy; ~ **stockée** ÉN RENOUV stored energy; ~ **thermique** ÉN RENOUV, POLL ATM thermal energy; ~ **thermique des mers** (*ETM*) ÉN RENOUV ocean thermal energy conversion (*OTEC*); ~ **thermique d'origine solaire** ÉN RENOUV solar thermal energy; ~ **thermique résiduelle** ÉN RENOUV residual thermal energy; ~ **des vagues** ÉN RENOUV wave energy

enfichage *m* DIST EAU, POLL EAU *canalisation* plugging
enflammer *vt* DÉCHETS, ÉCHANT, POLL EAU ignite
enfoncement: ~ **d'une pente** *m* DIST EAU slope recess
enfoncer *vt* DIST EAU break open
enfonçer: **s'~** *v réfl* DIST EAU sink
enfoui: ~ **dans le sous-sol** *adj* DÉCHETS, SOLS landfilled
enfouir: ~ **dans le sous-sol** *vt* DÉCHETS, SOLS landfill
enfouissement: ~ **dans le sol** *m* DÉCHETS, SOLS land burial; ~ **de déchets dangereux** *m* DÉCHETS underground landfill for hazardous waste; ~ **de déchets spéciaux** *m* DÉCHETS underground landfill for special waste; ~ **sanitaire** *m* DÉCHETS sanitary landfill
enfumage *f* POLL ATM fumigation
enfumer *vt* POLL ATM fumigate
engagement: ~ **délibéré** *m* POL ENV voluntary commitment
engin: ~ **de compactage** *m* DÉCHETS, SOLS compacting device, compacting machine, compaction equipment; ~ **compacteur** *m* DÉCHETS, SOLS compacting device, compacting machine, compaction equipment; ~ **de fonçage** *m* DIST EAU pile-driver; ~ **de levage** *m* SÉCURITÉ hoist; ~ **de nettoiement de la voie publique** *m* DÉCHETS street cleaner; ~ **de récupération** *m* DIST EAU recovery device, POLL MAR collection device; ~ **de remorquage** *m* POLL MAR towing gear
engins: ~ **de terrassement** *m pl* SOLS earth-moving machinery
engorgement *m* DIST EAU, POLL EAU *encrassement* fouling, clogging, *obstruction* stoppage; ~ **par l'eau** POLL EAU waterlogging
engorger *vt* DIST EAU, POLL EAU *encrasser* foul, clog, *obstruer* stop
engrais *m* POLL EAU, SOLS fertilizer; ~ **azoté** POLL EAU, SOLS nitrogenous fertilizer; ~ **chimique** POLL EAU, SOLS synthetic fertilizer; ~ **minéral azoté** POLL EAU, SOLS anorganic nitrogen fertilizer
engraissement *m* SOLS fertilization
engraisser *vt* SOLS fertilize
engrenage: ~ **à air comprimé** *m* DIST EAU pneumatic driving gear; ~ **épicycloïdal** *m* ÉN RENOUV planetary gear; ~ **planétaire** *m* ÉN RENOUV planetary gear; ~ **pneumatique** *m* DIST EAU pneumatic driving gear; ~ **de pompe** *m* DIST EAU, POLL EAU pump gear, pump-driving gear
enlaidissement: ~ **du milieu rural** *m* POL ENV disfigurement of rural environment
enlèvement *m* DÉCHETS refuse collection, waste collection; ~ **des déchets volumineux** DÉCHETS bulk collection; ~ **des ordures** (*cf élimination des ordures*) DÉCHETS scavenging, garbage collection (*AmE*)
enquête *f* ÉCHANT, GESTION investigation, investigative action, POL ENV survey, SÉCURITÉ investigation, investigative action; ~ **de comportement** POL ENV attitude survey; ~ **à domicile** POL ENV home interview survey; ~ **menée dans le cadre de l'étude d'impact** POL ENV environmental impact assessment investigation (*EIA investigation*), environment impact investigation; ~ **par sondage** POL ENV survey method; ~ **pédologique** SOLS soil survey; ~ **sanitaire** POL ENV health survey
enquêter: ~ **sur** *vt* ÉCHANT, GESTION, SÉCURITÉ investigate
enregistrement *m* DIST EAU, ÉCHANT, POL ENV, POLL ATM registration, recording; ~ **en continu** ÉCHANT continuous recording; ~ **de données** ÉCHANT data logging
enregistrer *vt* DIST EAU, ÉCHANT, POL ENV, POLL ATM register, record
enregistreur: **~-bloqueur** *m* ÉCHANT calibration storer; ~ **en courant continu** *m* ÉCHANT direct voltage recorder; ~ **d'étalonnage** *m* ÉCHANT calibration storer; ~ **de niveau** *m* POLL SON level recorder; ~ **séquentiel de données** *m* ÉCHANT data logger
enrichissement *m* POLL EAU, SOLS enrichment
enrobage *m* DÉCHETS, DIST EAU *déchets nucléaires* encapsulation; ~ **en grains** DÉCHETS grain encapsulation
enrobé *adj* POLL SON encapsulated
enrober *vt* DÉCHETS, DIST EAU *déchets nucleaires*, POLL SON encapsulate
ensachage *m* DÉCHETS bag packaging
ensemble: ~ **des émissions polluantes** *m* ÉCHANT total pollutant emissions; ~ **de mesurage** *m* ÉCHANT metering unit
ensemencement *m* ÉCHANT seeding
ensilage *m* SOLS ensilage, *agriculture* silage
ensiler *vt* DÉCHETS convey into silos
ensoleillement *m* ÉN RENOUV sunlight; ~ **global** POLL ATM global insolation
entartrage *m* DIST EAU furring, scale formation
entartrer *vi* DIST EAU fur, fur up, scale, scale up
entassement *m* DÉCHETS embankment
entérovirus *m* POLL EAU, SÉCURITÉ enterovirus
enthalpie *f* POLL ATM enthalpy; ~ **de combustion** POLL ATM combustion enthalpy; ~ **de fusion** POLL ATM enthalpy of fusion; ~ **de vaporisation** POLL ATM enthalpy of vaporization
entonnoir[1]: ~ **de décantation** *m* ÉCHANT separating funnel; ~ **de séparation** *m* ÉCHANT separating funnel
entonnoir[2]: **en** ~ *loc* POLL EAU funnel-shaped
entraîné *adj* POLL ATM, POLL EAU, SOLS entrained
entraînement[1] *m* DÉCHETS, DIST EAU removal, ÉCHANT washing away, ÉN RENOUV removal, POLL ATM entrainment, POLL EAU, SOLS entrainment, removal; ~ **complémentaire** DÉCHETS additional entrainment; ~ **de gaz** POLL ATM gas entrainment; ~ **des phosphates** POLL EAU phosphate stripping; ~ **de pompe** DIST EAU pump gear; ~ **sous les nuages** POLL ATM below-cloud scavenging
entraînement[2]: **à** ~ **électrique** *loc* ÉN RENOUV electrically-driven; **à** ~ **photovoltaïque** *loc* ÉN RENOUV photovoltaic-driven
entraîner *vt* ÉCHANT wash away, POLL ATM lead, entrain, POLL EAU, SOLS entrain; ~ **au lavage** ÉCHANT wash away
entrée *f* ÉCHANT input; ~ **des gaz** POLL ATM gas inlet
entreprise: ~ **d'élimination des déchets** *f* DÉCHETS, GESTION waste disposal company, waste disposal firm; ~ **de fourniture d'énergie solaire** *f* ÉN RENOUV solar energy enterprise; ~ **de gestion de déchets** *f* DÉCHETS, GESTION waste management company, waste management firm; ~ **viable** *f* GESTION sustainable business
entretien *m* GÉNÉRAL maintenance; ~ **permanent** GÉNÉRAL maintenance
envasement *m* POLL EAU silting up
enveloppe *f* DÉCHETS envelope, POLL ATM, POLL SON shell; ~ **de bulles d'air** DÉCHETS air bubble wrap; ~ **externe** POLL SON outer shell; ~ **de protection** SOLS shroud; ~ **souple** POLL SON flexible shell
environnement *m* GÉNÉRAL environment, natural environment; ~ **biosphérique** POL ENV biosphere environment; ~ **marin** POL ENV marine environment; ~ **physico-chimique** DIST EAU, POLL ATM physio-chemical environment; ~ **peu sûr** SÉCURITÉ unsafe environmental condition
environnemental *adj* GÉNÉRAL environmental
enzymatique *adj* POLL ATM, POLL EAU enzymatic
enzyme *f* POLL ATM, POLL EAU enzyme
éolienne *f* ÉN RENOUV wind turbine; ~ **à axe horizontal** ÉN RENOUV horizontal axis wind turbine (*HAWT*); ~ **à axe vertical** ÉN RENOUV vertical axis wind turbine (*VAWT*); ~ **isolée** ÉN RENOUV free-standing tower; ~ **monopale** ÉN RENOUV single-blade machine; ~ **multiple** ÉN RENOUV multiple-blade machine
EPA *abrév* (*agence de protection de l'environnement*) GESTION, POL ENV EPA (*environmental protection agency*)
épaisseur *f* DÉCHETS thickness; ~ **de couchage** DÉCHETS coating thickness; ~ **de couche** ÉCHANT, POLL SON layer

thickness; ~ **de la couche filtrante** POLL EAU filter bed depth; ~ **de décharge** DÉCHETS landfill thickness; ~ **de l'enduit** DÉCHETS *emballage* coating thickness; ~ **de paroi** DÉCHETS, DIST EAU wall depth; ~ **de particules** POLL ATM particulate thickness

épaissi *adj* DIST EAU, POLL EAU thickened; ~ **mécaniquement** POLL EAU mechanically-thickened

épaissir[1] *vt* DIST EAU, POLL EAU thicken

épaissir[2]: **s'~** *v réfl* DIST EAU, POLL EAU thicken

épaississant *m* DIST EAU, POLL EAU *agent* thickener; ~ **final** POLL EAU final thickener

épaississement *m* DIST EAU, POLL ATM, POLL EAU thickening; ~ **des boues** POLL EAU sludge thickening; ~ **gravitaire en cuve** POLL EAU gravity thickening in containers; ~ **gravitaire en filtre à manche** POLL EAU gravity thickening in filter bags; ~ **par flottation** POLL EAU flotation thickening; ~ **par gravité** POLL EAU gravity thickening; ~ **préalable** POLL EAU prethickening

épaississeur *m* DIST EAU, POLL EAU thickener; ~ **par gravité** POLL EAU gravity thickener; ~ **primaire** POLL EAU primary thickener

épanchements: ~ **d'hydrocarbures** *m pl* POLL EAU oil spillage

épandage *m* SOLS landfilling, manure application, manuring, spreading

épandre *vt* DIST EAU, SOLS, POLL EAU spread

épave *f* DÉCHETS *véhicule*, POLL MAR *bateau* wreck

éperon *m* POLL MAR fender

épilimnion *m* POLL EAU epilimnion

épisode *m* SOLS episode; ~ **de pollution atmosphérique** POLL ATM air-pollution episode

épluchures: ~ **de légumes** *f pl* DÉCHETS vegetable waste

épreuve: **à l'~ du feu** *loc* SÉCURITÉ fire-resistant; **à l'~ de l'humidité** *loc* DÉCHETS, DIST EAU moisture-proof

éprouvé *adj* ÉN RENOUV, POLL ATM, SOLS field-proven; ~ **en pratique** ÉN RENOUV, POLL ATM, SOLS field-proven

éprouvette: ~ **à gaz** *f* ÉCHANT gas-sampling tube; ~ **graduée** *f* ÉCHANT measuring jar; ~ **lambda** *f* ÉCHANT lambda probe

épuisable *adj* ÉN RENOUV exhaustible

épuisement *m* GÉNÉRAL drainage, *à la pompe* pumping, pumping out; ~ **du sol** SOLS soil depletion, soil exhaustion

épuiser *vt* GÉNÉRAL drain, *à la pompe* pump, pump out; ~ **par combustion** POLL ATM burn out

épurateur *m* DIST EAU purifier, SÉCURITÉ cleaner; ~ **catalytique** POLL ATM *gaz* catalytic waste gas cleaner; ~ **à double cycle** POLL ATM double-cycle scrubber; ~ **de fumées** DÉCHETS flue-gas scrubber; ~ **de gaz** POLL ATM gas purifier, waste-gas purifier; ~ **mixte** POLL ATM combination scrubber; ~ **d'oxydes de soufre** POLL ATM sulfur-oxides scrubber (*AmE*), sulphur-oxides scrubber (*BrE*); ~ **par voie sèche** POLL ATM dry scrubber; ~ **semi-sec** POLL ATM semidry scrubber; ~ **thermique** POLL ATM thermal air cleaner

épuration *f* DIST EAU sewage treatment, scrubbing, *des matières huileuses dans des séparateurs* removal of oil by separators, GÉNÉRAL treatment, POLL ATM cleaning, *gaz* scrubbing, POLL EAU purification; ~ **bactériologique** DÉCHETS bacteriological purification; ~ **catalytique de l'air par voie biologique** POLL ATM biologically-catalytic waste air cleaning; ~ **de l'eau** DIST EAU, POLL EAU water purification; ~ **des eaux** DIST EAU, POLL EAU water purification; ~ **des eaux fluviales** DIST EAU, POLL EAU river-water purification; ~ **des eaux résiduaires** DIST EAU, POLL EAU sewage purification; ~ **des eaux résiduaires industrielles** DIST EAU, POLL EAU industrial wastewater purification; ~ **des eaux usées** POLL EAU wastewater control, wastewater purification; ~ **des effluents gazeux sans eaux usées** DÉCHETS, POLL ATM wastewater-free flue-gas cleaning; ~ **des fumées** DÉCHETS, POLL ATM flue-gas washing; ~ **de gaz** POLL ATM gas purification; ~ **des gaz brûlés** DÉCHETS, POLL ATM flue-gas cleaning, waste-gas cleaning; ~ **des gaz d'échappement** POLL ATM exhaust waste-gas cleaning; ~ **des gaz par voie sèche** DÉCHETS, POLL ATM dry gas cleaning; ~ **des gaz perdus** POLL ATM waste-gas cleaning; ~ **intermédiaire** POLL EAU intermediate purification; ~ **des lixiviats de décharge** DÉCHETS landfill leachate purification; ~ **mécanique des eaux usées** POLL EAU mechanical wastewater treatment; ~ **naturelle** SOLS natural purification; ~ **par voie humide en boucle fermée** POLL ATM closed-loop wet scrubbing; ~ **préliminaire** DIST EAU preliminary purification; ~ **primaire** DIST EAU preliminary treatment; ~ **spécifique d'un secteur d'activité** POLL EAU industry-specific purification; ~ **supplémentaire** POLL ATM supplementary purification

épurer *vt* DÉCHETS treat, DIST EAU treat, scrub, GESTION treat, POLL ATM clean, *gaz* scrub, POLL EAU purify; ~ **par lavage** SOLS elutriate

équation: ~ **de Bernoulli** *f* ÉCHANT *dynamique des fluides* Bernoulli's equation

équilibre: ~ **biogéologique** *m* POL ENV bio-geo-equilibrium; ~ **biologique** *m* POL ENV biological balance; ~ **écologique** *m* POL ENV ecological balance; ~ **hydrobiologique** *m* POL ENV hydrobiological equilibrium; ~ **de rayonnement** *m* POLL ATM radiation equilibrium; ~ **de réaction** *m* POLL ATM reaction equilibrium

équipement *m* GÉNÉRAL equipment; ~ **de contrôle intégré** SOLS built-in monitoring equipment; ~ **de conversion** DÉCHETS conversion machinery; ~ **de dépollution par les hydrocarbures** POLL MAR oil spill combating equipment; ~ **d'épuration** SÉCURITÉ cleaning equipment; ~ **de mesure intégré** SOLS built-in measuring equipment; ~ **normalisé** SÉCURITÉ standardized equipment; ~ **de régulation intégré** SOLS built-in regulating equipment; ~ **de sécurité** SÉCURITÉ safety-related equipment

équipements: ~ **urbains** *m pl* POL ENV urban capital stocks

équiper *vt* ÉCHANT, SÉCURITÉ equip

équivalent: ~ **de dose collective** *m* ÉCHANT, POLL ATM group collective dose; ~ **en eau** *m* DIST EAU water equivalent; ~ **habitant** *m* DÉCHETS, POLL ATM, POLL EAU inhabitant equivalent; ~ **par million** *m* POLL ATM equivalent per million (*EPM*); ~ **de toxicité** *m* ÉCHANT toxicity equivalent

ergonomique *adj* DÉCHETS, SÉCURITÉ ergonomic

ériger *vt* DÉCHETS erect

érosif *adj* POLL ATM erosive

érosion *f* DIST EAU, SOLS erosion; ~ **accélérée** SOLS accelerated erosion; ~ **anthropique** SOLS anthropogenic erosion, man-caused erosion; ~ **éolienne** SOLS wind erosion; ~ **mécanique** SOLS mechanical erosion; ~ **en nappe** SOLS sheet erosion; ~ **par les vagues** POLL MAR, SOLS wave erosion; ~ **pluviale** SOLS rainfall erosion; ~ **provenant des mouvements de masse** SOLS slip erosion; ~ **ravinante** SOLS ravine erosion; ~ **en rigoles** SOLS rill erosion; ~ **des sols** SOLS soil erosion

erreur: ~ **d'analyse** *f* ÉCHANT analytical error; ~ **discrétionnaire** *f* POL ENV discretionary error; ~ **d'exploitation** *f* GESTION operational error; ~ **de gestion** *f* GESTION mismanagement; ~ **d'indication** *f* ÉCHANT indication error; ~ **de lecture** *f* ÉCHANT reading error; ~ **de manœuvre** *f* ÉCHANT, SÉCURITÉ operating error; ~ **de mesure** *f* ÉCHANT measurement error, measuring error; ~ **de relevé** *f* ÉCHANT reading error

escherichia: ~ **coli** *f* POLL EAU, SÉCURITÉ escherichia coli

espace: ~ **annulaire** *m* POLL ATM annulus, *condenseur* annular gap, SOLS *condenseur* annular gap; ~ **interstitiel** *m* POLL EAU void space; ~ **miné** *m* DÉCHETS mined space; ~ **universel** *m* POLL ATM universal space; ~ **de vente** *m* GESTION outlet; ~ **vert** *m* POL ENV green open space

espèce: ~ **benthique** *f* POLL MAR bottom-living communities; ~ **indicatrice** *f* DIST EAU, POLL EAU, SOLS indicator species; ~ **marine** *f* POLL MAR marine species; ~ **menacée** *f* POL ENV *de disparition* endangered species; ~ **sauvage** *f* POL ENV wild species; ~ **de végétales résistantes aux insectes** *f* SOLS insect-resisting crop strains

essai *m* ÉCHANT test, assay; ~ **acoustique** POLL SON acoustical test; ~ **d'adsorption** ÉCHANT adsorption test; ~ **d'approximation** ÉCHANT approximation test; ~ **arbitraire** ÉCHANT arbitrary test; ~ **en aveugle** ÉCHANT blind test; ~ **bactériologique** ÉCHANT bacteria test; ~ **biologique** ÉCHANT, POL ENV bioassay; ~ **à blanc** ÉCHANT blank test; ~ **au bleu de méthylène** DÉCHETS, ÉCHANT methylene blue test; ~ **de capacité en charge** ÉCHANT load capacity test; ~ **climatique** DÉCHETS climatic test; ~ **aux colibacilles** ÉCHANT coli test; ~ **comparatif** ÉCHANT comparative test; ~ **de compression** DÉCHETS, ÉCHANT compression test; ~ **de corrosion** ÉCHANT corrosion test; ~ **de criblage** ÉCHANT screening test; ~ **de décantation** ÉCHANT settling test; ~ **de décoloration** ÉCHANT stripping test; ~ **à la déshydrogénase** ÉCHANT dehydrogenase test; ~ **de dosage immunologique** ÉCHANT immunoassay test; ~ **d'élution** ÉCHANT elution test; ~ **d'eutrophisation** ÉCHANT algae test; ~ **d'extraction** ÉCHANT extraction test; ~ **de fermentation** ÉCHANT fermentation test; ~ **de filtration** ÉCHANT filter test; ~ **en flacon** ÉCHANT bottle test; ~ **de floculation** ÉCHANT flocculation test, jar test; ~ **de fuite** DÉCHETS leakage test; ~ **de houle** POLL MAR swell test; ~ **d'humidité** DÉCHETS moisture test; ~ **d'hygiénisation** DÉCHETS accelerated storage test; ~ **immunologique enzymatique** ÉCHANT enzyme mediated immunoassay test (*EMIT*); ~ **d'incinération** ÉCHANT incineration test; ~ **d'incombustibilité** SÉCURITÉ noncombustibility test; ~ **en laboratoire** ÉCHANT laboratory test; ~ **de lavage** ÉCHANT stripping test; ~ **de lessivage** DÉCHETS, ÉCHANT leachability test, leaching test, SOLS leaching test; ~ **de lessivage dynamique** DÉCHETS dynamic leachability test, dynamic leaching test; ~ **de lixiviabilité dynamique** ÉCHANT dynamic leachability test; ~ **de lixiviation** DÉCHETS, ÉCHANT, SOLS leaching test; ~ **de lixiviation statique** DÉCHETS, ÉCHANT, SOLS static leaching test, static leachability test; ~ **microbiologique** ÉCHANT microbiological testing; ~ **de mise à la masse** DÉCHETS bonding test; ~ **olfactif** ÉCHANT odor test (*AmE*), odour test (*BrE*); ~ **par bactérie luminescente** ÉCHANT luminescent bacteria test; ~ **par métabolites** ÉCHANT metabolites test; ~ **pilote** ÉCHANT pilot test; ~ **de pliage** DÉCHETS folding test; ~ **de pompage** DIST EAU, ÉCHANT, POLL EAU pumping test, pump test; ~ **préalable** ÉCHANT preliminary test, pretest; ~ **préliminaire** ÉCHANT preliminary trial, pretest; ~ **de séparabilité** ÉCHANT separability test; ~ **sur les poissons** ÉCHANT fish test; ~ **sur le terrain** ÉCHANT field trial, field test; ~ **témoin** ÉCHANT blank test; ~ **titrimétrique** ÉCHANT titrimetric test; ~ **de toxicité** ÉCHANT toxicity test; ~ **de toxicité sur daphnie** ÉCHANT daphnia test; ~ **très sensible** ÉCHANT highly-sensitive test; ~ **de vieillissement** DÉCHETS ageing test; ~ **visuel** ÉCHANT visual test

essayer *vt* ÉCHANT test

essence: ~ **sans plomb** *f* DIST EAU lead-free gasoline (*AmE*), lead-free petrol (*BrE*), unleaded gasoline (*AmE*), unleaded petrol (*BrE*)

essorage *m* DÉCHETS wringing, GÉNÉRAL drainage, draining, POLL ATM, POLL EAU extraction operation

essorer *vt* DÉCHETS wring, GÉNÉRAL drain

essoreuse *f* DÉCHETS wringer, POLL EAU extraction centrifuge, extraction vessel

esthétique: ~ **industrielle** *f* POL ENV industrial design

estimable *adj* ÉCHANT estimable

estimation: ~ **des risques** *f* GESTION risk estimation

étage: **à un** ~ *loc* POLL EAU *dispositif, installation* single-stage; **à** ~ **unique** *loc* POLL EAU single-stage

étages ~ **de séparation** *m pl* DIST EAU separation stages

étain *m* DÉCHETS tin

étalement: ~ **du panache** *m* POLL ATM plume shred

étalonnage *m* ÉCHANT calibration; ~ **d'appareil** ÉCHANT apparatus calibration; ~ **au centième** ÉCHANT hundred points calibration; ~ **logarithmique** ÉCHANT logarithmic calibration; ~ **de modèle** ÉCHANT model calibration

étalonné *adj* ÉCHANT calibrated

étalonner *vt* ÉCHANT calibrate

étalonneur *m* ÉCHANT calibrator; ~ **de gaz d'essai** ÉCHANT test gas calibrator

étamer *vt* DÉCHETS tin-plate

étanche *adj* GÉNÉRAL *résistant à l'eau* water-resistant, watertight, liquid-proof, tight, *à l'humidité* moisture-proof, moisture-resistant, damp-proof, *scellé* sealed, *imperméable* impermeable, impervious; ~ **à l'air** POLL ATM airtight; ~ **à l'eau** DÉCHETS, DIST EAU watertight, tight, POLL EAU watertight, POLL MAR watertight, tight; ~ **au gaz** POLL ATM gastight; ~ **aux graisses** DÉCHETS greaseproof; ~ **à l'humidité** DÉCHETS damp-proof, moisture-proof; ~ **à la lumière** DÉCHETS light-proof; ~ **à la poussière** SÉCURITÉ dust-proof

étanchéité: ~ **en plastique** *f* DIST EAU plastic sealing; ~ **souterraine** *f* DÉCHETS, DIST EAU underground sealing

étancher *vt* GÉNÉRAL seal, POLL SON waterproof

étang *m* DIST EAU, POLL EAU pond; ~ **d'épuration** DIST EAU, ÉN RENOUV, POLL EAU lagoon; ~ **d'oxydation** DIST EAU, POLL EAU oxidation pond, stabilization pond; ~ **de stabilisation** DIST EAU, POLL EAU sewage lagoon

étapes: **par** ~ *loc* ÉCHANT step-by-step

état[1]: ~ **absorbé** *m* POLL ATM absorbed state; ~ **de l'environnement** *m* POL ENV environmental conditions; ~ **gazeux d'agrégation** *m* POLL ATM gaseous state of aggregation; ~ **marécageux** *m* DÉCHETS marshiness; ~ **physique** *m* POLL ATM physical condition; ~ **solide** *m* POLL ATM solid state of aggregation, state of aggregation

état[2]: **à l'**~ **de vapeur** *loc* POLL ATM vaporous

éteindre *vt* DIST EAU *chaud* quench, slake, POLL ATM slake

étendue *f* ÉCHANT extent, *mesure* range; ~ **de mesure** ÉCHANT measuring range, field of measurement

éthanol *m* DÉCHETS ethanol

éthique: ~ **de l'environnement** *f* POL ENV environmental ethics

éthylène *m* POLL ATM ethylene

éthylènediamine *f* ÉCHANT *acide* ethylenediamine

étiage *m* DIST EAU low water, minimum low water

étiquetage *m* DÉCHETS, DIST EAU labeling (*AmE*), labelling (*BrE*); ~ **acoustique** POLL SON *sur machines* noise labeling (*AmE*), noise labelling (*BrE*); ~ **par soufflage d'air** DÉCHETS air blast labeling (*AmE*), air blast labelling (*BrE*)

étiqueté *adj* DÉCHETS labeled (*AmE*), labelled (*BrE*)

étiqueteuse *f* DÉCHETS labeling machine (*AmE*), labelling machine (*BrE*)

étiquette *f* DÉCHETS label; ~ **adhésive facile à décoller** DÉCHETS easy peel-off self-adhesive label; ~ **en bande** DÉCHETS band label; ~ **de contrôle** DÉCHETS control tag; ~ **décalquée à chaud** DÉCHETS hot transfer label; ~ **gommée** DÉCHETS gummed label; ~ **imprimée par transfert thermique** DÉCHETS heat transfer label; ~ **de polypropylène orienté** DÉCHETS oriented polypropylene label; ~ **de repérage** DÉCHETS marking label; ~ **thermocollée** DÉCHETS heat-seal label; ~ **volante** DÉCHETS hangtag

ETM *abrév* (*énergie thermique des mers*) ÉN RENOUV OTEC (*ocean thermal energy conversion*)

étouffer *vt* POLL SON deaden, muffle

étoupe *f* DIST EAU packing

étroite: ~ **surveillance** *f* SÉCURITÉ close supervision

étude: ~ **coût-avantage** *f* GESTION, POL ENV cost-benefit study; ~ **écologique** *f* ÉCHANT, POL ENV ecological analysis; ~ **de faisabilité** *f* ÉN RENOUV, GESTION, POL ENV feasibility study; ~ **d'impact** *f* GESTION, POL ENV impact study; ~ **d'impact sur l'environnement** *f* GESTION, POL ENV environmental impact assessment (*EIA*), environmental impact statement (*EIS*); **~ de la pollution** *f* ÉCHANT, GESTION, POL ENV pollution research; **~ radiologique** *f* ÉCHANT radiological survey; **~ de rentabilité** *f* GESTION, POL ENV cost-benefit study; **~ sur banc d'essai** *f* ÉCHANT, GESTION, POL ENV bench-scale study

étudier *vt* ÉCHANT, GESTION, SÉCURITÉ survey, examine

eutrophe *adj* DIST EAU, POLL EAU eutrophic

eutrophisation *f* DIST EAU, POLL EAU eutrophication

évacuateur *m* DÉCHETS, DIST EAU discharger; **~ de crue** DIST EAU flood spillway; **~ de secours** DIST EAU emergency spillway

évacuation *f* GÉNÉRAL disposal, evacuation, discharge, removal, *drainage* drainage, draining, draining off; **~ d'adsorbants** POLL ATM adsorbent outlet; **~ des boues** DIST EAU, POLL EAU de-sludging, sludge disposal; **~ de cheminée** POLL ATM chimney discharge; **~ des déchets** DÉCHETS residue disposal, waste disposal; **~ définitive** DÉCHETS final dumping; **~ directe des effluents industriels** POLL EAU direct industrial wastewater discharge; **~ des eaux usées** DIST EAU, POLL EAU sewerage, wastewater disposal; **~ des eaux usées urbains** DIST EAU, POLL EAU municipal wastewater discharge; **~ des gaz** SOLS venting; **~ des gaz de décharge** DÉCHETS landfill gas drainage; **~ de gaz épuré** POLL ATM clean-gas outlet; **~ des lixiviats** DÉCHETS, DIST EAU leachate drainage; **~ des lixiviats avec filtre à gravier** DÉCHETS, DIST EAU leachate drainage with filtering rubble; **~ de sols contaminés** SOLS removal of contaminated soils; **~ souterraine** DÉCHETS underground disposal; **~ souterraine des eaux usées** POLL EAU underground wastewater disposal; **~ de la station municipale d'épuration des eaux** POLL EAU discharge of municipal wastewater sludge plant; **~ tubulaire** DIST EAU tubular drainage

évacuer[1] *vt* GÉNÉRAL dispose of, evacuate, discharge, remove, *drainer* drain, drain off; **~ par lavage** GÉNÉRAL wash out; **~ par pompage** DÉCHETS, DIST EAU, POLL EAU pump out

évacuer[2]: **~ les boues de** *vi* DIST EAU, POLL EAU de-sludge

évaluation *f* GÉNÉRAL evaluation, assessment; **~ du bruit** ÉCHANT, POLL SON noise rating; **~ des coûts** GESTION cost appraisal; **~ du cycle de vie** GESTION life-cycle assessment, life-cycle analysis (*LCA*), POL ENV life-cycle analysis (*LCA*); **~ d'échantillon** ÉCHANT sample assessment; **~ écotoxicologique** GESTION, SÉCURITÉ ecotoxicological assessment; **~ de l'environnement** GESTION eco-rating; **~ de l'étude d'impact** POL ENV environmental impact assessment evaluation; **~ de faisabilité** ÉN RENOUV, GESTION, POLL ATM feasibility assessment; **~ préliminaire** DIST EAU, ÉCHANT, POL ENV, SÉCURITÉ, SOLS first assessment, preliminary assessment; **~ provisoire** ÉCHANT provisional evaluation; **~ qualitative** ÉCHANT qualitative evaluation; **~ quantitative** ÉCHANT quantitative evaluation; **~ des ressources** POL ENV survey of resources; **~ des risques** GESTION, POL ENV, POLL MAR, SÉCURITÉ danger assessment, risk assessment, SOLS risk assessment; **~ sans échantillon normalisé** ÉCHANT evaluation without standard samples; **~ de site** ÉCHANT site assessment; **~ spatiale** ÉCHANT spatial evaluation; **~ statistique** ÉCHANT statistical evaluation; **~ sur échantillons normalisés** ÉCHANT evaluation using standard samples; **~ sur site** ÉCHANT site assessment; **~ technologique** POL ENV technology assessment; **~ temporelle** ÉCHANT temporal evaluation; **~ toxicologique** GESTION, POL ENV toxicological assessment; **~ d'une vue aérienne** ÉCHANT aerial view evaluation

évaluer *vt* GÉNÉRAL evaluate, assess; **~ qualitativement** ÉCHANT assess qualitatively; **~ quantitativement** ÉCHANT assess quantitatively

évapodistillation *f* ÉCHANT steam distillation

évaporateur: **~ de circulation** *m* DIST EAU circulation evaporator; **~ instantané** *m* POLL EAU flash evaporator; **~ à mélangeur** *m* POLL EAU mixing device evaporator; **~ pelliculaire** *m* POLL EAU thin-film evaporator; **~ à plusieurs étages** *m* POLL EAU multistage evaporator; **~ rotatif** *m* DIST EAU, POLL EAU rotary evaporator; **~ sous vide** *m* DIST EAU vacuum evaporator; **~ à vide** *m* DIST EAU vacuum evaporator

évaporation *f* DIST EAU, ÉCHANT, POLL ATM evaporation; **~ annuelle** DIST EAU annual evaporation; **~ d'eau** DIST EAU, POLL ATM water evaporation; **~ efficace** DIST EAU effective evaporation; **~ d'éther** ÉCHANT ether evaporation

évaporer[1] *vt* DIST EAU, ÉCHANT, POLL ATM evaporate, steam out

évaporer[2]: **s'~** *v réfl* DIST EAU, ÉCHANT evaporate, ÉN RENOUV vaporize, POLL ATM evaporate, vaporize

évapotranspiration *f* DIST EAU evapotranspiration

évasement *m* DÉCHETS, POLL ATM, POLL EAU flare, flaring, splay, splaying

évaser *vt* DÉCHETS *lumière*, POLL ATM, POLL EAU *tuyau* flare, splay

évent *m* DIST EAU, POLL ATM vent

évier *m* DIST EAU sink; **~ à écoulement tourbillonnaire** POLL ATM eddy sink

évolution: **~ climatique** *f* POLL ATM climate changes; **~ écologique** *f* POL ENV ecological structural change

exactitude: **~ de mesure** *f* ÉCHANT measuring accuracy, accuracy

examen: **~ de l'eau** *m* ÉCHANT water examination

examiner *vt* ÉCHANT, SÉCURITÉ examine

excavateur *m* DÉCHETS, SOLS excavator

excavation *f* DÉCHETS, SOLS excavation; **~ de séparation** SOLS separating cut; **~ de tranchées** SOLS trenching

excavé *adj* DÉCHETS, SOLS excavated

excédent: **~ d'énergie électrique** *m* ÉN RENOUV excess power supply

exclusion *f* POL ENV exclusion

exempt: **~ de cendre** *adj* DIST EAU, POLL ATM ash-free; **~ de métaux lourds** *adj* POLL ATM, SOLS free of heavy metal; **~ de polluants** *adj* POLL ATM, SOLS pollutant-free; **~ de solvent** *adj* POLL ATM solvent-free

exemption: **~ fiscale** *f* POL ENV tax exemption; **~ de redevance** *f* POL ENV charge exemption

exhaure *f* GÉNÉRAL *à la pompe* pumping, pumping out

exhaussement *m* DIST EAU *du lit d'un cours d'eau* aggradation

exigence: **~ de conformité aux critères techniques de sécurité** *f* SÉCURITÉ technical safety demand; **~ de qualité** *f* SOLS quality requirement

exode: **~ rural** *m* POL ENV rural migration

exoenzyme *f* POLL ATM exoenzyme

exonération: **~ fiscale** *f* POL ENV tax relief

exosphère *f* POL ENV exosphere

exothermique *adj* POLL ATM, POLL EAU *réaction* exothermic

expérimenter *vt* ÉCHANT test

exploitant *m* POL ENV operator

exploitation *f* GÉNÉRAL operation; **~ autonome** ÉCHANT off-line operation; **~ commerciale** GESTION commercial operation; **~ compatible avec l'environnement** GESTION environmentally-compatible operation; **~ en continu** ÉCHANT on-line operation; **~ courante** POL ENV normal operation; **~ d'une décharge** DÉCHETS landfill operation; **~ de l'énergie géothermique** DIST EAU, ÉN RENOUV, POLL ATM geothermal energy exploitation; **~ de l'énergie des vagues selon le principe de la colonne oscillante** ÉN RENOUV oscillating water column operation (*OWC operation*); **~ éolienne** ÉN RENOUV wind farm; **~ éolienne en mer** ÉN RENOUV offshore wind farm; **~ expérimentale** DIST EAU, GESTION, POL ENV trial operation; **~ forestière** POL ENV, SOLS forest logging; **~ industrielle** POL ENV commercial operation; **~ des océans** ÉN RENOUV ocean harvesting; **~ des ressources du fond des mers** POL ENV exploitation of sea-bed resources; **~ simultanée** ÉCHANT simultaneous operation; ~ **d'un site** DIST EAU site operation; ~ **sous licence** POL ENV operating licence (*BrE*), operating license (*AmE*); ~ **en temps réel** ÉCHANT real-time operation

exploiter *vt* ÉN RENOUV *rivière, ressources* harness

explorateur: ~ **manuel** *m* DIST EAU, POLL ATM hand scanner

exploratoire *adj* ÉCHANT exploratory

explosibilité *f* POLL ATM explosibility

explosif[1] *adj* DÉCHETS, POLL ATM, SÉCURITÉ explosive

explosif[2] *m* DÉCHETS, POLL ATM, SÉCURITÉ explosive

explosion: ~ **de poussière** *f* POLL ATM dust explosion

exportateur: ~ **net** *m* DIST EAU net donator

exposition *f* SÉCURITÉ exposure; ~ **à l'amiante** SÉCURITÉ asbestos exposure; ~ **au bruit** POLL SON noise exposure; ~ **au bruit journalière** POLL SON daily noise exposure; ~ **aux fumées** POLL ATM, SÉCURITÉ exposure to fumes; ~ **maximale** SÉCURITÉ maximum exposure; ~ **quotidienne totale** SÉCURITÉ daily total exposure; ~ **au vent** DÉCHETS, POLL ATM wind exposure

expulser *vt* POLL ATM *gaz* scavenge

extérieur: ~ **à la décharge** *adj* DÉCHETS landfill-external

externalité *f* POL ENV external effect, externality

externe *adj* POL ENV external

extincteur *m* SOLS quencher; ~ **d'incendie** SÉCURITÉ fire extinguisher

extinction *f* ÉCHANT, POLL ATM extinction; ~ **à la chaux** POLL ATM lime slaking; ~ **d'espèces végétales** SOLS extinction of plant species

extracteur *m* DÉCHETS, SOLS extractor; ~ **de cendres** DIST EAU ash remover; ~ **de ferrailles** DÉCHETS scrap extractor; ~ **mélangeur décantateur** POLL EAU mixer-settler extractor; ~ **à rejets** DIST EAU refuse extractor; ~ **à vis** DIST EAU screw feeder

extraction *f* DIST EAU, ÉCHANT, POLL ATM extraction, removal; ~ **des boues** POLL EAU sludge removal; ~ **à contre-courant** SOLS counter-current stripping, counter-current extraction; ~ **des cyanures** POLL EAU cyanide removal; ~ **de déchets** DÉCHETS waste removal; ~ **d'énergie** ÉN RENOUV energy extraction; ~ **des ferrailles** DÉCHETS scrap metal separation; ~ **de fluide supercritique** ÉCHANT supercritical-fluid extraction (*SFE*); ~ **des gaz pollués** POLL ATM waste gas extraction; ~ **du gypse** POLL ATM gypsum separation; ~ **d'huile par séparation** DIST EAU removal of oil by separators; ~ **liquide-liquide** ÉCHANT liquid-liquid extraction; ~ **liquide-solide** ÉCHANT liquid-solid extraction; ~ **de matières premières** ÉN RENOUV, GESTION raw material extraction; ~ **de matières solides en suspension par décantation** DIST EAU removal of suspended solids by sedimentation; ~ **des métaux lourds** DIST EAU, POLL ATM, POLL EAU heavy metals removal; ~ **des oxydes d'azote** POLL ATM nitrogen oxides removal; ~ **par dissolution** SOLS dissolving away; ~ **par tube de Soxhlet** ÉCHANT Soxhlet extraction; ~ **par ultrasons** ÉCHANT ultrasonic extraction; ~ **de particules solides en suspension par décantation** POLL ATM removal of suspended solids by sedimentation; ~ **en phase solide** ÉCHANT solid-phase extraction (*SPE*); ~ **de poussière** POLL ATM dust extraction; ~ **des scories** DIST EAU, POLL ATM slag removal; ~ **séquentielle par phases spécifiques** DÉCHETS, DIST EAU phase-specific sequential extraction; ~ **solide-liquide** ÉCHANT solid-liquid extraction; ~ **des substances nutritives** POLL EAU nutrient removal; ~ **des substances organiques dissoutes** DIST EAU removal of dissolved organic substances; ~ **sur place** SOLS in-situ extraction, on-site extraction; ~ **totale** POLL ATM total removal; ~ **à la vapeur** SOLS steam stripping; ~ **à la vapeur sous vide** SOLS steam-vacuum stripping; ~ **des vapeurs** SOLS vapor extraction (*AmE*), vapour extraction (*BrE*); ~ **des vapeurs du sol** SOLS soil vapor extraction (*AmE*), soil vapour extraction (*BrE*)

extraire *vt* GÉNÉRAL extract, remove; ~ **par dissolution** SOLS dissolve away; ~ **par distillation** SOLS extract by distillation

extrait[1] *adj* GÉNÉRAL extracted, removed

extrait[2] *m* GÉNÉRAL extract; ~ **sec** POLL EAU dry residue

extrémité: ~ **de tuyauterie** *f* DIST EAU pipe end

extrudeuse: ~ **hélicoïdale** *f* DÉCHETS, DIST EAU cylindrical screw extruder; ~ **de laboratoire** *f* DÉCHETS laboratory extruder; ~ **à plastique** *f* DÉCHETS, DIST EAU plastics extruder; ~ **de regranulation** *f* DIST EAU regranulating extruder; ~ **à vis** *f* DÉCHETS, DIST EAU screw extruder, cylindrical screw extruder; ~ **à vis sans fin** *f* DÉCHETS, DIST EAU roller die extruder; ~ **à vis unique** *f* DÉCHETS, DIST EAU single-screw extruder

extrusion *f* DÉCHETS, DIST EAU extrusion, extruding; ~ **en couches multiples** DÉCHETS co-extrusion; ~ **directe** DÉCHETS, DIST EAU direct extrusion; **~-soufflage** DÉCHETS extrusion blow molding (*AmE*), extrusion blow moulding (*BrE*)

exutoire *m* POLL EAU outfall, outlet

F

F *abrév* (*fluor*) SOLS F (*fluorine*)

face: **~ supérieure** *f* DÉCHETS upper surface; **~ utilisable** *f* DÉCHETS operating face

façonnage: **~, débardage et stockage du bois** *m* ÉN RENOUV cutting, transporting and storing of wood

facteur: **~ alpha** *m* DIST EAU alpha factor; **~ bêta** *m* DIST EAU beta factor; **~ de bioaccumulation** *m* POLL ATM bioaccumulation factor; **~ climatique** *m* POLL ATM climate factor; **~ de conversion** *m* ÉCHANT conversion factor; **~ de coût** *m* GESTION cost factor; **~ écologique** *m* DIST EAU, POL ENV, SOLS ecological factor; **~ d'entraînement** *m* POLL ATM entrainment factor; **~ de marche** *m* GESTION onstream factor; **~ Q** *m* DIST EAU, POLL ATM quality factor; **~ de qualité** *m* DIST EAU, POLL ATM quality factor; **~ de risque** *m* GESTION risk factor; **~ de saturation en oxygène** *m* POLL EAU oxygen saturation factor; **~ de sécurité** *m* SÉCURITÉ safety factor; **~ volumétrique** *m* ÉCHANT volumetric factor

faible[1]: **de ~ acidité** *adj* SOLS mildly acidic; **~ en polluants** *adj* POLL ATM, SOLS low-pollutant

faible[2]: **~ charge en gaz brut** *f* DÉCHETS low raw-gas load; **~ concentration toxique** *f* SÉCURITÉ low toxic concentration; **~ dose létale** *f* SÉCURITÉ low lethal dose; **~ dose toxique** *f* SÉCURITÉ low toxic dose; **~ perméabilité du sol** *f* SOLS low soil permeability

faible[3]: **à ~ charge** *loc* POLL EAU *pompe* low-rate; **à ~ débit** *loc* POLL EAU *pompe* low-rate

faiblement: **~ radioactif** *adj* DÉCHETS low-level radioactive

faille *f* SOLS fault

faire[1]: **~ cailler** *vt* GÉNÉRAL coagulate; **~ circuler** *vt* POLL ATM circulate; **~ écouler** *vt* DÉCHETS, DIST EAU, POLL EAU *les eaux* drain off; **~ entrer par pression** *vt* DIST EAU press in; **~ évaporer** *vt* DIST EAU, ÉCHANT evaporate; **~ fermenter** *vt* DÉCHETS, ÉN RENOUV, POLL EAU ferment; **~ fondre** *vt* POLL EAU *métallurgie* melt; **~ le plein de** *vt* DIST EAU replenish; **~ pression** *vi* POL ENV lobby; **~ recirculer** *vt* DIST EAU recirculate; **~ vaciller** *vt* DÉCHETS *lumière*, POLL EAU flare

faire[2]: **~ de l'eau** *vi* DIST EAU water

faisabilité *f* ÉN RENOUV, GESTION, POL ENV, POLL ATM feasibility; **~ économique** POL ENV economic feasibility

faisceau *m* DÉCHETS *combustible nucléaire* bundle; **~ d'électrons** POLL ATM electron beam; **~ de lamelles** POLL ATM lamella bundle; **~ laser** DIST EAU, ÉCHANT laser beam; **~ de tubes** POLL ATM tube bundle

fardeau *m* POLL ATM burden

fardeleuse *f* DÉCHETS bottle packing machine; **~-ficeleuse** DÉCHETS bundle-tying machine

fatigue *f* ÉN RENOUV, POLL ATM, POLL SON, SOLS fatigue; **~ sensorielle** ÉCHANT sensory fatigue

faune: **~ aquatique** *f* POLL MAR aquatic life; **~ et flore marines** *f pl* POLL MAR marine biota

fausse: **~ manœuvre** *f* DIST EAU operational error

faux: **~-plafond** *m* POLL SON ceiling panel

favorable: **~ au recyclage** *adj* DIST EAU recycling-friendly; **~ au vent** *adj* ÉN RENOUV wind-favorable (*AmE*), wind-favourable (*BrE*)

Fe *abrév* (*fer*) GÉNÉRAL Fe (*iron*)

fécal *adj* POLL EAU faecal (*BrE*), fecal (*AmE*)

fèces *f pl* POLL EAU faecal matter (*BrE*), fecal matter (*AmE*)

feeder: **~ principal** *m* DIST EAU *électricité* trunk main

fenêtre: **~ d'habitabilité** *f* GESTION eco-window; **~ insonorisante** *f* POLL SON sound-protecting window; **~ lambda** *f* POLL ATM lambda window; **~ d'observation** *f* ÉCHANT inspection glass

fente: **~ de collage** *f* DÉCHETS glue joint

fer *m* (*Fe*) GÉNÉRAL iron (*Fe*): **~ blanc** DÉCHETS tin; **~ à marquer à chaud** DÉCHETS branding iron

fermentable *adj* DÉCHETS, ÉN RENOUV, POLL EAU fermentable

fermentatif *adj* DÉCHETS, ÉN RENOUV, POLL EAU fermentative

fermentation *f* DÉCHETS, ÉN RENOUV, POLL EAU fermentation; **~ acide** DÉCHETS acid fermentation; **~ alcaline** DÉCHETS alkaline fermentation; **~ contrôlée** DÉCHETS controlled fermentation; **~ lente** DÉCHETS slow fermentation; **~ méthanique** DÉCHETS methane fermentation; **~ méthanique instable** DÉCHETS unstable methane fermentation; **~ méthanogène** DÉCHETS methane fermentation; **~ des ordures** DÉCHETS fermentation of refuse; **~ rapide** DÉCHETS rapid fermentation

fermenter *vi* DÉCHETS, ÉN RENOUV, POLL EAU ferment

fermentescible *adj* DÉCHETS, ÉN RENOUV, POLL EAU fermentable

fermeture *f* DÉCHETS closure; **~ de cerclage** DÉCHETS strapping seal; **~ pour emballage thermoformé** DÉCHETS blister seal; **~ hermétique** DÉCHETS hermetic closure, hermetic seal; **~ hermétique collée** DÉCHETS bonded seal; **~ inviolable** SÉCURITÉ tamper-proof closure; **~ à membrane** DÉCHETS diaphragm sealing; **~ de sécurité** DÉCHETS guarantee closure; **~ de sécurité pour enfants** DÉCHETS child-resistant closure (*CRC*)

ferraille *f* DÉCHETS junk iron, scrap, scrap iron, ferrous scrap, metal scrap, scrap metal

ferrailleur *m* DÉCHETS scrap metal dealer, scrap metal merchant

ferreux *adj* ÉCHANT ferrous, iron-containing

ferromagnétique *adj* ÉCHANT ferromagnetic

fertilisation *f* SOLS fertilization; **~ excessive** POLL EAU, SOLS overfertilization

feu: **~ de forêt** *m* POLL ATM, SOLS forest fire

feuille: **~ de doublure** *f* DÉCHETS liner sheet; **~ de garde** *f* DÉCHETS lining paper; **~ isolante** *f* DÉCHETS insulating sheet; **~ métallique** *f* DÉCHETS metal foil; **~ de papier** *f* DÉCHETS paper laminate; **~ de plastique** *f* DÉCHETS plastic sheet; **~ plastique** *f* DÉCHETS plastic sheeting, plastic sheet

feuilletage *m* DÉCHETS lamination

feutre *m* POLL ATM, POLL SON felt; **~ bitumé** DIST EAU bitumen felt; **~ cousu** POLL ATM needled felt; **~ cousu à fibre fine** POLL ATM fine-fibered needled felt (*AmE*), fine-fibred needled felt (*BrE*); **~ de filtre** POLL ATM filter felt

feux: **~ de brousse** *m pl* POLL ATM, SOLS bush fires

fiabilité *f* GÉNÉRAL reliability

fibre *f* DÉCHETS, POLL ATM, POLL EAU fiber (*AmE*), fibre (*BrE*); **~ d'acier** POLL ATM steel fiber (*AmE*), steel fibre (*BrE*); **~ artificielle** DÉCHETS, DIST EAU, POLL ATM man-made fiber (*AmE*), man-made fibre (*BrE*); **~ de cellulose** DÉCHETS, DIST EAU cellulose fiber (*AmE*), cellulose fibre (*BrE*); **~ cellulosique** DÉCHETS, DIST EAU cellulose fiber (*AmE*), cellulose fibre (*BrE*); **~ d'emballage** DÉCHETS wood wool; **~ minérale** DIST EAU, POLL ATM, POLL SON mineral fiber (*AmE*), mineral fibre (*BrE*); **~ papetière** DÉCHETS, DIST EAU paper fiber (*AmE*), paper fibre (*BrE*); **~ à papier** DÉCHETS, DIST EAU paper fiber (*AmE*), paper fibre (*BrE*); **~ de récupération** DIST EAU secondary fiber (*AmE*), secondary fibre (*BrE*); **~ de synthèse** DÉCHETS, DIST EAU, POLL ATM synthetic fiber (*AmE*), synthetic fibre (*BrE*); **~ synthétique**

DÉCHETS, DIST EAU, POLL ATM man-made fiber (*AmE*), man-made fibre (*BrE*), synthetic fiber (*AmE*), synthetic fibre (*BrE*); ~ **de verre** DÉCHETS, POLL ATM, POLL SON fiberglass (*AmE*), fibreglass (*BrE*), glass fiber (*AmE*), glass fibre (*BrE*); ~ **de verre renforcée** DÉCHETS, DIST EAU, POLL ATM, POLL SON reinforced fiberglass (*AmE*), reinforced fibreglass (*BrE*)

fibres *f pl* DÉCHETS, POLL ATM, POLL EAU fibres, fibrous constituents

fibreux *adj* DÉCHETS, POLL ATM, POLL EAU fibrous

fiche: ~ **déchet** *f* DÉCHETS, POL ENV waste description sheet

fidélité *f* ÉCHANT precision

fil *m* POLL ATM yarn; ~ **d'agrafage** DÉCHETS stapling wire; ~ **torsadé** POLL ATM wire

filasse *f* POLL ATM yarn

filière *f* DÉCHETS *nucléaire* line; ~ **classique de fourniture d'énergie** ÉN RENOUV conventional energy supply system; ~ **classique de nettoyage du charbon** *m* POLL ATM conventional coal cleaning; ~ **de conversion** ÉN RENOUV, POL ENV conversion technique; ~ **d'élimination** DÉCHETS disposal pathway, disposal route; ~ **d'élimination de déchets** DÉCHETS disposal pathway, disposal route; ~ **énergétique** ÉN RENOUV energy chain; ~ **d'extrusion** DÉCHETS extrusion die, extrusion line

film[1]: ~ **d'acétate** *m* DÉCHETS acetate film; ~ **d'acétate de cellulose** *m* DÉCHETS cellulose acetate film; ~ **adhésif** *m* DÉCHETS adhesive film, cling film; ~ **biologique** *m* POLL EAU *épuration* filter film; ~ **à bulles** *m* DÉCHETS bubble film; ~ **à écoulement continu** *m* POLL ATM continuous flowing film; ~ **d'étiquetage** *m* DÉCHETS label film; ~ **étirable** *m* DÉCHETS stretch film; ~ **fixe** *m* POLL EAU fixed film; ~ **de polyéthylène téréphtalate** *f* DIST EAU polyethylene terephtalate film; ~ **poussiéreux** *m* POLL ATM dust-laden film; ~ **ruisselant** *m* POLL ATM trickling film; ~ **de séparation** *m* DÉCHETS barrier film

film[2]: **à ~ mince** *loc* DÉCHETS, POLL EAU thin-film

filtrabilité *f* GÉNÉRAL filterability

filtrable *adj* GÉNÉRAL filterable, filtrable

filtrage *m* GÉNÉRAL filtration, screening; ~ **des données mesurées** ÉCHANT measuring-data filtering; ~ **par compression** ÉCHANT compression filtration; ~ **des produits de fusion** DÉCHETS melt filtration

filtrant *adj* GÉNÉRAL filterable, filtrable

filtrat *m* GÉNÉRAL filtrate; ~ **lent** DIST EAU slow process filtrate

filtrateur *m* DIST EAU, POLL EAU filtering tank

filtration *f* GÉNÉRAL *passage à travers un filtre* filtration, filtering, screening, *écoulement* filtration, filtering, percolation, *extraction* filtration, filtering, leaching; ~ **biologique** POLL ATM biological filtration; ~ **des boues** DIST EAU sludge filtration; ~ **en fin de cycle** POLL EAU final filtration; ~ **finale** POLL ATM final filtering, POLL EAU final filtration; ~ **à forte charge** POLL EAU high-rate filtration; ~ **à haut rendement** DIST EAU, POLL ATM high-performance filtration; ~ **haute performance** DIST EAU, POLL ATM high-performance filtration; ~ **intrusive** DIST EAU intrusive filtration; ~ **inverse** DIST EAU reverse filtration; ~ **lente** DIST EAU, POLL EAU slow filtration; ~ **lente sur sable** DIST EAU, POLL EAU slow process sand filtration, slow sand filtration; ~ **mécanique** DÉCHETS mechanical filtration; ~ **de particules** POLL EAU particle filtration; ~ **préliminaire** POLL EAU preliminary filtration; ~ **en profondeur** POLL ATM, POLL EAU deep-bed filtration, depth filtration; ~ **rapide** DIST EAU, POLL EAU rapid filtration; ~ **rapide sur sable** DIST EAU, POLL EAU rapid sand filtration; ~ **à sec** POLL EAU dry filtration; ~ **sous vide** DÉCHETS, DIST EAU, POLL EAU vacuum filtration; ~ **sur diatomées** POLL EAU diatomaceous earth filtration; ~ **sur lit profond** POLL ATM, POLL EAU deep-bed filtration, depth filtration; ~ **sur membrane** ÉCHANT, POLL EAU membrane filtration; ~ **sur précouche** DIST EAU precoat filtration; ~ **en surface** DÉCHETS, POLL ATM, POLL EAU *traitement des boues* surface filtration; ~ **en surface sur membranes** DIST EAU, POLL EAU surface filtration with membranes; ~ **volumique** POLL EAU volume filtration

filtre *m* DÉCHETS, DIST EAU, ÉCHANT, POLL ATM, POLL EAU filter, POLL MAR strainer; ~ **adsorbant** POLL ATM adsorber filter; ~ **à air** POLL ATM, SÉCURITÉ air filter; ~ **à air à contre-courant** POLL ATM reverse air flow filter; ~ **à air sec** POLL ATM dry-air filter; ~ **aspirant** POLL ATM exhauster filter, suction filter; ~ **d'aspiration** DÉCHETS suction filter; ~ **bactérien** POLL ATM bacteria filter; ~ **biologique** POLL ATM biological filter; ~ **de cabine** SÉCURITÉ cab filter; ~ **à cartouche** POLL ATM cartridge filter; ~ **à cartouches pour granulats** POLL ATM granulate tubular filter; ~ **céramique** POLL ATM ceramic filter; ~ **à charbon actif** DIST EAU, POLL EAU activated-carbon filter; ~ **chargé** POLL ATM loaded filter; ~ **de chimisorption** POLL ATM chemisorption filter; ~ **à coke actif** DÉCHETS activated-coke filter; ~ **compact** POLL ATM compact filter; ~ **à compost** POLL ATM compost filter; ~ **à contre-courant** POLL ATM reverse-jet filter, POLL EAU reverse-flow filter; ~ **à couche fibreuse** POLL ATM fiber layer filter (*AmE*), fibre layer filter (*BrE*); ~ **à courant ascendant** DIST EAU upflow filter; ~ **de dépoussiérage** POLL ATM dust filter; ~ **de dérivation** POLL ATM bypass filter; ~ **à diaphragme** DIST EAU, ÉCHANT, POLL EAU membrane filter; ~ **à eau** DIST EAU water filter; ~ **pour échantillon gazeux** ÉCHANT sample gas filter; ~ **électrostatique de dépoussiérage par voie humide** POLL ATM wet electrostatic filter; ~ **électrostatique sec** POLL ATM dry electrostatic filter; ~ **externe** ÉCHANT outer filter; ~ **en fibre de verre** ÉCHANT glass-fiber filter (*AmE*), glass-fibre filter (*BrE*); ~ **final** POLL ATM final filter; ~ **à flux ascendant** DIST EAU upflow filter; ~ **à fumées** POLL ATM smoke filter; ~ **à garnissage** POLL ATM packed-bed filter; ~ **à garnissage à contre-courant** POLL ATM reverse-flow packed-bed filter; ~ **de Gauss** ÉCHANT Gaussian filter; ~ **à gaz chauds** POLL ATM hot-gas filter; ~ **à gravier** POLL ATM gravel-bed filter; ~ **gris** ÉCHANT gray filter (*AmE*), grey filter (*BrE*); ~ **grossier** POLL ATM coarse filter; ~ **à haute couche** POLL EAU deep-bed filter; ~ **haute performance** DIST EAU, POLL ATM high-performance filter; ~ **haute pression** POLL EAU high-pressure filter; ~ **à injection directe** POLL ATM direct-pulse filter; ~ **interférentiel** ÉCHANT interference filter; ~ **à jet** POLL ATM jet filter; ~ **à jet nettoyé à air comprimé** POLL ATM compressed-air-cleaned jet filter; ~ **à joint gonflable** POLL ATM blow-ring filter; ~ **lent à sable** DIST EAU, POLL EAU *purification de l'eau* slow sand filter; ~ **à lit de boues** POLL EAU sludge-blanket filter; ~ **à lit de granulats** POLL ATM granular-bed filter; ~ **à lit profond** POLL ATM deep-bed filter; ~ **à mailles serrées** POLL ATM compact filter; ~ **à manche** POLL ATM filter bag; ~ **à matelas fibreux** POLL ATM fibrous-mat filter; ~ **pour matière fondue** DÉCHETS melt filter; ~ **microporeux** DIST EAU, ÉCHANT, POLL EAU membrane filter; ~ **moléculaire** DIST EAU, ÉCHANT, POLL EAU membrane filter; ~ **multicouche** POLL EAU multilayer filter; ~ **nettoyé par jet pulsant** POLL ATM pulse-jet cleaned filter; **~-papier** ÉCHANT paper filter; ~ **à particules** POLL ATM particulate matter filter; ~ **à particules submicroniques** POLL ATM submicron particulate filter; ~ **passe-bande** ÉCHANT band-pass filter; ~ **percolateur** DIST EAU, POLL EAU percolating filter; ~ **à plis** POLL EAU folded filter; ~ **plissé** POLL EAU folded filter; ~ **à poche** POLL ATM pocket filter; ~ **à poche à secousses** POLL ATM shaker baghouse; ~ **à poussière** POLL ATM dust filter; **~-presse** DÉCHETS, DIST EAU, POLL EAU, SOLS filter press; **~-presse à cadres et plateaux** POLL EAU frame filter press; **~-presse à chambre** DÉCHETS, SOLS chamber filter press; **~-presse à plateaux** DÉCHETS, SOLS chamber filter press; ~ **à pression** DIST EAU, POLL ATM pressure filter; ~ **rapide à sable** DIST EAU, POLL EAU rapid sand filter; ~ **à recirculation** POLL ATM recirculating filter; ~ **respiratoire** SÉCURITÉ respiratory filter; ~ **rotatif** DIST EAU, POLL EAU drum filter, rotary filter; **~ à sable** DIST EAU, POLL EAU sand filter; **~ à sable à lit mobile** POLL EAU moving-bed sand filter; **~ à sac** POLL ATM, SOLS bag filter; **~ à sac compartimenté** POLL ATM multichamber bag filter; **~ sélectif** ÉCHANT selective filter; **~ pour sol** DIST EAU soil filter; **~ sous pression** DIST EAU, POLL ATM pressure filter; **~ sous vide** DÉCHETS, DIST EAU, POLL EAU vacuum filter; **~ de stockage** POLL ATM storage filter; **~ à succion** DIST EAU suction strainer; **~ de surface** DÉCHETS, POLL ATM, POLL EAU surface filter; **~ à tambour** DIST EAU drum filter, rotary filter, POLL ATM revolving filter, POLL EAU drum filter, rotary filter; **~ à tambour sous vide** POLL EAU rotary vacuum filter; **~ terminal** POLL EAU final filter; **~ en tissu** POLL ATM fabric filter; **~ tubulaire de dégazage** SOLS degasification filter tube; **~ tubulaire pour granulats** POLL ATM granulate tubular filter; **~ tubulaire à manche** POLL ATM tubular filter bag; **~ ultraviolet** POLL ATM, SÉCURITÉ ultraviolet filter; **~ à vide** DÉCHETS suction filter, vacuum filter, DIST EAU vacuum filter, POLL ATM vacuum belt filter, POLL EAU vacuum filter; **~ volumique** POLL EAU volume filter

filtré *adj* GÉNÉRAL filtered

filtrer[1] *vt* GÉNÉRAL *passer à travers un filtre* filter, *extraire* filter, leach, screen

filtrer[2] *vi* GÉNÉRAL leach

fin *adj* DIST EAU, POLL ATM, SOLS fine

final *adj* SOLS final

financement *m* GESTION, POL ENV funding; **~ des activités liées à l'environnement** GESTION, POL ENV environmental sponsoring

financer *vt* GESTION, POL ENV finance

fines[1] *f pl* POLL ATM *charbon* fine dust; **~ de catalyseur** POLL ATM catalyst fines

fines[2]: **à ~ particules** *loc* DÉCHETS fine-particled

fioul *m* POLL ATM fuel oil

fiscalement: **~ neutre** *loc* POL ENV revenue-neutral

fixateur *m* POLL ATM, SOLS fixative

fixatif *m* POLL ATM, SOLS fixative

fixation: **~ de l'azote** *f* POLL EAU nitrogen fixation

fixe *adj* DÉCHETS *matériel, source*, POLL ATM stationary

flacon *m* ÉCHANT flask; **~ à distillats** ÉCHANT distillate flask; **~ de Drechsel** ÉCHANT *lavage des gaz* Drechsel gas-washing bottle; **~ d'échantillonnage** ÉCHANT sampling bottle; **~ de Friedrich** ÉCHANT *lavage des gaz* Friedrich's gas-washing bottle; **~ de laboratoire à col large** ÉCHANT wide-mouthed laboratory bottle; **~ de lavage** ÉCHANT wash bottle, washing bottle; **~ mélangeur** ÉCHANT volumetric flask; **~ à mesurer la densité** ÉCHANT density bottle; **~ en verre** ÉCHANT glass bottle

flamme *f* ÉCHANT, POLL ATM flame

flammèche *f* POLL ATM flying spark

fléaux: **~ de la pollution** *m pl* POL ENV pollution plagues

fleur: **~ d'eau** *f* DIST EAU, POLL EAU, POLL MAR plankton bloom

fleuve *m* DIST EAU river

flexible *adj* DÉCHETS, POLL MAR, POLL SON flexible

floc *m* DIST EAU, POLL EAU floc; **~ zoogléal** POLL EAU *épuration biologique des eaux d'égout* zoogleal layer

flocon: **~ de filtration** *m* POLL EAU filtrable floc; **~ fin** *m* POLL EAU fine floc

floconnage *m* DÉCHETS *peinture* floc spraying

floconneux *adj* POLL ATM flaky

floculant *m* DIST EAU, POLL EAU flocculant

floculat *m* DIST EAU, POLL EAU flocculate

floculateur *m* DIST EAU, POLL EAU flocculator, flocculator basin; **~ à mélange hydraulique** POLL EAU hydraulically-mixed flocculator; **~ à mélangeur cylindrique** POLL EAU flocculator with cylindrical mixer; **~ à pales mécaniques** POLL EAU flocculator with mechanically-operated paddles

floculation *f* DIST EAU, POLL EAU flocculation; **~ péricinétique** DIST EAU perikinetic flocculation

floculer *vi* DIST EAU, POLL EAU flocculate

floculeux *adj* DIST EAU, POLL EAU flocculent

flot *m* DIST EAU flow

flottage *m* POLL EAU *bois* floating

flottant *adj* DIST EAU, POLL MAR, POLL EAU floating

flottation *f* DÉCHETS *épuration des eaux* flotation, POLL EAU *traitement des eaux usées* flotation, floating, POLL MAR flotation; **~ avec aération immergée** POLL EAU flotation with submerged aerator; **~ avec filtration intégrée** POLL EAU flotation with integrated filtration; **~ gravimétrique** DIST EAU sink-float flotation; **~ par air dissous** POLL EAU pressure flotation; **~ par écumage** SOLS foam flotation; **~ par insufflation d'air** POLL EAU dissolved air flotation; **~ sous vide** POLL EAU vacuum flotation

flotte *f* POL ENV fleet; **~ captive** POL ENV dedicated fleet

fluctuation: **~ climatique** *f* POLL ATM climate fluctuation; **~ de concentration** *f* POLL ATM concentration fluctuation; **~ de débit** *f* POLL ATM throughput fluctuation, thruput fluctuation (*AmE*); **~ temporelle** *f* POLL ATM temporal fluctuation

fluide: **~ caloporteur** *m* ÉN RENOUV heat transfer medium liquid coolant, storage medium; **~ caloporteur en phase gazeuse** *m* ÉN RENOUV gaseous working medium; **~ hydraulique** *m* ÉN RENOUV hydraulic fluid; **~ de mesure** *m* ÉCHANT measuring fluid; **~ moteur** *m* DIST EAU pump fluid; **~ réfrigérant** *m* DIST EAU cooling medium

fluidifiant *m* DÉCHETS, DIST EAU plasticizer

fluidiser *vt* POLL ATM, SOLS fluidize

fluor *m* (*F*) SOLS fluorine (*F*)

fluoration *f* DIST EAU fluorination; **~ de l'eau potable** DIST EAU fluoridation of drinking water

fluorescence *f* ÉCHANT fluorescence; **~ atomique** ÉCHANT atomic fluorescence

fluorescent *adj* ÉCHANT fluorescent

fluorimétrie *f* ÉCHANT fluorimetry

fluorimétrique *adj* ÉCHANT *dosage* fluorimetric

fluorure *m* ÉCHANT fluoride; **~ d'hydrogène** POLL ATM hydrogen fluoride

flux *m* DIST EAU *de la marée* flow, ÉN RENOUV flood tide; **~ de chaleur** POLL ATM heat flow; **~ de condensats** POLL ATM condensate flow; **~ de déchets** DÉCHETS waste stream, flow of waste, DIST EAU residue stream; **~ de déchets urbains** DIST EAU municipal waste stream (*MWS*); **~ de dilution** POLL ATM bypass flow; **~ d'énergie inépuisable** ÉN RENOUV inexhaustible energy stream; **~ entrant** DIST EAU, POLL EAU influent; **~ de gaz** POLL ATM gas flow, gas stream; **~ de gaz brûlés** POLL ATM flue-gas stream; **~ de gaz chargé** POLL ATM charged-gas stream; **~ de gaz chargé de particules** POLL ATM particulate-laden gas stream; **~ gazeux contaminant** POLL ATM contaminant gas stream; **~ de gaz à grande vitesse** POLL ATM high-velocity gas stream; **~ de gaz inerte** POLL ATM inert-gas flow; **~ de gaz pollué** POLL ATM waste-gas flow; **~ de gaz principal** POLL ATM main gas stream; **~ de gaz traité** POLL ATM process-gas flow; **~ de matériaux industriels** POLL ATM process materials stream; **~ de matières énergétiques** ÉN RENOUV energy material stream; **~ non radial** POLL ATM, POLL EAU nonradial flow; **~ radial** POLL ATM, POLL EAU radial flow; **~ thermique** POLL ATM heat flow

focalisation *f* ÉN RENOUV focusing; **~ isoélectrique capillaire** ÉCHANT capillary isoelectric focusing (*CIEF*)

foire: **~ aux produits de l'environnement** *f* GESTION environmental trade fair

fonction[1] *f* SÉCURITÉ function; **~ d'étalonnage d'une analyse** ÉCHANT analysis calibration function; **~ de protection** POLL ATM protective function; **~ de service** GESTION *d'un produit* user-related function; **~ du sol** SOLS soil function; **~ de stockage** ÉN RENOUV storage function

fonction[2]: **en ~ de la consommation** *loc* GÉNÉRAL use-related; **en ~ d'un incident** *loc* ÉCHANT incident-dependent; **en ~ du marché** *loc* GESTION market-dependent; **en ~ du**

site *loc* ÉN RENOUV site-dependent; **en ~ du temps** *loc* ÉCHANT *réaction, phénomène,* ÉN RENOUV time-dependent; **en ~ d'un usage** *loc* GÉNÉRAL use-related

fonctionnement[1]: **~ en continu** *m* ÉCHANT on-line operation; **~ d'une pompe alimentée par un château d'eau** *m* DIST EAU pump operation via water tower

fonctionnement[2]: **à ~ hydraulique** *loc* DIST EAU, ÉN RENOUV, POLL EAU hydraulically-operated

fond: **~ conique** *m* POLL ATM conical base; **~ de cuve** *m* POLL EAU tank bottom; **~ de décharge** *m* DÉCHETS landfill bottom; **~ distributeur** *m* POLL ATM distributing bottom; **~ de lit bactérien** *m* POLL EAU trickling filter bottom; **~ de lit fluidisé** *m* POLL ATM fluidized-bed bottom; **~ marin** *m* POLL MAR ocean bed, ocean floor; **~ océanique** *m* POLL MAR ocean bed, ocean floor; **~ perforé** *m* POLL ATM *tamis* sieve plate; **~ perforé à maille fine** *m* POLL ATM *tamis* close-meshed sieve plate; **~ de réacteur** *m* POLL EAU reactor bottom; **~ de rivière** *m* POLL EAU river floor; **~ de sac** *m* POLL ATM bag bottom; **~ vaseux** *m* POLL EAU muddy ground

fondamental *adj* GÉNÉRAL basic

fondant *m* DIST EAU *soudage* fluxing agent

fondation: **~ d'une décharge** *f* DÉCHETS landfill foundation

fondeuse: **~ en continu** *f* DIST EAU continuous casting machine

fondre[1] *vt* DÉCHETS, DIST EAU, POLL EAU melt

fondre[2] *vi* DÉCHETS, DIST EAU, POLL EAU melt

fondrière *f* SOLS bog

fonds: **~ pour l'environnement** *m* POL ENV environment fund

Fonds: **~ pour l'environnement mondial** *m* POL ENV Global Environment Facility (*GEF*)

fongicide: **~ métallique** *m* SOLS metal fungicide

fonte *f* DÉCHETS, DIST EAU, POLL EAU melting; **~ de ferraille** DÉCHETS scrap smelting

forage *m* ÉCHANT sounding boring, ÉN RENOUV, SOLS drilling; **~ d'injection** DIST EAU, SOLS injection well; **~ par injection** SOLS injection drilling; **~ à la tarière** SOLS hand augering; **~ au trépan-benne** SOLS grab drilling

force: **~ d'adhésion** *f* POLL ATM adhesive force; **~ d'attraction** *f* ÉN RENOUV, POLL EAU force of attraction; **~ d'auto-épuration** *f* DIST EAU, SOLS self-purification force; **~ centrifuge** *f* ÉN RENOUV, POLL ATM centrifugal force; **~-chaleur** *f* DÉCHETS, POLL ATM combined heat and power (*CHP*); **~ de cisaillage** *f* POLL ATM, POLL EAU shearing force; **~ de cisaillement** *f* POLL ATM, POLL EAU shearing force; **~ de conversion** *f* POLL ATM conversion force; **~ de Coriolis** *f* POLL ATM Coriolis force; **~ de dispersion** *f* POLL ATM dispersion force; **~ d'écoulement** *f* POLL ATM flow force; **~ électrique** *f* POLL ATM electrical force; **~ de friction** *m* POLL ATM frictional force; **~ de gravitation** *f* DIST EAU, ÉN RENOUV, POLL EAU gravitational force; **~ hydraulique** *f* ÉN RENOUV water power; **~ d'inertie** *f* POLL ATM inertial force; **~ d'interaction** *f* POLL ATM interaction force; **~ moléculaire** *f* POLL ATM molecular force; **~ de pressage** *f* DIST EAU pressure force; **~ de propulsion** *f* POLL ATM propelling force; **~ relative** *f* POLL ATM relative force; **~ de répulsion** *f* DIST EAU, POLL EAU repulsive power; **~ de répulsion électrostatique** *f* DIST EAU, POLL EAU electrostatic repulsive force; **~ de serrage** *f* DÉCHETS, DIST EAU clamping force; **~ utile** *f* ÉN RENOUV *éolienne* lift; **~ de valence** *f* POLL ATM, POLL EAU valence force; **~ de van der Waals** *f* DIST EAU, POLL ATM *attraction moléculaire* van der Waals force

forer *vi* DIST EAU, ÉN RENOUV, SOLS drill, *pour extraire de l'eau* bore

foret *m* ÉN RENOUV, SOLS drill

forêt: **~ à usage énergétique** *f* ÉN RENOUV energy forestry

foreuse *f* DÉCHETS bore machine

formage *m* DÉCHETS forming

formaldéhyde *m* POLL ATM formaldehyde

formation: **~ d'azote** *f* POLL ATM nitrogen formation; **~ de boue** *f* DIST EAU sludge formation; **~ de brouillard** *f* POLL ATM fog formation; **~ de condensats** *f* POLL ATM condensate formation; **~ en couches** *f* POLL ATM layer formation; **~ de déchets** *f* DÉCHETS waste formation; **~ d'un dépôt** *f* POLL ATM deposit build-up; **~ d'écume** *f* POLL EAU scum formation; **~ à l'environnement** *f* POL ENV environmental education; **~ excessive de dépôt** *f* POLL ATM excessive deposit build-up; **~ de floc** *f* POLL EAU floc formation; **~ de fumée** *f* POLL ATM smoke build-up; **~ de gaz** *f* DÉCHETS gas build-up, gas evolution, gas formation; **~ de gaz de décharge** *f* DÉCHETS landfill gas build-up, landfill gas development; **~ de gaz à effet de serre** *f* POLL ATM greenhouse gas build-up; **~ de gypse** *f* POLL ATM gypsum formation; **~ d'humus** *f* DIST EAU, POLL EAU humus formation; **~ d'une masse humique** *f* DIST EAU, POLL EAU humic mass formation; **~ de méthane** *f* DÉCHETS methane build-up; **~ de mousse** *f* DÉCHETS foam formation; **~ d'ozone** *f* POLL ATM ozone formation; **~ de poussière** *f* POLL ATM dust build-up; **~ de radicaux** *f* POLL ATM radical formation; **~ stratifiée** *f* POLL ATM layer formation

forme[1]: **~ d'énergie consommable** *f* ÉN RENOUV usable energy form; **~ d'énergie renouvelable** *f* ÉN RENOUV renewable-energy form; **~ du littoral** *f* ÉN RENOUV coastline form

forme[2]: **en ~ de poche** *loc* POLL ATM pocket-shaped; **en ~ de sac** *loc* POLL ATM bag-shaped; **en ~ de Venturi** *loc* POLL ATM Venturi-shaped

fort: **~ pouvoir calorifique** *m* POLL EAU high calorific value

forte[1]: **~ charge en gaz brut** *f* DÉCHETS high raw gas load; **~ concentration** *f* DIST EAU, ÉCHANT, POLL ATM high concentration; **~ masse volumique apparente après décantation** *f* ÉCHANT high-settled apparent density

forte[2]: **à ~ intensité** *loc* POLL EAU high-rate; **à ~ intensité énergétique** *loc* DIST EAU, ÉN RENOUV, POLL ATM *procédé* energy-intensive; **à ~ intensité de main-d'œuvre** *loc* DIST EAU, POLL ATM, POLL MAR, POLL SON, SOLS labor-intensive (*AmE*), labour-intensive (*BrE*); **à ~ intensité odorante** *loc* POLL ATM odor-intensive (*AmE*), odour-intensive (*BrE*); **à ~ intensité sonore** *loc* POLL MAR, POLL SON noise-intensive; **à ~ odeur** *loc* SOLS odor-intensive (*AmE*), odour-intensive (*BrE*); **à ~ pression** *loc* ÉCHANT, POLL ATM, POLL EAU, POLL MAR high-pressure

fortement: **~ absorbable** *loc* POLL ATM, POLL SON highly-absorbable; **~ contaminé** *adj* DÉCHETS, POLL EAU, SOLS highly-contaminated; **~ pollué** *adj* DÉCHETS, POLL EAU, SOLS highly-polluted; **~ radioactif** *adj* ÉN RENOUV highly-radioactive

fossé *m* DIST EAU ditch, POLL MAR *relief sous-marin* trench; **~ de drainage** DIST EAU drainage ditch; **~ d'infiltration** DIST EAU leaching trench; **~ d'oxydation** DÉCHETS oxidation ditch; **~ périphérique** SOLS peripheral ditch; **~ de réception** DIST EAU catchwater drain

fosse *f* DIST EAU, POLL EAU, SOLS pit; **~ d'aisance** POLL EAU cesspool; **~ à boue fraîche** DIST EAU fresh sludge sump, fresh sludge pit; **~ à boues** POLL EAU sludge hopper; **~ de décantation** POLL EAU cesspit; **~ Dortmund** POLL EAU Dortmund tank; **~ Emscher** POLL EAU *épuration des eaux* Emscher-type settling tank; **~ Imhoff** POLL EAU Imhoff tank; **~ à lisier** ÉN RENOUV slurry pit; **~ de sédimentation** DIST EAU settling basin; **~ septique** DIST EAU, POLL EAU digestion tank, septic tank

four: **~ à bande** *m* SOLS band oven; **~ chauffé** *m* ÉCHANT heated kiln; **~ à colonne chauffante** *m* ÉCHANT heated-column oven; **~ de contact** *m* POLL ATM contact reactor; **~ à cuve** *m* POLL EAU shaft furnace; **~ cyclone** *m* SOLS cyclone furnace; **~ à dôme** *m* POLL ATM cupola blast heater; **~ fixe à lit fluidisé** *m* DÉCHETS, POLL ATM, POLL EAU, SOLS stationary fluidized-bed oven; **~ de frittage** *m* SOLS sinter oven; **~ en graphite** *m* ÉCHANT graphite furnace (*GF*); **~ à infrarouge** *m* DIST EAU, ÉCHANT infrared furnace, infrared kiln; **~ à lit fixe** *m* POLL EAU fixed-bed furnace; **~ à lit fluidisé** *m*

DÉCHETS, POLL ATM, POLL EAU, SOLS fluidized-bed furnace, fluidized-bed oven; ~ **pour infrarouge** *m* infrared through-type furnace; ~ **à résistance** *m* ÉCHANT resistance kiln; ~ **rotatif** *m* DIST EAU, POLL ATM, POLL EAU, SOLS rotary furnace, rotary kiln; ~ **rotatif à chauffage direct** *m* SOLS direct-heated rotary kiln; ~ **rotatif à chauffage indirect** *m* SOLS indirect-heated rotary kiln; ~ **sécheur rotatif** *m* POLL EAU rotary kiln; ~ **à soles étagées** *m* POLL EAU tiered drier; ~ **à tambour rotatif** *m* POLL EAU rotary drum furnace; ~ **tunnel** *m* DÉCHETS heating tunnel

fourneau: ~ **à cuve** *m* DIST EAU, POLL ATM shaft furnace

fournir *vt* DIST EAU, ÉN RENOUV, POLL ATM, POLL EAU supply

fournisseur *m* DIST EAU, ÉN RENOUV, POLL ATM, POLL EAU supplier, deliverer

fourniture *f* DIST EAU, ÉN RENOUV, POLL ATM, POLL EAU supply; ~ **collective d'eau** DIST EAU public water supply; ~ **collective d'eau potable** DIST EAU public drinking-water supply; ~ **d'eau** DIST EAU water supply; ~ **d'eau en milieu rural** DIST EAU rural water supply; ~ **d'eau dans le secteur résidentiel** DIST EAU domestic water supply; ~ **d'électricité** ÉN RENOUV power supply; ~ **d'énergie** ÉN RENOUV energy supply; ~ **d'énergie classique** ÉN RENOUV conventional energy supply; ~ **d'énergie à partir de sources renouvelables** ÉN RENOUV energy supply based on renewable energy sources; ~ **d'énergie renouvelable** ÉN RENOUV renewable energy supply; ~ **d'hydroélectricité** ÉN RENOUV hydropower supply; ~ **de nutriments** POLL ATM nutriment supply; ~ **sans pression** DIST EAU unpressurized supply

fourreau *m* DIST EAU *tuyauterie* pipe sleeve, POLL EAU *d'une pompe* priming pipe

foyer: ~ **avec grille à chaîne à soufflage sous grille** *m* DIST EAU underblast chain grate stoker; ~ **à étage** *m* POLL ATM two-stage fired heater

fraction *f* GÉNÉRAL fraction; ~ **humide des déchets** DÉCHETS wet fraction of waste; ~ **de matériaux de récupération** DIST EAU secondary resource fraction; ~ **de poussière** POLL ATM dust fraction; ~ **roulante** DÉCHETS rolling fraction; ~ **surdimensionnée** DIST EAU oversize fraction; ~ **tamisable** DIST EAU sieve-through fraction; ~ **de tamisage** DIST EAU sieve fraction; ~ **triée** DÉCHETS sorted fraction; ~ **utilisable** DIST EAU usable fraction

fractionnement *m* GÉNÉRAL fractionation, fraction separation

fractionner *vt* GÉNÉRAL fractionate, break up

fracturation *f* SOLS fracturing

fracturer *vt* SÉCURITÉ *coffre, serrure* break open

fragile *adj* POL ENV *paquets et colis* handle with care

fragilité *f* POLL ATM brittleness

fragmenter *vt* POLL ATM break, *chimie* break down

fragmentographie: ~ **de masse** *f* ÉCHANT mass fragmentography

fragments: ~ **de floc** *m pl* POLL EAU floc fragments

frange: ~ **littorale** *f* POLL MAR littoral fringe

frappage *m* POLL ATM knocking, rapping; ~ **des électrodes** POLL ATM *dépoussiérage* rapping of electrodes

frapper *vt* POLL ATM knock

frein: ~ **de rotor** *m* ÉN RENOUV rotor brake

fréon: ~ **12** *m* POLL ATM freon 12

fréquence *f* GÉNÉRAL frequency; ~ **d'analyse** ÉCHANT analysis frequency; ~ **de courant triphasé** ÉN RENOUV rotary-current frequency; ~ **cumulée** ÉCHANT cumulative frequency; ~ **de détection** ÉCHANT detection frequency; ~ **d'échantillonnage** ÉCHANT sampling frequency; ~ **électrique** ÉN RENOUV power frequency; ~ **de transfert** DÉCHETS transfer frequency; ~ **des vagues** ÉN RENOUV wave frequency

friabilité *f* POLL ATM brittleness

friche: ~ **industrielle** *f* DÉCHETS scrapped plant

friction *f* GÉNÉRAL, POLL ATM friction; ~ **de particules** POLL ATM particle friction

frigorigène *m* DIST EAU refrigerant

frittage *m* DÉCHETS, POLL ATM calcination, SOLS sintering

fritter *vt* DÉCHETS, POLL ATM calcinate, SOLS sinter

front: ~ **d'air** *m* POLL ATM air front; ~ **de filtration** *m* DIST EAU filtration front; ~ **de taille** *m* DÉCHETS *mines* working face

frontal *m* SÉCURITÉ front-end display

frotter *vt* POLL ATM scrub

fuel-oil *m* POLL ATM fuel oil

fuir *vi* GÉNÉRAL leak

fuite *f* DÉCHETS, DIST EAU, POLL EAU, POLL MAR, SÉCURITÉ leak, leakage, seepage, SOLS seepage; ~ **acoustique** POLL SON acoustical leak; ~ **de gaz** POLL ATM gas leak; ~ **de substances nocives** SOLS escape of harmful substances; ~ **sur une tuyauterie** POLL EAU pipe bursting; ~ **de tuyauterie** POLL EAU pipe leakage

fumard: ~ **acide** *m* POLL ATM acid smog

fumée *f* POLL ATM, SÉCURITÉ smoke; ~ **d'échappement** POLL ATM exhaust smoke; ~ **équivalente standard** POLL ATM equivalent standard smoke

fumées *f pl* POLL ATM, POLL EAU, SÉCURITÉ emissions, fumes

fumerolle *f* ÉN RENOUV fumarole

fumeron *m* POLL ATM smut

fumigation *f* POLL ATM fumigation

fumimètre *m* ÉCHANT smoke density meter

fusée: ~ **sonde** *f* ÉCHANT rocket probe

fusion *f* DÉCHETS, DIST EAU melting, POLL ATM fusion, ÉN RENOUV, POLL EAU melting; ~ **de ferraille** DÉCHETS scrap smelting

fût *m* DÉCHETS *pour déchets nucléaires* canister, POLL EAU *de stockage*, POLL MAR drum; ~ **allégé** DÉCHETS, DIST EAU *en contre-plaqué* fiber drum (*AmE*), fibre drum (*BrE*); ~ **central d'alimentation** DIST EAU center feed column (*AmE*), centre feed column (*BrE*), influent pipe

G

gabare *f* POLL MAR lighter
gain: **~ d'énergie** *m* ÉN RENOUV energy gain
gaine *f* DIST EAU cover; **~ pour bouteille** DÉCHETS bottle jacket
galerie: **~ filtrante** *f* DIST EAU, POLL EAU filter gallery, infiltration gallery
galvanisation *f* DÉCHETS electroplating
galvaniser *vt* DÉCHETS electroplate
galvanomètre: **~ enregistreur** *m* ÉCHANT galvanometer
gamme *f* ÉCHANT *colorimétrie* range
garantie *f* SÉCURITÉ safeguard
garde: **~ forestier** *m* GESTION, SÉCURITÉ forest ranger
garni *adj* DIST EAU *colonne* packed
garnir *vt* POLL ATM lag, pack; **~ d'un revêtement étanche** SOLS lag, line
garnissage *m* POLL ATM lagging, packing; **~ en PVC** DIST EAU PVC insert fitment
garniture *f* DIST EAU armature, packing; **~ d'étanchéité** DIST EAU packing; **~ intérieure** DÉCHETS interior packing; **~ mobile** POLL ATM moveable fittings; **~ de presse-étoupe** DIST EAU gland packing, packing; **~ de tuyauterie** DIST EAU pipe fitting
gasohol *m* ÉN RENOUV gasohol
gaspillage: **~ d'énergie** *m* GÉNÉRAL waste of energy
gaspiller *vt* GÉNÉRAL waste
gâteau: **~ de boue** *m* POLL EAU sludge cake; **~ de filtration** *m* POLL ATM filter cake
gaz[1] *m* GÉNÉRAL gas; **~ acide** POLL ATM acid gas; **~ acide inorganique** POLL ATM inorganic acid gas; **~ acide organique** POLL ATM organic acid gas; **~ adsorbé** POLL ATM adsorbed gas; **~ d'allumage** POLL ATM ignition gas; **~ asphyxiant** DÉCHETS, DIST EAU, POLL ATM asphyxiating gas; **~ atmosphérique** POLL ATM atmospheric gas; **~ auxiliaire** ÉCHANT auxiliary gas; **~ de balayage** POLL ATM scavenging gas; **~ brûlable** POLL ATM burnable gas; **~ brûlé** DIST EAU, POLL ATM, POLL EAU exhaust gas, flue gas; **~ de calibration** POLL ATM *mesure* calibrating gas; **~ de carter** POLL ATM *véhicule à moteur* blowby gas; **~ chargé de particules** POLL ATM particle-laden waste gas; **~ chargé de poussière** POLL ATM dust-laden gas; **~ chaud** POLL ATM, SOLS hot gas; **~ chaud d'échantillonnage** ÉCHANT hot sampling gas; **~ de cheminée** POLL ATM stack gas; **~ de comburant** DIST EAU, POLL ATM combustion gas, exhaust gas, flue gas; **~ combustible** DÉCHETS, POLL ATM fuel gas; **~ de combustion** POLL ATM waste gas; **~ corrosif** POLL ATM corrosive gas; **~ de craquage** POLL ATM reformed gas; **~ de curage** POLL EAU fermentation gas, sludge gas; **~ de décharge** POLL ATM landfill gas; **~ dépoussiéré** POLL ATM dust-free gas, dust-removed dirty gas, scrubbed gas; **~ de digesteur** POLL EAU digester gas; **~ de digestion** DIST EAU digester gas, POLL EAU fermentation gas; **~ de distillation** POLL ATM distillation gas; **~ échantillon** ÉCHANT sample gas; **~ d'échantillonnage** ÉCHANT sampling gas; **~ d'échantillonnage pollué** ÉCHANT contaminated sampling gas; **~ d'échappement** DÉCHETS offgas, DIST EAU exhaust gas, offgas, POLL ATM exhaust, exhaust gas, offgas, waste gas; **~ d'échappement diesel** POLL ATM diesel exhaust gas; **~ d'échappement de véhicule à moteur** POLL ATM motor vehicle exhaust gas; **~ à effet de serre** POLL ATM greenhouse gas; **~ épuré** DÉCHETS, POLL ATM clean gas, cleaned gas; **~ d'essai** ÉCHANT test gas; **~ d'étalonnage** ÉCHANT, POLL ATM *mesure* calibration gas, calibrating gas; **~ à l'état de trace** POLL ATM trace gas; **~ à l'état de trace ayant un effet sur le climat** POLL ATM climate-relevant trace gas; **~ à l'état de trace dans l'atmosphère** POLL ATM atmospheric trace gas; **~ explosif** POLL ATM explosive gas; **~ d'extraction** ÉCHANT extraction gas; **~ de fermentation** DÉCHETS fermentation gas; **~ de fumée** POLL ATM flue gas; **~ de fumier** DÉCHETS manure gas; **~ halogéné** POLL ATM halogenated gas; **~ de houille** POLL ATM coal gas; **~ incolore** POLL ATM colorless gas (*AmE*), colourless gas (*BrE*); **~ industriel** POLL ATM manufactured gas; **~ industriel pollué** POLL ATM polluted process gas, process waste gas; **~ inerte** POLL ATM inert gas; **~ inflammable** POLL ATM flame gas; **~ inodore** POLL ATM odorless gas (*AmE*), odourless gas (*BrE*); **~ inorganique** POLL ATM inorganic gas; **~ lacrymogène** POLL ATM tear gas; **~ lavé** DIST EAU scrubbed gas; **~ mélangé** POLL ATM mixed gas; **~ de mesure** ÉCHANT measuring gas; **~ migrateur** DÉCHETS migrating gas; **~ naturel** DIST EAU natural gas; **~ naturel comprimé** POLL ATM compressed natural gas (*CNG*); **~ naturel liquéfié** (*GNL*) POLL ATM liquefied natural gas (*LNG*); **~ naturel de synthèse** ÉN RENOUV synthetic natural gas (*SNG*); **~ naturel véhicule** ÉN RENOUV natural gas for vehicles; **~ nocif** POLL ATM harmful gas; **~ odorant à traiter** POLL ATM odor-laden dirty gas (*AmE*), odour-laden dirty gas (*BrE*); **~ organique** POLL ATM organic gas; **~ parfait** POLL ATM ideal gas, perfect gas; **~ de pétrole liquéfié** (*GPL*) ÉN RENOUV, POLL ATM liquefied petroleum gas (*LPG*); **~ pollué** POLL ATM dirty gas; **~ porteur** ÉCHANT, POLL ATM carrier gas; **~ poussiéreux** POLL ATM dust-laden gas; **~ préalablement nettoyé** POLL ATM precleaned gas; **~ propre** POLL ATM clean gas; **~ de pyrolyse** DIST EAU pyrolysis gas; **~ de queue** POLL ATM tail gas; **~ radioactif** POLL ATM radioactive gas; **~ de raffinerie** POLL ATM refinery gas; **~ rare radioactif** POLL ATM radioactive rare gas; **~ de recyclage** POLL ATM recycling gas; **~ réducteur** POLL ATM reducing gas; **~ de référence** ÉCHANT reference gas, calibrating gas; **~ reformé** POLL ATM reformed gas; **~ de régénération** POLL ATM regenerative gas; **~ rejeté** POLL ATM waste gas; **~ résiduaire** POLL ATM residual gas, waste gas; **~ résiduel** DÉCHETS offgas, DIST EAU, POLL ATM offgas, residual gas; **~ riche désorbé** POLL ATM desorbed rich gas; **~ de rinçage** POLL ATM backflushing gas; **~ saturé** ÉCHANT, POLL ATM saturated gas; **~ secondaire** ÉCHANT auxiliary gas; **~ du sol** SOLS soil gas; **~ soluble** POLL ATM soluble gas; **~ sous forme de particules** POLL ATM particulate gas; **~ de stockage** DÉCHETS storage gas; **~ sulfureux** POLL ATM sulfurous gas (*AmE*), sulphurous gas (*BrE*); **~ de tête d'un dépoussiéreur par voie humide** POLL ATM wet-scrubbed waste top gas; **~ à traiter** POLL ATM dirty gas; **~ vecteur** ÉCHANT, POLL ATM carrier gas; **~ de ville** POLL ATM town gas; **~ de zéro** ÉCHANT calibration gas
gaz[2]: **~ et fumées** *loc* SÉCURITÉ gases and fumes
gazéificateur *m* ÉN RENOUV, POLL ATM gasifier
gazéification *f* ÉN RENOUV, POLL ATM gasification; **~ à haute température** DÉCHETS, DIST EAU, POLL EAU high-temperature gasification
gazéifier *vt* ÉN RENOUV, POLL ATM gasify
gazeux *adj* GÉNÉRAL gaseous
gazogène *m* DÉCHETS gas generator
gazomètre *m* DÉCHETS, ÉCHANT gas holder, gasometer, gas container, POLL EAU gas container, gasometer, SÉCURITÉ gas container

gazométrie *f* DÉCHETS, ÉCHANT gasometry
gazométrique *adj* DÉCHETS, ÉCHANT gasometric
gel: **~ de silice** *m* POLL ATM, POLL EAU silica gel
gélatine *f* DÉCHETS gelatine
gélifiant *m* POLL MAR gelling agent
GEMS *abrév* (*Système mondial de surveillance continue de l'environnement*) ÉCHANT, POL ENV GEMS (*Global Environment Monitoring System*)
gêne *f* GESTION discomfort; **~ olfactive** DÉCHETS, POLL ATM, POLL EAU odor annoyance (*AmE*), odour annoyance (*BrE*), odor nuisance (*AmE*), odour nuisance (*BrE*)
générateur: **~ asynchrone** *m* ÉN RENOUV induction generator; **~ d'ozone** *m* ÉCHANT ozone generator; **~ de vapeur** *m* (*GV*) DIST EAU, ÉN RENOUV steam generator; **~ de vapeur alimenté par des combustibles fossiles** *m* ÉN RENOUV fossil fuel-fired steam generator; **~ de vapeur chauffé à l'énergie solaire** *m* ÉN RENOUV solar-heated steam generator
génératrice *f* ÉN RENOUV generator; **~ diesel** ÉN RENOUV diesel generator
génétique *adj* SÉCURITÉ genetic
génie[1]: **de ~ civil** *adj* DÉCHETS, ÉN RENOUV structurally engineered
génie[2]: **~ civil** *m* GÉNÉRAL *ponts et chaussées* civil engineering, *bâtiments* structural engineering; **~ photovoltaïque** m (*PV*) ÉN RENOUV *discipline* photovoltaics (*PV*); **~ de protection de l'environnement** *m* GESTION environmental and safety engineering; **~ sanitaire** *m* GESTION, POL ENV sanitary engineering; **~ de la sécurité** *m* SÉCURITÉ safety engineering; **~ urbain** *m* GESTION, POL ENV municipal engineering
géothermie *f* ÉN RENOUV geothermics
géothermique *adj* ÉN RENOUV geothermal, geothermic
géré: **~ informatiquement** *adj* ÉCHANT software-controlled
germe: **~ de cristallisation** *m* POLL ATM crystallization germ; **~ pathogène** *m* POLL EAU pathogenic germ
gestion *f* GESTION management; **~ des crues** DIST EAU, ÉN RENOUV, GESTION, POLL MAR flood control; **~ de déchets** DÉCHETS, GESTION, POL ENV waste management; **~ de l'eau** DIST EAU, GESTION water management; **~ de l'eau de recirculation** DIST EAU, GESTION, POLL EAU recirculating water management; **~ des eaux superficielles** DIST EAU, GESTION surface water management; **~ des eaux urbaines** DIST EAU, GESTION urban water management; **~ de l'énergie** GESTION, ÉN RENOUV energy management; **~ de l'environnement** GESTION, POL ENV environmental administration, environmental management; **~ intégrée des déchets** DÉCHETS, GESTION integrated waste management (*IWM*); **~ à orientation écologique** GESTION environmentally-oriented management; **~ des ressources naturelles** GESTION natural resources management; **~ des risques** GESTION, SÉCURITÉ risk management; **~ tenant compte de l'environnement** GESTION ecology-oriented management; **~ visant à limiter la production de déchets** DÉCHETS, GESTION waste-economical planning
geyser *m* ÉN RENOUV geyser
GIE *abrév* (*groupe d'intérêt économique*) GESTION economic interest group
gisement *m* ÉN RENOUV deposit, POL ENV stock, SOLS deposit; **~ géothermique** ÉN RENOUV geothermal deposit
glacer *vt* DÉCHETS *papier* glaze
glaise *f* DIST EAU clay, loam
glissement: **~ de terrain** *m* SOLS landslide, landslip
gluant *adj* POLL ATM viscid, sticky
glutineux *adj* POLL ATM glutinous
glycolyse *f* DÉCHETS glycolysis
GNL *abrév* (*gaz naturel liquéfié*) POLL ATM LNG (*liquefied natural gas*)
goémon *m* POLL MAR seaweed
gonflable *adj* POLL ATM, POLL MAR inflatable
gonflement *m* POLL ATM *d'une chambre à air* inflation, POLL EAU *des boues* bulking, swelling, POLL MAR *d'un bateau* inflation
gonfler *vt* POLL ATM, POLL MAR inflate
goudron *m* POLL MAR tar
gouge *f* ÉCHANT scooper
goulotte *f* DIST EAU channel, trough, POLL ATM downspout; **~ basculante** POLL EAU tipping trough; **~ de chargement** DÉCHETS loading worm
gouttelette *f* POLL ATM droplet, liquid droplet; **~ de brume** POLL ATM mist drop; **~ cible** POLL ATM target droplet; **~ d'huile** POLL EAU oil droplet; **~ isolée** POLL ATM individual droplet; **~ de lavage** POLL ATM scrubbing liquid droplet; **~ de liquide entraînée** POLL ATM entrained liquid droplet; **~ de liquide en suspension** POLL ATM suspended liquid droplet
gouvernement *m* POL ENV government
GPL *abrév* (*gaz de pétrole liquéfié*) ÉN RENOUV, POLL ATM LPG (*liquefied petroleum gas*)
gradient: **~ barométrique** *m* SOLS pressure gradient; **~ géothermique** *m* ÉN RENOUV geothermal gradient; **~ de pression** *m* ÉCHANT pressure difference; **~ de pression hydrostatique** *m* DIST EAU, POLL EAU hydrostatic pressure difference; **~ de pression partielle** *m* ÉCHANT partial differential pressure, partial pressure difference; **~ sec adiabatique** *m* POLL ATM dry adiabatic lapse rate; **~ de température** *m* DIST EAU, POLL ATM, POLL EAU, POLL SON temperature gradient; **~ thermique adiabatique à sec** *m* POLL ATM dry adiabatic lapse rate; **~ thermométrique** *m* DIST EAU, POLL ATM, POLL EAU, POLL SON temperature gradient
graduation *f* ÉCHANT graduating, graduation, POLL ATM scaling
graduer *vt* ÉCHANT graduate, POLL ATM scale
grain *m* POLL ATM *catalyseur* pellet
grains: **~ de sable** *m pl* POLL ATM grit
graisin *m* DÉCHETS *recyclage du verre* cullet
graminée: **~ énergétique** *f* ÉN RENOUV grass for energy purposes
grand[1]: **~ teint** *adj* DÉCHETS *textile* colorfast (*AmE*), colourfast (*BrE*)
grand[2]: **~ ensemble** *m* POL ENV high-density housing
grande: **~ agglomération** *f* POL ENV conurbation; **~ sécurité** *f* SÉCURITÉ high security
granulat *m* DIST EAU, POLL ATM granulate; **~ de jets et débris de coulée** DIST EAU returns granulate
granulateur *m* DÉCHETS, DIST EAU granulator
granulation *f* DÉCHETS, DIST EAU, POLL EAU, SOLS granulation, granulating, pelletization, pelletizing
granulatoire *m* DÉCHETS granulating machine
granulatrice *f* DIST EAU pellet mill
granule *f* DÉCHETS granule
granulé *m* ÉN RENOUV pellet; **~ de paille** ÉN RENOUV straw pellet
granuler *vt* DÉCHETS, DIST EAU granulate, pelletize
granulométrie *f* DIST EAU, POLL ATM, POLL EAU particle size distribution, particle size separation, SOLS grain size distribution; **~ de la poussière** POLL ATM dust fineness
grappin *m* DÉCHETS clamshell, DIST EAU grab
grau *m* DIST EAU inlet
gravats *m pl* DÉCHETS, ÉN RENOUV building rubble, construction waste, demolition waste; **~ à faible teneur en chaux** DÉCHETS, DIST EAU low-lime content rubble
gravier *m* DÉCHETS, POLL ATM, POLL EAU *filtre à gravier* rubble, grit; **~ de construction routière** DÉCHETS road construction rubble; **~ filtrant** POLL EAU filter gravel
gravimétrique *adj* ÉCHANT gravimetric
gravité *f* GÉNÉRAL gravity
gray *m* (*gy*) ÉCHANT *radioprotection* gray (*Gy*)
grenailler *vt* DÉCHETS granulate
grenailleuse *f* DÉCHETS granulating machine

grenouillère *f* DIST EAU rose

grille *f* DIST EAU grate, screen, POLL ATM grate, screen, screening machine; **~ à barreaux** POLL ATM sieve grate; **~ basculante** DIST EAU rocking grate; **~ centrifuge autonettoyante** POLL EAU self-cleaning centrifugal screen; **~ de crible** POLL ATM sieve grate; **~ d'échantillonnage** SOLS sampling grid; **~ fixe horizontale** POLL EAU horizontal bar screen; **~ d'incinération** DÉCHETS incineration grate; **~ manuelle** POLL EAU manually-operated screen; **~ mécanique** DÉCHETS, DIST EAU, SOLS traveling grate (*AmE*), travelling grate (*BrE*); **~ de mélange** DÉCHETS mixing gate; **~ mobile** DIST EAU traveling grate (*AmE*), travelling grate (*BrE*); **~ à nettoyage automatique** POLL EAU mechanically-cleaned bar screen; **~ de retenue** POLL ATM retaining grate; **~ roulante** DIST EAU rolling grate; **~ sans fin** DÉCHETS, DIST EAU, SOLS traveling grate (*AmE*), travelling grate (*BrE*); **~ sèche** POLL EAU dry screen; **~ support** POLL ATM supporting grate

gros: **à ~ grain** *loc* POLL ATM, SOLS coarse

grossier *adj* POLL ATM, SOLS coarse

groupe: **~ cible** *m* POL ENV target group; **~ critique** *m* ÉCHANT critical group; **~ écologiste** *m* POL ENV environmental group; **~ fixe** *m* POLL EAU fixed group; **~ haute tension** *m* POLL ATM high-voltage aggregate; **~ d'intérêt économique** *m* (*GIE*) GESTION economic interest group; **~ de pression** *m* POL ENV lobby; **~ de pression écologiste** *m* POL ENV environmental lobby; **~ de protection de l'environnement** *m* POL ENV environmental group; **~ de travail chargé des questions d'environnement** *m* POL ENV environmentally-political working group; **~ turbo-alternateur d'éolienne** *m* ÉN RENOUV wind turbine generator (*WTG*)

grue *f* ÉN RENOUV crane; **~ pour conteneur** ÉN RENOUV container crane

guideau *m* POLL EAU guide wall

GV *abrév* (*générateur de vapeur*) DIST EAU steam generator

Gy *abrév* (*gray*) ÉCHANT *radioprotection* Gy (*gray*)

gypse *m* POLL ATM, SOLS gypsum, natural gypsum; **~ brut** POLL ATM raw gypsum; **~ réutilisable** POLL ATM reusable gypsum; **~ en suspension** POLL ATM gypsum suspension

gyromètre *m* ÉCHANT rotameter

H

H *abrév* (*hydrogène*) DÉCHETS, POLL ATM, POLL EAU H (*hydrogen*)
habillage: **~ antistatique** *m* SÉCURITÉ *contre les agressions chimiques* antistatic protective clothing
habitat *m* POL ENV habitat; **~ sous-marin** POLL MAR underwater habitat
habitudes: **~ de consommation** *f pl* GESTION consumer behavior (*AmE*), consumer behaviour (*BrE*), customer behavior (*AmE*), customer behaviour (*BrE*)
hall: **~ de stockage** *m* DÉCHETS tank room
halocline *m* POLL EAU halocline
halogénure *m* POLL ATM halide
haloïde *m* POLL ATM halide
halon *m* POLL ATM halon
hangar: **~ de débarquement** *m* POLL MAR shore reception facility; **~ de déchargement** *m* DIST EAU unloading hall
HAP *abrév* (*hydrocarbure aromatique polycyclique*) DIST EAU, POLL ATM PAH (*polycyclic aromatic hydrocarbon*)
harmonisation *f* POL ENV harmonization; **~ de la protection de l'environnement** POL ENV harmonization of environmental protection, harmonized environmental protection
harmonisé *adj* POL ENV harmonized
harmoniser *vt* POL ENV harmonize
hasardeux *adj* SÉCURITÉ unsafe
haut *m* POLL ATM top; **~ de l'eau** DIST EAU high water
haute[1]: **~ mer** *f* DIST EAU, POLL MAR high water; **~ sécurité** *f* SÉCURITÉ high security
haute[2]: **de ~ performance** *loc* DIST EAU, ÉCHANT, POLL ATM high-performance; **à ~ pression** *loc* ÉCHANT, POLL ATM, POLL EAU, POLL MAR high-pressure; **de ~ qualité** *loc* DIST EAU high-grade; **sous ~ pression** *loc* ÉCHANT, POLL ATM, POLL EAU, POLL MAR high-pressure; **à ~ température** *loc* DÉCHETS, DIST EAU, ÉN RENOUV, SOLS high-temperature
hautement: **~ toxique** *loc* DÉCHETS, POLL EAU, SOLS highly-toxic
hautes: **~ eaux** *f pl* DIST EAU high water
hauteur *f* DIST EAU lift, GÉNÉRAL height; **~ d'absorption** POLL ATM absorber head; **~ d'anémomètre** POLL ATM anemometer height; **~ d'aspiration** POLL EAU suction head, suction lift; **~ d'aspiration statique** POLL EAU static suction head, static suction lift; **~ de cheminée** POLL ATM chimney height, stack height; **~ de chute** DÉCHETS height of fall; **~ de chute totale** POLL EAU *barrage* total head; **~ de chute variable** DIST EAU changing head; **~ de couche filtrante** POLL ATM filter bed height; **~ de décharge** DÉCHETS dumping height; **~ dynamique** DIST EAU, POLL EAU *hydraulique* velocity head; **~ d'élévation** DIST EAU lift; **~ équivalente de perte de charge** POLL EAU friction head; **~ de garnissage** POLL ATM packed-bed height; **~ géométrique de refoulement** DIST EAU, POLL EAU discharge head; **~ manométrique** DIST EAU delivery head, discharge head, lift, pressure head, POLL ATM head, POLL EAU delivery head, discharge head, static head; **~ manométrique d'aspiration** POLL EAU manometric suction head; **~ manométrique de refoulement** POLL EAU manometric discharge head; **~ des marées** ÉN RENOUV tidal range; **~ nette positive d'aspiration** POLL EAU net positive suction height (*NPSH*); **~ piézométrique** POLL EAU static head; **~ de refoulement** DIST EAU, POLL EAU delivery head, discharge head; **~ de source** POLL ATM *de pollution* source height; **~ de tête** POLL ATM top height; **~ totale de cheminée** POLL ATM overall chimney height, overall stack height; **~ de tour** ÉN RENOUV tower height; **~ utile de cheminée** POLL ATM effective stack height
HCB *abrév* (*hexachlorobenzène*) SOLS HCB (*hexachlorobenzene*)
HCFC *abrév* (*hydrochlorofluorocarbure*) POLL ATM HCFC (*hydrochlorofluorocarbon*)
HCH *abrév* (*hexachlorocyclohexane*) SOLS HCH (*hexachlorocyclohexane*)
He *abrév* (*hélium*) POLL ATM He (*helium*)
hélice *f* DIST EAU, POLL MAR propeller; **~ à pas variable** POLL MAR variable-pitch propeller
hélicoïdal *adj* POLL ATM helical
héliostat *m* ÉN RENOUV heliostat, solar collector; **~ orientable** ÉN RENOUV tracking reflector
hélium *m* (*He*) POLL ATM helium (*He*)
hémioxyde: **~ d'azote** *m* (N_2O) POLL ATM nitrous oxide (N_2O)
heptane *m* POLL ATM heptane
heptaoxyde: **~ de soufre** *m* POLL ATM sulfur heptoxide (*AmE*), sulphur heptoxide (*BrE*)
herbage *m* SOLS grassland
herbicide *m* SOLS herbicide
hermétique *adj* GÉNÉRAL airtight, watertight, sealed
herse: **~ de tranquillisation** *f* DIST EAU stilling picket fence
hétérotrophe[1] *adj* DÉCHETS, POLL EAU heterotrophic
hétérotrophe[2] *m* DÉCHETS, POLL EAU heterotroph; **~ facultatif** POLL EAU facultative heterotroph
heurter: **se ~** *v réfl* POLL ATM collide, impinge
hexachlorobenzène *m* (*HCB*) SOLS hexachlorobenzene (*HCB*)
hexachlorocyclohexane *m* (*HCH*) SOLS hexachlorocyclohexane (*HCH*)
hexane *m* POLL ATM hexane
Hg *abrév* (*mercure*) POLL EAU Hg (*mercury*)
HNO_3 *abrév* (*acide nitrique*) POLL ATM, POLL ATM, SOLS HNO_3 (*nitric acid*)
homogène *adj* DÉCHETS, DIST EAU, POLL ATM, SOLS homogeneous
homogénéisation *f* DÉCHETS, DIST EAU, POLL ATM, SOLS homogenization
homogénéiser *vt* DÉCHETS, DIST EAU, POLL ATM, SOLS homogenize, mix thoroughly
homologation: **~ d'une installation** *f* POL ENV plant authorization
homopolaire *adj* POLL EAU homopolar
homopolymère *m* DÉCHETS homopolymer
horizon: **~ éluvial** *m* SOLS eluvial horizon; **~ de sol** *m* SOLS soil horizon
horizon-A *m* SOLS A-horizon
horizon-B *m* SOLS B-horizon
horizon-C *m* SOLS C-horizon
hotte *f* POLL ATM *d'un système de captage* hood; **~ de captage des buées** ÉN RENOUV de-mister unit
houille *f* POLL ATM coal; **~ blanche** ÉN RENOUV water power; **~ fuligineuse** POLL ATM soot coal
houle *f* ÉN RENOUV, POLL MAR swell
housse *f* DIST EAU jacket
housses: **~ et revêtements de palettes en polyéthylène** *m pl* DÉCHETS polyethylene pallet covers and liners
H_2SO_3 *abrév* (*acide sulfureux*) POLL ATM, POLL EAU, SOLS H_2SO_3 (*sulfurous acid AmE, sulphurous acid BrE*)
H_2SO_4 *abrév* (*acide sulfurique*) POLL ATM, POLL EAU, SOLS H_2SO_4 (*sulfuric acid AmE, sulphuric acid BrE*)

hublot *m* DÉCHETS inspection window
huile *f* GÉNÉRAL oil; ~ **alimentaire** ÉN RENOUV edible oil; ~ **de colza** POLL ATM rapeseed oil; ~ **combustible produite à partir des boues** POLL EAU oil from sludge (*OFS*); ~ **de coupe** DIST EAU, ÉN RENOUV cutting oil; ~ **désagrégée** POLL MAR weathered oil; ~ **émulsionnée** POLL EAU emulsified oil; ~ **épuisée** POLL EAU spent oil; ~ **lourde** POLL ATM heavy fuel; ~ **persistante** DIST EAU persistent oil; ~ **de pyrolyse** DIST EAU pyrolysis oil; ~ **de récupération** DIST EAU recovered oil; ~ **recyclée** DIST EAU recovered oil; ~ **résiduaire** DIST EAU used oil; ~ **résiduelle** DIST EAU residual oil; ~ **de synthèse** ÉN RENOUV synthetic oil; ~ **thermique** ÉN RENOUV thermal oil; ~ **usagée** DÉCHETS, SOLS waste oil; ~ **usée** DÉCHETS waste oil, DIST EAU used oil, SOLS waste oil
huileux *adj* ÉN RENOUV oil-containing, oily
humain *adj* ÉN RENOUV man-made
humide *adj* GÉNÉRAL humid, moist, damp, wet
humidifiable *adj* POLL ATM, SOLS wettable
humidificateur *m* POLL ATM humidifier, air humidifier
humidification *f* POLL ATM humidification, watering
humidifié *adj* POLL ATM, SOLS humidified, wetted
humidifier *vt* POLL ATM, SOLS humidify, wet
humidifuge *adj* DÉCHETS moisture-repellent
humidité *f* GÉNÉRAL humidity, moisture, dampness; ~ **de l'air** POLL ATM air humidity; ~ **des déchets** DÉCHETS waste dampness; ~ **d'une installation** DÉCHETS moisture of installation; ~ **relative** ÉCHANT relative humidity (*rH*); ~ **résiduelle** DIST EAU, ÉCHANT residual moisture
humification *f* DIST EAU, POLL EAU humification, humus formation
humus *m* DIST EAU, ÉN RENOUV, POLL EAU humus
hydraulique *adj* GÉNÉRAL hydraulic
hydrocarbure *m* DIST EAU hydrocarbon (*HC*), GÉNÉRAL oil, POLL ATM, POLL MAR hydrocarbon (*HC*); ~ **aliphatique** POLL ATM aliphatic hydrocarbon; ~ **aromatique** DIST EAU, POLL ATM aromatic hydrocarbon; ~ **aromatique polycyclique** (*HAP*) DIST EAU, POLL ATM polycyclic aromatic hydrocarbon (*PAH*); ~ **chloré** POLL ATM chlorinated hydrocarbon; ~ **chloré semi-volatil** POLL ATM, SOLS semivolatile chlorinated hydrocarbon; ~ **chloré volatile** POLL ATM, SOLS volatile chlorinated hydrocarbon; ~ **halogéné** DIST EAU, POLL ATM halogenated hydrocarbon; ~ **halogéné volatil** DIST EAU, POLL ATM volatile halogenated hydrocarbon; ~ **insaturé** POLL ATM unsaturated hydrocarbon; ~ **persistant** DÉCHETS persistent oil; ~ **polynucléaire aromatique** DIST EAU, POLL ATM polycyclic aromatic hydrocarbon (*PAH*); ~ **résistant aux hautes températures** ÉN RENOUV high-temperature-resistant oil; ~ **saturé** POLL ATM saturated hydrocarbon
hydrochlorofluorocarbure *m* (*HCFC*) POLL ATM hydrochlorofluorocarbon (*HCFC*)
hydrocinétique *adj* POLL EAU hydrokinetic
hydrocyclone *m* DÉCHETS, POLL ATM, POLL EAU, SOLS hydrocyclone
hydrodésulfuration *f* DIST EAU hydrodesulfurization (*AmE*), hydrodesulphurization (*BrE*)
hydrodynamique *adj* SOLS hydrodynamic
hydroélectricité *f* ÉN RENOUV hydroelectricity, hydropower
hydrofuge *adj* DÉCHETS, DIST EAU, POLL EAU moisture-proof, water-repellent, water-resistant
hydrogénation *f* DÉCHETS, DIST EAU, ÉN RENOUV, POLL ATM hydrogenation
hydrogène *m* (*H*) DÉCHETS, POLL ATM, POLL EAU hydrogen (*H*); ~ **fluoré** POLL ATM fluorohydrogen; ~ **sulfuré** DÉCHETS, DIST EAU hydrogen sulfide (*AmE*), hydrogen sulphide (*BrE*)
hydrogéner *vt* DÉCHETS, DIST EAU, ÉN RENOUV, POLL ATM hydrogenate
hydrogéologie *f* DIST EAU, SOLS hydrogeology
hydrogramme *m* DIST EAU hydrograph
hydrojecteur *m* DIST EAU jet pump
hydrologie *f* DIST EAU, POLL EAU hydrology; ~ **karstique** DIST EAU karst hydrology
hydrolyse *f* GÉNÉRAL hydrolysis; ~ **continue** SOLS continuous hydrolysis; ~ **discontinue** SOLS discontinuous hydrolysis
hydrolysé *adj* GÉNÉRAL hydrolysed (*BrE*), hydrolyzed (*AmE*)
hydrolyser *vt* GÉNÉRAL hydrolyse (*BrE*), hydrolyze (*AmE*)
hydromètre *m* DIST EAU, POLL EAU hydrometer
hydrométrie *f* DIST EAU, POLL EAU hydrometry
hydrophile *adj* POLL ATM, POLL EAU hydrophilic
hydrophobe *adj* POLL ATM, POLL EAU hydrophobic
hydrosphère *f* DIST EAU, SOLS hydrosphere
hydrostatique[1] *adj* DIST EAU, POLL EAU hydrostatic
hydrostatique[2] *f* DIST EAU, POLL EAU hydrostatics
hydrothermique *adj* ÉN RENOUV hydrothermal
hydrotimétrie *f* DIST EAU hardness determination
hygiène *f* SÉCURITÉ health; ~ **de l'environnement** POL ENV, SÉCURITÉ environmental health; ~ **du milieu** POL ENV, SÉCURITÉ environmental health; ~ **publique** DIST EAU, POL ENV, SÉCURITÉ public health, sanitation
hygiénique *adj* DIST EAU, SÉCURITÉ hygienic, sanitary
hygiénisation *f* DÉCHETS accelerated composting
hygiéniste *m* GESTION, POLL EAU hygienist
hygromètre *m* GÉNÉRAL hygrometer; ~ **capillaire** ÉCHANT hair hygrometer; ~ **à cheveu** ÉCHANT hair hygrometer; ~ **à point de gelée** ÉCHANT, POLL ATM freezing point hygrometer; ~ **à point de rosée** ÉCHANT dew point hygrometer
hygrométrie *f* GÉNÉRAL hygrometry, moisture measurement; ~ **électrique** ÉCHANT electrical moisture measurement
hygroscope *m* ÉCHANT, POLL ATM hygroscope
hygroscopicité *f* ÉCHANT, POLL ATM hygroscopicity
hygroscopique *adj* ÉCHANT, POLL ATM hygroscopic
hypersonique *adj* POLL SON hypersonic
hypochloration *f* POLL EAU hypochlorination
hypochlorite *f* DIST EAU chlorate
hypolimnion *m* POLL EAU *limnologie* hypolimnion

I

identification *f* ÉCHANT, SÉCURITÉ *de risques, dangers* identification; **~ de contenu** SÉCURITÉ identification of contents; **~ par code barre caché** DÉCHETS hidden bar code identification
identifier *vt* ÉCHANT, SÉCURITÉ identify
ignifuge *adj* DÉCHETS *peinture, revêtement*, SÉCURITÉ fireproof, fire-retardant
illicite *adj* POL ENV unlawful, illegal
imagerie: **~ détaillée** *f* ÉCHANT detailed mapping
imbibition *f* DIST EAU, ÉCHANT, POLL ATM, POLL EAU absorption
imbouchable *adj* DIST EAU, POLL EAU *pompe* nonclogging
imbroyé *adj* DÉCHETS unground
imbrûlés *m pl* DÉCHETS, DIST EAU combustion residue, burnt-out aggregate
immatriculation *f* DIST EAU registration
immédiat *adj* ÉCHANT on-the-spot
immergé *adj* DIST EAU, POLL EAU, POLL MAR underwater
immersion *f* DÉCHETS, POLL EAU, POLL MAR ocean disposal
immission: **~ acoustique** *f* POLL SON noise immission; **~ d'ozone** *f* POLL ATM ozone immission; **~ sonore** *f* POLL SON sound immission
immobile *adj* DÉCHETS stationary
immobilisation *f* DÉCHETS, ÉCHANT, POLL EAU, SOLS immobilization, fixation, fixing
immobilisé *adj* DÉCHETS, ÉCHANT, POLL EAU, SOLS immobilized, fixed
immobiliser *vt* DÉCHETS, ÉCHANT, POLL EAU, SOLS immobilize, fix
immondices *m pl* DÉCHETS litter
immuno-essai *m* ÉCHANT immunoassay
immunologie: **~ microbienne** *f* DIST EAU, SOLS microbial immunology
immunologique *adj* DIST EAU, SOLS immunological
impact *m* ÉCHANT, POLL ATM impact, impaction, impingement; **~ de particule** POLL ATM particle impact; **~ sanitaire** POL ENV health impact; **~ sur l'environnement** GÉNÉRAL environmental impact
impacteur *m* ÉCHANT, POLL ATM impinger, impactor
impédance: **~ d'électrode** *f* ÉCHANT electrode impedance
imperméabilisation *f* GÉNÉRAL sealing; **~ de la base** SOLS base sealing; **~ combinée de la base** SOLS combined base sealing; **~ des flancs** SOLS flanking sealing; **~ de fond** SOLS bottom sealing; **~ des talus** SOLS slope sealing
imperméabilisé *adj* GÉNÉRAL sealed
imperméabiliser *vt* GÉNÉRAL seal
imperméable *adj* DÉCHETS, DIST EAU, SÉCURITÉ, SOLS impermeable, impervious
implantation *f* ÉN RENOUV, SOLS plantation, setting
impluvium *m* DIST EAU rain area, precipitation area
importateur: **~ net** *m* DIST EAU net receiver
importation: **~ d'énergie** *f* ÉN RENOUV energy import
importer *vt* ÉN RENOUV *énergie* import
imposer *vt* POL ENV oblige
imposition: **~ différenciée** *f* POL ENV differential taxation; **~ d'une limite d'émissions** *f* POL ENV, POLL ATM emissions limitation
impossible: **~ à refondre** *adj* DÉCHETS, DIST EAU nonremeltable
impôt *m* POL ENV tax; **~ dégressif** POL ENV regressive tax; **~ neutre** POL ENV neutral tax; **~ progressif** POL ENV progressive tax; **~ réglementaire** POL ENV regulatory tax; **~ sur les déchets** POL ENV waste tax; **~ sur l'environnement** POL ENV eco-tax, green tax; **~ sur la protection climatique** POL ENV climate-protection tax; **~ vert** POL ENV green tax
imprégnation *f* DÉCHETS, ÉCHANT, POLL ATM impregnation; **~ catalytique** POLL ATM catalytic impregnation
imprégner *vt* DÉCHETS, ÉCHANT, POLL ATM impregnate
impulsion: **~ étalon** *f* ÉCHANT calibration pulse; **~ photonique** *f* ÉCHANT light-induced pulse
imputable: **~ aux activités humaines** *adj* POLL ATM, SOLS human-induced; **~ aux émissions** *adj* POLL ATM emissions-related
imputrescible *adj* DÉCHETS, DIST EAU, POLL EAU nonputrescible
incendie: **~ de forêt** *m* POLL ATM, SOLS forest fire
incertitude: **~ de mesure** *f* ÉCHANT uncertainty of measurement
inchangé *adj* ÉCHANT unchanged
incidence: **~ de l'ozone** *f* POLL ATM ozone impact
incinérateur *m* DÉCHETS, POLL ATM, SÉCURITÉ incinerator; **~ catalytique** POLL ATM catalytic incinerator; **~ de déchets** DÉCHETS garbage incinerator (*AmE*), refuse incinerator, waste incinerator; **~ à four rotatif** POLL ATM rotary-kiln incinerator; **~ à foyers multiples** DÉCHETS multiple-hearth incinerator; **~ pour gaz de fumées** POLL ATM fume incinerator; **~ d'ordures** DÉCHETS garbage incinerator (*AmE*), refuse incinerator, waste incinerator; **~ à ordures** DÉCHETS garbage incinerator (*AmE*), refuse incinerator; **~ thermique** POLL ATM thermal incinerator
incinération *f* DÉCHETS, POLL ATM, SÉCURITÉ incineration; **~ avec régénération** POLL ATM regenerative incineration; **~ des boues d'eaux résiduaires** POLL EAU wastewater sludge incineration; **~ des boues résiduaires** POLL EAU wastewater incineration; **~ des boues résiduaires dans des fours à charbon** POLL EAU incineration of wastewater sludge in coal furnaces; **~ des boues résiduaires dans des fours à ciment** POLL EAU incineration of wastewater sludge in cement furnaces; **~ des boues résiduaires dans des fours à lignite** POLL EAU incineration of wastewater sludge in lignite furnaces; **~ en boulettes** SOLS pellet incineration; **~ catalytique** POLL ATM catalytic incineration; **~ de déchets** DÉCHETS waste incineration, refuse incineration, garbage incineration (*AmE*), trash burning (*AmE*); **~ de déchets dangereux** DÉCHETS hazardous waste incineration; **~ de déchets en mer** DÉCHETS waste incineration at sea; **~ des eaux usées avec consommation de chaleur** POLL EAU wastewater incineration with heat use; **~ des eaux usées sans consommation de chaleur** POLL EAU wastewater incineration without heat use; **~ de gaz pollués** POLL ATM waste gas incineration; **~ à haute température** POLL ATM, POLL EAU high-temperature incineration; **~ hors site** SOLS off-site incineration; **~ en lit fluidisé** POLL EAU, SOLS fluidized-bed incineration; **~ d'ordures** DÉCHETS waste incineration, refuse incineration, garbage incineration (*AmE*), trash burning (*AmE*); **~ des ordures ménagères** DÉCHETS domestic refuse incineration, domestic waste burning, domestic waste incineration; **~ par étages** POLL ATM stages incineration; **~ sans récupération d'énergie** DÉCHETS incineration without energy recovery; **~ sans suie** POLL ATM soot-free incineration; **~ sur place** SOLS on-site incineration; **~ à température moyenne** DIST EAU, POLL ATM, POLL EAU mid-temperature incineration
incinéré *adj* DÉCHETS, POLL ATM, SÉCURITÉ incinerated

incinérer *vt* DÉCHETS, POLL ATM, SÉCURITÉ incinerate
incitation *f* GESTION, POL ENV incentive; **~ à l'économie de marché** GESTION, POL ENV market-economy incentive; **~ financière** GESTION, POL ENV financial incentive; **~ fiscale** GESTION, POL ENV tax incentive; **~ au recyclage** GESTION, POL ENV recycling incentive
incolore *adj* POLL ATM colorless (*AmE*), colourless (*BrE*)
incombustibilité *f* DÉCHETS, DIST EAU, POLL ATM, SÉCURITÉ incombustibility
incombustible *adj* DÉCHETS, DIST EAU, POLL ATM, SÉCURITÉ fireproof, incombustible
incomplet *adj* POLL ATM incomplete
incompressible *adj* POLL ATM incompressible, compression-proof
incorporé *adj* SOLS built-in
incorporer *vt* SOLS build in
incrustation *f* DIST EAU incrustation, scale, scaling
incubation *f* ÉCHANT incubation
indemnisation *f* POL ENV, SÉCURITÉ compensation
indemniser *vt* POL ENV, SÉCURITÉ compensate
indemnité *f* POL ENV, SÉCURITÉ compensation fee
indépendant *adj* POLL EAU separate
indicateur *m* GÉNÉRAL indicator, *isotopique, radioactif* tracer; **~ biologique** DIST EAU, ÉCHANT, SOLS biological indicator, bioindicator; **~ de couleur** ÉCHANT color indicator (*AmE*), colour indicator (*BrE*); **~ écologique** DIST EAU, POLL EAU, SOLS indicator organism, indicator species; **~ de fuite** SÉCURITÉ leakage indicator system; **~ de grisou** ÉCHANT, SÉCURITÉ fire damp indicator, gas indicator; **~ de niveau** ÉCHANT level indicator; **~ de niveau acoustique** ÉCHANT, POLL SON noise level indicator; **~ de niveau acoustique pondéré** ÉCHANT, POLL SON weighted noise level indicator; **~ de niveau d'eau** DIST EAU water gage (*AmE*), water gauge (*BrE*), water indicator, water level indicator; **~ de réaction** ÉCHANT reaction indicator; **~ de seuil acoustique** ÉCHANT, POLL SON noise limit indicator
indication: **~ de niveau sonore** *m* POLL SON *sur des matériels* noise labeling (*AmE*), noise labelling (*BrE*)
indice: **~ antidétonant** *m* POLL ATM antiknock value; **~ de boue** *m* POLL EAU sludge index; **~ de bruit** *m* POLL SON noise-guiding value; **~ de cétane** *m* POLL ATM cetane number; **~ de CO** *m* POLL ATM CO value; **~ de dureté** *m* POLL EAU degree of hardness; **~ d'émission** *m* POLL ATM emissions value; **~ d'émission des véhicules** *m* POLL ATM emissions value of cars; **~ estival d'ozone** *m* POLL ATM summertime ozone value; **~ foliaire** *m* POLL ATM leaf-area index; **~ de fumée noire** *m* POLL ATM black smoke number; **~ de Mohlmann** *m* POLL EAU Mohlmann index; **~ du niveau d'émission** *m* POLL ATM, POLL SON emissions level value; **~ de noircissement** *m* ÉCHANT smoke shade method, POLL ATM blackening index; **~ de noircissement par filtration sur papier** *m* POLL ATM smoke spot number; **~ de noircissement par opacité d'un panache de fumée** *m* POLL ATM plume spot number; **~ d'oxydes d'azote** *m* POLL ATM NO_x value; **~ d'octane** *m* POLL ATM octane number; **~ de pollution** *m* POL ENV, POLL ATM, POLL EAU, POLL MAR, POLL SON pollution index; **~ de pollution atmosphérique** *m* POLL ATM air pollution index; **~ de pollution gazeuse acide** *m* POLL ATM gaseous acid pollution index; **~ de pondération des bruits** *m* ÉCHANT, POLL SON noise rating index; **~ de qualité de l'air ambiant** *m* POLL ATM ambient-air quality value; **~ de qualité de transmission** *m* POLL ATM transmission performance rating; **~ de réduction du bruit** *m* POLL SON sound reduction index; **~ de réfraction** *m* DIST EAU, ÉCHANT refractive index; **~ de saturation** *m* DIST EAU saturation index; **~ sonore de référence** *m* POLL MAR, POLL SON noise standard value; **~ de turbidité** *m* ÉCHANT turbidity value; **~ de volume des boues** *m* (*IVB*) DIST EAU, POLL EAU sludge density index (*SDI*)
indices: **~ de la qualité de l'air** *m pl* ÉCHANT, POLL ATM air quality data
indirect *adj* GÉNÉRAL indirect
induit: **~ par le bruit** *loc* POLL MAR, POLL SON, SÉCURITÉ noise-induced
industrie: **~ de fabrication d'emballages** *f* DÉCHETS, GESTION packaging producing industry; **~ de fabrication de matériaux d'emballage** *f* DÉCHETS, GESTION packaging material producing industry; **~ de fourniture de matériel auxiliaire** *f* GESTION auxiliary material delivering industry; **~ mobile** *f* POL ENV footloose industry; **~ polluante** *f* POLL EAU polluting industry; **~ de recyclage du papier** *f* DÉCHETS paper-converting industry; **~ de transformation des plastiques** *f* DÉCHETS plastics processing industry
industriel *adj* GÉNÉRAL industrial
inépuisable *adj* ÉN RENOUV inexhaustible
inertage *m* DÉCHETS, DIST EAU, POLL ATM inerting
inertie *f* DÉCHETS, DIST EAU, POLL ATM inertia
inertiel *adj* DÉCHETS, DIST EAU, POLL ATM inertial
inertiser *vt* POLL EAU *déchets radioactifs* render inert
infiltration *f* DÉCHETS, DIST EAU, POLL EAU, POLL MAR infiltration, percolation, seepage, SÉCURITÉ seepage, SOLS infiltration, percolation, seepage; **~ dans les couches** DIST EAU intrusive filtration; **~ de gaz** POLL ATM gas permeation; **~ maximale** DIST EAU peak infiltration; **~ naturelle** DIST EAU, SOLS natural infiltration, natural seepage; **~ par percolation** DIST EAU percolative infiltration; **~ permanente** DIST EAU permanent infiltration
infiltrer: **s'~** *v réfl* GÉNÉRAL infiltrate, percolate, seep; **s'~ dans** *v réfl* GÉNÉRAL permeate
inflammabilité *f* DÉCHETS, POLL ATM, SÉCURITÉ inflammability, flammability, ignitability
inflammable *adj* DÉCHETS, POLL ATM, SÉCURITÉ inflammable, flammable, ignitable
inflation *f* POLL ATM, POLL MAR inflation
influx *m* DIST EAU, POLL EAU inflow
information: **~ confidentielle** *f* POL ENV confidential information; **~ du constructeur** *f* POL ENV firm-specific information; **~ d'entreprise** *f* POL ENV firm-specific information; **~ sur l'environnement** *f* POL ENV environmental information
informatisé *adj* ÉCHANT computerized
infraction: **~ à la loi** *f* POL ENV breach of law
infrarouge[1] *adj* (*IR*) ÉCHANT infrared (*IR*), ultrared (*UR*)
infrarouge[2] *m* (*IR*) ÉCHANT infrared (*IR*), ultrared (*UR*); **~ non dispersif** ÉCHANT nondispersive infrared (*NDIR*); **~ à transformée de Fourier** ÉCHANT Fourier-transform infrared (*FTIR*)
infrason *m* POLL SON infrasound
ingénierie: **~ du traitement de l'eau** *f* DIST EAU, POLL EAU water treatment engineering
ingénieur: **~ d'astreinte** *m* SÉCURITÉ duty engineer
ingérer *vt* DIST EAU ingest
inhibiteur *m* DIST EAU, POLL ATM inhibitor
ininflammable *adj* DÉCHETS, POLL ATM, SÉCURITÉ flameproof, flame-resistant, nonflammable, noninflammable
initiative: **~ de protection de l'environnement** *f* POL ENV environmental protection initiative
injecté: **~ automatiquement** *adj* ÉCHANT automatically injected
injecter *vt* ÉCHANT, POLL ATM inject, POLL ATM inject, blow into, POLL EAU inject
injecteur *m* ÉCHANT, POLL ATM, POLL EAU injector; **~ capillaire** ÉCHANT capillary injector; **~ capillaire haute densité** ÉCHANT packed-capillary injector; **~ chauffant** ÉCHANT heated injector; **~ en colonne** ÉCHANT *chromatographie* column injector; **~ d'échantillon** ÉCHANT sample injector; **~ haute densité** ÉCHANT packed injector; **~ pneumatique** POLL EAU pneumatic injector
injection[1] *f* ÉCHANT, POLL ATM, POLL EAU injection, com-

pressed-air injection; ~ **d'air secondaire** POLL ATM secondary-air injection; ~ **de déchets** DÉCHETS waste injection; ~ **directe** ÉCHANT *pompe, moteur,* SOLS direct injection; ~ **directe de liquide** ÉCHANT direct liquid introduction; ~ **d'eau** DIST EAU flushing; ~ **de gaz chaud** SOLS hot-gas injection; ~ **au jet** SOLS jet injection; ~ **de liquide** POLL ATM feeding-in of liquid; ~ **par lots** SOLS batch-bulk injection; ~ **en puits profond** POLL EAU deep-well injection

injection[2]: **à ~ automatique** *loc* ÉCHANT automatically injected

innocuité *f* DIST EAU, SOLS harmlessness

inodore *adj* DÉCHETS, DIST EAU, ÉCHANT, POLL ATM odourless (*BrE*), odorless (*AmE*)

inoffensif *adj* POLL ATM innocuous

inondation *f* DIST EAU, POLL EAU flooding, inundation, flood

inorganique *adj* DIST EAU, POLL ATM, POLL EAU inorganic

insaturé *adj* DÉCHETS, DIST EAU, POLL ATM, SOLS unsaturated

insecticide *m* DIST EAU, SOLS insecticide

insécurité *f* SÉCURITÉ unsafeness

insensibilité: ~ **à la désagrégation** *f* SOLS insensitivity to weathering

insensible *adj* POLL ATM insensitive

insert *m* DÉCHETS *plastique* insert

insipide *adj* POLL ATM tasteless

insolation *f* ÉN RENOUV insolation, sunlight

insoluble *adj* GÉNÉRAL insoluble; ~ **dans l'eau** DÉCHETS, DIST EAU, POLL ATM, POLL EAU water-insoluble

insonore *adj* POLL MAR, POLL SON, SÉCURITÉ noiseless, soundless

insonorisant *adj* POLL MAR, POLL SON, SÉCURITÉ soundproofing, noise-excluding, sound-protecting, noise-absorbing

insonorisation *f* POLL MAR, POLL SON, SÉCURITÉ soundproofing, acoustic insulation; ~ **des compresseurs** POLL SON, SÉCURITÉ noise protection for compressors; ~ **des locaux** POLL SON room lining; ~ **d'une pièce** POLL SON room lining

insonorisé *adj* POLL MAR, POLL SON, SÉCURITÉ noiseless, soundproof

insonoriser *vt* POLL MAR, POLL SON, SÉCURITÉ soundproof

inspecter *vt* GÉNÉRAL check

inspecteur *m* GESTION, SÉCURITÉ supervisor

inspection *f* GÉNÉRAL inspection, checking; ~ **de l'environnement** POL ENV environmental inspectorate; ~ **par des experts** POL ENV inspection by experts; ~ **rapide** DÉCHETS, DIST EAU high-speed inspection

instabilité *f* DÉCHETS instability

installation *f* GÉNÉRAL installation, plant, facility; ~ **adaptée au site** ÉN RENOUV site-specific installation; ~ **d'aspiration** POLL ATM suction device; ~ **d'assainissement par conteneurs** SOLS container remediation installation; ~ **de broyage** DIST EAU, ÉN RENOUV crusher unit, crushing installation; ~ **de broyage primaire** DIST EAU crushing plant; ~ **de chargement** DÉCHETS charging facility; ~ **de chauffage** ÉN RENOUV heating installation; ~ **de chauffage solaire** ÉN RENOUV solar heating plant; ~ **chauffée au bois** ÉN RENOUV wood firing plant; ~ **de clarification** DÉCHETS clarification plant; ~ **de clarification des eaux usées** POLL EAU wastewater clarification plant; ~ **Claus** POLL ATM *obtention de soufre* Claus plant; ~ **à combustion catalytique** POLL ATM catalytic combustion plant; ~ **de compostage** DÉCHETS, ÉN RENOUV composting plant; ~ **de compostage en conteneur** DÉCHETS, ÉN RENOUV container composting plant; ~ **de conditionnement d'air** SÉCURITÉ air conditioning installation; ~ **de couchage** DÉCHETS *industrie papetière* coating machine; ~ **en couches minces** DÉCHETS thin-layer installation; ~ **de craquage catalytique** POLL ATM catalytic cracking unit; ~ **de criblage** DIST EAU sieve installation; ~ **de débarquement** POLL MAR shore reception facility; ~ **de déchiquetage des véhicules** DÉCHETS car fragmentation plant; ~ **déclassée** POLL ATM antiquated plant; ~ **de dégazage** DIST EAU, ÉN RENOUV degassing facility; ~ **de dépoussiérage** POLL ATM, SÉCURITÉ dust extraction plant, dust removal installation, dust removal plant, precipitation unit; ~ **de dépoussiérage par voie sèche** POLL ATM, SÉCURITÉ dry dust-removal installation; ~ **de désencrage** POLL EAU de-inking unit; ~ **de dessalement** DIST EAU, POLL EAU desalination plant, desalting plant; ~ **de désulfuration** DIST EAU, POLL ATM sulfur recovery plant (*AmE*), sulphur recovery plant (*BrE*); ~ **de désulfuration des gaz par voie humide** POLL ATM wet flue-gas desulfurization plant (*AmE*), wet flue-gas desulphurization plant (*BrE*); ~ **à l'échelle industrielle** DÉCHETS, ÉN RENOUV, POLL ATM, SOLS commercial-scale plant; ~ **d'éclairage** ÉN RENOUV lighting installation; ~ **d'emballage** DÉCHETS packing station; ~ **d'épuration des gaz brûlés** SÉCURITÉ flue-gas cleaning installation; ~ **d'épuration des gaz d'échappement** SÉCURITÉ exhaust cleaning installation; ~ **d'évacuation des gaz d'échappement** SÉCURITÉ exhaust vent installation; ~ **d'évacuation des rejets de fumées et de chaleur** SÉCURITÉ smoke and heat exhaust installation; ~ **existant dans l'industrie** ÉN RENOUV commercially-available installation; ~ **d'extraction de poussière** POLL ATM dust extraction plant; ~ **de faible puissance** ÉN RENOUV kilowatt plant; ~ **de filtration** POLL EAU filter unit; ~ **de filtres à sac** POLL ATM baghouse; ~ **à grande échelle** ÉN RENOUV large-scale installation; ~ **à grille mobile** DIST EAU traveling-grate installation (*AmE*), travelling-grate installation (*BrE*); ~ **à grilles roulantes** DIST EAU rolling-grate installation; ~ **de lavage du sable** POLL EAU grit washer; ~ **à lit fluidisé pour le recyclage du plastique** DÉCHETS plastics recycling fluidized-bed plant; ~ **de méthanisation** DÉCHETS, POLL ATM boiler plant; ~ **moyenne** ÉN RENOUV medium-scale installation; ~ **neuve** POL ENV, POLL ATM new plant; ~ **de panneaux insonorisants** POLL SON acoustic paneling (*AmE*), acoustic panelling (*BrE*); ~ **à petite échelle** ÉN RENOUV small-scale installation; ~ **pilote** GESTION pilot plant; ~ **de plastification** DÉCHETS plasticizing unit; ~ **de pompage** POLL EAU pumping unit; ~ **de pompe à chaleur** ÉN RENOUV heat pump plant; ~ **de production combinée de chaleur et d'electricité** ÉN RENOUV, POLL ATM combined heat and power station (*CHP station*); ~ **propre à une décharge particulière** SOLS site-specific installation; ~ **de puissance** ÉN RENOUV megawatt plant; ~ **de puits solaire** ÉN RENOUV solar-driven well installation; ~ **de réception** POLL MAR *aménagement portuaire* shore reception facility; ~ **de récupération des vapeurs d'essence** DIST EAU, POLL ATM gasoline vapor recovery plant (*AmE*), petrol vapour recovery plant (*BrE*); ~ **de recyclage des gaz** DÉCHETS gas reuse plant; ~ **de réflecteurs paraboliques** ÉN RENOUV parabolic dish installation; ~ **de réflecteurs solaires** ÉN RENOUV solar dish installation; ~ **réglementée** POL ENV controlled plant; ~ **relocalisable** SOLS relocatable plant; ~ **de retraitement** DIST EAU processing unit; ~ **de rideaux d'air pour portes ouvertes** SÉCURITÉ air curtain installation for open doors; ~ **du site considéré** POLL ATM site-specific installation; ~ **solaire de chauffage central** ÉN RENOUV central solar heating plant; ~ **spécifique du site** POLL SON site-specific installation; ~ **de stockage** DÉCHETS, POLL ATM storage facility; ~ **subordonnée à la délivrance d'une licence** POL ENV plant requiring licence (*BrE*), plant requiring license (*AmE*); ~ **subordonnée à la délivrance d'un permis** POL ENV plant requiring permit; ~ **subordonnée à la délivrance d'un permis d'exploitation** POL ENV plant requiring authorization; ~ **subordonnée à l'octroi d'un permis d'exploitation** POL ENV plant requiring permit; ~ **de surface** DÉCHETS surface installation; ~ **de surface amont** DÉCHETS upward surface installation; ~ **de surface aval** DÉCHETS downward surface installation; ~ **de surface horizontale** DÉCHETS horizontal surface installation; ~ **de surface en pente ascendante**

DÉCHETS upward surface installation; ~ **de surface en pente descendante** DÉCHETS downward surface installation; ~ **de torchage** DÉCHETS flaring installation; ~ **d'une tour solaire** ÉN RENOUV solar tower installation; ~ **de traitement de déchets** DÉCHETS waste processing plant; ~ **en tranchées** DÉCHETS *type de décharge* tipped-edge installation; ~ **de tri des plastiques** DÉCHETS plastics sorting plant; ~ **de tulipage** DÉCHETS flaring installation

installer *vt* GÉNÉRAL install; ~ **dans le sous-sol** DIST EAU install underground; ~ **sous terre** SOLS install underground; ~ **en surface** DIST EAU install above ground

instance *f* POL ENV institution; ~ **délivrant une autorisation** POL ENV licensing authority; ~ **délivrant un permis** POL ENV licensing authority

Institut: ~ **allemand de normalisation** *m* (*DIN*) ÉCHANT German Institute for Standards (*DIN*)

institution *f* POL ENV institution

instructions: ~ **techniques sur la gestion des déchets** *f pl* DÉCHETS technical instruction on waste management

instrument: ~ **auto-testé** *m* ÉCHANT instrument self-test; ~ **de décision** *m* POL ENV policy instrument; ~ **d'échantillonnage** *m* ÉCHANT sampling instrument; ~ **de l'économie de marché** *m* GESTION, POL ENV market-economy instrument; ~ **économique** *m* POL ENV economic instrument; ~ **à étendues de mesure multiples** *m* ÉCHANT multirange instrument; ~ **de gestion de l'environnement** *m* POL ENV environmental policy instrument; ~ **indicateur** *m* ÉCHANT indicating instrument; ~ **juridique** *m* POL ENV legal instrument; ~ **de mesure** *m* ÉCHANT measuring instrument; ~ **de mesure de chimiluminescence** *m* ÉCHANT chemiluminescence measuring instrument; ~ **de mesure de poche** *m* ÉCHANT pocket measuring instrument; ~ **de mesure de type paramagnétique** *m* ÉCHANT measuring instrument based on paramagnetism; ~ **de mesure universel** *m* ÉCHANT universal measurement instrument (*UMI*); ~ **peu contraignant** *m* POL ENV soft instrument; ~ **radiométrique** *m* ÉCHANT radiation measuring instrument; ~ **à test automatique** *m* ÉCHANT instrument self-test

instrumentation *f* POL ENV instrumentation

insuffisance: ~ **d'air** *f* POLL ATM air deficiency; ~ **d'oxygène** *f* ÉCHANT oxygen deficiency; ~ **de saturation** *f* POLL ATM saturation deficit

intégrateur *m* ÉCHANT integrator

intégré *adj* GÉNÉRAL integrated

intempéries *f pl* POLL ATM, POLL EAU harsh weather

intensimètre *m* ÉCHANT *acoustique*, POLL SON sound intensity meter

intensité *f* ÉCHANT, ÉN RENOUV, POLL ATM intensity; ~ **acoustique** POLL MAR, POLL SON sound intensity; ~ **de couleur** ÉCHANT color depth (*AmE*), colour depth (*BrE*); ~ **énergétique** ÉN RENOUV energy intensity; ~ **lumineuse** ÉCHANT luminous intensity; ~ **de rayonnement** ÉCHANT, ÉN RENOUV radiation intensity; ~ **du rayonnement UV** POLL ATM UV radiation intensity; ~ **sonore** POLL SON loudness; ~ **de teinte** ÉCHANT color depth (*AmE*), colour depth (*BrE*); ~ **du traitement** SOLS treatment intensity

interactif *adj* ÉCHANT dialog-controlled (*AmE*), dialogue-controlled (*BrE*)

intercepter *vt* POLL ATM, POLL EAU *cours d'eau* intercept

intercepteur *m* DÉCHETS, DIST EAU, POLL EAU *égout* intercepting sewer

interception *f* POLL ATM, POLL EAU *d'un cours d'eau* interception; ~ **directe** POLL ATM direct interception

interchangeable *adj* ÉCHANT interchangeable

interdiction *f* POL ENV, POLL ATM prohibition, ban; ~ **du bruit** POLL SON noise ban; ~ **de mise en décharge des boues d'eaux résiduaires** POLL EAU wastewater sludge deposition ban; ~ **d'utilisation** DIST EAU, POL ENV use prohibition

interdire *vt* POL ENV, POLL ATM prohibit, ban

intérêt: ~ **couru** *m* GESTION interest profit; ~ **échu** *m* GESTION interest earned; ~ **moratoire** *m* GESTION default interest

interface *f* ÉCHANT interface; ~ **air-mer** POLL ATM, POLL MAR air-sea interface; ~ **hydrocarbures-eau** POLL MAR oil-water interface; ~ **de thermo-injection** ÉCHANT thermospray interface

interférence *f* ÉCHANT interference; ~ **anthropique dangereuse pour le système climatique** POLL ATM dangerous anthropogenic interference with the climate system

interférogramme *m* ÉCHANT interferogram

interféromètre: ~ **à gaz** *m* ÉCHANT gas interferometer; ~ **de Jamin** *m* ÉCHANT Jamin interferometer; ~ **de Mach-Zehnder** *m* ÉCHANT Mach-Zehnder interferometer; ~ **de Michelson** *m* ÉCHANT Michelson interferometer

interférométrie *f* ÉCHANT interferometry

intérieur: ~ **de la terre** *m* ÉN RENOUV earth's interior

intermédiaire *adj* GÉNÉRAL intermediate

intermittent *adj* DÉCHETS, DIST EAU, ÉCHANT *signal*, POLL ATM, POLL EAU intermittent

internalisation *f* POL ENV internalization; ~ **des effets externes** POL ENV internalization of external effects

interne *adj* DÉCHETS, DIST EAU, POLL ATM, POLL EAU internal

interphase *f* POLL ATM interphase

interprétation: ~ **des données** *f* ÉCHANT data interpretation

interstitiel *adj* DIST EAU interstitial

intervalle *m* ÉCHANT interval, lag time; ~ **de frappage** POLL ATM rapping interval; ~ **de mesurage** POLL SON span; ~ **de temps** ÉCHANT time interval

intoxication *f* DIST EAU poisoning

intrants *m pl* GESTION inputs

introduction: ~ **d'échantillon** *f* ÉCHANT sample introduction

intrusion *f* DÉCHETS, POLL SON, SÉCURITÉ intrusion; ~ **sonore** POLL SON noise intrusion

intumescent *adj* SÉCURITÉ intumescent

inutilisable *adj* GÉNÉRAL unusable

inventaire: ~ **du cycle de vie** *m* POLL ATM, POLL EAU life-cycle inventory; ~ **des déchets** *m* DÉCHETS waste inventory; ~ **des émissions** *m* ÉCHANT, POLL ATM emissions inventory; ~ **européen des nouvelles substances chimiques** *m* POL ENV European inventory of new chemical substances (*EINCS*); ~ **européen des substances chimiques commercialisées** *m* POL ENV European inventory of existing chemical substances (*EINECS*); ~ **forestier** *m* GESTION forest inventory; ~ **des polluants** *m* SÉCURITÉ pollutants inventory; ~ **des ressources** *m* GESTION resources survey; ~ **des substances dangereuses** *m* POL ENV, POLL ATM, SÉCURITÉ inventory of hazardous substances, register of hazardous substances; ~ **de la végétation** *m* GESTION, SOLS vegetational survey

inversion: ~ **atmosphérique** *f* POLL ATM inversion atmospheric conditions; ~ **de charge** *f* POLL ATM charge reversal; ~ **de température** *f* POLL ATM temperature inversion

investigation *f* ÉCHANT, SÉCURITÉ investigation

investissement: ~ **en matière de protection de l'environnement** *m* POL ENV environmental protection investment

invisible *adj* POLL ATM invisible

iodométrique *adj* ÉCHANT iodometric

ion *m* GÉNÉRAL ion; ~ **gazeux** POLL ATM gas ion; ~ **non résonant** ÉCHANT nonresonant ion; ~ **positif** DIST EAU positive ion; ~ **résonant** ÉCHANT resonant ion; ~ **sodium** DIST EAU sodium ion

ionique *adj* GÉNÉRAL ionic

ionisant *adj* GÉNÉRAL ionizing

ionisation *f* GÉNÉRAL ionization; ~ **de champ** ÉCHANT field ionization (*FI*); ~ **chimique négative** ÉCHANT negative chemical ionization (*NCI*); ~ **chimique positive** ÉCHANT positive chemical ionization (*PCI*); ~ **chimique sous pression atmosphérique** ÉCHANT atmospheric pressure chemical ionization (*APCI*); ~ **de flamme** ÉCHANT flame ionization; ~ **au laser** ÉCHANT laser-enhanced ionization (*LEI*); ~ **par choc** POLL ATM impact ionization; ~ **par collision** POLL ATM collision ionization; ~ **par impact** POLL ATM impact ionization; ~ **par impact électronique** ÉCHANT electron impact ionization

ionisé *adj* GÉNÉRAL ionized

ioniser *vt* GÉNÉRAL ionize

ionogène *adj* ÉCHANT, POLL EAU ionogenic

ionosphère *f* POLL ATM ionosphere

IR *abrév* (*infrarouge*) ÉCHANT IR (*infrared*), UR (*ultrared*)

iridescence *f* POLL MAR iridescence

irradiation: ~ **par rayonnement gamma** *f* POLL EAU *nucléaire* gamma-ray irradiation; ~ **des produits alimentaires** *f* DÉCHETS irradiation of food; ~ **de surface** *f* SOLS surface irradiation; ~ **ultrasonique** *f* SOLS ultrasound irradiation; ~ **aux ultraviolets** *f* DIST EAU, POLL EAU ultraviolet irradiation

irradié *adj* GÉNÉRAL irradiated

irradier *vt* GÉNÉRAL irradiate

irréversible *adj* GÉNÉRAL *processus* irreversible

irrigation *f* DIST EAU, POLL EAU, SOLS irrigation; ~ **avec des eaux recyclées** POLL EAU wastewater irrigation; ~ **en fossé** DIST EAU ditch irrigation; ~ **par aspersion** SOLS overhead irrigation; ~ **par infiltration** DIST EAU flood irrigation; ~ **par submersion** DIST EAU flood irrigation; ~ **à la raie** DIST EAU ditch irrigation; ~ **de surface** SOLS surface irrigation

irritant[1] *adj* SÉCURITÉ irritant

irritant[2] *m* SÉCURITÉ irritant; ~ **cutané** SÉCURITÉ skin irritant; ~ **oculaire** SÉCURITÉ eye irritant; ~ **pulmonaire** SÉCURITÉ lung irritant

ISO *abrév* (*Organisation internationale de normalisation*) ÉCHANT, POL ENV ISO (*International Standards Organization*)

isocinétique *adj* ÉCHANT, POLL ATM isokinetic

isolateur *m* GÉNÉRAL insulator, isolator; ~ **élastique** SÉCURITÉ *contre les vibrations* resilient isolator; ~ **de vibrations** POLL SON vibration isolator

isolation[1] *f* GÉNÉRAL isolation, insulation; ~ **acoustique** POLL SON acoustic insulation, sound insulation, sound isolation; ~ **acoustique intégrée** POLL SON structure-borne noise-proofing; ~ **calorifuge** DÉCHETS, SÉCURITÉ heat insulation; ~ **contre le bruit aérien** POLL SON airborne-noise proofing; ~ **contre le bruit et les vibrations** POLL SON, SÉCURITÉ insulation against sound and vibration; ~ **contre les vibrations** POLL SON vibration isolation; ~ **de protection** SOLS protective insulation

isolation[2]: **à ~ thermique** *loc* ÉCHANT thermally-isolated

isolé *adj* GÉNÉRAL isolated, insulated; ~ **thermiquement** ÉCHANT thermally-isolated

isolement *m* GÉNÉRAL isolation, insulation

isoler *vt* GÉNÉRAL isolate, insulate

isotherme[1] *adj* ÉCHANT, POLL ATM isothermal

isotherme[2] *f* ÉCHANT, POLL ATM isotherm, isothermal curve; ~ **d'absorption** POLL ATM absorption isotherm; ~ **d'adsorption** DIST EAU, POLL ATM, POLL EAU adsorption isotherm; ~ **de sorption** POLL ATM sorption isotherm

itération *f* DIST EAU, POLL ATM iteration

IVB *abrév* (*indice de volume des boues*) DIST EAU, POLL EAU SDI (*sludge density index*)

J

jachère *f* ÉN RENOUV, SOLS fallow
jauge *f* ÉCHANT measuring gage (*AmE*), measuring gauge (*BrE*), measuring rod; **~ de vide** ÉCHANT vacuum gage (*AmE*), vacuum gauge (*BrE*)
jaugeage *m* ÉCHANT gaging (*AmE*), gauging (*BrE*), calibration
jauger *vt* ÉCHANT gage (*AmE*), gauge (*BrE*)
javellisation *f* DIST EAU chlorination
javellisé *adj* DIST EAU chlorinated
javelliser *vt* DIST EAU chorinate
jet *m* DIST EAU, POLL ATM, SOLS jet; **~ d'air** POLL ATM air jet; **~ diphasique** POLL ATM two-phase jet; **~ d'eau** DIST EAU, POLL EAU water jet; **~ de liquide sous pression** POLL ATM pressurized liquid stream; **~ à phase unique** POLL ATM single-phase jet; **~ propulsé** POLL ATM propelled jet
jetable *adj* DÉCHETS disposable
jeter: **se ~** *v réfl* POLL EAU *affluent dans un fleuve* flow
jets: **~ et débris de coulée** *m pl* DIST EAU *génie industriel* returns
jeu *m* DIST EAU, ÉCHANT set
JO *abrév* (*journal officiel*) POL ENV OJ (*official journal*)
joint *m* DIST EAU bond; **~ en plastique** DÉCHETS plastic sealing; **~ torique** ÉCHANT *étanchéité* O-ring
jouissance *f* GÉNÉRAL *d'un bien* use
journal: **~ officiel** *m* (*JO*) POL ENV official journal (*OJ*)
juguler *vt* POLL MAR curb
jumbo *m* ÉCHANT wagon drill
juridique *adj* POL ENV juridical, lawful, legal
jurisprudence *f* POL ENV case law
jury: **~ en matière d'environnement** *m* POL ENV environmental jury

K

kilowatt *m* (*kW*) ÉN RENOUV kilowatt (*kW*); **~-heure** (*kWh*) ÉN RENOUV kilowatt-hour (*kWh*)
Kn *abrév* (*nombre de Knudsen*) POLL ATM Kn (*Knudsen's number*)
Kr *abrév* (*krypton*) POLL ATM Kr (*krypton*)
krypton *m* (*Kr*) POLL ATM krypton (*Kr*)
kW *abrév* (*kilowatt*) ÉN RENOUV kW (*kilowatt*)
kWh *abrév* (*kilowatt-heure*) ÉN RENOUV kWh (*kilowatt-hour*)
kWt *abrév* (*chaleur mesurée en kilowatt-heure*) ÉN RENOUV kWt (*kilowatt-hour heat*)

L

labile *adj* POLL ATM *météorologie* labile
labo *m* ÉCHANT lab
laborantin *m* ÉCHANT laboratory technician
laboratoire *m* ÉCHANT laboratory; **~ d'essai** ÉCHANT testing laboratory; **~ spécialisé** ÉCHANT specialist laboratory
labour: **~ profond** *m* SOLS deep ploughing (*BrE*), deep plowing (*AmE*)
lac *m* DIST EAU lake; **~ acide naturel** DIST EAU naturally acidic lake; **~ artificiel** DIST EAU artificial lake; **~ de barrage** DIST EAU artificial lake, impounding reservoir; **~ en transition** DIST EAU transition lake
lagunage *m* DIST EAU, ÉN RENOUV, POLL EAU lagooning
lagune *f* DIST EAU, ÉN RENOUV lagoon, POLL EAU lagoon, sludge lagoon; **~ de décantation** POLL EAU settling lagoon; **~ facultative** POLL EAU facultative lagoon
laine *f* POLL ATM wool; **~ de bois** DÉCHETS wood wool; **~ minérale** POLL SON mineral wool; **~ de roche** POLL SON rock wool; **~ de verre** DÉCHETS glass wool, POLL SON fiberglass fleece (*AmE*), fibreglass fleece (*BrE*)
lais *m* POLL EAU silt
lait: **~ de chaux** *m* POLL ATM lime slurry
laitier *m* DIST EAU slag; **~ d'incinération de déchets** DÉCHETS waste incineration slag; **~ vitrifié** SOLS vitrified slag
lame[1] *f* POLL ATM blade; **~ amovible** POLL ATM moveable blade; **~ déversante** DIST EAU overflow nappe
lame[2]: **à ~ unique** *loc* POLL SON single-leaf
lamelle *f* POLL ATM, POLL EAU *décantation lamellaire* lamella
laminaire *adj* DIST EAU, POLL ATM, SOLS laminar
lampe: **~ basse tension** *f* ÉCHANT low-voltage lamp; **~ de détecteur** *f* ÉCHANT detector lamp; **~ fluorescente** *f* ÉCHANT fluorescent lamp; **~ à incandescence** *f* ÉCHANT incandescent lamp; **~ interchangeable** *f* ÉCHANT interchangeable lamp; **~ à lumière ultraviolette** *f* ÉCHANT ultraviolet lamp
lance *f* POLL ATM lance; **~ à air comprimé** SOLS compressed-air lance; **~ d'arrosage** DÉCHETS nozzle, DIST EAU, POLL EAU water hose; **~ à incendie** DIST EAU fire hose; **~ d'injection** SOLS injection lance
lanterne *f* DIST EAU rose
laquage *m* DÉCHETS, DIST EAU enameling (*AmE*), enamelling (*BrE*)
large: **au ~** *adj* POLL MAR off-shore
largeur: **~ de bande effective** *f* ÉCHANT half-width value; **~ à mi-hauteur** *f* ÉCHANT half-width value; **~ de spectre à mi-hauteur** *f* ÉCHANT spectral half-width value
laser: **~ de décapage** *m* DÉCHETS compact laser for efficient ablation (*CLEAN*); **~ à diode** *m* ÉCHANT diode laser; **~ hélium-néon** *m* ÉCHANT helium-neon laser
lavage *m* GÉNÉRAL washing, washing away, washing out, washout, *d'un gaz* scrubbing, cleaning; **~ acide** POLL ATM acid scrubbing; **~ à l'air** DÉCHETS air stripping; **~ alcalin** POLL ATM caustic scrubbing; **~ atmosphérique** POLL ATM atmospheric scrubbing, below-cloud scavenging; **~ basique** POLL ATM caustic scrubbing; **~ basse pression** POLL ATM low-pressure cleaning; **~ biologique des gaz** POLL ATM biological gas scrubbing; **~ à la chaux** POLL ATM lime scrubbing; **~ à co-courant** DIST EAU, POLL ATM co-current washing; **~ à contre-courant** DIST EAU, POLL ATM backwashing, countercurrent washing, reverse-flow cleaning; **~ discontinu** POLL ATM intermittent cleaning; **~ à l'eau** POLL ATM water scrubbing; **~ à l'eau chaude** POLL MAR hot-water washing; **~ de filtre** POLL ATM filter washing; **~ des fumées** POLL ATM smoke scrubbing, flue-gas cleaning; **~ de gaz** POLL ATM gas scrubbing; **~ des gaz brûlés** DIST EAU waste gas cleaning; **~ du gaz combustible** DÉCHETS fuel gas cleaning; **~ des gaz par oxydation chimique** POLL ATM chemical-oxidative gas scrubbing; **~ des gaz perdus** DIST EAU waste-gas cleaning; **~ haute pression** POLL ATM high-pressure cleaning; **~ moyenne pression** POLL ATM medium-pressure cleaning; **~ à l'oxyde de magnésium** POLL ATM magnesium oxide scrubbing; **~ par filtration** POLL EAU filter washing; **~ par grenaillage** POLL ATM shot cleaning system; **~ par la pluie** DIST EAU, POLL EAU, SOLS washout; **~ par voie humide** POLL ATM *d'un gaz* wet scrubbing; **~ en parallèle** DIST EAU, POLL ATM co-current washing; **~ périodique** DIST EAU periodic flushing; **~ du sable** POLL EAU grit washing; **~ du sol et des sédiments** SOLS soil and sediment washing
laver *vt* GÉNÉRAL wash, wash away, wash out, *gaz* scrub
laveur: **~ à barbotage** *m* POLL ATM bubble washer; **~ biologique couplé** *m* POLL ATM coupled biological scrubber; **~ à chocs** *m* POLL ATM *dépoussiérage* impingement plate scrubber; **~ co-courant** *m* DIST EAU, POLL ATM co-current scrubber; **~ à col annulaire** *m* POLL ATM annular gap scrubber; **~ à colonne garnie** *m* DÉCHETS packed-column scrubber; **~ à contre-courant** *m* DIST EAU, POLL ATM cross-flow scrubber, *dépoussiérage hydraulique* reverse-flow scrubber; **~ à cycle unique** *m* POLL ATM single-cycle scrubber; **~ à double cycle** *m* POLL ATM double-cycle scrubber; **~ à écoulement parallèle** *m* POLL ATM co-current flow scrubber; **~ à écoulement radial** *m* POLL ATM radial-flow scrubber; **~ à écoulement turbulent** *m* POLL ATM eddy-flow scrubber; **~ à étage** *m* POLL ATM two-stage gas scrubber; **~ à garnissage** *m* DÉCHETS packed-tower scrubber, POLL ATM packed-bed scrubber; **~ haute performance** *m* POLL ATM high-performance scrubber; **~ hydraulique** *m* POLL ATM *dépoussiérage* wet scrubber; **~ mixte** *m* POLL ATM hybrid scrubber; **~ à mousse** *m* POLL ATM *dépoussiérage* foam scrubber; **~ d'oxydes d'azote** *m* POLL ATM nitrogen oxides scrubber; **~ par voie humide avec ionisation** *m* POLL ATM *dépoussiérage* ionizing wet scrubber (*IWS*); **~ à plateaux** *m* POLL ATM plate precipitator; **~ à plateaux à plaques d'impact** *m* POLL ATM *dépoussiérage* impingement plate scrubber; **~ préformé à pulvérisation** *m* POLL ATM preformed spray scrubber; **~ à pulvérisation gravitaire** *m* POLL ATM *dépoussiérage* gravity spray tower; **~ de sable** *m* POLL EAU grit washer; **~ de sol** *m* SOLS soil washer; **~ statique** *m* POLL ATM packed-tower scrubber; **~ de type annulaire** *m* POLL ATM annular-type scrubber; **~ à Venturi** *m* POLL ATM *dépoussiérage* Venturi scrubber; **~ à Venturi basse pression** *m* POLL ATM *dépoussiérage* low-pressure Venturi scrubber; **~ à Venturi haute pression** *m* POLL ATM *dépoussiérage* high-pressure Venturi scrubber
lecture *f* ÉCHANT reading; **~ directe** ÉCHANT *appareillage de mesure* direct reading; **~ simultanée** ÉCHANT simultaneous read-out; **~ des valeurs mesurées** ÉCHANT read-out of measured values
légende *f* DÉCHETS *étiquetage* caption
législateur *m* POL ENV legislator
législation *f* POL ENV legislation; **~ régissant l'ordre public** POL ENV public order law; **~ relative à la protection de l'environnement** POL ENV environmental protection legislation; **~ sur l'environnement** POL ENV environmental legislation; **~ sur la protection contre le bruit** POLL MAR, POLL SON noise protection law

lentille *f* ÉCHANT, ÉN RENOUV *optique* lens; **~ électronique** ÉCHANT electron lens; **~ de Fresnel** ÉN RENOUV *capteur solaire* Fresnel lens
lésion: **~ auditive** *f* POLL SON hearing damage
lessivable *adj* DÉCHETS, DIST EAU, POLL EAU, SOLS leachable
lessivage *m* DÉCHETS, DIST EAU, POLL EAU, SOLS leaching, lixiviation, scouring, *d'une décharge* washout; **~ microbien** SOLS microbial leaching; **~ rapide** DIST EAU flushing; **~ des terres cultivées** SOLS agricultural runoff
lessivat *m* DÉCHETS, DIST EAU, POLL EAU, SOLS leachate; **~ non dilué** DÉCHETS, DIST EAU undiluted leachate
lessive: **~ en poudre sans phosphate** *f* DIST EAU, POLL EAU phosphate-free washing-up powder (*BrE*); **~ résiduaire** *f* DÉCHETS spent lye, waste lye, DIST EAU spent lye, POLL EAU spent lye, waste lye
lessiver *vt* DÉCHETS, DIST EAU leach, lixiviate, scour, GÉNÉRAL *décharge* wash out, POLL EAU, SOLS leach, lixiviate, scour
létal *adj* SÉCURITÉ lethal
levage *m* DIST EAU lifting
levée *f* DIST EAU lift
LFC *abrév* (*lit fluidisé circulant*) POLL ATM, SOLS CBC (*circulating fluidized bed*)
liaison *f* DIST EAU, POLL ATM bond, bonding; **~ conjuguée** ÉCHANT *chimie* conjugate compound
liaisonner *vt* DÉCHETS, POLL ATM bond
liant *m* DÉCHETS bonding agent, binding agent, binder, DIST EAU, ÉN RENOUV, SOLS binder, binding agent; **~ de décharge** DÉCHETS landfill binder
libération *f* POLL SON release
libérer *vt* POLL EAU, POLL SON, SOLS release
libre[1] *adj* DÉCHETS freestanding
libre[2]: **~ parcours moyen** *m* POLL ATM mean free path
licence[1] *f* POL ENV *d'exploitation d'un procédé* licence (*BrE*), license (*AmE*)
licence[2]: **sous ~** *loc* POL ENV licensed
licite *adj* POL ENV lawful, legal
lidar *m* ÉCHANT LIDAR (*light detection and ranging*)
lié: **~ à la décharge** *adj* SOLS site-related; **~ à un élément** *adj* SOLS component-specific; **~ à l'énergie** *adj* ÉN RENOUV, POLL ATM, SOLS energy-related; **~ à l'environnement** *adj* DÉCHETS, DIST EAU, GESTION, POL ENV, SOLS environmental; **~ à l'objet** *adj* POLL MAR, POLL SON object-related; **~ à la population** *adj* POL ENV population-related; **~ aux sources** *adj* SOLS source-related
lier *vt* DÉCHETS bond, POLL ATM bind, bond, POLL EAU link
lieu: **~ d'analyse** *m* ÉCHANT analysis location; **~ de décharge** *m* DÉCHETS, DIST EAU landfill; **~ d'émission** *m* DIST EAU point source; **~ de mesure** *m* ÉCHANT measuring location; **~ de travail** *m* SÉCURITÉ workplace
ligne[1] *f* DÉCHETS, DIST EAU line, ÉCHANT peak; **~ aérodynamique** POLL ATM streamline; **~ de base** ÉCHANT *chromatographie* basic line; **~ de charge** POLL MAR *d'un navire* water line; **~ directrice sur l'environnement** POL ENV environmental guideline; **~ d'écoulement** POLL ATM flow line; **~ des effluents gazeux** POLL ATM flue-gas side; **~ d'emballage** DÉCHETS packaging line; **~ d'extrusion** DÉCHETS extrusion line; **~ focale** ÉN RENOUV *énergie solaire* focal line; **~ haute tension** ÉN RENOUV high-voltage line; **~ de partage des eaux** DIST EAU divide; **~ de prélèvement** ÉCHANT, POLL ATM sampling line; **~ de production en continu** DÉCHETS continuous production line; **~ de production de fermetures** DÉCHETS closure production line; **~ de soutirage des boues** POLL EAU return-sludge elevator; **~ de transfert** ÉCHANT, POLL ATM transfer line
ligne[2]: **en ~** *loc* ÉCHANT on-line
limitation: **~ des substances dangereuses pour la santé** *f* GESTION, POL ENV, SÉCURITÉ control of substances hazardous to health
limite: **~ d'acceptabilité** *f* POL ENV borderline acceptability; **~ de bruit** *f* POLL SON noise limit; **~ de détection** *f* ÉCHANT, POLL ATM lower limit of detectability, detection limit; **~ de dose professionnelle** *f* SÉCURITÉ occupational dose limit; **~ d'eau douce** *f* POLL EAU, POLL MAR freshwater limit; **~ d'émissions** *f* ÉCHANT, GESTION, POLL ATM, POL ENV emissions limit; **~ d'émissions de polluants** *f* ÉCHANT, GESTION, POLL ATM, POL ENV pollutant release limit; **~ d'émissions des véhicules** *f* ÉCHANT, GESTION, POLL ATM, POL ENV emissions limit for cars; **~ d'essai** *f* ÉCHANT testing limit; **~ d'exposition à l'amiante** *f* SÉCURITÉ asbestos exposure limit; **~ d'exposition à court terme** *f* ÉCHANT short-term exposure limit (*STEL*); **~ d'exposition aux substances dangereuses** *f* POL ENV, SÉCURITÉ exposure limit to hazardous substances; **~ imposée sur les émissions** *f* POL ENV, POLL ATM emissions limiting; **~ d'inflammabilité** *f* ÉCHANT inflammability limit; **~ maximale d'exposition** *f* SÉCURITÉ maximum exposure limit; **~ de résistance à la détonation** *f* POLL ATM knock limit; **~ de risque** *f* SÉCURITÉ risk limit; **~ du système** *f* GESTION system boundary
limiter *vt* GÉNÉRAL limit
limnimètre *m* ÉCHANT level indicator
limnologie *f* ÉCHANT limnology
limon *m* DIST EAU, POLL EAU silt, sludge
limpidité: **~ de l'eau** *f* DIST EAU, POLL EAU water clearness
lipophile *adj* POLL EAU lipophilic
lipophobe *adj* POLL EAU lipophobic
liposolubilité *f* POLL ATM liposolubility
liquéfaction *f* DIST EAU, ÉN RENOUV, POLL ATM liquefaction
liquéfier *vt* DIST EAU, ÉN RENOUV, POLL ATM liquefy
liqueur: **~ de boues** *f* DIST EAU, POLL EAU sludge liquor; **~ de lavage** *f* DIST EAU scrubbing liquor; **~ mixte** *f* POLL EAU mixed liquor; **~ surnageante** *f* DIST EAU sludge liquor, supernatant liquor
liquide[1] *adj* GÉNÉRAL liquid
liquide[2] *m* GÉNÉRAL liquid; **~ d'absorption** POLL ATM absorption liquid; **~ boueux** DIST EAU sludge-laden liquid; **~ de circulation** POLL ATM circulating liquid; **~ de départ** ÉCHANT separating liquid; **~ émulsifiant** SÉCURITÉ emulsifying liquid; **~ entraîné** POLL ATM entrained liquid; **~ d'étanchéité** ÉCHANT sealing liquid; **~ inciné́ré** POLL EAU incinerated liquid; **~ inflammable** SÉCURITÉ inflammable liquid, flammable liquid; **~ d'injection** SOLS injection fluid; **~ manométrique** ÉCHANT manometer liquid; **~ nauséabond** DIST EAU, POLL EAU evil-smelling liquid; **~ non miscible** POLL EAU immiscible liquid; **~ pompable** POLL ATM pumpable liquid; **~ de refroidissement** POLL ATM cooling liquid; **~ sous pression** POLL ATM pressurized liquid; **~ surnageant** DIST EAU sludge liquor, supernatant liquor; **~ visqueux** DIST EAU, POLL ATM, POLL EAU viscous liquid
lisible *adj* ÉCHANT readable
lisier *m* DIST EAU, POLL EAU liquid manure, slurry
lisière *f* POLL MAR skirt
lissage *m* DIST EAU polishing; **~ de surface** SOLS surface smoothing
lisser *vt* DIST EAU polish, SOLS smooth
liste: **~ des concentrations de substances nocives établie par la Hollande** *f* SOLS Dutch list; **~ noire** *f* POL ENV blacklist; **~ des priorités** *f* POL ENV priority list
lit *m* DIST EAU river bed; **~ adsorbant** POLL ATM adsorbent bed; **~ alluvial** POLL EAU alluvial bed; **~ bactérien** POLL EAU bacteria bed, biological filter, trickling filter; **~ bactérien à débit moyen** POLL EAU standard-rate trickling filter; **~ bactérien à faible charge** POLL EAU low-rate trickling filter; **~ bactérien à forte charge** POLL EAU high-rate trickling filter; **~ bactérien intermédiaire** POLL EAU intermediate trickling filter; **~ bactérien ouvert** POLL EAU open trickling filter; **~ bactérien secondaire** POLL EAU secondary trickling filter; **~ bactérien traditionnel** POLL EAU tower-type trickling filter; **~ bas** POLL ATM lower bed; **~ biologique** SOLS biological bed; **~ de boues** POLL EAU

mud blanket, sludge blanket; ~ **catalytique** POLL ATM catalyst bed; ~ **catalytique oxydant** POLL ATM oxidizing catalyst bed; ~ **à chargement frontal** DÉCHETS frontally-packed bed; ~ **filtrant** DIST EAU, POLL EAU filter bed; ~ **de filtration** DIST EAU, POLL EAU filter bed; ~ **de filtration lente sur sable** DIST EAU slow process sand filter bed; ~ **fixe** POLL ATM, POLL EAU fixed bed; ~ **fixe laminaire** SOLS laminar fixed bed; ~ **fixe suspendu** POLL EAU packed suspended bed (*PSB*); ~ **fluidisé** DIST EAU, POLL ATM, POLL EAU, SOLS fluidized bed, moving bed; ~ **fluidisé à charbons actifs** POLL ATM activated-carbon fluidized bed; ~ **fluidisé circulant** (*LFC*) POLL ATM, SOLS circulating fluidized bed (*CBC*); ~ **fluidisé fixe** POLL ATM, SOLS stationary fluidized bed; ~ **fluidisé multi-étage** POLL ATM multistage fluidized bed; ~ **à garnissage** POLL ATM packed bed; ~ **de hautes eaux** DIST EAU flood plain; ~ **majeur** DIST EAU flood plain; ~ **mobile** DIST EAU, POLL ATM, POLL EAU, SOLS moving bed; ~ **mobile à charbons actifs** POLL ATM activated-carbon moving bed; ~ **de mousse** POLL ATM *type de séparateur humide* foam bed; ~ **percolateur** DIST EAU biological filter, sprinkling filter, POLL ATM biological ground filter, DIST EAU, POLL EAU percolating filter; ~ **de pierre** POLL EAU rock bed; ~ **de réduction catalytique** POLL ATM reducing catalyst bed; ~ **de rivière** POLL EAU river bed; ~ **de séchage** DÉCHETS, POLL EAU drying bed; ~ **de séchage des boues** DIST EAU, POLL EAU sludge drying bed; ~ **supérieur** POLL ATM upper bed

lithosphère *f* DIST EAU lithosphere

litière *f* DÉCHETS litter; ~ **du sol** SOLS *humus* litter

littoral *m* POLL MAR shoreline

livraison: ~ **de produits pharmaceutiques** *f* DÉCHETS chemicals delivery

livreur *m* ÉN RENOUV deliverer

lixiviabilité *f* DÉCHETS, DIST EAU, POLL EAU, SOLS leachability

lixiviat *m* DÉCHETS, DIST EAU, POLL EAU, SOLS leachate; ~ **de décharge** DÉCHETS landfill leachate; ~ **non dilué** DÉCHETS, DIST EAU undiluted leachate; ~ **de pente** DIST EAU slope leachate

lixiviation *f* DÉCHETS, DIST EAU, POLL EAU, SOLS lixiviation, leaching

lixivier *vt* DÉCHETS, DIST EAU, POLL EAU, SOLS lixiviate, leach

local[1] *adj* GÉNÉRAL local, area-related, site-related, site-specific

local[2]: ~ **des pompes** *m* DIST EAU pump house

localement: ~ **représentatif** *adj* ÉCHANT locally-representative

localisation *f* SOLS localization; ~ **d'une décharge** DÉCHETS landfill location

location *f* POL ENV leasing

logarithme: ~ **de Briggs** *m* POLL ATM Briggs' logarithm

logiciel *m* ÉCHANT software; ~ **d'évaluation** ÉCHANT evaluation software

logistique: ~ **d'évacuation** *f* DIST EAU, GESTION, POLL ATM logistics of disposal; ~ **respectueuse de l'environnement** *f* GESTION environmentally-friendly logistics

loi: ~ **antipollution** *f* POL ENV antipollution law, Control of Pollution Act (*BrE*) (*CPA*); ~ **de Boyle-Mariotte** *f* ÉCHANT *gaz parfait* Boyle's law, Mariotte's law; ~ **cadre** *f* GESTION, POL ENV basic law; ~ **de Fick** *f* POLL ATM *diffusion moléculaire* Fick's law; ~ **de Henry** *f* POLL ATM, POLL EAU, SOLS Henry's law; ~ **de Henry et de Dalton** *f* DIST EAU Henry-Dalton law; ~ **de Lambert-Beer** *f* ÉCHANT *photométrie* Lambert-Beer law; ~ **de Planck** *f* POLL ATM Planck's law; ~ **de pression partielle de Dalton** *f* POLL ATM Dalton's law of partial pressure; ~ **de Stokes** *f* ÉCHANT *spectroscopie Raman*, POLL ATM *viscosité* Stokes's law; ~ **sur la dépollution de l'eau** *f* POLL MAR clean water act; ~ **sur l'élimination des carcasses animales** *f* DÉCHETS animal carcass disposal act; ~ **sur l'élimination de déchets** *f* DÉCHETS, GESTION, POL ENV waste disposal act; ~ **sur les épizooties** *f* DÉCHETS epizootic diseases act; ~ **sur la gestion de déchets** *f* DÉCHETS, GESTION, POL ENV waste management act; ~ **sur les huiles usagées et les corps gras de récupération** *f* DÉCHETS, POL ENV, SOLS waste oil act; ~ **sur les huiles usées** *f* DÉCHETS, POL ENV, SOLS waste oil act; ~ **sur la prévention et la gestion de déchets** *f* DÉCHETS, POL ENV waste avoidance and management act; ~ **sur la protection des espèces végétales** *f* DÉCHETS plant protection act; ~ **sur la protection des installations** *f* DIST EAU, POL ENV plant protection act; ~ **sur les rejets d'hydrocarbures** *f* DÉCHETS, POL ENV, SOLS waste oil act

lois: ~ **antipollution** *f pl* POL ENV antipollution laws; ~ **sur la protection de l'environnement** *f pl* POL ENV environmental protection legislation

lombricompostage *m* SOLS worm composting

long: ~ **à mettre en œuvre** *loc* GESTION time-intensive; **à ~ terme** *loc* GÉNÉRAL long-term

longévité *f* POLL ATM, SOLS longevity

longue: **de ~ durée** *loc* GESTION time-intensive; **à ~ durée de fonctionnement** *loc* ÉCHANT long-running

longueur: ~ **du front d'onde** *f* ÉN RENOUV wavefront length; ~ **d'onde** *f* ÉCHANT wavelength; ~ **d'onde de référence** *f* ÉCHANT reference wavelength; ~ **de sac** *f* POLL ATM bag length

lot *m* SOLS batch

louche: ~ **de dosage** *f* ÉCHANT measuring scoop

lubrifiant: ~ **pour déchets** *m* DÉCHETS waste lubricant; ~ **perdu** *m* DÉCHETS waste lubricant; ~ **résiduel** *m* DÉCHETS waste lubricant

luire *vi* POLL MAR lighten

lumière: ~ **ultraviolette** *f* ÉCHANT ultraviolet light

lunettes: ~ **de protection contre les rayons X** *f pl* SÉCURITÉ X-ray protective glasses

lutte *f* POLL ATM abatement, control; ~ **contre les adventices** SOLS weed control; ~ **contre le bruit** POLL SON, SÉCURITÉ noise abatement, noise check, noise control; ~ **contre la corrosion** ÉN RENOUV corrosion prevention; ~ **contre les déchets** DÉCHETS waste avoidance; ~ **contre la dégradation des sols** GESTION, SOLS soil degradation control; ~ **contre les émissions** POLL ATM, SOLS emissions control; ~ **contre les émissions à l'échappement** POLL ATM exhaust emissions control; ~ **contre la formation de mousse** DÉCHETS foam abatement; ~ **contre les inondations** DIST EAU, ÉN RENOUV, GESTION, POLL MAR flood control; ~ **contre les mauvaises herbes** SOLS weed control; ~ **contre les odeurs** DIST EAU, POLL ATM odor control (*AmE*), odour control (*BrE*); ~ **contre la pollution** POL ENV, POLL ATM, POLL EAU pollution control; ~ **contre la pollution atmosphérique** POLL ATM air pollution control, prevention of atmospheric pollution; ~ **contre la pollution des mers** POLL MAR control of marine pollution; ~ **contre la pollution par les hydrocarbures** POLL MAR oil pollution control; ~ **contre la production de déchets secondaires** DIST EAU secondary waste avoidance; ~ **contre la production de déchets à la source** DIST EAU residue avoidance at source; ~ **contre la prolifération des algues** POLL MAR algae control; ~ **contre les émissions de NO_x** DÉCHETS flue-gas NO_x abatement; ~ **intégrée contre la pollution** GESTION, POL ENV integrated pollution control (*IPC*); ~ **à long terme** POLL ATM, POLL SON *contre la pollution* long-term control

lyolyse *f* ÉCHANT lyolysis, solvolysis

lyophilisateur: ~ **sous vide** *m* DÉCHETS vacuum freeze-dryer

lyophilisation *f* ÉCHANT lyophilization, freeze-drying

lyophiliser *vt* ÉCHANT lyophilize, freeze-dry

lyse *m* ÉCHANT lysis

lysimètre *m* ÉCHANT lysimeter; ~ **flottant** ÉCHANT weighable lysimeter; ~ **non-pesable** ÉCHANT nonweighable lysimeter

M

macération *f* DÉCHETS maceration
macérer *vt* DÉCHETS macerate
mâchefer *m* DÉCHETS clinker, slag
machine: **~ à alimenter** *f* DÉCHETS feeding machine; **~ aspirante** *f* ÉCHANT suction pump; **~ à border** *f* DÉCHETS flanging machine; **~ à boucher les bouteilles** *f* DÉCHETS bottle corking machine; **~ à conditionner** *f* DÉCHETS parcelling machine; **~ à contrecoller** *f* DÉCHETS laminating machine; **~ de désemboîtement** *f* DÉCHETS denesting machine; **~ à doubler** *f* DÉCHETS laminating machine; **~ à dresser et fermer les cartons** *f* DÉCHETS carton erector and closer; **~ pour emballage moulant** *f* DÉCHETS blister packaging machine; **~ d'emballage sous vide** *f* DÉCHETS vacuum packaging machine; **~ à emballer les boîtes** *f* DÉCHETS can packing machine; **~ à empaqueter** *f* DÉCHETS parceling machine (*AmE*), parcelling machine (*BrE*); **~ à encaisser** *f* DÉCHETS case loader; **~ à encoder les étiquettes** *f* DÉCHETS label coding machine; **~ d'épuisement** *f* DIST EAU, POLL EAU draining engine, pumping engine; **~ d'exhaure** *f* DIST EAU, POLL EAU draining engine, pumping engine; **~ de fabrication des sacs** *f* DÉCHETS bag making machine; **~ à fermer les sacs** *f* DÉCHETS bag closing machine; **~ à fermer les caisses** *f* DÉCHETS case sealing machine; **~ de fermeture sous vide** *f* DÉCHETS vacuum closing machine; **~ à former, remplir et fermer les cartons caisses** *f* DÉCHETS case erecting, filling and closing machine; **~ à gainer les fils** *f* DÉCHETS wrapping machine; **~ à imprégner** *f* DÉCHETS impregnating machine; **~ insonorisée** *f* POLL SON low-noise machine; **~ de moulage par étirage-soufflage** *f* DÉCHETS extrusion stretch-blow molding machine (*AmE*), extrusion stretch-blow moulding machine (*BrE*); **~ de moulage par injection** *f* DÉCHETS injection molding machine (*AmE*), injection moulding machine (*BrE*), injection molding press (*AmE*), injection moulding press (*BrE*); **~ de moulage par injection en plusieurs éléments** *f* DÉCHETS multicomponent injection molding machine (*AmE*), multicomponent injection moulding machine (*BrE*); **~ à remplir** *f* DÉCHETS filling machine; **~ à remplir et fermer** *f* DÉCHETS filling and sealing machine; **~ de remplissage** *f* DÉCHETS feeding machine, filling machine; **~ de remplissage et de fermeture** *f* DÉCHETS filling and sealing machine; **~ de remplissage gravitaire** *f* DÉCHETS gravity filling machine; **~ de remplissage pondéral** *f* DÉCHETS weight filling machine; **~ de remplissage sous vide** *f* DÉCHETS vacuum filling machine; **~ de remplissage volumétrique** *f* DÉCHETS density filling machine; **~ à sceller les boîtes** *f* DÉCHETS can closing machine; **~ de thermoformage pour conditionnement sous vide** *f* DÉCHETS form, fill and seal machine; **~ de thermoformage sous vide** *f* DÉCHETS vacuum thermoforming machine; **~ de tri et d'inspection des produits pharmaceutiques** *f* DÉCHETS tablet sorting and inspection machine
mâchoire: **~ à balles** *f* DÉCHETS bale clamp
macro: **~-élément** *m* DIST EAU, POLL EAU macronutrient
macroclimat *m* POLL ATM macroclimate
macroconfinement *m* SOLS macroencapsulation
macrodéchet *m* DÉCHETS macrowaste
macrofloc: **~ décantable** *m* POLL EAU settleable macrofloc
macropore *m* POLL ATM, POLL EAU macropore
magasin: **~ de stockage** *m* DÉCHETS storeroom
magma *m* ÉN RENOUV magma
magnésium *m* (*Mg*) GÉNÉRAL magnesium (*Mg*)
magnétique *adj* GÉNÉRAL magnetic
maigre *adj* DIST EAU grease-free
maillage: **~ urbain** *m* POL ENV urban mesh
maille[1] *f* POLL EAU *tamis* mesh
maille[2]: **à ~ fine** *loc* DIST EAU close-meshed; **à ~ serrée** *loc* POLL ATM close-meshed
mailles: **à grandes ~** *loc* POLL EAU *filtre* wide-meshed
maintenance *f* GÉNÉRAL maintenance; **~ correctrice** GESTION corrective maintenance
maintenir: **~ fermé** *loc* ÉCHANT keep sealed
maîtrise: **~ des crues** *f* DIST EAU, ÉN RENOUV, GESTION, POLL MAR flood control; **~ des délais** *f* GESTION scheduling control; **~ de l'énergie** *f* ÉN RENOUV, GESTION energy management; **~ de la qualité** *f* DIST EAU, GESTION, POLL ATM quality control; **~ de la valeur** *f* GESTION value management
maladie: **~ d'origine hydrique** *f* POLL EAU, SÉCURITÉ waterborne disease; **~ professionnelle** *f* SÉCURITÉ occupational disease; **~ transmise par l'eau** *f* POLL EAU, SÉCURITÉ waterborne disease
malaxer *vt* DÉCHETS, DIST EAU knead
malaxeur *m* DÉCHETS *plastique*, DIST EAU kneader; **~ humide** DÉCHETS wet mixer; **~ à pales** DÉCHETS paddle mixer
malodorant *adj* POLL EAU foul-smelling
management: **~ de l'environnement** *m* GESTION environmental management
manche: **~ d'air** *f* POLL ATM air conduction cylinder; **~ de chargement** *f* DÉCHETS loading hose; **~ gonflable** *f* POLL MAR *d'un récupérateur* inflation cuff; **~ à incendie** *f* DIST EAU fire hose
manchon *m* DIST EAU socket, *tuyauterie* pipe sleeve, POLL EAU socket; **~ d'air** POLL ATM air conduction cylinder; **~ pour bouteille** DÉCHETS bottle sleeve; **~ d'expédition** DÉCHETS mailing sleeve; **~ de raccord** DIST EAU pipe joint; **~-raccord** DIST EAU pipe joint
manganèse *m* (*Mn*) POLL ATM manganese (*Mn*)
maniabilité *f* DÉCHETS *d'un engin*, ÉCHANT, POLL MAR, POLL SON handleability, controllability
manille *f* POLL MAR shackle
manioc *m* ÉN RENOUV cassava
manocontacteur *m* POLL ATM manometric switch
manomètre *m* DIST EAU, ÉCHANT, POLL ATM, POLL EAU manometer, pressure gage (*AmE*), pressure gauge (*BrE*); **~ différentiel** POLL ATM differential manometer, differential pressure gage (*AmE*), differential pressure gauge (*BrE*); **~ de réacteur** POLL ATM reactor pressure sensor
manométrique *adj* DIST EAU, ÉCHANT, POLL ATM, POLL EAU *hauteur* manometric
manque: **~ d'air** *m* POLL ATM air deficiency; **~ d'espace** *m* POL ENV urban congestion
manteau *m* ÉN RENOUV earth's interior
manuel[1] *adj* ÉCHANT hand-driven, hand-operated, manual
manuel[2] *m* ÉCHANT handbook; **~ qualité** GESTION quality manual
manutention: **~ des briquettes** *f* DÉCHETS briquette handling; **~ des emballages consignés** *f* DÉCHETS handling of returnables
maquette *f* ÉCHANT scale model, mock-up
maraîchage *m* SOLS market gardening (*BrE*), truck farming (*AmE*)
marais *m* DIST EAU, SOLS marsh

marcage *m* DIST EAU intake

marchandise: ~ **en vrac** *f* DÉCHETS bulk goods

marché *m* GESTION, POL ENV market; ~ **de la gestion de déchets** GESTION waste management market; ~ **immature** GESTION immature market; ~ **des technologies de l'environnement** GESTION environmental technology market

marécage *m* DIST EAU, SOLS marsh

marée *f* ÉN RENOUV, POLL MAR tide, ebb and flood; ~ **basse** DIST EAU low water, POLL MAR low tide, low-tide water, low water; ~ **descendante** ÉN RENOUV, POLL MAR ebb tide; ~ **haute** DIST EAU high water, POLL MAR high tide, high-tide water, high water; ~ **montante** ÉN RENOUV, POLL MAR flood tide; ~ **noire** POLL MAR oil slick, oil spill, oil spillage; ~ **rouge** POLL MAR red tide; ~ **de vives eaux** ÉN RENOUV, POLL MAR spring tide

marémoteur *adj* ÉN RENOUV tidal

marin *adj* POLL MAR sea-based

marketing: ~ **écologique** *m* GESTION ecology-oriented marketing

marne *f* DIST EAU marl

marque: ~ **déposée** *f* GESTION licensing mark; ~ **de niveau d'eau** *f* DIST EAU water mark

marqueuse *f* DÉCHETS marking machine

masque: ~ **anti-poussière** *m* SÉCURITÉ dust mask; ~ **à gaz** *m* SÉCURITÉ gas mask, respirator; ~ **respiratoire** *m* SÉCURITÉ respirator, respiratory mask

masse[1]: ~ **d'air** *f* POLL ATM air mass; ~ **anaérobie** *f* DIST EAU anaerobic mass; ~ **atmosphérique** *f* POLL ATM atmospheric mass; ~ **bactérienne** *f* DIST EAU bacterial mass; ~ **centrifuge** *f* ÉN RENOUV centrifugal mass; ~ **en contact** *f* POLL EAU contact mass; ~ **cuite** *f* DÉCHETS *industrie sucrière* fill mass; ~ **de déblaiement** *f* SOLS excavated mass; ~ **de déchets** *f* DÉCHETS waste mass, waste body; ~ **en décomposition** *f* DÉCHETS, DIST EAU rotting body; ~ **d'eau** *f* ÉN RENOUV water mass; ~ **filtrante** *f* DIST EAU, POLL EAU filter bed; ~ **humique** *f* POLL EAU humic mass; ~ **molaire** *f* ÉCHANT molar mass; ~ **moléculaire** *f* ÉCHANT, POLL EAU molecular weight; ~ **d'ordures** *f* DÉCHETS refuse mass; ~ **de pourrissage** *f* DÉCHETS, DIST EAU rotting body; ~ **sèche** *f* POLL EAU dry mass; ~ **sèche organique** *f* POLL EAU organic dry mass; ~ **volumique** *f* DIST EAU, ÉCHANT density; ~ **volumique apparente** *f* ÉCHANT bulk density, settled apparent density; ~ **volumique de gaz** *f* ÉCHANT gas density

masse[2]: **de ~** *loc* ÉCHANT mass; **en ~** *loc* ÉCHANT mass

massicot *m* DÉCHETS, ÉN RENOUV *papier* cutting shredder

mât *m* DÉCHETS strut

matelas: ~ **fibreux** *m* POLL ATM fibrous mat

matériau: ~ **absorbant** *m* POLL SON sound-absorbing material; ~ **amortisseur du son** *m* POLL SON sound-deadening material; ~ **de ballast** *m* DÉCHETS ballast material; ~ **de barrage** *m* DÉCHETS, DIST EAU barrier material; ~ **de base inorganique** *m* POLL ATM inorganic base material; ~ **biodégradable** *m* DÉCHETS, DIST EAU, POLL EAU biodegradable matter; ~ **catalyseur** *m* POLL ATM catalyst material; ~ **céramique** *m* POLL ATM ceramic material; ~ **combustible** *m* DIST EAU, POLL ATM, SÉCURITÉ combustible material; ~ **composite béton-polymère** *m* DIST EAU polymer-concrete composite material; ~ **composite isolant** *m* POLL SON insulating compound; ~ **de couverture** *m* DÉCHETS, ÉN RENOUV covering material; ~ **de couverture temporaire** *m* DÉCHETS interim covering material; ~ **de déchets** *m* DÉCHETS waste stuff; ~ **de doublure** *m* DÉCHETS lining compound; ~ **échangeur** *m* POLL EAU exchanger material; ~ **échangeur de chaleur** *m* DÉCHETS, ÉN RENOUV, POLL ATM heat exchanger material; ~ **d'emballage** *m* DÉCHETS packaging material; ~ **d'emballage anticolmatage** *m* DÉCHETS nonclogging packing material; ~ **d'emballage plastique** *m* DÉCHETS plastic packing material; ~ **d'emballage recyclé** *m* DÉCHETS packaging secondary resource; ~ **excavé** *m* DÉCHETS excavated material; ~ **fibreux** *m* POLL ATM fiber material (*AmE*), fibre material (*BrE*); ~ **filtrant** *m* POLL ATM, POLL EAU filter material, filter medium, filtering agent; ~ **industriel** *m* DIST EAU, POLL ATM process material; ~ **inerte** *m* POLL ATM inert material; ~ **inflammable** *m* SÉCURITÉ inflammable material, flammable material; ~ **infusible** *m* DÉCHETS nonmeltable material; ~ **d'injection** *m* SOLS grouting material; ~ **d'injection thixotrope** *m* SOLS thixotrope grouting material; ~ **insonorisant** *m* POLL SON sound insulation material, sound-absorbing material; ~ **isolant** *m* DÉCHETS, POLL SON insulating material; ~ **d'isolation** *m* DÉCHETS insulating compound; ~ **noir absorbant la chaleur** *m* ÉN RENOUV dark heat-absorbing material; ~ **d'operculage** *m* DÉCHETS cap sealing compound; ~ **porteur** *m* ÉCHANT, POLL ATM carrier material; ~ **primaire** *m* DÉCHETS primary material; ~ **produit en grandes quantités** *m* DÉCHETS bulk good; ~ **rebroyé** *m* DÉCHETS reground material; ~ **de rebut** *m* DÉCHETS waste material; ~ **récupérable** *m* DÉCHETS recoverable material; ~ **de récupération** *m* DÉCHETS secondary material; ~ **recyclable** *m* DÉCHETS recoverable material; ~ **recyclé** *m* DÉCHETS recycled material; ~ **de remblai** *m* DÉCHETS, SOLS fill material, landfill material; ~ **remeulé** *m* DÉCHETS reground material; ~ **de renfort** *m* DIST EAU reinforcing material; ~ **secondaire** *m* DÉCHETS secondary material; ~ **sédimentable** *m* POLL ATM sedimentable material; ~ **du sol** *m* SOLS soil component, soil constituent; ~ **du sol naturel** *m* DÉCHETS, SOLS natural soil material; ~ **solide** *m* DÉCHETS solid material; ~ **solidifié** *m* DÉCHETS solidified material; ~ **surdimensionné** *m* DÉCHETS oversize material; ~ **de synthèse** *m* DÉCHETS synthetic material; ~ **synthétique** *m* DÉCHETS synthetic material; ~ **thermoplastique** *m* POLL ATM thermoplastic material; ~ **thermosensible** *m* DÉCHETS heat-sensitive material; ~ **usagé** *m* DÉCHETS used material; ~ **vecteur** *m* ÉCHANT, POLL ATM carrier material; ~ **vierge** *m* DÉCHETS virgin material; ~ **en vrac** *m* DÉCHETS, POLL ATM bulk good, bulk material

matériel[1] *adj* DÉCHETS, DIST EAU, POLL ATM material, physical

matériel[2] *m* GÉNÉRAL equipment; ~ **d'agrafage** DÉCHETS stapling equipment; ~ **d'alimentation** DÉCHETS feeder equipment; ~ **d'analyse par voie sèche** ÉCHANT dry equipment; ~ **de banderolage** DÉCHETS strapping equipment; ~ **de construction insonorisé** POLL SON low-noise building machinery; ~ **de déshydratation des boues** POLL EAU sludge dewatering equipment; ~ **de désintégration** DÉCHETS repulping equipment; ~ **de dispersion** POLL MAR dispersing equipment; ~ **d'emballage alimentaire** DÉCHETS food wrapping equipment; ~ **d'enrubannage** DÉCHETS strapping equipment; ~ **d'épuisement** POLL EAU pumping plant; ~ **d'épuration et de désodorisation** SÉCURITÉ air purification and deodorization equipment; ~ **d'exhaure** POLL EAU pumping plant; ~ **d'exploitation** POLL ATM operating material; ~ **d'extraction des scories** DIST EAU slag removal equipment; ~ **d'extrusion de films** DÉCHETS film extrusion equipment; ~ **de finition en ligne** DÉCHETS in-line finishing equipment; ~ **insonorisé** POLL SON low-noise equipment; ~ **d'inspection** DÉCHETS, DIST EAU inspection equipment; ~ **de marquage** DÉCHETS marking equipment; ~ **de mesure** DÉCHETS, ÉCHANT measuring equipment, metering equipment; ~ **de mesure du bruit et des vibrations** POLL MAR, POLL SON, SÉCURITÉ noise and vibration measuring equipment; ~ **modifié du sol** DÉCHETS modified soil material; ~ **de nettoyage** SÉCURITÉ cleaning equipment; ~ **normalisé** SÉCURITÉ standardized equipment; ~ **d'operculage** DÉCHETS cap sealing equipment; ~ **de protection** SÉCURITÉ protective equipment; ~ **respiratoire** SÉCURITÉ respiratory equipment; ~ **de tamisage** DIST EAU screening equipment; ~ **de terrassement peu bruyant** POLL

SON low-noise building machinery; ~ **de thermocollage** DÉCHETS heat-sealing equipment; ~ **de transport** ÉN RENOUV conveying equipment

matière: ~ **allochtone** *f* SOLS allochthonous matter; ~ **autochtone** *f* SOLS autochthonous matter; ~ **dangereuse** *f* SÉCURITÉ dangerous material; ~ **décantée** *f* POLL EAU settling material, sinking material; ~ **dissoute** *f* POLL EAU dissolved substance; ~ **épaississante** *f* DIST EAU, POLL EAU thickener; ~ **fécale** *f* POLL EAU faecal matter (*BrE*), fecal matter (*AmE*); ~ **fissile** *f* ÉN RENOUV *nucléaire* core material; ~ **flottante** *f* DÉCHETS floating matter; ~ **indigestible** *f* DIST EAU roughage; ~ **indigestible séparable** *f* DIST EAU separable roughage; ~ **inerte** *f* DÉCHETS inert material; ~ **minérale** *f* DIST EAU mineral matter; ~ **à mouler par injection** *f* DÉCHETS injection molding compound (*AmE*), injection moulding compound (*BrE*); ~ **moulue** *f* DIST EAU, POLL ATM ground stock; ~ **organique** *f* POLL ATM, POLL EAU organic matter; ~ **organique dissoute** *f* DIST EAU, ÉCHANT, POLL ATM dissolved organic matter, POLL EAU dissolved organic matter, solubilized organics; ~ **organique non biodégradable** *f* POLL EAU nonbiodegradable organics; ~ **organique précipitable** *f* POLL EAU precipitable organics; ~ **organique solide en suspension** *f* POLL EAU suspended organic solid; ~ **particulaire** *f* POLL EAU particulate material; ~ **plastique acrylique** *f* DÉCHETS acrylic plastic; ~ **précieuse** *f* DÉCHETS valuable substance; ~ **première** *f* DÉCHETS, ÉN RENOUV raw material; ~ **première amylacée** *f* ÉN RENOUV starchy raw material; ~ **première cellulosique** *f* ÉN RENOUV cellulose raw material; ~ **première destinée à la production d'énergie** *f* ÉN RENOUV energy raw material; ~ **première pour papier** *f* DÉCHETS paper raw material; ~ **première de récupération** *f* DÉCHETS, ÉN RENOUV secondary raw material; ~ **première renouvelable** *f* ÉN RENOUV renewable raw material; ~ **première résiduelle** *f* DÉCHETS, ÉN RENOUV residue raw material, waste raw material; ~ **putrescente** *f* POLL EAU putrescent material; ~ **putrescible** *f* POLL EAU putrescible matter; ~ **sèche** *f* POLL EAU dry mass; ~ **solide décantable** *f* POLL EAU settleable solid; ~ **solide grossière** *f* POLL EAU ground coarse solids; ~ **solide sous forme particulaire** *f* POLL ATM particulate solid matter; ~ **solide en suspension** *f* POLL EAU suspended solid; ~ **solide en suspension et décantable** *f* POLL EAU suspended and settleable solid; ~ **solide total dissoute** *f* DIST EAU total dissolved solids (*TDS*); ~ **sous forme de particules submicroniques** *f* POLL ATM submicron particulate matter; ~ **sous forme de particules en suspension dans l'air** *f* POLL ATM airborne particulate matter; ~ **en suspension** *f* POLL EAU suspended matter, suspended solid; ~ **turbide** *f* POLL EAU turbid matter; ~ **vierge** *f* DÉCHETS virgin material

matières: ~ **solides en suspension dans une solution** *f pl* POLL EAU mixed liquor suspended solids (*MLSS*)

matrice: ~ **électrolytique** *f* POLL ATM electrolyte matrix; ~ **de plastique** *f* DÉCHETS, DIST EAU plastic mold (*AmE*), plastic mould (*BrE*); ~ **du plastique** *f* DÉCHETS, DIST EAU plastic matrix; ~ **du sol** *f* DIST EAU, SOLS soil matrix

matrices: ~ **moulées à partir de plastique recyclé** *f pl* DIST EAU molded parts using recycled plastics (*AmE*), moulded parts using recycled plastics (*BrE*)

maturation *f* DÉCHETS, POLL EAU, SOLS *conditionnement des boues* ripening; ~ **du compost** DÉCHETS, SOLS compost maturing

mauvaise: ~ **utilisation des terres** *f* GESTION, SOLS land misuse

maximal *adj* GÉNÉRAL maximum, peak

maximum: ~ **printanier des retombées** *m* POLL ATM spring maximum of fallout

mazout *m* ÉN RENOUV fuel oil

méandre *m* DIST EAU bend

mécanique[1] *adj* GÉNÉRAL mechanical

mécanique[2]: ~ **du sol** *f* SOLS soil mechanics

mécanisme: ~ **de chargement** *m* DÉCHETS loading mechanism; ~ **de commande pneumatique** *m* POLL EAU pneumatic driving gear; ~ **de décharge** *m* DÉCHETS dumping mechanism; ~ **de l'économie de marché** *m* POL ENV market-economy instrument; ~ **d'entraînement de grille** *m* DÉCHETS grate drive; ~ **d'orientation** *m* ÉN RENOUV tracking mechanism; ~ **de pointage** *m* ÉN RENOUV *de la course du soleil* tracking mechanism; ~ **de poursuite** *m* ÉN RENOUV *de la course du soleil* tracking mechanism; ~ **de secouage** *m* POLL ATM *dépoussiérage* shaker mechanism; ~ **de suivi des vents** *m* ÉN RENOUV wind direction tracking mechanism; ~ **de surveillance** *m* POL ENV monitoring instrument

mécanismes: ~ **du marché** *m pl* GESTION market forces

mèche: ~ **à bois à simple spirale** *f* DÉCHETS gimlet bit

médium *m* DÉCHETS medium

mégalopole *f* GESTION, POL ENV megalopolis (*AmE*)

mégawatt *m* (*MW*) ÉN RENOUV megawatt (*MW*)

meilleure: ~ **technique disponible** *f* DIST EAU, POL ENV, POLL ATM, SOLS best available technology (*BAT*); ~ **technique disponible n'entraînant pas des coûts excessifs** *f* DIST EAU, POL ENV, POLL ATM, SOLS best available technology not entailing excessive costs (*BATNEEC*); ~ **technique existante** *f* DIST EAU, POL ENV, POLL ATM, SOLS best available technology (*BAT*); ~ **technique existante n'entraînant pas des coûts excessifs** *f* DIST EAU, POL ENV, POLL ATM, SOLS best available technology not entailing excessive costs (*BATNEEC*); ~ **technologie existante** *f* DIST EAU, POL ENV, POLL ATM, SOLS best available technology (*BAT*); ~ **technologie existante n'entraînant pas des coûts excessifs** *f* DIST EAU, POL ENV, POLL ATM, SOLS best available technology not entailing excessive costs (*BATNEEC*)

mélange *m* GÉNÉRAL mixture, commixture, mixing; ~ **à des agrégats** SOLS mixing with aggregates; ~ **de boues fécales à l'état solide** POLL EAU faecal solids sludge composition (*BrE*), fecal solids sludge composition (*AmE*); ~ **ciment-argile** SOLS cement-clay mixture; ~ **de déchets** DÉCHETS waste mixture; ~ **déchets-boues** DÉCHETS waste-sludge mixture; ~ **déchets-charbon** DÉCHETS waste-coal mixture; ~ **détonant** POLL ATM explosive mixture; ~ **eau-ciment** SOLS cement-water mixture; ~ **explosif** POLL ATM explosive mixture; ~ **de fibres** POLL ATM fiber mixture (*AmE*), fibre mixture (*BrE*); ~ **gazeux** ÉCHANT gas mixture; ~ **intégral** POLL ATM thorough mixture; ~ **intime eau-boue** DIST EAU intimate water-sludge mixture; ~ **liquide-poussière** POLL ATM dust-liquid mixture; ~ **de poussière** POLL ATM dust mixture; ~ **solvant à base de benzène, de toluol et de xylène** POLL EAU solvent mix from benzene, toluol and xylene

mélangé *adj* GÉNÉRAL mixed

mélanger *vt* GÉNÉRAL mix, blend; ~ **avec des matériaux vierges** DÉCHETS *des plastiques recyclés* blend with virgin materials; ~ **intégralement** ÉCHANT mix thoroughly; ~ **par secousses** SOLS shake in

mélangeur[1] *adj* GÉNÉRAL mixing

mélangeur[2] *m* DÉCHETS, POLL EAU compounder, mixer, mixing device; ~ **à ailettes** DIST EAU paddle mixer; ~ **de gaz** ÉCHANT *chimie* gas mixer; ~ **à hélice** POLL EAU impeller-type mixer; ~ **primaire** DÉCHETS premixer; ~ **de produits de fusion** DÉCHETS melt mixer; ~ **rapide** DÉCHETS high-speed mixer; ~ **à soc** DÉCHETS, DIST EAU ploughshare mixer (*BrE*), plowshare mixer (*AmE*); ~ **statique** POLL EAU static mixer

membrane *f* ÉCHANT, POLL ATM, POLL EAU membrane; ~ **échangeuse d'ions** POLL EAU ion-exchanger membrane; ~ **à imprégnation** POLL ATM permeation membrane; ~ **microporeuse** POLL EAU microporous membrane; ~ **en polyamide** POLL EAU polyamide membrane; ~ **en polycarbonate** POLL EAU polycarbonate membrane; ~ **en polyester** POLL EAU elution; ~ **poreuse** DIST EAU porous membrane, POLL ATM permeation membrane, POLL EAU porous membrane; ~ **sélective d'ions** POLL EAU ion-selective membrane; ~ **semi-perméable** DIST EAU, POLL EAU semipermeable membrane

menace: ~ **écologique** *f* POL ENV ecological menace, ecological threat; ~ **pour l'environnement** *f* SÉCURITÉ unsafe environmental condition

menacé *adj* POL ENV endangered

mer[1] *f* POLL MAR sea; ~ **fermée** POLL ATM, POLL MAR enclosed sea, landlocked sea

mer[2]: **en** ~ *loc* POLL MAR offshore, sea-based

mercaptan *m* DÉCHETS, POLL ATM mercaptan, thiol; ~ **éthylique** DÉCHETS ethylmercaptane

mercatique: ~ **à orientation écologique** *f* GESTION environmentally-orientated marketing

mercure *m* (*Hg*) POLL EAU mercury (*Hg*)

mercuriel *adj* POLL EAU mercurial

mésophile *adj* POLL EAU *biologie, écologie* mesophilic

mésopore *m* POLL ATM mesopore

mésosphère *f* POLL ATM mesosphere

mésotrophie *f* POLL EAU mesotrophic water

mesurabilité *f* ÉCHANT measurability

mesurable *adj* ÉCHANT measurable

mesurage *m* ÉCHANT measurement, measuring; ~ **en continu** ÉCHANT on-line measurement, continuous measuring; ~ **discontinu** ÉCHANT discontinuous measuring

mesure *f* ÉCHANT *action* measurement, *grandeur* measure, measurement, measured value, GESTION, POL ENV *démarche* measure; ~ **à l'air libre** ÉCHANT open-air measurement; ~ **d'alcalinité** ÉCHANT alkalinity measurement; ~ **après fermeture** DIST EAU post-closure measure; ~ **atmosphérique** ÉCHANT ambient-air measuring; ~ **en aval** GESTION end-of-the-pipe measure; ~ **axée sur les sources** SOLS source-oriented measure; ~ **du bruit** ÉCHANT, POLL SON noise measurement, loudness measurement; ~ **du bruit ambiant** POLL SON, SÉCURITÉ environmental noise measurement; ~ **de comparaison** ÉCHANT comparative measurement; ~ **comparée** ÉCHANT comparative measurement; ~ **de la concentration d'oxygène** ÉCHANT, POLL ATM measurement of oxygen concentration; ~ **de la concentration de poussière** ÉCHANT, POLL ATM measurement of dust concentration; ~ **de confinement** SOLS encapsulation measure; ~ **contre la pollution** GESTION, POL ENV antipollution measure; ~ **de contrôle** SOLS control, measure; ~ **correctrice provisoire** SOLS temporary remediation; ~ **à court terme** GESTION, POL ENV short-term measure; ~ **de débit** ÉCHANT measurement of flow rate; ~ **de débit de gaz** ÉCHANT measurement of gas-flow rate; ~ **défectueuse** ÉCHANT faulty measurement; ~ **destinée à éviter la production de déchets primaires** DÉCHETS primary waste avoidance; ~ **discontinue** ÉCHANT discontinuous measuring; ~ **électrique de l'humidité** ÉCHANT electrical moisture measurement; ~ **électrochimique de l'ozone** ÉCHANT electrochemical ozone measurement; ~ **d'étanchéité** DIST EAU sealing measure; ~ **en fin de cycle** GESTION end-of-the-pipe measure; ~ **des gaz de décharge** ÉCHANT landfill gas measurement; ~ **de l'humidité** ÉCHANT humidity measurement; ~ **hydraulique parallèle** SOLS accompanying hydraulic measure; ~ **hygrométrique** ÉCHANT humidity measurement; ~ **d'intensité de la lumière diffusée** ÉCHANT measurement of intensity of scattered light; ~ **à long terme** GESTION, POL ENV long-term measure; ~ **de la lumière** ÉCHANT light measuring; ~ **de la lumière diffuse** ÉCHANT diffused-light measurement; ~ **de la lumière parasite** ÉCHANT diffused-light measurement; ~ **de lutte** POLL ATM *contre la pollution* abatement measure; ~ **de lutte contre les déchets dangereux** DÉCHETS hazardous waste avoidance; ~ **manométrique** ÉCHANT manometric measurement; ~ **olfactométrique** ÉCHANT, POLL ATM olfactometric measure; ~ **par ultrasons** ÉCHANT measuring with ultrasound, ultrasonic measurement; ~ **de pH** ÉCHANT pH-measurement; ~ **de la pollution atmosphérique** POLL ATM, SÉCURITÉ air pollution measurement; ~ **ponctuelle** ÉCHANT spot measurement; ~ **potentielle** ÉCHANT potential measurement; ~ **de précaution** SÉCURITÉ precautionary measure; ~ **de la précipitation de poussière** ÉCHANT, POLL ATM measurement of dust precipitation; ~ **de précision** ÉCHANT precision measurement; ~ **de pression acoustique** ÉCHANT sound pressure measuring; ~ **de pression de vapeur** ÉCHANT vapor pressure measurement (*AmE*), vapour pressure measurement (*BrE*); ~ **de prévention du bruit** POLL MAR, POLL SON noise prevention measure; ~ **de prévention contre les déchets** DÉCHETS waste avoidance; ~ **de prévention et de lutte intégrées contre la pollution** GESTION integrated prevention and pollution control (*IPPC*); ~ **préventive** GÉNÉRAL preventive measure; ~ **préventive applicable à l'environnement** POL ENV preventive environmental policy; ~ **préventive contre les déchets primaires** DÉCHETS primary-waste avoidance; ~ **préventive contre la production de déchets secondaires** DÉCHETS secondary-waste avoidance; ~ **préventive de protection de l'environnement** POL ENV preventive environmental protection; ~ **prise en cas de marée noire** POLL MAR oil spill response; ~ **de protection** POLL ATM, POLL MAR, SÉCURITÉ protective measure; ~ **de protection contre le bruit** POLL MAR, POLL SON noise protection measure; ~ **de protection de l'environnement** POL ENV environmental protection measure, environmental protection policy; ~ **protective** GESTION, POL ENV protective measure; ~ **de la qualité de l'air** ÉCHANT, POLL ATM air quality measurement; ~ **de la quantité de traces** ÉCHANT measurement of trace amounts; ~ **récursive** ÉCHANT repeated measurement; ~ **régionale** POL ENV regional policy; ~ **relative à l'environnement** POL ENV environmental policy, environmental policy measure; ~ **de sécurité** SÉCURITÉ security measure, safety measure; ~ **de sécurité axée sur les effets** SOLS effect-oriented security measure; ~ **simultanée** ÉCHANT simultaneous measurement; ~ **de sûreté** SÉCURITÉ, SOLS safeguard; ~ **de surveillance** SÉCURITÉ monitoring measure; ~ **technique de sécurité** SÉCURITÉ technical safety measure; ~ **de température** ÉCHANT temperature measurement; ~ **ultrasonique** ÉCHANT ultrasonic measurement; ~ **unique** ÉCHANT single measurement; ~ **d'urgence** POLL ATM, SÉCURITÉ, SOLS emergency measure; ~ **de volume de gaz** ÉCHANT measurement of gas quantity

mesurer *vt* DIST EAU, ÉCHANT, POLL ATM measure, *avec un compteur* meter; ~ **avec une pipette** ÉCHANT measure with a pipette; ~ **directement** ÉCHANT measure directly; ~ **par méthode photométrique** ÉCHANT measure photometrically

mesures: ~ **foncières** *f pl* POL ENV land policy

mesureur *m* DÉCHETS metering machine

métabolisme *m* DIST EAU, POLL ATM, POLL EAU metabolism

métabolite *m* DIST EAU, POLL ATM, POLL EAU metabolite

métal: ~ **catalytique actif** *m* POLL ATM catalytic active metal; ~ **fondu** *m* SÉCURITÉ molten metal; ~ **liquide** *m* DIST EAU liquid metal; ~ **lourd** *m* DÉCHETS, POLL ATM, POLL EAU heavy metal; ~ **lourd volatil** *m* DÉCHETS, POLL ATM, POLL EAU volatile heavy metal; ~ **noble** *m* DIST EAU, POLL ATM noble metal; ~ **non toxique** *m* POLL ATM, POLL EAU, SÉCURITÉ, SOLS nontoxic metal; ~ **en traces** *m* SOLS trace metal

métalimnion *m* POLL EAU metalimnion

métallifère *adj* POLL EAU *minerai* metalliferous

métallisation *f* DÉCHETS metal coating

météorologie *f* GESTION, POLL ATM meteorology

méthane *m* DÉCHETS, DIST EAU, POLL ATM, POLL EAU methane

méthanogénese *f* POLL EAU methangenesis

méthanogène *adj* DÉCHETS, DIST EAU, POLL ATM, POLL EAU methane-producing, methanogenic

méthanolyse *f* DÉCHETS methanolysis

méthode *f* GÉNÉRAL method, procedure; ~ **de l'alvéole**

DÉCHETS cell method; ~ **d'analyse** ÉCHANT analysis method; ~ **d'analyse paramagnétique** ÉCHANT paramagnetic analytical method; ~ **d'application** POLL ATM, SOLS application method; ~ **à l'avancement** DÉCHETS tip filling method; ~ **Bacharach** POLL ATM Bacharach method; ~ **calorimétrique à la pararosaniline** ÉCHANT calorimetric pararosaniline method; ~ **du casier** DÉCHETS cell method; ~ **CHEMFIX®** SOLS CHEMFIX® method; ~ **chimique par voie humide** ÉCHANT wet chemical method; ~ **complexométrique** ÉCHANT *volumétrie* complexometric method; ~ **de compostage** DÉCHETS composting technique; ~ **conductimétrique** ÉCHANT conductometric method; ~ **de désorption purge and trap** ÉCHANT purge-and-trap method; ~ **de détection** ÉCHANT detection method; ~ **de détection qualitative au microscope** ÉCHANT microscopic qualitative detection method; ~ **de détermination** ÉCHANT determination method; ~ **de détermination ampèremétrique** ÉCHANT amperometric determination method; ~ **de détermination polarographique** ÉCHANT polarographic determination method; ~ **d'échantillonnage** ÉCHANT sampling method; ~ **économisant l'énergie** POLL ATM energy-saving method; ~ **électrochimique** ÉCHANT electrochemical method; ~ **d'étalonnage** ÉCHANT calibration mode; ~ **d'extraction par stripping** ÉCHANT stripping-purging technique; ~ **forte consommatrice de temps** POLL ATM, POLL MAR, POLL SON, SOLS time-intensive method; ~ **à forte intensité de main-d'œuvre** DIST EAU, POLL MAR, POLL SON, SOLS labour-intensive method; ~ **gazométrique** ÉCHANT gasometric method; ~ **gravimétrique** ÉCHANT gravimetric method; ~ **de Griess Saltzmann** POLL ATM Griess Saltzmann method; ~ **d'impact électronique** ÉCHANT electron impact method; ~ **intégrée** POL ENV integrated approach; ~ **iodométrique** ÉCHANT iodometric method; ~ **de longue durée** POLL ATM, POLL MAR, POLL SON, SOLS time-intensive method; ~ **longue à mettre en œuvre** POLL ATM, POLL MAR, POLL SON, SOLS time-intensive method; ~ **manométrique dynamique** ÉCHANT dynamic headspace technique (*DHS*); ~ **mécanique d'épaississement des boues** POLL EAU mechanical sludge-thickening method; ~ **de mesure** ÉCHANT measuring method; ~ **de mesure directe** ÉCHANT direct measurement method; ~ **de mesure indirecte** ÉCHANT indirect measurement method; ~ **mise en œuvre in situ** SOLS in-situ method; ~ **mobile de mesure en laboratoire** ÉCHANT mobile laboratory-measuring method; ~ **mobile de mesure sur le terrain** ÉCHANT mobile field-measuring method; ~ **naturelle de déshydratation** POLL EAU natural dewatering system; ~ **par courants contrariés** ÉCHANT impinge method; ~ **à la pararosaniline** ÉCHANT *calorimétrie* pararosaniline method; ~ **perfectionnée d'épuration du charbon** POLL ATM advanced coal cleaning; ~ **peu consommatrice d'énergie** POLL ATM energy-saving method; ~ **de production laissant peu de résidus** DÉCHETS low-residue production method; ~ **de production sans déchets** DÉCHETS residue-free production method; ~ **de reconnaissance des formes** DIST EAU shape recognition method; ~ **reconnue** POL ENV generally-accepted technical procedure; ~ **Ringelmann** POLL ATM Ringelmann method; ~ **de saturation des gaz** ÉCHANT gas saturation method; ~ **SEALOSAFE®** SOLS SEALOSAFE® method; ~ **spectroscopique** ÉCHANT spectroscopic method; ~ **sur colonne** ÉCHANT *chromatographie* column method; ~ **de surface** DÉCHETS surface method; ~ **de tamisage par veine d'air** ÉCHANT air-jet screening method; ~ **de titration** ÉCHANT titration method; ~ **de traitement coûteuse** GESTION, POLL ATM cost-intensive method of treatment; ~ **des tranchées** DÉCHETS trench method; ~ **de travail peu bruyante** POLL SON low-noise working method; ~ **UCT** (*méthode de l'Université du Cape Town*) POLL EAU UCT method (*University of Cape Town method*); ~ **de l'Université du Cape Town** (*méthode UCT*) POLL EAU University of Cape Town method (*UCT method*); ~ **de vibration** ÉCHANT vibration method; ~ **volumétrique statique** ÉCHANT static headspace method

métropole: ~ **d'équilibre** *f* POL ENV metropolitan growth center (*AmE*), metropolitan growth centre (*BrE*)

mettre *vt* GÉNÉRAL put; ~ **en culture** ÉN RENOUV *terre*, SOLS bring into cultivation; ~ **en décharge** DÉCHETS, DIST EAU, POLL EAU, SOLS landfill; ~ **en œuvre** POL ENV implement; ~ **au point** GESTION *procédé* develop; ~ **au rebut** GÉNÉRAL dispose of; ~ **en silo** DÉCHETS, ÉN RENOUV convey into silos; ~ **en valeur** ÉN RENOUV harness

meuble *adj* SOLS loose

meulage *m* DÉCHETS, POLL EAU grinding

meuler *vt* DÉCHETS, POLL EAU grind

meuleuse *f* POLL EAU grinder

Mg *abrév* (*magnésium*) GÉNÉRAL Mg (*magnesium*)

micelle *f* POLL ATM finest dust

microbe *m* POLL ATM, POLL EAU, SOLS microbe, germ; ~ **pathogène** DIST EAU pathogenic microbe

microbien *adj* POLL ATM, POLL EAU, SOLS microbial

microbiologie *f* ÉCHANT, SÉCURITÉ, SOLS microbiology

microbiologique *adj* ÉCHANT, SÉCURITÉ, SOLS microbiological

microcentrale *f* ÉN RENOUV *hydroélectrique* kilowatt plant

microclimat *m* POLL ATM microclimate

microcoagulation *f* POLL EAU microcoagulation

microcolorimétrie *f* ÉCHANT microcolorimetry

microconfinement *m* SOLS microencapsulation

microcoulométrique *adj* ÉCHANT microcoulometric

microfiltration *f* POLL EAU microfiltration

microfiltre *m* POLL EAU microfilter

microflore *f* POLL ATM, SOLS microflora

micromilieu *m* POLL ATM, SOLS microenvironment

micromiroir: ~ **numérique** *m* ÉCHANT digital micromirror device (*DMD*)

micro-organisme *m* DIST EAU, POLL ATM, POLL EAU, SOLS microorganism; ~ **facultatif** DIST EAU, POLL ATM, POLL EAU, SOLS facultative microorganism; ~ **immobilisé** DIST EAU, POLL ATM, POLL EAU, SOLS immobilized microorganism; ~ **mésophile** DIST EAU, POLL ATM, POLL EAU, POLL MAR mesophilic microorganism; ~ **pathogène** DIST EAU, POLL EAU pathogenic microorganism; ~ **psychrophile** POLL EAU psychrophilic microorganism; ~ **revivifiable** POLL EAU, POLL MAR viable microorganism; ~ **thermophile** POLL EAU, POLL MAR thermophilic microorganism

microphone *m* ÉCHANT microphone; ~ **atténuateur de bruit** POLL SON antinoise microphone; ~ **à électret** ÉCHANT electret microphone; ~ **piézo-électrique** ÉCHANT piezo-electric microphone

micropipette *f* ÉCHANT micropipette

micropore *m* POLL ATM, POLL EAU micropore

microporeux *adj* POLL ATM, POLL EAU *pédologie, géologie, biologie* microporous

microprocesseur *m* ÉCHANT microprocessor

microscope *m* ÉCHANT microscope; ~ **à balayage électronique** ÉCHANT scanning electron microscope (*SEM*); ~ **à effet tunnel** ÉCHANT scanning tunnel microscope; ~ **électronique** ÉCHANT electron microscope; ~ **optique** ÉCHANT light-optical microscope

microscopie: ~ **à balayage électronique** *f* ÉCHANT scanning electron microscopy (*SEM*); ~ **électronique** *f* ÉCHANT electron microscopy; ~ **électronique à transmission** *f* ÉCHANT transmission electron microscopy (*TEM*); ~ **infrarouge à transformée de Fourier** *f* ÉCHANT Fourier-transform infrared microscopy

microscopique *adj* ÉCHANT, SÉCURITÉ microscopic

microtamisage *m* POLL EAU *traitement des eaux* microscreening

micro-usiné *adj* ÉCHANT micromachined

migration: ~ **dans les sols** *f* SOLS leaching; ~ **de gaz** *f*

DÉCHETS gas migration; ~ **horizontale des gaz** *f* DÉCHETS, SOLS horizontal gas migration; ~ **ionique** *f* SOLS ionic migration; ~ **de substances nocives** *f* SOLS migration of harmful substances
migrer *vi* DÉCHETS, DIST EAU, SOLS migrate
milieu *m* GÉNÉRAL environment, medium; ~ **ambiant** SOLS environmental medium; ~ **aménagé** DIST EAU planned environment; ~ **animal** SOLS animal environment; ~ **de culture** ÉCHANT culture medium; ~ **de dispersion** POLL ATM dispersion medium; ~ **écologique** POL ENV ecological conditions; ~ **d'essai** ÉCHANT test environment; ~ **gazeux** DIST EAU, POLL ATM gaseous medium; ~ **géologique** ÉN RENOUV geological conditions; ~ **hostile** GESTION, POLL ATM, POLL EAU, SOLS hostile environment; ~ **marin** POLL MAR marine environment; ~ **marin côtier** POLL MAR onshore marine environment; ~ **de mesure** ÉCHANT measurement conditions; ~ **naturel** DIST EAU, GESTION, POLL ATM, POLL EAU, SOLS natural environment; ~ **pédologique** SOLS soil environment; ~ **pélagique** POLL MAR pelagic environment; ~ **physico-chimique** DIST EAU, POLL ATM, POLL EAU physiochemical environment; ~ **rural** GESTION, POL ENV rural environment; ~ **peu sûr** SÉCURITÉ unsafe environmental condition; ~ **végétal** SOLS plant environment; ~ **zoologique** POLL ATM, SOLS animal environment
millivolt *m* (*mV*) ÉCHANT millivolt (*mV*)
minerai *m* GÉNÉRAL ore; ~ **de fer** GÉNÉRAL iron ore
minéral *m* GÉNÉRAL mineral
minéralisateur *adj* DÉCHETS, POLL ATM, POLL EAU, SOLS mineralizing
minéralisation *f* DÉCHETS, POLL ATM, POLL EAU, SOLS mineralization
minéraliser *vt* DÉCHETS, POLL ATM, POLL EAU, SOLS mineralize
minicentrale *f* ÉN RENOUV *hydroélectrique* kilowatt plant
ministère *m* POL ENV department, ministry; ~ **de l'environnement** POL ENV department of the environment, environment ministry
ministériel *adj* POL ENV ministerial
ministre *m* POL ENV minister; ~ **de l'environnement** POL ENV environment minister, minister for the environment
miroir *m* ÉN RENOUV mirror; ~ **cannelé** ÉCHANT grooved mirror; ~ **parabolique** ÉN RENOUV parabolic dish; ~ **renversé** ÉCHANT tilted mirror; ~ **à vide** ÉN RENOUV *miroir solaire réfléchissant* mirror-evacuated tube
mis: ~ **en danger** *loc* SÉCURITÉ endangered; ~ **en décharge** *loc* DÉCHETS, DIST EAU, POLL EAU, SOLS landfilled; ~ **en diffusion générale** *loc* SÉCURITÉ derestricted; ~ **à l'épreuve** *loc* ÉN RENOUV, POLL ATM, SOLS field-proven
miscibilité *f* DÉCHETS, DIST EAU, POLL EAU miscibility
mise: ~ **à l'air** *f* DIST EAU, POLL ATM vent; ~ **en balles** *f* ÉN RENOUV *de la paille* baling; ~ **en caisse** *f* DÉCHETS case packing; ~ **en charge** *f* DÉCHETS loading; ~ **en charge électrique** *f* POLL ATM field charging; ~ **en décharge** *f* DÉCHETS dumping landfill disposal, landfilling; ~ **en décharge à ciel ouvert** *f* DÉCHETS above-ground landfilling; ~ **en décharge conjointe** *f* DÉCHETS co-disposal landfill; ~ **en décharge contrôlée** *f* DÉCHETS sanitary landfilling; ~ **en décharge de déchets** *f* DÉCHETS waste landfilling, residue landfilling; ~ **en décharge de déchets dangereux** *f* DÉCHETS hazardous waste landfilling; ~ **en décharge souterraine** *f* DÉCHETS underground landfilling; ~ **en eau** *f* DIST EAU filling, flooding, POLL ATM watering, POLL EAU flooding; ~ **en œuvre** *f* POL ENV implementation; ~ **en œuvre de la précipitation** *f* POLL EAU precipitation operation; ~ **en péril de l'environnement** *f* POL ENV environmental endangerment; ~ **au point** *f* GESTION development; ~ **au rebut** *f* GÉNÉRAL disposal; ~ **en valeur de l'environnement** *f* GESTION environmental improvement; ~ **en valeur des sols** *f* SOLS soil reclamation, land reclamation; ~ **en valeur des terres** *f* SOLS soil reclamation, land reclamation; ~ **en vigueur** *f* POL ENV enforcement
mitraille *f* DÉCHETS scrap iron, junk iron, ferrous scrap, metal srap; ~ **pour paqueter** DÉCHETS baling scrap
mixte *adj* GÉNÉRAL mixed
Mn *abrév* (*manganèse*) POLL ATM Mn (*manganese*)
Mo *abrév* (*molybdène*) POLL ATM Mo (*molybdenum*)
mobile *adj* GÉNÉRAL mobile
mobilier: ~ **urbain** *m* POL ENV street furniture
mobilisation *f* DÉCHETS, DIST EAU, SOLS mobilization
mobilisé *adj* DÉCHETS, DIST EAU, SOLS mobilized
mobiliser *vt* DÉCHETS, DIST EAU mobilize, ÉN RENOUV *ressources* harness, SOLS mobilize
mobilité: ~ **ionique** *f* ÉCHANT ionic mobility
mode *m* GÉNÉRAL method; ~ **d'application** SÉCURITÉ method of application; ~ **de balayage** ÉCHANT scanning method; ~ **de collecte** POLL ATM collection behavior (*AmE*), collection behaviour (*BrE*); ~ **de consommation tenant compte de l'environnement** GESTION environmentally-aware buying, environmentally-aware purchasing; ~ **d'échantillonnage** ÉCHANT manner of sampling; ~ **d'élimination des déchets acceptable pour l'environnement** DÉCHETS environmentally-acceptable waste disposal; ~ **d'élimination de déchets neutre pour l'environnement** DÉCHETS environmentally-neutral disposal; ~ **d'emploi** DÉCHETS instructions for use, DIST EAU, POL ENV use recommendation; ~ **opératoire** GÉNÉRAL procedure; ~ **de stockage en environnement contrôlé** DÉCHETS controlled environment storage system
modèle: ~ **climatique mathématique** *m* POLL ATM mathematical climate model; ~ **climatique mondial** *m* POLL ATM global climate model; ~ **climatique planétaire** *m* POLL ATM global climate model; ~ **des eaux souterraines** *m* DIST EAU, ÉCHANT, GESTION groundwater model; ~ **d'écoulement** *m* DIST EAU flow pattern; ~ **de Gauss** *m* ÉCHANT Gaussian model; ~ **à grille numérique** *m* ÉCHANT numerical grid model; ~ **d'observation des polluants atmosphériques** *m* GESTION airborne pollutant observation model; ~ **de prévision** *m* ÉCHANT forecast model; ~ **spatial** *m* POLL ATM spatial pattern; ~ **de transport** *m* DIST EAU *polluants* transport model
modélisation: ~ **de la qualité de l'eau** *f* GESTION modeling of water quality (*AmE*), modelling of water quality (*BrE*)
modem: ~ **RTC** *m* SÉCURITÉ STN-modem
modernisation *f* GÉNÉRAL modernization, retrofitting, upgrading; ~ **d'installations anciennes** DÉCHETS, DIST EAU, POLL EAU modernization of old plants
moderniser *vt* GÉNÉRAL modernize, retrofit
modificateur: ~ **de tension superficielle** *m* POLL MAR surface-tension modifier
modification: ~ **écologique structurelle** *f* POL ENV ecological structural change
modulaire *adj* DÉCHETS, ÉCHANT modular
module: ~ **d'alimentation en air propre amovible** *m* POLL ATM moveable clean air feed; ~ **de catalyseur** *m* POLL ATM catalyst module; ~ **de cellules photovoltaïques** *m* ÉN RENOUV solar cell module; ~ **de cellules photovoltaïques en couche mince** *m* ÉN RENOUV thin-film module; ~ **à enroulement spirale** *m* POLL EAU *osmose inverse* spiral module; ~ **photovoltaïque** *m* ÉN RENOUV *centrale solaire* photovoltaic module; ~ **photovoltaïque couplé au réseau** *m* ÉN RENOUV *centrale solaire* grid-coupled photovoltaic module; ~ **à plaques** *m* POLL EAU plate module; ~ **en plastique** *m* POLL EAU plastic module; ~ **solaire** *m* ÉN RENOUV *centrale solaire* solar module; ~ **tubulaire** *m* POLL EAU tubular module
moisissure *f* GÉNÉRAL mold (*AmE*), mould (*BrE*)
moissonner *vt* ÉN RENOUV harvest
moissonneuse *f* ÉN RENOUV *céréales* harvester; **~-batteuse** ÉN RENOUV *céréales* harvester
mol *abrév* (*mole*) ÉCHANT mol (*mole*)
mole *f* (*mol*) ÉCHANT mole (*mol*)
moléculaire *adj* GÉNÉRAL molecular

molécule *f* ÉCHANT, POLL ATM, POLL EAU, SOLS molecule; **~ de gaz** POLL ATM gas molecule; **~ de gaz ionisé** POLL ATM ionized gas molecule; **~ de gaz neutre** POLL ATM neutral gas molecule
molybdène *m* (*Mo*) POLL ATM molybdenum (*Mo*)
mondial *adj* ÉN RENOUV, POLL ATM global
moniteur *m* ÉCHANT monitor; **~ longue distance** ÉCHANT long-path monitor
monobloc *adj* POLL ATM compact
monocristallin *adj* ÉN RENOUV *structure* single-crystalline
mono-étage *adj* POLL ATM, DIST EAU, POLL EAU single-stage
monolithe *m* POLL ATM monolith; **~ céramique** POLL ATM ceramic monolith
monolithique *adj* POLL ATM monolithic
monomère *m* DÉCHETS, DIST EAU monomer
monoxyde: **~ d'azote** *m* (*NO*) DIST EAU, POLL ATM nitric oxide (*NO*); **~ de carbone** *m* (*CO*) POLL ATM carbon monoxide (*CO*); **~ de diazote** *m* (N_2O) POLL ATM nitrous oxide (N_2O); **~ de soufre** *m* (*SO*) POLL ATM sulfur monoxide (*AmE*) (*SO*), sulphur monoxide (*BrE*) (*SO*)
montage *m* DIST EAU assembly; **~ antivibration** POLL SON antivibration mounting; **~ de mesure** ÉCHANT measuring set-up; **~ résilient** POLL SON resilient mounting; **~ tandem** POLL ATM tandem mounting
montre: **~ Seger** *f* ÉCHANT Seger cone
morbidité *f* POL ENV, POLL EAU morbidity
mort *adj* POLL EAU dead, still
mortalité *f* POL ENV, POLL EAU mortality
morte: **~-eau** *f* ÉN RENOUV neap tide
moteur[1]: **~ d'absorption** *m* POLL ATM absorber motor; **~ à air chaud** *m* ÉN RENOUV, POLL ATM Stirling engine (*BrE*), Stirling motor (*AmE*); **~ à air dilaté** *m* ÉN RENOUV, POLL ATM Stirling engine (*BrE*), Stirling motor (*AmE*); **~ à allumage par étincelles** *m* POLL ATM spark ignition engine (*BrE*), spark ignition motor (*AmE*); **~ à combustion externe** *m* ÉN RENOUV, POLL ATM external combustion engine; **~ à combustion interne** *m* DÉCHETS, POLL ATM, POLL EAU internal combustion engine; **~ à deux temps** *m* POLL ATM, POLL EAU two-cycle engine (*AmE*), two-stroke engine (*BrE*); **~ diesel** *m* DÉCHETS, POLL ATM, POLL EAU diesel engine; **~ électrique** *m* POLL EAU electrical motor; **~ d'épuisement** *m* DIST EAU, POLL EAU draining engine, pumping engine; **~ à essence** *m* POLL ATM spark ignition engine (*BrE*), spark ignition motor (*AmE*); **~ à induction à cage d'écureuil** *m* ÉN RENOUV, POLL EAU squirrel-cage induction motor; **~ à induction à rotor bobiné** *m* POLL EAU wound-rotor induction motor; **~ d'intégration** *m* ÉCHANT integrating motor; **~ à mélange pauvre** *m* POLL ATM lean-mix engine; **~ à mélange pauvre avec catalyseur d'oxydation** *m* POLL ATM lean-mix engine with oxidation catalyst; **~ à piston alternatif** *m* POLL ATM reciprocating piston engine; **~ de pompage** *m* DIST EAU water-drawing engine; **~ de pompe** *m* DIST EAU, POLL EAU pump-driving gear; **~ à pompes** *m* POLL EAU pumping engine; **~ à quatre temps** *m* DÉCHETS, POLL ATM, POLL EAU four-cycle engine (*AmE*), four-stroke engine (*BrE*); **~ silencieux** *m* POLL MAR, POLL SON, SÉCURITÉ noiseless motor; **~ de soutirage d'eau** *m* DIST EAU water-drawing engine; **~ Stirling** *m* ÉN RENOUV, POLL ATM Stirling engine (*BrE*), Stirling motor (*AmE*); **~ à vapeur à cycle de Rankine** *m* POLL ATM Rankine cycle steam engine; **~ à vitesse variable** *m* POLL EAU variable-speed motor
moteur[2]: **à ~** *loc* POLL EAU motorized
motopompe *f* DIST EAU, POLL ATM, POLL MAR motor pump
motorisé *adj* POLL EAU motorized
mouillabilité *f* POLL ATM, SOLS wettability, wetting
mouillable *adj* POLL ATM, SOLS wettable
mouillage: **~ de particule** *m* POLL ATM particle wetting
mouillant *adj* POLL ATM, SOLS wetting
mouillé *adj* POLL ATM, SOLS wet, wetted
moulage: **~ par air comprimé** *m* DÉCHETS blow molding (*AmE*), blow moulding (*BrE*); **~ par compression** *m* DÉCHETS compression molding (*AmE*), compression moulding (*BrE*); **~ par injection** *m* DÉCHETS injection molding (*AmE*), injection moulding (*BrE*); **~ par injection en couches alternées** *m* DÉCHETS sandwich injection molding (*AmE*), sandwich injection moulding (*BrE*); **~ par injection stratifiée** *m* DÉCHETS sandwich injection molding (*AmE*), sandwich injection moulding (*BrE*); **~ par injection-réaction en moule fermé** *m* DÉCHETS reinforced reaction injection molding (*AmE*) (*R-RIM*), reinforced reaction injection moulding (*BrE*) (*R-RIM*); **~ par injection-réaction en moule ouvert** *m* DÉCHETS structural reaction injection molding (*AmE*) (*S-RIM*), structural reaction injection moulding (*BrE*) (*S-RIM*); **~ par pressage basse pression avec injection de résine** *m* DÉCHETS resin transfer molding (*AmE*) (*RTM*), resin transfer moulding (*BrE*) (*RTM*); **~ par soufflage** *m* DÉCHETS blow molding (*AmE*), blow moulding (*BrE*); **~ de plastique** *m* DÉCHETS plastic molding (*AmE*), plastic moulding (*BrE*), plastics molding (*AmE*), plastics moulding (*BrE*)
moule *m* GÉNÉRAL mold (*AmE*), mould (*BrE*)
moulin: **~ à vent** *m* ÉN RENOUV windmill
moussage *m* POLL ATM foaming, frothing
mousse *f* POLL EAU, POLL MAR froth, scum, foam; **~ cloquée** DÉCHETS *plastique* close-cell foam; **~ de polyester** DÉCHETS, DIST EAU polyester foam
mouvement: **~ de l'air** *m* POLL ATM air motion; **~ brownien** *m* POLL ATM *état gazeux* Brownian movement; **~ circulaire** *m* POLL ATM circular movement; **~ de la mer** *m* ÉN RENOUV motion of the sea; **~ rotatif** *m* POLL ATM rotation movement; **~ transfrontière des déchets** *m* DÉCHETS transboundary waste movement; **~ des vagues** *m* ÉN RENOUV wave movement
moyenne *f* ÉCHANT mean value; **~ annuelle** ÉCHANT annual mean value; **~ horaire** ÉCHANT hourly mean value (*HMV*); **~ journalière** ÉCHANT daily average value, daily mean value; **~ mensuelle** ÉCHANT monthly mean value; **~ à pondération temporelle** ÉCHANT time interval; **~ pondérée dans le temps** ÉCHANT time-weighed average (*TWA*); **~ semi-horaire** ÉCHANT half-hourly average value, half-hourly mean value
moyens: **~ de mesure** *m pl* ÉCHANT measuring means
mullite *f* POLL ATM mullite
multicellulaire *adj* DIST EAU multistage
multicomposant *adj* ÉCHANT multicomponent
multicouche *adj* POLL SON multisheet
multi-élément *adj* ÉCHANT multicomponent
multi-étage *adj* GÉNÉRAL multistage
multifeuille *adj* POLL SON multisheet
multilatéral *adj* POL ENV multilateral
multiplicateur: **~ photoélectronique** *m* ÉCHANT photoelectron multiplier
mur: **~ antibruit** *m* POLL SON noise-insulating barrier; **~ antibruit constitué de déchets** *m* POLL SON noise protection wall made from waste; **~ de soutien** *m* POLL SON retaining wall
mûrissement *m* POLL EAU *conditionnement des boues* ripening; **~ du compost** DÉCHETS, SOLS compost ripening
mutagène *m* SÉCURITÉ mutagen
mutations: **~ urbaines** *f pl* POL ENV urban change
mV *abrév* (*millivolt*) ÉCHANT mV (*millivolt*)
MW *abrév* (*mégawatt*) ÉN RENOUV MW (*megawatt*)
mycète *m* POLL ATM, POLL EAU fungus
mycotoxine *f* POLL EAU mycotoxin

N

N *abrév* ÉCHANT (*newton*) N (*newton*), POLL ATM, POLL EAU (*azote*) N (*nitrogen*)
Na *abrév* (*sodium*) POLL ATM Na (*sodium*)
nanofiltration *f* DIST EAU nanofiltration
nanogramme *m* (*ng*) ÉCHANT nanogram (*ng*), nanogramme (*ng*)
nanomètre *m* (*nm*) ÉCHANT nanometre (*BrE*) (*nm*), nanometer (*AmE*) (*nm*)
naphtalène: **~ polychloré** *m* (*PCN*) SOLS polychlorinated naphthalene (*PCN*)
nappe: **~ aérée** *f* DIST EAU ventilated nappe; **~ aquifère** *f* DIST EAU aquifer; **~ artésienne** *f* DIST EAU artesian groundwater; **~ artificielle** *f* DIST EAU artificial groundwater; **~ captive** *f* DIST EAU confined groundwater; **~ d'eaux usées** *f* POLL EAU wastewater bed; **~ d'hydrocarbures** *f* DIST EAU, POLL MAR hydrocarbon slick, oil patch; **~ libre** *f* DIST EAU unconfined groundwater; **~ perchée** *f* DIST EAU perched groundwater, apparent water table; **~ de pétrole** *f* POLL MAR oil slick; **~ phréatique** *f* DIST EAU underground water, aquifer, water table; **~ phréatique décarbonisée** *f* DIST EAU decarbonized groundwater; **~ réalimentée artificiellement** *f* DIST EAU artificially-replenished groundwater; **~ rechargée artificiellement** *f* DIST EAU artificially-recharged groundwater; **~ superficielle** *f* DIST EAU unconfined groundwater; **~ suspendue** *f* DIST EAU perched groundwater
natte *f* POLL SON matting
nature *f* ÉN RENOUV nature; **~ du sol** SOLS soil condition
naturel *adj* DIST EAU, ÉN RENOUV natural
nauséabond *adj* POLL EAU foul-smelling
navigabilité: **~ en haute mer** *f* POLL MAR seaworthiness
navire: **~ de délestage** *m* POLL MAR lightening vessel; **~ dépollueur** *m* POLL MAR oil pollution combating ship
Ne *abrév* (*néon*) POLL ATM Ne (*neon*)
nébuliser: **se ~** *v réfl* POLL ATM shear
nébulosité *f* POLL ATM cloud cover; **~ partielle** POLL ATM cloud amount
nécrophage *m* SOLS scavenger
négociable *adj* POL ENV tradeable
négociant: **~ en ferrailles** *m* DÉCHETS scrap metal merchant
négociation *f* POL ENV negotiation; **~ bilatérale** POL ENV bilateral negotiation; **~ internationale** POL ENV international negotiation; **~ multilatérale** POL ENV multilateral negotiation
neige: **~ acide** *f* POLL ATM, SOLS acid snow
néon *m* (*Ne*) POLL ATM neon (*Ne*)
néoprène *m* DÉCHETS neoprene
néphélomètre *m* ÉCHANT nephelometer
néphélométrie *f* ÉCHANT nephelometry
néphélométrique *adj* ÉCHANT nephelometric
nessler *m* ÉCHANT *chimie* Nessler's solution
nettoyage[1] *m* GÉNÉRAL cleaning; **~ à air comprimé** POLL ATM compressed-air cleaning; **~ automatique** POLL EAU mechanical cleaning; **~ continu** POLL ATM continuous cleaning; **~ à contre-courant** POLL ATM reverse-flow cleaning; **~ au jet** DIST EAU spray washing; **~ au jet d'eau haute pression** POLL MAR high-pressure water blasting; **~ au jet pulsant** POLL ATM pulse-jet cleaning; **~ du littoral** POLL MAR shoreline clean-up; **~ par absorption** POLL ATM absorbent cleaning; **~ par voie humide** POLL EAU wet cleaning; **~ périodique** POLL ATM periodic cleaning; **~ préalable** POLL ATM precleaning; **~ à la vapeur** POLL MAR steam cleaning
nettoyage[2]: **à ~ automatique** *loc* POLL EAU mechanically-cleaned
nettoyé *adj* GÉNÉRAL cleaned; **~ à l'air comprimé** POLL ATM compressed-air-cleaned
nettoyer *vt* GÉNÉRAL clean, clean up
neutralisant *adj* GÉNÉRAL neutralizing
neutralisateur *m* GÉNÉRAL neutralizer
neutralisation *f* GÉNÉRAL neutralizing, neutralization; **~ des acides** POLL ATM acid neutralizing; **~ de déchets** DÉCHETS waste neutralization; **~ des eaux résiduaires** DIST EAU, POLL EAU *acides, alcalines* neutralization of wastewater; **~ du goût** DIST EAU taste control; **~ des odeurs** DIST EAU, POLL ATM, POLL EAU odor control (*AmE*), odour control (*BrE*)
neutraliser *vt* GÉNÉRAL neutralize
neutralité: **~ fiscale** *f* POL ENV revenue neutrality
neutre *adj* GÉNÉRAL neutral; **~ à l'air ambiant** DÉCHETS ambient-air neutral; **~ fiscalement** POL ENV income-neutral
newton *m* (*N*) ÉCHANT newton (*N*)
ng *abrév* (*nanogramme*) ÉCHANT ng (*nanogram*, *nanogramme*)
Ni *abrév* (*nickel*) POLL ATM, SOLS Ni (*nickel*)
nickel *m* (*Ni*) POLL ATM, SOLS nickel (*Ni*)
nitrate *m* GÉNÉRAL nitrate; **~ de péroxyacétyle** DIST EAU, POLL ATM peroxoacetyl nitrate (*PAN*)
nitrifiant *adj* DÉCHETS, DIST EAU, POLL EAU nitrogen-fixing
nitrification *f* DÉCHETS, DIST EAU, POLL EAU nitrification; **~-dénitrification** DIST EAU, POLL EAU nitrification-denitrification
nitrifier *vt* DÉCHETS, DIST EAU, POLL EAU nitrify
nitrite *m* POLL EAU, SOLS nitrite
nitromètre *m* ÉCHANT nitrometer
nitrosobactérie *f* POLL EAU nitrosomas bacteria
niveau *m* DIST EAU level, plane, ÉCHANT, POLL MAR, POLL SON level; **~ d'acidité** POLL ATM acidity level; **~ d'alarme** SÉCURITÉ danger level; **~ d'alerte** SÉCURITÉ warning level; **~ d'aspiration** SOLS suction level; **~ de bruit** POLL SON noise level, sound level; **~ de bruit direct** POLL SON direct noise level, direct sound level; **~ de bruit d'un événement isolé** POLL SON single-event noise level, single-event sound level; **~ de bruit moyen** POLL SON average noise level, average sound level; **~ de bruit perçu** POLL SON perceived noise level (*PNdB*); **~ de bruit pondéré** POLL SON weighted noise level, weighted sound level; **~ critique d'eau** DIST EAU critical water level; **~ de dose létale 0** SÉCURITÉ lethal dose level 0 (*LDL0*); **~ d'eau** DIST EAU, ÉN RENOUV, POLL MAR water level; **~ d'eau rehaussé par remblai** DÉCHETS, DIST EAU banked-up water level; **~ d'eau du réservoir** ÉN RENOUV chamber water level; **~ à effet nul** SÉCURITÉ no-effect level (*NEL*); **~ efficace** ÉCHANT, SÉCURITÉ effective level; **~ efficace franc** SÉCURITÉ frank effect level (*FEL*); **~ efficace observé** SÉCURITÉ observed effective level (*OEL*); **~ efficace le plus faible observé** SÉCURITÉ lowest observed effect level; **~ d'émissions** POLL ATM, POLL SON emissions level; **~ d'émissions dans la rue** POLL ATM, POLL SON street emissions level; **~ d'énergie** DIST EAU power level; **~ d'étiage** DIST EAU low-water level; **~ d'évaluation** ÉCHANT, POLL SON assessment level; **~ de fond** POLL ATM, POLL SON background level; **~ de gêne acoustique** POLL SON discomfort noise; **~ hydrostatique apparent** DIST EAU apparent water table; **~ instantané de nuisance** DIST EAU instantaneous concentration; **~ d'intensité sonore** POLL SON loudness level; **~ maximal en consécutif** POLL SON

maximum level in consecutive; ~ **de la mer** POLL MAR sea level; ~ **moyen annuel** DIST EAU annual mean water level; ~ **d'ozone** POLL ATM ozone level; ~ **de perturbation** POLL SON, SÉCURITÉ disturbance level; ~ **de perturbation observé le plus faible** POLL SON, SÉCURITÉ lowest observed adverse effect level (*LOAEL*); ~ **de pollution** POLL EAU pollution level; ~ **de pondération des bruits** ÉCHANT, POLL SON noise rating level; ~ **de pression acoustique** POLL SON sound pressure level (*SPL*); ~ **de pression acoustique pondéré en A** POLL SON A-weighted sound pressure level; ~ **de puissance** DIST EAU power level; ~ **de puissance sonore** POLL SON sound power level; ~ **de référence** POLL ATM, POLL SON background level; ~ **de remplissage** DÉCHETS fill level; ~ **de saturation** DIST EAU saturation level; ~ **seuil** POLL SON tolerance level (*TL*); ~ **sonore** POLL SON sound level; ~ **sonore direct** POLL SON direct noise level, direct sound level

niveleuse *f* POLL MAR grader

nm *abrév* (*nanomètre*) ÉCHANT nm (*nanometer AmE, nanometre BrE*)

NO *abrév* (*oxyde nitrique*) DIST EAU, POLL ATM NO (*nitric oxide*)

NO_2 *abrév* (*dioxyde d'azote*) POLL ATM, POLL EAU, SÉCURITÉ NO_2 (*nitrogen dioxide*)

N_2O *abrév* DIST EAU (*anhydride hypo-azoteux*) N_2O (*nitrous oxide*), POLL ATM (*oxyde nitreux, monoxyde de diazote, hémioxyde d'azote*) N_2O (*nitrous oxide*)

N_2O_5 *abrév* (*pentoxyde d'azote, pentoxyde de diazote*) POLL ATM N_2O_5 (*nitrogen pentoxide, dinitrogen pentoxide*)

NO_x *abrév* (*oxydes d'azote*) DÉCHETS NO_x (*nitrogen oxides*)

nocif *adj* GÉNÉRAL noxious, harmful

nocivité *f* SÉCURITÉ noxiousness, harmfulness; ~ **liminale** SÉCURITÉ threshold of harmfulness

noir: ~ **ériochrome** *m* ÉCHANT eriochrome black

noircissement *m* POLL ATM blackening

nombre: ~ **de germes coliformes** *m* POLL EAU bacterial coliform count; ~ **Hazen** *m* POLL EAU Hazen number; ~ **de Knudsen** *m* (*Kn*) POLL ATM Knudsen's number (*Kn*); ~ **de Reynolds** *m* (*Re*) DIST EAU, POLL ATM Reynolds number (*Re*); ~ **de Ringelmann** *m* POLL ATM Ringelmann number; ~ **de tours** *m* ÉN RENOUV number of revolutions

nominal *adj* DIST EAU rated

non: ~**-acide** *adj* GÉNÉRAL acid-free; ~**-aéré** *adj* POLL EAU nonaerated; ~**-amorti** *adj* POLL SON undamped; ~**-biodégradable** *adj* DÉCHETS, POLL EAU, POLL MAR, SOLS nonbiodegradable; ~**-broyé** *adj* DÉCHETS unground; ~**-catalytique** *adj* POLL ATM noncatalytic; ~**-chargé** *adj* POLL ATM, POLL EAU uncharged; ~**-compacté** *adj* DÉCHETS, DIST EAU uncompacted; ~**-connecté** *adj* ÉCHANT off-line; ~**-contrôlé** *adj* GÉNÉRAL uncontrolled; ~**-corrosif** *adj* POLL ATM noncorrosive; ~**-cristallin** *adj* POLL ATM noncrystalline; ~**-délayé** *adj* POLL MAR undiluted, neat; ~**-désaéré** *adj* POLL EAU underaerated; ~**-dilué** *adj* POLL MAR undiluted, neat; ~**-dissous** *adj* DIST EAU, POLL EAU undissolved; ~**-économique** *adj* GESTION, POL ENV uneconomical; ~**-engorgeable** *adj* DIST EAU, POLL EAU *pompe* nonclogging; ~**-épaissi** *adj* POLL EAU unthickened; ~**-épuré** *adj* DÉCHETS, POLL ATM uncleaned; ~**-éteint** *adj* POLL EAU unslaked; ~**-filtré** *adj* ÉCHANT, POLL ATM, POLL EAU unfiltered; ~**-ionique** *adj* POLL ATM, POLL EAU nonionic; ~**-ionogène** *adj* POLL EAU nonionogenic; ~**-jugulé** *adj* GÉNÉRAL uncontrolled; ~**-lixiviable** *adj* SOLS nonleachable; ~**-maîtrisé** *adj* GÉNÉRAL uncontrolled; ~**-miscible** *adj* POLL EAU immiscible; ~**-mouillant** *adj* POLL ATM nonwetting; ~**-nocif pour l'eau** *adj* POLL EAU not harmful to water, not hazardous to water; ~**-pathogène** *adj* DIST EAU, POLL EAU pathogen-free; ~**-perturbé** *adj* ÉCHANT undisturbed; ~**-pollué** *adj* POLL ATM, SOLS pollutant-free, unpolluted; ~**-poreux** *adj* POLL EAU nonporous; ~**-pressurisé** *adj* DIST EAU unpressurized; ~**-récupérable** *adj* DÉCHETS, DIST EAU nonrecoverable; ~**-remanié** *adj* ÉCHANT *échantillon géologique* undisturbed; ~**-rentable** *adj* GESTION, POL ENV uneconomical, unprofitable; ~**-revêtu** *adj* DIST EAU, POLL ATM uncoated; ~**-saturé** *adj* DÉCHETS, DIST EAU, POLL ATM, SOLS unsaturated; ~**-spécifique** *adj* ÉCHANT nonspecific; ~**-spécifique d'un gaz** *adj* ÉCHANT gas-nonspecific; ~**-tissé** *adj* POLL ATM nonwoven, unwoven; ~**-toxique** *adj* POLL ATM, POLL EAU, SÉCURITÉ, SOLS nontoxic; ~**-traité** *adj* DÉCHETS uncleaned, unprocessed, untreated, DIST EAU unprocessed, untreated, POLL ATM uncleaned, POLL EAU untreated; ~**-trié** *adj* DÉCHETS unsorted; ~**-volatile** *adj* POLL ATM nonvolatile; ~**-vulcanisé** *adj* DÉCHETS, DIST EAU nonvulcanized

non-conformité *f* GESTION nonconformity

normalisé *adj* SÉCURITÉ standardized

norme: ~ **d'émission** *f* ÉCHANT, POL ENV, POLL ATM emission standard; ~ **d'environnement** *f* DIST EAU environmental quality standard; ~ **d'exploitation** *f* POL ENV operating standard; ~ **industrielle** *f* POL ENV industrial standard; ~ **de qualité de l'air** *f* POLL ATM air quality standard; ~ **de qualité de l'air ambiant** *f* POLL ATM ambient-air quality standard; ~ **de qualité de l'environnement** *f* POL ENV environmental quality standard; ~ **de rejet** *f* DIST EAU discharge standard; ~ **de sécurité** *f* SÉCURITÉ safety standard

normes: ~ **de production** *f pl* POL ENV product standards; ~ **de protection de l'environnement** *f pl* POL ENV environmental protection standards

notification *f* GESTION notification, reporting, POLL MAR notification; ~ **des émissions** DIST EAU, POL ENV emissions statement

nourrir *vt* DIST EAU replenish

nouveau: ~ **dépôt** *m* SOLS redeposition

nouvelle: ~ **installation** *f* GÉNÉRAL new plant; ~ **technique** *f* GÉNÉRAL emerging technology, new technology; ~ **technologie** *f* GÉNÉRAL emerging technology, new technology; ~ **usine** *f* POL ENV, POLL ATM new plant

noyage *m* DIST EAU, POLL EAU flooding

noyau: ~ **de condensation** *m* DÉCHETS, POLL ATM condensation nucleus

noyé *adj* DIST EAU, POLL MAR underwater

noyer *vt* DIST EAU, POLL EAU flood

nuage: ~ **de gaz brûlés** *m* POLL ATM flue-gas cloud; ~ **toxique** *m* POLL ATM toxic cloud

nucléaire *adj* GÉNÉRAL nuclear

nuire: ~ **à** *vt* GÉNÉRAL harm

nuisance: ~ **olfactive** *f* DÉCHETS, POLL ATM, POLL EAU odor annoyance (*AmE*), odour annoyance (*BrE*), odor nuisance (*AmE*), odour nuisance (*BrE*)

nuisible *adj* GÉNÉRAL harmful, noxious; ~ **pour l'environnement** POL ENV, POLL ATM, POLL MAR, POLL SON, SOLS harmful to the environment

numération: ~ **bactérienne** *f* DIST EAU bacterial count

nutriment *m* DIST EAU, POLL EAU, POLL MAR nutriment

nylon *m* DÉCHETS, POLL ATM nylon

O

O *abrév* (*oxygène*) ÉN RENOUV, POLL ATM O (*oxygen*)
O_3 *abrév* (*ozone*) POLL ATM O_3 (*ozone*)
objectif: **~ d'assainissement** *m* SOLS remediation target value; **~ de gestion** *m* POL ENV policy objective; **~ de protection de l'environnement** *m* POL ENV environmental protection aim
objet: **~ d'analyse** *m* ÉCHANT analysis aim; **~ de mesure** *m* ÉCHANT measuring purpose
obligation *f* POL ENV obligation; **~ de recycler** POL ENV obligation to recycle; **~ de réutiliser les matériaux résiduels** POL ENV obligation to reuse residue materials; **~ de surveillance** POL ENV monitoring obligation
obligations: **~ de l'exploitant** *f pl* POL ENV operator duties
obliger *vt* POL ENV oblige
obscurité: **~ atmosphérique** *f* POLL ATM atmospheric obscurity
observable *adj* ÉCHANT observable
obsolescence: **~ intégrée** *f* DÉCHETS built-in obsolescence
obstacle *m* SÉCURITÉ hindrance; **~ au recyclage** DÉCHETS, DIST EAU recycling hindrance
obstruction *f* DIST EAU, POLL EAU blockage, stoppage, POLL SON obstruction; **~ à la propagation du bruit** POLL SON obstruction of noise propagation
obtention: **~ de biogaz par fermentation** *f* ÉN RENOUV fermentation to biogas; **~ de produits finals** *f* POLL ATM end product generation
obturation *f* DIST EAU, POLL EAU clogging
obturé *adj* DIST EAU, POLL EAU clogged
obturer *vt* DIST EAU, POLL EAU clog
occupation: **~ des sols** *f* DIST EAU, SOLS land use
océanique *adj* POLL MAR oceanic
octroi *m* POL ENV granting; **~ d'agrément** POL ENV leasing of certificate
oculaire *m* ÉCHANT *appareillage optique* ocular
OD *abrév* (*oxygène dissous*) DIST EAU, POLL ATM DO (*dissolved oxygen*)
odeur[1] *f* GÉNÉRAL odor (*AmE*), odour (*BrE*), smell; **~ âcre** DIST EAU, POLL ATM acrid odor (*AmE*), acrid odour (*BrE*); **~ limite** DIST EAU, POLL EAU threshold odor (*AmE*), threshold odour (*BrE*)
odeur[2]: **à ~ forte** *loc* POLL ATM, POLL EAU, POLL MAR odor-intensive (*AmE*), odour-intensive (*BrE*)
odorant *adj* POLL ATM, POLL EAU, POLL MAR odor-laden (*AmE*), odour-laden (*BrE*), odorous
odorimétrie *f* ÉCHANT odorimetry
odoriphore *adj* POLL ATM *chimie* odoriphore
OEB *abrév* (*Office Européen des Brevets*) ÉCHANT EPO (*European Patent Office*)
office: **~ de presse** *m* POL ENV press office
Office: **~ Européen des Brevets** *m* (*OEB*) ÉCHANT European Patent Office (*EPO*)
oléagineux *adj* ÉN RENOUV oil-containing
oléophile *adj* POLL EAU, POLL MAR oil-absorbing, oleophilic
olfactométrie *f* ÉCHANT, POLL ATM olfactometry
oligo: **~-élément** *m* DIST EAU, POLL EAU micronutrient, trace element
OM *abrév* (*ordures ménagères*) DÉCHETS household refuse, household waste
onde: **~ sonore** *f* POLL SON sound wave
opacimètre *m* ÉCHANT opacimeter, smoke density meter
opaque *adj* POLL EAU opaque
opération: **~ de cisaillage** *f* DIST EAU *du floc* shearing action; **~ de collecte** *f* POLL ATM collection operation; **~ de combustion** *f* POLL ATM combustion operation; **~ de démarrage** *f* POLL ATM starting operation; **~ d'emboutissage** *f* DIST EAU pressing operation; **~ d'extraction** *f* POLL ATM, POLL EAU extraction operation, removal operation; **~ de lavage** *f* POLL ATM washing operation; **~ de nettoyage** *f* POLL ATM cleaning operation; **~ de pourrissage** *f* DIST EAU rotting operation; **~ de précipitation** *f* POLL EAU precipitation operation; **~ de pressage** *f* DIST EAU pressing operation; **~ de raclage** *f* POLL EAU scraping operation; **~ de recyclage** *f* DÉCHETS, DIST EAU recycling operation; **~ de tri des matériaux** *f* DÉCHETS material-separation operation
operculage *m* DÉCHETS cap sealing
opercule: **~ en aluminium mince** *m* DÉCHETS foil seal
optimisation: **~ des emballages** *f* DÉCHETS packaging optimization
optique *adj* ÉCHANT optical
optoélectronique *adj* DÉCHETS, DIST EAU optoelectronic
orbite *f* POLL ATM orbit; **~ des particules** POLL ATM particle orbit; **~ périphériques de particules** POLL ATM periphery particle orbit
ordinateur: **~ de bord** *m* ÉCHANT on-board computer; **~ embarqué** *m* ÉCHANT on-board computer
ordures *f pl* DÉCHETS refuse, waste; **~ ménagères** (*OM*) DÉCHETS household refuse, household waste, ÉN RENOUV consumer waste
organe: **~ d'arrêt** *m* DIST EAU shut-off device; **~ compétent pour un secteur** *m* POL ENV sectoral authority
organique *adj* POLL ATM, POLL EAU organic
organisation: **~ chargée de l'environnement** *f* POL ENV environmental organization
Organisation: **~ internationale de normalisation** *f* (*ISO*) ÉCHANT, POL ENV International Standards Organization (*ISO*)
organisme *m* GÉNÉRAL institution, *être vivant* organism; **~ aérobie** POLL ATM, POLL EAU aerobic organism; **~ anaérobie** POLL ATM, POLL EAU anaerobic organism; **~ coliforme** DIST EAU coliform organism; **~ se développant dans les eaux usées** POLL EAU wastewater organism; **~ hétérotrophe** DÉCHETS, POLL EAU *écologie* heterotroph; **~ pélagique** POLL EAU pelagic organism; **~ témoin** DIST EAU, POLL ATM, POLL EAU, SOLS indicator organism
organochloré *adj* DIST EAU organochlorinated
organoleptique *adj* DIST EAU organoleptic
orientation[1] *f* POLL ATM *effluents gazeux* direction
orientation[2]: **à ~ écologique** *loc* GESTION, POL ENV environmentally-oriented
orienteur *m* ÉN RENOUV tracker
orifice: **~ d'admission** *m* POLL ATM entry inlet; **~ d'admission d'eau** *m* DIST EAU water inlet; **~ d'aspiration** *m* DÉCHETS suction port; **~ de charge** *m* DÉCHETS fill hole; **~ de chargement** *m* DÉCHETS filling hole; **~ de mesure** *m* ÉCHANT measuring nozzle; **~ en paroi mince** *m* ÉCHANT sharp-edge orifice; **~ de rejet de poussière** *m* POLL ATM dust discharge opening; **~ de ventilation** *m* DIST EAU, POLL ATM vent
origine: **d'~ humaine** *loc* POLL ATM, SOLS anthropogenically-caused, human-induced; **d'~ solaire** *loc* ÉN RENOUV solar-supported
orlon *m* DÉCHETS, POLL ATM orlon
oscillateur *m* POLL SON oscillator

oscillation: ~ **acoustique** *f* POLL SON acoustic oscillation; ~ **de pression** *f* DIST EAU pressure oscillation
oscilloscope *m* ÉCHANT oscilloscope
osmophore *m* POLL ATM odoriphore, osmophore
osmose *f* DIST EAU, POLL EAU osmosis; ~ **inverse** DIST EAU, POLL EAU reverse osmosis (*RO*); ~ **inverse avec évaporation des concentrats** DIST EAU, POLL EAU reverse osmosis with concentrate evaporation
osmotique *adj* DIST EAU, POLL EAU osmotic
ouate *m* DÉCHETS cotton, wadding
outils: ~ **d'information** *m pl* POL ENV information instruments
ouvrage: ~ **d'arrivée** *m* POLL EAU inlet structure; ~ **de crue** *m* DIST EAU flood control works; ~ **de pompage des boues** *m* POLL EAU sludge pump works; ~ **de prise d'eau** *m* DIST EAU, POLL EAU water intake structure; ~ **de retenue** *m* DIST EAU dam; ~ **de trop-plein** *m* DIST EAU overflow structure
ouvrages: ~ **publics** *m pl* POLL EAU public works
ouvrir *vt* GÉNÉRAL open; ~ **par effraction** SÉCURITÉ break open
oxydabilité *f* ÉCHANT, POLL ATM, SOLS oxidizability
oxydable *adj* ÉCHANT, POLL ATM, SOLS oxidizable
oxydant[1] *adj* POLL ATM, POLL EAU, SOLS oxidative, oxidizing; ~ **acide** POLL ATM acid-oxidative
oxydant[2] *m* POLL ATM, POLL EAU, SOLS oxidant, oxidizer; ~ **alcalin** POLL ATM alkaline oxidant; ~ **photochimique** POLL ATM photochemical oxidant
oxydation *f* ÉCHANT oxidation, POLL ATM trickling, oxidation, POLL EAU, SOLS oxidation; ~ **aérobie** DÉCHETS aerobic oxidation; ~ **catalytique** POLL ATM, SOLS catalytic oxidation; ~ **chimique** DÉCHETS chemical oxidation; ~ **due à des polluants** POLL ATM pollutant oxidation; ~ **humide** DÉCHETS, POLL EAU, SOLS wet oxidation; ~ **par voie humide** DÉCHETS, POLL EAU, SOLS wet oxidation, wet air oxidation (*WAO*); ~ **par voie humide à basse pression** POLL EAU low-pressure wet oxidation; ~ **par voie humide sous haute pression** POLL EAU high-pressure wet oxidation; ~ **en phase gazeuse** POLL EAU gas-phase oxidation; ~ **photocatalytique** DIST EAU photocatalytic oxidation; ~ **supercritique** SOLS supercritical oxidation; ~ **thermique** POLL ATM thermal oxidation; ~ **aux ultraviolets** DIST EAU ultraviolet oxidation
oxyde: ~ **d'aluminium activé** *m* POLL ATM activated aluminium oxide (*BrE*), activated aluminum oxide (*AmE*); ~ **de calcium** *m* (*CaO*) POLL ATM, SOLS calcium oxide (*CaO*); ~ **nitreux** *m* (N_2O) POLL ATM nitrous oxide (N_2O); ~ **nitrique** *m* (*NO*) DIST EAU, POLL ATM nitric oxide (*NO*); ~ **de soufre** *m* (*SO*) DIST EAU, POLL ATM sulfur oxide (*AmE*) (*SO*), sulphur oxide (*BrE*) (*SO*)
oxyder[1] *vt* ÉCHANT, POLL ATM, POLL EAU, SOLS oxidize
oxyder[2] *vi* ÉCHANT, POLL ATM, POLL EAU, SOLS oxidize
oxyder[3]: **s'~** *v réfl* ÉCHANT, POLL ATM, POLL EAU, SOLS oxidize
oxydes: ~ **d'azote** *m pl* (NO_x) DÉCHETS nitrogen oxides (NO_x); ~ **de soufre** *m pl* (SO_x) DIST EAU, POLL ATM sulfur oxides (*AmE*) (SO_x), sulphur oxides (*BrE*) (SO_x)
oxydimétrique *adj* ÉCHANT oxidimetric
oxygène *m* (*O*) ÉN RENOUV, POLL ATM oxygen (*O*); ~ **atmosphérique** POLL ATM atmospheric oxygen; ~ **dissous** (*OD*) DIST EAU, POLL ATM dissolved oxygen (*DO*); ~ **en excès** POLL ATM excess oxygen; ~ **pur** POLL ATM pure oxygen; ~ **résiduel** POLL ATM residual oxygen
oxygéné *adj* POLL EAU oxygenated, oxygenic
oxygènemètre *m* ÉCHANT, POLL ATM oxygen meter, oxygen measuring device
oxygéner *vt* POLL EAU oxygenate
oxysulfure: ~ **de carbone** *m* POLL ATM carbonyl sulfide (*AmE*), carbonyl sulphide (*BrE*)
ozone *m* (O_3) POLL ATM ozone (O_3); ~ **atmosphérique** POLL ATM atmospheric ozone; ~ **résiduel** POLL ATM residual ozone; ~ **stratosphérique** POLL ATM stratospheric ozone; ~ **troposphérique** POLL ATM tropospheric ozone
ozoneur *m* DIST EAU ozone generator, POLL ATM ozone equipment
ozonisation *f* DIST EAU, POLL EAU, SOLS ozonization; ~ **préliminaire** POLL EAU preliminary ozonization
ozoniser *vt* DIST EAU, POLL EAU, SOLS ozonize
ozoniseur *m* DIST EAU, POLL EAU, SOLS ozonizer
ozonométrie *f* ÉCHANT ozone measuring

P

p *abrév* (*pascal*) ÉCHANT p (*pascal*)
P *abrév* (*phosphore*) POLL ATM, POLL EAU P (*phosphorus*)
paiement: **~ compensatoire** *m* GESTION compensatory payment
paille: **~ de céréales** *f* ÉN RENOUV grain straw
paillis *m* ÉN RENOUV mulch
pale *f* POLL EAU paddle; **~ de déflecteur** POLL ATM deflector vane; **~ d'éolienne** ÉN RENOUV wind turbine blade; **~ de rotor** ÉN RENOUV rotor blade; **~ de rotor à pas fixe** ÉN RENOUV fixed-pitch rotor blade; **~ de rotor à pas réglable** ÉN RENOUV variable-pitch rotor blade
palette: **~ expansible** *f* DÉCHETS expandable pallet; **~ perdue** *f* DÉCHETS nonreturnable pallet, one-way pallet; **~ plate** *f* DÉCHETS flat pallet; **~ à quatre entrées** *f* DÉCHETS four-way pallet; **~ réutilisable** *f* DÉCHETS retunable pallet; **~ transportable traditionnelle** *f* DÉCHETS, ÉN RENOUV conventional transportable pallet
paliers: **par ~** *loc* ÉCHANT step-by-step
palissade *f* DÉCHETS fence
palladium *m* (*Pd*) POLL ATM palladium (*Pd*)
palper *vt* ÉCHANT sense
palpeur: **~ pyrométrique** *m* ÉCHANT temperature sensor
paludisme *m* SÉCURITÉ marsh fever
PAN *abrév* (*nitrate de péroxyacétyle, péroxiacétylnitrate*) DIST EAU, POLL ATM PAN (*peroxoacetyl nitrate*)
panache *m* POLL ATM plume; **~ en boucle** POLL ATM looping smoke plume; **~ en cloche** POLL ATM Gaussian plume; **~ d'effluents gazeux** POLL ATM waste gas plume; **~ de fumée** POLL ATM smoke plume; **~ de fumée en cône** POLL ATM coning smoke plume; **~ de fumée étalé** POLL ATM nonconing smoke plume; **~ de fumée en éventail** POLL ATM fanning smoke plume; **~ de gaz de combustion** POLL ATM flue-gas plume; **~ invisible** POLL ATM invisible plume; **~ nauséabond** POLL ATM malodorous plume; **~ de vapeur** DÉCHETS, DIST EAU, POLL ATM vapor plume (*AmE*), vapour plume (*BrE*); **~ de vapeur d'eau** POLL ATM water vapor plume (*AmE*), water vapour plume (*BrE*); **~ visible** POLL ATM visible plume
panne *f* POLL ATM *d'électricité* breakdown, SOLS *de machine* failure
panneau *m* DÉCHETS, DIST EAU panel, ÉN RENOUV chipboard, POLL MAR, POLL SON, SÉCURITÉ panel; **~ absorbant perforé** POLL SON perforated panel absorber; **~ à bouton** POLL SON knobbed panel; **~ d'extrémité** DÉCHETS end panel; **~ en fibre minérale** POLL MAR, POLL SON mineral fiber panel (*AmE*), mineral fibre panel (*BrE*); **~ insonorisant** POLL SON acoustic panel; **~ insonorisant en mousse** SÉCURITÉ sound-absorbing foam panel; **~ d'insonorisation** POLL SON sound-absorbing panel; **~ d'isolation** POLL SON insulating board; **~ léger** POLL SON lightweight panel; **~ de mousse** POLL SON foam panel; **~ de particules** DÉCHETS, ÉN RENOUV, POLL SON chipboard; **~ perforé** POLL SON perforated panel; **~ phono-absorbant** POLL SON sound-absorbing panel; **~ rainuré** POLL SON slotted panel; **~ résonateur** POLL SON panel resonator; **~ stratifié** POLL SON sandwich panel; **~ de verre** ÉN RENOUV *capteur solaire* glass pane
papier: **~ anticorrosion** *m* DÉCHETS corrosion-preventive paper; **~ brillant** *m* DÉCHETS high-gloss paper; **~ à carreaux** *m* DÉCHETS *emballage* cartridge paper; **~ ciré** *m* DÉCHETS wax paper; **~ à couchage synthétique** *m* DÉCHETS *emballage* coated synthetic paper; **~ couché** *m* DÉCHETS coated paper; **~ crêpé** *m* DÉCHETS crepe paper; **~ d'emballage** *m* DÉCHETS packaging paper, wrapping paper, bag paper; **~-filtre** *m* ÉCHANT filter paper; **~ filtre imprégné** *m* ÉCHANT impregnated filter paper; **~ glacé** *m* DÉCHETS glazed paper; **~ glacé par voie mécanique** *m* DÉCHETS machine-glazed paper; **~ gommé** *m* DÉCHETS gummed paper; **~ goudron** *m* DÉCHETS tar paper; **~ goudronné** *m* DÉCHETS bitumen-coated paper; **~ gradué** *m* DÉCHETS, ÉCHANT indicator paper; **~ imperméable aux graisses** *m* DÉCHETS grease-resistant paper; **~ imprégné** *m* DÉCHETS impregnated paper; **~ indicateur** *m* ÉCHANT test paper; **~ ingraissable** *m* DÉCHETS grease-resistant paper; **~ Joseph** *m* ÉCHANT filter paper; **~ kraft** *m* DÉCHETS kraft paper; **~ kraft pour sacherie** *m* DÉCHETS kraft sack paper; **~ de mise en cartes** *m* DÉCHETS *emballage* cartridge paper; **~ non acide** *m* DÉCHETS acid-free paper; **~ ondulé** *m* DÉCHETS corrugated paper; **~ parchemin** *m* DÉCHETS parchment paper; **~ parcheminé** *m* DÉCHETS greaseproof paper; **~ peint insonorisant** *m* POLL SON sound-absorbing wallpaper; **~ réactif** *m* ÉCHANT test paper; **~ recyclé** *m* DIST EAU recycled paper, recycling paper; **~ sulfurisé** *m* DÉCHETS greaseproof paper, parchment paper; **~ thermocollable** *m* DÉCHETS heat-sealable paper
paquet: **~ d'eau** *m* DIST EAU feed; **~ familial** *m* DÉCHETS family packet; **~ plat** *m* DÉCHETS flat pack
parabolique *adj* ÉN RENOUV parabolic
paraboloïde: **~ réfléchi** *m* ÉN RENOUV reflected paraboloid
parallèle *adj* ÉCHANT *fonctionnement* parallel
paramagnétique *adj* ÉCHANT paramagnetic
paramagnétisme *m* ÉCHANT paramagnetism
paramètre *m* DIST EAU, ÉCHANT, SOLS parameter; **~ chimique** SOLS chemical parameter; **~ cumulatif** ÉCHANT cumulative parameter; **~ de débit** ÉCHANT flow parameter; **~ de groupe** ÉCHANT group parameter; **~ physico-pédologique** SOLS soil-physical parameter; **~ physique** SOLS physical parameter; **~ du système** ÉCHANT system parameter
paramètres: **~ spécifique du site** *m pl* ÉN RENOUV site-related criterion
paramétrisation *f* DIST EAU, ÉCHANT, SOLS parameter determination, parameterization
parasite *m* DÉCHETS, DIST EAU vermin
parc: **~ automobile** *m* POL ENV car population; **~ d'éoliennes** *m* ÉN RENOUV wind farm, wind park; **~ à ferrailles** *m* DÉCHETS scrap yard; **~ national** *m* POL ENV national park; **~ naturel** *m* POL ENV nature park
pare: **~-écume** *m* POLL EAU scum baffle, scum board, scum remover; **~-étincelles** *m* POLL ATM spark arrester; **~-feu** *m* DIST EAU fire wall
paroi[1]: **~ absorbante** *f* POLL ATM absorber wall; **~ anti-bruit** *f* POLL MAR, POLL SON noise protection wall; **~ avant** *f* DÉCHETS working face; **~ de caisson** *f* POLL ATM housing wall; **~ de cheminée** *f* POLL ATM stack wall; **~ de cuve** *f* POLL ATM vessel wall; **~ de cyclone** *f* POLL ATM cyclone wall; **~ double** *f* POLL SON compound wall; **~ étanche** *f* POLL SON sealed wall; **~ de filtre** *f* POLL ATM filter wall; **~ insonorisante** *f* POLL MAR, POLL SON, SÉCURITÉ noise insulation wall, sound insulation wall; **~ interne de cyclone** *f* POLL ATM inner cyclone wall; **~ isolante** *f* POLL SON screening wall; **~ mise à la terre** *f* POLL ATM earthed wall; **~ phono-absorbante** *f* POLL SON sound-absorbing wall; **~ de protection** *f* POLL SON shielding wall;

~ **tamisante** *f* POLL SON screening wall; ~ **du trou de sonde** *f* SOLS borehole wall; ~ **tubulaire** *f* DIST EAU tube wall

paroi[2]: **à ~ mince** *loc* ÉCHANT thin-walled

part[1]: ~ **utilisable** *f* DIST EAU *de l'hydrosphère* usable portion

part[2]: **à ~** *loc* POLL EAU separate

partenaire: ~ **d'environnement** *m* GESTION, POL ENV environmental partner; ~ **environnemental** *m* GESTION, POL ENV environmental partner

participation: ~ **des habitants** *f* POL ENV citizens' participation, public participation; ~ **de la population** *f* POL ENV citizens' participation, public participation

particulaire *adj* POLL ATM particulate

particule *f* GÉNÉRAL particle; ~ **acide** POLL ATM acid particle, acidic particle; ~ **d'adsorbant** POLL ATM adsorbent particle; ~ **à bords aigus** POLL ATM sharp-edged particle; ~ **chargée** POLL ATM charged particle; ~ **colloïdale** DIST EAU, POLL ATM colloidal particle; ~ **déstabilisée** POLL EAU destabilized particle; ~ **émise** POLL ATM emitted particle; ~ **explosible** POLL ATM explosible particulate; ~ **fine** POLL ATM fine particle; ~ **grossière** POLL ATM coarse-particle; ~ **de gypse** POLL ATM gypsum particle; ~ **ionisante** POLL ATM ionizing particle; ~ **liquide** POLL ATM, POLL EAU liquid particle; ~ **non chargée** POLL ATM uncharged particle; ~ **de poussière** POLL ATM dust particle; ~ **de poussière collectée** POLL ATM collected dust particle; ~ **réactive** POLL ATM reactive particle; ~ **restabilisée** POLL EAU restabilized particle; ~ **de sol** SOLS soil particle; ~ **de sol entraînée** SOLS entrained soil particle; ~ **solide** DIST EAU, POLL ATM, POLL EAU solid particulate, solid particle; ~ **solide entraînée** POLL ATM entrained solid particle; ~ **solide en suspension** ÉCHANT, POLL ATM suspended solid particle; ~ **en suspension** POLL ATM particulate matter, suspended particle, suspended particulate matter; ~ **volatile** POLL ATM volatile fraction

particules[1]: ~ **solides en suspension dans une solution** *f pl* POLL EAU mixed liquor suspended solids (*MLSS*)

particules[2]: **à ~ grossières** *loc* DÉCHETS, DIST EAU, POLL ATM coarse-particled

partiel *adj* GÉNÉRAL partial

partiellement: ~ **automatisé** *adj* ÉCHANT partially-automated; ~ **halogéné** *adj* POLL ATM partially-halogenated; ~ **soluble dans l'eau** *adj* DIST EAU partially water soluble; ~ **stabilisé** *adj* ÉCHANT partially-stabilized

parties: ~ **par milliard** *loc* ÉCHANT parts per billion (*ppb*); ~ **par million** *loc* (*ppm*) ÉCHANT parts per million (*ppm*); ~ **par quadrillion** *loc* ÉCHANT parts per quadrillion (*ppq*); ~ **par trillion** *loc* ÉCHANT parts per trillion (*ppt*)

pas: ~ **à pas** *loc* ÉCHANT step-by-step

pascal *m* (*p*) ÉCHANT pascal (*p*)

passage: ~ **annulaire** *m* POLL ATM, SOLS annular gap; ~ **par dérivation** *m* POLL SON flanking path; ~ **latéral** *m* POLL SON flanking path; ~ **dans le sol** *m* DIST EAU soil passage

passer[1]: ~ **au filtre** *vt* DIST EAU, ÉCHANT, POLL EAU filter; ~ **à travers** *vt* DIST EAU percolate through; ~ **à travers de** *vt* ÉCHANT run through

passer[2]: ~ **à travers** *vi* POLL EAU percolate

passif[1] *adj* GÉNÉRAL passive

passif[2] *m* POL ENV *comptable, financier* liability

pasteurisation *f* POLL EAU pasteurization

pastillage *m* DÉCHETS, DIST EAU, POLL EAU, SOLS pelletization, pelletizing

pastille *f* POLL ATM, POLL EAU *combustible nucléaire* pellet

pastiller *vt* DÉCHETS, DIST EAU, POLL EAU, SOLS pelletize

pastilleur *m* DÉCHETS, DIST EAU pelletizer

pastilleuse *f* DIST EAU pellet mill, ÉCHANT tablet press

pâte: ~ **au bisulfite** *f* DIST EAU sulfite cellulose (*AmE*), sulphite cellulose (*BrE*); ~ **de bois** *f* DÉCHETS wood pulp; ~ **brute** *f* DÉCHETS *papier* crude pulp; ~ **chimique** *f* DÉCHETS *fabrication du papier* chemical pulp; ~ **de déchets** *f* DÉCHETS waste pulp; ~ **papetière** *f* DÉCHETS, DIST EAU paper pulp; ~ **mécanique** *f* DÉCHETS mechanical wood pulp; ~ **à papier** *f* DÉCHETS, DIST EAU paper pulp, pulp; ~ **de récupération** *f* DIST EAU recovered pulp

pâteux *adj* DÉCHETS, DIST EAU pasty

pathogène *adj* DIST EAU, POLL EAU pathogenic

patrimoine: ~ **naturel** *m* POL ENV natural environment

patrouilleur: ~ **des mers** *m* POLL MAR patrol boat

pauvre: ~ **en azote** *adj* POLL ATM low-nitrogen

pays: ~ **phare** *m* POL ENV leader country

paysage *m* POL ENV landscape

Pb *abrév* (*plomb*) GÉNÉRAL Pb (*lead*)

PC *abrév* (*polycarbonate*) DIST EAU, POLL EAU PC (*polycarbonate*)

PCDF *abrév* (*dibenzofuranne polychloré*) DÉCHETS, SOLS PCDF (*polychlorinated dibenzofuran*)

PCN *abrév* POLL EAU (*pouvoir calorifique net*) NCV (*net calorific value*), SOLS (*naphthalène polychloré*) PCN (*polychlorinated naphthale*)

PCP *abrév* (*pentachlorophénol*) POLL ATM, SOLS PCP (*pentachlorophenol*)

PCS *abrév* (*pouvoir calorifique supérieur*) DIST EAU, POLL ATM UCV (*upper calorific value*)

PCT *abrév* (*terphényl polychloré*) SOLS PCT (*polychlorinated terphenyl*)

Pd *abrév* (*palladium*) POLL ATM Pd (*palladium*)

PE *abrév* (*polyéthylène*) DÉCHETS, DIST EAU PE (*polyethylene*)

peau: ~ **d'étanchéité** *f* DÉCHETS liner

pédogenèse *f* SOLS pedogenesis, soil formation

pédologie *f* SOLS soil science, pedology

peinture: ~ **sans solvant** *f* DIST EAU solvent-free paint

pélagique *adj* POLL MAR pelagic

pelle *f* DIST EAU, POLL MAR scoop, shovel; ~ **à écumes** POLL EAU scum scraper

pelletiser *vt* DÉCHETS, DIST EAU, POLL EAU, SOLS pelletize

pelliculage *m* DÉCHETS film lamination

pellicule[1]: ~ **adhésive** *f* DÉCHETS adhesive film; ~ **brillante** *f* DÉCHETS *laminage* high-gloss foil; ~ **de colle** *f* DÉCHETS glue film; ~ **de condensats** *f* POLL ATM condensate film; ~ **d'eau** *f* POLL ATM water film; ~**-écran** *f* DÉCHETS barrier film; ~ **d'emballage à usage alimentaire** *f* DÉCHETS food-grade packaging film; ~ **extensible** *f* DÉCHETS stretch film; ~ **filtrante** *f* POLL ATM filter film; ~ **avec extrusion en couches multiples** *f* DÉCHETS co-extruded film; ~ **d'hydrocarbures** *f* POLL MAR oil film; ~ **liquide** *f* POLL ATM liquid film; ~ **photographique** *f* ÉCHANT photographic film; ~ **de poussière** *f* POLL ATM dust-laden film; ~ **de protection de surface** *f* DÉCHETS surface protection film; ~ **de surface** *f* POLL EAU surface film

pellicule[2]: **à ~ fine** *loc* DÉCHETS, POLL EAU thin-film

pellistor: ~ **passif** *m* ÉCHANT passive pellistor

pénétration *f* POLL ATM penetration; ~ **d'eau de mer** POLL MAR sea water intrusion; ~ **intempestive** DÉCHETS, SÉCURITÉ intrusion; ~ **intempestive de bruits** POLL SON noise intrusion; ~ **de poussière** POLL ATM dust breakthrough

pénétrer *vt* DIST EAU percolate through

pentachlorophénol *m* (*PCP*) POLL ATM, SOLS pentachlorophenol (*PCP*)

pente *f* DIST EAU slope; ~ **de décharge** DÉCHETS landfill slope

pentoxyde: ~ **d'azote** *m* (N_2O_5) POLL ATM nitrogen pentoxide (N_2O_5), dinitrogen pentoxide (N_2O_5); ~ **de diazote** *m* (N_2O_5) POLL ATM dinitrogen pentoxide (N_2O_5), nitrogen pentoxide (N_2O_5)

pénurie: ~ **de combustible** *f* POLL ATM lack of fuel; ~ **d'espace** *f* POL ENV urban congestion

PER *abrév* (*perchloroéthylène*) POLL ATM PER (*perchloroethylene*)

percé *adj* DIST EAU perforated

perception: ~ **de l'intensité sonore** *f* POLL SON loudness perception

percer *vt* DÉCHETS, DIST EAU *tuyau* perforate

perchloroéthylène *m* (*PER*) POLL ATM perchloroethylene (*PER*)
percolation *f* DIST EAU, POLL EAU percolation
percoler *vi* DIST EAU, POLL EAU percolate, percolate through
perdu *adj* DÉCHETS disposable
péréquation *f* POL ENV equalization
perforabilité *f* DIST EAU puncturability
perforation *f* DIST EAU pitting
perforé *adj* DÉCHETS, DIST EAU perforated
perforer *vt* DÉCHETS *tuyau*, DIST EAU *tuyau* perforate
performance: ~ **environnementale** *f* ÉCHANT environmental performance; ~ **sur le terrain** *f* ÉCHANT field performance
péricinétique *adj* DIST EAU perikinetic
périmètre: ~ **de protection** *m* POL ENV protected-area boundary
période: ~ **d'analyse** *f* ÉCHANT analysis time; ~ **de brouillard photochimique** *f* POLL ATM smog period; ~ **de collecte** *f* ÉCHANT collection period; ~ **de demi-vie** *f* POLL ATM radioactive half-life; ~ **de faible ensoleillement** *f* ÉN RENOUV weak-sun period; ~ **de fort ensoleillement** *f* ÉN RENOUV strong-sun period; ~ **de mesure** *f* ÉCHANT measuring period; ~ **radioactive** *f* POLL ATM radioactive half-life; ~ **sèche** *f* DIST EAU rain-free period
périodique *adj* ÉCHANT, ÉN RENOUV, POLL ATM periodic
périphérie *f* POLL SON periphery
perméabilité *f* GÉNÉRAL permeability; ~ **à l'air** SOLS air permeability; ~ **aux graisses** DÉCHETS permeability to grease; ~ **de membrane** POLL EAU membrane permeability; ~ **du sol** DIST EAU, SOLS soil permeability; ~ **sonore** POLL SON sound permeability
perméable *adj* GÉNÉRAL permeable, pervious; ~ **au gaz** DÉCHETS gas-permeable
perméamètre *m* ÉCHANT permeameter
perméance *f* POLL ATM penetration, transmission
perméat *m* GÉNÉRAL permeate
perméation: ~ **gazeuse** *f* POLL ATM gas permeation
permis *m* DIST EAU, POL ENV permission, permit; ~ **applicable à l'environnement** POL ENV environmental licence (*BrE*), environmental license (*AmE*); ~ **de construire** POL ENV building permit, construction licence (*BrE*), construction license (*AmE*), construction permission; ~ **d'exploitation** DIST EAU, POL ENV operating authorization, operating permission, plant permit; ~ **d'exploitation conforme au droit de l'eau** DIST EAU, POL ENV operating authorization, operating permission, plant permit according to law on water; ~ **d'exploitation conforme au droit nucléaire** DIST EAU, POL ENV plant permit according to law on nuclear power; ~ **d'exploitation conforme au droit sur les déchets** DIST EAU, POL ENV plant permit according to law on waste; ~ **d'exploitation conforme au droit sur la protection de l'air** DIST EAU, POL ENV plant permit according to law on ambient air protection; ~ **d'exploitation conforme à la législation sur les déchets** DIST EAU, POL ENV plant permit according to law on waste; ~ **d'exploitation d'une installation** POL ENV plant licence (*BrE*), plant license (*AmE*); ~ **d'exploitation pilote** POL ENV model operation permission
péroxiacétylnitrate *m* (*PAN*) DIST EAU, POLL ATM peroxoacetyl nitrate
péroxyde: ~ **d'azote** *m* DIST EAU, POLL ATM nitrogen peroxide; ~ **d'hydrogène** *m* POLL ATM, POLL EAU hydrogen peroxide
persistance *f* POLL ATM retentivity
personnel: ~ **de tri** *m* DÉCHETS sorting personnel
perspectives: ~ **d'utilisation** *f pl* ÉN RENOUV future usability
perte *f* DIST EAU leak, leakage; ~ **d'acuité auditive induite par le bruit** POLL SON, SÉCURITÉ noise-induced hearing damage, noise-induced hearing impairment; ~ **à l'allumage** POLL ATM ignition loss; ~ **par calcination** DIST EAU ignition loss; ~ **de charge** POLL ATM, POLL EAU head loss; ~ **de charge de grille** POLL EAU screen loss of head; ~ **de dégrillage** POLL EAU screen loss; ~ **d'eau** DIST EAU water loss; ~ **d'énergie** DIST EAU, ÉN RENOUV energy loss, POLL ATM energy loss, waste of energy; ~ **par évaporation** DIST EAU, POLL EAU evaporation loss; ~ **au feu** DIST EAU, POLL EAU ignition loss, loss on ignition; ~ **hydraulique** DIST EAU, POLL EAU *fuite* hydraulic loss; ~ **interne au réseau de distribution d'eau** DIST EAU water loss within distribution system; ~ **mécanique** POLL EAU mechanical loss; ~ **de pression** POLL ATM, POLL EAU head loss, pressure loss; ~ **de pression dans les conduites** DIST EAU pressure loss of pipeline; ~ **de pression dans le réseau de distribution** DIST EAU pressure loss in distribution network; ~ **sonore par transmission** POLL SON sound transmission loss; ~ **thermique** POLL EAU heat loss; ~ **thermique par dissipation** DÉCHETS heat radiation loss; ~ **par transmission** POLL SON transmission loss; ~ **volumétrique** DIST EAU, POLL EAU volumetric loss
pertinent: ~ **pour l'environnement** *adj* POLL ATM, POLL SON, SOLS environmentally-relevant
pertuis *m* DIST EAU sluice; ~ **d'entrée** DIST EAU inlet; ~ **de fond** DIST EAU bottom culvert
perturbation *f* POLL ATM, POLL MAR, SOLS stress; ~ **de l'écosystème** POLL ATM, POLL MAR, SOLS ecosystem stress; ~ **de l'environnement** POLL ATM, SOLS environmental stress; ~ **du milieu** POLL ATM, SOLS environmental stress
perturber *vt* POLL SON damage
pesage[1] *m* DÉCHETS, ÉCHANT weighing
pesage[2]: ~ **et mise en flein** *loc* DÉCHETS weighing and punnet filling
peser *vt* DÉCHETS, ÉCHANT weigh, weigh out; ~ **par voie analytique** ÉCHANT weigh out analytically
pesticide *m* DIST EAU, POLL EAU pesticide; ~ **organochloré** POLL EAU *base de chlore organique* organochlorine pesticide; ~ **tenace** POLL ATM persistent pesticide
pétrifier *vt* DÉCHETS, POLL ATM, POLL EAU, SOLS mineralize, petrify
pétrin: ~ **malaxeur** *m* SOLS pug mill
pétrir *vt* DÉCHETS, DIST EAU knead
pétrole *m* GÉNÉRAL oil, POLL EAU mineral oil; ~ **brut** DIST EAU, POLL ATM, POLL MAR crude oil; ~ **coagulé** POLL MAR oil clots
pétrolier *m* POLL MAR tanker
pH[1] *abrév* (*potentiel hydrogène*) GÉNÉRAL pH-value, pH (*potential of hydrogen*)
pH[2]: ~**-mètre** *m* ÉCHANT pH-meter; ~**-mètre industriel** *m* ÉCHANT commercially-available pH-measuring device
phase[1]: **en ~ vapeur** *loc* POLL ATM vaporous
phase[2]: ~ **d'agrégation liquide** *f* POLL ATM liquid state of aggregation; ~ **dispersée** *f* ÉCHANT, POLL ATM disperse phase; ~ **de dispersion** *f* ÉCHANT, POLL ATM disperse phase; ~ **gazeuse** *f* POLL ATM gas phase; ~ **liquide** *f* POLL ATM liquid phase; ~ **liquide en suspension** *f* POLL ATM suspended liquid phase; ~ **de transition** *f* POL ENV transition phase
phénol *m* DIST EAU, POLL EAU phenol
phénomène: ~ **acoustique** *m* POLL SON sound occurrence; ~ **atmosphérique** *m* POLL ATM atmospheric phenomenon
phéromone *f* SOLS pheromone
phone *m* ÉCHANT phon
phonemètre *m* ÉCHANT phonmeter
phono: ~**-conducteur** *adj* POLL SON sound-conducting
phonocapteur *m* POLL SON soundbox
phosphate *m* DIST EAU, POLL EAU phosphate; ~ **de calcium** POLL EAU phosphate of calcium
phosphaté *adj* DIST EAU, POLL EAU phosphate-containing
phosphore *m* (*P*) POLL ATM, POLL EAU phosphorus (*P*)
photo[1]: ~**-autotrophe** *adj* POLL EAU phototrophic
photo[2]: ~**-oxydant** *m* DIST EAU, POLL ATM photooxidant; ~**-oxydation** *f* DIST EAU, POLL ATM photooxidation
photobiologique *adj* DIST EAU photobiological
photocapteur *m* DIST EAU photosensor
photocatalyseur *m* DIST EAU photocatalyst

photocatalytique *adj* DIST EAU photocatalytic
photochimique *adj* DIST EAU, POLL ATM photochemical
photodégradable *adj* DIST EAU *plastique* photodegradable
photodégradation *f* DIST EAU photodegradation
photolyse *f* DIST EAU, ÉCHANT, SOLS photolysis
photolytique *adj* DIST EAU, ÉCHANT, SOLS photolytic
photomécanique *adj* DIST EAU photomechanical
photomètre *m* ÉCHANT photometer; **~ à double faisceau** ÉCHANT double-beam photometer; **~ de flamme** ÉCHANT flame photometer; **~ longue distance** ÉCHANT long-distance photometer
photométrie: **~ de flamme** *f* ÉCHANT flame photometry; **~ ultraviolette** *f* ÉCHANT ultraviolet photometry
photométrique: **~ de flame** *adj* ÉCHANT flame-photometric
photophorèse *f* POLL ATM photophoresis
photosynthèse *f* ÉN RENOUV, POLL EAU photosynthesis
phototrophe *adj* POLL EAU phototrophic
photovoltaïque *adj* ÉN RENOUV photovoltaic, photovoltaic-operated
physico: **~-chimique** *adj* GÉNÉRAL chemical-physical, physical-chemical, physio-chemical; **~-pédologique** *adj* SOLS soil-physical
physio: **~-chimique** *adj* DIST EAU physio-chemical
physique *adj* DÉCHETS, DIST EAU, POLL ATM physical
physisorption *f* POLL ATM physisorption
phytoécologie *f* GESTION, POL ENV plant ecology
phytopathologie *f* SOLS plant diseases
phytoplancton *m* DIST EAU plant plankton, phytoplankton
phytosanitaire *adj* SÉCURITÉ phytosanitary
phytotoxine *f* SÉCURITÉ phytotoxin
pic *m* DIST EAU line, ÉCHANT, POLL SON peak; **~ de bruit** POLL SON noise peak; **~ chromatographique** ÉCHANT chromatographic peak
pièce: **~ de filtre** *f* DIST EAU, POLL ATM, SÉCURITÉ filter element; **~ justificative** *f* GESTION voucher
pièces: **~ anatomiques** *f pl* DÉCHETS *déchets hospitaliers* human waste; **~ détachées de conteneur en acier recyclables** *f pl* DÉCHETS steel reusable CKD container
piédroit: **~ pour effluents** *m* POLL EAU effluent side walls
piège *m* DIST EAU separator, sink, ÉCHANT trap, POLL ATM, POLL EAU, POLL SON trap, baffle; **~ acoustique** POLL SON sound trap; **~ à gouttelettes** POLL ATM droplet trap; **~ ionique** ÉCHANT ion trap; **~ ionique quadripolaire** ÉCHANT *spectrométrie de masse* quadrupole ion trap; **~ à poussière** POLL ATM dust trap; **~ à sable** POLL EAU sand trap; **~ thermique atmosphérique** POLL ATM atmospheric heat trapping
piéger *vt* POLL MAR entrap
pierraille *f* DIST EAU rubble
piézomètre *m* ÉCHANT piezometer
pile: **~ atomique** *f* DIST EAU atomic reactor; **~ à combustible** *f* ÉN RENOUV, POLL ATM fuel cell; **~ à combustible à l'acide phosphorique** *f* POLL ATM phosphoric acid fuel cell (*PAFC*); **~ à combustible à carbonate fondu** *f* POLL ATM molten-carbonate fuel cell (*MCFA*); **~ à combustible à oxyde solide** *f* POLL ATM solid oxide fuel cell (*SOFC*)
piles: **à ~** *loc* GÉNÉRAL battery-operated
pilier *m* DÉCHETS strut
pilonneuse *f* DIST EAU ramming machine
pilot: **~ ballasté** *m* DIST EAU ballasted pile
pilotage *m* DÉCHETS, DIST EAU pile-driving
piloter *vt* DÉCHETS, DIST EAU control, pile-drive, ÉCHANT *essai*, POLL ATM control
pince: **~ d'agrafage** *f* DÉCHETS *emballage* stapling plier; **~ à plomber** *f* DÉCHETS lead-sealing plier
pipe *vt* DIST EAU, GESTION, POLL EAU canalize
pipeline *m* DIST EAU, POLL EAU pipeline
pipette: **~ d'absorption** *f* ÉCHANT absorption pipette; **~ d'Andreasen** *f* ÉCHANT Andreasen's pipette; **~ d'échantillonnage de gaz** *f* ÉCHANT gas-sampling pipette; **~ graduée** *f* ÉCHANT measuring pipette
piquage *m* DIST EAU *corrosion* branch connection, pitting
pissette *f* ÉCHANT wash bottle, washing bottle
piste: **~ cyclable** *f* POL ENV cycle path, cycle track
pistolet: **~ d'étiquetage** *m* DÉCHETS labeller; **~ pulvérisateur** *m* POLL MAR spray gun
piston: **~ déplaceur** *m* POLL ATM displacer; **~ hydraulique** *m* DÉCHETS hydraulic ram; **~ plongeur** *m* DIST EAU pump plunger, press ram; **~ de presse** *m* DIST EAU press plunger
pistonphone *m* ÉCHANT pistonphone
place: **sur ~** *loc* SOLS on-site
placoplâtre *m* POLL SON gypsum plaster board
plafond: **~ alvéolé** *m* POLL SON panel-louvered ceiling (*AmE*), panel-louvred ceiling (*BrE*); **~ étanche** *m* POLL SON sealed ceiling; **~ hermétique** *m* POLL SON sealed ceiling; **~ insonorisant** *m* POLL SON sound-absorbing ceiling, acoustical ceiling; **~ suspendu** *m* POLL SON suspended ceiling
plage *f* ÉCHANT *mesure* range; **~ d'audition** POLL SON range of audibility; **~ de fréquence** POLL MAR, POLL SON frequency range; **~ hypersonique** POLL SON hypersonic range; **~ d'indicateur** ÉCHANT indicator range; **~ de mesure** ÉCHANT measuring range; **~ de puissance** ÉN RENOUV range of capacity; **~ de température** POLL ATM temperature range; **~ de variation du pH** POLL EAU pH-range
plaine: **~ abyssale** *f* POLL MAR oceanic abyss; **~ alluviale** *f* DIST EAU alluvial plain; **~ d'inondation** *f* DIST EAU flood plain
plaintes: **~ relatives à la sécurité** *f pl* SÉCURITÉ safety complaints
plan *m* DIST EAU plane; **~ d'affectation des sols** POL ENV zoning plan; **~ d'aménagement** POL ENV building development plan; **~ de charge thermique** POLL ATM heat-load plan; **~ de circulation** POL ENV traffic plan; **~ directeur d'urbanisme** POL ENV master plan; **~ d'eau** DIST EAU *au-dessus d'un déversoir* flow; **~ d'eau rehaussé** DÉCHETS, DIST EAU banked-up water level; **~ d'eau surélevé** DÉCHETS, DIST EAU banked-up water level; **~ d'échantillonnage** ÉCHANT sampling plan; **~ d'élimination des déchets** DÉCHETS, GESTION, POL ENV waste disposal plan; **~ d'exécution** GESTION construction drawing; **~ incliné** DÉCHETS inclined table; **~ d'intervention** POLL MAR contingency plan; **~ de masse** GESTION *d'un aménagement, d'une installation* general layout; **~ de mesure** ÉCHANT measurement plan; **~ d'occupation des sols** DIST EAU, SOLS land-use plan, zoning map; **~ de situation** GESTION site plan; **~ d'urgence** SÉCURITÉ emergency plan
planchéiage *m* POLL SON planking
plancher: **~ flottant** *m* POLL SON floating floor; **~ mobile** *m* DIST EAU sliding floor
plancton *m* DIST EAU, POLL EAU, POLL MAR plankton; **~ aérien** POLL ATM air plankton
planétaire *adj* ÉN RENOUV, POLL ATM global
planification: **~ des coûts** *f* GESTION cost planning; **~ de la gestion de déchets** *f* GESTION waste management planning; **~ en matière d'environnement** *f* POL ENV environmental planning; **~ de la mise en décharge** *f* DÉCHETS waste landfill planning
planning: **~ conforme à l'exécution** *m* GESTION as-built schedule; **~ d'effectif** *m* GESTION manpower schedule; **~ des études** *m* GESTION engineering schedule; **~ préliminaire** *m* GESTION conceptual schedule; **~ prévisionnel** *m* GESTION forecast schedule; **~ de réalisation** *m* GESTION intermediate construction schedule
plant *m* ÉN RENOUV *d'une espèce* cutting
plantation *f* ÉN RENOUV plantation
plante[1] *f* GÉNÉRAL plant
plante[2]: **~ chlorophyllienne** *f* POLL ATM chlorophyll-containing plant
plantes: **~ amylacées** *f pl* ÉN RENOUV starchy plants;

~ **oléagineuses** *f pl* ÉN RENOUV oil-containing plants, plant oils; ~ **palustres** *f pl* DÉCHETS marsh plants
planteuse *f* ÉN RENOUV planting machine, *tubercules, arbres* harvester
plaque: ~ **de couverture** *f* POLL ATM cover plate; ~ **de déviation** *f* POLL SON baffle board; ~ **d'huile** *f* POLL MAR oil slick; ~ **d'insonorisation** *f* POLL SON acoustic plate; ~ **d'interception magnétique** *f* DÉCHETS magnetic interception plate; ~ **d'isolation sans amiante** *f* SÉCURITÉ asbestos-free insulating plate; ~ **de métal** *f* POLL ATM, POLL SON metal plate, sheet metal; ~ **mise à la terre** *f* POLL ATM *dépoussiéreur à deux étages* earthed wall; ~ **perforée** *f* DIST EAU, POLL ATM perforated plate; ~ **protectrice** *f* POLL SON baffle board; ~ **souple** *f* POLL SON flexible plate; ~ **spirale** *f* POLL ATM spiral plate; ~ **support de tube** *f* POLL ATM tube support plate
plaquette *f* ÉN RENOUV pellet, chip; ~ **de bois** ÉN RENOUV wood chip, wood pellet
plastifiant *m* DÉCHETS, DIST EAU plasticator
plastification *f* DÉCHETS, DIST EAU plasticizing; ~ **à chaud** DÉCHETS thermoplastic solidification
plastifié *adj* DÉCHETS, DIST EAU plasticized
plastifier *vt* DÉCHETS, DIST EAU plasticize
plastique[1] *adj* GÉNÉRAL plastic
plastique[2] *m* DÉCHETS, DIST EAU plastics; ~ **armé** DIST EAU reinforced plastics; ~ **biodégradable** DÉCHETS, POLL EAU biodegradable plastic; ~ **déchiqueté** DIST EAU shredded plastics; ~ **propre** DÉCHETS clean plastic; ~ **recyclé** DIST EAU recycled plastic; ~ **renforcé par de la fibre de verre** DÉCHETS glass-fiber reinforced plastics (*AmE*), glass-fibre reinforced plastics (*BrE*); ~ **souillé** DÉCHETS, DIST EAU contaminated plastics, soiled plastics; ~ **stratifié au verre textile** DÉCHETS glass-fiber laminate (*AmE*), glass-fibre laminate (*BrE*)
plastiques *m pl* DÉCHETS *ensemble de ces matières* plastics; ~ **mêlés** DÉCHETS commingled plastics
plasturgie *f* DÉCHETS plastics processing, plastics processing industry, DIST EAU plastics processing
plat[1] *adj* DÉCHETS, DIST EAU, SOLS flat
plat[2] *m* DÉCHETS, DIST EAU, SOLS flat
plate: ~**-forme** *f* DÉCHETS *de véhicule de décharge* tipping table; ~**-forme écologique** *f* POL ENV ecological station
plateau *m* DÉCHETS, POLL ATM, POLL SON plate, tray, cap; ~ **alimentaire** DÉCHETS food tray; ~ **alvéolé** DÉCHETS *emballage* compartmented tray; ~ **d'assemblage** DÉCHETS collating transit tray; ~ **de catalyse** POLL ATM catalyst plate; ~ **souple** POLL SON flexible plate; ~ **de tête** POLL ATM head plate
platine *m* (*Pt*) GÉNÉRAL platinum (*Pt*)
plâtre *m* POLL SON plaster
plein: ~ **de l'eau** *m* DIST EAU high water; **trop-~** *m* POLL EAU overflow, overflow pipe; **trop-~ des eaux pluviales** *m* DIST EAU storm overflow; **trop-~ de secours** *m* POLL EAU emergency overflow
plomb *m* (*Pb*) GÉNÉRAL lead (*Pb*)
plombifère *adj* DIST EAU plumbiferous
pluie: ~ **propre** *f* POLL ATM clean rain; ~ **au sol** *f* POLL ATM throughfall; ~ **torrentielle** *f* SOLS pouring rain
pluies: ~ **acides** *f pl* POLL ATM, SOLS acid rain, acidic rain
plusieurs: **à** ~ **étages** *loc* POLL EAU multistage; **à** ~ **étendues de mesure** *loc* ÉCHANT multirange
pluviomètre *m* DIST EAU, ÉCHANT pluviometer, rain gage (*BrE*), rain gauge (*BrE*), precipitation gage (*AmE*), precipitation gauge (*BrE*)
pluviométrie *f* DIST EAU, ÉCHANT pluviometry
pneu: ~ **rechapé** *m* DIST EAU retreaded tire (*AmE*), retreaded tyre (*BrE*); ~ **usé** *m* DIST EAU used tire (*AmE*), used tyre (*BrE*)
pneumatique *adj* GÉNÉRAL pneumatic
PNUE *abrév* (*Programme des Nations Unies pour l'Environnement*) POL ENV UNEP (*United Nations Environmental Programme*)
poêle *m* POLL ATM stove heater; ~ **à mazout** POLL ATM thermal oil heater
poids: ~ **net** *m* DÉCHETS net weight; ~ **à vide** *m* DÉCHETS tare weight; ~ **en volume** *m* DÉCHETS volume weight
poignée *f* DÉCHETS carrying handle
point: ~ **de bifurcation** *m* POLL ATM bifurcation point; ~ **de collecte** *m* DÉCHETS collection point; ~ **de départ** *m* ÉCHANT starting point; ~ **d'ébullition** *m* POLL ATM boiling point; ~ **d'échantillonnage** *m* ÉCHANT sampling point; ~ **d'émission** *m* POLL ATM emissions point; ~ **d'entraînement** *m* POLL ATM entrainment point; ~ **de focalisation central** *m* ÉN RENOUV central focal point; ~ **de fusion** *m* DIST EAU melting point; ~ **d'inflammation** *m* POLL ATM ignition point, kindling point; ~ **de mesure** *m* ÉCHANT *des eaux souterraines* measuring point; ~ **de neutralisation** *m* ÉCHANT neutralization point; ~ **de prélèvement** *m* ÉCHANT sampling station, SOLS sample point; ~ **de puisage** *m* DIST EAU point of withdrawal; ~ **du réseau** *m* ÉCHANT grid point; ~ **de rosée** *m* POLL ATM dew point; ~ **de transition** *m* ÉCHANT transition point, transition temperature; ~ **de trouble** *m* DIST EAU cloud point; ~ **vert** *m* DIST EAU, GESTION green dot
pointe[1]: **de** ~ *loc* GÉNÉRAL peak
pointe[2]: ~ **de consommation** *f* DIST EAU peak consumption; ~ **de la demande** *f* DIST EAU peak demand; ~ **de la demande d'eau** *f* DIST EAU peak water demand; ~ **journalière** *f* POLL EAU peak of the day
pointeur *m* ÉN RENOUV tracker
poison *m* DÉCHETS, DIST EAU, POLL ATM, SÉCURITÉ poison; ~ **catalytique** POLL ATM catalyst poison
polarimètre: ~ **à lames** *m* ÉCHANT foil polarometer
polarisation *f* ÉCHANT *par fluorescence* polarization
polarité *f* POLL ATM polarity
polarographe *m* ÉCHANT polarograph
polarographie *f* ÉCHANT polarography; ~ **à mode impulsionnel** ÉCHANT pulse polarography
polarographique *adj* ÉCHANT polarographic
polder *m* DIST EAU polder
pôle: ~ **d'équilibre** *m* POL ENV regional growth centre
polissage *m* DIST EAU, POLL EAU *d'effluents* polishing
politique: ~ **de l'eau** *f* GESTION, POL ENV water management policy; ~ **de l'environnement** *f* POL ENV environmental policy; ~ **qualité** *f* GESTION quality policy
polluant[1]: **peu** ~ *loc* POLL ATM, SOLS low-pollutant, low-polluting
polluant[2] *m* GÉNÉRAL contaminant, pollutant; ~ **anthropique** POLL ATM, SOLS anthropogenic pollutant; ~ **atmosphérique** POLL ATM air pollutant; ~ **dangereux** POLL EAU critical pollutant; ~ **de l'eau** POLL EAU water pollutant; ~ **gazeux** POLL ATM, POLL EAU gaseous contaminant, gaseous pollutant; ~ **gazeux acide** POLL ATM acid gas pollutant; ~ **gazeux combustible** POLL ATM combustible gaseous pollutant; ~ **industriel** POLL ATM, POLL EAU industrial pollutant; ~ **inorganique** POLL ATM inorganic contaminant; ~ **naturel** POLL ATM, SOLS natural pollutant; ~ **organique** POLL ATM organic contaminant; ~ **particulaire** POLL ATM particulate pollutant; ~ **en phase vapeur** POLL ATM vaporous pollutant; ~ **potentiel** POLL EAU potential pollutant; ~ **précurseur** POLL ATM precursor pollutant; ~ **des sols** DIST EAU, SOLS land pollutant; ~ **du sol** SOLS ground level contaminant; ~ **de source diffuse** POLL ATM, SOLS area source pollutant; ~ **de source ponctuelle** POLL ATM point source pollutant; ~ **sous forme de particule** POLL ATM particulate pollutant; ~ **de surface** POLL ATM, POLL EAU surface pollutant; ~ **thermoadoucissant** DÉCHETS thermo-softening contaminant
pollué[1] *adj* GÉNÉRAL contaminant-laden, contaminated, polluted; ~ **par le bruit** POLL MAR noise-contaminated

pollué[2] *m* POL ENV pollutee
polluer *vt* GÉNÉRAL pollute, contaminate
pollueur *m* POL ENV environmental polluter, polluter
pollution *f* GÉNÉRAL contamination, pollution, POL ENV pollution burden, POLL ATM, SÉCURITÉ *chimique*, SOLS contamination, pollution; **~ acide** POLL ATM, SOLS acid pollution; **~ acoustique** POLL SON noise pollution, sound pollution; **~ de l'air** POLL ATM air pollution; **~ de l'air à l'intérieur des locaux** POLL ATM indoor air pollution; **~ alimentaire** SÉCURITÉ food pollution; **~ des aliments** POLL ATM, POLL EAU, SÉCURITÉ food contamination; **~ anthropique** SÉCURITÉ anthropic pollution; **~ atmosphérique** POLL ATM air pollution, atmospheric pollution; **~ atmosphérique particulaire** POLL ATM particulate air pollution; **~ atmosphérique par des gaz** POLL ATM gaseous air pollution; **~ atmosphérique ponctuelle** POLL ATM air pollution incident; **~ atmosphérique transfrontière** POLL ATM transboundary air pollution; **~ continue** SÉCURITÉ persistent deterioration; **~ des côtes** POLL MAR coastal pollution, shore pollution; **~ des cours d'eau** DIST EAU river pollution; **~ diffuse** POL ENV non-point pollution, POLL ATM area pollution; **~ due à l'homme** SÉCURITÉ human pollution; **~ de l'eau** POLL EAU water pollution; **~ des eaux** DIST EAU water pollution; **~ des eaux côtières** POLL MAR pollution of coastal waters; **~ des eaux douces** POLL EAU freshwater pollution; **~ des eaux intérieures** POLL EAU pollution of inland waters; **~ des eaux souterraines** DÉCHETS groundwater pollution; **~ de fond** POL ENV background pollution; **~ par les hydrocarbures** POLL MAR oil contamination; **~ industrielle** POLL ATM, POLL EAU industrial pollution; **~ marine** POLL MAR marine pollution, sea-based pollution; **~ des mers** POLL MAR marine pollution; **~ du milieu marin** POLL MAR pollution of the sea; **~ de la nappe phréatique** POLL EAU water table pollution; **~ des océans** POLL MAR ocean pollution; **~ de l'ozone stratosphérique** POLL ATM stratospheric ozone pollution; **~ par les aérosols** POLL ATM aerosol pollution; **~ par le bruit** SÉCURITÉ noise pollution; **~ par le mercure** SÉCURITÉ mercury pollution; **~ par l'ozone troposphérique** POLL ATM tropospheric ozone pollution; **~ pétrolière** POLL MAR oil pollution; **~ radioactive** DIST EAU, POLL ATM, SÉCURITÉ radioactive pollution; **~ de référence** POLL ATM background pollution; **~ résiduelle** DIST EAU, POLL EAU residual pollution; **~ du sol** DIST EAU land pollution, SOLS land pollution, soil pollution; **~ solide** DIST EAU material pollution; **~ sonore** POLL SON noise contamination; **~ sous forme de particules** DIST EAU, POLL ATM particulate pollution; **~ thermique** DIST EAU, POLL ATM heat pollution, thermal pollution
pollutogramme *m* POL ENV pollution chart
polyacrylonitrile *m* POLL ATM polyacrylnitrile (*PAN*)
polyaddition *f* DIST EAU polyaddition
polyamide *m* DIST EAU, POLL EAU polyamide (*PA*); **~ aliphatique** POLL ATM aliphatic polyamide; **~ aromatique** POLL ATM aromatic polyamide
polycarbonate *m* (*PC*) DIST EAU, POLL EAU polycarbonate (*PC*)
polychlorure: **~ de vinyle** *m* (*PVC*) DIST EAU, SOLS polyvinyl chloride (*PVC*)
polycondensation *f* DIST EAU polycondensation
polycristallin *adj* ÉN RENOUV polycrystalline
polyélectrolyte *m* DIST EAU, ÉCHANT polyelectrolyte
polyester *m* GÉNÉRAL polyester (*PES*); **~ non saturé** DÉCHETS unsaturated polyester
polyéthylène *m* (*PE*) DÉCHETS, DIST EAU polyethylene (*PE*); **~ basse densité** DÉCHETS low-density polyethylene; **~ haute densité** (*polyéthylène HD*) DÉCHETS *matières plastiques* high-density polyethylene; **~ HD** (*polyéthylène haute densité*) DÉCHETS *matières plastiques* high-density polyethylene
polymère *m* DÉCHETS, DIST EAU, POLL ATM, POLL EAU polymer; **~ de condensation** DÉCHETS condensation polymer; **~ non-ionogène** POLL EAU nonionogenic polymer; **~ de synthèse** POLL EAU synthetic polymer
polymérisation *f* DÉCHETS, DIST EAU, POLL ATM, POLL EAU polymerization
polymètre *m* ÉCHANT polymeter
polyoléfine *f* DÉCHETS polyolefin (*PO*)
polypropylène *m* DIST EAU, POLL ATM polypropylene (*PP*)
polysaccharide *m* DIST EAU polysaccharide
polysaprobe *adj* POLL EAU polysaprobic
polystyrène *m* DIST EAU polystyrene (*PS*); **~ expansé** DÉCHETS expanded polystyrene (*ep*)
polytéraphtalate: **~ d'éthylène** *m* DÉCHETS polyethylene teraphtalate (*PET*)
polytétrafluoréthylène *m* POLL ATM polytetrafluoroethylene (*PTFE*)
polyuréthanne *m* DIST EAU polyurethane (*PUR*)
pompable *adj* DIST EAU, POLL ATM, POLL EAU pumpable
pompage *m* GÉNÉRAL pumping, pumping out
pompe *f* GÉNÉRAL pump; **~ alimentaire** POLL ATM feed pump, DIST EAU supply pump; **~ alimentaire de chaudière** DÉCHETS, POLL ATM boiler feedwater pump; **~ d'amorçage** DIST EAU priming pump; **~ aspirante** DIST EAU, ÉCHANT, POLL EAU lift pump, suction pump; **~ aspirante et foulante** DIST EAU, POLL EAU lift-and-force pump; **~ aspirante et refoulante** DIST EAU, POLL EAU lift-and-force pump; **~ d'assèchement** DIST EAU drain pump, POLL MAR stripping pump; **~ basse pression** (*pompe BP*) POLL EAU low-pressure pump; **~ BP** (*pompe basse pression*) POLL EAU low-pressure pump; **~ centrifuge à amorçage automatique** DIST EAU, POLL EAU self-priming centrifugal pump; **~ à chaleur** ÉN RENOUV heat pump; **~ à chaleur électrique** ÉN RENOUV electrically-driven heat pump; **~ à chaleur entraînée par des combustibles fossiles** ÉN RENOUV fossil-fuel-driven heat pump; **~ à chapelet** DIST EAU, POLL EAU paternoster pump; **~ de circulation** DÉCHETS, POLL ATM circulating pump, circulation pump; **~ à débit constant** DIST EAU constant speed pump; **~ demi-rotative** DIST EAU semirotary pump; **~ de dénoyage** DÉCHETS, DIST EAU dewatering pump; **~ de dérivation** POLL ATM bypass pump; **~ à diaphragme** DIST EAU, ÉCHANT, POLL EAU, POLL MAR diaphragm pump; **~ doseuse** DÉCHETS, DIST EAU, POLL EAU dosing pump, metering pump; **~ à eau** DIST EAU, ÉN RENOUV water pump; **~ d'eau brute** DIST EAU raw water pump; **~ d'échantillonnage** ÉCHANT sample pump, sampling pump; **~ à éjecteur** POLL EAU jet pump; **~ électromagnétique** DIST EAU, POLL EAU electromagnetic pump; **~ élévatoire** DIST EAU, POLL EAU lift pump; **~ à émulsion** POLL EAU mammoth pump; **~ à énergie cinétique** POLL EAU kinetic-energy pump; **~ d'épuisement** DÉCHETS, DIST EAU, POLL EAU dewatering pump, drainage pump, pump sump, pumping plant; **~ d'exhaure** DÉCHETS, DIST EAU, POLL EAU dewatering pump; **~ à flux radial** POLL EAU radial-flow pump; **~ de forage** DIST EAU borehole pump; **~ foulante** DIST EAU, POLL EAU plunger pump; **~ à godets** DIST EAU, POLL EAU paternoster pump; **~ haute pression** POLL EAU high-pressure pump; **~ à hélice** DIST EAU propeller pump; **~ hélico-centrifuge** POLL EAU mixed-flow pump; **~ imbouchable** POLL EAU nonclogging pump; **~ à jet** DIST EAU jet pump; **~ à membrane** DIST EAU, ÉCHANT, POLL EAU, POLL MAR diaphragm pump; **~ à mouvement alternatif** DIST EAU reciprocating pump; **~ multi-cellulaire** POLL EAU vane pump; **~ non engorgeable** POLL EAU nonclogging pump; **~ parallèle** DIST EAU, POLL EAU parallel pump; **~ périphérique** DIST EAU peripheral pump; **~ périphérique à un étage** DIST EAU single-stage peripheral pump; **~ péristaltique** POLL EAU peristaltic pump; **~ à pignons** POLL EAU mixed-flow pump; **~ à piston** DIST EAU, POLL EAU piston pump, reciprocating pump; **~ à piston plongeur**

DIST EAU, POLL EAU plunger pump; ~ **à plongeur** DIST EAU, POLL EAU plunger pump; ~ **de pression** DIST EAU pressure pump; ~ **primaire** POLL EAU preliminary pump; ~ **de recirculation** POLL EAU recirculating pump; ~ **refoulante** DIST EAU, POLL EAU plunger pump; ~ **de régénération** POLL EAU regenerative pump; ~ **de reprise** DIST EAU, POLL EAU lift pump; ~ **rotative** DIST EAU, POLL EAU rotary pump; ~ **rotative à ailettes** POLL EAU vane pump; ~ **semi-rotative** DIST EAU semirotary pump; ~ **soulevante** DIST EAU, POLL EAU lift pump; ~ **de station d'épuration** POLL EAU wastewater pump; ~ **submersible** ÉCHANT submersible pump; ~ **suceuse** ÉCHANT suction pump; ~ **tourbillonnaire** POLL EAU peripheral pump; ~ **tourbillonnaire monocellulaire** POLL EAU single-stage peripheral pump; ~ **tourbillonnaire à plusieurs étages** POLL EAU multistage peripheral pump; ~ **de transfert** POLL MAR transfer pump, SOLS conveyor pump; ~ **de vidange** POLL EAU flushing pump; ~ **à vide** ÉCHANT, ÉN RENOUV, POLL ATM, POLL MAR vacuum pump; ~ **à vis** DIST EAU, POLL EAU, POLL MAR screw pump, screw-type pump; ~ **à vitesse variable** POLL EAU variable-speed pump; ~ **volumétrique** DIST EAU, POLL EAU positive displacement pump, metering pump

pompé *adj* GÉNÉRAL pumped

pomper *vt* GÉNÉRAL pump, pump out; ~ **et traiter** DIST EAU pump and treat

pondéral: ~ **en continu** *m* ÉN RENOUV continuous motion weight filling

pondération: ~ **du bruit** *f* POLL SON noise rating, noise weighting

pont: ~ **porteur** *m* POLL ATM bearing bridge; ~ **racleur** *m* POLL EAU scraper bridge; ~ **de Wheatstone** *m* ÉCHANT measuring bridge, Wheatstone bridge

pore *m* DIST EAU, POLL EAU *de filtre* pore; ~ **submicronique** POLL ATM submicropore

poreux *adj* DIST EAU, POLL ATM, POLL EAU porous

poromètre *m* ÉCHANT porometer

porosité *f* DIST EAU, POLL ATM, POLL EAU porosity; ~ **de couche filtrante** POLL ATM filter bed porosity; ~ **efficace** DIST EAU *du sol* absorptive capacity

portance *f* DIST EAU, ÉN RENOUV *éolienne* lift

portatif *adj* ÉCHANT hand-held, portable

porte[1] *f* DIST EAU *d'un cours d'eau, d'un déversoir* discharge; ~ **insonorisante** POLL MAR, POLL SON, SÉCURITÉ sound-absorbing door; ~ **phono-absorbante** SÉCURITÉ sound-absorbing door

porte[2]: ~**-bouteille** *m* DÉCHETS bottle carrier; ~**-échantillons** *m* ÉCHANT sample holder; ~**-éprouvette** *m* ÉCHANT sample holder; ~**-parole** *m* POL ENV spokesperson; ~**-parole de l'environnement** *m* POL ENV environment spokesperson, environmental representative; ~**-parole d'une installation** *m* DIST EAU, GESTION, POL ENV plant representative; ~**-sacs** *m* DÉCHETS bag holder; ~**-vent** *m* POLL ATM blast connection

portée *f* ÉCHANT extent, *mesure* range; ~ **du bruit** POLL MAR, POLL SON influence zone of noise

porter: ~ **atteinte à** *vt* GÉNÉRAL damage, harm; ~ **à ébullition** *vt* ÉCHANT boil; ~ **à fusion** *vt* POLL EAU *métallurgie* melt

porteur *m* ÉCHANT, POLL ATM carrier

portillon *m* ÉCHANT gate electrode

pose *f* SOLS laying

positif *adj* DIST EAU, ÉCHANT, POLL ATM, POLL EAU positive

position: ~ **dominante en matière d'environnement** *f* POL ENV environmental leading position; ~ **de filtre** *f* POLL ATM filtering position; ~ **de mesure** *f* ÉCHANT measuring position; ~ **de nettoyage** *f* POLL ATM cleaning position

possibilité: ~ **d'application** *f* DIST EAU, POLL ATM, POLL SON, SÉCURITÉ application possibility; ~ **de collecte des gaz** *f* ÉCHANT gas collectability; ~ **d'épuisement** *f* ÉN RENOUV exhaustibility; ~ **de recueillir les gaz** *f* ÉCHANT gas collectability; ~ **de récupération des gaz** *f* ÉCHANT gas collectability; ~ **de réutilisation** *f* DIST EAU reusability; ~ **d'utilisation** *f* GÉNÉRAL usability, ÉN RENOUV use possibility; ~ **de vente** *f* GESTION marketability

postchloration *f* DIST EAU postchlorination

post-coagulation *f* DIST EAU post-coagulation

postcombustion *f* DÉCHETS, POLL ATM, SOLS afterburning; ~ **des gaz d'échappement** DÉCHETS exhaust waste-gas afterburning

post-filtration *f* POLL EAU post-filtration

poste: ~ **d'aspiration ponctuel** *m* SÉCURITÉ point vacuum-cleaning system; ~ **de broyage** *m* DÉCHETS crusher unit; ~ **de chloration** *m* DIST EAU chlorination unit; ~ **de classification** *m* DÉCHETS *répartition de déchets* classification station; ~ **de collecte** *m* DÉCHETS collection station; ~ **de commande** *m* DÉCHETS, POLL ATM control center (*AmE*), control centre (*BrE*), control unit; ~ **de commande central** *m* POLL ATM central control room; ~ **de conditionnement** *m* DÉCHETS packing station; ~ **de conduite** *m* SÉCURITÉ cab; ~ **de décrassage** *m* POLL EAU slag-removing plant; ~ **de dégazage** *m* DIST EAU, ÉN RENOUV degassing facility; ~ **de déshydratation des boues d'eaux résiduaires** *m* POLL EAU wastewater sludge dewatering equipment; ~ **de dessablage** *m* DIST EAU tumbling station; ~ **de déversement** *m* DÉCHETS tumbling station; ~ **d'eau chaude** *m* POLL ATM hot-water heater; ~ **d'emballage** *m* DÉCHETS packaging station, packing station; ~ **d'empaquetage** *m* DÉCHETS packer unit; ~ **d'entraînement** *m* DIST EAU drive unit; ~ **et installation relocalisable** *f* SOLS relocatable plant; ~ **d'extrusion** *m* DÉCHETS extrusion plant; ~ **d'extrusion-granulation** *m* DIST EAU *pressage et dessication des boues* regranulating extruder; ~ **d'incendie** *m* DIST EAU hydrant; ~ **d'injection** *m* SOLS injection plant; ~ **de mesurage** *m* ÉCHANT measuring station; ~ **de mesure mobile** *m* ÉCHANT mobile measuring unit; ~ **mobile** *m* SOLS mobile unit; ~ **de pesage** *m* DÉCHETS weigh office; ~ **de pompage** *m* DIST EAU, POLL EAU pump house, pump set, pump station; ~ **de réchauffage** *m* DÉCHETS, POLL ATM afterburner housing; ~ **de reconnaissance automatique** *m* DÉCHETS, DIST EAU automatic-recognition unit; ~ **de réglage** *m* POLL ATM regulating unit; ~ **de réglage du rapport air-carburant** *m* POLL ATM fuel-air ratio control unit; ~ **de réglage du rapport air-combustible** *m* POLL ATM fuel-air ratio control unit; ~ **de régulation** *m* DIST EAU, POLL ATM, SOLS regulating unit; ~ **de relèvement** *m* POLL EAU pumping unit; ~ **relocalisable** *m* SOLS relocatable unit; ~ **de télécommande** *m* DIST EAU, POLL EAU remote-control unit; ~ **de tonnelage** *m* DIST EAU tumbling station; ~ **de transit** *m* DÉCHETS transfer station; ~ **de travail** *m* SÉCURITÉ workplace, work station; ~ **de tri de déchets** *m* DÉCHETS, DIST EAU waste-sorting plant; ~ **de visualisation** *m* ÉCHANT display unit

post-ozonation *f* DIST EAU post-ozonation

postprécipitation *f* POLL EAU post-precipitation

pot: ~ **catalytique** *m* POLL ATM catalytic muffler; ~ **catalytique en circuit ouvert** *m* POLL ATM open-loop catalytic converter; ~ **catalytique à trois voies** *m* POLL ATM three-way catalytic converter (*TWC*); ~ **de chauffage** *m* POLL EAU *plastiques* heating cylinder; ~ **de détente des gaz** *m* POLL SON exhaust silencing; ~ **d'échappement** *m* POLL ATM, POLL SON exhaust muffler (*AmE*), exhaust silencer (*BrE*)

potabiliser *vt* DIST EAU, ÉN RENOUV *eau* treat

potable *adj* DIST EAU drinkable, potable

potasse *f* POLL EAU potash

potentiel[1] *adj* GÉNÉRAL potential

potentiel[2] *m* GESTION, POL ENV potential; ~ **d'adsorption** POLL ATM adsorption potential; ~ **agricole** SOLS agricultural potential; ~ **d'appauvrissement de la couche d'ozone** POL ENV, POLL ATM ozone depletion potential (*ODP*); ~ **d'approvisionnement** ÉN RENOUV supply potential; ~ **de bioaccumulation** POLL ATM bioaccumulation potential; ~ **biotique** SOLS biotic potential; ~ **de charge chlorée** POLL ATM chlorido load potential, chlorine-loading potential (*CLP*); ~ **de décomposition** SOLS decomposition potential; ~ **économique** ÉN RENOUV economic potential; ~ **d'électrode** ÉCHANT electrode potential; ~ **d'émissions** DÉCHETS emissions potential; ~ **énergétique** ÉN RENOUV energy potential, potential energy; ~ **d'énergie éolienne en mer** ÉN RENOUV offshore potential of wind energy; ~ **de formation de gaz** DÉCHETS, ÉCHANT gas build-up potential, gas formation potential; ~ **géothermique** ÉN RENOUV geothermal energy potential; ~ **hydrogène** (*pH*) DÉCHETS, DIST EAU, ÉCHANT, POLL ATM, POLL EAU, SOLS pH-value, potential of hydrogen (*pH*); ~ **d'ionisation** ÉCHANT ionization potential; ~ **marin** ÉN RENOUV offshore potential; ~ **d'oxydo-réduction** ÉCHANT, POLL EAU oxidation-reduction potential (*ORP*), redox potential; ~ **de production** ÉN RENOUV production potential; ~ **de recyclage** DIST EAU recycling potential; ~ **de risques** GESTION, SÉCURITÉ risk potential, danger potential; ~ **technique** ÉN RENOUV technical potential; ~ **théorique** ÉN RENOUV theoretic potential; ~ **valorisable** ÉN RENOUV usable potential

potentiomètre *m* ÉCHANT potentiometer

potentiométrie *f* ÉCHANT potentiometry

potentiométrique *adj* ÉCHANT potentiometric

potentiostat *m* ÉCHANT potentiostat

potentiostatique *adj* ÉCHANT potentiostatic

poubelle *f* DÉCHETS bin, dustbin (*BrE*), garbage can (*AmE*), refuse can, rubbish bin (*BrE*), trash can (*AmE*)

poudre: ~ **d'origine animale** *f* POLL ATM animal dust

pourcentage: ~ **d'eau de rétention** *m* DIST EAU share of retained water

pourcentile *m* ÉCHANT percentile value

pourrissage *m* ÉN RENOUV souring

pourriture *f* DÉCHETS, DIST EAU, POLL EAU rot; ~ **accélérée** POLL EAU rapid rotting; ~ **alvéolaire** DÉCHETS *du bois* pocket rot; ~ **fibreuse** DÉCHETS *du bois* string rot; ~ **molle** DÉCHETS *du bois* soft rot

pousse *f* ÉN RENOUV shoot

poussée *f* POLL MAR thrust

poussiérage *m* POLL ATM *papier* dusting

poussière *f* DÉCHETS, DIST EAU, POLL ATM, SÉCURITÉ dust, grit; ~ **abrasive** POLL ATM abrasive dust; ~ **adhésive** POLL ATM sticky dust; ~ **agglomérée** POLL ATM agglomerated dust; ~ **de catalyseur** POLL ATM catalyst dust; ~ **de charbon** POLL ATM coal dust; ~ **chimiquement agressive** POLL ATM chemically-aggressive dust; ~ **collante** POLL ATM sticky dust; ~ **collectée** POLL ATM collected dust; ~ **combustible** POLL ATM combustible dust; ~ **contaminée** POLL ATM contaminated dust; ~ **contenant du plomb** DIST EAU lead-containing dust; ~ **due aux activités humaines** POLL ATM man-made dust; ~ **de filtre** DÉCHETS, POLL ATM filter dust; ~ **humidifiable** POLL ATM wettable dust; ~ **industrielle** POLL ATM plant dust; ~ **inflammable** POLL ATM flammable dust; ~ **métallique** POLL ATM metallic dust; ~ **microscopique** SÉCURITÉ microscopic dust; ~ **minérale** POLL ATM mineral dust; ~ **mouillable** POLL ATM wettable dust; ~ **polluée** POLL ATM polluted dust; ~ **pyrophore** POLL ATM pyrophoric dust; ~ **radioactive** POLL ATM radioactive dust; ~ **réutilisable** POLL ATM reusable dust; ~ **de silice** SÉCURITÉ silica dust; ~ **de sol** SOLS soil dust; ~ **submicronique** POLL ATM finest dust; ~ **en suspension dans l'air** SÉCURITÉ airborne dust; ~ **totale** ÉCHANT, POLL ATM total dust; ~ **toxique** POLL ATM toxic dust; ~ **d'usine** POLL ATM plant dust

poussiéreux *adj* DIST EAU, POLL ATM, SÉCURITÉ dust-laden

pouvoir: ~ **adhérent** *m* ÉCHANT adhesive force; ~ **d'adsorption** *m* POLL ATM adsorption power; ~ **de bioaccumulation** *m* SOLS bioaccumulation potential; ~ **calorifique** *m* DÉCHETS, ÉCHANT, POLL ATM calorific value; ~ **calorifique faible** *m* DÉCHETS low calorific value; ~ **calorifique inférieur** *m* DÉCHETS lower calorific value; ~ **calorifique moyen** *m* POLL EAU medium calorific value; ~ **calorifique naturel** *m* DIST EAU net calorific value; ~ **calorifique net** *m* (*P.C.N.*) POLL EAU *combustible* net calorific value (*NCV*); ~ **calorifique supérieur** *m* (*PCS*) DIST EAU, POLL ATM upper calorific value (*UCV*); ~ **de dégradation** *m* POLL EAU decomposition capacity; ~ **de dilution** *m* POLL EAU dilution capacity; ~ **discrétionnaire** *m* POL ENV discretionary decision; ~ **dispersant** *m* DIST EAU dispersing power; ~ **encrassant** *m* DIST EAU fouling potential; ~ **d'entraînement** *m* POLL EAU flushing strength; ~ **d'épuration naturelle** *f* SOLS natural purification force; ~ **de protection** *m* POLL MAR protection capacity; ~ **de rinçage** *m* POLL EAU flushing strength; ~ **statutaire** *m* POL ENV statutory power; ~ **tampon** *m* SOLS buffer capacity; ~ **toxique** *m* SOLS toxicity potential

pouvoirs: ~ **publics** *m pl* POL ENV government; ~ **statutaires** *m pl* POL ENV statutory powers

pouzzolane *f* SOLS puzzolana

ppm *abrév* (*partie par million*) ÉCHANT ppm (*parts per million*)

pratiques: ~ **commerciales à orientation écologique** *f pl* GESTION environmentally-orientated marketing

pré-aération *f* POLL EAU preaeration

préamplificateur *m* ÉCHANT preamplifier

prébroyage *m* DIST EAU prior crushing

précaution *f* SÉCURITÉ precaution

précautions: ~ **contre la poussière** *f pl* SÉCURITÉ precautions against dust

précédant *adj* POLL EAU preceding

précédent *m* POL ENV *juridique* precedence case

préchauffage *m* GÉNÉRAL preheating; ~ **des gaz** POLL ATM *dépoussiérage* waste gas preheating; ~ **du gaz** POLL ATM, POLL EAU gas preheater

préchauffer *vt* GÉNÉRAL preheat

préchauffeur *m* GÉNÉRAL preheater; ~ **d'air rotatif** POLL ATM *dépoussiérage* rotating gas preheater

préchloration *f* DIST EAU prechlorination

précipitant *m* GÉNÉRAL precipitant

précipitateur: ~ **électrostatique** *m* DÉCHETS, DIST EAU, POLL ATM electrofilter, *dépoussiérage électrique des gaz* electrostatic precipitator (*ESP*); ~ **de grand rendement** *m* POLL ATM high-performance precipitator

précipitation *f* GÉNÉRAL deposition, precipitation; ~ **d'aérosols** POLL ATM aerosol precipitation; ~ **atmosphérique** DIST EAU, POLL ATM atmospheric precipitation, rainfall; ~ **à la chaux** POLL EAU precipitation with lime; ~ **chimique** DÉCHETS, POLL ATM chemical precipitation; ~ **directe** SOLS throughput, thruput (*AmE*); ~ **donnée** POLL ATM precipitation event; ~ **des métaux lourds** POLL EAU heavy metals precipitation; ~ **du phosphore** POLL EAU phosphorous precipitation; ~ **de poussières** POLL ATM grit precipitation; ~ **au sel ferrique** POLL EAU precipitation with ferric salt; ~ **aux sels d'aluminium** POLL EAU precipitation with aluminium salts; ~ **de sulfure** DIST EAU, POLL EAU sulfide precipitation (*AmE*), sulphide precipitation (*BrE*)

précipitations: ~ **acides** *f pl* POLL ATM acid precipitation

précipité *m* GÉNÉRAL precipitate; ~ **de phosphate** POLL EAU phosphate precipitate

précipiter[1] *vt* GÉNÉRAL precipitate

précipiter[2] *vi* DIST EAU *fluides* separate

précision: ~ **de lecture** *f* ÉCHANT reading accuracy; ~ **de mesure** *f* ÉCHANT measuring accuracy

préclarification *f* POLL EAU preclarification

préclassé *adj* DIST EAU, SOLS presorted

préclasser *vt* DIST EAU, SOLS presort

préconcentration *f* ÉCHANT preconcentration

préconcentré *m* ÉCHANT preconcentrate

précouche: ~ **de filtration** *f* POLL ATM precoat layer

précurseur: ~ **acide** *m* SOLS acidic precursor

prédation *f* POL ENV predation

prédénitrification *f* POLL EAU pre-denitrification

préfiltrage *m* ÉCHANT, POLL ATM, POLL EAU prefiltering
préfiltration *f* ÉCHANT, POLL ATM, POLL EAU prefiltration
préfiltre *m* ÉCHANT, POLL ATM, POLL EAU prefilter, primary filter
préjudice *m* POLL ATM impairment
préjudices: ~ **portés à l'environnement** *m pl* POL ENV environmental crime
préjudiciable *adj* POLL ATM *à l'environnement, à la qualité de l'air* harmful, SÉCURITÉ injurious, harmful
prélavage *m* DIST EAU prewashing, POLL ATM prescrubbing, precleaning
prélaver *vt* DIST EAU prewash, POLL ATM prescrub
prélaveur *m* DIST EAU prewasher, POLL ATM prescrubber
prélèvement *m* ÉCHANT sampling; ~ **automatique** ÉCHANT automatic sampling; ~ **d'eau** ÉCHANT water sample; ~ **d'eau du sol** ÉCHANT soil water sampling; ~ **d'eau stérile** ÉCHANT germ-free wastewater sample; ~ **d'eau à titre exploratoire** ÉCHANT, POLL EAU exploratory water sample; ~ **d'eaux usées après épuration** ÉCHANT clear wastewater sample; ~ **d'eaux usées neutres** ÉCHANT neutral wastewater sample; ~ **d'eaux usées troubles** ÉCHANT turbid wastewater sample; ~ **d'eaux usées turbides** ÉCHANT turbid wastewater sample; ~ **d'échantillon** ÉCHANT sampling, sample taking; ~ **d'échantillons atmosphériques** ÉCHANT air sampling; ~ **d'échantillons de sol** SOLS soil sampling; ~ **élémentaire** ÉCHANT increment; ~ **de gaz** ÉCHANT gas sample; ~ **de gaz pollués** POLL ATM waste gas probe; ~ **instantané** ÉCHANT spot sample; ~ **obligatoire** POL ENV levy; ~ **sanitaire** SÉCURITÉ health-related sampling
prélèvements *m pl* DIST EAU tapping
prémélangeur *m* DIST EAU premixer
prémoulage *m* DIST EAU premolding (*AmE*), premoulding (*BrE*)
préparation: ~ **d'échantillonnage** *f* ÉCHANT sampling preparation; ~ **d'échantillons** *f* ÉCHANT sample processing; ~ **des éluats** *f* ÉCHANT, POLL EAU eluate preparation; ~ **des résidus de bacs à graisse** *f* DÉCHETS waste oil preparation; ~ **du sol** *f* ÉN RENOUV, SOLS, tillage, tilling process; ~ **de vieux papiers** *f* DÉCHETS wastepaper preparation
préparer *vt* POL ENV draft; ~ **un projet d'accord cadre** POL ENV draft a framework agreement
préprogrammable *adj* ÉCHANT preprogrammable
préprogrammé *adj* ÉCHANT preprogrammed
préréfrigérant *m* POLL ATM precooler
préséchage: ~ **des boues résiduaires** *m* POLL EAU pre-drying of wastewater sludge
présentation: ~ **de rapports sur l'environnement** *f* POL ENV environmental reporting
préservation: ~ **des espèces** *f* POL ENV conservation of species; ~ **de la faune et de la flore sauvage** *f* POL ENV wildlife conservation; ~ **de la nature** *f* POL ENV conservation of nature
préserver *vt* GÉNÉRAL conserve
président *m* POL ENV *de séance* chairperson
pressage *m* DIST EAU pressing, ÉN RENOUV *paille* baling
presse[1]: ~**-balles** *m* DIST EAU packing press; ~**-conteneurs** *m* DIST EAU container press
presse[2]: ~ **à agglomérer** *f* DÉCHETS briquette press; ~ **à balles** *f* DÉCHETS baling press; ~ **à balles horizontale** *f* DÉCHETS channel ball press; ~ **à capsuler** *f* DÉCHETS capping press; ~ **à comprimés** *f* ÉCHANT tablet press; ~ **pour conteneurs** *f* DÉCHETS container press; ~ **en continu** *f* DIST EAU continuous press; ~ **à découper** *f* DIST EAU cutting press; ~ **à double bande** *f* DÉCHETS, DIST EAU double-belt press; ~ **à empaqueter** *f* DIST EAU packing press; ~ **à ferraille** *f* DÉCHETS scrap baler, scrap baling press, scrap press; ~ **à filer les tubes** *f* DÉCHETS extrusion machine for tubes; ~ **à impression** *f* DÉCHETS imprinter; ~ **à imprimer** *f* DÉCHETS imprinter; ~ **de moulage par injection** *f* DÉCHETS injection molding machine (*AmE*), injection moulding machine (*BrE*), injection molding press (*AmE*), injection moulding press (*BrE*); ~ **de moulage par injection-soufflage** *f* DÉCHETS injection blow-molding machine (*AmE*), injection blow-moulding machine (*BrE*); ~ **à mouler par injection** *f* DÉCHETS injection molding press (*AmE*), injection moulding press (*BrE*); ~ **pour vieux papiers** *f* DÉCHETS wastepaper compressing press
pressée *f* DIST EAU *filtre-presse* pressing
presser *vt* DIST EAU press
pression *f* GÉNÉRAL pressure; ~ **acoustique** POLL SON sound pressure; ~ **acoustique effective** POLL SON effective sound pressure; ~ **acoustique de référence** POLL SON reference sound pressure; ~ **d'admission** POLL EAU intake pressure; ~ **atmosphérique** POLL ATM atmospheric pressure; ~ **démographique** POL ENV population pressure; ~ **différentielle** POLL ATM differential pressure; ~ **dynamique** ÉCHANT dynamic pressure; ~ **d'eau** POLL EAU static head; ~ **de l'eau** DIST EAU water pressure; ~ **d'éclatement** DÉCHETS *emballage* bursting pressure; ~ **d'épuration** POLL ATM cleaning pressure; ~ **de filtration** POLL ATM filtration pressure; ~ **fiscale** POL ENV tax burden; ~ **hydraulique** POLL EAU hydraulic pressure; ~ **hydrostatique** DIST EAU, POLL EAU hydrostatic pressure; ~ **d'injection** SOLS grouting pressure, injection pressure; ~ **interne** POLL ATM internal pressure; ~ **manométrique totale** POLL EAU *barrage* total head; ~ **de moulage par injection** DÉCHETS injection molding pressure (*AmE*), injection moulding pressure (*BrE*); ~ **normale** ÉCHANT normal pressure; ~ **osmotique** POLL EAU osmotic pressure; ~ **partielle** ÉCHANT partial pressure; ~ **de rayonnement acoustique** POLL SON sound radiation pressure; ~ **de refoulement** DIST EAU delivery pressure; ~ **des rejets gazeux** POLL ATM waste gas pressure; ~ **de saturation** POLL ATM saturation pressure; ~ **statique** POLL ATM static pressure; ~ **de vapeur** DIST EAU, POLL EAU, SOLS steam pressure, vapor pressure (*AmE*), vapour pressure (*BrE*); ~ **du vent** POLL ATM velocity pressure
prétraitement *m* GÉNÉRAL pretreatment; ~ **de déchets** DÉCHETS waste pretreatment; ~ **physico-chimique** DÉCHETS chemical-physical pretreatment; ~ **pyrolytique** DIST EAU pyrolytic pretreatment
prétraiter *vt* DÉCHETS, DIST EAU, POLL ATM, SOLS pretreat
prétrié *adj* DIST EAU, SOLS presorted
prétrier *vt* DIST EAU, SOLS presort
préventif *adj* GÉNÉRAL preventive
prévention *f* GÉNÉRAL prevention; ~ **du bruit** POLL MAR, POLL SON noise prevention; ~ **des crues** DIST EAU flood prevention; ~ **des émissions** SOLS emissions avoidance; ~ **et lutte intégrées contre la pollution** GESTION integrated prevention and pollution control (*IPPC*); ~ **intégrée de la pollution** GESTION integrated pollution prevention; ~ **de la pollution** POL ENV pollution prevention; ~ **de la pollution sonore** POLL SON, SÉCURITÉ noise pollution prevention; ~ **des prises au feu inopinées d'atmosphères inflammables** SÉCURITÉ prevention of inadvertent ignition of flammable atmospheres by RF radiation; ~ **des risques** GESTION, SÉCURITÉ danger prevention, hazard prevention
prévision *f* ÉCHANT, GESTION, POL ENV forecast, forecasting; ~ **des catastrophes naturelles** POL ENV forecasting of natural disasters; ~ **d'exposition au bruit** POLL SON noise exposure forecast (*NEF*); ~ **de la pollution atmosphérique** POLL ATM air pollution forecast
prévoir *vt* ÉCHANT, GESTION, POL ENV forecast
primage *m* DIST EAU, POLL EAU *vapeur* priming
primaire *adj* POLL ATM, SOLS primary
principe: ~ **d'Archimède** *m* DIST EAU Archimedes' principle; ~ **des barrières multiples** *m* DÉCHETS, DIST EAU multibarrier principle; ~ **de conception ergonomique** *m* SÉCURITÉ ergonomic design principle; ~ **de détection** *m* ÉCHANT

detection principle; ~ **du droit d'accès à l'information** *m* POL ENV right-to-know principle; ~ **des droits acquis** *m* POL ENV grandfathering principle; ~ **de Huygens-Fresnel** *m* ÉCHANT *diffraction* Huygen-Fresnel principle; ~ **de mesure** *m* ÉCHANT measuring principle; ~ **pollueur-payeur** *m* DIST EAU, POL ENV *défini par l'OCDE* polluter pays principle; ~ **de prévention de la pollution** *m* POL ENV environmental prevention principle; ~ **de recondance des barrières** *m* DÉCHETS, DIST EAU multibarrier principle; ~ **de subsidiarité** *m* POL ENV subsidiarity principle; ~ **du Venturi** *m* POLL ATM Venturi principle; ~ **du vide** *m* ÉCHANT vacuum principle

priorité *f* POL ENV priority

prise *f* DÉCHETS caking, DIST EAU, ÉCHANT *durcissement* setting, POLL ATM *d'un matériau* caking; ~ **d'air** POLL ATM air intake; ~ **en compte** POL ENV internalization; ~ **en compte des coûts externes** POL ENV internalization of external costs; ~ **en compte des effets externes** POL ENV internalization of external effects, internalization of externality; ~ **en compte de l'environnement** GESTION, POL ENV environmental awareness; ~ **en compte du risque** POL ENV risk mitigation; ~ **d'eau** DIST EAU, POLL EAU water intake; ~ **d'eau souterraine** DIST EAU underground hydrant; ~ **d'essai** ÉCHANT test portion

prisme: ~ **d'inversion** *m* ÉCHANT inverting prism

prix: ~ **attribué dans le domaine de l'environnement** *m* POL ENV environmental prize; ~ **du marché** *m* GESTION, POL ENV market price; ~ **de référence** *m* POL ENV shadow price

problème: ~ **alimentaire mondial** *m* POL ENV world food problem; ~ **dû aux effluents gazeux** *m* POLL ATM waste gas problem; ~ **de formation de gaz de décharge** *m* DÉCHETS landfill gas problem

procédé *m* GÉNÉRAL process; ~ **d'absorption par atomisation** POLL ATM spray absorption process; ~ **d'absorption-filtration-adsorption** POLL ATM absorption-filtration-adsorption process; ~ **aux acides nitrique et sulfurique** POLL ATM nitric and sulfuric acid process (*AmE*), nitric and sulphuric acid process (*BrE*); ~ **ADSOX** POLL ATM ADSOX process; ~ **au charbon actif** POLL ATM activated-carbon process; ~ **CHEMFIX®** DÉCHETS CHEMFIX® process; ~ **Claus** POLL ATM Claus process; ~ **à coke actif** POLL ATM activated-coke process; ~ **de compostage en chambre de pourrissage** DIST EAU rotting-chamber composting process; ~ **de conditionnement** DÉCHETS, DIST EAU conditioning process; ~ **Cottrel** POLL ATM *dépoussiérage* Cottrel process; ~ **cryogénique** DÉCHETS, DIST EAU cryogenic process; ~ **Dano** DÉCHETS *traitement des déchets* Dano process; ~ **Degussa-Plinke** POLL ATM Degussa-Plinke process; ~ **de déphosphoration** POLL EAU Phoredox method; ~ **de dépoussiérage par voie humide** POLL ATM wet cleaning process; ~ **de destruction** DÉCHETS, SOLS destruction process; ~ **de désulfuration direct** POLL ATM direct desulfurization process (*AmE*), direct desulphurization process (*BrE*); ~ **de désulfuration par voie sèche** POLL ATM dry desulfurization process (*AmE*), dry desulphurization process (*BrE*); ~ **de désulfuration par voie humide** POLL ATM wet desulfurization process (*AmE*), wet desulphurization process (*BrE*); ~ **d'écumage** SOLS foam process; ~ **d'épuration** GÉNÉRAL treatment process; ~ **d'épuration des eaux usées** POLL EAU wastewater purification process; ~ **d'épuration par boues activées** DÉCHETS, DIST EAU, POLL ATM activated-sludge process; ~ **d'épuration par voie humide** POLL ATM wet cleaning process; ~ **d'épuration par voie sèche** POLL ATM *gaz* dry-cleaning process; ~ **exothermique** POLL EAU exothermic process; ~ **à faible charge** POLL EAU low-rate process; ~ **de filtration lent** DIST EAU slow filter process; ~ **à filtre ruisselant** POLL ATM trickling filter process; ~ **à forte charge** POLL EAU high-rate process; ~ **de fusion** ÉN RENOUV *énergie nucléaire* fusion process; ~ **industriel** DIST EAU industrial process; ~ **de lavage** POLL ATM washing process; ~ **par lit bactérien** POLL EAU trickling filter process; ~ **à lit fluidisé** SOLS fluidized-bed process; ~ **mettant en œuvre la dispersion par des réactifs chimiques** DIST EAU process of using dispersion through chemical reactions; ~ **mettant en œuvre un lit bactérien** POLL EAU trickling filter process; ~ **mettant en œuvre la vapeur** ÉN RENOUV steam process; ~ **microbiologique** SOLS microbiological process; ~ **mixte associant pyrolyse et incinération** POLL EAU *incinération de boues d'eaux usées* process combining pyrolysis and incineration; ~ **de moulage sous vide** DÉCHETS air mold process (*AmE*), air mould process (*BrE*); ~ **optoélectronique** DÉCHETS optoelectronic process; ~ **d'oxydo-réduction** POLL ATM oxidation-reduction process; ~ **par apport d'additifs secs** POLL ATM dry additive process; ~ **par boues activées** DÉCHETS, DIST EAU, POLL ATM *épuration des eaux* activated-sludge process; ~ **par boues activées avec alimentation étagée** POLL EAU multistage activated sludge process; ~ **par faisceau d'électrons** POLL ATM electron beam process; ~ **PETRIFIX** DÉCHETS PETRIFIX process, DIST EAU PETRIFIX process; ~ **Phoredox** POLL EAU *traitement biologique des effluents urbains* Phoredox method; ~ **photomécanique** DIST EAU photomechanical process; ~ **physico-chimique** DIST EAU, POLL EAU physical-chemical process, physio-chemical process; ~ **plasmatique haute température** SOLS high-temperature plasma process; ~ **de pourrissage** DIST EAU rotting process; ~ **de préparation** DIST EAU preparation process; ~ **de putréfaction-filtration** POLL EAU filtration-putrefaction process; ~ **radiographique** DIST EAU radiographic process; ~ **de recyclage** DIST EAU recycling process; ~ **de recyclage de matières premières** ÉN RENOUV raw material recycling process; ~ **de réduction catalytique** POLL ATM catalytic reduction process; ~ **SEALOSAFE®** DIST EAU SEALOSAFE® process; ~ **semi-humide de traitement des gaz** POLL ATM partially-dry waste-gas treatment; ~ **de séparation** POLL ATM separation process; ~ **de séparation par membrane** POLL EAU membrane separation process; ~ **de solidification** DIST EAU solidification process; ~ **de sorption biologique** POLL ATM *épuration des gaz* biological sorption process; ~ **Steuler** POLL ATM Steuler process; ~ **Sulzer-Ariqua** POLL ATM Sulzer-Ariqua process; ~ **de traitement** GÉNÉRAL treatment process; ~ **de vitrification** DÉCHETS vitrification process; ~ **Walther** POLL ATM *désulfuration* Walther process; ~ **Wellmann-Lord** POLL ATM *désulfuration des fumées* Wellmann-Lord process

procéder: ~ **au traitement final** *vi* DIST EAU treat finally

procédés: ~ **chimiques de base** *m pl* DIST EAU unit processes

procédure *f* GÉNÉRAL procedure; ~ **d'analyse** ÉCHANT analytical procedure; ~ **d'appel** POL ENV appeal procedure; ~ **d'application** POL ENV implementation procedure; ~ **d'autorisation** DIST EAU, POL ENV licensing process, permit process; ~ **d'essai dans l'environnement** GESTION environmental testing procedure; ~ **d'étalonnage** ÉCHANT calibration process; ~ **de l'étude d'impact** POL ENV environmental impact assessment procedure, EIA procedure; ~ **d'évaluation d'aménagement** DIST EAU, POL ENV planning assessment procedure; ~ **globale** POL ENV comprehensive procedure; ~ **de relais** DÉCHETS changeover procedure; ~ **simplifiée** POL ENV *juridique* simplified procedure; ~ **soumise à une étude d'impact** POL ENV EIA-dependent procedure

processus *m* GÉNÉRAL process; ~ **irréversible** POLL ATM, POLL EAU irreversible process; ~ **métabolique microbiologique** POLL ATM microbiological metabolic process; ~ **naturel de sédimentation** POLL EAU natural sedimentation process; ~ **de réaction** DIST EAU reaction process; ~ **réversible** POLL ATM, POLL EAU reversible process

producteur *m* DIST EAU producer; ~ **de déchets** DÉCHETS waste generator, waste producer

production *f* ÉCHANT output, ÉN RENOUV generation, SÉCUR-

ITÉ throughput, thruput (*AmE*); ~ **de biogaz** DIST EAU, POLL EAU digester gas yield; ~ **de bruit** POLL SON noise generation; ~ **de chaleur** DIST EAU, ÉN RENOUV heat generation; ~ **de chauffage urbain** DÉCHETS district heating generation; ~ **combinée** POLL ATM cogeneration; ~ **combinée de chaleur et d'électricité** DÉCHETS, POLL ATM combined heat and power (*CHP*); ~ **de déchets** DÉCHETS waste production, waste generation; ~ **d'eau chaude** POLL ATM warm-water generation; ~ **d'eaux résiduaires** POLL EAU waste-water production; ~ **d'eaux usées** POLL ATM, POLL EAU wastewater yield; ~ **d'électricité** DÉCHETS, DIST EAU, ÉN RENOUV electricity generation, power generation; ~ **d'électricité d'origine géothermique** ÉN RENOUV geothermal electrical power generation; ~ **d'électricité à partir des courants marins** ÉN RENOUV ocean current power recovery; ~ **d'énergie** ÉN RENOUV energy generation, energy output; ~ **d'énergie hydraulique** ÉN RENOUV hydro-power generation; ~ **d'énergie à partir de sources renouvelables** ÉN RENOUV energy generation from renewable energy sources; ~ **d'énergie photovoltaïque** ÉN RENOUV photovoltaic energy generation; ~ **d'éthanol** ÉN RENOUV ethanol generation; ~ **de gaz de décharge** DÉCHETS landfill gas production; ~ **de gaz naturel** DIST EAU gas recovery; ~ **de gaz de synthèse** DÉCHETS synthesis gas generation; ~ **de gouttelettes** POLL ATM droplet generation; ~ **de gypse** POLL ATM gypsum generation, gypsum yield; ~ **de lixiviats** DÉCHETS leachate production; ~ **de mâchefer** DIST EAU slag generation; ~ **de matières premières** ÉN RENOUV, GESTION production of raw material; ~ **d'odeurs** POLL ATM, POLL EAU odor formation (*AmE*), odour formation (*BrE*); ~ **d'ozone** POLL ATM ozone generation; ~ **primaire de papier** DIST EAU primary production of paper; ~ **de scories** DIST EAU slag generation; ~ **de son** POLL SON sound generation

productivité: ~ **énergétique** *f* ÉN RENOUV energy productivity

produire *vt* POLL SON *son*, ÉN RENOUV *électricité, vapeur* generate

produit: ~ **d'addition** *m* DIST EAU additive; ~ **blanc** *m* POLL MAR white product; ~ **de carbonisation** *m* DÉCHETS char; ~ **chimique** *m* GÉNÉRAL chemical; ~ **de combustion** *m* POLL ATM combustion product; ~ **de combustion gazeuse** *m* POLL ATM gaseous combustion product; ~ **compatible avec l'environnement** *m* GESTION environmentally-compatible product; ~ **congelé** *m* DÉCHETS frozen product; ~ **contenant des polluants** *m* DIST EAU product containing pollutants; ~ **de couchage** *m* DÉCHETS *revêtement des papiers* coating compound; ~ **de cristallisation** *m* DÉCHETS crystallizate; ~ **dangereux** *m* DÉCHETS dangerous good; ~ **de décantation** *m* POLL EAU settling material, sinking material, sinking matter; ~ **des déchets** *m* DÉCHETS, POLL EAU waste product; ~ **défoliant** *m* SOLS defoliant; ~ **de dégradation toxique** *m* DÉCHETS *dégradation des ordures ménagères en décharge* toxic degradation product; ~ **dérivé utilisable** *m* DÉCHETS, DIST EAU, GESTION, POLL ATM usable by-product; ~ **désémulsifiant** *m* DIST EAU, POLL MAR, POLL EAU demulsifying product; ~ **désinfectant** *m* DIST EAU, POLL EAU, SÉCURITÉ disinfectant; ~ **dispersant** *m* DIST EAU dispersing agent; ~ **de distillation** *m* DIST EAU distillate; ~ **de filtrage** *m* DIST EAU, POLL EAU filtrate; ~ **filtré** *m* DIST EAU, POLL EAU filtrate; ~ **final** *m* POLL ATM end product; ~ **final solidifié** *m* POLL ATM solidified end product; ~ **fini** *m* DÉCHETS, DIST EAU finished product; ~ **de fission** *m* DÉCHETS fission product; ~ **à flux libre** *m* DÉCHETS free-flow product; ~ **inflammable** *m* POLL ATM, SÉCURITÉ combustible, flammable product; ~ **jetable** *m* DÉCHETS throwaway product; ~ **à jeter** *m* DÉCHETS throwaway product; ~ **du métabolisme** *m* DIST EAU, POLL ATM, POLL EAU metabolite; ~ **à orientation écologique** *m* GESTION environmentally-oriented product; ~ **passé** *m* DIST EAU, POLL EAU filtrate; ~ **perdu** *m* DÉCHETS throwaway product; ~ **pétrolier** *m* DIST EAU petroleum product; ~ **phytosanitaire** *m* DIST EAU, POLL EAU pesticide; ~ **se prêtant au recyclage** *m* DIST EAU recycling-friendly product; ~ **de pyrolyse** *m* DIST EAU pyrolysis product; ~ **raffiné** *m* GÉNÉRAL refined product; ~ **de réaction** *m* POLL ATM reaction product; ~ **de réduction** *m* DIST EAU reducing product; ~ **réfrigéré** *m* DÉCHETS chilled good; ~ **de rembourrage** *m* DÉCHETS cushioning product; ~ **respectueux de l'environnement** *m* GESTION environmentally-friendly product; ~ **solidifié** *m* DIST EAU solidified product; ~ **tensioactif** *m* DIST EAU surfactant; ~ **toxique** *m* POLL ATM, SOLS toxicant; ~ **usagé** *m* DIST EAU used product; ~ **vert** *m* GESTION green product

produits: ~ **alimentaires conditionnés** *m pl* DÉCHETS parceled goods (*AmE*), parcelled goods (*BrE*); ~ **résiduels utilisables** *m pl* DÉCHETS usable waste products

professionnel *adj* SÉCURITÉ occupational

profil: ~ **aérodynamique** *m* POLL ATM streamline; ~ **d'emballage** *m* DÉCHETS packaging profile; ~ **spatial** *m* DIST EAU spatial pattern

profilage *m* DÉCHETS forming; ~ **de décharge** DÉCHETS landfill design

profileuse *f* POLL MAR grader

profond: **peu** ~ *adj* DIST EAU *puits* shallow

profondeur: ~ **de champ** *f* ÉCHANT visibility depth; ~ **d'écran** *f* SOLS curtain depth; ~ **de mur** *f* SOLS wall depth

programmable *adj* ÉCHANT programmable

programme: ~ **d'élimination des déchets** *m* DÉCHETS, GESTION, POL ENV waste disposal plan; ~ **d'environnement** *m* POL ENV environmental program (*AmE*), environmental programme (*BrE*); ~ **de fiscalité de l'environnement** *m* POL ENV eco-tax package; ~ **général d'assainissement** *m* DIST EAU master plan for sewage removal; ~ **de mesure** *m* ÉCHANT measurement program (*AmE*), measurement programme (*BrE*); ~ **pilote** *m* GESTION, POL ENV pilot program (*AmE*), pilot programme (*BrE*); ~ **de sélection** *m* ÉN RENOUV breeding program (*AmE*), breeding programme (*BrE*)

Programme: ~ **des Nations Unies pour l'Environnement** *m* (*PNUE*) POL ENV United Nations Environmental Programme (*UNEP*)

programmer *vt* ÉCHANT program

programmes: ~ **de recherches sur l'atmosphère** *m pl* POL ENV atmospheric research programmes (*BrE*), atmospheric research programs (*AmE*)

progressif *adj* POL ENV progressive

projection: ~ **d'eau haute pression** *f* SOLS high-pressure water jet method; ~ **établie par les pouvoirs publics** *f* POL ENV governmental projection

projet: ~ **d'accord cadre** *m* POL ENV framework agreement; ~ **d'aménagement d'un bassin fluvial** *m* POL ENV river-basin management plan; ~ **pilote** *m* GESTION, POL ENV pilot project

prolifération: ~ **algale** *f* POLL EAU algae bloom, algae growth; ~ **d'algues dans les lacs** *f* POLL EAU lake bloom; ~ **bactérienne** *f* DIST EAU bacterial growth; ~ **d'algues rouges** *f* POLL EAU red-tide bloom; ~ **de plancton** *f* DIST EAU, POLL EAU, POLL MAR plankton bloom

promoteur *m* POLL ATM promoter

propagation *f* GÉNÉRAL propagation, spreading; ~ **des bruits à l'extérieur** POLL SON outdoor sound propagation; ~ **de polluants** DIST EAU, POLL ATM, POLL EAU, SOLS propagation of pollutants; ~ **sonore** POLL SON noise propagation, spread of noise

prophylaxie *f* POL ENV prophylaxis

proportion: ~ **des eaux de surface** *f* DIST EAU share of surface waters

proposition *f* POL ENV proposal

propre *adj* GÉNÉRAL clean, POLL MAR neat; ~ **à un type** GÉNÉRAL type-specific

propriétés: **~ contaminantes** *f pl* POLL ATM contaminant properties; **~ de décantation** *f pl* DIST EAU settling characteristics, settling properties; **~ d'une eau** *f pl* DIST EAU water properties; **~ étanches** *f pl* DIST EAU sealing properties; **~ insonorisantes** *f pl* POLL SON sound-absorbing properties; **~ isolantes** *f pl* DÉCHETS, POLL SON insulating properties; **~ lixiviantes** *f pl* DIST EAU leaching properties; **~ du sol** *f pl* SOLS soil properties
propulseur *m* POLL ATM propellant
propulsion: **~ électrique** *f* ÉN RENOUV electric propulsion
protecteur *adj* GÉNÉRAL protective
protection[1]: **de ~** *loc* POLL MAR, POLL SON, SÉCURITÉ protective
protection[2]: **~ acoustique passive** *f* POLL SON passive sound protection; **~ anti-bruit** *f* POLL SON antinoise encapsulation; **~ antidéflagrante** *f* ÉCHANT, POLL ATM explosion protection; **~ anti-incendie** *f* SÉCURITÉ fire safety; **~ antirouille** *f* DIST EAU rust protection; **~ auditive** *f* POLL SON hearing protection; **~ cathodique** *f* DIST EAU cathodic protection; **~ climatique** *f* SÉCURITÉ climatic protection; **~ contre le bruit** *f* POLL SON noise protection; **~ contre l'érosion** *f* SOLS erosion protection; **~ contre l'explosion** *f* ÉCHANT, POLL ATM explosion protection; **~ contre les manipulations** *f* SÉCURITÉ tamper-proofing; **~ contre les rayonnements** *f* POLL ATM radiation protection; **~ contre les risques d'explosion** *f* SÉCURITÉ explosion safety; **~ contre la rouille** *f* DIST EAU rust protection; **~ des cultures** *f* SOLS crop protection; **~ des données** *f* ÉCHANT data security; **~ des eaux souterraines** *f* DÉCHETS groundwater protection; **~ de l'environnement** *f* POL ENV environmental protection; **~ de l'environnement à l'échelle nationale** *f* POL ENV national environmental protection; **~ de l'environnement urbain** *f* POL ENV municipal environmental protection; **~ incendie** *f* DIST EAU fire protection; **~ intégrée de l'environnement** *f* GESTION, POL ENV integrated environmental protection; **~ du littoral** *f* DIST EAU shore protection; **~ préventive de l'environnement** *f* DIST EAU preventative environmental protection; **~ des produits alimentaires** *f* POL ENV food protection; **~ respiratoire** *f* SÉCURITÉ breathing protection; **~ des ressources** *f* DIST EAU protection of resources, resource conservation; **~ du voisinage** *f* POL ENV protection of neighbors (*AmE*), protection of neighbours (*BrE*); **~ d'une zone de décharge** *f* DÉCHETS landfill area conservation
protégé: **~ contre le lessivage** *loc* SOLS safe against leaching; **~ de la lixiviation** *loc* DIST EAU safe against leaching
protège: **~-oreilles** *m* POLL SON ear protector
protéger *vt* SÉCURITÉ protect
protocole *m* DIST EAU, POL ENV protocol
protoplasme *m* POLL EAU protoplasm
protozoaire *m* POLL EAU protozoa
provision: **~ pour impôt** *f* POL ENV tax provision
psychromètre *m* ÉCHANT psychrometer
psychrophile *adj* POLL EAU psychrophilic
Pt *abrév* (*platine*) GÉNÉRAL Pt (*platinum*)
publication *f* POL ENV publication
puisard *m* POLL EAU cesspool, sump; **~ à boues** POLL EAU sludge sump; **~ de pompe** POLL ATM pump sump
puiser *vt* DIST EAU, ÉCHANT *de l'eau* draw; **~ avec une pompe** GÉNÉRAL pump, pump out
puisette *f* DIST EAU scoop
puissance *f* GÉNÉRAL capacity, power; **~ acoustique** POLL SON sound power; **~ d'une centrale solaire** ÉN RENOUV solar farm capacity; **~ d'une exploitation éolienne** ÉN RENOUV wind farm capacity; **~ fournie** ÉN RENOUV power output; **~ nominale** ÉN RENOUV rated capacity; **~ d'un parc d'éoliennes** ÉN RENOUV wind park capacity; **~ totale** ÉN RENOUV total capacity
puits *m* DIST EAU pit, sink, well, SOLS well; **~ absorbant** DIST EAU cesspool, inverted well, POLL EAU absorbing well; **~ d'acheminement** SOLS conveying well; **~ d'admission** DIST EAU, SOLS entrance shaft; **~ artésien** DIST EAU artesian well; **~ de boues** POLL EAU sludge hopper; **~ central** POLL ATM central well; **~ de chargement** DÉCHETS loading pit; **~ de collecte** DÉCHETS, DIST EAU collector shaft, collecting pit; **~ de décharge** DIST EAU relief well; **~ de décompression** DIST EAU relief well; **~ d'épuisement** DIST EAU pumping shaft; **~ d'évaporation sous pression** SOLS under-pressure evaporator well; **~ d'excavation** SOLS excavation well; **~ d'exhaure** DIST EAU pumping shaft; **~ filtrant** DIST EAU, POLL EAU filtering well, recharge well; **~ de filtration vertical** DIST EAU vertical filtering well; **~ de forage** DIST EAU bored well, borehole well; **~ foré** DIST EAU bored well; **~ de gaz** DÉCHETS gas shaft; **~ de gaz vertical** DÉCHETS, DIST EAU vertical gas well; **~ humide** DIST EAU wet well; **~ d'infiltration** DIST EAU, SOLS infiltration well, percolation well; **~ d'injection** DIST EAU, SOLS injection well, pressure well; **~ isolant de protection** SOLS protective insulation well; **~ d'observation** DÉCHETS observation well; **~ de percolation** DIST EAU percolation well; **~ de percolation vertical** DIST EAU vertical percolation well; **~ de pompage** DIST EAU pump well; **~ de pompe** DIST EAU, POLL EAU pump shaft; **~ profond** POLL EAU deep well; **~ de régénération** SOLS regeneration pit; **~ de retenue** SOLS retaining well; **~ sec** SOLS dry well; **~ de surveillance** DIST EAU, POLL ATM monitoring well; **~ de surveillance des eaux souterraines** ÉCHANT groundwater monitoring well; **~ tubé** DIST EAU bored well; **~ vertical** DIST EAU vertical well; **~ sous vide** SOLS vacuum well
pulvérisateur *m* DIST EAU, POLL ATM, SOLS atomizer, pulverizer; **~ ultrasonique** ÉCHANT ultrasound pulverizer
pulvérisation *f* DÉCHETS comminution, DIST EAU comminution, pulverizing, POLL ATM spraying; **~ d'agent dispersant** POLL MAR dispersant spraying; **~ d'eau** POLL ATM water spraying; **~ tournante** POLL ATM rotary atomization
pulvériser *vt* DIST EAU, POLL ATM pulverize, spray, atomize, POLL EAU spray, SOLS atomize
pulvimètre *m* ÉCHANT dust counter
pureté: **~ de l'air** *f* POLL ATM air purity; **~ de l'eau** *f* DIST EAU, POLL EAU water clearness; **~ du gaz final** *f* POLL ATM end-gas purity
purge *f* GÉNÉRAL drainage; **~ de l'air** DIST EAU air blow-down, vent, POLL ATM vent; **~ des gaz** DÉCHETS gas drainage; **~ tangentielle des gaz** DÉCHETS horizontal gas drainage
purgeoir *m* DIST EAU, POLL EAU filtering tank
purger *vt* DIST EAU drain, blow off, ÉCHANT purge
purification *f* DIST EAU, POLL EAU clarification, purification; **~ de l'eau** DIST EAU, POLL EAU water purification; **~ mécanique** POLL EAU mechanical purification; **~ naturelle** POLL EAU natural purification; **~ préliminaire** DIST EAU preliminary purification
purifier *vt* DIST EAU, POLL EAU clarify, purify
purin *m* DIST EAU, POLL EAU liquid manure, slurry
putréfaction *f* DIST EAU, POLL EAU decay, putrefaction, rot
putréfiant *adj* DIST EAU, POLL EAU putrefactive
putrescibilité *f* DIST EAU, POLL EAU putrescibility
putrescible *adj* DIST EAU, POLL EAU putrescible
PV *abrév* (*génie photovoltaïque*) ÉN RENOUV *discipline* PV (*photovoltaics*)
PVC[1] *abrév* (*polychlorure de vinyle*) DIST EAU, SOLS PVC (*polyvinyl chloride*)
PVC[2]: **~ rigide** *m* DIST EAU rigid PVC
pycnomètre *m* ÉCHANT density bottle, pyknometer
pycnométrique *adj* ÉCHANT pyknometric
pyramide: **~ des biomasses** *f* POL ENV biomass pyramid; **~ écologique** *f* DIST EAU, POL ENV ecological pyramid
pyranomètre *m* ÉCHANT pyranometer, solarimeter
pyrhéliomètre *m* ÉCHANT pyrheliometer
pyrolysat *m* DIST EAU pyrolysis product
pyrolyse *f* GÉNÉRAL pyrolysis; **~ de décharge contaminée** SOLS contaminated site pyrolysis; **~ sur lit fluidisé** DÉCHETS

fluidized-bed pyrolysis; ~ **des ordures** POLL EAU waste pyrolysis; ~ **de pneus usés** DIST EAU used tire pyrolysis (*AmE*), used tyre pyrolysis (*BrE*)

pyrolyser *vt* GÉNÉRAL pyrolyse (*BrE*), pyrolyze (*AmE*)

pyrolytique *adj* GÉNÉRAL pyrolytic

pyromètre: ~ **optique** *m* ÉCHANT *mesure thermique* radiation pyrometer

pyrophore[1] *adj* POLL ATM pyrophoric

pyrophore[2] *m* POLL ATM pyrophoric material

Q

quadripôle *m* ÉCHANT quadrupole

qualitatif *adj* ÉCHANT qualitative

qualité[1]: **de ~** *loc* DIST EAU high-grade; **de ~ microscopique** *loc* ÉCHANT microscopic qualitative

qualité[2]: **~ de l'air** *f* POLL ATM, SÉCURITÉ air quality; **~ de collecte** *f* POLL ATM collection quality; **~ des déchets** *f* DÉCHETS waste quality; **~ de décolmatage** *f* POLL ATM removal quality; **~ de l'eau** *f* DIST EAU, POLL EAU water quality; **~ d'eau brute** *f* DIST EAU raw water quality; **~ d'eau potable** *f* DIST EAU drinking water quality; **~ des eaux fluviales** *f* DIST EAU river water quality; **~ des eaux usées** *f* POLL EAU wastewater quality; **~ des effluents** *f* POLL EAU effluent quality; **~ de l'environnement** *f* POL ENV environmental quality; **~ de ferraille** *f* DÉCHETS scrap quality; **~ du filtrat** *f* POLL EAU filtrate quality; **~ globale de l'air** *f* POLL ATM total air quality; **~ de matériaux de récupération** *f* DIST EAU secondary resource quality; **~ de papier** *f* DÉCHETS, DIST EAU paper quality; **~ du sol** *f* SOLS soil quality; **~ totale de l'air** *f* POLL ATM total air quality

qualités: **~ requises** *f pl* POLL ATM quality requirements

quantification *f* ÉCHANT quantification

quantifier *vt* ÉCHANT quantify

quantitatif *adj* ÉCHANT quantitative

quantité: **~ analysée** *f* ÉCHANT analytical quantity; **~ d'eau tombée** *f* DIST EAU, POLL ATM rainfall; **~ d'échantillon minimale** *f* ÉCHANT minimum sample quantity; **~ mesurée** *f* ÉCHANT measuring quantity; **~ à mesurer** *f* ÉCHANT quantity to be measured; **~ minimale détectable** *f* ÉCHANT, POLL EAU lower limit of detectability; **~ passée** *f* POLL EAU throughput, thruput (*AmE*); **~ de polluants** *f* ÉCHANT pollutant amount; **~ recyclée** *f* DIST EAU recycling quantity; **~ totale de gaz** *f* ÉCHANT, POLL ATM total gas quantity

quasi: **~-simultané** *adj* ÉCHANT quasi-simultaneous

quistor *m* ÉCHANT *spectrométrie de masse* quadrupole ion trap

quota: **~ fiscal** *m* POL ENV tax quota

quotidien *adj* POLL ATM, SOLS daily

quotient: **~ olfactif** *m* ÉCHANT odor quotient (*AmE*), odour quotient (*BrE*)

R

rabats *m pl* DÉCHETS folding sides
raccord *m* POLL ATM connecting branch; **~ de pompe** DIST EAU, POLL EAU pump connection; **~ rapide** DIST EAU quick-fit connection; **~ de tuyauterie** DIST EAU pipe joint
raccordement: **~ au réseau** *m* ÉN RENOUV grid connection; **~ au secteur** *m* ÉN RENOUV mains connection
raclage: **~ du chapeau** *m* POLL EAU scum removal
racle *f* DÉCHETS doctor blade
racler *vt* POLL EAU, POLL MAR scrape
racleur *m* DIST EAU, POLL ATM, POLL EAU scraper; **~ à balancier** POLL EAU swinging scraper; **~ à chaîne** POLL EAU flight scraper; **~ d'écumes** POLL EAU scum scraper; **~ de fond** DIST EAU bottom scraper; **~ longitudinal** POLL EAU longitudinal scraper; **~ à sable** POLL EAU grit scraper; **~ de surface** DIST EAU surface skimmer
racleuse *f* POLL EAU, POLL MAR licker, scraper
rad *abrév* (*dose de rayonnement absorbé*) ÉCHANT rad (*radiation absorbed*)
radar: **~ à laser** *m* ÉCHANT laser radar
radical: **~ libre** *m* POLL ATM free radical
radier *m* DIST EAU invert
radioactif *adj* GÉNÉRAL radioactive
radioactivité: **~ naturelle** *f* POLL ATM, POLL EAU natural radioactivity; **~ d'origine tellurique** *f* POLL ATM earth radiation
radiographe *m* DIST EAU, ÉCHANT radiograph
radiographique *adj* DIST EAU, ÉCHANT radiographic
radiomessagerie *f* SÉCURITÉ radiopaging
radiomètre *m* ÉCHANT radiometer; **~ acoustique** ÉCHANT sound radiometer
radiométrique *adj* ÉCHANT radiometric
radiopolarographie *f* ÉCHANT radiopolarography
radioprotection *f* POLL ATM, SÉCURITÉ radiation protection
raffiné *adj* DIST EAU, POLL ATM, POLL MAR refined
raffiner *vt* DIST EAU, POLL ATM, POLL MAR refine
raidir *vt* DIST EAU, POLL EAU reinforce
raie: **~ spectrale** *f* ÉCHANT spectral line
raies: **~ de Fraunhofer** *f pl* ÉCHANT *raies sombres du spectre solaire* Fraunhofer lines
ralentisseur *m* DIST EAU retarder
ramassage *m* DÉCHETS collection, DIST EAU pick-up; **~ de ferraille** DÉCHETS scrap collection; **~ des ordures** DÉCHETS refuse collection, waste collection; **~ des ordures ménagères** DÉCHETS domestic waste collection
ramasser *vt* DÉCHETS, DIST EAU collect
ramasseuse: **~ aspiratrice** *f* DÉCHETS suction collector
ramification *f* DIST EAU branching
rampe: **~ d'accès** *f* DÉCHETS *décharge* access ramp; **~ à déversement** *f* DÉCHETS tipping ramp; **~ de pulvérisation** *f* POLL MAR spray boom
rangeur: **~ de bouteilles** *m* DÉCHETS bottle unscrambler
rapatrier *vt* SÉCURITÉ *informations, données* relay back
rapport: **~ d'absorption du sodium** *m* (*RAS*) DIST EAU sodium absorption ratio; **~ air-carburant** *m* POLL ATM fuel-air ratio; **~ air-combustible** *m* POLL ATM fuel-air ratio; **~ air-membrane** *m* POLL ATM air-to-cloth ratio; **~ du bruit à la superficie** *m* POLL MAR, POLL SON concentration of noise to area; **~ bruit-surface** *m* POLL MAR, POLL SON concentration of noise to area; **~ carbone-azote** *m* ÉCHANT carbon-nitrogen ratio (*C-N ratio*); **~ carbone-azote-phosphore** *m* ÉCHANT carbon-nitrogen-phosphorous ratio (*C-N-P ratio*); **~ coût-avantage** *m* GESTION cost-benefit ratio; **~ coût-durée de service** *m* GESTION cost-service life ratio; **~ DCO-DBO** *m* ÉCHANT COD-BOD ratio; **~ dose-effet** *m* DIST EAU, ÉCHANT dose-response relationship; **~ sur l'environnement** *m* POL ENV environmental report; **~ de l'étude d'impact** *m* POL ENV environmental impact assessment report; **~ molaire** *m* POLL ATM mol ratio
rare *adj* ÉN RENOUV, POL ENV, POLL ATM, SOLS scarce
raréfaction: **~ des ressources en oxygène** *f* ÉCHANT, POLL ATM, POLL EAU oxygen depletion
RAS *abrév* (*rapport d'absorption du sodium*) DIST EAU sodium absorption ratio
rassembler *vt* POLL ATM mass together
râteau *m* DIST EAU rake
ratifier *vt* POL ENV ratify
ravinement *m* SOLS gully erosion, ravine erosion
rayon: **~ de cyclone** *m* POLL ATM cyclone radius
rayonnement *m* GÉNÉRAL radiation; **~ acoustique** POLL SON sound radiation; **~ du bruit** POLL SON noise radiation; **~ de chaleur à onde longue** POLL ATM long-wave heat radiation; **~ cosmique** POLL ATM cosmic radiation; **~ cosmique primaire** POLL ATM primary cosmic radiation; **~ cosmique secondaire** POLL ATM secondary cosmic radiation; **~ diffus** ÉN RENOUV diffuse radiation; **~ direct** ÉN RENOUV direct radiation; **~ électromagnétique** POLL ATM electromagnetic radiation; **~ incident à onde courte** POLL ATM short-wave incoming radiation; **~ infrarouge** DIST EAU, POLL ATM infrared radiation; **~ ionisant** DIST EAU, POLL ATM, SOLS ionizing radiation; **~ laser** SÉCURITÉ laser radiation; **~ à onde longue réfléchi par la terre** POLL ATM long-wave reradiation from the earth; **~ planétaire** POLL ATM global radiation; **~ solaire** ÉN RENOUV, POLL ATM solar radiation, sunlight radiation; **~ solaire cosmique** POLL ATM cosmic solar radiation; **~ solaire global** ÉN RENOUV global solar radiation; **~ solaire à onde courte** POLL ATM short-wave sunlight; **~ terrestre** POLL ATM earth radiation; **~ thermique** ÉN RENOUV heat radiation; **~ ultraviolet** DIST EAU ultraviolet irradiation, POLL ATM ultraviolet radiation, ultraviolet irradiation; **~ ultraviolet-C** (*rayonnement UV-C*) DIST EAU ultraviolet-C radiation (*UV-C radiation*); **~ UV-C** (*rayonnement ultraviolet-C*) DIST EAU UV-C radiation (*ultraviolet-C radiation*)
rayonner *vt* POLL ATM *chaleur*, POLL SON radiate
rayons: **~ ultraviolets** *m pl* SÉCURITÉ ultraviolet rays; **~ X** *m pl* SÉCURITÉ X-rays
raz: **~-de-marée** *m* POLL MAR tidal wave
Re *abrév* (*nombre de Reynolds*) POLL ATM Re (*Reynolds number*)
réacteur *m* DÉCHETS, DIST EAU reactor, POLL ATM reactor, *chimie* reaction tower, POLL EAU, SOLS reactor; **~ de biofiltration** POLL ATM tower biofilter; **~ biologique à membrane** POLL ATM membrane biological reactor; **~ boues-huile** POLL EAU sludge-to-oil reactor system (*STORS*); **~ en cascade à lit fixe** POLL EAU fixed-bed cascade reactor; **~ à catalyse** POLL ATM reactor with catalyst; **~ catalytique** POLL ATM catalytic reactor; **~ chimique** POLL ATM phase-break vessel; **~ à collecteur d'échappement** POLL ATM exhaust manifold reactor; **~ de conditionnement** POLL ATM conditioning reactor; **~ de décontamination** SOLS decontamination reactor; **~ discontinu séquentiel** SOLS sequencing batch reactor (*SBR*); **~ à écoulement de type piston** POLL ATM hydrolift reactor; **~ haute pression** ÉN RENOUV high-pressure reac-

tor; ~ **à lit fixe** POLL EAU, SOLS fixed-bed reactor; ~ **à lit fixe à charbons actifs** POLL ATM activated carbon fixed-bed reactor; ~ **à lit fixe à haut rendement** DIST EAU high-performance fixed-bed reactor; ~ **à lit fixe haute performance** DIST EAU high-performance fixed-bed reactor; ~ **à lit fluidisé** DÉCHETS, POLL ATM, POLL EAU fluidized-bed reactor; ~ **à lit fluidisé circulant** POLL ATM circulating fluidized-bed reactor; ~ **à lit mobile** POLL ATM traveling-bed reactor (*AmE*), travelling-bed reactor (*BrE*); ~ **à lit mobile à admission tangentielle** POLL ATM transverse-entry traveling-bed reactor (*AmE*), transverse-entry travelling-bed reactor (*BrE*); ~ **à matériau en vrac** DIST EAU bulk material reactor; ~ **à membrane** POLL ATM membrane reactor; ~ **nucléaire** POLL EAU nuclear reactor; ~ **à puits profond** POLL EAU deep-shaft reactor; ~ **à pulvérisation** POLL ATM spray reactor; ~ **de réduction catalytique sélective** POLL ATM selective catalytic reduction reactor; ~ **séquentiel à alimentation discontinue** POLL EAU sequencing batch reactor (*SBR*); ~ **séquentiel par cuvées** POLL EAU sequencing batch reactor (*SBR*); ~ **solaire** DIST EAU solar reactor; ~ **à solvants** DIST EAU solvents reactor; ~ **thermique** POLL ATM thermal reactor; ~ **de trempe** POLL ATM quenching tower

réactif[1] *adj* DIST EAU, POLL ATM reactive

réactif[2] *m* ÉCHANT, POLL ATM, SOLS reactant, reagent, reacting agent; ~ **chimique** POLL ATM chemical reagent; ~ **liquide** ÉCHANT liquid-reacting agent; ~ **de régénération** GÉNÉRAL regenerant, POLL ATM regenerative reagent; ~ **solide** ÉCHANT solid reacting agent

réaction *f* GÉNÉRAL reaction; ~ **après-coup** POLL ATM post-reaction; ~ **catalytique** POLL ATM catalytic reaction; ~ **colorée** ÉCHANT color reaction (*AmE*), colour reaction (*BrE*); ~ **de combustion** POLL ATM combustion reaction; ~ **de conversion** POLL ATM conversion reaction; ~ **à la détection** ÉCHANT detection reaction; ~ **différée** POLL ATM post-reaction; ~ **endothermique** DIST EAU *énergie solaire*, POLL ATM endothermic reaction; ~ **exothermique** POLL ATM exothermic reaction; ~ **photobiologique** DIST EAU photobiological reaction; ~ **photochimique** DIST EAU photochemical reaction; ~ **physico-chimique** POLL ATM physio-chemical reaction; ~ **de reformage** POLL ATM reforming reaction; ~ **secondaire** POLL ATM side reaction; ~ **tampon** POLL ATM buffer reaction

réactivation *f* DIST EAU, POLL ATM, POLL EAU, SOLS reactivation

réactivé *adj* DIST EAU, POLL ATM, POLL EAU, SOLS reactivated

réactiver *vt* DIST EAU, POLL ATM, POLL EAU, SOLS reactivate

réactivité *f* DIST EAU, POLL ATM, POLL EAU, SOLS reactivity

réadaptation *f* GÉNÉRAL retrofitting

réadapter *vt* GÉNÉRAL retrofit

réalimentation *f* DIST EAU recharge; ~ **d'une nappe artificielle** DIST EAU artificial groundwater recharge; ~ **naturelle d'eau souterraine** DIST EAU natural groundwater recharge

réalimenté: ~ **artificiellement** *adj* DIST EAU artificially replenished

réaliser *vt* ÉCHANT *tests* conduct

réaménager *vt* POLL ATM retrofit

réaspirer: ~ **par pompage** *vt* DIST EAU, POLL EAU pump back

reboisement *m* POLL ATM reforestation

rebondir *vi* POLL ATM rebound

rebroyer *vt* DIST EAU *déchets* regrind

rebut *m* DÉCHETS, POLL ATM junk

recapter *vt* POLL ATM recapture

recarbonatation: ~ **en deux étapes** *f* POLL EAU two-stage recarbonation

recensement: ~ **des ressources de la terre** *m* GESTION earth resources survey

réceptacle *m* ÉCHANT receptacle

récepteur *m* ÉN RENOUV receiver, POLL ATM receptor, POLL EAU sink; ~ **de lumière** ÉCHANT light receiver; ~ **solaire central** ÉN RENOUV solar tower

recettes: ~ **dégagées par la fiscalité de l'environnement** *f pl* POL ENV eco-tax revenue; ~ **fiscales** *f pl* POL ENV revenue; ~ **perçues au titre des éco-taxes** *f pl* POL ENV revenue from eco-taxes; ~ **perçues au titre de la taxe sur l'énergie** *f pl* POL ENV energy tax revenue

rechappage *m* DÉCHETS *pneus* remolding (*AmE*), remoulding (*BrE*)

réchauffage *m* POLL ATM reheating; ~ **des gaz de combustion** POLL ATM flue-gas reheating; ~ **de gaz épuré** POLL ATM clean-gas reheating

réchauffement: ~ **dû à l'effet de serre** *m* POLL ATM greenhouse warming; ~ **planétaire** *m* POLL ATM global warming; ~ **de la planète** *m* POLL ATM global warming

réchauffer *vt* ÉCHANT, ÉN RENOUV, POLL ATM, POLL EAU reheat, heat up

réchauffeur *m* POLL EAU, POLL ATM preheater; ~ **d'air** DÉCHETS, POLL ATM air preheater; ~ **d'air à ailettes** POLL ATM finned tube air heater; ~ **d'air de régénération** POLL ATM regenerative air preheater; ~ **d'air rotatif** POLL ATM rotary air heater; ~ **de gaz** POLL ATM, POLL EAU gas preheater; ~ **de gaz de régénération** POLL ATM regenerative gas preheater; ~ **de vapeur** POLL ATM steam air preheater

recherche[1]: **de** ~ *loc* ÉCHANT exploratory

recherche[2] *f* ÉCHANT, POL ENV, SÉCURITÉ investigation; ~ **appliquée** ÉCHANT applied research; ~ **en matière d'environnement** POL ENV environmental research; ~ **océanique** POL ENV ocean research; ~ **sur la pollution** ÉCHANT pollution research

rechercher *vt* ÉCHANT, POL ENV, SÉCURITÉ investigate

recherches: ~ **sur le milieu** *f pl* POL ENV environmental research

récipient: ~ **d'aérosols** *m* DÉCHETS aerosol container; ~ **consigné** *m* DIST EAU returnable container; ~ **doseur** *m* DÉCHETS unit dose container; ~ **gradué** *m* ÉCHANT measuring container; ~ **en verre** *m* DÉCHETS glass jar

recirculation *f* DIST EAU, POLL ATM, POLL EAU recirculation; ~ **des gaz de combustion** POLL ATM flue-gas recirculation; ~ **des gaz d'échappement** POLL ATM exhaust-gas recirculation (*EGR*)

récolte *f* ÉN RENOUV harvesting; ~ **de troncs entiers** ÉN RENOUV shoot harvesting

récolter *vt* ÉN RENOUV harvest

récolteuse *f* ÉN RENOUV harvester

recontaminer *vt* POLL EAU recontaminate

reconvertir *vt* POLL ATM reconvert

rectificateur *m* POLL ATM, POLL EAU rectifier

rectification *f* POLL ATM, POLL EAU rectification

recueillir *vt* POLL MAR collect

récupérable *adj* DIST EAU, ÉCHANT, POLL ATM collectable, recoverable

récupérateur *m* DIST EAU recovery device, POLL ATM recuperator, POLL MAR skimmer; ~ **autopropulsé** POLL MAR self-propelled skimmer; ~ **à bande transporteuse** DIST EAU conveyor belt skimmer; ~ **à cage tubulaire** POLL ATM tube-cage recuperator; ~ **à corde flottante** POLL MAR rope skimmer; ~ **à cordon oléophile** POLL MAR oleophilic belt skimmer; ~ **à déversoir** POLL MAR weir boom, weir skimmer; ~ **à disque** POLL MAR disc skimmer (*BrE*), disk skimmer (*AmE*); ~ **dynamique** POLL MAR dynamic skimmer; ~ **de rayonnement** POLL ATM radiation recuperator; ~ **à seuil** POLL MAR weir boom, weir skimmer; ~ **à tambour** POLL MAR *pollution marine* drum skimmer; ~ **à tourbillon** POLL MAR vortex skimmer; ~ **en U** POLL ATM channel recuperator

récupération[1] *f* DÉCHETS, DIST EAU, POLL ATM, POLL MAR reclamation, recovery, collection; ~ **des boues** POLL EAU sludge recovery; ~ **des cendres volantes** POLL ATM fly-ash recovery; ~ **de chaleur** DÉCHETS, DIST EAU, POLL ATM heat recovery; ~ **de chaleur dissipée** DÉCHETS, DIST EAU, POLL ATM, POLL EAU waste heat recovery; ~ **de la chaleur dissipée par les déchets** DÉCHETS, DIST EAU, POLL ATM, POLL EAU waste heat recovery; ~ **de déchets** DÉCHETS waste recovery; ~ **de déchets non mélangés** DÉCHETS unmixed waste recovery; ~ **d'eau** DIST EAU water recovery; ~ **d'eau de mer** DIST EAU sea-water recovery; ~ **d'eau potable** DIST EAU *à partir des eaux de surface* drinking water recovery; ~ **d'énergie** DIST EAU, POLL ATM, POLL EAU energy recovery; ~ **de gaz** POLL ATM gas recovery; ~ **des huiles usagées** DÉCHETS waste oil recovery; ~ **des huiles usées** DÉCHETS waste oil recovery; ~ **du mâchefer** DIST EAU slag recovery; ~ **de matériaux** DÉCHETS material recovery; ~ **de mercure** DIST EAU mercury recovery; ~ **par type** DIST EAU recovery by type; ~ **physique** DÉCHETS physical recovery; ~ **de poussière** POLL ATM dust recovery; ~ **de produit** POLL ATM product recovery; ~ **des produits d'infiltration** POLL EAU permeate recovery; ~ **de ressources** DIST EAU resource recovery; ~ **des scories** DIST EAU slag recovery; ~ **sélective des substances dangereuses** DIST EAU unmixed recovery of harmful substances; ~ **de solvants** POLL ATM solvent recovery; ~ **des sous-produits** DÉCHETS, DIST EAU, POLL ATM by-product recovery; ~ **des substances dangereuses sans mélange** DIST EAU unmixed recovery of harmful substances; ~ **thermique** DÉCHETS thermal recovery

récupération[2]: **à** ~ *loc* POLL ATM recuperative; **de** ~ *loc* POLL ATM recuperative

récupéré *adj* GÉNÉRAL reclaimed, recovered

récupérer *vt* GÉNÉRAL collect, reclaim, recover; ~ **sous forme liquide** DIST EAU recover in liquid form

récurage *m* DIST EAU scour

récurer *vt* DIST EAU scour

recyclabilité *f* DIST EAU recyclability, POLL ATM recoverability

recyclable *adj* DIST EAU recoverable, recyclable, POLL ATM recoverable

recyclage *m* DÉCHETS, DIST EAU, POLL ATM, POLL EAU recirculation, recycling; ~ **de l'asphalte** DIST EAU asphalt recycling; ~ **biologique** DÉCHETS biological recycling; ~ **des boues** POLL EAU recycling of sludge, sludge recycling; ~ **des cendres volantes** DIST EAU recycling of fly ash; ~ **de chaleur de récupération** POLL ATM recuperative heat recovery; ~ **de la charge** DÉCHETS feedstock recycling; ~ **de compost inoculé** DIST EAU recycling of inoculated compost; ~ **de déchets** DIST EAU recycling of waste; ~ **de l'eau** POLL EAU pipe-to-pipe recycling; ~ **des eaux usées** POLL EAU wastewater reuse; ~ **des gaz de carter** POLL ATM positive crankcase ventilation (*PCV*); ~ **des gaz de tête** POLL ATM top-gas recycling; ~ **interne** POLL EAU internal recirculation; ~ **des matériaux** DÉCHETS, POLL EAU materials recycling; ~ **de matières premières** DIST EAU raw material recycling; ~ **mécanique** DÉCHETS, POLL EAU mechanical recycling; ~ **multimatériaux** DÉCHETS multimaterial recycling; ~ **de papier carton** DIST EAU recycling of paper cartons; ~ **des perméats** POLL EAU permeate recovery; ~ **physique** POLL EAU dilution; ~ **du plastique** DÉCHETS plastics recycling; ~ **primaire** DÉCHETS primary recycling; ~ **des produits de base** DÉCHETS feedstock recycling; ~ **quaternaire** DÉCHETS quaternary recycling; ~ **secondaire** DÉCHETS secondary recycling; ~ **tertiaire** DÉCHETS tertiary recycling; ~ **du verre** DÉCHETS glass recycling; ~ **de vieux papiers** DÉCHETS, DIST EAU wastepaper recycling

recyclé *adj* DIST EAU recirculated, recovered, recycled, POLL EAU recirculated

recycler *vt* DIST EAU recirculate, recover, recycle, POLL EAU recirculate; ~ **sous forme liquide** DIST EAU recover in liquid form

redéposer *vt* SOLS redeposit

redevance *f* GESTION, POL ENV charge; ~ **d'autorisation** GESTION licence charge (*BrE*), license charge (*AmE*); ~ **d'environnement** POL ENV environmental tax; ~ **d'épuration des eaux usées** POLL EAU wastewater treatment fee; ~ **imposée aux usagers de l'environnement** POL ENV environmental user tax; ~ **incitative** POL ENV incentive charge; ~ **d'indemnisation** POL ENV compensation charge; ~ **de péréquation des charges** POL ENV cost equalizing charge; ~ **sur le bruit** POL ENV, POLL SON noise charge; ~ **sur les déchets** POL ENV waste charge; ~ **sur l'eau** POL ENV water penny; ~ **sur les effluents** POL ENV effluent charge; ~ **sur les émissions** POL ENV emissions charge; ~ **sur la pollution** POL ENV pollution charge; ~ **sur le prélèvement d'eaux souterraines** POL ENV groundwater withdrawal charge, groundwater withdrawal tax; ~ **sur produit** POL ENV product charge; ~ **d'utilisation obligatoire** POL ENV mandatory user charge

redissolution *f* POLL EAU *du phosphore* redissolution

redresseur *m* POLL ATM, POLL EAU rectifier

réducteur[1] *adj* DIST EAU, ÉCHANT, POLL ATM, POLL EAU reducing; ~ **de bruit** POLL SON noise-reducing

réducteur[2] *m* POLL EAU reducing agent

réduction *f* GÉNÉRAL reduction, abatement; ~ **du bruit** POLL SON, SÉCURITÉ noise abatement, noise reduction; ~ **du bruit aérien** POLL SON airborne noise reduction; ~ **catalytique** POLL ATM catalytic reduction; ~ **catalytique par charbon actif** POLL ATM activated-carbon catalytic reduction (*ACCR*); ~ **catalytique sélective** DIST EAU, POLL ATM selective catalytic reduction (*SCR*); ~ **de la consommation** DIST EAU, POL ENV use reduction; ~ **de déchets** DÉCHETS waste reduction; ~ **d'échantillon** ÉCHANT sample diminution; ~ **des émissions** DÉCHETS, POLL ATM, SOLS emissions control, emissions reduction; ~ **des émissions de dioxyde de soufre** POLL ATM sulfur dioxide reduction (*AmE*), sulphur dioxide reduction (*BrE*); ~ **des émissions au minimum** POL ENV, POLL ATM emissions minimalization; ~ **des émissions de CO_2** *f pl* POLL ATM CO_2 emissions reduction; ~ **des fumées** POLL ATM smoke abatement; ~ **d'impôt** POL ENV tax reduction; ~ **intégrée du bruit** POLL SON structure-borne noise reduction; ~ **au minimum** DÉCHETS, DIST EAU minimization; ~ **de niveau** POLL SON level reduction; ~ **des oxydes d'azote** POLL ATM nitrogen oxides abatement; ~ **de la qualité** POLL ATM impairment; ~ **de redevance** POL ENV charge reduction; ~ **sélective non catalytique** DIST EAU selective non-catalytic reduction (*SNCR*); ~ **d'utilisation** DIST EAU, POL ENV use reduction; ~ **de volume** DÉCHETS volume reduction

réduire *vt* GÉNÉRAL reduce, abate; ~ **au minimum** POLL ATM minimize

réemploi *m* DÉCHETS, DIST EAU, POL ENV reuse; ~ **de ferraille** DÉCHETS scrap reuse; ~ **des terres abandonnées** POL ENV reuse of derelict land

réemployer *vt* DÉCHETS, DIST EAU, POL ENV reuse

réétalonnage *m* ÉCHANT recalibration

réétiquetage: ~ **des boîtes** *m* DÉCHETS can re-labeling (*AmE*), can re-labelling (*BrE*)

refiltration *f* DIST EAU, POLL ATM refiltering

REFIOM *abrév* (*résidus d'épuration des fumées d'incineration des ordures ménagères*) DÉCHETS garbage incineration refuse

réfléchi *adj* ÉN RENOUV, POLL ATM, POLL SON reflected

réfléchir *vt* ÉN RENOUV *rayonnement solaire*, POLL ATM, POLL SON reflect

réflecteur *m* ÉN RENOUV reflector; ~ **parabolique** ÉN RENOUV parabolic dish, *énergie solaire* parabolic reflector; ~ **de poursuite** ÉN RENOUV tracking reflector

réflectomètre *m* ÉCHANT reflectometer

refléter *vt* POLL ATM reflect

réflexion *f* ÉCHANT, POLL ATM, POLL SON reflection; ~ **atmosphérique** POLL ATM atmospheric reflection; ~ **du son** POLL SON sound reflection

refondre *vt* DIST EAU remelt

réforme *f* POL ENV reform, modernization; ~ **de la fiscalité de**

l'environnement POL ENV green tax reform, eco-tax reform; **~ de l'imposition sur l'environnement fiscalement neutre** POL ENV income-neutral eco-tax reform
réformer *vt* POL ENV modernize, reform
refoulement *m* DIST EAU delivery; **~ de boue en nappe** POLL EAU *égout* upflow sludge blanket process; **~ des gaz de combustion** POLL ATM waste gas back pressure
réfractaire *adj* DÉCHETS, DIST EAU, SÉCURITÉ heat-proof, heat-insulating, *matériau* heat-resistant, *brique* fireproof
réfractomètre: **~ interférentiel** *m* ÉCHANT interference refractometer; **~ de Rayleigh** *m* ÉCHANT Rayleigh refractometer
réfrigérant *m* DIST EAU, ÉCHANT, POLL ATM cooler, coolant, cooling medium, refrigerant; **~ complémentaire** POLL ATM aftercooler
réfrigération *f* GÉNÉRAL cooling; **~ de cuves de fermentation** POLL ATM *brasserie* attemperation
réfrigérer *vt* GÉNÉRAL cool
refroidir *vt* GÉNÉRAL cool
refroidissement *m* GÉNÉRAL cooling; **~ des gaz** POLL ATM gas cooling; **~ des gaz de combustion** POLL ATM flue-gas cooling; **~ de gaz pollué** POLL ATM dirty-gas cooling; **~ par évaporation** POLL ATM evaporation cooling
refroidisseur *m* GÉNÉRAL cooler; **~ complémentaire** POLL ATM aftercooler; **~ de filtre** POLL ATM filter cooler; **~ de gaz** POLL ATM gas cooler; **~ de gaz de mesure** ÉCHANT measuring gas cooler
refus *m pl* DIST EAU, POLL EAU residues; **~ bruts** DÉCHETS crude refuse; **~ de compostage** DÉCHETS composting reject; **~ de criblage** DIST EAU sieve residue; **~ de dégrillage** DIST EAU, POLL EAU screenings; **~ de fines** DÉCHETS fine refuse; **~ flottants** DÉCHETS floating refuse; **~ de forte granulométrie** DÉCHETS coarse refuse; **~ de grille** DIST EAU, POLL EAU screenings; **~ de taille moyenne** DÉCHETS medium-sized refuse
regard *m* DIST EAU pit, inspection hole, ÉCHANT inspection glass; **~ de service** DIST EAU monitoring well; **~ de visite** DIST EAU inspection chamber
régénérant *adj* GÉNÉRAL regenerative
régénérat *m* DÉCHETS *d'essence, d'huile* by-product
régénérateur *m* POLL ATM regenerator
régénération *f* GÉNÉRAL regeneration; **~ catalytique** POLL ATM catalyst regeneration; **~ catalytique par voie biologique** POLL ATM biologically-catalytic regeneration; **~ de décharge contaminée** SOLS contaminated site regeneration; **~ de filtre** POLL ATM filter regeneration; **~ par un agent de lavage** POLL ATM washing agent regeneration; **~ d'un solvant** SOLS solvent regeneration; **~ thermique** POLL EAU thermal regeneration
régénéré *adj* DIST EAU, POLL ATM, POLL EAU, SOLS regenerated, reclaimed
régénérer *vt* GÉNÉRAL reclaim, regenerate; **~ par voie microbiologique** POLL ATM regenerate microbiologically
régime: **~ des eaux** *m* DIST EAU water regime; **~ des peines applicables en matière d'environnement** *m* POL ENV environmental criminal law; **~ d'une pompe** *m* DIST EAU, POLL EAU pump speed; **~ des vents** *m* ÉN RENOUV wind conditions
région: **~ benthique** *f* DIST EAU benthic region; **~ désertique** *f* ÉN RENOUV desert region, SOLS wilderness area; **~ émettrice** *f* POLL ATM source area; **~ ensoleillée** *f* ÉN RENOUV sunny region; **~ inculte** *f* SOLS uncultivated area; **~ non aménagée** *f* POL ENV wild land; **~ non-cultivée** *f* SOLS uncultivated area; **~ protégée** *f* POL ENV protected area; **~ réceptrice** *f* POLL ATM receptor region; **~ semi-aride** *f* SOLS semi-arid zone; **~ source** *f* POLL ATM source area; **~ de steppe** *f* ÉN RENOUV steppe region; **~ ventée** *f* ÉN RENOUV wind-favorable region (*AmE*), wind-favourable region (*BrE*), windy region
registre: **~ d'échantillonnage** *m* ÉCHANT sample record; **~ des émissions** *m* DIST EAU, POL ENV emissions register; **~ d'étalonnage** *m* ÉCHANT calibration record; **~ de mise en décharge** *m* POL ENV landfill register; **~ mondial des ressources génétiques** *m* GESTION world register of genetic resources; **~ sonore** *m* POLL SON noise register; **~ des substances dangereuses** *m* DIST EAU, POL ENV harmful substance register, register of hazardous substances
Registre: **~ international des substances chimiques potentiellement toxiques** *m* (*RISCPT*) POL ENV International Register of Potentially Toxic Chemicals (*IRPTC*)
réglage[1] *m* DIST EAU *débit* regulation; **~ isoélectrique capillaire** ÉCHANT capillary isoelectric focusing (*CIEF*)
réglage[2]: **à ~ de température** *loc* ÉCHANT temperature-controlled
règlement *m* DIST EAU regulation
réglementation *f* POL ENV, SÉCURITÉ regulation; **~ applicable à l'élimination des déchets** POL ENV waste disposal regulation; **~ en cas d'accident** SÉCURITÉ regulation in case of accident; **~ de l'environnement** GESTION, POL ENV environmental administration; **~ relative à l'octroi de permis** POL ENV licensing law; **~ de sécurité** SÉCURITÉ safety regulations; **~ statutaire** POL ENV, SÉCURITÉ statutory regulation; **~ sur la protection contre le bruit** SOLS technical regulation on noise protection; **~ temporaire** POL ENV transition regulation; **~ transitoire** POL ENV transition regulation; **~ d'utilisation** POL ENV use regulation
règles: **~ de prévention de la pollution** *f pl* POL ENV environmental prevention rules
regranulation *f* DIST EAU regranulation
regranuler *vt* DIST EAU regranulate
régressif *adj* POLL ATM involutional
régularisation *f* DIST EAU regularization, POLL EAU *d'un cours d'eau* equalization; **~ d'un cours d'eau** DIST EAU river training
régulateur *m* ÉCHANT *de tension*, POLL ATM regulator; **~ de débit massique** ÉCHANT mass-flow controller, mass-flow regulator; **~ à dépression** POLL ATM *moteur automobile* suction-type governor; **~ de pression** ÉCHANT, POLL ATM pressure regulator
régulation *f* DIST EAU *débit*, ÉCHANT *processus* regulation; **~ des crues** DIST EAU, ÉN RENOUV, GESTION, POLL MAR flood control; **~ du débit de sortie** POLL EAU outlet flow control; **~ de la sensibilité de mesure** ÉCHANT measurement-sensitivity control
réguler *vt* ÉN RENOUV *pollution* fight
réhabilitation *f* GÉNÉRAL modernization, reclamation, refurbishment, remediation, retrofitting; **~ d'installations anciennes** DÉCHETS, DIST EAU, POLL EAU modernization of old plants
réhabilité *adj* GÉNÉRAL reclaimed, retrofitted, refurbished, modernized
réhabiliter *vt* GÉNÉRAL retrofit, refurbish, reclaim, modernize
rehaussement: **~ du niveau d'eau par remblai** *m* DIST EAU banking-up of water
réimplantation *f* SOLS relocation
réinjecter: **~ dans un circuit** *vt* DIST EAU, POLL EAU recirculate
réinjection *f* DIST EAU recharge
rejaillir *vi* POLL ATM rebound
rejet *m* DIST EAU discharge, effluent release, effluent, POLL ATM release, POLL EAU discharge, effluent release, effluent, POLL MAR discharge, spill; **~ à l'air libre** POLL EAU *système d'aération* air release; **~ atmosphérique chauffé** POLL ATM heated waste air; **~ atmosphérique dégraissé** POLL ATM fat-free waste air; **~ de cendres** DIST EAU ash discharge; **~ de déchets** DÉCHETS waste discharge; **~ définitif** DÉCHETS final dumping; **~ de drague** POLL EAU, POLL MAR dredging spoil; **~ d'eaux usées** POLL EAU wastewater discharge; **~ des eaux usées en mer** POLL EAU wastewater disposal in the ocean; **~ des feuilles défectueuses** DÉCHETS faulty sheet ejection; **~ de gaz** ÉCHANT gas release; **~ illicite**

DÉCHETS, POL ENV, POLL EAU illegal dumping; **~ illicite d'acides usagés** POLL MAR illegal dumping of spent acid; **~ indirect** POLL EAU *des effluents industriels* indirect discharge; **~ intentionnel** DIST EAU intentional discharge; **~ liquide** POLL ATM liquid discharge; **~ en mer** DÉCHETS, POLL EAU, POLL MAR dumping at sea, ocean disposal, ocean dumping, marine dumping; **~ de polluants** POLL ATM release of pollutants; **~ de poussières** POLL ATM dust discharge; **~ de pression** POLL ATM pressure release; **~ sauvage** DÉCHETS, DIST EAU, POL ENV, POLL EAU uncontrolled dumping, illegal dumping, uncontrolled disposal, phantom dump, uncontrolled tipping; **~ de scories** DIST EAU slag discharge; **~ en surface** DÉCHETS surface dumping; **~ thermique** DÉCHETS, POLL ATM exhaust heat, thermal discharge

rejeter *vt* GÉNÉRAL discharge, dump; **~ en mer** DÉCHETS, POLL EAU, POLL MAR dump at sea, dump into the ocean

relatif: **~ aux mesures d'environnement** *adj* POL ENV environmentally-political; **~ à une méthode de combustion** *adj* POLL ATM combustion-related; **~ à l'objet** *adj* POLL MAR, POLL SON object-related noise; **~ à la politique de l'environnement** *adj* POL ENV eco-political

relevé *m* ÉCHANT reading

rémanence *f* SOLS persistence

remblai *m* DÉCHETS, DIST EAU embankment, SOLS backfill; **~ de terre** DÉCHETS earth bank

remblaiement *m* SOLS aggradation

remblais: **~ et déblais** *m pl* DÉCHETS cuts and fills

remblayage *m* SOLS backfilling, landfilling

remblayé *adj* SOLS backfilled, landfilled

remblayer *vt* SOLS backfill, landfill

rembourrage *m* DÉCHETS wadding, POLL SON padding; **~ à bulles d'air** DÉCHETS air-bubble cushioning; **~ isolant** POLL SON isolation pad; **~ de protection auditive** POLL SON hearing-protection padding

remembrement *m* POL ENV replotting; **~ rural** POL ENV regrouping of lands

remettre: **~ en culture** *vt* POLL EAU, SOLS recultivate; **~ en état** *vt* GÉNÉRAL retrofit; **~ à neuf** *vt* ÉN RENOUV, POLL ATM renew; **~ en végétation** *vt* SOLS revegetate

remise: **~ en culture** *f* POLL EAU, SOLS recultivation; **~ en état** *f* DIST EAU, POLL EAU, POLL SON, SÉCURITÉ retrofitting, SOLS reclamation; **~ en mouvement** *f* DIST EAU *d'un produit chimique* remobilization; **~ en place** *f* SOLS redisposal; **~ en suspension** *f* DIST EAU resuspension; **~ en valeur** *f* DIST EAU *des terres, des eaux* reclamation

remontée *f* DIST EAU recovery

remorque *f* POLL MAR towline; **~ de camion** DÉCHETS truck trailer

remplacement: **~ de filtre** *m* POLL ATM filter replacement

remplissage *m* GÉNÉRAL loading, filling; **~ direct d'une nappe** DIST EAU direct groundwater recharge; **~ de fond** DÉCHETS bottom filling; **~ intégral** DÉCHETS complete fill; **~ en continu** ÉN RENOUV continuous motion weight filling; **~ par injection** DÉCHETS injection filling; **~ de réservoir** POLL MAR refuelling

remplisseuse: **~ de cartons** *f* DÉCHETS carton filler

rendement *m* DÉCHETS, DIST EAU, ÉCHANT, ÉN RENOUV effectiveness, efficiency, throughput, thruput (*AmE*), GESTION output, POLL ATM, POLL MAR effectiveness, efficiency, throughput, thruput (*AmE*); **~ de l'adjuvant de filtration** DÉCHETS, DIST EAU, POLL ATM, POLL EAU filter aid efficiency; **~ d'adsorption** DÉCHETS, POLL ATM, SOLS adsorption efficiency; **~ de boues** POLL EAU sludge yield; **~ de catalyse** POLL ATM catalyst efficiency; **~ de cisaillement** POLL ATM shearing efficiency; **~ de collecte** POLL ATM collection efficiency; **~ de conversion** ÉN RENOUV conversion efficiency; **~ de déchets** DIST EAU residue yield; **~ de décolmatage** POLL ATM removal efficiency; **~ de dépoussiérage** POLL ATM dust-removal efficiency; **~ d'échangeur de chaleur** DÉCHETS, ÉN RENOUV, POLL ATM heat-exchanger efficiency; **~ d'échantillonnage** ÉCHANT sample throughput, sample thruput (*AmE*); **~ énergétique** ÉN RENOUV, POLL ATM, SOLS energy efficiency, energy output, energy yield; **~ énergétique efficace** POL ENV effective energy use; **~ d'épuration** SOLS clean-up efficiency; **~ de filtration** POLL ATM filter efficiency; **~ industriel** ÉN RENOUV process effectiveness; **~ d'une pompe** DIST EAU, POLL EAU pump efficiency; **~ de procédé** ÉN RENOUV process effectiveness; **~ de séparation** POLL ATM separating efficiency; **~ de stockage** ÉN RENOUV storage efficiency; **~ thermique** POLL ATM thermal output; **~ total** ÉN RENOUV total efficiency; **~ total de boue** POLL EAU total sludge yield; **~ du transfert d'oxygène** POLL EAU oxygen transfer efficiency

renforcé: **~ à la fibre de verre** *adj* POLL SON fiberglass-reinforced (*AmE*), fibreglass-reinforced (*BrE*); **~ par du nylon** *adj* DÉCHETS nylon-reinforced

renforcement: **~ des coins** *m* ÉN RENOUV *carton* corner reinforcement

renforcer *vt* DIST EAU, POLL EAU reinforce

renfort: **~ adhésif à l'eau** *m* DÉCHETS water-based backing adhesive; **~ cartonné** *m* DÉCHETS cardboard backing

reniflard *m* DIST EAU snifter valve

renouvelable *adj* ÉN RENOUV, POLL ATM renewable

renouveler *vt* ÉN RENOUV, POLL ATM renew

renouvellement: **~ d'air** *m* POLL ATM air exchange; **~ interfacial** *m* POLL ATM interfacial renewal

rénovation *f* DIST EAU, POLL ATM, POLL EAU, SÉCURITÉ retrofitting; **~ urbaine** POL ENV urban improvement

rénover *vt* GÉNÉRAL retrofit

rentabilité *f* GESTION, POL ENV economic efficiency, profitability

rentable *adj* GESTION, POL ENV profitable

rentrer: **~ par dévidoir** *vt* POLL MAR reel in

renvoi *m* DIST EAU return

réorientation *f* POLL ATM direction change

réoxydation *f* DIST EAU *d'une substance chimique*, POLL ATM reoxidation

réoxygénation *f* DIST EAU *d'une substance chimique* reoxidation, POLL EAU reaeration

répandre: **se ~** *v réfl* DIST EAU, POLL EAU, POLL MAR spread

réparabilité *f* DIST EAU reparability

répartition *f* DIST EAU breaking up; **~ de chargement** POLL ATM loading distribution; **~ des gouttes** POLL ATM drop distribution; **~ granulométrique** ÉCHANT grain-size distribution; **~ granulométrique des particules de poussière** ÉCHANT dust particle size distribution; **~ sectorielle** POL ENV sectoral division; **~ spatiale** DIST EAU, POLL ATM spatial distribution; **~ verticale** POLL ATM, SOLS vertical distribution

répercussion: **~ d'une autorisation** *f* DIST EAU, POL ENV authorization impact, licensing impact, permit impact; **~ sur l'environnement** *f* DIST EAU, ÉN RENOUV, POL ENV, POLL ATM, POLL MAR, POLL SON, SOLS environmental effect, environmental impact

repère: **~ d'étalonnage** *m* ÉCHANT calibration mark

répertoire: **~ des sites contaminés** *m* SOLS contaminated site register

répertorier *vt* POL ENV *substances dangereuses* classify

répétabilité *f* ÉCHANT repeatability

replastification *f* DIST EAU replastification

repomper *vt* DIST EAU, POLL EAU pump back

report: **~ de crédit** *m* GESTION carry-over; **~ d'imposition** *m* POL ENV tax shift

reprendre: **~ par pompage** *vt* DIST EAU, POLL EAU pump back

réprimer *vt* POLL SON curb

reprise *f* DÉCHETS *mode de collecte des emballages usagés* bring system, DIST EAU recovery, return; **~ des boues**

DÉCHETS sludge collection; ~ **de décharge** DIST EAU recovery of landfill space

reproductibilité *f* ÉCHANT reproducibility

répulsion: ~ **électrostatique** *f* DIST EAU, POLL EAU electrostatic repulsion

réseau *m* GÉNÉRAL network, system; ~ **alimentaire marin** ÉN RENOUV marine food web; ~ **d'alimentation en énergie renouvelable** ÉN RENOUV renewable energy supply system; ~ **d'assainissement** DIST EAU sewerage system; ~ **d'assainissement classique** DÉCHETS conventional sewerage system; ~ **d'assainissement séparatif** DIST EAU separate sewerage system; ~ **d'assainissement unitaire** DIST EAU combined sewerage system; ~ **de canalisation** DÉCHETS piping system; ~ **de canaux défectueux** SOLS defect canal system; ~ **de collecte des eaux usées** POLL EAU wastewater collection system; ~ **de distribution d'eau** DIST EAU water distribution system; ~ **de distribution d'eau potable** DIST EAU drinking-water distribution system; ~ **de données** ÉCHANT data network; ~ **d'égouts** DIST EAU sewer system; ~ **électrique** ÉN RENOUV electricity grid, power grid; ~ **d'évacuation** POLL EAU discharge system; ~ **de fourniture d'énergie** ÉN RENOUV energy supply grid; ~ **maillé et de distribution** DIST EAU meshed mains and supply network; ~ **séparatif** POLL EAU separate sewerage system; ~ **de surveillance** ÉCHANT monitoring network; ~ **de transport** ÉN RENOUV *électricité haute tension* power grid; ~ **de tuyauterie** DIST EAU piping system

réserve *f* POLL MAR stand-by; ~ **de chasse** POL ENV game reserve, hunting reserve; ~ **d'eau** DIST EAU water reserve; ~ **naturelle** POL ENV natural reserve, natural sanctuary; ~ **vidangeable** DIST EAU live storage

réserves: ~ **énergétiques** *f pl* ÉN RENOUV energy reserves

réservoir *m* DÉCHETS, DIST EAU tank, reservoir, ÉN RENOUV reservoir, POLL ATM tank, reservoir, POLL EAU tank, reservoir, cistern; ~ **de boues activées à forte charge** POLL EAU high-rate activated sludge tank; ~ **à carburant** POLL ATM fuel tank; ~ **de collecte** POLL EAU receiver tank; ~ **de collecte des eaux usées** DIST EAU, POLL EAU wastewater collection tank; ~ **à eau** POLL EAU cistern; ~ **d'eau brute** DIST EAU, POLL EAU untreated water tank; ~ **d'eau de lavage** POLL ATM wash-water tank; ~ **d'eau non-traitée** DIST EAU, POLL EAU untreated water tank; ~ **d'eau surélevé** DIST EAU elevated water reservoir; ~ **d'eaux pluviales** DIST EAU rainwater tank; ~ **fixe** POLL ATM stationary tank; ~ **de gaz** DÉCHETS, ÉCHANT, POLL EAU, SÉCURITÉ gas container; ~ **de gaz à effet de serre** POLL ATM reservoir of greenhouse gases; ~ **géothermique** ÉN RENOUV geothermal reservoir; ~ **de liquide de lavage** POLL ATM scrubbing liquid tank; ~ **des lixiviats de décharge** DÉCHETS, POLL EAU landfill leachate tank; ~ **à ordures** DIST EAU refuse bunker; ~ **de préparation** POLL EAU *de produits chimiques* batching tank; ~ **récepteur** POLL EAU receiving tank; ~ **de retenue d'eau** DIST EAU, ÉN RENOUV, POLL EAU reservoir; ~ **de stockage** POLL EAU storage basin; ~ **de surface** POLL EAU surface reservoir; ~ **de trop-plein** DIST EAU overflow tank

résidu *m* GÉNÉRAL residue; ~ **d'allumage** ÉCHANT residue on ignition; ~ **d'aluminium** DÉCHETS aluminium scrap (*BrE*), aluminum scrap (*AmE*); ~ **de calcination** ÉCHANT ignition residue; ~ **charbonneux** DÉCHETS char; ~ **de combustion** POLL ATM incineration ashes; ~ **contenant des polluants** DIST EAU residue containing pollutants; ~ **de digestion** DÉCHETS digestion deposit; ~ **de distillation** DIST EAU distillation residue; ~ **d'évaporation** ÉCHANT residue from evaporation; ~ **d'extraction** DÉCHETS extract residue; ~ **de filtration** ÉCHANT, POLL ATM filtration residue, filter residue; ~ **de flottation** DÉCHETS floating refuse; ~ **de houille** POLL ATM colliery waste; ~ **d'incinération** DÉCHETS, POLL EAU incineration residue; ~ **d'inflammation** POLL EAU ignition residue; ~ **d'installation de traitement des eaux d'égout** DIST EAU sewage treatment plant residue; ~ **de mise en balles** DÉCHETS baling scrap; ~ **des opérations minières** DÉCHETS, POLL EAU mining waste; ~ **des pesticides** POLL EAU pesticide residue; ~ **phytosanitaire** POLL EAU pesticide residue; ~ **de plomb** DÉCHETS scrap lead; ~ **de production** DIST EAU production residue; ~ **de raffinage** DIST EAU residue; ~ **sec** POLL EAU dry residue; ~ **de solvant** DIST EAU, SOLS solvent residue; ~ **de tamisage** DIST EAU sieve residue; ~ **urbain** DÉCHETS town refuse

résidus *m pl* GÉNÉRAL residue; ~ **éliminés** DÉCHETS, DIST EAU, POLL EAU, SOLS disposed-of residues; ~ **agricoles** POLL EAU agricultural waste; ~ **d'épuration des fumées d'incineration des ordures ménagères** (*REFIOM*) DÉCHETS garbage incineration refuse; ~ **évacués** DÉCHETS, DIST EAU, POLL EAU, SOLS disposed-of residues; ~ **formant des boues** DIST EAU sludge-forming residue; ~ **incombustibles** DÉCHETS, POLL ATM noncombustible residue, incombustible residue; ~ **inutilisables** DIST EAU unusable residue; ~ **de lavage** SOLS washings; ~ **liquides** POLL ATM liquid residue; ~ **métalliques** DÉCHETS scrap metal; ~ **non recyclables** DÉCHETS nonreusable residue wastes; ~ **non réutilisables** DÉCHETS nonreusable residue wastes; ~ **non-traités** DÉCHETS untreated refuse; ~ **de PVC** DIST EAU PVC residue; ~ **de récolte** DÉCHETS, ÉN RENOUV crop waste; ~ **respirables** SÉCURITÉ respirable residues; ~ **solides** DIST EAU solid residue; ~ **totaux** ÉCHANT total residue; **traités** DÉCHETS, DIST EAU processed waste

résilient *adj* POLL SON, SÉCURITÉ resilient

résine *f* DÉCHETS, DIST EAU, POLL EAU resin; ~ **acrylique** DÉCHETS acrylic resin; ~ **à base d'acétone** DÉCHETS acetone resin; ~ **échangeuse ionique** DIST EAU, POLL EAU, SOLS ion-exchange resin; ~ **échangeuse d'ions sélective** DIST EAU selective ion-exchange resin; ~ **époxyde** DIST EAU epoxy resin; ~ **phénolique** DIST EAU, POLL EAU phenolic resin; ~ **de polyester non saturée** DÉCHETS unsaturated polyester resin; ~ **saturée** POLL EAU saturated resin; ~ **thermodurcie renforcée à la fibre d'aramide** DIST EAU aramide fiber-reinforced thermoset (*AmE*), aramide fibre-reinforced thermoset (*BrE*)

résistance *f* GÉNÉRAL resistance; ~ **à l'abrasion** POLL ATM abrasion resistance; ~ **aux acides** POLL ATM, SOLS acid resistance; ~ **à l'attrition** POLL ATM attrition resistance; ~ **de canalisation** DIST EAU, POLL EAU pipeline resistance; ~ **chimique** POLL ATM chemical resistance; ~ **au cisaillement** POLL EAU *floc* shearing resistance; ~ **à la compression** DÉCHETS, DIST EAU, ÉCHANT, SOLS compression strength; ~ **d'une conduite** DIST EAU, POLL EAU pipeline resistance; ~ **à la corrosion** DIST EAU corrosion resistance; ~ **d'une couche** POLL ATM layer strength; ~ **à la crevaison** DIST EAU puncture resistance; ~ **à la déchirure** POLL ATM, POLL MAR tear strength, resistance to tearing; ~ **à la détonation** POLL ATM knock resistance; ~ **aux écarts de température** POLL ATM temperature resistance; ~ **à l'éclatement** DÉCHETS *emballage* bursting strength; ~ **à l'écrasement** SOLS soil-bearing capacity; ~ **à effet couronne** POLL ATM corona resistance; ~ **électrique** ÉCHANT electric resistance; ~ **de filtre** POLL ATM filter resistance; ~ **hydrodynamique** POLL ATM drag; ~ **aux intempéries** SOLS weathering resistance; ~ **au laminage** DÉCHETS lamination strength; ~ **aux maladies** SOLS resistance to diseases; ~ **de mesure** ÉCHANT measuring resistance, measuring resistor; ~ **à l'oxydation** DIST EAU resistance to oxidation; ~ **aux parasites** SOLS resistance to pests; ~ **à la perforation** DIST EAU puncture resistance; ~ **au pliage** DÉCHETS folding strength; ~ **aux polluants** POLL ATM, SOLS resistance to pollutants; ~ **de référence** ÉCHANT reference resistor; ~ **du sol aux agressions** SOLS stress capacity of soil; ~ **à la traction** POLL ATM tensile strength; ~ **aux ultraviolets** DIST EAU ultraviolet resistance; ~ **à l'usure** SOLS insensitivity to weathering

résistant *adj* DIST EAU, ÉCHANT, POLL ATM, POLL MAR resistant,

SOLS firm, resistant; ~ **aux acides** DÉCHETS, POLL ATM acid-proof, acid-resistant; ~ **à la chaleur** DÉCHETS, DIST EAU, SÉCURITÉ heat-proof, heat-insulating, *matériau* heat-resistant; ~ **aux insectes** DÉCHETS insect-proof; ~ **aux polluants** POLL ATM, SOLS pollutant-resistant; ~ **à l'eau** DÉCHETS, DIST EAU, POLL EAU water resistant; ~ **à l'usure** DÉCHETS, POLL ATM wear-resistant

résistivité *f* POLL ATM resistivity

résolution *f* ÉCHANT *d'un appareil de mesure* resolution; ~ **spatiale** DIST EAU, POLL ATM spatial resolution; ~ **temporelle** DIST EAU, POLL ATM temporal resolution

résonance *f* POLL MAR, POLL SON resonance; ~ **de spin électronique** (*RSE*) ÉCHANT electron spin resonance (*ESR*)

résonateur *m* POLL SON resonator; ~ **de Helmholtz** POLL SON Helmholtz resonator; ~ **de quart d'onde** POLL SON quarter-wave resonator

respect: ~ **de l'environnement** *m* GESTION, POL ENV environmental friendliness

respectueux: ~ **de l'environnement** *adj* GESTION, POL ENV environmentally-aware, environmentally-friendly

respirable *adj* SÉCURITÉ respirable

respiration: ~ **endogène** *f* DIST EAU endogenous respiration; ~ **du sulfate** *f* DIST EAU sulfate respiration (*AmE*), sulphate respiration (*BrE*); ~ **tissulaire** *f* DIST EAU exogenous respiration

responsabilité *f* GESTION, POL ENV *juridique* liability; ~ **civile** GESTION third-party liability; ~ **conjointe et solidaire** GESTION joint and several liability; ~ **délictuelle** GESTION tort liability; ~ **à l'égard de l'environnement** GESTION, POL ENV environmental responsibility, environmental commitment; ~ **en matière de dégradation de l'environnement** GESTION environment impairment liability (*EIL*); ~ **en matière d'environnement** GESTION, POL ENV environmental liability, environmental responsibility; ~ **objective** GESTION, POL ENV absolute liability, strict liability; ~ **pénale** GESTION criminal liability; ~ **résultant d'une faute** GESTION fault liability; ~ **surérogatoire** GESTION excess liability

responsable: ~ **de délits à l'encontre de l'environnement** *m* POL ENV environmental criminal; ~ **de la gestion de déchets** *m* DÉCHETS, GESTION, POL ENV waste management officer; ~ **de mouvement écologique** *m* POL ENV environmental leader; ~ **de la qualité de l'eau** *m* GESTION, POL ENV, POLL EAU water pollution control officer

ressaut *m* DIST EAU hydraulic jump

ressources *f pl* GÉNÉRAL resources; ~ **aquatiques** ÉN RENOUV aquatic resources; ~ **atmosphériques** ÉN RENOUV atmospheric resources; ~ **biologiques** ÉN RENOUV biological resources; ~ **biotiques** ÉN RENOUV biotic resources; ~ **en eau** DIST EAU, ÉN RENOUV water resources; ~ **en eaux souterraines** DIST EAU groundwater resources; ~ **énergétiques** DIST EAU, ÉN RENOUV, SOLS energy resources; ~ **énergétiques des courants marins** ÉN RENOUV ocean current resources; ~ **de l'énergie éolienne** ÉN RENOUV wind power resources; ~ **en énergie marémotrice** ÉN RENOUV tidal energy resources; ~ **en énergie solaire** ÉN RENOUV solar energy resources; ~ **de l'énergie des vagues** ÉN RENOUV wave energy resources; ~ **géothermiques** ÉN RENOUV geothermal energy resources, geothermal resources; ~ **hydriques** DIST EAU, ÉN RENOUV water resources; ~ **hydroélectriques** ÉN RENOUV hydropower resources; ~ **hydrologiques** DIST EAU, ÉN RENOUV water resources; ~ **en matières premières** DIST EAU raw material resources; ~ **du milieu peu abondantes** SOLS scarce environmental resources; ~ **minérales** ÉN RENOUV mineral resources; ~ **non renouvelables** ÉN RENOUV nonrenewable resources; ~ **rares dans l'environnement** ÉN RENOUV, POLL ATM scarce environmental resources; ~ **renouvelables** ÉN RENOUV renewable resources; ~ **du sous-sol** ÉN RENOUV subsoil resources; ~ **végétales** ÉN RENOUV plant resources; ~ **vivantes** ÉN RENOUV living resources

restabilisation *f* DIST EAU, POLL EAU restabilization

restabilisé *adj* DIST EAU, POLL EAU restabilized

restabiliser *vt* DIST EAU, POLL EAU restabilize

restauration: ~ **des terres** *f* SOLS land reclamation

restaurer *vt* SOLS reclaim, restore; ~ **le couvert végétal** SOLS revegetate

restrictions: ~ **écologiques** *f pl* POL ENV ecological restrictions

résultat: ~ **d'analyse** *m* ÉCHANT analysis data, analysis result, analytical finding, analytical result; ~ **de laboratoire** *m* ÉCHANT laboratory result; ~ **de mesurage** *m* ÉCHANT measurement result, measuring result; ~ **de mesure** *m* ÉCHANT measurement result

résurgent *adj* DIST EAU resurgent

rétablissement *m* DIST EAU recovery

retard *m* POLL ATM lagging

retardateur *m* DIST EAU retarder

rétention *f* POLL ATM holding

retenue *f* DIST EAU impounding reservoir, *d'eau* reservoir, ÉN RENOUV storage reservoir, *barrage* reservoir, POLL EAU reservoir; ~ **hydroélectrique** ÉN RENOUV storage hydroelectric plant

réticule *m* ÉCHANT *optique* cross lines

réticulé *adj* ÉCHANT cross-linked, grid-shaped, ÉN RENOUV, POLL ATM cross-linked

réticuler *vt* ÉCHANT, ÉN RENOUV, POLL ATM cross-link

retirer: **se** ~ *v réfl* DIST EAU *mer* ebb

retitrage *m* ÉCHANT retitration

retombées *f pl* DIST EAU deposition, POLL ATM fall-out, POLL EAU *nucléaires* deposition, SOLS fall-out; ~ **acides** POLL ATM acid fallout; ~ **acides humides** POLL ATM wet acidic fallout; ~ **acides sèches** POLL ATM dry acidic fallout, SOLS dry acid fallout; ~ **atmosphériques** POLL ATM atmospheric fallout; ~ **entraînées par la pluie** POLL ATM rainout; ~ **globales** DIST EAU bulk deposition; ~ **polluantes** POLL ATM contamination fallout; ~ **de polluants** DIST EAU pollutant deposition; ~ **pulvrulaires** POLL ATM dust fall; ~ **radioactives** DIST EAU, POLL ATM radioactive fall-out

retour *m* DIST EAU, POLL EAU return; ~ **des boues** POLL EAU sludge return, sludge return flow; ~ **d'eau** DIST EAU backflow; ~ **des eaux de filtration et des eaux résiduaires** POLL EAU filtration and residue water return

retournement: ~ **du sol** *m* SOLS soil break-up

retrait *m* DIST EAU, POL ENV repeal; ~ **d'autorisation** POL ENV repeal of authorization; ~ **de licence** POL ENV repeal of licence (*BrE*), repeal of license (*AmE*); ~ **d'un permis** DIST EAU repeal of licence (*BrE*), repeal of license (*AmE*); ~ **de permis** POL ENV repeal of permit

retraité *adj* DÉCHETS, DIST EAU reprocessed

retraitement *m* DÉCHETS, DIST EAU processing, reprocessing; ~ **de déchets** DÉCHETS waste reprocessing

retraiter *vt* DÉCHETS, DIST EAU reprocess, POLL ATM retreat

rétroaction *f* DIST EAU feedback

rétrobalayage *m* ÉCHANT backscattering

rétrodiffusion *f* ÉCHANT backscattering; ~ **de laser infrarouge** ÉCHANT infrared laser backscattering

rétrolavage *m* DIST EAU, POLL ATM backwashing; ~ **basse pression** POLL ATM low-pressure backwashing; ~ **en colonne amont** ÉCHANT pre-column back-flush operation

rétroprojecteur *m* GESTION back-projected screen

rétrotitrage *m* ÉCHANT back titration

réunir *vt* DIST EAU *eaux* collect

réutilisation *f* GÉNÉRAL reuse; ~ **des déchets** DÉCHETS, ÉN RENOUV reuse of waste; ~ **des eaux après épuration** POLL EAU wastewater reuse

réutiliser *vt* GÉNÉRAL reuse

revenus: ~ **de la fiscalité de l'environnement** *m pl* POL ENV eco-tax revenue

réverbérant *adj* POLL SON reverberant

réverbération *f* POLL SON reverberation

réversible *adj* POLL ATM, POLL EAU reversible

revêtement *m* DÉCHETS, DIST EAU, POLL ATM, POLL EAU coating, lining, POLL SON cladding, lining, SOLS sheeting; ~ **anti-bruit en caoutchouc mou** POLL SON, SÉCURITÉ antinoise soft rubber lining; ~ **en asphalte** DÉCHETS, DIST EAU asphalt sealing; ~ **à base de solvants** DIST EAU solvent-based coating; ~ **du bassin de retenue** POLL EAU reservoir lining; ~ **catalytique** POLL ATM catalytic coating; ~ **de cheminée** POLL ATM stack lining; ~ **de conduite** POLL SON duct lining; ~ **d'imperméabilisation** SOLS sealing sheeting; ~ **insonorisant** POLL SON sound-absorbing lining, sound-absorbing wall lining, acoustic lining; ~ **insonorisant en caoutchouc mou** POLL SON, SÉCURITÉ anti-noise soft rubber lining; ~ **naturel** DÉCHETS natural lining; ~ **en papier pour caisses** DÉCHETS case lining paper; ~ **d'une pièce** POLL SON room lining; ~ **de plastique** SOLS plastic liner; ~ **en plastique** DIST EAU plastic liner; ~ **de protection détachable** DÉCHETS peelable protective coating; ~ **de protection pelable** DÉCHETS peelable protective coating; ~ **résistant à l'usure** POLL ATM wear-resistant lining; ~ **de retenue** DIST EAU reservoir lining; ~ **de surface** DÉCHETS, POLL SON lining of surface, surface lining; ~ **synthétique** DÉCHETS synthetic lining; ~ **thermoscellé** DÉCHETS heat-seal coating

revêtir *vt* DÉCHETS, DIST EAU, ÉCHANT, POLL ATM coat, line, POLL SON line

revêtu *adj* DÉCHETS, DIST EAU, ÉCHANT, POLL ATM coated, lined, POLL SON lined; ~ **d'absorbant** POLL SON absorbent-lined; ~ **mécaniquement** DÉCHETS machine-coated; ~ **par émulsion** DÉCHETS, DIST EAU emulsion-coated

revitalisation *f* SOLS revitalization

révocation *f* DIST EAU, POL ENV repeal

Rh *abrév* (*rhodium*) POLL ATM Rh (*rhodium*)

rhéomètre *m* ÉCHANT rheometer

rhodium *m* (*Rh*) POLL ATM rhodium (*Rh*)

riblons *m pl* DÉCHETS scrap iron

riche: ~ **en soufre** *adj* POLL ATM sulfur-rich (*AmE*), sulphur-rich (*BrE*)

rideau: ~ **étanche** *m* SOLS retention wall; ~ **étroit** *m* DIST EAU narrow curtain; ~ **de gouttes** *m* POLL ATM drop film; ~ **liquide** *m* POLL ATM drop film; ~ **de palplanches** *m* DIST EAU sheet piling

rigole *f* DIST EAU, POLL EAU ditch, gutter; ~ **d'assèchement** DIST EAU drainage ditch; ~ **de décharge** POLL EAU effluent trough; ~ **de drainage** DIST EAU drainage ditch; ~ **d'évacuation** POLL EAU discharge gutter; ~ **de filtration du sable** POLL EAU sand filtration ditch; ~ **de forme parabolique** ÉN RENOUV parabolically-formed gutter; ~ **de grille** POLL EAU screen gutter

rinçage: ~ **du sol** *m* SOLS soil flushing

rincer *vt* DÉCHETS *bouteille* rinse, DIST EAU scour, spray, POLL EAU spray, SOLS rinse

rinceuse *f* DÉCHETS bottle washing machine, bottle rinsing machine

RISCPT *abrév* (*Registre international des substances chimiques potentiellement toxiques*) POL ENV IRPTC (*International Register of Potentially Toxic Chemicals*)

risque *m* GESTION, SÉCURITÉ hazard, risk; ~ **d'accumulation** POL ENV, SÉCURITÉ accumulation hazard; ~ **d'admission de gaz** DÉCHETS gas entrance possibility; ~ **biologique** SÉCURITÉ biological hazard; ~ **de bouchage** DÉCHETS, POLL ATM, POLL EAU clogging risk, risk of clogging; ~ **de brouillard photochimique** POLL ATM smog risk; ~ **de calaminage** POLL ATM scaling risk; ~ **de cancer** SÉCURITÉ cancer risk; ~ **chimique** SÉCURITÉ chemical hazard; ~ **chronique pour la santé** SÉCURITÉ chronic health risk; ~ **climatique** SÉCURITÉ climatic hazard; ~ **de colmatage** DÉCHETS, DIST EAU, POLL ATM clogging risk, risk of clogging; ~ **de colmatage par le fer** DIST EAU risk of iron clogging; ~ **de corrosion** POLL ATM corrosion risk; ~ **de dégradation** SOLS degradation hazard; ~ **de dépôt d'eaux usées** POLL EAU possibility of wastewater deposition; ~ **pour les eaux souterraines** DIST EAU groundwater risk; ~ **électrique** SÉCURITÉ electrical hazard; ~ **d'empaillassonnage** POLL EAU *aération des boues* risk of clogging, *de membranes* interlocking risk; ~ **d'encrassement** POLL ATM risk of clogging; ~ **d'erreur** ÉCHANT possibility of error; ~ **d'explosion** DÉCHETS, ÉCHANT explosion risk; ~ **d'incendie** ÉN RENOUV, SÉCURITÉ danger of fire, fire risk, fire hazard; ~ **d'interférence radio** SÉCURITÉ radiowave hazard; ~ **lié au rayonnement laser** SÉCURITÉ laser radiation hazard; ~ **mécanique** SÉCURITÉ mechanical hazard; ~ **microbiologique** SÉCURITÉ microbiological hazard; ~ **pour la nappe phréatique** DIST EAU groundwater risk; ~ **pathogène** SÉCURITÉ pathogenic hazard; ~ **potentiel** SÉCURITÉ potential danger; ~ **potentiel pour la sécurité** SÉCURITÉ potential safety risk; ~ **de rayonnement** SÉCURITÉ radiation hazard; ~ **de rémanence** SOLS persistence potential; ~ **restant** GESTION remaining risk; ~ **sanitaire** SÉCURITÉ health hazard, health risk; ~ **sanitaire manifeste** SÉCURITÉ clear health risk; ~ **pour la sécurité** SÉCURITÉ safety risk; ~ **statistique d'erreur** ÉCHANT statistical possibility of error; ~ **systématique d'erreur** ÉCHANT systematic possibility of error; ~ **total** SÉCURITÉ total risk; ~ **ultrasonique** SÉCURITÉ ultrasonic hazard

rive *f* DIST EAU *d'une rivière* bank

riverain *adj* DIST EAU riparian

rivière *f* DIST EAU river; ~ **capturée** DIST EAU beheaded river; ~ **côtière** DIST EAU coastal river

robinet: ~ **d'amorçage** *m* DIST EAU priming cock; ~ **d'arrosage** *m* DIST EAU jet cock; ~ **à deux eaux** *m* DIST EAU two-way cock; ~ **à deux faces** *m* DIST EAU twin cock; ~ **à deux voies** *m* DIST EAU two-way cock; ~ **d'échantillonnage de gaz** *m* ÉCHANT gas-sampling valve; ~ **d'essai** *m* DIST EAU try cock; ~ **de hauteur d'eau** *m* DIST EAU, POLL EAU gage cock (*AmE*), gauge cock (*BrE*), try cock; ~ **de jauge** *m* DIST EAU, POLL EAU gage cock (*AmE*), gauge cock (*BrE*), try cock; ~ **de lavage** *m* POLL ATM scrubbing vane; ~ **à membrane** *m* POLL ATM, POLL EAU diaphragm valve; ~ **de regard** *m* DIST EAU try cock; ~ **de vidange** *m* DIST EAU purging cock

robinetterie *f* DIST EAU valves and fittings

robustesse *f* DIST EAU robustness

roches: ~ **chaudes et sèches** *f pl* ÉN RENOUV *géothermie* hot dry rock (*HDR*)

rognures: ~ **de papier** *f pl* DÉCHETS paper chips

rompre *vt* POLL ATM break up

rondin *m* ÉN RENOUV billet

rose: ~ **des vents** *f* POLL ATM wind rose

rotatif *adj* POLL ATM rotating

rotation: ~ **des cultures** *f* SOLS crop rotation; ~ **terrestre** *f* ÉN RENOUV earth's rotation

rotor *m* ÉN RENOUV, POLL ATM rotor; ~ **à aubes** DIST EAU, POLL EAU pump impeller; ~ **Darius** ÉN RENOUV *éolienne à axe vertical* Darius rotor; ~ **d'émulsion** POLL EAU mammoth rotor; ~ **à nombre de tours fixe** ÉN RENOUV rotors with fixed number of revolutions; ~ **à nombre de tours variable** ÉN RENOUV rotors with variable number of revolutions; ~ **Savonius** ÉN RENOUV *éolienne à axe vertical* Savonius rotor

roue: ~ **à aube** *f* POLL ATM impeller; ~ **à eau** *f* ÉN RENOUV waterwheel; ~ **mobile** *f* POLL ATM impeller

roulant *adj* DIST EAU rolling

rouleau: ~ **mouilleur** *m* ÉN RENOUV *emballage* damping roll; ~ **de séparation magnétique** *m* POLL ATM magnetic separation roller

RSE *abrév* (*résonance de spin électronique*) ÉCHANT ESR (*electron spin resonance*)

Ru *abrév* (*ruthénium*) POLL ATM Ru (*ruthenium*)

ruban: ~ **d'acier** *m* DÉCHETS steel band strapping; ~ **adhésif**

m DÉCHETS adhesive tape, masking tape; ~ **thermoadhésif** *m* DÉCHETS heat-sealing tape

rubaneuse *f* DÉCHETS wrapping machine

ruisseler *vi* DIST EAU flow, run off, trickle, POLL ATM, POLL MAR leach, run off

ruissellement *m* DIST EAU flow, run-off, trickle, trickling, POLL ATM, POLL MAR leaching, run-off; ~ **acide** DIST EAU, SOLS acid run-off; ~ **de surface** DIST EAU surface run-off

rupture *f* DIST EAU, POLL ATM, POLL EAU rupture; ~ **de conduite** POLL EAU pipe break; ~ **de l'équilibre écologique** POL ENV disruption of the ecological balance; ~ **du floc** POLL EAU rupture of floc; ~ **de tuyauterie** DIST EAU pipe break, pipe breakage, pipe bursting

ruthénium *m* (*Ru*) POLL ATM ruthenium (*Ru*)

rythme: ~ **de frappage** *m* POLL ATM rapping interval; ~ **d'implantation** *m* DIST EAU settlement speed

S

S *abrév* (*soufre*) POLL EAU, SOLS S (*sulfur AmE, sulphur BrE*)
sabin *m* DIST EAU, ÉCHANT, POLL MAR sabin
sable *m* DÉCHETS, DIST EAU sand, POLL EAU grit; **~ flottant** POLL MAR quicksand; **~ mouvant** POLL MAR quicksand; **~ vert** DIST EAU greensand
sablière *f* POLL EAU sandpit; **~ encaissée** POLL EAU deep sand trap
sac *m* DÉCHETS, DIST EAU, POLL ATM bag; **~ doublé** DÉCHETS lined bag; **~ d'échantillonnage de gaz** ÉCHANT gas-sampling bag; **~ à fermeture éclair** DÉCHETS zip-lock bag; **~ de filtre tubulaire** POLL ATM tubular filter bag; **~ insérable** DÉCHETS insertable sack; **~ de jute** DÉCHETS jute sacking; **~ à ordures** DÉCHETS refuse sack, garbage bag (*AmE*), rubbish bag (*BrE*); **~ en papier** DÉCHETS, DIST EAU paper bag, paper sack; **~ à plusieurs épaisseurs** DÉCHETS multiwall sack; **~ poubelle** DÉCHETS garbage bag (*AmE*), rubbish bag (*BrE*), waste bag, refuse sack; **~ à provisions** DÉCHETS carrier bag
sacherie: **~ en jute** *f* DÉCHETS jute sacking
sachet *m* DÉCHETS bag, DIST EAU small bag
saison: **~ humide** *f* POLL ATM wet period; **~ sèche** *f* DIST EAU rain-free period
salant *adj* DIST EAU *marais* saline
salin *adj* DIST EAU saline
salinisation *f* DIST EAU salination
salinité *f* DIST EAU salinity; **~ absolue** DIST EAU absolute salinity; **~ excessive** DIST EAU oversalting; **~ pratique** DIST EAU practical salinity
salir *vt* DÉCHETS, ÉCHANT, POLL ATM, POLL MAR foul
salissure *f* DÉCHETS, ÉCHANT, POLL ATM, POLL MAR fouling
salle: **~ de commande** *f* DIST EAU, POLL ATM control room; **~ de commande centrale** *f* POLL ATM central control room; **~ de pompage** *f* DIST EAU pump room; **~ des pompes** *f* DIST EAU pump room; **~ de stockage** *f* DÉCHETS storage chamber; **~ de transit** *f* DÉCHETS transfer room
salubrité *f* SÉCURITÉ sanitary conditions
sangle *f* POLL MAR webbing
sanitaire *adj* DIST EAU sanitary, SÉCURITÉ health-related, sanitary
sans: **~ acide** *adj* POLL ATM acid-free; **~ agent pathogène** *adj* DIST EAU, POLL EAU pathogen-free; **~ amiante** *adj* SÉCURITÉ asbestos-free; **~ aucune activité biologique** *adj* SOLS biologically dead; **~ bois** *adj* DÉCHETS wood-free; **~ cendre** *adj* DIST EAU, POLL ATM ash-free; **~ déchet** *adj* DÉCHETS waste-free, residue-free; **~ eau** *adj* DIST EAU, POLL ATM water-free; **~ eaux usées** *adj* POLL ATM wastewater-free; **~ écho** *adj* POLL SON *chambre* anechoic; **~ fumée** *adj* SÉCURITÉ smokeless; **~ graisse** *adj* DIST EAU, POLL ATM fat-free, grease-free; **~ incidence négative pour la couche d'ozone** *adj* POLL ATM ozone-friendly; **~ matières indigestibles** *adj* DIST EAU roughage-free; **~ mélange préalable** *adj* POLL EAU *brûleur* nonaerated; **~ odeur** *adj* DÉCHETS, ÉCHANT, POLL ATM odourless (*BrE*), odorless (*AmE*); **~ phosphate** *adj* DIST EAU, POLL EAU phosphate-free; **~ poussière** *adj* POLL ATM dust-free; **~ pression** *adj* DIST EAU unpressurized; **~ résidu** *adj* DÉCHETS, POLL ATM residue-free, free of residues; **~ solvant** *adj* DIST EAU, POLL ATM solvent-free; **~ soufre** *adj* POLL ATM sulfur-free (*AmE*), sulphur-free (*BrE*); **~ stockage** *adj* ÉN RENOUV storage-free; **~ suie** *adj* POLL ATM soot-free
santé *f* SÉCURITÉ health; **~ publique** POL ENV public health
sarment: **~ de vigne** *m* ÉN RENOUV vine shoot
sas: **~ à air** *m* SÉCURITÉ airlock
sassement *m* DIST EAU, ÉCHANT screening
sasser *vt* DIST EAU, ÉCHANT screen
satellite: **~ extraterrestre** *m* ÉCHANT extraterrestrial-operated satellite; **~ géostationnaire** *m* POL ENV geostationary satellite; **~ d'observation de la terre** *m* POL ENV earth observation satellite
saturation *f* GÉNÉRAL saturation; **~ en oxygène** POLL EAU oxygen saturation; **~ du sol** SOLS soil saturation
saturé *adj* DIST EAU, ÉCHANT, POLL ATM, SOLS saturated; **~ en eau** DÉCHETS, DIST EAU, POLL EAU water-saturated
saturer *vt* DIST EAU, ÉCHANT, POLL ATM, SOLS saturate
saumâtre *adj* DIST EAU brackish, *eau* saline
saumure *f* POLL EAU brine
sauvegarde *f* DIST EAU safekeeping; **~ de la nature** POL ENV nature conservation; **~ du paysage** POL ENV landscape protection; **~ des ressources en eau** POL ENV water resources protection
Sb *abrév* (*antimoine*) DIST EAU Sb (*antimony*)
scanner: **~ manuel** *m* DIST EAU, POLL ATM hand scanner
scellé *adj* DÉCHETS, DIST EAU encapsulated, GÉNÉRAL sealed
scellement *m* DÉCHETS, DIST EAU encapsulation, GÉNÉRAL sealing; **~ au bitume** DÉCHETS, DIST EAU asphalt sealing; **~ du fond** DÉCHETS base sealing; **~ minéral** DÉCHETS mineral sealing; **~ mixte** DÉCHETS combination sealing; **~ sous vide en chambre** DÉCHETS chamber-type vacuum sealing
sceller *vt* DÉCHETS, DIST EAU encapsulate, GÉNÉRAL seal
science: **~ de l'eau** *f* DIST EAU hydroscience
sciences: **~ de la terre** *f pl* POL ENV earth sciences, geosciences
scintillomètre *m* ÉCHANT scintillation counter
sciure *f* ÉN RENOUV sawdust
scories *f pl* DÉCHETS, DIST EAU slag; **~ d'incinération** DÉCHETS incineration slag; **~ non-traitées** DÉCHETS untreated slag; **~ salines** DÉCHETS, DIST EAU salt slag; **~ traitées** DÉCHETS, DIST EAU processed slag; **~ d'une usine d'incinération** DÉCHETS slag from waste incineration plant; **~ volcaniques** DÉCHETS, POLL EAU lava slag
séchage *m* DÉCHETS drying, DIST EAU airing, drying, dewatering, ÉCHANT, POLL ATM drying, POLL EAU drying, dewatering, SOLS drying; **~ des boues** POLL EAU sludge drying; **~ à chaud** POLL EAU heat drying; **~ du gypse** POLL ATM gypsum drying; **~ indirect** POLL EAU indirect drying; **~ en lit fluidisé** POLL ATM, POLL EAU fluidized-bed drying; **~ par infrarouge** ÉCHANT infrared drying; **~ par contact direct** POLL EAU direct drying; **~ rapide** DIST EAU quick drying; **~ sous pression** POLL EAU *des boues résiduaires* press drying; **~ sur couches minces** POLL EAU thin-layer drying; **~ thermique** POLL EAU *des boues* thermal drying
sécher *vt* DÉCHETS dry, DIST EAU dry, air, dewater, ÉCHANT, POLL ATM dry, POLL EAU dry, dewater, SOLS dry
sécheresse *f* DIST EAU drought
sécheur: **~ à étages** *m* POLL EAU tiered dryer; **~ en lit fluidisé** *m* DIST EAU, POLL ATM fluidized-bed dryer; **~ à lit mobile** *m* POLL EAU moving-bed dryer; **~-pulvérisateur** *m* POLL EAU spray dryer
séchoir *m* ÉCHANT drying closet; **~ à bande** POLL EAU suspension dryer; **~-éclair** POLL ATM flash dryer; **~ en lit fluidisé** DIST EAU, POLL EAU fluidized-bed dryer; **~ par rayonnement** DIST EAU radiation dryer; **~ rotatif** POLL EAU, SOLS drum-type dryer; **~ à tambour** POLL EAU, SOLS drum-

type dryer; ~ **à tapis** POLL EAU suspension dryer; ~ **à tissu flottant** POLL ATM suspension dryer
secondaire *adj* GÉNÉRAL secondary
secousse *f* POLL ATM shaking; ~ **sismique artificielle** SOLS man-made earth tremor; ~ **tellurique** SOLS earth tremor
secrétariat *m* POL ENV secretariat
secteur *m* DIST EAU *électrique* service mains; ~ **d'approvisionnement en matières premières** GESTION raw material delivering industry; ~ **de l'environnement** POL ENV environmental sector; ~ **fournisseur de maitères premières** GESTION raw material delivering industry
secteurs: ~ **classiques du recyclage** *m pl* GESTION classical recycling sectors; ~ **traditionnels du recyclage** *m pl* GESTION classical recycling sectors
section: ~ **de convection** *f* POLL ATM convection section; ~ **cylindrique** *f* POLL ATM cylindrical section; ~ **d'injection** *f* SOLS grouting section, injection section; ~ **de mesure** *f* ÉCHANT measuring section
sécuritaire *adj* SÉCURITÉ safety-related
sécurité *f* SÉCURITÉ safety, security; ~ **de l'environnement** SÉCURITÉ environmental safety; ~ **incendie** SÉCURITÉ fire safety; ~ **à long terme** DÉCHETS long-term security; ~ **des produits** SÉCURITÉ product safety; ~ **du travail** SÉCURITÉ safety of the work place, work safety; ~ **d'utilisation** SÉCURITÉ use safety
sédiment *m* GÉNÉRAL precipitate, sediment; ~ **benthique** DIST EAU benthic deposit; ~ **marin pélagique** POLL MAR deep-sea sediment
sédimentable *adj* DIST EAU settleable, sedimentable, POLL ATM sedimentable, POLL EAU settleable, sedimentable, POLL MAR sedimentable
sédimentation *f* GÉNÉRAL sedimentation; ~ **naturelle** POLL EAU natural sedimentation; ~ **par la force centrifuge** ÉCHANT centrifugal force sedimentation; ~ **par gravité** ÉCHANT gravity sedimentation
sédimenter *vt* DIST EAU, POLL ATM, POLL EAU, POLL MAR sediment
segment: ~ **gazeux** *m* POLL ATM gas compartment
ségrégation *f* DIST EAU segregation, POLL EAU demixing
séisme: ~ **artificiel** *m* SOLS man-made earthquake
sel: ~ **ferrique** *m* POLL EAU ferric salt; ~ **inorganique** *m* DIST EAU inorganic salt; ~ **mélangé** *m* DIST EAU mixed salt; ~ **résiduaire** *m* DIST EAU waste salts
sélectif *adj* POLL EAU *électrode sélective* selective; ~ **d'ions** ÉCHANT *membrane* ion-selective; ~ **de masse** ÉCHANT mass-selective
sélection *f* ÉN RENOUV *sélection variétale* breeding
sélectionner *vt* ÉCHANT select
selle: ~ **de Berl** *f* POLL ATM *colonne à garnissage* berl saddle; ~ **Intalox** *f* POLL ATM *colonne à garnissage* Intalox saddle
sels: ~ **métalliques** *m pl* POLL EAU metallic salts
semelle *f* DIST EAU sill
semiconducteur *m* ÉN RENOUV semiconductor
semi-continu *adj* ÉCHANT partially-continuous
semi-solide *adj* DIST EAU semisolid
semi-mobile *adj* ÉCHANT semiportable
semi-perméable *adj* DIST EAU, POLL EAU semipermeable
semi-remorque *f* DIST EAU semitrailer
sens[1]: **dans le ~ du vent** *adv* POLL ATM downwind
sens[2] *m* DIST EAU, POLL MAR *d'un courant* drift; ~ **de filtration** POLL ATM filtration direction
sensibilisation: ~ **à l'environnement** *f* GESTION, POL ENV environmental awareness
sensibilisé: ~ **à l'environnement** *adj* GESTION, POL ENV environmentally-aware
sensibilité *f* ÉCHANT sensitivity; ~ **au bruit** POLL SON sensitivity to noise; ~ **de détection** ÉCHANT detection sensitivity; ~-**limite de détection** ÉCHANT, POLL EAU lower limit of detectability; ~ **magnétique** ÉCHANT magnetic susceptibility; ~ **de mesurage** ÉCHANT measuring sensitivity; ~ **aux odeurs** ÉCHANT odor sensitivity (*AmE*), odour sensitivity (*BrE*); ~ **olfactive** ÉCHANT odor sensitivity (*AmE*), odour sensitivity (*BrE*)
sensible *adj* ÉCHANT, POLL ATM sensitive; ~ **aux ions** ÉCHANT ion-sensitive; ~ **au pH** DIST EAU, POLL ATM, POLL EAU pH-sensitive; **très** ~ ÉCHANT highly-sensitive
sensotrode *m* ÉCHANT measuring sonde, measuring tube
séparable *adj* DIST EAU, POLL EAU dissociable, separable
séparateur *m* DÉCHETS, DIST EAU separator, ÉCHANT trap, POLL ATM separator, POLL EAU separator, single-stage digester, POLL MAR separator; ~ **d'aérosol à col sonique** POLL ATM ring-jet aerosol precipitator; ~ **balistique** DÉCHETS ballistic sorter; ~ **à buse** POLL ATM nozzle separator; ~ **de cendres grossières** POLL ATM coarse-ash separator; ~ **centrifuge** POLL ATM *séparation mécanique* centrifugal-force precipitator; ~ **à contre-courant** POLL ATM cross-flow air separator; ~ **électromagnétique** DÉCHETS electromagnetic separator; ~ **électrostatique** DÉCHETS, DIST EAU, POLL ATM *dépoussiérage électrique des gaz* electrostatic precipitator (*ESP*); ~ **de faisceau** ÉCHANT *optique laser* beamsplitter; ~ **de fractionnement** DÉCHETS emulsifying agent; ~ **de graisse** DÉCHETS, DIST EAU, POLL EAU grease separator, grease trap; ~ **gravimétrique** DIST EAU gravity separator; ~ **gravitaire** DÉCHETS, POLL ATM *dépoussiéreur mécanique*, POLL EAU gravity separator; ~ **d'huile de coalescence à forte charge** POLL EAU high-rate coalescence oil separator; ~ **huile-eau** POLL EAU oil-water separator; ~ **hydrocarbures/eau** POLL MAR separator of oily waters; ~ **à impacts** DIST EAU impinger; ~ **lamellaire** POLL ATM lamellar precipitator; ~ **laveur** DÉCHETS Venturi scrubber; ~-**laveur à plateaux** POLL ATM plate-column scrubber; ~ **magnétique** SOLS magnetic separator; ~ **magnétique de faible puissance** SOLS low-intensity magnet separator; ~ **magnétique de grande puissance** SOLS high-intensity magnet separator; ~ **de matières solides** DIST EAU, POLL EAU solids separator; ~ **de métaux** DÉCHETS metal separator; ~ **par filtration** POLL ATM filtering precipitator; ~ **par voie humide** POLL EAU wet classifier; ~ **de particules secondaires** POLL ATM secondary particle separator; ~ **de phase** POLL ATM phase-break vessel; ~ **pneumatique** DÉCHETS, POLL ATM air classifier, air separator; ~ **de très fines gouttelettes** POLL ATM finest-drop precipitator
séparation *f* DIST EAU segregation; ~ **balistique** *m* DÉCHETS ballistic separation; ~ **centrifuge** POLL ATM centrifugal precipitation, centrifugal separation; ~ **à contre-courant** ÉN RENOUV counter-flow air separation, cross-flow air separation; ~ **électromécanique** DÉCHETS, DIST EAU electromechanical separation; ~ **d'une émulsion** DIST EAU, POLL EAU emulsion separation; ~ **des fines** POLL ATM fine-dust precipitation; ~ **de gouttelettes** POLL ATM droplet separation; ~ **d'huile** DIST EAU, POLL EAU, SOLS oil separation; ~ **huile-eau** POLL EAU oil-water separation; ~ **magnétique** DÉCHETS, POLL ATM, POLL EAU, SOLS magnetic separation; ~ **magnétique en continu** POLL ATM *de ferraillage* permanent magnetic separating chute; ~ **manuelle** DÉCHETS manual separation; ~ **manuelle assistée mécaniquement** DÉCHETS mechanically-assisted hand separation; ~ **de matière solide** POLL ATM solid-matter separation; ~ **des matières sèches** POLL EAU dry-mass separation; ~ **mécanisée** DÉCHETS mechanical separation; ~ **par chromatographie en phase gazeuse** ÉCHANT gas chromatographic separation; ~ **par cyclone** POLL ATM cyclone separation; ~ **par décantation** ÉCHANT elutriation; ~ **par gravité** DÉCHETS, DIST EAU, POLL ATM, POLL EAU, SOLS gravity separation; ~ **par liquide dense** DÉCHETS heavy-medium separation; ~ **par sédimentation** POLL EAU settling out; ~ **par voie humide** POLL ATM *dépoussiérage* wet precipitation; ~ **de particules** POLL ATM particulate separation; ~ **de phases** DÉCHETS, DIST EAU, ÉCHANT, POLL ATM phase

separation; ~ **des plastiques** DIST EAU plastic separation; ~ **préliminaire** DIST EAU prior separation

séparé *adj* POLL EAU separate

séparer[1] *vt* DIST EAU segregate, separate, POLL ATM *les matières solides des gaz*, POLL EAU, SOLS separate; ~ **par filtration** DÉCHETS, DIST EAU, ÉCHANT, POLL ATM filter out; ~ **par voie électrolytique** DIST EAU separate electrolytically

séparer[2]: **se ~ par précipitation** *v réfl* GÉNÉRAL precipitate

septique *adj* DIST EAU, POLL EAU septic

séquentiel *adj* POLL ATM tiered, POLL EAU sequential, tiered

serpentin: ~ **de chargement** *m* DÉCHETS loading worm

serrage *m* DÉCHETS wraparound

service *m* POL ENV *d'une entreprise* department; ~ **contractuel d'emballage thermoformé** DÉCHETS contract blister-packaging service; ~ **des eaux** DIST EAU, POL ENV water authority; ~ **public** GESTION public utility; ~ **de ramassage** DÉCHETS *ordures ménagères* collection service; ~ **de ramassage des ordures** DIST EAU refuse collection service

services: ~ **collectifs** *m pl* POL ENV community facilities

servocommande: ~ **de pompe** *f* POLL EAU pump controls

seuil *m* DIST EAU sill, *d'un barrage, d'un déversoir* crest, POL ENV, SÉCURITÉ threshold; ~ **acoustique** ÉCHANT noise limit; ~ **d'audition** POLL SON threshold of audibility; ~ **de concentration atmosphérique** ÉCHANT ambient-air concentration limit; ~ **de concentration olfactive** POLL EAU threshold odor concentration (*AmE*), threshold odour concentration (*BrE*); ~ **de détectabilité** POLL ATM lower limit of detectability; ~ **de détection** ÉCHANT, POLL ATM lower limit of detectability; ~ **de détermination** ÉCHANT determination limit; ~ **d'émission de gaz de décharge** DÉCHETS landfill gas limit; ~ **d'émissions envisageable** POLL ATM lowest achievable emission rate; ~ **d'explosibilité** ÉCHANT, POLL ATM explosibility limit; ~ **d'explosion** SÉCURITÉ lower explosion limit; ~ **seuil de détectabilité** ÉCHANT lower level of detectability; ~ **d'inflammabilité** POLL ATM limit of inflammability; ~ **de nocivité** SÉCURITÉ threshold of harmfulness; ~ **olfactif** DIST EAU, ÉCHANT, POLL EAU odour threshold (*BrE*), odor threshold (*AmE*); ~ **de 90%** ÉCHANT 90% threshold value range; ~ **de risque** SÉCURITÉ risk limit; ~ **de toxicité** SÉCURITÉ threshold of toxicity

signal: ~ **d'alarme** *m* ÉCHANT emergency signal, SÉCURITÉ warning signal; ~ **de mesure** *m* ÉCHANT measuring signal

silencieux[1] *adj* POLL MAR, POLL SON, SÉCURITÉ noiseless

silencieux[2] *m* POLL ATM exhaust muffler (*AmE*), exhaust silencer (*BrE*), POLL SON sound-control system, attenuator, exhaust muffler (*AmE*), exhaust silencer (*BrE*), muffler (*AmE*), silencer (*BrE*), noise muffler (*AmE*), sound silencer (*BrE*); ~ **absorbant les résonances** POLL SON resonance-absorption silencer (*BrE*), resonance-absorption muffler (*AmE*); ~ **actif** POLL SON active sound-control system; ~ **de chambre de détente des gaz** POLL SON expansion chamber silencer (*BrE*), expansion chamber muffler (*AmE*); ~ **à chicanes** POLL SON baffle silencer (*BrE*), baffle muffler (*AmE*); ~ **d'évent** POLL SON vent silencer (*BrE*), vent muffler (*AmE*); ~ **réactif** POLL SON, SÉCURITÉ reactive silencer (*BrE*), reactive muffler (*AmE*)

sillage *m* POLL MAR wake; ~ **sonore** POLL SON noise carpet

silo *m* DIST EAU silo, ÉN RENOUV clamp, POLL ATM silo; ~ **à additifs** POLL ATM additives silo; ~ **à cendres de chambre de combustion** POLL ATM combustion chamber ash silo; ~ **à cendres volantes** POLL ATM fly-ash silo; ~ **de gypse** POLL ATM gypsum silo; ~ **à ordures** DIST EAU refuse bunker; ~ **à poussières** POLL ATM dust silo; ~ **de stockage temporaire** POLL ATM temporary storage silo

simili: ~**-sol** *m* SOLS soil-like material

simulation: ~ **bassins hydrographiques** *f pl* GESTION simulation river systems

siphon *m* DIST EAU, POLL EAU siphon, trap; ~ **à boues** POLL EAU sludge siphon; ~ **de dépôt de graisse** DÉCHETS, DIST EAU, POLL EAU *plomberie* grease trap, grease separator

siphonner *vt* DIST EAU, POLL EAU siphon, siphon off

site[1]: **du ~** *loc* GÉNÉRAL site-related, site-specific; **sur ~** *loc* DÉCHETS, DIST EAU, ÉCHANT on-site

site[2] *m* DÉCHETS, DIST EAU, POLL ATM site; ~ **de décharge** DÉCHETS tipping site, landfill, DIST EAU landfill; ~ **de décharge illicite** DÉCHETS illegal dump site; ~ **d'élimination des ordures** DIST EAU refuse disposal site; ~ **industriell** SOLS industrial site; ~ **de mise en décharge** POLL MAR landfill site; ~ **problématique** DIST EAU problem site; ~ **de stockage** DÉCHETS storage site

situation: ~ **écologique** *m* POL ENV ecological conditions

SM-CLHP *abrév* (*spectrométrie de masse-chromatographie en phase liquide haute performance*) ÉCHANT HPLC-MS (*high-performance liquid chromatography-mass spectrometry*)

SME *abrév* (*système de management environnemental*) GESTION EMS (*environmental management system*)

smog *m* POLL ATM smog; ~ **acide** POLL ATM acid smog; ~ **photochimique** POLL ATM photochemical smog

SO *abrév* DIST EAU (*oxyde de soufre*) SO (*sulfur oxide AmE, sulphur oxide BrE*), POLL ATM (*monoxyde de soufre*) SO (*sulfur monoxide AmE, sulphur monoxide BrE*)

SO_2 *abrév* DIST EAU (*anhydride sulfureux*) SO_2 (*sulfur anhydride AmE, sulphur anhydride BrE*), POLL ATM (*dioxyde de soufre*) SO_2 (*sulfur dioxide AmE, sulphur dioxide BrE*)

SO_3 *abrév* POLL ATM (*trioxyde de soufre*) SO_3 (*sulfur trioxide AmE, sulphur trioxide BrE*), DIST EAU (*anhydride sulphurique*) SO_3 (*sulfuric anhydride AmE, sulphuric anhydride BrE*)

SO_4 *abrév* (*tétraoxyde de soufre*) POLL ATM SO_4 (*sulfur tetroxide AmE, sulphur tetroxide BrE*)

SO_x *abrév* (*oxydes de soufre*) DIST EAU, POLL ATM SO_x (*sulfur oxides AmE, sulphur oxides BrE*)

société: ~ **de traitement de déchets** *f* DÉCHETS, GESTION waste disposal firm, waste disposal company

sodium *m* (*Na*) POLL ATM sodium (*Na*)

sol *m* DÉCHETS, DIST EAU, SOLS ground, soil; ~ **boueux** SOLS soil slurry; ~ **brut** SOLS raw soil; ~ **de couverture** DÉCHETS *décharge* top soil, DIST EAU, ÉN RENOUV covering soil; ~ **décontaminé** SOLS decontaminated soil; ~ **excavé** DÉCHETS, SOLS excavated soil; ~ **extrait** SOLS extracted soil; ~ **gorgé d'eau** SOLS waterlogged soil; ~ **humifère** SOLS humus-containing soil; ~ **d'origine** SOLS original soil; ~ **pollué** SOLS polluted ground; ~ **solidifié** SOLS solidified soil

solaire *adj* ÉN RENOUV solar, solar-operated

solarimètre *m* ÉCHANT solarimeter

sole *f* DIST EAU sill

solides: ~ **décantés** *m pl* POLL EAU settled solids; ~ **filtrants** *m pl* POLL EAU filterable solids

solidificateur *m* POLL MAR solidifier

solidification *f* DIST EAU, POLL ATM, SOLS solidification; ~ **avec des polymères organiques** DIST EAU solidification with organic polymers; ~ **in situ** SOLS in-situ solidification

solidifié *adj* DÉCHETS, DIST EAU, POLL ATM, SOLS solidified

solidifier *vt* DÉCHETS, DIST EAU, POLL ATM, SOLS solidify

solubilisation *f* DIST EAU, POLL ATM, POLL EAU, SOLS solubilization

solubilisé *adj* DIST EAU, POLL ATM, POLL EAU, SOLS solubilized

solubiliser *vt* DIST EAU, POLL ATM, POLL EAU, SOLS solubilize

solubilité *f* DIST EAU, POLL ATM, POLL EAU, SOLS solubility, solvability; ~ **des gaz** POLL ATM gas solubility

soluble *adj* DIST EAU, POLL ATM, POLL EAU, SOLS soluble; ~ **dans l'eau** DÉCHETS, DIST EAU, POLL ATM, POLL EAU water-soluble

soluté *m* POLL ATM solute

solution *f* DIST EAU, ÉCHANT, POLL EAU solution; ~ **d'absorption** ÉCHANT absorption solution; ~ **analytique** ÉCHANT analytical solution; ~ **diluée** ÉCHANT diluted solution; ~ **d'essai** ÉCHANT test solution; ~ **d'étalonnage**

ÉCHANT calibration solution; ~ **indicatrice** ÉCHANT indicator solution; ~ **initiale** ÉCHANT parent solution; ~ **de lavage** POLL ATM washing solution; ~ **la meilleure pour l'environnement** POL ENV best practical environmental option (*BPEO*); ~ **mère** ÉCHANT parent solution; ~ **de Nessler** ÉCHANT Nessler's solution; ~ **de réactifs** ÉCHANT reagent solution; ~ **tampon** ÉCHANT buffer solution

solvant *m* GÉNÉRAL solvent; ~ **actif** DÉCHETS active solvent; ~ **chargé** POLL ATM loaded solvent; ~ **combustible** POLL ATM combustible solvent; ~ **d'épuisement** ÉCHANT, SOLS extraction solvent; ~ **d'extraction** ÉCHANT, SOLS extraction solvent, extractant; ~ **nauséabond** POLL ATM odor-intensive solvent (*AmE*), odour-intensive solvent (*BrE*); ~ **organique** POLL ATM organic solvent; ~ **organique halogéné** DIST EAU halogenated organic solvent; ~ **organique non halogéné** DIST EAU halogen-free organic solvent; ~ **en phase vapeur** POLL ATM vaporous solvent; ~ **régénéré** POLL ATM regenerated solvent; ~ **de séparation** SOLS separation solvent

solvatation *f* POLL EAU solvation

solvaté *adj* POLL EAU solvated

solvater *vt* POLL EAU solvate

solvolyse *f* DIST EAU solvolysis

sommet: ~ **sur l'environnement** *m* POL ENV environmental summit; ~ **d'une tour solaire** *m* ÉN RENOUV solar tower top

son *m* DIST EAU, ÉCHANT, POLL SON sound; ~ **résiduel** POLL SON residual sound; ~ **secondaire** POLL SON secondary sound; ~ **supersonique** POLL SON supersonic sound

sondage *m* ÉCHANT sound boring, sounding boring, ÉN RENOUV drilling; ~ **de pression** ÉCHANT pressure sounding; ~**-tarière** *m* SOLS hand augering; ~ **au trépan-benne** SOLS grab boring

sonde *f* DIST EAU detector, probe, ÉCHANT measuring gage (*AmE*), measuring gauge (*BrE*); ~ **acoustique** ÉCHANT sound probe; ~ **électrochimique à ozone** ÉCHANT electrochemical ozone sonde; ~ **de mesure** ÉCHANT measuring sonde; ~ **microphonique** ÉCHANT probe microphone; ~ **de prélèvement** ÉCHANT sampling probe; ~ **de température** ÉCHANT temperature probe; ~ **de vitesse et de pression** ÉCHANT pressure velocity probe

sonder *vt* ÉCHANT sound, sense, ÉN RENOUV drill

sondeuse *f* DÉCHETS bore machine; ~ **portée** ÉCHANT wagon drill

sone *m* ÉCHANT sone

sonique *adj* DIST EAU, POLL SON sonic

sonomètre *m* ÉCHANT, POLL MAR, POLL SON, SÉCURITÉ noise meter, sound-exposure meter, sound-level meter (*SLM*); ~ **à impulsion** ÉCHANT impulse sound-level meter; ~ **modulaire** ÉCHANT module sound-level meter; ~ **de précision** ÉCHANT precision sound-level meter

sorbant *m* POLL ATM, POLL MAR sorbent; ~ **de désulfuration** POLL ATM desulfurization sorbent (*AmE*), desulphurization sorbent (*BrE*)

sorbat *m* POLL ATM, POLL MAR sorbate

sorption *f* DIST EAU, POLL ATM, POLL EAU sorption

sorte: ~ **de papier** *f* DÉCHETS paper grade

sortie *f* DIST EAU outflow, POLL ATM outlet, POLL EAU outflow; ~ **d'air de refroidissement** POLL ATM cooling air outlet; ~ **de boue** POLL ATM sludge outlet; ~ **de chambre de combustion** POLL ATM combustion chamber outlet; ~ **de cheminée** POLL ATM stack orifice; ~ **d'eau** DIST EAU water outlet; ~ **d'eau de lavage** DIST EAU washout; ~ **des eaux usées** POLL EAU wastewater outflow; ~ **de gaz** POLL ATM gas outlet; ~ **de gaz épuré** POLL ATM clean-gas out; ~ **de poussière** POLL ATM dust outlet; ~ **de réfrigérant** POLL ATM coolant out, coolant outlet; ~ **de vidange** DIST EAU scouring outlet

soubassement: ~ **non-fini** *m* DIST EAU unfinished support

soude *f* DIST EAU, POLL EAU soda, sodium hydroxide; ~ **calcinée** DIST EAU soda ash

soufflage *m* POLL ATM blowing through

soufflante: ~ **centrifuge** *f* POLL ATM centrifugal fan; ~ **des gaz de combustion** *f* POLL ATM waste gas blower; ~ **sous grille** *f* DIST EAU underblast fan

souffler *vt* POLL ATM blow, inflate

soufflerie *f* DÉCHETS blower; ~ **à air comprimé** POLL ATM compressed-air-blast

soufflet *m* DÉCHETS bottom fold

soufre *m* (*S*) POLL EAU, SOLS sulfur (*AmE*) (*S*), sulphur (*BrE*) (*S*); ~ **atmosphérique** POLL ATM atmospheric sulfur (*AmE*), atmospheric sulphur (*BrE*)

souiller *vt* DÉCHETS, ÉCHANT, POLL MAR foul, soil

souillure *f* DÉCHETS, ÉCHANT, POLL MAR fouling, soiling

soumis: ~ **à autorisation** *adj* POL ENV requiring authorization; ~ **à des essais en laboratoire** *adj* ÉCHANT laboratory-tested; ~ **à étalonnage** *adj* ÉCHANT subject to calibration; ~ **à une la pollution acide** *adj* SOLS acid-stressed

soumission: ~ **de dossier à l'inspection publique** *f* POL ENV presentation of documents for public inspection

soupape: ~ **d'échappement** *f* POLL ATM exhaust valve; ~ **de sûreté** *f* DIST EAU safety valve

souple *adj* DÉCHETS, POLL MAR, POLL SON flexible

source[1] *adj* DIST EAU, POLL ATM, POLL MAR, POLL SON source-related; **de la** ~ DIST EAU, POLL MAR, POLL SON source-related; ~ **de déchets** DÉCHETS waste-generating

source[2] *f* DIST EAU, POLL ATM source, emission source; ~ **alignée** DIST EAU, POLL EAU line source; ~ **anthropique** POLL ATM anthropogenic sources; ~ **artésienne** DIST EAU artesian spring; ~ **de bruit** POLL MAR, POLL SON noise source; ~ **de contact** DIST EAU contact spring; ~ **de dépression** DIST EAU depression spring; ~ **de déversement** DIST EAU depression spring; ~ **diffuse** POLL ATM, SOLS area source; ~ **d'eau chaude** ÉN RENOUV geyser, hot-water source; ~ **d'électricité** ÉN RENOUV electricity source, power source; ~ **éloignée** DIST EAU, POLL ATM distant source; ~ **émettrice** DIST EAU emission source; ~ **d'émission** DIST EAU, POL ENV, POLL ATM emission source; ~ **d'émission indirecte** POLL ATM indirect emission source; ~ **d'émissions fixe** POLL ATM stationary emission source; ~ **d'énergie** ÉN RENOUV energy source; ~ **d'énergie fossile** ÉN RENOUV fossil energy source; ~ **d'énergie inépuisable** ÉN RENOUV inexhaustible energy source; ~ **d'énergie non polluante** ÉN RENOUV clean-energy source; ~ **d'énergie nucléaire** ÉN RENOUV nuclear energy source; ~ **d'énergie primaire** ÉN RENOUV primary energy source; ~ **d'énergie propre** ÉN RENOUV clean energy source; ~ **d'énergie renouvelable** ÉN RENOUV, POLL ATM renewable energy source; ~ **d'énergie renouvelable autre que solaire** ÉN RENOUV nonsolar-based renewable energy source; ~ **d'énergie renouvelable d'origine solaire** ÉN RENOUV solar-based renewable energy source; ~ **d'énergie secondaire** ÉN RENOUV secondary energy source; ~ **étrangère** POLL ATM foreign source; ~ **de faille** DIST EAU fault spring; ~ **hydrothermale souterraine** POLL MAR deep ocean geyser, hydrothermal vent; ~ **industrielle** POLL ATM industrial source, *polluants* commercial source; ~ **intermittente** DIST EAU intermittent spring; ~ **interne** DIST EAU, POLL ATM internal source; ~ **d'ions** ÉCHANT ion source; ~ **karstique** DIST EAU karstic spring; ~ **linéaire** DIST EAU, POLL ATM *de neutrons* line source; ~ **locale d'émission** POLL ATM area emission source, local emission source; ~ **majeure** DIST EAU, POLL ATM major source; ~ **de métaux lourds** POLL ATM, SOLS heavy metals source; ~ **minérale** DIST EAU mineral spring; ~ **mobile** DIST EAU transportation source; ~ **mobile d'émission** POLL ATM mobile emission source; ~ **naturelle** POLL ATM natural source; ~ **naturelle d'énergie** ÉN RENOUV natural energy source; ~ **olfactive** POLL ATM odor source (*AmE*), odour source (*AmE*), source of odor (*AmE*), source of odour (*BrE*); ~ **de pollution** DIST EAU, POL ENV, POLL ATM, SOLS pollution source; ~ **ponctuelle** DIST EAU, POLL ATM, SOLS point source; ~ **principale** DIST EAU, POLL ATM major source; ~ **de revenus** POL ENV revenue source; ~ **salée** DIST EAU saline spring; ~ **saline** DIST EAU brine spring, saline spring; ~ **saumâtre** DIST EAU saline spring; ~ **sonore** POLL MAR, POLL SON noise source; ~ **thermique** ÉN RENOUV heat source, thermal spring; ~ **traditionnelle d'énergie** ÉN RENOUV traditional energy source

sous-biez *m* DIST EAU backwater

sous-bois *m* SOLS undergrowth

sous-cellule *f* DÉCHETS subcell

sous-écoulement: ~ **de tamis** *m* DIST EAU screen underflow

sous-indice *m* POL ENV sub-index

sous-marin *adj* DIST EAU, POLL MAR underwater

sous-micron *m* POLL ATM submicron

sous-produit *m* DÉCHETS, DIST EAU, GESTION, POLL ATM by-product; ~ **de déchet** DÉCHETS usable waste by-product; ~ **de récupération** DIST EAU recovery by-product; ~ **utilisable** DÉCHETS, DIST EAU, GESTION, POLL EAU usable by-product; ~ **valorisable** DÉCHETS, DIST EAU, GESTION, POLL ATM usable by-product

sous-refroidir *vt* POLL ATM subcool

sous-sol: ~ **de décharge** *m* DÉCHETS landfill underground; ~ **plat** *m* SOLS flat subsurface; ~ **résistant** *m* SOLS firm subsurface

sous-verse *m* DIST EAU *flottateur* underflow

soute *f* DIST EAU, POLL ATM bunker tank, pit

souterrain *adj* DÉCHETS underground

soutirer *vt* POLL EAU, SOLS draw off

spécialiste: ~ **de l'environnement** *m* GESTION, POL ENV environmentalist

spécification: ~ **de contenu** *f* DÉCHETS *suivi de déchets*, SÉCURITÉ contents declaration, identification of contents; ~ **du type** *f* DÉCHETS type specification

spécifique: ~ **d'un composant** *adj* ÉCHANT component-specific; ~ **d'un contaminant** *adj* POLL ATM contaminant-specific; ~ **de l'installation** *adj* GÉNÉRAL plant-specific; ~ **d'une phase** *adj* DIST EAU phase-specific; ~ **d'un polluant** *adj* POLL ATM, SOLS pollutant-specific; ~ **de la population** *adj* POL ENV population-specific; ~ **d'un type** *adj* GÉNÉRAL type-specific

spécimen *m* ÉCHANT specimen

spectre: ~ **d'absorption** *m* ÉCHANT absorption spectrum; ~ **acoustique** *m* POLL SON sound spectrum; ~ **d'émission** *m* ÉCHANT, POLL ATM emissions spectrum; ~ **de fluorescence** *m* ÉCHANT fluorescence spectrum; ~ **de fréquence** *m* POLL SON frequency spectrum; ~ **de l'infrarouge** *m* ÉCHANT infrared spectrum; ~ **de masse** *m* ÉCHANT mass spectrum; ~ **de pression acoustique** *m* POLL SON sound-pressure spectrum; ~ **de référence** *m* ÉCHANT reference spectrum; ~ **de transmission** *m* ÉCHANT transmission spectrum

spectrographe: ~ **de masse** *m* ÉCHANT mass spectrograph

spectrographie: ~ **de masse** *f* ÉCHANT mass spectrography

spectromètre *m* ÉCHANT spectrometer; ~ **d'émission** ÉCHANT emissions spectrometer; ~ **imageur de l'ozone total** (*TOMS*) ÉCHANT, POLL ATM total ozone mapping spectrometer (*TOMS*); ~ **infrarouge à transformée de Fourier** ÉCHANT Fourier transform infrared spectrometer; ~ **à ionisation** ÉCHANT ionization spectrometer; ~ **de masse** ÉCHANT mass spectrometer (*MS*); ~ **de masse à impact électronique** ÉCHANT electron impact mass spectrometer (*EI/MS*); ~ **de masse quadripolaire** ÉCHANT quadrupole mass spectrometer; ~ **de masse à secteur magnétique** ÉCHANT magnetic sector mass spectrometer; ~ **de mobilité ionique** ÉCHANT ion-mobility spectrometer; ~ **à rayonnement ultraviolet et visible** ÉCHANT ultraviolet and visible spectrometer

spectrométrie *f* ÉCHANT spectrometry; ~ **d'absorption atomique** ÉCHANT atomic absorption spectrometry (*AAS*); ~ **d'absorption atomique d'un four en graphite** ÉCHANT graphite furnace atomic absorption spectrometry; ~ **d'absorption sur jet** ÉCHANT flow injection analysis spectrometry (*FIAS*); ~ **d'émission atomique** ÉCHANT *dosage des métaux en solution* atomic emission spectrometry (*AES*); ~ **d'émission optique de plasma avec couplage inductif** ÉCHANT inductively-coupled plasma-optical emission spectrometry (*ICP-OES*); ~ **de fluorescence atomique** ÉCHANT atomic fluorescence spectrometry (*AFS*); ~ **de fluorescence atomique avec excitation laser** ÉCHANT laser-excited atomic fluorescence spectrometry (*LEAFS*); ~ **infrarouge** ÉCHANT infrared spectrometry; ~ **infrarouge non dispersive** ÉCHANT non-dispersive infrared spectrometry (*NDIR spectrometry*); ~ **infrarouge à transformée de Fourier** ÉCHANT Fourier-transform infrared spectrometry (*FTIR spectrometry*); ~ **de masse** ÉCHANT mass spectrometry; ~ **de masse par réaction ion-molécule** ÉCHANT ion-molecule reaction mass spectrometry (*CI-MS*); ~ **de masse au laser** ÉCHANT laser mass spectrometry (*LAMS*); ~ **de masse par ionisation chimique** ÉCHANT chemical ionization mass spectrometry (*CI-MS*); ~ **de masse de plasma avec couplage inductif** ÉCHANT inductively-coupled plasma-mass spectrometry (*ICP-MS*); ~ **de masse-chromatographie liquide à faisceau de particules** ÉCHANT particle beam liquid chromatography-mass spectrometry (*PB LC-MS*); ~ **de masse-chromatographie en phase gazeuse avec pyrolyse** ÉCHANT pyrolysis-gas chromatography-mass spectrometry (*PY-GC-MS*); ~ **de masse-chromatographie en phase liquide haute performance** (*SM-CLHP*) ÉCHANT high-performance liquid chromatography-mass spectrometry (*HPLC-MS*); ~ **de mobilité des ions** ÉCHANT ion-mobility spectrometry (*IMS*); ~ **moléculaire** ÉCHANT molecular spectrometry

spectrométrique *adj* ÉCHANT spectrometric

spectrophotomètre *m* ÉCHANT spectrophotometer; ~ **d'absorption atomique** ÉCHANT atomic absorption spectrophotometer; ~ **d'enregistrement** ÉCHANT recording spectrophotometer; ~ **à infrarouge** ÉCHANT infrared spectrophotometer; ~ **portable** ÉCHANT portable spectrophotometer

spectrophotométrie: ~ **ultraviolette et visible** *f* ÉCHANT ultraviolet and visible spectrophotometry

spectroscope: ~ **de masse** *m* ÉCHANT mass spectroscope; ~ **portatif** *m* ÉCHANT hand-held spectroscope

spectroscopie *f* ÉCHANT spectroscopy; ~ **d'absorption atomique avec atomisation par flammes** ÉCHANT flame atomic absorption spectroscopy (*flame-AAS*); ~ **de l'infrarouge proche** DÉCHETS near infrared spectroscopy; ~ **laser** ÉCHANT laser spectroscopy; ~ **laser à effet Raman** ÉCHANT laser Raman spectroscopy; ~ **de masse** ÉCHANT mass spectroscopy; ~ **de masse par émission ionique secondaire** ÉCHANT secondary ion mass spectroscopy (*SIMS*); ~ **par fluorescence** ÉCHANT fluorescent spectroscopy; ~ **photoacoustique** ÉCHANT photoacoustic spectroscopy (*PAS*); ~ **photoélectronique** ÉCHANT photoelectron spectroscopy (*PES*); ~ **Raman** ÉCHANT Raman spectroscopy; ~ **de réflexion** ÉCHANT reflection spectroscopy; ~ **ultraviolette** ÉCHANT ultraviolet spectroscopy; ~ **ultraviolette et visible** ÉCHANT ultraviolet and visible spectroscopy

spectroscopique *adj* ÉCHANT spectroscopic

spiral *adj* POLL ATM helical

spirale: **en** ~ *loc* POLL ATM spiraling (*AmE*), spiralling (*BrE*)

spiromètre *m* ÉCHANT rotameter

spray *m* POLL ATM spray

sprinkler *m* DIST EAU *incendie*, POLL EAU *traitement des eaux usées* sprinkler

stabilisant *m* GÉNÉRAL stabilizer

stabilisation *f* GÉNÉRAL stabilization; ~ **biologique** SOLS biological stabilization; ~ **des boues** POLL EAU sludge stabilization; ~ **enzymatique** POLL EAU enzymatic stabilization; ~ **du pH** POLL EAU pH-stabilization; ~ **physique**

DÉCHETS, DIST EAU physical stabilization; ~ **thermophile aérobie en deux étapes** POLL EAU two-stage aerobic thermophilic stabilization

stabilisé *adj* GÉNÉRAL stabilized

stabiliser *vt* GÉNÉRAL stabilize

stabilité *f* GÉNÉRAL stability; ~ **de forme** POLL ATM form stability; ~ **du procédé** POLL EAU process stability; ~ **de température** ÉCHANT temperature stability; ~ **thermique** DÉCHETS, DIST EAU heat stability; ~ **au vieillissement** POLL ATM ageing stability

stable *adj* GÉNÉRAL stable; ~ **en température** ÉCHANT temperature-stable

stade: ~ **d'absorption** *m* POLL ATM absorption stage; ~ **d'adsorption** *m* POLL ATM adsorption stage; ~ **de catalyse** *m* POLL ATM catalyst step; ~ **de décomposition** *m* DÉCHETS stage of decomposition; ~ **expérimental** *m* ÉN RENOUV test phase; ~ **de filtration** *m* POLL ATM filter stage; ~ **de lavage** *m* POLL ATM scrubbing stage; ~ **de mise en décharge** *m* DÉCHETS landfill phase; ~ **de projet** *m* ÉN RENOUV project phase

stagnation *f* POLL EAU stagnation

staphylocoques *m pl* POLL EAU stapylococci

station: ~ **de collecte** *f* DÉCHETS collection station; ~ **d'épuration** *f* DÉCHETS clarification plant, treatment plant, DIST EAU, POLL EAU treatment plant, treatment unit, purification plant, sewage plant, wastewater purification plant; ~ **d'épuration de l'eau d'infiltration** *f* POLL EAU percolating water treatment plant; ~ **d'épuration des eaux usées** *f* POLL EAU wastewater treatment plant; ~ **d'épuration des eaux usées urbaines** *f* POLL EAU municipal wastewater works; ~ **d'infiltration souterraine** *f* DIST EAU underground infiltration facility; ~ **de mesure** *f* ÉCHANT measuring station; ~ **de mesure de la pollution de fond** *f* POL ENV background station; ~ **de pompage** *f* DIST EAU, POLL EAU pump station, pumping plant, pumping station, POLL MAR pumping unit; ~ **de pompage amont** *f* POLL EAU influent pump works; ~ **de pompage mobile** *f* POLL EAU mobile pumping unit; ~ **de traitement de déchets** *f* DÉCHETS waste treatment plant; ~ **de traitement des eaux usées** *f* ÉCHANT, POLL EAU sewage treatment plant, wastewater monitoring station; ~ **de transit** *f* DÉCHETS transfer station

stationnaire *adj* DÉCHETS, ÉCHANT, POLL ATM stationary

statique *adj* GÉNÉRAL static

statistique *adj* ÉCHANT statistical

statistiques: ~ **sur l'environnement** *f pl* POL ENV environmental statistics

statutaire *adj* POL ENV, SÉCURITÉ statutory

statut: ~ **d'une société** *m* GESTION bye-law

stérile *adj* ÉCHANT, POLL EAU germ-free, sterile

stériles *m pl* DÉCHETS tailings, waste washings

stérilisation *f* ÉCHANT POLL EAU sterilization

stockable *adj* ÉN RENOUV, POLL EAU, POLL MAR, SÉCURITÉ storable

stockage *m* ÉN RENOUV, POLL EAU, POLL MAR, SÉCURITÉ *de matières dangereuses* storage; ~ **de chaleur** ÉN RENOUV heat storage; ~ **de chaleur saisonnier** ÉN RENOUV seasonal heat storage; ~ **à court terme** ÉN RENOUV short-term storage; ~ **de courte durée** ÉN RENOUV short-term storage; ~ **de déchets** DÉCHETS waste storage; ~ **définitif** DÉCHETS final storage, *déchets nucléaires* ultimate storage; ~ **des données** ÉCHANT data storage; ~ **des données brutes** ÉCHANT raw-data storage; ~ **d'échantillon** ÉCHANT sample storage; ~ **d'énergie** ÉN RENOUV energy storage; ~ **intermédiaire** DÉCHETS intermediate storage; ~ **à long terme** DÉCHETS, ÉN RENOUV long-term storage; ~ **permanent** DIST EAU permanent storage; ~ **permanent des déchets** DÉCHETS permanent waste storage; ~ **des produits finis** DÉCHETS finished goods store; ~ **en silos** DIST EAU silo storage; ~ **souterrain** DIST EAU underground storage; ~ **temporaire** DÉCHETS temporary storage

stocké *adj* DÉCHETS, ÉCHANT, ÉN RENOUV, SOLS stored

stocker *vt* DÉCHETS, ÉCHANT, ÉN RENOUV *rayonnement solaire*, SOLS store; ~ **l'énergie** ÉN RENOUV store energy; ~ **provisoirement** DÉCHETS, SOLS store intermediately; ~ **temporairement** DÉCHETS, SOLS store intermediately

stœchiométrie *f* ÉCHANT, POLL ATM, POLL EAU stoichiometry

stœchiométrique *adj* ÉCHANT, POLL ATM, POLL EAU stoichiometric

strate: ~ **de déchets** *f* DÉCHETS waste layer

stratégie: ~ **de sécurité** *f* SÉCURITÉ safety strategy; ~ **de sûreté** *f* SÉCURITÉ safety strategy

stratification *f* DÉCHETS laminating; ~ **neutre des températures** POLL ATM neutral temperature layering; ~ **stable des températures** POLL ATM stable temperature layering

stratosphère *f* POLL ATM stratosphere

stratosphérique *adj* POLL ATM stratospheric

streptocoques *m pl* POLL EAU streptococci

stripage *m* POLL EAU stripping; ~ **à la vapeur** POLL EAU *élimination de l'ammoniac* steam stripping

stroboscope *m* ÉCHANT stroboscope

structure *f* GÉNÉRAL structure; ~ **amovible intégrée** POLL ATM moveable built-in structure; ~ **des coûts** GESTION cost structure; ~ **en étoile** POL ENV star-shaped procedure; ~ **intégrée** POLL ATM built-in structure; ~ **rotative intégrée** POLL ATM rotating built-in structure; ~ **stratifiée** ÉN RENOUV sandwich structure

subdivision *f* POL ENV subdivision

sublétal *adj* SÉCURITÉ sub-lethal

submersion *f* DIST EAU, POLL EAU inundation

subordonné: ~ **à l'octroi d'une licence** *adj* POL ENV requiring a licence

subsidence *f* SOLS subsidence

subsidiarité *f* POL ENV subsidiarity

substance *f* GÉNÉRAL substance; ~ **adsorbée** POLL ATM adsorbed substance; ~ **biodégradable** DÉCHETS, DIST EAU, POLL EAU biodegradable substance; ~ **carcinogène** SÉCURITÉ carcinogenic substance; ~ **corrosive** SÉCURITÉ corrosive substance; ~ **dangereuse** SÉCURITÉ hazardous substance, harmful substance, dangerous substance; ~ **de déchets liquides** POLL ATM liquid waste substance; ~ **dissoute** POLL EAU dissolved substance; ~ **émise** POLL ATM emitted substance; ~ **étalonnée** ÉCHANT calibrated substance; ~ **miscible** DÉCHETS, DIST EAU, POLL EAU miscible substance; ~ **nauséabonde** POLL ATM, POLL EAU odor-intensive substance (*AmE*), odour-intensive substance (*BrE*), malodorous substance; ~ **nocive** SÉCURITÉ harmful substance; ~ **non-dissoute** DIST EAU, POLL EAU undissolved substance; ~ **à odeur désagréable** POLL ATM malodorous substance; ~ **à odeur gênante** POLL ATM malodorous substance; ~ **odoriférante** POLL ATM, POLL EAU odor-forming substance (*AmE*), odour-forming substance (*BrE*), odoriferous substance; ~ **organique nocive** POLL EAU organic harmful substance; ~ **oxydable** POLL ATM, SOLS oxidizable substance; ~ **oxydante** POLL ATM, SÉCURITÉ oxidizing substance; ~ **perdue** DÉCHETS waste substance; ~ **radioactive** DIST EAU, POLL ATM, SÉCURITÉ radioactive substance; ~ **réductrice** POLL ATM reducing substance; ~ **réfractaire** POLL EAU refractory substance; ~ **rejetée** DÉCHETS waste substance; ~ **séparable** POLL EAU dissociable substance; ~ **solide en suspension** POLL EAU floating solid; ~ **tensioactive** SOLS surface-active substance; ~ **titrimétrique** ÉCHANT titrimetric substance; ~ **toxique** DÉCHETS, POLL ATM, SÉCURITÉ, SOLS toxic substance

substances: ~ **dangereuses** *f pl* DIST EAU hazardous goods

substitution: ~ **de combustible** *f* POLL ATM fuel replacement; ~ **d'énergie finale** *f* ÉN RENOUV final energy substitution;

~ **à l'énergie primaire** *f* ÉN RENOUV primary energy substitution

substrat *m* DIST EAU substrate; ~ **organique** POLL ATM organic base material

subvention *f* POL ENV subsidy; ~ **préjudiciable à l'environnement** POL ENV environmentally-harmful subsidy

subventionner *vt* POL ENV subsidize

succion *f* DIST EAU, POLL ATM, POLL EAU suction

suie *f* POLL ATM soot

suintement *m* DÉCHETS leaching, POLL EAU seeping

suinter *vi* DÉCHETS leach, POLL EAU seep

suivi *m* GÉNÉRAL monitoring; ~ **d'une décharge** DÉCHETS landfill progress; ~ **des émissions** ÉCHANT emissions monitoring; ~ **de fonctionnement** SÉCURITÉ function monitoring; ~ **à long terme** ÉCHANT, POLL MAR, POLL SON long-term monitoring; ~ **des normes de mise en vigueur** POL ENV standards implementation monitoring; ~ **de produit** DIST EAU product stewardship

sulfate *m* POLL ATM, SOLS sulfate (*AmE*), sulphate (*BrE*); ~ **d'aluminium** DIST EAU, POLL EAU aluminum sulfate (*AmE*), aluminium sulphate (*BrE*); ~ **de calcium semi-hydraté** POLL ATM calcium sulfate hemihydrated (*AmE*), calcium sulphate hemihydrated (*BrE*); ~ **de fer** DIST EAU, POLL EAU iron sulfate (*AmE*), iron sulphate (*BrE*); ~ **de magnésium** DIST EAU magnesium sulfate (*AmE*), magnesium sulphate (*BrE*); ~ **de strontium** POLL ATM strontium sulfate (*AmE*), strontium sulphate (*BrE*)

sulfure *m* POLL ATM sulfide (*AmE*), sulphide (*BrE*); ~ **d'hydrogène** DÉCHETS, DIST EAU, POLL ATM hydrogen sulfide (*AmE*), hydrogen sulphide (*BrE*)

sulfureux *adj* POLL ATM sulfurous (*AmE*), sulphurous (*BrE*)

superficie *f* GÉNÉRAL surface

support *m* POLL ATM carrier layer, carrier medium, support body; ~ **de catalyse** POLL ATM catalyst carrier; ~ **céramique** POLL ATM ceramic carrier; ~ **non-façonné** DÉCHETS unfinished support

supporter *vt* SÉCURITÉ *blessures* sustain

suppression: ~ **de bruit** *f* POLL SON noise suppression

supprimer *vt* GÉNÉRAL remove, POLL MAR, POLL SON suppress

sûr *adj* SÉCURITÉ safe; **peu** ~ SÉCURITÉ unsafe

surcharge *f* POLL ATM overload; ~ **pondérale** DÉCHETS excess weight

surchauffe *f* POLL ATM overheating

surchauffeur *m* DIST EAU, ÉN RENOUV, POLL ATM, POLL EAU superheater

surchloration *f* POLL EAU superchlorination

surdité: ~ **induite par le bruit** *f* POLL SON, SÉCURITÉ noise-induced deafness

surdosage *m* POLL ATM, SÉCURITÉ overdosage, overdosing

surdose *f* SÉCURITÉ overdose (*OD*)

suréchantillonnage *m* ÉCHANT oversampling

suremballage *m* DÉCHETS over-packaging, overwrapping

suremballer *vt* DÉCHETS over-package, overwrap

suremballeuse *f* DÉCHETS exterior packaging machine

sûreté *f* SÉCURITÉ safety, security; ~ **des données** ÉCHANT data security; ~ **d'évacuation** DÉCHETS disposal safety; ~ **d'exploitation** SÉCURITÉ operations safety, safe operation dependability; ~ **de fonctionnement** SÉCURITÉ function safety; ~ **de l'installation** SÉCURITÉ plant safety; ~ **de localisation d'une décharge** DÉCHETS landfill location security; ~ **à long terme** DÉCHETS long-term security; ~ **nucléaire** SÉCURITÉ nuclear safety; ~ **technique** SÉCURITÉ technical safety; ~ **de transport** SÉCURITÉ transport safety; ~ **d'utilisation** SÉCURITÉ use safety

surexploitation *f* SOLS overexploitation; ~ **des fonds de pêche** POLL MAR overfishing

surface[1]: **de** ~ *loc* DÉCHETS, DIST EAU above ground

surface[2] *f* GÉNÉRAL surface, surface area; ~ **active de catalyseur** POLL ATM active catalyst surface; ~ **adsorbante** POLL ATM adsorbent surface; ~ **catalytique** POLL ATM catalyst surface; ~ **de cloison** POLL SON partition surface; ~ **construite** SOLS built-up surface; ~ **de contact** POLL ATM contact surface; ~ **de décharge** DÉCHETS landfill surface; ~ **de l'eau** ÉN RENOUV water surface; ~ **équipotentielle terrestre** DIST EAU equipotential surface; ~ **externe** DÉCHETS exterior surface; ~ **fibreuse** POLL ATM fiber surface (*AmE*), fibre surface (*BrE*); ~ **filtrante** POLL EAU filter area; ~ **de filtration** POLL ATM filter surface; ~ **liquide** POLL ATM liquid surface; ~ **de lit bactérien** POLL EAU trickling film surface; ~ **luisante** POLL MAR oil slick; ~ **de matériau filtrant** POLL ATM filter medium surface; ~ **de membrane** ÉCHANT membrane surface; ~ **de mesure** ÉCHANT measuring surface; ~ **non revêtue** DIST EAU, POLL ATM uncoated surface; ~ **de particule** POLL ATM particle surface; ~ **porteuse** POLL ATM carrier surface; ~ **de réaction** POLL ATM reaction surface; ~ **revêtue** POLL ATM coated surface; ~ **de séparation** POLL ATM *chimie* boundary layer; ~ **spécifique** POLL EAU specific surface area; ~ **de stockage** DÉCHETS storage area; ~ **supérieure** DIST EAU upper surface; ~ **terrestre** ÉN RENOUV, POLL ATM earth's surface; ~ **utile** DÉCHETS useful area

surfactant *m* GÉNÉRAL surface-active agent, surfactant

surfertilisation *f* POLL EAU, SOLS overfertilization

surnageant *m* POLL EAU supernatant liquor

surpâturage *m* SOLS overgrazing

surpêche *f* POLL MAR overfishing

surpeuplement *m* POL ENV overpopulation

surpopulation *f* POL ENV overpopulation

surpression *f* POLL EAU, POLL SON excess pressure

sursaturation: ~ **en oxygène** *f* POLL EAU oxygen supersaturation

sursaturé *adj* DIST EAU, POLL ATM supersaturated

surveillance *f* GÉNÉRAL supervision, surveillance, monitoring; ~ **après fermeture** SOLS post-closure monitoring; ~ **automatique** POL ENV self-monitoring; ~ **du bruit** POLL MAR, POLL SON noise monitoring; ~ **en continu des émissions** POLL ATM emissions-related monitoring; ~ **à court terme** ÉCHANT, POLL ATM, POLL MAR, POLL SON short-term monitoring; ~ **des émissions** ÉCHANT emissions monitoring; ~ **de l'environnement** ÉCHANT environmental monitoring; ~ **surveillance** SÉCURITÉ close supervision; ~ **de fonctionnement** SÉCURITÉ function monitoring; ~ **à long terme** ÉCHANT, POLL ATM, POLL MAR, POLL SON long-term monitoring; ~ **officielle** POL ENV official supervision; ~ **des produits** POL ENV product stewardship; ~ **de la qualité de l'eau** DIST EAU, POLL EAU water quality monitoring; ~ **du site après fermeture** DÉCHETS monitoring after site closure; ~ **des sols** GESTION soil monitoring

surveiller *vt* GÉNÉRAL monitor

suspendu *adj* GÉNÉRAL suspended

suspension[1]: **en** ~ *loc* GÉNÉRAL suspended

suspension[2] *f* DÉCHETS suspension; ~ **d'additifs** POLL ATM additive suspension; ~ **d'appui** SOLS supporting suspension; ~ **d'autorisation** POL ENV reservation of permission; ~ **de calcaire** POLL ATM limestone suspension; ~ **de charbon actif** ÉCHANT activated-carbon suspension; ~ **chargée** POLL ATM loaded suspension; ~ **de chaux éteinte** POLL ATM slaked lime suspension; ~ **colloïdale** DIST EAU colloidal suspension; ~ **épaisse** DIST EAU slurry; ~ **d'une licence** POL ENV repeal of licence (*BrE*), repeal of license (*AmE*); ~ **de permis** POL ENV reservation of permission

sylviculture: ~ **viable** *f* POL ENV sustainable forestry

symbiose *f* DIST EAU symbiosis

symbole: ~ **de danger** *m* SÉCURITÉ danger symbol

synchrone *adj* ÉCHANT synchronous

synergie *f* POLL ATM synergy

synthèse: **de** ~ *loc* POLL EAU synthetic

synthétique *adj* POLL EAU synthetic

système *m* ÉCHANT, GESTION, POL ENV system; ~ **d'aération** DIST EAU aeration system; ~ **d'aération de fûts** DÉCHETS barrel-venting system; ~ **d'agglomération de pâte** DIST EAU pulp-molding system (*AmE*), pulp-moulding system (*BrE*); ~ **d'alerte** SÉCURITÉ warning system; ~ **anti-fumées** POLL ATM smoke-control system; ~ **d'assemblage** DÉCHETS *emballage* collating system; ~ **d'assurance-qualité** GESTION quality-assurance system; ~ **de barrage** DÉCHETS barrier system; ~ **de broyage** DÉCHETS crushing system; ~ **de certification par tierce partie** GESTION third-party certification system; ~ **de chaudière à chaleur de récupération** , DÉCHETS DIST EAU, POLL ATM, POLL EAU waste heat boiler system; ~ **de chauffage** ÉN RENOUV heating system; ~ **de chauffage urbain alimenté par l'énergie solaire** ÉN RENOUV solar-supported district heating system; ~ **en circuit fermé** DÉCHETS closed-loop system; ~ **de classification des risques** SÉCURITÉ Hazard-Ranking System (*HRS*); ~ **climatique** POLL ATM climate system; ~ **de collecte** DÉCHETS collection system; ~ **de combustion** POLL ATM combustion system; ~ **de combustion contrôlée** POLL ATM controlled combustion system (*CCS*); ~ **de commande** ÉN RENOUV control system; ~ **de consigne** POL ENV deposit fee system; ~ **de conteneur** DIST EAU container system; ~ **de contournement** ÉCHANT, POLL ATM bypass system; ~ **de contrôle** ÉCHANT monitoring system; ~ **de conversion de l'énergie éolienne** ÉN RENOUV wind energy conversion system (*WECS*); ~ **couplant éolienne et groupe électrogène** ÉN RENOUV wind-diesel system; ~ **détachable** DÉCHETS peelable system; ~ **de détection des déversements d'hydrocarbures** POLL MAR oil-spill identification system; ~ **à dispersion de colloïdes** POLL ATM colloid-disperse system; ~ **de données** ÉCHANT *chromatographiques* data system; ~ **dual** DÉCHETS, GESTION dual system; ~ **d'échangeur de chaleur intégré** POLL ATM integrated heat exchanger system; ~ **d'échantillonnage** ÉCHANT sampling system; ~ **d'éco-bilan en parties doubles** GESTION double eco-balance system; ~ **écologique naturel** POLL MAR, SOLS natural ecological system; ~ **d'écotaxes** POL ENV eco-tax system; ~ **d'emballage par thermoformage** DÉCHETS thermoforming packaging system; ~ **d'encollage à froid** DÉCHETS cold gluing system; ~ **d'épuration des eaux d'égout** DIST EAU sewage scheme; ~ **d'épuration utilisant des produits végétaux** DIST EAU purification system using plants; ~ **d'étiquetage des conteneurs en plastique recyclable** DÉCHETS, DIST EAU labeling system for recyclable plastic containers (*AmE*), labelling system for recyclable plastic containers (*BrE*); ~ **d'excavation** DÉCHETS excavation system, extraction system; ~ **d'extraction** DÉCHETS, DIST EAU *des déchets* excavation system, extraction system; ~ **de filtre classique** POLL ATM conventional filter system; ~ **à grappin** DIST EAU grab system; ~ **d'imperméabilisation combinée de la base** SOLS combined base-sealing system; ~ **d'information sur l'environnement** GESTION, POL ENV environmental information system; ~ **d'injection** ÉCHANT injection system; ~ **d'insonorisation** POLL MAR, POLL SON noise-protection system; ~ **de lavage** POLL ATM washing system; ~ **de lutte contre les émissions** POLL ATM, SOLS emissions control system; ~ **de lutte contre les émissions à l'échappement** POLL ATM exhaust emissions control system; ~ **de management environnemental** (*SME*) GESTION environmental management system (*EMS*); ~ **de manutention par bande transporteuse** ÉN RENOUV conveyor handling system; ~ **à matériau unique** DIST EAU single-material system; ~ **de moulage de pâte** DIST EAU pulp-molding system (*AmE*), pulp-moulding system (*BrE*); ~ **ouvert** DÉCHETS open system; ~ **partiellement intégré** DIST EAU partially-integrated system; ~ **pelable** DÉCHETS peelable system; ~ **photovoltaïque** ÉN RENOUV photovoltaic system; ~ **de protection contre le bruit** POLL MAR, POLL SON noise-protection system; ~ **de protection contre le bruit fortement absorbant** DIST EAU high-absorbing noise-protection system; ~ **de protection respiratoire** SÉCURITÉ breathing-protection system, respiratory-protection system; ~ **de recirculation** POLL EAU recirculation system; ~ **de récupération** DIST EAU recovery system; ~ **à récupération de chaleur** DÉCHETS, DIST EAU, POLL ATM, POLL EAU waste heat system; ~ **de refroidissement** DIST EAU, ÉN RENOUV, POLL ATM cooling system; ~ **de remplissage par injection** DÉCHETS blow fill seal system; ~ **sanitaire** DIST EAU sanitation; ~ **de sécurité** SÉCURITÉ security system; ~ **stator-rotor** POLL ATM stator-rotor system; ~ **de surveillance** ÉCHANT monitoring system; ~ **à tambours** SOLS drum system; ~ **de télécommande** DIST EAU, POLL EAU telecontrol system, remote-control system; ~ **à tours** DÉCHETS tower system; ~ **de transport de films sous vide** DÉCHETS vacuum film transport system

Système: ~ **mondial de surveillance continue de l'environnement** *m* (*GEMS*) ÉCHANT, POL ENV Global Environment Monitoring System (*GEMS*)

T

Ta *abrév* (*tantale*) DIST EAU Ta (*tantalum*)
table: **~ à secousses** *f* DÉCHETS jarring table
tache: **~ d'huile** *f* POLL MAR oil slick; **~ de mesure** *f* ÉCHANT measuring operation
tâche: **~ de mesure** *f* ÉCHANT measurement task
taille: **~ d'échantillon** *f* ÉCHANT sample size; **~ de gouttelette** *f* POLL ATM droplet size; **~ de maille** *f* POLL EAU mesh size; **~ de particule** *f* DÉCHETS, POLL ATM particle size; **~ de pore** *f* POLL ATM pore size; **~ de tamis** *f* DIST EAU sieve size
taillis *m* ÉN RENOUV coppice
talus *m* DIST EAU embankment, slope; **~ en terre** DÉCHETS earth bank
tambour *m* DÉCHETS drum, DIST EAU revolving drum, POLL ATM, POLL MAR, SOLS drum; **~ de câble** DÉCHETS *emballage* cable drum; **~ de compostage** DÉCHETS composting drum; **~ de criblage** DIST EAU drum sieve; **~ de cuisson** SOLS firing drum; **~ métallique** DÉCHETS metal drum; **~ de refroidissement** SOLS cooling drum; **~ rotatif** POLL EAU *traitement des eaux de consommation* drum screen, SOLS revolving drum; **~ de séchage** SOLS rotary dryer; **~ sécheur** POLL EAU, SOLS drum-type dryer; **~ de séparation magnétique en continu** DÉCHETS, DIST EAU permanent magnetic-separating drum
tamis *m* GÉNÉRAL screen, sieve, screening machine, fine screen, strainer; **~ de lavage** DÉCHETS, SOLS jigger; **~ moléculaire** ÉCHANT, POLL ATM, POLL EAU, SOLS molecular sieve; **~ rotatif** POLL EAU mixing tank; **~ à secousse** DÉCHETS, SOLS vibrating screen; **~ à tambour** POLL EAU drum screen; **~ vibrant** DÉCHETS vibrator screen, vibrating screen, DIST EAU, POLL ATM, POLL EAU vibrating sieve, shaking screen, SOLS vibrator screen, vibrating screen; **~ vibrant sans balancier** DÉCHETS unbalance-type vibrating screen; **~ à vibrations** DÉCHETS, SOLS vibrating screen, vibrator screen
tamisage *m* DÉCHETS, DIST EAU, ÉCHANT screening, GÉNÉRAL sieving, POLL ATM, POLL EAU, POLL MAR, SOLS screening; **~ de contrôle** ÉCHANT test sieving; **~ par effet de champ** ÉCHANT field screening; **~ par voie humide** POLL EAU wet screening; **~ secondaire** DIST EAU secondary sieving
tamiser *vt* DIST EAU, ÉCHANT, POLL ATM, POLL MAR screen, sieve, sift, POLL SON muffle, SOLS screen, sieve, sift
tampon *m* DIST EAU *d'égout* manhole cover, ÉCHANT, POLL ATM buffer
tamponner *vt* ÉCHANT, POLL ATM buffer
tantale *m* (*Ta*) DIST EAU tantalum (*Ta*)
tapis: **~ d'alimentation d'une presse** *m* DIST EAU feeder belt; **~ éjecteur** *m* SOLS ejection conveyor; **~ végétal** *m* SOLS plant cover, vegetation cover
tarage: **~ sur zéro** *m* ÉCHANT null balance
tare *f* DÉCHETS tare
taré *adj* DIST EAU calibrated, rated
tarière *f* SOLS auger
tarif: **~ marchandise** *m* DÉCHETS freight rate
tarification: **~ routière** *f* POL ENV road pricing
tarir *vi* DIST EAU run dry
tarissement *m* DIST EAU drying up
tartre *m* DIST EAU deposit, scale, boiler scale, POLL EAU, SOLS boiler scale
tas: **~ de compost trapézoïdal** *m* DIST EAU trapezoid compost heap
tasse *f* DÉCHETS cup
tassé *adj* DÉCHETS, DIST EAU, SOLS compacted
tassement *m* DÉCHETS, DIST EAU, SOLS compacting, compaction; **~ primaire** SOLS primary settlement; **~ secondaire** SOLS secondary settlement
taux: **~ d'absorption** *m* DIST EAU rate of absorption; **~ d'application de papier recyclé** *m* DIST EAU recycled paper application rate; **~ d'assainissement** *m* SOLS remediation rate; **~ de brûlage** *m* POLL ATM *métallurgie des poudres* burn-off rate; **~ de charge habituel** *m* POL ENV normal rate of charge; **~ de combustion** *m* POLL ATM burn-off rate; **~ de conversion** *m* POLL ATM conversion rate, transformation rate; **~ de dépôt** *m* SOLS deposition rate; **~ d'élimination** *m* POLL EAU removal efficiency; **~ d'émission** *m* POLL ATM emissions rate; **~ d'épaississement** *m* POLL EAU degree of thickening; **~ d'épuration** *m* SOLS clean-up rate; **~ d'évaporation** *m* DÉCHETS evaporation rate; **~ de filtration** *m* ÉCHANT, POLL ATM, POLL EAU filtering rate, filtration rate; **~ de floculation** *m* POLL EAU flocculation rate; **~ d'humidité** *m* DÉCHETS, DIST EAU, POLL EAU water content, moisture content; **~ d'hygrométrie** *m* POLL EAU *caractéristique de l'air* water content; **~ d'imposition** *m* POL ENV rate of charge, tax rate; **~ d'imposition différencié** *m* POL ENV differential tax rate; **~ de lavage** *m* POLL ATM washout rate; **~ de lixiviation** *m* DÉCHETS leachate yield; **~ de récupération** *m* DIST EAU recovery rate; **~ de recyclage** *m* DIST EAU recycling rate; **~ réduit d'imposition** *m* POL ENV reduced rate of charge; **~ de reflux** *m* DIST EAU reflux ratio; **~ de reprise** *m* DIST EAU return rate; **~ de restitution** *m* DIST EAU return rate; **~ de retour** *m* DIST EAU reflux ratio; **~ de retour des matériaux de rebut** *m* DÉCHETS waste material return rate; **~ de transfert** *m* POLL EAU transfer rate; **~ de transformation** *m* POLL ATM transformation rate; **~ d'utilisation** *m* DIST EAU, POLL EAU rate of utilization; **~ d'utilisation des matériaux de rebut** *m* DÉCHETS waste material use rate; **~ de vides** *m* POLL ATM voids fraction
taxe *f* POL ENV levy, tax; **~ d'environnement** POL ENV environmental tax; **~ de gestion de l'eau** POL ENV water control charge; **~ intérieure sur les produits pétroliers** (*TIPP*) POL ENV domestic tax on oil products; **~ réglementaire** POL ENV regulatory tax; **~ sur l'abus d'engrais** POL ENV excess-manure tax; **~ sur le captage d'eaux souterraines** POL ENV groundwater extraction tax; **~ sur les déchets** POL ENV waste tax; **~ sur l'élimination des déchets** POL ENV waste disposal tax; **~ sur les émissions** POL ENV emissions tax; **~ sur l'énergie** POL ENV energy tax; **~ sur la ferraille** POL ENV scrapping fee; **~ sur les huiles minérales** POL ENV mineral-oil tax; **~ sur la pollution** POL ENV effluent charge; **~ sur la pollution de l'eau** POL ENV water pollution tax; **~ sur la protection du climat** POL ENV climate-protection tax; **~ sur le stockage des déchets ménagers et assimilés** POL ENV landfill charge; **~ à la valeur ajoutée appliquée à l'environnement** POL ENV green VAT; **~ visant les émissions des véhicules** POL ENV emissions-targeted automobile tax
TCDD *abrév* (*tétrachlorodibenzoparadioxine*) SOLS TCDD (*tetrachlorodibenzoparadioxin*)
technicien: **~ de laboratoire** *m* ÉCHANT laboratory technician
technique[1] *adj* GÉNÉRAL technical
technique[2] *f* GÉNÉRAL technique, technology; **~ classique** DÉCHETS, POLL ATM, POLL SON, SOLS conventional technol-

ogy; ~ **classique de chauffage** ÉN RENOUV conventional heating technology; ~ **de construction** POLL ATM structural engineering technology; ~ **de décontamination** SOLS decontamination technique; ~ **de dépollution** POLL MAR clean-up technique; ~ **de dépoussiérage** POLL ATM dust-removal technology; ~ **énergétique classique** ÉN RENOUV conventional energy technology; ~ **existante** DIST EAU, POLL ATM available technology; ~ **de fixation** DÉCHETS fixation technique; ~ **de génie civil** ÉN RENOUV structural engineering technology; ~ **de granulation** ÉCHANT pellet technique; ~ **de mesure** ÉCHANT measurement technique; ~ **de mesure et d'analyse** ÉCHANT analysis-measuring technique; ~ **de mesure du bruit** POLL SON noise measurement technology; ~ **de minéralisation** DÉCHETS mineralization technique; ~ **de mise en décharge** DÉCHETS landfill technique, landfilling practice; ~ **de moulage par transfert** DÉCHETS transfer mold technique (*AmE*), transfer mould technique (*BrE*); ~ **de moulage de pâte** DIST EAU pulp-molding system (*AmE*), pulp-moulding system (*BrE*); ~ **de pelletisation** ÉCHANT pellet technique; ~ **de protection contre le bruit** POLL MAR, POLL SON noise-protection technology; ~ **du radier à dents de scie** POLL EAU ridge-and-furrow technique; ~ **de récupération classique** DÉCHETS conventional recovery technology; ~ **de recyclage** ÉN RENOUV recycling technique; ~ **de recyclage de chaleur** ÉN RENOUV regenerative-heating technology; ~ **de remplissage de moule** DÉCHETS, DIST EAU mold-filling technique (*AmE*), mould-filling technique (*BrE*); ~ **sanitaire** DIST EAU sanitary engineering; ~ **de séparation** DIST EAU separation technology; ~ **de solidification** DIST EAU solidification technique; ~ **de stockage** ÉN RENOUV storage technology; ~ **de traitement des eaux usées** POLL EAU wastewater technology; ~ **du vrac** DÉCHETS *d'essence, d'huile* bulk solids handling

techniques: ~ **informatiques CG-SM** *f pl* ÉCHANT GC-MS-COMP techniques; ~ **innovantes** *f pl* ÉN RENOUV, POLL ATM innovative technologies

technologie *f* GÉNÉRAL technology; ~ **d'application** DIST EAU, POLL ATM, SÉCURITÉ application technology; ~ **en bout de chaîne** POLL ATM, POLL EAU end-of-pipe technology; ~ **classique** DÉCHETS, POLL ATM, POLL SON, SOLS conventional technology; ~ **couplant l'énergie solaire à l'hydrogène** ÉN RENOUV solar-hydrogen technology; ~ **économisant l'énergie** DÉCHETS, POLL ATM energy-saving technology; ~ **énergétique** ÉN RENOUV, POLL ATM energy technology; ~ **de l'environnement** ÉN RENOUV, GESTION, POLL ATM, POLL SON, SOLS environmental technology; ~ **existante** DIST EAU, POLL ATM available technology; ~ **en fin de canalisation** POLL ATM, POLL EAU end-of-pipe technology; ~ **de génie civil** POLL ATM structural-engineering technology; ~ **d'insonorisation** POLL MAR, POLL SON noise-protection technology; ~ **mettant en œuvre une énergie renouvelable** ÉN RENOUV renewable-energy technology; ~ **naissante** GÉNÉRAL emerging technology; ~ **ne produisant pas de déchets** DÉCHETS, DIST EAU, POLL ATM nonwaste technology (*NWT*); ~ **non polluante** DÉCHETS, DIST EAU, GESTION, POLL ATM clean technology, nonwaste technology (*NWT*); ~ **peu consommatrice d'énergie** DÉCHETS, POLL ATM energy-saving technology; ~ **peu polluante** DIST EAU, POLL ATM low-waste technology; ~ **produisant peu de déchets** DIST EAU, POLL ATM low-waste technology; ~ **propre** DÉCHETS, GESTION, POLL ATM clean technology; ~ **renouvelable** ÉN RENOUV renewable technology; ~ **de retraitement** DIST EAU reprocessing technology; ~ **solaire** ÉN RENOUV solar technology

technologies: ~ **existantes** *f pl* ÉN RENOUV, POLL ATM established technologies; ~ **innovantes** *f pl* ÉN RENOUV, POLL ATM innovative technologies; ~ **novatrices** *f pl* ÉN RENOUV, POLL ATM innovative technologies; ~ **reconnues** *f pl* ÉN RENOUV, POLL ATM established technologies

téléalimentation: ~ **en eau** *f* DIST EAU distant water supply

télédétection *f* ÉCHANT, POLL MAR remote sensing

télémesure *f* ÉCHANT remote sensing, telemetering; ~ **de dispersion** ÉCHANT dispersion telemetering; ~ **de dispersion de nuages** ÉCHANT scattering cloud telemetering; ~ **par laser** ÉCHANT remote laser measuring

télémétrie *f* ÉCHANT telemetry; ~ **par ultrasons** ÉCHANT ultrasound telemetering

télépsychromètre: ~ **d'Assmann** *m* ÉCHANT *psychromètre à ventilation* aspiration psychrometer

télescope *m* ÉCHANT telescope; ~ **récepteur** ÉCHANT receiving telescope

température[1] *f* GÉNÉRAL temperature; ~ **d'absorption** POLL ATM absorption temperature; ~ **d'adsorption** POLL ATM adsorption temperature; ~ **de l'air résiduaire** POLL ATM waste air temperature; ~ **ambiante** ÉCHANT room temperature; ~ **atmosphérique** POLL ATM atmospheric temperature; ~ **de combustion** POLL ATM combustion temperature; ~ **critique** POLL ATM critical temperature; ~ **de décomposition** DÉCHETS decomposition temperature; ~ **de l'eau de mer superficielle** ÉN RENOUV surface water temperature; ~ **des eaux résiduaires** POLL EAU wastewater temperature; ~ **des effluents gazeux** POLL ATM waste gas temperature; ~ **finale en chambre de combustion** POLL ATM combustion chamber final temperature; ~ **de formage** DÉCHETS forming temperature; ~ **de gaz** POLL ATM gas temperature; ~ **de gaz pollués** POLL ATM waste gas temperature; ~ **humide** ÉCHANT wet-bulb temperature; ~ **d'inflammation** POLL ATM ignition temperature; ~ **de point de rosée** POLL ATM dew-point temperature; ~ **des rejets gazeux** DÉCHETS waste gas temperature; ~ **de saturation** POLL ATM saturation temperature; ~ **sèche** ÉCHANT dry-bulb temperature; ~ **de thermoscellage** DÉCHETS heat-seal temperature; ~ **de transition** ÉCHANT transition point, transition temperature; ~ **du volume d'air** POLL ATM air-charge temperature

température[2]: **à ~ élevée** *loc* DÉCHETS, DIST EAU, ÉN RENOUV, SOLS high-temperature; **à ~ moyenne** *loc* ÉN RENOUV mid-temperature; ~ **et pression normales** *loc* ÉCHANT standard temperature and pressure (*STP*)

temps[1] *m* ÉN RENOUV, POLL ATM weather; ~ **d'amorçage** POLL EAU *siphon* initiation time; ~ **d'attente** ÉCHANT waiting time; ~ **de contact** DÉCHETS contact time; ~ **de décantation** DIST EAU settling time; ~ **de démarrage** POLL EAU *exploitation* initiation time; ~ **de dépôt** DIST EAU settling time; ~ **de désorption** POLL ATM desorption time; ~ **de digestion** DIST EAU, POLL EAU digestion period, digestion time; ~ **d'épuration** POLL ATM duration of cleaning; ~ **de filtration** POLL ATM, POLL EAU filtering time; ~ **d'impulsion** POLL ATM pulse stroke; ~ **létal moyen** ÉCHANT *radioprotection*, POLL ATM median lethal time; ~ **de mesure** ÉCHANT measuring time; ~ **de pose** ÉCHANT exposure time; ~ **de pourrissage** DIST EAU rotting time; ~ **de prise** DIST EAU setting time; ~ **de réaction** POLL ATM, POLL EAU reaction time; ~ **de réaction secondaire** POLL EAU secondary reaction time; ~ **de récupération énergétique** ÉN RENOUV energy pay-back time; ~ **de rétention** POLL ATM, POLL EAU retention time; ~ **de réverbération** POLL SON reverberation time; ~ **sec** POLL ATM dry weather; ~ **de sédimentation** POLL EAU settling time, sludge digestion, *chimie* detention time; ~ **de séjour** DIST EAU, POLL ATM, POLL EAU *d'un produit dans un milieu* dwell time, residence time; ~ **de séjour des boues** POLL EAU sludge residence time (*SRT*); ~ **de séjour dans l'atmosphère** POLL ATM *substance* atmospheric lifetime; ~ **de séjour des déchets** DIST EAU residence time of waste; ~ **de séjour des matières solides** DIST EAU residence time of solids

temps[2]: **en ~ réel** *loc* ÉCHANT in real time

tendance: ~ **à écumer** *f* POLL EAU frothing tendency; ~ **au**

moussage *f* POLL EAU foaming tendency; ~ **spatiale** *f* POLL ATM spatial trend
teneur: ~ **acide** *f* POLL ATM acid content; ~ **en cendres** *f* DIST EAU, POLL ATM ash content; ~ **en contaminants** *f* SOLS contaminant content; ~ **en eau** *f* DÉCHETS, DIST EAU, ÉCHANT, POLL EAU water content, moisture content; ~ **en eau condensée** *f* DÉCHETS, POLL ATM condensed water content; ~ **énergétique** *f* ÉN RENOUV energy content; ~ **en humidité** *f* ÉCHANT, POLL ATM humidity content, moisture content; ~ **hygrométrique** *f* ÉN RENOUV moisture content; ~ **en matériaux de récupération** *f* DIST EAU secondary resource content; ~ **en matières sèches** *f* POLL EAU dry-mass content; ~ **en matières solides** *f* DIST EAU, POLL EAU solids contents; ~ **maximale en émissions** *f* DIST EAU maximum emission concentration; ~ **en métaux lourds** *f* POLL ATM, SOLS heavy-metal content; ~ **en oxygène** *f* ÉCHANT, ÉN RENOUV oxygen content; ~ **en ozone** *f* POLL ATM ozone content; ~ **en polluants** *f* DIST EAU, POLL ATM, SOLS pollutant content; ~ **en poussière** *f* DIST EAU dust content; ~ **en poussière résiduelle** *f* POLL ATM residual dust content; ~ **en soufre** *f* POLL ATM sulfur content (*AmE*), sulphur content (*BrE*); ~ **totale en poussières** *f* POLL ATM total dust content
tenir *vt* GÉNÉRAL keep; ~ **à l'abri de la lumière** ÉCHANT keep in the dark; ~ **au frais** ÉCHANT keep cool
tensioactif[1] *m* DIST EAU, POLL ATM, POLL EAU, SOLS surface active agent, surfactant
tensioactif[2] *adj* POLL ATM, POLL EAU, SOLS surface-active
tensiomètre *m* ÉCHANT dynamometer, tensiometer
tenue: ~ **de la couleur** *f* DIST EAU colour fastness; ~ **à long terme** *f* DÉCHETS, DIST EAU, POLL ATM long-term behaviour (*BrE*), long-term behavior (*AmE*)
tératogène *adj* SÉCURITÉ teratogen
terminal: ~ **de transport de marchandises** *m* DÉCHETS freight terminal
terne *adj* DIST EAU cloudy
terphényle: ~ **polychloré** *m* (*TPC*) POLL ATM, SOLS polychlorinated terphenyl (*PCT*)
terrain: ~ **aquifère** *m* DIST EAU, SOLS water-bearing soil; ~ **de collecte** *m* POLL ATM collection field; ~ **marécageux** *m* SOLS marshland, wetland; ~ **réhabilité** *m* DIST EAU reclaimed area; ~ **remis en valeur** *m* DIST EAU reclaimed area; ~ **vague** *m* POL ENV waste land
terrains: ~ **bâtis** *m pl* POL ENV land-built areas; ~ **d'épandage** *m pl* POLL EAU spreading ground
terre[1]: **à** ~ *loc* ÉN RENOUV land-based
terre[2] *f* DÉCHETS ground; ~ **acide** SOLS acid earth; ~ **dégradée** SOLS degraded land; ~ **à diatomées** DIST EAU diatomaceous earth; ~ **à foulon** POLL ATM Fuller's earth; ~ **en friche** ÉN RENOUV fallow land; ~ **gorgée d'eau** SOLS waterlogged belt; ~ **inculte** SOLS waste land; ~ **labourable** SOLS arable land
Terre *f* POL ENV Mother earth
terreau *m* SOLS soil, compost
terrestre *adj* ÉN RENOUV land-based
test: ~ **de dosage immunologique** *m* ÉCHANT immunoassay test
testé: ~ **en laboratoire** *adj* ÉCHANT laboratory-tested
tête *f* POLL ATM head, top; ~ **de boudineuse** DÉCHETS extrusion head; ~ **d'écrémage** POLL MAR skimming head; ~ **d'extrusion** DÉCHETS extrusion head; ~ **de galet** DÉCHETS *granulateur à extrusion* granulating head
tétrachlorodibenzoparadioxine *f* (*TCDD*) SOLS tetrachlorodibenzoparadioxin (*TCDD*)
tétraoxyde: ~ **de soufre** *m* (SO_4) POLL ATM sulfur tetroxide (*AmE*) (SO_4), sulphur tetroxide (*BrE*) (SO_4)
texture: ~ **du sol** *f* SOLS soil texture
thermique *adj* ÉN RENOUV thermal; ~ **solaire** ÉN RENOUV solar-thermal
thermiquement: ~ **stabilisé** *adj* POLL EAU thermally-stabilized
thermistor *m* ÉCHANT thermistor
thermocline *f* POLL EAU thermocline
thermocollage *m* DÉCHETS heat sealing
thermocolleuse: ~ **sous vide** *f* DÉCHETS vacuum heat-sealer
thermocouple *m* POLL ATM thermocouple
thermodésorption: ~ **en phase gazeuse** *f* ÉCHANT GC thermodesorption
thermodurci *adj* DÉCHETS, DIST EAU thermoset
thermodurcissable *adj* DÉCHETS, DIST EAU thermosetting
thermodurcissage *m* DÉCHETS, DIST EAU thermosetting
thermogravimétrie *f* ÉCHANT thermogravimetry
thermohygrographe *m* ÉCHANT thermohygrograph
thermolyse *f* DÉCHETS thermolysis
thermomagnétique *adj* ÉCHANT thermomagnetic
thermomètre *m* ÉCHANT thermometer; ~ **à colonne de liquide** ÉCHANT liquid spring thermometer; ~ **à dilatation** ÉCHANT expansion thermometer; ~ **à dilatation de liquide dans une gaine de verre** ÉCHANT liquid-in-glass thermometer; ~ **industriel** ÉCHANT commercially-available thermometer; ~ **à infrarouge** ÉCHANT infrared thermometer; ~ **à pression de vapeur** ÉCHANT steam pressure thermometer; ~ **à résistance** ÉCHANT, POLL ATM resistance thermometer
thermophile *adj* POLL EAU thermophilic; ~**-mésophile** POLL EAU thermophilic-mesophilic
thermoplastification *f* DÉCHETS thermoplastic solidification
thermoplastique[1] *adj* DÉCHETS thermoplastic
thermoplastique[2]: ~ **expansible** *m* DÉCHETS expandable thermoplastics; ~ **renforcé à la fibre de verre courte** *m* DIST EAU short-glass-fiber-reinforced thermoplast (*AmE*), short-glass-fibre-reinforced thermoplast (*BrE*)
thermorésistance *f* POLL ATM temperature resistance
thermorésistant *adj* POLL ATM temperature-resistant
thermorétractable *adj* DÉCHETS thermoshrinkable
thermoscelleuse *f* DÉCHETS heat-sealing machine; ~ **par impulsion** DÉCHETS impulse heat-sealer; ~ **sous vide** DÉCHETS vacuum heat-sealer, vacuum sealing machine
thermosoudage *m* DÉCHETS heat sealing, heat welding
thermosphère *f* POLL ATM thermosphere
thermostabilité *f* POLL ATM, SOLS thermostability
thermostable *adj* POLL ATM, SOLS thermostable, temperature-resistant
thermostat *m* ÉCHANT thermostat
thermostatique *adj* ÉCHANT thermostatic
thiol *m* POLL ATM thiol
TIPP *abrév* (*taxe intérieure sur les produits pétroliers*) POL ENV domestic tax on oil products
tirage: ~ **forcé** *m* POLL ATM induced draft (*AmE*), induced draught (*BrE*) (*ID*)
tissu: ~ **fibreux** *m* POLL ATM fiber material (*AmE*), fibre material (*BrE*); ~ **filtrant** *m* POLL ATM, POLL EAU filter fabric, pump-driving gear; ~ **imprégné** *m* DÉCHETS impregnated fabric; ~ **industriel** *m* POL ENV industrial fabric; ~ **non-tissé** *m* POLL ATM nonwoven fabric; ~ **urbain** *m* POL ENV urban fabric; ~ **de verre** *m* DÉCHETS glass fabric
titrage *m* ÉCHANT titration; ~ **complexométrique** ÉCHANT complexometric titration; ~ **coulométrique** ÉCHANT coulometric titration; ~ **microcoulométrique** ÉCHANT microcoulometric titration; ~ **oxydimétrique** ÉCHANT oxidimetric titration; ~ **en phase gazeuse** ÉCHANT gas-phase titration (*GPT*); ~ **potentiométrique** ÉCHANT potentiometric titration; ~ **turbidimétrique** ÉCHANT turbidimetric titration
titre *m* ÉCHANT titer (*AmE*), titre (*BrE*); ~ **chimique de l'eau** DIST EAU chemical water quality; ~ **de l'eau** DIST EAU, POLL EAU water quality; ~ **de l'eau brute** DIST EAU, POLL EAU untreated water quality; ~ **de l'eau non-traitée** DIST EAU, POLL EAU untreated water quality; ~ **hydrotimétrique** DIST

EAU hardness; ~ **hydrotimétrique calcique** DIST EAU calcium hardness; ~ **hydrotimétrique magnésien** DIST EAU magnesium hardness
titrer *vt* ÉCHANT titrate
titrimètre: ~ **de phase gazeuse** *m* DIST EAU gas-phase titrator, ÉCHANT gas-phase titrator, titrimeter
titrimétrique *adj* ÉCHANT titrimetric
toile: ~ **de filtre** *f* DIST EAU, ÉCHANT, POLL ATM, POLL EAU filter cloth, filter belt; ~ **de filtre-presse** *f* DIST EAU, POLL ATM, POLL EAU filter cloth; ~ **métallique** *f* POLL EAU woven-wire cloth
toit: ~ **de verre** *m* ÉN RENOUV *maison solaire* glass roof
toiture *f* POLL ATM *caisson d'électrofiltre* roofing; ~ **en verre** ÉN RENOUV glass roofing
tôle *f* DIST EAU, POLL EAU plate; ~ **insonorisante** POLL SON acoustic sheet; ~ **perforée** DIST EAU *tamisage*, POLL EAU perforated plate
tolérable *adj* SOLS tolerable
tolérance *f* SOLS tolerance; ~ **à l'acidité** SOLS acid tolerance
toluène *m* DÉCHETS toluene
tomodensitomètre *m* ÉCHANT *radiographie* scanner
TOMS *abrév* (*spectromètre imageur de l'ozone total*) ÉCHANT, POLL ATM TOMS (*total ozone mapping spectrometer*)
tonnage *m* DÉCHETS throughput, thruput (*AmE*)
torchage *m* DÉCHETS *industrie pétrolière* flaring
torchère *f* POLL ATM waste-gas burner
tordre *vt* DÉCHETS wring
total *m* POLL ATM total; ~ **des matières solides en suspension** POLL EAU total suspended solids (*TSS*), total undissolved solids
totalement: ~ **halogéné** *adj* POLL ATM fully-halogenated
tour *f* ÉN RENOUV tower; ~ **d'absorption** POLL ATM absorption tower, *industrie pétrolière* absorber; ~ **collectrice** ÉN RENOUV *solaire* receiving tower; ~ **de contact** POLL ATM contact tower; ~ **de débenzolage** POLL ATM *pétrole* scrubbing tower; ~ **de déchiquetage** DÉCHETS shredding tower; ~ **de déshydratation** POLL ATM drying tower; ~ **de fermentation** POLL EAU fermentation tower; ~ **de lavage** POLL ATM scrubbing tower; ~ **de pourrissage** DIST EAU rotting tower; ~ **à pulvérisation** POLL ATM spray tower; ~ **de réaction catalytique** POLL ATM catalytic tower; ~ **de relavage** POLL ATM rescrubber; ~ **de séchage** POLL ATM drying tower; ~ **solaire** ÉN RENOUV *centrale de type tour* solar tower
tourbillon *m* DIST EAU eddy, POLL ATM vortex; ~ **dextrogyre** DIST EAU outer eddy, POLL ATM outer vortex; ~ **interne** DIST EAU internal eddy; ~ **lévogyre** POLL ATM inner vortex; ~ **sinistrogyre** DIST EAU internal eddy
tourbillonnement *m* POLL ATM whirling
tourisme: ~ **des déchets** *m* DÉCHETS, POL ENV waste tourism; ~ **écologique** *m* GESTION eco-tourism, soft tourism; ~ **vert** *m* GESTION eco-tourism, green tourism
tournoyer *vi* DIST EAU spin
tout: ~**-à-l'égout** *m* DÉCHETS, DIST EAU, POLL EAU main sewer, main drain
toxicité *f* GÉNÉRAL toxicity; ~ **des eaux usées** POLL EAU wastewater toxicity; ~ **liminale** SÉCURITÉ threshold of toxicity; ~ **pour les poissons** POLL EAU fish toxicity
toxicologie *f* SÉCURITÉ toxicology
toxine *f* POLL ATM, POLL MAR toxin
toxique *adj* GÉNÉRAL toxic; **très** ~ DÉCHETS, POLL EAU, SOLS highly-toxic
TPC *abrév* (*terphényle polychloré*) POLL ATM, SOLS PCT (*polychlorinated terphenyl*)
traçabilité *f* ÉCHANT traceability
tracer *vt* ÉCHANT record, *chromatogramme* plot
traces *f pl* DÉCHETS trace concentration, ÉCHANT traces, trace concentration, SOLS trace concentration
traceur *m* ÉCHANT, POLL ATM plotter, *marquage de liquide* tracer; ~ **biochimique** DIST EAU, ÉCHANT biochemical tracer; ~ **de pollution** ÉCHANT pollution tracer
tracteur: ~ **de chargement** *m* SOLS loading tractor
train: ~ **d'incinérateur** *m* DÉCHETS incineration train; ~ **de roulement** *m* ÉN RENOUV drive train
traînée *f* ÉN RENOUV drag; ~ **d'eaux usées** POLL EAU wastewater trail
traité *adj* GÉNÉRAL treated; ~ **par lots** POLL ATM batch-processed
traitement *m* GÉNÉRAL treatment, processing; ~ **de l'agent d'extraction** SOLS extracting-agent processing, extraction-agent processing; ~ **alterné par boues activées en deux étapes** POLL EAU alternated two-stage activated sludge process; ~ **à l'ammoniac** POLL ATM ammonia process; ~ **anti-corrosion** ÉN RENOUV corrosion prevention; ~ **anti-corrosion préalable** DIST EAU prerotting treatment; ~ **anti-corrosion préventif** DIST EAU prerotting treatment; ~ **anti-rouille** POLL EAU post-rotting treatment; ~ **anti-rouille curatif** DIST EAU post-rotting treatment; ~ **anti-rouille principal** DIST EAU principal rotting treatment; ~ **bactériologique** DÉCHETS bacteriological treatment; ~ **biologique** DÉCHETS, DIST EAU biological treatment; ~ **biologique des effluents gazeux** POLL ATM biological waste air treatment; ~ **biotechnologique** POLL ATM biotechnological treatment; ~ **des boues** POLL EAU, SOLS sludge treatment, sludge processing; ~ **des boues d'eaux résiduaires** POLL EAU wastewater sludge treatment; ~ **des boues fécales solides** POLL EAU faecal solids sludge treatment (*BrE*), fecal solids sludge treatment (*AmE*); ~ **au charbon actif** DÉCHETS, POLL ATM, POLL EAU activated-carbon treatment; ~ **chimique atmosphérique** POLL ATM atmospheric chemical process; ~ **des condensats** POLL EAU *échange d'ions* condensate polishing; ~ **curatif de la corrosion** DIST EAU post-rotting treatment; ~ **de déchets** DÉCHETS waste processing, waste treatment; ~ **des déchets dangereux** DÉCHETS hazardous waste treatment; ~ **différencié** POLL EAU separate treatment; ~ **des données** ÉCHANT data processing; ~ **de l'eau potable** DIST EAU drinking water treatment; ~ **de l'eau de rinçage** SOLS rinsing-water processing; ~ **des eaux** DIST EAU, POLL EAU water treatment; ~ **des eaux d'égout** DIST EAU sewage treatment; ~ **des eaux pluviales** POLL EAU storm water treatment; ~ **des eaux de surface** DIST EAU, POLL EAU surface water treatment; ~ **des eaux usées** POLL EAU wastewater treatment; ~ **d'échantillon** ÉCHANT sample treatment; ~ **d'échantillons** ÉCHANT sample processing; ~ **de fin de cycle** DÉCHETS final treatment; ~ **final** DÉCHETS final treatment; ~ **des gaz de décharge** DIST EAU processing of landfill gas; ~ **des gaz par voie humide** POLL ATM wet waste gas treatment; ~ **des gaz par voie sèche** POLL ATM dry waste-gas treatment; ~ **des gaz riches** POLL ATM rich gas processing; ~ **des lixiviats** DÉCHETS leachate treatment; ~ **de matières premières** GESTION processing of raw material; ~ **par un agent de lavage** POLL ATM washing-agent processing; ~ **par boues activées en deux étapes** POLL EAU two-stage activated sludge process; ~ **par boues activées en deux étapes avec prédénitrification** POLL EAU two-stage activated sludge process with pre-denitrification; ~ **par boues activées en une seule étape** POLL EAU single-stage activated sludge process; ~ **par charbon actif** DÉCHETS, POLL ATM, POLL EAU activated-carbon treatment; ~ **par lots** POLL ATM batch operation, batch process; ~ **à part** POLL EAU separate treatment; ~ **photochimique** POLL EAU photochemical treatment; ~ **physico-chimique** POLL ATM, POLL EAU, SOLS physical-chemical process, physio-chemical treatment; ~ **physique de l'eau** DIST EAU physical water treatment; ~ **des plastiques** DIST EAU plastics processing; ~ **postérieur** SOLS post-treatment; ~ **préalable** GÉNÉRAL pretreatment; ~ **préliminaire** GÉNÉRAL pretreatment; ~ **primaire des eaux résiduaires** POLL EAU primary

wastewater treatment; ~ **secondaire** POLL ATM secondary treatment; ~ **secondaire des eaux résiduaires** POLL EAU secondary wastewater treatment; ~ **secondaire des effluents gazeux** POLL ATM secondary waste-gas treatment; ~ **séquentiel** ÉCHANT *informatique* off-line operation; ~ **du signal en temps réel** ÉCHANT real-time signal processing; ~ **du sol** SOLS soil processing; ~ **du solvant d'extraction** SOLS extractant processing; ~ **tertiaire des eaux usées** POLL EAU tertiary wastewater treatment; ~ **thermique** DÉCHETS, POLL EAU thermal process, thermal treatment; ~ **thermique des déchets** DÉCHETS thermal waste treatment; ~ **unitaire** POLL EAU unit processes; ~ **d'urgence** SÉCURITÉ, SOLS emergency treatment; ~ **de la valeur mesurée** ÉCHANT measured-value processing; ~ **à la vapeur** DIST EAU vapor treatment (*AmE*), vapour treatment (*BrE*)

traiter *vt* DÉCHETS treat, DIST EAU treat, process, ÉN RENOUV process, POLL ATM, POLL EAU treat, process, SOLS process; ~ **par lots** DÉCHETS batch; ~ **par pompage** DIST EAU pump and treat; ~ **sur lit fluidisé** POLL ATM, SOLS fluidize

tranche: ~ **vidangeable** *f* DIST EAU *aménagement hydraulique* live storage

tranchée *f* SOLS cut; ~ **d'égout** DIST EAU sewer trench

trancheuse *f* SOLS ditcher

transbordement *m* DIST EAU transshipment

transducteur: ~ **acoustique à résonance** *m* POLL SON resonance transducer; ~ **de mesure** *m* ÉCHANT measuring transducer

transférer *vt* POL ENV shift, POLL ATM, SOLS transfer

transfert: ~ **de chaleur** *m* DIST EAU, POLL EAU heat transfer; ~ **de données** *m* ÉCHANT data transfer; ~ **de gaz** *m* POLL EAU gas transfer; ~ **de masse** *m* POLL ATM mass transfer; ~ **d'oxygène** *m* POLL EAU oxygen transfer; ~ **thermique** *m* DIST EAU, POLL EAU heat transfer; ~ **thermique par conduction** *m* POLL ATM heat transfer via conduction; ~ **thermique par convection** *m* POLL ATM heat transfer via convection

transformateur *m* ÉN RENOUV transformer

transformation *f* DÉCHETS, DIST EAU, ÉN RENOUV processing; ~ **de déchets** DÉCHETS waste processing; ~ **des ferrailles** DÉCHETS scrap processing; ~ **des plastiques** DIST EAU plastics processing

transformer *vt* DÉCHETS, DIST EAU, ÉN RENOUV process

transfrontière *adj* POLL ATM transboundary

transition *f* POL ENV transition; ~ **thermique** POLL ATM heat transition

transmettre *vt* ÉCHANT, ÉN RENOUV transmit; ~ **la puissance** ÉN RENOUV transmit power

transmission *f* ÉCHANT, ÉN RENOUV, POLL ATM, POLL SON transmission; ~ **du bruit aérien** POLL SON airborne-noise transmission; ~ **électrique** POLL ATM electric drive; ~ **hydraulique** ÉN RENOUV hydraulic transmission; ~ **indirecte** POLL SON *acoustique* flanking transmission; ~ **intégrée du bruit** POLL SON structure-borne noise transmission; ~ **par dérivation** POLL SON *acoustique* flanking transmission; ~ **pneumatique** DIST EAU pneumatic driving gear; ~ **de polluants** POLL ATM pollutant transmission; ~ **de prélèvement** ÉCHANT sampling transmission; ~ **sonore** POLL SON sound transmission

transpiration *f* POLL ATM transpiration

transport *m* DÉCHETS, DIST EAU, SOLS transport; ~ **d'air brut** POLL ATM raw-air conveyance; ~ **en camion** DÉCHETS truck transport; ~ **d'énergie** ÉN RENOUV energy transport; ~ **ferroviaire** DÉCHETS rail transport (*BrE*), railroad transport (*AmE*); ~ **maritime** DÉCHETS carriage by sea; ~ **par chemin de fer** DÉCHETS rail transport (*BrE*), railroad transport (*AmE*); ~ **par rail** DÉCHETS rail transport (*BrE*), railroad transport (*AmE*); ~ **pneumatique** DIST EAU pneumatic transport; ~ **routier** DÉCHETS road transport, truck transport; ~ **en vrac** DÉCHETS bulk transport

transportable *adj* GÉNÉRAL transportable

transporter *vt* GÉNÉRAL transport

transporteur *m* DÉCHETS, DIST EAU, ÉN RENOUV conveyor; ~ **à palettes métalliques** DIST EAU apron conveyor; ~ **à secousses non équilibré** DÉCHETS unbalance-type vibratory conveyor; ~ **à rouleaux par gravité** DÉCHETS gravity roller conveyor; ~ **à secousses** DIST EAU vibrating conveyor; ~ **vibrant** DIST EAU vibrating conveyor; ~ **à vis** DIST EAU screw conveyor; ~ **à vis sans fin** POLL EAU screw conveyor

transrouleur *m* DÉCHETS gravity roller

transvasement *m* GÉNÉRAL decantation

transvaser *vt* GÉNÉRAL decant

trappe: ~ **de fond** *f* DÉCHETS bottom flap

travaux: ~ **publics** *m pl* DIST EAU public works

traverser *vt* ÉCHANT, POLL ATM run through, pass through, penetrate, flow through

trémail *m* ÉN RENOUV, POLL MAR, POLL EAU weir

trémie *f* GÉNÉRAL hopper; ~ **d'alimentation** DÉCHETS, SOLS feed hopper, feeding hopper; ~ **d'alimentation avec volet mobile** DÉCHETS feed hopper with flap gate; ~ **de chargement** DÉCHETS, SOLS charging hopper, feed hopper, loading hopper, feeding hopper; ~ **de décharge** POLL EAU discharge hopper; ~ **de décharge des remblais** DIST EAU refuse extractor; ~ **de déchargement** DÉCHETS, DIST EAU unloading hopper; ~ **de dépoussiérage** POLL ATM dust-collecting hopper; ~ **à obturateur** SOLS lock hopper; ~ **à poussière** POLL ATM dust-collecting hopper; ~ **de réception** DIST EAU receiving bunker; ~ **de récupération** DÉCHETS collection hopper; ~**-tampon** SOLS surge hopper

tremper *vt* DIST EAU quench, SOLS soak

trépan: ~**-benne** *m* SOLS drilling grab; ~**-benne de forage au câble** *m* SOLS cable-guided drilling grab

treuil *m* POLL MAR winch, SÉCURITÉ hoist

tri *m* DÉCHETS sorting, DIST EAU, ÉCHANT, SOLS classification; ~ **automatique** DÉCHETS, DIST EAU automatic sorting; ~ **balistique** DÉCHETS ballistic sorting; ~ **de déchets** DÉCHETS sorting of refuse, waste sorting, waste segregation, DIST EAU waste sorting, waste segregation; ~ **électronique** DÉCHETS, DIST EAU electronic sorting; ~ **de ferraille** DÉCHETS scrap sorting; ~ **final** DÉCHETS final sorting; ~ **manuel** DÉCHETS hand sorting, manual sorting; ~ **par couleur** DÉCHETS separation by color (*AmE*), separation by colour (*BrE*); ~ **par densité** DÉCHETS sorting based on density; ~ **par jet d'air** DÉCHETS air-stream sorting; ~ **par voie humide** DÉCHETS wet sorting; ~ **des plastiques par couleur** DÉCHETS color sorting of plastics (*AmE*), colour sorting of plastics (*BrE*); ~ **préalable** DIST EAU, SOLS presorting; ~ **à sec** DÉCHETS, DIST EAU dry sorting; ~ **semi-humide** DIST EAU semi-wet sorting

triage *m* DÉCHETS sorting

tribunal: ~ **chargé des questions d'environnement** *m* POL ENV environmental court

tribune *f* POL ENV forum

trichloréthylène *m* POLL ATM trichloroethylene (*TRI*)

trichlorofluorométhane *m* POLL ATM trichlorofluoromethane

trié *adj* DIST EAU checked, separated, DÉCHETS sorted, ÉCHANT, SOLS classified

trier *vt* DÉCHETS sort, separate ÉCHANT, SOLS classify; ~ **par couleur** DIST EAU *verre* separate by color (*AmE*), separate by colour (*BrE*); ~ **par taille** DÉCHETS sort according to size

trieur: ~ **balistique** *m* DÉCHETS ballistic sorter; ~ **magnétique** *m* DÉCHETS, DIST EAU *séparateur* magnetic separator; ~ **optique** *m* DÉCHETS optical sorter; ~ **pneumatique** *m* DÉCHETS pneumatic sorter

trieuse *f* DÉCHETS sorting machine

triglycéride *m* POLL EAU triglyceride

trioxyde: ~ **de soufre** *m* (SO_3) POLL ATM sulfur trioxide (*AmE*) (SO_3), sulphur trioxide (*BrE*) (SO_3)

triturateur *m* DIST EAU *papier* pulper

triturer *vt* DIST EAU grind, pulverize, *papier* pulp

trompe: ~ **soufflante** *f* DIST EAU, POLL EAU water jet pump
trop ~**-plein** *m* DIST EAU, POLL ATM overflow, overflow pipe
tropopause *f* POLL ATM tropopause
troposphère *f* POLL ATM troposphere
troposphérique *adj* POLL ATM tropospheric
trou: ~ **d'homme** *m* DIST EAU manhole; ~ **d'ozone** *m* POLL ATM ozone hole; ~ **de sondage** *m* ÉCHANT *géologie* borehole; ~ **de visite** *m* DIST EAU manhole
trouble *adj* DIST EAU, POLL EAU cloudy, turbid; **très** ~ POLL EAU very cloudy
tube: ~ **d'admission cylindrique** *m* POLL ATM cylindrical inlet tube; ~ **d'adsorption** *m* ÉCHANT adsorption tube; ~ **avec coude central** *m* DÉCHETS center-folding tubing (*AmE*), centre-folding tubing (*BrE*); ~ **à buse** *m* POLL ATM nozzle tube; ~ **capillaire** *m* ÉCHANT capillary tube; ~ **en carton** *m* DÉCHETS cardboard tube; ~ **cathodique** *m* SÉCURITÉ cathode ray tube (*CRT*); ~ **de combustion** *m* ÉCHANT combustion tube; ~ **de compensation** *m* DÉCHETS compensating tube; ~ **cyclone** *m* POLL ATM cyclone tube; ~ **détecteur** *m* ÉCHANT detector tube; ~ **détecteur automatique** *m* ÉCHANT automatic detector tube; ~ **détecteur à long terme** *m* ÉCHANT long-term detector tube; ~ **à essai** *m* ÉCHANT test tube; ~ **indicateur colorimétrique** *m* ÉCHANT colorimetric indicator tube; ~ **moniteur** *m* ÉCHANT monitor tube; ~ **à ondes stationnaires** *m* POLL SON standing-wave tube; ~ **de photomultiplicateur** *m* ÉCHANT photomultiplier tube (*PMT*); ~ **de Pitot** *m* ÉCHANT pitostatic tube; ~ **de prélèvement** *m* ÉCHANT sampling tube; ~ **à réaction** *m* ÉCHANT reaction tube; ~ **souple** *m* DÉCHETS collapsible tube; ~ **de Soxhlet** *m* ÉCHANT Soxhlet extractor; ~ **Venturi** *m* ÉCHANT, POLL ATM Venturi tube; ~ **en verre** *m* ÉCHANT glass tube
tubercule *m* ÉN RENOUV tuber
tubulaire *adj* POLL ATM tubular
tubulure *f* POLL ATM nozzle pipe; ~ **d'admission** DIST EAU *pompe* inlet branch, inlet piping; ~ **de chaudière** DÉCHETS boiler tube; ~ **d'évacuation** POLL ATM tailpipe; ~ **de réacteur** POLL ATM reactor tube; ~ **de refoulement** DIST EAU delivery pipe, head pipe; ~ **de sortie** DIST EAU *pompe* outlet branch, outlet piping
tulipage *m* DÉCHETS *plastique* flaring
tunnel: ~ **d'insonorisation** *m* POLL MAR, POLL SON noise-protection tunnel; ~ **de protection contre le bruit** *m* POLL MAR, POLL SON noise-protection tunnel
turbide *adj* DIST EAU, POLL EAU turbid
turbidimètre *m* ÉCHANT, POLL ATM, POLL EAU turbidimeter
turbidimétrie *f* ÉCHANT, POLL ATM, POLL EAU turbidimetry
turbidimétrique *adj* ÉCHANT, POLL ATM, POLL EAU turbidimetric
turbidité *f* ÉCHANT, POLL ATM, POLL EAU turbidity; ~ **des eaux** POLL EAU turbidity; ~ **des effluents gazeux** ÉCHANT waste gas turbidity; ~ **normale** ÉCHANT standard turbidity
turbine *f* DÉCHETS, ÉN RENOUV, POLL ATM, POLL EAU turbine; ~ **éolienne** ÉN RENOUV wind turbine; ~ **à gaz** POLL EAU gas turbine; ~ **à gaz à cycle combiné** POLL ATM combined-cycle gas turbine (*CCGT*); ~ **hydraulique** ÉN RENOUV water turbine; ~ **Kaplan** ÉN RENOUV *hydroélectricité* Kaplan turbine; ~ **à vapeur** ÉN RENOUV steam turbine
turbo: ~**-aérateur** *m* POLL EAU turbo-aerator; ~**-alternateur** *m* ÉN RENOUV turbogenerator; ~**-alternateur d'éolienne** *m* ÉN RENOUV wind turbine generator (*WTG*)
turbocompresseur *m* POLL EAU turbocompressor
turbolaveur *m* DÉCHETS *régénération des plastiques* turbowasher
turbomachine *f* DIST EAU, POLL EAU turbomachine
turbopompe *f* DIST EAU turbopump
turbulence *f* DÉCHETS, DIST EAU, POLL ATM turbulence
turbulent *adj* POLL EAU turbulent
tuyau *m* DIST EAU *flexible* hose, *rigide* pipe, tube; ~ **d'aération** DIST EAU pipe aerator; ~ **d'amenée** DIST EAU supply pipe; ~ **en amiante ciment** DIST EAU asbestos cement pipe; ~ **d'amorçage** DIST EAU priming pipe; ~ **d'assainissement** DIST EAU soil pipe; ~ **de décharge** DIST EAU, POLL EAU discharge pipe; ~ **de descente** POLL EAU drainpipe, downpipe, downspout; ~ **distributeur** DIST EAU distributing pipe; ~ **d'eau** DIST EAU water pipe; ~ **d'eaux pluviales** POLL ATM downspout, POLL EAU downpipe; ~ **en fonte** DIST EAU cast-iron pipe; ~ **à fourche** POLL ATM breeches pipe; ~ **en grès** POLL EAU stoneware pipe, vitrified clay pipe; ~ **d'incendie** SÉCURITÉ fire hose; ~ **de lavage** DÉCHETS cleaning pipe; ~ **en plastique** ÉN RENOUV plastic pipe; ~ **de projection d'eau haute pression** SOLS high-pressure water jet pipe; ~ **en PVC** DIST EAU PVC pipe; ~ **de raccordement** SOLS connecting pipe; ~ **de refoulement** DIST EAU delivery pipe; ~ **de sortie** POLL ATM tailpipe; ~ **souple** POLL MAR flexible hose; ~ **de trop-plein** DIST EAU overflow pipe
tuyauterie: ~ **d'aspiration** *f* POLL EAU suction piping; ~ **d'évacuation** *f* DIST EAU, POLL EAU discharge piping; ~ **métallique** *f* POLL ATM metal pipe; ~ **en PVC** *f* POLL EAU PVC pipe; ~ **de refoulement** *f* DIST EAU, POLL EAU discharge piping; ~ **sous pression** *f* DIST EAU pressure pipe
tuyère: ~ **d'aspiration** *f* POLL EAU suction nozzle; ~ **d'épuration** *f* POLL ATM cleaning nozzle; ~ **de filtre** *f* POLL EAU filter nozzle; ~ **de pulvérisation** *f* POLL ATM spray nozzle; ~ **de Venturi** *f* POLL ATM Venturi nozzle
TVA: ~ **appliquée à l'environnement** *f* POL ENV green VAT; ~ **verte** *f* POL ENV green VAT
type: ~ **de bruit** *m* POLL SON noise type; ~ **de certificat** *m* POL ENV certificate concept; ~ **de contaminant** *m* DÉCHETS, DIST EAU, POLL ATM, SOLS contaminant type; ~ **de contamination** *m* ÉCHANT contamination type; ~ **de décharge** *m* DÉCHETS landfill type, SOLS site type; ~ **de déchets** *m* DÉCHETS waste type; ~ **d'éco-bilan** *m* GESTION eco-balance design; ~ **de filtre** *m* POLL ATM filter type; ~ **de gestion de déchets** *m* GESTION waste-management concept; ~ **de polluant** *m* POLL ATM, SOLS contaminant type, pollutant type; ~ **de poussière** *m* POLL ATM dust type

U

ultrafiltration *f* GÉNÉRAL ultrafiltration; **~ discontinue** POLL EAU discontinuous ultrafiltration
ultramicroanalyse *f* ÉCHANT ultratrace analysis
ultrason *m* ÉCHANT, POLL SON, SOLS ultrasound
ultrasonique *adj* ÉCHANT, SÉCURITÉ ultrasonic
ultraviolet[1] *adj* (*UV*) GÉNÉRAL ultraviolet (*UV*); **~ et visible** (*UV/VIS*) ÉCHANT ultraviolet and visible (*UV/VIS*)
ultraviolet[2] *m* (*UV*) GÉNÉRAL ultraviolet (*UV*)
unité: **~ analytique** *f* ÉCHANT analytical unit; **~ calendaire** *f* DÉCHETS calender unit; **~ de désulfuration Claus** *f* POLL ATM Claus plant; **~ d'échantillonnage** *f* ÉCHANT sampling unit; **~ d'étalonnage** *f* ÉCHANT calibration unit; **~ gros bétail** *f* ÉN RENOUV livestock unit; **~ de mesure** *f* ÉCHANT measuring unit; **~ néphélométrique** *f* ÉCHANT nephelometry turbidity unit (*NTU*); **~ de reconnaissance automatique** *f* DÉCHETS, DIST EAU automatic-recognition unit; **~ de turbidité formazine** *f* ÉCHANT formazin turbidity unit (*FTU*)
urbanisation *f* GESTION, POL ENV urbanization; **~ anarchique** POL ENV urban sprawl; **~ débridée** GESTION uncontrolled urbanization
urbanisme *m* POL ENV urban planning
urée *f* POLL EAU urea
usage *m* GÉNÉRAL use; **~ de l'environnement** POL ENV environmental use
usagé *adj* GÉNÉRAL used
usine[1]: **d'~** *loc* GÉNÉRAL plant-specific, plant-related
usine[2] *f* GÉNÉRAL, plant; **~ d'adoucissement** DIST EAU softening plant; **~ alimentée en biocombustible** DÉCHETS waste-fired plant; **~ alimentée au gaz** DÉCHETS gas-use plant; **~ brûlant de la paille** ÉN RENOUV straw-burning plant; **~ de clarification collective** POLL EAU group clarification plant; **~ de clarification à grande échelle** POLL EAU large-scale clarification plant; **~ de conditionnement des boues d'eaux résiduaires** POLL EAU wastewater sludge conditioning plant; **~ de conversion de l'énergie thermique des mers** (*usine ETM*) ÉN RENOUV ocean thermal energy conversion plant (*OTEC plant*); **~ de conversion de l'énergie thermique des mers en circuit fermé** ÉN RENOUV closed cycle OTEC plant; **~ de conversion de l'énergie thermique des mers à cycle ouvert** ÉN RENOUV open-cycle plant; **~ de dessalement** DIST EAU desalination plant; **~ de dessalement d'eau de mer** DIST EAU salt-water plant; **~ de destruction des carcasses animales** DÉCHETS offal destruction plant; **~ de désulfuration des gaz par voie humide** POLL ATM wet flue-gas desulfurization plant (*AmE*), wet flue-gas desulphurization plant (*BrE*); **~ équipée de station d'épuration** DÉCHETS clarification-system-using plant; **~ ETM** (*usine de conversion de l'énergie thermique des mers*) ÉN RENOUV OTEC plant (*ocean thermal energy conversion plant*); **~ existante** POL ENV existing plant; **~ fonctionnant au gaz** DÉCHETS gas-use plant; **~ géothermique** ÉN RENOUV geothermal heating plant; **~ hydroélectrique** ÉN RENOUV hydroelectric plant, hydropower plant; **~ d'incinération** POLL ATM incineration plant, incinerator plant; **~ d'incinération des boues d'eaux résiduaires** POLL EAU wastewater sludge incineration plant; **~ d'incinération catalytique** POLL ATM catalytic incineration plant; **~ d'incinération de déchets** DÉCHETS waste incineration plant; **~ d'incinération des déchets dangereux** DÉCHETS hazardous waste incineration plant; **~ d'incinération en lit fluidisé** POLL EAU fluidized-bed incineration plant; **~ d'incinération des ordures** DIST EAU refuse incineration plant; **~ marémotrice** ÉN RENOUV tidal barrage, tidal power plant; **~ neuve** POL ENV, POLL ATM new plant; **~ pilote** GESTION pilot plant; **~ de préparation** DIST EAU preparation plant; **~ de récupération** DIST EAU reclamation plant, resource recovery plant; **~ de récupération des solvants** DIST EAU solvent recovery plant, solvent recovery unit; **~ de recyclage** DIST EAU recycling plant; **~ de recyclage du plastique** DÉCHETS, DIST EAU plastics recycling plant; **~ de régénération des huiles usées** DÉCHETS, DIST EAU oil regeneration plant; **~ de retraitement** DIST EAU processing plant; **~ de séparation** POLL ATM, SÉCURITÉ precipitation unit; **~ source de déchets** DÉCHETS waste generating factory; **~ de stockage** DÉCHETS storage facility; **~ de traitement** GÉNÉRAL treatment plant; **~ de traitement de déchets** DÉCHETS, DIST EAU waste treatment plant; **~ de tri** DÉCHETS sorting plant; **~ de tri de déchets** DÉCHETS, DIST EAU waste sorting plant; **~ de tri des déchets** DIST EAU refuse separation plant; **~ de valorisation de déchets** DÉCHETS waste utilization plant
utilisabilité *f* GÉNÉRAL usability
utilisable *adj* GÉNÉRAL *déchets* usable
utilisation *f* GÉNÉRAL use; **~ dans l'environnement** POL ENV environmental use; **~ durable des sols** SOLS sustainable soil use; **~ des eaux pluviales** DIST EAU rainwater utilization; **~ de l'énergie des courants marins** ÉN RENOUV use of ocean currents; **~ de l'énergie éolienne** ÉN RENOUV wind power use; **~ d'engrais** SOLS fertilizer use; **~ de l'espace** POL ENV space utilization; **~ rationnelle des terres** SOLS proper use of land; **~ des ressources** POL ENV resources use; **~ du sol** SOLS soil use
UV *abrév* (*ultraviolet*) GÉNÉRAL UV (*ultraviolet*)
UV/VIS *abrév* (*ultraviolet et visible*) ÉCHANT UV/VIS (*ultraviolet and visible*)

V

V *abrév* (*vanadium*) POLL ATM V (*vanadium*)
vague *f* ÉN RENOUV, POLL MAR wave; **~ de pollution atmosphérique** POLL ATM air-pollution episode
vagueur *m* POLL ATM agitating arm
valence *f* ÉCHANT, POLL ATM, POLL EAU valence; **~ ionique** POLL EAU ionic valence
valeur: **~ acoustique de référence** *f* POLL MAR, POLL SON noise standard value; **~ auditive de référence** *f* ÉCHANT audibility reference value; **~ à blanc** *f* ÉCHANT blank value; **~ brute mesurée** *f* ÉCHANT raw-measured value; **~ d'essai** *f* ÉCHANT test value; **~ étalonnée** *f* ÉCHANT calibrated value; **~ fertilisante** *f* POLL EAU *agriculture* fertilizing value; **~ horaire** *f* ÉCHANT hourly value; **~ de l'humidité relative** *f* ÉCHANT rH-value; **~ limite** *f* POL ENV limit value, threshold limit value, (*TLV*), threshold value, POLL ATM limit value, threshold limit value (*TLV*), POLL SON threshold limit value (*TLV*), SÉCURITÉ limit value, threshold limit value (*TLV*), threshold value; **~ limite admissible à l'air libre** *f* POLL ATM threshold limit value in the free environment; **~ limite de bruit** *f* POLL SON noise-limiting value; **~ limite d'émission acoustique** *f* POLL MAR, POLL SON noise limit, noise limit value; **~ limite d'exposition à l'air libre** *f* POLL ATM threshold limit value in the free environment; **~ limite de gaz d'échappement** *f* POL ENV exhaust-gas limit value; **~ manométrique** *f* ÉCHANT manometer value; **~ mesurée** *f* ÉCHANT measured value; **~ moyenne** *f* ÉCHANT average value; **~ numérique** *f* ÉCHANT numerical value; **~ obtenue par analyse** *f* ÉCHANT analytical finding; **~ de l'ozone** *f* POLL ATM ozone value; **~ de pH** *f* DÉCHETS, DIST EAU, ÉCHANT, POLL ATM, POLL EAU, SOLS pH-value; **~ de référence** *f* ÉCHANT reference spectrum; **~ de référence de pression acoustique** *f* ÉCHANT sound-pressure reference value; **~ des retombées** *f* DIST EAU, SOLS deposition value; **~ seuil** *f* POLL ATM, POL ENV, POLL SON threshold limit value (*TLV*), SÉCURITÉ threshold limit value (*TLV*), threshold value; **~ seuil à l'air libre** *f* POLL SON threshold limit value in the free environment; **~ seuil dans l'environnement** *f* POL ENV threshold limit value in the free environment; **~ seuil en milieu professionnel** *f* DIST EAU occupational threshold limit value (*occupational TLV*); **~ seuil sur le lieu de travail** *f* SÉCURITÉ threshold limit value at the work place
valeurs: **~ limites intégrées** *f pl* DIST EAU, POL ENV integrated limit values
validation: **~ externe** *f* GESTION external validation
valorisation *f* GÉNÉRAL reuse, valorization, utilization, recovery, usability, upgrading; **~ de la bioénergie** ÉN RENOUV use of bioenergy; **~ des boues** POLL EAU sludge recovery; **~ des boues humides** POLL EAU wet sludge utilization; **~ des boues traitées comme combustible** POLL EAU use of treated sludge as fuel; **~ de chaleur de récupération** DÉCHETS, DIST EAU, POLL ATM, POLL EAU waste heat use waste heat utilization; **~ de chaleur résiduelle** DÉCHETS, DIST EAU, POLL ATM, POLL EAU waste heat use, waste heat utilization; **~ de déchets** DÉCHETS waste utilization; **~ des eaux usées** POLL EAU wastewater utilization; **~ énergétique des déchets** DÉCHETS energy production from waste; **~ de l'énergie éolienne** ÉN RENOUV use of wind energy; **~ de l'énergie géothermique** ÉN RENOUV use of geothermal energy; **~ de l'énergie marémotrice** ÉN RENOUV use of tidal energy; **~ de l'énergie des vagues** ÉN RENOUV use of wave energy, wave energy use; **~ des gaz de décharge** DÉCHETS landfill gas use; **~ des gaz de digestion** DIST EAU, POLL EAU digester gas utilization; **~ des rejets thermiques** DÉCHETS, DIST EAU, POLL ATM, POLL EAU waste heat utilization, waste heat use, heat use; **~ thermique** POLL EAU *des déchets* thermal use
valoriser *vt* GÉNÉRAL use, utilize
valve: **~ d'aérosol** *f* DÉCHETS aerosol valve
vanadium *m* (*V*) POLL ATM vanadium (*V*)
vannage *m* ÉN RENOUV sluiceway
vanne *f* DIST EAU, POLL EAU slide gate, valve, penstock, sluice; **~ anti-reflux** DIST EAU check valve; **~ d'arrêt** DIST EAU check valve, POLL ATM shutoff valve; **~ d'arrêt à contre-pression** POLL ATM bypass stop valve; **~ d'arrêt et de contournement** POLL ATM bypass stop valve; **~ de décharge** POLL EAU discharge sluice; **~ d'extraction des boues** DIST EAU sludge draw-off valve; **~ à glissières** DIST EAU penstock; **~ à membrane** POLL ATM, POLL EAU diaphragm valve; **~ murale** DIST EAU penstock; **~ régulatrice de débit** DIST EAU flow-regulating valve; **~-rideau** DIST EAU penstock; **~ de secours** POLL EAU emergency gate
vapeur *f* ÉN RENOUV, POLL ATM, SOLS steam, vapor (*AmE*), vapour (*BrE*); **~ à basse pression** POLL ATM low-pressure steam; **~ contaminante** POLL ATM contaminant vapor (*AmE*), contaminant vapour (*BrE*); **~ d'eau** ÉN RENOUV, POLL ATM steam, water vapor (*AmE*), water vapour (*BrE*), SOLS steam; **~ d'échappement** POLL EAU exhaust vapor (*AmE*), exhaust vapour (*BrE*); **~ extraite** SOLS extracted vapor (*AmE*), extracted vapour (*BrE*); **~ industrielle** ÉN RENOUV process steam; **~ inflammable** SÉCURITÉ flammable vapor (*AmE*), flammable vapour (*BrE*); **~ polluante** POLL ATM contaminant vapor (*AmE*), contaminant vapour (*BrE*); **~ résiduelle** DIST EAU waste steam; **~ du sol** SOLS soil vapor (*AmE*), soil vapour (*BrE*); **~ surchauffée** DIST EAU, ÉN RENOUV, POLL ATM superheated steam
vapeurs *f pl* POLL EAU fumes
vaporisateur *m* DIST EAU, ÉN RENOUV, POLL ATM, SOLS vaporizer, atomizer; **~ à buse** POLL ATM nozzle atomizer
vaporisation *f* DIST EAU, ÉCHANT, ÉN RENOUV, POLL ATM evaporation, vaporization, atomization, POLL EAU, POLL MAR spraying; **~ d'eau** POLL ATM water spraying; **~ par une buse** POLL ATM nozzle atomization; **~ par injection** ÉCHANT injection spraying; **~ de solvants** POLL ATM solvent vaporization
vaporiser *vt* DIST EAU, ÉCHANT, ÉN RENOUV, POLL ATM atomize, spray, evaporate, vaporize, POLL EAU, POLL MAR spray
varech *m* ÉN RENOUV, POLL MAR seaweed
variabilité: **~ spatiale** *f* POLL ATM spatial variability
variable[1] *adj* POLL ATM variable
variable[2]: **~ mesurable** *f* ÉCHANT measurable variable
variation: **~ temporelle** *f* POLL ATM temporal variation
variations: **~ climatiques** *f pl* POLL ATM climatic variations
variété: **~ résistante aux pesticides** *f* SOLS pest-resistant variety
vase[1]: **~ d'absorption** *m* ÉCHANT absorption vessel; **~ à clarifier** *m* POLL EAU sludge dewatering tank; **~ de décantation** *m* ÉCHANT sedimentation glass; **~ Dewar** *m* ÉCHANT Dewar vessel; **~ d'Erlenmeyer** *m* ÉCHANT Erlenmeyer flask; **~ d'expansion** *m* POLL ATM expander; **~ gradué** *m* ÉCHANT measuring vessel; **~ gradué cylindrique** *m* ÉCHANT cylindrical graduated measuring glass

vase[2] *f* DIST EAU, POLL EAU mud, silt, sludge; ~ **de fond** POLL EAU bottom deposit
vasque *f* DIST EAU, ÉCHANT, POLL EAU basin
vecteur *m* ÉN RENOUV, ÉCHANT, POLL ATM working medium, carrier, carrier medium; ~ **de charge** POLL ATM charge carrier; ~ **en pahse gazeuse** ÉN RENOUV gaseous working medium; ~ **métallique** POLL ATM metal carrier; ~ **du milieu** POLL ATM environmental medium
véhicule: ~ **à alimentation photovoltaïque** *m* ÉN RENOUV photovoltaic-powered vehicle; ~ **de compactage** *m* DÉCHETS compactor vehicle; ~ **de compression** *m* DÉCHETS, DIST EAU compression vehicle; ~ **électrique** *m* POLL ATM battery-operated vehicle, electric car; ~ **en fin de vie** *m* DÉCHETS end-of-life vehicle; ~ **de lavage des rues** *m* DÉCHETS street-cleaning vehicle; ~ **de mesure** *m* ÉCHANT measurement vehicle; ~ **de nettoyage des chaussées** *m* DÉCHETS street-cleaning vehicle; ~ **peu polluant** *m* POLL ATM low-emissions vehicle; ~ **de ramassage** *m* DÉCHETS, DIST EAU collection vehicle; ~ **tout terrain pour décharge** *m* DÉCHETS landfill-suitable vehicle
vélocimètre *m* ÉCHANT speedometer
vélocité: ~ **de l'air** *f* POLL ATM air velocity
vent[1]: **sous le** ~ *loc* POLL ATM downwind
vent[2] *m* ÉN RENOUV, POLL ATM wind; ~ **arrière** POLL ATM down wind; ~ **descendant** POLL ATM down wind; ~ **en poupe** POLL ATM down wind; ~ **de sable** POLL ATM airborne sand
vente: **en** ~ **dans le commerce** *loc* ÉCHANT commercially available
venté *adj* ÉN RENOUV wind-favorable (*AmE*), wind-favourable (*BrE*), windy
ventilateur *m* DÉCHETS ventilator, DIST EAU, POLL ATM fan, jet aerator, ventilator, POLL EAU fan, jet aerator, SÉCURITÉ ventilator; ~ **d'air résiduel** POLL ATM waste air fan; ~ **aspirant** POLL ATM *chaudière* exhaust fan; ~ **auxiliaire** DIST EAU auxiliary fan; ~ **axial** POLL ATM axial fan; ~ **axial à réaction** POLL ATM reaction-type axial fan; ~ **de combustion** DÉCHETS combustion air fan; ~ **de dépoussiérage** SÉCURITÉ dust exhaust fan; ~ **d'effluents gazeux** POLL ATM waste gas blower; ~ **d'évents principal** POLL ATM main waste-air blower; ~ **à fonctionnement périodique** DÉCHETS chopper ventilator; ~ **de gaz brûlés** POLL ATM flue-gas blower; ~ **de gaz de refroidissement** POLL ATM cooling-gas fan; ~ **de gaz de régénération** POLL ATM regenerative-gas fan; ~ **de surpression** POLL ATM booster fan; ~ **à tirage forcé** DÉCHETS, POLL ATM forced draft fan (*AmE*), forced draught fan (*BrE*), induced draft fan (*AmE*) (*ID fan*), induced draught fan (*BrE*) (*ID fan*); ~ **d'usine de désulfuration des gaz** POLL ATM flue-gas desulfurization plant blower (*AmE*), flue-gas desulphurization plant blower (*BrE*)
ventilation *f* DÉCHETS, DIST EAU, POLL ATM, SÉCURITÉ aeration, ventilation; ~ **de carter** POLL ATM crankcase ventilation; ~ **forcée** POLL EAU forced aeration; ~ **par aspiration** POLL EAU forced aeration
ventiler *vt* DÉCHETS, DIST EAU, POLL ATM, SÉCURITÉ aerate, ventilate
Venturi *m* ÉCHANT Venturi tube
Venturimètre *m* POLL ATM Venturi meter
venue: ~ **d'eau** *f* DIST EAU water inflow
ver *m* POLL EAU, SOLS worm; ~ **des boues** POLL EAU sludge worm
vérifiable *adj* ÉCHANT verifiable
vérification *f* GÉNÉRAL checking; ~ **mécanique de la qualité de la production** DÉCHETS machine-version verification of production quality; ~ **du poids** DÉCHETS *à l'admission en décharge* check weighing
vérifier *vt* GÉNÉRAL check
vermine *f* DÉCHETS, DIST EAU vermin
vernis: ~ **résistant à l'acide** *m* DÉCHETS acid-proof varnish
vernissage *m* DÉCHETS, DIST EAU enameling (*AmE*), enamelling (*BrE*)
vernisser *vt* DÉCHETS *papier* glaze
verre: ~ **cassé** *m* DÉCHETS cullet; ~ **gradué** *m* ÉCHANT measuring jar; ~ **insonorisant** *m* POLL MAR, POLL SON, SÉCURITÉ sound-absorbing glass; ~ **isolant** *m* POLL SON insulating glass; ~ **perdu** *m* DÉCHETS waste glass; ~ **protecteur contre les rayons X** *m* SÉCURITÉ X-ray protective glass; ~ **recyclé** *m* DIST EAU recycled glass; ~ **textile** *m* DÉCHETS glass fabric; ~ **usagé** *m* DÉCHETS waste glass, DIST EAU used glass; ~ **à vitres** *m* DIST EAU sheet glass
vert *adj* GESTION, POL ENV green
Verts: **les** ~ *m pl* POL ENV the Greens
vésicule *f* POLL ATM liquid droplet; ~ **de lavage** POLL ATM scrubbing liquid droplet
vêtement: ~ **de protection** *m* SÉCURITÉ protective clothing, protective suit; ~ **de protection contre les agressions chimiques** *m* SÉCURITÉ chemically-resistant protective suit; ~ **recyclé** *m* DIST EAU recycled clothing
viabilité *f* POL ENV sustainability; ~ **écologique** DIST EAU ecological viability
viable *adj* POL ENV sustainable
vibrant *adj* GÉNÉRAL vibrating
vibration *f* GÉNÉRAL vibration; ~ **acoustique** POLL SON acoustic vibration, sound vibration; ~ **des bâtiments** POLL SON building vibration; ~ **de construction** POLL SON building vibration
vibrer *vi* GÉNÉRAL vibrate
vibrion *m* POLL EAU vibrion
vibromètre *m* ÉCHANT vibrometer
vice: ~ **apparent** *m* GESTION conspicuous defect; ~ **caché** *m* GESTION latent defect
vicier *vt* GESTION *contrat* invalidate
victime: ~ **de la pollution** *f* POL ENV pollutee
vidage *m* DÉCHETS emptying; ~ **intégré** DÉCHETS enclosed emptying; ~ **réglementé** DÉCHETS *de soutes* controlled emptying
vidange *f* GÉNÉRAL emptying, draining off; ~ **de fond** DIST EAU bottom outlet; ~ **périodique** DÉCHETS, DIST EAU periodic flushing; ~ **de la tuyauterie** POLL EAU flushing of pipe network
vidanger *vt* GÉNÉRAL drain, drain off
vidanges *f pl* DÉCHETS empties
vide *m* ÉCHANT, POLL EAU, POLL MAR, SOLS void, vacuum; **~-cave** DIST EAU pump sump; **~-ordures** DIST EAU refuse chute; ~ **à retardement** POLL ATM time-delayed vacuum
vider[1] *vt* GÉNÉRAL drain
vider[2]: **se** ~ *v réfl* DIST EAU, POLL EAU *eau d'une cuve* pump out, run off
vie *f* GÉNÉRAL life; ~ **en milieu marin** POLL MAR marine life
vieillissement *m* DÉCHETS ageing; ~ **artificiel** DÉCHETS, DIST EAU artificial ageing
vieux: ~ **papiers** *m pl* DÉCHETS waste paper; ~ **plomb** *m* DÉCHETS scrap lead
vif *adj* POLL EAU unslaked
vigueur: **en** ~ *loc* POL ENV *légistation* in force
virulicide *adj* DIST EAU viricidal
virus *m* POLL EAU virus
vis: ~ **d'Archimède** *f* DIST EAU Archimedean screw; ~ **filetée** *f* DÉCHETS continuous thread (*ct*); ~ **mélangeuse** *f* POLL EAU mixing worm; ~ **des rejets** *f* DIST EAU refuse worm; ~ **de relèvement** *f* DIST EAU, POLL EAU, POLL MAR screw-type pump; ~ **sans fin** *f* POLL ATM *extraction des poussières dans un électrofiltre* screw conveyor; ~ **transporteuse** *f* SOLS screw conveyer; ~ **vibrante** *f* SOLS vibrating worm
viscosimètre *m* ÉCHANT viscosimeter, viscosity meter; ~ **aval** ÉCHANT flowing-out viscosimeter; ~ **à chute de bille** ÉCHANT falling-sphere viscosimeter; ~ **à cônes tournants** ÉCHANT drag-torque viscosimeter; ~ **de Poiseuil** ÉCHANT capillary viscosimeter; ~ **rotatif** ÉCHANT rotameter viscosi-

meter; ~ **de sortie** ÉCHANT flowing-out viscosimeter; ~ **à tube capillaire** ÉCHANT capillary viscosimeter; ~ **de vibration** ÉCHANT vibration viscosimeter
viscosité *f* DIST EAU, POLL ATM, POLL EAU viscosity; ~ **cinématique** DIST EAU kinematic viscosity; ~ **dynamique** DIST EAU dynamic viscosity; ~ **turbulente** POLL ATM *météorologie et mécanique des fluides* eddy viscosity
visible *adj* ÉCHANT, POLL ATM visible
visite *f* DIST EAU inspection
visqueux *adj* DIST EAU, POLL ATM, POLL EAU viscid, viscous
vitesse: ~ **d'absorption** *f* DIST EAU rate of absorption; ~ **d'admission** *f* POLL ATM inlet velocity; ~ **d'adsorption** *f* POLL ATM adsorption velocity; ~ **de l'air** *f* POLL ATM air speed; ~ **ascensionnelle** *f* POLL ATM rate of ascent; ~ **de crête** *f* DIST EAU peak velocity; ~ **de décantation** *f* DIST EAU settling velocity, POLL ATM settling speed, POLL EAU settling velocity; ~ **de décomposition** *f* DÉCHETS, POLL EAU rate of decomposition; ~ **de dépôt** *f* DIST EAU, POLL ATM deposition velocity; ~ **de dérive** *f* POLL ATM drift velocity; ~ **d'échappement** *f* ÉCHANT speed of escape; ~ **d'écoulement** *f* ÉCHANT, POLL EAU velocity of flow, flow velocity; ~ **d'écoulement laminaire** *f* POLL ATM laminar-flow velocity; ~ **d'établissement** *f* DIST EAU settlement speed; ~ **de filtration** *f* POLL ATM filtration velocity; ~ **de fonctionnement** *f* ÉN RENOUV operating speed; ~ **de formation des gaz** *f* DÉCHETS gas build-up rate; ~ **du gaz** *f* POLL ATM gas velocity; ~ **de gaz résiduels** *f* POLL ATM waste gas velocity; ~ **d'imprégnation** *f* POLL ATM permeation velocity; ~ **de métabolisation** *f* POLL EAU metabolic rate; ~ **de migration** *f* ÉCHANT *gaz*, POLL ATM velocity of migration; ~ **de montée** *f* POLL ATM rate of ascent; ~ **nominale** *f* POLL ATM, POLL EAU face velocity; ~ **périphérique** *f* POLL ATM circumferential velocity; ~ **de perméation** *f* POLL ATM permeation velocity; ~ **de pointe** *f* DIST EAU peak velocity; ~ **de propagation de flamme** *f* POLL ATM rate of flame propagation; ~ **de réaction** *f* DIST EAU, POLL ATM, POLL EAU reaction rate, reaction velocity; ~ **de rotation** *f* DIST EAU, POLL ATM rotating velocity, rotation velocity; ~ **de sédimentation** *f* DIST EAU, POLL EAU deposition velocity, settling velocity; ~ **de séparation** *f* DIST EAU separation rate; ~ **du son** *f* POLL SON speed of sound; ~ **de soufflage** *f* POLL ATM blower stream velocity; ~ **spatiale** *f* POLL ATM space velocity; ~ **du vent** *f* ÉN RENOUV, POLL ATM wind velocity; ~ **de ventilation** *f* POLL ATM blower stream velocity; ~ **de vidange** *f* POLL EAU flushing speed
vitrage: ~ **transparent** *m* ÉN RENOUV *collecteur photovoltaïque* glazing
vitrer *vt* DÉCHETS glaze
vitrification *f* DÉCHETS *déchets nucléaires*, SOLS vitrification; ~ **des cendres** DÉCHETS, DIST EAU ash vitrification; ~ **par cyclone** SOLS cyclone vitrification; ~ **in situ** SOLS in-situ vitrification (*ISV*)
vitrifié *adj* DÉCHETS, SOLS vitrified
vitrifier *vt* DÉCHETS, SOLS vitrify
vocation: ~ **agricole** *f* SOLS agricultural potential
voie: ~ **d'exposition** *f* SÉCURITÉ *à un polluant* exposure path
voies: ~ **d'accès à une décharge** *f pl* DÉCHETS landfill roads; ~ **d'écoulement** *f pl* POLL ATM flow lanes
voile *m* POLL ATM haze, mist; ~ **de boue** POLL EAU top level of sludge; ~ **de carde** DÉCHETS *textile* deflector, fibrous web; ~ **étroit** DÉCHETS narrow curtain; ~ **de pollution** POLL ATM veil of pollution
voirie *f* POL ENV road system
volatil *adj* POLL ATM, POLL EAU, SOLS volatile
volatilisation *f* DÉCHETS, SOLS volatilization
volatiliser[1] *vt* DÉCHETS, SOLS volatilize
volatiliser[2]: **se** ~ *v réfl* DIST EAU, ÉCHANT evaporate
volet *m* DIST EAU screen; ~ **acoustique** POLL SON acoustic louver (*AmE*), acoustic louvre (*BrE*)
voligeage *m* POLL SON planking
voltampèremétrie *f* ÉCHANT *inverse* voltammetry
voltampèremétrique *adj* ÉCHANT voltammetric
volume *m* GÉNÉRAL volume; ~ **admis** POLL EAU intake volume; ~ **d'air** POLL ATM air charge; ~ **d'air effectif** POLL ATM actual air volume; ~ **analysé** ÉCHANT analysis volume; ~ **de biogaz** DIST EAU, POLL EAU digester gas quantity; ~ **de boues** POLL EAU sludge volume; ~ **des boues d'eaux résiduaires** POLL EAU wastewater sludge amount; ~ **des boues excédentaires** POLL EAU excess sludge volume; ~ **de catalyseur** POLL ATM catalyst volume; ~ **de compost** SOLS compost heap; ~ **concentré d'échantillon** ÉCHANT concentrated amount of sample; ~ **de décharge** DÉCHETS landfill volume; ~ **de déchets** DÉCHETS waste volume; ~ **de déchets dangereux** DÉCHETS volume of hazardous waste; ~ **d'eau** DIST EAU volume of water; ~ **d'eau recyclable** DIST EAU recoverable volume of water; ~ **d'eaux usées** POLL EAU wastewater volume; ~ **d'échantillon** ÉCHANT sample amount, sample volume; ~ **de fret** DÉCHETS, DIST EAU transport volume; ~ **de gaz** POLL ATM amount of gas; ~ **des gaz** ÉCHANT gas volume; ~ **de gaz produit par une décharge** DÉCHETS landfill gas quantity; ~ **ingéré** POLL EAU intake volume; ~ **molaire** ÉCHANT molar volume; ~ **d'ordures** DIST EAU refuse volume; ~ **partiel de gaz** ÉCHANT partial gas quantity; ~ **de polluants** ÉCHANT pollutant amount; ~ **de pore** DIST EAU pore volume; ~ **des pores** POLL ATM pore space, pore volume; ~ **de poussière** POLL ATM dust content; ~ **de poussière contenu dans les effluents gazeux** POLL ATM dust content in flue gas; ~ **de précipités** POLL EAU quantity of precipitate; ~ **de rejets gazeux** POLL ATM waste gas volume; ~ **sonore** POLL SON loudness; ~ **de stockage** DÉCHETS, ÉN RENOUV storage space, storage volume; ~ **total** ÉCHANT total volume; ~ **total de gaz** ÉCHANT, POLL ATM total gas quantity; ~ **total de poussière** ÉCHANT, POLL ATM total dust; ~ **transporté** DÉCHETS, DIST EAU transport volume; ~ **de vide** POLL ATM void volume
voluménomètre: ~ **de pilonnage** *m* ÉCHANT tamping volume meter
volumètre *m* ÉCHANT volume unit meter (*VU meter*)
volumétrique *adj* ÉCHANT volumetric
volumineux *adj* DÉCHETS *déchets* bulky
vortex *m* POLL ATM vortex
voyage: ~ **à vide** *m* DÉCHETS empty trip
vrac: **en** ~ *loc* DÉCHETS, DIST EAU, POLL ATM bulk, in batch quantities
vulcanisé *adj* DÉCHETS vulcanized
vulcaniser *vt* DÉCHETS vulcanize
vulnérabilité *f* POL ENV *d'un écosystème* vulnerability
vumètre *m* ÉCHANT volume unit meter (*VU meter*)

W

wagon: **~ citerne** *m* POLL MAR tanker; **~ basculant** *m* DÉCHETS dumper truck, dumper car

wall: **~ crête** *m* ÉN RENOUV peak wall

Water: **~, Energy, Land, Materials, Manpower** *loc* (*WELMM*) GESTION Water, Energy, Land, Materials, Manpower (*WELMM*)

WELMM *abrév* (*Water, Energy, Land, Materials, Manpower*) GESTION WELMM (*Water, Energy, Land, Materials, Manpower*)

X Y Z

Xe *abrév* (*xénon*) POLL ATM Xe (*xenon*)
xénobiotique *adj* SOLS xenobiotic
xénon *m* (*Xe*) POLL ATM xenon (*Xe*)
xéromorphe *adj* SOLS xeromorphic
xérophyle *adj* SOLS xerophilous, xerophil
xérotherme *adj* POLL ATM xerothermic, xerothermal

yardang *m* SOLS yardang
Yb *abrév* (*ytterbium*) SOLS Yb (*ytterbium*)
ylem *m* SOLS ylem
ylure *m* ÉCHANT ylide
ytterbium *m* (*Yb*) SOLS ytterbium (*Yb*)

zéolite *f* POLL ATM zeolite
zéolitique *adj* POLL ATM zeolitic
zéro: ~ **émission de CO_2** *m* POLL ATM *objectif de dépollution* zero CO_2 emissions
zinc *m* (*Zn*) POLL ATM, POLL EAU, SOLS zinc (*Zn*)
Zn *abrév* (*zinc*) POLL ATM, POLL EAU, SOLS Zn (*zinc*)
zonage *m* DIST EAU, SOLS zoning; ~ **des terres** SOLS land zoning
zone: ~ **absorbant le son** *f* POLL SON sound-absorbent zone; ~ **d'acidité** *f* SOLS acidic area; ~ **d'adsorption** *f* POLL ATM adsorption zone; ~ **d'aération** *f* SOLS zone of aeration; ~ **d'air pur** *f* POLL ATM area of pure air; ~ **aménagée** *f* DÉCHETS constructed area, ÉN RENOUV installation area, POL ENV developed area; ~ **aphotique** *f* POLL EAU aphotic zone; ~ **aride** *f* SOLS arid zone; ~ **d'aspersion** *f* POLL ATM spraying zone; ~ **de balancement des marées** *f* POLL MAR tidal action zone; ~ **bâtie** *f* POL ENV built-up area; ~ **de brouillard photochimique** *f* POLL ATM smog area; ~ **bruyante** *f* POLL SON noise area; ~ **de chargement** *f* POLL ATM charging zone; ~ **de chauffe** *f* POLL ATM heating zone; ~ **climatique** *f* POLL ATM climatic zone; ~ **de compactage** *f* DÉCHETS compaction zone; ~ **construite** *f* DÉCHETS built-up surface; ~ **côtière** *f* POLL MAR near-shore area; ~ **de cristallisation** *f* POLL ATM crystallization zone; ~ **dangereuse** *f* SÉCURITÉ danger zone; ~ **de décantation** *f* POLL EAU settling zone; ~ **de décharge** *f* DÉCHETS, DIST EAU disposal site, disposal zone, landfill area; ~ **de dénitrification** *f* POLL EAU denitrification zone; ~ **déshéritée** *f* POL ENV deprived area; ~ **de déshydratation** *f* POLL ATM drying zone; ~ **de désorption** *f* POLL ATM desorption zone; ~ **de drainage** *f* DIST EAU drainage area; ~ **euphotique** *f* POLL EAU euphotic zone; ~ **d'évacuation** *f* DÉCHETS, DIST EAU disposal site, disposal zone; ~ **facultative** *f* POLL EAU facultative zone; ~ **de floculation** *f* POLL EAU flocculation zone; ~ **à forte densité de population** *f* DIST EAU densely-populated area; ~ **de fracture des plaques continentales** *f* ÉN RENOUV fracture zone of continental plates; ~ **d'incinération** *f* POLL ATM, POLL EAU incineration zone; ~ **d'infiltration** *f* DIST EAU area of water filtration; ~ **d'influence** *f* DIST EAU area of influence; ~ **d'influence du bruit** *f* POLL MAR, POLL SON influence zone of noise; ~ **de l'infrarouge** *f* ÉCHANT infrared range, infrared region; ~ **d'installation** *f* ÉN RENOUV installation area; ~ **intercotidale** *f* POLL MAR intertidal zone; ~ **interdite** *f* POL ENV forbidden area; ~ **intertidale** *f* POLL MAR intertidal zone; ~ **ionisante** *f* POLL ATM ionizing zone; ~ **d'ionisation** *f* POLL ATM ionizing zone; ~ **latérale** *f* DIST EAU side area; ~ **de lavage** *f* POLL ATM scrubbing zone; ~ **à lit mobile** *f* POLL ATM moving-bed zone; ~ **littorale** *f* POLL EAU littoral zone; ~ **morte** *f* POLL MAR dead zone; ~ **de nitrification** *f* DIST EAU, POLL EAU nitrification zone; ~ **non saturée** *f* DIST EAU unsaturated zone; ~ **d'oxydation** *f* POLL ATM oxidation zone; ~ **périphérique** *f* POLL MAR fringe area; ~ **perméable** *f* POLL EAU, SOLS pervious area; ~ **de pluviosité** *f* DIST EAU rainfall area; ~ **polluée** *f* POLL ATM, POLL EAU polluted area; ~ **de précipitation** *f* DIST EAU precipitation area, rain area, rainfall area; ~ **de protection** *f* SÉCURITÉ protection zone; ~ **de protection contre le bruit** *f* POLL MAR, POLL SON noise-protection zone; ~ **protégée** *f* POL ENV trust area; ~ **de pulvérisation** *f* POLL ATM spraying zone; ~ **de réaction** *f* POLL EAU reaction zone; ~ **de récifs** *f* POLL MAR reef area; ~ **de réduction du bruit** *f* POLL SON noise-abatement zone; ~ **de refroidissement** *f* POLL ATM cooling zone; ~ **de régénération** *f* DIST EAU *sols* recovery area; ~ **remise en état** *f* SOLS reclaimed area; ~ **de repos** *f* POL ENV amenity area; ~ **de saturation** *f* SOLS zone of saturation; ~ **saturée** *f* DIST EAU saturated zone; ~ **de séchage** *f* POLL ATM drying zone; ~ **de silence** *f* POLL MAR, POLL SON noise-abatement zone, noise-protection area, noise-protection zone; ~ **sinistrée** *f* POL ENV damaged area; ~ **de sol** *f* SOLS soil area; ~ **de sol non saturée** *f* SOLS unsaturated soil zone; ~ **de sol saturé** *f* SOLS saturated soil zone; ~ **source** *f* POLL ATM source area; ~ **sous grille** *f* DIST EAU underblast zone; ~ **sous pression** *f* DIST EAU pressure zone; ~ **spectrale** *f* ÉCHANT spectral region; ~ **spectroscopique** *f* ÉCHANT spectroscopic spectral region; ~ **de stockage** *f* SOLS storage area; ~ **subtidale** *f* POLL MAR subtidal zone; ~ **de sûreté** *f* SÉCURITÉ safety zone; ~ **surpeuplée** *f* DIST EAU overcrowded area; ~ **tampon** *f* POLL SON buffer zone; ~ **de transition** *f* POLL EAU, POLL MAR transition zone; ~ **de trempe** *f* POLL ATM quench zone; ~ **utile** *f* DIST EAU useful area
zooplancton *m* POLL EAU zooplankton

ENGLISH–FRENCH
ANGLAIS–FRANÇAIS

A

AAS *abbr* (*atomic absorption spectrometry*) ANALYSIS spectrométrie d'absorption atomique *f*
abate *vt* GEN réduire
abatement *n* GEN lutte *f*, réduction *f*; ~ **measure** AIR mesure de lutte *f*
abattoir: ~ **waste** *n* WASTE déchets d'abattoir *m pl*
abiotic: ~ **degradation** *n* WATER POLL dégradation abiotique *f*
above[1]: ~ **ground** *adj* WASTE *landfill* à ciel ouvert, de surface, WATER SUPP de surface
above[2]: ~**-ground landfilling** *n* WASTE mise en décharge à ciel ouvert *f*
abrasion *n* AIR, WATER SUPP abrasion *f*; ~ **resistance** AIR résistance à l'abrasion *f*
abrasive[1] *adj* AIR, WATER SUPP abrasif
abrasive[2]: ~ **dust** *n* AIR poussière abrasive *f*
ABS *abbr* (*acrylonitrile-butadiene-styrene*) WASTE ABS (*acrylonitrile-butadiène-styrène*)
absolute: ~ **liability** *n* MANAGE responsabilité objective *f*; ~ **salinity** *n* WATER SUPP salinité absolue *f*
absorb *vt* GEN absorber
absorbable *adj* GEN absorbable
absorbate *n* AIR absorbat *m*
absorbed[1] *adj* GEN absorbé
absorbed[2]: ~ **dose** *n* SAFETY dose absorbée *f*; ~ **state** *n* AIR état absorbé *m*
absorbency *n* GEN capacité d'absorption *f*, absorptance *f*
absorbent[1] *adj* GEN absorbant; ~**-lined** NOISE revêtu d'absorbant
absorbent[2] *n* GEN absorbant *m*; ~ **cleaning** AIR nettoyage par absorption *m*; ~**-material filler** NOISE charge de matériau absorbant *f*
absorber *n* GEN absorbeur *m*; ~ **head** AIR hauteur d'absorption *f*; ~ **motor** AIR moteur d'absorption *m*; ~ **sump** AIR carter de colonne d'absorption *m*; ~ **tank** AIR cuve d'absorption *f*; ~ **wall** AIR *bubble-cap tray tower* paroi absorbante *f*
absorbing: ~ **well** *n* WATER POLL puits absorbant *m*
absorption *n* GEN absorption *f*, imbibition *f*; ~ **apparatus without internal fittings** AIR appareil d'absorption sans aménagement interne *m*, appareil d'absorption nu *m*; ~ **band** ANALYSIS bande d'absorption *f*; ~ **capacity** AIR capacité d'absorption *f*; ~ **coefficient** AIR, NOISE coefficient d'absorption *m*; ~ **column** AIR *scrubber column* colonne d'absorption *f*; ~**-filtration-adsorption process** AIR procédé d'absorption-filtration-adsorption *m*; ~ **isotherm** AIR isotherme d'absorption *f*; ~ **liquid** AIR liquide d'absorption *m*; ~ **pipette** ANALYSIS pipette d'absorption *f*; ~ **solution** ANALYSIS solution d'absorption *f*; ~ **spectrum** ANALYSIS *of infrared spectrophotometer* spectre d'absorption *m*; ~ **stage** AIR stade d'absorption *m*; ~ **temperature** AIR température d'absorption *f*; ~ **tower** AIR tour d'absorption *f*; ~ **vessel** ANALYSIS vase d'absorption *m*
absorptive: ~ **attenuator** *n* NOISE atténuateur d'absorption *m*; ~ **capacity** *n* WASTE capacité d'absorption *f*, WATER SUPP *of well* capacité d'absorption *f*, porosité efficace *f*
absorptivity *n* AIR absorptivité *f*
abstraction *n* WATER POLL, WATER SUPP captage *m*
abundance: ~ **coefficient** *n* ANALYSIS coefficient d'abondance *m*
AC *abbr* (*alternating current*) RENEW ENERG CA (*courant alternatif*)
acaricide *n* AIR acaricide *m*
acaride *n* AIR acarien *m*
accelerate *vt* AIR *particle* accélérer
accelerated[1] *adj* AIR accéléré
accelerated[2]: ~ **composting** *n* WASTE compostage accéléré *m*, hygiénisation *f*; ~ **electron** *n* AIR *electron beam process* électron accéléré *m*; ~ **erosion** *n* SOIL érosion accélérée *f*; ~ **storage test** *n* WASTE essai d'hygiénisation *m*
acceleration *n* AIR accélération *f*
accelerometer *n* ANALYSIS accéléromètre *m*
acceptable: ~ **daily intake** *n* (*ADI*) SAFETY dose journalière admissible *f* (*DJA*)
acceptance: ~ **capacity** *n* MANAGE, POLICY capacité de prise en charge *f*, capacité de prise en tolérance *f*
access: ~ **ramp** *n* WASTE rampe d'accès *f*
acclimate *vt* AIR, WATER POLL acclimater
acclimation *n* AIR, WATER POLL acclimatation *f*
acclimatization *n* AIR, WATER POLL acclimatation *f*
acclimatize *vt* AIR, WATER POLL acclimater
accompanying: ~ **hydraulic measure** *n* SOIL *soil vapour extraction* mesure hydraulique parallèle *f*
ACCR *abbr* (*activated-carbon catalytic reduction*) AIR réduction catalytique par charbon actif *f*
accretion *n* WATER SUPP *of sediment* accrétion *f*
accumulate *vt* GEN accumuler
accumulation *n* GEN accumulation *f*; ~ **of contaminants** AIR, SOIL accumulation de contaminants *f*; ~ **of harmful substances** AIR, SOIL accumulation de substances nocives *f*; ~ **hazard** POLICY, SAFETY risque d'accumulation *m*; ~ **of pollutants** AIR, SOIL accumulation de polluants *f*
accuracy *n* ANALYSIS *of measurement* exactitude de mesure *f*
acetate: ~ **adhesive** *n* WASTE adhésif à base d'acétate *m*; ~ **film** *n* WASTE film d'acétate *m*; ~ **glue** *n* WASTE colle à base d'acétate *f*
acetone *n* AIR, WASTE *landfill gas component* acétone *f*; ~ **resin** WASTE résine à base d'acétone *f*
acid[1] *adj* GEN acide; ~**-free** AIR sans acide, non-acide; ~**-oxidative** AIR oxydant acide; ~**-proof** AIR résistant aux acides; ~**-resistant** AIR, WASTE résistant aux acides; ~**-stressed** SOIL soumis à une pollution acide
acid[2] *n* GEN *additive* acide *m*; ~ **aerosol** AIR aérosol acide *m*; ~ **concentration** AIR *state* concentration d'acides *f*; ~ **content** AIR teneur acide *f*; ~ **deposit** AIR dépôt acide *m*; ~ **deposition** AIR dépôt acide *m*; ~ **earth** SOIL terre acide *f*; ~ **fallout** AIR retombées acides *f pl*; ~ **fermentation** WASTE *landfill gas phase* fermentation acide *f*; ~ **fog** AIR brouillard acide *m*; ~ **former** AIR, SOIL, WATER SUPP acidifiant *m*; ~**-free paper** WASTE papier non acide *m*; ~ **gas** AIR gaz acide *m*; ~ **gas pollutant** AIR polluant gazeux acide *m*; ~ **loading** AIR concentration acide *f*; ~ **neutralizing** AIR neutralisation des acides *f*; ~**-neutralizing capacity** AIR capacité de neutralisation des acides *f*; ~ **particle** AIR particule acide *f*; ~ **pollution** AIR, SOIL pollution acide *f*; ~ **precipitation** AIR précipitations acides *f pl*; ~**-proof varnish** WASTE vernis résistant à l'acide *m*; ~ **rain** AIR, SOIL pluies acides *f pl*; ~ **resistance** AIR, SOIL résistance aux acides *f*; ~ **run-off** SOIL, WATER SUPP ruissellement acide *m*; ~ **scrubbing** AIR lavage acide *m*; ~ **smog** AIR brouillard photochimique acide *m*, fumard acide *m*, smog acide *m*; ~ **snow** SOIL neige acide *f*; ~ **storm water** WATER POLL eaux pluviales acides *f pl*; ~ **stress** SOIL agression acide *f*; ~ **tolerance** SOIL *of soil* tolérance à l'acidité *f*; ~ **waste gas** AIR, WASTE effluent

gazeux acide *m*; ~ **wastewater** WASTE, WATER POLL eaux résiduaires acides *f pl*

acidic[1] *adj* GEN acide

acidic[2]: ~ **area** *n* SOIL zone d'acidité *f*; ~ **particle** *n* AIR particule acide *f*; ~ **precursor** *n* SOIL précurseur acide *m*; ~ **rain** *n* AIR, SOIL pluies acides *f pl*

acidification *n* GEN acidification *f*

acidify *vt* GEN acidifier

acidifying *adj* GEN acidifiant

acidity *n* GEN acidité *f*; ~ **level** AIR, SOIL niveau d'acidité *m*, degré d'acidité *m*

acknowledge *vt* SAFETY *alarm* acquitter

acoustic[1] *adj* NOISE acoustique

acoustic[2]: ~ **baffle** *n* NOISE écran acoustique *m*, déflecteur *m*; ~ **barrier** *n* NOISE *on roads* écran acoustique *m*; ~ **board** *n* NOISE carton d'insonorisation *m*; ~ **booth** *n* NOISE cabine acoustique *f*; ~ **capacity** *n* NOISE capacité acoustique *f*; ~ **coupler** *n* ANALYSIS, NOISE coupleur acoustique *m*; ~ **emission analysis** *n* ANALYSIS analyse d'émissions acoustiques *f*; ~ **emissions** *n pl* (*AE*) NOISE émissions acoustiques *f pl*; ~ **enclosure** *n* NOISE enceinte acoustique *f*; ~ **insulation** *n* MARINE, NOISE, SAFETY isolation acoustique *f*, insonorisation *f*; ~ **lining** *n* NOISE revêtement insonorisant *m*; ~ **louver** *AmE*, ~ **louvre** *BrE* *n* NOISE volet acoustique *m*; ~ **oscillation** *n* NOISE oscillation acoustique *f*; ~ **panel** *n* NOISE panneau insonorisant *m*; ~ **paneling** *AmE*, ~ **panelling** *BrE* *n* NOISE installation de panneaux insonorisants *f*; ~ **plaster** *n* NOISE enduit insonorisant *m*; ~ **plate** *n* NOISE *for airborne noise damping* plaque d'insonorisation *f*; ~ **screen** *n* NOISE écran acoustique *m*; ~ **sheet** *n* NOISE tôle insonorisante *f*; ~ **shield** *n* NOISE écran acoustique *m*; ~ **shielding** *n* NOISE blindage acoustique *m*; ~ **slab** *n* NOISE dalle insonorisante *f*; ~ **vibration** *n* NOISE vibration acoustique *f*

acoustical: ~ **ceiling** *n* NOISE plafond insonorisant *m*; ~ **emissions** *n pl* NOISE émissions acoustiques *f pl*; ~ **leak** *n* NOISE fuite acoustique *f*; ~ **test** *n* NOISE essai acoustique *m*; ~ **tile** *n* NOISE carreau insonorisant *m*

acrid: ~ **odor** *AmE*, ~ **odour** *BrE* *n* AIR, WATER SUPP odeur âcre *f*

acrylic: ~ **plastic** *n* WASTE *Plexiglas®* matière plastique acrylique *f*; ~ **resin** *n* WASTE résine acrylique *f*; ~ **rubber** *n* WASTE caoutchouc acrylique *m*

acrylonitrile: **~-butadiene-styrene** *n* (*ABS*) WASTE acrylonitrile-butadiène-styrène *f* (*ABS*)

actinometer *n* ANALYSIS actinomètre *m*

activated[1] *adj* GEN activé

activated[2]: ~ **alumina** *n* AIR *adsorbent* alumine activée *f*; ~ **aluminium oxide** *BrE*, ~ **aluminum oxide** *AmE* *n* AIR *adsorbent* oxyde d'aluminium activé *m*; ~ **carbon** *n* AIR *adsorbent*, SOIL, WATER POLL, WATER SUPP charbon actif *m*; **~-carbon absorption** *n* SOIL absorption sur carbone actif *f*; **~-carbon adsorber** *n* AIR adsorbeur à charbon actif *m*; **~-carbon adsorption** *n* AIR adsorption sur charbon actif *f*; **~-carbon canister** *n* AIR, WATER POLL absorbeur de charbon actif *m*; **~-carbon catalytic reduction** *n* (*ACCR*) AIR réduction catalytique par charbon actif *f*; **~-carbon column** *n* ANALYSIS colonne de charbon actif *f*; **~-carbon filter** *n* WATER POLL, WATER SUPP filtre à charbon actif *m*; ~ **carbon fixed-bed reactor** *n* AIR réacteur à lit fixe à charbons actifs *m*; **~-carbon fluidized bed** *n* AIR lit fluidisé à charbons actifs *m*; **~-carbon moving bed** *n* AIR lit mobile à charbons actifs *m*; **~-carbon process** *n* AIR procédé au charbon actif *m*; **~-carbon suspension** *n* ANALYSIS suspension de charbon actif *f*; **~-carbon treatment** *n* AIR, WASTE, WATER POLL traitement au charbon actif *m*, traitement par charbon actif *m*; ~ **coke** *n* AIR, WASTE coke actif *m*; **~-coke adsorber** *n* AIR, WASTE adsorbeur au coke actif *m*; **~-coke filter** *n* WASTE filtre à coke actif *m*; **~-coke process** *n* AIR procédé à coke actif *m*; ~ **sludge** *n* AIR, WATER POLL, WATER SUPP boues activées *f pl*; **~-sludge process** *n* AIR *biowasher*, WASTE, WATER SUPP procédé par boues activées *m*, procédé d'épuration par boues activées *m*; **~-sludge tank** *n* AIR *biowasher* cuve à boues activées *f*

activation *n* GEN activation *f*; ~ **analysis** ANALYSIS *heavy metal detection* analyse par activation *f*; ~ **energy** AIR énergie d'activation *f*

active[1] *adj* GEN actif

active[2]: ~ **catalyst surface** *n* AIR surface active de catalyseur *f*; ~ **matrix liquid-crystal display** *n* (*AMLCD*) ANALYSIS écran à cristaux liquides à matrice active *m*; ~ **solvent** *n* WASTE solvant actif *m*; ~ **sound-control system** *n* NOISE silencieux actif *m*; ~ **water** *n* WATER SUPP eau corrosive *f*

actual: ~ **air requirement** *n* AIR besoin d'air effectif *m*; ~ **air volume** *n* AIR volume d'air effectif *m*

acute: ~ **effect** *n* AIR, WATER SUPP effet aigu *m*

acyclic *adj* AIR acyclique

additional: ~ **display** *n* ANALYSIS *measuring instruments* affichage complémentaire *m*; ~ **entrainment** *n* WASTE entraînement complémentaire *m*

additive *n* AIR *used sorbents*, SOIL, WASTE additif *m*, WATER SUPP additif *m*, produit d'addition *m*; ~ **suspension** AIR suspension d'additifs *f*

additives: ~ **silo** *n* AIR silo à additifs *m*

adhere *vi* AIR *particles*, WASTE adhérer; ~ **by suction** AIR adhérer par aspiration

adherence *n* AIR *of particles*, WASTE adhérence *f*

adhesion *n* AIR, WASTE adhésion *f*

adhesive[1] *adj* WATER POLL adhésif

adhesive[2] *n* AIR, WASTE adhésif *m*; ~ **agent** AIR agent adhésif *m*; ~ **film** WASTE film adhésif *m*, pellicule adhésive *f*; ~ **force** AIR force d'adhésion *f*, ANALYSIS *chromatography* adhérence *f*, pouvoir adhérent *m*; ~ **tape** WASTE ruban adhésif *m*

adhesiveness *n* WATER POLL adhésivité *f*

ADI *abbr* (*acceptable daily intake*) SAFETY DJA (*dose journalière admissible*)

adiabatic *adj* AIR adiabatique

adsorb *vt* GEN adsorber

adsorbable: ~ **organic halogen compounds** *n pl* (*AOX*) WATER SUPP composés organo-halogénés adsorbables *m pl*

adsorbate *n* GEN adsorbat *m*

adsorbed[1] *adj* GEN adsorbé

adsorbed[2]: ~ **component** *n* AIR élément adsorbé *m*; ~ **gas** *n* AIR gaz adsorbé *m*; ~ **substance** *n* AIR substance adsorbée *f*

adsorbent[1] *adj* GEN adsorbant

adsorbent[2] *n* GEN adsorbant *m*; ~ **bed** AIR lit adsorbant *m*; ~ **containing carbon** AIR adsorbant au carbone *m*; ~ **density** AIR densité d'adsorbants *f*; ~ **dosage** AIR dosage d'adsorbants *m*; ~ **inlet** AIR admission d'adsorbants *f*; ~ **outlet** AIR évacuation d'adsorbants *f*; ~ **particle** AIR particule d'adsorbant *f*; ~ **surface** AIR surface adsorbante *f*

adsorber *n* GEN adsorbeur *m*; ~ **filter** AIR filtre adsorbant *m*

adsorption *n* GEN adsorption *f*; ~ **capacity** AIR capacité d'adsorption *f*; ~ **chromatography** ANALYSIS chromatographie d'adsorption *f*; ~ **column** ANALYSIS colonne d'adsorption *f*; ~ **duration** AIR durée d'adsorption *f*; ~ **efficiency** AIR, SOIL, WASTE rendement d'adsorption *m*; ~ **isotherm** AIR, WATER POLL, WATER SUPP isotherme d'adsorption *f*; ~ **potential** AIR potentiel d'adsorption *m*; ~ **power** AIR pouvoir d'adsorption *m*; ~ **stage** AIR stade d'adsorption *m*; ~ **temperature** AIR température d'adsorption *f*; ~ **test** ANALYSIS essai d'adsorption *m*; ~ **tube** ANALYSIS tube d'adsorption *m*; ~ **velocity** AIR vitesse d'adsorption *f*; ~ **zone** AIR zone d'adsorption *f*

adsorptive *adj* GEN adsorbant

ADSOX: ~ **process** *n* AIR *adsorption* procédé ADSOX *m*

advanced: ~ **coal cleaning** *n* AIR méthode perfectionnée d'épuration du charbon *m*

AE[1] *abbr* (*acoustic emissions*) NOISE émissions acoustiques *f pl*

AE[2]: ~ **analysis** *n* ANALYSIS analyse d'émissions acoustiques *f*

AED *abbr* (*atomic emission detection*) ANALYSIS détection d'émissions atomiques *f*
aerate *vt* GEN aérer
aerated: ~ **pile composting** *n* (*APC*) SOIL compostage aéré en tas *m*
aeration *n* GEN aération *f*, carbonatation *f*, consommation d'air *f*, ventilation *f*; ~ **basin** WATER POLL bassin d'aération *m*; ~ **system** WATER SUPP système d'aération *m*; ~ **tank** WATER POLL bassin d'aération *m*
aerator *n* WATER SUPP aérateur *m*
aerial: ~ **conveyor** *n* WASTE bande transporteuse aérienne *f*; ~ **view evaluation** *n* ANALYSIS évaluation d'une vue aérienne *f*
aerobic[1] *adj* WASTE, WATER SUPP aérobie
aerobic[2]: ~ **digestion** *n* WATER SUPP digestion aérobie *f*; ~ **organism** *n* WATER POLL organisme aérobie *m*; ~ **oxidation** *n* WASTE *landfill gas phase* oxydation aérobie *f*
aerodynamic: ~ **diameter** *n* AIR, ANALYSIS diamètre aérodynamique *m*
aerometer *n* ANALYSIS aéromètre *m*
aerosol *n* AIR, WASTE, WATER SUPP aérosol *m*; ~ **cap** WASTE capsule d'aérosol *f*; ~ **container** WASTE boîtier d'aérosol *m*, récipient d'aérosols *m*; ~ **dispenser** WASTE atomiseur *m*, bombe aérosol *f*; ~ **pollution** AIR pollution par les aérosols *f*; ~ **precipitation** AIR précipitation d'aérosols *f*; ~ **valve** WASTE valve d'aérosol *f*
AES *abbr* (*atomic emission spectrometry*) ANALYSIS spectrométrie d'émission atomique *f*
afforestation *n* POLICY, SOIL boisement *m*
AFNOR *abbr* (*French Standards Association*) ANALYSIS, POLICY AFNOR (*Association Française de Normalisation*)
AFS *abbr* (*atomic fluorescence spectrometry*) ANALYSIS spectrométrie de fluorescence atomique *f*
afterburner *n* AIR, WASTE brûleur de postcombustion *m*, chambre de postcombustion *f*; ~ **chamber** AIR, WASTE chambre de postcombustion *f*; ~ **housing** AIR, WASTE chambre de postcombustion *f*, poste de réchauffage *m*
afterburning *n* AIR, SOIL, WASTE postcombustion *f*; ~ **chamber** AIR, WASTE chambre de postcombustion *f*
aftercooler *n* AIR réfrigérant complémentaire *m*, refroidisseur complémentaire *m*
aged: ~ **catalyst** *n* AIR catalyseur usagé *m*
ageing[1]: ~**-resistant** *adj* WASTE anti-vieillissement
ageing[2] *n* WASTE vieillissement *m*; ~ **stability** AIR stabilité au vieillissement *f*; ~ **test** WASTE essai de vieillissement *m*
agent: ~ **orange** *n* AIR agent orange *m*
agglomerate[1] *n* AIR, SOIL, WASTE, WATER SUPP agglomérat *m*
agglomerate[2] *vt* AIR, SOIL, WASTE, WATER SUPP agglomérer
agglomerated: ~ **dust** *n* AIR *in bag filter* poussière agglomérée *f*
agglomeration *n* AIR, SOIL, WASTE, WATER SUPP agglomération *f*; ~ **capacity** AIR, ANALYSIS capacité d'agglomération *f*
agglomerator *n* WASTE agglomérateur *m*
agglutination *n* AIR, WASTE, WATER SUPP agglutination *f*
aggradation *n* SOIL remblaiement *m*, WATER SUPP alluvionnement *m*, exhaussement *m*
aggradational: ~ **deposit** *n* WATER SUPP couche alluvionnaire *f*
aggregate *n* AIR, SOIL, WATER SUPP agrégat *m*
aggressive: ~ **gas component** *n* AIR élément gazeux agressif *m*; ~ **water** *n* WATER SUPP eau agressive *f*
agitate *vt* WASTE brasser
agitating: ~ **arm** *n* AIR agitateur *m*, vagueur *m*
agitation *n* WASTE brassage *m*
agitator *n* SOIL agitateur *m*, WASTE agitateur *m*, brasseur *m*
agricultural: ~ **potential** *n* SOIL potentiel agricole *m*, vocation agricole *f*; ~ **runoff** *n* SOIL lessivage des terres cultivées *m*; ~ **waste** *n* WATER POLL résidus agricoles *m pl*
agro: ~**-ecology** *n* MANAGE, POLICY agro-écologie *f*
A-horizon *n* SOIL horizon-A *m*
air[1]: ~ **ambient** *n* AIR air ambiant *m*; ~ **blast labeling** *AmE*, ~ **blast labelling** *BrE n* WASTE étiquetage par soufflage d'air *m*; ~ **blow-down** *n* WATER SUPP purge de l'air *f*; ~ **breathing apparatus** *n* SAFETY appareil respiratoire *m*; ~**-bubble cushioning** *n* WASTE rembourrage à bulles d'air *m*; ~ **bubble wrap** *n* WASTE enveloppe de bulles d'air *f*; ~ **charge** *n* AIR volume d'air *m*; ~**-charge temperature** *n* AIR température du volume d'air *f*; ~ **classification** *n* AIR classification pneumatique *f*; ~ **classifier** *n* AIR, WASTE classificateur pneumatique *m*, séparateur *m*; ~ **compressor** *n* AIR compresseur d'air *m*; ~ **conditioning installation** *n* SAFETY installation de conditionnement d'air *f*; ~ **conduction cylinder** *n* AIR manche d'air *f*, manchon d'air *m*; ~ **curtain installation for open doors** *n* SAFETY installation de rideaux d'air pour portes ouvertes *f*; ~ **cushion** *n* NOISE coussin d'air *m*; ~ **decontamination** *n* AIR décontamination de l'air *f*; ~ **deficiency** *n* AIR insuffisance d'air *f*, manque d'air *m*; ~ **dehumidifier** *n* AIR déshumidificateur *m*; ~ **distribution** *n* AIR *filters* distribution d'air *f*; ~ **dosage** *n* AIR dosage d'air *m*; ~ **emissions** *n pl* AIR émissions atmosphériques *f pl*; ~ **exchange** *n* AIR échange atmosphérique *m*, renouvellement d'air *m*; ~ **filter** *n* AIR, SAFETY filtre à air *m*; ~ **friction damping** *n* ANALYSIS *in measuring instruments* amortissement par friction d'air *m*; ~ **front** *n* AIR front d'air *m*; ~ **humidifier** *n* AIR humidificateur *m*; ~ **humidity** *n* AIR humidité de l'air *f*; ~ **intake** *n* AIR admission d'air *f*, prise d'air *f*; ~ **jet** *n* AIR jet d'air *m*; ~**-jet screening method** *n* ANALYSIS *dust analysis* méthode de tamisage par veine d'air *f*; ~ **layer** *n* AIR couche d'air *f*; ~ **mass** *n* AIR masse d'air *f*; ~ **mold process** *AmE see air mould process BrE*; ~ **motion** *n* AIR mouvement de l'air *m*; ~ **mould process** *n BrE* WASTE procédé de moulage sous vide *m*; ~ **permeability** *n* SOIL *in extraction of soil atmosphere* perméabilité à l'air *f*; ~ **plankton** *n* AIR plancton aérien *m*; ~ **poisoning** *n* AIR empoisonnement de l'air *m*; ~ **pollutant** *n* AIR polluant atmosphérique *m*; ~ **pollution** *n* AIR pollution atmosphérique *f*, pollution de l'air *f*; ~ **pollution control** *n* AIR lutte contre la pollution atmosphérique *f*; ~ **pollution emissions** *n pl* AIR émissions atmosphériques polluantes *f pl*; ~ **pollution episode** *n* AIR épisode de pollution atmosphérique *m*, vague de pollution atmosphérique *f*; ~ **pollution forecast** *n* AIR prévision de la pollution atmosphérique *f*; ~ **pollution incident** *n* AIR pollution atmosphérique ponctuelle *f*; ~ **pollution index** *n* AIR indice de pollution atmosphérique *m*; ~ **pollution measurement** *n* AIR, ANALYSIS mesure de la pollution atmosphérique *f*; ~ **preheater** *n* AIR, WASTE réchauffeur d'air *m*; ~ **purification and deodorization equipment** *n* AIR matériel d'épuration et de désodorisation *m*; ~ **purifier** *n* AIR appareil d'assainissement de l'air *m*; ~ **purity** *n* AIR pureté de l'air *f*; ~ **quality** *n* AIR qualité de l'air *f*; ~ **quality criteria** *n pl* AIR critères de qualité atmosphérique *m pl*; ~ **quality data** *n pl* AIR, ANALYSIS données relatives à la qualité de l'air *f pl*, indices de la qualité de l'air *m pl*; ~ **quality measurement** *n* AIR, ANALYSIS mesure de la qualité de l'air *f*; ~ **quality standard** *n* AIR norme de qualité de l'air *f*; ~ **release** *n* WATER POLL *aeration system* rejet à l'air libre *m*; ~ **sampling** *n* ANALYSIS prélèvement d'échantillons atmosphériques *m*; ~**-sea interface** *n* AIR, MARINE interface air-mer *f*; ~ **separation plant** *n* WASTE classificateur pneumatique *m*; ~ **separator** *n* AIR, WASTE séparateur pneumatique *m*; ~ **shed** *n* AIR bassin atmosphérique *m*; ~ **slide** *n* WASTE aéroglissière *f*; ~ **sparging** *n* SOIL aspersion à l'air *f*; ~ **speed** *n* AIR vitesse de l'air *f*; ~**-stream sorting** *n* WASTE tri par jet d'air *m*; ~ **stripping** *n* WASTE lavage à l'air *m*; ~ **supply** *n* AIR *thermal combustion* apport d'air *m*; ~**-to-cloth ratio** *n* AIR *fabric filtration* rapport air-membrane *m*; ~ **unfit for respiration** *n* AIR, SAFETY air impropre à la respiration *m*, air irrespirable *m*; ~ **velocity** *n* AIR vitesse de l'air *f*

air[2] *vt* WATER SUPP aérer, sécher
airborne[1] *adj* AIR aérien, atmosphérique
airborne[2]: ~ **dust** *n* AIR, SAFETY poussière en suspension dans l'air *f*; ~ **dust concentration** *n* SAFETY concentration de poussières en suspension dans l'air *f*; ~ **noise** *n* NOISE, SAFETY bruit aérien *m*; ~ **noise proofing** *n* NOISE isolation contre le bruit aérien *f*; ~ **noise reduction** *n* NOISE *air absorption* réduction du bruit aérien *f*; ~ **noise transmission** *n* NOISE transmission du bruit aérien *f*; ~ **particulate matter** *n* AIR matière sous forme de particules en suspension dans l'air *f*; ~ **pollutant observation model** *n* MANAGE modèle d'observation des polluants atmosphériques *m*; ~ **sand** *n* AIR vent de sable *m*
airflow *n* AIR *mass force separator* débit d'air *m*; ~ **sensor** AIR capteur de débit d'air *m*, débitmètre *m*
airing *n* WATER SUPP séchage *m*
airlock *n* SAFETY sas à air *m*
airtight *adj* AIR étanche à l'air, hermétique
Al *abbr* (*aluminium BrE, aluminum AmE*) SOIL, WATER SUPP Al (*aluminium*)
albedo *n* AIR albédo *m*
albumen *n* AIR *enzyme* albumine *f*
alcohol *n* AIR, WATER POLL alcool *m*
alga *n* WATER SUPP algue *f*
algae *n* MARINE, WATER SUPP algues *f pl*; ~ **bloom** WATER POLL prolifération algale *f*; ~ **control** MARINE, WATER POLL lutte contre la prolifération des algues *f*; ~ **growth** WATER POLL prolifération algale *f*; ~ **test** ANALYSIS essai d'eutrophisation *m*
alicyclic[1] *adj* AIR alicyclique
alicyclic[2]: ~ **compound** *n* AIR composé alicyclique *m*
aliphatic[1] *adj* AIR, WASTE aliphatique
aliphatic[2]: ~ **compound** *n* AIR composé aliphatique *m*; ~ **hydrocarbon** *n* AIR hydrocarbure aliphatique *m*; ~ **polyamide** *n* AIR *filter fibre material* polyamide aliphatique *m*
alkaline[1] *adj* AIR, SOIL, WATER POLL, WATER SUPP alcalin, basique
alkaline[2]: ~ **fermentation** *n* WASTE fermentation alcaline *f*; ~ **oxidant** *n* AIR oxydant alcalin *m*; ~ **temporary hardness** *n* WATER SUPP dureté alcaline *f*
alkalinity *n* AIR, SOIL, WATER POLL, WATER SUPP alcalinité *f*; ~ **measurement** ANALYSIS mesure d'alcalinité *f*
alkane *n* AIR *saturated hydrocarbons* alcane *m*
alkene *n* AIR *unsaturated hydrocarbons* alcène *m*
alkylene *n* AIR alkylène *m*
allochthonous[1] *adj* WATER SUPP allochtone
allochthonous[2]: ~ **matter** *n* SOIL matière allochtone *f*
all-purpose: ~ **adhesive** *n* WASTE adhésif universel *m*
alluvial: ~ **bed** *n* WATER POLL lit alluvial *m*, WATER SUPP canal alluvial *m*; ~ **cone** *n* WATER SUPP cône d'éboulis *m*, cône d'alluvions *m*; ~ **deposit** *n* WASTE alluvions *f pl*, dépôt alluvial *m*, dépôt alluvionnaire *m*, WATER SUPP alluvion *m*, dépôt alluvial *m*, dépôt alluvionnaire *m*; ~ **plain** *n* WATER SUPP plaine alluviale *f*; ~ **silt** *n* WATER POLL boues alluviales *f pl*
alpha: ~ **factor** *n* WATER SUPP facteur alpha *m*
alternate *vt* AIR alterner, ANALYSIS affaiblir
alternated: ~ **two-stage activated sludge process** *n* WATER POLL traitement alterné par boues activées en deux étapes *m*
alternating: ~ **current** *n* RENEW ENERG courant alternatif *m*
aluminium *n BrE* (*Al*) SOIL, WATER SUPP aluminium *m* (*Al*); ~ **scrap** *BrE* WASTE résidu d'aluminium *m*; ~ **sulphate** *BrE* WATER POLL, WATER SUPP sulfate d'aluminium *m*
aluminum *AmE see aluminium BrE*
ambient[1]: ~**-air neutral** *adj* WASTE neutre à l'air ambiant
ambient[2]: ~ **air** *n* AIR air ambiant *m*; ~**-air analysis** *n* ANALYSIS analyse de l'air ambiant *f*, analyse du milieu atmosphérique *f*; ~**-air concentration** *n* AIR concentration atmosphérique *f*; ~**-air concentration limit** *n* ANALYSIS seuil de concentration atmosphérique *m*; ~**-air measuring** *n* ANALYSIS mesure atmosphérique *f*; ~**-air quality analyser** *BrE*, ~**-air quality analyzer** *AmE n* ANALYSIS analyseur de la qualité atmosphérique *m*; ~**-air quality standard** *n* AIR norme de qualité de l'air ambiant *f*; ~**-air quality value** *n* AIR indice de qualité de l'air ambiant *m*; ~ **atmosphere** *n* AIR atmosphère ambiante *f*; ~ **noise** *n* NOISE bruit ambiant *m*
amelioration *n* SOIL amélioration *f*
amenity: ~ **area** *n* POLICY zone de repos *f*; ~ **rights** *n pl* POLICY droit à un environnement vivable *m*
AMLCD *abbr* (*active matrix liquid-crystal display*) ANALYSIS écran à cristaux liquides à matrice active *m*
ammeter *n* ANALYSIS ampèremètre *m*
ammonia *n* AIR, SOIL, WASTE, WATER SUPP ammoniac *m*, ammoniaque *m*; ~ **process** AIR *Walther* traitement à l'ammoniac *m*
ammoniation *n* WATER POLL ammonisation *f*
ammonifying: ~ **bacteria** *n pl* WATER POLL bactéries ammonifiantes *f pl*
amorphous[1] *adj* AIR amorphe
amorphous[2]: ~ **adsorbent** *n* AIR adsorbant amorphe *m*
amount: ~ **of gas** *n* AIR volume de gaz *m*
amperometric[1] *adj* ANALYSIS ampèremétrique
amperometric[2]: ~ **determination method** *n* ANALYSIS méthode de détermination ampèremétrique *f*
amperometry *n* ANALYSIS *voltametric process* ampèremétrie *f*
amplifier *n* ANALYSIS amplificateur *m*
ampoule *n* WASTE ampoule *f*
anaerobic[1] *adj* WATER SUPP *sewage* anaérobie
anaerobic[2]: ~ **bacteria** *n pl* WATER POLL bactéries anaérobies *f pl*; ~ **digestion** *n* WATER POLL, WATER SUPP digestion anaérobie *f*; ~ **mass** *n* WATER SUPP masse anaérobie *f*; ~ **organism** *n* WATER POLL organisme anaérobie *m*
analog: ~**-controlled** *adj* ANALYSIS *programming* à commande analogique
analysable *adj BrE* ANALYSIS analysable
analyse *vt BrE* ANALYSIS analyser; ~ **oxidimetrically** *BrE* ANALYSIS analyser par oxydimétrie
analyser *n BrE* ANALYSIS analyseur *m*
analysis *n* ANALYSIS analyse *f*; ~ **aim** ANALYSIS objet d'analyse *m*; ~ **calibration function** ANALYSIS fonction d'étalonnage d'une analyse *f*; ~ **data** ANALYSIS résultat d'analyse *m*; ~ **frequency** ANALYSIS fréquence d'analyse *f*; ~ **of iron in wastewater** ANALYSIS détermination de la concentration de fer dans les eaux usées *f*, dosage de la concentration de fer dans les eaux usées *m*; ~ **location** ANALYSIS lieu d'analyse *m*; ~**-measuring technique** ANALYSIS technique de mesure et d'analyse *f*; ~ **method** ANALYSIS méthode d'analyse *f*; ~ **of residue** ANALYSIS analyse des déchets *f*; ~ **result** ANALYSIS résultat d'analyse *m*; ~ **time** ANALYSIS durée d'analyse *f*, période d'analyse *f*; ~ **volume** ANALYSIS volume analysé *m*; ~ **of weakest points** ANALYSIS analyse des points faibles *f*
analytical[1] *adj* ANALYSIS analytique
analytical[2]: ~ **error** *n* ANALYSIS erreur d'analyse *f*; ~ **finding** *n* ANALYSIS résultat d'analyse *m*, valeur obtenue par analyse *f*; ~ **procedure** *n* ANALYSIS procédure d'analyse *f*; ~ **quality control** *n* ANALYSIS contrôle-qualité d'analyse *m*; ~ **quantity** *n* ANALYSIS quantité analysée *f*; ~ **result** *n* ANALYSIS résultat d'analyse *m*; ~ **sample** *n* ANALYSIS échantillon d'analyse *m*; ~ **scale** *n* ANALYSIS échelle d'analyse *f*; ~ **solution** *n* ANALYSIS solution analytique *f*; ~ **unit** *n* ANALYSIS unité analytique *f*
analytically: ~**-pure** *adj* ANALYSIS analytiquement pur
analyzable *AmE see analysable BrE*
analyze *AmE see analyse BrE*
analyzer *AmE see analyser BrE*
anatomical: ~ **waste** *n* WASTE déchets anatomiques *m pl*
Andreasen: ~**'s pipette** *n* ANALYSIS pipette d'Andreasen *f*
anechoic *adj* NOISE *room* sans écho

anemometer *n* ANALYSIS anémomètre *m*; ~ **height** AIR *stack* hauteur d'anémomètre *f*
anhydrite *n* AIR *gypsum form* anhydrite *m*
anhydrous *adj* WATER SUPP anhydre
animal: ~ **carcass disposal act** *n* WASTE loi sur l'élimination des carcasses animales *f*; ~ **dung** *n* RENEW ENERG déjections animales *f*; ~ **dust** *n* AIR poudre d'origine animale *f*; ~ **ecology** *n* MANAGE, POLICY écologie animale *f*; ~ **environment** *n* AIR milieu zoologique *m*, SOIL milieu animal *m*, milieu zoologique *m*; ~ **waste** *n* WASTE déchets d'origine animale *m*; ~ **wastes** *n pl* WASTE déchets animaux *m pl*
anion: ~ **exchange** *n* SOIL échange anionique *m*; ~**-exchange chromatography** *n* ANALYSIS chromatographie d'échange anionique *f*
anisokinetic: ~ **sampling** *n* ANALYSIS échantillonnage anisocinétique *m*
annual: ~ **evaporation** *n* WATER SUPP évaporation annuelle *f*; ~ **flood** *n* WATER SUPP crue annuelle *f*; ~ **flow** *n* WATER SUPP débit annuel *m*; ~ **load** *n* WATER SUPP charge annuelle *f*; ~ **mean value** *n* ANALYSIS moyenne annuelle *f*; ~ **mean water level** *n* WATER SUPP niveau moyen annuel *m*; ~ **runoff** *n* WATER SUPP écoulement annuel *m*
annular: ~ **gap** *n* AIR, SOIL col annulaire *m*, espace annulaire *m*, passage annulaire *m*; ~ **gap scrubber** *n* AIR laveur à col annulaire *m*; ~**-type scrubber** *n* AIR laveur de type annulaire *m*
annulus *n* AIR couronne *f*, espace annulaire *m*
anode *n* AIR, SOIL anode *f*
anorganic: ~ **nitrogen fertilizer** *n* SOIL *conditioner*, WATER POLL engrais minéral azoté *m*
anthropic: ~ **pollution** *n* SAFETY pollution anthropique *f*
anthropogenic[1] *adj* AIR, SOIL anthropique
anthropogenic[2]: ~ **acidification** *n* SOIL acidification anthropique *f*; ~ **climate change** *n* AIR changement climatique d'origine humaine *m*; ~ **ecosystem** *n* AIR, SOIL écosystème anthropique *m*; ~ **erosion** *n* SOIL érosion anthropique *f*; ~ **greenhouse effect** *n* AIR effet de serre anthropique *m*; ~ **pollutant** *n* AIR, SOIL polluant anthropique *m*; ~ **sources** *n pl* AIR source anthropique *f*
anthropogenically[1]: ~ **caused** *adj* AIR, SOIL anthropique, d'origine humaine
anthropogenically[2]: ~**-caused soil contamination** *n* SOIL contamination du sol imputable aux activités humaines *f*
anthroposphere *n* SOIL anthroposphère *f*
antidust: ~ **agent** *n* AIR agent antipoussière *m*
antifoaming: ~ **agent** *n* WATER SUPP antimousse *m*, agent antimousse *m*
antiknock: ~ **additive** *n* AIR additif antidétonant *m*; ~ **agent** *n* AIR agent antidétonant *m*; ~ **fuel** *n* AIR combustible antidétonant *m*; ~ **value** *n* AIR indice antidétonant *m*
antimony *n* (*Sb*) WATER SUPP antimoine *m* (*Sb*)
antinoise[1] *adj* NOISE antibruit
antinoise[2]: ~ **encapsulation** *n* NOISE protection anti-bruit *f*; ~ **hood** *n* NOISE capot insonorisant *m*; ~ **microphone** *n* NOISE microphone atténuateur de bruit *m*; ~ **soft rubber lining** *n* NOISE, SAFETY revêtement anti-bruit en caoutchouc mou *m*, revêtement insonorisant en caoutchouc mou *m*
antipollution: ~ **laws** *n pl* POLICY lois antipollution *f pl*; ~ **measure** *n* MANAGE, POLICY mesure contre la pollution *f*
antiquated: ~ **plant** *n* AIR installation déclassée *f*
antistatic: ~ **protective clothing** *n* SAFETY *against chemical aggression* habillage antistatique *m*
antivibration: ~ **mounting** *n* NOISE montage antivibration *m*
AOX[1] *abbr* (*adsorbable organic halogen compounds*) WATER SUPP composés organo-halogénés adsorbables *m pl*
AOX[2]: ~ **analysis** *n* ANALYSIS analyse des composés organo-halogénés adsorbables *f*
APC *abbr* (*aerated pile composting*) SOIL compostage aéré en tas *m*
APCI *abbr* (*atmospheric pressure chemical ionization*) ANALYSIS ionisation chimique sous pression atmosphérique *f*
aphotic: ~ **zone** *n* WATER POLL zone aphotique *f*
apparatus: ~ **calibration** *n* ANALYSIS étalonnage d'appareil *m*
apparent: ~ **water table** *n* WATER SUPP nappe perchée *f*, niveau hydrostatique apparent *m*
appeal: ~ **authorization** *n* WATER SUPP autorisation d'appel *f*; ~ **deadline** *n* POLICY date limite d'appel *f*; ~ **procedure** *n* POLICY procédure d'appel *f*
appearance: ~ **of waste** *n* WATER SUPP aspect des déchets *m*
applicability *n* AIR, RENEW ENERG, WATER SUPP *of process* applicabilité *f*
application *n* GEN application *f*; ~ **method** AIR, SOIL méthode d'application *f*; ~ **possibility** AIR, NOISE, SAFETY, WATER SUPP *of process* débouché possible *m*, possibilité d'application *f*; ~ **technology** AIR, SAFETY, WATER SUPP technologie d'application *f*
applied: ~ **research** *n* ANALYSIS recherche appliquée *f*
apply *vt* GEN appliquer
apportion *vt* ANALYSIS attribuer, doser
approximation: ~ **test** *n* ANALYSIS essai d'approximation *m*
apron: ~ **conveyor** *n* WASTE convoyeur à plateaux *m*; WATER SUPP convoyeur à plateaux *m*, transporteur à palettes métalliques *m*
aquaculture *n* WATER SUPP aquaculture *f*, aquiculture *f*
aquatic: ~ **life** *n* MARINE faune aquatique *f*; ~ **resources** *n pl* RENEW ENERG ressources aquatiques *f pl*
aqueduct *n* WATER SUPP aqueduc *m*
aqueous *adj* WATER SUPP aqueux
aquiclude *n* WATER SUPP aquiclude *f*
aquifer *n* WATER SUPP aquifère *m*, couche aquifère *f*, couche perméable *f*, nappe aquifère *f*, nappe phréatique *f*
aquiferous *adj* WATER SUPP aquifère
aquifuge *n* WATER SUPP aquifuge *m*, couche imperméable *f*
Ar *abbr* (*argon*) AIR, WASTE, WATER SUPP Ar (*argon*)
arable: ~ **land** *n* SOIL terre labourable *f*
aramide: ~ **fiber-reinforced thermoset** *AmE*, ~ **fibre-reinforced thermoset** *BrE* *n* WATER SUPP résine thermodurcie renforcée à la fibre d'aramide *f*
arbitrary: ~ **analysis** *n* ANALYSIS analyse arbitraire *f*; ~ **test** *n* ANALYSIS essai arbitraire *m*
arc: ~**-over** *n* AIR amorçage d'arc *m*
Archimedean: ~ **screw** *n* WATER SUPP vis d'Archimède *f*
Archimedes': ~ **principle** *n* WATER SUPP principe d'Archimède *m*
area[1]: ~**-related** *adj* AIR, NOISE, SOIL, WATER SUPP local
area[2]: ~ **emission source** *n* AIR source locale d'émission *f*; ~ **of influence** *n* WATER SUPP zone d'influence *f*; ~ **pollution** *n* AIR pollution diffuse *f*; ~ **of pure air** *n* AIR zone d'air pur *f*; ~**-related noise** *n* NOISE bruit localisé *m*; ~**-related water demand** *n* WATER SUPP demande locale d'eau *f*; ~ **source** *n* AIR, SOIL source diffuse *f*; ~ **source pollutant** *n* AIR, SOIL polluant de source diffuse *m*; ~ **of water filtration** *n* WATER SUPP zone d'infiltration *f*
argon *n* (*Ar*) AIR, WASTE, WATER SUPP argon *m* (*Ar*)
arid: ~ **zone** *n* SOIL zone aride *f*
armament: ~ **repository site** *n* SOIL dépôt d'armes *m*
armature *n* WATER SUPP garniture *f*
aromatic[1] *adj* AIR, WATER SUPP aromatique
aromatic[2]: ~ **compounds** *n pl* AIR composés aromatiques *m pl*; ~ **hydrocarbon** *n* AIR, WATER SUPP hydrocarbure aromatique *m*; ~ **polyamide** *n* AIR *filter fibre material* polyamide aromatique *m*
arsenic *n* (*As*) SOIL arsenic *m* (*As*)
artesian[1]: *adj* WATER SUPP artésien
artesian[2]: ~ **groundwater** *n* WATER SUPP nappe artésienne *f*; ~ **spring** *n* WATER SUPP source artésienne *f*; ~ **water** *n* WATER SUPP eau artésienne *f*; ~ **well** *n* WATER SUPP puits artésien *m*
artificial[1] *adj* WATER SUPP artificiel

artificial[2]: ~ **acidification** *n* SOIL acidification artificielle *f*; ~ **ageing** *n* WASTE, WATER SUPP vieillissement artificiel *m*; ~ **groundwater** *n* WATER SUPP nappe artificielle *f*; ~ **groundwater recharge** *n* WATER SUPP réalimentation d'une nappe artificielle *f*; ~ **lake** *n* WATER SUPP lac artificiel *m*, lac de barrage *m*
artificially[1]: ~ **replenished** *adj* WATER SUPP réalimenté artificiellement
artificially[2]: ~-**recharged groundwater** *n* WATER SUPP nappe rechargée artificiellement *f*; ~-**replenished groundwater** *n* WATER SUPP nappe réalimentée artificiellement *f*
As *abbr* (*arsenic*) SOIL As (*arsenic*)
asbestos[1]: ~-**free** *adj* SAFETY sans amiante
asbestos[2] *n* AIR, SAFETY amiante *f*; ~ **cement pipe** WATER SUPP tuyau en amiante ciment *m*; ~ **exposure** SAFETY exposition à l'amiante *f*; ~ **exposure limit** SAFETY limite d'exposition à l'amiante *f*; ~-**free insulating plate** SAFETY plaque d'isolation sans amiante *f*; ~ **removal** POLICY désamiantage *m*, WASTE élimination de l'amiante *f*
asbestosis *n* AIR asbestose *m*
as-built: ~ **schedule** *n* MANAGE planning conforme à l'exécution *m*
aseptic *adj* WASTE aseptique
ash[1]: ~-**free** *adj* AIR, WATER SUPP exempt de cendre, sans cendre
ash[2] *n* AIR, WASTE, WATER POLL, WATER SUPP cendre *f*; ~ **content** AIR, WATER SUPP teneur en cendres *f*; ~ **discharge** WATER SUPP rejet de cendres *m*; ~ **removal** WATER SUPP *waste incineration plant* élimination des cendres *f*; ~ **remover** WATER SUPP extracteur de cendres *m*; ~ **vitrification** WASTE, WATER SUPP vitrification des cendres *f*
asphalt: ~ **recycling** *n* WATER SUPP recyclage de l'asphalte *m*; ~ **sealing** *n* WASTE, WATER SUPP asphaltisation *f*, bitumage *f*, revêtement en asphalte *m*, scellement au bitume *m*
asphyxiating: ~ **gas** *n* AIR, WASTE, WATER SUPP gaz asphyxiant *m*
aspiration: ~ **psychrometer** *n* ANALYSIS télépsychromètre d'Assmann *m*
assay *n* ANALYSIS *sample* échantillon prélevé *m*, *test* essai *m*
assembling: ~ **unit** *n* WASTE, WATER SUPP chaîne de montage *f*
assembly *n* WATER SUPP montage *m*
assess *vt* GEN évaluer; ~ **qualitatively** ANALYSIS évaluer qualitativement; ~ **quantitatively** ANALYSIS évaluer quantitativement
assessed: ~ **sample** *n* ANALYSIS échantillon évalué *m*
assessment *n* GEN évaluation *f*; ~ **criteria** *n pl* AIR, NOISE, SOIL critères d'évaluation *m pl*; ~ **level** ANALYSIS, NOISE niveau d'évaluation *m*
assimilable: ~ **organic carbon** *n* WATER POLL carbone organique assimilable *m*
assimilative: ~ **capacity** *n* WASTE, WATER SUPP capacité d'assimilation *f*
association: ~ **colloid** *n* WATER POLL colloïde micellaire *m*
atmosphere *n* AIR atmosphère *f*
atmospheric[1] *adj* AIR atmosphérique
atmospheric[2]: ~ **acidity** *n* AIR acidité atmosphérique *f*; ~ **boundary layer** *n* AIR couche limite atmosphérique *f*; ~ **chemical process** *n* AIR traitement chimique atmosphérique *m*; ~ **chemistry** *n* AIR chimie atmosphérique *f*; ~ **concentration** *n* AIR concentration atmosphérique *f*; ~ **conditions** *n pl* AIR conditions atmosphériques *f pl*; ~ **fallout** *n* AIR retombées atmosphériques *f pl*; ~ **fumes** *n pl* AIR effluent atmosphérique *m*; ~ **gas** *n* AIR gaz atmosphérique *m*; ~ **heat trapping** *n* AIR piège thermique atmosphérique *m*; ~ **inversion layer** *n* AIR couche d'inversion atmosphérique *f*; ~ **lifetime** *n* AIR durée de vie dans l'atmosphère *f*, temps de séjour dans l'atmosphère *m*; ~ **load** *n* AIR concentration dans l'atmosphère *f*; ~ **loading** *n* AIR concentration dans l'atmosphère *f*; ~ **mass** *n* AIR masse atmosphérique *f*; ~ **obscurity** *n* AIR obscurité atmosphérique *f*; ~ **oxygen** *n* AIR oxygène atmosphérique *m*; ~ **ozone** *n* AIR ozone atmosphérique *m*; ~ **ozone concentration** *n* AIR concentration d'ozone dans l'atmosphère *f*; ~ **phenomenon** *n* AIR phénomène atmosphérique *m*; ~ **pollution** *n* AIR pollution atmosphérique *f*; ~ **precipitation** *n* AIR, WATER SUPP précipitation atmosphérique *f*; ~ **pressure** *n* AIR pression atmosphérique *f*; ~ **pressure chemical ionization** *n* (*APCI*) ANALYSIS ionisation chimique sous pression atmosphérique *f*; ~ **reflection** *n* AIR réflexion atmosphérique *f*; ~ **research programmes** *n pl BrE*, ~ **research programs** *AmE* POLICY programmes de recherches sur l'atmosphère *m pl*; ~ **resources** *n pl* RENEW ENERG ressources atmosphériques *f pl*; ~ **scrubbing** *n* AIR lavage atmosphérique *m*; ~ **sulfur** *AmE*, ~ **sulphur** *BrE n* AIR soufre atmosphérique *m*; ~ **temperature** *n* AIR température atmosphérique *f*; ~ **trace gas** *n* AIR gaz à l'état de trace dans l'atmosphère *m pl*
atomic: ~ **absorption spectrometry** *n* (*AAS*) ANALYSIS spectrométrie d'absorption atomique *f*; ~ **absorption spectrophotometer** *n* ANALYSIS spectrophotomètre d'absorption atomique *m*; ~ **emission detection** *n* (*AED*) ANALYSIS détection d'émissions atomiques *f*; ~ **emission detector** *n* ANALYSIS détecteur d'émissions atomiques *m*; ~ **emission spectrometry** *n* (*AES*) ANALYSIS spectrométrie d'émission atomique *f*; ~ **fluorescence** *n* ANALYSIS fluorescence atomique *f*; ~ **fluorescence spectrometry** *n* (*AFS*) ANALYSIS spectrométrie de fluorescence atomique *f*; ~ **reactor** *n* WATER SUPP pile atomique *f*
atomization *n* AIR, SOIL, WATER SUPP atomisation *f*, pulvérisation *f*, vaporisation *f*
atomize *vt* AIR, SOIL, WATER SUPP atomiser, pulvériser, vaporiser
atomizer *n* AIR, SOIL, WATER SUPP atomiseur *m*, pulvérisateur *m*, vaporisateur *m*; ~ **nozzle** AIR buse d'atomiseur *f*
attached: ~ **shell** *n* NOISE *surface lining* anneau correspondant *m*
attemperation *n* AIR contrôle de température *m*, réfrigération de cuves de fermentation *f*
attenuate *vt* ANALYSIS affaiblir, *chemically* atténuer, NOISE amortir, atténuer
attenuated *adj* ANALYSIS affaibli
attenuation *n* ANALYSIS affaiblissement *m*, *chemical* atténuation *f*, NOISE amortissement *m*, atténuation *f*
attenuator *n* NOISE atténuateur *m*, silencieux *m*; ~ **pad** NOISE coussinet amortisseur *m*
attitude: ~ **survey** *n* POLICY enquête de comportement *f*
attrition *n* AIR attrition *f*; ~ **resistance** AIR résistance à l'attrition *f*
audibility *n* NOISE audibilité *f*; ~ **reference value** ANALYSIS valeur auditive de référence *f*
audible *adj* NOISE audible
audiometer *n* ANALYSIS, SAFETY audiomètre *m*
auditee *n* MANAGE audité *m*
auditor *n* MANAGE auditeur *m*
auger *n* SOIL *bioremediation* tarière *f*
authorization *n* POLICY, WATER SUPP autorisation *f*; ~ **application** WATER SUPP demande d'autorisation *f*; ~ **impact** WATER SUPP répercussion d'une autorisation *f*
authorize *vt* POLICY autoriser
autochthonous: ~ **matter** *n* SOIL matière autochtone *f*
autoecology *n* MANAGE, POLICY autoécologie *f*, écologie de l'individu *f*
automatic: ~ **burner** *n* WASTE brûleur automatique *m*; ~ **detector tube** *n* ANALYSIS tube détecteur automatique *m*; ~-**recognition unit** *n* WASTE, WATER SUPP poste de reconnaissance automatique *m*, unité de reconnaissance automatique *f*; ~ **sampling** *n* ANALYSIS échantillonnage automatique *m*, prélèvement automatique *m*; ~ **sorting** *n* WASTE, WATER SUPP tri automatique *m*

automatically: ~ **injected** *adj* ANALYSIS à injection automatique, injecté automatiquement
autosampler *n* ANALYSIS auto-échantillonneur *m*, dispositif d'auto-prélèvement *m*
autosampling *n* ANALYSIS auto-échantillonnage *m*
autothermic *adj* AIR autotherme
autotrophic: ~ **bacteria** *n pl* WATER POLL bactéries autotrophes *f pl*
auxiliary: ~ **air** *n* AIR air d'appoint *m*; ~ **burner** *n* AIR brûleur d'appoint *m*, WATER SUPP brûleur auxiliaire *m*; ~ **fan** *n* WATER SUPP ventilateur auxiliaire *m*; ~ **gas** *n* ANALYSIS gaz auxiliaire *m*, gaz secondaire *m*; ~ **material delivering industry** *n* MANAGE industrie de fourniture de matériel auxiliaire *f*
available: ~ **technology** *n* GEN technique existante *f*, technologie existante *f*
average: ~ **monthly dose** *n* AIR, ANALYSIS, SAFETY dose moyenne mensuelle *f*; ~ **noise level** *n* NOISE niveau de bruit moyen *m*; ~ **sample** *n* ANALYSIS échantillon moyen *m*; ~ **sound level** *n* NOISE niveau de bruit moyen *m*; ~ **value** *n* ANALYSIS valeur moyenne *f*
Avogadro: ~**'s constant** *n* AIR constante d'Avogadro *f*
A-weighted: ~ **sound pressure level** *n* NOISE niveau de pression acoustique pondéré en A *m*
axial[1] *adj* AIR axial
axial[2]: ~**-entry cyclone** *n* AIR cyclone à admission axiale *m*; ~ **fan** *n* AIR ventilateur axial *m*
azotometer *n* ANALYSIS azotomètre *m*

B

Bacharach: ~ **method** *n* AIR méthode Bacharach *f*; ~ **scale** *n* ANALYSIS *dust measurement* échelle de Bacharach *f*

back: ~ **pressure** *n* AIR contre-pression *f*; **~-projected screen** *n* RENEW ENERG rétroprojecteur *m*; ~ **titration** *n* ANALYSIS rétrotitrage *m*

backfill[1] *n* SOIL remblai *m*

backfill[2] *vt* SOIL remblayer, WASTE combler

backfilled *adj* SOIL remblayé, WASTE comblé

backfilling *n* SOIL remblayage *m*, WASTE comblement *m*

backflow *n* WATER SUPP retour d'eau *m*

backflush: ~ **screen changer** *n* WASTE changeur de grille de rinçage *m*, WATER SUPP dispositif de renouvellement de tamis à contre-courant *m*

backflushing: ~ **gas** *n* AIR gaz de rinçage *m*

background: ~ **concentration** *n* AIR concentration de référence *f*; ~ **level** *n* AIR, NOISE niveau de fond *m*, niveau de référence *m*; ~ **noise** *n* NOISE bruit de fond *m*; ~ **ozone concentration** *n* AIR concentration d'ozone de référence *f*; ~ **pollution** *n* AIR pollution de référence *f*, POLICY pollution de fond *f*; ~ **sample** *n* ANALYSIS échantillon de référence *m*; ~ **station** *n* POLICY station de mesure de la pollution de fond *f*

backscattering *n* ANALYSIS *ozone detection* rétrobalayage *m*, rétrodiffusion *f*

backward: **~-curved blade compressor** *n* AIR compresseur à lame rétroincurvée *m*

backwash: ~ **water** *n* AIR, SOIL, WATER SUPP eau de lavage *f*, eau de rinçage *f*

backwashing *n* AIR lavage à contre-courant *m*, WATER SUPP *treatment* lavage à contre-courant *m*, rétrolavage *m*

backwater *n* WATER SUPP *stagnant* bras mort *m*, *driven back* eau de remous *f*, *of canal* sous-biez *m*

bacteria *n pl* AIR *biological filter*, SOIL, WATER POLL bactéries *f pl*; ~ **activity** *n* AIR, SOIL, WATER POLL *biological filter* activité bactérienne *f*; ~ **bed** *n* WATER POLL lit bactérien *m*; ~ **filter** *n* AIR filtre bactérien *m*; ~ **test** *n* ANALYSIS essai bactériologique *m*

bacterial: ~ **coliform count** *n* WATER POLL nombre de germes coliformes *m*; ~ **count** *n* WATER SUPP numération bactérienne *f*; ~ **growth** *n* WATER SUPP prolifération bactérienne *f*; ~ **mass** *n* WATER SUPP masse bactérienne *f*

bacteriological: ~ **purification** *n* WASTE épuration bactériologique *f*; ~ **treatment** *n* WASTE traitement bactériologique *m*

baffle *n* AIR chicane *f*, piège *m*, ANALYSIS piège *m*, NOISE, WATER POLL chicane *f*, piège *m*; ~ **board** NOISE plaque de déviation *f*, plaque protectrice *f*; ~ **chamber** AIR chambre à chicanes *f*; ~ **collector** NOISE collecteur à déflecteur *m*, dépoussiéreur à chicane *m*; ~ **muffler** *AmE* (*cf baffle silencer BrE*) NOISE silencieux à chicanes *m*; ~ **silencer** *BrE* (*cf baffle muffler AmE*) NOISE silencieux à chicanes *m*

bag[1]: **~-shaped** *adj* AIR *filter* en forme de sac

bag[2] *n* AIR sac *m*, WASTE *large* sac *m*, *small* sachet *m*; ~ **bottom** AIR *of bag filter* fond de sac *m*; ~ **closing machine** WASTE machine à fermer *f*; ~ **column** AIR colonne à garnissage *f*, colonne à poche *f*; ~ **filter** AIR, SOIL filtre à sac *m*; ~ **holder** WASTE porte-sacs *m*; ~ **length** AIR longueur de sac *f*; ~ **making machine** WASTE machine de fabrication des sacs *f*; ~ **packaging** WASTE ensachage *m*; ~ **paper** WASTE papier d'emballage *m*

bagasse *n* RENEW ENERG bagasse *f*

baghouse *n* AIR installation de filtres à sac *f*

bale: ~ **clamp** *n* WASTE mâchoire à balles *f*; ~ **landfill** *n* WASTE, WATER SUPP décharge de balles *f*; ~ **storage belt** *n* WASTE bande transporteuse pour le stockage en balles *f*

baling *n* RENEW ENERG mise en balles *f*, pressage *m*; ~ **press** WASTE presse à balles *f*; ~ **scrap** WASTE mitraille pour paqueter *f*, résidu de mise en balles *m*

ballast: ~ **material** *n* WASTE matériau de ballast *m*

ballasted: ~ **pile** *n* WATER SUPP pilot ballasté *m*

ballistic[1] *adj* WASTE balistique

ballistic[2]: ~ **separation** *n* WASTE séparation balistique *f*; ~ **sorter** *n* WASTE trieur balistique *m*, séparateur balistique *m*; ~ **sorting** *n* WASTE tri balistique *m*

balloon: ~ **sonde** *n* ANALYSIS ballon-sonde *m*

bamboo: ~ **reed** *n* RENEW ENERG canne de Provence *f*

ban[1] *n* AIR, POLICY interdiction

ban[2] *vt* AIR, POLICY interdire

band: ~ **conveyor** *n* RENEW ENERG, WASTE bande transporteuse *f*; ~ **label** *n* WASTE étiquette en bande *f*; ~ **oven** *n* SOIL *pellet incineration* four à bande *m*; ~ **sealer** *n* WASTE banderoleuse *f*; ~ **sealing** *n* WASTE banderolage *m*

bandpass: ~ **filter** *n* ANALYSIS filtre passe-bande *m*

bank *n* WASTE *recycling* conteneur de collecte *m*, WATER SUPP *of river, lake* rive *f*, berge *f*, bord *m*, *of pond* rive *f*, *sea bed* banc *m*

banked: **~-up water level** *n* WASTE, WATER SUPP niveau d'eau rehaussé par remblai *m*, plan d'eau rehaussé *m*, plan d'eau surélevé *m*

banking: **~-up of water** *n* WATER SUPP rehaussement du niveau d'eau par remblai *m*

bar: ~ **coding** *n* WASTE, WATER SUPP *for identification* apposition d'un code à barres *f*; ~ **screen** *n* WATER POLL dégrilleur *m*

barometer *n* ANALYSIS baromètre *m*

barrage *n* RENEW ENERG barrage *m*

barrel: **~-venting system** *n* WASTE système d'aération de fûts *m*

barrier *n* WASTE barrière *f*; ~ **film** WASTE film de séparation *m*, pellicule-écran *f*; ~ **layer** WASTE, WATER SUPP couche de barrage *f*; ~ **material** WASTE, WATER SUPP matériau de barrage *m*; ~ **packaging** WASTE, WATER SUPP emballage de protection *m*, blindage de protection *m*; ~ **system** WASTE système de barrage *m*

base *n* SOIL *additive* base *f*; ~ **cation** AIR cation basique *m*; ~ **cup** WASTE coupelle *f*, embase *f*; **~-metal catalyst** AIR catalyseur métallique basique *m*; ~ **sealing** SOIL imperméabilisation de la base *f*, WASTE scellement du fond *m*

basic[1] *adj* GEN de base, fondamental

basic[2]: ~ **absorbent** *n* AIR absorbant basique *m*; ~ **law** *n* MANAGE, POLICY loi cadre *f*; ~ **line** *n* ANALYSIS ligne de base *f*

basin *n* ANALYSIS, WATER POLL, WATER SUPP bassin *m*, cuvette *f*, vasque *f*

basis: ~ **for charges** *n* POLICY assiette des redevances *f*

BAT *abbr* (*best available technology*) AIR, POLICY, SOIL, WATER SUPP meilleure technique disponible *f*, meilleure technique existante *f*, meilleure technologie existante *f*

batch[1]: **~-processed** *adj* AIR traité par lots

batch[2]: **in ~ quantities** *adv* AIR en vrac, WATER POLL par cuvées, WATER SUPP en vrac

batch[3] *n* SOIL lot *m*; **~-bulk injection** SOIL injection par lots *f*; ~ **distillation** SOIL distillation discontinue *f*; ~ **distillation tank** SOIL cuve de distillation discontinue *f*; ~ **operation** AIR traitement par lots *m*; ~ **process** AIR traitement par lots *m*

batch[4] *vt* WASTE traiter par lots

batching: **~ tank** *n* WATER POLL réservoir de préparation *m*

bathing: **~ place** *n* POLICY baignade *f*

BATNEEC *abbr* (*best available technology not entailing excessive costs*) AIR, POLICY, WATER SUPP meilleure technique disponible n'entraînant pas des coûts excessifs *f*, meilleure technique existante n'entraînant pas des coûts excessifs *f*, meilleure technologie existante n'entraînant pas des coûts excessifs *f*

battery[1]: **~-operated** *adj* GEN à piles; **~-powered** *adj* GEN alimenté par des piles

battery[2] *n* WASTE batterie *f*; **~-operated vehicle** AIR véhicule électrique *m*; **~-powered gas chromatograph** ANALYSIS chromatographie gazeuse alimentée par des piles *f*

bauxite *n* AIR *adsorbent*, WATER SUPP *for aluminium production* bauxite *f*

Be *abbr* (*beryllium*) AIR Be (*béryllium*)

beamsplitter *n* ANALYSIS diviseur de faisceau *m*, séparateur de faisceau *m*

bearing: **~ bridge** *n* AIR *packed-bed filter* pont porteur *m*

beat *vt* WATER SUPP battre

bed: **~ load** *n* WATER SUPP charriage de sédiment *m*

beheaded: **~ river** *n* WATER SUPP rivière capturée *f*

below: **~-cloud scavenging** *n* AIR entraînement sous les nuages *m*, lavage atmosphérique *m*

belt: **~ conveyor** *n* WASTE, WATER SUPP convoyeur à bande *m*

bench: **~-scale study** *n* ANALYSIS, MANAGE, POLICY étude sur banc d'essai *f*

bend *n* WATER SUPP *of river* méandre *m*

benthic: **~ community** *n* MARINE, WATER POLL communauté benthique *f*; **~ deposit** *n* WATER SUPP sédiment benthique *m*; **~ region** *n* WATER SUPP région benthique *f*

benthos *n* MARINE, WATER SUPP, WATER POLL benthos *m*

benzene *n* AIR benzène *m*

benzofuran *n* SOIL benzofurane *m*

berl: **~ saddle** *n* AIR *packing of packed-tower scrubber* selle de Berl *f*

berm: **~ ditch** *n* WATER SUPP caniveau *m*, contre-fossé *m*

Bernoulli: **~'s equation** *n* ANALYSIS équation de Bernoulli *f*

beryllium *n* (*Be*) AIR béryllium *m* (*Be*)

best: **~ available technology** *n* (*BAT*) AIR, POLICY, SOIL, WATER SUPP meilleure technique disponible *f*, meilleure technique existante *f*, meilleure technologie existante *f*; **~ available technology not entailing excessive costs** *n* (*BATNEEC*) AIR, POLICY, SOIL, WATER SUPP meilleure technique disponible n'entraînant pas des coûts excessifs *f*, meilleure technique existante n'entraînant pas des coûts excessifs *f*, meilleure technologie existante n'entraînant pas des coûts excessifs *f*; **~ practical environmental option** *n* (*BPEO*) POLICY solution la meilleure pour l'environnement *f*

beta: **~ factor** *n* WATER SUPP facteur bêta *m*; **~-ray absorption sampler** *n* ANALYSIS échantillonneur d'absorption par rayonnement bêta *m*

beverage: **~ container** *n* WASTE conteneur de boisson *m*

B-horizon *n* SOIL horizon-B *m*

bicycle: **~ path** *n* POLICY bande cyclable *f*

bifurcation: **~ point** *n* AIR *splitting of plume stack* point de bifurcation *m*

bilateral[1] *adj* POLICY bilatéral

bilateral[2]: **~ environmental agreement** *n* POLICY accord bilatéral en matière d'environnement *m*; **~ negotiation** *n* POLICY négociation bilatérale *f*

bilge: **~ water** *n* MARINE eau de cale *f*

bilharzia *n* WATER POLL bilharziose *f*

billet *n* RENEW ENERG rondin *m*

binary *adj* ANALYSIS *measuring gas* binaire

bind *vt* AIR *particles with washing liquid* lier, WASTE, WATER SUPP colmater

binder *n* RENEW ENERG, SOIL, WASTE, WATER SUPP liant *m*

binding *n* WASTE, WATER SUPP colmatage *m*; **~ agent** AIR agglutinant, RENEW ENERG, SOIL liant *m*, WASTE agglutinant *m*, liant *m*, WATER SUPP agglutinant *m*, agglomérant *m*, liant *m*, agent de liaison *m*

bioaccumulate *vt* AIR accumuler par voie biologique

bioaccumulation *n* AIR, SOIL bioaccumulation *f*; **~ capacity** AIR capacité de bioaccumulation *f*; **~ factor** AIR facteur de bioaccumulation *m*; **~ potential** AIR potentiel de bioaccumulation *m*, SOIL pouvoir de bioaccumulation *m*

bioaccumulative *adj* AIR, SOIL bioaccumulatif

bioassay *n* ANALYSIS, POLICY essai biologique *m*, bioessai *m*

bioaugmentation *n* SOIL bioaugmentation *f*

biocenosis *AmE see biocoenosis BrE*

biochemical: **~ oxygen demand** *n* (*BOD*) AIR, WASTE, WATER SUPP demande biochimique en oxygène *f* (*DBO*); **~ oxygen demand in five days** *n* (BOD_5) AIR, WASTE, WATER SUPP demande biochimique d'oxygène en cinq jours *f* (DBO_5); **~ tracer** *n* ANALYSIS, WATER SUPP traceur biochimique *m*

biocide *n* SOIL biocide *m*

biocoenosis *n BrE* AIR biocénose *f*

biodegradability *n* AIR, WASTE, WATER POLL, WATER SUPP biodégradabilité *f*

biodegradable[1] *adj* AIR, WASTE, WATER POLL, WATER SUPP biodégradable

biodegradable[2]: **~ matter** *n* WASTE, WATER POLL, WATER SUPP matériau biodégradable *m*; **~ plastic** *n* WASTE, WATER POLL plastique biodégradable *m*; **~ substance** *n* WASTE, WATER POLL, WATER SUPP substance biodégradable *f*; **~ waste** *n* WASTE, WATER POLL, WATER SUPP déchets biodégradables *m pl*

biodegradation *n* AIR *biological gas scrubbing* biodégradation *f*, WASTE, WATER POLL, WATER SUPP biodégradation *f*, dégradation biologique *f*

biodisc *n BrE* WATER POLL disque biologique *m*

biodisk *AmE see biodisc BrE*

biofilm *n* SOIL biofilm *m*

biofilter *n* AIR, WASTE, WATER SUPP biofiltre *m*; **~ effect** WASTE, WATER SUPP effet de biofiltre *m*

biofiltration *n* AIR, WASTE, WATER SUPP biofiltration *f*

biogas *n* WASTE, WATER SUPP biogaz *m*

biogeocenosis *AmE see biogeocoenosis BrE*

bio-geochemical: **~ cycle** *n* AIR cycle externe *m*, WATER POLL cycle biogéochimique *m*

biogeocoenosis *n BrE* AIR biogéocénose *f*

bio-geo-equilibrium *n* POLICY équilibre biogéologique *m*

bioindicator *n* ANALYSIS indicateur biologique *m*, SOIL, WATER SUPP bioindicateur *m*, indicateur biologique *m*

biological[1] *adj* GEN biologique

biological[2]: **~ agent** *n* WATER SUPP agent biologique *m*, agent antimicrobien *m*; **~ balance** *n* POLICY équilibre biologique *m*; **~ bed** *n* SOIL *injected with bacteria* lit biologique *m*; **~ catalyst** *n* AIR catalyseur biologique *m*; **~ damage** *n* SAFETY dommage biologique *m*; **~ degradation** *n* WASTE, WATER SUPP dégradation biologique *f*; **~ denitrification** *n* WATER SUPP dénitrification biologique *f*; **~ elimination** *n* SOIL élimination biologique *f*; **~ energy conversion** *n* WASTE, WATER SUPP conversion de l'énergie par voie biologique *f*, conversion de la bioénergie *f*; **~ filter** *n* AIR filtre biologique *m*, WATER POLL lit bactérien *m*, WATER SUPP lit percolateur *m*; **~ filtration** *n* AIR biofiltration *f*, *biological sorption process* filtration biologique *f*, WATER SUPP biofiltration *f*; **~ gas scrubbing** *n* AIR *biological sorption process* lavage biologique des gaz *m*; **~ ground filter** *n* AIR lit percolateur *m*; **~ harm** *n* SAFETY dommage biologique *m*; **~ hazard** *n* SAFETY risque biologique *m*; **~ indicator** *n* ANALYSIS, SOIL, WATER SUPP indicateur biologique *m*; **~ oxygen demand** *n* (*BOD*) AIR, WASTE, WATER SUPP demande biologique en oxygène *f* (*DBO*); **~ oxygen demand in five days** *n* (BOD_5) AIR, WASTE, WATER SUPP demande biologique d'oxygène en cinq jours *f* (DBO_5); **~ recycling** *n* WASTE recyclage biologique *m*; ~ **resources** *n pl* RENEW ENERG ressources biologiques *f pl*; ~ **sorption process** *n* AIR procédé de sorption biologique *m*; ~ **stabilization** *n* SOIL stabilisation biologique *f*; ~ **treatment** *n* WASTE, WATER SUPP traitement biologique *m*; ~ **waste** *n* WASTE déchets biologiques *m pl*; ~ **waste air treatment** *n* AIR traitement biologique des effluents gazeux *m*; ~ **water cycle** *n* WATER POLL cycle biologique de l'eau *m*; ~ **water quality determination** *n* ANALYSIS détermination biologique de la qualité de l'eau *f*

biologically[1]: ~ **catalytic** *adj* AIR catalytique par voie biologique; ~ **dead** *adj* SOIL sans aucune activité biologique

biologically[2]: ~**-catalytic regeneration** *n* AIR régénération catalytique par voie biologique *f*; ~**-catalytic waste air cleaning** *n* AIR épuration catalytique de l'air par voie biologique *f*

biomass *n* SOIL, WATER POLL, WATER SUPP biomasse *f*; ~ **pyramid** POLICY pyramide des biomasses *f*

bio-medical: ~ **waste** *n* WASTE déchets biomédicaux *m pl*

biomonitoring *n* ANALYSIS contrôle biologique *m*, MANAGE, POLICY biosurveillance *f*

bio-refuse *n* WASTE déchets d'origine biologique *m pl*

bio-remediation *n* SOIL *for soil containing organic compounds* assainissement biologique *m*

biosensor *n* ANALYSIS biocapteur *m*

biosphere *n* POLICY biosphère *f*; ~ **environment** POLICY environnement biosphérique *m*

biostabilizer *n* AIR, WASTE, WATER SUPP biostabilisateur *m*

biota *n* POLICY biote *m*

biotechnological[1] *adj* AIR, POLICY, RENEW ENERG WATER POLL biotechnologique

biotechnological[2]: ~ **treatment** *n* AIR traitement biotechnologique *m*

biotechnology *n* AIR, POLICY, RENEW ENERG, WATER POLL biotechnologie *f*

biotic[1] *adj* RENEW ENERG, WATER SUPP biotique

biotic[2]: ~ **potential** *n* MANAGE, POLICY capacité biotique *f*, SOIL potentiel biotique *m*; ~ **resources** *n pl* RENEW ENERG ressources biotiques *f pl*

biotope *n* POLICY biotope *m*

biowasher *n* AIR *biological sorption process* biolaveur *m*

bitumen: ~**-coated paper** *n* WASTE papier goudronné *m*; ~ **felt** *n* WATER SUPP feutre bitumé *m*

black: ~ **body** *n* AIR, ANALYSIS corps noir *m*; ~ **smoke number** *n* AIR indice de fumée noire *m*; ~ **water** *n* WATER POLL eau noire *f*

blackening *n* AIR *of building* noircissement *m*; ~ **index** AIR indice de noircissement *m*

blacklist *n* POLICY liste noire *f*

blade *n* AIR lame *f*, ailette *f*

blank: ~ **test** *n* ANALYSIS essai à blanc *m*, essai témoin *m*; ~ **value** *n* ANALYSIS valeur à blanc *f*

blast: ~ **connection** *n* AIR porte-vent *m*

bleach *vt* WATER POLL, WATER SUPP décolorer

blend *vt* ANALYSIS, SOIL, WASTE *plastics*, WATER POLL, WATER SUPP mélanger; ~ **with virgin materials** WASTE *plastics* mélanger avec des matériaux vierges

blind: ~ **drainage area** *n* WATER SUPP bassin fermé *m*; ~ **test** *n* ANALYSIS essai en aveugle *m*

blister: ~ **pack** *n* WASTE emballage sous film *m*, emballage thermoformé *m*; ~ **packaging machine** *n* WASTE machine pour emballage moulant *f*; ~ **seal** *n* WASTE fermeture pour emballage thermoformé *f*

block: ~ **polymer** *n* WASTE copolymère bloc *m*

blockage *n* AIR *catalyst* blocage *m*, WATER POLL, WATER SUPP obstruction *f*

blow[1]: ~ **fill seal system** *n* WASTE système de remplissage par injection *m*; ~ **molding** *AmE*, ~ **moulding** *BrE* *n* WASTE moulage par air comprimé *m*, moulage par soufflage *m*

blow[2] *vt* AIR souffler; ~ **into** AIR injecter; ~ **off** WATER SUPP purger

blowby: ~ **gas** *n* AIR gaz de carter *m*

blower *n* WASTE soufflerie *f*; ~ **stream velocity** AIR *catalyst* vitesse de soufflage *f*, vitesse de ventilation *f*

blowing: ~ **agent** *n* WASTE agent gonflant *m*; ~ **through** *n* AIR soufflage *m*

blown: ~ **bottle** *n* WASTE bouteille soufflée *f*

blowring: ~ **filter** *n* AIR filtre à joint gonflable *m*

blue: ~ **angel** *n* POLICY *eco-label* ange bleu *m*

BOD[1] *abbr* AIR, WASTE, WATER SUPP (*biological oxygen demand*) DBO (*demande biologique en oxygène*), (*biochemical oxygen demand*) DBO (*demande biochimique en oxygène*)

BOD[2]: ~ **determination** *n* ANALYSIS calcul de la DBO *m*; ~ **determination in wastewater** *n* ANALYSIS calcul de la DBO des eaux usées *m*

BOD_5[1] *abbr* AIR, WASTE, WATER SUPP (*biological oxygen demand in five days*) DBO_5 (*demande biologique d'oxygène en cinq jours*), (*biochemical oxygen demand in five days*) DBO_5 (*demande biochimique d'oxygène en cinq jours*)

BOD_5[2]: ~ **decomposition** *n* WATER SUPP décomposition mesurée par la DBO_5 *f*; ~ **determination** *n* ANALYSIS calcul de la DBO_5 *m*

bodying: ~ **agent** *n* RENEW ENERG agent épaississant *m*

bog *n* SOIL fondrière *f*

boil *vt* ANALYSIS bouillir, porter à ébullition

boiled: ~ **sample** *n* ANALYSIS échantillon porté à ébullition *m*

boiler *n* GEN chaudière *f*; ~ **ash** WASTE cendre de chaudière *f*; ~ **feedwater** AIR eau alimentaire de chaudière *f*; ~ **feed-water pump** AIR *full- and partial-load*, WASTE pompe alimentaire de chaudière *f*; ~ **plant** AIR, WASTE *landfill gas use* installation de méthanisation *f*; ~ **scale** SOIL, WATER POLL, WATER SUPP tartre *m*, calamine *f*; ~ **tube** WASTE tubulure de chaudière *f*; ~ **water** AIR eau de chaudière *f*

boiling *n* ANALYSIS ébullition *f*; ~ **analysis** ANALYSIS ébulliométrie *f*; ~ **point** AIR point d'ébullition *m*; ~ **vessel** AIR bouilloire *f*, cuve à ébullition *f*

bomb: ~ **calorimeter** *n* ANALYSIS bombe calorimétrique *f*

bond[1] *n* AIR liaison *f*, WATER SUPP *of fuel and can* joint *m*, liaison *f*

bond[2] *vt* AIR, WASTE liaisonner, lier

bonded: ~ **seal** *n* WASTE fermeture hermétique collée *f*

bonding *n* AIR, WATER SUPP *particles on surface* liaison *f*; ~ **agent** WASTE adhésif *m*, liant *m*; ~ **test** WASTE essai de mise à la masse *m*

booster: ~ **fan** *n* AIR ventilateur de surpression *m*

borderline: ~ **acceptability** *n* POLICY limite d'acceptabilité *f*

bore[1]: ~ **machine** *n* WASTE foreuse *f*, sondeuse *f*

bore[2] *vi* RENEW ENERG, WATER SUPP *for water* forer

bored: ~ **well** *n* WATER SUPP puits de forage *m*, puits tubé *m*, puits foré *m*

borehole *n* ANALYSIS trou de sondage *m*; ~ **pump** WATER SUPP pompe de forage *f*; ~ **wall** SOIL paroi du trou de sonde *f*; ~ **well** WATER SUPP puits de forage *m*

bottle: ~ **bank** *n* WASTE conteneur pour verre usagé *m*; ~ **cap** *n* WASTE capsule *f*; ~ **carrier** *n* WASTE porte-bouteille *m*; ~ **corking machine** *n* WASTE machine à boucher les bouteilles *f*; ~ **crate** *n* WASTE caisse à bouteilles *f*; ~ **deposit** *n* WASTE consigne de bouteille *f*; ~ **jacket** *n* WASTE gaine pour bouteille *f*; ~ **leak detector** *n* WASTE détecteur de fuite sur bouteille *m*; ~ **packing machine** *n* WASTE fardeleuse *f*; ~ **rinsing machine** *n* WASTE rinceuse *f*; ~ **sealing machine** *n* WASTE encapsuleuse *f*; ~ **sleeve** *n* WASTE manchon pour bouteille *m*; ~ **stopper** *n* WASTE bouchon de bouteille *m*; ~ **test** *n* ANALYSIS essai en flacon *m*; ~ **unscrambler** *n* WASTE rangeur de bouteilles *m*; ~ **washing machine** *n* WASTE rinceuse *f*

bottling: ~ **line** *n* WASTE chaîne de mise en bouteille *f*; ~ **machine** *n* WASTE embouteilleuse *f*

bottom: ~ **coder** *n* WASTE dispositif de codage de fond *m*;

~ **culvert** *n* WATER SUPP pertuis de fond *m*; ~ **deposit** *n* WATER POLL vase de fond *f*; ~ **deposit sampler** *n* ANALYSIS échantillonneur de dépôt de sol *m*; ~ **drainage** *n* WASTE drainage de fond *m*; ~ **filling** *n* WASTE remplissage de fond *m*; ~ **flap** *n* WASTE trappe de fond *f*; ~ **fold** *n* WASTE soufflet *m*; ~**-living communities** *n pl* MARINE espèce benthique *f*; ~ **outlet** *n* WATER SUPP vidange de fond *f*; ~ **scraper** *n* WATER SUPP racleur de fond *m*; ~ **sealing** *n* SOIL imperméabilisation de fond *f*

boundary: ~ **layer** *n* AIR couche limite *f*, surface de séparation *f*; ~ **layer flow** *n* AIR écoulement sur la couche limite *m*, écoulement sur la surface de séparation *m*

boxing: ~ **machine** *n* WASTE encaisseuse *f*

Boyle: ~**'s law** *n* ANALYSIS loi de Boyle-Mariotte *f*

BPEO *abbr* (*best practical environmental option*) POLICY solution la meilleure pour l'environnement *f*

brace *n* AIR *stringer* bride de liaison *f*

brackish *adj* WATER SUPP saumâtre

branch: ~ **connection** *n* WATER SUPP piquage *m*; ~ **sewer** *n* WATER POLL collecteur secondaire *m*

branching *n* WATER SUPP branchement *m*, embranchement *m*, ramification *f*

branding: ~ **iron** *n* WASTE fer à marquer à chaud *m*

breach: ~ **of law** *n* POLICY infraction à la loi *f*

break[1] *vt* AIR *liquid into fine droplets* fragmenter, WASTE concasser; ~ **down** AIR décomposer, dissocier, fragmenter, SOIL, WASTE, WATER POLL, WATER SUPP décomposer; ~ **open** SAFETY *safe, lock* fracturer, ouvrir par effraction, WATER SUPP enfoncer; ~ **up** AIR fractionner, rompre, WATER POLL fractionner

break[2]: ~ **down** *vi* AIR, WASTE, WATER POLL, WATER SUPP se décomposer

breakdown *n* AIR panne *f*, WASTE, WATER POLL, WATER SUPP *of organic substances* décomposition *f*

breaker *n* SOIL, WASTE concasseur *m*

breaking: ~ **load** *n* WASTE charge de rupture *f*; ~ **up** *n* WATER SUPP répartition *f*

breakpoint: ~ **chlorination** *n* WATER POLL chloration au point critique *f*

breathable *adj* AIR, SAFETY *air* respirable

breathing: ~ **capacity** *n* AIR capacité respiratoire *f*; ~ **protection** *n* SAFETY protection respiratoire *f*; ~**-protection system** *n* SAFETY système de protection respiratoire *m*

breeches: ~ **pipe** *n* AIR tuyau à fourche *m*

breeding *n* RENEW ENERG sélection *f*; ~ **program** *AmE*, ~ **programme** *BrE* RENEW ENERG programme de sélection *m*

Briggs': ~ **logarithm** *n* AIR logarithme de Briggs *m*

brine *n* WATER POLL saumure *f*; ~ **spring** WATER SUPP source saline *f*

bring[1]: ~ **system** *n* WASTE reprise *f*

bring[2]: ~ **into cultivation** *phr* RENEW ENERG, SOIL mettre en culture

briquette *n* WASTE briquette *f*; ~ **handling** WASTE manutention des briquettes *f*; ~ **pression** WASTE presse à agglomérer *f*

British: ~ **Standards Institution** *n* (*BSI*) ANALYSIS British Standards Institution *m* (*BSI*)

brittleness *n* AIR *filter materials* fragilité *f*, friabilité *f*

broken: ~ **glass** *n* WATER SUPP calcin *m*

bromide *n* WATER SUPP bromure *m*

brown: ~ **coal power station** *n* AIR centrale thermique au lignite *f*

Brownian: ~ **movement** *n* AIR *of particles* mouvement brownien *m*

BSI *abbr* (*British Standards Institution*) ANALYSIS BSI (*British Standards Institution*)

bubble *n* WATER POLL bulle *f*; ~ **cap** AIR calotte à barbotage *f*; ~**-cap column** AIR colonne à plateaux de barbotage *f*; ~**-cap tray column** AIR colonne à plateaux de barbotage *f*; ~**-cap tray tower** AIR colonne à plateaux de barbotage *f*; ~ **column** AIR colonne à bulles *f*, colonne à plateaux de barbotage *f*; ~ **film** WASTE film à bulles *m*; ~ **pack** WASTE emballage alvéolé *m*, emballage coque *m*; ~ **washer** AIR laveur à barbotage *m*

bubbler *n* ANALYSIS barboteur *m*

bubbling *n* ANALYSIS, WATER POLL barbotage *m*; ~ **aeration** WATER SUPP aération par barbotage *f*

bucket: ~ **conveyor** *n* WASTE, WATER SUPP convoyeur à godets *m*

buffer[1] *n* AIR, ANALYSIS tampon *m*; ~ **capacity** SOIL pouvoir tampon *m*; ~ **effect** AIR, ANALYSIS effet tampon *m*; ~ **reaction** AIR réaction tampon *f*; ~ **solution** ANALYSIS solution tampon *f*; ~ **tank** AIR cuve tampon *f*; ~ **zone** NOISE zone tampon *f*

buffer[2] *vt* AIR, ANALYSIS amortir, tamponner

buffering *n* AIR, ANALYSIS amortissement *m*; ~ **action** AIR, ANALYSIS action d'amortissement *f*, effet tampon *m*; ~ **effect** AIR, ANALYSIS effet tampon *m*

build: ~ **in** *vt* SOIL incorporer

building: ~ **density** *n* POLICY densité de construction *f*; ~ **development plan** *n* POLICY plan d'aménagement *m*; ~ **permit** *n* POLICY permis de construire *m*; ~ **rubble** *n* WASTE gravats *m pl*, blocaille *f*; ~ **vibration** *n* NOISE vibration de construction *f*, vibration des bâtiments *f*

built[1]: ~**-in** *adj* SOIL incorporé

built[2]: ~**-in measuring equipment** *n* SOIL équipement de mesure intégré *m*; ~**-in monitoring equipment** *n* SOIL équipement de contrôle intégré *m*; ~**-in obsolescence** *n* WASTE obsolescence intégrée *f*; ~**-in regulating equipment** *n* SOIL équipement de régulation intégré *m*; ~**-in structure** *n* AIR structure intégrée *f*; ~**-up area** *n* POLICY zone bâtie *f*; ~**-up surface** *n* SOIL surface construite *f*, WASTE zone construite *f*

bulk[1] *adj* AIR, WASTE, WATER SUPP en vrac

bulk[2]: ~ **collection** *n* WASTE enlèvement des déchets volumineux *m*, collecte des déchets volumineux *f*, collecte en vrac *f*; ~ **density** *n* ANALYSIS masse volumique apparente *f*, WASTE densité apparente *f*, densité en vrac *f*; ~ **deposition** *n* WATER SUPP dépôt global *m*, dépôt brut *m*, dépôt en vrac *m*, retombées globales *f pl*; ~ **good** *n* AIR matériau en vrac *m*, WASTE matériau en vrac *m*, matériau produit en grandes quantités *m*; ~ **goods** *n pl* WASTE marchandise en vrac *f*; ~ **material** *n* AIR *packed-bed filter*, WASTE matériau en vrac *m*; ~ **material catalyst** *n* AIR catalyseur en vrac *m*; ~ **material reactor** *n* WATER SUPP réacteur à matériau en vrac *m*; ~ **solids handling** *n* WASTE technique du vrac *f*; ~ **transport** *n* WASTE transport en vrac *m*

bulking *n* WATER POLL *of sludge* gonflement *m*; ~ **agent** SOIL agent de gonflement *m*

bulky[1] *adj* WASTE encombrant, volumineux

bulky[2]: ~ **waste** *n* WASTE, WATER SUPP déchets encombrants *m pl*, déchets monstres *m pl*

bulldozer *n* WASTE bouteur *m*

bundle *n* WASTE botte *f*, faisceau *m*; ~**-tying machine** WASTE fardeleuse-ficeleuse *f*

bunker: ~ **tank** *n* AIR, WATER SUPP soute *f*

Bunsen: ~ **burner** *n* ANALYSIS bec Bunsen *m*

burden *n* AIR charge *f*, fardeau *m*

burette *n* ANALYSIS burette *f*

burn[1]: ~**-off rate** *n* AIR taux de brûlage *m*, taux de combustion *m*

burn[2] *vt* WASTE, WATER POLL brûler; ~ **out** AIR *flue gas* épuiser par combustion, SOIL brûler totalement

burnable: ~ **gas** *n* AIR gaz brûlable *m*

burner *n* AIR *thermal and catalytic combustion*, WASTE, WATER POLL, WATER SUPP brûleur *m*

burning: ~ **out** *n* AIR *of pollutants in thermal combustion* dégagement *m*

burnout *n* AIR, WASTE arrêt par épuisement *m*, caléfaction *f*

burnt: **~-out aggregate** *n* WASTE, WATER SUPP imbrûlés *m pl*

bursting: **~ pressure** *n* WASTE pression d'éclatement *f*; **~ strength** *n* WASTE résistance à l'éclatement *f*

bush: **~ fires** *n pl* AIR, SOIL feux de brousse *m pl*; **~ killer** *n* SOIL, WATER POLL débroussaillant *m*

butadiene *n* AIR butadiène *m*

butane *n* AIR butane *m*

bye-law *n pl* MANAGE statut d'une société *m*

bypass *n* AIR *protection against overheating* dérivation *f*, ANALYSIS bypass *m*, dérivation *f*; **~ chimney** AIR cheminée de dérivation *f*; **~ duct** AIR conduite de dérivation *f*; **~ filter** AIR filtre de dérivation *m*; **~ flow** AIR flux de dilution *m*; **~-pressure sensor** AIR capteur de pression de contournement *m*; **~ pump** AIR pompe de dérivation *f*; **~ stack** AIR cheminée de dérivation *f*; **~ stop valve** AIR vanne d'arrêt à contre-pression *f*, vanne d'arrêt et de contournement *f*; **~ system** AIR, ANALYSIS système de contournement *m*

by-product *n* AIR *of combustion*, MANAGE, WASTE, WATER SUPP sous-produit *m*, WASTE régénérat *m*; **~ recovery** AIR, WASTE, WATER SUPP récupération des sous-produits *f*

C

C *abbr* (*carbon*) AIR, ANALYSIS C (*carbone*)
Ca *abbr* (*calcium*) AIR, SOIL, WATER SUPP Ca (*calcium*)
CA *abbr* (*controlled atmosphere*) WASTE, WATER SUPP atmosphère contrôlée *f*
cab *n* SAFETY *building machinery* cabine de conduite *f*, poste de conduite *m*; **~ enclosure** SAFETY cage de cabine *f*, coque de cabine *f*; **~ filter** SAFETY filtre de cabine *m*; **~ with integrated activated-carbon filter unit** SAFETY cabine avec filtre à charbons actifs intégré *f*
cable: **~ drum** *n* WASTE tambour de câble *m*; **~-guided drilling grab** *n* SOIL trépan-benne de forage au câble *m*; **~ reel** *n* WASTE bobine de câble *f*
cadmium *n* (*Cd*) SOIL cadmium *m* (*Cd*)
caisson *n* WATER POLL caisson *m*, caisson de plongée *m*
caking *n* AIR, WASTE *of bag filter* agglomération *f*, collage *m*, prise *f*
calcinate *vt* AIR, WASTE, calciner, fritter
calcination *n* AIR, WASTE calcination *f*, frittage *m*
calcium *n* (*Ca*) AIR, SOIL, WATER SUPP calcium *m* (*Ca*); **~ hardness** WATER SUPP titre hydrotimétrique calcique *m*; **~ oxide** (*CaO*) AIR, SOIL *quicklime* oxyde de calcium *m*; **~ sulfate hemihydrated** *AmE*, **~ sulphate hemihydrated** *BrE* AIR *gypsum form* sulfate de calcium semi-hydraté *m*
calculable *adj* ANALYSIS calculable
calculation *n* ANALYSIS calcul *m*
calender: **~ unit** *n* WASTE unité calendaire *f*
calibrate *vt* ANALYSIS étalonner
calibrated[1] *adj* ANALYSIS calibré, étalonné, taré
calibrated[2]: **~ substance** *n* ANALYSIS substance étalonnée *f*; **~ value** *n* ANALYSIS valeur étalonnée *f*; **~ watershed** *n* SOIL *agriculture* bassin jaugé *m*
calibrating: **~ gas** *n* AIR gaz d'étalonnage *m*, gaz de référence *m*, ANALYSIS gaz d'étalonnage *m*
calibration *n* ANALYSIS étalonnage *m*, calibrage *m*, jaugeage *m*; **~ buffer** ANALYSIS *pH-meter* correcteur d'étalonnage *m*; **~ curve** ANALYSIS courbe d'étalonnage *f*; **~ data** *n pl* ANALYSIS caractéristiques d'étalonnage *f pl*; **~ gas** AIR gaz d'étalonnage *m*, ANALYSIS gaz d'étalonnage *m*, gaz de zéro *m*; **~ gas bottle** ANALYSIS bouteille de gaz de référence *f*; **~ mark** ANALYSIS repère d'étalonnage *m*; **~ mode** ANALYSIS méthode d'étalonnage *f*; **~ process** ANALYSIS procédure d'étalonnage *f*; **~ pulse** ANALYSIS impulsion étalon *f*; **~ record** ANALYSIS dossier d'étalonnage *m*, registre d'étalonnage *m*; **~ solution** ANALYSIS solution d'étalonnage *f*; **~ storer** ANALYSIS enregistreur d'étalonnage *m*, enregistreur-bloqueur *m*; **~ unit** ANALYSIS banc d'étalonnage *m*, unité d'étalonnage *f*
calibrator *n* ANALYSIS étalonneur *m*
calorific: **~ value** *n* AIR, ANALYSIS, WASTE *of material* pouvoir calorifique *m*
calorimeter *n* ANALYSIS calorimètre *m*
calorimetric: **~ pararosaniline method** *n* ANALYSIS *gas measurement* méthode calorimétrique à la pararosaniline *f*
calorimetry *n* ANALYSIS calorimétrie *f*
campaign *n* POLICY campagne *f*
can *n* (*cf tin BrE*) ANALYSIS bidon *m*, WASTE boîte *f*, boîte de conserve *f*, boîte en fer blanc *f*; **~ closing machine** WASTE machine à sceller les boîtes *f*; **~-filling line** WASTE chaîne de remplissage des boîtes *f*; **~ packing machine** WASTE machine à emballer les boîtes *f*; **~ relabeling** *AmE*, **~ relabelling** *BrE* WASTE réétiquetage des boîtes *m*; **~ sealing compound** WASTE élément de sertissage *m*
canal *n* WATER POLL, WATER SUPP canal *m*; **~ junction** WATER SUPP embranchement de canal *m*
canalization *n* MANAGE, WATER POLL, WATER SUPP canalisation *f*
canalize *vt* MANAGE, WATER POLL, WATER SUPP canaliser
cancer: **~ risk** *n* SAFETY risque de cancer *m*
canister *n* WASTE bidon *m*, fût *m*
canopy *n* RENEW ENERG, SOIL couvert forestier *m*
CaO *abbr* (*calcium oxide*) AIR, SOIL CaO (*oxyde de calcium*)
caoutchouc *n* WASTE caoutchouc *m*
cap[1] *n* AIR, NOISE plateau *m*, WASTE *of bottle* capsule *f*, plateau *m*; **~ sealing** WASTE operculage *m*; **~ sealing compound** WASTE matériau d'operculage *m*; **~ sealing equipment** WASTE matériel d'operculage *m*
cap[2] *vt* WASTE capsuler
CAP *abbr* (*controlled-atmosphere packaging*) WASTE conditionnement sous atmosphère contrôlée *m*
capacity *n* GEN puissance *f*, capacité *f*, débit *m*
capillarity *n* SOIL, WATER POLL capillarité *f*
capillary *n* ANALYSIS capillaire *m*; **~ column** ANALYSIS colonne capillaire *f*; **~ electrophoresis** (*CE*) ANALYSIS électrophorèse capillaire *f*; **~ electrophoresis-mass spectrometry** (*CE-MS*) ANALYSIS électrophorèse capillaire couplée à la spectrométrie de masse *f*; **~ flow meter** ANALYSIS débitmètre capillaire *m*; **~ gas chromatography** ANALYSIS chromatographie en phase gazeuse par capillarité *f*; **~ gas chromatography mass spectrometer** (*GC-MS*) ANALYSIS chromatographie gazeuse par capillarité couplée à un spectromètre de masse *f* (*CG-SM*); **~ gel electrophoresis** (*CGE*) ANALYSIS électrophorèse capillaire sur gel *f*; **~ injector** ANALYSIS injecteur capillaire *m*; **~ iso-electric focusing** (*CIEF*) ANALYSIS focalisation isoélectrique capillaire *f*, réglage isoélectrique capillaire *m*; **~ tube** ANALYSIS tube capillaire *m*; **~ viscosimeter** ANALYSIS viscosimètre à tube capillaire *m*, viscosimètre de Poiseuil *m*; **~ water** SOIL, WATER POLL, WATER SUPP eau capillaire *f*, eau de rétention capillaire *f*; **~ zone electrophoresis** (*CZE*) ANALYSIS électrophorèse de zone capillaire *f*
capnograph *n* ANALYSIS *for continuous measurement of dust concentration* capnographe *m*
capper *n* WASTE capsuleuse *f*
capping *n* WASTE capsulation *f*; **~ press** WASTE presse à capsuler *f*
caption *n* WASTE légende *f*
captive: **~ balloon** *n* AIR ballon captif *m*
capture *vt* AIR *harmful substance* capter, capturer, MANAGE, NOISE capter
car: **~ fragmentation plant** *n* WASTE installation de déchiquetage des véhicules *f*; **~ pool** *n* POLICY co-voiturage *m*; **~ population** *n* POLICY parc automobile *m*
carbide *n* AIR, WATER POLL carbure *m*
carbon *n* (*C*) AIR, ANALYSIS carbone *m* (*C*); **~ cycle** AIR cycle du carbone *m*; **~ dioxide** AIR (CO_2) dioxyde de carbone *m* (CO_2); **~ dioxide budget** AIR bilan de dioxyde de carbone *m*; **~ dioxide greenhouse effect** AIR effet de serre induit par le dioxyde de carbone *m*; **~ monoxide** (*CO*) AIR monoxyde de carbone *m* (*CO*); **~ monoxide detector** ANALYSIS détecteur de monoxyde de carbone *m*; **~-nitrogen ratio** (*C-N ratio*) ANALYSIS rapport carbone-azote *m*; **~-nitrogen-phosphorous ratio** (*C-N-P ratio*) ANALYSIS rapport carbone-azote-phosphore *m*

carbonate: ~ **hardness** *n* WATER POLL dureté carbonatée *f*; ~ **removal** *n* WATER SUPP décarbonisation *f*
carbonic: ~ **anhydride** *n* (CO_2) WATER SUPP anhydride carbonique *m* (CO_2)
carboniferous *adj* AIR carbonifère
carbonyl: ~ **sulfide** *AmE*, ~ **sulphide** *BrE n* AIR oxysulfure de carbone *m*
carburetor *AmE see carburettor BrE*
carburettor *n BrE* AIR carburateur *m*; ~ **emissions** *n pl BrE* AIR *hydrocarbon* émissions de carburateur *f pl*
carcinogenic[1] *adj* SAFETY carcinogène, cancérigène
carcinogenic[2]: ~ **substance** *n* SAFETY substance carcinogène *f*
cardboard: ~ **backing** *n* WASTE renfort cartonné *m*; ~ **machine** *n* WASTE cartonneuse *f*; ~ **packaging** *n* WASTE emballage en carton *m*; ~ **tray** *n* WASTE barquette en carton *f*; ~ **tube** *n* WASTE tube en carton *m*
carded: ~ **packaging** *n* WASTE conditionnement sous carte *m*
carousel: ~ **extractor** *n* WASTE carrousel d'extraction *m*
carriage: ~ **by sea** *n* WASTE transport maritime *m*
carrier *n* AIR, ANALYSIS porteur *m*, vecteur *m*; ~ **bag** WASTE sac à provisions *m*; ~ **box** WASTE boîte à provisions *f*; ~ **gas** AIR, ANALYSIS gaz porteur *m*, gaz vecteur *m*; ~ **layer** AIR *catalyser* support *m*; ~ **material** AIR, ANALYSIS matériau porteur *m*, matériau vecteur *m*, SOIL base support *f*; ~ **medium** AIR support *m*, vecteur *m*; ~ **surface** AIR *catalyst* surface porteuse *f*
carry[1]: ~-**over** *n* MANAGE report de crédit *m*
carry[2]: ~ **out** *vt* AIR, MANAGE effectuer
carrying: ~ **capacity** *n* WASTE *of pipe* capacité de transport *f*, WATER POLL, WATER SUPP capacité limite *f*, charge biotique utile *f*; ~ **handle** *n* WASTE poignée *f*
carton *n* WASTE carton *m*; ~ **erector and closer** WASTE machine à dresser et fermer les cartons *f*; ~ **filler** WASTE remplisseuse de cartons *f*
cartoner *n* WASTE encartonneuse *f*
cartoning: ~ **machine** *n* WASTE encartonneuse *f*
cartridge: ~ **filter** *n* AIR filtre à cartouche *m*; ~ **paper** *n* WASTE papier à carreaux *m*, papier de mise en cartes *m*
cascade: ~ **acidification** *n* SOIL acidification en cascade *f*; ~ **impactor** *n* ANALYSIS échantillonneur à chocs en cascade *m*, conimètre à impact en cascade *m*
case: ~ **erecting, filling and closing machine** *n* WASTE machine à former, remplir et fermer les cartons caisses *f*; ~ **law** *n* POLICY jurisprudence *f*; ~ **lining paper** *n* WASTE revêtement en papier pour caisses *m*; ~ **loader** *n* WASTE machine à encaisser *f*; ~ **packing** *n* WASTE mise en caisse *f*; ~ **sealing machine** *n* WASTE machine à fermer les caisses *f*
cassava *n* RENEW ENERG manioc *m*
cast: ~-**iron pipe** *n* WATER SUPP tuyau en fonte *m*
caster: ~-**equipped container** *n* WASTE conteneur sur roulettes *m*
catalyse *vt BrE* AIR, SOIL catalyser
catalysis *n* AIR, SOIL catalyse *f*
catalyst *n* AIR, ANALYSIS, SOIL catalyseur *m*; ~ **bed** AIR lit catalytique *m*; ~ **carrier** AIR support de catalyse *m*; ~ **charge** AIR charge catalytique *f*; ~ **container** AIR bac catalytique *m*; ~ **contamination** AIR contamination du catalyseur *f*, empoisonnement du catalyseur *m*; ~ **deactivation** AIR désactivation du catalyseur *f*; ~ **degradation** AIR dégradation du catalyseur *f*; ~ **deterioration** AIR détérioration du catalyseur *f*; ~ **dust** AIR poussière de catalyseur *f*; ~ **efficiency** AIR rendement de catalyse *m*; ~ **element** AIR élément du catalyseur *m*; ~ **fines** *n pl* AIR fines de catalyseur *f pl*; ~ **lifetime** AIR durée de vie du catalyseur *f*; ~ **material** AIR matériau catalyseur *m*; ~ **module** AIR module de catalyseur *m*; ~ **plate** AIR plateau de catalyse *m*; ~ **poison** AIR poison catalytique *m*; ~ **poisoning** AIR empoisonnement du catalyseur *m*; ~ **regeneration** AIR régénération catalytique *f*; ~ **step** AIR stade de catalyse *m*; ~ **surface** AIR surface catalytique *f*; ~ **volume** AIR volume de catalyseur *m*
catalytic[1] *adj* AIR, SOIL catalytique
catalytic[2]: ~ **acceleration** *n* AIR accélération catalytique *f*; ~ **active layer** *n* AIR couche active de catalyseur *f*; ~ **active metal** *n* AIR métal catalytique actif *m*; ~ **activity** *n* AIR activité catalytique *f*; ~ **coating** *n* AIR *of filter material* revêtement catalytique *m*; ~ **combustion** *n* AIR combustion catalytique *f*; ~ **combustion plant** *n* AIR installation à combustion catalytique *f*; ~ **converter** *n* AIR convertisseur catalytique *m*; ~ **converter closed loop** *n* AIR convertisseur catalytique en circuit fermé *m*; ~ **converter open loop** *n* AIR convertisseur catalytique en circuit ouvert *m*; ~ **cracker** *n* AIR craqueur catalytique *m*; ~ **cracking** *n* AIR craquage catalytique *m*; ~ **cracking unit** *n* AIR installation de craquage catalytique *f*; ~ **impregnation** *n* AIR imprégnation catalytique *f*; ~ **incineration** *n* AIR incinération catalytique *f*; ~ **incineration plant** *n* AIR usine d'incinération catalytique *f*; ~ **incinerator** *n* AIR incinérateur catalytique *m*; ~ **muffler** *n* AIR pot catalytique *m*; ~ **oxidation** *n* AIR, SOIL oxydation catalytique *f*; ~ **reaction** *n* AIR réaction catalytique *f*; ~ **reactor** *n* AIR réacteur catalytique *m*; ~ **reduction** *n* AIR réduction catalytique *f*; ~ **reduction process** *n* AIR procédé de réduction catalytique *m*; ~ **tower** *n* AIR tour de réaction catalytique *f*; ~ **waste gas cleaner** *n* AIR épurateur catalytique *m*
catalyze *AmE see catalyse BrE*
catch: ~ **crop** *n* RENEW ENERG culture dérobée *f*; ~ **feeder** *n* WATER SUPP canal d'irrigation *m*; ~ **tank** *n* WASTE bac récepteur *m*, cuve de captage *f*
catchment *n* WATER POLL captation *f*, WATER SUPP captage *m*
catchwater: ~ **drain** *n* WATER POLL canalisation de drainage *f*, WATER SUPP fossé de réception *m*
caterpillar *n* SOIL, WASTE, WATER SUPP chenille *f*
cathode *n* SOIL cathode *f*; ~ **ray tube** (*CRT*) SAFETY tube cathodique *m*
cathodic: ~ **protection** *n* WATER SUPP protection cathodique *f*
cation *n* AIR, SOIL, WATER POLL, WATER SUPP cation *m*; ~ **exchange** SOIL, WATER POLL, WATER SUPP échange cationique *m*; ~ **exchange capacity** (*CEC*) SOIL capacité d'échange cationique *f* (*CEC*); ~ **exchanger** SOIL échangeur cationique *m*
caustic[1] *adj* SOIL, WATER POLL caustique
caustic[2]: ~ **scrubbing** *n* AIR lavage alcalin *m*, lavage basique *m*
causticity *n* SOIL, WATER POLL causticité *f*
cavitation *n* WATER SUPP cavitation *f*
cavity: ~ **absorber** *n* NOISE absorbeur à cavités *m*
CBC *abbr* (*circulating fluidized bed*) AIR, SOIL LFC (*lit fluidisé circulant*)
CC *abbr* (*closed cycle*) RENEW ENERG circuit fermé *m*
CCGT *abbr* (*combined-cycle gas turbine*) AIR turbine à gaz à cycle combiné *f*
CCS *abbr* (*controlled combustion system*) AIR système de combustion contrôlée *m*
Cd *abbr* (*cadmium*) SOIL Cd (*cadmium*)
CD *abbr* (*chlorine demand*) WATER SUPP demande de chlore *f*
CE *abbr* (*capillary electrophoresis*) ANALYSIS électrophorèse capillaire *f*
CEC *abbr* (*cation exchange capacity*) SOIL CEC (*capacité d'échange cationique*)
ceiling: ~ **panel** *n* NOISE faux-plafond *m*
cell: ~ **method** *n* WASTE méthode de l'alvéole *f*, méthode du casier *f*; ~ **stack** *n* AIR *fuel cell* empilement de cellules *m*
cellophane *n* WASTE cellophane *f*
cellular: ~ **wheel sluice** *n* AIR *of cyclone* écluse à roue cellulaire *f*
Celluloid® *n* WASTE, WATER SUPP Celluloïd® *m*
cellulose *n* RENEW ENERG, WASTE, WATER POLL, WATER SUPP cellulose *f*; ~ **acetate** WATER POLL, WATER SUPP acétate de

cellulose *m*; ~ **acetate film** WASTE film d'acétate de cellulose *m*; ~ **fiber** *AmE*, ~ **fibre** *BrE* WASTE, WATER SUPP fibre cellulosique *f*, fibre de cellulose *f*; ~ **raw material** RENEW ENERG *biomass conversion base material* matière première cellulosique *f*; ~ **waste** WASTE déchets cellulosiques *m pl*
cement *n* SOIL *binding agent* ciment *m*; **~-clay mixture** SOIL *slurry* mélange ciment-argile *m*; **~-water mixture** SOIL *slurry* mélange eau-ciment *m*
CE-MS *abbr* (*capillary electrophoresis-mass spectrometry*) ANALYSIS électrophorèse capillaire couplée à la spectrométrie de masse *f*
CEN *abbr* (*European Committee for Standardization*) ANALYSIS, POLICY CEN (*Comité Européen de Normalisation*)
center *AmE see centre BrE*
central: ~ **control room** *n* AIR poste de commande central *m*, salle de commande centrale *f*; ~ **focal point** *n* RENEW ENERG point de focalisation central *m*; ~ **landfill** *n* WASTE décharge centralisée *f*; ~ **pipe** *n* WASTE *degassing shaft* conduite centrale *f*; ~ **solar heating plant** *n* RENEW ENERG installation solaire de chauffage central *f*; ~ **vacuum-cleaning system** *n* SAFETY aspirateur central *m*; ~ **well** *n* AIR puits central *m*
centre: ~ **feed column** *n BrE* WATER SUPP fût central d'alimentation *m*; **~-folding tubing** *n BrE* WASTE tube avec coude central *m*
centrifugal[1] *adj* AIR, SOIL, WASTE, WATER POLL centrifuge
centrifugal[2]: ~ **acceleration** *n* AIR accélération centrifuge *f*; ~ **collector** *n* AIR collecteur centrifuge *m*; ~ **compressor** *n* AIR compresseur centrifuge *m*; ~ **fan** *n* AIR soufflante centrifuge *f*; ~ **force** *n* AIR, RENEW ENERG *tidal power* force centrifuge *f*; ~ **force activity** *n* AIR activité par force centrifuge *f*; ~ **force precipitator** *n* AIR *cyclone* séparateur centrifuge *m*; ~ **force sedimentation** *n* ANALYSIS *dust measurement dust* sédimentation par la force centrifuge *f*; ~ **mass** *n* RENEW ENERG masse centrifuge *f*; ~ **precipitation** *n* AIR séparation centrifuge *f*; ~ **separation** *n* AIR séparation centrifuge *f*
centrifugation *n* AIR, SOIL, WASTE, WATER POLL centrifugation *f*
centrifuge[1] *n* AIR, SOIL, WASTE, WATER POLL, centrifugeuse *f*; ~ **pre-storage tank** AIR cuve de pré-stockage centrifuge *f*
centrifuge[2] *vt* AIR, SOIL, WASTE, WATER POLL centrifuger
centrifuging *n* AIR, SOIL, WASTE, WATER POLL centrifugation *f*
ceramic: ~ **body** *n* AIR *of catalyst* corps céramique *m*; ~ **carrier** *n* AIR *of catalyst* support céramique *m*; ~ **catalyst** *n* AIR catalyseur céramique *m*; **~-coated body** *n* AIR corps à revêtement céramique *m*; ~ **filter** *n* AIR filtre céramique *m*; **~-lined combustion chamber** *n* AIR chambre de combustion à revêtement céramique *f*; ~ **material** *n* AIR *filter material* matériau céramique *m*; ~ **monolith** *n* AIR *as catalyst* monolithe céramique *m*; ~ **pellet** *n* SOIL bille de céramique *f*
ceramization *n* SOIL *of pellets* céramisation *f*
certificate: ~ **of compliance** *n* MANAGE certificat de conformité *m*; ~ **concept** *n* POLICY type de certificat *m*; ~ **of inspection** *n* POLICY certificat d'inspection *m*; ~ **of origin** *n* MANAGE certificat d'origine *m*
certification: ~ **of conformity** *n* MANAGE, POLICY certification de conformité *f*
cesspit *n* WATER POLL fosse de décantation *f*
cesspool *n* WATER POLL fosse d'aisance *f*, puisard *m*, WATER SUPP puits absorbant *m*
cetane: ~ **number** *n* AIR indice de cétane *m*
CFC[1] *abbr* (*chlorofluorocarbon*) AIR, WASTE, WATER SUPP CFC (*chlorofluorocarbone*)
CFC[2]: **~-processed waste** *n* WASTE déchets traités aux chlorofluorocarbones *m pl*
CGE *abbr* (*capillary gel electrophoresis*) ANALYSIS électrophorèse capillaire sur gel *f*
chain: ~ **conveyor** *n* AIR *packed-bed filter* convoyeur à chaîne *m*
chairperson *n* POLICY président *m*
chamber *n* AIR, WASTE, WATER SUPP chambre *f*; ~ **filter press** SOIL, WASTE filtre-presse à chambre *m*, filtre-presse à plateaux *m*; **~-type vacuum sealing** WASTE scellement sous vide en chambre *m*; ~ **water level** RENEW ENERG *wave power plant* niveau d'eau du réservoir *m*
chaneling *AmE see channelling BrE*
changeover: ~ **procedure** *n* WASTE procédure de relais *f*
changing: ~ **head** *n* WATER SUPP hauteur de chute variable *f*
channel *n* WATER POLL canal *m*, WATER SUPP canal *m*, goulotte *f*; ~ **ball press** WASTE presse à balles horizontale *f*; ~ **recuperator** AIR conduit de récupération *m*, récupérateur en U *m*
channelling *n BrE* AIR effet de canalisation *m*
char *n* WASTE produit de carbonisation *m*, résidu charbonneux *m*
charge *n* MANAGE, POLICY redevance *f*, SOIL *of kiln* charge *f*; ~ **carrier** AIR *electrostatic precipitator* vecteur de charge *m*; ~ **exemption** POLICY exemption de redevance *f*; ~ **reduction** POLICY réduction de redevance *f*; ~ **reversal** AIR *particulates in electrostatic precipitator* inversion de charge *f*
charged[1] *adj* AIR chargé; ~ **with solar power** RENEW ENERG alimenté par l'énergie solaire
charged[2]: **~-gas stream** *n* AIR flux de gaz chargé *m*; ~ **particle** *n* AIR *electrostatic precipitator* particule chargée *f*
charges: ~ **quota** *n* POLICY assiette des redevances *f*; ~ **tariff** *n* POLICY barème des redevances *m*
charging: ~ **chamber** *n* WASTE chambre de chargement *f*; **~-electrode wire** *n* AIR *in plate-wire electrostatic precipitators* électrode émissive *f*, électrode d'émission *f*; ~ **facility** *n* WASTE installation de chargement *f*; ~ **hopper** *n* WASTE trémie de chargement *f*; ~ **material** *n* AIR charge *f*; ~ **zone** *n* AIR zone de chargement *f*
check[1] *n* GEN contrôle *m*; ~ **valve** WATER SUPP clapet de retenue *m*, vanne anti-reflux *f*, vanne d'arrêt *f*; ~ **weighing** WASTE vérification du poids *f*
check[2] *vt* GEN contrôler, inspecter, vérifier
checking *n* GEN contrôle *m*, inspection *f*, vérification *f*; ~ **apparatus** WASTE appareil de tri *m*
chelate *n* SOIL *additive* chélate *m*
CHEMFIX®: ~ **method** *n* SOIL *American solidification process* méthode CHEMFIX® *f*; ~ **process** *n* WASTE procédé CHEMFIX® *m*
chemical[1] *adj* GEN chimique; **~-physical** AIR, SOIL, WASTE, WATER POLL, WATER SUPP physico-chimique
chemical[2] *n* GEN produit chimique *m*; ~ **adsorption** AIR adsorption chimique *f*; ~ **aggressivity** WASTE, WATER SUPP agressivité chimique *f*; ~ **dosimeter** ANALYSIS dosimètre chimique *m*; ~ **hazard** SAFETY risque chimique *m*; ~ **ionization mass spectrometry** (*CI-MS*) ANALYSIS spectrométrie de masse par ionisation chimique *f*; ~ **oxidation** WASTE oxydation chimique *f*; **~-oxidative gas scrubbing** AIR lavage des gaz par oxydation chimique *m*; ~ **oxygen demand** (*COD*) AIR, WASTE, WATER POLL, WATER SUPP demande chimique en oxygène *f* (*DCO*); ~ **parameter** SOIL paramètre chimique *m*; **~-physical pretreatment** WASTE prétraitement physico-chimique *m*; ~ **precipitation** AIR, WASTE *sewage* précipitation chimique *f*; ~ **pulp** WASTE pâte chimique *f*; ~ **reagent** AIR réactif chimique *m*; ~ **resistance** AIR *filter materials* résistance chimique *f*; ~ **waste** WASTE déchets d'origine chimique *m pl*; ~ **water quality** WATER SUPP titre chimique de l'eau *m*
chemically[1]: ~ **aggressive** *adj* AIR chimiquement agressif; ~ **resistant** *adj* AIR, SAFETY, WASTE chimiquement résistant
chemically[2]: **~-aggressive dust** *n* AIR poussière chimiquement agressive *f*; **~-resistant protective suit** *n* SAFETY vêtement de protection contre les agressions chimiques *m*
chemicals: ~ **delivery** *n* WASTE livraison de produits pharmaceutiques *f*

chemiluminescence: ~ **detection** *n* (*CLD*) ANALYSIS détection par chimiluminescence *f*; ~ **measuring instrument** *n* ANALYSIS *for detection of ozone* instrument de mesure de chimiluminescence *m*
chemisorb *vt* AIR absorber chimiquement
chemisorption *n* AIR chimisorption *f*; ~ **filter** AIR filtre de chimisorption *m*
chemistry *n* GEN chimie *f*
chemolysis *n* WASTE *feedstock recycling* chimiolyse *f*
chevron: ~ **baffle** *n* AIR baffle à chevrons *m*
child: ~**-proof closure** *n* (*CPC*) SAFETY bouchon de sécurité *m*; ~**-resistant closure** *n* (*CRC*) SAFETY fermeture de sécurité pour enfants *f*; ~**-resistant packaging** *n* SAFETY emballage de sécurité *m*
chill: ~**-permanent adhesive** *n* WASTE colle permanente résistante à la réfrigération *f*
chimney *n* AIR, RENEW ENERG cheminée *f*; ~ **discharge** AIR évacuation de cheminée *f*; ~ **emissions** *n pl* AIR émissions en sortie de cheminée *f pl*; ~ **height** AIR hauteur de cheminée *f*
chip *n* RENEW ENERG plaquette *f*
chipboard *n* NOISE, RENEW ENERG, WASTE carton gris *m*, panneau de particules *m*
chipping *n* RENEW ENERG déchiquetage *m*
chlorate *n* WATER SUPP hypochlorite *m*
chloride *n* AIR, SOIL chlorure *m*
chlorido: ~ **load potential** *n* (*CLP*) AIR potentiel de charge chlorée *m*
chlorinate *vt* AIR, SOIL, WASTE chlorer, WATER SUPP javelliser
chlorinated[1] *adj* AIR, SOIL, WASTE chloré, WATER SUPP javellisé
chlorinated[2]: ~ **hydrocarbon** *n* AIR hydrocarbure chloré *m*; ~ **waste** *n* WASTE déchets chlorés *m pl*
chlorination *n* AIR, SOIL, WASTE chloration *f*, WATER SUPP javellisation *f*; ~ **unit** WATER SUPP poste de chloration *m*
chlorine *n* (*Cl*) AIR *gaseous oxidant*, SOIL, WASTE chlore *m* (*Cl*); ~ **demand** (*CD*) WATER SUPP demande de chlore *f*; ~**-loading potential** (*CLP*) AIR potentiel de charge chlorée *m*
chlorobenzene *n* AIR chlorobenzène *m*
chlorofluorocarbon *n* (*CFC*) AIR, WASTE, WATER SUPP chlorofluorocarbone *m* (*CFC*)
chlorophenol *n* AIR chlorophénol *m*
chlorophyll *n* RENEW ENERG chlorophylle *f*; ~**-containing plant** AIR plante chlorophyllienne *f*
chopper: ~ **ventilator** *n* WASTE ventilateur à fonctionnement périodique *m*
C-horizon *n* SOIL horizon-C *m*
CHP[1] *abbr* (*combined heat and power*) AIR, WASTE force-chaleur *f*, production combinée de chaleur et d'électricité *f*
CHP[2]: ~ **station** *n* AIR, RENEW ENERG, WASTE installation de production combinée de chaleur et d'électricité *f*, centrale force-chaleur *f*
chromatogram *n* ANALYSIS chromatogramme *m*
chromatograph *n* ANALYSIS chromatographe *m*
chromatographic[1] *adj* ANALYSIS chromatographique
chromatographic[2]: ~ **column** *n* ANALYSIS colonne chromatographique *f*; ~ **peak** *n* ANALYSIS pic chromatographique *m*
chromium *n* (*Cr*) AIR, SOIL chrome *m* (*Cr*)
chromogen *n* ANALYSIS chromogène *m*
chronic[1] *adj* SAFETY chronique
chronic[2]: ~ **effect** *n* SAFETY effet chronique *m*; ~ **health risk** *n* SAFETY risque chronique pour la santé *m*; ~ **poisoning** *n* SAFETY empoisonnement chronique *m*
chronopotentiometry *n* ANALYSIS chronopotentiométrie *f*
chunk: ~ **activated carbon** *n* AIR *adsorbent* charbon actif en bâtonnets *m*
CIEF *abbr* (*capillary isoelectric focusing*) ANALYSIS focalisation isoélectrique capillaire *f*, réglage isoélectrique capillaire *m*
CI-MS *abbr* (*chemical ionization mass spectrometry*) ANALYSIS spectrométrie de masse par ionisation chimique *f*
circular: ~ **broom** *n* WASTE, WATER SUPP balai circulaire *m*; ~ **main** *n* WATER SUPP collecteur circulaire *m*; ~ **movement** *n* AIR mouvement circulaire *m*
circulate *vt* AIR faire circuler
circulating: ~ **fluidized bed** *n* (*CBC*) AIR, SOIL lit fluidisé circulant *m* (*LFC*); ~ **fluidized-bed reactor** *n* AIR réacteur à lit fluidisé circulant *m*; ~ **liquid** *n* AIR liquide de circulation *m*; ~ **pump** *n* AIR, WASTE pompe de circulation *f*; ~ **water** *n* AIR eau de circulation *f*
circulation *n* AIR, WATER POLL circulation *f*; ~ **evaporator** WATER SUPP évaporateur de circulation *m*; ~ **pump** AIR, WASTE pompe de circulation *f*
circumferential: ~ **velocity** *n* AIR vitesse périphérique *f*
cistern *n* WATER POLL, WASTE citerne *f*, réservoir *m*, réservoir à eau *m*, WATER SUPP bâche *f*, cuve *f*
citizens': ~ **participation** *n* POLICY participation de la population *f*, participation des habitants *f*
civil: ~ **engineering** *n* GEN génie civil *m*; ~ **law** *n* POLICY droit civil *m*
Cl *abbr* (*chlorine*) AIR, SOIL, WASTE Cl (*chlore*)
cladding *n* NOISE revêtement *m*
clam: ~ **pack** *n* WASTE dispositif de serrage *m*
clamp *n* RENEW ENERG silo *m*
clamping: ~ **force** *n* WASTE *sandwich moulding machine*, WATER SUPP force de serrage *f*
clamshell *n* WASTE grappin *m*; ~ **blister** WASTE blister double coque *m*
clarification *n* AIR décantation *f*, déposition *f*, WATER POLL débourbage *m*, décantation *f*, purification *f*, WATER SUPP débourbage *m*, purification *f*; ~ **plant** WASTE installation de clarification *f*, station d'épuration *f*; ~**-system-using plant** WASTE usine équipée de station d'épuration *f*
clarifier *n* WATER POLL *substance* agent pathogène *m*, *vessel* décanteur *m*
clarify *vt* AIR décanter, déposer, WATER POLL clarifier, débourber, décanter, purifier, WATER SUPP débourber, purifier
clarifying: ~ **tank** *n* WATER POLL bassin d'épuration *m*
class: ~ **of accuracy** *n* ANALYSIS classe de précision *f*; ~ **of pollution** *n* POLICY classe de pollution *f*
classical: ~ **recycling sectors** *n pl* MANAGE *metal, scrap, paper* secteurs classiques du recyclage *m pl*, secteurs traditionnels du recyclage *m pl*
classifiable *adj* ANALYSIS classifiable
classification *n* ANALYSIS, SOIL, WASTE classification *f*; ~ **station** WASTE poste de classification *m*
classifier *n* AIR, ANALYSIS, SOIL, WASTE classificateur *m*
classify *vt* ANALYSIS, SOIL, WASTE classifier, trier, POLICY *harmful substances* répertorier
Claus: ~ **plant** *n* AIR installation Claus *f*, unité de désulfuration Claus *f*; ~ **process** *n* AIR procédé Claus *m*
clay *n* WATER SUPP glaise *f*; ~ **barrier** WASTE barrière argileuse *f*; ~ **layer** WASTE couche argileuse *f*; ~ **mineral layer** SOIL couche de minéral argileux *f*
CLD *abbr* (*chemiluminescence detection*) ANALYSIS détection par chimiluminescence *f*
clean[1] *adj* GEN propre
clean[2]: ~ **air** *n* AIR air pur *m*; ~**-air device** *n* SAFETY dispositif d'épuration de l'air *m*; ~**-air suction** *n* AIR aspiration d'air pur *f*; ~**-energy source** *n* RENEW ENERG source d'énergie non polluante *f*, source d'énergie propre *f*; ~ **gas** *n* AIR, WASTE gaz épuré *m*, gaz propre *m*; ~**-gas canal** *n* AIR conduite de gaz épuré *f*; ~**-gas compartment** *n* AIR compartiment des gaz propres *m*; ~**-gas concentration** *n* AIR concentration de gaz propre *f*; ~**-gas dust content** *n* AIR concentration de poussière dans un gaz propre *f*; ~**-gas lid** *n* AIR couche de gaz propres *f*; ~**-gas out** *n* AIR sortie de gaz épuré *f*; ~**-gas outlet** *n* AIR évacuation de gaz épuré *f*;

~-gas reheating *n* AIR réchauffage de gaz épuré *m*; **~ plastic** *n* WASTE plastique propre *m*; **~ rain** *n* AIR pluie propre *f*; **~ technology** *n* AIR, MANAGE, WASTE technologie non polluante *f*, technologie propre *f*; **~-up** *n* MARINE dépollution *f*, SOIL assainissement *m*; **~-up efficiency** *n* SOIL rendement d'épuration *m*; **~-up rate** *n* SOIL taux d'épuration *m*; **~-up technique** *n* MARINE technique de dépollution *f*; **~ water** *n* WATER SUPP eau propre *f*, eau pure *f*; **~ water act** *n* MARINE loi sur la dépollution de l'eau *f*

clean[3] *vt* AIR *gas* épurer, nettoyer, *filter* décolmater, WASTE balayer, nettoyer, WATER SUPP nettoyer, *heat exchanger* décalaminer; **~ out** WATER SUPP curer, débourber; **~ up** AIR nettoyer, MARINE dépolluer, SOIL assainir, WASTE, WATER SUPP nettoyer

CLEAN *abbr* (*compact laser for efficient ablation*) WASTE laser de décapage *m*

cleaned[1] *adj* AIR *gas* épuré, nettoyé, *filter* décolmaté, WASTE balayé, nettoyé, WATER SUPP nettoyé, *heat exchanger* décalaminé

cleaned[2]: **~ gas** *n* AIR, WASTE gaz épuré *m*; **~ waste air** *n* AIR air résiduel épuré *m*

cleaner *n* SAFETY épurateur *m*

cleaning *n* AIR *of gas* épuration *f*, nettoyage *m*, *of filter* décolmatage *m*, WASTE balayage *m*, nettoyage *m*, WATER SUPP nettoyage *m*, *of heat exchanger* décalaminage *m*; **~ capacity** AIR capacité d'épuration *f*; **~ equipment** SAFETY équipement d'épuration *m*, matériel de nettoyage *m*; **~ nozzle** AIR tuyère d'épuration *f*; **~ operation** AIR opération de nettoyage *f*; **~ out** WATER SUPP curage *m*, débourbage *m*; **~ pipe** WASTE tuyau de lavage *m*; **~ position** AIR *of bag filter* position de nettoyage *f*; **~ pressure** AIR pression d'épuration *f*; **~-up costs** *n pl* MANAGE coûts de dépollution *m pl*

cleanse *vt* WATER POLL assainir, curer

cleansing *n* WATER POLL assainissement *m*, curage *m*, débourbage *m*

clear: **~-cutting** *n* SOIL déboisement *m*; **~ health risk** *n* SAFETY risque sanitaire manifeste *m*; **~ wastewater sample** *n* ANALYSIS prélèvement d'eaux usées après épuration *m*; **~ water** *n* WATER SUPP eau claire *f*, eau limpide *f*

climate *n* AIR climat *m*; **~ changes** *n pl* AIR *global warming* évolution climatique *f*, changement climatique *m*; **~ deterioration** AIR dégradation du climat *f*; **~ factor** AIR facteur climatique *m*; **~ fluctuation** AIR fluctuation climatique *f*; **~-protection tax** POLICY impôt sur la protection climatique *m*, taxe sur la protection du climat *f*; **~-relevant trace gas** AIR gaz à l'état de trace ayant un effet sur le climat *m*; **~ system** AIR système climatique *m*

climatic[1] *adj* AIR climatique

climatic[2]: **~ chamber** *n* AIR, WASTE chambre climatique *f*; **~ conditions** *n pl* AIR, WASTE conditions climatiques *f pl*; **~ data** *n pl* MANAGE, POLICY données climatiques *f pl*; **~ hazard** *n* SAFETY risque climatique *m*; **~ protection** *n* SAFETY protection climatique *f*; **~ stress** *n* AIR conditions climatiques éprouvantes *f pl*, contrainte climatique *f*; **~ test** *n* WASTE essai climatique *m*; **~ variations** *n pl* AIR variations climatiques *f pl*; **~ zone** *n* AIR zone climatique *f*

cling: **~ film** *n* WASTE film adhésif *m*

clinker *n* WASTE mâchefer *m*, WATER SUPP brique vitrifiée *f*

clip[1]: **~ band** *n* WASTE bande de serrage par pince *f*; **~-on carrier** *n* WASTE agrippe-col *m*

clip[2] *vt* WASTE attacher avec une pince, agrafer

clipping *n* WASTE agrafage *m*

clog *vt* AIR colmater, encrasser, WASTE colmater, WATER POLL, WATER SUPP boucher, colmater, engorger, obturer

clogged *adj* AIR encrassé, GEN colmaté, WATER POLL, WATER SUPP bouché, obturé

clogging *n* AIR colmatage *m*, encrassement *m*, WATER POLL, WATER SUPP bouchage *m*, colmatage *m*, engorgement *m*, obturation *f*; **~ risk** WASTE risque de bouchage *m*, risque de colmatage *m*

close[1]: **~-coupled** *adj* SOIL *of soil remediation machinery* attenant; **~-meshed** *adj* AIR à maille serrée, WATER SUPP *grate* à maille fine

close[2]: **~-cell foam** *n* WASTE mousse cloquée *f*; **~-meshed sieve plate** *n* AIR fond perforé à maille fine *m*; **~ supervision** *n* SAFETY étroite surveillance *f*

closed: **~ canal** *n* WATER SUPP canal fermé *m*; **~ cycle** *n* (*CC*) RENEW ENERG circuit fermé *m*; **~ cycle OTEC plant** *n* RENEW ENERG usine de conversion de l'énergie thermique des mers en circuit fermé *f*; **~-loop circulation** *n* AIR circuit en boucle fermée *m*; **~-loop system** *n* WASTE système en circuit fermé *m*; **~-loop wet scrubbing** *n* AIR épuration par voie humide en boucle fermée *f*; **~-pit landfill** *n* WASTE décharge couverte *f*; **~ Rankine cycle** *n* RENEW ENERG *ocean thermal energy* cycle de Rankine fermé *m*; **~ system** *n* WASTE circuit fermé *m*

closure *n* WASTE fermeture *f*; **~ production line** WASTE ligne de production de fermetures *f*

cloud: **~ amount** *n* AIR *of stack* nébulosité partielle *f*; **~ cover** *n* AIR nébulosité *f*; **~ point** *n* WATER SUPP point de trouble *m*

cloudy *adj* WATER POLL, WATER SUPP terne, trouble; **very ~** WATER POLL, WATER SUPP *degree of turbidity* très trouble

CLP *abbr* (*chlorine-loading potential*) AIR potentiel de charge chlorée *m*

cluster *n* WATER SUPP agglomérat *m*

C-N: **~ ratio** *n* (*carbon-nitrogen ratio*) ANALYSIS rapport carbone-azote *m*

CNG *abbr* (*compressed natural gas*) AIR gaz naturel comprimé *m*

C-N-P: **~ ratio** *n* (*carbon-nitrogen-phosphorous ratio*) ANALYSIS rapport carbone-azote-phosphore *m*

Co *abbr* (*cobalt*) AIR, SOIL Co (*cobalt*)

CO[1] *abbr* (*carbon monoxide*) AIR CO (*monoxyde de carbone*)

CO[2]: **~ emissions** *n pl* AIR émissions de CO *f pl*; **~ value** *n* AIR indice de CO *m*

CO_2[1] *abbr* AIR (*carbon dioxide*) CO_2 (*dioxyde de carbone*), WATER SUPP (*carbonic anhydride*) CO_2 (*anhydride carbonique*)

CO_2[2]: **~ concentration** *n* AIR concentration de CO_2 *f*; **~ emissions reduction** *n* AIR réduction des émissions de CO_2 *f*

coagulant *n* GEN agent de précipitation *m*, coagulant *m*

coagulate[1] *vt* GEN coaguler, conglutiner, faire cailler

coagulate[2] *vi* GEN cailler, se coaguler, conglutiner

coagulation *n* GEN coagulation *f*, caillement *m*, floculation *f*, conglutination *f*

coal: **~ brick** *n* RENEW ENERG briquette de charbon *f*; **~ desulfurization** *AmE*, **~ desulphurization** *BrE* *n* AIR désulfuration du charbon *f*; **~ dust** *n* AIR poussière de charbon *f*; **~ gas** *n* AIR gaz de houille *m*; **~ power station** *n* AIR centrale thermique au charbon *f*

coalesce *vi* AIR, WATER SUPP s'associer, se combiner

coalescence *n* AIR, WATER SUPP coalescence *f*

coarse[1] *adj* AIR, SOIL à gros grain, grossier; **~-particle** AIR, WASTE, WATER SUPP à particules grossières; **~-particled** AIR, WASTE, WATER SUPP à particules grossières

coarse[2]: **~-ash separator** *n* AIR séparateur de cendres grossières *m*; **~ crusher** *n* WASTE broyeur grossier *m*, concasseur primaire *m*; **~ filter** *n* AIR filtre grossier *m*; **~ particle** *n* AIR particule grossière; **~ refuse** *n* WASTE refus de forte granulométrie *m pl*

coastal: **~ deposit** *n* WATER SUPP dépôt côtier *m*; **~ pollution** *n* MARINE pollution des côtes *f*; **~ river** *n* WATER SUPP rivière côtière *f*

coastline: **~ form** *n* RENEW ENERG forme du littoral *f*

coat *vt* AIR, WASTE, WATER SUPP revêtir

coated[1] *adj* AIR revêtu

coated[2]: **~ paper** *n* WASTE papier couché *m*; **~ surface** *n* AIR

surface revêtue *f*; ~ **synthetic paper** *n* WASTE papier à couchage synthétique *m*
coating *n* AIR revêtement *m*, WASTE couchage *m*, revêtement *m*, WATER SUPP revêtement *m*; ~ **compound** WASTE produit de couchage *m*; ~ **machine** WASTE installation de couchage *f*; ~ **thickness** WASTE épaisseur de couchage *f*, épaisseur de l'enduit *f*
cobalt *n* (*Co*) AIR, SOIL cobalt *m* (*Co*)
COC *abbr* (*conventional oxidation catalyst*) AIR catalyseur d'oxydation classique *m*
co-current: ~ **flow** *n* AIR *waste gas flow* écoulement parallèle *m*; ~ **flow scrubber** *n* AIR laveur à écoulement parallèle *m*; ~ **scrubber** *n* AIR, WATER SUPP laveur co-courant *m*; ~ **washing** *n* AIR, WATER SUPP lavage à co-courant *m*, lavage en parallèle *m*
COD[1] *abbr* (*chemical oxygen demand*) AIR, WASTE, WATER POLL, WATER SUPP DCO (*demande chimique en oxygène*)
COD[2]: ~**-BOD ratio** *n* ANALYSIS rapport DCO-DBO *m*; ~ **determination** *n* ANALYSIS calcul de la DCO *m*
co-deposition *n* WASTE dépôt commun *m*
coding: ~ **of measured values** *n* ANALYSIS codage des valeurs mesurées *m*, encodage des valeurs mesurées *m*
co-disposal *n* WASTE élimination commune *f*; ~ **landfill** WASTE mise en décharge conjointe *f*
coefficient: ~ **of haze** *n* AIR coefficient de brume *m*; ~ **of permeability** *n* AIR, SOIL, WASTE, WATER SUPP coefficient de perméabilité *m*
co-extruded: ~ **film** *n* WASTE pellicule avec extrusion en couches multiples *f*
co-extrusion *n* WASTE *plastics* extrusion en couches multiples *f*
coffer *n* NOISE caisson *m*, WATER SUPP coffre *m*
cofferdam *n* RENEW ENERG *tidal power plant* bâtardeau *m*
cogeneration *n* AIR *waste heat* production combinée *f*
cohesion *n* AIR cohésion *f*
coil *n* WASTE bobine *f*
co-injection *n* WASTE *plastics* co-injection *f*
coke *vt* RENEW ENERG encrasser
coking *n* RENEW ENERG encrassement *m*
cold: ~ **gluing system** *n* WASTE système d'encollage à froid *m*; ~ **seawater** *n* RENEW ENERG eau de mer froide *f*
coli: ~ **bacillus** *n* WATER SUPP *indicator germ* colibacille *m*; ~ **test** *n* ANALYSIS essai aux colibacilles *m*
colibacterium *n* WATER SUPP colibacille *m*
coliform *n* WATER SUPP coliforme *m*; ~ **bacteria** *n pl* ANALYSIS *indicator* bactéries coliformes *f pl*; ~ **organism** WATER SUPP organisme coliforme *m*
collapse *n* AIR *of bag in filter* affaissement *m*
collapsible: ~ **case** *n* WASTE caisse pliable *f*; ~ **and reusable packaging system** *n* WASTE emballage pliable et réutilisable *m*; ~ **tube** *n* WASTE tube souple *m*
collating: ~ **device** *n* AIR *catalytic incineration* dispositif d'interclassement *m*; ~ **system** *n* WASTE système d'assemblage *m*; ~ **transit tray** *n* WASTE plateau d'assemblage *m*
collect *vt* AIR *dust* collecter, capter, ANALYSIS *samples* collectur, récupérer MARINE collecter, recueillir, WASTE collecter, récupérer, ramasser, WATER SUPP *water* capter, réunir
collectable *adj* ANALYSIS récupérable
collected[1] *adj* AIR, ANALYSIS, MARINE, WASTE collecté
collected[2]: ~ **dust** *n* AIR poussière collectée *f*; ~ **dust particle** *n* AIR particule de poussière collectée *f*
collecting: ~ **electrode** *n* ANALYSIS, WASTE électrode collectrice *f*, électrode de dépôt *f*; ~ **main** *n* AIR barillet *m*, collecteur *m*; ~ **pit** *n* WASTE, WATER SUPP puits de collecte *m*; ~ **receptacle** *n* ANALYSIS bac de collecte *m*
collection *n* AIR *dust* captation *f*, MARINE récupération *f*, WASTE collecte *f*, récupération *f*, ramassage *m*, WATER SUPP *of water* captage *m*; ~ **behavior** *AmE*, ~ **behaviour** *BrE* AIR mode de collecte *m*; ~ **costs** *n pl* MANAGE coûts de collecte *m pl*; ~ **device** MARINE engin de récupération *m*; ~ **efficiency** AIR rendement de collecte *m*; ~ **field** AIR terrain de collecte *m*; ~ **hopper** WASTE trémie de récupération *f*; ~ **operation** AIR opération de collecte *f*; ~ **period** ANALYSIS *sampling* campagne *f*, période de collecte *f*; ~ **point** WASTE point de collecte *m*; ~ **quality** AIR qualité de collecte *f*; ~ **service** WASTE service de ramassage *m*; ~ **of solar heat** RENEW ENERG captation d'énergie solaire *f*; ~ **station** WASTE poste de collecte *m*, station de collecte *f*; ~ **system** WASTE système de collecte *m*; ~ **vehicle** WASTE, WATER SUPP véhicule de ramassage *m*
collector *n* WASTE collecteur *m*; ~ **field** RENEW ENERG *solar farm* champ d'héliostats *m*, champ de capteurs *m*; ~ **shaft** WASTE, WATER SUPP puits de collecte *m*
collide *vi* AIR, SAFETY entrer en collision, se heurter
colliery: ~ **waste** *n* AIR résidu de houille *m*, WASTE déchets miniers *m pl*
collimator *n* ANALYSIS collimateur *m*
collision *n* AIR, SAFETY, WATER SUPP collision *f*; ~ **ionization** AIR ionisation par collision *f*
colloid *n* AIR, WATER SUPP, WATER POLL colloïde *m*; ~**-disperse system** AIR système à dispersion de colloïdes *m*
colloidal[1] *adj* AIR, WATER POLL, WATER SUPP colloïdal
colloidal[2]: ~ **concentration** *n* WATER SUPP concentration colloïdale *f*; ~ **particle** *n* AIR, WATER SUPP particule colloïdale *f*; ~ **suspension** *n* WATER SUPP suspension colloïdale *f*
colluvium *n* SOIL colluvion *f*
colony: ~ **count** *n* WATER POLL comptage des colonies *m*
color *AmE see colour BrE*
colored *AmE see coloured BrE*
colorfast *AmE see colourfast BrE*
colorimeter *n* ANALYSIS colorimètre *m*
colorimetric[1] *adj* ANALYSIS colorimétrique
colorimetric[2]: ~ **detection** *n* ANALYSIS détection colorimétrique *f*; ~ **determination** *n* ANALYSIS détermination colorimétrique *f*; ~ **indicator tube** *n* ANALYSIS tube indicateur colorimétrique *m*; ~ **pH-value determination** *n* ANALYSIS détermination colorimétrique du pH *f*
colorimetry *n* ANALYSIS colorimétrie *f*
coloring *AmE see colouring BrE*
colorless *AmE see colourless BrE*
colour: ~ **analysis** *n BrE* ANALYSIS analyse colorimétrique *f*, analyse de la couleur *f*; ~ **change** *n BrE* ANALYSIS changement de couleur; ~ **depth** *n BrE* ANALYSIS intensité de couleur *f*, intensité de teinte *f*; ~ **designation** *n BrE* ANALYSIS désignation de la couleur *f*; ~ **fastness** *n BrE* WATER SUPP *glass recycling* tenue de la couleur *f*; ~ **indicator** *n BrE* ANALYSIS indicateur de couleur *m*; ~ **reaction** *n BrE* ANALYSIS *absorption solution* réaction colorée *f*; ~ **reduction** *n BrE* WATER POLL abattement de la couleur *m*; ~ **removal** *n BrE* WATER POLL abattement de la couleur *m*; ~ **sorting of plastics** *n BrE* WASTE tri des plastiques par couleur *m*
coloured: ~ **strapping** *n BrE* WASTE banderolage de couleur *m*
colourfast *adj BrE* WASTE grand teint
colouring *n BrE* ANALYSIS coloration *f*
colourless[1] *adj BrE* AIR incolore
colourless[2]: ~ **gas** *n BrE* AIR gaz incolore *m*
column: ~ **chromatography** *n* ANALYSIS chromatographie sur colonne *f*; ~ **injector** *n* ANALYSIS injecteur en colonne *m*; ~ **method** *n* ANALYSIS méthode sur colonne *f*
combination: ~ **pH-electrode** *n* ANALYSIS *pH-meter* électrode combinée de mesure du pH *f*; ~ **scrubber** *n* AIR épurateur mixte *m*; ~ **sealing** *n* WASTE *of plastic sheets and mineral layer* couverture alternée *f*
combined[1] *adj* SOIL combiné
combined[2]: ~ **base sealing** *n* SOIL imperméabilisation

combinée de la base *f*; ~ **base sealing system** *n* SOIL système d'imperméabilisation combinée de la base *m*; ~**-cycle gas turbine** *n* (*CCGT*) AIR turbine à gaz à cycle combiné *f*; ~ **heat and power** *n* (*CHP*) AIR, WASTE force-chaleur *f*, production combinée de chaleur et d'électricité *f*; ~ **heat and power station** *n* AIR, RENEW ENERG centrale force-chaleur *f*, installation de production combinée de chaleur et d'électricité *f*; ~ **packaging** *n* WASTE emballage composé *m*; ~ **sewerage system** *n* WATER SUPP réseau d'assainissement unitaire *m*

combustibility *n* GEN combustibilité *f*

combustible[1] *adj* GEN combustible

combustible[2] *n* AIR, SAFETY produit inflammable *m*; ~ **dust** AIR poussière combustible *f*; ~ **fossil fuel** AIR, RENEW ENERG combustible fossile combustible *m*; ~ **gaseous pollutant** AIR polluant gazeux combustible *m*; ~ **material** AIR, SAFETY, WATER SUPP matériau combustible *m*; ~ **solvent** AIR solvant combustible *m*; ~ **waste** WATER SUPP déchets combustibles *m pl*

combustion[1]: ~**-related** *adj* AIR relatif à une méthode de combustion

combustion[2] *n* AIR, RENEW ENERG, SAFETY combustion *f*; ~ **air** AIR *thermal and catalytic combustion* air de combustion *m*; ~ **air fan** WASTE ventilateur de combustion *m*; ~ **chamber** AIR, SOIL, WASTE *furnace* chambre de combustion *f*; ~ **chamber ash silo** AIR silo à cendres de chambre de combustion *m*; ~ **chamber final temperature** AIR *thermal combustion* température finale en chambre de combustion *f*; ~ **chamber outlet** AIR sortie de chambre de combustion *f*; ~ **enthalpy** AIR enthalpie de combustion *f*; ~ **gas** AIR, WATER SUPP gaz de comburant *m*; ~ **operation** AIR opération de combustion *f*; ~ **product** AIR produit de combustion *m*; ~ **reaction** AIR réaction de combustion *f*; ~ **residue** WASTE, WATER SUPP imbrûlés *m pl*; ~ **system** AIR *thermal combustion* système de combustion *m*; ~ **temperature** AIR température de combustion *f*; ~ **tube** ANALYSIS *nitrogen determination* tube de combustion *m*; ~ **waste gas** AIR gaz de combustion *m*

commercial[1]: ~ **global energy demand** *n* RENEW ENERG demande globale d'énergie dans l'industrie *f*, demande mondiale d'énergie dans l'industrie *f*; ~ **operation** *n* MANAGE *of plant* exploitation commerciale *f*, POLICY *of plant* exploitation industrielle *f*; ~ **packaging** *n* WASTE *reused for transport by last user* emballage industriel *m*; ~**-scale application** *n* AIR, RENEW ENERG application à l'échelle industrielle *f*; ~**-scale plant** *n* AIR, RENEW ENERG, SOIL, WASTE installation à l'échelle industrielle *f*; ~**-scale solar pond** *n* RENEW ENERG bassin solaire à l'échelle industrielle *m*; ~ **source** *n* AIR source industrielle *f*; ~ **waste** *n* WASTE déchets des commerces *m pl*, déchets industriels *m pl*

commercial[2]: **on a ~ scale** *phr* AIR, RENEW ENERG, SOIL, WASTE à l'échelle industrielle

commercially[1]: ~ **available** *adj* ANALYSIS, MANAGE, MARINE commercialisé, en vente dans le commerce, RENEW ENERG commercialisé

commercially[2]: ~**-available installation** *n* RENEW ENERG installation existant dans l'industrie *f*; ~**-available measuring device** *n* ANALYSIS appareil de mesure industriel *m*; ~**-available pH-measuring device** *n* ANALYSIS pH-mètre industriel *m*; ~**-available thermometer** *n* ANALYSIS thermomètre industriel *m*

commingled: ~ **plastics** *n pl* WASTE plastiques mêlés *m pl*

comminution *n* WASTE, WATER SUPP broyage *m*, pulvérisation *f*

comminutor *n* WASTE, WATER SUPP *sewage* broyeur *m*, dilacérateur *m*

Commission: ~ **on Sustainable Development** *n* (*CSD*) POLICY Commission du Développement Durable *f*

commixture *n* WASTE mélange *m*

commodity: ~ **group** *n* WASTE catégorie de biens *f*

community: ~ **facilities** *n pl* NOISE POLICY services collectifs *m pl*

compact[1] *adj* AIR monobloc

compact[2]: ~ **absorber** *n* NOISE absorbeur compact *m*; ~ **filter** *n* AIR *in container* filtre à mailles serrées *m*, filtre compact *m*; ~ **flocculation plant** *n* WATER SUPP banc de compactage des floculants *m*; ~ **gas scrubber** *n* AIR dépoussiéreur monobloc *m*; ~ **heat exchanger** *n* AIR échangeur de chaleur compact *m*; ~ **laser for efficient ablation** *n* (*CLEAN*) WASTE laser de décapage *m*

compact[3] *vt* SOIL, WASTE, WATER SUPP compacter

compacted *adj* SOIL, WASTE compacté, tassé, WATER SUPP compacté

compacting *n* SOIL, WASTE compactage *m*, tassement *m*; ~ **device** SOIL, WASTE compacteur *m*, engin compacteur *m*, engin de compactage *m*; ~ **machine** SOIL, WASTE compacteur *m*, engin compacteur *m*, engin de compactage *m*

compaction *n* SOIL, WASTE, WATER SUPP compactage *m*, tassement *m*; ~ **equipment** SOIL, WASTE compacteur *m*, engin compacteur *m*, engin de compactage *m*; ~ **landfill** WASTE *dominant landfill type* décharge compactée *f*; ~ **zone** WASTE zone de compactage *f*

compactor *n* SOIL, WASTE *device* compacteur *m*; ~ **vehicle** SOIL, WASTE camion tasseur *m*, véhicule de compactage *m*

comparative: ~ **measurement** *n* ANALYSIS mesure comparée *f*, mesure de comparaison *f*; ~ **test** *n* ANALYSIS essai comparatif *m*

compartment: ~ **case** *n* WASTE caisse compartimentée *f*

compartmented: ~ **insert** *n* WASTE compartiment emboîtable *m*; ~ **tray** *n* WASTE plateau alvéolé *m*

compatibility *n* AIR, POLICY, WASTE, WATER SUPP compatibilité *f*

compatibilizer *n* WASTE, WATER SUPP compatibilisateur *m*

compensate *vt* POLICY, SAFETY *for injuries received* compenser, dédommager, indemniser

compensating: ~ **tube** *n* WASTE tube de compensation *m*

compensation *n* POLICY, SAFETY dédommagement *m*, indemnisation *f*, *for injuries received* compensation *f*; ~ **charge** POLICY redevance d'indemnisation *f*; ~ **fee** POLICY, SAFETY indemnité *f*

compensatory: ~ **payment** *n* MANAGE paiement compensatoire *m*

competent: ~ **authority** *n* POLICY autorité compétente *f*

competitiveness *n* MANAGE, POLICY, RENEW ENERG *of energy sources* compétitivité *f*

complete: ~ **enclosure** *n* NOISE enceinte intégrale *f*; ~ **fill** *n* WASTE remplissage intégral *m*

completed: ~ **activity** *n* POLICY activité achevée *f*

complexing: ~ **agent** *n* WATER POLL agent complexant *m*

complexometric[1] *adj* ANALYSIS complexométrique

complexometric[2]: ~ **determination** *n* ANALYSIS détermination complexométrique *f*; ~ **method** *n* ANALYSIS méthode complexométrique *f*; ~ **titration** *n* ANALYSIS titrage complexométrique *m*

component[1]: ~**-specific** *adj* ANALYSIS spécifique d'un composant, SOIL lié à un élément

component[2] *n* AIR, ANALYSIS, SOIL élément *m*, composant *m*; ~**-specific gas analyser** *BrE*, ~**-specific gas analyzer** *AmE* ANALYSIS analyseur de gaz spécifique d'un composant *m*

composite[1] *adj* WASTE, WATER SUPP composite

composite[2]: ~ **container** *n* WASTE conteneur en matériau composite *m*; ~ **packaging** *n* WASTE emballage composite *m*; ~ **sample** *n* ANALYSIS échantillon composite *m*; ~ **waste** *n* WASTE déchets composites *m pl*

compost *n* SOIL compost brut *m*, terreau *m*, WASTE, WATER SUPP compost brut *m*; ~ **aeration** SOIL, WASTE aération du compost *f*; ~ **container** WATER SUPP bac de compostage *m*; ~ **filter** AIR *biological filter* filtre à compost *m*; ~ **heap** SOIL

volume de compost *m*; ~ **maturing** SOIL, WASTE maturation du compost *f*; ~ **ripening** SOIL, WASTE mûrissement du compost *m*; ~**-screening machine** SOIL, WASTE crible pour compost *m*
compostable[1] *adj* SOIL, WASTE, WATER SUPP compostable
compostable[2]: ~ **waste** *n* WASTE déchets compostables *m pl*
composting *n* SOIL, WASTE, WATER SUPP *sewage* compostage *m*; ~ **drum** WASTE tambour de compostage *m*; ~ **plant** RENEW ENERG, WASTE installation de compostage *f*; ~ **reject** WASTE refus de compostage *m pl*; ~ **technique** WASTE méthode de compostage *f*
compound: ~ **wall** *n* NOISE paroi double *f*
compounder *n* WASTE, WATER POLL mélangeur *m*
comprehensive: ~ **procedure** *n* POLICY procédure globale *f*
compress *vt* RENEW ENERG, SOIL, WASTE compresser, comprimer, WATER POLL, WATER SUPP comprimer
compressed[1]: ~**-air-cleaned** *adj* AIR nettoyé à l'air comprimé
compressed[2]: ~ **air** *n* AIR air comprimé *m*; ~**-air blast** *n* AIR *bag filter* soufflerie à air comprimé *f*; ~**-air-cleaned jet filter** *n* AIR filtre à jet nettoyé à air comprimé *m*; ~**-air cleaning** *n* AIR nettoyage à air comprimé *m*; ~**-air consumption** *n* AIR consommation d'air comprimé *f*; ~**-air industrial breathing apparatus** *n* SAFETY appareil respiratoire industriel à air comprimé *m*; ~**-air injection** *n* AIR injection d'air comprimé *f*; ~**-air lance** *n* SOIL *soil vapour extraction* lance à air comprimé *f*; ~**-air receiver** *n* AIR collecteur d'air comprimé *m*; ~**-air storage tank** *n* AIR *pulse jet filter* cuve de stockage à air comprimé *f*; ~ **natural gas** *n* (*CNG*) AIR gaz naturel comprimé *m*
compressible: ~ **waste** *n* WASTE déchets compressibles *m pl*
compression[1]: ~**-proof** *adj* AIR incompressible
compression[2] *n* RENEW ENERG, SOIL, WASTE, WATER SUPP compression *f*; ~ **damage** WASTE dégât par compression *m*; ~ **filtration** ANALYSIS filtrage par compression *m*; ~ **load** ANALYSIS, WASTE charge de compression *f*; ~ **molding** *AmE*, ~ **moulding** *BrE* WASTE moulage par compression *m*; ~ **strength** ANALYSIS, SOIL, WASTE, WATER SUPP résistance à la compression *f*; ~ **test** ANALYSIS, WASTE essai de compression *m*; ~ **test machine** ANALYSIS, WASTE banc d'essai de compression *m*; ~ **vehicle** WASTE, WATER SUPP véhicule de compression *m*
compressor *n* AIR, RENEW ENERG *heat pump* compresseur *m*; ~ **casing** AIR carter compresseur *m*
computable *adj* ANALYSIS calculable
computation *n* ANALYSIS calcul *m*
computer: ~**-controlled catalyst** *n* AIR catalyseur géré par ordinateur *m*; ~**-controlled catalytic converter** *n* AIR convertisseur catalytique à commande informatique *m*
computerized[1] *adj* ANALYSIS informatisé
computerized[2]: ~ **analyser** *BrE*, ~ **analyzer** *AmE* *n* ANALYSIS analyseur informatique *m*
concentrate[1] *n* AIR, SOIL, WASTE concentré *m*
concentrate[2] *vt* AIR, RENEW ENERG *rays*, SOIL, WASTE, WATER SUPP concentrer
concentrated[1] *adj* GEN concentré
concentrated[2]: ~ **amount of sample** *n* ANALYSIS concentration d'échantillon *f*, volume concentré d'échantillon *m*; ~ **solar energy** *n* RENEW ENERG énergie solaire concentrée *f*
concentrating: ~ **collector** *n* RENEW ENERG *parabolic* collecteur solaire à concentration *m*
concentration *n* GEN concentration *f*; ~ **fluctuation** AIR *of pollutants* fluctuation de concentration *f*; ~ **of hazardous substances** AIR concentration de substances dangereuses *f*; ~ **of noise to area** MARINE, NOISE rapport bruit-surface *m*, rapport du bruit à la superficie *m*
conceptual: ~ **schedule** *n* MANAGE planning préliminaire *m*
concrete: ~ **ring of tubings** *n* WASTE anneau en béton *m*
condensate *n* AIR, SOIL, WATER POLL condensat *m*; ~ **film** AIR pellicule de condensats *f*; ~ **flow** AIR flux de condensats *m*; ~ **formation** AIR formation de condensats *f*; ~ **polishing** WATER POLL traitement des condensats *m*
condensation *n* GEN condensation *f*; ~ **aerosol** AIR aérosol de condensation *m*; ~**-force wet electrostatic precipitator** AIR dépoussiéreur électrostatique par condensation *m*; ~**-force wet ESP** AIR dépoussiéreur électrostatique par condensation *m*; ~ **nucleus** AIR, WASTE noyau de condensation *m*; ~ **nucleus counter** AIR, WASTE compteur de noyaux de condensation *m*; ~ **polymer** WASTE polymère de condensation *m*; ~ **scrubber** AIR dépoussiéreur par condensation *m*
condense *vt* GEN condenser
condensed: ~ **water** *n* AIR, WASTE eau condensée *f*; ~ **water content** *n* AIR, WASTE teneur en eau condensée *f*
condenser *n* GEN condenseur *m*
condition *vt* AIR, WASTE, WATER SUPP conditionner
conditioning *n* AIR, WASTE, WATER SUPP conditionnement *m*; ~ **process** WASTE, WATER SUPP procédé de conditionnement *m*; ~ **reactor** AIR réacteur de conditionnement *m*
conditions *n pl* GEN conditions *f pl*; ~ **for authorization** POLICY conditions d'autorisation *f pl*; ~ **for licensing** POLICY conditions d'octroi de licence *f pl*
conduct *vt* AIR conduire, ANALYSIS *test* réaliser, MANAGE, NOISE *structure-borne sound* conduire
conductibility *n* AIR, ANALYSIS conductibilité *f*
conductivity *n* ANALYSIS, SOIL, WATER SUPP conductivité *f*; ~ **measuring instrument** ANALYSIS conductivimètre *m*; ~ **meter** ANALYSIS conductivimètre *m*
conductometer *n* ANALYSIS conductimètre *m*
conductometric[1] *adj* ANALYSIS conductimétrique
conductometric[2]: ~ **analysis** *n* ANALYSIS analyse conductimétrique *f*; ~ **method** *n* ANALYSIS méthode conductimétrique *f*
conduit *n* AIR conduite *f*, WATER SUPP conduite *f*, canalisation *f*
cone *n* ANALYSIS, AIR *of cyclone*, WATER SUPP cône *m*
confidential: ~ **document** *n* POLICY document confidentiel *m*; ~ **information** *n* POLICY information confidentielle *f*
confine *vt* WATER SUPP confiner
confined[1] *adj* WATER SUPP confiné
confined[2]: ~ **aquifer** *n* WATER SUPP aquifère captif *m*; ~ **groundwater** *n* WATER SUPP nappe captive *f*
confluence *n* WATER SUPP confluent *m*
conglutination *n* GEN conglutination *f*
conical[1] *adj* AIR conique
conical[2]: ~ **base** *n* AIR *of cyclone* fond conique *m*
conimeter *n* ANALYSIS conimètre *m*
coning: ~ **smoke plume** *n* AIR panache de fumée en cône *m*
conjugate: ~ **compound** *n* ANALYSIS liaison conjuguée *f*
connate: ~ **water** *n* WATER POLL eau connée *f*
connecting: ~ **branch** *n* AIR raccord *m*; ~ **pipe** *n* SOIL tuyau de raccordement *m*
connection *n* WATER SUPP couplage *m*
conservation *n* GEN conservation *f*; ~ **of nature** POLICY préservation de la nature *f*; ~ **of species** POLICY préservation des espèces *f*
conserve *vt* GEN conserver, préserver, économiser; ~ **energy** MANAGE, RENEW ENERG, WASTE économiser l'énergie
consistency *n* ANALYSIS cohérence *f*, compatibilité *f*
conspicuous: ~ **defect** *n* MANAGE vice apparent *m*
constant: ~ **speed pump** *n* WATER SUPP pompe à débit constant *f*; ~ **volume sampling** *n* (*CVS*) AIR, ANALYSIS échantillonnage à volume constant *m*
constituent *n* AIR élément constitutif *m*, composant *m*, ANALYSIS composant *m*, SOIL composant *m*, constituant *m*
constrained: ~ **layer** *n* NOISE *vibration damping* couche comprimée *f*
constructed: ~ **area** *n* WASTE *of landfill* zone aménagée *f*; ~ **layer** *n* WASTE *of landfill* couche construite *f*
construction: ~ **drawing** *n* MANAGE plan d'exécution *m*;

~ **licence** *BrE*, ~ **license** *AmE* *n* POLICY permis de construire *m*; ~ **noise** *n* NOISE bruit de construction *m*; ~ **permission** *n* POLICY permis de construire *m*; ~ **waste** *n* WASTE déblais *m pl*, déchets de démolition *m pl*, gravats *m pl*

consume *vt* MANAGE, RENEW ENERG, WATER SUPP consommer

consumer *n* MANAGE consommateur *m*; ~ **behavior** *AmE*, ~ **behaviour** *BrE* MANAGE comportement du consommateur *m*, habitudes de consommation *f pl*; ~ **goods** *n pl* WASTE biens de consommation *m pl*; ~ **waste** RENEW ENERG, WATER SUPP ordures ménagères *f pl*

consumption *n* GEN consommation *f*; ~ **water** WATER POLL, WATER SUPP eau de consommation *f*

contact *n* AIR contact *m*; ~ **adhesive** WASTE adhésif de contact *m*, colle de contact *f*; ~ **catalysis** AIR catalyse de contact *f*; ~ **condenser** AIR condenseur par contact *m*; ~ **mass** WATER POLL masse en contact *f*; ~ **reactor** AIR four de contact *m*; ~ **spring** WATER SUPP source de contact *f*; ~ **surface** AIR surface de contact *f*; ~ **time** WASTE temps de contact *m*; ~ **tower** AIR tour de contact *f*

contained: ~ **energy** *n* RENEW ENERG énergie confinée *f*, énergie contenue *f*

container *n* AIR, SOIL, WASTE conteneur *m*; ~ **board box** WASTE boîte en carton caisse *f*; ~ **composting** RENEW ENERG compostage en conteneur *m*, compostage en fût *m*; ~ **composting plant** RENEW ENERG, WASTE installation de compostage en conteneur *f*; ~ **crane** RENEW ENERG grue pour conteneur *f*; ~ **press** WASTE presse pour conteneurs *f*, presse-conteneurs *m*; ~ **remediation installation** SOIL installation d'assainissement par conteneurs *f*; ~ **slide** WATER SUPP châssis de conteneur *m*; ~ **system** WATER SUPP système de conteneur *m*

containerizable *adj* RENEW ENERG, SOIL, WASTE conteneurisable

containerization *n* RENEW ENERG, SOIL, WASTE conteneurisation *f*

containerize *vt* RENEW ENERG, SOIL, WASTE conteneuriser

containerized *adj* RENEW ENERG, SOIL, WASTE conteneurisé

contaminant[1]: ~**-laden** *adj* GEN contaminé, pollué; ~**-specific** *adj* AIR, SOIL spécifique d'un contaminant

contaminant[2] *n* GEN contaminant *m*, polluant *m*; ~ **bound to soil** SOIL contaminant du sol *m*; ~ **concentrate** SOIL contaminant concentré *m*; ~ **content** SOIL teneur en contaminants *f*; ~ **degradation** AIR dégradation des contaminants *f*; ~ **emissions** *n pl* AIR émissions de contaminants *f pl*; ~ **gas stream** AIR flux gazeux contaminant *m*; ~ **properties** *n pl* AIR propriétés contaminantes *f pl*; ~ **type** AIR, SOIL, WASTE, WATER SUPP type de contaminant *m*; ~ **vapor** *AmE*, ~ **vapour** *BrE* AIR vapeur contaminante *f*

contaminate *vt* GEN contaminer, polluer

contaminated[1] *adj* GEN contaminé, pollué

contaminated[2]: ~ **dust** *n* AIR poussière contaminée *f*; ~ **plastics** *n* WASTE, WATER SUPP plastique souillé *m*; ~ **sampling gas** *n* ANALYSIS gaz d'échantillonnage pollué *m*; ~ **site** *n* SOIL décharge contaminée *f*; ~ **site pyrolysis** *n* SOIL pyrolyse de décharge contaminée *f*; ~ **site regeneration** *n* SOIL régénération de décharge contaminée *f*; ~ **site register** *n* SOIL répertoire des sites contaminés *m*

contamination *n* GEN contamination *f*, pollution *f*; ~ **fallout** AIR retombées polluantes *f pl*; ~ **meter** WATER SUPP compteur de contamination *m*; ~ **type** ANALYSIS type de contamination *m*

contents: ~ **declaration** *n* SAFETY, WASTE spécification de contenu *f*

contingency: ~ **plan** *n* GEN dispositif d'urgence *m*, plan d'intervention *m*

continuous[1] *adj* ANALYSIS *processes* continu

continuous[2]: ~ **casting machine** *n* WATER SUPP fondeuse en continu *f*; ~ **cleaning** *n* AIR *filter bag* nettoyage continu *m*; ~ **flowing film** *n* AIR film à écoulement continu *m*; ~ **hydrolysis** *n* SOIL hydrolyse continue *f*; ~ **measuring** *n* ANALYSIS mesurage en continu *m*; ~ **motion weight filling** *n* RENEW ENERG pondéral en continu *m*; ~ **press** *n* WATER SUPP presse en continu *f*; ~ **press line** *n* WASTE *for production from waste material* chaîne de presse en continu *f*; ~ **production line** *n* WASTE ligne de production en continu *f*; ~ **recording** *n* ANALYSIS enregistrement en continu *m*; ~ **sampling** *n* ANALYSIS échantillonnage en continu *m*; ~ **thread** *n* (*ct*) WASTE vis filetée *f*; ~**-thread cap** *n* WASTE capsule à vis préfiletée *f*; ~**-thread closure** *n* WASTE capsule à vis préfiletée *f*

contract: ~ **blister-packaging service** *n* WASTE service contractuel d'emballage thermoformé *m*; ~ **packaging** *n* WASTE emballage contractuel *m*

contravene *vt* POLICY, SAFETY *law, rule* contrevenir à

control[1] *n* AIR lutte *f*, GEN contrôle *m*, SOIL mesure de contrôle *f*; ~ **center** *AmE*, ~ **centre** *BrE* AIR poste de commande *m*, WASTE centre de contrôle *m*, poste de commande *m*; ~ **device** WATER SUPP dispositif de commande *m*; ~ **of marine pollution** MARINE lutte contre la pollution des mers *f*; ~ **room** AIR, WATER SUPP salle de commande *f*; ~ **shaft** RENEW ENERG, WASTE arbre de commande *m*; ~ **of substances hazardous to health** MANAGE, POLICY, SAFETY limitation des substances dangereuses pour la santé *f*; ~ **system** RENEW ENERG système de commande *m*; ~ **tag** WASTE étiquette de contrôle *f*; ~ **unit** AIR, WASTE poste de commande *m*

control[2] *vt* AIR *toxicity* contrôler, ANALYSIS commander, contrôler, piloter, MARINE commander, NOISE, RENEW ENERG contrôler, WASTE commander, contrôler, piloter, WATER SUPP piloter

Control: ~ **of Pollution Act** *n* *BrE* (*CPA*) POLICY loi antipollution *f*

controllability *n* AIR contrôlabilité *f*, ANALYSIS, MARINE, NOISE, WASTE *of landfill* maniabilité *f*

controlled: ~ **atmosphere** *n* (*CA*) WASTE, WATER SUPP atmosphère contrôlée *f*; ~**-atmosphere packaging** *n* (*CAP*) WASTE conditionnement sous atmosphère contrôlée *m*; ~ **combustion system** *n* (*CCS*) AIR système de combustion contrôlée *m*; ~ **dumping** *n* WATER SUPP décharge contrôlée *f*, dépôt contrôlé *m*; ~ **dust content chamber** *n* AIR enceinte à repoussièrement contrôlé *f*; ~ **emptying** *n* WASTE vidage réglementé *m*; ~ **environment storage system** *n* WASTE mode de stockage en environnement contrôlé *m*; ~ **fermentation** *n* WASTE fermentation contrôlée *f*; ~ **high-level landfill** *n* WASTE décharge contrôlée pour déchets spéciaux *f*; ~ **plant** *n* POLICY installation réglementée *f*; ~ **tipping** *n* WATER SUPP décharge contrôlée *f*

controlling: ~ **loop** *n* RENEW ENERG, WATER SUPP boucle de commande *f*, boucle de contrôle *f*

controls *n pl* WATER SUPP asservissement *m*

conurbation *n* MANAGE, POLICY grande agglomération *f*, conurbation *f*

convection *n* AIR, ANALYSIS, RENEW ENERG, WATER POLL convection *f*; ~ **heat exchanger** AIR échangeur de chaleur par convection *m*; ~ **section** AIR section de convection *f*

convention *n* AIR, POLICY convention *f*

conventional[1] *adj* AIR, NOISE, RENEW ENERG, SOIL classique

conventional[2]: ~ **above-ground multi-component landfill** *n* WASTE décharge classique à ciel ouvert et usages multiples *f*; ~ **coal cleaning** *n* AIR filière classique de nettoyage du charbon *f*; ~ **energy supply** *n* RENEW ENERG approvisionnement d'énergie classique *m*, fourniture d'énergie classique *f*; ~ **energy supply system** *n* RENEW ENERG filière classique de fourniture d'énergie *f*; ~ **energy technology** *n* RENEW ENERG technique énergétique classique *f*; ~ **filter system** *n* AIR système de filtre classique *m*; ~ **heating technology** *n* RENEW ENERG technique de chauffage classique *f*; ~ **oxidation catalyst** *n* (*COC*) AIR catalyseur d'oxydation classique *m*; ~ **recovery technology**

n WASTE technique de récupération classique *f*; ~ **sewerage system** *n* WASTE réseau d'assainissement classique *m*; ~ **technology** *n* AIR, NOISE, SOIL, WASTE technologie classique *f*, technique classique *f*; ~ **transportable pallet** *n* RENEW ENERG, WASTE palette transportable traditionnelle *f*
conversion *n* GEN conversion *f*; ~ **efficiency** RENEW ENERG rendement de conversion *m*; ~ **factor** ANALYSIS facteur de conversion *m*; ~ **force** AIR force de conversion *f*; ~ **into usable energy form** RENEW ENERG conversion sous forme d'énergie consommable *f*; ~ **machinery** WASTE équipement de conversion *m*; ~ **rate** AIR taux de conversion *m*; ~ **reaction** AIR *fuel cell* réaction de conversion *f*; ~ **technique** POLICY, RENEW ENERG filière de conversion *f*
convert *vt* GEN convertir; ~ **to electricity** RENEW ENERG convertir en électricité; ~ **to heat** RENEW ENERG convertir en chaleur
converter *n* GEN convertisseur *m*
convey: ~ **into silos** *vt* RENEW ENERG, WASTE ensiler, mettre en silo
conveying: ~ **equipment** *n* RENEW ENERG matériel de transport *m*; ~ **well** *n* SOIL puits d'acheminement *m*
conveyor *n* RENEW ENERG transporteur *m*, WASTE convoyeur *m*, WATER SUPP convoyeur *m*, transporteur *m*; ~ **belt** RENEW ENERG, WASTE bande d'amenée *f*, bande transporteuse *f*; ~ **belt skimmer** WATER SUPP récupérateur à bande transporteuse *m*; ~ **handling system** RENEW ENERG système de manutention par bande transporteuse *m*; ~ **pump** SOIL pompe de transfert *f*
cool *vt* GEN refroidir, réfrigérer
coolant *n* AIR réfrigérant *m*; ~ **flow rate** AIR débit de réfrigérant *m*; ~ **inlet** AIR admission de réfrigérant *f*; ~ **out** AIR sortie de réfrigérant *f*; ~ **outlet** AIR sortie de réfrigérant *f*
cooler *n* AIR, ANALYSIS, WATER SUPP *of ancillary system* réfrigérant *m*, refroidisseur *m*
cooling *n* GEN réfrigération *f*, refroidissement *m*; ~ **air** AIR air de refroidissement *m*; ~**-air inlet** AIR admission d'air de refroidissement *f*; ~ **air outlet** AIR sortie d'air de refroidissement *f*; ~ **drum** SOIL tambour de refroidissement *m*; ~**-gas fan** AIR ventilateur de gaz de refroidissement *m*; ~ **liquid** AIR liquide de refroidissement *m*; ~ **medium** AIR réfrigérant *m*, WATER SUPP fluide réfrigérant *m*; ~ **pond** WATER POLL, WATER SUPP bassin de refroidissement *m*, bassin de réfrigération *m*; ~ **system** AIR circuit de refroidissement *m*, dispositif de réfrigération *m*, système de refroidissement *m*, RENEW ENERG, WATER SUPP circuit de réfrigération *m*, circuit de refroidissement *m*, dispositif de réfrigération *m*, système de refroidissement *m*; ~ **tank** AIR cuve de refroidissement *f*; ~ **tower** RENEW ENERG aéroréfrigérant *m*; ~ **tower exhaust air** AIR effluent d'aéroréfrigérant *m*; ~ **water** AIR eau de refroidissement *f*; ~**-water loop** RENEW ENERG branche froide *f*; ~ **zone** AIR zone de refroidissement *f*
copolymer *n* WASTE, WATER SUPP copolymère *m*
copolymerization *n* WASTE, WATER SUPP copolymérisation *f*
copper *n* (*Cu*) AIR, SOIL, WATER SUPP cuivre *m* (*Cu*); ~**-indium diselenide cell** RENEW ENERG cellule au diséléniure cuivre-indium *f*
coppice *n* RENEW ENERG taillis *m*
cordierite *n* AIR *ceramic carrier* cordiérite *f*
core *n* RENEW ENERG âme *f*; ~ **material** RENEW ENERG matière fissile *f*
Coriolis: ~ **force** *n* AIR force de Coriolis *f*
corking: ~ **machine** *n* WASTE bouche-bouteille *f*; ~ **plug** *n* WASTE bouchon *m*
corner: ~ **reinforcement** *n* WASTE renforcement des coins *m*; ~ **stapling** *n* WASTE agrafage sur coin *m*; ~ **stapling machine** *n* WASTE agrafeuse sur coin *f*
corona *n* AIR couronne *f*; ~ **discharge** AIR décharge lumineuse *f*; ~ **effect** AIR effet couronne *m*; ~ **resistance** AIR résistance à effet couronne *f*
corral *vt* MARINE *pollution* encercler
corrective: ~ **maintenance** *n* MANAGE maintenance correctrice *f*
corresponding: ~ **sampling** *n* ANALYSIS échantillonnage correspondant *m*
corrosion[1]: ~**-preventive** *adj* RENEW ENERG anticorrosion; ~**-resistant** *adj* AIR anticorrosion
corrosion[2] *n* AIR, SAFETY, WASTE, WATER SUPP corrosion *f*; ~ **inhibitor** WATER SUPP agent anticorrosion *m*; ~ **prevention** RENEW ENERG lutte contre la corrosion *f*, WATER SUPP traitement anticorrosion *m*; ~**-preventive paper** WASTE papier anticorrosion *m*; ~ **resistance** WATER SUPP résistance à la corrosion *f*; ~ **risk** AIR risque de corrosion *m*; ~ **test** ANALYSIS essai de corrosion *m*
corrosive[1] *adj* AIR, SAFETY, WASTE, WATER SUPP corrosif
corrosive[2]: ~ **gas** *n* AIR gaz corrosif *m*; ~ **hazardous waste** *n* RENEW ENERG, WASTE déchets dangereux corrosifs *m pl*; ~ **substance** *n* SAFETY substance corrosive *f*
corrosiveness *n* AIR, SAFETY, WASTE, WATER SUPP corrosivité *f*
corrugated: ~ **board** *n* WASTE carton ondulé *m*; ~ **board box** *n* WASTE boîte en carton ondulé *f*; ~ **fiberboard** *AmE*, ~ **fibreboard** *BrE* *n* WASTE carton ondulé *m*; ~ **paper** *n* WASTE papier ondulé *m*
cosmic[1] *adj* AIR cosmique
cosmic[2]: ~ **radiation** *n* AIR rayonnement cosmique *m*; ~ **solar radiation** *n* AIR rayonnement solaire cosmique *m*
cost[1]: ~**-effective** *adj* MANAGE, POLICY économique, efficace par rapport au coût; ~**-intensive** *adj* MANAGE coûteux
cost[2]: ~ **analysis** *n* MANAGE analyse des coûts *f*; ~ **appraisal** *n* MANAGE évaluation des coûts *f*; ~**-benefit analysis** *n* MANAGE analyse coût-avantage *f*; ~**-benefit ratio** *n* MANAGE rapport coût-avantage *m*; ~**-benefit study** *n* MANAGE étude coût-avantage *f*, étude de rentabilité *f*; ~ **calculation** *n* MANAGE calcul des coûts *m*; ~ **effectiveness** *n* MANAGE efficacité par rapport au coût *f*; ~ **equalizing charge** *n* POLICY redevance de péréquation des charges *f*; ~ **factor** *n* MANAGE facteur de coût *m*; ~**-intensive method of treatment** *n* MANAGE méthode de traitement coûteuse *f*; ~ **planning** *n* MANAGE planification des coûts *f*; ~**-service life ratio** *n* MANAGE rapport coût-durée de service *m*; ~ **structure** *n* MANAGE structure des coûts *f*
cotton *n* WASTE ouate *m*
Cottrel: ~ **process** *n* AIR *electrostatic gas cleaning process* procédé Cottrel *m*
coulometer *n* ANALYSIS coulomètre *m*
coulometric[1] *adj* ANALYSIS coulométrique
coulometric[2]: ~ **analyser** *BrE*, ~ **analyzer** *AmE* *n* ANALYSIS analyseur coulométrique *m*; ~ **titration** *n* ANALYSIS titrage coulométrique *m*
coulometry *n* ANALYSIS coulométrie *f*
Council: ~ **Directive** *n* MANAGE, POLICY *EU* Directive du Conseil *f*; ~ **of environmental ministers** *n* POLICY Conseil des ministres de l'environnement *m*
counter: ~**-current extraction** *n* SOIL extraction à contre-courant *f*; ~**-current stripping** *n* SOIL extraction à contre-courant *f*; ~**-flow air separation** *n* RENEW ENERG séparation à contre-courant *f*; ~**-current washing** *n* AIR, WATER SUPP lavage à contre-courant *m*
coupled: ~ **biological scrubber** *n* AIR laveur biologique couplé *m*
coupling *n* AIR, ANALYSIS couplage *m*
course *n* WATER SUPP *of a river* cours *m*
cover[1] *n* RENEW ENERG couche *f*, couche de revêtement *f*, WASTE couverture *f*, WATER SUPP couverture *f*, gaine *f*; ~ **plate** AIR *of bag filter* plaque de couverture *f*
cover[2] *vt* RENEW ENERG couvrir; ~ **energy demand** RENEW ENERG couvrir la demande d'énergie
coverage: ~ **density** *n* POLICY densité d'occupation des sols *f*; ~ **ratio** *n* POLICY coefficient d'emprise au sol *m*
covering *n* NOISE, RENEW ENERG, WASTE, WATER SUPP

couverture *f*; ~ **layer** SOIL couche de revêtement *f*, WASTE couche de couverture *f*; ~ **material** RENEW ENERG, WASTE matériau de couverture *m*; ~ **sheet** NOISE couche de couverture *f*; ~ **soil** SOIL, WATER SUPP sol de couverture *m*
cowling *n* NOISE capuchonnement de ventilateur *m*
CPA *abbr BrE* (*Control of Pollution Act*) POLICY loi antipollution *f*
CPC *abbr* (*child-proof closure*) SAFETY bouchon de sécurité *m*
Cr *abbr* (*chromium*) AIR, SOIL Cr (*chrome*)
cradle: **~-to-grave analysis** *n* MANAGE, POLICY analyse de bout en bout *f*, analyse du cycle de vie *f*, analyse sur la totalité du cycle *f*
crane *n* RENEW ENERG grue *f*
crankcase *n* AIR carter *m*; ~ **emissions** *n pl* AIR *hydrocarbon* émissions de carter *f pl*; ~ **ventilation** AIR ventilation de carter *f*
crate *n* WASTE caisse à claire-voies *f*, casier *m*
CRC *abbr* (*child-resistant closure*) SAFETY fermeture de sécurité pour enfants *f*
crepe: ~ **paper** *n* WASTE papier crêpé *m*
crest *n* WATER SUPP *of dam* couronnement *m*, *of weir* crête *f*, seuil *m*
criminal: ~ **liability** *n* MANAGE responsabilité pénale *f*
criteria *n pl* GEN critères *m pl*
critical: ~ **group** *n* ANALYSIS groupe critique *m*; ~ **load** *n* SOIL charge critique *f*; ~ **pollutant** *n* WATER POLL polluant dangereux *m*; ~ **temperature** *n* AIR température critique *f*; ~ **water level** *n* WATER SUPP niveau critique d'eau *m*
crop: ~ **biomass** *n* SOIL biomasse végétale *f*; ~ **ecology** *n* MANAGE, POLICY écologie des cultures agricoles *f*; ~ **protection** *n* SOIL protection des cultures *f*; ~ **rotation** *n* SOIL rotation des cultures *f*; ~ **waste** *n* RENEW ENERG, WASTE, WATER SUPP déchets agricoles *m pl*, résidus de récolte *m pl*
cross[1]: **~-linked** *adj* AIR, RENEW ENERG réticulé
cross[2]: ~ **flow** *n* AIR *of waste gas* contre-courant *m*; **~-flow air separation** *n* AIR, RENEW ENERG séparation à contre-courant *f*; **~-flow scrubber** *n* AIR, WATER SUPP laveur à contre-courant *m*; **~-flow separator** *n* AIR, RENEW ENERG séparateur à contre-courant *m*; ~ **lines** *n pl* ANALYSIS *interferential pit gas tester* réticule *m*; **~-section sampling** *n* ANALYSIS échantillonnage en coupe *m*
cross[3]: **~-link** *vt* AIR *particle*, ANALYSIS réticuler
crosstalk *n* NOISE *transfer of airborne noise* diaphonie *f*
crown: ~ **cap** *n* WASTE capsule-couronne *f*; ~ **closure** *n* WASTE capsule-couronne *f*; ~ **cork** *n* WASTE capsule-couronne *f*
CRT *abbr* (*cathode ray tube*) SAFETY tube cathodique *m*
crude: ~ **oil** *n* AIR *auxiliary fuel*, MARINE, WATER SUPP pétrole brut *m*; ~ **pulp** *n* WASTE pâte brute *f*; ~ **refuse** *n* WASTE refus bruts *m pl*
crumble *vi* SOIL s'effriter
crush *vt* SOIL, WASTE, WATER SUPP écraser, broyer, concasser
crushed[1] *adj* SOIL, WASTE, WATER SUPP écrasé, broyé, concassé
crushed[2]: ~ **waste** *n* WASTE, RENEW ENERG déchets broyés *m pl*; **~-waste landfill** *n* WASTE décharge d'ordures broyées *f*
crusher *n* WASTE broyeur *m*, concasseur *m*, WATER SUPP broyeur *m*; ~ **unit** RENEW ENERG, WASTE, WATER SUPP poste de broyage *m*, installation de broyage *f*
crushing *n* SOIL, WASTE, WATER SUPP écrasement *m*, broyage *m*, concassage *m*; ~ **installation** RENEW ENERG, WASTE, WATER SUPP installation de broyage *f*; ~ **plant** WATER SUPP installation de broyage primaire *f*; ~ **system** WASTE système de broyage *m*
cryogenic: ~ **crushing** *n* WASTE *of solid waste* cryobroyage *m*; ~ **process** *n* WASTE, WATER SUPP procédé cryogénique *m*
cryogrinding *n* WASTE cryobroyage *m*
crystalline *adj* RENEW ENERG cristallin
crystallizate *n* WASTE produit de cristallisation *m*
crystallization *n* AIR cristallisation *f*; ~ **germ** AIR germe de cristallisation *m*; ~ **zone** AIR zone de cristallisation *f*
crystallize *vt* AIR cristalliser, se cristalliser
crystallography *n* ANALYSIS cristallographie *f*
CSD *abbr* (*Commission on Sustainable Development*) POLICY Commission du Développement Durable *f*
ct *abbr* (*continuous thread*) WASTE vis filetée *f*
Cu *abbr* (*copper*) AIR, SOIL, WATER SUPP Cu (*cuivre*)
cullet *n* WASTE calcin *m*, graisin *m*, verre cassé *m*
cultivation *n* ANALYSIS, SOIL culture *f*
culture: ~ **medium** *n* ANALYSIS milieu de culture *m*
culvert *n* WATER POLL, WATER SUPP canal *m*
cumulative: ~ **dose** *n* SAFETY dose cumulée *f*; ~ **frequency** *n* ANALYSIS fréquence cumulée *f*; ~ **parameter** *n* ANALYSIS paramètre cumulatif *m*; ~ **toxic effect** *n* MARINE, SAFETY, WATER SUPP effet toxique cumulatif *m*
cup *n* WASTE coupe *f*, coupelle *f*, tasse *f*; ~ **anemometer** ANALYSIS anémomètre à coupelles *m*
cupola: ~ **blast heater** *n* AIR four à dôme *m*
curb *vt* MARINE juguler, NOISE réprimer
curl *n* RENEW ENERG courbure *f*
current *n* AIR, RENEW ENERG, WATER POLL, WATER SUPP courant *m*; ~ **meter** WATER SUPP courantomètre *m*
curtain: ~ **boom** *n* MARINE barrage rideau *m*; ~ **depth** *n* SOIL profondeur d'écran *f*
cushion *vt* NOISE amortir, WATER SUPP affaiblir
cushioned *adj* WATER SUPP affaibli
cushioning *n* NOISE amortissement *m*, WATER SUPP affaiblissement *m*; ~ **product** WASTE produit de rembourrage *m*
customer *n* MANAGE client *m*; ~ **behavior** *AmE*, ~ **behaviour** *BrE* MANAGE comportement du client *m*, habitudes de consommation *f pl*
cut *n* SOIL *in subterraneous curtain method* tranchée *f*
cuts: ~ **and fills** *n pl* WASTE remblais et déblais *m pl*
cutter *n* SOIL désagrégateur *m*
cutting *n* RENEW ENERG éclaircie *f*, coupe *f*, plant *m*; ~ **oil** RENEW ENERG, WATER SUPP huile de coupe *f*; ~ **press** WATER SUPP presse à découper *f*; ~ **shredder** RENEW ENERG, WASTE déchiqueteuse *f*, massicot *m*; **~, transporting and storing of wood** RENEW ENERG façonnage, débardage et stockage du bois *m*
CVS *abbr* (*constant volume sampling*) AIR, ANALYSIS échantillonnage à volume constant *m*
cyanide[1]: **~-containing** *adj* WATER POLL cyanuré
cyanide[2] *n* SOIL, WATER POLL cyanure *m*; **~-containing sludge** WATER POLL boues cyanurées *f*; ~ **removal** WATER POLL extraction des cyanures *f*
cycle: ~ **path** *n* POLICY piste cyclable *f*; ~ **track** *n* POLICY piste cyclable *f*
cyclone *n* AIR, SOIL, WATER SUPP cyclone *m*; ~ **burning chamber** WASTE chambre de combustion cyclone *f*; ~ **furnace** SOIL four cyclone *m*; ~ **housing** AIR corps de cyclone *m*; ~ **radius** AIR rayon de cyclone *m*; ~ **separation** AIR séparation par cyclone *f*; ~ **separator** AIR cyclone de dévésiculeur *m*; ~ **tube** AIR tube cyclone *m*; ~ **vitrification** SOIL vitrification par cyclone *f*; ~ **wall** AIR paroi de cyclone *f*
cylinder: ~ **broom** *n* WASTE balai cylindrique *m*, brosse cylindrique *f*, WATER SUPP balai cylindrique *m*
cylindrical[1] *adj* AIR cylindrique
cylindrical[2]: ~ **cyclone** *n* AIR cyclone cylindrique *m*; ~ **graduated measuring glass** *n* ANALYSIS vase gradué cylindrique *m*; ~ **inlet tube** *n* AIR *of cyclone* tube d'admission cylindrique *m*; ~ **screw extruder** *n* WASTE, WATER SUPP extrudeuse à vis *f*, extrudeuse hélicoïdale *f*; ~ **section** *n* AIR *of cyclone* section cylindrique *f*
cytotoxicity *n* SAFETY cytotoxicité *f*
CZE *abbr* (*capillary zone electrophoresis*) ANALYSIS électrophorèse de zone capillaire *f*

D

Dacron® *n AmE* (*cf Terylene BrE*) AIR *filter material* Dacron® *m*
daily[1] *adj* AIR, ANALYSIS, SAFETY, WATER SUPP quotidien
daily[2]: **~ average value** *n* ANALYSIS moyenne journalière *f*; **~ consumption** *n* WATER SUPP consommation journalière *f*; **~ cover** *n* WASTE couverture journalière *f*; **~ mean value** *n* ANALYSIS moyenne journalière *f*; **~ noise exposure** *n* NOISE exposition au bruit journalière *f*; **~ processing capacity** *n* AIR, SOIL, WATER POLL, WATER SUPP capacité de traitement journalier *f*; **~ total exposure** *n* SAFETY exposition quotidienne totale *f*; **~ water flow** *n* WATER SUPP débit journalier *m*
Dalton: **~'s law of partial pressure** *n* AIR loi de pression partielle de Dalton *f*
dam *n* RENEW ENERG barrage *m*, WATER SUPP barrage de retenue *m*, ouvrage de retenue *m*
damage[1] *n* AIR destruction *f*, dégradation *f*; **~ to health** SAFETY atteinte à la santé *f*; **~ to vegetation** RENEW ENERG atteinte portée à la végétation *f*, dégradation de la végétation *f*
damage[2] *vt* GEN endommager, porter atteinte à, NOISE altérer, perturber
damaged *adj* GEN endommagé, NOISE altéré; **~ area** *n* POLICY zone sinistrée *f*
damp[1]: **~-proof** *adj* GEN étanche, étanche à l'humidité
damp[2]: **~ heat** *n* AIR chaleur humide *f*
damp[3] *vt* AIR, NOISE amortir
damped: **~ layer** *n* NOISE couche d'amortissement *f*
dampen *vt* WATER SUPP affaiblir
dampened *adj* WATER SUPP affaibli
damper *n* AIR, NOISE amortisseur *m*
damping *n* AIR, NOISE amortissement *m*; **~ roll** RENEW ENERG rouleau mouilleur *m*
dampness *n* GEN humidité *f*
danger: **~ assessment** *n* SAFETY évaluation des risques *f*; **~ of fire** *n* RENEW ENERG risque d'incendie *m*; **~ level** *n* SAFETY niveau d'alarme *m*; **~ potential** *n* MANAGE, SAFETY potentiel de risques *m*; **~ prevention** *n* SAFETY prévention des risques *f*; **~ symbol** *n* SAFETY symbole de danger *m*; **~ zone** *n* SAFETY zone dangereuse *f*
dangerous[1] *adj* AIR, SAFETY dangereux
dangerous[2]: **~ anthropogenic interference with the climate system** *n* AIR interférence anthropique dangereuse pour le système climatique *f*; **~ good** *n* SAFETY, WASTE produit dangereux *m*; **~ load** *n* SAFETY, WASTE, WATER SUPP charge dangereuse *f*; **~ material** *n* SAFETY matière dangereuse *f*; **~ substance** *n* SAFETY substance dangereuse *f*
Dano: **~ process** *n* WASTE procédé Dano *m*
daphnia: **~ test** *n* ANALYSIS essai de toxicité sur daphnie *m*
Darius: **~ rotor** *n* RENEW ENERG *vertical axle* rotor Darius *m*
dark: **~ heat-absorbing material** *n* RENEW ENERG *solar pond* matériau noir absorbant la chaleur *m*
data: **~ acquisition** *n* ANALYSIS acquisition des données *f*; **~ analysis** *n* ANALYSIS analyse des données *f*; **~ interpretation** *n* ANALYSIS interprétation des données *f*; **~ logger** *n* ANALYSIS collecteur de données *m*, enregistreur séquentiel de données *m*; **~ logging** *n* ANALYSIS enregistrement de données *m*; **~ network** *n* ANALYSIS réseau de données *m*; **~ processing** *n* ANALYSIS traitement des données *m*; **~ security** *n* ANALYSIS protection des données *f*, sûreté des données *f*; **~ storage** *n* ANALYSIS stockage des données *m*; **~ system** *n* ANALYSIS système de données *m*; **~ transfer** *n* ANALYSIS transfert de données *m*
database *n* ANALYSIS base de données *f*
dB *abbr* (*decibel*) ANALYSIS, NOISE dB (*décibel*)
DC *abbr* (*direct current*) RENEW ENERG CC (*courant continu*)
DC/AC: **~ converter** *n* RENEW ENERG convertisseur CC/CA *m*
DDT *abbr* (*dichlorodiphenyltrichloroethane*) SOIL DDT (*dichloro-diphényl-trichloroéthane*)
deactivation *n* AIR, RENEW ENERG désactivation *f*
dead[1] *adj* WATER POLL mort
dead[2]: **~ bacteria** *n pl* WATER POLL bactéries mortes *f pl*; **~ channel** *n* WATER SUPP bras mort *m*; **~ man's switch** *n* SAFETY dispositif homme mort *m*; **~ zone** *n* MARINE zone morte *f*
deaden *vt* NOISE étouffer, amortir
deadening *n* NOISE amortissement *m*
de-aerate *vt* WATER POLL, WATER SUPP désaérer, dégazer
de-aeration *n* WATER POLL, WATER SUPP désaération *f*, dégazage *m*
de-aerator *n* WATER POLL, WATER SUPP désaérateur *m*, dégazeur *m*
debris *n* AIR, WASTE débris
debt: **~ for nature swaps** *n* POLICY conversion de dette en investissements écologiques *f*
decant *vt* GEN décanter, transvaser
decantation *n* GEN décantage *m*, décantation *f*, transvasement *m*
decanter *n* GEN *device* décanteur *m*
decanting: **~ centrifuge** *n* RENEW ENERG, WATER POLL décanteur centrifuge *m*
decarbonate *vt* WATER SUPP décarbonater
decarbonated *adj* WATER SUPP décarbonaté
decarbonator *n* WATER SUPP éliminateur de gaz carbonique *m*
decarbonization *n* WATER SUPP décalaminage *m*, décarbonisation *f*, décarburation *f*; **~ plant** WATER SUPP banc de décarbonisation *m*; **~ with slaked lime** WATER SUPP décalaminage à la chaux éteinte *m*
decarbonize *vt* WATER SUPP décalaminer, décarboniser, décarburer
decarbonized[1] *adj* WATER SUPP décalaminé, décarbonisé, décarburé
decarbonized[2]: **~ groundwater** *n* WATER SUPP nappe phréatique décarbonisée *f*
decay *n* WASTE, WATER POLL décomposition *f*, putréfaction *f*, *of radiactive material* désintégration radioactive, WATER SUPP putréfaction *f*; **~ of organic matter** WASTE, WATER POLL décomposition des matières organiques *f*; **~ tank** WATER POLL cuve de désactivation *f*, cuve de fermentation *f*
deceleration *n* AIR *filter medium* décélération *f*
decentralized: **~ photoelectric use** *n* RENEW ENERG application photoélectrique décentralisée *f*
dechlorinate *vt* SOIL, WATER SUPP déchlorer, éliminer le chlore
dechlorination *n* SOIL, WATER SUPP déchloration *f*
decibel *n* (*dB*) ANALYSIS, NOISE décibel *m* (*dB*)
decision *n* POLICY, WATER SUPP décision *f*
declaration: **~ of public utility** *n* POLICY déclaration d'utilité publique *f* (*DUP*)
decoding: **~ of measured value** *n* ANALYSIS décodage de la valeur mesurée *m*
decoloration *n* ANALYSIS décoloration *f*
decompose *vi* SOIL, WASTE se décomposer, WATER POLL, WA-

TER SUPP se décomposer, se dégrader; ~ **oxidatively** WATER POLL se décomposer par oxydation
decomposition *n* SOIL décomposition *f*, WATER POLL *anaerobic and aerobic* décomposition *f*, dégradation *f*, WATER SUPP décomposition *f*, désagrégation *f*, WASTE décomposition *f*; ~ **capacity** WATER POLL pouvoir de dégradation *m*; ~ **of harmful substances** WATER POLL dégradation de substances dangereuses *f*; ~ **potential** SOIL potentiel de décomposition *m*; ~ **temperature** WASTE température de décomposition *f*
decontaminability *n* SOIL décontaminabilité *f*
decontaminate *vt* GEN décontaminer, dépolluer
decontaminated[1] *adj* GEN décontaminé, dépollué
decontaminated[2]: ~ **soil** *n* SOIL sol décontaminé *m*
decontamination *n* GEN décontamination *f*, dépollution *f*; ~ **reactor** SOIL réacteur de décontamination *m*; ~ **technique** SOIL technique de décontamination *f*
dedicated: ~ **fleet** *n* POLICY flotte captive *f*
de-dust *vt* AIR, WATER SUPP dépoussiérer
de-duster *n* AIR, WATER SUPP dépoussiéreur *m*
de-dusting *n* AIR, WATER SUPP dépoussiérage *m*
deep: ~**-bed filter** *n* AIR *separator* filtre à lit profond *m*, WATER POLL filtre à haute couche *m*; ~**-bed filtration** *n* AIR, WATER POLL filtration en profondeur *f*, filtration sur lit profond *f*; ~ **ocean geyser** *n* MARINE source hydrothermale souterraine *f*; ~ **ploughing** *BrE*, ~ **plowing** *AmE* *n* SOIL labour profond *m*; ~ **sand trap** *n* WATER POLL sablière encaissée *f*; ~**-sea sediment** *n* MARINE sédiment marin pélagique *m*; ~**-shaft reactor** *n* WATER POLL réacteur à puits profond *m*; ~ **well** *n* WATER POLL puits profond *m*; ~**-well injection** *n* WATER POLL injection en puits profond *f*
default: ~ **interest** *n* MANAGE intérêt moratoire *m*
defect: ~ **canal system** *n* SOIL *source of soil degradation* réseau de canaux défectueux *m*
deferrization *n* RENEW ENERG déferraillage *m*, WATER SUPP déferrisation *f*
deflector *n* AIR déflecteur *m*, WASTE voile de carde *m*; ~ **separator** AIR dépoussiéreur à déflecteur *m*; ~ **vane** AIR pale de déflecteur *f*
defoliant *n* SOIL, WATER POLL défoliant *m*, produit défoliant *m*
defoliate *vt* SOIL, WATER POLL défolier
defoliating: ~ **agent** *n* SOIL agent défoliant *m*
defoliation *n* SOIL, WATER POLL défoliation *f*
deforestation *n* AIR, SOIL déboisement *m*, déforestation *f*
degas *vt* GEN dégazer
degasification *n* GEN dégazage *m*; ~ **filter tube** SOIL filtre tubulaire de dégazage *m*
degasify *vt* GEN dégazer
degasser *n* GEN dégazeur *m*
degassing *n* GEN dégazage *m*; ~ **chamber** RENEW ENERG, WASTE, WATER SUPP chambre de dégazage *f*; ~ **facility** RENEW ENERG, WASTE, WATER SUPP installation de dégazage *f*, poste de dégazage *m*; ~ **with gabion** RENEW ENERG dégazage en gabion *m*; ~ **with gas wells** RENEW ENERG dégazage en puits *m*; ~ **pipe** RENEW ENERG conduite de dégazage *f*; ~ **with sonde** RENEW ENERG dégazage à la sonde *m*
degradability *n* GEN dégradabilité *f*
degradable *adj* GEN dégradable
degradation *n* AIR dégradation *f*, WATER POLL dégradation *f*, détérioration *f*; ~ **hazard** SOIL risque de dégradation *m*; ~ **performance** WATER POLL dégradation *f*
degrade *vt* GEN dégrader
degraded[1] *adj* AIR, SOIL, WATER POLL, WATER SUPP dégradé
degraded[2]: ~ **land** *n* SOIL terre dégradée *f*
degreaser *n* WASTE dégraisseur *m*
degreasing: ~ **compound** *n* WASTE composition dégraissante *f*
degree: ~ **of absorption** *n* AIR degré d'absorption *m*; ~ **of contamination** *n* SOIL degré de contamination *m*; ~ **of decontamination** *n* SOIL degré de décontamination *m*; ~ **of degradability** *n* WATER POLL degré de dégradabilité *m*; ~ **of degradation** *n* WATER POLL degré de dégradation *m*; ~ **of desulfurization** *AmE*, ~ **of desulphurization** *BrE* *n* AIR degré de désulfuration *m*; ~ **of dewatering** *n* WATER POLL degré de déshydratation *m*; ~ **of dissociation** *n* ANALYSIS degré de dissociation *m*; ~ **of dust removal** *n* AIR degré de dépoussiérage *m*; ~ **of exchange** *n* WATER SUPP degré d'échange *m*; ~ **of fermentation** *n* WATER SUPP degré de fermentation *m*; ~ **of fraction removal** *n* AIR degré de fractionnement *m*; ~ **of hardness** *n* ANALYSIS degré de dureté *m*, WATER POLL indice de dureté *m*, WATER SUPP degré de dureté *m*; ~ **of humification** *n* WATER POLL *composting* degré d'humification *m*; ~ **of nitrogen oxides removal** *n* AIR degré d'extraction des oxydes d'azote *m*; ~**-of-fraction collection** *n* AIR collecte par degré de fractionnement *f*; ~ **of ozone depletion** *n* AIR degré d'appauvrissement de la couche d'ozone *m*; ~ **of pollution** *n* AIR, MARINE, SOIL, WATER POLL, WATER SUPP degré de pollution *m*; ~ **of purification** *n* WATER POLL degré d'épuration *m*; ~ **of purity** *n* WATER SUPP degré de pureté *m*; ~ **of separation** *n* AIR, WATER POLL degré de séparation *m*; ~ **of stabilization** *n* WATER POLL degré de stabilisation *m*; ~ **of thickening** *n* WATER POLL *dewatering* taux d'épaississement *m*
degrit *vt* WATER POLL dessabler
degritter *n* WATER POLL dessableur *m*
Degussa-Plinke: ~ **process** *n* AIR procédé Degussa-Plinke *m*
dehalogenate *vt* SOIL déshalogéner
dehalogenation *n* SOIL déshalogénation *f*
dehumidification *n* ANALYSIS déshumidification *f*; ~ **effect** ANALYSIS effet desséchant *m*
dehumidify *vt* ANALYSIS déshumidifier
dehydrate *vt* AIR déshydrater
dehydration *n* AIR déshydratation *f*
dehydrogenase: ~ **test** *n* ANALYSIS essai à la déshydrogénase *m*
de-ink *vt* WATER POLL désencrer
de-inking *n* WATER POLL désencrage *m*; ~ **by flotation** WATER POLL désencrage par flottation *m*; ~ **sludge** WATER POLL boues de désencrage *f pl*; ~ **unit** WATER POLL installation de désencrage *f*
deionization *n* WATER POLL, WATER SUPP désionisation *f*
deionize *vt* WATER POLL, WATER SUPP désioniser
deionized[1] *adj* WATER POLL, WATER SUPP désionisé
deionized[2]: ~ **water** *n* WATER POLL, WATER SUPP eau désionisée *f*
de-labeling *AmE*, **de-labelling** *BrE* *n* WASTE désétiquetage *m*
deleterious: ~ **effect** *n* AIR, SOIL, WATER POLL effet délétère *m*
deliberate: ~ **dumping** *n* WATER POLL déversement délibéré *m*
deliverer *n* RENEW ENERG fournisseur *m*, livreur *m*
delivery *n* WATER POLL, WATER SUPP *of pump* approvisionnement *m*, *of water* débit *m*, *of pump* refoulement *m*; ~ **channel** WATER POLL canal d'amenée *m*, WATER SUPP canal distributeur *m*, canalisation d'amenée *f*; ~ **head** WATER POLL, WATER SUPP hauteur de refoulement *f*, hauteur manométrique *f*; ~ **main** WATER SUPP conduite de refoulement *f*; ~ **pipe** WATER SUPP conduite de refoulement *f*, tubulure de refoulement *f*, tuyau de refoulement *m*; ~ **pressure** WATER SUPP pression de refoulement *f*; ~ **race** WATER SUPP canal distributeur *m*; ~ **time** WASTE, WATER SUPP délai d'acheminement *m*
demineralization *n* WATER POLL, WATER SUPP *removal of minerals* déminéralisation *f*, *removal of salt* dessalage *m*, dessalement *m*
demineralize *vt* WATER POLL, WATER SUPP *remove minerals from* déminéraliser, *remove salt from* dessaler
demineralized[1] *adj* WATER POLL, WATER SUPP *minerals removed* déminéralisé, *salts removed* dessalé
demineralized[2]: ~ **water** *n* WATER POLL, WATER SUPP eau déminéralisée *f*
demister *n* AIR dispositif antibuée *m*, *ozone* chambre de désaturation *f*; ~ **unit** AIR dispositif antibuée *m*, RENEW

ENERG désembueur *m*, dispositif de désembuage *m*, hotte de captage des buées *f*
demixing *n* WATER POLL ségrégation *f*
demolition: **~ waste** *n* RENEW ENERG, WASTE déchets de démolition *m pl*, gravats *m pl*, déblais *m pl*
demonstration: **~ power plant** *n* RENEW ENERG centrale prototype de production d'électricité *f*
demulsification *n* MARINE, WATER POLL, WATER SUPP désémulsification *f*
demulsifier *n* MARINE, WATER POLL, WATER SUPP désémulsifiant *m*
demulsify *vt* MARINE, WATER POLL, WATER SUPP désémulsifier
demulsifying: **~ product** *n* MARINE, WATER POLL, WATER SUPP produit désémulsifiant *m*
denesting: **~ machine** *n* WASTE machine de désemboîtement *f*
denitrate *vt* GEN dénitrifier, dénitrurer
denitration *n* GEN dénitrification *f*
denitrification *n* GEN dénitrification *f*; **~ in circulating fluidized bed** AIR dénitration sur lit fluidisé circulant *f*; **~ zone** WATER POLL zone de dénitrification *f*; **in situ ~** WATER POLL dénitrification in situ *f*
denitrify *vt* GEN dénitrifier
denitrifying: **~ bacteria** *n pl* WATER POLL bactéries dénitrifiantes *f pl*
denox[1] *abbr* (*NO_x removal*) AIR, SOIL, WATER SUPP dénitrification *f* (*élimination de NO_x*)
denox[2] *n* AIR, SOIL, WATER SUPP dénitration *f*; **~ catalyst** AIR catalyseur dénitrifiant *m*; **~ in circulating fluidized bed** AIR dénitrification sur lit fluidisé circulant *f*
denoxer *n* AIR dispositif de dénitration *m*
densely: **~-populated area** *n* WATER SUPP zone à forte densité de population *f*
densifier *n* RENEW ENERG, WASTE *for continuous material feeding* densificateur *m*
densify *vt* SOIL *into pellets*, WASTE densifier
densimeter *n* ANALYSIS *density determination of liquids* densimètre *m*
density *n* ANALYSIS, WATER SUPP densité *f*, densité absolue *f*, masse volumique *f*; **~ bottle** ANALYSIS flacon à mesurer la densité *m*, pycnomètre *m*; **~ current** WATER POLL, WATER SUPP courant de densité *m*; **~ determination** ANALYSIS calcul de masse volumique *m*; **~ filling machine** WASTE machine de remplissage volumétrique *f*; **~ of measurement stations** ANALYSIS densité des stations de mesure *f*; **~ of volume charge** AIR *electrostatic precipitator* densité de charge volumique *f*
deodorization *n* AIR, WATER POLL, WATER SUPP désodorisation *f*
deodorize *vt* AIR, WATER POLL, WATER SUPP désodoriser
de-oil *vt* WATER POLL déshuiler
de-oiling *n* WATER POLL déshuilage *m*
deoxygenate *vt* WATER POLL, WATER SUPP désoxygéner
deoxygenated: **~ water** *n* WATER POLL, WATER SUPP eau désoxygénée *f*
deoxygenation *n* WATER POLL, WATER SUPP désoxygénation *f*
depalletization *n* RENEW ENERG, WASTE, WATER SUPP dépalettisation *f*
depalletize *vt* RENEW ENERG, WASTE, WATER SUPP dépalettiser
depalletizer *n* RENEW ENERG, WASTE dépalettisateur *m*
department *n* POLICY *administrative unit* ministère *m*, service *m*; **~ of the environment** POLICY ministère de l'environnement *m*
dependent: **~ ecosystem** *n* AIR, MARINE, SOIL écosystème dépendant *m*, écosystème tributaire *m*
dephenolate *vt* WATER POLL déphénoler
dephenolating *n* WATER POLL déphénolage *f*
dephenolize *vt* WATER POLL déphénoler
dephenolizing *n* WATER POLL déphénolage *f*
deplate *vt* WATER SUPP déchromer, dénickeler
depletion: **~ time** *n* WATER POLL délai d'épuisement *m*, délai d'appauvrissement *m*
depolarization *n* ANALYSIS dépolarisation *f*
depolarize *vt* ANALYSIS dépolariser
depolarizer *n* ANALYSIS dépolariseur *m*
depolarizing *adj* ANALYSIS dépolarisant
depolymerization *n* RENEW ENERG, WATER SUPP dépolymérisation *f*
depolymerize *vt* RENEW ENERG, WATER SUPP dépolymériser
depopulate *vt* POLICY dépeupler
depopulation: **~ of countryside** *n* POLICY dépeuplement des campagnes *m*
deposit[1] *n* AIR, SOIL, WATER POLL, WATER SUPP *scale* dépôt *m*, tartre *m*; **~ bottle** WASTE bouteille consignée *f*; **~ build-up** AIR, SOIL, WATER POLL, WATER SUPP constitution d'un dépôt *f*, formation d'un dépôt *f*; **~ fee system** POLICY système de consigne *m*; **~ gage** *AmE*, **~ gauge** *BrE* AIR collecteur de précipitations atmosphériques *m*
deposit[2] *vt* WATER POLL, WATER SUPP déposer
deposit[3] *vi* WATER POLL, WATER SUPP se déposer
deposition *n* AIR, SAFETY précipitation *f*, SOIL, WATER POLL, WATER SUPP précipitation *f*, dépôt *m*, retombées *f pl*; **~ of cardboard** SOIL, WATER POLL, WATER SUPP dépôt de carton *m*; **~ rate** SOIL taux de dépôt *m*; **~ of silt** SOIL, WATER SUPP dépôt de limon *m*; **~ value** SOIL, WATER SUPP valeur des retombées *f*; **~ velocity** AIR vitesse de dépôt *f*, WATER SUPP vitesse de dépôt *f*, vitesse de sédimentation *f*; **~ of waste-water sludge on grassland** WATER POLL dépôt de boues d'eaux résiduaires sur des herbages *m*; **~ of wastewater sludge on ground used for forage** WATER POLL dépôt de boues d'épuration sur des terres à fourrage *m*
depot: **~ container** *n* SOIL conteneur de stockage *m*, WASTE conteneur de dépôt *m*, conteneur de parc *m*
depression: **~ cone** *n* WATER SUPP cône d'appel *m*, cône de dépression *m*; **~ spring** *n* WATER SUPP source de dépression *f*, source de déversement *f*
deprived: **~ area** *n* POLICY zone déshéritée *f*
depth: **~ filtration** *n* WATER POLL filtration en profondeur *f*
derestricted *adj* MANAGE mis en diffusion générale
derrick *n* RENEW ENERG *tidal power plant* derrick *m*
desalinate *vt* WATER SUPP dessaler
desalination *n* WATER SUPP dessalage *m*, dessalement *m*; **~ plant** WATER POLL, WATER SUPP installation de dessalement *f*, usine de dessalement *f*
desalinization *n* WATER SUPP dessalage *m*, dessalement *m*
desalinize *vt* WATER SUPP dessaler
desalt *vt* WATER SUPP dessaler
desalting: **~ plant** *n* WATER POLL, WATER SUPP installation de dessalement *f*, usine de dessalement
descale *vt* WATER POLL, WATER SUPP décalaminer, décrasser
descaling *n* WATER POLL, WATER SUPP décalaminage *m*, décrassage *m*
desert: **~ formation** *n* SOIL désertification *f*; **~ region** *n* RENEW ENERG *solar farm installation* région désertique *f*
desertification *n* SOIL désertification *f*
desiccator *n* ANALYSIS dessiccateur *m*
design *n* GEN dessin *m*, conception *f*; **~ flow** WATER SUPP débit nominal *m*; **~ to cost** MANAGE conception à coût objectif *f*
designation *n* ANALYSIS désignation *f*
deslag *vt* WATER POLL, WATER SUPP décrasser
deslagging *n* WATER POLL, WATER SUPP décrassage *m*
de-sludge *vt* WATER POLL, WATER SUPP évacuer les boues de, débourber
de-sludger *n* WATER POLL, WATER SUPP débourbeur *m*
de-sludging *n* WATER POLL, WATER SUPP évacuation des boues *f*, débourbage *m*
desonox *abbr* (*SO_2/NO_xremoval*) AIR desonox (*élimination de SO_2/NO_x*)
desorb *vt* AIR, ANALYSIS, SOIL, WATER POLL désorber; **~ thermally** ANALYSIS désorber par voie thermique

desorbed[1] *adj* AIR, ANALYSIS, SOIL, WATER POLL désorbé
desorbed[2]: ~ **absorbent** *n* AIR absorbant désorbé *m*; ~ **activated coke** *n* AIR coke actif désorbé *m*; ~ **rich gas** *n* AIR gaz riche désorbé *m*
desorbent *n* AIR désorbant *m*
desorber: ~ **feed** *n* AIR charge de désorbeur *f*
desorption *n* AIR, ANALYSIS, SOIL, WATER POLL désorption *f*; ~ **air** AIR air de désorption *m*; ~-**air heat exchanger** AIR échangeur d'air de désorption *m*; ~ **column** AIR colonne de désorption *f*; ~ **time** AIR temps de désorption *m*; ~ **zone** AIR zone de désorption *f*
desox[1] *abbr* (*desulphurization*) AIR, WATER SUPP desox *f* (*désulfuration*)
desox[2]: ~ **in circulating fluidized bed** *n* AIR désulfuration sur lit fluidisé circulant *f*
destabilization *n* WATER POLL *of colloids* déstabilisation *f*
destabilize *vt* WATER POLL *colloids* déstabiliser
destabilized[1] *adj* WATER POLL déstabilisé
destabilized[2]: ~ **particle** *n* WATER POLL *flocculation* particule déstabilisée *f*
de-stack *vt* WASTE désempiler
de-stacking *n* WASTE désempilement *m*
destroy *vt* AIR *ozone layer, pollutant*, SOIL *harmful substance*, WATER POLL détruire
destruction *n* AIR, SOIL, WATER POLL destruction *f*; ~ **of harmful substances** AIR, SOIL, WATER POLL destruction de substances dangereuses *f*; ~ **process** AIR, SOIL, WASTE, WATER POLL procédé de destruction *m*; ~ **and removal efficiency** (*DRE*) SOIL efficacité de destruction et d'élimination *f*
desulfurization *AmE see desulphurization BrE*
desulfurize *AmE see desulphurize BrE*
desulphurization *n BrE* (*desox*) AIR, WATER SUPP désulfuration *f*; ~ **in circulating fluidized bed** *BrE* AIR désulfuration sur lit fluidisé circulant *f*; ~ **with magnesium oxide** *BrE* AIR désulfuration à l'oxyde de magnésium *f*; ~ **of process gases** *BrE* AIR désulfuration des gaz industriels *f*; ~ **sorbent** *BrE* AIR sorbant de désulfuration *m*
desulphurize *vt BrE* AIR, WATER SUPP désulfurer
detailed: ~ **mapping** *n* ANALYSIS cartographie détaillée *f*, imagerie détaillée *f*
detect *vt* ANALYSIS détecter
detectability *n* ANALYSIS détectabilité *f*
detectable *adj* ANALYSIS détectable
detection *n* ANALYSIS détection *f*; ~ **frequency** ANALYSIS fréquence de détection *f*; ~ **limit** ANALYSIS limite de détection *f*; ~ **method** ANALYSIS méthode de détection *f*; ~ **principle** ANALYSIS principe de détection *m*; ~ **reaction** ANALYSIS réaction à la détection *f*; ~ **sensitivity** ANALYSIS sensibilité de détection *f*
detector *n* ANALYSIS détecteur *m*, WATER SUPP sonde *f*; ~ **lamp** ANALYSIS lampe de détecteur *f*; ~ **tube** ANALYSIS *gas* tube détecteur *m*
detention: ~ **time** *n* WATER POLL temps de sédimentation *m*
detergent[1] *adj* SOIL, WATER POLL détergent, détersif
detergent[2] *n* SOIL, WATER POLL détergent *m*, détersif *m*
deteriorate *vi* AIR, SOIL détériorer, se détériorer
deterioration *n* AIR, SOIL détérioration *f*; ~ **of the environment** POLICY dégradation de l'environnement *f*; ~ **of fresh waters** WATER POLL dégradation des ressources en eau douce *f*; ~ **of natural resources** AIR, POLICY, SOIL, WATER POLL dégradation des ressources naturelles *f*
determination *n* ANALYSIS *fixing* détermination *f*, *calculation* calcul *m*, dosage *m*; ~ **of acid capacity in wastewater** ANALYSIS détermination de la capacité de neutralisation des acides des eaux usées *f*; ~ **of ammonium content in wastewater** ANALYSIS détermination de la concentration d'ammonium dans les eaux usées *f*, dosage de la concentration d'ammonium dans les eaux usées *m*; ~ **component** ANALYSIS élément entrant dans la détermination *m*; ~ **of concentration of nitrogen dioxide** ANALYSIS détermination de la concentration de dioxyde d'azote *f*; ~ **of content** ANALYSIS calcul de concentration *m*, calcul de teneur *m*; ~ **of dew point** ANALYSIS détermination du point de rosée *f*; ~ **following sample distillation** ANALYSIS détermination après distillation sur échantillon *f*, détermination consécutive à la distillation d'échantillon *f*; ~ **of gaseous acid air pollution** ANALYSIS détermination de la pollution atmosphérique acide *f*; ~ **limit** ANALYSIS seuil de détermination *m*; ~ **of mass concentration of sulfur dioxide in ambient air** *AmE*, ~ **of mass concentration of sulphur dioxide in ambient air** *BrE* SAFETY détermination de la concentration massique de dioxyde de soufre dans l'air ambiant *f*, dosage de la concentration massique de dioxyde de soufre dans l'air ambiant *m*; ~ **method** ANALYSIS méthode de détermination *f*; ~ **of natrium content in wastewater** ANALYSIS calcul de la teneur en sodium des eaux usées *m*; ~ **of particulate lead and lead compounds** SAFETY détermination des particules et des composés de plomb *f*, dosage des particules et des composés de plomb *m*; ~ **of settleable and undissolved matter in wastewater** ANALYSIS détermination des matières décantables et non dissoutes dans les eaux usées *f*, dosage des matières décantables et non dissoutes dans les eaux usées *m*; ~ **of total hardness in wastewater** ANALYSIS détermination de la dureté totale des eaux usées *f*; ~ **of wastewater density** ANALYSIS détermination de la masse volumique des eaux usées *f*; ~ **of wastewater pH-value** ANALYSIS calcul du pH des eaux usées *m*; ~ **of water demand** WATER SUPP détermination de la demande en eau *f*
de-tin *vt* WATER POLL désétamer
de-tinning *n* WATER POLL désétamage *m*
detoxification *n* AIR, SOIL, WATER POLL, WATER SUPP détoxication *f*; ~ **under batch conditions** WATER POLL détoxication par lots *f*
detoxify *vt* AIR, SOIL, WATER POLL, WATER SUPP détoxiquer
detritor *n* WASTE, WATER POLL bassin de dessablement *m*, dessableur *m*
detritus *n* WASTE, WATER POLL détritus *m pl*; ~ **feeder** WASTE détritiphage *m*, détritivore *m*
develop *vt* MANAGE *process* développer, mettre au point, POLICY aménager, développer
developed: ~ **area** *n* POLICY zone aménagée *f*
development *n* MANAGE *of process* développement *m*, mise au point *f*, POLICY développement *m*
deviate *vi* AIR *gas flow* dévier
device *n* GEN dispositif *m*; ~ **for measured value transmission** ANALYSIS dispositif de transmission des valeurs mesurées *m*
dew: ~ **point** *n* AIR point de rosée *m*; ~ **point hygrometer** *n* ANALYSIS hygromètre à point de rosée *m*; ~ **point temperature** *n* AIR température de point de rosée *f*
Dewar: ~ **vessel** *n* ANALYSIS vase Dewar *m*
dewater *vt* SOIL dénoyer, drainer, WATER POLL, WATER SUPP assécher, déshydrater, sécher
dewaterability *n* AIR déshydratabilité *f*, WATER POLL déshydratabilité *f*, drainabilité *f*
dewaterable: ~ **crystal** *n* AIR cristal déshydratable *m*
dewatered[1] *adj* WATER POLL, WATER SUPP asséché, déshydraté, séché
dewatered[2]: ~ **sludge** *n* WATER POLL boues déshydratées *f pl*
dewatering *n* SOIL dénoyage *m*, drainage *m*, WATER POLL, WATER SUPP assèchement *m*, déshydratation *f*, séchage *m*; ~ **pump** WASTE pompe d'épuisement *f*, pompe d'exhaure *f*, pompe de dénoyage *f*, WATER POLL pompe d'épuisement *f*, pompe d'exhaure *f*, WATER SUPP pompe d'épuisement *f*, pompe d'exhaure *f*, pompe de dénoyage *f*; ~ **via freezing** WATER POLL déshydratation par congélation *f*
dezincification *n* WATER SUPP dézingage *m*
dezincify *vt* WATER SUPP dézincifier

DHS *abbr* (*dynamic headspace technique*) ANALYSIS méthode manométrique dynamique *f*

diagnosis *n* ANALYSIS diagnostic *m*

dial: ~ **gage** *AmE*, ~ **gauge** *BrE* *n* ANALYSIS *measuring* comparateur à cadran *m*; ~-**type timer** *n* ANALYSIS *gas meter* chronométreur à cadran *m*

dialog[1]: ~-**controlled** *AmE see dialogue-controlled BrE*

dialog[2]: ~ **display** *AmE see dialogue display BrE*

dialogue[1]: ~-**controlled** *adj BrE* ANALYSIS *programming* interactif

dialogue[2]: ~ **display** *n BrE* ANALYSIS *on measuring instrument* affichage conversationnel *m*

dialysis *n* ANALYSIS, WATER POLL dialyse *f*

diamagnetic *adj* ANALYSIS diamagnétique

diaphragm: ~ **pump** *n* ANALYSIS, MARINE, WATER POLL, WATER SUPP pompe à diaphragme *f*, pompe à membrane *f*; ~ **sealing** *n* WASTE fermeture à membrane *f*; ~ **valve** *n* AIR, WATER POLL robinet à membrane *m*, vanne à membrane *f*

diatomaceous: ~ **earth** *n* WATER SUPP terre à diatomées *f*; ~ **earth filtration** *n* WATER POLL filtration sur diatomées *f*

dichlorodifluoromethane *n* AIR *propellant* dichlorodifluorométhane *m*

dichlorodiphenyltrichloroethane *n* (*DDT*) SOIL dichlorodiphényl-trichloroéthane *m* (*DDT*)

dichloromethane *n* AIR *halogenated hydrocarbon* dichlorométhane *m*

diesel: ~ **engine** *n* AIR, WASTE, WATER POLL moteur diesel *m*; ~ **exhaust gas** *n* AIR gaz d'échappement diesel *m*; ~ **generator** *n* RENEW ENERG génératrice diesel *f*

differential: ~ **gas chromatogram** *n* ANALYSIS chromatogramme différentiel en phase gazeuse *m*; ~ **manometer** *n* AIR manomètre différentiel *m*; ~ **pressure** *n* AIR *in filter layer* pression différentielle *f*; ~ **pressure gage** *AmE*, ~ **pressure gauge** *BrE* *n* AIR manomètre différentiel *m*; ~ **scanning calorimetry** *n* (*DSC*) ANALYSIS calorimétrie différentielle *f*; ~ **tax rate** *n* POLICY taux d'imposition différencié *m*; ~ **taxation** *n* POLICY imposition différenciée *f*; ~ **thermal analysis** *n* (*DTA*) ANALYSIS analyse thermique différentielle *f* (*ATD*)

diffraction *n* ANALYSIS diffraction *f*; ~ **of light** ANALYSIS diffraction de la lumière *f*

diffuse[1]: ~ **field** *n* NOISE champ diffus *m*; ~ **radiation** *n* RENEW ENERG rayonnement diffus *m*

diffuse[2] *vt* AIR diffuser

diffused: ~-**air aeration** *n* WATER POLL aération par air surpressé *f*, aération par diffusion *f*; ~-**air tank** *n* WATER SUPP ballon d'air surpressé *m*; ~-**light measurement** *n* ANALYSIS *turbidity determination* mesure de la lumière diffuse *f*, mesure de la lumière parasite *f*; ~-**light sensor** *n* ANALYSIS capteur de lumière diffuse *m*, capteur de lumière parasite *m*

diffuser *n* GEN diffuseur *m*

diffusion *n* GEN diffusion *f*; ~ **analysis** ANALYSIS analyse de diffusion *f*; ~ **charging** AIR charge de diffusion *f*

digested: ~ **sludge** *n* WATER POLL boues digérées *f*, boues putréfiées *f*

digester *n* WASTE, WATER POLL digesteur *m*; ~ **disruption** WATER POLL destruction d'un digesteur *f*; ~ **gas** WATER SUPP biogaz *m*, gaz de digestion *m*; ~ **gas quantity** WATER POLL, WATER SUPP volume de biogaz *m*; ~ **gas utilization** WATER POLL, WATER SUPP valorisation des gaz de digestion *f*; ~ **gas yield** WATER SUPP, WATER POLL production de biogaz *f*

digestion *n* WATER POLL, WATER SUPP digestion *f*; ~ **with added composting** WATER POLL, WATER SUPP digestion avec apport de compost *f*; ~ **deposit** WASTE résidu de digestion *m*; ~ **period** WATER POLL, WATER SUPP durée de la digestion *f*, temps de digestion *m*; ~ **tank** WATER POLL, WATER SUPP chambre de digestion *f*, fosse septique *f*; ~ **time** WATER POLL, WATER SUPP temps de digestion *m*

digestor: ~ **gas** *n* WATER POLL gaz de digesteur *m*

digital: ~ **micromirror device** *n* (*DMD*) ANALYSIS micromiroir numérique *m*

dilute *vt* AIR, ANALYSIS, WASTE diluer, délayer

diluted[1] *adj* AIR, ANALYSIS, WASTE dilué, délayé

diluted[2]: ~ **sample** *n* ANALYSIS échantillon dilué *m*; ~ **solution** *n* ANALYSIS solution diluée *f*

dilution *n* AIR, ANALYSIS, WASTE dilution *f*, délayage *m*, WATER POLL dilution *f*, recyclage physique *m*, WATER SUPP dilution *f*; ~ **analysis** ANALYSIS analyse de dilution *f*, analyse par dilution *f*; ~ **capacity** WATER POLL *of waters* pouvoir de dilution *m*; ~ **sample** ANALYSIS échantillon dilué *m*

DIN *abbr* (*German Institute for Standards*) ANALYSIS DIN (*Institut allemand de normalisation*)

dinitrogen: ~ **pentoxide** *n* (N_2O_5) AIR pentoxyde de diazote *m* (N_2O_5), pentoxyde d'azote *m* (N_2O_5)

diode: ~ **array detector** *n* ANALYSIS détecteur à diodes en série *m*; ~ **laser** *n* ANALYSIS *measurement of gaseous pollutants* laser à diode *m*

dioxin *n* SOIL dioxine *f*

diplomacy *n* POLICY diplomatie *f*

direct[1] *adj* AIR, RENEW ENERG, WATER POLL direct

direct[2]: ~ **air sampler** *n* ANALYSIS échantillonneur à courant direct *m*; ~ **chipping** *n* RENEW ENERG déchiquetage immédiat en copeaux *m*; ~-**contact condenser** *n* AIR condenseur à contact direct *m*; ~ **current** *n* (*DC*) RENEW ENERG courant continu *m* (*CC*); ~ **current voltage field** *n* AIR *electrostatic precipitator* champ de courant continu *m*; ~ **desulfurization process** *AmE*, ~ **desulphurization process** *BrE* *n* AIR procédé de désulfuration direct *m*; ~ **discharger** *n* WATER POLL déchargeur automatique *m*, déversoir direct *m*; ~ **drying** *n* WATER POLL *suspension, drum or moving-bed dryer* séchage par contact direct *m*; ~ **extrusion** *n* WASTE, WATER SUPP extrusion directe *f*; ~ **groundwater recharge** *n* WATER SUPP remplissage direct d'une nappe *m*; ~-**heated rotary kiln** *n* SOIL four rotatif à chauffage direct *m*; ~ **heating** *n* SOIL chauffage direct *m*; ~ **heating by radiation** *n* SOIL *thermal destruction* chauffage direct par rayonnement *m*; ~ **industrial wastewater discharge** *n* WATER POLL évacuation directe des effluents industriels *f*; ~ **injection** *n* ANALYSIS, SOIL injection directe *f*; ~ **interception** *n* AIR interception directe *f*; ~ **liquid introduction** *n* ANALYSIS *chromatography* injection directe de liquide *f*; ~ **measurement method** *n* ANALYSIS méthode de mesure directe *f*; ~ **noise level** *n* NOISE niveau de bruit direct *m*, niveau sonore direct *m*; ~ **poisoning** *n* SAFETY empoisonnement immédiat *m*; ~ **pulse filter** *n* AIR filtre à injection directe *m*; ~ **radiation** *n* RENEW ENERG rayonnement direct *m*; ~ **reading** *n* ANALYSIS *instrumentation* lecture directe *f*; ~ **sound level** *n* NOISE niveau de bruit direct *m*, niveau sonore direct *m*; ~ **suction skimmer** *n* MARINE écrémeuse à aspiration directe *f*; ~ **voltage recorder** *n* ANALYSIS enregistreur en courant continu *m*

direction *n* AIR *of flue gas* orientation *f*, direction *f*, NOISE, WATER SUPP direction *f*; ~ **change** AIR *cyclone* changement de direction *m*, réorientation *f*; ~ AIR orientation des effluents gazeux *f*

directive *n* POLICY *legal, EU* directive *f*; ~ **on major industrial hazards** POLICY directive sur les principaux risques industriels *f*

directorate *n* POLICY *administrative unit* direction *f*

dirt: ~-**gas inlet** *n* AIR admission de gaz pollué *f*

dirty: ~ **gas** *n* AIR gaz à traiter *m*, gaz pollué *m*; ~-**gas canal** *n* AIR conduite de gaz pollué *f*; ~-**gas compartment** *n* AIR compartiment des gaz à traiter *m*; ~-**gas concentration** *n* AIR concentration de gaz pollué *f*; ~-**gas conditioning** *n* AIR conditionnement de gaz pollué *m*; ~-**gas cooling** *n* AIR refroidissement de gaz pollué *m*; ~-**gas dust content** *n* AIR concentration de poussière dans un gaz pollué *f*; ~-**gas dust loading** *n* AIR charge pulvimétrique d'un gaz pollué *f*; ~-**gas**

flow *n* AIR débit de gaz pollué *m*; **~-gas load** *n* AIR charge de gaz pollué *f*; **~-gas loading** *n* AIR charge de gaz pollué *f*; **~-gas volume flow** *n* AIR débit volumique de gaz pollué *m*; **~ oil** *n* AIR, WATER POLL cambouis *m*
disaggregate *vi* SOIL s'effriter
disc: **~ aerator** *n BrE* WATER POLL diffuseur à disques *m*; **~ skimmer** *n BrE* MARINE écrémeuse à disque *f*, récupérateur à disque *m*
discharge[1] *n* AIR émission *f*, évacuation *f*, MARINE évacuation *f*, rejet *m*, SOIL dépôt *m*, évacuation *f*, WASTE évacuation *f*, WATER POLL écoulement *m*, évacuation *f*, déchargement *m*, déversement *m*, rejet *m*, WATER SUPP rejet *m*, écoulement *m*, émission *f*, évacuation *f*, décharge *f*, déversement *m*, *of pump, compressor, blower* débit *m*, *of stream, weir* porte *f*; **~ canal** WATER POLL canal de déversement *m*; **~ capacity** WATER POLL débit d'évacuation *m*, débit de rejet *m*; **~ conditions** *n pl* AIR *of stack* conditions de rejet *f pl*; **~ conveyor** WASTE, WATER SUPP convoyeur de décharge *m*; **~ gutter** WATER POLL canal de décharge *m*, rigole d'évacuation *f*; **~ head** WATER POLL, WATER SUPP *of pump* hauteur de refoulement *f*, hauteur géométrique de refoulement *f*, hauteur manométrique *f*; **~ hopper** WATER POLL déversoir de trop-plein *m*, trémie de décharge *f*; **~ of municipal wastewater sludge plant** WATER POLL évacuation de la station municipale d'épuration des eaux *f*; **~ nozzle** WATER POLL, WATER SUPP *flowing out* buse d'évacuation *f*, *flowing in* buse d'alimentation *f*; **~ pipe** WATER POLL, WATER SUPP conduite d'évacuation *f*, tuyau de décharge *m*; **~ piping** WATER POLL, WATER SUPP tuyauterie d'évacuation *f*, tuyauterie de refoulement *f*; **~ sluice** WATER POLL vanne de décharge *f*; **~ standard** WATER SUPP norme de rejet *f*; **~ system** WATER POLL *for wastewater* réseau d'évacuation *m*; **~ worm** WASTE conduite d'égout *f*
discharge[2] *vt* AIR évacuer, rejeter, MARINE, SOIL déverser, évacuer, WASTE évacuer, WATER POLL *wastewater pump* déverser, évacuer, rejeter, WATER SUPP évacuer, rejeter
discharge[3] *vi* WATER SUPP se déverser
discharger *n* WASTE évacuateur *m*, WATER POLL déversoir *m*, WATER SUPP évacuateur *m*
discolor *AmE see discolour BrE*
discoloration *n* ANALYSIS décoloration *f*
discolour *vt BrE* ANALYSIS décolorer
discomfort *n* MANAGE gêne *f*; **~ noise** NOISE niveau de gêne acoustique *m*
discontinuous[1] *adj* GEN *process* discontinu
discontinuous[2]: **~ fixed-bed adsorber** *n* AIR, WATER POLL adsorbeur à lit fixe discontinu *m*; **~ hydrolysis** *n* SOIL hydrolyse discontinue *f*; **~ measuring** *n* ANALYSIS mesurage discontinu *m*, mesure discontinue *f*; **~ ultrafiltration** *n* WATER POLL ultrafiltration discontinue *f*
discretionary: **~ decision** *n* POLICY pouvoir discrétionnaire *m*; **~ error** *n* POLICY erreur discrétionnaire *f*
disfigurement: **~ of rural environment** *n* POLICY enlaidissement du milieu rural *m*
disinfect *vt* SAFETY, WATER POLL, WATER SUPP désinfecter
disinfectant *n* SAFETY, WATER POLL, WATER SUPP désinfectant *m*, produit désinfectant *m*
disinfection *n* SAFETY, WATER POLL, WATER SUPP désinfection *f*; **~ effect** WATER POLL effet désinfectant *m*
disintegrate[1] *vt* SOIL, WASTE, WATER SUPP désintégrer
disintegrate[2] *vi* SOIL, WASTE, WATER SUPP se désintégrer
disintegration *n* SOIL, WASTE, WATER SUPP désintégration *f*
disintegrator *n* AIR désintégrateur *m*, RENEW ENERG, WATER SUPP dilacérateur *m*
disinterment: **~ of waste** *n* WASTE affleurement de déchets *m*
disk *AmE see disc BrE*
dislocation *n* WATER SUPP dislocation *f*
dispersant *n* GEN agent dispersant *m*, dispersant *m*; **~ agent** AIR, ANALYSIS, MARINE, SOIL, WATER POLL dispersant *m*; **~ spraying** MARINE pulvérisation d'agent dispersant *f*
disperse[1]: **~ phase** *n* AIR, ANALYSIS *in liquid* phase de dispersion *f*, phase dispersée *f*
disperse[2] *vt* GEN disperser
dispersed *adj* GEN dispersé
dispersing: **~ agent** *n* GEN agent dispersant *m*; **~ equipment** *n* MARINE matériel de dispersion *m*; **~ power** *n* WATER SUPP pouvoir dispersant *m*
dispersion *n* GEN dispersion *f*; **~ aerosol** AIR aérosol de dispersion *m*; **~ behavior** *AmE*, **~ behaviour** *BrE* SOIL, WATER SUPP comportement de dispersion *m*; **~ colloid** WATER POLL colloïde de dispersion *m*; **~ force** AIR *van der Waals forces* force de dispersion *f*; **~ medium** AIR milieu de dispersion *m*; **~ particle counter** ANALYSIS compteur de particules de dispersion *m*; **~ telemetering** ANALYSIS télémesure de dispersion *f*
displacement *n* WATER SUPP déplacement *m*, dislocation *f*; **~ transducer** ANALYSIS capteur de déplacement *m*
displacer *n* AIR piston déplaceur *m*
display: **~ unit** *n* ANALYSIS écran d'affichage *m*, console de visualisation *f*, poste de visualisation *m*
disposable[1] *adj* WASTE jetable, perdu
disposable[2]: **~ bottle** *n* WASTE bouteille jetable *f*; **~ container** *n* WASTE conteneur jetable *m*
disposal[1]: **~-oriented** *adj* WASTE axé sur l'élimination
disposal[2] *n* GEN élimination *f*, évacuation *f*, *scrapping* mise au rebut *f*; **~ channel** WATER SUPP canal d'évacuation *m*; **~ cost** MANAGE coût d'élimination *m*; **~ pathway** WASTE filière d'élimination *f*, filière d'élimination de déchets *f*; **~ of plastics** SOIL, WASTE élimination des plastiques *f*; **~ route** WASTE filière d'élimination *f*, filière d'élimination de déchets *f*; **~ safety** WASTE sûreté d'évacuation *f*; **~ site** WASTE, WATER SUPP zone d'évacuation *f*, zone de décharge *f*; **~ zone** WASTE zone d'évacuation *f*, zone de décharge *f*
dispose: **~ of** *vt* GEN *scrap* mettre au rebut, éliminer, évacuer
disposed: **~-of residues** *n pl* SOIL, WASTE, WATER POLL, WATER SUPP déchets éliminés *m pl*, résidus éliminés *m pl*, résidus évacués *m pl*
disruption: **~ of the ecological balance** *n* POLICY rupture de l'équilibre écologique *f*
dissimilation *n* WATER POLL *biological denitrification process* dissimilation *f*
dissociable[1] *adj* ANALYSIS dissociable, WATER POLL, WATER SUPP dissociable, séparable
dissociable[2]: **~ substance** *n* WATER POLL substance séparable *f*
dissociation *n* ANALYSIS, WATER POLL dissociation *f*
dissolution *n* AIR, ANALYSIS, WATER POLL, WATER SUPP dissolution *f*
dissolve *vt* AIR *gas into liquid*, ANALYSIS, WATER POLL, WATER SUPP dissoudre; **~ away** SOIL extraire par dissolution; **~ printed color from fiber** *AmE*, **~ printed colour from fibre** *BrE* WATER SUPP dissoudre l'encre d'impression de fibre
dissolved[1] *adj* AIR, ANALYSIS, WATER POLL, WATER SUPP dissous
dissolved[2]: **~ air flotation** *n* WATER POLL flottation par insufflation d'air *f*; **~ inorganic carbon** *n* AIR, ANALYSIS, WATER SUPP carbone inorganique dissous *m*; **~ organic carbon** *n* (*DOC*) AIR, ANALYSIS, WATER SUPP carbone organique dissous *m* (*COD*); **~ organic matter** *n* AIR, ANALYSIS, WATER POLL, WATER SUPP matière organique dissoute *f*; **~ oxygen** *n* (*DO*) AIR, WATER SUPP oxygène dissous *m* (*OD*); **~ substance** *n* WATER POLL substance dissoute *f*, matière dissoute *f*
dissolving: **~ away** *n* SOIL *of contaminant* extraction par dissolution *f*
distant: **~ source** *n* AIR, WATER SUPP source éloignée *f*; **~ water supply** *n* WATER SUPP alimentation en eau à distance *f*, téléalimentation en eau *f*
distil[1] *vt BrE* ANALYSIS, SOIL, WATER POLL, WATER SUPP distiller

distil[2] *vi BrE* ANALYSIS, SOIL, WATER POLL, WATER SUPP se distiller
distill *AmE see distil BrE*
distillate *n* ANALYSIS distillat *m*, WATER SUPP produit de distillation *m*; ~ **flask** ANALYSIS flacon à distillats *m*
distillation *n* ANALYSIS, SOIL, WATER POLL, WATER SUPP *process* distillation *f*, *product* distillat *m*; ~ **column** WATER SUPP colonne de distillation *f*; ~ **gas** AIR gaz de distillation *m*; ~ **residue** WATER SUPP résidu de distillation *m*
distilled: ~ **water** *n* ANALYSIS *use in sampling* eau distillée *f*
distillery: ~ **wastewater** *n* WASTE, WATER POLL eaux résiduaires de distillerie *f pl*
distilling: ~ **column** *n* ANALYSIS colonne de distillation *f*
distinguishable *adj* ANALYSIS discernable
distributing: ~ **bottom** *n* AIR *of wet precipitator* fond distributeur *m*; ~ **pipe** *n* WATER SUPP conduit principal *m*, tuyau distributeur *m*
distribution *n* GEN distribution
distributor *n* WASTE distributeur *m*; ~ **worm** WASTE conduite de distribution *f*
district: ~ **heating generation** *n* WASTE *from waste incineration* production de chauffage urbain *f*
disturbance: ~ **level** *n* NOISE niveau de perturbation *m*
disturbed: ~ **sample** *n* ANALYSIS échantillon perturbé *m*
disturbing: ~ **noise** *n* NOISE bruit perturbant *m*
ditch *n* WATER SUPP fossé *m*, rigole *f*; ~ **irrigation** WATER SUPP irrigation à la raie *f*, irrigation en fossé *f*
ditcher *n* SOIL *for installing retention wall* trancheuse *f*
diversity *n* MANAGE, POLICY *of genetic resources* diversité *f*
divide *n* WATER SUPP ligne de partage des eaux *f*
division *n* POLICY *administrative unit* direction *f*
DMD *abbr* (*digital micromirror device*) ANALYSIS micromiroir numérique *m*
DO *abbr* (*dissolved oxygen*) AIR, WATER SUPP OD (*oxygène dissous*)
DOC *abbr* (*dissolved organic carbon*) AIR, ANALYSIS, WATER SUPP COD (*carbone organique dissous*)
doctor: ~ **blade** *n* WASTE racle *f*
dolomite: ~ **meal** *n* AIR *additive* charge de dolomie *f*
domestic: ~ **emissions** *n pl* AIR émissions nationales *f pl*; ~**-like industrial waste** *n* WASTE déchets industriels assimilables aux ordures ménagères *m pl*, déchets industriels banals *m pl* (*DIB*); ~ **refuse incineration** *n* WASTE incinération des ordures ménagères *f*; ~ **tax on oil products** *n* POLICY taxe intérieure sur les produits pétroliers *f* (*TIPP*); ~ **waste** *n* WASTE déchets domestiques *m pl*, déchets ménagers *m pl*; ~ **waste analysis** *n* ANALYSIS analyse des ordures ménagères *f*; ~ **waste burning** *n* WASTE incinération des ordures ménagères *f*; ~ **waste collection** *n* WASTE ramassage des ordures ménagères *m*; ~ **waste composition** *n* WASTE composition des ordures ménagères *f*; ~ **waste incineration** *n* WASTE incinération des ordures ménagères *f*; ~ **wastewater** *n* WATER POLL eaux usées ménagères *f pl*; ~ **water supply** *n* WATER SUPP approvisionnement en eau dans le secteur résidentiel *m*, fourniture d'eau dans le secteur résidentiel *f*
dope *vt* AIR, RENEW ENERG doper
doping *n* AIR, RENEW ENERG dopage *m*
Dortmund: ~ **tank** *n* WATER POLL décanteur Dortmund *m*, fosse Dortmund *f*
dosage *n* ANALYSIS, WASTE, WATER POLL dosage *m*; ~ **device** ANALYSIS doseur *m*
dose[1] *n* ANALYSIS, WASTE, WATER POLL, WATER SUPP dose *f*; ~ **response** ANALYSIS, WATER SUPP dose effet *f*; ~**-response relationship** ANALYSIS, WATER SUPP rapport dose-effet *m*
dose[2] *vt* ANALYSIS, WASTE, WATER POLL, WATER SUPP doser
dosimeter *n* ANALYSIS dosimètre *m*
dosing *n* ANALYSIS, WASTE, WATER POLL, WATER SUPP dosage *m*; ~ **apparatus** WASTE doseur *m*; ~ **feeder** WASTE chargeur-doseur *m*; ~ **machine** WATER SUPP doseuse *f*; ~ **pump** WASTE, WATER POLL, WATER SUPP pompe doseuse *f*; ~ **scale** AIR échelle de dosage *f*
double[1]: ~**-cycle** *adj* AIR à double cycle; ~**-leaf** *adj* NOISE à double feuille; ~**-pipe** *adj* AIR à double tube; ~**-sheet** *adj* NOISE *encapsulation* à double couche, à double feuille
double[2]: ~**-beam photometer** *n* ANALYSIS photomètre à double faisceau *m*; ~**-belt press** *n* WASTE, WATER SUPP presse à double bande *f*; ~**-bucket collector** *n* WATER SUPP collecteur à deux augets *m*; ~ **counter-current filtration** *n* WATER SUPP double filtration à contre-courant *f*; ~**-cycle scrubber** *n* AIR épurateur à double cycle *m*, laveur à double cycle *m*; ~**-detector gas chromatography** *n* ANALYSIS chromatographie en phase gazeuse à double détecteur *f*; ~ **eco-balance system** *n* MANAGE système d'éco-bilan en parties doubles *m*; ~**-pipe condenser** *n* AIR condenseur à double tube *m*; ~**-pipe heat exchanger** *n* AIR échangeur de chaleur à tubes doubles *m*, échangeur de chaleur à tubes jumelés *m*; ~ **wall** *n* NOISE *for level reduction* double paroi *f*
down: ~ **wind** *n* AIR vent arrière *m*, vent en poupe *m*, vent descendant *m*
downpipe *n* WATER POLL descente *f*, tuyau d'eaux pluviales *m*, tuyau de descente *m*
downspout *n* AIR couloir *m*, goulotte *f*, tuyau d'eaux pluviales *m*, WATER POLL descente *f*, tuyau de descente *m*
downstream[1] *adj* AIR, WATER SUPP d'aval
downstream[2] *adv* WATER SUPP en aval
downward: ~ **surface installation** *n* WASTE installation de surface aval *f*, installation de surface en pente descendante *f*
downwind[1] *adj* AIR dans le sens du vent, sous le vent
downwind[2]: ~ **power plant** *n* RENEW ENERG centrale sous le vent *f*
dracone *n* MARINE citerne souple flottante *f*
draft *vt* POLICY préparer; ~ **a framework agreement** POLICY préparer un projet d'accord cadre
drag *n* AIR résistance hydrodynamique *f*, RENEW ENERG *of wind turbine* traînée *f*; ~ **chain** WASTE chaîne d'attelage *f*, WATER SUPP chaîne à drague *f*; ~**-torque viscosimeter** ANALYSIS viscosimètre à cônes tournants *m*
drain[1]: ~ **pump** *n* WATER SUPP pompe d'assèchement *f*; ~ **water** *n* WASTE eau de purge *f*, eau de vidange *f*
drain[2] *vt* GEN assécher, drainer, vider, *seep* écouler, décharger, *liquid from soil* dénoyer, évacuer, *dry up* épuiser, *by centrifugal force* essorer; ~ **off** WASTE, WATER POLL, WATER SUPP *liquid* évacuer, faire écouler, vidanger
drainage *n* GEN *act of draining* purge *f*, assèchement *m*, drainage *m*, *seepage* écoulement *m*, déchargement *m*, *of liquid from soil* dénoyage *m*, évacuation *f*, *drying-up* épuisement *m*, *by centrifugal force* essorage *m*; ~ **area** WATER SUPP bassin hydrographique *m*, bassin versant *m*, bassin hydrologique *m*, zone de drainage *f*; ~ **basin** WATER SUPP bassin versant *m*, bassin hydrologique *m*, bassin hydrographique *m*; ~ **channel** WATER SUPP canal d'épuisement *m*, chenal de déversement *m*, canal d'assèchement *m*, canal d'évacuation *m*; ~ **ditch** WATER SUPP fossé de drainage *m*, collecteur-fossé *m*, rigole de drainage *f*, rigole d'assèchement *f*; ~ **layer** WASTE couche drainante *f*; ~ **pump** WASTE, WATER POLL, WATER SUPP pompe d'épuisement *f*
draining *n* GEN *of liquid from soil* dénoyage *m*, évacuation *f*, *by centrifugal force* essorage *m*; ~ **ditch** WATER SUPP émissaire *m*; ~ **engine** WATER POLL, WATER SUPP machine d'épuisement *f*, machine d'exhaure *f*, moteur d'épuisement *m*; ~ **off** GEN évacuation *f*, vidange *f*
drainpipe *n* WATER POLL descente *f*, tuyau de descente *m*; ~ **spout** WATER POLL descente de vidange *f*
draw *vt* MARINE *ship* caler, WATER SUPP *water* puiser; ~ **off** SOIL, WATER POLL soutirer
drawing *n* SOIL aspiration *f*
DRE *abbr* (*destruction and removal efficiency*) SOIL efficacité de destruction et d'élimination *f*

Drechsel: ~ **gas-washing bottle** *n* ANALYSIS flacon de Drechsel *m*
dredge *vt* MARINE, WATER POLL draguer
dredger *n* MARINE, WATER POLL dragueuse *f*
dredging: ~ **spoil** *n* MARINE, WATER POLL rejet de drague *m*
dried: ~ **sludge** *n* WATER POLL boues déshydratées *f pl*, boues séchées *f pl*
drift[1] *n* AIR *deviation from course* dérive *f*, MARINE, WATER SUPP *direction of current* direction *f*, sens *m*, *deviation from course* dérive *f*; ~ **velocity** AIR vitesse de dérive *f*
drift[2] *vt* MARINE, WATER SUPP dériver
drill[1] *n* RENEW ENERG, SOIL foret *m*
drill[2] *vt* RENEW ENERG forer, sonder, SOIL forer
drilling *n* RENEW ENERG forage *m*, sondage *m*, SOIL forage *m*; **~-core analysis** ANALYSIS analyse de carotte *f*; ~ **grab** SOIL trépan-benne *m*; ~ **shaft** SOIL corps de sonde *m*
drinkable *adj* WATER SUPP potable, consommable
drinking: ~ **water** *n* WATER POLL, WATER SUPP eau potable *f*, eau de consommation *f*; ~ **water analysis** *n* ANALYSIS analyse de l'eau potable *f*; ~ **water consumption** *n* WATER SUPP consommation d'eau potable *f*; ~ **water distribution system** *n* WATER SUPP réseau de distribution d'eau potable *m*; ~ **water quality** *n* WATER SUPP qualité d'eau potable *f*; ~ **water quality control** *n* WATER POLL contrôle de la qualité de l'eau de consommation *m*; ~ **water recovery** *n* WATER SUPP *from surface waters* récupération d'eau potable *f*; ~ **water softening** *n* WATER SUPP adoucissement de l'eau potable *m*; ~ **water supply** *n* WATER SUPP alimentation en eau potable *f*, distribution d'eau potable *f*; ~ **water treatment** *n* WATER SUPP traitement de l'eau potable *m*
drive: ~ **train** *n* RENEW ENERG train de roulement *m*; ~ **unit** *n* WATER SUPP poste d'entraînement *m*
drop[1]: **~-laden** *adj* AIR chargé de gouttes
drop[2]: ~ **distribution** *n* AIR répartition des gouttes *f*; ~ **film** *n* AIR rideau de gouttes *m*, rideau liquide *m*; ~ **inlet** *n* WATER POLL *to manhole* bouche d'égout *f*
droplet *n* AIR gouttelette *f*; ~ **diameter** AIR diamètre de gouttelette *m*; ~ **dispersion** AIR dispersion de gouttelettes *f*; ~ **generation** AIR production de gouttelettes *f*; ~ **separation** AIR séparation de gouttelettes *f*; ~ **separator** AIR dévésiculeur *m*; ~ **size** AIR taille de gouttelette *f*; ~ **trap** AIR piège à gouttelettes *m*
dropping: ~ **below dew point** *n* AIR baisse au-dessous du point de rosée *f*
drought *n* WATER SUPP sécheresse *f*
drowned: ~ **weir** *n* WATER SUPP déversoir noyé *m*, déversoir à nappe noyée *m*
drum *n* AIR *of waste heat boiler* tambour *m*, MARINE tambour *m*, baril *m*, fût *m*, bidon *m*, SOIL *in soil washing process* tambour *m*, WATER POLL fût *m*; ~ **filter** WATER POLL, WATER SUPP filtre à tambour *m*, filtre rotatif *m*; ~ **screen** WATER POLL tambour rotatif *m*, tamis à tambour *m*; ~ **sieve** WATER SUPP tambour de criblage *m*; ~ **skimmer** MARINE récupérateur à tambour *m*; ~ **system** SOIL système à tambours *m*; **~-type dryer** SOIL, WATER POLL tambour sécheur *m*, séchoir à tambour *m*, séchoir rotatif *m*
dry[1]: ~ **absorption** *n* AIR absorption à sec *f*; ~ **acid deposit** *n* AIR, SOIL dépôt acide sec *m*; ~ **acid fallout** *n* SOIL retombées acides sèches *f pl*; ~ **acidic fallout** *n* AIR retombées acides sèches *f pl*; ~ **additive process** *n* AIR procédé par apport d'additifs secs *m*; ~ **adiabatic lapse rate** *n* AIR gradient thermique adiabatique à sec *m*, gradient sec adiabatique *m*; ~ **adsorption** *n* AIR adsorption par voie sèche *f*; ~ **air** *n* AIR air sec *m*; **~-air filter** *n* AIR filtre à air sec *m*; **~-bulb temperature** *n* ANALYSIS température sèche *f*; **~-cleaning process** *n* AIR *of dry electrostatic precipitator through rapping* procédé d'épuration par voie sèche *m*; ~ **deposition** *n* AIR, SOIL dépôt sec *m*; ~ **desulfurization process** *AmE*, ~ **desulphurization process** *BrE* *n* AIR procédé de désulfuration par voie sèche *m*; ~ **distillation** *n* SOIL distillation sèche *f*; ~ **dust-removal installation** *n* AIR, SAFETY installation de dépoussiérage par voie sèche *f*; ~ **electrostatic filter** *n* AIR filtre électrostatique sec *m*; ~ **equipment** *n* ANALYSIS matériel d'analyse par voie sèche *m*; ~ **farming** *n* POLICY, SOIL culture sèche *f*; ~ **filtration** *n* WATER POLL filtration à sec *f*; ~ **gas cleaning** *n* AIR, WASTE épuration des gaz par voie sèche *f*; ~ **grinding** *n* WASTE broyage sec *m*; ~ **heat** *n* AIR chaleur sèche *f*; ~ **mass** *n* WATER POLL masse sèche *f*, matière sèche *f*; **~-mass concentration** *n* WATER POLL concentration en matières sèches *f*; **~-mass content** *n* WATER POLL teneur en matières sèches *f*; **~-mass separation** *n* WATER POLL séparation des matières sèches *f*; ~ **residue** *n* WATER POLL extrait sec *m*, résidu sec *m*; ~ **screen** *n* WATER POLL grille sèche *f*; ~ **scrubber** *n* AIR épurateur par voie sèche *m*, dépoussiéreur par voie sèche *m*; ~ **scrubbing** *n* AIR dépoussiéreur par voie sèche *m*; ~ **sludge** *n* WATER POLL boues sèches *f pl*; **~-sludge disposal site** *n* WATER POLL décharge pour boues sèches *f*; ~ **sorting** *n* WASTE, WATER SUPP tri à sec *m*; ~ **waste-gas treatment** *n* AIR traitement des gaz par voie sèche *m*; ~ **weather** *n* AIR temps sec *m*; ~ **well** *n* SOIL puits sec *m*
dry[2] *vt* AIR déshydrater, sécher, ANALYSIS, SOIL, WASTE, WATER POLL, WATER SUPP sécher
drying *n* AIR déshydratation *f*, séchage *m*, ANALYSIS, SOIL, WASTE *in waste incineration* séchage *m*, WATER POLL, WATER SUPP séchage *m*; ~ **agent** ANALYSIS déshydratant *m*, dessiccateur *m*; ~ **bed** WASTE, WATER POLL lit de séchage *m*; ~ **closet** ANALYSIS séchoir *m*; ~ **medium** AIR agent déshydratant *m*, dessiccateur *m*; ~ **tower** AIR tour de déshydratation *f*, tour de séchage *f*; ~ **up** WATER SUPP *of stream* tarissement *m*; ~ **zone** AIR zone de déshydratation *f*, zone de séchage *f*
DSC *abbr* (*differential scanning calorimetry*) ANALYSIS calorimétrie différentielle *f*
DTA *abbr* (*differential thermal analysis*) ANALYSIS ATD (*analyse thermique différentielle*)
dual: **~-bed bulk material catalyst** *n* AIR catalyseur massique sur lits superposés *m*; **~-bed catalyst** *n* AIR catalyseur à lit double *m*; **~-bed monolithic catalyst** *n* AIR catalyseur monolithique à lit double *m*; ~ **system** *n* MANAGE, WASTE *waste management concept* système dual *m*
duct *n* AIR conduit *m*; ~ **lining** NOISE revêtement de conduite *m*
dummy *n* WASTE emballage vide *m*; ~ **activity** POLICY activité fictive *f*
dump[1] *n* WASTE décharge *f*, WATER POLL, WATER SUPP dépôt *m*; ~ **for bulky waste** WASTE décharge de déchets encombrants *f*
dump[2] *vt* MARINE *liquid* déverser, *solids* rejeter, WASTE déposer; ~ **at sea** MARINE, WASTE, WATER POLL rejeter en mer; ~ **into the ocean** MARINE, WASTE, WATER POLL rejeter en mer
dumper *n* WASTE camion benne *m*; ~ **car** WASTE wagon basculant *m*; ~ **truck** WASTE wagon basculant *m*
dumping *n* WASTE décharge *f*, dépôt *m*; ~ **at sea** MARINE, WASTE, WATER POLL rejet en mer *m*; ~ **height** WASTE hauteur de décharge *f*; ~ **landfill disposal** WASTE mise en décharge *f*; ~ **mechanism** WASTE mécanisme de décharge *m*
duplex: ~ **board** *n* WASTE carton deux couches *m*
durable *adj* GEN durable
duration: ~ **of cleaning** *n* AIR *of filter bag* temps d'épuration *m*
duromer *n* SOIL duromère *f*
dust[1]: **~-free** *adj* AIR dépoussiéré, sans poussière; **~-laden** *adj* AIR chargé de poussières, poussiéreux, SAFETY, WATER SUPP poussiéreux; **~-proof** *adj* SAFETY étanche à la poussière; **~-removed** *adj* AIR dépoussiéré
dust[2] *n* AIR *containing solid particulates*, SAFETY, WASTE, WATER SUPP poussière *f*; **~-binding agent** AIR *filters* agent antipoussière *m*; ~ **breakthrough** AIR pénétration de poussière *f*; ~ **build-up** AIR formation de poussière *f*;

~ **burden** AIR charge de pollution *f*, charge de poussière *f*, charge pulvimétrique *f*; ~ **catalyst** AIR catalyseur de poussière *m*; ~ **catcher** AIR collecteur de poussière *m*; ~ **coating** AIR couche de poussière *f*; ~**-collecting hopper** AIR trémie à poussière *f*, trémie de dépoussiérage *f*; ~ **collector** ANALYSIS dépoussiéreur *m*, WATER SUPP capteur de poussière *m*; ~ **combustion** AIR combustion de poussière *f*; ~ **component** AIR élément constitutif de la poussière *m*; ~ **concentration** AIR concentration de poussière *f*; ~ **content** AIR volume de poussière *m*, WATER SUPP teneur en poussière *f*; ~ **content in flue gas** AIR volume de poussière contenu dans les effluents gazeux *m*; ~ **control** AIR dépoussiérage *m*; ~ **counter** ANALYSIS détecteur de poussières *m*, pulvimètre *m*; ~ **cover** SAFETY couche de poussière *f*; ~ **deposit** AIR, SOIL dépôt de poussière *m*, dépôt poussiéreux *m*; ~ **discharge** AIR rejet de poussières *m*; ~ **discharge opening** AIR *of cyclone* orifice de rejet de poussière *m*, gueulard *m*; ~ **dispersion** AIR dispersion de poussière *f*; ~ **emissions** *n pl* AIR émissions de poussière *f pl*, émissions pulvérulentes *f pl*; ~ **exhaust fan** SAFETY ventilateur de dépoussiérage *m*; ~ **explosion** AIR explosion de poussière *f*; ~ **extraction** AIR extraction de poussière *f*; ~ **extraction plant** AIR installation d'extraction de poussière *f*, installation de dépoussiérage *f*, SAFETY installation de dépoussiérage *f*; ~ **extractor** AIR dépoussiéreur *m*; ~ **fall** AIR retombées pulvérulaires *f pl*; ~ **filter** AIR filtre à poussière *m*, filtre de dépoussiérage *m*; ~ **fineness** AIR granulométrie de la poussière *f*; ~ **fraction** AIR fraction de poussière *f*; ~**-free gas** AIR gaz dépoussiéré *m*; ~**-free waste air** AIR effluent gazeux dépoussiéré *m*; ~**-laden air** AIR, SAFETY air poussiéreux *m*; ~**-laden film** AIR film poussiéreux *m*, pellicule de poussière *f*; ~**-laden flue gas** AIR effluent gazeux poussiéreux *m*; ~**-laden gas** AIR gaz chargé de poussière *m*, gaz poussiéreux *m*; ~**-liquid mixture** AIR *to separate from gas* mélange liquide-poussière *m*; ~ **loading** AIR concentration de poussière *f*; ~ **mask** SAFETY masque anti-poussière *m*; ~ **mixture** AIR mélange de poussière *m*; ~ **outlet** AIR sortie de poussière *f*; ~ **particle** AIR particule de poussière *f*; ~ **particle size distribution** ANALYSIS répartition granulométrique des particules de poussière *f*; ~**-particulate layer** AIR *electrostatic precipitator* couche de particules de poussière *f*; ~ **receiver** AIR collecteur de poussière *m*; ~ **recovery** AIR récupération de poussière *f*; ~ **removal** AIR dépoussiérage *m*; ~ **removal efficiency** AIR rendement de dépoussiérage *m*; ~ **removal installation** AIR, SAFETY installation de dépoussiérage *f*; ~ **removal plant** AIR, SAFETY installation de dépoussiérage *f*; ~ **removal technology** AIR technique de dépoussiérage *f*; ~**-removed dirty gas** AIR gaz dépoussiéré *m*; ~ **remover** AIR dépoussiéreur *m*; ~ **scale** ANALYSIS échelle pulvimétrique *f*; ~ **separator** AIR, SAFETY *cleaning equipment* dépoussiéreur *m*; ~ **shield** AIR cache-poussières *m*; ~ **silo** AIR silo à poussières *m*; ~**-transporting worm** AIR conduite d'évents *f*; ~ **trap** AIR piège à poussière *m*; ~ **type** AIR type de poussière *m*

dust[3] *vt* AIR, WATER SUPP dépoussiérer

dustbin *n BrE* (*cf trash can AmE, garbage can AmE*) WASTE boîte à ordures *f*, poubelle *f*; ~ **lorry** *BrE* (*cf dustcart AmE*) WASTE benne à ordures *f*

dustcart *n AmE* (*cf dustbin lorry BrE*) WASTE benne à ordures *f*

dusting *n* AIR empoussiérage *m*, poussiérage *m*

Dutch: ~ **list** *n* SOIL *of harmful substance concentrations* liste des concentrations de substances nocives établie par la Hollande *f*

duty: ~ **engineer** *n* SAFETY cadre d'astreinte *m*, ingénieur d'astreinte *m*; ~ **of prevention** *n* POLICY devoir de prévention *m*

dwell: ~ **time** *n* AIR, WATER POLL temps de séjour *m*

dyke *n* RENEW ENERG *tidal power* digue *f*

dynamic[1] *adj* ANALYSIS, MARINE, WASTE, WATER SUPP dynamique

dynamic[2]: ~ **headspace technique** *n* (*DHS*) ANALYSIS méthode manométrique dynamique *f*; ~ **leachability test** *n* ANALYSIS essai de lixiviabilité dynamique *m*, WASTE essai de lessivage dynamique *m*; ~ **leaching test** *n* WASTE essai de lessivage dynamique *m*; ~ **pressure** *n* ANALYSIS pression dynamique *f*; ~ **skimmer** *n* MARINE récupérateur dynamique *m*; ~ **viscosity** *n* WATER SUPP viscosité dynamique *f*

dynamometer *n* ANALYSIS tensiomètre *m*, dynamomètre *m*

E

EA *abbr* (*elemental analysis*) ANALYSIS analyse élémentaire *f*
ear: ~ **defenders** *n pl* NOISE casque *m*; ~ **protector** *n* NOISE protège-oreilles *m*; ~ **protectors** *n pl* NOISE casque *m*
earplugs *n pl* NOISE bouchons antibruit *m pl*
earth: ~ **bank** *n* WASTE *dump* talus en terre *m*, remblai de terre *m*; ~**-moving machinery** *n* SOIL engins de terrassement *m pl*; ~ **observation satellite** *n* POLICY satellite d'observation de la terre *m*; ~ **radiation** *n* AIR rayonnement terrestre *m*, radioactivité d'origine tellurique *f*; ~ **resources survey** *n* MANAGE recensement des ressources de la terre *m*; ~ **sciences** *n* POLICY sciences de la terre *f pl*; ~ **tremor** *n* SOIL secousse tellurique *f*
earthed: ~ **wall** *n* AIR *collection anode* paroi mise à la terre *f*, plaque mise à la terre *f*
earth: ~**'s atmosphere** *n* AIR atmosphère terrestre *f*; ~**'s crust** *n* RENEW ENERG *geothermal energy* croûte terrestre *f*; ~**'s interior** *n* RENEW ENERG *geothermal energy* intérieur de la terre *m*, manteau *m*; ~**'s rotation** *n* RENEW ENERG *source of tidal power* rotation terrestre *f*; ~**'s sunbelt** *n* RENEW ENERG *optimal site of solar energy* ceinture de van Allen *f*; ~**'s surface** *n* AIR, RENEW ENERG surface terrestre *f*
easy: ~ **peel-off self-adhesive label** *n* WASTE étiquette adhésive facile à décoller *f*
ebb[1]: ~ **and flood** *n* MARINE, RENEW ENERG *tidal power* marée *f*; ~ **tide** *n* MARINE, RENEW ENERG *tidal power plant* marée descendante *f*
ebb[2] *vi* WATER SUPP *sea* se retirer
ECD *abbr* (*electron capture detector*) ANALYSIS détecteur de capture d'électrons *m*
eco-audit *n* MANAGE, POLICY, WATER SUPP *of environmental impacts* éco-audit *m*, audit de l'environnement *m*
eco-balance *n* MANAGE, POLICY éco-bilan *m*, bilan écologique *m*, WATER SUPP éco-bilan *m*, bilan écologique *m*; ~ **design** MANAGE type d'éco-bilan *m*
eco-climate *n* MANAGE, POLICY éco-climat *m*
eco-controlling *n* MANAGE éco-régulation *f*
eco-label *n* MANAGE, POLICY, WASTE éco-étiquette *f*
eco-labeling *AmE see eco-labelling BrE*
eco-labelling *n BrE* MANAGE, WASTE, POLICY éco-étiquetage *m*
ecological[1] *adj* GEN écologique
ecological[2]: ~ **account** *n* MANAGE comptabilité écologique *f*; ~ **analysis** *n* ANALYSIS, POLICY étude écologique *f*; ~ **balance** *n* POLICY équilibre écologique *m*; ~ **buying guideline** *n* MANAGE directive d'achat écologique *f*; ~ **collapse** *n* AIR, MANAGE, POLICY catastrophe écologique *f*; ~ **conditions** *n pl* POLICY milieu écologique *m*, situation écologique *f*; ~ **damage** *n* AIR, SOIL dégradation écologique *f*; ~ **database** *n* MANAGE base de données écologiques *f*; ~ **factor** *n* POLICY, SOIL, WATER SUPP facteur écologique *m*; ~ **menace** *n* POLICY menace écologique *f*; ~ **pyramid** *n* POLICY, WATER SUPP pyramide écologique *f*; ~ **restrictions** *n pl* POLICY restrictions écologiques *f pl*; ~ **station** *n* POLICY plate-forme écologique *f*; ~ **structural change** *n* POLICY modification écologique structurelle *f*, évolution écologique *f*; ~ **threat** *n* POLICY menace écologique *f*; ~ **viability** *n* WATER SUPP viabilité écologique *f*
ecologically: ~**-relevant** *adj* GEN écologique
ecologist *n* MANAGE, POLICY écologiste *m*
ecology[1]: ~**-oriented** *adj* MANAGE, POLICY axé sur l'environnement, écologique
ecology[2] *n* GEN écologie *f*; ~**-oriented management** MANAGE gestion tenant compte de l'environnement *f*; ~**-oriented marketing** MANAGE marketing écologique *m*
economic[1] *adj* MANAGE, POLICY économique
economic[2]: ~ **efficiency** *n* POLICY *of eco-taxes* rentabilité *f*; ~ **feasibility** *n* POLICY *of environmental policy measures* faisabilité économique *f*; ~ **instrument** *n* POLICY instrument économique *m*; ~ **interest group** *n* MANAGE groupe d'intérêt économique *m* (*GIE*); ~ **potential** *n* RENEW ENERG *of sources* potentiel économique *m*
economy: ~**-size pack** *n* WASTE emballage économique *m*; ~ **of space** *n* WASTE économie d'espace *f*
eco-political *adj* POLICY relatif à la politique de l'environnement
eco-quality: ~ **criteria** *n* MANAGE critères de qualité écologique *m pl*, critères de qualité de l'environnement *m pl*
eco-rating *n* MANAGE évaluation de l'environnement *f*
eco-retailer *n* MANAGE éco-détaillant *m*
ecosphere *n* AIR, MARINE, SOIL écosphère *f*
ecosystem *n* GEN écosystème *m*; ~ **dynamics** *n pl* POLICY dynamique des écosystèmes *f*; ~ **stress** AIR, MARINE, SOIL perturbation de l'écosystème *f*
eco-tax *n* POLICY écotaxe *f*, impôt sur l'environnement *m*; ~ **package** POLICY programme de fiscalité de l'environnement *m*; ~ **reform** POLICY réforme de la fiscalité de l'environnement *f*; ~ **revenue** POLICY revenus de la fiscalité de l'environnement *m pl*, recettes dégagées par la fiscalité de l'environnement *f pl*; ~ **system** POLICY barème fiscal applicable à l'environnement *m*, système d'éco-taxes *m*
eco-tourism *n* MANAGE tourisme écologique *m*, tourisme vert *m*
ecotoxicity *n* WATER POLL écotoxicité *f*
ecotoxicological: ~ **assessment** *n* MANAGE, SAFETY évaluation écotoxicologique *f*
ecotoxicology *n* POLICY, SAFETY, WATER POLL écotoxicologie *f*
eco-window *n* MANAGE fenêtre d'habitabilité *f*, encart écologique *m*
eddy *n* WATER SUPP tourbillon *m*; ~ **flow** AIR *cyclone*, WATER SUPP écoulement turbulent *m*; ~**-flow scrubber** AIR laveur à écoulement turbulent *m*; ~ **sink** AIR évier à écoulement tourbillonnaire *m*; ~ **viscosity** AIR viscosité turbulente *f*; ~ **viscosity coefficient** AIR coefficient de viscosité turbulente *m*
edible: ~ **oil** *n* RENEW ENERG huile alimentaire *f*
EDTA *abbr* (*ethylenediamine tetracetic acid*) ANALYSIS acide éthylènediamine-tétracétique *m*
eductor: ~**-basin cleaner** *n* WASTE camion-vidange *m*
effect[1]: ~**-oriented** *adj* SOIL axé sur les effets
effect[2]: ~**-oriented security measure** *n* SOIL mesure de sécurité axée sur les effets *f*
effective: ~ **energy use** *n* POLICY, RENEW ENERG consommation effective d'énergie *f*, rendement énergétique efficace *m*; ~ **evaporation** *n* WATER SUPP évaporation efficace *f*; ~ **level** *n* ANALYSIS, SAFETY niveau efficace *m*; ~ **sound pressure** *n* NOISE pression acoustique effective *f*; ~ **stack height** *n* AIR hauteur utile de cheminée *f*
effectiveness *n* GEN efficacité *f*, rendement *m*
efficiency *n* GEN efficacité *f*, rendement *m*; ~ **analysis** ANALYSIS analyse de rendement *f*
efficient: ~ **packaging** *n* WASTE emballage efficace *m*
effluent *n* WATER POLL, WATER SUPP effluent *m*, rejet *m*; ~ **channel** WATER POLL conduite de décharge *f*; ~ **charge**

POLICY taxe sur la pollution *f*, redevance sur les effluents *f*; **~ from nuclear power stations** AIR effluent des centrales nucléaires *m*; **~ quality** WATER POLL qualité des effluents *f*; **~ sewer** WATER POLL conduite de décharge *f*; **~ side walls** WATER POLL piédroit pour effluents *m*; **~ sludge** *n pl* WATER POLL boues d'effluents *f pl*; **~ trough** WATER POLL rigole de décharge *f*; **~ weir** WATER POLL déversoir de sortie *m*

EGR *abbr* (*exhaust gas recirculation*) AIR recirculation des gaz d'échappement *f*

EIA[1] *abbr* (*environmental impact assessment*) MANAGE, POLICY étude d'impact sur l'environnement *f*

EIA[2]: **~-dependent procedure** *n* POLICY procédure soumise à une étude d'impact *f*; **~ evaluation** *n* POLICY évaluation de l'étude d'impact *f*; **~ investigation** *n* POLICY enquête menée dans le cadre de l'étude d'impact *f*, évaluation de l'étude d'impact *f*; **~ procedure** *n* POLICY procédure de l'étude d'impact *f*; **~ report** *n* POLICY compte-rendu de l'étude d'impact *m*

EIL *abbr* (*environment impairment liability*) MANAGE responsabilité en matière de dégradation de l'environnement *f*

EI/MS *abbr* (*electron impact mass spectrometer*) ANALYSIS spectromètre de masse à impact électronique *m*

EINCS *abbr* (*European inventory of new chemical substances*) POLICY inventaire européen des nouvelles substances chimiques *m*

EINECS *abbr* (*European inventory of existing chemical substances*) POLICY inventaire européen des substances chimiques commercialisées *m*

EIS *abbr* (*environmental impact statement*) MANAGE, POLICY étude d'impact sur l'environnement *f*, dossier d'impact sur l'environnement *m*

ejection: **~ conveyor** *n* SOIL tapis éjecteur *m*

EKC *abbr* (*electrokinetic capillary chromatography*) ANALYSIS chromatographie capillaire électrocinétique *f*

EL *abbr* (*electroluminescent display*) ANALYSIS écran luminescent *m*

elastic: **~ temporary layer** *n* AIR couche temporaire élastique *f*

elastomer *n* NOISE *resilient material*, WASTE élastomère *m*

elbow *n* AIR *separator, scrubber* coude *m*

electret: **~ microphone** *n* ANALYSIS *noise measurement* microphone à électret *m*

electric[1] *adj* GEN électrique

electric[2]: **~ car** *n* AIR véhicule électrique *m*; **~ drive** *n* AIR transmission électrique *f*; **~ energy** *n* RENEW ENERG énergie électrique *f*; **~ field** *n* AIR, RENEW ENERG champ électrique *m*; **~ measuring apparatus** *n* SAFETY appareil de mesure électrique *m*; **~ propulsion** *n* RENEW ENERG propulsion électrique *f*; **~ resistance** *n* ANALYSIS résistance électrique *f*

electrical[1] *adj* GEN électrique

electrical[2]: **~ dust-removal installation** *n* SAFETY dépoussiéreur électrique *m*; **~ energy** *n* RENEW ENERG énergie électrique *f*; **~ field** *n* AIR, RENEW ENERG *electrostatic precipitator* champ électrique *m*; **~ force** *n* AIR force électrique *f*; **~ hazard** *n* SAFETY risque électrique *m*; **~ moisture measurement** *n* ANALYSIS hygrométrie électrique *f*, mesure électrique de l'humidité *f*; **~ motor** *n* WATER POLL moteur électrique *m*; **~ power** *n* AIR électricité *f*, RENEW ENERG, WATER SUPP électricité *f*, énergie électrique *f*

electrically[1]: **~-driven** *adj* RENEW ENERG à entraînement électrique

electrically[2]: **~-driven heat pump** *n* RENEW ENERG pompe à chaleur électrique *f*

electricity *n* AIR, RENEW ENERG, WATER SUPP électricité *f*; **~ generation** RENEW ENERG, WASTE *from incineration*, WATER SUPP production d'électricité *f*; **~ grid** RENEW ENERG réseau électrique *m*; **~ source** RENEW ENERG source d'électricité *f*

electroacoustic *adj* NOISE électroacoustique

electroanalytic *adj* ANALYSIS électroanalytique

electrochemical[1] *adj* ANALYSIS électrochimique

electrochemical[2]: **~ detector** *n pl* ANALYSIS *gas measuring device* détecteur électrochimique *m*; **~ method** *n* ANALYSIS méthode électrochimique *f*; **~ ozone measurement** *n* ANALYSIS mesure électrochimique de l'ozone *f*; **~ ozone sonde** *n* ANALYSIS sonde électrochimique à ozone *f*; **~ stripping analysis** *n* (*ESA*) ANALYSIS analyse par séparation électrochimique *f*

electrode *n* ANALYSIS, WATER SUPP électrode *f*; **~ cell** ANALYSIS *process conductivity meter* cellule d'électrode *f*; **~ impedance** ANALYSIS impédance d'électrode *f*; **~ potential** ANALYSIS *pH-meter span* potentiel d'électrode *m*; **~ sensor** ANALYSIS capteur à électrode *m*

electrodialysis *n* ANALYSIS, WATER POLL, WATER SUPP électrodialyse *f*

electrofilter *n* AIR, WASTE, WATER SUPP électrofiltre *m*, précipitateur électrostatique *m*

electroflotation *n* WATER POLL électroflottation *f*

electrokinetic[1] *adj* ANALYSIS, SOIL électrocinétique

electrokinetic[2]: **~ capillary chromatography** *n* (*EKC*) ANALYSIS chromatographie capillaire électrocinétique *f*; **~ remediation** *n* SOIL *in-situ* assainissement électrocinétique *m*

electroluminescent: **~ display** *n* (*EL*) ANALYSIS écran luminescent *m*

electrolysis *n* SOIL, WATER POLL, WATER SUPP électrolyse *f*; **~ cell** ANALYSIS cellule d'électrolyse *f*; **~ installation** RENEW ENERG banc électrolytique *m*

electrolyte *n* ANALYSIS, WATER SUPP électrolyte *m*; **~ matrix** AIR *fuel cell* matrice électrolytique *f*

electrolytic: **~ concentration** *n* WATER POLL concentration électrolytique *f*

electromagnetic[1] *adj* AIR, WASTE, WATER POLL, WATER SUPP électromagnétique

electromagnetic[2]: **~ pump** *n* WATER POLL, WATER SUPP *for liquid metals* pompe électromagnétique *f*; **~ radiation** *n* AIR rayonnement électromagnétique *m*; **~ separator** *n* WASTE séparateur électromagnétique *m*

electromechanical: **~ separation** *n* WASTE, WATER SUPP séparation électromécanique *f*

electrometric[1] *adj* ANALYSIS électrométrique

electrometric[2]: **~ pH-value determination** *n* ANALYSIS calcul électrométrique du pH *m*

electron *n* AIR, ANALYSIS, SOIL électron *m*; **~ accelerator** AIR accélérateur d'électrons *m*; **~ amelioration** SOIL amendement aux électrons *m*; **~ beam** AIR faisceau d'électrons *m*; **~ beam process** AIR *flue-gas cleaning* procédé par faisceau d'électrons *m*; **~ capture** ANALYSIS capture d'électrons *f*; **~-capture detector** (*ECD*) ANALYSIS détecteur de capture d'électrons *m*; **~ impact ionization** ANALYSIS ionisation par impact électronique *f*; **~ impact mass spectrometer** (*EI/MS*) ANALYSIS spectromètre de masse à impact électronique *m*; **~ impact method** ANALYSIS méthode d'impact électronique *f*; **~ lens** *n pl* ANALYSIS *quadrupole ion trap* lentille électronique *f*; **~ microscope** ANALYSIS *analysis of dust* microscope électronique *m*; **~ microscopy** ANALYSIS microscopie électronique *f*; **~ spin resonance** (*ESR*) ANALYSIS résonance de spin électronique *f* (*RSE*)

electronic[1] *adj* ANALYSIS *measuring devices*, WASTE, WATER SUPP électronique

electronic[2]: **~ scrap** *n* WASTE déchets électroniques *m pl*; **~ sorting** *n* WASTE, WATER SUPP tri électronique *m*

electro-osmosis *n* SOIL électro-osmose *f*

electro-osmotic *adj* SOIL électro-osmotique

electrophoresis *n* ANALYSIS, SOIL électrophorèse *f*

electrophoretic *adj* ANALYSIS, SOIL électrophorétique

electroplate *vt* WASTE galvaniser

electroplating *n* WASTE galvanisation *f*

electrostatic[1] *adj* AIR, WASTE, WATER POLL, WATER SUPP électrostatique

electrostatic[2]: **~ attraction** *n* AIR attraction électrostatique *f*; **~ precipitator** *n* (*ESP*) AIR, WASTE, WATER SUPP électrofiltre *m*, précipitateur électrostatique *m*, séparateur électrostatique *m*, dépoussiéreur électrostatique *m*; **~ repulsion** *n* WATER POLL, WATER SUPP répulsion électrostatique *f*; **~ repulsive force** *n* WATER POLL *colloidal particle*, WATER SUPP force de répulsion électrostatique *f*

electrostatics *n pl* AIR, WASTE, WATER POLL, WATER SUPP électrostatique *f*

element *n* AIR, SOIL élément *m*; **~-specific analysis** ANALYSIS analyse par élément *f*, analyse spécifique *f*

elemental: **~ analysis** *n* (*EA*) ANALYSIS analyse élémentaire *f*

elevated: **~ water reservoir** *n* WATER SUPP réservoir d'eau surélevé *m*

elimination: **~ of contaminants** *n* WATER SUPP décontamination *f*

eluant *n* ANALYSIS, WATER POLL éluant *m*

eluate *n* ANALYSIS, WATER POLL éluat *m*; **~ control** ANALYSIS, WATER POLL contrôle des éluats *m*; **~ preparation** ANALYSIS, WATER POLL préparation des éluats *f*

elute *vt* ANALYSIS, WATER POLL éluer

elution *n* ANALYSIS, WATER POLL élution *f*; **~ chromatography** ANALYSIS chromatographie par élution *f*; **~ test** ANALYSIS essai d'élution *m*

elutriate *vt* SOIL épurer par lavage

elutriation *n* AIR, ANALYSIS, WASTE, WATER POLL élutriation *f*, séparation par décantation *f*

elutriator *n* AIR, ANALYSIS, WASTE, WATER POLL appareil d'élutriation *m*

eluvial: **~ horizon** *n* SOIL horizon éluvial *m*

embankment *n* WASTE endigage *m*, entassement *m*, remblai *m*, WATER SUPP digue *f*, remblai *m*, talus *m*

emergency: **~ bypass** *n* WATER POLL dérivation de secours *f*; **~ gate** *n* WATER POLL vanne de secours *f*; **~ measure** *n* AIR, SAFETY, SOIL mesure d'urgence *f*; **~ overflow** *n* WATER POLL trop-plein de secours *m*; **~ plan** *n* SAFETY plan d'urgence *m*; **~ signal** *n* ANALYSIS signal d'alarme *m*; **~ spillway** *n* WATER SUPP évacuateur de secours *m*, canal de décharge *m*; **~ treatment** *n* SAFETY, SOIL traitement d'urgence *m*

emerging: **~ technology** *n* GEN nouvelle technique *f*, technologie naissante *f*, nouvelle technologie *f*

emission *n* AIR, WATER SUPP émission *f*; **~ source** AIR, POLICY, WATER SUPP source *f*, source émettrice *f*, source d'émission *f*; **~ standard** AIR, ANALYSIS, POLICY norme d'émission *f*

emissions[1]: **~-related** *adj* AIR imputable aux émissions

emissions[2] *n pl* AIR, NOISE, SAFETY *pollution* émissions *f pl*, fumées *f pl*, WATER POLL *from wastewater sludge incineration* émissions *f pl*, fumées *f pl*; **~ analysis** *n* ANALYSIS analyse des émissions *f*; **~ avoidance** *n* SOIL prévention des émissions *f*; **~ banking** *n* POLICY constitution d'une base de données sur les émissions *f*; **~ barrier** *n* WASTE écran anti-émission *m*; **~ charge** *n* POLICY redevance sur les émissions *f*; **~ control** *n* AIR, SOIL lutte contre les émissions *f*, réduction des émissions *f*; **~ control system** *n* AIR, SOIL système de lutte contre les émissions *m*; **~ credit** *n* POLICY crédit d'émissions *m*; **~ data** AIR, ANALYSIS caractéristiques des émissions *f pl*; **~ inventory** *n* AIR, ANALYSIS *air quality measurement* inventaire des émissions *m*; **~ level** *n* AIR, NOISE *of sound* niveau d'émissions *m*; **~ level value** *n* AIR, NOISE indice du niveau d'émission *m*; **~ limit** *n* AIR, ANALYSIS, MANAGE, POLICY limite d'émissions *f*; **~ limit for cars** *n* AIR, ANALYSIS, MANAGE, POLICY limite d'émissions des véhicules *f*; **~ limitation** *n* AIR, POLICY imposition d'une limite d'émissions *f*; **~ limiting** *n* AIR, ANALYSIS, MANAGE, POLICY limite imposée sur les émissions *f*; **~ minimalization** *n* AIR, POLICY réduction des émissions au minimum *f*; **~ monitoring** *n* ANALYSIS suivi des émissions *m*, surveillance des émissions *f*; **~ point** *n* AIR point d'émission *m*; **~ potential** *n* WASTE potentiel d'émissions *m*; **~ rate** *n* AIR taux d'émission *m*; **~ reduction** *n* AIR, WASTE réduction des émissions *f*; **~ register** *n* POLICY, WATER SUPP registre des émissions *m*; **~-related monitoring** *n* AIR surveillance en continu des émissions *f*; **~ requirements** POLICY conditions applicables aux émissions *f pl*; **~ spectrometer** *n* ANALYSIS spectromètre d'émission *m*; **~ spectrum** *n* AIR, ANALYSIS spectre d'émission *m*; **~ statement** *n* POLICY, WATER SUPP notification des émissions *f*; **~-targeted automobile tax** *n* POLICY taxe visant les émissions des véhicules *f*; **~ tax** *n* POLICY taxe sur les émissions *f*; **~ value** *n* AIR indice d'émission *m*; **~ value of cars** *n* AIR indice d'émission des véhicules *m*

emissivity *n* AIR émissivité *f*

emit *vt* AIR, NOISE, WATER SUPP émettre

EMIT *abbr* (*enzyme mediated immunoassay test*) ANALYSIS essai immunologique enzymatique *m*

emitted[1] *adj* AIR, NOISE, WATER SUPP émis

emitted[2]: **~ particle** *n* AIR particule émise *f*; **~ substance** *n* AIR substance émise *f*

emitter *n* AIR, NOISE, WATER SUPP émetteur *m*

emplaced: **~ waste** *n* WASTE déchets mis en place *m pl*

empties *n pl* WASTE emballages vides *m pl*, vidanges *f pl*

empty: **~ trip** *n* WASTE *transport* voyage à vide *m*

emptying *n* WASTE *of container* dépotage *m*, vidage *m*, WATER POLL *of wastewater canals* vidange *f*

EMS *abbr* (*environmental management system*) MANAGE SME (*système de management environnemental*)

Emscher: **~-type settling tank** *n* WATER POLL fosse Emscher *f*

emulsified[1] *adj* WASTE, WATER POLL, WATER SUPP émulsifié, émulsionné

emulsified[2]: **~ oil** *n* WATER POLL huile émulsionnée *f*

emulsifier *n* WASTE, WATER POLL, WATER SUPP émulsifiant *m*, émulsionnant *m*

emulsify *vt* WASTE, WATER POLL, WATER SUPP émulsifier, émulsionner

emulsifying: **~ agent** *n* SOIL agent émulsifiant *m*, WASTE séparateur de fractionnement *m*; **~ liquid** *n* SAFETY liquide émulsifiant *m*

emulsion[1]: **~-coated** *adj* WASTE, WATER SUPP couché par émulsion, revêtu par émulsion

emulsion[2] *n* WASTE, WATER POLL, WATER SUPP émulsion *f*; **~ adhesive** WASTE colle en émulsion *f*; **~ breaker** MARINE, WATER POLL désémulsifiant *m*, WATER SUPP briseur d'émulsion *m*, casseur d'émulsion *m*, désémulsifiant *m*; **~ composition** WATER POLL composition d'une émulsion *f*; **~ separation** WATER POLL, WATER SUPP séparation d'une émulsion *f*

enamel: **~ board** *n* WASTE carton glacé *m*

enameling *AmE see enamelling BrE*

enamelling *n BrE* WASTE, WATER SUPP émaillage *m*, laquage *m*, vernissage *m*

encapsulate *vt* NOISE enrober, SOIL confiner, WASTE confiner, enrober, sceller, WATER SUPP *nuclear waste* enrober, sceller

encapsulated *adj* NOISE enrobé, SOIL confiné, WASTE scellé, confiné, WATER SUPP *nuclear waste* enrobé, scellé

encapsulation *n* NOISE enceinte acoustique *f*, WASTE, WATER SUPP enrobage *m*, scellement *m*, confinement *m*; **~ measure** SOIL mesure de confinement *f*; **~ using inert materials** SOIL encapsulage avec des matériaux inertes *m*

enclosed: **~ emptying** *n* WASTE vidage intégré *m*; **~ sea** *n* AIR, MARINE mer fermée *f*

enclosure *n* NOISE *to prevent sound propagation* enceinte *f*

encoding *n* SOIL, WASTE codage *m*, encodage *m*

end: **~-frame member** *n* SOIL barre pour cadre d'extrémité *f*; **~-gas purity** *n* AIR pureté du gaz final *f*; **~-of-life vehicle** *n* WASTE véhicule en fin de vie *m*; **~-of-line packaging** *n* WASTE emballage en fin de chaîne de production *m*; **~-of-pipe technology** *n* AIR, WATER POLL *waste water* technologie en bout de chaîne *f*, technologie en fin de canalisation *f*; **~-of-**

the-pipe measure *n* MANAGE mesure en aval *f*, mesure en fin de cycle *f*; ~ **panel** *n* WASTE panneau d'extrémité *m*; ~ **product** *n* AIR produit final *m*; ~ **product generation** *n* AIR obtention de produits finals *f*

endanger *vt* SAFETY mettre en danger

endangered[1] *adj* POLICY menacé SAFETY mis en danger

endangered[2]: ~ **species** *n pl* POLICY espèce menacée *f*

endogenous: ~ **catabolism** *n* WATER POLL catabolisme endogène *m*; ~ **respiration** *n* WATER SUPP respiration endogène *f*

endothermic[1] *adj* AIR, WATER SUPP endothermique

endothermic[2]: ~ **reaction** *n* AIR, WATER SUPP *solar power* réaction endothermique *f*

energetic *adj* AIR, RENEW ENERG, SOIL, WATER SUPP énergétique

energy[1]: ~**-consumptive** *adj* AIR, RENEW ENERG consommateur d'énergie; ~**-intensive** *adj* AIR, RENEW ENERG, WATER SUPP à forte intensité énergétique; ~**-related** *adj* AIR, RENEW ENERG, SOIL énergétique, lié à l'énergie; ~**-saving** *adj* MANAGE, RENEW ENERG, WASTE économisant l'énergie

energy[2] *n* AIR, RENEW ENERG, SOIL, WATER SUPP énergie *f*; ~ **balance** AIR, RENEW ENERG, WASTE, WATER SUPP bilan énergétique *m*; ~ **barrier** WATER POLL, WATER SUPP écran énergétique *m*; ~ **budget** AIR, RENEW ENERG, WASTE, WATER SUPP bilan énergétique *m*; ~ **chain** RENEW ENERG filière énergétique *f*; ~ **conservation** MANAGE, RENEW ENERG, WASTE économies d'énergie *f pl*; ~ **consumption** AIR, POLICY, RENEW ENERG, WASTE consommation d'énergie *f*; ~ **content** RENEW ENERG teneur énergétique *f*; ~ **conversion** AIR, RENEW ENERG conversion énergétique *f*; ~ **costs** *n pl* RENEW ENERG coûts énergétiques *m pl*; ~ **cycle** RENEW ENERG cycle de l'énergie *m*; ~ **demand** RENEW ENERG demande d'énergie *f*; ~ **density** RENEW ENERG, WATER SUPP densité énergétique *f*; ~ **efficiency** AIR, RENEW ENERG, SOIL rendement énergétique *m*; ~ **expenditure** AIR *of processes*, RENEW ENERG, WATER SUPP consommation d'énergie *f*, dépense d'énergie *f*; ~ **extraction** RENEW ENERG extraction d'énergie *f*; ~ **farming** RENEW ENERG *of renewable raw materials from biomass* agri-énergie *f*; ~ **forestry** RENEW ENERG forêt à usage énergétique *f*; ~ **gain** RENEW ENERG gain d'énergie *m*; ~ **generation** RENEW ENERG production d'énergie *f*; ~ **generation from renewable energy sources** RENEW ENERG production d'énergie à partir de sources renouvelables *f*; ~ **import** RENEW ENERG importation d'énergie *f*; ~ **intensity** RENEW ENERG intensité énergétique *f*; ~ **loss** AIR perte d'énergie *f*, RENEW ENERG dissipation d'énergie *f*, perte d'énergie *f*; ~ **management** MANAGE gestion de l'énergie *f*, maîtrise de l'énergie *f*, RENEW ENERG gestion de l'énergie *f*, maîtrise de l'énergie *f*; ~ **material stream** RENEW ENERG flux de matières énergétiques *m*; ~ **output** RENEW ENERG production d'énergie *f*, rendement énergétique *m*; ~ **pay-back time** RENEW ENERG délai d'amortissement d'un investissement énergétique *m*, temps de récupération énergétique *m*; ~ **potential** RENEW ENERG *of sources* potentiel énergétique *m*; ~ **production from waste** WASTE valorisation énergétique des déchets *f*; ~ **productivity** RENEW ENERG productivité énergétique *f*; ~ **purpose** RENEW ENERG application énergétique *f*; ~ **raw material** RENEW ENERG matière première destinée à la production d'énergie *f*; ~ **recovery** AIR, WATER POLL, WATER SUPP récupération d'énergie *f*; ~ **reserves** *n pl* RENEW ENERG réserves énergétiques *f pl*; ~ **resources** *n pl* RENEW ENERG, SOIL, WATER SUPP ressources énergétiques *f pl*; ~ **saving** AIR, WASTE économies d'énergie *f pl*; ~**-saving method** AIR méthode économisant l'énergie *f*, méthode peu consommatrice d'énergie *f*; ~**-saving technology** AIR, WASTE technologie peu consommatrice d'énergie *f*, technologie économisant l'énergie *f*; ~ **source** RENEW ENERG source d'énergie *f*; ~ **storage** RENEW ENERG stockage d'énergie *m*; ~ **supply** AIR approvisionnement en énergie *m*, RENEW ENERG fourniture d'énergie *f*; ~ **supply based on renewable energy sources** RENEW ENERG fourniture d'énergie à partir de sources renouvelables *f*; ~ **supply grid** RENEW ENERG réseau de fourniture d'énergie *m*; ~ **tax** POLICY taxe sur l'énergie *f*; ~ **tax revenue** POLICY recettes perçues au titre de la taxe sur l'énergie *f pl*; ~ **technology** AIR, RENEW ENERG technologie énergétique *f*; ~ **transport** RENEW ENERG *from sources* transport d'énergie *m*; ~ **use** AIR, RENEW ENERG, WASTE application énergétique *f*, consommation d'énergie *f*; ~ **yield** RENEW ENERG rendement énergétique *m*

enforcement *n* POLICY *of legislation* mise en vigueur *f*

engineering: ~ **schedule** *n* MANAGE planning des études *m*

enrichment *n* SOIL, WATER POLL enrichissement *m*

ensilage *n* SOIL ensilage *m*

enterovirus *n* SAFETY, WATER POLL entérovirus *m*

enthalpy *n* AIR enthalpie *f*; ~ **of fusion** AIR enthalpie de fusion *f*; ~ **of vaporization** AIR enthalpie de vaporisation *f*

entrain *vt* AIR, SOIL, WATER POLL entraîner

entrained[1] *adj* AIR, SOIL, WATER POLL entraîné

entrained[2]: ~ **liquid** *n* AIR liquide entraîné *m*; ~ **liquid droplet** *n* AIR gouttelette de liquide entraînée *f*; ~ **soil particle** *n* SOIL particule de sol entraînée *f*; ~ **solid particle** *n* AIR particule solide entraînée *f*

entrainment *n* AIR, SOIL WATER POLL entraînement *m*; ~ **factor** AIR facteur d'entraînement *m*; ~ **point** AIR point d'entraînement *m*

entrance: ~ **shaft** *n* SOIL, WATER SUPP arbre d'admission *m*, puits d'admission *m*

entrap *vt* MARINE piéger

entrapped: ~ **air** *n* WASTE bulle d'air *f*

entry: ~ **inlet** *n* AIR *of cyclone* orifice d'admission *m*

envelope *n* WASTE enveloppe *f*

environment *n* GEN environnement *m*, milieu *m*; ~ **atlas** POLICY atlas de l'environnement *m*; ~ **committee** POLICY *of parliament* comité de l'environnement *m*; ~ **fund** POLICY fonds pour l'environnement *m*; ~ **impact investigation** POLICY enquête menée dans le cadre de l'étude d'impact *f*; ~ **impairment liability** (*EIL*) MANAGE responsabilité en matière de dégradation de l'environnement *f*; ~ **minister** POLICY ministre de l'environnement *m*; ~ **ministry** POLICY ministère de l'environnement *m*; ~ **spokesperson** POLICY *of parliamentary party* porte-parole de l'environnement *m*

environmental[1] *adj* GEN environnemental, lié à l'environnement; ~**-protection-oriented** MANAGE, POLICY axé sur la protection de l'environnement

environmental[2]: ~ **activist** *n* MANAGE, POLICY écologiste *m*; ~ **activity** *n* MANAGE, POLICY activité environnementale *f*, activité liée à l'environnement *f*; ~ **administration** *n* MANAGE, POLICY gestion de l'environnement *f*, réglementation de l'environnement *f*; ~ **advice** *n* POLICY conseil en matière d'environnement *m*; ~ **adviser** *n* POLICY conseiller à l'environnement *m*; ~ **agreement** *n* POLICY accord relatif à l'environnement *m*; ~ **analysis** *n* ANALYSIS, MANAGE analyse de l'environnement *f*; ~ **audit** *n* MANAGE, POLICY, WATER SUPP audit de l'environnement *m*; ~ **awareness** *n* MANAGE, POLICY prise en compte de l'environnement *f*, sensibilisation à l'environnement *f*; ~ **code** *n* POLICY code de l'environnement *m*; ~ **collapse** *n* SOIL dégradation du milieu *f*; ~ **commitment** *n* POLICY responsabilité à l'égard de l'environnement *f*; ~ **communication** *n* POLICY communication en matière d'environnement *f*; ~ **compatibility** *n* AIR, POLICY, WATER SUPP compatibilité avec l'environnement *f*, compatibilité avec le milieu; ~ **conditions** *n pl* POLICY état de l'environnement *m*; ~ **counseling** *AmE*, ~ **counselling** *BrE* *n* POLICY conseil en matière d'environnement *m*; ~ **court** *n* POLICY tribunal chargé des questions d'environnement *m*; **~ crime** *n* POLICY délits relatifs à l'environnement *m pl*, préjudices portés à l'environnement *m pl*; **~ criminal** *n* POLICY responsable de délits à l'encontre de l'environnement *m*; **~ criminal law** *n* POLICY droit pénal en matière d'environnement *m*, régime des peines applicables en matière d'environnement *m*; **~ criminality** *n* POLICY délits à l'encontre de l'environnement *m pl*; **~ damage** *n* AIR, POLICY, SOIL atteinte à l'environnement *f*, dégradation de l'environnement *f*, dégradation du milieu *f*; **~ directive** *n* POLICY directive applicable à l'environnement *f*; **~ disaster** *n* AIR, MANAGE, POLICY, SOIL catastrophe écologique *f*; **~ economics** *n pl* POLICY économie de l'environnement *f*; **~ economist** *n* POLICY économiste spécialisé dans les questions d'environnement *m*; **~ education** *n* POLICY formation à l'environnement *f*; **~ effect** *n* AIR, MARINE, NOISE, SOIL, WATER POLL effet sur l'environnement *m*, répercussion sur l'environnement *f*; **~ endangerment** *n* POLICY mise en péril de l'environnement *f*; **~ ethics** *n pl* POLICY éthique de l'environnement *f*; **~ expert opinion** *n* POLICY avis d'expert en matière d'environnement *m*; **~ friendliness** *n* MANAGE, POLICY respect de l'environnement *m*; **~ group** *n* POLICY groupe écologiste *m*, groupe de protection de l'environnement *m*; **~ guideline** *n* POLICY ligne directrice sur l'environnement *f*; **~ health** *n* POLICY, SAFETY hygiène de l'environnement *f*, hygiène du milieu *f*; **~ idea** *n* MANAGE conception de l'environnement *f*; **~ impact** *n* AIR, MARINE, NOISE, POLICY, RENEW ENERG, SOIL, WATER SUPP impact sur l'environnement *m*, répercussion sur l'environnement *f*; **~ impact assessment** *n* (*EIA*) MANAGE, POLICY étude d'impact sur l'environnement *f*; **~ impact assessment investigation** *n* POLICY étude menée dans le cadre de l'étude d'impact *f*, évaluation de l'étude d'impact *f*; **~ impact assessment procedure** *n* POLICY procédure de l'étude d'impact *f*; **~ impact assessment report** *n* POLICY compte-rendu de l'étude d'impact *m*, rapport de l'étude d'impact *m*; **~ impact statement** *n* (*EIS*) MANAGE, POLICY étude d'impact sur l'environnement *f*, dossier d'impact sur l'environnement *m*; **~ improvement** *n* MANAGE mise en valeur de l'environnement *f*; **~ information** *n* POLICY information sur l'environnement *f*; **~ information collection** *n* POLICY collecte d'informations sur l'environnement *f*; **~ information system** *n* MANAGE, POLICY système d'information sur l'environnement *m*; **~ inspectorate** *n* POLICY inspection de l'environnement *f*; **~ issues** *n pl* POLICY questions relatives à l'environnement *f pl*; **~ jury** *n* POLICY jury en matière d'environnement *m*; **~ law** *n* POLICY *specific ruling* droit de l'environnement *m*; **~ leader** *n* POLICY responsable de mouvement écologique *m*; **~ leading position** *n* POLICY position dominante en matière d'environnement *f*; **~ legislation** *n* POLICY législation sur l'environnement *f*; **~ liability** *n* POLICY responsabilité en matière d'environnement *f*; **~ licence** *n BrE*, **~ license** *n AmE* POLICY permis applicable à l'environnement *m*; **~ lobby** *n* POLICY groupe de pression écologiste *m*; **~ management** *n* MANAGE, POLICY gestion de l'environnement *f*, management de l'environnement *m*; **~ management system** *n* (*EMS*) MANAGE système de management environnemental *m* (*SME*); **~ management system audit** *n* MANAGE audit du système de management environnemental *m*; **~ medium** *n* AIR agent du milieu *m*, vecteur du milieu *m*, POLICY milieu *m*, SOIL milieu ambiant *m*; **~ medium compartment** *n* SOIL *subsoil, groundwater* couche de milieu ambiant *f*; **~ monitoring** *n* ANALYSIS surveillance de l'environnement *f*; **~ noise** *n* NOISE, SAFETY bruit ambiant *m*; **~ noise measurement** *n* NOISE, SAFETY mesure du bruit ambiant *f*; **~ organization** *n* POLICY organisation chargée de l'environnement *f*; **~ partner** *n* MANAGE, POLICY partenaire environnemental *m*, partenaire d'environnement *m*; **~ performance** *n* ANALYSIS performance environnementale *f*; **~ planning** *n* POLICY aménagement de l'environnement *m*, planification en matière d'environnement *f*; **~ policy** *n* POLICY mesure relative à l'environnement *f*, politique de l'environnement *f*; **~ policy instrument** *n* POLICY instrument de gestion de l'environnement *m*; **~ policy measure** *n* POLICY mesure relative à l'environnement *f*; **~ polluter** *n* POLICY pollueur *m*; **~ prevention principle** *n* POLICY principe de prévention de la pollution *m*; **~ prevention rules** *n pl* POLICY règles de prévention de la pollution *f pl*; **~ prize** *n* POLICY prix attribué dans le domaine de l'environnement *m*; **~ profile analysis** *n* POLICY analyse de profil environnemental *f*; **~ program** *n AmE*, **~ programme** *BrE n* POLICY programme d'environnement *m*; **~ protection** *n* POLICY protection de l'environnement *f*; **~ protection agency** *n* (*EPA*) MANAGE, POLICY agence de protection de l'environnement *f* (*EPA*); **~ protection aim** *n* POLICY objectif de protection de l'environnement *m*; **~ protection initiative** *n* POLICY initiative de protection de l'environnement *f*; **~ protection investment** *n* POLICY investissement en matière de protection de l'environnement *m*; **~ protection legislation** *n* POLICY législation relative à la protection de l'environnement *f*, lois sur la protection de l'environnement *f pl*; **~ protection measure** *n* POLICY mesure de protection de l'environnement *f*; **~ protection policy** *n* POLICY mesure de protection de l'environnement *f*; **~ protection standards** *n pl* POLICY normes de protection de l'environnement *f pl*; **~ quality** *n* POLICY qualité de l'environnement *f*; **~ quality standard** *n* POLICY norme de qualité de l'environnement *f*, WATER SUPP norme d'environnement *f*; **~ report** *n* POLICY rapport sur l'environnement *m*; **~ reporting** *n* POLICY présentation de rapports sur l'environnement *f*; **~ representative** *n* POLICY porte-parole de l'environnement *m*; **~ research** *n* POLICY recherche en matière d'environnement *f*, recherches sur le milieu *f pl*; **~ responsibility** *n* MANAGE, POLICY responsabilité à l'égard de l'environnement *f*, responsabilité en matière d'environnement *f*; **~ safety** *n* SAFETY sécurité de l'environnement *f*; **~ and safety engineering** *n* MANAGE génie de protection de l'environnement *m*; **~ sanitation** *n* POLICY assainissement du milieu *m*; **~ sector** *n* POLICY secteur de l'environnement *m*; **~ sponsoring** *n* MANAGE, POLICY financement des activités liées à l'environnement *m*; **~ statistics** *n pl* POLICY statistiques sur l'environnement *f pl*; **~ stress** *n* AIR, SOIL contrainte s'exerçant sur l'environnement *f*, perturbation de l'environnement *f*, perturbation du milieu *f*; **~ summit** *n* POLICY sommet sur l'environnement *m*; **~ tax** *n* POLICY redevance d'environnement *f*, taxe d'environnement *f*; **~ technology** *n* AIR, MANAGE, NOISE, RENEW ENERG, SOIL technologie de l'environnement *f*; **~ technology market** *n* MANAGE marché des technologies de l'environnement *m*; **~ testing procedure** *n* MANAGE procédure d'essai dans l'environnement *f*; **~ trade fair** *n* MANAGE foire aux produits de l'environnement *f*; **~ use** *n* POLICY usage de l'environnement *m*, utilisation dans l'environnement *f*; **~ use rights** *n pl* POLICY droits de jouissance de l'environnement *m pl*, droits d'utilisation de l'environnement *m pl*; **~ user tax** *n* POLICY redevance imposée aux usagers de l'environnement *f*

environmentalism *n* MANAGE, POLICY écologisme *m*

environmentalist *n* MANAGE, POLICY écologiste *m*, spécialiste de l'environnement *m*

environmentally[1]: **~-acceptable** *adj* WASTE, WATER SUPP acceptable pour l'environnement; **~-aware** *adj* MANAGE, POLICY respectueux de l'environnement, sensibilisé à l'environnement; **~-compatible** *adj* MANAGE, POLICY compatible avec l'environnement; **~-friendly** *adj* MANAGE, POLICY respectueux de l'environnement; **~-oriented** *adj* MANAGE, POLICY axé sur l'environnement, à orientation écologique, écologique; **~-political** *adj* POLICY relatif aux mesures d'environnement; **~-relevant** *adj* AIR, NOISE, SOIL pertinent pour l'environnement; **~-sustainable** *adj* MANAGE, POLICY écologiquement durable

environmentally[2]: **~-acceptable waste disposal** *n* WASTE mode d'élimination des déchets acceptable pour l'environnement *m*; **~-aware buying** *n* MANAGE mode de consommation tenant compte de l'environnement *m*; **~-aware purchasing** *n* MANAGE mode de consommation tenant compte de l'environnement *m*; **~-compatible development** *n* POLICY développement compatible avec l'environnement *m*; **~-compatible operation** *n* MANAGE exploitation compatible avec l'environnement *f*; **~-compatible product** *n* MANAGE produit compatible avec l'environnement *m*; **~-friendly logistics** *n* MANAGE logistique respectueuse de l'environnement *f*; **~-friendly product** *n* MANAGE produit respectueux de l'environnement *m*; **~-harmful subsidy** *n* POLICY subvention préjudiciable à l'environnement *f*; **~-neutral disposal** *n* WASTE mode d'élimination de déchets neutre pour l'environnement *m*; **~-oriented management** *n* MANAGE gestion à orientation écologique *f*; **~-oriented marketing** *n* MANAGE mercatique à orientation écologique *f*, pratiques commerciales à orientation écologique *f pl*; **~-oriented product** *n* MANAGE produit à orientation écologique *m*; **~-political working group** *n* POLICY groupe de travail chargé des questions d'environnement *m*

enzymatic[1] *adj* AIR, WATER POLL enzymatique

enzymatic[2]: **~ stabilization** *n* WATER POLL stabilisation enzymatique *f*

enzyme *n* AIR *biological scrubber*, WATER POLL enzyme *f*; **~ mediated immunoassay test** (*EMIT*) ANALYSIS essai immunologique enzymatique *m*

EOCL *abbr* (*extractable organic chloride*) AIR chlorure organique extractible *m pl*

ep *abbr* (*expanded polystyrene*) WASTE polystyrène expansé *m*

EPA *abbr* (*environmental protection agency*) MANAGE, POLICY EPA (*agence de protection de l'environnement*)

epilimnion *n* WATER POLL épilimnion *m*

episode *n* SOIL épisode *m*

epizootic: **~ diseases act** *n* WASTE loi sur les épizooties *f*

EPM *abbr* (*equivalent per million*) AIR équivalent par million *m*

EPO *abbr* (*European Patent Office*) ANALYSIS OEB (*Office Européen des Brevets*)

epoxy: **~ resin** *n* WATER SUPP résine époxyde *f*

equalization *n* POLICY péréquation *f*, WATER POLL égalisation *f*, régularisation *f*

equalizing: **~ tank** *n* WATER SUPP bassin de compensation *m*, bassin de stabilisation *m*

equilibrium: **~ constant** *n* AIR constante d'équilibre *f*; **~ load** *n* AIR *adsorption* charge d'équilibre *f*

equip *vt* ANALYSIS *with measuring instruments* SAFETY équiper

equipment *n* GEN équipement *m*, matériel *m*

equipotential: **~ surface** *n* WATER SUPP surface équipotentielle terrestre *f*

equivalent: **~ diameter** *n* AIR diamètre équivalent *m*; **~ per million** *n* (*EPM*) AIR équivalent par million *m*; **~ standard smoke** *n* AIR fumée équivalente standard *f*

erect *vt* WASTE ériger

ergonomic[1] *adj* SAFETY, WASTE ergonomique

ergonomic[2]: **~ design principle** *n* SAFETY principe de conception ergonomique *m*

eriochrome: **~ black** *n* ANALYSIS noir ériochrome *m*

Erlenmeyer: **~ flask** *n* ANALYSIS vase d'Erlenmeyer *m*

erosion *n* SOIL, WATER SUPP érosion *f*; **~ protection** SOIL protection contre l'érosion *f*

erosive *adj* AIR *dust property* érosif

ESA *abbr* (*electrochemical stripping analysis*) ANALYSIS analyse par séparation électrochimique *f*

escape[1]: **~ of harmful substances** *n* SOIL *into soil* fuite de substances nocives *f*, dégagement de substances nocives *m*

escape[2] *vi* ANALYSIS échapper, s'échapper

escherichia: **~ coli** *n* SAFETY, WATER POLL escherichia coli *f*

ESP *abbr* (*electrostatic precipitator*) AIR, WASTE, WATER SUPP électrofiltre *m*, dépoussiéreur électrostatique *m*, précipitateur électrostatique *m*, séparateur électrostatique *m*

ESR *abbr* (*electron spin resonance*) ANALYSIS RSE (*résonance de spin électronique*)

established: **~ technologies** *n pl* AIR technologies existantes *f pl*, technologies reconnues *f pl*, RENEW ENERG technologies existantes *f pl*, technologies reconnues *f pl*

estimable *adj* ANALYSIS estimable

ethanol *n* WASTE *landfill gas component* éthanol *m*; **~ generation** RENEW ENERG *biomass conversion* production d'éthanol *f*

ether: **~ evaporation** *n* ANALYSIS *dew point hygrometer* évaporation d'éther *f*

ethylene *n* AIR éthylène *m*

ethylenediamine *n* ANALYSIS *acid* éthylènediamine *f*; **~ tetracetic acid** (*EDTA*) ANALYSIS acide éthylènediamine-tétracétique *m*

ethylmercaptane *n* WASTE *landfill gas component* mercaptan éthylique *m*

euphotic: **~ zone** *n* WATER POLL zone euphotique *f*

European: **~ Committee for Standardization** *n* (*CEN*) ANALYSIS, POLICY Comité Européen de Normalisation *m* (*CEN*); **~ inventory of existing chemical substances** *n* (*EINECS*) POLICY inventaire européen des substances chimiques commercialisées *m*; **~ inventory of new chemical substances** *n* (*EINCS*) POLICY inventaire européen des nouvelles substances chimiques *m*; **~ Patent Office** *n* (*EPO*) ANALYSIS Office Européen des Brevets *f* (*OEB*)

eutrophic *adj* WATER POLL, WATER SUPP eutrophe

eutrophication *n* WATER POLL, WATER SUPP eutrophisation *f*

evacuate *vt* GEN évacuer

evacuation *n* GEN évacuation *f*

evaluate *vt* GEN évaluer

evaluation *n* GEN évaluation *f*; **~ criteria** *n pl* POLICY *environmental policy* critères d'évaluation *m pl*; **~ software** ANALYSIS logiciel d'évaluation *m*; **~ using standard samples** ANALYSIS évaluation sur échantillons normalisés *f*; **~ without standard samples** ANALYSIS évaluation sans échantillon normalisé *f*

evaporate[1] *vt* AIR évaporer, ANALYSIS, WATER SUPP évaporer, faire évaporer, vaporiser

evaporate[2] *vi* AIR s'évaporer, ANALYSIS, WATER SUPP s'évaporer, se volatiliser

evaporation *n* AIR, ANALYSIS, WATER SUPP évaporation *f*, vaporisation *f*; **~ cooling** AIR refroidissement par évaporation *m*; **~ crystallizer** AIR cristallisoir d'évaporation *m*; **~ loss** WATER POLL, WATER SUPP perte par évaporation *f*; **~ pan** WATER SUPP bac d'évaporation *m*; **~ rate** WASTE taux d'évaporation *m*

evaporative: **~ emissions** *n pl* AIR émissions d'évaporation *f pl*

evaporator: **~ distillation** *n* SOIL distillation à l'évaporateur *f*

evapotranspiration *n* WATER SUPP évapotranspiration *f*

evil: **~-smelling liquid** *n* WATER POLL, WATER SUPP liquide nauséabond *m*

examine *vt* ANALYSIS étudier, examiner, MANAGE étudier, SAFETY étudier, examiner

excavated[1] *adj* SOIL, WASTE excavé

excavated[2]: **~ mass** *n* SOIL masse de déblaiement *f*; **~ material** *n* SOIL déblais *m pl*, WASTE *inert material for mono-landfill* matériau excavé *m*; **~ soil** *n* SOIL, WASTE *waste type* sol excavé *m*

excavation *n* SOIL, WASTE excavation *f*; **~ system** WASTE,

WATER SUPP système d'excavation *m*, système d'extraction *m*; ~ **well** SOIL puits d'excavation *m*
excavator *n* SOIL, WASTE excavateur *m*
excess: ~ **air** *n* AIR air en excès *m*; ~ **biomass** *n* WATER POLL biomasse excédentaire *f*; ~ **fuel** *n* AIR combustible en excès *m*; ~ **liability** *n* MANAGE responsabilité surérogatoire *f*; ~ **load** *n* SOIL charge excessive *f*; ~ **manure tax** *n* POLICY taxe sur l'abus d'engrais *f*; ~ **oxygen** *n* AIR oxygène en excès *m*; ~ **power supply** *n* RENEW ENERG excédent d'énergie électrique *m*; ~ **pressure** *n* NOISE, WATER POLL surpression *f*; ~ **sludge** *n* WATER POLL boues en excès *f pl*; ~ **sludge volume** *n* WATER POLL volume des boues excédentaires *m*; ~ **weight** *n* WASTE surcharge pondérale *f*
excessive: ~ **deposit build-up** *n* AIR formation excessive de dépôt *f*
exchange[1] *n* GEN échange *m*
exchange[2] *vt* AIR, WATER POLL échanger
exchangeable[1] *adj* AIR, WATER POLL échangeable
exchangeable[2]: ~ **cation** *n* AIR cation échangeable *m*
exchanger: ~ **material** *n* WATER POLL matériau échangeur *m*
excise: ~ **duty** *n* MANAGE, POLICY droit de consommation *m*
exclusion *n* POLICY *of public participation* exclusion *f*
exhaust *n* AIR gaz d'échappement *m*; ~ **cleaning installation** SAFETY installation d'épuration des gaz d'échappement *f*; ~ **emissions** *n pl* AIR émissions à l'échappement *f pl*; ~ **emissions control** AIR lutte contre les émissions à l'échappement *f*; ~ **emissions control system** AIR système de lutte contre les émissions à l'échappement *m*; ~ **fan** AIR ventilateur aspirant *m*; ~ **gas** AIR, WASTE, WATER POLL, WATER SUPP effluent gazeux *m*, gaz brûlé *m*, gaz d'échappement *m*, gaz de comburant *m*; ~-**gas flow** ANALYSIS débit d'effluents gazeux *m*; ~-**gas limit value** POLICY valeur limite de gaz d'échappement *f*; ~ **gas recirculation** (*EGR*) AIR recirculation des gaz d'échappement *f*; ~ **heat** AIR rejet thermique *m*; ~ **housing** AIR chambre de reprise de gaz *m*, corps d'échappement *m*; ~ **manifold** AIR collecteur d'échappement *m*; ~ **manifold catalyst** AIR catalyseur à collecteur d'échappement *m*; ~ **manifold reactor** AIR réacteur à collecteur d'échappement *m*; ~ **muffler** *AmE* (*cf exhaust silencer BrE*) AIR, NOISE pot d'échappement *m*, silencieux *m*; ~ **noise** NOISE bruit à l'échappement *m*; ~ **silencer** *BrE* (*cf exhaust muffler AmE*) AIR, NOISE pot d'échappement *m*, silencieux *m*; ~ **silencing** NOISE pot de détente des gaz *m*; ~ **smoke** AIR fumée d'échappement *f*; ~ **valve** AIR soupape d'échappement *f*; ~ **vapor** *AmE*, ~ **vapour** *BrE* WATER POLL vapeur d'échappement *f*; ~ **vent installation** SAFETY installation d'évacuation des gaz d'échappement *f*; ~ **waste-gas afterburning** WASTE postcombustion des gaz d'échappement *f*; ~ **waste-gas cleaning** AIR épuration des gaz d'échappement *f*; ~ **waste-gas desulfurization** *AmE*, ~ **waste-gas desulphurization** *BrE* WASTE désulfuration des gaz d'échappement *f*
exhauster: ~ **filter** *n* AIR filtre aspirant *m*
exhaustibility *n* RENEW ENERG *sources* possibilité d'épuisement *f*
exhaustible *adj* RENEW ENERG épuisable
existing: ~ **plant** *n* POLICY usine existante *f*
exoelectron: ~ **dosimeter** *n* ANALYSIS dosimètre exoélectronique *m*
exoenzyme *n* AIR exoenzyme *f*
exogenous: ~ **respiration** *n* WATER SUPP respiration tissulaire *f*
exosphere *n* POLICY exosphère *f*
exothermic[1] *adj* AIR, WATER POLL exothermique
exothermic[2]: ~ **process** *n* WATER POLL *wastewater incineration* procédé exothermique *m*; ~ **reaction** *n* AIR réaction exothermique *f*
expandable: ~ **pallet** *n* WASTE *throwaway palette* palette expansible *f*; ~ **thermoplastics** *n* WASTE thermoplastique expansible *m*
expanded: ~ **polystyrene** *n* (*ep*) WASTE polystyrène expansé *m*; ~ **polythene packaging** *n* WASTE emballage en polyéthylène expansé *m*
expander *n* AIR *Venturi tube* vase d'expansion *m*
expansion *n* WATER SUPP détente *f*; ~ **chamber muffler** *AmE* (*cf expansion chamber silencer BrE*) NOISE silencieux de chambre de détente des gaz *m*; ~ **chamber silencer** *BrE* (*cf expansion chamber muffler AmE*) NOISE silencieux de chambre de détente des gaz *m*; ~ **thermometer** ANALYSIS thermomètre à dilatation *m*
experimental: ~ **basin** *n* WATER SUPP bassin versant expérimental *m*
explanatory: ~ **clause** *n* POLICY *of environmental label* clause explicative *f*
exploitation: ~ **of sea-bed resources** *n* POLICY exploitation des ressources du fond des mers *f*
exploratory[1] *adj* ANALYSIS de recherche, exploratoire
exploratory[2]: ~ **sample** *n* ANALYSIS échantillon de recherche *m*; ~ **water sample** *n* ANALYSIS, WATER POLL prélèvement d'eau à titre exploratoire *m*
explosibility *n* AIR, ANALYSIS explosibilité *f*; ~ **limit** AIR, ANALYSIS seuil d'explosibilité *m*
explosible: ~ **particulate** *n* AIR particule explosible *f*
explosion[1]: ~-**proof** *adj* WASTE antidéflagrant
explosion[2]: ~ **protection** *n* AIR, ANALYSIS protection antidéflagrante *f*, protection contre l'explosion *f*; ~ **risk** *n* ANALYSIS, WASTE risque d'explosion *m*; ~ **safety** *n* SAFETY protection contre les risques d'explosion *f*; ~ **suppression device** *n* AIR *of filter apparatus* dispositif antiexplosion *m*; ~ **warning device** *n* ANALYSIS alarme antidéflagrante *f*
explosive[1] *adj* AIR, SAFETY *as danger symbol*, WASTE explosif
explosive[2] *n* AIR, SAFETY, WASTE explosif *m*; ~ **atmosphere** SAFETY atmosphère présentant des risques d'explosion *f*; ~ **gas** AIR gaz explosif *m*; ~ **mixture** AIR *gases* mélange détonant *m*, mélange explosif *m*; ~ **waste** WASTE déchets explosifs *m pl*
exposure *n* SAFETY exposition *f*; ~ **limit to hazardous substances** POLICY, SAFETY limite d'exposition aux substances dangereuses *f*; ~ **path** SAFETY voie d'exposition *f*; ~ **time** ANALYSIS durée d'exposition *f*, temps de pose *m*; ~ **to fumes** AIR, SAFETY exposition aux fumées *f*
extended: ~ **aeration** *n* AIR, WATER SUPP aération prolongée *f*; ~-**period aeration** *n* WATER POLL aération sur une période prolongée *f*
extent *n* ANALYSIS *of sample* étendue *f*, portée *f*
exterior: ~ **packaging machine** *n* WASTE suremballeuse *f*; ~ **surface** *n* WASTE surface externe *f*
external[1] *adj* POLICY externe
external[2]: ~ **combustion engine** *n* AIR, RENEW ENERG moteur à combustion externe *m*; ~ **costs** *n pl* POLICY coûts externes *m pl*; ~ **effect** *n* POLICY effet externe *m*, externalité *f*; ~ **input** *n* AIR, WATER SUPP apport extérieur *m*; ~ **validation** *n* MANAGE validation externe *f*; ~ **water** *n* WATER POLL eau étrangère *f*
externality *n* POLICY effet externe *m*, externalité *f*
extinction *n* AIR, ANALYSIS, SOIL extinction *f*; ~ **coefficient** AIR, ANALYSIS coefficient d'extinction *m*; ~ **of plant species** SOIL extinction d'espèces végétales *f*
extract[1] *n* GEN extrait *m*; ~ **residue** WASTE résidu d'extraction *m*
extract[2] *vt* GEN extraire; ~ **by distillation** SOIL extraire par distillation
extractable: ~ **organic chloride** *n* (*EOCL*) AIR chlorure organique extractible *m*
extractant *n* ANALYSIS, SOIL solvant d'extraction *m*; ~ **processing** SOIL traitement du solvant d'extraction *m*
extracted[1] *adj* GEN extrait
extracted[2]: ~ **component** *n* WATER POLL élément extrait *m*;

~ **soil** *n* SOIL sol extrait *m*; ~ **vapor** *AmE*, ~ **vapour** *BrE* *n* SOIL vapeur extraite *f*

extracting: ~**-agent processing** *n* SOIL traitement de l'agent d'extraction *m*

extraction *n* GEN extraction *f*; ~ **agent** SOIL agent d'extraction *m*; ~**-agent processing** SOIL traitement de l'agent d'extraction *m*; ~ **centrifuge** WATER POLL centrifugeuse d'extraction *f*, essoreuse *f*; ~ **gas** ANALYSIS gaz d'extraction *m*; ~ **operation** AIR, WATER POLL opération d'extraction *f*, essorage *m*; ~ **solvent** ANALYSIS, SOIL solvant d'épuisement *m*, solvant d'extraction *m*; ~ **system** WASTE *grinder*, WATER SUPP système d'excavation *m*, système d'extraction *m*; ~ **tank** SOIL cuve d'extraction *f*; ~ **test** ANALYSIS essai d'extraction *m*; ~ **vessel** WATER POLL cuve d'extraction *f*, essoreuse *f*

extractor *n* SOIL, WASTE extracteur *m*, WATER SUPP colonne d'extraction *f*

extraneous: ~ **noise** *n* NOISE bruit extérieur *m*

extraterrestrial: ~**-operated satellite** *n* ANALYSIS satellite extraterrestre *m*

extruding *n* WASTE, WATER SUPP extrusion *f*

extrusion *n* WASTE, WATER SUPP extrusion *f*; ~ **blow molding** *AmE*, ~ **blow moulding** *BrE* WASTE extrusion-soufflage *f*; ~ **die** WASTE filière d'extrusion *f*; ~ **head** WASTE tête d'extrusion *f*, tête de boudineuse *f*; ~ **line** WASTE filière d'extrusion *f*, ligne d'extrusion *f*; ~ **machine for tubes** WASTE presse à filer les tubes *f*; ~ **plant** WASTE *for processing used plastics* poste d'extrusion *m*; ~ **stretch-blow molding machine** *AmE*, ~ **stretch-blow moulding machine** *BrE* WASTE machine de moulage par étirage-soufflage *f*

eye: ~ **irritant** *n* SAFETY irritant oculaire *m*

eyelet *n* WASTE cosse *f*

F

F *abbr* (*fluorine*) SOIL F (*fluor*)
fabric: **~ filter** *n* AIR *baghouse* filtre en tissu *m*
face: **~ velocity** *n* AIR, WATER POLL vitesse nominale *f*
facility *n* GEN installation *f*
factory: **~ farming** *n* POLICY, SOIL agriculture industrielle *f*
facultative: **~ anaerobic bacteria** *n pl* WATER POLL bactéries pouvant se développer en milieu anaérobie *f pl*, bactéries anaérobies facultatives *f pl*; **~ heterotroph** *n* WATER POLL hétérotrophe facultatif *m*; **~ lagoon** *n* WATER POLL lagune facultative *f*; **~ microorganism** *n* AIR, SOIL, WATER POLL, WATER SUPP micro-organisme facultatif *m*; **~ zone** *n* WATER POLL zone facultative *f*
faecal[1] *adj BrE* WATER POLL fécal
faecal[2]: **~ matter** *n BrE* WATER POLL fèces *f pl*, matière fécale *f*; **~ solids sludge** *n BrE* WATER POLL boues fécales solides *f pl*; **~ solids sludge composition** *n BrE* WATER POLL mélange de boues fécales à l'état solide *m*; **~ solids sludge treatment** *n BrE* WATER POLL traitement des boues fécales solides *m*; **~ wastewater** *n BrE* WATER POLL eaux gadoueuses *f pl*, eaux usées d'origine fécale *f pl*
failure *n* SOIL panne *f*
fall *n* WATER SUPP baisse *f*, *between two water levels* chute *f*; **~-out** AIR, SOIL retombées *f pl*
falling: **~-film absorber** *n* AIR absorbeur de film ruisselant *m*, absorbeur de film tombant *m*; **~-sphere viscosimeter** *n* ANALYSIS viscosimètre à chute de bille *m*; **~ stage** *n* WATER SUPP décrue *f*
fallow *n* RENEW ENERG, SOIL champ en friche *m*, jachère *f*; **~ land** RENEW ENERG terre en friche *f*
family: **~ packet** *n* WASTE paquet familial *m*
fan *n* AIR ventilateur *m*; **~ noise** NOISE bruit de ventilateur *m*; **~ shroud** NOISE *casing* capuchonnement de ventilateur *m*
fanning: **~ smoke plume** *n* AIR panache de fumée en éventail *m*
far: **~-ultraviolet absorbance detector** *n* (*FUVAD*) ANALYSIS détecteur d'absorption dans l'ultraviolet lointain *m*
farm: **~ waste** *n* WASTE déchets d'exploitation agricole *m pl*
fat[1]: **~-free** *adj* AIR dégraissé, sans graisse
fat[2]: **~-free waste air** *n* AIR rejet atmosphérique dégraissé *m*
fatal: **~ accident** *n* SAFETY accident mortel *m*
fatigue *n* AIR, NOISE, RENEW ENERG, SOIL *of materials* fatigue *f*
fatty: **~ acid** *n* WATER POLL acide gras *m*; **~ alcohol** *n* WATER POLL alcool gras *m*
fault *n* SOIL faille *f*; **~ liability** MANAGE responsabilité résultant d'une faute *f*; **~ spring** WATER SUPP source de faille *f*
faulty: **~ measurement** *n* ANALYSIS mesure défectueuse *f*; **~ sheet ejection** *n* WASTE rejet des feuilles défectueuses *m*
FBC *abbr* (*fluidized-bed combustion*) AIR combustion sur lit fluidisé *f*
FD *abbr* (*field desorption*) ANALYSIS désorption par champ électrique *f*
Fe *abbr* (*iron*) GEN Fe (*fer*)
feasibility *n* AIR, MANAGE, POLICY, RENEW ENERG faisabilité *f*; **~ assessment** AIR, MANAGE, RENEW ENERG évaluation de faisabilité *f*; **~ study** MANAGE, POLICY, RENEW ENERG étude de faisabilité *f*
fecal *AmE see faecal BrE*
FED *abbr* (*field emission display*) ANALYSIS écran à émission de champ *m*, écran à micropointes *m*
feed[1] *n* WATER SUPP alimentation *f*, paquet d'eau *m*; **~ hopper** WASTE trémie d'alimentation *f*, trémie de chargement *f*; **~ hopper with flap gate** WASTE trémie d'alimentation avec volet mobile *f*; **~ pump** AIR pompe alimentaire *f*; **~ tank** AIR, WASTE bâche alimentaire *f*, cuve d'alimentation *f*
feed[2] *vt* GEN alimenter; **~ in** SOIL *process water* alimenter; **~ tangentially** AIR alimenter tangentiellement
feedback *n* WATER SUPP rétroaction *f*
feeder: **~ belt** *n* WATER SUPP courroie d'alimentation *f*, tapis d'alimentation d'une presse *m*; **~ broom** *n* WASTE balai d'alimentation *m*; **~ equipment** *n* WASTE matériel d'alimentation *m*; **~ worm** *n* SOIL distributeur à vis *m*
feeding *n* WATER POLL alimentation *f*; **~-chain trough conveyor** WASTE, WATER SUPP convoyeur en auge avec chaîne d'alimentation *m*; **~ hopper** SOIL trémie d'alimentation *f*, trémie de chargement *f*; **~-in of liquid** AIR *absorber* injection de liquide *f*; **~ machine** WASTE machine à alimenter *f*, machine de remplissage *f*
feedlot *n* RENEW ENERG élevage hors sol *m*
feedstock: **~ recycling** *n* WASTE recyclage des produits de base *m*, *plastics* recyclage de la charge *m*
FEL *abbr* (*frank effect level*) SAFETY niveau efficace franc *m*
felt *n* AIR, NOISE feutre *m*
fence *n* WASTE barrière *f*, clôture *f*, palissade *f*; **~ boom** MARINE barrage-barrière *m*
fender *n* MARINE éperon *m*
ferment[1] *vt* RENEW ENERG, WASTE, WATER POLL faire fermenter
ferment[2] *vi* RENEW ENERG *biomass conversion*, WASTE, WATER POLL fermenter
fermentable *adj* RENEW ENERG, WASTE, WATER POLL fermentable, fermentescible
fermentation *n* RENEW ENERG *biomass conversion*, WASTE, WATER POLL *sewage* fermentation *f*; **~ gas** WASTE gaz de fermentation *m*, WATER POLL biogaz *m*, gaz de curage *m*, gaz de digestion *m*; **~ of refuse** WASTE fermentation des ordures *f*; **~ test** ANALYSIS essai de fermentation *m*; **~ to biogas** RENEW ENERG obtention de biogaz par fermentation *f*; **~ tower** WATER POLL tour de fermentation *f*
fermentative *adj* RENEW ENERG, WASTE, WATER POLL fermentatif
fermented: **~ waste** *n* WASTE déchets fermentés *m pl*
ferric: **~ salt** *n* WATER POLL sel ferrique *m*
ferromagnetic *adj* ANALYSIS ferromagnétique
ferrous[1] *adj* ANALYSIS ferreux
ferrous[2]: **~ scrap** *n* WASTE déchets de fer *m pl*, ferraille *f*, mitraille *f*
fertilization *n* SOIL engraissement *m*, fertilisation *f*
fertilize *vt* SOIL engraisser
fertilizer *n* SOIL, WATER POLL engrais *m*; **~ use** SOIL utilisation d'engrais *f*
fertilizing: **~ value** *n* WATER POLL valeur fertilisante *f*
FGD *abbr* (*flue-gas desulfurization AmE, flue-gas desulphurization BrE*) AIR désulfuration des gaz de combustion *f*
FI *abbr* (*field ionization*) ANALYSIS ionisation de champ *f*
FIA *abbr* (*fluorescent indicator adsorption*) ANALYSIS adsorption à l'indicateur fluorescent *f*
FIAS *abbr* (*flow injection analysis spectrometry*) ANALYSIS spectrométrie d'absorption sur jet *f*
fiber *AmE see fibre BrE*
fiberglass *AmE see fibreglassBrE*
fibre *n BrE* AIR, WASTE, WATER SUPP fibre *f*; **~ composition** *BrE* AIR, WASTE composition de fibre *f*; **~ diameter** *BrE* AIR diamètre de fibre *m*; **~ drum** *BrE* WASTE, WATER SUPP *for radioactive wastes* fût allégé *m*; **~ layer** *BrE* AIR *filter*

medium couche de fibre *f*; ~ **layer filter** *BrE* AIR filtre à couche fibreuse *m*; ~ **material** *BrE* AIR *for filter* matériau fibreux *m*, tissu fibreux *m*; ~ **mixture** *BrE* AIR mélange de fibres *m*; ~ **surface** *BrE* AIR *filtering separators* surface fibreuse *f*

fibreglass[1]: ~**-reinforced** *adj BrE* NOISE renforcé à la fibre de verre

fibreglass[2] *n BrE* AIR *filter medium*, NOISE, WASTE, WATER SUPP fibre de verre *f*; ~ **fleece** *BrE* NOISE *absorption material* laine de verre *f*

fibrous[1] *adj* AIR *particle description*, WASTE, WATER POLL fibreux

fibrous[2]: ~ **constituents** *n pl* WATER POLL fibres *f pl*; ~ **mat** *n* AIR matelas fibreux *m*; ~ **mat filter** *n* AIR filtre à matelas fibreux *m*; ~ **web** *n* WASTE *sealing* voile de carde *m*

Fick: ~**'s law** *n* AIR *of diffusion* loi de Fick *f*

FID *abbr* (*flame ionization detector*) ANALYSIS détecteur à ionisation de flamme *m*

field[1]: ~**-proven** *adj* AIR, RENEW ENERG, SOIL *processes* éprouvé, éprouvé en pratique, mis à l'épreuve

field[2]: ~ **analysis** *n* ANALYSIS, SOIL analyse sur le terrain *f*; ~ **capacity** *n* SOIL capacité de champ *f*; ~ **capacity of landfill** *n* WASTE capacité effective d'une décharge *f*; ~ **charging** *n* AIR mise en charge électrique *f*; ~ **data** *n pl* ANALYSIS données recueillies sur le terrain *f pl*; ~ **desorption** *n* (*FD*) ANALYSIS désorption par champ électrique *f*; ~ **emission display** *n* (*FED*) ANALYSIS écran à émission de champ *m*, écran à micropointes *m*; ~ **food crop** *n* POLICY, SOIL culture vivrière en plein champ *f*; ~ **ionization** *n* (*FI*) ANALYSIS ionisation de champ *f*; ~ **of measurement** *n* ANALYSIS étendue de mesure *f*; ~ **performance** *n* ANALYSIS performance sur le terrain *f*; ~ **screening** *n* ANALYSIS tamisage par effet de champ *m*; ~ **strength** *n* AIR *electrostatic precipitator* champ de charge *m*; ~ **test** *n* ANALYSIS essai sur le terrain *m*; ~ **trial** *n* ANALYSIS essai sur le terrain *m*

50%: ~ **inhibiting concentration** *n* SAFETY, WATER POLL concentration inhibitrice a 50% *f* (CI_{50})

fight *vt* RENEW ENERG *pollution* réguler

fill: ~ **hole** *n* WASTE orifice de charge *m*; ~ **level** *n* WASTE niveau de remplissage *m*; ~ **mass** *n* WASTE charge de remplissage *f*, masse cuite *f*; ~ **material** *n* SOIL *from decontaminated soils*, WASTE matériau de remblai *m*

filler *n* SOIL appareil de remplissage *m*

filling *n* ANALYSIS, SOIL, WASTE remplissage *m*, WATER SUPP mise en eau *f*, remplissage *m*; ~ **hole** WASTE orifice de chargement *m*; ~ **machine** WASTE machine de remplissage *f*, machine à remplir *f*; ~ **and sealing machine** WASTE machine de remplissage et de fermeture *f*, machine à remplir et fermer *f*

film: ~ **badge** *n* ANALYSIS *radiation dosimeter* dosifilm *m*, dosimètre photographique individuel *m*; ~ **cartoning** *n* WASTE encartonnage par pellicules *m*; ~ **coating** *n* WASTE couchage *m*; ~ **extrusion equipment** *n* WASTE matériel d'extrusion de films *m*; ~ **lamination** *n* WASTE pelliculage *m*; ~ **wrap** *n* WASTE cellophane *f*; ~ **wrapping machine** *n* WASTE cellophaneuse *f*

filter[1] *n* AIR filtre à air *m*, WATER POLL, WATER SUPP filtre *m*; ~ **aid** WATER POLL, WATER SUPP adjuvant de filtration *m*; ~ **aid efficiency** AIR, WASTE, WATER POLL, WATER SUPP rendement de l'adjuvant de filtration *m*; ~ **area** WATER POLL surface filtrante *f*; ~ **bag** AIR filtre à manche *m*; ~ **bed** WATER POLL, WATER SUPP bassin de filtration *m*, bassin filtrant *m*, couche filtrante *f*, lit de filtration *m*, lit filtrant *m*, masse filtrante *f*; ~ **bed depth** WATER POLL épaisseur de couche filtrante *f*; ~ **bed height** AIR hauteur de couche filtrante *f*; ~ **bed porosity** AIR porosité de couche filtrante *f*; ~ **belt** ANALYSIS toile de filtre *f*; ~ **cage** AIR cage filtrante *f*; ~ **cake** AIR gâteau de filtration *m*; ~ **cloth** AIR, WATER POLL, WATER SUPP toile de filtre *f*, toile de filtre-presse *f*; ~ **column** AIR colonne filtrante *f*; ~ **compartment** AIR compartiment de filtre *m*; ~ **cooler** AIR refroidisseur de filtre *m*; ~ **dimensioning** AIR dimensionnement de filtre *m*; ~ **dust** AIR, WASTE *fly ash* poussière de filtre *f*; ~ **efficacy** AIR efficacité de filtration *f*; ~ **efficiency** AIR rendement de filtration *m*; ~ **element** AIR, SAFETY, WATER POLL élément filtrant *m*, pièce de filtre *f*; ~ **fabric** AIR tissu filtrant *m*; ~ **felt** AIR feutre de filtre *m*; ~ **field** WATER SUPP champ de filtration *m*; ~ **film** AIR *biowasher* pellicule filtrante *f*, WATER POLL film biologique *m*; ~ **gallery** WATER POLL, WATER SUPP galerie filtrante *f*; ~ **gravel** WATER POLL gravier filtrant *m*; ~ **housing** AIR corps de filtre *m*; ~ **layer** AIR couche filtrante *f*; ~ **material** AIR, WATER POLL matériau filtrant *m*; ~ **material circulation** AIR circulation sur matériau filtrant *f*; ~ **medium** AIR, WATER POLL agent filtrant *m*, matériau filtrant *m*; ~ **medium surface** AIR surface de matériau filtrant *f*; ~ **nozzle** WATER POLL tuyère de filtre *f*; ~ **paper** ANALYSIS papier Joseph *m*, papier-filtre *m*; ~ **plant** SAFETY banc de filtration *m*, batterie de filtres *f*; ~ **plant for dusts and fibres** SAFETY *textile industry* batterie de filtres pour fibres et poussières *f*; ~ **press** SOIL, WASTE, WATER POLL, WATER SUPP filtre-presse *m*; ~ **regeneration** AIR régénération de filtre *f*; ~ **replacement** AIR remplacement de filtre *m*; ~ **residue** AIR résidu de filtration *m*; ~ **resistance** AIR résistance de filtre *f*; ~ **stage** AIR stade de filtration *m*; ~ **surface** AIR surface de filtration *f*; ~ **test** ANALYSIS essai de filtration *m*; ~ **type** AIR type de filtre *m*; ~ **unit** AIR *bag filters* batterie de filtres *f*, WATER POLL installation de filtration *f*; ~ **wall** AIR paroi de filtre *f*; ~ **washing** AIR lavage de filtre *m*, WATER POLL lavage par filtration *m*

filter[2] *vt* AIR filtrer, ANALYSIS, WATER POLL, WATER SUPP clarifier par filtre, filtrer, passer au filtre; ~ **out** AIR, ANALYSIS, WASTE, WATER SUPP séparer par filtration

filterability *n* AIR, ANALYSIS, WATER POLL, WATER SUPP filtrabilité *f*

filterable[1] *adj* AIR, ANALYSIS, WATER POLL, WATER SUPP filtrable

filterable[2]: ~ **solids** *n pl* WATER POLL solides filtrants *m pl*

filtered[1] *adj* AIR, ANALYSIS, WATER POLL, WATER SUPP filtré

filtered[2]: ~ **sample** *n* ANALYSIS échantillon filtré *m*

filtering *n* AIR, ANALYSIS, WATER POLL, WATER SUPP filtration *f*; ~ **agent** AIR, WATER POLL agent filtrant *m*, matériau filtrant *m*; ~ **operation** AIR action de filtration *f*; ~ **position** AIR *bag filter* position de filtre *f*; ~ **precipitator** AIR séparateur par filtration *m*; ~ **rate** AIR, ANALYSIS, WATER POLL taux de filtration *m*; ~ **tank** WATER POLL, WATER SUPP filtrateur *m*, purgeoir *m*, cuve de filtration *f*; ~ **time** AIR, WATER POLL délai de filtration *m*, temps de filtration *m*; ~ **unit** AIR, SAFETY, WATER POLL, WATER SUPP élément filtrant *m*, batterie de filtres *f*, banc de filtration *m*; ~ **well** WATER POLL, WATER SUPP puits filtrant *m*

filtrable[1] *adj* AIR, ANALYSIS, WATER POLL, WATER SUPP filtrant

filtrable[2]: ~ **floc** *n* WATER POLL flocon de filtration *m*

filtrate *n* AIR, WATER POLL, WATER SUPP filtrat *m*, produit de filtrage *m*, produit filtré *m*, produit passé *m*; ~ **contamination** WATER POLL contamination des filtrats *f*; ~ **flow** WATER POLL écoulement de filtrats *m*; ~ **quality** WATER POLL qualité du filtrat *f*

filtration *n* AIR, ANALYSIS, WATER POLL, WATER SUPP filtrage *m*, filtration *f*; ~ **capacity** AIR capacité de filtration *f*; ~ **curve** AIR courbe de filtration *f*; ~ **cycle** AIR cycle de filtration *m*; ~ **direction** AIR sens de filtration *m*; ~ **effect** AIR effet filtrant *m*; ~ **front** WATER SUPP front de filtration *m*; ~ **plant** AIR banc de filtration *m*; ~ **pressure** AIR pression de filtration *f*; ~**-putrefaction process** WATER POLL procédé de putréfaction-filtration *m*; ~ **rate** AIR, ANALYSIS, WATER POLL taux de filtration *m*; ~ **residue** ANALYSIS résidu de filtration *m*; ~ **and residue water return** WATER POLL retour des eaux de filtration et des eaux résiduaires *m*; ~ **sludge** WATER POLL boues de filtration *f pl*; ~ **vat** WATER POLL bac de filtration *m*, cuve de filtration *f*; ~ **velocity** AIR vitesse de filtration *f*

final[1] *adj* SOIL final

final[2]: ~ **clarification** *n* WATER POLL décantation finale *f*; ~ **cover** *n* WASTE *of landfill* couverture définitive *f*; ~ **covering** *n* WASTE *of landfill* couverture définitive *f*; ~ **crusher** *n* WASTE broyeur final *m*; ~ **dumping** *n* WASTE *of radioactive waste* rejet définitif *m*, évacuation définitive *f*; ~ **energy** *n* RENEW ENERG énergie finale *f*; ~ **energy consumption** *n* RENEW ENERG consommation d'énergie finale *f*; ~ **energy substitution** *n* RENEW ENERG substitution d'énergie finale *f*; ~ **filter** *n* AIR filtre final *m*, WATER POLL filtre terminal *m*; ~ **filtering** *n* AIR filtration finale *f*; ~ **filtration** *n* WATER POLL filtration finale *f*, filtration en fin de cycle *f*; ~ **remediation** *n* SOIL assainissement final *m*; ~ **sedimentation tank** *n* WATER POLL bassin de décantation finale *m*, décanteur final *m*; ~ **settling tank** *n* WATER POLL bassin de décantation finale *m*, décanteur final *m*; ~ **sorting** *n* WASTE tri final *m*; ~ **storage** *n* WASTE stockage définitif *m*; ~ **thickener** *n* WATER POLL épaississant final *m*; ~ **treatment** *n* WASTE traitement final *m*, traitement de fin de cycle *m*

finance *vt* MANAGE, POLICY financer

financial: ~ **incentive** *n* MANAGE, POLICY incitation financière *f*

fine[1] *adj* AIR, SOIL, WATER SUPP *free from impurity* fin; ~-**particled** WASTE à fines particules

fine[2] *n* POLICY amende *f*; ~-**bubble aeration** WATER POLL carbonatation à bulles fines *f*; ~ **dust** AIR fines *f pl*; ~-**dust precipitation** AIR séparation des fines *f*; ~-**fibered needled felt** *AmE*, ~-**fibred needled felt** *BrE* AIR feutre cousu à fibre fine *m*; ~ **floc** WATER POLL flocon fin *m*; ~ **particle** AIR particule fine *f*; ~ **refuse** WASTE *grain size 0 to 8 mm* refus de fines *m pl*; ~ **screen** WATER POLL tamis *m*

finest: ~-**drop precipitator** *n* AIR séparateur de très fines gouttelettes *m*; ~ **dust** *n* AIR micelle *f*, poussière submicronique *f*

finished: ~ **goods store** *n* WASTE stockage des produits finis *m*; ~ **product** *n* WASTE, WATER SUPP produit fini *m*

finned: ~ **tube air heater** *n* AIR réchauffeur d'air à ailettes *m*

fire[1]: ~-**resistant** *adj* SAFETY à l'épreuve du feu; ~-**retardant** *adj* SAFETY ignifuge

fire[2]: ~ **damp indicator** *n* ANALYSIS, SAFETY indicateur de grisou *m*; ~ **extinguisher** *n* SAFETY extincteur d'incendie *m*; ~ **hazard** *n* SAFETY risque d'incendie *m*; ~ **hose** *n* SAFETY tuyau d'incendie *m*, WATER SUPP lance à incendie *f*, manche à incendie *f*; ~ **hydrant** *n* WATER SUPP bouche d'incendie *f*; ~ **monitor** *n* MARINE, SAFETY détecteur d'incendie *m*; ~ **protection** *n* WATER SUPP protection incendie *f*; ~ **risk** *n* SAFETY risque d'incendie *m*; ~ **safety** *n* SAFETY protection anti-incendie *f*, sécurité incendie *f*; ~ **wall** *n* WATER SUPP pare-feu *m*

fireproof *adj* AIR incombustible, SAFETY, WASTE coupe-feu, ignifuge, incombustible, réfractaire, WATER SUPP incombustible

firewood *n* RENEW ENERG *biomass conversion base material* bois de chauffe *m*

firing: ~ **drum** *n* SOIL tambour de cuisson *m*

firm[1] *adj* SOIL résistant

firm[2]: ~-**specific documents** *n pl* POLICY documentation à usage interne *f*, documentation d'entreprise *f*; ~-**specific information** *n* POLICY information d'entreprise *f*, information du constructeur *f*; ~ **subsurface** *n* SOIL sous-sol résistant *m*

first: ~ **assessment** *n* GEN évaluation préliminaire *f*

fish: ~ **test** *n* ANALYSIS essai sur les poissons *m*; ~ **toxicity** *n* WATER POLL toxicité pour les poissons *f*

fission: ~ **power** *n* RENEW ENERG énergie de fission *f*; ~ **product** *n* WASTE produit de fission *m*

fix *vt* ANALYSIS, SOIL, WASTE, WATER POLL immobiliser

fixation *n* ANALYSIS, SOIL, WASTE, WATER POLL immobilisation *f*; ~ **technique** WASTE technique de fixation *f*

fixative *n* AIR, SOIL fixateur *m*, fixatif *m*

fixed[1] *adj* ANALYSIS, SOIL, WASTE, WATER POLL immobilisé

fixed[2]: ~ **bed** *n* AIR, WATER POLL lit fixe *m*; ~-**bed adsorber** *n* AIR, WATER POLL adsorbeur à lit fixe *m*; ~-**bed cascade reactor** *n* WATER POLL réacteur en cascade à lit fixe *m*; ~-**bed catalyst** *n* AIR catalyseur sur lit fixe *m*; ~-**bed electrolysis** *n* WATER POLL électrolyse sur lit fixe *f*; ~-**bed furnace** *n* WATER POLL *incineration of wastewater sludge* four à lit fixe *m*; ~-**bed ion exchanger** *n* WATER POLL échangeur d'ions à lit fixe *m*; ~-**bed reactor** *n* SOIL, WATER POLL réacteur à lit fixe *m*; ~ **carbon** *n* RENEW ENERG *biomass* carbone fixe *m*; ~ **catalyst** *n* AIR catalyseur fixe *m*; ~ **film** *n* WATER POLL *attached growth* film fixe *m*; ~ **group** *n* WATER POLL *ion exchanger* groupe fixe *m*; ~-**pitch rotor blade** *n* RENEW ENERG *wind turbine* pale de rotor à pas fixe *f*

fixing *n* SOIL immobilisation *f*

flaking *n* WATER POLL *slow process filter* écaillage *m*, arrachage *m*

flaky *adj* AIR *particle description* écailleux, floconneux

flame[1]: ~-**photometric** *adj* ANALYSIS photométrique de flame; ~-**resistant** *adj* AIR, SAFETY, WASTE ininflammable

flame[2] *n* AIR, ANALYSIS flamme *f*; ~-**AAS** (*flame atomic absorption spectroscopy*) ANALYSIS spectroscopie d'absorption atomique avec atomisation par flammes *f*; ~ **atomic absorption spectroscopy** (*flame-AAS*) ANALYSIS spectroscopie d'absorption atomique avec atomisation par flammes *f*; ~ **coloration** ANALYSIS coloration de flamme *f*; ~ **gas** AIR gaz inflammable *m*; ~ **ionization** ANALYSIS ionisation de flamme *f*; ~ **ionization detector** (*FID*) ANALYSIS détecteur à ionisation de flamme *m*; ~ **photometer** ANALYSIS photomètre de flamme *m*; ~ **photometric detection** (*FPD*) ANALYSIS détection par photométrie de flamme *f*; ~ **photometric determination** ANALYSIS détermination par photométrie de flamme *f*, dosage par photométrie de flamme *m*; ~ **photometry** ANALYSIS photométrie de flamme *f*

flameproof *adj* AIR, SAFETY, WASTE ininflammable

flammability *n* AIR, SAFETY, WASTE inflammabilité *f*

flammable[1] *adj* AIR, SAFETY, WASTE inflammable

flammable[2]: ~ **dust** *n* AIR poussière inflammable *f*; ~ **liquid** *n* SAFETY liquide inflammable *m*; ~ **material** *n* SAFETY matériau inflammable *m*; ~ **product** *n* AIR, SAFETY produit inflammable *m*; ~ **vapor** *AmE*, ~ **vapour** *BrE* *n* SAFETY vapeur inflammable *f*; ~ **waste** *n* WASTE déchets inflammables *m pl*

flange *n* AIR, MARINE bride *f*

flanged[1] *adj* AIR, MARINE à brides

flanged[2]: ~ **edge** *n* WASTE bord bridé *m*, bord tombé *m*; ~ **heat exchanger** *n* AIR échangeur de chaleur à ailettes *m*

flanging: ~ **machine** *n* WASTE machine à border *f*

flanking: ~ **path** *n* NOISE passage latéral *m*, passage par dérivation *m*; ~ **sealing** *n* SOIL imperméabilisation des flancs *f*; ~ **transmission** *n* NOISE transmission indirecte *f*, transmission par dérivation *f*

flare[1] *n* AIR, WASTE, WATER POLL évasement *m*

flare[2] *vt* AIR évaser, WASTE *light*, WATER POLL *pipe* évaser, brûler, faire vaciller

flaring *n* AIR évasement *m*, WASTE évasement *m*, torchage *m*, tulipage *m*, WATER POLL évasement *m*; ~ **installation** WASTE installation de torchage *f*, installation de tulipage *f*

flash: ~ **distillation** *n* WATER SUPP distillation instantanée *f*; ~ **dryer** *n* AIR séchoir-éclair *m*; ~ **evaporator** *n* WATER POLL évaporateur instantané *m*; ~ **point determination** *n* ANALYSIS détermination du point éclair *f*

flask *n* ANALYSIS flacon *m*

flat[1] *adj* SOIL, WASTE, WATER SUPP plat

flat[2] *n* WATER SUPP plat *m*; ~-**bed biofilter** AIR *biological sorption process* biofiltre à fond plat *m*; ~ **pack** WASTE paquet plat *m*; ~ **pallet** WASTE palette plate *f*; ~ **panel display** ANALYSIS écran plat *m*; ~-**plate collector** RENEW ENERG capteur plan *m*, collecteur plan *m*; ~ **subsurface** SOIL sous-sol plat *m*; ~-**top chain conveyor** WASTE chaîne à

palettes *f*, convoyeur à charnières *m*, WATER SUPP convoyeur à charnières *m*
fleet *n* POLICY flotte *f*
flexible[1] *adj* MARINE, NOISE, WASTE flexible, souple
flexible[2]: ~ **hose** *n* MARINE tuyau souple *m*; ~ **package** *n* WASTE emballage souple *m*; ~ **plate** *n* NOISE plaque souple *f*, plateau souple *m*; **~-plate absorber** *n* NOISE absorbeur à plateau souple *m*; ~ **shell** *n* NOISE enveloppe souple *f*
flight: ~ **scraper** *n* WATER POLL racleur à chaîne *m*
floating[1] *adj* MARINE, WATER POLL, WATER SUPP flottant
floating[2] *n* MARINE, WATER POLL, WATER SUPP flottage *m*, flottation *f*; ~ **flexible tank** MARINE citerne souple flottante *f*; ~ **floor** NOISE *resilient layer* plancher flottant *m*; ~ **matter** WASTE matière flottante *f*; ~ **OTEC plant** RENEW ENERG centrale ETM flottante *f*; ~ **refuse** WASTE résidu de flottation *m*, refus flottants *m pl*; ~ **sample** ANALYSIS échantillon flottant *m*; ~ **solid** WATER POLL substance solide en suspension *f*
floc *n* WATER POLL floc *m*; ~ **formation** WATER POLL formation de floc *f*; ~ **fragments** *n pl* WATER POLL fragments de floc *m pl*; ~ **spraying** WASTE floconnage *m*
flocculant *n* WATER POLL, WATER SUPP floculant *m*
flocculate[1] *n* WATER POLL, WATER SUPP floculat *m*
flocculate[2] *vi* WATER POLL, WATER SUPP floculer
flocculation *n* WATER POLL, WATER SUPP floculation *f*; ~ **aid** WATER SUPP adjuvant de floculation *m*; ~ **rate** WATER POLL taux de floculation *m*; ~ **tank** WATER POLL, WATER SUPP bassin de floculation *m*, cuve de floculation *f*; ~ **test** ANALYSIS essai de floculation *m*; ~ **zone** WATER POLL zone de floculation *f*
flocculator *n* WATER POLL, WATER SUPP bassin de floculation *m*, floculateur *m*; ~ **basin** WATER POLL, WATER SUPP bassin de floculation *m*, floculateur *m*; ~ **with cylindrical mixer** WATER POLL floculateur à mélangeur cylindrique *m*; ~ **with mechanically-operated paddles** WATER POLL floculateur à pales mécaniques *m*
flocculent *adj* WATER POLL, WATER SUPP floculeux
flood[1] *n* WATER POLL, WATER SUPP crue *f*, inondation *f*; ~ **control** MANAGE, MARINE, RENEW ENERG, WATER SUPP gestion des crues *f*, maîtrise des crues *f*, régulation des crues *f*, lutte contre les inondations *f*; ~ **control works** *n pl* WATER SUPP ouvrage de crue *m*; ~ **irrigation** WATER SUPP irrigation par infiltration *f*, irrigation par submersion *f*; ~ **plain** WATER SUPP lit de hautes eaux *m*, lit majeur *m*, plaine d'inondation *f*; ~ **prevention** WATER SUPP prévention des crues *f*; ~ **spillway** WATER SUPP évacuateur de crue *m*; ~ **tide** MARINE marée montante *f*, RENEW ENERG *tidal power plant* flux *m*, marée montante *f*
flood[2] *vt* WATER POLL, WATER SUPP noyer
flooding *n* WATER POLL, WATER SUPP inondation *f*, noyage *m*, mise en eau *f*
flotation *n* MARINE, WATER POLL, WATER SUPP flottation *f*; ~ **agent** SOIL, WATER POLL agent de flottation *m*; ~ **cell** SOIL compartiment de flottation *m*; ~ **chamber** MARINE chambre de flottation *f*; ~ **with integrated filtration** WATER POLL flottation avec filtration intégrée *f*; ~ **plant** WATER POLL banc de flottation *m*; ~ **with submerged aerator** WATER POLL flottation avec aération immergée *f*; ~ **tank** WATER POLL cuve de flottation *f*; ~ **thickening** WATER POLL épaississement par flottation *m*
flow[1] *n* AIR écoulement *m*, WATER POLL courant *m*, WATER SUPP *motion* écoulement *m*, ruissellement *m*, *of river, tide, current* cours *m*, flot *m*, flux *m*, courant *m*, *of water above weir* plan d'eau *m*; ~ **deviation** AIR détournement d'écoulement *m*; **~-duration curve** WATER POLL courbe de débit *f*; ~ **field** AIR *cyclone* champ de densité *m*; ~ **force** AIR force d'écoulement *f*; ~ **injection analysis** (*FIA*) ANALYSIS analyse de débit d'injection *f*; ~ **injection analysis spectrometry** (*FIAS*) ANALYSIS spectrométrie d'absorption sur jet *f*; ~ **lanes** *n pl* AIR *of gas in electrostatic precipitator* voies d'écoulement *f pl*; ~ **line** AIR *carrier medium* bec de déversement *m*, ligne d'écoulement *f*; **~-measuring flume** WATER POLL canal de mesure *m*; ~ **meter** ANALYSIS, WATER POLL, WATER SUPP débitmètre *m*; ~ **parameter** ANALYSIS paramètre de débit *m*; ~ **pattern** WATER SUPP diagramme d'écoulement *m*, modèle d'écoulement *m*; ~ **rate** AIR, ANALYSIS, WATER SUPP débit *m*; ~ **rate measuring device** ANALYSIS débitmètre *m*; **~-regulating valve** WATER SUPP vanne régulatrice de débit *f*; ~ **sample** AIR, ANALYSIS échantillon de débit *m*; ~ **sensor** AIR capteur de débit *m*; ~ **velocity** WATER POLL vitesse d'écoulement *f*; ~ **of waste** WASTE flux de déchets *m*
flow[2]: ~ **through** *vt* AIR traverser, WATER POLL circuler
flow[3] *vi* AIR, RENEW ENERG couler, WATER POLL *tributary into river* se jeter, *river, stream* s'écouler, couler, WATER SUPP *river* s'écouler, couler, ruisseler; ~ **over** ANALYSIS déborder
flowing[1] *adj* AIR, WATER SUPP coulant
flowing[2]: **~-out viscosimeter** *n* ANALYSIS viscosimètre aval *m*, viscosimètre de sortie *m*
flowmeter *n* ANALYSIS débitmètre *m*
flue *n* AIR conduite de cheminée *f*; **~-dust catcher** AIR collecteur de cendres volantes *m*, capuchon à cendres volantes *m*; ~ **gas** AIR gaz brûlé *m*, gaz de comburant *m*, gaz de fumée *m*; **~-gas blower** AIR ventilateur de gaz brûlés *m*; **~-gas cleaning** AIR épuration des gaz brûlés *f*, lavage des fumées *m*, WASTE épuration des gaz brûlés *f*; **~-gas cleaning installation** SAFETY installation d'épuration des gaz brûlés *f*; **~-gas cloud** AIR nuage de gaz brûlés *m*; **~-gas cooling** AIR refroidissement des gaz de combustion *m*; **~-gas desulfurization** *AmE*, **~-gas desulphurization** *BrE* (*FGD*) AIR désulfuration des gaz de combustion *f*; **~-gas desulphurization plant blower** *BrE* AIR ventilateur d'usine de désulfuration des gaz *m*; **~-gas heat exchanger** AIR échangeur de chaleur pour gaz de combustion *m*; **~-gas NO_x abatement** WASTE lutte contre les émissions de NO_x *f*; **~-gas plume** AIR panache de gaz de combustion *m*; **~-gas recirculation** AIR recirculation des gaz de combustion *f*; **~-gas reheating** AIR réchauffage des gaz de combustion *m*; **~-gas scrubber** WASTE épurateur de fumées *m*; **~-gas side** AIR ligne des effluents gazeux *f*; **~-gas stream** AIR flux de gaz brûlés *m*; **~-gas tester** ANALYSIS analyseur de fumées *m*; **~-gas throughput**, **~-gas thruput** *AmE* AIR débit des gaz brûlés *m*, débit des gaz de combustion *m*; **~-gas to flue-gas heat exchanger** WASTE échangeur de chaleur fumée-fumée *m*; **~-gas washing** AIR, WASTE épuration des fumées *f*
fluid: ~ **catalyst** *n* AIR catalyseur fluide *m*
fluidize *vt* AIR, SOIL fluidiser, traiter sur lit fluidisé
fluidized: ~ **bed** *n* AIR *adsorption process*, SOIL, WATER POLL, WATER SUPP bain fluidisé *m*, couche fluidifiée *f*, lit fluidisé *m*; **~-bed absorber** *n* AIR, WATER POLL absorbeur à lit fluidisé *m*; **~-bed adsorber** *n* AIR, WATER POLL adsorbeur à lit fluidisé *m*; **~-bed bottom** *n* AIR fond de lit fluidisé *m*; **~-bed combustion** *n* (*FBC*) AIR combustion sur lit fluidisé *f*; **~-bed dryer** *n* AIR, WATER POLL sécheur en lit fluidisé *m*, séchoir en lit fluidisé *m*; **~-bed drying** *n* AIR, WATER POLL séchage en lit fluidisé *m*; **~-bed furnace** *n* AIR, SOIL, WASTE, WATER POLL four à lit fluidisé *m*; **~-bed incineration** *n* SOIL, WATER POLL incinération en lit fluidisé *f*; **~-bed incineration plant** *n* WATER POLL usine d'incinération en lit fluidisé *f*; **~-bed oven** *n* AIR, SOIL, WASTE, WATER POLL four à lit fluidisé *m*; **~-bed process** *n* SOIL *for fine-grained contaminated soil* procédé à lit fluidisé *m*; **~-bed pyrolysis** *n* WASTE pyrolyse sur lit fluidisé *f*; **~-bed reactor** *n* AIR, WASTE, WATER POLL réacteur à lit fluidisé *m*; ~ **solid system** *n* WATER SUPP couche turbulente *f*
flume *n* WATER SUPP caniveau *m*, *artificial open channel* canal de dérivation *m*, canal de prise *m*, canal en bois *m*, canal en planches *m*, *turbine chamber* chambre d'eau *f*; ~ **flushing** WATER POLL chasse du caniveau *f*
fluorescence *n* ANALYSIS fluorescence *f*; ~ **analysis** ANALYSIS

analyse par fluorescence *f*; ~ **spectrum** ANALYSIS *infrared spectrometer* spectre de fluorescence *m*
fluorescent[1] *adj* ANALYSIS fluorescent
fluorescent[2]: ~ **detection** *n* ANALYSIS détection fluorescente *f*; ~ **indicator adsorption** *n* (*FIA*) ANALYSIS adsorption à l'indicateur fluorescent *f*; ~ **lamp** *n* ANALYSIS lampe fluorescente *f*; ~ **spectroscopy** *n* ANALYSIS spectroscopie par fluorescence *f*
fluoridation: ~ **of drinking water** *n* WATER SUPP fluoration de l'eau potable *f*
fluoride *n* ANALYSIS fluorure *m*; ~ **analysis in wastewater** ANALYSIS analyse des fluorures dans les eaux usées *f*
fluorimetric *adj* ANALYSIS fluorimétrique
fluorimetry *n* ANALYSIS *heavy metal detection* fluorimétrie *f*
fluorination *n* WATER SUPP fluoration *f*
fluorine *n* (*F*) SOIL fluor *m* (*F*)
fluorohydrogen *n* AIR hydrogène fluoré *m*
flush *vt* WATER SUPP curer
flushing *n* MARINE chasse *f*, WATER SUPP lessivage rapide *m*, *water-flush drilling* curage *m*, injection d'eau *f*; ~ **of pipe network** WATER POLL vidange de la tuyauterie *f*; ~ **pump** WATER POLL pompe de vidange *f*; ~ **speed** WATER POLL vitesse de vidange *f*; ~ **strength** WATER POLL pouvoir d'entraînement *m*, pouvoir de rinçage *m*
fluvial: ~ **alluvium** *n* WATER SUPP alluvions fluviales *f pl*
fluxing: ~ **agent** *n* WATER SUPP fondant *m*
fly: ~ **ash** *n* AIR *containing solid particulates*, WASTE cendres volantes *f pl*; ~**-ash conditioning** *n* AIR conditionnement des cendres volantes *m*; ~**-ash recovery** *n* AIR récupération des cendres volantes *f*; ~**-ash silo** *n* AIR silo à cendres volantes *m*; ~ **tipping** *n* WASTE, WATER SUPP décharge sauvage *f*
flying: ~ **spark** *n* AIR flammèche *f*
foam *n* MARINE, WATER POLL écume *f*, mousse *f*; ~ **abatement** WASTE lutte contre la formation de mousse *f*; ~ **bed** AIR lit de mousse *m*; ~ **flotation** SOIL flottation par écumage *f*; ~ **formation** WASTE formation de mousse *f*; ~ **panel** NOISE panneau de mousse *m*; ~ **process** SOIL procédé d'écumage *m*; ~ **scrubber** AIR laveur à mousse *m*
foaming *n* MARINE, WATER POLL moussage *m*; ~ **tendency** WATER POLL tendance au moussage *f*
focal: ~ **line** *n* RENEW ENERG *on solar energy plant reflector* axe focal *m*, ligne focale *f*
focusing *n* RENEW ENERG *of sunlight* focalisation *f*
fog *n* AIR brouillard *m*; ~ **formation** AIR formation de brouillard *f*
foil: ~ **polarometer** *n* ANALYSIS *oxygen measurement* polarimètre à lames *m*; ~ **seal** *n* WASTE opercule en aluminium mince *m*
folded: ~**-bottom box** *n* WASTE boîte à fond plié *f*; ~ **filter** *n* WATER POLL filtre à plis *m*, filtre plissé *m*
folding: ~ **carton** *n* WASTE carton pliant *m*, carton à rabats *m*; ~ **sides** *n pl* WASTE rabats *m pl*; ~ **strength** *n* WASTE résistance au pliage *f*; ~ **test** *n* WASTE essai de pliage *m*
food: ~ **chain** *n* POLICY chaîne alimentaire *f*; ~ **contamination** *n* AIR, SAFETY, WATER POLL contamination des aliments *f*, pollution des aliments *f*; ~**-grade packaging film** *n* WASTE pellicule d'emballage à usage alimentaire *f*; ~ **packaging** *n* WASTE emballage à usage alimentaire *m*; ~ **pollution** *n* SAFETY pollution alimentaire *f*; ~ **protection** *n* POLICY protection des produits alimentaires *f*; ~ **tray** *n* WASTE plateau alimentaire *m*; ~ **waste** *n* WASTE déchets alimentaires *m pl*; ~ **wrapping equipment** *n* WASTE matériel d'emballage alimentaire *m*
footloose: ~ **industry** *n* POLICY industrie mobile *f*
forbidden: ~ **area** *n* POLICY zone interdite *f*
force[1]: **in** ~ *phr* POLICY *legislation* en vigueur
force[2]: ~ **of attraction** *n* RENEW ENERG, WATER POLL force d'attraction *f*
forced: ~ **aeration** *n* SOIL aération forcée *f*, WATER POLL ventilation forcée *f*, ventilation par aspiration *f*; ~ **degassing** *n* WASTE dégazage forcé *m*; ~ **draft fan** *AmE*, ~ **draught fan** *BrE* *n* AIR, WASTE ventilateur à tirage forcé *m*
forecast[1] *n* ANALYSIS, MANAGE, POLICY prévision *f*; ~ **model** ANALYSIS modèle de prévision *m*; ~ **schedule** MANAGE planning prévisionnel *m*
forecast[2] *vt* ANALYSIS, MANAGE, POLICY prévoir
forecasting *n* ANALYSIS, MANAGE, POLICY prévision *f*; ~ **of natural disasters** POLICY prévision des catastrophes naturelles *f*
foreign: ~ **emissions** *n pl* AIR émissions de source étrangère *f pl*; ~ **source** *n* AIR source étrangère *f*
forest: ~ **biomass** *n* SOIL biomasse forestière *f*; ~ **cover** *n* RENEW ENERG, SOIL couvert forestier *m*; ~ **fire** *n* AIR, SOIL feu de forêt *m*, incendie de forêt *m*; ~ **inventory** *n* MANAGE inventaire forestier *m*; ~ **logging** *n* POLICY, SOIL exploitation forestière *f*; ~ **ranger** *n* MANAGE, SAFETY garde forestier *m*
fork: ~**-lift truck** *n* WASTE chariot élévateur à fourche *m*, chariot élévateur automoteur *m*
form: ~, **fill and seal machine** *n* WASTE machine de thermoformage pour conditionnement sous vide *f*; ~ **stability** *n* AIR *of filter* stabilité de forme *f*
formaldehyde *n* AIR formaldéhyde *m*
formazin: ~ **turbidity unit** *n* (*FTU*) ANALYSIS unité de turbidité formazine *f*
forming *n* WASTE emboutissage *m*, formage *m*, profilage *m*; ~ **temperature** WASTE température de formage *f*
forum *n* POLICY tribune *f*
forward: ~**-curved blade compressor** *n* AIR compresseur à lame incurvée vers l'avant *m*; ~ **scattering of radiation** *n* ANALYSIS *heavy metal detection* diffusion du rayonnement vers l'avant *f*
fossil[1]: ~**-fuel-driven** *adj* RENEW ENERG à combustibles fossiles, alimenté par des combustibles fossiles; ~**-fuel-fired** *adj* RENEW ENERG à combustibles fossiles, alimenté par des combustibles fossiles
fossil[2]: ~ **energy source** *n* RENEW ENERG source d'énergie fossile *f*; ~**-fuel** *n* AIR, RENEW ENERG combustible fossile *m*; ~**-fuel combustion** *n* AIR combustion de combustible fossile *f*; ~**-fuel-driven heat pump** *n* RENEW ENERG pompe à chaleur entraînée par des combustibles fossiles *f*; ~**-fuel-fired steam generator** *n* RENEW ENERG générateur de vapeur alimenté par des combustibles fossiles *m*
foul[1]: ~**-smelling** *adj* WATER POLL malodorant, nauséabond
foul[2] *vt* AIR salir, ANALYSIS, MARINE, WASTE encrasser, souiller, salir, WATER POLL, WATER SUPP boucher, encrasser, engorger
fouled *adj* WATER POLL, WATER SUPP bouché
fouling *n* AIR salissure *f*, ANALYSIS *of sample*, MARINE, WASTE encrassement *m* salissure *f*, souillure *f*, WATER POLL, WATER SUPP bouchage *m*, encrassement *m*, engorgement *m*; ~ **potential** WATER SUPP pouvoir encrassant *m*
four: ~**-cycle engine** *AmE*, ~**-stroke engine** *BrE* *n* AIR, WASTE, WATER POLL moteur à quatre temps *m*; ~**-way pallet** *n* WASTE palette à quatre entrées *f*
Fourier: ~**-transform infrared** *n* (*FTIR*) ANALYSIS infrarouge à transformée de Fourier *m*; ~**-transform infrared microscopy** *n* ANALYSIS microscopie infrarouge à transformée de Fourier *f*; ~ **transform infrared spectrometer** *n* ANALYSIS spectromètre infrarouge à transformée de Fourier *m*; ~**-transform infrared spectrometry** *n* ANALYSIS spectrométrie infrarouge à transformée de Fourier *f*
FPD *abbr* (*flame photometric detection*) ANALYSIS détection par photométrie de flamme *f*
fraction *n* GEN fraction *f*; ~ **separation** AIR, ANALYSIS, WASTE fractionnement *m*, concassage *m*, égrenage *m*; ~ **separator** WASTE concaisseur *m*, égrenuer *m*
fractionate *vt* GEN fractionner
fractionation *n* GEN fractionnement *m*
fracture: ~ **zone of continental plates** *n* RENEW ENERG zone de fracture des plaques continentales *f*

fracturing *n* SOIL *pneumatic* fracturation *f*
frame: ~ **filter press** *n* WATER POLL filtre-presse à cadres et plateaux *m*
framework: ~ **agreement** *n* POLICY projet d'accord cadre *m*
frank: ~ **effect level** *n* (*FEL*) SAFETY niveau efficace franc *m*
Fraunhofer: ~ **lines** *n pl* ANALYSIS *diffraction* raies de Fraunhofer *f pl*
free[1]: ~ **of heavy metal** *adj* AIR, SOIL exempt de métaux lourds; ~ **of residues** *adj* WASTE sans résidu
free[2]: ~ **carbon dioxide** *n* WATER SUPP dioxyde de carbone libre *m*; ~ **cyanide** *n* WATER POLL cyanure libre *m*; ~ **electron** *n* AIR électron libre *m*; ~ **fall** *n* WASTE chute libre *f*; ~**-fall weir** *n* WATER POLL déversoir dénoyé *m*; ~ **field** *n* NOISE champ infini *m*, champ libre *m*; ~**-flow product** *n* WASTE produit à flux libre *m*; ~ **radical** *n* AIR *electron beam process* radical libre *m*; ~**-standing tower** *n* RENEW ENERG *wind turbine* éolienne isolée *f*
free[3]: ~ **from dust** *vt* AIR dépoussiérer
freestanding *adj* WASTE isolé, libre
freeze[1]: ~**-drying** *n* ANALYSIS lyophilisation *f*; ~**-grinding** *n* WASTE cryobroyage *m*
freeze[2]: ~**-dry** *vt* ANALYSIS, WASTE lyophiliser
freezing: ~ **point hygrometer** *n* AIR, ANALYSIS hygromètre à point de gelée *m*
freight: ~ **rate** *n* WASTE tarif marchandise *m*; ~ **terminal** *n* WASTE terminal de transport de marchandises *m*
French: ~ **Standards Association** *n* (*AFNOR*) ANALYSIS, POLICY Association Française de Normalisation *f* (*AFNOR*)
freon: ~ **12** *n* AIR *propellant* fréon 12 *m*
frequency *n* GEN fréquence *f*; ~ **analyser** *BrE*, ~ **analyzer** *AmE* ANALYSIS analyseur de fréquence *m*; ~ **distribution** ANALYSIS distribution de fréquence *f*; ~ **range** MARINE, NOISE plage de fréquence *f*; ~ **spectrum** NOISE spectre de fréquence *m*
fresh: ~ **coke** *n* AIR coke frais *m*; ~ **compost** *n* WASTE compost frais *m*; ~ **sludge pit** *n* WATER SUPP fosse à boue fraîche *f*; ~ **sludge sump** *n* WATER SUPP fosse à boue fraîche *f*
freshness: ~ **seal** *n* WASTE emballage fraîcheur *m*
freshwater *n* GEN eau douce *f*; ~ **conservation** WATER SUPP conservation de l'eau douce *f*; ~ **introduction** WASTE apport d'eau douce *m*; ~ **limit** MARINE, WATER POLL limite d'eau douce *f*; ~ **pollution** WATER POLL pollution des eaux douces *f*; ~ **supply** AIR approvisionnement en eau douce *m*
Fresnel: ~ **lens** *n* RENEW ENERG *solar collector* lentille de Fresnel *f*
friction: ~ **head** *n* WATER POLL hauteur équivalente de perte de charge *f*; ~ **layer** *n* AIR couche de frottement *f*
frictional: ~ **force** *n* AIR force de friction *f*
Friedrich: ~**'s gas-washing bottle** *n* ANALYSIS flacon de Friedrich *m*
Friends: ~ **Of The Earth** *n* POLICY Amis de la Terre *m pl*
fringe: ~ **area** *n* MARINE zone périphérique *f*
front: ~**-end display** *n* ANALYSIS frontal *m*; ~**-end loader** *n* MARINE chargeur frontal *m*
frontally: ~**-packed bed** *n* WASTE lit à chargement frontal *m*
frost: ~**-preventive agent** *n* WASTE antigélif *m*
froth *n* MARINE, WATER POLL écume *f*, mousse *f*
frothing *n* AIR moussage *m*; ~ **tendency** WATER POLL tendance à écumer *f*
frozen: ~ **product** *n* WASTE produit congelé *m*
FTIR *abbr* (*Fourier-transform infrared*) ANALYSIS infrarouge à transformée de Fourier *m*
FTU *abbr* (*formazin turbidity unit*) ANALYSIS unité de turbidité formazine *f*
fuel *n* AIR, RENEW ENERG combustible *m*; ~**-air ratio** AIR rapport air-carburant *m*, rapport air-combustible *m*; ~**-air ratio control unit** AIR poste de réglage du rapport air-carburant *m*, poste de réglage du rapport air-combustible *m*; ~ **cell** AIR, RENEW ENERG pile à combustible *f*; ~ **combustion** AIR combustion de combustible *f*; ~ **from biomass** RENEW ENERG biocombustible *m*, combustible obtenu à partir de la biomasse *m*; ~ **gas** AIR, WASTE *from waste gasification* gaz combustible *m*; ~ **gas cleaning** WASTE lavage du gaz combustible *m*; ~ **oil** AIR carburant *m*, *auxiliary fuel* fioul *m*, fuel-oil *m*, RENEW ENERG mazout *m*; ~ **replacement** AIR substitution de combustible *f*; ~ **tank** AIR cuve à combustible *f*, réservoir à carburant *m*; ~ **tank emissions** *n pl* AIR *hydrocarbon* émissions du réservoir de carburant *f pl*
fugitive: ~ **emissions** *n pl* AIR émissions fugaces *f pl*
full: ~ **analysis** *n* ANALYSIS analyse intégrale *f*; ~ **gluing** *n* WASTE collage intégral *m*
Fuller: ~**'s earth** *n* AIR *adsorbent* terre à foulon *f*
fully[1]: ~**-halogenated** *adj* AIR totalement halogéné
fully[2]: ~**-halogenated CFC** *n* AIR CFC totalement halogéné *m*
fumarole *n* RENEW ENERG *geothermal energy* fumerolle *f*
fume: ~ **incinerator** *n* AIR *afterburner* incinérateur pour gaz de fumées *m*; ~ **inlet** *n* AIR admission de fumées *f*
fumes *n pl* AIR *containing solid particulates*, SAFETY *from paints, lacquers, adhesives*, WATER POLL fumées *f pl*, vapeurs *f pl*
fumigate *vt* AIR enfumer
fumigation *n* AIR enfumage *f*, fumigation *f*
function *n* SAFETY fonction *f*; ~ **monitoring** SAFETY suivi de fonctionnement *m*, surveillance de fonctionnement *f*; ~ **safety** SAFETY sûreté de fonctionnement *f*
funding *n* MANAGE financement *m*, POLICY consolidation *f*, financement *m*
fungus *n* AIR, WATER POLL champignon *m*, mycète *m*
funnel[1]: ~**-shaped** *adj* WATER POLL en entonnoir
funnel[2]: ~**-shaped settling chamber** *n* WATER POLL bassin de décantation en entonnoir *m*
fur[1] *vi* WATER SUPP entartrer
fur[2]: ~ **up** *vi* WATER SUPP entartrer
furnace: ~ **boiler** *n* WASTE chaudière à chambre de combustion *f*
furring *n* WATER SUPP entartrage *m*
fusion *n* AIR, RENEW ENERG fusion *f*; ~ **process** RENEW ENERG *solar energy* procédé de fusion *m*
future: ~ **usability** *n* RENEW ENERG *of energy sources* perspectives d'utilisation *f pl*
FUVAD *abbr* (*far-ultraviolet absorbance detector*) ANALYSIS détecteur d'absorption dans l'ultraviolet lointain *m*

G

GAC *abbr* (*granulated activated carbon*) AIR, WATER POLL charbon actif en grains *m*
gage *AmE see gauge BrE*
gaging *AmE see gauging BrE*
galvanic: ~ **sludge** *n* WATER POLL boues galvaniques *f pl*
galvanometer *n* ANALYSIS galvanomètre enregistreur *m*
game: ~ **reserve** *n* POLICY réserve de chasse *f*
gamma: ~-**ray irradiation** *n* WATER POLL irradiation par rayonnement gamma *f*
garbage *n AmE* (*cf rubbish BrE*) WASTE déchets *m pl*, ordures *f pl*, ordures ménagères *f pl* (*OM*); ~ **bag** *AmE* WASTE sac poubelle *m*, sac à ordures *m*; ~ **can** *AmE* (*cf dustbin BrE, rubbish bin BrE*) WASTE boîte à ordures *f*, poubelle *f*; ~ **collection** *AmE* WASTE enlèvement des ordures *m*; ~ **disposal** *AmE* WASTE élimination des ordures *f*, élimination des déchets *f*; ~ **incineration** *AmE* WASTE incinération de déchets *f*, incinération d'ordures *f*; ~ **incineration refuse** *AmE* WASTE résidus d'épuration des fumées d'incinération des ordures ménagères *m pl* (*REFIOM*); ~ **incinerator** *AmE* WASTE incinérateur à ordures *m*, incinérateur d'ordures *m*, incinérateur de déchets *m*; ~ **truck** WASTE benne à ordures *f*
garden: ~ **waste** *n* WASTE déchets de jardin *m pl*, déchets de jardinage *m pl*
gas[1]: ~-**chromatographic** *adj* ANALYSIS chromatographique en phase gazeuse; ~-**nonspecific** *adj* ANALYSIS *measurement principle* non-spécifique d'un gaz; ~-**permeable** *adj* WASTE perméable au gaz; ~-**tight** *adj* AIR étanche au gaz
gas[2] *n* GEN gaz *m*; ~ **absorbent** AIR absorbant gazeux *m*; ~ **absorption** AIR absorption gazeuse *f*; ~ **acceleration** AIR *Venturi nozzle* accélération de la vitesse du gaz *f*; ~ **analyser** *BrE*, ~ **analyzer** *AmE* ANALYSIS analyseur de gaz *m*; ~ **bubble** WATER POLL bulle de gaz *f*; ~ **build-up** WASTE formation de gaz *f*; ~ **build-up potential** ANALYSIS, WASTE potentiel de formation de gaz *m*; ~ **build-up rate** WASTE vitesse de formation des gaz *f*; ~ **burette** ANALYSIS burette à gaz *f*; ~ **capture** AIR, NOISE captage de gaz *m*; ~ **chromatograph** (*GC*) ANALYSIS chromatographe en phase gazeuse *m* (*CPG*); ~ **chromatograph nitrogen-phosphorus detection** (*GC-NPD*) ANALYSIS détection d'azote et de phosphore par chromatographie en phase gazeuse *f*; ~ **chromatographic separation** ANALYSIS séparation par chromatographie en phase gazeuse *f*; ~ **chromatography** (*GC*) ANALYSIS chromatographie en phase gazeuse *f* (*CPG*); ~ **chromatography coupled with flame ionization detectors** ANALYSIS chromatographie en phase gazeuse couplée à des détecteurs d'ionisation de flamme *f*; ~ **chromatography coupled with mass spectrometers** ANALYSIS chromatographie en phase gazeuse couplée à des spectromètres de masse *f*; ~ **chromatography detector** (*GCD*) ANALYSIS détecteur par chromatographie en phase gazeuse *m*; ~ **chromatography-electron capture detection** (*GC-ECD*) ANALYSIS détection de capture d'électrons par chromatographie en phase gazeuse *f*; ~ **chromatography-flame ionization detection** (*GC-FID*) ANALYSIS détection d'ionisation de flamme par chromatographie en phase gazeuse *f*; ~ **chromatography-flame photometric detection** (*GC-FPD*) ANALYSIS détection par photométrie de flamme-chromatographie en phase gazeuse *f*; ~ **chromatography-gas chromatography** (*GC-GC*) ANALYSIS chromatographie de partage gaz-gaz *f*; ~ **collectability** ANALYSIS *landfill gas* possibilité de collecte des gaz *f*, possibilité de récupération des gaz *f*, possibilité de recueillir les gaz *f*; ~ **collection outside of landfill** WASTE collecte des gaz hors décharge *f*; ~ **compartment** AIR compartiment des gaz *m*, segment gazeux *m*; ~ **component** AIR composant d'un gaz *m*; ~ **composition** AIR composition des gaz *f*; ~ **container** ANALYSIS, SAFETY, WASTE gazomètre *m*, réservoir de gaz *m*, WATER POLL cloche *f*, gazomètre *m*; ~ **cooler** AIR refroidisseur de gaz *m*; ~ **cooling** AIR refroidissement des gaz *m*; ~ **density** ANALYSIS masse volumique de gaz *f*; ~ **density determination** ANALYSIS détermination de la masse volumique des gaz *f*; ~ **desulfurization** *AmE*, ~ **desulphurization** *BrE* SAFETY, WATER SUPP désulfuration des gaz *f*; ~ **detector** ANALYSIS détecteur de gaz *m*; ~ **dispersion** AIR dispersion des gaz *f*; ~ **distribution** AIR distribution de gaz *f*; ~ **drainage** WASTE purge des gaz *f*; ~ **effluent** AIR, WASTE, WATER SUPP effluent gazeux *m*; ~ **emissions** *n pl* AIR, RENEW ENERG émissions de gaz *f pl*, émissions gazeuses *f pl*; ~ **entrainment** AIR entraînement de gaz *m*; ~ **entrance possibility** WASTE risque d'admission de gaz *m*; ~ **evolution** WASTE *of dump* formation de gaz *f*; ~ **flow** AIR flux de gaz *m*, ANALYSIS débit de gaz *m*; ~ **flow counter** ANALYSIS débitmètre de gaz *m*; ~ **flow meter** ANALYSIS débitmètre de gaz *m*; ~ **formation** WASTE formation de gaz *f*; ~ **formation potential** ANALYSIS, WASTE *landfill gas* potentiel de formation de gaz *m*; ~ **generator** WASTE gazogène *m*; ~ **heat exchanger** AIR échangeur de chaleur en phase gazeuse *m*; ~ **holder** ANALYSIS, WASTE gazomètre *m*; ~ **hood** WATER POLL cloche à gaz *f*, calotte de gaz *f*; ~ **indicator** ANALYSIS détecteur de gaz *m*, indicateur de grisou *m*, SAFETY indicateur de grisou *m*; ~ **inlet** AIR admission de gaz *f*, entrée des gaz *f*; ~ **interferometer** ANALYSIS interféromètre à gaz *m*; ~ **ion** AIR *in electrostatic precipitator corona field* ion gazeux *m*; ~ **leak** AIR fuite de gaz *f*; ~-**liquid partition chromatography** ANALYSIS chromatographie gaz-liquide *f*; ~ **mask** SAFETY masque à gaz *m*; ~ **meter** ANALYSIS compteur à gaz *m*; ~ **migration** WASTE *in underground* migration de gaz *f*; ~ **mixer** ANALYSIS mélangeur de gaz *m*; ~-**mixing chamber** ANALYSIS chambre de mélange des gaz *f*; ~ **mixture** ANALYSIS mélange gazeux *m*; ~ **molecule** AIR molécule de gaz *f*; ~ **outlet** AIR sortie de gaz *f*; ~ **partition chromatography** ANALYSIS chromatographie de partage en phase gazeuse *f*; ~ **permeation** AIR infiltration de gaz *f*, perméation gazeuse *f*; ~ **phase** AIR phase gazeuse *f*; ~-**phase oxidation** WATER POLL *wastewater incineration* oxydation en phase gazeuse *f*; ~-**phase titration** (*GPT*) ANALYSIS titrage en phase gazeuse *m*; ~-**phase titrator** ANALYSIS, WATER SUPP titrimètre de phase gazeuse *m*; ~ **preheater** AIR, WATER POLL préchauffage du gaz *m*, réchauffeur de gaz *m*; ~ **purification** AIR épuration de gaz *f*; ~ **purifier** AIR épurateur de gaz *m*; ~ **recovery** AIR récupération de gaz *f*, WATER SUPP production de gaz naturel *f*; ~ **release** ANALYSIS *from sample* échappement de gaz *m*, rejet de gaz *m*; ~ **reuse plant** WASTE installation de recyclage des gaz *f*; ~ **sample** ANALYSIS prélèvement de gaz *m*; ~-**sampling bag** ANALYSIS sac d'échantillonnage de gaz *m*; ~-**sampling pipette** ANALYSIS pipette d'échantillonnage de gaz *f*; ~-**sampling tube** ANALYSIS éprouvette à gaz *f*; ~-**sampling valve** ANALYSIS robinet d'échantillonnage de gaz *m*; ~ **saturation method** ANALYSIS *for vapour pressure determination* méthode de saturation des gaz *f*; ~ **scrubbing**

AIR lavage de gaz *m*; **~-sensitive electrode** ANALYSIS *in measuring cell* électrode sensible aux gaz *f*; **~ shaft** WASTE puits de gaz *m*; **~ soil pipe** SOIL canalisation d'alimentation de gaz *f*; **~-solid chromatography** (*GSC*) ANALYSIS chromatographie gaz-solide *f* (*CGS*); **~ solubility** AIR solubilité des gaz *f*; **~ storage tank** WATER POLL cuve de stockage du gaz *f*; **~ stream** AIR flux de gaz *m*; **~ suction** WASTE aspiration des gaz *f*; **~ temperature** AIR température de gaz *f*; **~ throughput**, **~ thruput** *AmE* AIR débit de gaz *m*; **~ trace analyser** *BrE*, **~ trace analyzer** *AmE* ANALYSIS analyseur de gaz à l'état de trace *m*; **~ transfer** WATER POLL transfert de gaz *m*; **~ turbine** WATER POLL turbine à gaz *f*; **~-use plant** WASTE usine alimentée au gaz *f*, usine fonctionnant au gaz *f*; **~ velocity** AIR vitesse du gaz *f*; **~ volume** ANALYSIS volume des gaz *m*; **~ washing bottle with sintered head** ANALYSIS barboteur à gaz avec tête frittée *m*; **~ well chain** WASTE chaîne de puits de gaz *f*

gaseous[1] *adj* GEN gazeux

gaseous[2]: **~ acid pollution index** *n* AIR indice de pollution gazeuse acide *m*; **~ air pollution** *n* AIR pollution atmosphérique par des gaz *f*; **~ combustion product** *n* AIR produit de combustion gazeuse *m*; **~ contaminant** *n* AIR contaminant gazeux *m*, polluant gazeux *m*, WATER POLL polluant gazeux *m*; **~ emissions** *n pl* AIR émissions gazeuses *f pl*; **~ fuel** *n* AIR combustible en phase gazeuse *m*; **~ medium** *n* AIR, WATER SUPP milieu gazeux *m*; **~ pollutant** *n* AIR, WATER POLL polluant gazeux *m*; **~ state of aggregation** *n* AIR état gazeux d'agrégation *m*; **~ waste** *n* WASTE, WATER SUPP déchets sous forme gazeuse *m pl*; **~ working medium** *n* RENEW ENERG *heat pump* fluide caloporteur en phase gazeuse *m*, vecteur en phase gazeuse *m*

gases: **~ and fumes** *n pl* SAFETY *hazards* gaz et fumées

gasification *n* AIR gazéification *f*

gasifier *n* RENEW ENERG *biomass conversion* gazéificateur *m*

gasify *vt* AIR gazéifier

gasohol *n* RENEW ENERG carburol *m*, gasohol *m*

gasoline: **~ vapor recovery plant** *n AmE* (*cf petrol vapour recovery plant BrE*) AIR installation de récupération des vapeurs d'essence *f*

gasometer *n* ANALYSIS gazomètre *m*

gasometric[1] *adj* ANALYSIS gazométrique

gasometric[2]: **~ method** *n* ANALYSIS méthode gazométrique *f*

gasometry *n* ANALYSIS gazométrie *f*

gate: **~ electrode** *n* ANALYSIS *quadrupole ion trap* portillon *m*

gauge[1]: **~ cock** *n BrE* WATER POLL, WATER SUPP robinet de hauteur d'eau *m*, robinet de jauge *m*; **~ wire** *n BrE* ANALYSIS calibre pour fils métalliques *m*

gauge[2] *vt BrE* ANALYSIS jauger

gauging *n BrE* ANALYSIS jaugeage *m*; **~ tank** *BrE* ANALYSIS bac jaugeur *m*

Gaussian: **~ filter** *n* ANALYSIS filtre de Gauss *m*; **~ model** *n* ANALYSIS modèle de Gauss *m*; **~ plume** *n* AIR panache en cloche *m*

gazoline: **~ stripping** *n* WASTE désessenciement *m*

GC[1] *abbr* ANALYSIS (*gas chromatograph*) CPG (*chromatographe en phase gazeuse*), (*gas chromatography*) CPG (*chromatographie en phase gazeuse*)

GC[2]: **~ thermodesorption** *n* ANALYSIS thermodésorption en phase gazeuse *f*

GCD *abbr* (*gas chromatography detector*) ANALYSIS détecteur par chromatographie en phase gazeuse *m*

GC-ECD *abbr* (*gas chromatography-electron capture detection*) ANALYSIS détection de capture d'électrons par chromatographie en phase gazeuse *f*

GC-FID *abbr* (*gas chromatography-flame ionization detection*) ANALYSIS détection d'ionisation de flamme par chromatographie en phase gazeuse *f*

GC-FPD *abbr* (*gas chromatography-flame photometric detection*) ANALYSIS détection par photométrie de flamme-chromatographie en phase gazeuse *f*

GC-GC *abbr* (*gas chromatography-gas chromatography*) ANALYSIS chromatographie de partage gaz-gaz *f*

GC-MS *abbr* ANALYSIS (*capillary gas chromatography mass spectrometer*) CG-SM (*chromatographie gazeuse par capillarité couplée à un spectromètre de masse*)

GC-MS-COMP: **~ techniques** *n pl* ANALYSIS techniques informatiques CG-SM *f pl*

GC-NPD *abbr* (*gas chromatograph nitrogen-phosphorus detection*) ANALYSIS détection d'azote et de phosphore par chromatographie en phase gazeuse *f*

gear: **~ drive** *n* WATER POLL commande à engrenage *f*, commande à roues dentées *f*

GEF *abbr* (*Global Environment Facility*) POLICY Fonds pour l'environnement mondial *m*

Geiger-Müller ~ counter *n* ANALYSIS compteur Geiger-Müller *m*

gel: **~ permeation chromatography** *n* (*GPC*) ANALYSIS chromatographie sur gel perméable *f*; **~ permeation chromatography-gas chromatography** *n* (*GPC-GC*) ANALYSIS chromatographie en phase gazeuse-chromatographie sur gel perméable *f*

gelatine *n* WASTE gélatine *f*; **~ capsule** WASTE capsule de gélatine *f*

gelling: **~ agent** *n* MARINE gélifiant *m*

GEMS *abbr* (*Global Environment Monitoring System*) ANALYSIS, POLICY GEMS (*Système mondial de surveillance continue de l'environnement*)

gene: **~ pool** *n* POLICY, SAFETY capital génétique *m*

general: **~ cargo** *n* WASTE conteneur d'usage courant *m*; **~ layout** *n* MANAGE plan de masse *m*

generally: **~-accepted technical procedure** *n* POLICY méthode reconnue *f*

generate *vt* NOISE *sound*, RENEW ENERG *electricity, steam* produire

generated: **~ electricity** *n* RENEW ENERG électricité produite *f*; **~ power** *n* RENEW ENERG électricité produite *f*

generation *n* NOISE *of sound*, RENEW ENERG *of electrical energy* production *f*

generator *n* RENEW ENERG génératrice *f*, alternateur *m*; **~ house** NOISE caisson d'alternateur *m*

genetic *adj* SAFETY génétique

genetically: **~ significant dose** *n* AIR, SAFETY dose génétiquement significative *f*

geological: **~ conditions** *n pl* RENEW ENERG milieu géologique *m*

geosciences *n pl* POLICY sciences de la terre *f pl*

geostationary: **~ satellite** *n* POLICY satellite géostationnaire *m*

geothermal[1] *adj* RENEW ENERG géothermique

geothermal[2]: **~ deposit** *n* RENEW ENERG gisement géothermique *m*; **~ electrical power generation** *n* RENEW ENERG production d'électricité d'origine géothermique *f*; **~ energy** *n* AIR, RENEW ENERG, WATER SUPP énergie géothermique *f*; **~ energy exploitation** *n* AIR, RENEW ENERG, WATER SUPP exploitation de l'énergie géothermique *f*; **~ energy potential** *n* RENEW ENERG potentiel géothermique *m*; **~ energy resources** *n pl* RENEW ENERG ressources géothermiques *f pl*; **~ gradient** *n* RENEW ENERG gradient géothermique *m*; **~ heating plant** *n* RENEW ENERG usine géothermique *f*; **~ power plant** *n* RENEW ENERG centrale géothermique *f*; **~ reservoir** *n* RENEW ENERG réservoir géothermique *m*; **~ resources** *n pl* RENEW ENERG ressources géothermiques *f pl*

geothermic *adj* RENEW ENERG géothermique

geothermics *n pl* RENEW ENERG géothermie *f*

germ[1] *n* AIR, SOIL, WATER POLL germe *m*; **~-free** *adj* ANALYSIS *sample* stérile

germ[2]: **~-free wastewater sample** *n* ANALYSIS prélèvement d'eau stérile *m*

German: ~ **Institute for Standards** *n* (*DIN*) ANALYSIS Institut allemand de normalisation *m* (*DIN*)
geyser *n* RENEW ENERG *geothermal energy* geyser *m*, source d'eau chaude *f*
GF *abbr* (*graphite furnace*) ANALYSIS four en graphite *m*
gimlet: ~ **bit** *n* WASTE mèche à bois à simple spirale *f*
glance: ~ **coal** *n* AIR anthracite *f*
gland: ~ **packing** *n* WATER SUPP garniture de presse-étoupe *f*
glass: ~ **bottle** *n* ANALYSIS *in sampling* flacon en verre *m*; ~ **container** *n* WASTE conteneur à verre *m*; ~ **fabric** *n* WASTE tissu de verre *m*, verre textile *m*; ~ **fiber** *AmE*, ~ **fibre** *BrE n* AIR, NOISE, WASTE fibre de verre *f*; ~-**fibre absorber** *n BrE* NOISE absorbeur en fibre de verre *m*; ~-**fibre filter** *n BrE* ANALYSIS filtre en fibre de verre *m*; ~-**fibre laminate** *n BrE* WASTE plastique stratifié au verre textile *m*; ~-**fibre reinforced plastics** *n pl BrE* WASTE plastique renforcé par de la fibre de verre *m*; ~ **jar** *n* WASTE récipient en verre *m*; ~ **pane** *n* RENEW ENERG *solar collector* panneau de verre *m*; ~ **recycling** *n* WASTE recyclage du verre *m*; ~ **roof** *n* RENEW ENERG *solar power plant* toit de verre *m*; ~ **roofing** *n* RENEW ENERG *solar power plant* toiture en verre *f*; ~ **tube** *n* ANALYSIS tube en verre *m*; ~ **wool** *n* WASTE laine de verre *f*
glaze *vt* WASTE *paper* glacer, vernisser, vitrer
glazed: ~ **board** *n* WASTE carton glacé *m*; ~ **paper** *n* WASTE papier glacé *m*
glazing *n* RENEW ENERG *solar collector* vitrage transparent *m*
GLC *abbr* (*ground level concentration*) SOIL concentration au niveau du sol *f*
global[1] *adj* AIR, RENEW ENERG mondial, planétaire
global[2]: ~ **climate model** *n* AIR modèle climatique mondial *m*, modèle climatique planétaire *m*; ~ **ecosystem** *n* AIR, MARINE, SOIL écosystème mondial *m*, écosystème planétaire *m*; ~ **emissions** *n pl* AIR émissions planétaires *f pl*; ~ **energy consumption** *n* RENEW ENERG consommation globale d'énergie *f*, consommation mondiale d'énergie *f*; ~ **energy demand** *n* RENEW ENERG demande globale d'énergie *f*, demande mondiale d'énergie *f*; ~ **energy supply** *n* RENEW ENERG approvisionnement mondial en énergie *m*; ~ **insolation** *n* AIR ensoleillement global *m*; ~ **irradiance** *n* AIR éclairement planétaire *m*; ~ **radiation** *n* AIR rayonnement planétaire *m*; ~ **solar radiation** *n* RENEW ENERG rayonnement solaire global *m*; ~ **sulfur budget** *AmE*, ~ **sulphur budget** *n BrE* AIR bilan des émissions mondiales de soufre *m*; ~ **warming** *n* AIR réchauffement planétaire *m*, réchauffement de la planète *m*; ~ **warming potential** *n* AIR, SOIL, WATER POLL effet de serre *m*
Global: ~ **Environment Facility** *n* (*GEF*) POLICY Fonds pour l'environnement mondial *m*; ~ **Environment Monitoring System** *n* (*GEMS*) ANALYSIS, POLICY Système mondial de surveillance continue de l'environnement *m* (*GEMS*)
GLP *abbr* (*good laboratory practices*) ANALYSIS bonnes pratiques de laboratoire *f pl*
glue *n* WASTE colle *f*; ~ **film** WASTE pellicule de colle *f*; ~ **joint** WASTE fente de collage *f*
gluing *n* WASTE collage *m*; ~ **device** WASTE colleur *m*
glutinous *adj* AIR *dust property* glutineux
glycolysis *n* WASTE glycolyse *f*
good: ~ **laboratory practices** *n pl* (*GLP*) ANALYSIS bonnes pratiques de laboratoire *f pl*
government *n* POLICY gouvernement *m*, pouvoirs publics *m pl*
governmental: ~ **projection** *n* POLICY projection établie par les pouvoirs publics *f*
GPC *abbr* (*gel permeation chromatography*) ANALYSIS chromatographie sur gel perméable *f*
GPC-GC *abbr* (*gel permeation chromatography-gas chromatography*) ANALYSIS chromatographie en phase gazeuse-chromatographie sur gel perméable *f*
GPT *abbr* (*gas-phase titration*) ANALYSIS titrage en phase gazeuse *m*
grab *n* WATER SUPP *of crane or refuelling machine* grappin *m*; ~ **boring** SOIL sondage au trépan-benne *m*; ~ **drilling** SOIL forage au trépan-benne *m*; ~ **sample** ANALYSIS échantillon pris au hasard *m*; ~ **system** WATER SUPP système à grappin *m*
grade[1]: ~ **efficiency curve** *n* AIR courbe de rendement de niveau *f*
grade[2] *vt* AIR cribler, ANALYSIS calibrer, cribler, MARINE cribler, WASTE calibrer, WATER SUPP calibrer, cribler
graded *adj* ANALYSIS, WASTE, WATER SUPP calibré
grader *n* MARINE niveleuse *f*, profileuse *f*
grading *n* ANALYSIS calibrage *m*, WASTE attribution d'une classe *f*, calibrage *m*, WATER SUPP criblage *m*
graduate *vt* ANALYSIS graduer
grain *n* RENEW ENERG céréales *f pl*; ~ **encapsulation** WASTE confinement en grains *m*, *of hazardous waste* enrobage en grains *m*; ~ **size** SOIL calibre de grain *m*; ~-**size analysis** ANALYSIS analyse granulométrique *f*; ~-**size distribution** ANALYSIS, SOIL répartition granulométrique *f*, granulométrie *f*; ~ **straw** RENEW ENERG paille de céréales *f*
grandfathering: ~ **principle** *n* POLICY principe des droits acquis *m*
granting *n* POLICY *of permit* octroi *m*
granular: ~-**bed filter** *n* AIR *of granulate layers* filtre à lit de granulats *m*; ~ **catalyst** *n* AIR catalyseur en grains *m*
granulate[1] *n* AIR *granular bed filter* granulat *m*; ~ **body** AIR *granular bed filter* corps granuleux *m*; ~ **layer** AIR *granular bed filter* couche de granulats *f*; ~ **tubular filter** AIR filtre à cartouches pour granulats *m*, filtre tubulaire pour granulats *m*
granulate[2] *vt* WASTE granuler, grenailler
granulated: ~ **activated carbon** *n* (*GAC*) AIR, WATER POLL charbon actif en grains *m*; ~ **layer** *n* AIR *filter medium* couche granuleuse *f*
granulating: ~ **head** *n* WASTE tête de galet *f*; ~ **machine** *n* WASTE granulatoire *m*, grenailleuse *f*
granulation *n* WASTE granulation *f*
granulator *n* WASTE granulateur *m*
granule *n* WASTE granule *f*
graphite: ~ **furnace** *n* (*GF*) ANALYSIS four en graphite *m*; ~ **furnace atomic absorption spectrometry** *n* ANALYSIS spectrométrie d'absorption atomique en four de graphite *f*
grass: ~ **cover** *n* RENEW ENERG, SOIL couvert herbacé *m*; ~ **for energy purposes** *n* RENEW ENERG graminée énergétique *f*
grassland *n* SOIL herbage *m*
grate *n* AIR, WATER SUPP grille *f*; ~ **drive** WASTE mécanisme d'entraînement de grille *m*; ~ **firing** AIR combustion sur grille *f*
gravel: ~-**bed filter** *n* AIR filtre à gravier *m*; ~ **filter layer** *n* WASTE couche de filtre à gravier *f*
gravimetric[1] *adj* ANALYSIS gravimétrique
gravimetric[2]: ~ **analysis** *n* ANALYSIS analyse gravimétrique *f*; ~ **determination** *n* ANALYSIS calcul gravimétrique *m*; ~ **gas analysis** *n* ANALYSIS analyse gravimétrique des gaz *f*; ~ **gas density determination** *n* ANALYSIS détermination de la masse volumique de gaz par gravimétrie *f*; ~ **method** *n* ANALYSIS méthode gravimétrique *f*
gravitational: ~ **force** *n* RENEW ENERG, WATER POLL *dewatering*, WATER SUPP force de gravitation *f*, attraction universelle *f*; ~ **settling** *n* AIR décantation gravitaire *f*; ~ **water** *n* SOIL eau gravitaire *f*
gravity[1]: ~-**fed** *adj* RENEW ENERG alimenté par gravité
gravity[2] *n* GEN gravité *f*; ~ **filling machine** WASTE machine de remplissage gravitaire *f*; ~ **flow** WATER SUPP écoulement gravitaire *m*; ~ **roller** WASTE transrouleur *m*; ~ **roller conveyor** WASTE transporteur à rouleaux par gravité *m*; ~ **sedimentation** ANALYSIS *measurement for dust* sédimentation par gravité *f*; ~ **separation** AIR, SOIL, WASTE, WATER POLL, WATER SUPP séparation par gravité *f*; ~ **separator** AIR, WASTE, WATER POLL, WATER SUPP séparateur gravitaire *m*, séparateur gravimétrique *m*; ~ **settler** AIR

décanteur gravitaire *m*; ~ **settling** AIR décantation gravitaire *f*; ~ **sewer** WASTE, WATER POLL, WATER SUPP égout gravitaire *m*; ~ **spray tower** AIR laveur à pulvérisation gravitaire *m*; ~ **thickener** WATER POLL épaississeur par gravité *m*; ~ **thickening** WATER POLL épaississement par gravité *m*; ~ **thickening in containers** WATER POLL épaississement gravitaire en cuve *m*; ~ **thickening in filter bags** WATER POLL épaississement gravitaire en filtre à manche *m*

gray *n* (*Gy*) ANALYSIS Gray (*Gy*); ~ **board** *AmE see grey board BrE*; ~ **container** *AmE see grey container BrE*; ~ **filter** *AmE see grey filter BrE*; ~ **water** *AmE see grey water BrE*

grease[1]: ~-**free** *adj* WATER SUPP dégraissé, maigre, sans graisse

grease[2]: ~-**resistant paper** *n* WASTE papier imperméable aux graisses *m*, papier ingraissable *m*; ~ **separator** *n* WASTE, WATER SUPP dégraisseur *m*, séparateur de graisse *m*; ~ **trap** *n* WASTE, WATER POLL, WATER SUPP dégraisseur *m*, siphon de dépôt de graisse *m*, séparateur de graisse *m*

greaseproof *adj* WASTE étanche aux graisses; ~ **paper** *n* WASTE papier parcheminé *m*, papier sulfurisé *m*

green[1] *adj* MANAGE, POLICY vert

green[2]: ~ **belt** *n* POLICY ceinture verte *f*; ~ **consumerism** *n* MANAGE consommation verte *f*; ~ **container** *n* WATER SUPP conteneur vert *m*; ~ **dot** *n* MANAGE, WATER SUPP point vert *m*; ~ **open space** *n* POLICY espace vert *m*; ~ **product** *n* MANAGE produit vert *m*; ~ **tax** *n* POLICY impôt sur l'environnement *m*, impôt vert *m*; ~ **tax reform** *n* POLICY réforme de la fiscalité de l'environnement *f*; ~ **tourism** *n* MANAGE tourisme vert *m*; ~ **VAT** *n* POLICY taxe à la valeur ajoutée appliquée à l'environnement *f*, TVA verte *f*, TVA appliquée à l'environnement *f*; ~ **waste** *n* WATER SUPP déchets verts *m pl*

greenhouse: ~ **effect** *n* AIR, SOIL, WATER POLL effet de serre *m*; ~ **gas** *n* AIR gaz à effet de serre *m*; ~ **gas build-up** *n* AIR formation de gaz à effet de serre *f*; ~ **warming** *n* AIR réchauffement dû à l'effet de serre *m*

Greens: **the** ~ *n pl* POLICY *political party* les Verts *m pl*

greensand *n* WATER SUPP sable vert *m*

grey: ~ **board** *n BrE* WASTE carton gris *m*, carton pour pancartes *m*; ~ **container** *n BrE* WASTE conteneur gris *m*; ~ **filter** *n BrE* ANALYSIS filtre gris *m*; ~ **water** *n BrE* WATER POLL eau grise *f*

grid[1]: ~-**coupled** *adj* RENEW ENERG couplé au réseau; ~-**shaped** *adj* ANALYSIS réticulé

grid[2]: ~ **connection** *n* RENEW ENERG couplage au réseau *m*, raccordement au réseau *m*; ~-**coupled photovoltaic module** *n* RENEW ENERG module photovoltaïque couplé au réseau *m*; ~ **crusher** *n* WASTE concasseur à grille *m*; ~ **point** *n* ANALYSIS point du réseau *m*

Griess: ~ **Saltzmann method** *n* AIR méthode de Griess Saltzmann *f*

grind *vt* ANALYSIS, SOIL broyer, WASTE broyer, meuler, concasser, WATER POLL, WATER SUPP broyer, triturer

grinder *n* WASTE concasseur *m*, WATER POLL meuleuse *f*

grinding *n* ANALYSIS, SOIL broyage *m*, WASTE meulage *m*, concassage *m*, égrenage *m*, *of refuse* broyage *m*, WATER POLL meulage *m*, WATER SUPP broyage *m*

grit *n* AIR *greater than 10*m* grains de sable *m pl*, gravier *m*, poussière *f*, SAFETY poussière *f*, WASTE, WATER POLL gravier *m*, poussière *f*, sable *m*; ~ **additive** WATER SUPP adjuvant de sablage *m*; ~ **can** WATER POLL collecteur de sable *m*; ~ **chamber** WASTE, WATER POLL bassin de dessablement *m*, chambre à sable *f*; ~ **precipitation** AIR précipitation de poussières *f*; ~ **removal** WATER POLL dégravoyage *m*; ~ **scraper** WATER POLL racleur à sable *m*; ~ **washer** WATER POLL installation de lavage du sable *f*, laveur de sable *m*; ~ **washing** WATER POLL lavage du sable *m*

grooved: ~ **mirror** *n* ANALYSIS miroir cannelé *m*

ground[1] *adj* ANALYSIS, SOIL, WASTE, WATER SUPP broyé

ground[2] *n* SOIL sol *m*, WASTE sol *m*, terre *f*, WATER SUPP sol *m*; ~ **coarse solids** *n pl* WATER POLL matière solide grossière *f*; ~ **container** POLICY conteneur enterré *m*; ~-**glass stopper** ANALYSIS *sampling bottle* bouchon de verre dépoli *m*, bouchon émeri *m*; ~ **level concentration** (*GLC*) SOIL concentration au niveau du sol *f*; ~ **level contaminant** SOIL polluant du sol *m*; ~ **stock** AIR, WATER SUPP matière moulue *f*

grounding *n* MARINE échouage *m*, échouement *m*

groundwater *n* ANALYSIS, MANAGE, WATER SUPP eaux souterraines *f pl*; ~ **contamination** SOIL, WASTE, WATER SUPP contamination des eaux souterraines *f*; ~ **extraction tax** POLICY taxe sur le captage d'eaux souterraines *f*; ~ **model** ANALYSIS, MANAGE modèle des eaux souterraines *m*; ~ **monitoring well** ANALYSIS puits de surveillance des eaux souterraines *m*; ~ **pollution** WASTE pollution des eaux souterraines *f*; ~ **protection** WASTE protection des eaux souterraines *f*; ~ **resources** *n pl* WATER SUPP ressources en eaux souterraines *f pl*; ~ **risk** WATER SUPP risque pour la nappe phréatique *m*, risque pour les eaux souterraines *m*; ~ **withdrawal charge** POLICY redevance sur le prélèvement d'eaux souterraines *f*; ~ **withdrawal tax** POLICY redevance sur le prélèvement d'eaux souterraines *f*

group: ~ **clarification plant** *n* WATER POLL usine de clarification collective *f*; ~ **collective dose** *n* AIR, ANALYSIS dose collective *f*, équivalent de dose collective *m*; ~ **parameter** *n* ANALYSIS paramètre de groupe *m*

grouting: ~ **material** *n* SOIL matériau d'injection *m*; ~ **pressure, injection pressure** *n* SOIL pression d'injection *f*; ~ **section** *n* SOIL section d'injection *f*

GSC *abbr* (*gas-solid chromatography*) ANALYSIS CGS (*chromatographie gaz-solide*)

guarantee: ~ **cap** *n* WASTE bouchon de sécurité *m*; ~ **closure** *n* WASTE fermeture de sécurité *f*

guide: ~ **vane** *n* AIR, RENEW ENERG, WATER SUPP *on turbines* aube directrice *f*; ~ **wall** *n* WATER POLL guideau *m*

guideline *n* POLICY directive *f*

guillotine *n* WASTE, WATER SUPP cisaille à guillotine *f*

gulley: ~ **sucker** *n* MARINE aspirateur à goulet *m*

gully *n* WATER POLL caniveau *m*; ~ **erosion** SOIL ravinement *m*; ~ **hole** WATER SUPP bouche d'égout *f*

gummed: ~ **edge** *n* WASTE bord gommé *m*; ~ **label** *n* WASTE étiquette gommée *f*; ~ **paper** *n* WASTE papier gommé *m*

gutter *n* WATER POLL caniveau *m*, WATER SUPP rigole *f*

Gy *abbr* (*gray*) ANALYSIS Gy (*gray*)

gypsum *n* AIR, SOIL *binding agent* gypse *m*; ~ **crystal** AIR cristal de gypse *m*; ~ **dewatering** AIR déshydratation du gypse *f*; ~ **drying** AIR séchage du gypse *m*; ~ **formation** AIR formation de gypse *f*; ~ **generation** AIR production de gypse *f*; ~ **particle** AIR particule de gypse *f*; ~ **plaster board** NOISE *example of single wall* placoplâtre *m*; ~ **separation** AIR extraction du gypse *f*; ~ **silo** AIR silo de gypse *m*; ~ **suspension** AIR gypse en suspension *m*; ~ **yield** AIR production de gypse *f*

H

H *abbr* (*hydrogen*) AIR, WASTE, WATER POLL H (*hydrogène*)
habitat *n* POLICY habitat *m*
habituation *n* AIR, WATER POLL accoutumance *f*
hair: ~ **hygrometer** *n* ANALYSIS hygromètre à cheveu *m*, hygromètre capillaire *m*
half: ~-**hourly average value** *n* ANALYSIS moyenne semi-horaire *f*; ~-**hourly mean value** *n* ANALYSIS moyenne semi-horaire *f*; ~-**width value** *n* ANALYSIS demi-largeur *f*, largeur à mi-hauteur *f*, largeur de bande effective *f*
halide *n* AIR haloïde *m*, halogénure *m*
halocline *n* WATER POLL halocline *m*
haloform: ~ **compounds** *n pl* WATER POLL composés haloformes *m pl*
halogen: ~-**free organic solvent** *n* WATER SUPP solvant organique non halogéné *m*
halogenated: ~ **gas** *n* AIR gaz halogéné *m*; ~ **hydrocarbon** *n* AIR, WATER SUPP hydrocarbure halogéné *m*; ~ **organic solvent** *n* WATER SUPP solvant organique halogéné *m*
halon *n* AIR halon *m*
hammermill *n* WATER SUPP broyeur à marteaux *m*; ~ **crusher** WATER SUPP broyeur-concasseur à marteaux *m*
hand[1]: ~-**driven** *adj* ANALYSIS à commande manuelle, manuel; ~-**held** *adj* ANALYSIS portatif; ~-**operated** *adj* ANALYSIS à commande manuelle, manuel
hand[2]: ~ **augering** *n* SOIL forage à la tarière *m*, sondage-tarière *m*; ~-**held centrifuge** *n* ANALYSIS centrifugeuse portative *f*; ~-**held spectroscope** *n* ANALYSIS spectroscope portatif *m*; ~ **scanner** *n* AIR, WATER SUPP explorateur manuel *m*, scanner manuel *m*; ~ **sorting** *n* WASTE tri manuel *m*
handbook *n* ANALYSIS manuel *m*
handle: ~ **with care** *phr* POLICY *on parcels and packets* fragile
handleability *n* ANALYSIS *of detectors*, MARINE, NOISE, WASTE maniabilité *f*
handling: ~ **of returnables** *n* WASTE manutention des emballages consignés *f*
hangtag *n* WASTE étiquette volante *f*
hard: ~ **coal** *n* AIR anthracite *f*; ~ **detergent** *n* WATER POLL détergent dur *m*; ~ **water** *n* WATER SUPP eau dure *f*, eau calcaire *f*
hardboard *n* WASTE carton fort *m*
hardness *n* WATER SUPP titre hydrotimétrique *m*; ~ **detection** ANALYSIS détection de dureté *f*; ~ **determination** ANALYSIS détermination de la dureté *f*, WATER SUPP hydrotimétrie *f*
harm *vt* GEN porter atteinte à, endommager, nuire à
harmed *adj* GEN endommagé
harmful: ~ **agent** *n* WATER POLL agent nocif *m*; ~ **effect** *n* AIR, SAFETY, WATER SUPP effet nocif *m*, effet nuisible *m*, effet préjudiciable *m*; ~ **emissions** *n pl* AIR émissions nocives *f pl*; ~ **gas** *n* AIR gaz nocif *m*; ~ **gas adsorption** *n* AIR adsorption de gaz nocif *f*; ~ **substance** *n* SAFETY substance nocive *f*, substance dangereuse *f*; ~ **substance concentration** *n* WATER POLL concentration de substances nocives *f*; ~ **substance register** *n* POLICY registre des substances dangereuses *m*; ~ **substances analysis** *n* ANALYSIS analyse de substances dangereuses *f*
harmfulness *n* SAFETY nocivité *f*
harmlessness *n* SOIL, WATER SUPP innocuité *f*
harmonization *n* POLICY *of environmental protection* harmonisation *f*
harmonize *vt* POLICY harmoniser
harmonized[1] *adj* POLICY harmonisé
harmonized[2]: ~ **environmental protection** *n* POLICY harmonisation de la protection de l'environnement *f*
harness *vt* RENEW ENERG *resources* aménager, exploiter, mettre en valeur, mobiliser
harsh: ~ **weather** *n* AIR, WATER POLL intempéries *f pl*
harvest *vt* RENEW ENERG moissonner, récolter
harvestable: ~ **wood** *n* RENEW ENERG bois récoltable *m*
harvester *n* RENEW ENERG moissonneuse *f*, moissonneuse-batteuse *f*, planteuse *f*, récolteuse *f*
harvesting *n* RENEW ENERG coupe *f*, récolte *f*
HAWT *abbr* (*horizontal axis wind turbine*) RENEW ENERG éolienne à axe horizontal *f*
hazard *n* SAFETY danger potentiel *m*, risque *m*; ~ **prevention** MANAGE, SAFETY prévention des risques *f*
hazardous[1] *adj* SAFETY dangereux
hazardous[2]: ~ **goods** *n pl* SAFETY substances dangereuses *f pl*; ~ **liquid waste** *n* WASTE, WATER POLL effluent liquide dangereux *m*; ~ **substance** *n* SAFETY substance dangereuse *f*; ~ **waste** *n* WASTE déchets dangereux *m pl*; ~ **waste avoidance** *n* WASTE mesure de lutte contre les déchets dangereux *f*; ~ **waste collection** *n* WASTE collecte des déchets dangereux *f*; ~ **waste concentrate** *n* WASTE concentré de déchets dangereux *m*; ~ **waste incineration** *n* WASTE incinération de déchets dangereux *f*; ~ **waste incineration plant** *n* WASTE usine d'incinération des déchets dangereux *f*; ~ **waste landfill** *n* WASTE décharge de déchets dangereux *f*; ~ **waste landfilling** *n* WASTE mise en décharge de déchets dangereux *f*; ~ **waste treatment** *n* WASTE traitement des déchets dangereux *m*; ~ **wastes** *n pl* WASTE déchets dangereux *m pl*
Hazard-Ranking: ~ **System** *n* (*HRS*) SAFETY système de classification des risques *m*
haze *n* AIR brume *f*, voile *m*, WASTE brume *f*; ~ **canopy** AIR calotte de brume *f*
Hazen: ~ **number** *n* WATER POLL nombre Hazen *m*
HC *abbr* (*hydrocarbon*) AIR, MARINE, WATER SUPP hydrocarbure *m*
HCB *abbr* (*hexachlorobenzene*) SOIL HCB (*hexachlorobenzène*)
HCFC *abbr* (*hydrochlorofluorocarbon*) AIR HCFC (*hydrochlorofluorocarbure*)
HCH *abbr* (*hexachlorocyclohexane*) SOIL HCH (*hexachlorocyclohexane*)
HDR *abbr* (*hot dry rock*) RENEW ENERG roches chaudes et sèches *f pl*
He *abbr* (*helium*) AIR He (*hélium*)
head *n* AIR *of reactor* hauteur manométrique *f*, tête *f*; ~ **loss** AIR *filter bag* perte de charge *f*, perte de pression *f*, WATER POLL perte de charge *f*, perte de pression *f*; ~ **pipe** WATER SUPP *of pump* tubulure de refoulement *f*; ~ **plate** AIR plateau de tête *m*
headphones *n pl* NOISE écouteurs *m pl*, casque *m*
headspace: ~ **analysis** *n* ANALYSIS analyse volumétrique *f*; ~ **gas chromatography** *n* (*HSGC*) ANALYSIS chromatographie volumétrique en phase gazeuse *f*, chromatographie de tête en phase gazeuse *f*; ~ **sampler** *n* ANALYSIS échantillonneur de tête *m*
health[1]: ~-**related** *adj* SAFETY sanitaire
health[2] *n* SAFETY hygiène *f*, santé *f*; ~ **hazard** SAFETY risque sanitaire *m*; ~ **impact** POLICY impact sanitaire *m*; ~-**related sampling** SAFETY prélèvement sanitaire *m*; ~ **risk** SAFETY risque sanitaire *m*; ~ **and safety requirements** *n pl* SAFETY

conditions d'hygiène et de sécurité *f pl*; ~ **survey** POLICY enquête sanitaire *f*

hearing: ~ **damage** *n* NOISE lésion auditive *f*; ~ **date** *n* POLICY date d'audition *f*; ~ **protection** *n* NOISE protection auditive *f*; ~-**protection padding** *n* NOISE rembourrage de protection auditive *m*; ~-**protection plugs** *n pl* NOISE bouchons de protection auditive *m pl*

heat[1]: ~-**insulating** *adj* SAFETY, WASTE réfractaire, calorifuge, résistant à la chaleur; ~-**proof** *adj* SAFETY, WASTE réfractaire, calorifuge, résistant à la chaleur; ~-**resistant** *adj* SAFETY, WASTE réfractaire, calorifuge, résistant à la chaleur

heat[2] *n* GEN chaleur *f*; ~ **accumulator** AIR accumulateur de chaleur *m*; ~ **balance** AIR, WATER POLL bilan thermique *m*; ~ **capacity** AIR capacité calorifique *f*; ~ **carrier medium** RENEW ENERG caloporteur *m*; ~ **conduction** AIR, ANALYSIS conduction thermique *f*; ~ **conductivity** AIR, ANALYSIS conductibilité thermique *f*, conductivité thermique *f*; ~ **content** RENEW ENERG, WATER SUPP charge thermique *f*; ~ **convection** AIR convection thermique *f*; ~ **drying** WATER POLL *sludge disposal* séchage à chaud *m*; ~ **emissions** *n pl* AIR émissions thermiques *f pl*; ~ **exchange** AIR, RENEW ENERG, WASTE échange thermique *m*; ~ **exchanger** AIR, RENEW ENERG, WASTE échangeur de chaleur *m*; ~-**exchanger efficiency** AIR, RENEW ENERG, WASTE rendement d'échangeur de chaleur *m*; ~ **exchanger material** AIR, RENEW ENERG, WASTE matériau échangeur de chaleur *m*; ~ **flow** AIR flux de chaleur *m*, flux thermique *m*; ~ **generation** RENEW ENERG, WATER SUPP production de chaleur *f*; ~ **insulation** SAFETY, WASTE calorifugeage *m*, isolation calorifuge *f*; ~-**load plan** AIR plan de charge thermique *m*; ~ **loss** WATER POLL *due to incomplete combustion* perte thermique *f*; ~ **pollution** AIR, WATER SUPP pollution thermique *f*; ~ **pump** RENEW ENERG pompe à chaleur *f*; ~ **pump plant** RENEW ENERG installation de pompe à chaleur *f*; ~ **radiation** RENEW ENERG rayonnement thermique *m*; ~ **radiation loss** WASTE perte thermique par dissipation *f*; ~ **of reaction** ANALYSIS chaleur de réaction *f*; ~ **recovery** AIR, WASTE, WATER SUPP récupération de chaleur *f*; ~-**seal coating** WASTE revêtement thermoscellé *m*; ~-**seal label** WASTE étiquette thermocollée *f*; ~-**seal temperature** WASTE température de thermoscellage *f*; ~-**sealable paper** WASTE papier thermocollable *m*; ~ **sealing** WASTE thermocollage *m*, thermosoudage *m*; ~-**sealing adhesive** WASTE adhésif thermocollant *m*; ~-**sealing equipment** WASTE matériel de thermocollage *m*; ~-**sealing machine** WASTE thermoscelleuse *f*; ~-**sealing tape** WASTE ruban thermoadhésif *m*; ~-**sensitive material** WASTE matériau thermosensible *m*; ~ **source** RENEW ENERG source thermique *f*; ~ **stability** WASTE, WATER SUPP stabilité thermique *f*; ~ **storage** RENEW ENERG stockage de chaleur *m*; ~ **transfer** WATER POLL, WATER SUPP transfert de chaleur *m*, transfert thermique *m*; ~ **transfer label** WASTE étiquette imprimée par transfert thermique *f*; ~ **transfer medium** RENEW ENERG *solar collector*, WATER POLL caloporteur *m*; ~ **transfer medium liquid coolant** RENEW ENERG fluide caloporteur *m*; ~ **transfer via conduction** AIR transfert thermique par conduction *m*; ~ **transfer via convection** AIR transfert thermique par convection *m*; ~ **transition** AIR transition thermique *f*; ~ **use** AIR consommation de chaleur *f*, WATER POLL *from stabilized sludge, exhaust gas* valorisation des rejets thermiques *f*; ~ **welding** WASTE thermosoudage *m*

heat[3] *vt* GEN chauffer; ~ **up** GEN chauffer, réchauffer

heated[1] *adj* GEN chauffé

heated[2]: ~ **column oven** *n* ANALYSIS four à colonne chauffante *m*; ~ **digester** *n* WATER POLL digesteur chauffé *m*; ~ **injector** *n* ANALYSIS *liquid or gas sampling* injecteur chauffant *m*; ~ **kiln** *n* ANALYSIS four chauffé *m*; ~ **waste air** *n* AIR rejet atmosphérique chauffé *m*

heating *n* GEN chauffage *m*; ~ **channel** WASTE conduite de chauffage *f*; ~ **cylinder** WATER POLL pot de chauffage *m*; ~ **installation** RENEW ENERG chaufferie *f*, installation de chauffage *f*; ~ **purpose** RENEW ENERG application de chauffage *f*; ~ **system** RENEW ENERG circuit de chauffage *m*, système de chauffage *m*; ~ **tunnel** WASTE four tunnel *m*; ~ **up** AIR, RENEW ENERG, SOIL, WASTE, WATER POLL chauffage *m*; ~ **zone** AIR zone de chauffe *f*

heavily: ~-**colored wastewater** *AmE*, ~-**coloured wastewater** *BrE* *n* WATER POLL eaux usées fortement colorées *f pl*; ~-**dyed wastewater** *n* WATER POLL eaux usées fortement teintées *f pl*

heavy: ~-**duty corrugated fiberboard** *AmE*, ~-**duty corrugated fibreboard** *BrE* *n* WASTE carton ondulé très résistant *m*; ~ **fuel** *n* AIR huile lourde *f*; ~-**medium container** *n* WASTE conteneur de liquide dense *m*; ~-**medium separation** *n* WASTE séparation par liquide dense *f*; ~ **metal** *n* AIR, WASTE, WATER POLL métal lourd *m*; ~-**metal analysis** *n* ANALYSIS analyse des métaux lourds *f*; ~-**metal concentration** *n* AIR, SOIL concentration de métaux lourds *f*; ~-**metal content** *n* AIR, SOIL teneur en métaux lourds *f*; ~-**metal determination** *n* ANALYSIS détermination des métaux lourds *f*, dosage des métaux lourds *m*; ~-**metal emissions** *n pl* AIR émissions de métaux lourds *f pl*; ~ **metals accumulation** *n* AIR, SOIL accumulation de métaux lourds *f*; ~ **metals circulation** *n* AIR, SOIL circulation de métaux lourds *f*; ~ **metals deposition** *n* AIR, SOIL dépôt de métaux lourds *m*; ~ **metals precipitation** *n* WATER POLL précipitation des métaux lourds *f*; ~ **metals removal** *n* AIR, WATER POLL, WATER SUPP extraction des métaux lourds *f*; ~ **metals source** *n* AIR, SOIL source de métaux lourds *f*; ~ **oil desulfurization** *AmE*, ~ **oil desulphurization** *BrE* *n* WASTE, WATER SUPP désulfuration des huiles lourdes *f*; ~ **water** *n* WATER POLL eau lourde *f*

height: ~ **of fall** *n* WASTE hauteur de chute *f*

helical *adj* AIR hélicoïdal, spiral

heliostat *n* RENEW ENERG héliostat *m*

helium *n* (*He*) AIR hélium *m* (*He*); ~-**neon laser** ANALYSIS laser hélium-néon *m*

helmet *n* SAFETY casque *m*

Helmholtz: ~ **resonator** *n* NOISE résonateur de Helmholtz *m*

hemp: ~ **rope** *n* WASTE corde en chanvre *f*

Henry-Dalton: ~ **law** *n* WATER SUPP loi de Henry et de Dalton *f*

Henry: ~'**s law** *n* AIR, SOIL, WATER POLL loi de Henry *f*

heptane *n* WASTE *landfill gas component* heptane *m*

herbicide *n* SOIL herbicide *m*

herder: ~ **effect** *n* MARINE effet repousseur *m*

herding: ~ **agent** *n* MARINE, WATER SUPP agent repousseur *m*

hermetic: ~ **closure** *n* WASTE fermeture hermétique *f*; ~ **seal** *n* WASTE fermeture hermétique *f*

heterogeneous: ~ **catalyst** *n* AIR catalyseur hétérogène *m*

heterotroph *n* WASTE, WATER POLL hétérotrophe *m*, organisme hétérotrophe *m*

heterotrophic[1] *adj* WASTE, WATER POLL hétérotrophe

heterotrophic[2]: ~ **bacteria** *n pl* WATER POLL bactéries hétérotrophes *f pl*; ~ **denitrification** *n* WATER POLL dénitrification hétérotrophe *f*

hexachlorobenzene *n* (*HCB*) SOIL hexachlorobenzène *m* (*HCB*)

hexachlorocyclohexane *n* (*HCH*) SOIL hexachlorocyclohexane *m* (*HCH*)

hexane *n* AIR hexane *m*

Hg *abbr* (*mercury*) WATER POLL Hg (*mercure*)

hidden: ~ **bar code identification** *n* WASTE identification par code barre caché *f*

high[1]: ~-**grade** *adj* WATER SUPP de qualité, de haute qualité; ~-**performance** *adj* AIR, ANALYSIS, WATER SUPP de haute performance; ~-**pressure** *adj* AIR, ANALYSIS, MARINE, WATER POLL à forte pression, à haute pression, sous haute pression; ~-**rate** *adj* WATER POLL à forte intensité; ~-**temperature** *adj*

RENEW ENERG, SOIL, WASTE, WATER SUPP à haute température, à température élevée

high[2]: **~-absorbing noise-protection system** *n* WATER SUPP système de protection contre le bruit fortement absorbant *m*; **~ barrier** *n* WATER SUPP écran renforcé *m*; **~ calorific value** *n* WATER POLL fort pouvoir calorifique *m*; **~ concentration** *n* AIR, ANALYSIS, WATER SUPP forte concentration *f*; **~-density housing** *n* POLICY grand ensemble *m*; **~-density polyethylene** *n* WASTE polyéthylène haute densité *m* (*polyéthylène HD*); **~-efficiency cyclone** *n* AIR cyclone à haut rendement *m*; **~-gloss foil** *n* WASTE pellicule brillante *f*; **~-gloss paper** *n* WASTE papier brillant *m*; **~-intensity magnet separator** *n* SOIL séparateur magnétique de grande puissance *m*, déferreur magnétique de grande puissance *m*; **~-level landfill** *n* WASTE décharge de haute activité *f*; **~-level radioactive waste** *n* AIR, WASTE, WATER POLL déchets radioactifs de haute activité *m pl*; **~-performance cyclone** *n* AIR cyclone haute performance *m*; **~-performance dust removal** *n* AIR dépoussiérage haute performance *m*; **~-performance filter** *n* AIR, WATER SUPP filtre haute performance *m*; **~-performance filtration** *n* AIR, WATER SUPP filtration à haut rendement *f*, filtration haute performance *f*; **~-performance fixed-bed reactor** *n* WATER SUPP réacteur à lit fixe à haut rendement *m*, réacteur à lit fixe haute performance *m*; **~-performance liquid chromatography** *n* (*HPLC*) ANALYSIS chromatographie en phase liquide haute performance *f* (*CLHP*), chromatographie liquide à haute performance *f* (*CLHP*); **~-performance liquid chromatography-mass spectrometry** *n* (*HPLC-MS*) ANALYSIS spectrométrie de masse-chromatographie en phase liquide haute performance *f* (*SM-CLHP*); **~-performance precipitator** *n* AIR dépoussiéreur haute performance *m*, précipitateur de grand rendement *m*; **~-performance scrubber** *n* AIR laveur haute performance *m*; **~-pressure cleaning** *n* AIR lavage haute pression *m*; **~-pressure filter** *n* WATER POLL filtre haute pression *m*; **~-pressure flushing** *n* MARINE chasse haute pression *f*; **~-pressure pump** *n* WATER POLL pompe haute pression *f*; **~-pressure radial flow compressor** *n* SOIL *for vapour extraction* compresseur radial haute pression *m*; **~-pressure reactor** *n* RENEW ENERG *biomass conversion* réacteur haute pression *m*; **~-pressure side** *n* AIR branche haute pression *f*, côté haute pression *m*; **~-pressure spray nozzle** *n* AIR buse de pulvérisation haute pression *f*; **~-pressure thin-layer chromatography** *n* ANALYSIS chromatographie en couche mince sous haute pression *f*, chromatographie HP en couche mince *f*; **~-pressure Venturi scrubber** *n* AIR laveur à Venturi haute pression *m*; **~-pressure water blasting** *n* MARINE nettoyage au jet d'eau haute pression *m*; **~-pressure water jet** *n* SOIL canon d'eau haute pression *m*; **~-pressure water jet method** *n* SOIL projection d'eau haute pression *f*; **~-pressure water jet pipe** *n* SOIL tuyau de projection d'eau haute pression *m*; **~-pressure wet oxidation** *n* WATER POLL oxydation par voie humide sous haute pression *f*; **~-rate activated sludge tank** *n* WATER POLL réservoir de boues activées à forte charge *m*; **~-rate coalescence oil separator** *n* WATER POLL séparateur d'huile de coalescence à forte charge *m*; **~-rate filtration** *n* WATER POLL filtration à forte charge *f*; **~-rate process** *n* WATER POLL *activated sludge* procédé à forte charge *m*; **~-rate trickling filter** *n* WATER POLL lit bactérien à forte charge *m*; **~ raw gas load** *n* WASTE forte charge en gaz brut *f*; **~ security** *n* SAFETY *of site* haute sécurité *f*, grande sécurité *f*; **~-security landfill** *n* WASTE décharge sous surveillance renforcée *f*; **~-settled apparent density** *n* ANALYSIS forte masse volumique apparente après décantation *f*; **~-speed inspection** *n* WASTE, WATER SUPP inspection rapide *f*; **~-speed mixer** *n* WASTE mélangeur rapide *m*; **~-temperature gasification** *n* WASTE, WATER POLL, WATER SUPP gazéification à haute température *f*; **~-temperature heat** *n* RENEW ENERG *from solar energy* chaleur haute température *f*; **~-temperature incineration** *n* AIR, WASTE, WATER POLL incinération à haute température *f*; **~-temperature plasma process** *n* SOIL procédé plasmatique haute température *m*; **~-temperature-resistant oil** *n* RENEW ENERG hydrocarbure résistant aux hautes températures *m*; **~ tide** *n* MARINE marée haute *f*; **~-tide water** *n* MARINE marée haute *f*; **~-velocity gas stream** *n* AIR flux de gaz à grande vitesse *m*; **~-voltage aggregate** *n* AIR groupe haute tension *m*; **~-voltage discharge electrode** *n* AIR, WATER SUPP électrode de décharge haute tension *f*; **~-voltage electrostatic precipitator** *n* AIR électrofiltre haute tension *m*, dépoussiéreur électrostatique haute tension *m*; **~-voltage generator** *n* AIR, RENEW ENERG alternateur haute tension *m*; **~-voltage line** *n* RENEW ENERG ligne haute tension *f*; **~-volume sampler** *n* ANALYSIS échantillonneur à grand volume *m*; **~ water** *n* MARINE marée haute *f*, WATER SUPP haute mer *f*, *river* hautes eaux *f pl*, *tidal* marée haute *f*, plein de l'eau *m*, haut de l'eau *m*

highly[1]: **~-absorbable** *adj* AIR, NOISE fortement absorbable; **~-absorbent** *adj* AIR, NOISE très absorbant; **~-contaminated** *adj* SOIL, WASTE, WATER POLL fortement contaminé; **~-polluted** *adj* SOIL, WASTE, WATER POLL fortement pollué; **~-radioactive** *adj* RENEW ENERG fortement radioactif; **~-sensitive** *adj* ANALYSIS très sensible; **~-toxic** *adj* SOIL, WASTE, WATER POLL hautement toxique, très toxique

highly[2]: **~-contaminated municipal wastewater** *n* WATER POLL eaux usées municipales fortement contaminées *f pl*; **~-polluted faecal solids sludge** *BrE*, **~-polluted fecal solids sludge** *AmE* *n* WATER POLL boues de matières fécales fortement polluées *f pl*; **~-polluted wastewater** *n pl* WATER POLL eaux usées fortement polluées *f pl*; **~-polluted waters** *n pl* WATER POLL eaux fortement polluées *f pl*; **~-sensitive detector** *n* ANALYSIS détecteur de grande sensibilité *m*; **~-sensitive scale** *n* ANALYSIS échelle très sensible *f*; **~-sensitive test** *n* ANALYSIS essai très sensible *m*; **~-toxic hazardous waste** *n* WASTE, WATER POLL déchets dangereux très toxiques *m pl*

hindered: **~ settling** *n* WATER POLL décantation retardée *f*

hindrance *n* SAFETY obstacle *m*

hinged: **~ lid** *n* WASTE, WATER SUPP couvercle à charnière *m*

historical: **~ data** *n pl* WATER SUPP données antérieures *f pl*, données historiques *f pl*

HMV *abbr* (*hourly mean value*) ANALYSIS moyenne horaire *f*

HNO_3 *abbr* (*nitric acid*) AIR, SOIL, WATER POLL HNO_3 (*acide nitrique*)

hoist *n* SAFETY engin de levage *m*, treuil *m*

holding *n* AIR rétention *f*; **~ capacity** WASTE capacité de rétention *f*

hollow: **~ cylinder** *n* AIR *rotating adsorber* cylindre creux *m*

home: **~ interview survey** *n* POLICY enquête à domicile *f*

homogeneous[1] *adj* AIR, SOIL, WASTE, WATER SUPP homogène

homogeneous[2]: **~ catalyst** *n* AIR catalyseur homogène *m*

homogenization *n* AIR, SOIL, WASTE, WATER SUPP homogénéisation *f*

homogenize *vt* AIR, SOIL, WASTE, WATER SUPP homogénéiser

homopolar *adj* WATER POLL homopolaire

homopolymer *n* WASTE homopolymère *m*

honeycomb: **~ catalyst** *n* AIR catalyseur poreux *m*; **~ element** *n* AIR *catalyst* élément alvéolaire *m*; **~-shaped channel** *n* AIR *catalyst* conduite alvéolée *f*

hood *n* AIR capot *m*, hotte *f*

hooping *n* WASTE cerclage *m*

hopper *n* WATER POLL trémie *f*; **~-bottom tank** WATER POLL bassin à fond pyramidal *m*; **~-furnace feed chute** WASTE chute de trémie d'alimentation *f*

horizontal: **~ axis wind turbine** *n* (*HAWT*) RENEW ENERG éolienne à axe horizontal *f*; **~ bar screen** *n* WATER POLL grille fixe horizontale *f*; **~ degassing** *n* AIR, WATER SUPP

dégazage horizontal *m*, dégazage tangentiel *m*; ~ **fixed-bed adsorber** *n* AIR, WATER POLL adsorbeur à lit fixe horizontal *m*; ~ **gas drainage** *n* WASTE purge tangentielle des gaz *f*; ~ **gas migration** *n* SOIL, WASTE migration horizontale des gaz *f*; ~ **moving-bed adsorber** *n* AIR, WATER POLL adsorbeur à lit mobile horizontal *m*; ~ **surface installation** *n* WASTE installation de surface horizontale *f*
hose *n* WATER SUPP tuyau *m*
hospital: ~ **waste** *n* WASTE déchets hospitaliers *m pl*
hostile: ~ **environment** *n* AIR, MANAGE, SOIL, WATER POLL milieu hostile *m*
hot: ~**-air radiation heating system** *n* SAFETY chauffage par rayonnement d'air chaud *m*; ~ **dry rock** *n* (*HDR*) RENEW ENERG roches chaudes et sèches *f pl*; ~ **gas** *n* AIR, SOIL gaz chaud *m*; ~**-gas compressor** *n* AIR compresseur de gaz chaud *m*; ~**-gas dust removal** *n* AIR *incinerator gas* dépoussiérage de gaz chauds *m*; ~**-gas filter** *n* AIR filtre à gaz chauds *m*; ~**-gas injection** *n* SOIL injection de gaz chaud *f*; ~ **sampling gas** *n* ANALYSIS gaz chaud d'échantillonnage *m*; ~**-soak loss** *n* AIR déperdition de chaleur du moteur arrêté *f*; ~ **transfer label** *n* WASTE étiquette décalquée à chaud *f*; ~ **tubing** *n* (*ht*) AIR branche chaude *f*; ~ **waste gas** *n* AIR effluent gazeux chaud *m*; ~**-water heater** *n* AIR *thermal combustion* poste d'eau chaude *m*; ~**-water source** *n* RENEW ENERG source d'eau chaude *f*; ~**-water washing** *n* MARINE lavage à l'eau chaude *m*; ~**-wire anemometer** *n* ANALYSIS anémomètre à résistance électrique *m*
hourly: ~ **capacity** *n* WASTE *incineration* capacité horaire *f*; ~ **mean value** *n* (*HMV*) ANALYSIS moyenne horaire *f*; ~ **value** *n* ANALYSIS valeur horaire *f*
household: ~ **refuse** *n* WASTE ordures ménagères *f pl* (*OM*); ~ **waste** *n* WASTE ordures ménagères *f pl* (*OM*); ~ **wastewater** *n* WATER POLL eaux usées ménagères *f pl*
housing *n* RENEW ENERG *solar collector* culasse *f*; ~ **wall** AIR paroi de caisson *f*
HPLC *abbr* (*high-performance liquid chromatography*) ANALYSIS CLHP (*chromatographie liquide à haute performance, chromatographie en phase liquide haute performance*)
HPLC-MS *abbr* (*high-performance liquid chromatography-mass spectrometry*) ANALYSIS SM-CLHP (*spectrométrie de masse-chromatographie en phase liquide haute performance*)
HRS *abbr* (*Hazard-Ranking System*) SAFETY système de classification des risques *m*
HSGC *abbr* (*headspace gas chromatography*) ANALYSIS chromatographie de tête en phase gazeuse *f*, chromatographie volumétrique en phase gazeuse *f*
H_2SO_3 *abbr* (*sulfurous acid AmE, sulphurous acid BrE*) AIR, SOIL, WATER POLL H_2SO_3 (*acide sulfureux*)
H_2SO_4 *abbr* (*sulfuric acid AmE, sulphuric acid BrE*) AIR, SOIL, WATER POLL H_2SO_4 (*acide sulfurique*)
ht *abbr* (*hot tubing*) AIR branche chaude *f*
human[1]: ~**-induced** *adj* AIR, SOIL anthropique, d'origine humaine, imputable aux activités humaines
human[2]: ~**-induced soil degradation** *n* SOIL dégradation des sols imputable aux activités humaines *f*; ~ **pollution** *n* SAFETY pollution due à l'homme *f*; ~ **waste** *n* WASTE pièces anatomiques *f pl*, déchets anatomiques *m pl*
humic: ~ **mass** *n* WATER POLL masse humique *f*; ~ **mass formation** *n* WATER POLL formation d'une masse humique *f*
humidify *vt* AIR humidifier
humidity *n* GEN humidité *f*; ~ **content** ANALYSIS *of gases* teneur en humidité *f*; ~ **content determination** ANALYSIS détermination de la teneur en humidité *f*; ~ **measurement** ANALYSIS mesure de l'humidité *f*, mesure hygrométrique *f*
humification *n* WATER POLL humification *f*
humus *n* WATER POLL, WATER SUPP humus *m*; ~**-containing soil** SOIL sol humifère *m*; ~ **formation** WATER POLL, WATER SUPP formation d'humus *f*, humification *f*; ~ **sludge** WATER POLL boue humique *f*; ~ **tank** WATER POLL, WATER SUPP bac à humus *m*
hundred: ~ **points calibration** *n* ANALYSIS étalonnage au centième *m*
hunting: ~ **reserve** *n* POLICY réserve de chasse *f*
Huygen-Fresnel: ~ **principle** *n* ANALYSIS principe de Huygens-Fresnel *m*
hybrid: ~ **scrubber** *n* AIR laveur mixte *m*
hydrant *n* WATER SUPP poste d'incendie *m*
hydraulic[1] *adj* GEN hydraulique
hydraulic[2]: ~ **classifier** *n* SOIL classificateur hydraulique *m*; ~ **cylinder** *n* WASTE cylindre hydraulique *m*; ~ **fluid** *n* RENEW ENERG fluide hydraulique *m*; ~ **jump** *n* WATER SUPP ressaut *m*; ~ **loading** *n* WATER POLL charge hydraulique *f*; ~ **loss** *n* WATER POLL, WATER SUPP perte hydraulique *f*; ~ **pressure** *n* WATER POLL pression hydraulique *f*; ~ **ram** *n* WASTE piston hydraulique *m*; ~ **transmission** *n* RENEW ENERG transmission hydraulique *f*
hydraulically[1]: ~**-controlled** *adj* RENEW ENERG, WATER POLL, WATER SUPP à commande hydraulique; ~**-operated** *adj* RENEW ENERG, WATER POLL, WATER SUPP à fonctionnement hydraulique
hydraulically[2]: ~**-controlled flap** *n* AIR clapet à commande hydraulique *m*; ~**-mixed flocculator** *n* WATER POLL floculateur à mélange hydraulique *m*
hydrobiological: ~ **equilibrium** *n* POLICY équilibre hydrobiologique *m*, WATER POLL bilan hydrobiologique *m*
hydrocarbon *n* (*HC*) AIR, MARINE, WATER SUPP hydrocarbure *m*; ~ **catalyst** AIR catalyseur d'hydrocarbures *m*; ~ **emissions** *n pl* AIR émissions d'hydrocarbures; ~ **slick** MARINE, WATER SUPP nappe d'hydrocarbures *f*
hydrochlorofluorocarbon *n* (*HCFC*) AIR hydrochlorofluorocarbure *m* (*HCFC*)
hydrocyclone *n* AIR, SOIL, WASTE, WATER POLL hydrocyclone *m*; ~ **battery** AIR batterie d'hydrocyclones *f*
hydrodesulfurization *AmE see hydrodesulphurization BrE*
hydrodesulphurization *n BrE* WATER SUPP hydrodésulfuration *f*
hydrodynamic[1] *adj* SOIL hydrodynamique
hydrodynamic[2]: ~ **behavior in subsoil** *AmE*, ~ **behaviour in subsoil** *BrE n* SOIL comportement hydrodynamique dans le sous-sol *m*
hydroelectric: ~ **dam** *n* RENEW ENERG barrage hydroélectrique *m*; ~ **plant** *n* RENEW ENERG, WATER SUPP centrale hydroélectrique *f*, usine hydroélectrique *f*
hydroelectricity *n* RENEW ENERG hydroélectricité
hydrogen *n* (*H*) AIR, WASTE *landfill gas component*, WATER SUPP hydrogène *m* (*H*); ~ **fluoride** AIR fluorure d'hydrogène *m*; ~ **peroxide** AIR *liquid oxidant*, WATER POLL *wastewater disinfection* eau oxygénée *f*, péroxyde d'hydrogène *m*; ~ **sulfide** *AmE*, ~ **sulphide** *BrE* AIR, WASTE, WATER SUPP acide sulfhydrique *m*, hydrogène sulfuré *m*, sulfure d'hydrogène *m*
hydrogenate *vt* AIR, RENEW ENERG, WASTE, WATER SUPP hydrogéner
hydrogenation *n* AIR, RENEW ENERG, WASTE, WATER SUPP hydrogénation *f*
hydrogeological: ~ **analysis** *n* ANALYSIS, SOIL, WATER SUPP analyse hydrogéologique *f*
hydrogeology *n* SOIL, WATER SUPP hydrogéologie *f*
hydrograph *n* WATER SUPP hydrogramme *m*
hydrokinetic *adj* WATER POLL hydrocinétique
hydrolift: ~ **reactor** *n* AIR réacteur à écoulement de type piston *m*
hydrolizing: ~ **tank** *n* AIR cuve à hydrolyse *f*
hydrologic: ~ **cycle** *n* WATER SUPP cycle hydrologique *m*
hydrological: ~ **balance** *n* WATER SUPP bilan hydrologique *m*
hydrology *n* WATER POLL, WATER SUPP hydrologie *f*
hydrolyse *vt BrE* GEN hydrolyser
hydrolysed[1] *adj BrE* GEN hydrolysé

hydrolysed[2]: ~ **raw sludge** *n BrE* WATER POLL boues brutes hydrolysées *f pl*
hydrolysis *n* GEN hydrolyse *f*
hydrolyze *AmE see hydrolyse BrE*
hydrolyzed *AmE see hydrolysed BrE*
hydrometer *n* WATER POLL, WATER SUPP hydromètre *m*
hydrometry *n* WATER POLL, WATER SUPP hydrométrie *f*
hydrophilic *adj* AIR, WATER POLL hydrophile
hydrophobic *adj* AIR, WATER POLL hydrophobe
hydropower *n* RENEW ENERG hydroélectricité *f*; ~ **generation** RENEW ENERG production d'énergie hydraulique *f*; ~ **plant** RENEW ENERG usine hydroélectrique *f*; ~ **resources** *n pl* RENEW ENERG ressources hydroélectriques *f pl*; ~ **supply** RENEW ENERG fourniture d'hydroélectricité *f*
hydroscience *n* WATER SUPP science de l'eau *f*
hydrosphere *n* SOIL, WATER SUPP hydrosphère *f*
hydrostatic[1] *adj* WATER POLL, WATER SUPP hydrostatique
hydrostatic[2]: ~ **barrier** *n* WATER POLL, WATER SUPP écran hydrostatique *m*; ~ **pressure** *n* WATER POLL, WATER SUPP pression hydrostatique *f*; ~ **pressure difference** *n* WATER POLL, WATER SUPP gradient de pression hydrostatique *m*
hydrostatics *n* WATER POLL, WATER SUPP hydrostatique *f*
hydrothermal[1] *adj* RENEW ENERG hydrothermique
hydrothermal[2]: ~ **vent** *n* MARINE source hydrothermale souterraine *f*
hydroxide: ~ **sludge** *n* WATER POLL boues d'hydroxyde *f pl*
hygienic *adj* SAFETY, WATER SUPP hygiénique
hygienist *n* MANAGE, WATER POLL hygiéniste *m*
hygrometer *n* GEN hygromètre *m*
hygrometry *n* GEN hygrométrie *f*
hygroscope *n* AIR, ANALYSIS hygroscope *m*
hygroscopic[1] *adj* AIR, ANALYSIS hygroscopique
hygroscopic[2]: ~ **water** *n* WATER SUPP eau hygroscopique *f*
hygroscopicity *n* AIR, ANALYSIS hygroscopicité *f*
hypersonic *n* NOISE hypersonique; ~ **range** NOISE plage hypersonique *f*
hypoacusis *n* NOISE hypoacousie *f*
hypochlorination *n* WATER POLL hypochloration *f*
hypolimnion *n* WATER POLL hypolimnion *m*

I

IC *abbr* (*ion-exchange chromatography*) ANALYSIS chromatographie d'échange d'ions *f*
ICP-MS *abbr* (*inductively-coupled plasma-mass spectrometry*) ANALYSIS spectrométrie de masse de plasma avec couplage inductif *f*
ICP-OES *abbr* (*inductively-coupled plasma-optical emission spectrometry*) ANALYSIS spectrométrie d'émission optique de plasma avec couplage inductif *f*
ID[1] *abbr* (*induced draft AmE, induced draught BrE*) AIR tirage forcé *m*
ID[2]: **~ fan** *n* AIR, WASTE ventilateur à tirage forcé *m*
ideal: **~ gas** *n* AIR gaz parfait *m*
identification *n* ANALYSIS, SAFETY *of hazards* identification *f*; **~ of contents** SAFETY, WASTE *of industrial gas containers* identification de contenu *f*, spécification de contenu *f*
identify *vt* ANALYSIS, SAFETY identifier
IE *abbr* (*ion exchange*) ANALYSIS, WATER POLL, WATER SUPP échange d'ions *m*, échange ionique *m*
ignitability *n* AIR, SAFETY, WASTE inflammabilité *f*
ignitable *adj* AIR, SAFETY, WASTE inflammable
ignite *vt* AIR allumer, ANALYSIS *plasma* enflammer, WASTE, WATER POLL allumer, enflammer
ignition: **~ gas** *n* AIR *thermal combustion* gaz d'allumage *m*; **~ loss** *n* AIR perte à l'allumage *f*, WATER POLL perte au feu *f*, WATER SUPP perte au feu *f*, perte par calcination *f*; **~ point** *n* AIR point d'inflammation *m*; **~ residue** *n* ANALYSIS résidu de calcination *m*, WATER POLL résidu d'inflammation *m*; **~ temperature** *n* AIR *thermal waste gas treatment* température d'inflammation *f*
illegal[1] *adj* POLICY illicite
illegal[2]: **~ dump site** *n* WASTE site de décharge illicite *m*; **~ dumping** *n* POLICY, WASTE, WATER SUPP dépôt sauvage *m*, rejet illicite *m*, déversement illégal *m*, rejet sauvage *m*; **~ dumping of spent acid** *n* MARINE rejet illicite d'acides usagés *m*; **~ waste disposal** *n* WASTE élimination illégale des déchets *f*, décharge sauvage *f*
image: **~ analysis** *n* ANALYSIS analyse d'image *f*
Imhoff: **~ cone** *n* ANALYSIS cône d'Imhoff *m*; **~ tank** *n* WATER POLL *Emscher tank* fosse Imhoff *f*
immature: **~ market** *n* MANAGE marché immature *m*
immersion: **~ stirrer** *n* ANALYSIS *pH-meter* agitateur à immersion *m*, barboteur *m*
immiscible[1] *adj* WATER POLL non miscible
immiscible[2]: **~ liquid** *n* WATER POLL liquide non-miscible *m*
immobilization *n* ANALYSIS, SOIL, WASTE, WATER POLL immobilisation *f*
immobilize *vt* ANALYSIS, SOIL, WASTE, WATER POLL immobiliser
immobilized[1] *adj* ANALYSIS, SOIL, WASTE, WATER POLL immobilisé
immobilized[2]: **~ microorganism** *n* AIR, SOIL, WATER POLL, WATER SUPP micro-organisme immobilisé *m*
immunoassay *n* ANALYSIS dosage immunologique *m*, immuno-essai *m*; **~ test** ANALYSIS essai de dosage immunologique *m*, test de dosage immunologique *m*
immunological *adj* SOIL, WATER SUPP immunologique
impact *n* AIR, ANALYSIS impact *m*; **~ analysis** MANAGE analyse d'impact *f*; **~ crusher** WASTE concasseur à mâchoires *m*, concasseur à percussion *m*; **~ ionization** AIR ionisation par choc *f*, ionisation par impact *f*; **~ study** MANAGE, POLICY étude d'impact *f*
impaction *n* AIR, ANALYSIS impact *m*
impactor *n* AIR, ANALYSIS échantillonneur à chocs *m*, conimètre à impact *m*, impacteur *m*; **~ nozzle** AIR buse d'impacteur *f*
impair *vt* GEN *damage* endommager, déteriorer, dégrader, altérer, *weaken* affaiblir
impaired *adj* GEN *damaged* endommagé, détérioré, dégradé, altéré, *weakened* affaibli
impairment *n* AIR dégradation *f*, préjudice *m*, réduction de la qualité *f*, MARINE dégradation *f*, NOISE altération *f*, SOIL dégradation *f*, WATER SUPP affaiblissement *m*
impeller *n* AIR roue à aube *f*, roue mobile *f*; **~ anemometer** ANALYSIS anémomètre à hélice *m*; **~-type mixer** WATER POLL mélangeur à hélice *m*
impermeable[1] *adj* GEN étanche, imperméable
impermeable[2]: **~ barrier** *n* SOIL écran imperméable *m*, couche imperméable *f*
impervious *adj* GEN étanche, imperméable
impinge[1]: **~ method** *n* ANALYSIS *measurement of air quality* méthode par courants contrariés *f*
impinge[2] *vi* AIR empiéter, entrer en collision, se heurter
impingement *n* AIR collision *f*, impact *m*, ANALYSIS impact *m*; **~ plate scrubber** AIR laveur à chocs *m*, laveur à plateaux à plaques d'impact *m*
impinger *n* AIR impacteur *m*, ANALYSIS conimètre à impact *m*, impacteur *m*, WATER SUPP séparateur à impacts *m*
implement *vt* POLICY appliquer, mettre en œuvre
implementation *n* POLICY application *f*, mise en œuvre *f*; **~ procedure** POLICY *in court* procédure d'application *f*
import *vt* GEN importer
impounding: **~ reservoir** *n* WATER SUPP bassin de retenue *m*, retenue *f*, *artificial* lac de barrage *m*
impregnate *vt* AIR, ANALYSIS, WASTE imprégner
impregnated: **~ fabric** *n* WASTE tissu imprégné *m*; **~ filter paper** *n* ANALYSIS papier filtre imprégné *m*; **~ paper** *n* WASTE papier impregné *m*
impregnating: **~ agent** *n* WASTE agent d'imprégnation *m*; **~ machine** *n* WASTE machine à imprégner *f*; **~ wax** *n* WASTE cire d'imprégnation *f*
impregnation *n* AIR, ANALYSIS, WASTE imprégnation *f*
imprinter *n* WASTE presse à impression *f*, presse à imprimer *f*
impulse: **~ heat-sealer** *n* WASTE thermoscelleuse par impulsion *f*; **~ noise** *n* NOISE bruit impulsionnel *m*; **~ sound-level meter** *n* ANALYSIS sonomètre à impulsion *m*
IMS *abbr* (*ion-mobility spectrometry*) ANALYSIS spectrométrie de mobilité des ions *f*
inactivation *n* WATER SUPP désactivation *f*
incandescent: **~ lamp** *n* ANALYSIS lampe à incandescence *f*
incentive *n* MANAGE, POLICY incitation *f*; **~ charge** POLICY redevance incitative *f*
incident: **~-dependent** *adj* ANALYSIS en fonction d'un incident
incinerate *vt* AIR, SAFETY, WASTE incinérer
incinerated[1] *adj* AIR, SAFETY, WASTE incinéré
incinerated[2]: **~ liquid** *n* WATER POLL liquide incinéré *m*
incineration *n* AIR, SAFETY, WASTE incinération *f*; **~ ash** WASTE cendres d'incinération *f pl*; **~ ashes** *n pl* AIR résidu de combustion *m*; **~ calorimeter** ANALYSIS calorimètre d'incinération *m*; **~ capacity** WASTE, WATER POLL capacité d'incinération *f*; **~ grate** WASTE grille d'incinération *f*; **~ plant** AIR usine d'incinération *f*; **~ residue** WASTE, WATER POLL résidu d'incinération *m*; **~ slag** WASTE scories d'incinération *f pl*; **~ stock** WATER SUPP charge d'incinération *f*; **~ test** ANALYSIS essai d'incinération *m*;

~ **train** WASTE train d'incinérateur *m*; ~ **of wastewater sludge in cement furnaces** WATER POLL incinération des boues résiduaires dans des fours à ciment *f*; ~ **of wastewater sludge in coal furnaces** WATER POLL incinération des boues résiduaires dans des fours à charbon *f*; ~ **of wastewater sludge in lignite furnaces** WATER POLL incinération des boues résiduaires dans des fours à lignite *f*; ~ **without energy recovery** WASTE incinération sans récupération d'énergie *f*; ~ **zone** AIR, WATER POLL zone d'incinération *f*

incinerator *n* AIR, SAFETY, WASTE incinérateur *m*; ~ **plant** WASTE usine d'incinération des déchets *f*

inclined: ~ **table** *n* WASTE plan incliné *m*

incombustibility *n* AIR, SAFETY, WASTE, WATER SUPP incombustibilité *f*

incombustible[1] *adj* AIR, SAFETY, WASTE, WATER SUPP incombustible

incombustible[2]: ~ **residue** *n* AIR résidus incombustibles *m pl*

income[1]: ~-**neutral** *adj* POLICY neutre fiscalement

income[2]: ~-**neutral eco-tax reform** *n* POLICY réforme de l'imposition sur l'environnement fiscalement neutre *f*

incomplete[1] *adj* AIR incomplet

incomplete[2]: ~ **burning** *n* AIR combustion incomplète *f*; ~ **combustion** *n* AIR combustion incomplète *f*

incompressible *adj* AIR incompressible

increment *n* ANALYSIS prélèvement élémentaire *m*

incremental: ~ **heating** *n* AIR, WATER SUPP chauffage progressif *m*

incrustation *n* WATER SUPP incrustation *f*

incubation *n* ANALYSIS incubation *f*

indicating: ~ **electrode** *n* ANALYSIS électrode indicatrice *f*; ~ **instrument** *n* ANALYSIS instrument indicateur *m*

indication: ~ **error** *n* ANALYSIS erreur d'indication *f*

indicator *n* GEN indicateur *m*; ~ **organism** SOIL, WATER POLL, WATER SUPP indicateur écologique *m*, organisme témoin *m*; ~ **paper** ANALYSIS, WASTE papier gradué *m*; ~ **range** ANALYSIS plage d'indicateur *f*; ~ **solution** ANALYSIS solution indicatrice *f*; ~ **species** SOIL, WATER POLL, WATER SUPP espèce indicatrice *f*, indicateur écologique *m*

indirect[1] *adj* GEN indirect

indirect[2]: ~ **condenser** *n* AIR condenseur indirect *m*; ~ **discharge** *n* WATER POLL rejet indirect *m*; ~ **discharger** *n* WATER POLL déversoir indirect *m*; ~ **drying** *n* WATER POLL séchage indirect *m*; ~ **emission source** *n* AIR source d'émission indirecte *f*; ~-**heated rotary kiln** *n* SOIL four rotatif à chauffage indirect *m*; ~ **heating** *n* SOIL chauffage indirect *m*; ~ **measurement method** *n* ANALYSIS méthode de mesure indirecte *f*

indiscriminate: ~ **dumping** *n* WASTE décharge inconsidérée *f*; ~ **waste disposal** *n* WASTE décharge sauvage *f*

individual: ~ **droplet** *n* AIR gouttelette isolée *f*; ~ **dust-removal apparatus** *n* SAFETY dépoussiéreur isolé *m*; ~ **sample** *n* ANALYSIS échantillon isolé *m*; ~ **substance analysis** *n* ANALYSIS analyse d'une substance isolée *f*

indoor: ~ **air pollution** *n* AIR pollution de l'air à l'intérieur des locaux *f*; ~ **shielding** *n* NOISE blindage interne *m*

induced: ~ **draft** *AmE*, ~ **draught** *BrE n* (*ID*) AIR tirage forcé *m*; ~ **draught fan** *n BrE* (*ID fan*) AIR, WASTE ventilateur à tirage forcé *m*

induction: ~ **generator** *n* RENEW ENERG *wind turbine* alternateur asynchrone *m*, générateur asynchrone *m*; ~ **heating** *n* AIR chauffage par induction *m*

inductively: ~-**coupled plasma-mass spectrometry** *n* (*ICP-MS*) ANALYSIS spectrométrie de masse de plasma avec couplage inductif *f*; ~-**coupled plasma-optical emission spectrometry** *n* (*ICP-OES*) ANALYSIS spectrométrie d'émission optique de plasma avec couplage inductif *f*

industrial[1] *adj* GEN industriel

industrial[2]: ~ **accident** *n* SAFETY accident du travail *m*; ~ **bulk-container system** *n* WASTE conteneur industriel pour vrac *m*; ~ **design** *n* POLICY esthétique industrielle *f*; ~ **effluent** *n* WASTE, WATER POLL, WATER SUPP effluent industriel *m*; ~ **emissions** *n pl* AIR émissions d'origine industrielle *f pl*; ~ **fabric** *n* POLICY tissu industriel *m*; ~ **insurance** *n* SAFETY assurance contre les risques professionnels *f*, assurance professionnelle *f*; ~ **landfill** *n* WASTE décharge industrielle *f*; ~ **noise** *n* NOISE bruit industriel *m*; ~ **outfall** *n* WATER POLL déversement industriel *m*; ~ **packing** *n* WASTE emballage industriel *m*; ~ **pollutant** *n* AIR, WATER POLL polluant industriel *m*; ~ **pollution** *n* AIR, WATER POLL pollution industrielle *f*; ~ **process** *n* WATER SUPP procédé industriel *m*; ~ **site** *n* SOIL site industriell *m*; ~ **source** *n* AIR source industrielle *f*; ~ **standard** *n* POLICY norme industrielle *f*; ~ **waste** *n* WASTE, WATER POLL, WATER SUPP déchets industriels *m pl*, effluent industriel *m*; ~ **waste follow-up form** *n* POLICY bordereau de suivi des déchets industriels *m*; ~ **wastewater** *n* WASTE, WATER POLL eaux résiduaires industrielles *f pl*; ~ **wastewater purification** *n* WATER POLL, WATER SUPP épuration des eaux résiduaires industrielles *f*; ~ **wastewater sludge** *n* WATER POLL boues résiduaires industrielles *f pl*; ~ **water** *n* WATER POLL eau à usage industriel *f*

industrial[3]: **on an ~ scale** *phr* AIR, RENEW ENERG à l'échelle industrielle

industry: ~-**specific purification** *n* WATER POLL épuration spécifique d'un secteur d'activité *f*

inert: ~ **gas** *n* AIR gaz inerte *m*; ~-**gas combustion chamber** *n* AIR chambre de combustion pour gaz inerte *f*; ~-**gas flow** *n* AIR flux de gaz inerte *m*; ~ **landfill** *n* WASTE décharge inerte *f*; ~ **material** *n* AIR matériau inerte *m*, WASTE matière inerte *f*; ~ **waste** *n* WASTE déchets inertes *m pl*

inertia *n* AIR, WASTE, WATER SUPP inertie *f*

inertial[1] *adj* AIR, WASTE, WATER SUPP inertiel

inertial[2]: ~ **collector** *n* AIR collecteur inertiel *m*; ~ **force** *n* AIR force d'inertie *f*; ~ **separator** *n* AIR dépoussiéreur à inertie *m*

inerting *n* AIR, WASTE *of harmful substances*, WATER SUPP inertage *m*

inexhaustibility *n* RENEW ENERG caractère inépuisable *m*

inexhaustible[1] *adj* RENEW ENERG inépuisable

inexhaustible[2]: ~ **energy source** *n* RENEW ENERG source d'énergie inépuisable *f*; ~ **energy stream** *n* RENEW ENERG flux d'énergie inépuisable *m*

infectious: ~ **waste** *n* WASTE déchets infectieux *m pl*

infiltration *n* SOIL, WASTE, WATER POLL, WATER SUPP infiltration *f*; ~ **basin** WATER POLL bassin d'infiltration *m*; ~ **gallery** WATER POLL, WATER SUPP galerie filtrante *f*; ~ **well** SOIL, WATER SUPP puits d'infiltration *m*

inflammability *n* AIR, SAFETY, WASTE inflammabilité *f*; ~ **limit** ANALYSIS limite d'inflammabilité *f*

inflammable[1] *adj* AIR, SAFETY, WASTE inflammable

inflammable[2]: ~ **liquid** *n* SAFETY liquide inflammable *m*; ~ **material** *n* SAFETY matériau inflammable *m*

inflatable *adj* AIR, MARINE gonflable

inflate *vt* AIR gonfler, souffler, MARINE gonfler

inflation *n* AIR, MARINE inflation *f*; ~ **cuff** MARINE manche gonflable *f*

inflow *n* WATER POLL, WATER SUPP admission *f*, influx *m*

influence: ~ **zone of noise** *n* MARINE, NOISE portée du bruit *f*, zone d'influence du bruit *f*

influent *n* WATER POLL, WATER SUPP affluent *m*, flux entrant *m*, apport d'eau *m*; ~ **concentration** WATER POLL concentration d'affluents *f*; ~ **pipe** WATER SUPP fût central d'alimentation *m*, colonne centrale d'arrivée *f*; ~ **pump works** WATER POLL station de pompage amont *f*

information: ~ **campaign** *n* POLICY campagne d'information *f*; ~ **instruments** *n pl* POLICY outils d'information *m pl*

infrared[1] *adj* (*IR*) ANALYSIS infrarouge (*IR*)

infrared[2] *n* (*IR*) ANALYSIS infrarouge *m* (*IR*); ~ **absorption measuring device** ANALYSIS absorptiomètre à infrarouge *m*; ~ **drying** ANALYSIS séchage à l'infrarouge *m*; ~ **furnace** ANALYSIS, WATER SUPP four à infrarouge *m*; ~ **gas analyser**

BrE, ~ **gas analyzer** *AmE* ANALYSIS analyseur de gaz à infrarouge *m*; ~ **kiln** ANALYSIS, WATER SUPP four à infrarouge *m*; ~ **laser backscattering** ANALYSIS rétrodiffusion de laser infrarouge *f*; ~ **optical absorption** ANALYSIS absorption optique des infrarouges *f*; ~ **radiation** AIR, WATER SUPP rayonnement infrarouge *m*; ~ **range** ANALYSIS zone de l'infrarouge *f*; ~ **region** ANALYSIS zone de l'infrarouge *f*; ~ **spectrometry** ANALYSIS spectrométrie infrarouge *f*; ~ **spectrophotometer** ANALYSIS spectrophotomètre à infrarouge *m*; ~ **spectrum** ANALYSIS spectre de l'infrarouge *m*; ~ **thermal destruction** SOIL destruction thermique à l'infrarouge *f*; ~ **thermometer** ANALYSIS thermomètre à infrarouge *m*; ~ **through-type furnace** ANALYSIS, WATER SUPP four par infrarouge *m*; ~ **volatile organic compound analyser** *BrE* (*IR-VOC analyser*), ~ **volatile organic compound analyzer** *AmE* (*IR-VOC analyzer*) ANALYSIS analyseur de composés organiques volatils à infrarouge *m*

infrasound *n* NOISE infrason *m*

infringer *n* POLICY contrevenant *m*

ingest *vt* WATER SUPP ingérer

inhabitant: ~ **equivalent** *n* AIR, WASTE, WATER POLL équivalent habitant *m*

inherited: ~ **trait** *n* POLICY, SAFETY caractère héréditaire *m*

inhibitor *n* AIR, WATER SUPP inhibiteur *m*

in-house: ~ **water recirculation system** *n* WATER POLL circuit interne de recyclage des eaux *m*

initial: ~ **adsorption at optimum polymer dosage** *n* WATER POLL *flocculation* adsorption initiale à dosage optimal de polymères *f*

initiation: ~ **time** *n* WATER POLL temps d'amorçage *m*, temps de démarrage *m*

inject *vt* GEN injecter

injection *n* GEN injection *f*; ~ **blow-molding machine** *AmE*, ~ **blow-moulding machine** *BrE* WASTE presse de moulage par injection-soufflage *f*; ~ **drilling** SOIL forage par injection *m*; ~ **filling** WASTE remplissage par injection *m*; ~ **fluid** SOIL liquide d'injection *m*; ~ **lance** SOIL lance d'injection *f*; ~ **molding** *AmE*, ~ **moulding** *BrE* WASTE moulage par injection *m*; ~ **moulding compound** *BrE* WASTE matière à mouler par injection *f*; ~ **moulding machine** *BrE* WASTE machine de moulage par injection *f*, presse de moulage par injection *f*; ~ **moulding press** *BrE* WASTE machine de moulage par injection *f*, presse à mouler par injection *f*, presse de moulage par injection *f*; ~ **moulding pressure** *BrE* WASTE pression de moulage par injection *f*; ~ **nozzle** AIR buse d'injection *f*; ~ **plant** SOIL *grouting* poste d'injection *m*; ~ **section** SOIL section d'injection *f*; ~ **spraying** ANALYSIS *dosing of gas, solutions* vaporisation par injection *f*; ~ **system** ANALYSIS système d'injection *m*; ~ **well** SOIL, WATER SUPP forage d'injection *m*, puits d'injection *m*

injector *n* AIR, ANALYSIS injecteur *m*, WATER POLL injecteur *m*, éjecteur *m*; ~ **action** AIR *bag filter* effet d'injection *m*

injurious *adj* SAFETY dommageable, préjudiciable

inkless: ~ **ink-jet system** *n* WASTE système à jet d'encre sans encre *m*

inland: ~ **waters** *n pl* WATER SUPP eaux intérieures *f pl*

inlet *n* AIR admission *f*, WATER POLL amenée *f*, crique *f*, WATER SUPP crique *f*, grau *m*, pertuis d'entrée *m*; ~ **branch** WATER SUPP *pump* tubulure d'admission *f*; ~ **dust burden** AIR charge de poussière à l'admission *f*; ~ **emissions** *n pl* AIR émissions à l'admission *f pl*; ~ **piping** WATER SUPP tubulure d'admission *f*; ~ **structure** WATER POLL ouvrage d'arrivée *m*, ouvrage d'amenée *m*; ~ **velocity** AIR *cyclone* vitesse d'admission *f*

in-line: ~ **contact coding** *n* WASTE codage par contact en ligne *m*; ~ **finishing equipment** *n* WASTE matériel de finition en ligne *m*

inner: ~ **cyclone wall** *n* AIR paroi interne de cyclone *f*; ~ **vortex** *n* AIR tourbillon lévogyre *m*

innocuous *adj* AIR anodin, inoffensif

innovative: ~ **technologies** *n pl* AIR, RENEW ENERG techniques innovantes *f pl*, technologies innovantes *f pl*, technologies novatrices *f pl*

inorganic[1] *adj* AIR, WATER POLL, WATER SUPP inorganique

inorganic[2]: ~ **acid gas** *n* AIR gaz acide inorganique *m*; ~ **base material** *n* AIR *biofilter* matériau de base inorganique *m*; ~ **contaminant** *n* AIR polluant inorganique *m*; ~ **gas** *n* AIR gaz inorganique *m*; ~ **salt** *n* WATER SUPP sel inorganique *m*

input *n* ANALYSIS données saisies *f pl*, entrée *f*; ~ **of rotor** RENEW ENERG *wind turbine* admission du rotor *f*

inputs *n* MANAGE *materials and energy* intrants *m pl*

insect[1]: ~**-proof** *adj* WASTE résistant aux insectes

insect[2]: ~**-resisting crop strains** *n pl* SOIL espèce de végétales résistantes aux insectes *f*

insecticide *n* SOIL, WATER SUPP insecticide *m*

insensitive *adj* AIR insensible

insensitivity: ~ **to weathering** *n* SOIL *liners* insensibilité à la désagrégation *f*, résistance à l'usure *f*

insert *n* WASTE insert *m*

insertable: ~ **sack** *n* WASTE sac insérable *m*

insertion: ~ **loss** *n* NOISE affaiblissement d'insertion *m*

inshore: ~ **water** *n* MARINE, POLICY eau littorale *f*

in-situ: ~ **bio-remediation** *n* SOIL *for hydrocarbons, chlorinated solvents, metals* assainissement biologique sur place *m*; ~ **electro-acoustic decontamination** *n* SOIL décontamination électroacoustique in situ *f*; ~ **extraction** *n* SOIL extraction sur place *f*; ~ **method** *n* SOIL méthode mise en œuvre in situ *f*; ~ **solidification** *n* SOIL solidification in situ *f*; ~ **thermal desorption** *n* SOIL désorption thermique in situ *f*; ~ **vitrification** *n* (*ISV*) SOIL vitrification in situ *f*

insolation *n* RENEW ENERG insolation *f*

insoluble *adj* GEN insoluble

inspection *n* GEN visite *f*, *of workplace* inspection *f*; ~ **by experts** POLICY inspection par des experts *f*; ~ **chamber** WATER SUPP regard de visite *m*; ~ **equipment** WASTE, WATER SUPP matériel d'inspection *m*; ~ **glass** ANALYSIS fenêtre d'observation *f*, regard *m*; ~ **hole** WATER SUPP regard *m*; ~ **window** WASTE hublot *m*

instability *n* WASTE instabilité *f*

install *vt* GEN installer; ~ **above ground** WATER SUPP installer en surface; ~ **underground** SOIL installer sous terre, WATER SUPP installer dans le sous-sol

installation *n* GEN installation *f*; ~ **area** RENEW ENERG *of solar farm* zone aménagée *f*, zone d'installation *f*

instant: ~ **flow meter** *n* WATER SUPP débitmètre instantané *m*

instantaneous: ~ **concentration** *n* WATER SUPP concentration instantanée *f*, niveau instantané de nuisance *m*

institution *n* POLICY instance *f*, institution *f*, organisme *m*

instructions: ~ **for opening** *n pl* WASTE consigne d'ouverture *f*; ~ **for use** *n pl* WASTE mode d'emploi *m*

instrument: ~ **self-test** *n* ANALYSIS instrument à test automatique *m*, instrument auto-testé *m*

instrumental: ~ **analysis** *n* ANALYSIS analyse instrumentale *f*

instrumentation *n* POLICY instrumentation *f*

insulate *vt* GEN isoler

insulated *adj* GEN isolé

insulating: ~ **board** *n* NOISE carton isolant *m*, panneau d'isolation *m*, WASTE carton isolant *m*; ~ **brick** *n* NOISE brique isolante *f*; ~ **compound** *n* NOISE composé isolant *m*, matériau composite isolant *m*, WASTE composé isolant *m*, matériau d'isolation *m*; ~ **glass** *n* NOISE verre isolant *m*; ~ **layer** *n* NOISE, WASTE couche d'isolation *f*, couche isolante *f*; ~ **material** *n* NOISE, WASTE matériau isolant *m*; ~ **properties** *n pl* NOISE, WASTE propriétés isolantes *f pl*; ~ **sheet** *n* WASTE feuille isolante *f*

insulation *n* GEN isolation *f*, isolement *m*; ~ **against sound and vibration** NOISE, SAFETY isolation contre le bruit et les vibrations *f*

insulator *n* GEN isolateur *m*

intake *n* AIR, WATER POLL consommation *f*, apport *m*, WATER SUPP marcage *m*, captage *m*, canal de prise d'eau *m*; **~ air** AIR air admis *m*; **~ main** WATER POLL, WATER SUPP canal d'amenée *m*, canal de décharge *m*; **~ manifold** AIR collecteur d'admission *m*; **~ pressure** WATER POLL pression d'admission *f*; **~ volume** WATER POLL volume admis *m*, volume ingéré *m*

Intalox: **~ saddle** *n* AIR *packed-tower scrubber* selle Intalox *f*

integral: **~ detector** *n* ANALYSIS *in chromatography* détecteur intégral *m*; **~ gas chromatogram** *n* ANALYSIS chromatogramme intégral d'un gaz *m*

integrated[1] *adj* GEN intégré

integrated[2]: **~ approach** *n* POLICY méthode intégrée *f*; **~ environmental protection** *n* MANAGE, POLICY protection intégrée de l'environnement *f*; **~ heat exchanger system** *n* AIR système d'échangeur de chaleur intégré *m*; **~ limit values** *n pl* POLICY, WATER SUPP valeurs limites intégrées *f pl*; **~ pollution control** *n* (*IPC*) MANAGE, POLICY lutte intégrée contre la pollution *f*; **~ pollution prevention** *n* MANAGE prévention intégrée de la pollution *f*; **~ prevention and pollution control** *n* (*IPPC*) MANAGE mesure de prévention et de lutte intégrées contre la pollution *f*, prévention et lutte intégrées contre la pollution *f*; **~ waste management** *n* (*IWM*) MANAGE, WASTE gestion intégrée des déchets *f*

integrating: **~ motor** *n* ANALYSIS moteur d'intégration *m*

integrator *n* ANALYSIS *for quick determination* intégrateur *m*

intensity *n* AIR, ANALYSIS, RENEW ENERG intensité *f*

intensive: **~ farming** *n* POLICY, SOIL agriculture intensive *f*

intentional: **~ discharge** *n* WATER SUPP rejet intentionnel *m*

interaction: **~ force** *n* AIR force d'interaction *f*

intercept *vt* AIR *particle*, WATER POLL *flow of water* intercepter

intercepting: **~ sewer** *n* WASTE, WATER POLL, WATER SUPP égout captant *m*, intercepteur *m*

interception *n* AIR *particle*, WATER POLL interception *f*

interchangeable[1] *adj* ANALYSIS interchangeable

interchangeable[2]: **~ detector** *n* ANALYSIS détecteur amovible *m*; **~ lamp** *n* ANALYSIS lampe interchangeable *f*

interest: **~ earned** *n* MANAGE intérêt échu *m*; **~ profit** *n* MANAGE intérêt couru *m*

interface *n* ANALYSIS interface *f*

interfacial: **~ renewal** *n* AIR renouvellement interfacial *m*

interference *n* ANALYSIS interférence *f*; **~ filter** ANALYSIS filtre interférentiel *m*; **~ refractometer** ANALYSIS réfractomètre interférentiel *m*

interferogram *n* ANALYSIS interférogramme *m*

interferometry *n* ANALYSIS interférométrie *f*

interim: **~ covering material** *n* WASTE matériau de couverture temporaire *m*; **~ tank** *n* SOIL cuve de stockage temporaire *f*

interior: **~ electrode** *n* ANALYSIS électrode interne *f*; **~ packing** *n* WASTE garniture intérieure *f*; **~-polished cyclone** *n* AIR cyclone à parois lisses *m*; **~ wrapping** *n* WASTE emballage intérieur *m*, emballage interne *m*

interlocking *n* WATER POLL *of membrane* empaillassonnage *m*, enchevêtrement *m*; **~ risk** WATER POLL *of membrane* risque d'empaillassonnage *m*

intermediate[1] *adj* GEN intermédiaire

intermediate[2]: **~ bulk container** *n* WASTE conteneur de stockage temporaire en vrac *m*; **~ construction schedule** *n* MANAGE planning de réalisation *m*; **~ cover** *n* WASTE couverture intermédiaire *f*; **~ layer** *n* NOISE *in encapsulation* couche intermédiaire *f*; **~ purification** *n* WATER POLL épuration intermédiaire *f*; **~ settling** *n* WATER POLL décantation intermédiaire *f*; **~ settling tank** *n* WATER POLL bassin de décantation intermédiaire *m*; **~ storage** *n* WASTE stockage intermédiaire *m*; **~ trickling filter** *n* WATER POLL lit bactérien intermédiaire *m*

intermittent[1] *adj* GEN discontinu, intermittent

intermittent[2]: **~ cleaning** *n* AIR *filter bag* lavage discontinu *m*; **~ denitrification** *n* WATER POLL dénitrification intermittente *f*; **~ spring** *n* WATER SUPP source intermittente *f*

internal[1] *adj* AIR, WASTE, WATER POLL, WATER SUPP interne

internal[2]: **~ combustion engine** *n* AIR, WASTE, WATER POLL *in large pumping stations* moteur à combustion interne *m*; **~ compactor** *n* WATER SUPP compacteur intégré *m*; **~ eddy** *n* WATER SUPP tourbillon interne *m*, tourbillon sinistrogyre *m*; **~ pressure** *n* AIR *bag filter* pression interne *f*; **~ recirculation** *n* WATER POLL recyclage interne *m*; **~ source** *n* AIR, WATER SUPP source interne *f*

internalization *n* POLICY internalisation *f*, prise en compte *f*; **~ of external costs** POLICY prise en compte des coûts externes *f*; **~ of external effects** POLICY internalisation des effets externes *f*, prise en compte des effets externes *f*; **~ of externality** POLICY prise en compte des effets externes *f*

international: **~ law** *n* POLICY droit international *m*; **~ negotiation** *n* POLICY négociation internationale *f*; **~ tradeable emissions rights** *n pl* POLICY droits d'émissions négociables à l'échelle internationale *m pl*

International: **~ Register of Potentially Toxic Chemicals** *n* (*IRPTC*) POLICY Registre international des substances chimiques potentiellement toxiques *m* (*RISCPT*); **~ Standards Organization** *n* (*ISO*) POLICY, ANALYSIS Organisation internationale de normalisation *f* (*ISO*)

interphase *n* AIR interphase *f*

interstitial[1] *adj* WATER SUPP interstitiel

interstitial[2]: **~ water** *n* WATER SUPP eau interstitielle *f*

intertidal: **~ zone** *n* MARINE zone intercotidale *f*, zone intertidale *f*

interval *n* ANALYSIS intervalle *m*

intimate: **~ water-sludge mixture** *n* WATER SUPP mélange intime eau-boue *m*

intrusion *n* NOISE intrusion, SAFETY *of polluted air*, WASTE intrusion *f*, pénétration intempestive *f*

intrusive: **~ filtration** *n* WATER SUPP filtration intrusive *f*, infiltration dans les couches *f*

intumescent *adj* SAFETY intumescent

inundation *n* WATER POLL, WATER SUPP crue *f*, inondation *f*, submersion *f*

invalidate *vt* MANAGE vicier

inventory: **~ of hazardous substances** *n* AIR, POLICY, SAFETY inventaire des substances dangereuses *m*

inversion: **~ atmospheric conditions** *n pl* AIR inversion atmosphérique *f*; **~ layer** *n* AIR couche d'inversion *f*

invert *n* WATER SUPP radier *m*

inverted: **~ well** *n* WATER SUPP puits absorbant *m*

inverting: **~ prism** *n* ANALYSIS *interferential pit gas tester* prisme d'inversion *m*

investigate *vt* ANALYSIS, MANAGE, SAFETY enquêter sur, rechercher

investigation *n* ANALYSIS, MANAGE, SAFETY enquête *f*, investigation *f*, recherche *f*

investigative: **~ action** *n* ANALYSIS, MANAGE, SAFETY enquête *f*

invisible[1] *adj* AIR invisible

invisible[2]: **~ plume** *n* AIR panache invisible *m*

in vitro: **~ analysis** *n* WATER SUPP analyse in vitro *f*

in vivo: **~ analysis** *n* WATER SUPP analyse in vivo *f*

involutional *adj* AIR régressif

iodometric[1] *adj* ANALYSIS iodométrique

iodometric[2]: **~ determination** *n* ANALYSIS détermination iodométrique *f*, dosage iodométrique *m*; **~ method** *n* ANALYSIS méthode iodométrique *f*

ion[1]: **~-selective** *adj* ANALYSIS sélectif d'ions; **~-sensitive** *adj* ANALYSIS sensible aux ions

ion[2] *n* GEN ion *m*; **~ activity** ANALYSIS activité ionique *f*; **~ budget** AIR, WATER SUPP bilan ionique *m*; **~ current** AIR courant ionique *m*, courant ionique *m*; **~ exchange** (*IE*) ANALYSIS, WATER POLL, WATER SUPP échange d'ions *m*, échange ionique *m*; **~-exchange capacity** WATER POLL capacité d'échange ionique *f*; **~-exchange**

chromatography (*IC*) ANALYSIS chromatographie d'échange d'ions *f*; **~-exchange resin** SOIL, WATER POLL, WATER SUPP résine échangeuse ionique *f*; **~-exchange softener** WATER SUPP adoucisseur par échange ionique *m*; **~-exchange softening** WATER SUPP adoucissement par échange ionique *m*; **~ exchanger** WATER POLL échangeur d'ions *m*; **~-exchanger column** WATER POLL colonne échangeuse d'ions *f*; **~-exchanger membrane** WATER POLL membrane échangeuse d'ions *f*; **~-mobility spectrometer** ANALYSIS spectromètre de mobilité ionique *m*; **~-mobility spectrometry** (*IMS*) ANALYSIS spectrométrie de mobilité des ions *f*; **~-molecule reaction mass spectrometry** ANALYSIS spectrométrie de masse avec réaction ion-molécule *f*; **~-selective electrode** (*ISE*) ANALYSIS électrode sélective d'ions *f*; **~-selective electrode with glass membrane** ANALYSIS électrode sélective d'ions avec membrane en verre *f*; **~-selective membrane** WATER POLL membrane sélective d'ions *f*; **~-sensitive electrode** ANALYSIS électrode sensible aux ions *f*; **~ source** ANALYSIS *quadrupole mass analyser* source d'ions *f*; **~ trap** ANALYSIS piège ionique *m*; **~-trap detector** ANALYSIS détecteur de piège ionique *m*

ionic[1] *adj* GEN ionique

ionic[2]: **~ concentration** *n* WATER POLL, WATER SUPP concentration ionique *f*; **~ migration** *n* SOIL migration ionique *f*; **~ mobility** *n* ANALYSIS mobilité ionique *f*; **~ valence** *n* WATER POLL valence ionique *f*

ionization *n* GEN ionisation *f*; **~ chamber** ANALYSIS *gas-filled detectors* chambre d'ionisation *f*; **~ potential** ANALYSIS potentiel d'ionisation *m*; **~ spectrometer** ANALYSIS spectromètre à ionisation *m*

ionize *vt* GEN ioniser

ionized[1] *adj* GEN ionisé

ionized[2]: **~ gas molecule** *n* AIR molécule de gaz ionisé *f*

ionizing[1] *adj* GEN ionisant

ionizing[2]: **~ particle** *n* AIR particule ionisante *f*; **~ radiation** *n* AIR, SOIL, WATER SUPP rayonnement ionisant *m*; **~ wet scrubber** *n* (*IWS*) AIR laveur par voie humide avec ionisation *m*; **~ zone** *n* AIR zone d'ionisation *f*, zone ionisante *f*

ionogenic *adj* ANALYSIS, WATER POLL ionogène

ionosphere *n* AIR ionosphère *f*

IPC *abbr* (*integrated pollution control*) MANAGE, POLICY lutte intégrée contre la pollution *f*

IPPC *abbr* (*integrated prevention and pollution control*) MANAGE mesure de prévention et de lutte intégrées contre la pollution *f*, prévention et lutte intégrées contre la pollution *f*

IR *abbr* (*infrared*) ANALYSIS IR (*infrarouge*)

iridescence *n* MARINE iridescence *f*

iron *n* (*Fe*) GEN fer *m* (*Fe*); **~ ore** GEN minerai de fer *m*; **~ sulfate** *AmE*, **~ sulphate** *BrE* WATER POLL, WATER SUPP sulfate de fer *m*

IRPTC *abbr* (*International Register of Potentially Toxic Chemicals*) POLICY RISCPT (*Registre international des substances chimiques potentiellement toxiques*)

irradiated[1] *adj* ANALYSIS, RENEW ENERG irradié

irradiated[2]: **~ energy** *n* RENEW ENERG énergie irradiée *f*

irradiation: **~ of food** *n* WASTE irradiation des produits alimentaires *f*

irreplaceables *n pl* WASTE biens irremplaçables *m pl*

irreversible[1] *adj* GEN irréversible

irreversible[2]: **~ process** *n* AIR *chemisorption*, WATER POLL *physisorption* processus irréversible *m*

irrigation *n* SOIL, WATER POLL *wastewater disposal*, WATER SUPP irrigation *f*; **~ channel** WATER SUPP canal d'irrigation *m*; **~ field** WASTE champ irrigué *m*; **~ water** WATER SUPP eau d'irrigation *f*

irritant[1] *adj* SAFETY *as danger symbol* irritant

irritant[2] *n* SAFETY irritant *m*

IR-VOC: **~ analyser** *n BrE* (*infrared volatile organic compound-analyser*), **~ analyzer** *n AmE* (*infrared volatile organic compound-analyzer*) ANALYSIS analyseur de composés organiques volatils à infrarouge *m*

ISE *abbr* (*ion-selective electrode*) ANALYSIS électrode sélective d'ions *f*

ISO *abbr* (*International Standards Organization*), ANALYSIS , POLICY ISO (*Organisation internationale de normalisation*)

isokinetic[1] *adj* AIR, ANALYSIS isocinétique

isokinetic[2]: **~ sampling** *n* ANALYSIS échantillonnage isocinétique *m*; **~ suction** *n* ANALYSIS aspiration isocinétique *f*

isolate *vt* GEN isoler

isolated[1] *adj* GEN isolé

isolated[2]: **~ thermal use** *n* RENEW ENERG consommation thermique isolée *f*

isolation *n* GEN isolation *f*, isolement *m*; **~ pad** NOISE rembourrage isolant *m*

isolator *n* GEN isolateur *m*

isotherm *n* AIR, ANALYSIS isotherme *f*

isothermal[1] *adj* AIR, ANALYSIS isotherme

isothermal[2]: **~ curve** *n* AIR isotherme *f*; **~ gas chromatography** *n* ANALYSIS chromatographie en phase gazeuse isotherme *f*

ISV *abbr* (*in-situ vitrification*) SOIL vitrification in situ *f*

iteration *n* AIR, WATER SUPP itération *f*

IWM *abbr* (*integrated waste management*) MANAGE, WASTE gestion intégrée des déchets *f*

IWS *abbr* (*ionizing wet scrubber*) AIR laveur par voie humide avec ionisation *m*

J

jacket *n* WATER SUPP chemise *f*, housse *f*
Jamin: ~ **interferometer** *n* ANALYSIS interféromètre de Jamin *m*
jar: ~ **test** *n* ANALYSIS *water* essai de floculation *m*
jarring: ~ **table** *n* WASTE table à secousses *f*
jaw: ~ **breaker** *n* WASTE concasseur à mâchoires *m*
jet *n* AIR, SOIL, WATER POLL, WATER SUPP *emission of fluid* jet *m*, *nozzle* ajutage *m*; ~ **aeration** WATER POLL aération avec injection d'air *f*, aération à jet *f*; ~ **aerator** WATER POLL ventilateur *m*; ~ **cock** WATER SUPP robinet d'arrosage *m*; ~ **filter** AIR filtre à jet *m*; ~ **injection** SOIL injection au jet *f*; ~ **pump** WATER POLL, WATER SUPP pompe à éjecteur *f*, pompe à jet *f*, hydrojecteur *m*
jig[1] *n* SOIL, WASTE crible *m*
jig[2] *vt* SOIL, WASTE cribler
jigger *n* SOIL, WASTE crible *m*, crible de lavage *m*, crible pulsant *m*, tamis de lavage *m*
jigging *n* SOIL, WASTE criblage *m*
job: ~ **instructions** *n pl* MANAGE cahier des charges *m*
joint: ~ **and several liability** *n* MANAGE responsabilité conjointe et solidaire *f*
Joule: ~ **heating effect** *n* SOIL *thermal desorption* effet Joule *m*
junk *n* AIR, WASTE rebut *m*; ~ **iron** WASTE ferraille *f*, débris de fer *m pl*, mitraille *f*; ~ **pile** WASTE dépotoir *m*
juridical *adj* POLICY juridique
jute: ~ **sacking** *n* WASTE sac de jute *m*, sacherie en jute *f*
juvenile: ~ **water** *n* WATER SUPP eau juvénile *f*

K

Kaplan: ~ **turbine** *n* RENEW ENERG *tidal power plant* turbine Kaplan *f*
karst: ~ **hydrology** *n* WATER SUPP hydrologie karstique *f*
karstic: ~ **spring** *n* WATER SUPP source karstique *f*
keep *vt* GEN tenir, garder; ~ **cool** ANALYSIS *sample* tenir au frais, conserver au frais; ~ **in the dark** ANALYSIS *sample* tenir à l'abri de la lumière, conserver à l'abri de la lumière; ~ **sealed** ANALYSIS *sample* maintenir fermé
kilowatt *n* (*kW*) RENEW ENERG kilowatt *m* (*kW*); **~-hour** (*kWh*) RENEW ENERG kilowatt-heure *m* (*kWh*); **~-hour heat** (*kWt*) RENEW ENERG chaleur mesurée en kilowatt-heure *f* (*kWt*); ~ **plant** RENEW ENERG installation de faible puissance *f*, microcentrale *f*, minicentrale *f*
kindling: ~ **point** *n* AIR point d'inflammation *m*
kinematic: ~ **viscosity** *n* WATER SUPP viscosité cinématique *f*
kinetic: ~ **energy** *n* AIR, RENEW ENERG *wind power* énergie cinétique *f*; **~-energy pump** *n* WATER POLL pompe à énergie cinétique *f*
Kjeldahl: ~ **nitrogen** *n* WATER SUPP azote Kjeldahl *m*
Kn *abbr* (*Knudsen's number*) AIR Kn (*nombre de Knudsen*)
knead *vt* WASTE, WATER SUPP malaxer, pétrir
kneader *n* WASTE, WATER SUPP malaxeur *m*
knobbed: ~ **panel** *n* NOISE panneau à bouton *m*
knock[1]: ~ **limit** *n* AIR limite de résistance à la détonation *f*; ~ **resistance** *n* AIR résistance à la détonation *f*
knock[2] *vt* AIR cogner, frapper
knocking *n* AIR détonation *f*, frappage *m*
Knudsen: **~'s number** *n* (*Kn*) AIR nombre de Knudsen *m* (*Kn*)
Kr *abbr* (*krypton*) AIR Kr (*krypton*)
kraft: ~ **paper** *n* WASTE papier kraft *m*; ~ **sack paper** *n* WASTE papier kraft pour sacherie *m*
krypton *n* (*Kr*) AIR krypton *m* (*Kr*)
kW *abbr* (*kilowatt*) RENEW ENERG kW (*kilowatt*)
kWh *abbr* (*kilowatt-hour*) RENEW ENERG kWh (*kilowatt-heure*)
kWt *abbr* (*kilowatt-hour heat*) RENEW ENERG kWt (*chaleur mesurée en kilowatt-heure*)

L

lab *n* ANALYSIS labo *m*

label *n* WASTE étiquette *f*; **~ coding machine** WASTE machine à encoder les étiquettes *f*; **~ dispenser** WASTE distributeur d'étiquettes *m*; **~ film** WASTE film d'étiquetage *f*, bande d'étiquetage *f*

labeled *AmE see labelled BrE*

labeling *AmE see labelling BrE*

labelled *adj BrE* WASTE *plastics* étiqueté

labeller *n* WASTE pistolet d'étiquetage *m*

labelling *n BrE* WASTE, WATER SUPP étiquetage *m*; **~ machine** *BrE* WASTE étiqueteuse *f*; **~ system for recyclable plastic containers** WASTE, WATER SUPP système d'étiquetage des conteneurs en plastique recyclable *m*

labile *adj* AIR *waste gas plume propagation class* labile

labor: **~-intensive** *AmE see labour-intensive BrE*

laboratory[1]: **~-tested** *adj* ANALYSIS soumis à des essais en laboratoire, testé en laboratoire

laboratory[2] *n* ANALYSIS laboratoire *m*; **~ extruder** WASTE extrudeuse de laboratoire *f*; **~ result** ANALYSIS résultat de laboratoire *m*; **~ sample** ANALYSIS échantillon de laboratoire *m*; **~ technician** ANALYSIS laborantin *m*, technicien de laboratoire *m*; **~ test** ANALYSIS essai en laboratoire *m*

labour[1]: **~-intensive** *adj BrE* GEN à forte intensité de main-d'œuvre

labour[2]: **~-intensive method** *n* MARINE, NOISE, SOIL, WATER SUPP méthode à forte intensité de main-d'œuvre *f*

lack: **~ of fuel** *n* AIR pénurie de combustible *f*; **~ of implementation** *n* POLICY *of competent authority* absence de mise en œuvre *f*; **~ of oxygen** *n* WATER POLL absence d'oxygène *f*

laden *adj* AIR chargé

lag[1]: **~ time** *n* ANALYSIS *between sampling* décalage *m*, intervalle *m*

lag[2] *vt* AIR garnir, SOIL garnir d'un revêtement étanche

lagging *n* AIR garnissage *m*, retard *m*

lagoon *n* RENEW ENERG, WATER POLL, WATER SUPP lagune *f*, étang d'épuration *m*

lagooning *n* RENEW ENERG, WATER POLL, WATER SUPP lagunage *m*

lake *n* WATER SUPP lac *m*; **~ bloom** WATER POLL prolifération d'algues dans les lacs *f*

lambda: **~ probe** *n* AIR *catalyst* échantillon lambda *m*, ANALYSIS éprouvette lambda *f*; **~ window** *n* AIR fenêtre lambda *f*

Lambert-Beer: **~ law** *n* ANALYSIS loi de Lambert-Beer *f*

lamella *n* AIR, WATER POLL *plates within settling tank* lamelle *f*; **~ bundle** AIR faisceau de lamelles *m*

lamellar: **~ heat exchanger** *n* AIR échangeur de chaleur lamellaire *m*; **~ precipitator** *n* AIR séparateur lamellaire *m*, décanteur lamellaire *m*

laminar[1] *adj* AIR, SOIL, WATER SUPP laminaire

laminar[2]: **~ boundary layer** *n* AIR *cyclone* couche limite laminaire *f*; **~ boundary layer flow** *n* AIR écoulement sur la couche limite laminaire *m*; **~ fixed bed** *n* SOIL lit fixe laminaire *m*; **~ flow** *n* AIR écoulement en régime laminaire *m*; **~-flow velocity** *n* AIR vitesse d'écoulement laminaire *f*

laminated: **~ pack** *n* WASTE emballage stratifié *m*

laminating *n* WASTE stratification *f*; **~ machine** WASTE machine à contrecoller *f*, machine à doubler *f*

lamination *n* WASTE feuilletage *m*, doublure *f*; **~ strength** WASTE résistance au laminage *f*

LAMS *abbr* (*laser mass spectrometry*) ANALYSIS spectrométrie de masse au laser *f*

lance *n* AIR lance *f*

land[1]: **~-based** *adj* RENEW ENERG à terre, terrestre

land[2]: **~ application** *n* WATER POLL application agricole *f*; **~-based OTEC plant** *n* RENEW ENERG centrale ETM terrestre *f*; **~-built areas** *n pl* POLICY terrains bâtis *m pl*; **~ burial** *n* SOIL, WASTE enfouissement dans le sol *m*; **~ clearing** *n* POLICY défrichement *m*; **~ degradation** *n* SOIL, WATER SUPP dégradation des terres *f*, dégradation des sols *f*, dévastation du sol *f*; **~ disturbance** *n* WATER SUPP dévastation du sol *f*; **~ misuse** MANAGE, SOIL mauvaise utilisation des terres *f*; **~ policy** *n* POLICY mesures foncières *f pl*; **~ pollutant** *n* SOIL, WATER SUPP polluant des sols *m*; **~ pollution** *n* SOIL pollution du sol *f*; **~ reclamation** *n* SOIL mise en valeur des sols *f*, mise en valeur des terres *f*, restauration des terres *f*; **~ use** *n* SOIL, WATER SUPP occupation des sols *f*; **~-use plan** *n* WATER SUPP plan d'occupation des sols *m*; **~ zoning** *n* SOIL zonage des terres *m*

landfill[1]: **~-external** *adj* WASTE extérieur à la décharge

landfill[2] *n* WASTE décharge *f*, lieu de décharge *m*, site de décharge *m*; **~ age** WASTE âge de décharge *m*; **~ area** WASTE zone de décharge *f*; **~ area conservation** WASTE protection d'une zone de décharge *f*; **~ binder** WASTE liant de décharge *m*; **~ body** SOIL corps de décharge *m*; **~ bottom** WASTE fond de décharge *m*; **~ building** WASTE aménagement d'une décharge *m*; **~ capacity** WASTE capacité de décharge *f*; **~ cell** WASTE alvéole de décharge *m*, casier de décharge *m*, compartiment de décharge *m*; **~ charge** POLICY taxe sur le stockage des déchets ménagers et assimilés *f*; **~ class** WASTE classe de décharge *f*; **~ compactor** WASTE compacteur de déchets *m*; **~ design** WASTE conception d'une décharge *f*, profitage de décharge *m*; **~ foundation** WASTE fondation d'une décharge *f*; **~ gas** AIR gaz de décharge *m*; **~ gas analysis** ANALYSIS analyse de gaz de décharge *f*; **~ gas build-up** WASTE formation de gaz de décharge *f*; **~ gas composition** WASTE composition des gaz de décharge *f*; **~ gas development** WASTE formation de gaz de décharge *f*; **~ gas drainage** WASTE évacuation des gaz de décharge *f*; **~ gas limit** WASTE seuil d'émission de gaz de décharge *m*; **~ gas measurement** ANALYSIS mesure des gaz de décharge *f*; **~ gas problem** WASTE problème de formation de gaz de décharge *m*; **~ gas production** WASTE production de gaz de décharge *f*; **~ gas quantity** WASTE volume de gaz produit par une décharge *m*; **~ gas use** WASTE valorisation des gaz de décharge *f*; **~ installation** WASTE aménagement d'une décharge *m*; **~ leachate** WASTE lixiviat de décharge *m*; **~ leachate purification** WASTE épuration des lixiviats de décharge *f*; **~ leachate tank** WASTE, WATER POLL réservoir des lixiviats de décharge *m*, cuve de stockage des lixiviats *f*; **~ location** WASTE localisation d'une décharge *f*; **~ location security** WASTE sûreté de localisation d'une décharge *f*; **~ material** SOIL, WASTE matériau de remblai *m*; **~ operation** WASTE exploitation d'une décharge *f*; **~ phase** WASTE stade de mise en décharge *m*; **~ progress** WASTE suivi d'une décharge *m*; **~ register** POLICY registre de mise en décharge *m*; **~ requirement** WASTE conditions de mise en décharge *f pl*; **~ roads** *n pl* WASTE voies d'accès à une décharge *f pl*; **~ site** MARINE site de mise en décharge *m*; **~ slope** WASTE pente de décharge *f*; **~ structure** WASTE aménagement intérieur d'une décharge *m*; **~-suitable vehicle** WASTE véhicule tout terrain pour décharge *m*; ~ **surface** WASTE aire de décharge *f*, surface de décharge *f*; ~ **technique** WASTE technique de mise en décharge *f*; ~ **thickness** WASTE épaisseur de décharge *f*; ~ **type** WASTE type de décharge *m*; ~ **underground** WASTE sous-sol de décharge *m*; ~ **volume** WASTE volume de décharge *m*

landfill[3] *vt* SOIL combler, remblayer, *residues from flue-gas cleaning* mettre en décharge, enfouir dans le sous-sol, WASTE combler, enfouir dans le sous-sol, mettre en décharge

landfilled *adj* SOIL remblayé, mis en décharge, enfoui dans le sous-sol, WASTE comblé, enfoui dans le sous-sol, mis en décharge

landfilling *n* SOIL épandage *m*, remblayage *m*, WASTE mise en décharge *f*; ~ **ability** WATER POLL *wastewater sludge ash* aptitude à la mise en décharge *f*; ~ **practice** WASTE technique de mise en décharge *f*

landlocked: ~ **sea** *n* AIR, MARINE mer fermée *f*

landscape *n* POLICY paysage *m*; ~ **architect** MANAGE architecte-paysagiste *m*; ~ **protection** POLICY sauvegarde du paysage *f*

landscaping *n* RENEW ENERG *effect of renewable energy technology* aménagement du paysage *m*

landslide *n* SOIL éboulement *m*, glissement de terrain *m*

landslip *n* SOIL éboulement *m*, glissement de terrain *m*

large[1]: ~**-scale** *adj* RENEW ENERG, WATER POLL à grande échelle

large[2]: ~ **area use** *n* RENEW ENERG application à grande échelle *f*, consommation à grande échelle *f*; ~ **area use of solar energy** *n* RENEW ENERG application de l'énergie solaire à grande échelle *f*; ~ **bubble aeration** *n* WATER POLL aération par grosses bulles *f*; ~**-scale clarification plant** *n* WATER POLL usine de clarification à grande échelle *f*; ~**-scale installation** *n* RENEW ENERG *wind, solar* installation à grande échelle *f*; ~**-size container** *n* WASTE conteneur de grandes dimensions *m*, conteneur grande contenance *m*

laser: ~ **beam** *n* ANALYSIS, WATER SUPP faisceau laser *m*; ~**-enhanced ionization** *n* (*LEI*) ANALYSIS ionisation au laser *f*; ~**-excited atomic fluorescence spectrometry** *n* (*LEAFS*) ANALYSIS spectrométrie de fluorescence atomique avec excitation laser *f*; ~ **mass spectrometry** *n* (*LAMS*) ANALYSIS spectrométrie de masse au laser *f*; ~ **microprobe mass analysis** *n* ANALYSIS analyse de masse avec microsonde laser *f*; ~ **radar** *n* ANALYSIS *measurement of pollutant concentration* radar à laser *m*; ~ **radiation** *n* SAFETY rayonnement laser *m*; ~ **radiation hazard** *n* SAFETY risque lié au rayonnement laser *m*; ~ **Raman spectroscopy** *n* ANALYSIS spectroscopie laser à effet Raman *f*; ~ **sampler** *n* ANALYSIS *solids sampling* échantillonneur laser *m*; ~ **spectroscopy** *n* ANALYSIS spectroscopie laser *f*

latent: ~ **defect** *n* MANAGE vice caché *m*; ~ **heat** *n* AIR chaleur latente *f*

lateral: ~ **sewer** *n* WASTE, WATER POLL, WATER SUPP égout latéral *m*; ~ **stapling** *n* WASTE agrafage latéral *m*

lava: ~ **slag** *n* WASTE, WATER POLL *trickling filter-packing material* scories volcaniques *f pl*

lawful *adj* POLICY juridique, licite

layer *n* GEN couche *f*; ~ **formation** AIR *on collecting electrodes of electrostatic precipitator* formation en couches *f*, formation stratifiée *f*; ~ **sampling** ANALYSIS échantillonnage laminaire *m*; ~ **strength** AIR *filters* résistance d'une couche *f*; ~ **thickness** ANALYSIS, NOISE épaisseur de couche *f*

layered *adj* ANALYSIS en couche

laying *n* SOIL *of liners* pose *f*

LBG *abbr* (*low-thermal-value biogas*) RENEW ENERG biogaz à faible pouvoir calorifique *m*

LC *abbr* (*lethal concentration*) AIR, ANALYSIS, SAFETY, WATER SUPP concentration létale (*CL*)

LCA *abbr* (*life-cycle analysis*) MANAGE, POLICY évaluation du cycle de vie *f*, analyse du cycle de vie *f*, analyse sur la totalité du cycle *f*

LCD[1] *abbr* (*liquid-crystal display*) ANALYSIS affichage à cristaux liquides *m*

LCD[2]: ~ **screen** *n* ANALYSIS écran d'affichage à cristaux liquides *m*

LC-EC *abbr* (*liquid chromatography-electrochemical detection*) ANALYSIS détection électrochimique par chromatographie en phase liquide *f*

LC-FL *abbr* (*liquid chromatography-fluorescence detection*) ANALYSIS détection de fluorescence par chromatographie en phase liquide *f*

LC-LC *abbr* (*liquid chromatography-liquid chromatography*) ANALYSIS chromatographie de partage liquide-liquide *f*

LC-UV: ~ **detection** *n* (*liquid chromatography-UV detection*) ANALYSIS détection d'UV par chromatographie en phase liquide *f*

LC_0 *abbr* (*lethal concentration zero*) SAFETY CL_0 (*concentration létale zéro*)

LC_{50} *abbr* (*lethal concentration fifty*) AIR, ANALYSIS, SAFETY, WATER SUPP CL_{50} (*concentration létale à 50%, concentration létale cinquante*)

LC_x *abbr* (*lethal concentration x*) SAFETY CL_x (*concentration létale x*)

LD *abbr* (*lethal dose*) AIR, ANALYSIS, SAFETY, WATER SUPP DL (*dose létale*)

LDL_0 *abbr* (*lethal dose level 0*) SAFETY niveau de dose létale 0 *m*

LD_{50} *abbr* (*median lethal dose*) SAFETY DL_{50} (*dose létale 50, dose semi-létale*)

LD_x *abbr* (*lethal dose x*) SAFETY DL_x (*dose létale x*), dose létale indéterminée *f*

leach[1] *n* SOIL, WASTE, WATER POLL, WATER SUPP discontinuité hydraulique *f*

leach[2] *vt* SOIL lixivier, WASTE lessiver, lixivier, *percolate* filtrer, WATER POLL, WATER SUPP lessiver, lixivier

leach[3] *vi* MARINE ruisseler, WASTE filtrer, suinter

leachability *n* SOIL, WASTE, WATER POLL, WATER SUPP lixiviabilité *f*; ~ **test** ANALYSIS, WASTE essai de lessivage *m*

leachable[1] *adj* SOIL, WASTE, WATER POLL, WATER SUPP lessivable

leachable[2]: ~ **waste** *n* WASTE déchets lessivables *m pl*

leachate *n* SOIL lixiviat *m*, WASTE *from storage area* lixiviat *m*, lessivat *m*, WATER POLL eau de lessivage *f*, lessivat *m*, WATER SUPP *from storage area* lixiviat *m*, lessivat *m*; ~ **behavior** *AmE*, ~ **behaviour** *BrE* SOIL, WASTE comportement des produits de lixiviation *m*, comportement des lixiviats *m*; ~ **collection** WASTE collecte des lixiviats *f*; ~ **composition** WASTE composition des eaux de ruissellement *f*, composition des lixiviats *f*; ~ **drainage** WASTE, WATER SUPP évacuation des lixiviats *f*, drainage des eaux de ruissellement *m*; ~ **drainage with filtering rubble** WASTE drainage des eaux de ruissellement sur une couche de gravier *m*, évacuation des lixiviats avec filtre à gravier *f*; ~ **production** WASTE production de lixiviats *f*; ~ **treatment** WASTE traitement des lixiviats *m*; ~ **yield** WASTE taux de lixiviation *m*

leaching *n* SOIL migration dans les sols *f*, lixiviation *f*, WASTE suintement *m*, lixiviation *f*, WATER POLL, WATER SUPP lessivage *m*, lixiviation *f*, ruissellement *m*; ~ **behavior** *AmE*, ~ **behaviour** *BrE* SOIL, WASTE, WATER SUPP comportement au ruissellement *m*, comportement au lessivage *m*; ~ **properties** *n pl* WATER SUPP propriétés lixiviantes *f pl*; ~ **test** ANALYSIS, SOIL, WASTE essai de lessivage *m*, essai de lixiviation *m*; ~ **trench** WATER SUPP fossé d'infiltration *m*

lead[1] *n* (*Pb*) GEN plomb *m* (*Pb*); ~**-acid battery** AIR accumulateur au plomb *m*; ~ **additive** WATER SUPP additif au plomb *m*; ~**-containing dust** WATER SUPP poussière contenant du plomb *f*; ~**-containing fuel** AIR carburant au plomb *m*; ~**-free fuel** AIR carburant sans plomb *m*; ~**-free gasoline** *AmE* (*cf lead-free petrol BrE*) WATER SUPP essence sans plomb *f*; **~-free petrol** *BrE* (*cf lead-free gasoline AmE*) WATER SUPP essence sans plomb *f*; **~ poisoning** SAFETY empoisonnement au plomb *m*; **~ seal** WASTE cachet en plomb *m*; **~-sealing plier** WASTE pince à plomber *f*

lead[2] *vt* AIR conduire, entraîner

leader: **~ country** *n* POLICY pays phare *m*

leaf: **~-area index** *n* AIR indice foliaire *m*

LEAFS *abbr* (*laser-excited atomic fluorescence spectrometry*) ANALYSIS spectrométrie de fluorescence atomique avec excitation laser *f*

leak[1] *n* WATER SUPP perte *f*, fuite *f*; **~ detection** AIR, ANALYSIS, SOIL, WASTE détection de fuite *f*; **~ detector** ANALYSIS détecteur de fuite *m*; **~ water** WATER SUPP eau de fuite *f*

leak[2] *vi* MARINE, SAFETY, WASTE, WATER POLL fuir

leakage *n* WATER SUPP fuite *f*, perte *f*, échappement *m*, coulage *m*, écoulement *m*; **~ indicator system** SAFETY indicateur de fuite *m*; **~ test** WASTE essai de fuite *m*; **~ warning** SAFETY alerte aux fuites *f*; **~ water** WASTE, WATER SUPP eau d'infiltration *f*

lean: **~ eco-balancing** *n* MANAGE, POLICY éco-bilan pauvre *m*; **~-mix engine** *n* AIR moteur à mélange pauvre *m*; **~-mix engine with oxidation catalyst** *n* AIR moteur à mélange pauvre avec catalyseur d'oxydation *m*

leasing *n* MANAGE affermage *m*, POLICY concession *f*, crédit-bail *m*, location *f*; **~ of certificate** POLICY octroi d'agrément *m*

LED[1] *abbr* (*light-emitting diode*) ANALYSIS DEL (*diode électroluminescente*)

LED[2]: **~ display** *n* ANALYSIS affichage à diode électroluminescente *m*

legal[1] *adj* POLICY juridique, licite

legal[2]: **~ instrument** *n* POLICY instrument juridique *m*; **~ requirements** *n pl* POLICY conditions juridiques *f pl*; **~ standards** *n pl* POLICY critères juridiques *m pl*

legislation *n* POLICY législation *f*

legislator *n* POLICY législateur *m*

LEI *abbr* (*laser-enhanced ionization*) ANALYSIS ionisation au laser *f*

lens *n* ANALYSIS, RENEW ENERG *solar energy plant* lentille *f*

lethal[1] *adj* SAFETY létal

lethal[2]: **~ concentration** *n* (*LC*) AIR, ANALYSIS, SAFETY, WATER SUPP concentration létale *f* (*CL*); **~ concentration fifty** *n* (LC_{50}) AIR, ANALYSIS, SAFETY, WATER SUPP concentration létale cinquante *f* (CL_{50}), concentration létale à 50% (CL_{50}); **~ concentration x** *n* (LC_x) SAFETY concentration létale x *f* (CL_x); **~ concentration zero** *n* (LC_0) SAFETY concentration létale zéro *f* (CL_0); **~ dose** *n* (*LD*) AIR, ANALYSIS, SAFETY, WATER SUPP dose létale *f* (*DL*); **~ dose fifty** *n* (LD_{50}) SAFETY dose létale 50 *f* (DL_{50}), dose semi-létale *f* (DL_{50}); **~ dose x** *n* (LD_x) SAFETY dose létale indéterminée *f*, dose létale x *f* (DL_x); **~ dose level 0** *n* (LDL_0) SAFETY niveau de dose létale 0 *m*; **~ effect** *n* SAFETY effet létal *m*

level *n* GEN niveau *m*; **~ indicator** ANALYSIS indicateur de niveau *m*, limnimètre *m*; **~ recorder** ANALYSIS *of water* contrôleur de niveau *m*, NOISE enregistreur de niveau *m*; **~ reduction** MARINE baisse de niveau *f*, NOISE réduction de niveau *f*, WATER SUPP baisse de niveau *f*

levy *n* POLICY droit *m*, prélèvement obligatoire *m*, taxe *f*

liability *n* POLICY passif *m*, responsabilité *f*

licence *n* *BrE* POLICY, WATER SUPP autorisation *f*, licence *f*; **~ charge** *BrE* MANAGE redevance d'autorisation *f*

license *AmE see licence BrE*

licensed *adj* POLICY sous licence

licensing *n* POLICY, WATER SUPP délivrance d'autorisation *f*; **~ application** POLICY demande d'autorisation *f*; **~ authority** POLICY instance délivrant un permis *f*, instance délivrant une autorisation *f*; **~ conference** POLICY, WATER SUPP conférence sur les procédures d'autorisation *f*, délibération sur l'octroi de licence *f*, délibération sur les procédures d'autorisation *f*; **~ impact** POLICY répercussion d'une autorisation *f*; **~ law** POLICY réglementation relative à l'octroi de permis *f*; **~ mark** MANAGE marque déposée *f*; **~ process** POLICY, WATER SUPP procédure d'autorisation *f*

licker *n* MARINE, WATER POLL écrémeuse *f*, racleuse *f*

lid *n* AIR *meteorology* couvercle *m*

LIDAR *abbr* (*light detection and ranging*) ANALYSIS lidar *m*

life *n* GEN durée de vie *f*, vie *f*; **~-cycle analysis** (*LCA*) MANAGE, POLICY évaluation du cycle de vie *f*, analyse du cycle de vie *f*, analyse sur la totalité du cycle *f*; **~-cycle assessment** MANAGE évaluation du cycle de vie *f*; **~-cycle cost** MANAGE coût global *m*; **~-cycle inventory** AIR inventaire du cycle de vie *m*, MANAGE bilan du cycle de vie *m*, WATER POLL inventaire du cycle de vie *m*

lifespan *n* AIR, ANALYSIS, SOIL, WASTE durée de vie *f*

lifetime *n* AIR, ANALYSIS, MANAGE, WASTE durée de vie

lift *n* RENEW ENERG *wind turbine* force utile *f*, portance *f*, WATER SUPP hauteur *f*, portance *f*, hauteur d'élévation *f*, *valve, control rod* hauteur manométrique *f*, levée *f*, course *f*, *canal locks* chute *f*; **~-and-force pump** WATER POLL, WATER SUPP pompe aspirante et foulante *f*, pompe aspirante et refoulante *f*; **~ pump** ANALYSIS, WATER POLL, WATER SUPP pompe aspirante *f*, pompe élévatoire *f*, pompe de reprise *f*, pompe soulevante *f*

lifting *n* WATER SUPP *by machine* levage *m*

light[1]: **~-proof** *adj* WASTE étanche à la lumière

light[2]: **~ detection and ranging** *n* (*LIDAR*) ANALYSIS lidar *m*; **~-emitting diode** *n* (*LED*) ANALYSIS diode électroluminescente *f* (*DEL*); **~-induced pulse** *n* ANALYSIS impulsion photonique *f*; **~ measuring** *n* ANALYSIS mesure de la lumière *f*; **~-optical microscope** *n* ANALYSIS microscope optique *m*; **~ receiver** *n* ANALYSIS récepteur de lumière *m*; **~ scattering** *n* ANALYSIS *dispersion* diffraction de la lumière *f*; **~-scattering detector** *n* ANALYSIS détecteur de dispersion de la lumière *m*; **~ sender** *n* ANALYSIS émetteur de lumière *m*

lighten *vi* GEN éclairer, alléger, luire

lightening *n* GEN éclairement *m*, allègement *m*, délestage *m*; **~ vessel** MARINE navire de délestage *m*

lighter *n* MARINE allège *f*, allumoir *m*, gabare *f*

lighting: **~ installation** *n* RENEW ENERG installation d'éclairage *f*

lightly: **~-compacted landfill** *n* WASTE décharge légèrement compactée *f*

lightweight: **~ panel** *n* NOISE panneau léger *m*

ligneous: **~ prunings** *n pl* RENEW ENERG émondes ligneuses *f pl*

lignite: **~ power station** *n* AIR, RENEW ENERG centrale thermique au lignite *f*

lime[1]: **~-aggressive** *adj* WATER POLL caustique

lime[2] *n* AIR, SOIL, WATER POLL, WATER SUPP chaux *f*; **~ carbonate** WATER POLL carbonate de chaux *m*; **~ scale** SOIL, WATER POLL calcaire *m*; **~ scrubbing** AIR *wet flue gas desulphurization* lavage à la chaux *m*; **~ slaking** AIR extinction à la chaux *f*; **~ slurry** AIR lait de chaux *m*; **~ softening** WATER SUPP décarbonisation à l'eau de chaux *f*

limestone: **~ ballast** *n* AIR ballast calcaire *m*; **~ meal** *n* AIR *additive* calcaire pulvérulent *m*; **~ suspension** *n* AIR suspension de calcaire *f*

limit[1]: **~ concentration** *n* AIR, SAFETY concentration limite *f*; **~ of inflammability** *n* AIR seuil d'inflammabilité *m*; **~ value** *n* POLICY, SAFETY valeur limite *f*; **~-value control** *n* ANALYSIS contrôle de valeur limite *m*

limit[2] *vt* GEN *dispersion of hazardous substances* limiter

limnology *n* ANALYSIS limnologie *f*

line[1] *n* GEN filière *f*, ligne *f*, pic *m*; **~ source** AIR, WATER SUPP source linéaire *f*, source alignée *f*

line[2] *vt* NOISE revêtir, SOIL *bioremediation pit* garnir d'un revêtement étanche

linear: **~ absorption coefficient** *n* ANALYSIS coefficient d'absorption linéaire *m*

lined[1] *adj* AIR, NOISE revêtu, doublé

lined[2]: ~ **bag** *n* WASTE sac doublé *m*
liner *n* WASTE *of pinch device* peau d'étanchéité *f*, WATER SUPP chemise *f*; ~ **bag** WASTE doublure *f*; ~ **sheet** WASTE feuille de doublure *f*
lining *n* AIR *of catalyst*, NOISE, WATER POLL *of basin, tank* revêtement *m*, chemisage *m*, WATER SUPP *of pump* chemise *f*; ~ **compound** WASTE matériau de doublure *m*; ~ **paper** WASTE doublure *f*, feuille de garde *f*; ~ **of surface** WASTE revêtement de surface *m*
link *vt* WATER POLL lier
lipophilic *adj* WATER POLL lipophile
lipophobic *adj* WATER POLL lipophobe
liposolubility *n* WATER POLL liposolubilité *f*
liquefaction *n* AIR, RENEW ENERG, WATER SUPP *biomass conversion* liquéfaction *f*
liquefied: ~ **natural gas** *n* (*LNG*) AIR gaz naturel liquéfié *m* (*GNL*); ~ **petroleum gas** *n* (*LPG*) AIR, RENEW ENERG gaz de pétrole liquéfié *m* (*GPL*)
liquefy *vt* AIR, RENEW ENERG, WATER SUPP liquéfier
liquid[1] *adj* GEN liquide; ~**-proof** GEN étanche
liquid[2] *n* GEN liquide *m*; ~ **chromatography detector** (*LCD*) ANALYSIS détecteur de chromatographie en phase liquide *m*; ~ **chromatography-electrochemical detection** (*LC-EC*) ANALYSIS détection électrochimique par chromatographie en phase liquide *f*; ~ **chromatography-fluorescence detection** (*LC-FL*) ANALYSIS détection de fluorescence par chromatographie en phase liquide *f*; ~ **chromatography-liquid chromatography** (*LC-LC*) ANALYSIS chromatographie de partage liquide-liquide *f*; ~ **chromatography-UV detection** (*LC-UV detection*) ANALYSIS détection d'UV par chromatographie en phase liquide *f*; ~ **column** ANALYSIS colonne de liquide *f*; ~ **composting** WATER POLL compostage liquide *m*; ~**-crystal display** (*LCD*) ANALYSIS affichage à cristaux liquides *m*; ~**-crystal display screen** ANALYSIS écran d'affichage à cristaux liquides *m*; ~ **discharge** AIR *absorber* rejet liquide *m*; ~ **distributor** AIR distributeur de liquide *m*; ~ **droplet** AIR gouttelette *f*, vésicule *f*; ~ **film** AIR pellicule liquide *f*; ~ **gel chromatography** ANALYSIS chromatographie en phase liquide sur gel *f*; ~**-in-glass thermometer** ANALYSIS thermomètre à dilatation de liquide dans une gaine de verre *m*; ~ **inlet** AIR *bubble-cap tray tower* admission de liquide *f*; ~ **layer** AIR couche liquide *f*; ~**-liquid extraction** ANALYSIS extraction liquide-liquide *f*; ~ **manure** WATER POLL, WATER SUPP lisier *m*, purin *m*; ~ **metal** WATER SUPP métal liquide *m*; ~ **particle** AIR, WATER POLL particule liquide *f*; ~ **phase** AIR phase liquide *f*; ~**-proof carton** WASTE carton étanche *m*; ~**-reacting agent** ANALYSIS réactif liquide *m*; ~ **residue** AIR résidus liquides *m pl*; ~ **sludge** WATER POLL boues liquides *f pl*; ~**-sludge conditioning** WATER POLL conditionnement des boues liquides *m*; ~**-solid extraction** ANALYSIS extraction liquide-solide *f*; ~ **spring thermometer** ANALYSIS thermomètre à colonne de liquide *m*; ~ **state of aggregation** AIR phase d'agrégation liquide *f*; ~ **surface** AIR surface liquide *f*; ~ **waste** WASTE, WATER POLL, WATER SUPP déchets liquides *m pl*, effluent liquide *m*; ~ **waste substance** AIR substance de déchets liquides *f*; ~ **working medium** RENEW ENERG *heat pump* caloporteur liquide *m*
lithographed: ~ **package** *n* WASTE emballage lithographié *m*
lithosphere *n* WATER SUPP lithosphère *f*
litter *n* SOIL litière du sol *f*, WASTE déchets sauvages *m pl*, immondices *m pl*, litière *f*
littoral: ~ **fringe** *n* MARINE frange littorale *f*, cordon littoral *m*; ~ **zone** *n* WATER POLL zone littorale *f*
live: ~ **bacteria** *n pl* WATER POLL bactéries vivantes *f pl*; ~ **storage** *n* WATER SUPP réserve vidangeable *f*, tranche vidangeable *f*
livestock: ~ **unit** *n* RENEW ENERG unité gros bétail *f*
living: ~ **community** *n* WATER SUPP communauté vivante *f*; ~ **resources** *n pl* RENEW ENERG ressources vivantes *f pl*
lixiviate *vt* SOIL, WASTE, WATER POLL, WATER SUPP lessiver, lixivier
lixiviation *n* SOIL, WASTE, WATER POLL, WATER SUPP lessivage *m*, lixiviation *f*
LNB *abbr* (*low NO_x burner, low nitrogen burner*) AIR brûleur à faibles émissions de NO_x *m*
LNG *abbr* (*liquefied natural gas*) AIR GNL (*gaz naturel liquéfié*)
load[1] *n* AIR, ANALYSIS, WATER POLL, WATER SUPP charge *f*; ~ **capacity test** ANALYSIS essai de capacité en charge *m*
load[2] *vt* AIR charger
loaded[1] *adj* AIR chargé
loaded[2]: ~ **activated coke** *n* AIR coke actif chargé *m*; ~ **adsorbent** *n* AIR adsorbant chargé *m*; ~ **filter** *n* AIR filtre chargé *m*; ~ **solvent** *n* AIR solvant chargé *m*; ~ **suspension** *n* AIR suspension chargée *f*; ~ **washing agent** *n* AIR agent de lavage chargé *m*
loading *n* GEN chargement *m*, remplissage *m*, *sewage* mise en charge *f*; ~ **capacity** AIR capacité de chargement *f*; ~ **distribution** AIR répartition de chargement *f*; ~ **of dry mass** WATER POLL concentration de masse sèche *f*; ~ **hopper** WASTE trémie de chargement *f*; ~ **hose** WASTE manche de chargement *f*; ~ **mechanism** WASTE mécanisme de chargement *m*; ~ **pit** WASTE puits de chargement *m*; ~ **tractor** SOIL tracteur de chargement *m*; ~ **worm** WASTE serpentin de chargement *m*, goulotte de chargement *f*
LOAEL *abbr* (*lowest observed adverse effect level*) SAFETY niveau de perturbation observé le plus faible *m*
loam *n* WATER SUPP glaise *f*
lobby[1] *n* POLICY groupe de pression *m*
lobby[2] *vi* POLICY faire pression
local[1] *adj* GEN local
local[2]: ~ **authorities** *n pl* MANAGE, POLICY collectivités locales *f pl*; ~ **emission source** *n* AIR source locale d'émission *f*; ~ **emissions** *n pl* AIR émissions locales *f pl*; ~ **government** *n* POLICY collectivité locale *f*; ~ **politicians** *n pl* POLICY élus locaux *m pl*
localization *n* SOIL *of contaminated site* localisation *f*
locally: ~**-representative** *adj* ANALYSIS *sample* localement représentatif
lock *n* WATER SUPP écluse *f*; ~ **hopper** SOIL trémie à obturateur *f*; ~ **side wall** WATER SUPP bajoyer *m*; ~ **weir** RENEW ENERG barrage *m*
LOEC *abbr* (*lowest observed effect concentration*) SAFETY concentration à effet le plus bas observé *f*, concentration à effet le plus faible observé *f*
logarithmic: ~ **calibration** *n* ANALYSIS étalonnage logarithmique *m*
logistics: ~ **of disposal** *n pl* AIR, MANAGE, WATER SUPP logistique d'évacuation *f*
long[1]: ~**-running** *adj* ANALYSIS *of gas* à longue durée de fonctionnement; ~**-term** *adj* ANALYSIS, POLICY à long terme
long[2]: ~**-distance photometer** *n* ANALYSIS photomètre longue distance *m*; ~**-life radioactive waste** *n* WASTE, WATER SUPP déchets radioactifs à vie longue *m pl*; ~**-path monitor** *n* ANALYSIS moniteur longue distance *m*; ~**-term behavior** *AmE*, ~**-term behaviour** *BrE* *n* GEN comportement à long terme *m*, tenue à long terme *f*; ~**-term control** *n* AIR, NOISE contrôle à long terme *f* lutte à long terme *m*; ~**-term detector tube** *n* ANALYSIS tube détecteur à long terme *m*; ~**-term dosimeter** *n* ANALYSIS dosimètre à long terme *m*; ~**-term durability** *n* AIR, MARINE, NOISE durabilité à long terme *f*; ~**-term effect** *n* AIR, MARINE, NOISE, POLICY effet à long terme *m*; ~**-term gas analysis** *n* ANALYSIS analyse de gaz à long terme *f*; ~**-term measure** *n* MANAGE, POLICY mesure à long terme *f*; ~**-term monitoring** *n* AIR, ANALYSIS, MARINE, NOISE suivi à long terme *m*, surveillance à long terme *f*; ~**-term security** *n* WASTE sûreté à long terme *f*,

sécurité à long terme *f*; **~-term storage** *n* RENEW ENERG, WASTE stockage à long terme *m*; **~-throw sprinkler** *n* WATER POLL asperseur longue portée *m*; **~-wave heat radiation** *n* AIR rayonnement de chaleur à onde longue *m*; **~-wave reradiation from the earth** *n* AIR rayonnement à onde longue réfléchi par la terre *m*

longevity *n* AIR *of CFCs*, SOIL *of contaminants* longévité *f*

longitudinal: **~ scraper** *n* WATER POLL racleur longitudinal *m*

loop *n* RENEW ENERG boucle *f*; **~-type bubble column** WATER POLL colonne de barbotage en boucle *f*

looping: **~ smoke plume** *n* AIR panache en boucle *m*

loose *adj* SOIL meuble

loosen *vt* SOIL, WATER SUPP ameublir

loosening: **~ up** *n* SOIL, WATER SUPP détente *f*

loss: **~ on ignition** *n* WATER POLL, WATER SUPP perte au feu *f*

loudness *n* NOISE intensité sonore *f*, volume sonore *m*; **~ level** NOISE niveau d'intensité sonore *m*; **~ measurement** ANALYSIS, NOISE mesure du bruit *f*; **~ perception** NOISE perception de l'intensité sonore *f*

low[1]: **~-energy** *adj* RENEW ENERG peu consommateur d'énergie; **~-level radioactive** *adj* WASTE faiblement radioactif; **~-nitrogen** *adj* AIR pauvre en azote; **~-noise** *adj* NOISE peu bruyant; **~-pollutant** *adj* AIR, SOIL faible en polluants, peu polluant; **~-polluting** *adj* AIR, SOIL peu polluant; **~-pressure** *adj* AIR à basse pression; **~-rate** *adj* WATER POLL *pump* à faible charge, à faible débit; **~-temperature** *adj* AIR, SOIL à basse température; **~-voltage** *adj* AIR à basse tension

low[2]: **~ calorific value** *n* WASTE pouvoir calorifique faible *m*; **~-contaminated municipal wastewater** *n* WATER SUPP, WATER POLL effluent urbain peu contaminé *m*; **~-density polyethylene** *n* WASTE *thermoplastic* polyéthylène basse densité *m*; **~-emissions vehicle** *n* AIR véhicule peu polluant *m*; **~-energy automobile** *n* AIR automobile à faible consommation *f*; **~-frequency absorber** *n* NOISE absorbeur basse fréquence *m*; **~-intensity magnet separator** *n* SOIL séparateur magnétique de faible puissance *m*; **~ lethal dose** *n* SAFETY faible dose létale *f*; **~-lime content rubble** *n* WASTE, WATER SUPP gravats à faible teneur en chaux *m pl*; **~ nitrogen burner** *n* (*LNB*) AIR brûleur à faibles émissions de NO_x *m*; **~ NO_x burner** *n* (*LNB*) AIR brûleur à faibles émissions de NO_x *m*; **~ NO_x combustor** *n* AIR brûleur à faibles émissions de NO_x *m*; **~-noise building machinery** *n* NOISE matériel de terrassement peu bruyant *m*, matériel de construction insonorisé *m*; **~-noise construction** *n* NOISE construction insonorisée *f*; **~-noise equipment** *n* NOISE matériel insonorisé *m*; **~-noise machine** *n* NOISE machine insonorisée *f*; **~-noise working method** *n* NOISE méthode de travail peu bruyante *f*; **~ pressure** *n* AIR basse pression *f*; **~-pressure backwashing** *n* AIR, MARINE, WATER POLL rétrolavage basse pression *m*; **~-pressure cleaning** *n* AIR lavage basse pression *m*; **~-pressure flushing** *n* WATER POLL chasse basse pression *f*; **~-pressure pump** *n* WATER POLL pompe basse pression *f*; **~-pressure side** *n* AIR branche basse pression *f*; **~-pressure steam** *n* AIR vapeur à basse pression *f*; **~-pressure supply** *n* WATER SUPP alimentation basse pression *f*; **~-pressure Venturi scrubber** *n* AIR laveur à Venturi basse pression *m*; **~-pressure wet oxidation** *n* WATER POLL oxydation par voie humide à basse pression *f*; **~-rate process** *n* WATER POLL *activated sludge* procédé à faible charge *m*; **~-rate trickling filter** *n* WATER POLL lit bactérien à faible charge *m*; **~ raw-gas load** *n* WASTE faible charge en gaz brut *f*; **~-residue production method** *n* WASTE méthode de production laissant peu de résidus *f*; **~ soil permeability** *n* SOIL faible perméabilité du sol *f*; **~ temperature** *n* RENEW ENERG, WATER POLL basse température *f*; **~-temperature carbonization wastewater** *n* WATER POLL effluent de carbonisation à basse température *m*; **~-temperature conversion** *n* WATER POLL *wastewater sludge incineration* conversion à basse température *f*; **~-temperature heat distribution** *n* RENEW ENERG distribution de chaleur basse température *f*; **~-temperature solar collector plant** *n* RENEW ENERG centrale solaire basse température *f*; **~-thermal-value biogas** *n* (*LBG*) RENEW ENERG biogaz à faible pouvoir calorifique *m*; **~ tide** *n* MARINE marée basse *f*; **~-tide water** *n* MARINE marée basse *f*; **~ toxic concentration** *n* SAFETY faible concentration toxique *f*; **~ toxic dose** *n* SAFETY faible dose toxique *f*; **~-voltage electrostatic precipitator** *n* AIR électrofiltre basse tension *m*; **~-voltage ESP** *n* AIR électrofiltre basse tension *m*; **~-voltage lamp** *n* ANALYSIS lampe basse tension *f*; **~-waste technology** *n* AIR, WATER SUPP technologie peu polluante *f*, technologie produisant peu de déchets *f*; **~ water** *n* MARINE marée basse *f*, WATER SUPP *in river, lake* étiage *m*, *tide* basses eaux *f pl*, marée basse *f*, basse mer *f*; **~-water discharge** *n* WATER SUPP débit d'étiage *m*; **~-water level** *n* WATER SUPP niveau d'étiage *m*

lower[1]: **~ bed** *n* AIR *catalyst* lit bas *m*; **~ calorific value** *n* WASTE (*LCV*) pouvoir calorifique inférieur *m*; **~ course** *n* WATER SUPP cours inférieur *m*; **~ explosion limit** *n* SAFETY seuil d'explosion *m*; **~-level administration** *n* POLICY échelon administratif inférieur *m*; **~ level of detectability** *n* ANALYSIS seuil inférieur de détectabilité *m*; **~ limit of detectability** *n* AIR, ANALYSIS limite de détection *f*, seuil de détection *m*, quantité minimale détectable *f*, sensibilité-limite de détection *f*

lower[2] *vt* AIR *temperature* baisser, diminuer, WATER SUPP *load* abaisser, descendre

lowest: **~ achievable emission rate** *n* AIR seuil d'émissions envisageable *m*; **~ observed adverse effect level** *n* (*LOAEL*) SAFETY niveau de perturbation observé le plus faible *m*; **~ observed effect concentration** *n* (*LOEC*) SAFETY concentration à effet le plus bas observé *f*, concentration à effet le plus faible observé *f*; **~ observed effect level** *n* SAFETY niveau efficace le plus faible observé *m*

LPG *abbr* (*liquefied petroleum gas*) AIR, RENEW ENERG GPL (*gaz de pétrole liquéfié*)

luminescent: **~ bacteria test** *n* ANALYSIS essai par bactérie luminescente *m*

luminous: **~ intensity** *n* ANALYSIS intensité lumineuse *f*

lung: **~ irritant** *n* SAFETY irritant pulmonaire *m*

lyolysis *n* ANALYSIS lyolyse *f*

lyophilization *n* ANALYSIS lyophilisation *f*

lyophilize *vt* ANALYSIS lyophiliser

lysimeter *n* ANALYSIS *leachate determination* lysimètre *m*

lysis *n* ANALYSIS lyse *m*

M

MAC[1] *abbr* (*maximum allowable concentration*) AIR, ANALYSIS, WATER SUPP *in the workplace* concentration maximale admissible *f*

MAC[2]: **~ in the free environment** *n* AIR, ANALYSIS, WATER SUPPconcentration maximale admissible à l'air libre *f*

maceration *n* WASTE *of waste* macération *f*

machine[1]: **~-coated** *adj* WASTE revêtu mécaniquement

machine[2]: **~ enclosure** *n* NOISE enceinte de matériel *f*; **~-glazed paper** *n* WASTE papier glacé par voie mécanique *m*; **~-made board** *n* WASTE carton de fabrication mécanique en continu *m*; **~-version verification of production quality** *n* WASTE vérification mécanique de la qualité de la production *f*

Mach-Zehnder: **~ interferometer** *n* ANALYSIS interféromètre de Mach-Zehnder *m*

macroclimate *n* AIR macroclimat *m*

macroencapsulation *n* SOIL macroconfinement *m*

macronutrient *n* WATER POLL, WATER SUPP macro-élément *m*

macropore *n* AIR *biological scrubber*, WATER POLL *activated carbon* macropore *m*

macroscopic: **~ biological water quality determination** *n* ANALYSIS détermination biomacroscopique de la qualité de l'eau *f*

macrowaste *n* WASTE macrodéchet *m*, déchet monstre *m*

magma *n* RENEW ENERG *geothermal energy resource* magma *m*

magnesium *n* (*Mg*) GEN magnésium *m* (*Mg*): **~ hardness** WATER SUPP titre hydrotimétrique magnésien *m*; **~ oxide scrubbing** AIR *wet flue-gas desulphurization* lavage à l'oxyde de magnésium *m*; **~ sulfate** *AmE*, **~ sulphate** *BrE* WATER SUPP sulfate de magnésium *m*

magnet *n* ANALYSIS, WASTE, WATER SUPP aimant *m*

magnetic[1] *adj* GEN magnétique

magnetic[2]: **~ coupling** *n* WATER POLL couplage magnétique *m*; **~ field** *n* ANALYSIS champ magnétique *m*; **~ interception plate** *n* WASTE plaque d'interception magnétique *f*; **~ oxygen analyser** *BrE*, **~ oxygen analyzer** *AmE* *n* ANALYSIS analyseur magnétique d'oxygène *m*; **~ sector mass spectrometer** *n* ANALYSIS spectromètre de masse à secteur magnétique *m*; **~ separation** *n* AIR, SOIL, WASTE, WATER POLL séparation magnetique *f*; **~ separation roller** *n* AIR rouleau de séparation magnétique *m*; **~ separator** *n* SOIL séparateur magnétique *m*, WASTE, WATER SUPP trieur magnétique *m*; **~ stirrer** *n* WATER POLL agitateur magnétique *m*; **~ susceptibility** *n* ANALYSIS sensibilité magnétique *f*

mailing: **~ sleeve** *n* WASTE manchon d'expédition *m*; **~ tube** *n* WASTE carton d'expédition cylindrique *m*

main: **~ analysis** *n* ANALYSIS analyse principale *f*; **~ catalyst** *n* AIR catalyseur principal *m*; **~ collector** *n* WASTE collecteur principal *m*; **~ combustion chamber** *n* AIR chambre de combustion principale *f*; **~ display** *n* ANALYSIS *on measuring instruments* affichage principal *m*; **~ drain** *n* WASTE, WATER POLL, WATER SUPP tout-à-l'égout *m*; **~ flow** *n* AIR écoulement principal *m*; **~ gas stream** *n* AIR flux de gaz principal *m*; **~ pipe** *n* WASTE, WATER SUPP conduite principale *f*; **~ sealing layer** *n* AIR, WASTE couche principale d'étanchéité *f*; **~ sewer** *n* WASTE, WATER POLL, WATER SUPP égout principal *m*, collecteur principal *m*, égout collecteur *m*, tout-à-l'égout *m*; **~ stream** *n* AIR courant principal *m*; **~ wash circulation** *n* AIR *main stream oxidation* circulation de lavage principal *f*; **~-wash circulation loop** *n* AIR *main stream oxidation* boucle de lavage principale *f*; **~ waste-air blower** *n* AIR ventilateur d'évents principal *m*

mains: **~ connection** *n* RENEW ENERG raccordement au secteur *m*

maintenance *n* GEN entretien *m*, entretien permanent *m*, maintenance *f*; **~ emissions** AIR émissions liées à l'entretien permanent *f pl*

major: **~ source** *n* AIR, WATER SUPP source principale *f*, source majeure *f*

make: **~-up fuel** *n* WATER POLL combustible d'appoint *m*; **~-up water** *n* AIR eau d'appoint *f*

malfunction *n* ANALYSIS, SAFETY dysfonctionnement *m*; **~ detection** SAFETY détection des dysfonctionnements *f*

malodorous: **~ plume** *n* AIR panache nauséabond *m*; **~ substance** *n* AIR, WATER POLL substance nauséabonde *f*, substance malodorante *f*, substance à odeur désagréable *f*, substance à odeur gênante *f*; **~ wastewater** *n* *BrE* AIR, WATER SUPP eaux usées nauséabondes *f pl*

mammoth: **~ pump** *n* WATER POLL pompe géante *f*; **~ rotor** *n* WATER POLL rotor d'émulsion *m*

man[1]: **~-made** *adj* RENEW ENERG humain, SOIL artificiel

man[2]: **~-caused erosion** *n* SOIL érosion anthropique *f*; **~-made dust** *n* AIR poussière due aux activités humaines *f*; **~-made earth tremor** *n* SOIL secousse sismique artificielle *f*; **~-made earthquake** *n* SOIL séisme artificiel *m*; **~-made fiber** *AmE*, **~-made fibre** *BrE* *n* AIR, WASTE, WATER SUPP fibre artificielle *f*, fibre synthétique *f*

management *n* MANAGE, POLICY gestion *f*, direction *f*

mandatory: **~ user charge** *n* POLICY redevance d'utilisation obligatoire *f*

manganese *n* (*Mn*) AIR manganèse *m* (*Mn*)

manhole *n* WATER SUPP trou d'homme *m*, trou de visite *m*; **~ cover** WATER SUPP tampon *m*

manner: **~ of sampling** *n* ANALYSIS mode d'échantillonnage *m*

manometer *n* AIR, ANALYSIS, WATER POLL, WATER SUPP manomètre *m*; **~ liquid** ANALYSIS liquide manométrique *m*; **~ value** ANALYSIS valeur manométrique *f*

manometric[1] *adj* AIR, ANALYSIS, WATER POLL, WATER SUPP manométrique

manometric[2]: **~ discharge head** *n* WATER POLL hauteur manométrique de refoulement *f*; **~ measurement** *n* ANALYSIS mesure manométrique *f*; **~ suction head** *n* WATER POLL hauteur manométrique d'aspiration *f*; **~ switch** *n* AIR commutateur manométrique *m*, manocontacteur *m*

manpower: **~ schedule** *n* MANAGE planning d'effectif *m*

manual[1] *adj* ANALYSIS manuel

manual[2]: **~ feed** *n* WASTE alimentation manuelle *f*; **~ separation** *n* WASTE séparation manuelle *f*; **~ sorting** *n* WASTE tri manuel *m*

manually: **~ operated screen** *n* WATER POLL grille manuelle *f*

manufactured: **~ gas** *n* AIR gaz industriel *m*

manure: **~ application** *n* SOIL épandage *m*; **~ gas** *n* WASTE gaz de fumier *m*

manuring *n* SOIL épandage *m*

margin: **~ gluer** *n* WASTE encolleuse sur marge *f*

marine: **~ biota** *n* MARINE faune et flore marines *f pl*; **~ climate** *n* MARINE climat marin *m*; **~ dredge** *n* MARINE drague de mer *f*; **~ dumping** *n* MARINE déversement en mer *m*, rejet en mer *m*; **~ ecology** *n* MANAGE, MARINE, POLICY écologie marine *f*; **~ ecosystem** *n* MARINE écosystème marin *m*; **~ environment** *n* MARINE milieu marin *m*, environnement marin *m*; **~ food chain** *n* MARINE chaîne alimentaire marine *f*; **~ food web** *n* RENEW ENERG réseau alimentaire marin *m*; ~ **life** *n* MARINE vie en milieu marin *f*; ~ **pollution** *n* MARINE pollution marine *f*, pollution des mers *f*; ~ **species** *n pl* MARINE espèce marine *f*

Mariotte: ~**'s law** *n* ANALYSIS loi de Boyle-Mariotte *f*

market[1]: ~**-dependent** *adj* MANAGE en fonction du marché

market[2] *n* MANAGE, POLICY marché *m*; ~ **economy** MANAGE, POLICY économie de marché *f*; ~**-economy incentive** MANAGE, POLICY incitation à l'économie de marché *f*; ~**-economy instrument** MANAGE, POLICY instrument de l'économie de marché *m*, mécanisme de l'économie de marché *m*; ~ **forces** *n pl* MANAGE mécanismes du marché *m pl*; ~ **gardening** *BrE* (*cf truck farming AmE*) SOIL maraîchage *m*; ~ **price** MANAGE, POLICY prix du marché *m*; ~ **waste** WASTE déchets de marché *m pl*

marketability *n* MANAGE *of recycled products* débouchés possibles *m pl*, possibilité de vente *f*

marketable *adj* MANAGE commercialisable

marketing *n* MANAGE *of recycled products* commercialisation *f*

marking: ~ **equipment** *n* WASTE matériel de marquage *m*; ~ **label** *n* WASTE étiquette de repérage *f*; ~ **machine** *n* WASTE marqueuse *f*

marl *n* WATER SUPP marne *f*

marly: ~ **clay** *n* WATER SUPP argile marneuse *f*

marsh *n* SOIL, WATER SUPP marais *m*, marécage *m*; ~ **fever** SAFETY paludisme *m*; ~ **plants** *n pl* WASTE plantes palustres *f pl*

marshiness *n* WASTE état marécageux *m*, caractère marécageux *m*

marshland *n* SOIL terrain marécageux *m*

masking: ~ **noise** *n* NOISE bruit masquant *m*, bruit recouvrant *m*; ~ **tape** *n* WASTE ruban adhésif *m*

masonry: ~ **dam** *n* RENEW ENERG barrage en maçonnerie *m*

mass[1] *adj* ANALYSIS de masse, en masse; ~**-selective** ANALYSIS sélectif de masse

mass[2]: ~ **concentration** *n* AIR *gas, dust* concentration en masse *f*; ~**-flow controller** *n* ANALYSIS régulateur de débit massique *m*; ~**-flow rate** *n* WATER SUPP débit-masse *m*; ~**-flow rate meter** *n* ANALYSIS débitmètre massique *m*; ~**-flow regulator** *n* ANALYSIS régulateur de débit massique *m*; ~ **fragmentography** *n* ANALYSIS fragmentographie de masse *f*; ~**-selective detector** *n* ANALYSIS détecteur sélectif de masse *m*; ~ **spectrograph** *n* ANALYSIS spectrographe de masse *m*; ~ **spectrography** *n* ANALYSIS *process type* spectrographie de masse *f*; ~ **spectrometer** *n* (*MS*) ANALYSIS spectromètre de masse *m*; ~ **spectrometry** *n* ANALYSIS spectrométrie de masse *f*; ~**-spectrometry detector** *n* (*MSD*) ANALYSIS détecteur de spectrométrie de masse *m*; ~ **spectroscope** *n* ANALYSIS spectroscope de masse *m*; ~ **spectroscopy** *n* ANALYSIS *process type* spectroscopie de masse *f*; ~ **spectrum** *n* ANALYSIS spectre de masse *m*; ~ **transfer** *n* AIR *biological gas scrubbing* transfert de masse *m*

mass[3]: ~ **together** *vt* AIR rassembler

master: ~ **plan** *n* POLICY plan directeur d'urbanisme *m*; ~ **plan for sewage removal** *n* WATER SUPP programme général d'assainissement *m*

MATC *abbr* (*maximum allowable toxicant concentration*) SAFETY concentration de produits toxiques maximale admissible *f*

material[1] *adj* AIR, WASTE matériel

material[2] *n* WATER SUPP bilan matières *m*; ~ **flow** WASTE débit matière *m*; ~**-flow analysis** WATER SUPP bilan matières *m*; ~ **pollution** WATER SUPP pollution solide *f*; ~ **recovery** WASTE récupération de matériaux *f*; ~**-separation operation** WASTE opération de tri des matériaux *f*

materials: ~ **balance** *n* MANAGE, POLICY, WATER SUPP bilan matières *m*; ~ **recycling** *n* WASTE *through crushing, cleaning, melting and forming*, WATER POLL recyclage des matériaux *m*

mathematical: ~ **climate model** *n* AIR modèle climatique mathématique *m*

matting *n* NOISE *surface lining* natte *f*

maturation: ~ **pond** *n* WATER POLL bassin de maturation *m*

maximum[1] *adj* GEN maximal

maximum[2]: ~ **allowable concentration** *n* (*MAC*) AIR, ANALYSIS, WATER SUPP concentration maximale admissible *f*; ~ **allowable concentration at the workplace** *n* SAFETY concentration maximale admissible sur le lieu de travail *f*; ~ **allowable concentration in the free environment** *n* AIR, ANALYSIS concentration maximale admissible à l'air libre *f*, concentration maximale admissible dans l'environnement *f*; ~ **allowable emissions concentration** *n* ANALYSIS concentration d'émissions maximale admissible *f*; ~ **allowable organic concentration** *n* ANALYSIS concentration organique maximale admissible *f*; ~ **allowable toxicant concentration** *n* (*MATC*) SAFETY concentration de produits toxiques maximale admissible *f*; ~ **credible accident** *n* SAFETY accident maximal de référence *m*; ~ **emission concentration** *n* AIR, WATER SUPP concentration maximale d'émissions *f*, teneur maximale en émissions *f*; ~ **exposure** *n* SAFETY exposition maximale *f*; ~ **exposure limit** *n* SAFETY *to hazardous substances* limite maximale d'exposition *f*; ~ **flow** *n* WATER SUPP débit maximal *m*; ~ **immission concentration** *n* WATER SUPP concentration maximale d'immissions *f*; ~ **level in consecutive** *n* NOISE *time intervals* niveau maximal en consécutif *m*; ~ **permissible dose** *n* (*MPD*) SAFETY dose maximale admissible *f* (*DMA*); ~ **tolerated dose** *n* (*MTD*) SAFETY dose maximale tolérée *f*

MBG *abbr* (*middle-thermal-value biogas*) RENEW ENERG biogaz à pouvoir calorifique moyen *m*

MCA *abbr* (*multicriteria analysis*) WATER SUPP analyse multicritère *f*

MCFA *abbr* (*molten-carbonate fuel cell*) AIR pile à combustible à carbonate fondu *f*

mean: ~ **daily flow** *n* WATER SUPP débit journalier moyen *m*; ~ **free path** *n* AIR libre parcours moyen *m*; ~ **value** *n* ANALYSIS moyenne *f*

measurability *n* ANALYSIS mesurabilité *f*

measurable[1] *adj* ANALYSIS mesurable

measurable[2]: ~ **variable** *n* ANALYSIS *infrared absorption, caloric conductibility* variable mesurable *f*

measure[1] *n* ANALYSIS, MANAGE, POLICY mesure *f*, SOIL mesure de contrôle *f*

measure[2] *vt* ANALYSIS *emissions* mesurer; ~ **directly** ANALYSIS mesurer directement; ~ **photometrically** ANALYSIS mesurer par méthode photométrique; ~ **with a pipette** ANALYSIS mesurer avec une pipette

measured: ~ **value** *n* ANALYSIS mesure *f*, valeur mesurée *f*; ~**-value distortion** *n* ANALYSIS distorsion de la valeur mesurée *f*; ~**-value processing** *n* ANALYSIS traitement de la valeur mesurée *m*

measurement *n* ANALYSIS mesure *f*, *of environmental conditions* mesurage *m*; ~ **component** ANALYSIS composant de mesure *m*; ~ **conditions** *n pl* ANALYSIS conditions de mesurage *f pl*, milieu de mesure *m*; ~ **of dust concentration** AIR, ANALYSIS mesure de la concentration de poussière *f*; ~ **of dust precipitation** AIR, ANALYSIS mesure de la précipitation de poussière *f*; ~ **error** ANALYSIS erreur de mesure *f*; ~ **of flow rate** ANALYSIS mesure de débit *f*; ~ **of gas quantity** ANALYSIS mesure de volume de gaz *f*; ~ **of gas-flow rate** ANALYSIS mesure de débit de gaz *f*; ~ **of intensity of scattered light** ANALYSIS mesure d'intensité de la lumière diffusée *f*; ~ **of oxygen concentration** AIR, ANALYSIS mesure de la concentration d'oxygène *f*; ~ **plan** ANALYSIS campagne de mesure *f*, plan de mesure *m*; ~ **prerequisites** *n pl* ANALYSIS conditions préliminaires de mesure *f pl*; ~ **program** *AmE*, ~ **programme** *BrE* ANALYSIS campagne de mesure *f*, programme de mesure *m*; ~ **result** ANALYSIS résultat de mesurage *m*, résultat de mesure *m*; ~**-sensitivity control** ANALYSIS régulation de la sensibilité de mesure *f*;

~ **task** ANALYSIS tâche de mesure *f*; ~ **technique** ANALYSIS technique de mesure *f*; ~ **of trace amounts** ANALYSIS mesure de la quantité de traces *f*; ~ **vehicle** ANALYSIS véhicule de mesure *m*

measuring *n* ANALYSIS mesurage *m*; ~ **accuracy** ANALYSIS exactitude de mesure *f*, précision de mesure *f*; ~ **amplifier** ANALYSIS amplificateur de mesure *m*; ~ **bridge** ANALYSIS pont de Wheatstone *m*; ~ **cell** ANALYSIS cellule de mesure *f*; ~ **chamber** ANALYSIS chambre de mesure *f*; ~ **conditions** ANALYSIS conditions de mesurage *f pl*; ~ **container** ANALYSIS récipient gradué *m*; ~ **cup** ANALYSIS coupelle de mesure *f*; ~ **cuvette** ANALYSIS cuvette de mesure *f*; ~-**data filtering** ANALYSIS filtrage des données mesurées *m*; ~-**data scanning** ANALYSIS balayage des données mesurées *m*; ~ **device** ANALYSIS dispositif de mesure *m*; ~ **electrode** ANALYSIS électrode de mesure *f*; ~ **element** ANALYSIS élément de mesure *m*; ~ **equipment** ANALYSIS, WASTE matériel de mesure *m*; ~ **error** ANALYSIS erreur de mesure *f*; ~ **fluid** ANALYSIS fluide de mesure *m*; ~ **gage** *AmE see measuring gauge BrE*; ~ **gas** ANALYSIS gaz de mesure *m*; ~ **gas cooler** ANALYSIS refroidisseur de gaz de mesure *m*; ~ **gauge** *BrE* ANALYSIS jauge *f*, sonde *f*; ~-**head cable** ANALYSIS câble de sonde *m*; ~ **instrument** ANALYSIS instrument de mesure *m*; ~ **instrument based on paramagnetism** ANALYSIS *measurement of oxygen concentration* instrument de mesure de type paramagnétique *m*; ~ **jar** ANALYSIS éprouvette graduée *f*, burette graduée *f*, verre gradué *m*; ~ **location** ANALYSIS lieu de mesure *m*; ~ **means** *n pl* ANALYSIS moyens de mesure *m pl*; ~ **method** ANALYSIS méthode de mesure *f*; ~ **nozzle** ANALYSIS orifice de mesure *m*; ~ **operation** ANALYSIS tâche de mesure *f*; ~ **period** ANALYSIS campagne de mesure *f*, période de mesure *f*; ~ **pipette** ANALYSIS pipette graduée *f*; ~ **point** ANALYSIS *of groundwater* point de mesure *m*; ~ **position** ANALYSIS position de mesure *f*; ~ **principle** ANALYSIS principe de mesure *m*; ~ **probe** ANALYSIS échantillon mesuré *m*; ~ **purpose** ANALYSIS objet de mesure *m*; ~ **quantity** ANALYSIS quantité mesurée *f*; ~ **range** ANALYSIS étendue de mesure *f*, plage de mesure *f*; ~-**range deviation** ANALYSIS écart de mesure *m*; ~ **report** ANALYSIS compte-rendu de mesure *m*; ~ **resistance** ANALYSIS résistance de mesure *f*; ~ **resistor** ANALYSIS résistance de mesure *f*; ~ **result** ANALYSIS résultat de mesurage *m*; ~ **rod** ANALYSIS jauge *f*; ~ **scoop** ANALYSIS cuiller de dosage *f*, louche de dosage *f*; ~ **section** ANALYSIS section de mesure *f*; ~ **sensitivity** ANALYSIS sensibilité de mesurage *f*; ~ **set-up** ANALYSIS montage de mesure *m*; ~ **shaft** ANALYSIS axe de mesure *m*; ~ **signal** ANALYSIS signal de mesure *m*; ~ **sonde** ANALYSIS sensotrode *m*, sonde de mesure *f*; ~ **station** ANALYSIS poste de mesurage *m*, station de mesure *f*; ~ **surface** ANALYSIS surface de mesure *f*; ~ **switch** ANALYSIS clé de mesure *f*; ~ **tank** ANALYSIS bac jaugeur *m*; ~ **time** ANALYSIS temps de mesure *m*; ~ **transducer** ANALYSIS *pH-measurement* convertisseur de mesure *m*, transducteur de mesure *m*; ~ **tube** ANALYSIS sensotrode *m*; ~ **with ultrasound** ANALYSIS mesure par ultrason *f*; ~ **unit** ANALYSIS unité de mesure *f*; ~ **vessel** ANALYSIS vase gradué *m*; ~ **weir** WATER SUPP déversoir de mesure *m*

mechanical[1] *adj* GEN mécanique

mechanical[2]: ~ **cleaning** *n* WATER POLL *of screen* nettoyage automatique *m*; ~ **collector** *n* AIR collecteur mécanique *m*; ~ **composting** *n* WASTE compostage mécanique *m*; ~ **energy** *n* RENEW ENERG énergie mécanique *f*; ~ **erosion** *n* SOIL érosion mécanique *f*; ~ **filtration** *n* WASTE filtration mécanique *f*; ~ **hazard** *n* SAFETY risque mécanique *m*; ~ **loss** *n* WATER POLL perte mécanique *f*; ~ **purification** *n* WATER POLL purification mécanique *f*; ~ **recycling** *n* WASTE, WATER POLL recyclage mécanique *m*; ~ **separation** *n* WASTE séparation mécanisée *f*; ~ **sludge-thickening method** *n* WATER POLL *vacuum filtration, centrifugation* méthode mécanique d'épaississement des boues *f*; ~ **softening** *n* WATER SUPP adoucissement mécanique *m*; ~ **sweeper** *n* WASTE balayeuse mécanique *f*, WATER SUPP balai mécanique *m*; ~ **wastewater treatment** *n* WATER POLL épuration mécanique des eaux usées *f*; ~ **wood pulp** *n* WASTE cellulose de bois *f*, pâte mécanique *f*

mechanically[1]: ~-**cleaned** *adj* WATER POLL à nettoyage automatique; ~-**dewatered** *adj* WATER POLL déshydraté mécaniquement; ~-**thickened** *adj* WATER POLL épaissi mécaniquement

mechanically[2]: ~-**assisted hand separation** *n* WASTE séparation manuelle assistée mécaniquement *t*; ~ **cleaned bar screen** *n* WATER POLL grille à nettoyage automatique *f*; ~-**dewatered sludge** *n* WATER POLL boues déshydratées mécaniquement *f pl*; ~-**thickened sludge** *n* WATER POLL boues épaissies mécaniquement *f pl*

median: ~ **lethal concentration** *n* AIR, ANALYSIS, WATER SUPP concentration létale à 50% *f* (CL_{50}); ~ **lethal dose** *n* (LD_{50}) SAFETY dose létale à 50% *f* (DL_{50}), dose semi-létale *f* (DL_{50}); ~ **lethal time** *n* AIR, ANALYSIS temps létal moyen *m*

medium[1]: ~-**scale** *adj* RENEW ENERG à échelle moyenne

medium[2] *n* SOIL , WASTE, WATER SUPP médium *m*, milieu *m*; ~ **calorific value** WATER POLL pouvoir calorifique moyen *m*; ~-**pressure cleaning** AIR lavage moyenne pression *m*; ~-**scale installation** RENEW ENERG *wind, solar* installation moyenne *f*; ~-**size bubble aeration** WATER POLL aération par bulles moyennes *f*; ~-**sized refuse** WASTE *grain size 8 to 40 mm* refus de taille moyenne *m pl*; ~-**specific component** ANALYSIS élément constitutif du milieu *m*, élément du milieu *m*

megalopolis *n AmE* MANAGE, POLICY mégalopole *f*

megawatt *n* (*MW*) RENEW ENERG mégawatt *m* (*MW*); ~ **plant** RENEW ENERG installation de puissance *f*

melt[1]: ~ **filter** *n* WASTE *plastics* filtre pour matière fondue *m*; ~ **filtration** *n* WASTE filtrage des produits de fusion *m*; ~ **mixer** *n* WASTE mélangeur de produits de fusion *m*

melt[2] *vt* WASTE fondre, WATER POLL, WATER SUPP porter à fusion, *snow, ice* fondre

melt[3] *vi* WASTE, WATER POLL, WATER SUPP fondre

melting *n* WASTE, WATER POLL, WATER SUPP fusion *f*, fonte *f*; ~ **point** WATER SUPP point de fusion *m*

membrane *n* AIR, ANALYSIS, WATER POLL membrane *f*; ~ **biological reactor** AIR réacteur biologique à membrane *m*; ~ **distillation** WATER POLL distillation par membrane *f*; ~ **filter** ANALYSIS, WATER POLL, WATER SUPP filtre à diaphragme *m*, filtre microporeux *m*, filtre moléculaire *m*; ~ **filtration** ANALYSIS, WATER POLL filtration sur membrane *f*; ~ **layer** WATER POLL, WATER SUPP couche de membrane *f*; ~ **permeability** WATER POLL perméabilité de membrane *f*; ~ **reactor** AIR réacteur à membrane *m*; ~ **separation process** WATER POLL procédé de séparation par membrane *m*; ~ **surface** ANALYSIS surface de membrane *f*

mercaptan *n* AIR, WASTE mercaptan *m*

mercurial[1] *adj* WATER POLL mercuriel

mercurial[2]: ~ **wastewater** *n* WATER POLL eaux usées mercurielles *f pl*

mercury *n* (*Hg*) WATER POLL mercure *m* (*Hg*); ~ **poisoning** WATER POLL empoisonnement par le mercure *m*; ~ **pollution** SAFETY pollution par le mercure *f*; ~ **recovery** WATER SUPP récupération de mercure *f*

mesh *n* WATER POLL maille *f*; ~ **size** WATER POLL *of screen* taille de maille *f*

meshed: ~ **mains and supply network** *n BrE* WATER SUPP réseau maillé et de distribution *m*

mesophilic[1] *adj* WATER POLL mésophile

mesophilic[2]: ~ **bacteria** *n pl* WATER POLL bactéries mésophiles *f pl*; ~ **digestion** *n* WATER POLL digestion mésophile *f*; ~ **microorganism** *n* AIR, MARINE, WATER POLL, WATER SUPP micro-organisme mésophile *m*

mesopore *n* AIR *biological scrubber* mésopore *m*

mesoscale: ~ **circulation** *n* AIR circulation à échelle intermédiaire *f*

mesosphere *n* AIR mésosphère *f*

mesotrophic: ~ **water** *n* WATER POLL mésotrophie *f*

metabolic: ~ **activity** *n* WATER POLL activité métabolique *f*; ~ **chain** *n* POLICY chaîne alimentaire *f*; ~ **rate** *n* WATER POLL vitesse de métabolisation *f*; ~ **waste** *n* WASTE déchets du métabolisme *m pl*

metabolism *n* AIR, WATER POLL, WATER SUPP métabolisme *m*

metabolite *n* AIR, WATER POLL, WATER SUPP métabolite *m*, produit du métabolisme *m*

metabolites: ~ **test** *n* ANALYSIS essai par métabolites *m*

metal: ~ **box** *n* WASTE boîte métallique *f*, coffret métallique *m*; ~ **can** *n* WASTE boîte *f*, boîte de conserve *f*; ~ **carrier** *n* AIR *catalyst* vecteur métallique *m*; ~ **coating** *n* WASTE métallisation *f*; ~ **complex** *n* SOIL complexe métallifère *m*; ~ **detector** *n* WASTE détecteur d'éclats *m*, détecteur de métaux *m*; ~ **drum** *n* WASTE tambour métallique *m*; ~ **foil** *n* WASTE feuille métallique *f*; ~ **fungicide** *n* SOIL fongicide métallique *m*; ~ **housing** *n* AIR corps métallique *m*; ~ **pipe** *n* AIR tuyauterie métallique *f*; ~ **plate** *n* AIR, NOISE plaque de métal *f*; ~ **scrap** *n* WASTE, WATER POLL, WATER SUPP déchets métalliques *m pl*, ferraille *f*, mitraille *f*; ~ **separator** *n* WASTE séparateur de métaux *m*; ~ **waste** *n* WASTE déchets métalliques *m pl*

metalimnion *n* WATER POLL métalimnion *m*

metallic: ~ **dust** *n* AIR poussière métallique *f*; ~ **salts** *n pl* WATER POLL sels métalliques *m pl*; ~ **sludge** *n* WATER POLL boues métalliques *f pl*

metalliferous[1] *adj* WATER POLL métallifère

metalliferous[2]: ~ **wastewater** *n* WASTE, WATER POLL eaux résiduaires métallifères *f pl*, eaux usées métallifères *f pl*

meteorology *n* AIR, MANAGE météorologie *f*

meter[1]: ~ **housing** *n* ANALYSIS boîtier de compteur *m*; ~ **terminal** *n* ANALYSIS borne *f*

meter[2] *vt* AIR, ANALYSIS, WATER SUPP mesurer

metereological: ~ **conditions** *n* AIR, RENEW ENERG conditions météorologiques *f pl*

metering: ~ **equipment** *n* ANALYSIS matériel de mesure *m*; ~ **machine** *n* WASTE mesureur *m*; ~ **pump** *n* WASTE pompe doseuse *f*, WATER POLL, WATER SUPP pompe doseuse *f*, pompe volumétrique *f*; ~ **unit** *n* ANALYSIS ensemble de mesurage *m*

methane[1]: ~**-producing** *adj* AIR, WASTE, WATER POLL, WATER SUPP méthanogène

methane[2] *n* AIR *trace gas*, WASTE, WATER POLL, WATER SUPP méthane *m*; ~ **build-up** WASTE formation de méthane *f*; ~ **digestion** WATER POLL digestion méthanique *f*; ~ **fermentation** WASTE fermentation méthanogène *f*, *sewage* fermentation méthanique *f*; ~ **former** WATER POLL agent méthanogène *m*

methangenesis *n* WATER POLL méthanogénese *f*

methanogenic[1] *adj* AIR, WASTE, WATER POLL, WATER SUPP méthanogène

methanogenic[2]: ~ **bacteria** *n pl* WATER POLL bactéries méthanogènes *f pl*

methanolysis *n* WASTE méthanolyse *f*

method *n* GEN méthode *f*, modc *m*; ~ **of application** SAFETY mode d'application *m*

methylene: ~ **blue test** *n* ANALYSIS, WASTE essai au bleu de méthylène *m*; ~ **chloride** *n* AIR chlorure de méthylène *m*

metropolitan: ~ **growth center** *AmE*, ~ **growth centre** *BrE* *n* POLICY métropole d'équilibre *f*

Mg *abbr* (*magnesium*) GEN Mg (*magnésium*)

Michelson: ~ **interferometer** *n* ANALYSIS interféromètre de Michelson *m*

microbe *n* AIR, SOIL, WATER POLL microbe *m*

microbial[1] *adj* AIR, SOIL, WATER POLL microbien

microbial[2]: ~ **activity** *n* SOIL activité microbienne *f*; ~ **decomposition** *n* AIR décomposition microbienne *f*; ~ **degradation** *n* SOIL décomposition microbienne *f*; ~ **immunology** *n* SOIL, WATER SUPP immunologie microbienne *f*; ~ **leaching** *n* SOIL lessivage microbien *m*

microbiological[1] *adj* ANALYSIS, SAFETY, SOIL microbiologique

microbiological[2]: ~ **hazard** *n* SAFETY risque microbiologique *m*; ~ **metabolic process** *n* AIR processus métabolique microbiologique *m*; ~ **process** *n* SOIL procédé microbiologique *m*; ~ **testing** *n* ANALYSIS essai microbiologique *m*

microbiology *n* ANALYSIS, SAFETY, SOIL microbiologie *f*

microclimate *n* AIR microclimat *m*

microclimatic: ~ **change** *n pl* AIR changement microclimatique *m*

microcoagulation *n* WATER POLL microcoagulation *f*

microcolorimetry *n* ANALYSIS microcolorimétrie *f*

microcoulometric[1] *adj* ANALYSIS microcoulométrique

microcoulometric[2]: ~ **titration** *n* ANALYSIS titrage microcoulométrique *m*

microencapsulation *n* SOIL microconfinement *m*

microenvironment *n* AIR, SOIL micromilieu *m*

microfilter *n* WATER POLL microfiltre *m*

microfiltration *n* WATER POLL microfiltration *f*

microflora *n* AIR, SOIL microflore *f*

micromachined *adj* ANALYSIS micro-usiné

micronutrient *n* WATER POLL, WATER SUPP élément oligodynamique *m*, oligo-élément *m*

microorganism *n* AIR *biological sorption process*, SOIL, WATER POLL, WATER SUPP micro-organisme *m*

microphone *n* ANALYSIS microphone *m*

micropipette *n* ANALYSIS *dosing of liquids* micropipette *f*

micropore *n* AIR *biological scrubber*, WATER POLL *activated carbon* micropore *m*

microporous[1] *adj* AIR, WATER POLL microporeux

microporous[2]: ~ **membrane** *n* WATER POLL membrane microporeuse *f*

microprocessor *n* ANALYSIS *built-in in analyser* microprocesseur *m*; ~**-controlled photoionization detector** ANALYSIS détecteur d'ionisation photoélectrique à microprocesseur *m*

microscope *n* ANALYSIS microscope *m*

microscopic[1] *adj* ANALYSIS, SAFETY microscopique; ~ **qualitative** ANALYSIS de qualité microscopique

microscopic[2]: ~ **dust** *n* SAFETY poussière microscopique *f*; ~ **qualitative detection method** *n* ANALYSIS méthode de détection qualitative au microscope *f*

microscreening *n* WATER POLL microtamisage *m*

mid[1]: ~**-temperature** *adj* RENEW ENERG à température moyenne

mid[2]: ~**-temperature heat** *n* RENEW ENERG *from solar energy* chaleur à température moyenne *f*; ~**-temperature incineration** *n* AIR, WASTE incinération à température moyenne *f*; ~**-thermal-value biogas** *n* (*MBG*) RENEW ENERG biogaz à pouvoir calorifique moyen *m*

Mie: ~ **scattering** *n* AIR, ANALYSIS diffusion de Mie *f*

migrate *vi* SOIL, WASTE *landfill gas*, WATER SUPP migrer

migrating: ~ **gas** *n* WASTE gaz migrateur *m*

migration: ~ **of harmful substances** *n* SOIL migration de substances nocives *f*

mildly: ~ **acidic** *adj* SOIL de faible acidité

millivolt *n* (*mV*) ANALYSIS millivolt *m* (*mV*)

mined: ~ **space** *n* WASTE espace miné *m*

mineral *n* GEN minéral *m*; ~**-aggregate landfill** WASTE décharge pour agrégats miniers *f*; ~ **dust** AIR poussière minérale *f*; ~ **fiber** *AmE*, ~ **fibre** *BrE* AIR, NOISE, WATER SUPP fibre minérale *f*; ~ **fibre absorber** *BrE* NOISE absorbeur en fibre minérale *m*; ~ **fibre panel** *BrE* MARINE, NOISE panneau en fibre minérale *m*; ~ **matter** WATER SUPP matière minérale *f*; ~ **oil** WATER POLL pétrole *m*; ~**-oil tax** POLICY taxe sur les huiles minérales *f*; ~ **resources** *n pl* RENEW ENERG ressources minérales *f pl*; ~ **sealing** WASTE scellement minéral *m*; ~ **spring** WATER SUPP source minérale *f*;

~ **waste** WASTE déchets miniers *m pl*; ~ **water** WATER SUPP eau minérale *f*; ~ **wool** NOISE laine minérale *f*
mineralization *n* AIR, SOIL, WASTE, WATER POLL minéralisation *f*; ~ **technique** WASTE technique de minéralisation *f*
mineralize *vt* AIR, SOIL, WASTE, WATER POLL minéraliser, pétrifier
mineralizing[1] *adj* AIR, SOIL, WASTE, WATER POLL minéralisateur
mineralizing[2] *n* AIR, SOIL, WASTE, WATER POLL minéralisateur
minimization *n* WASTE *of emissions*, WATER SUPP *of emissions* réduction au minimum *f*
minimize *vt* AIR *emissions* réduire au minimum
minimum: ~ **flow** *n* WATER SUPP débit minimal *m*; ~ **low water** *n* WATER SUPP étiage *m*; ~ **sample quantity** *n* ANALYSIS quantité d'échantillon minimale *f*
mining: ~ **waste** *n* WASTE, WATER POLL résidu des opérations minières *m*
minister *n* POLICY ministre *m*; ~ **for the environment** POLICY ministre de l'environnement *m*
ministerial[1] *adj* POLICY ministériel
ministerial[2]: ~ **statement** *n* POLICY déclaration ministérielle *f*
ministry *n* POLICY ministère *m*
mirror *n* RENEW ENERG *in solar tower* miroir *m*; ~**-evacuated tube** RENEW ENERG *non-concentrating collector* collecteur à vide *m*, miroir à vide *m*
miscibility *n* WASTE *of plastics*, WATER POLL, WATER SUPP miscibilité *f*
miscible: ~ **substance** *n* WASTE, WATER POLL, WATER SUPP substance miscible *f*
mismanagement *n* MANAGE erreur de gestion *f*
missing: ~ **data** *n* ANALYSIS *in eco-balance* données manquantes *f pl*
mist *n* AIR *containing liquid particulates* brume *f*, voile *m*, WASTE brume *f*; ~ **drop** AIR *from stack* gouttelette de brume *f*; ~ **eliminator** AIR désembueur *m*, dévésiculeur *m*, WASTE dévésiculeur *m*, dispositif d'élimination des buées *m*
mitigate *vt* POLICY atténuer
mitigation *n* POLICY atténuation *f*
mix *vt* GEN brasser, mélanger; ~ **thoroughly** ANALYSIS mélanger intégralement, SOIL homogénéiser
mixed[1] *adj* GEN mélangé, mixte
mixed[2]: ~ **adhesive** *n* WASTE adhésif mixte *m*; ~**-bed ion exchanger** *n* WATER POLL échangeur d'ions à lits mélangés *m*; ~ **canalization** *n* WATER POLL conduite mixte *f*; ~ **culture** *n* WATER POLL culture mélangée *f*; ~**-flow pump** *n* WATER POLL pompe à pignons *f*, pompe hélico-centrifuge *f*; ~ **gas** *n* AIR gaz mélangé *m*; ~ **liquor** *n* WATER POLL liqueur mixte *f*; ~ **liquor suspended solids** *n pl* (*MLSS*) WATER POLL matières solides en suspension dans une solution *f pl*, particules solides en suspension dans une solution *f pl*; ~ **primary sludge** *n* WATER POLL boues primaires mélangées *f pl*; ~ **salt** *n* WATER SUPP sel mélangé *m*; ~ **sludge** *n* WATER POLL boues mélangées *f pl*; ~ **waste** *n* WASTE déchets mélangés *m pl*, déchets mêlés *m pl*
mixer *n* WASTE, WATER POLL mélangeur *m*; ~**-settler extractor** WATER POLL extracteur mélangeur décantateur *m*
mixing[1] *adj* GEN mélangeur
mixing[2] *n* GEN mélange *m*, brassage *m*; ~ **with aggregates** SOIL mélange à des agrégats *m*; ~ **device** WASTE, WATER POLL mélangeur *m*; ~ **device evaporator** WATER POLL évaporateur à mélangeur *m*; ~ **gate** WASTE grille de mélange *f*; ~ **tank** WATER POLL bassin de contact *m*, tamis rotatif *m*, WATER SUPP cuve de mélange *f*; ~ **worm** WATER POLL vis mélangeuse *f*
mixture *n* GEN mélange *m*
MLSS *abbr* (*mixed liquor suspended solids*) WATER POLL matières solides en suspension dans une solution *f pl*, particules solides en suspension dans une solution *f pl*
Mn *abbr* (*manganese*) AIR Mn (*manganèse*)
Mo *abbr* (*molybdenum*) AIR Mo (*molybdène*)
mobile[1] *adj* GEN mobile
mobile[2]: ~ **component** *n* AIR, WATER SUPP élément mobile *m*; ~ **emission source** *n* AIR source mobile d'émission *f*; ~ **field-measuring method** *n* ANALYSIS méthode mobile de mesure sur le terrain *f*; ~ **laboratory-measuring method** *n* ANALYSIS méthode mobile de mesure en laboratoire *f*; ~ **measuring unit** *n* ANALYSIS poste de mesure mobile *m*; ~ **office** *n* POLICY bureau mobile *m*; ~ **pumping unit** *n* WATER POLL station de pompage mobile *f*; ~ **sampler** *n* ANALYSIS échantillonneur mobile *m*; ~ **unit** *n* SOIL *for soil treatment* poste mobile *m*
mobilization *n* SOIL *of heavy metals*, WASTE, WATER SUPP *of harmful substances* mobilisation *f*
mobilize *vt* SOIL, WASTE, WATER SUPP mobiliser
mobilized *adj* SOIL, WASTE, WATER SUPP mobilisé
mock: ~**-up** *n* ANALYSIS maquette *f*
model: ~ **calculation** *n* AIR calcul de modèle *m*; ~ **calibration** *n* ANALYSIS calage de modèle *m*, étalonnage de modèle *m*, WATER SUPP calibrage de modèle *m*; ~ **operation permission** *n* POLICY permis d'exploitation pilote *m*
modeling: ~ **of water quality** *AmE see modelling of water quality BrE*
modelling: ~ **of water quality** *n BrE* MANAGE modélisation de la qualité de l'eau *f*
moderately: ~ **acidic** *adj* SOIL d'acidité moyenne
modernization *n* POLICY *of tax system* actualisation *f*, modernisation *f*, réforme *f*, WASTE, WATER POLL, WATER SUPP *industrial site* modernisation *f*, réhabilitation *f*
modernize *vt* POLICY *tax system* actualiser, moderniser, réformer, WASTE, WATER POLL, WATER SUPP moderniser, réhabiliter
modernized *adj* GEN réhabilité, réformé
modified: ~ **soil material** *n* WASTE *landfill sealing* matériel modifié du sol *m*
modular[1] *adj* ANALYSIS, WASTE modulaire
modular[2]: ~ **design** *n* WASTE *of containers* conception modulaire *f*
module: ~ **sound-level meter** *n* ANALYSIS sonomètre modulaire *m*
Mohlmann: ~ **index** *n* WATER POLL indice de Mohlmann *m*
Mohr: ~ **balance** *n* ANALYSIS balance de Mohr *f*
moisture[1]: ~**-proof** *adj* GEN étanche, étanche à l'humidité, à l'épreuve de l'humidité, hydrofuge; ~**-repellent** *adj* GEN humidifuge; ~**-resistant** *adj* GEN étanche
moisture[2] *n* GEN humidité *f*; ~ **content** AIR, RENEW ENERG *biomass dust* teneur en humidité *f*, degré d'hygrométrie *m*, WASTE, WATER POLL, WATER SUPP taux d'humidité *m*, *textiles* teneur en eau *f*; ~ **determination** WASTE détermination de la teneur en humidité *f*; ~ **eliminator** AIR déshumidificateur *m*; ~ **of installation** WASTE humidité d'une installation *f*; ~ **measurement** AIR, WASTE, WATER POLL, WATER SUPP hygrométrie *f*; ~ **sensor** ANALYSIS capteur hygrométrique *m*; ~ **test** WASTE essai d'humidité *m*
mol *abbr* (*mole*) ANALYSIS mole *f* (*mole*); ~ **ratio** *n* AIR rapport molaire *m*
molar: ~ **mass** *n* ANALYSIS masse molaire *f*; ~ **volume** *n* ANALYSIS volume molaire *m*
mold *AmE see mould BrE*
molded *AmE see moulded BrE*
mole *n* (*mol*) ANALYSIS mole *f* (*mol*)
molecular[1] *adj* GEN moléculaire
molecular[2]: ~ **colloid** *n* WATER POLL colloïde moléculaire *m*; ~ **flow** *n* WATER SUPP écoulement moléculaire *m*; ~ **force** *n* AIR *adsorption* force moléculaire *f*; ~ **sieve** *n* AIR *adsorbent*, ANALYSIS, SOIL, WATER POLL *adsorbent* tamis moléculaire *m*; ~ **spectrometry** *n* ANALYSIS spectrométrie moléculaire *f*; ~ **weight** *n* ANALYSIS, WATER POLL masse moléculaire *f*
molecule *n* AIR, ANALYSIS, SOIL, WATER POLL molécule *f*
molten: ~**-carbonate fuel cell** *n* (*MCFA*) AIR pile à combustible à carbonate fondu *f*; ~ **metal** *n* SAFETY métal fondu *m*

molybdenum *n* (*Mo*) AIR molybdène *m* (*Mo*)
monitor[1] *n* ANALYSIS moniteur *m*; **~ tube** ANALYSIS *mass spectroscopy* tube moniteur *m*
monitor[2] *vt* GEN contrôler, surveiller
monitoring *n* GEN surveillance *f*; **~ after site closure** WASTE surveillance du site après fermeture *f*; **~ instrument** POLICY mécanisme de surveillance *m*; **~ measure** SAFETY mesure de surveillance *f*; **~ network** ANALYSIS réseau de surveillance *m*; **~ obligation** POLICY obligation de surveillance *f*; **~ system** ANALYSIS système de contrôle *m*, système de surveillance *m*; **~ well** AIR puits de surveillance *m*, WASTE regard de service *m*, WATER SUPP puits de surveillance *m*
mono-dump *n* WASTE décharge réservée à un type de déchets *f*, décharge spécialisée *f*, WATER POLL décharge spécialisée *f*
monolith *n* AIR *catalyst* monolithe *m*
monolithic *adj* AIR monolithique
monomer *n* WASTE, WATER SUPP monomère *m*
monthly: **~ mean value** *n* ANALYSIS moyenne mensuelle *f*
mooring: **~ buoy** *n* MARINE coffre *m*
morainic: **~ filter layer** *n* AIR, WASTE couche de filtre morainique *f*
morbidity *n* POLICY, WATER POLL morbidité *f*
mortality *n* POLICY, WATER POLL mortalité *f*
motion: **~ of the sea** *n* RENEW ENERG *tidal energy* mouvement de la mer *m*
motor: **~ pump** *n* AIR, MARINE, WATER SUPP motopompe *f*; **~ vehicle exhaust gas** *n* AIR gaz d'échappement de véhicule à moteur *m*
motorized *adj* WATER POLL à moteur, motorisé
mould *n BrE* GEN *fungi* moisissure *f*, *shape* moule *m*; **~-filling technique** *BrE* WASTE, WATER SUPP technique de remplissage de moule *f*
moulded: **~ pulp article** *n BrE* WASTE article en cellulose moulée *m*
moveable: **~ blade** *n* AIR lame amovible *f*; **~ built-in structure** *n* AIR *scrubbing area* structure amovible intégrée *f*; **~ clean air feed** *n* AIR module d'alimentation en air propre amovible *m*; **~ fittings** *n pl* AIR garniture mobile *f*
moving: **~ bed** *n* AIR, SOIL, WATER POLL, WATER SUPP lit fluidisé *m*, lit mobile *m*; **~-bed adsorber** *n* AIR, SOIL, WATER POLL adsorbeur à lit mobile *m*, adsorbeur à lit fluidisé *m*; **~-bed dryer** *n* WATER POLL sécheur à lit mobile *m*; **~-bed sand filter** *n* WATER POLL filtre à sable à lit mobile *m*; **~-bed zone** *n* AIR zone à lit mobile *f*
MPD *abbr* (*maximum permissible dose*) SAFETY DMA (*dose maximale admissible*)
MS *abbr* (*mass spectrometer*) ANALYSIS spectromètre de masse *m*
MSD *n* (*mass-spectrometry detector*) ANALYSIS détecteur de spectrométrie de masse *m*
MSW *abbr* (*municipal solid waste*) WASTE, WATER SUPP déchets urbains solides *m pl*
MTD *abbr* (*maximum-tolerated dose*) SAFETY dose maximale tolérée *f*
mud *n* WATER POLL vase *f*; **~ blanket** WATER POLL lit de boues *m*; **~ content** WATER POLL concentration de boues *f*
muddy: **~ ground** *n* WATER POLL fond vaseux *m*; **~ water** *n* WATER POLL eau chargée *f*, eau baneuse *f*
muffle *vt* NOISE étouffer, tamiser
muffler *n AmE* (*cf silencer BrE*) NOISE silencieux *m*
mulch *n* RENEW ENERG paillis *m*
mullite *n* AIR *ceramic carrier* mullite *f*
multibarrier: **~ principle** *n* WASTE, WATER SUPP principe des barrières multiples *m*, principe de recondance des barrières *m*
multichamber: **~ bag filter** *n* AIR filtre à sac compartimenté *m*; **~ container** *n* WASTE conteneur compartimenté *m*
multicomponent *adj* ANALYSIS *measuring devices, analyzers* multi-élément, multicomposant; **~ injection molding machine** *AmE*, **~ injection moulding machine** *BrE* *n* WASTE machine de moulage par injection en plusieurs éléments *f*
multicriteria: **~ analysis** *n* (*MCA*) ANALYSIS analyse multicritère *f*
multilateral[1] *adj* POLICY multilatéral
multilateral[2]: **~ negotiation** *n* POLICY négociation multilatérale *f*
multilayer: **~ container** *n* WASTE conteneur multicouche *m*; **~ filter** *n* WATER POLL filtre multicouche *m*
multimaterial: **~ recycling** *n* WASTE recyclage multimatériaux *m*
multipack *n* WASTE emballage à usages multiples *m*
multiple: **~-blade machine** *n* RENEW ENERG *wind turbine* éolienne multiple *f*; **~ board** *n* WASTE carton multicouche *m*; **~-hearth incinerator** *n* WASTE incinérateur à foyers multiples *m*
multirange[1] *adj* ANALYSIS à plusieurs étendues de mesure
multirange[2]: **~ instrument** *n* ANALYSIS instrument à étendues de mesure multiples *m*
multisheet *adj* NOISE *as with noise prevention walls* multicouche, multifeuille
multistage[1] *adj* GEN multi-étage, à plusieurs étages, multicellulaire
multistage[2]: **~ activated sludge process** *n* WATER POLL procédé par boues activées avec alimentation étagée *m*; **~ cyclone** *n* ANALYSIS cyclone multi-étages *m*; **~ evaporator** *n* WATER POLL évaporateur à plusieurs étages *m*; **~ fluidized bed** *n* AIR lit fluidisé multi-étage *m*; **~ fluidized-bed adsorber** *n* AIR adsorbeur à lit fluidisé multi-étage *m*; **~ peripheral pump** *n* WATER POLL pompe tourbillonnaire à plusieurs étages *f*
multitube: **~ cyclone** *n* AIR cyclone multitubulaire *m*
multiunit: **~ container** *n* WASTE conteneur compartimenté *m*, conteneur multicellulaire *m*; **~ cyclones** *n* AIR cyclones en série *m pl*
multi-wall: **~ sack** *n* WASTE sac à plusieurs épaisseurs *m*
municipal: **~ effluent** *n* WATER POLL, WATER SUPP effluent urbain *m*; **~ engineering** *n* MANAGE, POLICY génie urbain *m*; **~ environmental protection** *n* POLICY protection de l'environnement urbain *f*; **~ solid waste** *n* (*MSW*) WASTE, WATER SUPP déchets urbains solides *m pl*; **~ waste** *n* WASTE, WATER SUPP déchets urbains *m pl*; **~ waste landfill** *n* WASTE décharge municipale *f*; **~ waste stream** *n* (*MWS*) WATER SUPP flux de déchets urbains *m*; **~ wastewater** *n* WATER POLL eaux usées municipales *f pl*; **~ wastewater discharge** *n* WATER POLL, WATER SUPP évacuation des eaux usées urbains *f*; **~ wastewater sludge** *n* WATER POLL boues des eaux usées urbaines *f pl*; **~ wastewater works** *n* WATER POLL station d'épuration des eaux usées urbaines *f*
mutagen *n* SAFETY mutagène *m*
mV *abbr* (*millivolt*) ANALYSIS mV (*millivolt*)
MW *abbr* (*megawatt*) RENEW ENERG MW (*mégawatt*)
MWS *abbr* (*municipal waste stream*) WATER SUPP flux de déchets urbains *m*
mycotoxin *n* WATER POLL mycotoxine *f*

N

N *abbr* AIR (*nitrogen*) azote *m* (*N*), ANALYSIS (*newton*) newton *m* (*N*), WATER POLL azote *m* (*N*)
Na *abbr* (*sodium*) AIR Na (*sodium*)
nanofiltration *n* WATER SUPP nanofiltration *f*
nanogram *n* (*ng*) ANALYSIS nanogramme *m* (*ng*)
nanogramme *see nanogram*
nanometer *AmE see nanometre BrE*
nanometre *n BrE* (*nm*) ANALYSIS nanomètre *m* (*nm*)
narrow: ~ **curtain** *n* WASTE voile étroit *m*, WATER SUPP rideau étroit *m*
national: ~ **environmental protection** *n* POLICY protection de l'environnement à l'échelle nationale *f*; ~ **park** *n* POLICY parc national *m*
natural[1] *adj* RENEW ENERG, WATER SUPP naturel
natural[2]: ~ **acidification** *n* AIR, SOIL, WATER SUPP acidification naturelle *f*; ~ **convection** *n* AIR convection naturelle *f*; ~ **dewatering system** *n* WATER POLL *sand drying bed, sludge lagoon* méthode naturelle de déshydratation *f*; ~ **disaster** *n* POLICY catastrophe naturelle *f*; ~ **drainage** *n* WATER SUPP drainage naturel *m*; ~ **ecological system** *n* MARINE, SOIL système écologique naturel *m*; ~ **ecosystem** *n* AIR, MARINE, SOIL écosystème naturel *m*; ~ **energy balance** *n* RENEW ENERG bilan énergétique naturel *m*; ~ **energy source** *n* RENEW ENERG source naturelle d'énergie *f*; ~ **environment** *n* AIR, MANAGE, POLICY, SOIL, WATER POLL, WATER SUPP environnement *m*, patrimoine naturel *m*, milieu naturel *m*; ~ **gas** *n* WATER SUPP gaz naturel *m*; ~ **gas for vehicles** *n* RENEW ENERG gaz naturel véhicule *m*; ~ **greenhouse effect** *n* AIR effet de serre naturel *m*; ~ **groundwater recharge** *n* WATER SUPP réalimentation naturelle d'eau souterraine *f*; ~ **gypsum** *n* AIR, SOIL gypse *m*; ~ **infiltration** *n* SOIL, WATER SUPP infiltration naturelle *f*; ~ **lining** *n* WASTE revêtement naturel *m*; ~ **ozone content** *n* AIR concentration d'ozone dans l'environnement *f*; ~ **pollutant** *n* AIR, SOIL polluant naturel *m*; ~ **purification** *n* SOIL épuration naturelle *f*, WATER POLL purification naturelle *f*; ~ **purification force** *n* SOIL pouvoir d'épuration naturelle *m*; ~ **radioactivity** *n* AIR, WATER POLL radioactivité naturelle *f*; ~ **reserve** *n* POLICY réserve naturelle *f*; ~ **resources management** *n* MANAGE gestion des ressources naturelles *f*; ~ **sanctuary** *n* POLICY réserve naturelle *f*; ~ **sedimentation** *n* WATER POLL sédimentation naturelle *f*; ~ **sedimentation process** *n* WATER POLL processus naturel de sédimentation *m*; ~ **seepage** *n* SOIL, WATER SUPP infiltration naturelle *f*; ~ **self-purification** *n* WATER POLL, WATER SUPP auto-épuration naturelle *f*; ~ **soil material** *n* SOIL, WASTE *landfill sealing* matériau du sol naturel *m*; ~ **source** *n* AIR source naturelle *f*
naturally: ~ **acidic lake** *n* WATER SUPP lac acide naturel *m*
nature *n* RENEW ENERG nature *f*; ~ **conservation** POLICY sauvegarde de la nature *f*; ~ **park** POLICY parc naturel *m*
nautical: ~ **chart** *n* MARINE carte marine *f*
NC: ~-**curve** *n* (*noise criterion curve*) ANALYSIS courbe de bruit *f*
NCI *abbr* (*negative chemical ionization*) ANALYSIS ionisation chimique négative *f*
NCV *abbr* (*net calorific value*) WATER POLL PCN (*pouvoir calorifique net*)
NDIR *abbr* (*nondispersive infrared*) ANALYSIS infrarouge non dispersif *m*
NDIRA *abbr* (*nondispersive infrared absorption*) ANALYSIS absorption infrarouge non dispersive *f*
Ne *abbr* (*neon*) AIR Ne (*néon*)
neap: ~ **tide** *n* RENEW ENERG *tidal energy* morte-eau *f*
near: ~ **infrared spectroscopy** *n* WASTE *automatic sorting of plastics* spectroscopie de l'infrarouge proche *f*; ~-**shore area** *n* MARINE zone côtière *f*; ~-**size particle** *n* AIR élément limite *m*
neat *adj* MARINE non-délayé, non-dilué, propre
need: ~ **for remediation** *n* SOIL besoin d'assainissement *m*
needled: ~ **felt** *n* AIR *filter medium* feutre cousu *m*
NEF *abbr* (*noise exposure forecast*) NOISE prévision d'exposition au bruit *f*
negative: ~ **catalysis** *n* AIR catalyse négative *f*; ~ **chemical ionization** *n* (*NCI*) ANALYSIS ionisation chimique négative *f*; ~ **electrode** *n* AIR électrode négative *f*; ~ **pressure** *n* SOIL *suction of soil atmosphere* dépression *f*
negotiation *n* POLICY négociation *f*
NEL *abbr* (*no-effect level*) SAFETY niveau à effet nul *m*
neon *n* (*Ne*) AIR néon *m* (*Ne*)
neoprene *n* WASTE néoprène *m*
nephelometer *n* ANALYSIS néphélomètre *m*
nephelometric[1] *adj* ANALYSIS néphélométrique
nephelometric[2]: ~ **analysis** *n* ANALYSIS *of water* analyse néphélométrique *f*
nephelometry *n* ANALYSIS néphélométrie *f*; ~ **turbidity unit** (*NTU*) ANALYSIS unité néphélométrique *f*
Nessler: ~**'s solution** *n* ANALYSIS solution de Nessler *f*, nessler *m*
net: ~ **calorific value** *n* (*NCV*) WATER POLL pouvoir calorifique net *m* (*PCN*), WATER SUPP pouvoir calorifique naturel *m*; ~ **donator** *n* WATER SUPP exportateur net *m*; ~ **positive suction height** *n* (*NPSH*) WATER POLL hauteur nette positive d'aspiration *f*; ~ **receiver** *n* WATER SUPP importateur net *m*; ~ **weight** *n* WASTE poids net *m*
network *n* WATER SUPP réseau *m*; ~ **calculation** WATER SUPP calcul de réseau *m*
neutral[1] *adj* AIR *waste gas-plume propagation class*, POLICY neutre
neutral[2]: ~ **gas molecule** *n* AIR molécule de gaz neutre *f*; ~ **tax** *n* POLICY impôt neutre *m*; ~ **temperature layering** *n* AIR stratification neutre des températures *f*; ~ **wastewater sample** *n* ANALYSIS prélèvement d'eaux usées neutres *m*
neutralization *n* GEN *of detergents* neutralisation *f*; ~ **point** ANALYSIS point de neutralisation *m*; ~ **of wastewater** WATER POLL *acid, alkaline* neutralisation des eaux résiduaires *f*
neutralize *vt* GEN neutraliser
neutralizer *n* GEN neutralisateur *m*
neutralizing[1] *adj* GEN neutralisant
neutralizing[2]: ~ **agent** *n* AIR, SAFETY, WASTE agent neutralisateur *m*, agent neutralisant *m*
new: ~ **money** *n* MANAGE crédits de restructuration *m pl*; ~ **plant** *n* AIR, POLICY installation neuve *f*, nouvelle usine *f*, nouvelle installation *f*, usine neuve *f* WATER SUPP *created after 1975* centrale nouvelle *f*; ~ **technology** *n* GEN nouvelle technique *f*, technologie naissante *f*, nouvelle technologie *f*; ~ **water** *n* WATER SUPP eau neuve *f*
newton *n* (*N*) ANALYSIS newton *m* (*N*)
ng *abbr* (*nanogram, nanogramme*) ANALYSIS ng (*nanogramme*)
Ni *abbr* (*nickel*) AIR, SOIL Ni (*nickel*)
nickel *n* (*Ni*) AIR, SOIL nickel *m* (*Ni*)
90%: ~ **threshold value range** *n* ANALYSIS seuil de 90% *m*
NIPTS *abbr* (*noise-induced permanent threshold shift*) NOISE dépassement permanent de seuil induit par le bruit *m*

nitrate *n* AIR, SOIL, WATER POLL, WATER SUPP nitrate *m*; ~ **load** WATER POLL charge de nitrates *f*
nitric: ~ **acid** *n* (*HNO_3*) AIR, WATER SUPP acide nitrique *m* (*HNO_3*); ~ **bacteria** *n pl* WATER POLL bactéries nitriques *f pl*; ~ **oxide** *n* (*NO*) AIR, WATER SUPP oxyde nitrique *m* (*NO*), monoxyde d'azote *m* (*NO*); ~ **and sulfuric acid process** *AmE*, ~ **and sulphuric acid process** *BrE n* AIR procédé aux acides nitrique et sulfurique *m*
nitrification *n* WASTE, WATER POLL, WATER SUPP nitrification *f*; ~**-denitrification** WATER POLL, WATER SUPP nitrification-dénitrification *f*; ~ **zone** WATER POLL, WATER SUPP zone de nitrification *f*
nitrify *vt* WASTE, WATER POLL, WATER SUPP nitrifier
nitrifying: ~ **bacteria** *n pl* WATER POLL bactéries nitrifiantes *f pl*
nitrite *n* SOIL, WATER POLL nitrite *m*
nitrogen[1]: ~**-fixing** *adj* WATER POLL, WATER SUPP nitrifiant
nitrogen[2] *n* (*N*) AIR, WATER POLL azote *m* (*N*); ~ **compound** AIR, WATER POLL composé azoté *m*; ~ **cycle** AIR, WATER POLL cycle de l'azote *m*; ~ **dioxide** (*NO_2*) AIR, SAFETY, WATER POLL dioxyde d'azote *m* (*NO_2*); ~ **fixation** WATER POLL fixation de l'azote *f*; ~ **formation** AIR formation d'azote *f*; ~ **oxides** *n pl* (*NO_x*) WASTE oxydes d'azote *m pl* (*NO_x*); ~ **oxides abatement** AIR réduction des oxydes d'azote *f*; ~ **oxides removal** AIR extraction des oxydes d'azote *f*; ~ **oxides scrubber** AIR laveur d'oxydes d'azote *m*; ~ **pentoxide** (*N_2O_5*) AIR pentoxyde d'azote *m* (*N_2O_5*), pentoxyde de diazote *m* (*N_2O_5*); ~ **peroxide** AIR, WATER POLL péroxyde d'azote *m*; ~**-phosphorus detector** (*NPD*) ANALYSIS détecteur d'azote et de phosphore *m*
nitrogenous: ~ **fertilizer** *n* SOIL, WATER POLL engrais azoté *m*
nitrometer *n* ANALYSIS azotomètre *m*, nitromètre *m*
nitrosomas: ~ **bacteria** *n* WATER POLL *oxidizing ammonia to nitrate* nitrosobactérie *f*
nitrous: ~ **oxide** *n* (*N_2O*) AIR hémioxyde d'azote *m* (*N_2O*), monoxyde de diazote *m* (*N_2O*), oxyde nitreux *m* (*N_2O*), WATER SUPP anhydride hypo-azoteux *m* (*N_2O*)
nm *abbr* (*nanometer AmE, nanometre BrE*) ANALYSIS nm (*nanomètre*)
NO *abbr* (*nitric oxide*) AIR, WATER SUPP NO (*oxyde nitrique, monoxyde d'azote*)
NO_2 *abbr* (*nitrogen dioxide*) AIR, SAFETY, WATER POLL NO_2 (*dioxyde d'azote*)
N_2O *abbr* (*nitrous oxide*) AIR N_2O (*oxyde nitreux, hémioxyde d'azote, monoxyde de diazote*), WATER POLL N_2O (*anhydride hypo-azoteux*)
N_2O_5 *abbr* (*nitrogen pentoxide, dinitrogen pentoxide*) AIR N_2O_5 (*pentoxyde d'azote, pentoxyde de diazote*)
NO_x[1] *abbr* (*nitrogen oxides*) WASTE NO_x (*oxydes d'azote*)
NO_x[2]: ~ **emissions** *n pl* AIR émissions d'oxydes d'azote *f pl*; ~ **removal** *n* (*denox*) AIR, SOIL, WASTE élimination de NO_x (*denox*); ~ **value** *n* AIR indice d'oxydes d'azote *m*
NOAEL *abbr* (*no observed adverse-effect level*) SAFETY absence totale de perturbation observée *f*
noble: ~ **metal** *n* AIR, WATER SUPP métal noble *m*; ~ **metal catalyst** *n* AIR, WATER SUPP catalyseur à base de métaux nobles *m*
NOEC *abbr* (*no observed effect concentration*) SAFETY CSEO (*concentration sans effet observé*)
no-effect: ~ **level**; *n* (*NEL*) SAFETY niveau à effet nul *m*, absence d'effet *f*
noise[1]: ~**-absorbing** *adj* MARINE, NOISE, SAFETY insonorisant; ~**-contaminated** *adj* MARINE pollué par le bruit, NOISE contaminé par le bruit; ~**-excluding** *adj* MARINE, NOISE, SAFETY insonorisant; ~**-induced** *adj* MARINE, NOISE, SAFETY induit par le bruit; ~**-intensive** *adj* MARINE, NOISE à forte intensité sonore; ~**-reducing** *adj* NOISE réducteur de bruit
noise[2] *n* NOISE, SAFETY bruit *m*; ~ **abatement** NOISE, SAFETY lutte contre le bruit *f*, réduction du bruit *f*, assourdissement *m*; ~**-abatement zone** NOISE zone de réduction du bruit *f*, zone de silence *f*; ~**-absorbing structural element** NOISE élément structurel insonorisant *m*; ~ **ambient impairment** *n pl* NOISE détérioration du milieu sonore *f*; ~ **area** NOISE zone bruyante *f*; ~ **area map** NOISE cartographie acoustique *f*; ~ **at the workplace** MANAGE, NOISE bruit sur le lieu de travail *m*; ~ **ban** NOISE interdiction du bruit *f*; ~ **carpet** NOISE empreinte acoustique *f*, sillage sonore *m*; ~ **caused by air traffic** NOISE bruit de la navigation aérienne *m*; ~ **caused by rail traffic** NOISE bruit lié au trafic ferroviaire *m*; ~ **charge** NOISE, POLICY redevance sur le bruit *f*; ~ **check** NOISE, SAFETY lutte contre le bruit *f*; ~ **contamination** NOISE pollution sonore *f*; ~ **control** NOISE, SAFETY lutte contre le bruit *f*; ~ **criterion curve** (*NC-curve*) ANALYSIS courbe du bruit *f*; ~ **curve** ANALYSIS courbe de bruit *f*; ~ **dosimeter** ANALYSIS dosimètre acoustique *m*; ~ **emissions** *n pl* AIR, NOISE, RENEW ENERG *from wind turbines* émissions acoustiques *f pl*, émissions sonores *f pl*; ~ **emitter** NOISE émetteur de bruit *m*; ~ **exclusion helmet** NOISE, SAFETY casque antibruit *m*; ~ **exposure** NOISE exposition au bruit *f*; ~ **exposure forecast** (*NEF*) NOISE prévision d'exposition au bruit *f*; ~ **exposure meter** ANALYSIS, NOISE appareil de mesure de l'exposition au bruit *m*; ~ **generation** NOISE production de bruit *f*; ~ **guard** *n pl* NOISE capot de protection acoustique *m*, dispositif anti-bruit *m*; ~**-guiding value** NOISE indice de bruit *m*; ~ **immission** NOISE immission acoustique *f*; ~**-induced deafness** NOISE, SAFETY surdité induite par le bruit *f*; ~**-induced hearing damage** NOISE, SAFETY perte d'acuité auditive induite par le bruit *f*; ~**-induced hearing impairment** NOISE, SAFETY déficience auditive induite par le bruit *f*, perte d'acuité auditive induite par le bruit *f*; ~**-induced permanent threshold shift** (*NIPTS*) NOISE dépassement permanent de seuil induit par le bruit *m*; ~**-insulating barrier** NOISE *on streets* mur antibruit *m*; ~ **insulation wall** NOISE paroi insonorisante *f*; ~ **intrusion** NOISE intrusion sonore *f*, pénétration intempestive de bruits *f*; ~ **labeling** *AmE*, ~ **labelling** *BrE* NOISE étiquetage acoustique *m*, indication de niveau sonore *m*; ~ **level** NOISE niveau de bruit *m*; ~ **level analyser** *BrE*, ~ **level analyzer** *AmE* ANALYSIS, NOISE analyseur de niveau sonore *m*; ~ **level indicator** ANALYSIS, NOISE indicateur de niveau acoustique *m*; ~ **limit** ANALYSIS seuil acoustique *m*, MARINE valeur limite d'émission acoustique *f*, NOISE limite de bruit *f*; ~ **limit indicator** ANALYSIS, NOISE indicateur de seuil acoustique *m*; ~ **limit value** NOISE valeur limite d'émission acoustique *f*; ~**-limiting value** NOISE valeur limite de bruit *f*; ~ **map** NOISE carte acoustique *f*; ~ **measurement** ANALYSIS, NOISE mesure du bruit *f*; ~ **measurement technology** NOISE technique de mesure du bruit *f*; ~ **meter** ANALYSIS sonomètre *m*; ~ **monitor** ANALYSIS, MARINE, NOISE contrôleur de bruit *m*; ~ **monitoring** ANALYSIS contrôle du bruit *m*, MARINE, NOISE surveillance du bruit *f*; ~ **muffler** *AmE* (*cf sound silencer BrE*) NOISE silencieux *m*; ~ **nuisance effect** NOISE effet de nuisance sonore *m*; ~ **peak** NOISE pic de bruit *m*; ~ **pollution** NOISE pollution acoustique *f*, SAFETY pollution par le bruit *f*; ~ **pollution prevention** NOISE, SAFETY prévention de la pollution sonore *f*; ~ **prevention** MARINE, NOISE prévention du bruit *f*; ~ **prevention measure** MARINE, NOISE mesure de prévention du bruit *f*; ~ **propagation** NOISE propagation sonore *f*; ~ **protection** NOISE protection contre le bruit *f*; ~ **protection area** NOISE zone de silence *f*; ~ **protection booth** MARINE, NOISE, SAFETY cabine de protection contre le bruit *f*, cabine insonorisée *f*; ~ **protection for compressors** NOISE, SAFETY insonorisation des compresseurs *f*; ~ **protection law** MARINE, NOISE législation sur la protection contre le bruit *f*; ~ **protection measure** MARINE, NOISE mesure de protection contre le bruit *f*; ~ **protection system** MARINE, NOISE système d'insonorisation *m*, système de protection contre le bruit *m*; ~ **protection technology** MARINE, NOISE technologie d'insonorisation *f*, technique de protection

contre le bruit *f*; ~ **protection tunnel** MARINE, NOISE tunnel d'insonorisation *m*, tunnel de protection contre le bruit *m*; ~ **protection wall** MARINE, NOISE paroi anti-bruit *f*; ~ **protection wall made from waste** NOISE mur antibruit constitué de déchets *m*; ~ **protection zone** MARINE, NOISE zone de protection contre le bruit *f*, zone de silence *f*; ~ **radiation** NOISE rayonnement du bruit *m*; ~ **rating** ANALYSIS, NOISE évaluation du bruit *f*, pondération du bruit *f*; ~ **rating curve** (*NR-curve*) ANALYSIS, NOISE courbe d'évaluation de bruit *f*; ~ **rating index** ANALYSIS, NOISE indice de pondération des bruits *m*; ~ **rating level** ANALYSIS, NOISE niveau de pondération des bruits *m*; ~ **reduction** NOISE, SAFETY réduction du bruit *f*; ~ **register** NOISE registre sonore *m*; ~ **screen** NOISE écran acoustique *m*, écran anti-bruit *m*; ~ **shield** NOISE écran acoustique *m*, écran anti-bruit *m*, blindage acoustique *m*; ~ **source** MARINE, NOISE source sonore *f*, source de bruit *f*; ~ **standard value** MARINE, NOISE indice sonore de référence *m*, valeur acoustique de référence *f*; ~ **suppression** NOISE suppression de bruit *f*; ~ **type** NOISE type de bruit *m*; ~ **and vibration measuring equipment** MARINE, NOISE, SAFETY matériel de mesure du bruit et des vibrations *m*; ~ **weighting** NOISE pondération du bruit *f*

noise[3]: ~ **and vibration** *phr* MARINE, NOISE, SAFETY bruit et vibrations *m pl*

noiseless[1] *adj* MARINE, NOISE, SAFETY insonore, insonorisé, silencieux

noiseless[2]: ~ **motor** *n* MARINE, NOISE, SAFETY moteur silencieux *m*

nonaerated *adj* WATER POLL non-aéré, sans mélange préalable; ~ **oxidation pond** *n* WATER POLL bassin d'oxydation non-aéré *m*

nonalkaline: ~ **permanent hardness** *n* WATER SUPP dureté non alcaline *f*

nonbiodegradable[1] *adj* MARINE, SOIL, WASTE, WATER POLL non biodégradable

nonbiodegradable[2]: ~ **organics** *n pl* WATER POLL matière organique non biodégradable *f*; ~ **packaging** *n* WASTE emballage non-biodégradable *m*; ~ **waste** *n* WASTE, WATER SUPP déchets non biodégradables *m pl*

noncatalytic *adj* AIR non-catalytique

nonclogging[1] *adj* WASTE, WATER POLL, WATER SUPP anti-colmatage, imbouchable, non-engorgeable

nonclogging[2]: ~ **packing material** *n* WASTE matériau d'emballage anti-colmatage *m*; ~ **pump** *n* WATER POLL pompe non-engorgeable *f*, pompe imbouchable *f*

noncombustibility: ~ **test** *n* SAFETY essai d'incombustibilité *m*

noncombustible: ~ **residue** *n* WASTE résidus incombustibles *m pl*

noncompostable: ~ **waste** *n* WASTE, WATER SUPP déchets non compostables *m pl*

nonconcentrating: ~ **collector** *n* RENEW ENERG *mirror evacuated tube* collecteur sans concentration *m*

nonconformity *n* MANAGE non-conformité *f*

nonconing: ~ **smoke plume** *n* AIR panache de fumée étalé *m*

noncorrosive *adj* AIR non-corrosif

noncrystalline *adj* AIR non-cristallin

nondispersive: ~ **infrared** *n* (*NDIR*) ANALYSIS infrarouge non dispersif *m*; ~ **infrared absorber** *n* ANALYSIS absorbeur d'infrarouge non dispersif *m*; ~ **infrared absorption** *n* (*NDIRA*) ANALYSIS absorption infrarouge non dispersive *f*; ~ **infrared analyser** *BrE*, ~ **infrared analyzer** *AmE* *n* ANALYSIS analyseur d'infrarouge non dispersif *m*; ~ **infrared spectrometry** *n* ANALYSIS spectrométrie infrarouge non dispersive *f*

nonflammable *adj* AIR, SAFETY, WASTE ininflammable

nonhazardous: ~ **waste** *n* WASTE déchets banals *m pl*

noninflammable *adj* WASTE ininflammable

nonionic *adj* AIR, WATER POLL non-ionique

nonionogenic[1] *adj* WATER POLL non-ionogène

nonionogenic[2]: ~ **polymer** *n* WATER POLL polymère non-ionogène *m*

nonleachable *adj* SOIL non-lixiviable

nonmeltable: ~ **material** *n* WASTE matériau infusible *m*

non-point: ~ **pollution** *n* POLICY pollution diffuse *f*

nonporous *adj* WATER POLL non-poreux

nonputrescible *adj* WASTE, WATER POLL, WATER SUPP imputrescible

nonradial: ~ **flow** *n* AIR, WATER POLL *of settling tank* flux non radial *m*

nonrecoverable[1] *adj* WASTE, WATER SUPP non-récupérable

nonrecoverable[2]: ~ **waste** *n* WASTE déchets non recyclables *m pl*

nonregenerable: ~ **flue-gas desulfurization** *AmE*, ~ **flue-gas desulphurization** *BrE* *n* AIR désulfuration d'effluents gazeux non recyclables *f*

nonremeltable *adj* WASTE, WATER SUPP impossible à refondre

nonrenewable: ~ **resources** *n pl* RENEW ENERG ressources non renouvelables *f pl*

nonresonant: ~ **ion** *n* ANALYSIS *quadrupole mass analyser* ion non résonant *m*

nonreturnable: ~ **bottle** *n* WASTE bouteille non consignée *f*, bouteille à jeter *f*, bouteille en verre perdu *f*; ~ **container** *n* WASTE conteneur perdu *m*; ~ **packaging** *n* WASTE emballage perdu *m*; ~ **pallet** *n* WASTE palette perdue *f*

nonreusable: ~ **residue wastes** *n pl* WASTE résidus non recyclables *m pl*, résidus non réutilisables *m pl*

nonsolar: ~-**based renewable energy source** *n* RENEW ENERG *tidal and geothermal energy* source d'énergie renouvelable autre que solaire *f*

nonspecific *adj* ANALYSIS non-spécifique

nontoxic[1] *adj* AIR, SAFETY, SOIL, WATER POLL non-toxique

nontoxic[2]: ~ **metal** *n* AIR, SAFETY, SOIL, WATER POLL métal non toxique *m*

nonvolatile *adj* AIR non-volatile

nonvulcanized *adj* WASTE, WATER SUPP non-vulcanisé

nonwaste: ~ **technology** *n* (*NWT*) AIR, WASTE, WATER SUPP technologie ne produisant pas de déchets *f*, technologie non polluante *f*

nonweighable: ~ **lysimeter** *n* ANALYSIS lysimètre non-pesable *m*

nonwetting *adj* AIR non-mouillant

nonwoven[1] *adj* AIR non-tissé

nonwoven[2]: ~ **fabric** *n* AIR *filter fibre* tissu non-tissé *m*

normal: ~ **operation** *n* POLICY *of plant* exploitation courante *f*; ~ **pressure** *n* ANALYSIS pression normale *f*; ~ **rate of charge** *n* POLICY taux de charge habituel *m*

notification *n* MARINE notification *f*

notifier *n* POLICY déclarant *m*

noxious *adj* GEN nocif, nuisible

noxiousness *n* SAFETY nocivité *f*

nozzle *n* WASTE lance d'arrosage *f*, WATER SUPP ajutage *m*, lance d'arrosage *f*; ~ **atomization** AIR *spray absorption* vaporisation par une buse *f*; ~ **atomizer** AIR *spray absorber* atomiseur à buse *m*, vaporisateur à buse *m*; ~ **pipe** AIR ajoutoir *m*, tubulure *f*; ~ **separator** AIR séparateur à buse *m*; ~ **tube** AIR tube à buse *m*

NPD *abbr* (*nitrogen-phosphorus detector*) ANALYSIS détecteur d'azote et de phosphore *m*

NPSH *abbr* (*net positive suction height*) WATER POLL hauteur nette positive d'aspiration *f*

NR: ~-**curve** *n* (*noise rating curve*) ANALYSIS, NOISE courbe d'évaluation de bruit *f*

NTU *abbr* (*nephelometry turbidity unit*) ANALYSIS unité néphélométrique de turbidité *f*

nuclear[1] *adj* GEN nucléaire

nuclear[2]: ~ **energy source** *n* RENEW ENERG source d'énergie nucléaire *f*; ~ **reactor** *n* WATER POLL réacteur nucléaire *m*;

~ **safety** *n* SAFETY sûreté nucléaire *f*; ~ **waste** *n* WASTE déchets nucléaires *m pl*
null: ~ **balance** *n* ANALYSIS *nullification* compensation à zéro *f*, tarage sur zéro *m*
numerical: ~ **grid model** *n* ANALYSIS modèle à grille numérique *m*; ~ **value** *n* ANALYSIS valeur numérique *f*
nutrient *n* SOIL, WASTE, WATER POLL élément nutritif *m*; ~ **cycle** POLICY cycle nutritif *m*; ~ **removal** WATER POLL extraction des substances nutritives *f*; ~**-rich wastes** *n pl* WASTE déchets riches en éléments nutritifs *m pl*; ~**-solution tank** AIR *biological gas scrubbing* cuve pour nutriments en solution *f*
nutriment *n* MARINE, WATER POLL, WATER SUPP nutriment *m*; ~ **supply** AIR fourniture de nutriments *f*
NWT *abbr* (*nonwaste technology*) AIR, WASTE, WATER SUPP technologie ne produisant pas de déchets *f*, technologie non polluante *f*
nylon[1]: ~**-reinforced** *adj* WASTE renforcé par du nylon
nylon[2] *n* AIR *filter material*, WASTE nylon *m*

O

O *abbr* (*oxygen*) AIR, RENEW ENERG O (*oxygène*)
O₃ *abbr* (*ozone*) AIR O_3 (*ozone*)
object[1]: **~-related** *adj* MARINE, NOISE lié à l'objet, relatif à l'objet
object[2]: **~-related noise** *n* MARINE, NOISE *machines* bruit lié à l'objet *m*
obligation *n* POLICY obligation *f*; **~ to recycle** POLICY obligation de recycler *f*; **~ to reuse residue materials** POLICY obligation de réutiliser les matériaux résiduels *f*
oblige *vt* POLICY imposer, obliger
observable *adj* ANALYSIS observable
observation: **~ time** *n* ANALYSIS durée d'observation *f*; **~ well** *n* WASTE puits d'observation *m*
observed: **no ~ adverse-effect level** *n* (*NOAEL*) SAFETY absence totale de perturbation observée *f*, **no ~ effect concentration** *n* (*NOEC*) SAFETY concentration sans effet observé *f* (*CSEO*); **~ effective level** *n* (*OEL*) SAFETY niveau efficace observé *m*
obstruction: **~ of noise propagation** *n* NOISE obstruction à la propagation du bruit *f*
OC *abbr* (*open cycle*) RENEW ENERG cycle ouvert *m*
occupational[1] *adj* SAFETY professionnel
occupational[2]: **~ disease** *n* SAFETY maladie professionnelle *f*; **~ dose limit** *n* SAFETY limite de dose professionnelle *f*; **~ threshold limit value** *n* (*occupational TLV*) WATER SUPP valeur seuil en milieu professionnel *f*; **~ TLV** *n* (*occupational threshold limit value*) WATER SUPP valeur seuil en milieu professionnel *f*
ocean: **~ basin** *n* MARINE bassin océanique *m*; **~ bed** *n* MARINE fond océanique *m*, fond marin *m*; **~ current** *n* RENEW ENERG *energy source* courant marin *m*, courant océanique *m*; **~ current energy** *n* RENEW ENERG énergie des courants marins *f*; **~ current power** *n* RENEW ENERG électricité produite à partir des courants marins *f*; **~ current power plant** *n* RENEW ENERG centrale exploitant les courants marins *f*; **~ current power recovery** *n* RENEW ENERG production d'électricité à partir des courants marins *f*; **~ current resources** *n pl* RENEW ENERG ressources énergétiques des courants marins *f pl*; **~ disposal** *n* MARINE élimination en mer *f*, immersion *f*, rejet en mer *m*, WASTE, WATER POLL *of wastewater* immersion *f*, rejet en mer *m*; **~ dumping** *n* MARINE, WASTE, WATER POLL rejet en mer *m*; **~ energy** *n* RENEW ENERG énergie océanique *f*; **~ floor** *n* MARINE fond océanique *m*, fond marin *m*; **~ harvesting** *n* RENEW ENERG exploitation des océans *f*; **~ pollution** *n* MARINE pollution des océans *f*; **~ research** *n* POLICY recherche océanique *f*; **~ thermal energy conversion** *n* (*OTEC*) RENEW ENERG énergie thermique des mers *f* (*ETM*); **~ thermal energy conversion plant** *n* (*OTEC plant*) RENEW ENERG usine de conversion de l'énergie thermique des mers *f*; **~ water** *n* MARINE, RENEW ENERG, WATER SUPP eau de mer *f*
oceanic[1] *adj* MARINE océanique
oceanic[2]: **~ abyss** *n* MARINE plaine abyssale *f*
octane: **~ number** *n* AIR indice d'octane *m*
ocular *n* ANALYSIS *interferential pit gas tester* oculaire *m*
OD *abbr* ANALYSIS (*oxygen deficit*) déficit en oxygène *m*, SAFETY (*overdose*) dépassement de dose *m*, surdose *f*
odor *AmE see odour BrE*
odoriferous: **~ substance** *n* AIR, WATER POLL substance odoriférante *f*; **~ substance emissions** *n pl* AIR émissions de substances odoriférantes *f pl*
odorimetry *n* ANALYSIS odorimétrie *f*
odoriphore *n* AIR osmophore *m*, odoriphore
odorless *AmE see odourless BrE*
odorous *adj* AIR odorant
odour[1]: **~-intensive** *adj BrE* AIR à forte intensité odorante, à forte odeur, à odeur forte, très odorant; **~-laden** *adj BrE* AIR odorant
odour[2] *n BrE* AIR, ANALYSIS, WASTE, WATER POLL odeur *f*; **~ analysis** *BrE* ANALYSIS analyse olfactive *f*, analyse d'odeur *f*; **~ analysis in wastewater** *BrE* ANALYSIS, WASTE analyse de l'odeur des eaux usées *f*, analyse olfactive des eaux usées *f*; **~ annoyance** *BrE* AIR, WASTE, WATER POLL nuisance olfactive *f*, gêne olfactive *f*; **~ control** *BrE* AIR, WATER POLL, WATER SUPP désodorisation *f*, neutralisation des odeurs *f*; **~ determination** *BrE* ANALYSIS détermination de l'odeur *f*; **~ emissions** *n pl BrE* AIR, WATER POLL émissions d'odeurs *f pl*, émissions olfactives *f pl*; **~ formation** *BrE* AIR, WATER POLL apparition d'odeurs *f*, production d'odeurs *f*; **~-forming substance** *BrE* AIR, WATER POLL substance odoriférante *f*; **~-intensive** *BrE* AIR, WATER POLL à odeur forte; **~-intensive solvent** *BrE* WATER POLL solvant nauséabond *m*; **~-intensive substance** *BrE* AIR, WATER POLL substance nauséabonde *f*; **~-laden dirty gas** *BrE* AIR gaz odorant à traiter *m*; **~-laden waste air** *BrE* AIR effluent gazeux odorant *m*; **~ nuisance** *BrE* AIR, WASTE, WATER POLL nuisance olfactive *f*, gêne olfactive *f*; **~ quotient** *BrE* ANALYSIS quotient olfactif *m*; **~ removal** *BrE* AIR, WATER POLL, WATER SUPP désodorisation *f*; **~ sensitivity** *BrE* ANALYSIS sensibilité aux odeurs *f*, sensibilité olfactive *f*; **~-substance-laden air** *BrE* AIR air chargé en substances odoriférantes *m*; **~ test** *BrE* ANALYSIS essai olfactif *m*; **~ threshold** *BrE* AIR, ANALYSIS, WATER POLL, WATER SUPP seuil olfactif *m*
odourless[1] *adj BrE* AIR, ANALYSIS, WASTE, WATER SUPP inodore, sans odeur
odourless[2]: **~ gas** *n BrE* AIR gaz inodore *m*
ODP *abbr* (*ozone depletion potential*) AIR, POLICY potentiel d'appauvrissement de la couche d'ozone *m*
OEL *abbr* (*observed effective level*) SAFETY niveau efficace observé *m*
offgas *n* AIR, WASTE, WATER SUPP gaz d'échappement *m*, gaz résiduel *m*
office: **~ waste** *n* WASTE déchets administratifs *m pl*, déchets de bureau *m pl*
official: **~ journal** *n* (*OJ*) POLICY *of EU* journal officiel *m* (*JO*); **~ supervision** *n* POLICY surveillance officielle *f*
off-line[1] *adj* ANALYSIS autonome, non-connecté
off-line[2]: **~ operation** *n* ANALYSIS exploitation autonome *f*, traitement séquentiel *m*
off-shore *adj* MARINE en mer, au large
offshore: **~ potential** *n* RENEW ENERG potentiel marin *m*; **~ potential of wind energy** *n* RENEW ENERG potentiel d'énergie éolienne en mer *m*; **~ wind farm** *n* RENEW ENERG exploitation éolienne en mer *f*
off-site: **~ analysis** *n* ANALYSIS analyse hors site *f*; **~ disposal** *n* SOIL élimination hors site *f*; **~ incineration** *n* SOIL incinération hors site *f*
OFS *abbr* (*oil from sludge*) WATER POLL huile combustible produite à partir des boues *f*
oil[1]: **~-absorbing** *adj* MARINE, WATER POLL oléophile; **~-containing** *adj* RENEW ENERG huileux, oléagineux
oil[2] *n* GEN huile *f*, pétrole *m*, hydrocarbure *m*; **~ boom** MARINE

barrage flottant *m*, barrière flottante *f*; ~ **clots** *n pl* MARINE pétrole coagulé *m*; ~ **concentration** WATER POLL concentration d'huile *f*; ~**-containing plants** *n pl* RENEW ENERG *biomass conversion base material* plantes oléagineuses *f pl*; ~ **contamination** MARINE pollution par les hydrocarbures *f*; ~**-content meter** SOIL, WATER POLL appareil de mesure de la teneur en hydrocarbures *m*; ~ **dispersant** WATER POLL dispersant du pétrole *m*; ~ **droplet** WATER POLL gouttelette d'huile *f*; ~**-filled receiving tube** RENEW ENERG *solar power plant* collecteur tubulaire à bain d'huile *m*; ~ **film** MARINE pellicule d'hydrocarbures *f*; ~**-fired power plant** POLICY, RENEW ENERG centrale thermique au fioul *f*; ~ **from sludge** (*OFS*) WATER POLL huile combustible produite à partir des boues *f*; ~ **leakage** MARINE déversement d'hydrocarbures *m*; ~ **patch** MARINE, WATER SUPP nappe d'hydrocarbures *f*; ~**-polluted wastewater** WATER POLL eaux polluées par les hydrocarbures *f pl*; ~ **pollution** MARINE pollution pétrolière *f*; ~ **pollution combating ship** MARINE navire dépollueur *m*; ~ **pollution control** MARINE lutte contre la pollution par les hydrocarbures *f*; ~ **pollution fighter** MARINE, WATER POLL, WATER SUPP bâtiment de dépollution *m*; ~ **regeneration plant** WASTE, WATER SUPP usine de régénération des huiles usées *f*; ~ **removal** WATER POLL déshuilage *m*; ~ **separation** SOIL, WATER POLL, WATER SUPP séparation d'huile *f*; ~ **separator** SOIL déshuileur *m*, WATER POLL dispositif collecteur *m*; ~ **slick** MARINE marée noire *f*, nappe de pétrole *f*, plaque d'huile *f*, surface luisante *f*, tache d'huile *f*; ~ **spill** MARINE déversement d'hydrocarbures *m*, marée noire *f*; ~ **spill combating equipment** MARINE équipement de dépollution par les hydrocarbures *m*; ~**-spill identification system** MARINE système de détection des déversements d'hydrocarbures *m*; ~ **spill response** MARINE mesure prise en cas de marée noire *f*; ~ **spillage** MARINE déversement d'hydrocarbures *m*, marée noire *f*, WATER POLL épanchements d'hydrocarbures *m pl*; ~ **sump** SOIL, WATER POLL carter à huile *m*; ~ **trap** WATER POLL déshuileur *m*, dispositif collecteur *m*; ~ **waste** WATER SUPP déchets huileux *m pl*; ~**-water interface** MARINE interface hydrocarbures-eau *f*; ~**-water separation** WATER POLL séparation huile-eau *f*; ~**-water separator** WATER POLL séparateur huile-eau *m*

oily *adj* RENEW ENERG huileux

OJ *abbr* (*official journal*) POLICY JO (*journal officiel*)

oleophilic[1] *adj* MARINE, WATER POLL oléophile

oleophilic[2]: ~ **belt** *n* MARINE cordon oléophile *m*; ~ **belt skimmer** *n* MARINE récupérateur à cordon oléophile *m*

olfactometric: ~ **measure** *n* AIR, ANALYSIS mesure olfactométrique *f*

olfactometry *n* AIR, ANALYSIS olfactométrie *f*

on-board: ~ **computer** *n* ANALYSIS ordinateur de bord *m*, ordinateur embarqué *m*

one: ~**-way bottle** *n* WASTE bouteille non consignée *f*, bouteille perdue *f*; ~**-way container** *n* WASTE conteneur perdu *m*; ~**-way pack** *n* WASTE emballage perdu *m*; ~**-way pallet** *n* WASTE palette perdue *f*

on-line[1] *adj* ANALYSIS en ligne

on-line[2]: ~ **analyser** *BrE*, ~ **analyzer** *AmE n* ANALYSIS, WATER POLL analyseur en continu *m*; ~ **analysis** *n* ANALYSIS, WATER POLL analyse en continu *f*; ~ **measurement** *n* ANALYSIS mesurage en continu *m*; ~ **operation** *n* ANALYSIS exploitation en continu *f*, fonctionnement en continu *m*

onshore: ~ **marine environment** *n* MARINE milieu marin côtier *m*

on-site[1] *adj* ANALYSIS, SOIL,WASTE,WATER POLL, WATER SUPP, sur site, sur place

on-site[2]: ~ **analysis** *n* ANALYSIS, SOIL analyse sur site *f*; ~ **application** *n* SOIL application sur place *f*; ~ **extraction** *n* SOIL extraction sur place *f*; ~ **incineration** *n* SOIL incinération sur place *f*

onstream: ~ **factor** *n* MANAGE facteur de marche *m*

on-the-spot *adj* GEN immédiat, direct

ooze: ~ **mud** *n* WATER POLL boues grasses *f pl*

opacimeter *n* ANALYSIS opacimètre *m*

opaque *adj* WATER POLL *degree of turbidity* opaque

open: ~**-air measurement** *n* ANALYSIS mesure à l'air libre *f*; ~ **cycle** *n* (*OC*) RENEW ENERG cycle ouvert *m*; ~**-cycle plant** *n* RENEW ENERG *ocean thermal energy* centrale à cycle ouvert *f*, usine de conversion de l'énergie thermique des mers à cycle ouvert *f*; ~ **drain** *n* WATER POLL collecteur à ciel ouvert *m*; ~ **dump** *n* WASTE décharge à ciel ouvert *f*, décharge brute *f*, décharge non contrôlée *f*; ~**-loop catalyst** *n* AIR catalyseur en circuit ouvert *m*; ~**-loop catalytic converter** *n* AIR pot catalytique en circuit ouvert *m*; ~**-ocean ecosystem** *n* MARINE écosystème marin *m*; ~**-pit landfill** *n* WASTE décharge à ciel ouvert *f*; ~ **Rankine cycle** *n* RENEW ENERG *ocean thermal energy* cycle de Rankine ouvert *m*; ~ **sewer** *n* WASTE, WATER POLL, WATER SUPP égout à ciel ouvert *m*; ~ **system** *n* WASTE circuit ouvert *m*, système ouvert *m*; ~ **trickling filter** *n* WATER POLL lit bactérien ouvert *m*

operating: ~ **authorization** *n* POLICY, WATER SUPP permis d'exploitation *m*; ~ **costs** *n pl* POLICY, WATER SUPP coûts d'exploitation *m pl*; ~ **error** *n* ANALYSIS, SAFETY dysfonctionnement *m*, erreur de manœuvre *f*; ~ **face** *n* WASTE côté utilisable *m*, face utilisable *f*; ~ **licence** *BrE*, ~ **license** *AmE n* POLICY exploitation sous licence *f*; ~ **material** *n* AIR matériel d'exploitation *m*; ~ **permission** *n* POLICY, WATER SUPP permis d'exploitation *m*; ~ **speed** *n* RENEW ENERG *wind turbine* vitesse de fonctionnement *f*; ~ **standard** *n* POLICY norme d'exploitation *f*

operation *n* GEN exploitation *f*; ~ **time** RENEW ENERG durée de service *f*

operational: ~ **error** *n* MANAGE erreur d'exploitation *f*, WATER SUPP fausse manœuvre *f*

operations: ~ **safety** *n* SAFETY sûreté d'exploitation *f*

operative: ~ **range** *n* SOIL domaine d'activité *m*

operator *n* POLICY exploitant *m*; ~ **duties** *n pl* POLICY obligations de l'exploitant *f pl*

optical[1] *adj* ANALYSIS *analysers* optique

optical[2]: ~ **absorption** *n* ANALYSIS *in gases* absorption de lumière *f*, absorption optique *f*; ~ **sorter** *n* WASTE trieur optique *m*

optically: ~**-absorptive pollutant accumulation** *n* ANALYSIS accumulation de polluants absorbant la lumière *f*

optimum: ~ **polymer dosage** *n* WATER POLL dosage optimal de polymères *m*

optoelectronic[1] *adj* WASTE, WATER SUPP optoélectronique

optoelectronic[2]: ~ **process** *n* WASTE *binding* procédé optoélectronique *m*

ore *n* GEN minerai *m*

orbit *n* AIR *raw gas in cyclone* orbite *f*

organic[1] *adj* AIR, WATER POLL organique

organic[2]: ~ **acid gas** *n* AIR gaz acide organique *m*; ~ **base material** *n* AIR *biofilter* substrat organique *m*; ~ **biomass** *n* RENEW ENERG *biomass conversion* biomasse organique *f*; ~ **contaminant** *n* AIR polluant organique *m*; ~ **debris** *n pl* WASTE débris organiques *m pl*; ~ **dry mass** *n* WATER POLL masse sèche organique *f*; ~ **farming** *n* POLICY, SOIL agriculture biologique *f*; ~ **gas** *n* AIR gaz organique *m*; ~ **harmful substance** *n* WATER POLL substance organique nocive *f*; ~ **matter** *n* AIR, WATER POLL matière organique *f*; ~ **solvent** *n* AIR solvant organique *m*; ~ **waste** *n* WASTE déchets organiques *m pl*

organism *n* WATER POLL organisme *m*

organochlorinated *adj* WATER SUPP organochloré

organochlorine: ~ **pesticide** *n* WATER POLL pesticide organochloré *m*

organoleptic *adj* WATER SUPP organoleptique

oriented: ~ **polypropylene label** *n* WASTE étiquette de polypropylène orienté *f*

original: ~ **soil** *n* SOIL sol d'origine *m*

O-ring *n* ANALYSIS joint torique *m*
orlon *n* AIR *filter material* orlon *m*
ORP *abbr* (*oxidation-reduction potential*) ANALYSIS, WATER POLL potentiel d'oxydo-réduction *m*
Orsat: ~ **apparatus** *n* ANALYSIS appareil de mesure des fumées d'Orsat *m*; ~ **burette** *n* ANALYSIS burette d'Orsat *f*; ~ **gas analyser** *BrE*, ~ **gas analyzer** *AmE* *n* ANALYSIS analyseur de fumées type Orsat *m*
oscillating: ~ **conveyor** *n* WASTE convoyeur à secousses *m*; ~ **water column** *n* (*OWC*) RENEW ENERG colonne d'eau oscillante *f*; ~ **water column operation** *n* (*OWC operation*) RENEW ENERG exploitation de l'énergie des vagues selon le principe de la colonne oscillante *f*
oscillator *n* NOISE oscillateur *m*
oscilloscope *n* ANALYSIS *mass spectroscopy* oscilloscope *m*
osmophore *n* AIR osmophore *m*
osmosis *n* WATER POLL osmose *f*, WATER SUPP osmose *f*
osmotic[1] *adj* WATER POLL, WATER SUPP osmotique
osmotic[2]: ~ **pressure** *n* WATER POLL pression osmotique *f*
OTEC[1] *abbr* (*ocean thermal energy conversion*) RENEW ENERG ETM (*énergie thermique des mers*)
OTEC[2]: ~ **plant** *n* RENEW ENERG usine ETM *f*
outdoor: ~ **shielding** *n* NOISE écran externe *m*; ~ **sound propagation** *n* NOISE propagation des bruits à l'extérieur *f*
outer: ~ **air** *n* AIR air extérieur *m*; ~ **eddy** *n* WATER SUPP tourbillon dextrogyre *m*; ~ **electrode** *n* ANALYSIS électrode extérieure *f*; ~ **filter** *n* ANALYSIS filtre externe *m*; ~ **shell** *n* NOISE *in encapsulation* enveloppe externe *f*; ~ **vortex** *n* AIR tourbillon dextrogyre *m*
outfall *n* WATER POLL émissaire *m*, déversoir *m*, exutoire *m*; ~ **of sewage** WATER POLL déversement d'eaux usées *m*
outflow *n* WATER POLL, WATER SUPP écoulement *m*, débit sortant *m*, sortie *f*
outlet *n* AIR sortie *f*, MANAGE débouché *m*, espace de vente *m*, WATER POLL émissaire *m*, déversoir *m*, exutoire *m*; ~ **branch** WATER SUPP *pump* tubulure de sortie *f*; ~ **channel** WATER POLL conduite d'évacuation *f*; ~ **flow control** WATER POLL régulation du débit de sortie *f*; ~ **pipe** WATER POLL conduite d'évacuation *f*; ~ **piping** WATER SUPP *pump* tubulure de sortie *f*
output *n* ANALYSIS production *f*, MANAGE *from product life cycle* rendement *m*
overall: ~ **chimney height** *n* AIR hauteur totale de cheminée *f*; ~ **stack height** *n* AIR hauteur totale de cheminée *f*
overburning *n* SOIL combustion excessive *f*
overcrowded: ~ **area** *n* WATER SUPP zone surpeuplée *f*
overdosage *n* AIR, SAFETY surdosage *m*
overdose *n* (*OD*) ANALYSIS, SAFETY dépassement de dose *m*, surdose *f*
overdosing *n* AIR, SAFETY surdosage *m*
overexploitation *n* SOIL *of soil* surexploitation *f*
overfertilization *n* SOIL, WATER POLL fertilisation excessive *f*, surfertilisation *f*
overfishing *n* MARINE surexploitation des fonds de pêche *f*, surpêche *f*
overflow[1] *n* AIR, WATER SUPP débordement *m*, trop-plein *m*; ~ **nappe** WATER SUPP lame déversante *f*; ~ **pipe** AIR conduite de débordement *f*, WATER SUPP trop-plein *m*, tuyau de trop-plein *m*; ~ **rate** WATER SUPP débit de trop-plein *m*; ~ **structure** WATER SUPP ouvrage de trop-plein *m*; ~ **tank** WATER SUPP réservoir de trop-plein *m*; ~ **weir** AIR, WATER POLL déversoir de trop-plein *m*
overflow[2] *vi* AIR, WATER SUPP déborder, se déverser
overgrazing *n* SOIL surpâturage *m*
overhead: ~ **irrigation** *n* SOIL irrigation par aspersion *f*; ~ **magnet** *n* WASTE aimant aérien *m*; ~ **sound absorber** *n* NOISE absorbeur acoustique aérien *m*, absorbeur acoustique surélevé *m*
overheating *n* AIR surchauffe *f*
overload *n* AIR surcharge *f*
over-package *vt* WASTE suremballer
over-packaging *n* WASTE suremballage *m*
overpopulation *n* POLICY surpeuplement *m*, surpopulation *f*
oversalting *n* WATER SUPP salinité excessive *f*
oversampling *n* ANALYSIS suréchantillonnage *m*
oversize: ~ **fraction** *n* WATER SUPP fraction surdimensionnée *f*; ~ **material** *n* WASTE *in sorting* matériau surdimensionné *m*
overwrap *vt* WASTE suremballer
overwrapping *n* WASTE suremballage *m*
OWC[1] *abbr* (*oscillating water column*) RENEW ENERG colonne d'eau oscillante *f*
OWC[2]: ~ **operation** *n* RENEW ENERG *wave energy* exploitation de l'énergie des vagues selon le principe de la colonne oscillante *f*
oxbow *n* WATER SUPP bras mort *m*
oxidant *n* AIR, SOIL, WATER POLL comburant *m*, oxydant *m*
oxidation *n* AIR, ANALYSIS, SOIL oxydation *f*; ~ **air** AIR air oxydant *m*; ~ **ditch** WASTE *sewage* fossé d'oxydation *m*, WATER POLL chenal d'oxydation *m*; ~ **pond** WASTE bassin d'oxydation *m*, WATER POLL étang d'oxydation *m*, *maturation pond* bassin d'oxydation *m*, WATER SUPP étang d'oxydation *m*; ~**-reduction potential** (*ORP*) ANALYSIS, WATER POLL potentiel d'oxydo-réduction *m*; ~**-reduction process** (*ORP*) AIR *for waste air cleaning* procédé d'oxydo-réduction *m*; ~ **tank** AIR bassin d'oxydation *m*; ~ **zone** AIR zone d'oxydation *f*
oxidative[1] *adj* AIR, SOIL, WATER POLL oxydant
oxidative[2]: ~ **decomposition** *n* WATER POLL décomposition par oxydation *f*
oxidimetric[1] *adj* ANALYSIS oxydimétrique
oxidimetric[2]: ~ **determination** *n* ANALYSIS détermination oxydimétrique *f*, dosage oxydimétrique *m*; ~ **titration** *n* ANALYSIS titrage oxydimétrique *m*
oxidizability *n* AIR, ANALYSIS, SOIL oxydabilité *f*
oxidizable[1] *adj* AIR, SOIL oxydable
oxidizable[2]: ~ **substance** *n* AIR, SOIL substance oxydable *f*
oxidize[1] *vt* AIR, SOIL, WATER POLL oxyder
oxidize[2] *vi* AIR, SOIL, WATER POLL s'oxyder
oxidizer *n* AIR, SOIL, WATER POLL comburant *m*, oxydant *m*
oxidizing[1] *adj* AIR, SOIL, WATER POLL oxydant
oxidizing[2]: ~ **agent** *n* AIR, SOIL, WATER POLL agent oxydant *m*, comburant *m*; ~ **catalyst** *n* AIR catalyseur d'oxydation *m*; ~ **catalyst bed** *n* AIR lit catalytique oxydant *m*; ~ **substance** *n* AIR, SAFETY substance oxydante *f*
oxygen[1]: ~**-demanding** *adj* SOIL exigeant en oxygène; ~**-free** *adj* WATER SUPP anaérobie
oxygen[2] *n* (*O*) AIR, RENEW ENERG oxygène *m* (*O*); ~ **breathing apparatus** SAFETY appareil respiratoire à oxygène *m*; ~ **budget** AIR bilan d'oxygène *m*; ~ **concentration** ANALYSIS, WATER POLL concentration d'oxygène *f*, concentration en oxygène *f*; ~ **consumption** WATER POLL consommation d'oxygène *f*; ~ **content** ANALYSIS, RENEW ENERG *of biomass* teneur en oxygène *f*; ~ **deficiency** ANALYSIS insuffisance d'oxygène *f*, WATER POLL carence en oxygène *f*; ~ **deficit** (*OD*) ANALYSIS déficit en oxygène *m*; ~ **depletion** AIR, ANALYSIS, WATER POLL appauvrissement en oxygène *m*, raréfaction des ressources en oxygène *f*; ~ **flask** ANALYSIS bouteille d'oxygène *f*; ~ **introduction** RENEW ENERG *biomass conversion*, WATER SUPP apport d'oxygène *m*; ~ **measuring device** AIR, ANALYSIS oxygènemètre *m*; ~ **meter** AIR, ANALYSIS *catalyst* capteur d'oxygène *m*, oxygènemètre *m*; ~ **saturation** WATER POLL saturation en oxygène *f*; ~ **saturation factor** WATER POLL facteur de saturation en oxygène *m*; ~ **sensor** AIR, ANALYSIS capteur d'oxygène *m*; ~ **supersaturation** WATER POLL sursaturation en oxygène *f*; ~ **supply** AIR, WATER POLL alimentation en oxygène *f*, approvisionnement en oxygène *m*; ~ **transfer** WATER POLL transfert d'oxygène *m*; ~ **transfer capacity** WATER POLL *in clean water* capacité de transfert d'oxygène *f*;

~ transfer efficiency WATER POLL rendement du transfert d'oxygène *m*; **~ uptake** WATER POLL apport d'oxygène *m*
oxygenate *vt* WATER POLL oxygéner
oxygenated *adj* WATER POLL oxygéné
oxygenic *adj* WATER POLL oxygéné
ozone[1]: **~-damaging** *adj* AIR destructeur d'ozone; **~-friendly** *adj* AIR sans incidence négative pour la couche d'ozone
ozone[2] *n* (O_3) AIR ozone *m* (O_3); **~ analyser** *BrE*, **~ analyzer** *AmE* ANALYSIS analyseur d'ozone *m*; **~ budget** AIR bilan d'ozone *m*; **~ capillary** AIR capillaire à ozone *m*; **~ concentration** AIR concentration d'ozone *f*; **~ content** AIR teneur en ozone *f*; **~ damage** AIR destruction de l'ozone *f*; **~ deficit** AIR déficit en ozone *m*; **~ depletion potential** (*ODP*) AIR, POLICY potentiel d'appauvrissement de la couche d'ozone *m*; **~ destroyer** AIR destructeur d'ozone *m*; **~ destruction** AIR destruction de l'ozone *f*; **~ equipment** AIR ozoneur *m*; **~ formation** AIR formation d'ozone *f*; **~ generation** AIR production d'ozone *f*; **~ generator** ANALYSIS générateur d'ozone *m*, WATER SUPP ozoneur *m*; **~ hole** AIR trou d'ozone *m*; **~ immission** AIR immission d'ozone *f*; **~ impact** AIR incidence de l'ozone *f*; **~ layer** AIR couche d'ozone *f*; **~-layer depletion** AIR; appauvrissement de la couche d'ozone *m*, appauvrissement en ozone *m*; **~ layer destruction** AIR destruction de la couche d'ozone *f*; **~ level** AIR niveau d'ozone *m*; **~ loss** AIR diminution de la couche d'ozone *f*; **~ measuring** ANALYSIS ozonométrie *f*; **~ shield** AIR écran d'ozone *m*, bouclier d'ozone *m*; **~ shrinkage** AIR amenuisement de la couche d'ozone *m*; **~ smog** AIR brouillard d'ozone *m*; **~ value** AIR valeur de l'ozone *f*; **~ warning** AIR alerte à l'ozone *f*
ozonization *n* SOIL, WATER POLL, WATER SUPP ozonisation *f*
ozonize *vt* SOIL, WATER POLL, WATER SUPP ozoniser
ozonizer *n* SOIL, WATER POLL, WATER SUPP ozoniseur *m*

P

p *abbr* (*pascal*) ANALYSIS p (*pascal*)
P *abbr* (*phosphorus*) AIR, WATER POLL P (*phosphore*)
PA *abbr* (*polyamide*) WATER POLL, WATER SUPP polyamide *m*
PAC *abbr* (*powdered activated carbon*) WATER POLL, WATER SUPP charbon actif en poudre *m*
pack *vt* AIR pack, WASTE emballer
package[1] *n* WASTE emballage *m*
package[2] *vt* WASTE emballer
packaging: ~ **line** *n* WASTE chaîne d'emballage *m*, ligne d'emballage *f*; ~ **material** *n* WASTE matériau d'emballage *m*; ~ **material producing industry** *n* MANAGE, WASTE industrie de fabrication de matériaux d'emballage *f*; ~ **optimization** *n* WASTE optimisation des emballages *f*; ~ **paper** *n* WASTE papier d'emballage *m*; ~ **producing industry** *n* MANAGE, WASTE industrie de fabrication d'emballages *f*; ~ **profile** *n* WASTE profil d'emballage *m*; ~ **secondary resource** *n* WASTE matériau d'emballage recyclé *m*; ~ **station** *n* WASTE poste d'emballage *m*; ~ **waste** *n* WASTE, WATER SUPP déchets d'emballage *m pl*
packed[1] *adj* WASTE emballé, WATER SUPP garni
packed[2]: ~ **bed** *n* AIR lit à garnissage *m*; ~-**bed column** *n* AIR colonne à garnissage *f*; ~-**bed filter** *n* AIR filtre à garnissage *m*; ~-**bed height** *n* AIR hauteur de garnissage *f*; ~-**bed scrubber** *n* AIR laveur à garnissage *m*; ~-**capillary injector** *n* ANALYSIS injecteur capillaire haute densité *m*; ~ **column** *n* AIR, ANALYSIS, WATER SUPP colonne garnie *f*, colonne à corps de remplissage *f*, colonne à garnissage *f*, colonne garnie *f*; ~-**column scrubber** *n* WASTE *wet scrubber* laveur à colonne garnie *m*, dépoussiéreur à colonne garnie *m*; ~ **injector** *n* ANALYSIS injecteur haute densité *m*; ~ **suspended bed** *n* (*PSB*) WATER POLL lit fixe suspendu *m*; ~-**tower scrubber** *n* WASTE laveur à garnissage *m*, laveur statique *m*
packer *n* SOIL compacteur *m*; ~ **body** WASTE benne étanche à ordures *f*, WATER SUPP benne tasseuse *f*; ~ **unit** WASTE poste d'empaquetage *m*, WATER SUPP compacteur *m*
packing *n* AIR garnissage *m*, WASTE emballage *m*, WATER SUPP étoupe *f*, garniture *f*, garniture d'étanchéité *f*, garniture de presse-étoupe *f*; ~ **press** WATER SUPP presse à empaqueter *f*, presse-balles *m*; ~ **station** WASTE installation d'emballage *f*, poste d'amballage *m*, poste de conditionnement *m*
padding *n* NOISE rembourrage *m*
paddle *n* WATER POLL ailette *f*, pale *f*; ~ **blade** WATER POLL aube *f*; ~ **mixer** WASTE malaxeur à pales *m*, WATER SUPP mélangeur à ailettes *m*; ~-**wheel aeration** WATER POLL aération par roue à ailettes *f*
PAFC *abbr* (*phosphoric acid fuel cell*) AIR pile à combustible à l'acide phosphorique *f*
PAH *abbr* (*polycyclic aromatic hydrocarbon*) AIR, WATER SUPP HAP (*hydrocarbure aromatique polycyclique*), hydrocarbure polynucléaire aromatique *m*
paint: ~ **sludge** *n* WATER POLL, WATER SUPP boues de peinture *f pl*, boues de décapage *f pl*
pall: ~ **ring** *n* AIR *packing of packed-tower scrubbers* anneau de Pall *m*
palladium *n* (*Pd*) AIR palladium *m* (*Pd*)
pallet: ~ **hood** *n* WASTE chausse de palette *f*; ~ **shrink wrapping** *n* WASTE emballage rétractable pour palette *m*
PAN *abbr* AIR (*peroxoacetyl nitrate*) PAN (*nitrate de péroxyacétyle, péroxiacétylnitrate*), polyacrylnitrile *m*
panel *n* GEN panneau *m*; ~ **louvered ceiling** *AmE*, ~-**louvred ceiling** *BrE* NOISE plafond alvéolé *m*; ~ **resonator** NOISE panneau résonateur *m*

paper: ~ **bag** *n* WASTE, WATER SUPP sac en papier *m*; ~ **board** *n* WATER SUPP carton *m*; ~ **board waste** *n* WASTE, WATER SUPP *production-specific waste* déchets de carton *m pl*; ~ **chips** *n* WASTE rognures de papier *f pl*; ~ **chromatography** *n* ANALYSIS chromatographie sur papier *f*; ~ **collection** *n* WASTE collecte des papiers *f*; ~ **consumption** *n* WATER SUPP consommation de papier *f*; ~-**converting industry** *n* WASTE industrie de recyclage du papier *f*; ~ **fiber** *AmE*, ~ **fibre** *BrE* *n* WASTE, WATER SUPP fibre papetière *f*; ~ **filter** *n* ANALYSIS filtre-papier *m*; ~ **grade** *n* WASTE sorte de papier *f*; ~ **laminate** *n* WASTE feuille de papier *f*; ~ **liner** *n* WASTE couchage du papier *m*; ~ **pulp** *n* WASTE, WATER SUPP pâte à papier *f*, pâte papetière *f*; ~ **quality** *n* WASTE, WATER SUPP qualité de papier *f*; ~ **raw material** *n* WASTE matière première pour papier *f*; ~ **sack** *n* WASTE, WATER SUPP sac en papier *m*; ~ **wrapping** *n* WASTE emballage en papier *m*
parabolic[1] *adj* RENEW ENERG parabolique
parabolic[2]: ~ **dish** *n* RENEW ENERG miroir parabolique *m*, réflecteur parabolique *m*; ~ **dish installation** *n* RENEW ENERG installation de réflecteurs paraboliques *f*; ~ **reflector** *n* RENEW ENERG *solar power plant* réflecteur parabolique *m*
parabolically: ~-**formed gutter** *n* RENEW ENERG *collector* rigole de forme parabolique *f*
parallel[1] *adj* ANALYSIS *of measuring* parallèle
parallel[2]: ~ **pump** *n* WATER POLL, WATER SUPP pompe parallèle *f*
paramagnetic[1] *adj* ANALYSIS paramagnétique
paramagnetic[2]: ~ **analytical method** *n* ANALYSIS méthode d'analyse paramagnétique *f*
paramagnetism *n* ANALYSIS paramagnétisme *m*
parameter *n* ANALYSIS, SOIL, WATER SUPP paramètre *m*; ~ **determination** ANALYSIS, SOIL, WATER SUPP paramétrisation *f*
parameterization *n* ANALYSIS, SOIL, WATER SUPP paramétrisation *f*
pararosaniline: ~ **method** *n* ANALYSIS méthode à la pararosaniline *f*
parceled *AmE parcelled BrE*
parceling *AmE see parcelling BrE*
parcelled: ~ **goods** *n pl BrE* WASTE produits alimentaires conditionnés *m pl*
parcelling: ~ **machine** *n BrE* WASTE machine à empaqueter *f*, machine à conditionner *f*
parchment: ~ **paper** *n* WASTE papier parchemin *m*, papier sulfurisé *m*
parent: ~ **solution** *n* ANALYSIS solution initiale *f*, solution mère *f*
park: ~ **waste** *n* WASTE, WATER SUPP déchets horticoles *m pl*
partial[1] *adj* GEN partiel
partial[2]: ~ **authorization** *n* POLICY autorisation partielle *f*; ~ **current** *n* WATER POLL *of production water* courant partiel *m*; ~ **differential pressure** *n* AIR, WATER POLL gradient de pression partielle *m*; ~ **enclosure** *n* NOISE enceinte partielle *f*; ~ **gas quantity** *n* ANALYSIS volume partiel de gaz *m*; ~ **pressure** *n* ANALYSIS pression partielle *f*; ~ **pressure difference** *n* ANALYSIS différence de pression partielle *f*, gradient de pression partielle *m*; ~ **soil decontamination** *n* SOIL décontamination partielle du sol *f*; ~ **soil remediation** *n* SOIL assainissement partiel du sol *m*; ~ **stream** *n* AIR courant partiel *m*
partially[1]: ~-**automated** *adj* ANALYSIS partiellement

automatisé; **~-continuous** *adj* ANALYSIS semi-continu; **~-halogenated** *adj* AIR partiellement halogéné; **~-stabilized** *adj* ANALYSIS partiellement stabilisé; **~ water soluble** *adj* WATER SUPP partiellement soluble dans l'eau

partially[2]: **~-dry waste-gas treatment** *n* AIR procédé semi-humide de traitement des gaz *m*; **~-halogenated CFC** *n* AIR CFC partiellement halogéné *m*; **~-integrated system** *n* WATER SUPP système partiellement intégré *m*; **~-stabilized sludge** *n* WATER POLL boues partiellement stabilisées *f pl*

particle *n* GEN particule *f*; **~ beam liquid chromatography-mass spectrometry** (*PB LC-MS*) ANALYSIS spectrométrie de masse-chromatographie liquide à faisceau de particules *f*; **~ charging** AIR charge de particule *f*; **~ counter** ANALYSIS compteur de particules *m*; **~ counting device** ANALYSIS compteur de particules *m*; **~ diameter** WATER POLL diamètre de particule *m*; **~-drop agglomerate** AIR agglomérat goutte-particule *m*; **~ emissions** *n pl* AIR émissions de particules *f pl*; **~ filtration** WATER POLL filtration de particules *f*; **~ friction** AIR friction de particules *f*; **~ impact** AIR impact de particule *m*; **~-laden waste gas** AIR gaz chargé de particules *m*; **~ orbit** AIR *collection operation* orbite des particules *f*; **~ size** AIR taille de particule *f*, WASTE calibre *m*, dimension de particules *f*, taille de particule *f*, WATER SUPP calibre *m*; **~ size analyser** *BrE*, **~ size analyzer** *AmE* ANALYSIS analyseur granulométrique *m*; **~ size class** AIR, WATER POLL, WATER SUPP classe granulométrique *f*; **~ size distribution** AIR, WATER POLL, WATER SUPP granulométrie *f*; **~ size separation** AIR, WATER POLL, WATER SUPP granulométrie *f*; **~ surface** AIR *wet precipitator* surface de particule *f*; **~ wetting** AIR mouillage de particule *m*

particulate[1] *adj* AIR en particule, particulaire; **~-laden** AIR chargé de particules

particulate[2]: **~ air pollution** *n* AIR pollution atmosphérique particulaire *f*; **~ collection** *n* AIR, WASTE, WATER SUPP collecte de particules *f*; **~ emissions** *n pl* AIR émissions sous forme de particules *f pl*; **~ gas** *n* AIR gaz sous forme de particules *m*; **~-laden gas stream** *n* AIR flux de gaz chargé de particules *m*; **~ material** *n* WATER POLL matière particulaire *f*; **~ matter** *n* AIR particule en suspension *f*; **~ matter filter** *n* AIR filtre à particules *m*; **~ pollutant** *n* AIR polluant particulaire *m*, *in dust, fumes, mist, smoke, spray* polluant sous forme de particule *m*; **~ pollution** *n* AIR, WATER SUPP pollution sous forme de particules *f*; **~ separation** *n* AIR séparation de particules *f*; **~ size distribution** *n* AIR granulométrie *f*; **~ smoke emissions** *n pl* AIR émissions de fumées sous forme de particules *f pl*; **~ solid matter** *n* AIR matière solide sous forme particulaire *f*; **~ thickness** *n* AIR épaisseur de particules *f*

particulates: **~ monitoring** *n* ANALYSIS contrôle des particules *m*

partition *n* NOISE cloison *f*; **~ chromatography** ANALYSIS chromatographie de partage *f*; **~ surface** NOISE *walls, ceilings* surface de cloison *f*

parts: **~ per billion** *phr* (*ppb*) ANALYSIS parties par milliard; **~ per million** *phr* (*ppm*) ANALYSIS parties par million (*ppm*); **~ per quadrillion** *phr* (*ppq*) ANALYSIS parties par quadrillion; **~ per trillion** *phr* (*ppt*) ANALYSIS parties par trillion

PAS *abbr* (*photoacoustic spectroscopy*) ANALYSIS spectroscopie photoacoustique *f*

pascal *n* (*p*) ANALYSIS pascal *m* (*p*)

pass *vt* POLICY *law* adopter; **~ through** AIR traverser

passive[1] *adj* MARINE, NOISE, RENEW ENERG, SOIL passif

passive[2]: **~ matrix liquid-crystal display** *n* (*PMLCD*) ANALYSIS écran à cristaux liquides à matrice passive *m*; **~ sound protection** *n* NOISE protection acoustique passive *f*; **~ thermal solar energy use** *n* RENEW ENERG consommation passive d'énergie thermique d'origine solaire *f*

pasteurization *n* WATER POLL pasteurisation *f*

pasty *adj* WASTE, WATER SUPP pâteux

paternoster: **~ pump** *n* WATER POLL, WATER SUPP pompe à chapelet *f*, pompe à godets *f*

pathogen[1]: **~-free** *adj* WATER POLL, WATER SUPP non-pathogène, sans agent pathogène

pathogen[2] *n* WATER POLL, WATER SUPP agent pathogène *m*

pathogenic[1] *adj* WATER POLL, WATER SUPP pathogène

pathogenic[2]: **~ bacteria** *n pl* WATER POLL, WATER SUPP bactéries pathogènes *f pl*; **~ germ** *n* WATER POLL germe pathogène *m*; **~ hazard** *n* SAFETY risque pathogène *m*; **~ microbe** *n* WATER SUPP microbe pathogène *m*; **~ microorganism** *n* WATER POLL, WATER SUPP micro-organisme pathogène *m*

pathological: **~ waste** *n* WASTE, WATER POLL, WATER SUPP déchets contaminés *m pl*, déchets pathogènes *m pl*

patrol: **~ boat** *n* MARINE patrouilleur des mers *m*

pay: **~-back time** *n* RENEW ENERG délai d'amortissement *m*

Pb *abbr* (*lead*) GEN Pb (*plomb*)

PBL *abbr* (*planetary boundary layer*) AIR couche limite planétaire *f*

PB LC-MS *abbr* (*particle beam liquid chromatography-mass spectrometry*) ANALYSIS spectrométrie de masse-chromatographie liquide à faisceau de particules *f*

PC *abbr* (*polycarbonate*) WATER POLL, WATER SUPP PC (*polycarbonate*)

PCB[1] *abbr* (*polychlorinated biphenyl*) AIR, SOIL, WASTE BPC (*biphényle polychloré*)

PCB[2]: **~-capacitor** *n* WATER SUPP condensateur à PCB *m*

PCDF *abbr* (*polychlorinated dibenzofuran*) SOIL, WASTE PCDF (*dibenzofuranne polychloré*)

PCI *abbr* (*positive chemical ionization*) ANALYSIS ionisation chimique positive *f*

PCN *abbr* (*polychlorinated naphthale*) SOIL PCN (*naphthalène polychloré*)

PCP *abbr* (*pentachlorophenol*) AIR, SOIL PCP (*pentachlorophénol*)

PCT *abbr* (*polychlorinated terphenyl*) AIR, SOIL TPC (*terphényle polychloré*)

PCV *abbr* (*positive crankcase ventilation*) AIR recyclage des gaz de carter *m*

Pd *abbr* (*palladium*) AIR Pd (*palladium*)

Pd(II)/Cu(II): **~ catalyst** *n* AIR catalyseur Pd(II)/Cu(II) *m*

PDP *abbr* (*plasma display panel*) ANALYSIS écran à plasma *m*

PE *abbr* (*polyethylene*) WASTE, WATER SUPP PE (*polyéthylène*)

peak[1] *adj* GEN maximal, de pointe

peak[2] *n* ANALYSIS *in energy spectrum* ligne *f*, pic *m*, NOISE pic *m*; **~ concentration** AIR, ANALYSIS concentration maximale *f*; **~ consumption** WATER SUPP consommation maximale *f*, pointe de consommation *f*; **~ of the day** WATER POLL pointe journalière *f*; **~ demand** WATER SUPP consommation de pointe *f*, pointe de la demande *f*; **~ detection** ANALYSIS détection de pic *f*, détection de pointe *f*; **~ flow** WATER POLL débit de pointe *m*; **~ infiltration** WATER SUPP infiltration maximale *f*; **~ load** RENEW ENERG charge de pointe *f*; **~-load power plant** RENEW ENERG centrale de pointe *f*; **~ ozone damage** AIR destruction aiguë de l'ozone *f*; **~ velocity** WATER SUPP vitesse de crête *f*, vitesse de pointe *f*; **~ wall** (*Wp*) RENEW ENERG wall crête *m*; **~ water demand** WATER SUPP pointe de la demande d'eau *f*; **~ water flow** WATER SUPP débit de pointe *m*

PED *abbr* (*plasma emission detection*) ANALYSIS détection d'émissions de plasma *f*

pedogenesis *n* SOIL pédogenèse *f*

pedology *n* SOIL pédologie *f*

peel: **~-off wrapping** *n* WASTE emballage pelable *m*

peelable: **~ protective coating** *n* WASTE revêtement de protection détachable *m*, revêtement de protection pelable *m*; **~ system** *n* WASTE système détachable *m*, système pelable *m*

pelagic[1] *adj* MARINE pélagique
pelagic[2]: **~ environment** *n* MARINE milieu pélagique *m*; **~ organism** *n* WATER POLL organisme pélagique *m*
pellet *n* AIR grain *m*, pastille *f*, RENEW ENERG granulé *m*, plaquette *f*, SOIL *vitrification*, WASTE boulette *f*, WATER POLL *sludge cake* pastille *f*, boulette *f*; **~ catalyst** AIR catalyseur en grains *m*; **~ incineration** SOIL incinération en boulettes *f*; **~ layer** SOIL couche granuleuse *f*; **~ mill** WATER SUPP pastilleuse *f*, granulatrice *f*; **~ technique** ANALYSIS *in infrared spectroscopy* technique de granulation *f*, technique de pelletisation *f*
pelletization *n* SOIL, WASTE, WATER POLL, WATER SUPP pastillage *m*, bouletage *m*, granulation *f*
pelletize *vt* SOIL, WASTE, WATER POLL, WATER SUPP pastiller, bouleter, pelletiser
pelletizer *n* WASTE *machine*, WATER SUPP pastilleur *m*
pelletizing *n* SOIL, WASTE, WATER POLL, WATER SUPP pastillage *m*, bouletage *m*, granulation *f*
Peltier: **~ element** *n* ANALYSIS *for cooling* élément Peltier *m*
penal: **~ law** *n* POLICY droit pénal *m*
penetrate *vt* AIR traverser
penetrated: **~ by radiation** *adj* ANALYSIS irradié
penetration *n* AIR perméance *f*, *of dust particle* pénétration *f*
penstock *n* WATER SUPP canal de prise *m*, vanne *f*, *dams* conduite forcée *f*, *hydroelectric dams* vanne-rideau *f*, vanne à glissières *f*, *intake structure* vanne murale *f*
pentachlorophenol *n* (*PCP*) AIR, SOIL pentachlorophénol *m* (*PCP*)
per: **~ capita consumption** *n* WATER SUPP consommation par habitant *f*; **~ capita energy consumption** *n* RENEW ENERG consommation d'énergie par habitant *f*
PER *abbr* (*perchloroethylene*) AIR PER (*perchloroéthylène*)
perceived: **~ noise level** *n* (*PNdB*) NOISE niveau de bruit perçu *m*; **~ sound level** *n* NOISE niveau de bruit perçu *m*
percentile: **~ value** *n* ANALYSIS pourcentile *m*
perched: **~ groundwater** *n* WATER SUPP nappe perchée *f*, nappe suspendue *f*
perchloroethylene *n* (*PER*) AIR perchloroéthylène *m* (*PER*)
percolate[1]: **~ through** *vt* WATER POLL percoler, WATER SUPP pénétrer, passer à travers, percoler
percolate[2] *vi* WATER POLL passer à travers, percoler, s'infiltrer, WATER SUPP percoler
percolating: **~ filter** *n* WATER POLL, WATER SUPP filtre percolateur *m*, lit percolateur *m*; **~ water** *n* WASTE, WATER SUPP eau de percolation *f*, eau d'infiltration *f*; **~ water treatment plant** *n* WATER POLL station d'épuration de l'eau d'infiltration *f*
percolation *n* AIR, ANALYSIS filtration *f*, WASTE, WATER POLL, WATER SUPP filtration *f*, infiltration *f*, percolation *f*; **~ capacity** WATER SUPP capacité d'infiltration *f*; **~ well** SOIL puits d'infiltration *m*, WATER SUPP puits d'infiltration *m*, puits de percolation *m*
percolative: **~ infiltration** *n* WATER SUPP infiltration par percolation *f*
perfect: **~ gas** *n* AIR gaz parfait *m*
perforate *vt* WASTE, WATER SUPP *pipe* percer, perforer
perforated[1] *adj* WASTE, WATER SUPP *pipe* percé, perforé
perforated[2]: **~ panel** *n* NOISE panneau perforé *m*; **~ panel absorber** *n* NOISE panneau absorbant perforé *m*; **~ plate** *n* AIR plaque perforée *f*, WATER POLL *of screen* tôle perforée *f*, WATER SUPP tôle perforée *f*, *in fuel assembly* plaque perforée *f*
perikinetic[1] *adj* WATER SUPP péricinétique
perikinetic[2]: **~ flocculation** *n* WATER SUPP floculation péricinétique *f*
periodic[1] *adj* AIR, ANALYSIS, RENEW ENERG périodique
periodic[2]: **~ cleaning** *n* AIR *filter bag* nettoyage périodique *m*; **~ flushing** *n* WASTE vidange périodique *f*, WATER SUPP lavage périodique *m*, vidange périodique *f*
peripheral: **~ ditch** *n* SOIL fossé périphérique *m*; **~ pump** *n* WATER POLL pompe tourbillonnaire *f*, WATER SUPP pompe périphérique *f*
periphery *n* NOISE *areas to be lined* périphérie *f*; **~ particle orbit** AIR *collection operation* orbite périphériques de particules *f*
peristaltic: **~ pump** *n* WATER POLL pompe péristaltique *f*
permanent[1]: **~ magnetic** *adj* WATER SUPP à aimantation permanente
permanent[2]: **~ hardness** *n* WATER POLL, WATER SUPP dureté permanente *f*; **~ infiltration** *n* WATER SUPP infiltration permanente *f*; **~ magnet** *n* ANALYSIS, WASTE, WATER SUPP aimant permanent *m*; **~ magnetic-separating chute** *n* WASTE séparation magnétique en continu *f*; **~ magnetic separating drum** *n* WASTE, WATER SUPP tambour de séparation magnétique en continu *m*; **~ storage** *n* WATER SUPP *of waste* stockage permanent *m*; **~ threshold shift** *n* (*PTS*) NOISE déplacement permanent de seuil *m*; **~ waste storage** *n* WASTE stockage permanent des déchets *m*
permeability *n* GEN perméabilité *f*; **~ to grease** WASTE perméabilité aux graisses *f*
permeable *adj* GEN perméable
permeameter *n* ANALYSIS *permeability measuring instrument* perméamètre *m*
permeate[1] *n* GEN perméat *m*; **~ recovery** WATER POLL récupération des produits d'infiltration *f*, recyclage des perméats *m*
permeate[2] *vt* GEN s'infiltrer dans, impregner
permeation: **~ membrane** *n* AIR *reactor* membrane à imprégnation *f*, membrane poreuse *f*; **~ velocity** *n* AIR vitesse d'imprégnation *f*, vitesse de passage *f*
permission *n* POLICY, WATER SUPP autorisation *f*, permis *m*
permit[1] *n* POLICY autorisation *f*, permis *m*; **~ application** POLICY demande d'autorisation *f*, WATER SUPP demande de permis *f*; **~ impact** POLICY, WATER SUPP répercussion d'une autorisation *f*; **~ process** POLICY, WATER SUPP procédure d'autorisation *f*
permit[2] *vt* POLICY autoriser
peroxoacetyl: **~ nitrate** *n* (*PAN*) AIR, WATER SUPP nitrate de péroxyacétyle *m*, péroxiacétylnitrate *m* (*PAN*)
persistence *n* SOIL rémanence *f*; **~ potential** SOIL risque de rémanence *m*
persistent: **~ deterioration** *n* SAFETY pollution continue *f*, dégradation continue *f*; **~ oil** *n* WASTE hydrocarbure persistant *m*, WATER SUPP huile persistante *f*; **~ pesticide** *n* AIR pesticide tenace *m*
pervious[1] *adj* GEN perméable
pervious[2]: **~ area** *n* SOIL, WATER POLL zone perméable *f*
PES *abbr* ANALYSIS (*photoelectron spectroscopy*) spectroscopie photoélectronique *f*, GEN (*polyester*) polyester *m*
pest: **~ control** *n* MANAGE action phytosanitaire *f*
pesticide *n* WATER POLL, WATER SUPP pesticide *m*, produit phytosanitaire *m*; **~ residue** WATER POLL résidu des pesticides *m*, résidu phytosanitaire *m*; **~-resistant variety** SOIL variété résistante aux pesticides *f*
PET *abbr* (*polyethylene teraphtalate*) WASTE polytéraphtalate d'éthylène *m*
PETRIFIX: **~ process** *n* WASTE , WATER SUPP procédé PETRIFIX *m*
petrify *vt* AIR, SOIL, WASTE, WATER POLL pétrifier
petrol: **~ vapour recovery plant** *n* *BrE* (*cf gasoline vapor recovery plant AmE*) AIR installation de récupération des vapeurs d'essence *f*
petroleum: **~ product** *n* WATER SUPP produit pétrolier *m*
pH[1] *abbr* (*potential of hydrogen*) GEN pH (*potentiel hydrogène*)
pH[2]: **~-sensitive** *adj* AIR, WATER POLL, WATER SUPP sensible au pH
pH[3]: **~-depression** *n* AIR, ANALYSIS, WATER POLL baisse de pH *f*; **~-drop** *n* AIR, ANALYSIS, WATER POLL chute de pH *f*; **~-electrode** *n* ANALYSIS électrode de mesure du pH *f*; **~-measurement** *n* ANALYSIS mesure de pH *f*; **~-meter** *n*

ANALYSIS pH-mètre *m*; **~-paper strip** *n* ANALYSIS bande de papier de mesure du pH *f*; **~-range** *n* WATER POLL plage de variation du pH *f*; **~-stabilization** *n* WATER POLL stabilisation du pH *f*; **~-value** *n* AIR, ANALYSIS, SOIL, WASTE, WATER POLL, WATER SUPP potentiel hydrogène *m* (*pH*), valeur de pH *f*; **~-value increase** *n* SOIL augmentation du pH *f*

phantom: **~ dump** *n* WASTE décharge clandestine *f*, décharge sauvage *f*, WATER SUPP dépôt clandestin *m*, rejet sauvage *m*

pharmaceutical: **~ industry waste** *n* WASTE déchets de l'industrie pharmaceutique *m pl*; **~ waste** *n* WASTE déchets pharmaceutiques *m pl*

phase[1]: **~-specific** *adj* WATER SUPP spécifique d'une phase

phase[2]: **~ analysis** *n* ANALYSIS *of sample* analyse de phase *f*; **~-break vessel** *n* AIR cuve de séparation *f*, réacteur chimique *m*, séparateur de phase *m*; **~ separation** *n* AIR *solid from gaseous phase*, ANALYSIS, WASTE, WATER SUPP séparation de phases *f*; **~-specific sequential extraction** *n* WASTE, WATER SUPP *form of binding* extraction séquentielle par phases spécifiques *f*

phenol *n* WATER POLL, WATER SUPP phénol *m*

phenolic: **~ resin** *n* WATER POLL, WATER SUPP résine phénolique *f*

pheromone *n* SOIL phéromone *f*

phon *n* ANALYSIS *unit of noise level* phone *m*

phonmeter *n* ANALYSIS phonemètre *m*

Phoredox: **~ method** *n* WATER POLL *phosphorous removal* procédé Phoredox *m*, procédé de déphosphoration *m*

phosphate[1]: **~-containing** *adj* WATER POLL, WATER SUPP phosphaté; **~-free** *adj* WATER POLL, WATER SUPP sans phosphate

phosphate[2] *n* WATER POLL, WATER SUPP phosphate *m*; **~ of calcium** WATER POLL phosphate de calcium *m*; **~ elimination** WATER POLL déphosphatation *f*; **~-free washing-up powder** WATER POLL, WATER SUPP détergent sans phosphate *m*, lessive en poudre sans phosphate *f*; **~ precipitate** WATER POLL précipité de phosphate *m*; **~ removal** WATER POLL déphosphatation *f*; **~ sludge** WATER POLL boues phosphatées *f pl*; **~ stripping** WATER POLL entraînement des phosphates *m*

phosphoric: **~ acid fuel cell** *n* (*PAFC*) AIR pile à combustible à l'acide phosphorique *f*

phosphorous: **~ precipitation** *n* WATER POLL précipitation du phosphore *f*

phosphorus *n* (*P*) AIR, WATER POLL phosphore *m* (*P*); **~ cycle** AIR cycle du phosphore *m*; **~ removal** WATER POLL déphosphoration *f*

photoacoustic: **~ spectroscopy** *n* (*PAS*) ANALYSIS spectroscopie photoacoustique *f*

photobiological[1] *adj* WATER SUPP photobiologique

photobiological[2]: **~ reaction** *n* WATER SUPP réaction photobiologique *f*

photocatalyst *n* WATER SUPP photocatalyseur *m*

photocatalytic[1] *adj* WATER SUPP photocatalytique

photocatalytic[2]: **~ oxidation** *n* WATER SUPP oxydation photocatalytique *f*

photochemical[1] *adj* AIR, WATER SUPP photochimique

photochemical[2]: **~ oxidant** *n* AIR oxydant photochimique *m*; **~ reaction** *n* WATER SUPP réaction photochimique *f*; **~ smog** *n* AIR smog photochimique *m*, brouillard photochimique *m*; **~ treatment** *n* WATER POLL traitement photochimique *m*

photodegradable *adj* WATER SUPP photodégradable

photodegradation *n* WATER SUPP *plastics* photodégradation *f*

photoelectric: **~ current** *n* RENEW ENERG *solar cell* courant photoélectrique *m*; **~ light barrier** *n* WASTE écran photoélectrique *m*; **~ register control** *n* WASTE commande de registre photoélectrique *f*

photoelectron: **~ multiplier** *n* ANALYSIS multiplicateur photoélectronique *m*; **~ spectroscopy** *n* (*PES*) ANALYSIS spectroscopie photoélectronique *f*

photographic: **~ film** *n* ANALYSIS *film badge* pellicule photographique *f*; **~ infrared control** *n* ANALYSIS contrôle photographique infrarouge *m*

photoionization: **~ detector** *n* (*PID*) ANALYSIS détecteur d'ionisation photoélectrique *m*

photolysis *n* ANALYSIS, SOIL, WATER SUPP photolyse *f*

photolytic[1] *adj* ANALYSIS *converter*, SOIL, WATER SUPP photolytique

photolytic[2]: **~ degradation** *n* SOIL dégradation photolytique *f*; **~ destruction** *n* SOIL destruction photolytique *f*

photomechanical[1] *adj* WATER SUPP photomécanique

photomechanical[2]: **~ process** *n* WATER SUPP procédé photomécanique *m*

photometer *n* ANALYSIS photomètre *m*

photomultiplier: **~ tube** *n* (*PMT*) ANALYSIS tube de photomultiplicateur *m*

photooxidant *n* AIR, WATER SUPP photo-oxydant *m*

photooxidation *n* AIR, WATER SUPP photo-oxydation *f*

photophoresis *n* AIR photophorèse *f*

photosensor *n* WATER SUPP photocapteur *m*

photosynthesis *n* RENEW ENERG *biomass*, WATER POLL photosynthèse *f*

phototrophic *adj* WATER POLL *bacteria* photo-autotrophe, phototrophe

photovoltaic[1] *adj* RENEW ENERG photovoltaïque; **~-driven** RENEW ENERG à entraînement photovoltaïque; **~-operated** RENEW ENERG photovoltaïque; **~-powered** RENEW ENERG alimenté par l'énergie photovoltaïque

photovoltaic[2]: **~ cell** *n* RENEW ENERG cellule photovoltaïque *f*; **~ effect** *n* RENEW ENERG effet photovoltaïque *m*; **~ element** *n* RENEW ENERG élément photovoltaïque *m*; **~-energy generation** *n* RENEW ENERG conversion de l'énergie photovoltaïque *f*, production d'énergie photovoltaïque *f*; **~ module** *n* RENEW ENERG module photovoltaïque *m*; **~ power plant** *n* RENEW ENERG centrale photovoltaïque *f*; **~-powered vehicle** *n* RENEW ENERG véhicule à alimentation photovoltaïque *m*; **~ system** *n* RENEW ENERG système photovoltaïque *m*

photovoltaics *n pl* (*PV*) RENEW ENERG génie photovoltaïque *m* (*PV*)

physical[1] *adj* AIR, WASTE, WATER SUPP matériel, physique; **~-chemical** AIR, SOIL, WASTE, WATER POLL, WATER SUPP physico-chimique

physical[2]: **~ adsorption** *n* AIR adsorption physique *f*; **~ agent** *n* WATER SUPP agent physique *m*; **~-chemical process** *n* AIR, SOIL traitement physico-chimique *m*, WATER POLL procédé physico-chimique *m*, traitement physico-chimique *m*, WATER SUPP procédé physico-chimique *m*; **~ condition** *n* AIR état physique *m*; **~ parameter** *n* SOIL paramètre physique *m*; **~ recovery** *n* WASTE récupération physique *f*; **~ stabilization** *n* WASTE, WATER SUPP stabilisation physique *f*; **~ water treatment** *n* WATER SUPP traitement physique de l'eau *m*

physio-chemical[1] *adj* GEN physico-chimique

physio-chemical[2]: **~ environment** *n* AIR, WATER POLL, WATER SUPP environnement physico-chimique *m*, caractéristiques physico-chimiques du milieu *f pl*, milieu physico-chimique *m*; **~ phosphorus removal** *n* WATER POLL déphosphoration physico-chimique *f*; **~ process** *n* WATER POLL, WATER SUPP procédé physico-chimique *m*; **~ reaction** *n* AIR réaction physico-chimique *f*; **~ treatment** *n* WATER POLL traitement physico-chimique *m*

physiological: **~ effect** *n* SAFETY effet physiologique *m*

physisorption *n* AIR physisorption *f*

phytoplankton *n* WATER SUPP phytoplancton *m*

phytosanitary *adj* SAFETY phytosanitaire

phytotoxin *n* SAFETY phytotoxine *f*

pick: **~-up** *n* WASTE benne *f*, WATER SUPP ramassage *m*; **~-up mechanism** *n* WATER SUPP dispositif de collecte *m*

PID *abbr* (*photoionization detector*) ANALYSIS détecteur d'ionisation photoélectrique *m*
piezoelectric: ~ **microphone** *n* ANALYSIS microphone piézo-électrique *m*
piezometer *n* ANALYSIS piézomètre *m*
pigment: ~ **sludge** *n* WATER POLL boues de pigments *f pl*
pile[1]: ~-**driver** *n* WASTE bélier *m*, WATER SUPP bélier *m*, engin de fonçage *m*; ~-**driving** *n* WASTE, WATER SUPP battage de pilots *m*, pilotage *m*
pile[2]: ~-**drive** *vt* WASTE piloter
pile[3]: ~-**drive** *vi* WASTE, WATER SUPP battre les pieux, piloter
pilot: ~ **audit** *n* POLICY audit pilote *m*; ~ **plant** *n* MANAGE installation pilote *f*, usine pilote *f*; ~ **plant scale** *n* MANAGE échelle pilote *f*; ~ **program** *AmE*, ~ **programme** *n* MANAGE, POLICY programme pilote *m*; ~ **project** *n* MANAGE, POLICY projet pilote *m*; ~ **sample** *n* ANALYSIS échantillon pilote *m*; ~ **test** *n* ANALYSIS essai pilote *m*
pipe *n* WATER SUPP tuyau *m*, conduit *m*, conduite *f*; ~ **aerator** WATER SUPP tuyau d'aération *m*; ~ **break** WATER POLL rupture de conduite *f*, WATER SUPP rupture de tuyauterie *f*; ~ **breakage** WATER SUPP rupture de tuyauterie *f*; ~ **bridge** WATER POLL canalisation sur passerelle *f*; ~ **bursting** WATER POLL fuite sur une tuyauterie *f*, rupture de tuyauterie *f*; ~ **end** WATER SUPP extrémité de tuyauterie *f*; ~ **fitting** WATER SUPP garniture de tuyauterie *f*; ~ **joint** WATER SUPP manchon-raccord *m*, manchon de raccord *m*, raccord de tuyauterie *m*; ~ **leakage** WATER POLL fuite de tuyauterie *f*; ~ **sleeve** WATER SUPP fourreau *m*, manchon *m*
pipeline *n* WATER POLL, WATER SUPP canalisation *f*, conduite *f*, pipeline *m*; ~ **resistance** WATER POLL, WATER SUPP résistance d'une conduite *f*, résistance de canalisation *f*
piping: ~ **system** *n* WASTE réseau de canalisation *m*, WATER SUPP réseau de tuyauterie *m*
piston: ~ **capacity** *n* WATER SUPP *of plunger* cylindrée *f*; ~ **compressor** *n* WATER POLL, WATER SUPP compresseur à piston *m*; ~ **pump** *n* WATER POLL, WATER SUPP pompe à piston *f*
pistonphone *n* ANALYSIS *sound measuring* pistonphone *m*
pit *n* SOIL, WATER POLL, WATER SUPP bac *m*, fosse *f*, *mine* puits *m*, regard *m*, soute *f*
pitostatic: ~ **tube** *n* ANALYSIS tube de Pitot *m*
pitting *n* WATER SUPP perforation *f*, piquage *m*
pivoting: ~ **pan conveyor** *n* WASTE, WATER SUPP convoyeur à plateaux pivotants *m*
PIXE[1] *abbr* (*proton-induced x-ray emission*) ANALYSIS émission de rayons X induite par des protons *f*
PIXE[2]: ~ **analysis** *n* ANALYSIS analyse d'émission de rayons X induite par des protons *f*
Planck: ~**'s constant** *n* AIR constante de Planck *f*; ~**'s law** *n* AIR loi de Planck *f*
plane *n* WATER SUPP niveau *m*, plan *m*
planetary: ~ **boundary layer** *n* (*PBL*) AIR couche limite planétaire *f*; ~ **gear** *n* RENEW ENERG *wind turbine* engrenage épicycloïdal *m*, engrenage planétaire *m*
planking *n* NOISE *sound absorbing panels* planchéiage *m*, voligeage *m*
plankton *n* MARINE, WATER POLL, WATER SUPP plancton *m*; ~ **bloom** MARINE, WATER POLL, WATER SUPP fleur d'eau *f*, prolifération de plancton *f*
planned: ~ **environment** *n* WATER SUPP milieu aménagé *m*
planning *n* POLICY aménagement *m*; ~ **assessment decision** POLICY, WATER SUPP décision d'aménagement *f*; ~ **assessment procedure** POLICY, WATER SUPP procédure d'évaluation d'aménagement *f*
plant[1]: ~-**specific** *adj* GEN d'usine, spécifique de l'installation
plant[2] *n* GEN banc *m*, *factory* installation *f*, usine *f*, centrale *f*, *organism* plante *f*; ~ **authorization** POLICY homologation d'une installation *f*; ~ **cover** SOIL tapis végétal *m*; ~ **diseases** *n pl* SOIL phytopathologie *f*; ~ **dust** AIR poussière d'usine *f*, poussière industrielle *f*; ~ **ecology** MANAGE, POLICY écologie végétale *f*, phytoécologie *f*; ~ **environment** SOIL milieu végétal *m*; ~ **licence** *BrE*, ~ **license** *AmE* POLICY permis d'exploitation d'une installation *m*; ~ **nutrient** SOIL, WATER POLL élément nutritif végétal *m*; ~ **oils** *n pl* RENEW ENERG *end product of extraction* plantes oléagineuses *f pl*; ~ **permit** POLICY, WATER SUPP permis d'exploitation *m*; ~ **permit according to law on ambient air protection** POLICY, WATER SUPP permis d'exploitation conforme au droit sur la protection de l'air *m*; ~ **permit according to law on nuclear power** POLICY, WATER SUPP permis d'exploitation conforme au droit nucléaire *m*; ~ **permit according to law on waste** POLICY, WATER SUPP permis d'exploitation conforme à la législation sur les déchets *m*, permis d'exploitation conforme au droit sur les déchets *m*; ~ **permit according to law on water** POLICY, WATER SUPP permis d'exploitation conforme au droit de l'eau *m*; ~ **plankton** WATER SUPP phytoplancton *m*; ~ **protection act** POLICY loi sur la protection des installations *f*, WASTE loi sur la protection des espèces végétales *f*, WATER SUPP loi sur la protection des installations *f*; ~ **representative** MANAGE, POLICY, WATER SUPP porte-parole d'une installation *m*; ~ **requiring authorization** POLICY installation subordonnée à la délivrance d'un permis d'exploitation *f*; ~ **requiring licence** *BrE*, ~ **requiring license** *AmE* POLICY installation subordonnée à la délivrance d'une licence *f*; ~ **requiring permit** POLICY installation subordonnée à l'octroi d'un permis d'exploitation *f*, installation subordonnée à la délivrance d'un permis *f*; ~ **resources** *n pl* RENEW ENERG ressources végétales *f pl*; ~ **safety** SAFETY sûreté de l'installation *f*
plantation *n* RENEW ENERG, SOIL *planting* implantation *f*, *place* plantation *f*
planting: ~ **machine** *n* RENEW ENERG planteuse *f*
plasma: ~ **display panel** *n* (*PDP*) ANALYSIS écran à plasma *m*; ~ **emission detection** *n* (*PED*) ANALYSIS détection d'émissions de plasma *f*
plaster *n* NOISE *surface lining* plâtre *m*
plasti: ~-**foam packaging** *n* WASTE emballage en mousse de plastique *m*
plastic[1] *adj* GEN plastique
plastic[2]: ~ **bottle** *n* ANALYSIS *in sampling*, WATER SUPP bouteille en plastique *f*; ~ **container** *n* WASTE, WATER SUPP conteneur en plastique *m*, conteneur à plastiques *m*; ~ **liner** *n* SOIL, WATER SUPP revêtement de plastique *m*, revêtement en plastique *m*; ~ **matrix** *n* WASTE, WATER SUPP matrice du plastique *f*; ~ **module** *n* WATER POLL *trickling-filter packing material* module en plastique *m*; ~ **mold** *AmE see plastic mould BrE*; ~ **molding** *AmE see plastic moulding BrE*; ~ **mould** *BrE n* WASTE matrice de plastique *f*; ~ **moulding** *n BrE* WASTE moulage de plastique *m*; ~ **packing material** *n* WASTE matériau d'emballage plastique *m*; ~ **packaging** *n* WASTE emballage en plastique *m*; ~ **pipe** *n* RENEW ENERG *heat pump* tuyau en plastique *m*; ~ **sealing** *n* WASTE joint en plastique *m*, WATER SUPP étanchéité en plastique *f*; ~ **separation** *n* WATER SUPP séparation des plastiques *f*; ~ **sheet** *n* WASTE, WATER SUPP feuille de plastique *f*; ~ **waste** *n* WASTE déchets de plastique *m pl*
plasticator *n* WASTE, WATER SUPP plastifiant *m*
plasticize *vt* WASTE, WATER SUPP plastifier
plasticized *adj* WASTE, WATER SUPP plastifié
plasticizer *n* WASTE, WATER SUPP fluidifiant *m*
plasticizing *n* WASTE, WATER SUPP *plastics* plastification *f*; ~ **unit** WASTE installation de plastification *f*, WATER SUPP banc de plastification *m*
plastics *n pl* WASTE plastique *m*, plastiques *m pl*, WATER SUPP plastique *m*; ~ **application** *n* WATER SUPP application de plastique *f*; ~ **container** *n* WASTE conteneur en plastique *m*; ~ **extruder** *n* WASTE, WATER SUPP extrudeuse à plastique *f*; ~ **molding** *AmE*, ~ **moulding** *BrE n* WASTE moulage du plastique *m*; ~ **processing** *n* WASTE plasturgie *f*, WATER SUPP

plasturgie *f*, traitement des plastiques *m*, transformation des plastiques *f*; ~ **processing industry** *n* WASTE industrie de transformation des plastiques *f*, plasturgie *f*; ~ **recycling** *n* WASTE recyclage du plastique *m*; ~ **recycling fluidized-bed plant** *n* WASTE installation à lit fluidisé pour le recyclage du plastique *f*; ~ **recycling plant** *n* WASTE, WASTE usine de recyclage du plastique *f*; ~ **sheeting** *n* WASTE, WATER SUPP feuille plastique *f*; ~ **sorting plant** *n* WASTE installation de tri des plastiques *f*; ~ **waste** *n* WASTE déchets plastiques *m pl*

plate *n* AIR plateau *m*, NOISE, WASTE plateau *m*, WATER SUPP tôle *f*; ~ **catalyst** AIR catalyseur supporté *m*, catalyseur sur support *m*; ~ **column** AIR colonne à plateaux *f*; ~**-column scrubber** AIR séparateur-laveur à plateaux *m*; ~ **diffuser** WATER POLL diffuseur à plaques *m*; ~ **module** WATER POLL module à plaques *m*; ~ **precipitator** AIR laveur à plateaux *m*; ~ **tower** AIR colonne à plateaux *f*; ~**-wire electrostatic precipitator** AIR *dry electrofilter* électrofiltre à plaques *m*, dépoussiéreur à plaques *m*; ~**-wire ESP** AIR *dry electrofilter* électrofiltre à plaques *m*, dépoussiéreur à plaques *m*

platinum *n* (*Pt*) GEN platine *m* (*Pt*); ~ **alloy** AIR *catalyst* alliage au platine *m*

plenum: ~ **chamber** *n* NOISE *type of silencer* chambre de mélange d'air *f*

plot *vt* ANALYSIS *chromatogram* tracer

plotter *n* ANALYSIS traceur *m*

ploughshare: ~ **mixer** *n BrE* WASTE, WATER SUPP mélangeur à soc *m*

plowshare *AmE see ploughshare BrE*

plug[1] *n* WATER SUPP bonde *f*

plug[2] *vt* WATER POLL, WATER SUPP boucher

plugged *adj* WATER POLL, WATER SUPP bouché

plugging *n* WATER POLL *filter* bouchage *m*, enfichage *m*, WATER SUPP bouchage *m*, embrochage *m*, enfichage *m*

plumbiferous *adj* WATER SUPP plombifère

plume *n* AIR panache *m*; ~ **axis** AIR axe du panache *m*; ~ **rise** AIR ascension du panache *f*; ~ **shred** AIR étalement du panache *m*, effilochage du panache *m*; ~ **spot number** AIR indice de noircissement par opacité d'un panache de fumée *m*

plunger: ~ **pump** *n* WATER POLL, WATER SUPP pompe à piston plongeur *f*, pompe à plongeur *f*, pompe foulante *f*, pompe refoulante *f*

pluviometer *n* ANALYSIS, WATER SUPP pluviomètre *m*

pluviometry *n* ANALYSIS, WATER SUPP pluviométrie *f*

PMLCD *abbr* (*passive matrix liquid-crystal display*) ANALYSIS écran à cristaux liquides à matrice passive *m*

PMT *abbr* (*photomultiplier tube*) ANALYSIS tube de photomultiplicateur *m*

PNdB *abbr* (*perceived noise level*) NOISE niveau de bruit perçu *m*

pneumatic[1] *adj* GEN pneumatique

pneumatic[2]: ~ **classification** *n* WASTE, WATER SUPP classification pneumatique *f*; ~ **driving gear** *n* WATER POLL mécanisme de commande pneumatique *m*, WATER SUPP engrenage à air comprimé *m*, engrenage pneumatique *m*, transmission pneumatique *f*; ~ **flotation** *n* WATER SUPP brassage à l'air *m*; ~ **injector** *n* WATER POLL injecteur pneumatique *m*; ~ **sorter** *n* WASTE trieur pneumatique *m*, WATER SUPP classificateur pneumatique *m*; ~ **sound absorber** *n* MARINE, NOISE absorbeur acoustique pneumatique *m*; ~ **transport** *n* WATER SUPP transport pneumatique *m*

PO *abbr* (*polyolefin*) WASTE, WATER SUPP polyoléfine *f*

pocket[1]: ~**-shaped** *adj* AIR *filter* en forme de poche

pocket[2]: ~ **filter** *n* AIR filtre à poche *m*; ~ **measuring instrument** *n* ANALYSIS instrument de mesure de poche *m*; ~ **rot** *n* WASTE pourriture alvéolaire *f*

point: ~ **source** *n* AIR *pollution*, SOIL, WATER SUPP source ponctuelle *f*; ~ **source pollutant** *n* AIR polluant de source ponctuelle *m*; ~ **vacuum-cleaning system** *n* SAFETY poste d'aspiration ponctuel *m*; ~ **of withdrawal** *n* WATER SUPP point de puisage *m*

poison[1] *n* AIR, SAFETY, WASTE, WATER SUPP poison *m*

poison[2] *vt* SAFETY, WATER POLL, WATER SUPP empoisonner

poisoning *n* AIR, SAFETY empoisonnement *m*, WATER SUPP empoisonnement *m*, intoxication *f*

poisonous: ~ **waste** *n BrE* SAFETY, WASTE, WATER SUPP déchets toxiques *m pl*

polarity *n* AIR *positive or negative* polarité *f*

polarization *n* ANALYSIS polarisation *f*

polarograph *n* ANALYSIS polarographe *m*

polarographic[1] *adj* ANALYSIS *detectors* polarographique

polarographic[2]: ~ **determination method** *n* ANALYSIS méthode de détermination polarographique *f*

polarography *n* ANALYSIS *potentiometry* polarographie *f*

polder *n* WATER SUPP polder *m*

policy: ~ **instrument** *n* POLICY instrument de décision *m*; ~ **objective** *n* POLICY objectif de gestion *m*

polish *vt* WATER SUPP lisser

polishing *n* WATER POLL polissage *m*, WATER SUPP *plastics recycling* lissage *m*, polissage *m*

pollutant[1]: ~**-free** *adj* AIR, SOIL exempt de polluants, non-pollué; ~**-resistant** *adj* AIR, SOIL résistant aux polluants; ~**-specific** *adj* AIR, SOIL spécifique d'un polluant

pollutant[2] *n* GEN contaminant *m*, polluant *m*; ~ **amount** ANALYSIS quantité de polluants *f*, volume de polluants *m*; ~ **behavior** *AmE*, ~ **behaviour** *BrE* AIR, SOIL comportement des polluants *m*; ~ **concentration** AIR, ANALYSIS, WATER POLL concentration des polluants *f*; ~ **content** AIR, SOIL, WATER SUPP teneur en polluants *f*; ~ **conversion** AIR conversion de polluants *f*; ~ **deposition** AIR dépôt de polluants *m*, WATER SUPP dépôt de polluants *m*, retombées de polluants *f pl*; ~ **emissions** *n pl* AIR émissions de polluants *f pl*, émissions polluantes *f pl*; ~ **load** AIR charge de pollution *f*, charge polluante *f*; ~ **oxidation** AIR oxydation due à des polluants *f*; ~ **release limit** AIR, ANALYSIS, MANAGE, POLICY limite d'émissions de polluants *f*; ~ **transmission** AIR transmission de polluants *f*; ~ **type** AIR, SOIL type de polluant *m*

pollutants: ~ **inventory** *n* SAFETY inventaire des polluants *m*

pollute *vt* GEN contaminer, polluer

polluted[1] *adj* GEN pollué

polluted[2]: ~ **area** *n* AIR, WATER POLL zone polluée *f*; ~ **dust** *n* AIR poussière polluée *f*; ~ **ground** *n* SOIL sol pollué *m*; ~ **process gas** *n* AIR gaz industriel pollué *m*; ~ **storm water** *n* WATER POLL eau pluviale polluée *f*; ~ **water** *n* MARINE, WATER POLL eau polluée *f*; ~ **waters** *n pl* MARINE, WATER POLL eaux polluées *f pl*

pollutee *n* POLICY pollué *m*, victime de la pollution *f*

polluter *n* POLICY pollueur *m*; ~ **group** POLICY catégorie de pollueurs *f*; ~ **pays principle** POLICY, WATER SUPP principe pollueur-payeur *m*

polluting: ~ **agent** *n* SAFETY agent polluant *m*; ~ **industry** *n* WATER POLL industrie polluante *f*

pollution *n* GEN contamination *f*, pollution *f*; ~ **burden** AIR charge de pollution *f*, POLICY charge polluante *f*, pollution *f*, WATER SUPP charge polluante *f*; ~ **charge** POLICY redevance sur la pollution *f*; ~ **chart** POLICY pollutogramme *m*; ~ **of coastal waters** MARINE pollution des eaux côtières *f*; ~ **control** AIR, POLICY, WATER POLL, WATER SUPP lutte contre la pollution *f*, dépollution *f*; ~ **deposition** SOIL, WATER POLL dépôt polluant *m*; ~ **emitter** AIR, SOIL émetteur de polluants *m*; ~ **index** AIR, MARINE, NOISE, POLICY, WATER POLL indice de pollution *m*; ~ **of inland waters** WATER POLL pollution des eaux intérieures *f*; ~ **level** WATER POLL degré de pollution *m*, niveau de pollution *m*; ~ **plagues** *n pl* POLICY fléaux de la pollution *m pl*; ~ **prevention** POLICY prévention de la pollution *f*; ~ **research** ANALYSIS recherche sur la pollution *f*, étude de la pollution *f*; ~ **of the sea** MARINE pollution du milieu marin *f*; ~ **source** AIR, POLICY, SOIL, WATER SUPP

source de pollution *f*; ~ **tracer** ANALYSIS traceur de pollution *m*
polyacrylnitrile *n* (*PAN*) AIR *filter fibre material* polyacrylonitrile *m*
polyaddition *n* WATER SUPP polyaddition *f*
polyamide *n* (*PA*) WATER POLL, WATER SUPP polyamide *m*; ~ **membrane** WATER POLL membrane en polyamide *f*
polycarbonate *n* (*PC*) WATER POLL, WATER SUPP polycarbonate *m* (*PC*); ~ **membrane** WATER POLL membrane en polycarbonate *f*
polychlorinated: ~ **biphenyl** *n* (*PCB*) AIR, SOIL, WASTE biphényle polychloré *m* (*BPC*); ~ **dibenzofuran** *n* (*PCDF*) SOIL, WASTE dibenzofuranne polychloré *m* (*PCDF*); ~ **naphthalene** *n* (*PCN*) SOIL naphtalène polychloré *m* (*PCN*); ~ **terphenyl** *n* (*PCT*) AIR, SOIL terphényle polychloré *m* (*TPC*)
polycondensation *n* WATER SUPP polycondensation *f*
polycrystalline[1] *adj* RENEW ENERG polycristallin
polycrystalline[2]: ~ **silicon cell** *n* RENEW ENERG *photovoltaic* cellule au silicium polycristallin *f*
polycyclic: ~ **aromatic hydrocarbon** *n* (*PAH*) AIR, WATER SUPP hydrocarbure aromatique polycyclique *m* (*HAP*), hydrocarbure polynucléaire aromatique *m*
polyelectrolyte *n* ANALYSIS, WATER SUPP polyélectrolyte *m*
polyester *n* (*PES*) GEN polyester *m*; ~ **foam** WASTE, WATER SUPP mousse de polyester *f*
polyethylene[1] *n* (*PE*) WASTE, WATER SUPP polyéthylène (*PE*) *m*; ~ **container** WASTE, WATER SUPP conteneur en polyéthylène *m*; ~ **pallet covers and liners** WASTE housses et revêtements de palettes en polyéthylène *m pl*; ~ **terephthalate** (*PET*) WATER SUPP polytéraphtalate d'éthylène *m*; ~ **terephthalate bottle** (*bottle PETP*) WASTE bouteille en PETP *f*, bouteille en polyéthylène téréphthalate *f*; ~ **terephthalate film** WATER SUPP film de polyéthylène téréphthalate *m*
polymer *n* AIR, WASTE, WATER POLL, WATER SUPP polymère *m*; ~ **concrete** WASTE béton polymérisé *m*; ~**-concrete composite material** WATER SUPP *after treatment of filter dust* matériau composite béton-polymère *m*
polymerization *n* AIR, WASTE, WATER POLL, WATER SUPP polymérisation *f*
polymeter *n* ANALYSIS polymètre *m*; dispositif de mesure multi-usages *m*
polyolefin *n* (*PO*) WASTE polyoléfine *f*; ~ **container** WASTE, WATER SUPP conteneur en polyoléfine *m*
polypropylene *n* (*PP*) AIR, WATER SUPP *filter fibre material* polypropylène *m*
polysaccharide *n* WATER SUPP polysaccharide *m*
polysaprobic *adj* WATER POLL polysaprobe
polystyrene *n* (*PS*) WATER SUPP polystyrène *m*
polytetrafluoroethylene *n* (*PTFE*) AIR *filter fibre material* polytétrafluoréthylène *m*
polyurethane *n* (*PUR*) WATER SUPP polyuréthanne *m*
polyvinyl: ~ **chloride** *n* (*PVC*) SOIL, WATER SUPP polychlorure de vinyle *m* (*PVC*)
pond *n* WATER POLL étang *m*, bassin *m*, WATER SUPP étang *m*, bassin *m*, bief *m*
poorly: ~**-treated wastewater** *n* WATER POLL eaux usées peu épurées *f pl*
population[1]: ~**-related** *adj* POLICY démographique, lié à la population; ~**-specific** *adj* POLICY spécifique de la population
population[2]: ~ **dose** *n* AIR *nuclear*, SAFETY dose reçue par une population *f*; ~ **growth** *n* POLICY accroissement démographique *m*; ~ **pressure** *n* POLICY pression démographique *f*; ~**-related water consumption** *n* WATER SUPP consommation d'eau fonction de la population *f*; ~**-related water demand** *n* WATER SUPP demande d'eau fonction de la population *f*; ~**-specific water demand** *n* WATER SUPP demande d'eau spécifique de la population *f*
pore *n* WATER POLL, WATER SUPP pore *m*; ~ **diameter** AIR diamètre de pore *m*; ~ **size** AIR taille de pore *f*; ~ **space** AIR volume des pores *m*; ~ **volume** AIR, WATER SUPP volume des pores *m*
porometer *n* ANALYSIS poromètre *m*
porosity *n* AIR, WATER POLL, WATER SUPP porosité *f*
porous[1] *adj* AIR, WATER POLL, WATER SUPP poreux
porous[2]: ~ **activated carbon** *n* AIR charbon actif poreux *m*; ~ **adsorbent** *n* AIR adsorbant poreux *m*; ~ **cover** *n* WATER SUPP *of landfill* couverture poreuse *f*; ~ **membrane** *n* WATER POLL, WATER SUPP membrane poreuse *f*
portable[1] *adj* ANALYSIS *gas analyser* portatif, portable
portable[2]: ~ **digital colorimeter** *n* ANALYSIS colorimètre numérique portable *m*; ~ **gas chromatograph** *n* ANALYSIS chromatographe en phase gazeuse portable *m*; ~ **GC** *n* ANALYSIS CPG portable; ~ **hand-held analyser** *BrE*, ~ **hand-held analyzer** *AmE* *n* ANALYSIS analyseur portable portatif *m*; ~ **PID** *n* ANALYSIS détecteur d'ionisation photoélectrique portable *m*; ~ **spectrophotometer** *n* ANALYSIS spectrophotomètre portable *m*; ~ **X-ray fluorescence analyser** *BrE*, ~ **X-ray fluorescence analyzer** *AmE* *n* ANALYSIS analyseur de fluorescence X portable *m*
positive[1] *adj* AIR, ANALYSIS, WATER POLL, WATER SUPP positif
positive[2]: ~ **chemical ionization** *n* (*PCI*) ANALYSIS ionisation chimique positive *f*; ~ **crankcase ventilation** *n* (*PCV*) AIR recyclage des gaz de carter *m*; ~ **displacement pump** *n* WATER POLL, WATER SUPP *reciprocating pumps, Archimedean screw* pompe volumétrique *f*; ~ **electrode** *n* AIR électrode positive *f*; ~ **ion** *n* WATER SUPP ion positif *m*
possibility: ~ **of error** *n* ANALYSIS risque d'erreur *m*; ~ **of wastewater deposition** *n* WATER POLL risque de dépôt d'eaux usées *m*
post: ~**-closure expenditure** *n* WATER SUPP dépenses après fermeture *f pl*; ~**-closure measure** *n* WATER SUPP mesure après fermeture *f*; ~**-closure monitoring** *n* SOIL surveillance après fermeture *f*; ~**-coagulation** *n* WATER SUPP post-coagulation *f*; ~**-consumer plastics waste** *n* WATER SUPP déchets de plastique après consommation *m pl*; ~**-filtration** *n* WATER POLL post-filtration *f*; ~**-ozonation** *n* WATER SUPP post-ozonation *f*; ~**-precipitation** *n* WATER POLL postprécipitation *f*; ~**-reaction** *n* AIR réaction après-coup *f*, réaction différée *f*; ~**-rotting treatment** *n* WATER POLL traitement anti-rouille *m*, WATER SUPP *build-up phase* traitement anti-rouille curatif *m*, traitement curatif de la corrosion *m*; ~**-treatment** *n* SOIL traitement postérieur *m*
postchlorination *n* WATER SUPP postchloration *f*
poster: ~ **campaign** *n* POLICY campagne d'affichage *f*
potable *adj* WATER SUPP buvable, potable
potash *n* WATER POLL potasse *f*
potential[1] *adj* GEN potentiel
potential[2] *n* MANAGE, POLICY capacité *f*, potentiel *m*; ~ **acidity** SOIL acidité potentielle *f*; ~ **danger** SAFETY danger potentiel *m*, risque potentiel *m*; ~ **energy** RENEW ENERG potentiel énergétique *m*; ~ **of hydrogen** (*pH*) AIR, ANALYSIS, SOIL, WASTE, WATER POLL potentiel hydrogène (*pH*); ~ **measurement** ANALYSIS mesure potentielle *f*; ~ **pollutant** WATER POLL polluant potentiel *m*; ~ **safety risk** SAFETY risque potentiel pour la sécurité *m*
potentiometer *n* ANALYSIS potentiomètre *m*
potentiometric[1] *adj* ANALYSIS potentiométrique
potentiometric[2]: ~ **sensor** *n* ANALYSIS capteur potentiométrique *m*; ~ **titration** *n* ANALYSIS titrage potentiométrique *m*
potentiometry *n* ANALYSIS *polarography* potentiométrie *f*
potentiostat *n* ANALYSIS potentiostat *m*
potentiostatic[1] *adj* ANALYSIS potentiostatique
potentiostatic[2]: ~ **electrode sensor** *n* ANALYSIS capteur à électrode potentiostatique *m*
pourable *adj* AIR, WATER SUPP coulable, coulant
pouring: ~ **rain** *n* SOIL pluie torrentielle *f*

powdered: ~ **activated carbon** *n* (*PAC*) WATER POLL, WATER SUPP charbon actif en poudre *m*
power *n* AIR électricité *f*, GEN puissance *f*, RENEW ENERG, WATER SUPP électricité *f*; ~ **capacity** AIR capacité électrique *f*; ~ **frequency** RENEW ENERG fréquence électrique *f*; ~ **generation** RENEW ENERG, WASTE, WATER SUPP *via biogas* production d'électricité *f*; ~ **grid** RENEW ENERG réseau électrique *m*, réseau de transport *m*; ~ **level** WATER SUPP niveau d'énergie *m*, niveau de puissance *m*; ~ **output** RENEW ENERG puissance fournie *f*; ~ **plant** AIR centrale électrique *f*; ~ **source** RENEW ENERG source d'électricité *f*; ~ **station** RENEW ENERG centrale électrique *f*; ~ **supply** RENEW ENERG fourniture d'électricité *f*
powerhouse *n* RENEW ENERG centrale *f*
POX *abbr* (*purgeable organic halogen compounds*) WATER SUPP composés organiques halogénés éliminables *m pl*
PP *abbr* (*polypropylene*) AIR, WATER SUPP polypropylène *m*
ppb *abbr* (*parts per billion*) ANALYSIS parties par milliard
ppm *abbr* (*parts per million*) ANALYSIS ppm (*parties par million*)
ppq *abbr* (*parts per quadrillion*) ANALYSIS parties par quadrillion
ppt *abbr* (*parts per trillion*) ANALYSIS parties par trillion
practical: ~ **salinity** *n* WATER SUPP salinité pratique *f*
pre-acidification: ~ **alkalinity** *n* AIR alcalinité initiale *f*
preaeration *n* WATER POLL pré-aération *f*; ~ **tank** WATER POLL cuve de pré-aération *f*
preamplifier *n* ANALYSIS préamplificateur *m*
precaution *n* SAFETY précaution *f*
precautionary: ~ **measure** *n* SAFETY mesure de précaution *f*
precautions: ~ **against dust** *n* SAFETY précautions contre la poussière *f pl*
precedence: ~ **case** *n* POLICY précédent *m*
preceding *adj* WATER POLL précédant
prechlorination *n* WATER SUPP préchloration *f*
precipitable: ~ **organics** *n pl* WATER POLL matière organique précipitable *f*
precipitant *n* GEN précipitant *m*
precipitate[1] *n* GEN précipité *m*, sédiment *m*; ~ **feeding** WATER POLL alimentation en précipitats *f*
precipitate[2] *vt* GEN précipiter
precipitate[3] *vi* GEN se déposer, se séparer par précipitation
precipitation *n* GEN précipitation *f*; ~ **with aluminium salts** WATER POLL précipitation aux sels d'aluminium *f*; ~ **area** WATER SUPP zone de précipitation *f*, impluvium *m*; ~ **collector** AIR collecteur de précipitation *m*; ~ **event** AIR précipitation donnée *f*; ~ **with ferric salt** WATER POLL précipitation au sel ferrique *f*; ~ **gage** *AmE*, ~ **gauge** *BrE* ANALYSIS, WATER SUPP pluviomètre *m*; ~ **with lime** WATER POLL précipitation à la chaux *f*; ~ **operation** WATER POLL mise en œuvre de la précipitation *f*, opération de précipitation *f*; ~ **sludge** WATER POLL boues de précipitation *f pl*; ~ **unit** AIR, SAFETY installation de dépoussiérage *f*, usine de séparation *f*
precision *n* ANALYSIS fidélité *f*; ~ **measurement** ANALYSIS mesure de précision *f*; ~ **sound-level meter** ANALYSIS sonomètre de précision *m*
preclarification *n* WATER POLL préclarification *f*
precleaned: ~ **gas** *n* AIR gaz préalablement nettoyé *m*
precleaning *n* AIR nettoyage préalable *m*, prélavage *m*; ~ **basin** WATER SUPP bassin de prélavage *m*
precoat: ~ **filtration** *n* WATER SUPP filtration sur précouche *f*; ~ **layer** *n* AIR *bag filter* précouche de filtration *f*
pre-column: ~ **back-flush operation** *n* ANALYSIS rétrolavage en colonne amont *m*
precombustion: ~ **chamber** *n* AIR chambre de précombustion *f*
precombustor *n* SOIL appareil de précombustion *m*
preconcentrate *n* ANALYSIS préconcentré *m*
preconcentration *n* ANALYSIS préconcentration *f*
precooler *n* AIR préréfrigérant *m*
precursor: ~ **pollutant** *n* AIR polluant précurseur *m*
predation *n* POLICY prédation *f*
pre-denitrification *n* WATER POLL prédénitrification *f*
predictive: ~ **capability** *n* AIR capacité de prévision *f*
pre-drying: ~ **of wastewater sludge** *n* WATER POLL préséchage des boues résiduaires *m*
prefilter *n* AIR, ANALYSIS, WATER POLL préfiltre *m*
prefiltering *n* AIR, ANALYSIS, WATER POLL préfiltrage *m*
prefiltration *n* AIR, ANALYSIS, WATER POLL préfiltration *f*
preformed: ~ **spray scrubber** *n* AIR laveur préformé à pulvérisation *m*
preheat *vt* GEN préchauffer
preheated: ~ **combustion air** *n* AIR, SOIL air de combustion préchauffé *m*
preheater *n* GEN préchauffeur *m*
preheating *n* GEN préchauffage *m*
preliminary: ~ **analysis** *n* ANALYSIS analyse préliminaire *f*; ~ **assessment** *n* ANALYSIS, POLICY, SAFETY, SOIL, WATER SUPP évaluation préliminaire *f*; ~ **decision** *n* POLICY, WATER SUPP décision préliminaire *f*; ~ **dewatering** *n* WATER POLL déshydratation préliminaire *f*; ~ **dust removal** *n* AIR dépoussiérage préliminaire *m*; ~ **filtration** *n* WATER POLL filtration préliminaire *f*; ~ **ozonization** *n* WATER POLL ozonation préliminaire *f*; ~ **pump** *n* WATER POLL pompe primaire *f*; ~ **purification** *n* WATER SUPP épuration préliminaire *f*, purification préliminaire *f*; ~ **sample** *n* ANALYSIS échantillon préliminaire *m*; ~**-settled sludge** *n* WATER POLL boues de première décantation *f pl*; ~ **settlement** *n* WATER SUPP décantation préliminaire *f*; ~ **settling tank** *n* WATER POLL bassin de décantation préliminaire *m*; ~ **test** *n* ANALYSIS essai préalable *m*; ~ **treatment** *n* WATER SUPP épuration primaire *f*; ~ **trial** *n* ANALYSIS essai préliminaire *m*
premixer *n* WASTE mélangeur primaire *m*, WATER SUPP dispositif de prémélange *m*, prémélangeur *m*
premolding *AmE see premoulding BrE*
premoulding *n BrE* WATER SUPP prémoulage *m*
preparation: ~ **plant** *n* WATER SUPP usine de préparation *f*; ~ **process** *n* WATER SUPP procédé de préparation *m*
preprogrammable *adj* ANALYSIS *measuring instrument* préprogrammable
preprogrammed *adj* ANALYSIS préprogrammé
prerotting: ~ **treatment** *n* WATER SUPP *degradation and sanitation phase* traitement anti-corrosion préalable *m*, traitement anti-corrosion préventif *m*
prescrub *vt* AIR prélaver
prescrubber *n* AIR prélaveur *m*
prescrubbing *n* AIR prélavage *m*; ~ **tank** AIR *biological gas scrubbing* cuve de prélavage *f*
presedimentation *n* WATER SUPP décantation primaire *f*
presentation: ~ **of documents for public inspection** *phr* POLICY soumission de dossier à l'inspection publique *f*
preseparation: ~ **cyclone** *n* AIR cyclone de dépoussiérage préliminaire *m*
presort *vt* SOIL, WATER SUPP *excavated mass* préclasser, prétrier
presorted[1] *adj* SOIL, WATER SUPP préclassé, prétrié
presorted[2]: ~ **waste** *n* SOIL, WASTE, WATER SUPP déchets prétriés *m pl*
presorting *n* SOIL, WATER SUPP tri préalable *m*
press[1]: ~ **communication** *n* POLICY communiqué de presse *m*; ~ **drying** *n* WATER POLL *dewatering by pressure* séchage sous pression *m*; ~ **office** *n* POLICY office de presse *m*; ~ **plunger** *n* WATER SUPP piston de presse *m*; ~ **ram** *n* WATER SUPP piston plongeur *m*; ~ **release** *n* POLICY communiqué de presse *m*; ~ **stroke** *n* WATER SUPP course de presse *f*
press[2] *vt* SOIL, WASTE, WATER POLL comprimer, WATER SUPP comprimer, presser; ~ **down** SOIL, WASTE, WATER POLL, WATER SUPP comprimer; ~ **in** WATER SUPP faire entrer par

pression; ~ **together** SOIL, WASTE, WATER POLL, WATER SUPP comprimer

pressing *n* RENEW ENERG, SOIL, WASTE, WATER POLL compression *f*, WATER SUPP compression *f*, emboutissage *m*, pressée *f*, pressage *m*; ~ **operation** WATER SUPP opération d'emboutissage *f*, opération de pressage *f*

pressure *n* GEN pression *f*; ~ **chamber** AIR *ground filter* chambre de pression *f*; ~ **compensation** WATER SUPP compensation de pression *f*; ~ **difference** ANALYSIS différence de pression *f*, gradient de pression *m*; ~-**difference device** ANALYSIS appareil déprimogène *m*; ~ **drainage** WATER POLL écoulement sous pression *m*; ~ **drop** AIR *filter bag* baisse de pression *f*; ~ **energy** AIR énergie de pression *f*; ~ **filter** AIR, WATER SUPP filtre à pression *m*, filtre sous pression *m*; ~ **flotation** WATER POLL flottation par air dissous *f*; ~ **force** WATER SUPP force de pressage *f*; ~ **gage** *AmE*, ~ **gauge** *BrE* AIR, ANALYSIS, WATER POLL, WATER SUPP manomètre *m*; ~ **governor** AIR commande de pression *f*; ~ **gradient** SOIL gradient barométrique *m*; ~ **head** WATER SUPP hauteur manométrique *f*; ~ **increase** AIR augmentation de pression *f*; ~ **loss** AIR *filter bag*, WATER POLL perte de pression *f*; ~ **loss in distribution network** WATER SUPP perte de pression dans le réseau de distribution *f*; ~ **loss of pipeline** WATER SUPP perte de pression dans les conduites *f*; ~ **oscillation** WATER SUPP oscillation de pression *f*; ~ **pipe** WATER POLL conduite forcée *f*, WATER SUPP tuyauterie sous pression *f*; ~ **pump** WATER SUPP pompe de pression *f*; ~ **regulator** AIR, ANALYSIS *trace analyser* régulateur de pression *m*; ~ **release** AIR *of filtering apparatus* rejet de pression *m*; ~ **sensor** AIR capteur de pression *m*; ~ **sounding** ANALYSIS sondage de pression *m*; ~ **velocity probe** ANALYSIS sonde de vitesse et de pression *f*; ~ **well** SOIL, WATER SUPP puits d'injection *m*; ~ **zone** WATER SUPP zone sous pression *f*

pressurized: ~ **liquid** *n* AIR liquide sous pression *m*; ~ **liquid stream** *n* AIR jet de liquide sous pression *m*

prestressed: ~-**concrete pipe** *n* WATER POLL, WATER SUPP conduite en béton précontraint *f*

pretest *n* ANALYSIS essai préalable *m*, essai préliminaire *m*

prethickening *n* WATER POLL épaississement préalable *m*

pretreat *vt* AIR, SOIL, WASTE, WATER SUPP prétraiter

pretreatment *n* GEN traitement préliminaire *m*, prétraitement *m*, traitement préalable *m*

prevent *vt* MARINE, WATER SUPP empêcher

prevention *n* GEN prévention *f*; ~ **of atmospheric pollution** AIR lutte contre la pollution atmosphérique *f*; ~ **of inadvertent ignition of flammable atmospheres by RF radiation** SAFETY prévention des prises au feu inopinées d'atmosphères inflammables *f*

preventive[1] *adj* GEN préventif

preventive[2]: ~ **environmental policy** *n* POLICY mesure préventive applicable à l'environnement *f*; ~ **environmental protection** *n* POLICY, WATER SUPP protection préventive de l'environnement *f*; ~ **measure** *n* GEN mesure préventive *f*

prewash *vt* WATER SUPP prélaver

prewasher *n* WATER SUPP prélaveur *m*

prewashing *n* WATER SUPP prélavage *m*

primary[1] *adj* GEN primaire

primary[2]: ~ **air** *n* AIR air primaire *m*; ~ **aluminium** *BrE*, ~ **aluminum** *AmE* *n* WATER SUPP aluminium de première fusion *m*; ~ **catalyst** *n* AIR catalyseur primaire *m*; ~ **clarification** *n* WATER POLL clarification primaire *f*; ~ **combustion air** *n* AIR, WATER SUPP air de combustion primaire *m*; ~ **combustion chamber** *n* WATER SUPP chambre de combustion primaire *f*; ~ **cosmic radiation** *n* AIR rayonnement cosmique primaire *m*; ~ **crusher** *n* WATER SUPP broyeur primaire *m*; ~ **downpipe** *n* AIR *packed-bed filter* descente primaire *f*; ~ **energy balance** *n* RENEW ENERG bilan d'énergie primaire *m*; ~ **energy consumption** *n* RENEW ENERG consommation d'énergie primaire *f*; ~ **energy demand** *n* RENEW ENERG demande d'énergie primaire *f*; ~ **energy source** *n* RENEW ENERG source d'énergie primaire *f*; ~ **energy substitution** *n* RENEW ENERG substitution à l'énergie primaire *f*; ~ **filter** *n* AIR, ANALYSIS, WATER POLL préfiltre *m*; ~ **fuel** *n* WATER SUPP combustible primaire *m*; ~ **material** *n* WASTE matériau primaire *m*; ~ **production of paper** *n* WATER SUPP production primaire de papier *f*; ~ **recycling** *n* WASTE recyclage primaire *m*; ~ **settlement** *n* SOIL tassement primaire *m*, WATER SUPP décantation primaire *f*; ~ **settling basin** *n* WATER POLL décanteur primaire *m*; ~ **settling tank** *n* WATER POLL décanteur primaire *m*; ~ **sludge** *n* WATER POLL boues primaires *f pl*; ~ **thickener** *n* WATER POLL épaississeur primaire *m*; ~ **waste-avoidance** *n* WASTE mesure destinée à éviter la production de déchets primaires *f*, mesure préventive contre les déchets primaires *f*; ~ **wastewater treatment** *n* WATER POLL *physical process* traitement primaire des eaux résiduaires *m*

prime *vt* WATER POLL *pump* amorcer, WATER SUPP *pump* abreuver, amorcer

priming *n* WATER POLL, WATER SUPP amorçage *m*, *of tide* avance *f*, primage *m*; ~ **cock** WATER SUPP amorceur *m*, robinet d'amorçage *m*; ~ **pipe** WATER POLL fourreau *m*, WATER SUPP *of pump* conduite d'amorçage *f*, tuyau d'amorçage *m*; ~ **pump** WATER SUPP pompe d'amorçage *f*

principal: ~ **rotting treatment** *n* WATER SUPP *conversion phase* traitement anti-rouille principal *m*

prior: ~ **crushing** *n* WATER SUPP *of waste* broyage préliminaire *m*, prébroyage *m*; ~ **separation** *n* WATER SUPP *of harmful substances before landfilling* séparation préliminaire *f*

priority *n* POLICY priorité *f*; ~ **determination** POLICY choix des priorités *m*; ~ **list** POLICY *of measure* liste des priorités *f*

private: ~ **collection** *n* WASTE collecte privée *f*; ~ **disposal company** *n* WATER SUPP compagnie privée d'élimination des déchets *f*; ~ **drinking water consumption** *n* WATER SUPP consommation d'eau de boisson des ménages *f*

probe *n* WATER SUPP sonde *f*; ~ **microphone** ANALYSIS sonde microphonique *f*

problem: ~ **definition** *n* ANALYSIS définition du problème *f*; ~ **site** *n* WASTE *landfill* décharge point noir *f*, WATER SUPP site problématique *m*

procedure *n* GEN démarche *f*, méthode *f*, procédure *f*, mode opératoire *m*

process[1] *n* GEN procédé *m*, processus *m*; ~ **combining pyrolysis and incineration** WATER POLL *wastewater sludge incineration* procédé mixte associant pyrolyse et incinération *m*; ~ **conductivity meter** ANALYSIS conductivimètre industriel *m*; ~ **costs** *n pl* MANAGE coûts de traitement *m pl*; ~ **effectiveness** AIR, ANALYSIS efficacité d'un procédé *f*, RENEW ENERG rendement de procédé *m*, rendement industriel *m*, SOIL, WATER SUPP efficacité de traitement *f*; ~-**gas flow** AIR flux de gaz traité *m*; ~ **heat** RENEW ENERG chaleur industrielle *f*; ~ **material** AIR, WATER SUPP matériau industriel *m*; ~ **materials stream** AIR flux de matériaux industriels *m*; ~ **stability** WATER POLL stabilité du procédé *f*; ~ **steam** RENEW ENERG vapeur industrielle *f*; ~ **of using dispersion through chemical reactions** WATER SUPP *hazardous waste treatment* procédé mettant en œuvre la dispersion par des réactifs chimiques *m*; ~ **waste** WATER SUPP déchets industriels *m pl*; ~ **waste gas** AIR gaz industriel pollué *m*; ~ **water** WATER POLL, WATER SUPP eau de procédé *f*, eau industrielle *f*, eau de traitement *f*, eau de fabrication *f*

process[2] *vt* AIR, RENEW ENERG traiter, SOIL assainir, traiter, WATER SUPP traiter, transformer

processed: ~ **slag** *n* WASTE, WATER SUPP scories traitées *f pl*, résidus traités *m pl*; ~ **waste** *n* WASTE déchets traités *m pl*

processing *n* AIR, MANAGE traitement *m*, RENEW ENERG *of biomass* traitement *m*, transformation *f*, SOIL traitement *m*, assainissement *m*, WATER SUPP retraitement *m*; ~ **capacity**

AIR, SOIL capacité de traitement *f*, WATER SUPP capacité de retraitement *f*; **~ center for recyclable solid waste materials** *AmE*, **~ centre for recyclable solid waste materials** *BrE* WATER SUPP centre de retraitement des déchets solides *m*; **~ of landfill gas** WATER SUPP traitement des gaz de décharge *m*; **~ plant** WATER SUPP usine de retraitement *f*; **~ of raw material** MANAGE traitement de matières premières *m*; **~ speed** POLICY cadence de traitement *f*; **~ unit** WATER SUPP installation de retraitement *f*
producer *n* WATER SUPP producteur *m*
product: **~ charge** *n* POLICY redevance sur produit *f*; **~ containing pollutants** *n* WATER SUPP produit contenant des polluants *m*; **~ life-cycle** *n* MANAGE cycle de vie d'un produit *m*; **~-line analysis** *n* MANAGE analyse d'une ligne de produits *f*; **~ recovery** *n* AIR récupération de produit *f*; **~ safety** *n* SAFETY sécurité des produits *f*; **~ standards** *n pl* POLICY normes de production *f pl*; **~ stewardship** *n* POLICY *waste* surveillance des produits *f*, WATER SUPP bonne gestion de produit *f*, suivi de produit *m*
production: **~-consumption loop** *n* WATER SUPP boucle production-consommation *f*; **~ line** *n* WATER SUPP chaîne de production *f*; **~ potential** *n* RENEW ENERG potentiel de production *m*; **~ of raw material** *n* MANAGE, RENEW ENERG *biomass conversion* production de matières premières *f*; **~ residue** *n* WATER SUPP résidu de production *m*; **~-specific waste** *n* WASTE, WATER SUPP déchets spécifiques d'une production *m pl*; **~ water** *n* WATER SUPP eau de production *f*
profitability *n* MANAGE, POLICY rentabilité *f*
profitable *adj* MANAGE, POLICY rentable
program *vt* ANALYSIS programmer
programmable *adj* ANALYSIS programmable
progressive[1] *adj* POLICY progressif
progressive[2]: **~ tax** *n* POLICY impôt progressif *m*
prohibit *vt* AIR, POLICY interdire
prohibition *n* AIR, POLICY interdiction *f*
project: **~ phase** *n* RENEW ENERG *of installations* stade de projet *m*
promoter *n* AIR *catalyst* promoteur *m*
propagation *n* GEN propagation *f*; **~ class** AIR *waste gas plume of stack* classe de propagation *f*; **~ deadening** NOISE affaiblissement de la propagation *m*; **~ of pollutants** AIR, SOIL, WATER POLL, WATER SUPP propagation de polluants *f*
propellant *n* AIR agent moteur *m*, propulseur *m*
propelled: **~ jet** *n* AIR *annular gap nozzle* jet propulsé *m*
propeller *n* MARINE, WATER SUPP hélice *f*; **~ pump** WATER SUPP pompe à hélice *f*
propelling: **~ force** *n* AIR force de propulsion *f*
proper: **~ use of land** *n* SOIL utilisation rationnelle des terres *f*
properties: **~ of materials** *n* WASTE caractéristiques des matériaux *f pl*
prophylaxis *n* POLICY *prevention* prophylaxie *f*
propionic: **~ acid** *n* WATER POLL acide propionique *m*
proportional: **~ counter** *n* ANALYSIS *gas-filled detectors* compteur proportionnel *m*; **~ sampling** *n* ANALYSIS échantillonnage proportionnel *m*
proposal *n* POLICY proposition *f*
protect *vt* POLICY, SAFETY protéger
protected: **~ area** *n* POLICY région protégée *f*; **~-area boundary** *n* POLICY périmètre de protection *m*
protection: **~ capacity** *n* MARINE *from noise* pouvoir de protection *m*, NOISE capacité de protection *f*; **~ of neighbors** *AmE*, **~ of neighbours** *BrE* *n* POLICY protection du voisinage *f*; **~ of resources** *n* WATER SUPP protection des ressources *f*; **~ zone** *n* SAFETY zone de protection *f*
protective[1] *adj* GEN de protection, protecteur
protective[2]: **~-aerated cab** *n* SAFETY *of building machinery* cabine avec aération de protection *f*; **~ aeration** *n* SAFETY aération de sécurité *f*; **~ agent** *n* WATER SUPP agent protecteur *m*; **~ clothing** *n* SAFETY vêtement de protection *m*; **~ equipment** *n* SAFETY matériel de protection *m*; **~ function** *n* AIR *of ozone layer* fonction de protection *f*; **~ housing** *n* AIR *filter material* boîtier protecteur *m*, corps protecteur *m*; **~ insulation** *n* SOIL isolation de protection *f*; **~ insulation well** *n* SOIL puits isolant de protection *m*; **~ layer** *n* WATER SUPP couche protectrice *f*; **~ measure** *n* AIR, MANAGE, MARINE, POLICY, SAFETY mesure protective *f*, mesure de protection *f*; **~ shield** *n* AIR *against UV radiation*, SAFETY écran de protection *m*; **~ suit** *n* SAFETY vêtement de protection *m*
protein: **~ feed** *n* RENEW ENERG aliments protéiques *m pl*
protocol *n* POLICY, WATER SUPP protocole *m*
proton: **~-absorptive capacity** *n* AIR capacité d'absorption des protons *f*; **~-induced x-ray emission** *n* (*PIXE*) ANALYSIS émission de rayons X induite par des protons *f*; **~ induced X-ray emission analysis** *n* ANALYSIS analyse d'émission de rayons X induite par des protons *f*
protoplasm *n* WATER POLL protoplasme *m*
protozoa *n pl* WATER POLL protozaire *m*
provisional: **~ evaluation** *n* ANALYSIS évaluation provisoire *f*
proximate: **~ analysis** *n* ANALYSIS analyse immédiate *f*
proximity: **~ pick-up** *n* ANALYSIS capteur de proximité *m*; **~ sensor** *n* ANALYSIS capteur de proximité *m*
PS *abbr* (*polystyrene*) WATER SUPP polystyrène *m*
PSB *abbr* (*packed suspended bed*) WATER POLL lit fixe suspendu *m*
psychrometer *n* ANALYSIS psychromètre *m*
psychrophilic[1] *adj* WATER POLL psychrophile
psychrophilic[2]: **~ microorganism** *n* WATER POLL microorganisme psychrophile *m*
Pt *abbr* (*platinum*) GEN Pt (*platine*)
PTFE *abbr* (*polytetrafluoroethylene*) AIR polytétrafluoréthylène *m*
PTS *abbr* (*permanent threshold shift*) NOISE déplacement permanent du seuil *m*
public: **~ awareness campaign** *n* POLICY campagne de sensibilisation *f*; **~ canalization** *n* WATER POLL canalisation du réseau public *f*; **~ collection station** *n* WATER SUPP centre de collecte *m*, déchetterie *f*, dépôt public *m*; **~ drinking-water supply** *n* WATER SUPP approvisionnement collectif en eau potable *m*, fourniture collective d'eau potable *f*; **~ health** *n* POLICY santé publique *f*, SAFETY, WATER SUPP hygiène publique *f*; **~ hearing** *n* POLICY audition *f*; **~ hearing date** *n* POLICY date d'audition publique *f*; **~ order law** *n* POLICY législation régissant l'ordre public *f*; **~ participation** *n* POLICY participation des habitants *f*, participation de la population *f*; **~ utility** *n* MANAGE compagnie de service public *f*, service public *m*; **~ water supply** *n*, WATER SUPP distribution publique d'eau *f*, fourniture collective d'eau *f*; **~ works** *n* POLICY, WATER POLL, WATER SUPP ouvrages publics *m pl*, travaux publics *m pl*
publication *n* POLICY publication *f*
puff *n* AIR bouffée *f*
pug: **~ mill** *n* SOIL pétrin malaxeur *m*
pulp *n* WASTE, WATER SUPP pâte à papier *f*; **~-molding system** *AmE*, **~-moulding system** *BrE* WATER SUPP système d'agglomération de pâte *m*, système de moulage de pâte *m*, technique de moulage de pâte *f*; **~ wood** WASTE bois à pâte *m*
pulper *n* WATER SUPP désintégrateur *m*, triturateur *m*
pulse: **~-jet cleaned filter** *n* AIR filtre nettoyé par jet pulsant *m*; **~-jet cleaning** *n* AIR nettoyage au jet pulsant *m*; **~ polarography** *n* ANALYSIS polarographie à mode impulsionnel *f*; **~ stroke** *n* AIR temps d'impulsion *m*
pulverize *vt* AIR atomiser, pulvériser, MARINE, SOIL atomiser, WATER POLL atomiser, pulvériser, WATER SUPP atomiser, pulvériser, triturer
pulverizer *n* AIR pulvérisateur *m*, WATER SUPP appareil à pulvérisation d'eau *m*, pulvérisateur *m*

pulverizing *n* AIR, MARINE, SOIL, WATER POLL atomisation *f*, WATER SUPP atomisation *f*, pulvérisation *f*

pump[1] *n* GEN pompe *f*; ~ **barrel** WATER POLL, WATER SUPP barillet *m*, corps de pompe *m*, cylindre de pompe *m*; ~ **capacity** WATER POLL capacité de refoulement *f*; ~ **case** WATER POLL, WATER SUPP corps de pompe *m*; ~ **casing** WATER POLL carter de pompe *m*, corps de pompe *m*, WATER SUPP corps de pompe *m*; ~ **cavitation** WATER POLL cavitation d'une pompe *f*; ~ **characteristics** *n pl* WATER POLL, WATER SUPP caractéristiques d'une pompe *f pl*; ~ **connection** WATER POLL, WATER SUPP raccord de pompe *m*; ~ **controls** WATER POLL servocommande de pompe *f*, WATER SUPP commande de pompe *f*; ~ **cylinder** WATER POLL, WATER SUPP barillet *m*, corps de pompe *m*, cylindre de pompe *m*; ~**-driving gear** WATER POLL, WATER SUPP moteur de pompe *m*, engrenage de pompe *m*; ~ **efficiency** WATER POLL, WATER SUPP rendement d'une pompe *m*; ~ **fluid** WATER SUPP fluide moteur *m*; ~ **gear** WATER POLL armature de pompe *f*, WATER SUPP engrenage de pompe *m*, entraînement de pompe *m*; ~ **house** WATER POLL bâtiment des pompes *m*, poste de pompage *m*, WATER SUPP local des pompes *m*, poste de pompage *m*; ~ **impeller** WATER POLL, WATER SUPP rotor à aubes *m*; ~ **operation via water tower** WATER SUPP fonctionnement d'une pompe alimentée par un château d'eau *m*; ~ **plunger** WATER SUPP piston plongeur *m*; ~**-plunger shaft** WATER POLL, WATER SUPP arbre barillet *m*; ~ **room** WATER SUPP salle de pompage *f*, salle des pompes *f*; ~ **set** WATER POLL poste de pompage *m*, WATER SUPP batterie de pompes *f*, poste de pompage *m*; ~ **shaft** WATER POLL, WATER SUPP *well* puits de pompe *m*, *rod* arbre de pompe *m*; ~ **speed** WATER POLL, WATER SUPP régime d'une pompe *m*; ~ **station** WATER POLL poste de pompage *m*, WATER SUPP poste de pompage *m*, station de pompage *f*; ~ **sump** AIR puisard de pompe *m*, WASTE, WATER POLL pompe d'épuisement *f*, WATER SUPP pompe d'épuisement *f*, vide-cave *m*; ~ **test** ANALYSIS, WATER POLL, WATER SUPP essai de pompage *m*; ~ **water** WATER SUPP eau alimentaire *f*, eau de puits *f*; ~ **well** WATER SUPP puits de pompage *m*

pump[2] *vt* GEN pomper, épuiser, puiser avec une pompe, assécher; ~ **back** WATER POLL, WATER SUPP repomper, réaspirer par pompage, reprendre par pompage; ~ **out** WASTE, WATER POLL, WATER SUPP épuiser, évacuer par pompage, *mine* dénoyer, *water from tank* pomper, puiser avec une pompe, se vider, *empty* assécher; ~ **and treat** WATER SUPP pomper et traiter, traiter par pompage

pumpable[1] *adj* AIR, WATER POLL, WATER SUPP pompable

pumpable[2]: ~ **hazardous wastes** *n* WASTE, WATER POLL, WATER SUPP déchets dangereux pompables *m pl*; ~ **liquid** *n* AIR liquide pompable *m*; ~ **wastes** *n pl* WASTE, WATER SUPP déchets pompables *m pl*

pumped[1] *adj* GEN pompé

pumped[2]: ~**-water flow** *n* WATER SUPP débit d'eau pompée *m*

pumping *n* GEN épuisement *m*, assèchement *m*, exhaure *f*, pompage *m*; ~ **engine** WATER POLL machine d'épuisement *f*, machine d'exhaure *f*, moteur à pompes *m*, moteur d'épuisement *m*; ~ **out** WASTE, WATER POLL, WATER SUPP épuisement *m*, assèchement *m*, exhaure *f*, pompage *m*; ~ **plant** WASTE pompe d'épuisement *f*, WATER POLL *equipment* matériel d'épuisement *m*, matériel d'exhaure *m*, pompe d'épuisement *f*, WATER SUPP *equipment* appareil d'exhaure *m*, pompe d'épuisement *f*, *pumping station* station de pompage *f*; ~ **shaft** WATER SUPP puits d'épuisement *m*, puits d'exhaure *m*; ~ **station** WATER POLL, WATER SUPP station de pompage *f*; ~ **test** ANALYSIS, WATER POLL, WATER SUPP essai de pompage *m*; ~ **unit** MARINE station de pompage *f*, WATER POLL installation de pompage *f*, poste de relèvement *m*

puncturability *n* WATER SUPP perforabilité *f*

puncture: ~ **resistance** *n* WATER SUPP résistance à la crevaison *f*, résistance à la perforation *f*

PUR *abbr* (*polyurethane*) WATER SUPP polyuréthanne *m*

pure: ~ **oxygen** *n* AIR oxygène pur *m*; ~ **water** *n* WATER SUPP eau pure *f*

purge[1]: ~**-air inlet** *n* AIR *activated carbon filter* admission d'air purgé *f*; ~**-and-trap gas chromatography** *n* ANALYSIS chromatographie en phase gazeuse par piégeage *f*; ~**-and-trap method** *n* ANALYSIS méthode de désorption purge and trap *f*; ~**-and-trap sampler** *n* ANALYSIS échantillonneur de purge et de piégeage *m*

purge[2] *vt* ANALYSIS clarifier, purger, WATER SUPP clarifier

purgeable: **organic halogen compounds** *n* (*POX*) WATER SUPP composés organiques halogénés éliminables *m pl*

purging: ~ **cock** *n* WATER SUPP robinet de vidange *m*

purification *n* WATER POLL épuration *f*, purification *f*, WATER SUPP purification *f*; ~ **capacity** WATER POLL capacité d'auto-épuration *f*; ~ **plant** WATER POLL station d'épuration *f*; ~ **system using plants** WATER SUPP système d'épuration utilisant des produits végétaux *m*

purified: ~ **wastewater** *n* WATER POLL eaux usées épurées *f pl*

purifier *n* WATER SUPP épurateur *m*

purify *vt* WATER POLL clarifier, épurer, purifier, WATER SUPP clarifier, purifier

putrefaction *n* WATER POLL, WATER SUPP putréfaction *f*

putrefactive *adj* WATER POLL, WATER SUPP putréfiant

putrescent: ~ **material** *n* WATER POLL matière putrescente *f*

putrescibility *n* WATER POLL, WATER SUPP putrescibilité *f*

putrescible[1] *adj* WATER POLL, WATER SUPP putrescible

putrescible[2]: ~ **matter** *n* WATER POLL matière putrescible *f*; ~ **sludge** *n* WATER POLL boues putrescibles *f pl*; ~ **waste** *n* WASTE, WATER SUPP déchets putrescibles *m pl*

puzzolana *n* SOIL *binding agent* pouzzolane *f*

PV *abbr* (*photovoltaics*) RENEW ENERG PV (*genie photovoltaïque*)

PVC[1] *abbr* (*polyvinyl chloride*) SOIL, WATER SUPP PVC (*polychlorure de vinyle*)

PVC[2]: ~ **bottle** *n* WASTE bouteille en PVC *f*; ~ **insert fitment** *n* WATER SUPP garnissage en PVC *m*; ~ **pipe** *n* WATER POLL tuyauterie en PVC *f*, WATER SUPP tuyau en PVC *m*; ~ **residue** *n* WATER SUPP résidus de PVC *m pl*

PY-GC *abbr* (*pyrolysis gas chromatography*) ANALYSIS chromatographie en phase gazeuse avec pyrolyse *f*

PY-GC-MS *abbr* (*pyrolysis-gas chromatography-mass spectrometry*) ANALYSIS spectrométrie de masse-chromatographie en phase gazeuse avec pyrolyse *f*

pyknometer *n* ANALYSIS pycnomètre *m*

pyknometric *adj* ANALYSIS pycnométrique

pyranometer *n* ANALYSIS pyranomètre *m*

pyrheliometer *n* ANALYSIS pyrhéliomètre *m*

pyrolyse *vt BrE* GEN pyrolyser

pyrolysis *n* GEN pyrolyse *f*; ~ **gas** WATER SUPP gaz de pyrolyse *m*; ~ **gas chromatography** (*PY-GC*) ANALYSIS chromatographie en phase gazeuse avec pyrolyse *f*; ~**-gas chromatography-mass spectrometry** (*PY-GC-MS*) ANALYSIS spectrométrie de masse-chromatographie en phase gazeuse avec pyrolyse *f*; ~ **oil** WATER SUPP huile de pyrolyse *f*; ~ **product** WATER SUPP produit de pyrolyse *m*, pyrolysat *m*

pyrolytic[1] *adj* GEN pyrolytique

pyrolytic[2]: ~ **pretreatment** *n* WATER SUPP prétraitement pyrolytique *m*

pyrolyze *AmE see pyrolyse BrE*

pyrophoric[1] *adj* AIR pyrophore

pyrophoric[2]: ~ **dust** *n* AIR poussière pyrophore *f*; ~ **material** *n* AIR *dust property* pyrophore *m*

Q

QA *abbr* (*quality assurance*) MANAGE AQ (*assurance-qualité*)

quadrupole *n* ANALYSIS quadripôle *m*; **~ ion trap** ANALYSIS piège ionique quadripolaire *m*, quistor *m*; **~-mass analyser** *BrE*, **~-mass analyzer** *AmE* ANALYSIS analyseur de masse quadripolaire *m*; **~ mass spectrometer** ANALYSIS spectromètre de masse quadripolaire *m*

qualitative[1] *adj* ANALYSIS qualitatif

qualitative[2]: **~ analysis** *n* ANALYSIS analyse qualitative *f*; **~ detection** *n* ANALYSIS détection qualitative *f*; **~ determination** *n* ANALYSIS détermination qualitative *f*, dosage qualitatif *m*; **~ evaluation** *n* ANALYSIS évaluation qualitative *f*; **~ gas analysis** *n* ANALYSIS *hand-driven gas detectors* analyse qualitative des gaz *f*; **~ odor analysis** *AmE*, **~ odour analysis** *BrE n* ANALYSIS analyse olfactive qualitative *f*, analyse qualitative des odeurs *f*

qualitatively: **~ detect** *phr* ANALYSIS détecter qualitativement

quality: **~ assurance** *n* (*QA*) MANAGE assurance-qualité *f* (*AQ*); **~-assurance system** *n* MANAGE système d'assurance-qualité *m*; **~ control** *n* AIR, MANAGE, WATER SUPP contrôle-qualité *m*, maîtrise de la qualité *f*; **~ factor** *n* AIR, WATER SUPP facteur de qualité *m*, facteur Q *m*; **~ loop** *n* MANAGE boucle de la qualité *f*; **~ manual** *n* MANAGE manuel qualité *m*; **~ monitoring of ambient air** *n* SAFETY contrôle de la qualité de l'air ambiant *m*; **~ policy** *n* MANAGE politique qualité *f*; **~ requirement** *n* SOIL exigence de qualité *f*; **~ requirements** *n pl* AIR, MANAGE, MARINE, NOISE, SAFETY, WATER SUPP critères de qualité *m pl*

quantification *n* ANALYSIS quantification *f*

quantify *vt* ANALYSIS quantifier

quantitative[1] *adj* ANALYSIS quantitatif

quantitative[2]: **~ analysis** *n* ANALYSIS analyse quantitative *f*; **~ detection** *n* ANALYSIS détection quantitative *f*; **~ determination** *n* ANALYSIS détermination quantitative *f*, dosage quantitatif *m*; **~ evaluation** *n* ANALYSIS évaluation quantitative *f*; **~ gas analysis** *n* ANALYSIS *infrared absorption spectrometry* analyse quantitative des gaz *f*; **~ odor analysis** *AmE*, **~ odour analysis** *BrE n* ANALYSIS analyse olfactive quantitative *f*, analyse quantitative des odeurs *f*

quantity: **~ of precipitate** *n* WATER POLL volume de précipités *m*; **~ to be measured** *n* ANALYSIS quantité à mesurer *f*

quarter: **~-wave resonator** *n* NOISE résonateur de quart d'onde *m*

quasi: **~-simultaneous** *adj* ANALYSIS quasi-simultané

quaternary: **~ recycling** *n* WASTE recyclage quaternaire *m*

quench[1]: **~ tank** *n* AIR cuve de trempe *f*; **~ zone** *n* AIR zone de trempe *f*

quench[2] *vt* WATER SUPP *lime* éteindre, extruder à plat, tremper

quencher *n* AIR agent de trempe *m*, SOIL extincteur *m*; **~ suspension** AIR agent de trempe en suspension *m*; **~ tank** AIR cuve de trempe *f*

quenching: **~ tower** *n* AIR réacteur de trempe *m*

quick: **~ drying** *n* WATER SUPP séchage rapide *m*; **~-fit connection** *n* WATER SUPP raccord rapide *m*

quicklime *n* GEN chaux vive *f*

quicksand *n* MARINE sable flottant *m*, sable mouvant *m*

quiet: **~ water** *n* WATER SUPP eau calme *f*

R

rabble: ~ **arm** *n* WATER SUPP bras de raclage *m*
rad *abbr* (*radiation absorbed*) ANALYSIS rad (*dose de rayonnement absorbé*)
radial: ~ **flow** *n* AIR *cyclone* écoulement radial *m*, flux radial *m*, WATER POLL flux radial *m*; ~**-flow pump** *n* WATER POLL pompe à flux radial *f*; ~**-flow scrubber** *n* AIR laveur à écoulement radial *m*
radiate *vt* AIR *heat*, NOISE rayonner
radiation *n* GEN rayonnement *m*; ~ **absorbed** (*rad*) ANALYSIS dose de rayonnement absorbée *f* (*rad*); ~ **absorption** AIR *dust concentration*, ANALYSIS absorption de rayonnement *f*; ~ **detector** ANALYSIS *radiation monitoring* détecteur de rayonnement *m*; ~ **dose** AIR, SAFETY dose de rayonnement *f*; ~ **dryer** WATER SUPP séchoir par rayonnement *m*; ~ **emitter** WATER SUPP *UV disinfection* émetteur de rayonnement *m*; ~ **energy** RENEW ENERG *solar energy plant* énergie rayonnée *f*; ~ **equilibrium** AIR équilibre de rayonnement *m*; ~ **hazard** SAFETY risque de rayonnement *m*; ~ **intensity** ANALYSIS, RENEW ENERG intensité de rayonnement *f*; ~ **measuring instrument** ANALYSIS instrument radiométrique *m*; ~ **protection** AIR radioprotection *f*, protection contre les rayonnements *f*, SAFETY radioprotection *f*; ~ **pyrometer** ANALYSIS pyromètre optique *m*; ~ **recuperator** AIR récupérateur de rayonnement *m*; ~ **shield** SAFETY écran de protection contre les rayonnements *m*, bouclier de protection contre les rayonnements *m*; ~ **shielding** SAFETY blindage de protection contre les rayonnements *m*
radical: ~ **formation** *n* AIR *electron beam process* formation de radicaux *f*
radio: ~**-frequency heating** *n* SOIL *thermal desorption* chauffage à haute fréquence *m*
radioactive[1] *adj* GEN radioactif
radioactive[2]: ~ **burial ground** *n* WATER SUPP cimetière de déchets radioactifs *m*; ~ **contaminant** *n* AIR, WATER POLL contaminant radioactif *m*; ~ **decay** *n* RENEW ENERG *geothermal energy*, WATER POLL désintégration radioactive *f*; ~ **dust** *n* AIR poussière radioactive *f*; ~ **emissions** *n pl* AIR émissions radioactives *f pl*; ~ **fall-out** *n* AIR, WATER SUPP retombées radioactives *f pl*; ~ **gas** *n* AIR gaz radioactif *m*; ~ **graveyard** *n* WATER SUPP cimetière de déchets radioactifs *m*; ~ **half-life** *n* AIR période radioactive *f*, période de demi-vie *f*; ~ **pollution** *n* AIR, SAFETY, WATER POLL, WATER SUPP pollution radioactive *f*, contamination radioactive *f*; ~ **rare gas** *n* AIR gaz rare radioactif *m*; ~ **substance** *n* AIR, SAFETY, WATER SUPP substance radioactive *f*; ~ **waste** *n* WASTE, WATER SUPP déchets radioactifs *m pl*; ~ **wastewater** *n* WATER POLL effluent radioactif *m*
radioactivity: ~ **decay** *n* AIR décroissement de la radioactivité *m*
radiograph *n* ANALYSIS, WATER SUPP radiographe *m*
radiographic[1] *adj* ANALYSIS, WATER SUPP radiographique
radiographic[2]: ~ **process** *n* WATER SUPP *form of binding* procédé radiographique *m*
radiological: ~ **survey** *n* ANALYSIS étude radiologique *f*
radiometer *n* ANALYSIS radiomètre *m*
radiometric *adj* ANALYSIS radiométrique
radiopaging *n* SAFETY radiomessagerie *f*
radiopolarography *n* ANALYSIS radiopolarographie *f*
radiowave: ~ **hazard** *n* SAFETY risque d'interférence radio *m*
rail: ~ **transport** *n BrE* (*cf railroad transport AmE*) WASTE transport par chemin de fer *m*, transport par rail *m*, transport ferroviaire *m*
railroad: ~ **transport** *n AmE* (*cf rail transport BrE*) WASTE transport par chemin de fer *m*, transport par rail *m*, transport ferroviaire *m*
rain: ~ **area** *n* WATER SUPP impluvium *m*, zone de précipitation *f*; ~**-free period** *n* WATER SUPP période sèche *f*, saison sèche *f*; ~ **gage** *AmE*, ~ **gauge** *BrE n* ANALYSIS, WATER SUPP pluviomètre *m*
rainfall *n* AIR, WATER SUPP précipitation atmosphérique *f*, quantité d'eau tombée *f*; ~ **area** WATER SUPP zone de pluviosité *f*, zone de précipitation *f*; ~ **erosion** SOIL érosion pluviale *f*
rainout *n* AIR retombées entraînées par la pluie *f pl*
rainwater *n* WATER POLL, WATER SUPP eau de pluie *f*, eau météorique *f*, eau pluviale *f*, eau précipitée *f*; ~ **tank** WATER SUPP réservoir d'eaux pluviales *m*; ~ **utilization** WATER SUPP utilisation des eaux pluviales *f*
raising: ~ **of the earth's temperature** *n* AIR élévation de la température terrestre *f*
rake *n* WATER SUPP râteau *m*
Raman: ~ **spectroscopy** *n* ANALYSIS spectroscopie Raman *f*
ramming: ~ **machine** *n* WATER SUPP pilonneuse *f*
ramp: ~ **landfill** *n* WATER SUPP décharge inclinée *f*
random: ~ **noise** *n* NOISE bruit aléatoire *m*; ~ **sample** *n* ANALYSIS échantillon aléatoire *m*
range *n* ANALYSIS étendue *f*, gamme *f*, plage *f*, portée *f*; ~ **of audibility** NOISE plage d'audition *f*; ~ **of capacity** RENEW ENERG plage de puissance *f*
Rankine: ~ **cycle steam engine** *n* AIR moteur à vapeur à cycle de Rankine *m*
rapeseed: ~ **oil** *n* AIR *as fuel* huile de colza *f*
rapid: ~ **carbonate removal** *n* WATER SUPP décalaminage rapide *m*, décarbonisation rapide *f*; ~ **fermentation** *n* WASTE fermentation rapide *f*; ~ **filtration** *n* WATER POLL, WATER SUPP filtration rapide *f*; ~ **rotting** *n* WATER POLL pourriture accélérée *f*; ~ **sand filter** *n* WATER POLL, WATER SUPP filtre rapide à sable *m*; ~ **sand filtration** *n* WATER POLL, WATER SUPP filtration rapide sur sable *f*
rapper *n* AIR dispositif de frappage *m*
rapping *n* AIR frappage *m*; ~ **of electrodes** AIR frappage des électrodes *m*; ~ **interval** AIR *filter* intervalle de frappage *m*, rythme de frappage *m*
Raschig: ~ **ring** *n* AIR *packing of wet scrubber* anneau Raschig *m*
rate: ~ **of absorption** *n* WATER SUPP taux d'absorption *m*, vitesse d'absorption *f*; ~ **of ascent** *n* AIR *of gases* vitesse ascensionnelle *f*, vitesse de montée *f*; ~ **of charge** *n* POLICY taux d'imposition *m*; ~ **of decomposition** *n* WASTE, WATER POLL vitesse de décomposition *f*; ~ **of flame propagation** *n* AIR vitesse de propagation de flamme *f*; ~ **of flow** *n* ANALYSIS, WATER POLL, WATER SUPP débit *m*; ~ **of flow originating from groundwater discharge** *n* WATER SUPP débit tributaire du rejet d'eaux souterraines *m*; ~ **of utilization** *n* WATER POLL, WATER SUPP taux d'utilisation *m*
rated[1] *adj* WATER SUPP taré, nominal
rated[2]: ~ **capacity** *n* RENEW ENERG puissance nominale *f*, WATER POLL, WATER SUPP capacité nominale *f*; ~ **load** *n* WATER POLL, WATER SUPP charge nominale *f*
ratify *vt* POLICY ratifier
ravine: ~ **erosion** *n* SOIL érosion ravinante *f*, ravinement *m*
raw[1] *adj* WATER SUPP brut

raw[2]: **~-air conveyance** *n* AIR transport d'air brut *m*; **~ compost** *n* WATER SUPP compost brut *m*; **~ data** *n* ANALYSIS données brutes *f pl*; **~-data storage** *n* ANALYSIS stockage des données brutes *m*; **~ gypsum** *n* AIR gypse brut *m*; **~ material** *n* RENEW ENERG, WASTE matière première *f*; **~ material decomposition** *n* WATER SUPP décomposition de matière première *f*; **~ material delivering industry** *n* MANAGE secteur d'approvisionnement en matières premières *m*, secteur fournisseur de matières premières *m*; **~ material extraction** *n* MANAGE, RENEW ENERG extraction de matières premières *f*; **~ material recycling** *n* WATER SUPP recyclage de matières premières *m*; **~ material recycling process** *n* RENEW ENERG procédé de recyclage de matières premières *m*; **~ material resources** *n pl* WATER SUPP ressources en matières premières *f pl*; **~ measured value** *n* ANALYSIS valeur brute mesurée *f*; **~ sewage** *n* WATER POLL eaux usées brutes *f pl*; **~ sludge** *n* WATER POLL boues brutes *f pl*; **~ soil** *n* SOIL sol brut *m*; **~ waste gas** *n* AIR effluent gazeux non traité *m*; **~ wastewater** *n* WASTE, WATER POLL *untreated* eaux résiduaires brutes *f pl*; **~ water** *n* WATER POLL eau brute *f*, WATER SUPP eau brute *f*, eau crue *f*; **~-water main** *n* WATER SUPP collecteur d'eau brute *m*, conduite d'eau brute *f*; **~ water pump** *n* WATER SUPP pompe d'eau brute *f*; **~ water quality** *n* WATER SUPP qualité d'eau brute *f*

Rayleigh: **~ disk** *n* ANALYSIS *measurement of sound intensity* disque de Rayleigh *m*; **~ refractometer** *n* ANALYSIS réfractomètre de Rayleigh *m*; **~ scattering** *n* AIR, ANALYSIS diffusion de Rayleigh *f*

RBC *abbr* (*rotating biological contactor*) WATER POLL disque biologique *m*

Re *abbr* (*Reynolds number*) AIR, WATER SUPP Re (*nombre de Reynolds*)

reach *n* WATER SUPP bief *m*

reactant *n* AIR réactif *m*

reacting: **~ agent** *n* ANALYSIS réactif *m*

reaction *n* GEN réaction *f*; **~ chamber** AIR chambre de réaction *f*, ANALYSIS chambre de réaction *f*, colonne de réaction *f*; **~ equilibrium** AIR équilibre de réaction *m*; **~ indicator** ANALYSIS indicateur de réaction *m*; **~ process** WATER SUPP *in landfill* processus de réaction *m*; **~ product** AIR produit de réaction *m*; **~ rate** WATER SUPP vitesse de réaction *f*; **~ surface** AIR surface de réaction *f*; **~ tank** WATER POLL cuve de réaction *f*; **~ time** AIR, WATER POLL délai de réaction *m*, temps de réaction *m*; **~ tower** AIR réacteur *m*; **~ tube** ANALYSIS tube à réaction *m*; **~-type axial fan** AIR ventilateur axial à réaction *m*; **~ velocity** AIR, WATER POLL, WATER SUPP vitesse de réaction *f*; **~ zone** WATER POLL zone de réaction *f*

reactivate *vt* GEN réactiver

reactivated *adj* GEN réactivé

reactivation *n* GEN réactivation *f*

reactive[1] *adj* GEN réactif

reactive[2]: **~ attenuator** *n* NOISE atténuateur réactif *m*; **~ hazardous waste** *n* WASTE, WATER POLL, WATER SUPP déchets dangereux réactifs *m pl*; **~ muffler** *n AmE* (*cf reactive silencer BrE*) NOISE, SAFETY silencieux réactif *m*; **~ particle** *n* AIR particule réactive *f*; **~ silencer** *n BrE* (*cf reactive muffler AmE*) NOISE, SAFETY silencieux réactif *m*

reactivity *n* AIR, SOIL, WATER POLL, WATER SUPP réactivité *f*

reactor *n* AIR, WATER POLL réacteur *m*; **~ bottom** WATER POLL fond de réacteur *m*; **~ with catalyst** AIR réacteur à catalyse *m*; **~ pressure sensor** AIR manomètre de réacteur *m*; **~ tube** AIR *electron beam process* tubulure de réacteur *f*

read: **~-out display** *n* ANALYSIS affichage de lecture *m*; **~-out of measured values** *n* ANALYSIS lecture des valeurs mesurées *f*

readable *adj* ANALYSIS lisible

readied: **~ compost** *n* WATER SUPP compost mûri *m*

reading *n* ANALYSIS lecture *f*, relevé *m*; **~ accuracy** ANALYSIS précision de lecture *f*; **~ constancy** ANALYSIS constance de lecture *f*; **~ error** ANALYSIS erreur de lecture *f*, erreur de relevé *f*

reaeration *n* WATER POLL réoxygénation *f*

reagent *n* AIR, ANALYSIS, SOIL réactif *m*; **~ solution** ANALYSIS solution de réactifs *f*

real[1]: **~-time analyser** *BrE*, **~-time analyzer** *AmE n* ANALYSIS analyseur en temps réel *m*; **~-time chromatogram** *n* ANALYSIS chromatogramme en temps réel *m*; **~-time operation** *n* ANALYSIS exploitation en temps réel *f*; **~-time sampling** *n* ANALYSIS échantillonnage en temps réel *m*; **~-time signal processing** *n* ANALYSIS traitement du signal en temps réel *m*

real[2]: **in ~ time** *phr* ANALYSIS *instrumentation* en temps réel

rebound *vi* AIR *particles on fibre* rebondir, rejaillir

recalibration *n* ANALYSIS réétalonnage *m*

recapture *vt* AIR recapter

receiver *n* RENEW ENERG *solar power plant* collecteur *m*, récepteur *m*; **~ tank** WATER POLL réservoir de collecte *m*

receiving: **~ bunker** *n* WATER SUPP trémie de réception *f*; **~ chamber** *n* WATER POLL chambre réceptrice *f*; **~ tank** *n* WATER POLL réservoir récepteur *m*, collecteur *m*; **~ telescope** *n* ANALYSIS télescope récepteur *m*; **~ tower** *n* RENEW ENERG *solar power plant* tour collectrice *f*

receptacle *n* ANALYSIS réceptacle *m*

receptor *n* AIR récepteur *m*; **~ region** AIR région réceptrice *f*

rechargable: **~ storage battery** *n* RENEW ENERG batterie à accumulation rechargeable *f*

recharge *n* WATER SUPP réalimentation *f*, réinjection *f*; **~ basin** WATER SUPP bassin de réalimentation *m*; **~ well** WATER POLL, WATER SUPP puits filtrant *m*

reciprocal: **~ effect** *n* ANALYSIS effet réciproque *m*

reciprocating: **~ piston engine** *n* AIR moteur à piston alternatif *m*; **~ pump** *n* WATER POLL pompe à piston *f*, WATER SUPP pompe à mouvement alternatif *f*, pompe à piston *f*

recirculate *vt* WATER POLL, WATER SUPP faire recirculer, réinjecter dans un circuit, recycler

recirculated: **~ heating** *n* WATER POLL, WATER SUPP chaleur recyclée *f*

recirculating: **~ filter** *n* AIR filtre à recirculation *m*; **~ pump** *n* WATER POLL pompe de recirculation *f*; **~ washing agent** *n* AIR agent de lavage recirculant *m*; **~ water** *n* WATER POLL, WATER SUPP eau de recirculation *f*, eau de retour *f*; **~ water management** *n* MANAGE, WATER POLL, WATER SUPP gestion de l'eau de recirculation *f*

recirculation *n* WATER POLL, WATER SUPP recirculation *f*, recyclage *m*; **~ system** WATER POLL circuit de recirculation *m*, système de recirculation *m*

reclaim *vt* GEN régénérer, récupérer, réhabiliter, SOIL restaurer

reclaimed[1] *adj* GEN régénéré, récupéré, réhabilité

reclaimed[2]: **~ area** *n* SOIL zone remise en état *f*, WATER SUPP terrain réhabilité *m*, terrain remis en valeur *m*; **~ wastewater** *n* WATER POLL eaux usées régénérées *f pl*; **~ water** *n* WATER POLL eau régénérée *f*

reclamation *n* SOIL *of land* remise en état *f*, WATER SUPP récupération *f*, *of land or water* réhabilitation *f*, remise en valeur *f*; **~ plant** WATER SUPP usine de récupération *f*

recontaminate *vt* WATER POLL *treated wastewater* recontaminer, polluer de nouveau

reconvert *vt* AIR reconvertir

record *vt* AIR, ANALYSIS, POLICY, WATER SUPP enregistrer

recording *n* AIR, ANALYSIS, POLICY, WATER SUPP enregistrement *m*; **~ spectrophotometer** ANALYSIS spectrophotomètre d'enregistrement *m*

recover *vt* AIR, SOIL, WATER POLL, WATER SUPP récupérer, recycler; **~ in liquid form** WATER SUPP récupérer sous forme liquide, recycler sous forme liquide

recoverability *n* AIR recyclabilité

recoverable[1] *adj* AIR, WATER SUPP récupérable, recyclable

recoverable[2]: **~ material** *n* WASTE matériau récupérable *m*,

paper, cardboard, plastics, glass matériau recyclable *m*; ~ **volume of water** *n* WATER SUPP volume d'eau recyclable *m*; ~ **waste** *n* WATER SUPP déchets recyclables *m pl*
recovered[1] *adj* AIR, SOIL, WATER POLL, WATER SUPP récupéré, recyclé
recovered[2]: ~ **heat** *n* AIR, WATER SUPP chaleur récupérée *f*; ~ **oil** *n* WATER SUPP huile de récupération *f*, huile recyclée *f*; ~ **pulp** *n* WATER SUPP pâte de récupération *f*; ~ **wastewater** *n* WATER POLL eaux usées régénérées *f pl*; ~ **water** *n* WATER POLL eau régénérée *f*
recovery *n* AIR récupération *f*, *of dust* captation *f*, MARINE récupération *f*, RENEW ENERG valorisation *f*, WATER SUPP remontée *f*, rétablissement *m*, reprise *f*, récupération *f*; ~ **area** WATER SUPP zone de régénération *f*; ~ **boiler** AIR, WATER SUPP chaudière de récupération *f*; ~ **by type** WATER SUPP récupération par type *f*; ~ **by-product** WATER SUPP sous-produit de récupération *m*; ~ **device** AIR, MARINE dispositif de récupération *m*, WATER SUPP appareil de récupération *m*, dispositif de récupération *m*, engin de récupération *m*, récupérateur *m*; ~ **of landfill space** WATER SUPP reprise de décharge *f*; ~ **rate** WATER SUPP *of take-back system* taux de récupération *m*; ~ **system** WATER SUPP système de récupération *m*
recreation: ~ **area** *n* POLICY aire de loisirs *f*
rectangular: ~ **basin** *n* WATER POLL bassin rectangulaire *m*, clarificateur rectangulaire *m*; ~ **clarifier** *n* WATER POLL bassin rectangulaire *m*, clarificateur rectangulaire *m*; ~ **tank** *n* WATER POLL citerne rectangulaire *f*
rectification *n* AIR, WATER POLL rectification *f*
rectifier *n* AIR, WATER POLL rectificateur *m*, redresseur *m*
recultivate *vt* SOIL, WATER POLL remettre en culture
recultivation *n* SOIL, WATER POLL remise en culture *f*
recuperative[1] *adj* AIR à récupération, de récupération
recuperative[2]: ~ **heat exchanger** *n* AIR échangeur de chaleur à récupération *m*; ~ **heat recovery** *n* AIR recyclage de chaleur de récupération *m*
recuperator *n* AIR récupérateur *m*
recyclability *n* WATER SUPP recyclabilité *f*
recyclable[1] *adj* WASTE recyclable
recyclable[2]: ~ **battery** *n* WASTE batterie recyclable *f*
recycle *vt* WASTE recycler
recycled: ~ **bottle** *n* WASTE bouteille recyclée *f*; ~ **clothing** *n* WASTE vêtement recyclé *m*; ~ **glass** *n* WASTE verre recyclé *m*; ~-**glass application** *n* WASTE application de verre recyclé *f*; ~ **glass cullet** *n* WASTE calcin recyclé *m*; ~ **material** *n* WASTE matériau recyclé *m*; ~ **paper** *n* WASTE papier recyclé *m*; ~ **paper application** *n* WASTE application de papier recyclé *f*; ~ **paper application rate** *n* WASTE taux d'application de papier recyclé *m*; ~-**paper use** *n* WASTE consommation de papier recyclé *f*; ~ **plastic** *n* WASTE plastique recyclé *m*
recycling[1]: ~-**friendly** *adj* WASTE favorable au recyclage; ~-**oriented** *adj* WASTE axé sur le recyclage
recycling[2] *n* WASTE recyclage *m*; ~ **capacity** WASTE capacité de recyclage *f*; ~ **caterpillar** WASTE chenille de recyclage *f*; ~ **center** *AmE* WASTEcentre de recyclage *m*; ~ **centre** *BrE* WASTE centre de recyclage *m*; ~ **costs** *n pl* MANAGE coûts de recyclage *m pl*; ~ **economy** POLICY économie de recyclage *f*, activités de recyclage *f pl*, RENEW ENERG, WASTE économie de recyclage *f*; ~ **economy law** POLICY *in Germany*, WASTE droit du recyclage *m*; ~ **of fly ash** WASTE recyclage des cendres volantes *m*; ~-**friendly product** WASTE produit se prêtant au recyclage *m*; ~ **gas** AIR gaz de recyclage *m*; ~ **hindrance** WASTE obstacle au recyclage *m*; ~ **incentive** MANAGE, POLICY incitation au recyclage *f*; ~ **of inoculated compost** WASTE recyclage de compost inoculé *m*; ~ **operation** WASTE opération de recyclage *f*; ~ **paper** WASTE papier recyclé *m*; ~ **of paper cartons** WASTE recyclage de papier carton *m*; ~ **plant** WASTE usine de recyclage *f*; ~ **potential** WASTE potentiel de recyclage *m*; ~ **process** WASTE procédé de recyclage *m*; ~ **quantity** WASTE quantité recyclée *f*; ~ **rate** WASTE taux de recyclage *m*; ~ **of sludge** WASTE recyclage des boues *m*; ~ **technique** RENEW ENERG technique de recyclage *f*; ~ **of waste** WASTE recyclage de déchets *m*
red: ~ **algae** *n* WATER SUPP algues rouges *f pl*; ~ **mud** *n* WATER POLL boues rouges *f pl*; ~ **tide** *n* MARINE, WATER POLL marée rouge *f*; ~-**tide bloom** *n* MARINE, WATER POLL prolifération d'algues rouges *f*
redeposit *vt* SOIL *after treatment* redéposer
redeposition *n* SOIL *after treatment* nouveau dépôt *m*
redisposal *n* SOIL *of unpolluted arable soil* remise en place *f*
redissolution *n* WATER POLL redissolution *f*
redox: ~ **potential** *n* ANALYSIS, WATER POLL potentiel d'oxydo-réduction *m*
reduce *vt* ANALYSIS, NOISE, WATER POLL, WATER SUPP *waste* réduire; ~ **the pressure** WATER SUPP détendre la pression
reduced: ~ **rate of charge** *n* POLICY taux réduit d'imposition *m*
reducing[1] *adj* AIR, ANALYSIS, WATER POLL, WATER SUPP réducteur
reducing[2]: ~ **agent** *n* AIR, ANALYSIS agent réducteur *m*, WATER POLL agent réducteur *m*, réducteur *m*; ~ **catalyst bed** *n* AIR lit de réduction catalytique *m*; ~ **gas** *n* AIR gaz réducteur *m*; ~ **product** *n* WATER SUPP produit de réduction *m*; ~ **substance** *n* AIR substance réductrice *f*
reduction *n* WATER SUPP diminution *f*, réduction *f*; ~ **catalyst** AIR catalyseur de réduction *m*; ~ **of waste gas retention time** AIR diminution du temps de rétention des effluents gazeux *f*
reef *n* MARINE brisant *m*; ~ **area** MARINE zone de récifs *f*
reel: ~ **in** *vt* MARINE rentrer par dévidoir
reference: ~ **chamber** *n* ANALYSIS chambre de référence *f*; ~ **detector** *n* ANALYSIS *UV emissions* détecteur de référence *m*; ~ **electrode** *n* ANALYSIS *pH measurement* électrode de référence *f*; ~ **gas** *n* ANALYSIS gaz de référence *m*; ~ **resistor** *n* ANALYSIS résistance de référence *f*; ~ **sound pressure** *n* NOISE pression acoustique de référence *f*; ~ **spectrum** *n* ANALYSIS spectre de référence *m*, valeur de référence *f*; ~ **wavelength** *n* ANALYSIS longueur d'onde de référence *f*
refiltering *n* AIR, WATER SUPP refiltration *f*
refine *vt* AIR, MARINE, WATER SUPP raffiner
refined[1] *adj* AIR, MARINE, WATER SUPP raffiné
refined[2]: ~ **product** *n* GEN produit raffiné *m*
refinery: ~ **gas** *n* AIR gaz de raffinerie *m*; ~ **waste** *n* WASTE déchets de raffinerie *m pl*
refining *n* AIR, MARINE, WATER SUPP affinage *m*
reflect *vt* AIR réfléchir, refléter, NOISE *on the separating wall*, RENEW ENERG *solar radiation* réfléchir
reflected[1] *adj* AIR, NOISE, RENEW ENERG réfléchi
reflected[2]: ~ **paraboloid** *n* RENEW ENERG paraboloïde réfléchi *m*
reflection *n* AIR, ANALYSIS réflexion *f*; ~ **spectroscopy** ANALYSIS spectroscopie de réflexion *f*
reflectometer *n* ANALYSIS *measures volumetric soil water content* réflectomètre *m*
reflector *n* RENEW ENERG *solar tower* réflecteur *m*
reflux: ~ **ratio** *n* WATER SUPP taux de reflux *m*, taux de retour *m*
reforestation *n* AIR reboisement *m*
reform[1] *n* POLICY réforme *f*
reform[2] *vt* POLICY réformer
reformed: ~ **gas** *n* AIR gaz de craquage *m*, gaz reformé *m*
reforming: ~ **reaction** *n* AIR réaction de reformage *f*
refractive: ~ **index** *n* ANALYSIS *for detector*, WATER SUPP indice de réfraction *m*
refractory: ~ **substance** *n* WATER POLL substance réfractaire *f*
refrigerant *n* AIR *containing CFCs* réfrigérant *m*, WATER SUPP frigorigène *m*, réfrigérant *m*
refueling *AmE see refuelling BrE*
refuelling *n BrE* MARINE remplissage de réservoir *m*
refurbish *vt* GEN réhabiliter

refurbished *adj* GEN réhabilité
refurbishment *n* GEN réhabilitation *f*
refuse *n* WASTE déchets *m pl*, ordures *f pl*, détritus *m pl*; ~ **bunker** WASTE réservoir à ordures *m*, silo à ordures *m*; ~ **can** WASTE poubelle *f*, boîte à ordures *f*; ~ **cell** WASTE compartiment *m*; ~ **chute** WASTE vide-ordures *m*; ~ **collection** WASTE enlèvement *m*, ramassage des ordures *m*; ~ **collection vehicle** WASTE benne à ordures *f*; ~ **collection service** WASTE service de ramassage des ordures *m*; ~ **collector** WASTE camion de collecte des ordures *m*; ~ **composition** WASTE composition des ordures *f*; ~ **compost** WASTE compost d'ordures *m*; ~ **composting** WASTE compostage des ordures *m*; ~ **crusher** WASTE broyeur à ordures *m*; ~ **disposal** MANAGE, POLICY, WASTE élimination des déchets *f*, élimination des ordures *f*; ~ **disposal site** WASTE site d'élimination des ordures *m*; ~ **extractor** WASTE extracteur à rejets *m*, trémie de décharge des remblais *f*; ~ **glass** WASTE débris de verre *m pl*, calcin *m*; ~ **incineration** WASTE, WATER SUPP incinération de déchets *f*, incinération des ordures *f*; **incineration plant** WASTE usine d'incinération des ordures *f*; ~ **incinerator** WASTE incinérateur d'ordures *m*, incinérateur de déchets *m*, incinérateur à ordures *m*; ~ **lorry** *BrE* (*cf refuse truck AmE*) WASTE camion de collecte des ordures *m*; ~ **mass** WASTE masse d'ordures *f*; ~ **sack** WASTE sac à ordures *m*, sac poubelle *m*; ~ **sack collection** WASTE collecte de sacs poubelles *f*; ~ **separation plant** WASTE usine de tri des déchets *f*; ~ **truck** *AmE* (*cf refuse lorry BrE*) WASTE camion de collecte des ordures *m*; ~ **volume** WASTE volume d'ordures *m*; ~ **worm** WASTE vis des rejets *f*
regenerable: ~ **flue-gas desulfurization** *AmE*, ~ **flue-gas desulphurization** *BrE n* AIR désulfuration des effluents gazeux régénérables *f*
regenerant *n* GEN réactif de régénération *m*
regenerate *vt* GEN régénérer; ~ **microbiologically** AIR régénérer par voie microbiologique
regenerated[1]: ~ **activated carbon** *n* AIR charbon actif régénéré *m*; ~ **activated coke** *n* AIR coke actif régénéré *m*; ~ **adsorbent** *n* AIR adsorbant régénéré *m*; ~ **solvent** *n* AIR solvant régénéré *m*
regenerated[2] *adj* GEN régénéré
regeneration *n* GEN régénération *f*; ~ **basin** WATER POLL bassin de régénération *m*; ~ **capacity** WATER POLL capacité de régénération *f*; ~ **pit** SOIL puits de régénération *m*
regenerative[1] *adj* GEN régénérant
regenerative[2]: ~ **air preheater** *n* AIR réchauffeur d'air de régénération *m*; ~ **gas** *n* AIR gaz de régénération *m*; ~ **gas fan** *n* AIR ventilateur de gaz de régénération *m*; ~ **gas preheater** *n* AIR réchauffeur de gaz de régénération *m*; ~ **heat exchanger** *n* AIR échangeur de chaleur régénérée *m*; ~ **heating technology** *n* RENEW ENERG technique de recyclage de chaleur *f*; ~ **incineration** *n* AIR *exhaust gas* incinération avec régénération *f*; ~ **pump** *n* WATER POLL pompe de régénération *f*; ~ **reagent** *n* AIR réactif de régénération *m*
regenerator *n* AIR régénérateur *m*
regional: ~ **growth centre** *n* POLICY pôle d'équilibre *m*; ~ **policy** *n* POLICY mesure régionale *f*
register[1]: ~ **of hazardous substances** *n* AIR, POLICY, SAFETY inventaire des substances dangereuses *m*, WATER SUPP registre des substances dangereuses *m*
register[2] *vt* AIR, ANALYSIS, POLICY, WATER SUPP enregistrer
registration *n* AIR, ANALYSIS, POLICY, WATER SUPP déclaration *f*, enregistrement *m*, immatriculation *f*
regranulate *vt* WASTE regranuler
regranulating: ~ **extruder** *n* WASTE extrudeuse de regranulation *f*, poste d'extrusion-granulation *m*
regranulation *n* WASTE regranulation *f*
regression: ~ **curve** *n* ANALYSIS courbe de régression *f*
regressive[1] *adj* POLICY dégressif
regressive[2]: ~ **tax** *n* POLICY impôt dégressif *m*
regrind *vt* WATER SUPP *waste* rebroyer
regrinding *n* WATER SUPP *of waste* broyage secondaire *m*
reground: ~ **material** *n* WASTE matériau rebroyé *m*, matériau remeulé *m*
regrouping: ~ **of lands** *n* POLICY remembrement rural *m*
regularization *n* WATER SUPP régularisation *f*
regulated: ~ **deposition** *n* WATER SUPP dépôt réglementé *m*
regulating: ~ **unit** *n* AIR poste de réglage *m*, poste de régulation *m*, SOIL, WATER SUPP poste de régulation *m*
regulation *n* ANALYSIS régulation *f*, POLICY *legal* règlementation *f*, WATER SUPP *control* réglage *m*, régulation *f*, *rule* réglement *m*; ~ **in case of accident** SAFETY règlementation en cas d'accident *f*
regulator *n* ANALYSIS régulateur *m*
regulatory: ~ **tax** *n* POLICY impôt réglementaire *m*, taxe réglementaire *f*
rehabilitate *vt* SOIL assainir
rehabilitation *n* SOIL assainissement *m*
reheat *vt* GEN réchauffer
reheating *n* GEN réchauffage *m*
reinforce *vt* WATER POLL, WATER SUPP armer, raidir, renforcer
reinforced: ~-**concrete pipe** *n* WATER POLL, WATER SUPP conduite en béton armé *f*; ~ **fiberglass** *AmE*, ~ **fibreglass** *BrE n* AIR, NOISE, WASTE, WATER SUPP fibre de verre renforcée *f*; ~ **plastics** *n pl* WATER SUPP plastique armé *m*; ~ **reaction injection molding** *AmE*, ~ **reaction injection moulding** *BrE n* (*R-RIM*) WASTE moulage par injection-réaction en moule fermé *m*
reinforcing: ~ **material** *n* WASTE *plastics reprocessing* matériau de renfort *m*
relative: ~ **density** *n* WATER SUPP densité relative *f*; ~ **force** *n* AIR *of particle to drop* force relative *f*; ~ **humidity** *n* (*rH*) ANALYSIS humidité relative *f*
relay: ~ **back** *vt* SAFETY rapatrier
release[1] *n* AIR émission *f*, *of harmful substances* rejet *m*, NOISE décharge *f*, dégagement *m*, libération *f*; ~ **of pollutants** AIR rejet de polluants *m*
release[2] *vt* AIR émettre, NOISE, SOIL, WATER POLL *harmful substance* dégager, libérer
released: ~ **heat** *n* AIR, WATER POLL dégagement de chaleur *m*
reliability *n* GEN fiabilité *f*
relief: ~ **well** *n* WATER SUPP puits de décharge *m*, puits de décompression *m*
relocatable: ~ **plant** *n* SOIL installation relocalisable *f*; ~ **unit** *n* SOIL poste relocalisable *m*
relocation *n* SOIL réimplantation *f*
remaining: ~ **dust removal** *n* AIR élimination de la poussière résiduelle *f*, décolmatage *m*; ~ **risk** *n* MANAGE risque restant *m*
remediate *vt* SOIL assainir
remediation *n* AIR, NOISE, SAFETY réhabilitation *f*, SOIL assainissement *m*, WATER SUPP réhabilitation *f*; ~ **need** SOIL besoin d'assainissement *m*; ~ **rate** SOIL taux d'assainissement *m*; ~ **target value** SOIL objectif d'assainissement *m*
remelt *vt* WATER SUPP refondre
remobilization *n* WATER SUPP remise en mouvement *f*
remolding *AmE see remoulding BrE*
remote: ~-**control system** *n* WATER POLL, WATER SUPP système de télécommande *m*; ~-**control unit** *n* WATER POLL, WATER SUPP poste de télécommande *m*; ~ **laser measuring** *n* ANALYSIS télémesure par laser *f*; ~ **sensing** *n* ANALYSIS télédétection *f*, télémesure *f*, MARINE télédétection *f*
remoulding *n BrE* WASTE *pneus* rechappage *m*
removal *n* GEN extraction *f*, élimination *f*, évacuation *f*, entraînement *m*; ~ **by suction** WATER POLL élimination par aspiration *f*; ~ **of contaminated soils** SOIL évacuation de sols contaminés *f*; ~ **of dissolved organic substances** WATER SUPP extraction des substances organiques dissoutes

f; ~ **efficiency** AIR rendement de décolmatage *m*, WATER POLL taux d'élimination *m*; ~ **of oil by separators** WATER SUPP extraction d'huile par séparation *f*, épuration *f*; ~ **operation** AIR, WATER POLL opération d'extraction *f*; ~ **of organic matter** AIR, WATER SUPP élimination de matière organique *f*; ~ **of pollutants** AIR, WATER SUPP élimination de polluants *f*; ~ **quality** AIR qualité de décolmatage *f*; ~ **of sludge** WATER POLL déschlammage *m*; ~ **of suspended solids by sedimentation** AIR extraction de particules solides en suspension par décantation *f*, WATER SUPP extraction de matières solides en suspension par décantation *f*

remove *vt* GEN évacuer, *particles from gas* extraire, *solid material* éliminer, *take away* emporter, *suppress* supprimer; ~ **dust from** WATER SUPP dépoussiérer; ~ **nitrogen from** GEN dénitrurer

removed *adj* GEN extrait

render: ~ **inert** *vt* WATER POLL *radioactive waste* inertiser

renew *vt* AIR, RENEW ENERG remettre à neuf, renouveler

renewable[1] *adj* AIR, RENEW ENERG renouvelable

renewable[2]: ~ **energy** *n* RENEW ENERG énergie renouvelable *f*; ~-**energy form** *n* RENEW ENERG forme d'énergie renouvelable *f*; ~-**energy source** *n* RENEW ENERG source d'énergie renouvelable *f*; ~-**energy supply** *n* RENEW ENERG fourniture d'énergie renouvelable *f*; ~-**energy supply system** *n* RENEW ENERG réseau d'alimentation en énergie renouvelable *m*; ~-**energy technology** *n* RENEW ENERG technologie mettant en œuvre une énergie renouvelable *f*; ~ **raw material** *n pl* RENEW ENERG matière première renouvelable *f*; ~ **resources** *n pl* RENEW ENERG ressources renouvelables *f pl*; ~ **technology** *n* RENEW ENERG technologie renouvelable *f*

reoxidation *n* AIR réoxydation *f*, WATER SUPP réoxydation *f*, réoxygénation *f*

reparability *n* WATER SUPP *of landfill* réparabilité *f*

repeal[1] *n* AIR, POLICY, WATER SUPP abrogation *f*, POLICY, WATER SUPP révocation *f*, retrait *m*; ~ **of authorization** POLICY retrait d'autorisation *m*; ~ **of licence** *BrE*, ~ **of license** *AmE* POLICY retrait de licence *m*, suspension d'une licence *f*, WATER SUPP retrait d'un permis *m*; ~ **of permit** POLICY retrait de permis *m*

repeal[2] *vt* POLICY abroger, annuler

repeatability *n* ANALYSIS répétabilité *f*

repeated: ~ **measurement** *n* ANALYSIS mesure récursive *f*

repellency: ~ **effect** *n* AIR effet répulsif *m*

replastification *n* WATER SUPP *recycled asphalt* replastification *f*

replenish *vt* WATER SUPP faire le plein de, nourrir

replotting *n* POLICY remembrement *m*

reporting *n* MANAGE déclaration *f*, notification *f*

repository: ~ **for radioactive waste** *n* WASTE, WATER SUPP dépôt de déchets radioactifs *m*, cimetière de déchets radioactifs *m*

reprocess *vt* WASTE, WATER SUPP retraiter

reprocessed *adj* WASTE, WATER SUPP retraité

reprocessing *n* WASTE, WATER SUPP retraitement; ~ **costs** *n pl* WATER SUPP coûts de retraitement *m pl*; ~ **technology** WATER SUPP technologie de retraitement *f*

reproducibility *n* ANALYSIS reproductibilité *f*

repulpable: ~ **adhesive** *n* WASTE adhésif pouvant être refondu *m*

repulping: ~ **equipment** *n* WASTE matériel de désintégration *m*

repulsive: ~ **power** *n* WATER POLL, WATER SUPP *colloidal particle* force de répulsion *f*

requiring: ~ **authorization** *adj* POLICY soumis à autorisation; ~ **a licence** *adj* POLICY subordonné à l'octroi d'une licence

rescrubber *n* AIR tour de relavage *f*

reservation: ~ **of permission** *n* POLICY suspension d'autorisation *f*, suspension de permis *f*

reservoir *n* AIR réservoir *m*, RENEW ENERG, WATER POLL, WATER SUPP bassin *m*, réservoir de retenue d'eau *m*, réservoir *m*, retenue *f*, bassin de retenue *m*; ~ **cleaning** WATER SUPP curage de réservoir *m*; ~ **of greenhouse gases** AIR réservoir de gaz à effet de serre *m*; ~ **lining** WATER POLL revêtement du bassin de retenue *m*, WATER SUPP revêtement de retenue *m*; ~ **water** WATER SUPP eau de réservoir *f*, eau de retenue *f*

residence: ~ **time** *n* AIR, WATER SUPP temps de séjour *m*; ~ **time of solids** *n* WATER SUPP temps de séjour des matières solides *m*; ~ **time of waste** *n* WATER SUPP temps de séjour des déchets *m*

residual: ~ **activity** *n* AIR *catalyst* activité résiduelle *f*; ~ **dust content** *n* AIR teneur en poussière résiduelle *f*; ~ **emissions** *n pl* WATER POLL émissions résiduelles *f*; ~ **gas** *n* AIR gaz résiduel *m*, WATER SUPP gaz résiduaire *m*, gaz résiduel *m*; ~ **impact of soils with contaminants** *n* SOIL effet résiduel de sols contaminés *m*; ~ **moisture** *n* ANALYSIS, WATER SUPP humidité résiduelle *f*; ~ **noise** *n* NOISE bruit résiduel *m*; ~ **oil** *n* WATER SUPP huile résiduelle *f*; ~ **oxygen** *n* AIR oxygène résiduel *m*; ~ **ozone** *n* AIR ozone résiduel *m*; ~ **pollution** *n* WATER POLL, WATER SUPP pollution résiduelle *f*; ~ **solvent removal** *n* SOIL élimination des solvants résiduels *f*; ~ **sound** *n* NOISE son résiduel *m*; ~ **thermal energy** *n* RENEW ENERG énergie thermique résiduelle *f*

residue[1]: ~-**free** *adj* AIR *spray absorption process* sans résidu, WATER SUPP sans déchet

residue[2] *n* GEN résidu *m pl*, WATER SUPP résidu de raffinage *m*, débris *m pl*; ~ **avoidance at source** WATER SUPP lutte contre la production de déchets à la source *f*; ~ **concentration** WATER SUPP concentration de déchets *f*; ~ **concentration of pollutants** AIR concentration de résidus de polluants *f*; ~ **containing pollutants** WATER SUPP résidu contenant des polluants *m*; ~ **disposal** WASTE, WATER POLL, WATER SUPP élimination des déchets *f*, évacuation des déchets *f*; ~-**dust loading** AIR charge de poussière en résidus *f*; ~ **emissions** *n pl* AIR, WATER SUPP émissions résiduaires *f pl*; ~-**free production method** WASTE méthode de production sans déchets *f*; ~ **from evaporation** ANALYSIS résidu d'évaporation *m*; ~ **landfill** WATER SUPP décharge de déchets *f*; ~ **landfilling** WASTE mise en décharge de déchets *f*; ~ **load** AIR charge résiduelle *f*; ~ **on ignition** ANALYSIS résidu d'allumage *m*; ~ **raw material** RENEW ENERG *biomass conversion base material*, WASTE matière première résiduelle *f*; ~ **stream** WATER SUPP flux de déchets *m*; ~ **yield** WATER SUPP rendement de déchets *m*

residues *n pl* GEN résidus *m pl*, WATER POLL, WATER SUPP refus *m pl*

resilient[1] *adj* NOISE, SAFETY élastique, résilient

resilient[2]: ~ **isolator** *n* SAFETY isolateur élastique *m*; ~ **layer** *n* NOISE couche élastique *f*, couche résiliente *f*; ~ **mounting** *n* NOISE *of machines* montage résilient *m*

resin *n* WATER SUPP résine *f*; ~ **transfer molding** *AmE*, ~ **transfer moulding** *BrE* (*RTM*) WASTE moulage par pressage basse pression avec injection de résine *m*

resistance *n* GEN résistance *f*; ~ **capacity** ANALYSIS capacité de résistance *f*; ~ **kiln** ANALYSIS *AOX determination* four à résistance *m*; ~ **thermometer** AIR, ANALYSIS thermomètre à résistance *m*; ~ **to diseases** SOIL résistance aux maladies *f*; ~ **to oxidation** WATER SUPP *recycled materials* résistance à l'oxydation *f*; ~ **to pests** SOIL résistance aux parasites *f*; ~ **to pollutants** AIR, SOIL résistance aux polluants *f*; ~ **to tearing** MARINE résistance à la déchirure *f*

resistant *adj* GEN résistant

resistivity *n* AIR *dust* résistivité *f*

resolution *n* ANALYSIS résolution *f*

resonance *n* MARINE, NOISE résonance *f*; ~ **absorber** NOISE absorbeur de résonance *m*; ~ **absorption** NOISE absorption de résonance *f*; ~-**absorption muffler** *AmE* (*cf resonance-absorption silencer BrE*) NOISE silencieux absorbant les résonances *m*; ~-**absorption silencer** *BrE* (*cf resonance-*

absorption muffler AmE) NOISE silencieux absorbant les résonances *m*; ~ **screen** WATER SUPP crible à résonance *m*; ~ **transducer** NOISE transducteur acoustique à résonance *m*
resonant: ~ **ion** *n* ANALYSIS *quadrupole mass analyser* ion résonant *m*
resonator *n* NOISE résonateur *m*
resource: ~ **conservation** *n* WATER SUPP protection des ressources *f*, économies de ressources *f pl*; ~ **consumption** *n* WATER SUPP consommation des ressources *f*; ~ **recovery** *n* WATER SUPP récupération de ressources *f*; ~ **recovery plant** *n* WATER SUPP usine de récupération *f*
resources *n pl* GEN ressources *f pl*; ~ **survey** *n* MANAGE inventaire des ressources *m*; ~ **use** *n* POLICY utilisation des ressources *f*
respirable[1] *adj* AIR, SAFETY breathable, respirable
respirable[2]: ~ **residues** *n pl* SAFETY résidus respirables *m pl*
respirator *n* SAFETY masque à gaz *m*, masque respiratoire *m*
respiratory: ~ **equipment** *n* SAFETY matériel respiratoire *m*; ~ **filter** *n* SAFETY filtre respiratoire *m*; ~ **mask** *n* SAFETY masque respiratoire *m*; ~-**protection apparatus** *n* SAFETY appareil de protection respiratoire *m*; ~-**protection system** *n* SAFETY système de protection respiratoire *m*
responsible: ~ **authority** *n* POLICY autorité compétente *f*
restabilization *n* WATER POLL, WATER SUPP restabilisation *f*
restabilize *vt* WATER POLL, WATER SUPP restabiliser
restabilized[1] *adj* WATER POLL, WATER SUPP restabilisé
restabilized[2]: ~ **particle** *n* WATER POLL particule restabilisée *f*
restore *vt* SOIL restaurer
resurgent *adj* WATER SUPP résurgent
resuspension *n* WATER SUPP remise en suspension *f*
retailer *n* MANAGE détaillant *m*
retained: ~ **water** *n* WATER POLL, WATER SUPP eau capturée *f*, eau de rétention *f*
retaining: ~ **grate** *n* AIR grille de retenue *f*; ~ **wall** *n* NOISE mur de soutien *m*; ~ **well** *n* SOIL puits de retenue *m*
retarder *n* WATER SUPP ralentisseur *m*, retardateur *m*
retarding: ~ **agent** *n* WATER SUPP agent retardateur *m*
retention: ~ **cage** *n* AIR *fluidized-bed absorber* cage de rétention *f*; ~ **capacity** *n* WASTE capacité de rétention *f*; ~ **tank** *n* WATER POLL bassin de rétention *m*; ~ **time** *n* AIR, WATER POLL temps de rétention *m*; ~ **wall** *n* SOIL rideau étanche *m*
retentivity *n* AIR aimantation rémanente *f*, persistance *f*
retitration *n* ANALYSIS retitrage *m*
retreaded: ~ **tire** *AmE*, ~ **tyre** *BrE* *n* WATER SUPP pneu rechapé *m*
retreat *vt* AIR retraiter
retrofit *vt* GEN moderniser, réhabiliter, rénover, remettre en état, réadapter
retrofitted *adj* GEN réhabilité
retrofitting *n* GEN modernisation *f*, réhabilitation *f*, rénovation *f*, remise en état *f*, réadaptation *f*
return *n* WASTE renvoi *m*, reprise *f*, retour *m*; ~ **rate** WASTE taux de reprise *m*, taux de restitution *m*; ~ **sludge** WATER POLL boues de recyclage *f pl*, boues de retour *f pl*; ~-**sludge elevator** WATER POLL dispositif de reprise des boues *m*, ligne de soutirage des boues *f*; ~ **sludge pipe** WATER POLL conduite de boues de recyclage *f*
returnable: ~ **bottle** *n* WASTE bouteille consignée *f*; ~ **container** *n* WASTE récipient consigné *m*; ~ **pack** *n* WASTE emballage consigné *m*; ~ **packaging** *n* WASTE emballage consigné *m*, emballage repris *m*; ~ **pallet** *n* WASTE palette réutilisable *f*
returns *n pl* WASTE *industrial engineering* jets et débris de coulée *m pl*; ~ **granulate** *n* WASTE granulat de jets et débris de coulée *m*
reusability *n* WASTE possibilité de réutilisation *f*
reusable: ~ **box** *n* WATER SUPP boîte réutilisable *f*; ~ **dust** *n* AIR poussière réutilisable *f*; ~ **gypsum** *n* AIR *product of flue-gas cleaning* gypse réutilisable *m*; ~ **packaging** *n* WASTE emballage réutilisable *m*; ~ **waste products** *n pl* AIR, WASTE déchets réutilisables *m pl*
reuse[1] *n* AIR, POLICY *of collected dust* réutilisation *f*, réutiliser, WASTE réemploi *m*, valorisation *f*, WATER SUPP réemploi *m*, réutilisation *f*; ~ **of derelict land** POLICY réemploi des terres abandonnées *m*; ~ **of waste** RENEW ENERG, WASTE réutilisation des déchets *f*
reuse[2] *vt* WASTE, WATER SUPP réemployer, réutiliser
revegetate *vi* SOIL remettre en végétation, restaurer le couvert végétal
revenue[1]: ~-**neutral** *adj* POLICY fiscalement neutre
revenue[2] *n* POLICY *from taxes* recettes fiscales *f pl*; ~ **from eco-taxes** POLICY recettes perçues au titre des éco-taxes *f pl*; ~ **neutrality** POLICY neutralité fiscale *f*; ~ **source** POLICY source de revenus *f*
reverberant *adj* NOISE réverbérant
reverberation *n* NOISE réverbération *f*; ~ **chamber** NOISE chambre de réverbération *f*; ~ **time** NOISE temps de réverbération *m*
reverse: ~ **air flow filter** *n* AIR filtre à air à contre-courant *m*; ~ **filtration** *n* WATER SUPP filtration inverse *f*; ~ **flow** *n* AIR contre-courant *m*; ~-**flow cleaning** *n* AIR, WATER SUPP lavage à contre-courant *m*, nettoyage à contre-courant *m*; ~-**flow duct** *n* AIR *bag filter* conduite à contre-courant *f*; ~-**flow filter** *n* WATER POLL filtre à contre-courant *m*; ~-**flow packed-bed filter** *n* AIR filtre à garnissage à contre-courant *m*; ~-**flow precipitator** *n* AIR électrofiltre à contre-courant *m*, dépoussiéreur à contre-courant *m*; ~-**flow scrubber** *n* AIR laveur à contre-courant *m*; ~-**jet filter** *n* AIR filtre à contre-courant *m*; ~ **osmosis** *n* (*RO*) WATER POLL, WATER SUPP osmose inverse *f*; ~ **osmosis with concentrate evaporation** *n* WATER POLL, WATER SUPP osmose inverse avec évaporation des concentrats *f*; ~ **osmosis plant** *n* WATER SUPP banc d'osmose inverse *m*
reversible[1] *adj* AIR, WATER POLL réversible
reversible[2]: ~ **process** *n* AIR, WATER POLL *chemisorption* processus réversible *m*
revitalization *n* SOIL revitalisation *f*
revolving: ~ **drum** *n* SOIL tambour rotatif *m*, WATER SUPP tambour *m*; ~ **filter** *n* AIR *nuclear technology* filtre à tambour *m*
Reynolds: ~ **number** *n* (*Re*) AIR, WATER SUPP nombre de Reynolds *m* (*Re*)
rH[1] *abbr* (*relative humidity*) ANALYSIS humidité relative *f*
rH[2]: ~-**value** *n* ANALYSIS valeur de l'humidité relative *f*
Rh *abbr* (*rhodium*) AIR Rh (*rhodium*)
rheometer *n* ANALYSIS rhéomètre *m*
rhodium *n* (*Rh*) AIR rhodium *m* (*Rh*)
rich: ~-**gas processing** *n* AIR traitement des gaz riches *m*; ~-**gas tank** *n* AIR cuve de gaz riches *f*
ridge: ~-**and-furrow technique** *n* WATER POLL *land application of wastewater* technique du radier à dents de scie *f*
right: ~-**to-know principle** *n* POLICY *environmental information* principe du droit d'accès à l'information *m*
rigid: ~-**plate absorber** *n* NOISE absorbeur à plateaux rigides *m*; ~ **PVC** *n* WATER SUPP PVC rigide *m*
rill: ~ **erosion** *n* SOIL érosion en rigoles *f*
ring: ~ **adsorber** *n* AIR *fixed bed adsorber, layered in rings* adsorbeur à anneau *m*; ~-**jet aerosol precipitator** *n* AIR séparateur d'aérosol à col sonique *m*; ~-**jet element** *n* AIR élément à col sonique *m*; ~ **of tubings** *n* WATER SUPP bague de tuyauterie *f*
ringed: ~ **layer** *n* AIR couche annelée *f*
Ringelmann: ~ **method** *n* AIR méthode Ringelmann *f*; ~ **number** *n* AIR nombre de Ringelmann *m*; ~ **scale** *n* AIR, ANALYSIS *dust measurement* échelle de Ringelmann *f*
rinse *vt* SOIL, WASTE *bottle* rincer
rinsing: ~ **PMT case** *n* AIR cuve de rinçage des particules *f*; ~ **water** *n* SOIL, WATER SUPP eau de rinçage *f*; ~ **water processing** *n* SOIL traitement de l'eau de rinçage *m*

riparian *adj* WATER SUPP riverain
ripening *n* SOIL, WASTE maturation *f*, WATER POLL *sludge conditioning* mûrissement *m*, maturation *f*
rising: ~ **energy prices** *n* RENEW ENERG augmentation du prix de l'énergie *f*; ~ **flow rate** *n* WATER POLL débit croissant *m*; ~ **of pH-value** *n* WATER POLL augmentation de la valeur du pH *f*; ~ **of water level** *n* WATER SUPP élévation du niveau d'eau *f*
risk *n* GEN risque *m*; ~ **analysis** POLICY *Seveso directive*, WATER SUPP analyse de risque *f*; ~ **assessment** MANAGE, MARINE, POLICY, SAFETY, SOIL *contaminated sites* évaluation des risques *f*; ~ **of clogging** AIR risque d'encrassement *m*, risque de colmatage *m*, WATER POLL risque d'empaillassonnage *m*, risque de bouchage *m*, WATER SUPP risque de colmatage *m*; ~ **estimation** MANAGE *eco-audit* estimation des risques *f*; ~ **factor** MANAGE facteur de risque *m*; ~ **of iron clogging** WATER SUPP risque de colmatage par le fer *m*; ~ **limit** SAFETY limite de risque *f*, seuil de risque *m*; ~ **management** MANAGE, SAFETY gestion des risques *f*; ~ **mitigation** POLICY prise en compte du risque *f*; ~ **of plugging** AIR risque de bouchage *m*; ~ **potential** MANAGE potentiel de risques *m*
river *n* WATER SUPP rivière *f*, fleuve *m*, cours d'eau *m*; ~ **authority** WATER SUPP agence de bassin *f*; ~ **basin** WATER SUPP bassin fluvial *m*, bassin hydrographique *m*, bassin versant *m*; ~ **basin authority** POLICY agence de bassin *f*; ~ **basin management plan** POLICY projet d'aménagement d'un bassin fluvial *m*; ~ **bed** WATER POLL lit de rivière *m*, WATER SUPP lit *m*; ~ **discharge** WATER POLL déversement des eaux fluviales *m*; ~ **floor** WATER POLL fond de rivière *m*; ~ **pollution** WATER SUPP pollution des cours d'eau *f*; ~ **tapping** WATER SUPP captage de rivière *m*; ~ **training** MANAGE canalisation *f*, WATER SUPP régularisation d'un cours d'eau *f*; ~ **wall** WATER SUPP bajoyer *m*; ~ **water** WATER POLL, WATER SUPP eau de rivière *f*, eau fluviale *f*; ~-**water purification** WATER POLL, WATER SUPP épuration des eaux fluviales *f*; ~ **water quality** WATER SUPP qualité des eaux fluviales *f*; ~ **water tapping** WATER SUPP captage d'eaux fluviales *m*; ~ **works** WATER SUPP aménagement des cours d'eau *m*
RO *abbr* (*reverse osmosis*) WATER POLL, WATER SUPP osmose inverse *f*
road: ~ **construction rubble** *n* WASTE gravier de construction routière *m*; ~ **pricing** *n* POLICY tarification routière *f*; ~-**sweeping lorry** *n* *BrE* (*cf road-sweeping truck AmE*) WATER SUPP balayeuse *f*; ~ **sweepings** *n pl* WASTE détritus ramassés sur la voie publique *m pl*; ~-**sweeping truck** *n* *AmE* (*cf road-sweeping lorry BrE*) WATER SUPP balayeuse *f*; ~ **system** *n* POLICY voirie *f*; ~ **tanker** *n* MARINE camion citerne *m*; ~ **transport** *n* WASTE transport routier *m*
robustness *n* WATER SUPP robustesse *f*
rock: ~ **bed** *n* WATER POLL lit de pierre *m*, WATER SUPP banc rocheux *m*; ~ **wool** *n* NOISE laine de roche *f*
rocket: ~ **probe** *n* ANALYSIS fusée sonde *f*
rocking: ~ **grate** *n* WATER SUPP grille basculante *f*
roll: ~ **body** *n* SOIL cylindre *m*; ~ **crusher** *n* WASTE concasseur à cylindre *m*; ~-**out waste container** *n* WASTE conteneur de déchets amovible *m*, conteneur de déchets de grande capacité *m*
roller: ~ **die extruder** *n* WASTE, WATER SUPP extrudeuse à vis sans fin *f*
rolling[1] *adj* WATER SUPP roulant
rolling[2]: ~ **fraction** *n* WASTE fraction roulante *f*; ~ **grate** *n* WATER SUPP grille roulante *f*; ~-**grate installation** *n* WATER SUPP installation à grilles roulantes *f*
roofing *n* AIR *of filter* toiture *f*
room: ~ **acoustics** *n pl* NOISE acoustique des locaux *f*; ~ **lining** *n* NOISE insonorisation d'une pièce *f*, revêtement d'une pièce *m*, *insulation* insonorisation des locaux *f*; ~ **temperature** *n* ANALYSIS température ambiante *f*
rope: ~ **skimmer** *n* MARINE écrémeur à corde flottante *m*, récupérateur à corde flottante *m*
rose *n* WATER SUPP crépine *f*, crépine d'aspiration *f*, grenouillère *f*, lanterne *f*
rot *n* WASTE pourriture *f*, WATER POLL, WATER SUPP pourriture *f*, putréfaction *f*; ~ **landfill** WATER SUPP décharge en décomposition *f*
rotameter *n* ANALYSIS débitmètre à flotteur *m*, gyromètre *m*, spiromètre *m*; ~ **viscosimeter** ANALYSIS viscosimètre rotatif *m*
rotary: ~ **air heater** *n* AIR réchauffeur d'air rotatif *m*; ~ **atomization** *n* AIR *spray absorption* pulvérisation tournante *f*; ~-**current frequency** *n* RENEW ENERG *wind turbine* fréquence de courant triphasé *f*; ~ **distributor** *n* WATER POLL *trickling filtration* distributeur rotatif *m*; ~ **dryer** *n* SOIL tambour de séchage *m*; ~ **drum furnace** *n* WATER POLL *incineration* four à tambour rotatif *m*; ~ **evaporator** *n* WATER POLL, WATER SUPP évaporateur rotatif *m*; ~ **filter** *n* WATER POLL, WATER SUPP filtre à tambour *m*, filtre rotatif *m*; ~ **furnace** *n* AIR, SOIL, WATER POLL, WATER SUPP four rotatif *m*; ~ **kiln** *n* AIR, SOIL four rotatif *m*, WATER POLL four rotatif *m*, four sécheur rotatif *m*, WATER SUPP four rotatif *m*; ~-**kiln incinerator** *n* AIR incinérateur à four rotatif *m*; ~-**piston compressor** *n* AIR compresseur à piston rotatif *m*; ~ **pump** *n* WATER POLL, WATER SUPP pompe rotative *f*; ~ **vacuum filter** *n* WATER POLL filtre à tambour sous vide *m*; ~ **valve** *n* WASTE écluse rotative *f*
rotatable: ~ **arm** *n* MARINE bras tournant *m*
rotating[1] *adj* AIR rotatif
rotating[2]: ~ **adsorber** *n* AIR adsorbeur rotatif *m*; ~ **biological contactor** *n* (*RBC*) WATER POLL disque biologique *m*; ~ **built-in structure** *n* AIR *scrubbing area* structure rotative intégrée *f*; ~-**disk biological contactor** *n* (*RBC*) SOIL colonne biologique à disques rotatifs *f*; ~ **gas preheater** *n* AIR préchauffeur d'air rotatif *m*; ~ **sprayer** *n* WATER SUPP *sewage* arroseur rotatif *m*; ~ **velocity** *n* WATER SUPP vitesse de rotation *f*
rotation: ~ **movement** *n* AIR *cyclone* mouvement rotatif *m*; ~ **velocity** *n* AIR vitesse de rotation *f*
rotor *n* AIR, RENEW ENERG *wind turbine* rotor *m*; ~ **blade** RENEW ENERG pale de rotor *f*; ~ **brake** RENEW ENERG frein de rotor *m*; ~ **hub** RENEW ENERG arbre de rotor *m*; ~ **hub with adjustable blade mechanism** RENEW ENERG arbre de rotor équipé d'un mécanisme de réglage des pales *m*; ~ **with fixed number of revolutions** RENEW ENERG rotor à nombre de tours fixe *m*; ~ **with variable number of revolutions** RENEW ENERG rotor à nombre de tours variable *m*
rotting: ~ **body** *n* WASTE masse de pourrissage *f*, masse en décomposition *f*; ~ **chamber** *n* WASTE chambre de pourrissage *f*; ~-**chamber composting process** *n* WASTE procédé de compostage en chambre de pourrissage *m*; ~ **container** *n* WASTE cuve de pourrissage *m*; ~ **hall** *n* WASTE chambre de pourrissage *f*; ~ **operation** *n* WASTE opération de pourrissage *f*; ~ **process** *n* WASTE procédé de pourrissage *m*; ~ **standstill** *n* WASTE arrêt de pourrissage *m*; ~ **time** *n* WASTE temps de pourrissage *m*; ~ **tower** *n* WASTE tour de pourrissage *f*
roughage[1]: ~-**free** *adj* WATER SUPP sans matières indigestibles
roughage[2] *n* WATER SUPP matière indigestible *f*
roughing: ~ **tank** *n* WATER POLL dégrossisseur *m*
round: ~-**bottomed flask** *n* ANALYSIS *in gas density determination* ballon rond *m*
routine: ~ **analysis** *n* ANALYSIS analyse de routine *f*
R-RIM *abbr* (*reinforced reaction unjection moulding BrE*, *reinforced reaction injection molding AmE*) WASTE moulage par injection-réaction en moule fermé *m*
RTM *abbr* (*resin transfer moulding BrE*, *resin transfer molding AmE*) WASTE moulage par pressage basse pression avec injection de résine *m*
Ru *abbr* (*ruthenium*) AIR Ru (*ruthénium*)

rubber *n* WASTE caoutchouc *m*; ~ **waste** WASTE déchets de caoutchouc *m pl*
rubbish *n* AIR, WASTE débris; ~ **bag** *BrE* (*cf garbage bag AmE*) WASTE sac à ordures *m*, sac poubelle *m*; ~ **bin** *BrE* (*cf garbage can AmE, trash can AmE*) WASTE boîte à ordures *f*, poubelle *f*
rubble[1]: ~-**filled** *adj* WATER SUPP comblé par du gravier
rubble[2] *n* AIR *ground filter*, WASTE, WATER POLL gravier *m*, WATER SUPP éboulis *m*, pierraille *f*
run *vi* WATER POLL, WATER SUPP *into sea* affluer, couler, déboucher; ~ **aground** MARINE échouer; ~ **dry** WATER SUPP s'assécher, tarir; ~ **off** WATER SUPP *precipitation* ruisseler, *water from tank* couler, se vider; ~ **out** WATER SUPP s'écouler; ~ **over** WATER SUPP déborder
running: ~ **water** *n* WATER SUPP eau courante *f*, eau vive *f*
runoff *n* AIR *rainwater*, MARINE ruissellement *m*, WATER SUPP eau de ruissellement *f*, écoulement *m*, ruissellement *m*; ~ **capacity** WATER SUPP capacité de ruissellement *f*; ~ **coefficient** WATER POLL coefficient de ruissellement *m*, WATER SUPP coefficient d'écoulement *m*; ~ **cycle** WATER SUPP cycle d'écoulement *m*; ~ **duration** WATER POLL, WATER SUPP durée d'écoulement *f*, durée de ruissellement *f*
rupture *n* AIR, WATER SUPP *filter bag* rupture *f*; ~ **of floc** WATER POLL rupture du floc *f*
rural: ~ **development** *n* POLICY aménagement rural *m*; ~ **environment** *n* MANAGE, POLICY milieu rural *m*; ~ **migration** *n* POLICY exode rural *m*; ~ **water supply** *n* WATER SUPP fourniture d'eau en milieu rural *f*, distribution d'eau en milieu rural *f*
rust[1]: ~-**preventive** *adj* WATER SUPP antirouille; ~-**resistant** *adj* WATER SUPP antirouille
rust[2]: ~ **protection** *n* WATER SUPP protection contre la rouille *f*, protection antirouille *f*
ruthenium *n* (*Ru*) AIR ruthénium *m* (*Ru*)

S

S *abbr* (*sulfur AmE, sulphur BrE*) SOIL, WATER POLL S (*soufre*)
sabin *n* ANALYSIS, MARINE, WATER SUPP *unit of sound absorption* sabin *m*
safe[1] *adj* SAFETY sûr; **~ against leaching** SOIL protégé contre le lessivage, WATER SUPP protégé de la lixiviation
safe[2]: **~ disposal** *n* SAFETY, WATER SUPP élimination inoffensive *f*, élimination en toute sécurité *f*; **~ operation dependability** *n* SAFETY sûreté d'exploitation *f*
safeguard *n* SAFETY, SOIL mesure de sûreté *f*
safekeeping *n* SAFETY garantie *f*, WATER SUPP sauvegarde *f*
safety[1]: **~-related** *adj* SAFETY sécuritaire
safety[2] *n* SAFETY sécurité *f*, sûreté *f*; **~ analysis** SAFETY analyse de sûreté *f*; **~ check** SAFETY contrôle de sécurité *m*; **~ code** SAFETY code de sécurité *m*; **~ complaints** *n pl* SAFETY plaintes relatives à la sécurité *f pl*; **~ engineering** SAFETY génie de la sécurité *m*; **~ factor** SAFETY facteur de sécurité *m*; **~ helmet** NOISE, SAFETY casque de protection *m*; **~ measure** SAFETY mesure de sécurité *f*; **~ regulations** *n pl* SAFETY réglementation de sécurité *f*; **~-related equipment** SAFETY équipement de sécurité *m*; **~ requirements** *n pl* SAFETY critères de sécurité *m pl*; **~ risk** SAFETY risque pour la sécurité *m*; **~ standard** SAFETY norme de sécurité *f*; **~ strategy** SAFETY stratégie de sécurité *f*, stratégie de sûreté *f*; **~ valve** WATER SUPP soupape de sûreté *f*; **~ of the work place** SAFETY sécurité du travail *f*; **~ zone** SAFETY zone de sûreté *f*
salination *n* WATER SUPP salinisation *f*
saline[1] *adj* WATER SUPP salin, saumâtre, salant
saline[2]: **~ concentration** *n* RENEW ENERG *for solar pond* concentration saline *f*; **~ spring** *n* WATER SUPP source salée *f*, source saline *f*, source saumâtre *f*; **~ water** *n* WATER SUPP eau saumâtre *f*, eau salée *f*; **~-water conversion** *n* WATER SUPP dessalage *m*, dessalement *m*
salinity *n* WATER SUPP salinité *f*
salt: **~ aerosol** *n* AIR aérosol salin *m*; **~ slag** *n* WASTE, WATER SUPP *from conventional aluminium recycling* scories salines *f pl*; **~ water** *n* WATER SUPP eau salée *f*, eau saumâtre *f*; **~-water plant** *n* WATER SUPP usine de dessalement d'eau de mer *f*
sample[1] *n* ANALYSIS échantillon *m*; **~ amount** ANALYSIS volume d'échantillon *m*; **~ analysis** ANALYSIS analyse d'échantillon *f*, analyse sur échantillon *f*; **~ assessment** ANALYSIS évaluation d'échantillon *f*; **~ calibration** ANALYSIS calibrage d'échantillon *m*; **~ collection** ANALYSIS collecte d'échantillons *f*; **~ components** *n pl* ANALYSIS éléments constitutifs d'échantillon *m pl*, éléments d'échantillonnage *m pl*; **~ composition** ANALYSIS composition d'échantillon *f*; **~ container** ANALYSIS bac d'échantillonnage *m*, conteneur d'échantillonnage *m*; **~ diminution** ANALYSIS réduction d'échantillon *f*; **~ dispenser** ANALYSIS distributeur d'échantillon *m*; **~ dosing** ANALYSIS dosage d'échantillon *m*; **~ gas** ANALYSIS gaz échantillon *m*; **~ gas capillary** ANALYSIS capillaire pour prélèvement gazeux *m*; **~ gas filter** ANALYSIS filtre pour échantillon gazeux *m*; **~ holder** ANALYSIS *for holding test tube* porte-échantillons *m*, porte-éprouvette *m*; **~ injector** ANALYSIS injecteur d'échantillon *m*; **~ introduction** ANALYSIS introduction d'échantillon *f*; **~ point** SOIL point de prélèvement *m*; **~ processing** ANALYSIS préparation d'échantillons *f*, traitement d'échantillons *m*; **~ pump** ANALYSIS pompe d'échantillonnage *f*; **~ record** ANALYSIS registre d'échantillonnage *m*; **~ size** ANALYSIS *statistically* taille d'échantillon *f*; **~ storage** ANALYSIS stockage d'échantillon *m*; **~ taking** ANALYSIS prélèvement d'échantillon *m*; **~ throughput**, **~ thruput** *AmE* ANALYSIS rendement d'échantillonnage *m*; **~ treatment** ANALYSIS traitement d'échantillon *m*; **~ volume** ANALYSIS volume d'échantillon *m*
sample[2] *vt* ANALYSIS échantillonner, prélever un échantillon
sampler *n* ANALYSIS échantillonneur *m*; **~ for flow-proportional samples** ANALYSIS échantillonneur à prélèvement proportionnel au débit *m*
sampling *n* ANALYSIS échantillonnage *m*, prélèvement *m*, prélèvement d'échantillon *m*; **~ aid** ANALYSIS accessoire d'échantillonnage *m*, adjuvant d'échantillonnage *m*; **~ bottle** ANALYSIS flacon d'échantillonnage *m*; **~ capacity** ANALYSIS capacité d'échantillonnage *f*; **~ cell** ANALYSIS cellule d'échantillonnage *f*; **~ condensation** ANALYSIS condensation d'échantillonnage *f*; **~ device** ANALYSIS dispositif de prélèvement d'échantillon *m*; **~ frequency** ANALYSIS fréquence d'échantillonnage *f*; **~ gas** ANALYSIS gaz d'échantillonnage *m*; **~ grid** SOIL grille d'échantillonnage *f*; **~ instrument** ANALYSIS instrument d'échantillonnage *m*; **~ line** AIR ligne de prélèvement *f*, ANALYSIS conduite d'échantillonnage *f*, ligne de prélèvement *f*; **~ method** ANALYSIS méthode d'échantillonnage *f*; **~ plan** ANALYSIS campagne d'échantillonnage *f*, plan d'échantillonnage *m*; **~ point** ANALYSIS point d'échantillonnage *m*; **~ preparation** ANALYSIS préparation d'échantillonnage *f*; **~ probe** ANALYSIS sonde de prélèvement *f*; **~ pump** ANALYSIS pompe d'échantillonnage *f*; **~ station** ANALYSIS *for water control* point de prélèvement *m*; **~ system** ANALYSIS système d'échantillonnage *m*; **~ transmission** ANALYSIS transmission de prélèvement *f*; **~ tube** ANALYSIS canne d'échantillonnage *f*, tube de prélèvement *m*; **~ unit** ANALYSIS unité d'échantillonnage *f*; **~ vessel** ANALYSIS cuve d'échantillonnage *f*
sand *n* WASTE, WATER SUPP sable *m*; **~ deposition** WATER SUPP dépôt de sable *m*; **~ filter** WATER POLL, WATER SUPP filtre à sable *m*; **~ filtration ditch** WATER POLL rigole de filtration du sable *f*; **~ layer** WATER SUPP *in filtering tank* couche de sable *f*; **~ trap** ANALYSIS échantillonneur de sable *m*, WATER POLL échantillonneur de sable *m*, piège à sable *m*
sandpit *n* WATER POLL sablière *f*
sandwich: **~ construction** *n* NOISE construction en stratifié isolant *f*; **~ injection molding** *AmE*, **~ injection moulding** *BrE* *n* WASTE moulage par injection en couches alternées *m*, moulage par injection stratifiée *m*; **~ panel** *n* NOISE *vibration damping* panneau stratifié *m*; **~ structure** *n* RENEW ENERG structure stratifiée *f*
sanitary[1] *adj* SAFETY, WATER SUPP hygiénique, sanitaire
sanitary[2]: **~ conditions** *n pl* SAFETY salubrité *f*; **~ engineering** *n* MANAGE, POLICY génie sanitaire *m*, WATER SUPP technique sanitaire *f*; **~ landfill** *n* WASTE décharge contrôlée *f*, enfouissement sanitaire *m*; **~ landfilling** *n* WASTE mise en décharge contrôlée *f*; **~ wastewater** *n* WATER POLL eaux usées sanitaires *f pl*
sanitation *n* SAFETY hygiène publique *f*, WATER SUPP *science* hygiène publique *f*, *system* système sanitaire *m*
saturate *vt* GEN saturer
saturated[1] *adj* GEN saturé
saturated[2]: **~ activated carbon** *n* WATER POLL charbon actif saturé *m*; **~ gas** *n* AIR, ANALYSIS gaz saturé *m*; **~ hydrocarbon** *n* AIR hydrocarbure saturé *m*; **~ resin** *n*

WATER POLL *ion exchanger* résine saturée *f*; ~ **soil zone** *n* SOIL zone de sol saturé *f*; ~ **zone** *n* WATER SUPP zone saturée *f*
saturation *n* GEN saturation *f*; ~ **concentration** WATER SUPP charge de saturation *f*; ~ **deficit** AIR insuffisance de saturation *f*; ~ **index** WATER SUPP indice de saturation *m*; ~ **level** WATER SUPP niveau de saturation *m*; ~ **load** AIR *adsorption* charge de saturation *f*; ~ **pressure** AIR pression de saturation *f*; ~ **temperature** AIR température de saturation *f*
save *vt* RENEW ENERG économiser; ~ **energy** MANAGE, RENEW ENERG, WASTE économiser l'énergie
Savonius: ~ **rotor** *n* RENEW ENERG *vertical axle* rotor Savonius *m*
sawdust *n* RENEW ENERG sciure *f*
Sb *abbr* (*antimony*) WATER SUPP Sb (*antimoine*)
SBR *abbr* (*sequencing batch reactor*) SOIL réacteur discontinu séquentiel *m*, WATER POLL réacteur séquentiel à alimentation discontinue *m*, réacteur séquentiel par cuvées *m*
SC *abbr* (*standard conditions*) ANALYSIS conditions normales *f pl*
scale[1]: ~-**like** *adj* AIR écailleux
scale[2] *n* WATER SUPP calamine *f*, incrustation *f*, tartre *m*; ~ **formation** WATER SUPP entartrage *m*; ~-**forming properties** *n pl* WATER POLL caractère entartrant *m*; ~ **model** ANALYSIS maquette *f*
scale[3] *vi* WATER SUPP entartrer; ~ **up** WATER SUPP entartrer
scales *n pl* ANALYSIS balance *f*, WATER SUPP déchets *m pl*
scaling *n* AIR *formation of scale* calaminage *m*, *marking with units of measurement* graduation *f*, WATER SUPP incrustation *f*; ~ **risk** AIR risque de calaminage *m*
scanner *n* ANALYSIS analyseur à balayage *m*, tomodensitomètre *m*
scanning: ~ **electron microscope** *n* (*SEM*) ANALYSIS microscope à balayage électronique *m*; ~ **electron microscopy** *n* (*SEM*) ANALYSIS microscopie à balayage électronique *f*; ~ **method** *n* ANALYSIS mode de balayage *m*; ~ **tunnel microscope** *n* ANALYSIS microscope à effet tunnel *m*
scarce[1] *adj* AIR, POLICY, RENEW ENERG, SOIL rare, peu abondant
scarce[2]: ~ **environmental resources** *n pl* AIR, RENEW ENERG ressources rares dans l'environnement *f pl*, SOIL ressources du milieu peu abondantes *f pl*
scatter *vt* ANALYSIS diffuser
scattering *n* ANALYSIS diffusion *f*, dispersion *f*; ~ **cloud telemetering** ANALYSIS télémesure de dispersion de nuages *f*
scavenge *vt* AIR *gas* balayer, décrasser, expulser, WASTE balayer
scavenger *n* AIR agent d'épuration *m*, décrassant *m*, SOIL agent d'épuration *m*, nécrophage *m*, WASTE détritiphage *m*
scavenging *n* AIR balayage *m*, WASTE balayage *m*, enlèvement des ordures *m*; ~ **gas** AIR gaz de balayage *m*
SCC *abbr* (*secondary combustion chamber*) WATER SUPP chambre de combustion secondaire *f*
scheduled: ~ **activity** *n* POLICY activité datée *f*, activité programmée *f*
scheduling: ~ **control** *n* MANAGE maîtrise des délais *f*
schistosomiasis *n* WATER POLL bilharziose *f*
scintillation: ~ **counter** *n* ANALYSIS scintillomètre *m*
scoop *n* MARINE écope *f*, pelle *f*, WASTE benne de ramassage *f*, WATER SUPP écope *f*, pelle *f*, puisette *f*
scooper *n* ANALYSIS gouge *f*
scour[1] *n* WATER SUPP *of dam* curage *m*, dégorgement *m*, *wash of water* décapage *m*, récurage *m*
scour[2] *vt* SOIL, WASTE, WATER POLL lessiver, WATER SUPP affouiller, curer, dégorger, lessiver, décaper, récurer, rincer
scouring *n* SOIL, WASTE, WATER POLL lessivage *m*, WATER SUPP chasse *f*, curage *m*, décapage *m*, dégraissage *m*, dérochage *m*, lessivage *m*; ~ **outlet** WATER SUPP sortie de vidange *f*
scowing *n* WATER SUPP décolmatage *m*
SCR *abbr* (*selective catalytic reduction*) AIR, WATER SUPP réduction catalytique sélective *f*
scrap *n* WASTE déchets *m pl*, *metal* ferraille *f*, débris *m pl*; ~ **application** WASTE application de ferraille *f*; ~ **bale** WASTE balle de ferraille *f*; ~ **baler** WASTE presse à ferraille *f*; ~ **baling press** WASTE presse à ferraille *f*; ~ **collection** WASTE collecte de ferraille *f*, ramassage de ferraille *m*; ~ **extractor** WASTE extracteur de ferrailles *m*; ~ **iron** WASTE ferraille *f*, mitraille *f*, riblons *m pl*; ~ **lead** WASTE vieux plomb *m*, *residue* résidu de plomb *m*; ~ **metal** WASTE ferraille *f*, résidus métalliques *m pl*; ~ **metal dealer** WASTE ferrailleur *m*; ~ **metal merchant** WASTE ferrailleur *m*, négociant en ferrailles *m*; ~ **metal separation** WASTE extraction des ferrailles *f*; ~ **motorcar** WASTE carcasse de voiture *f*; ~ **press** WASTE presse à ferraille *f*; ~ **processing** WASTE transformation des ferrailles *f*; ~ **quality** WASTE qualité de ferraille *f*; ~ **reuse** WASTE réemploi de ferraille *m*; ~ **shear** WASTE cisaillage de ferraille *m*; ~ **smelting** WASTE fonte de ferraille *f*, fusion de ferraille *f*; ~ **sorting** WASTE tri de ferraille *m*; ~ **yard** WASTE parc à ferrailles *m*
scrape *vt* MARINE, WATER POLL curer, racler
scraper *n* AIR, MARINE, WATER POLL racleur *m*, WATER SUPP cureuse *f*, racleuse *f*, racleur *m*; ~ **bridge** WATER POLL pont racleur *m*
scraping *n* MARINE curage *m*; ~ **operation** WATER POLL opération de raclage *f*
scrapped: ~ **plant** *n* WASTE friche industrielle *f*
scrapping: ~ **fee** *n* POLICY *automotive* taxe sur la ferraille *f*
screen[1] *n* GEN crible *m*, grille *f*, NOISE écran *m*; ~ **analysis** ANALYSIS analyse granulométrique *f*; ~-**changing device** WATER SUPP dispositif de changement de grille *m*; ~ **cleaner** WATER POLL dégrilleur *m*; ~ **cleaning** WATER POLL dégrillage *m*; ~ **gutter** WATER POLL rigole de grille *f*; ~ **loss** WATER POLL perte de dégrillage *f*; ~ **loss of head** WATER POLL perte de charge de grille *f*; ~ **overflow** WATER SUPP débordement de tamis *m*; ~ **underflow** WATER SUPP sous-écoulement de tamis *m*
screen[2] *vt* GEN cribler, filtrer, *remove screen* dégriller
screening *n* GEN criblage *m*, filtrage *m*, *removal of screen* dégrillage; ~ **equipment** WATER SUPP matériel de tamisage *m*; ~ **machine** AIR crible *m*, grille *f*, tamis *m*; ~ **magnet** WATER SUPP crible magnétique *m*; ~ **test** ANALYSIS essai de criblage *m*; ~ **wall** NOISE paroi isolante *f*, paroi tamisante *f*
screenings *n pl* WATER POLL, WATER SUPP détritus de criblage *m pl*, refus de dégrillage *m pl*, refus de grille *m pl*; ~ **bale press** *n* WATER POLL compacteur de refus de dégrillage *m*
screw: ~ **compressor** *n* WATER SUPP compresseur à vis *m*; ~ **conveyer** *n* SOIL vis transporteuse *f*; ~ **conveyor** *n* AIR vis sans fin *f*, WASTE convoyeur à vis sans fin *m*, WATER POLL transporteur à vis sans fin *m*, WATER SUPP convoyeur à vis sans fin *m*, transporteur à vis *m*; ~ **extruder** *n* WASTE, WATER SUPP extrudeuse à vis *f*; ~ **feeder** *n* WATER SUPP extracteur à vis *m*; ~ **pump** *n* MARINE, WATER POLL, WATER SUPP pompe à vis *f*; ~-**type pump** *n* MARINE, WATER POLL, WATER SUPP pompe à vis *f*, vis de relèvement *f*
scrub *vt* AIR *gases* épurer, frotter, *remove dust* dépoussiérer, GEN laver, WATER SUPP épurer
scrubbed: ~ **gas** *n* AIR gaz dépoussiéré *m*, WATER SUPP gaz lavé *m*
scrubber *n* AIR dépoussiéreur *m*; ~ **clogging** AIR colmatage de dépoussiéreur *m*; ~ **plugging** AIR bouchage de dépoussiéreur *m*
scrubbing *n* AIR épuration *f*, lavage *m*, SOIL, WASTE lavage *m*, WATER SUPP lavage *m*, épuration *f*; ~-**area design** AIR conception de la zone de lavage *f*; ~ **liquid droplet** AIR gouttelette de lavage *f*, vésicule de lavage *f*; ~ **liquid jet** AIR aspersion de liquide de lavage *f*; ~ **liquid tank** AIR réservoir de liquide de lavage *m*; ~ **liquor** WATER SUPP liqueur de lavage *f*; ~ **stage** AIR stade de lavage *m*; ~ **tower** AIR tour de

débenzolage *f*, tour de lavage *f*; ~ **vane** AIR robinet de lavage *m*; ~ **zone** AIR zone de lavage *f*

SCS *abbr* (*stationary container system*) WASTE, WATER SUPP conteneur fixe *m*

scum *n* MARINE, WATER POLL écume *f*, chapeau *m*, mousse *f*; ~ **baffle** WATER POLL pare-écume *m*; ~ **board** WATER POLL pare-écume *m*; ~ **formation** WATER POLL formation d'écume *f*; ~ **removal** WATER POLL raclage du chapeau *m*; ~ **remover** WATER POLL pare-écume *m*; ~ **scraper** WATER POLL pelle à écumes *f*, racleur d'écumes *m*; ~ **skimmer** MARINE, WATER POLL, WATER SUPP écrémeur *m*, écumeuse *f*, écumoire *f*

SDA *abbr* (*spray-dry absorber*) AIR absorbeur à pulvérisation par voie sèche *m*, absorbeur de séchage par pulvérisation *m*

SDI *abbr* (*sludge density index*) WATER POLL, WATER SUPP IVB (*indice de volume des boues*)

sea[1]: ~**-based** *adj* MARINE en mer, marin

sea[2] *n* MARINE mer *f*; ~**-based pollution** MARINE pollution marine *f*; ~ **chart** MARINE carte marine *f*; ~ **level** MARINE niveau de la mer *m*; ~ **water intrusion** MARINE pénétration d'eau de mer *f*; ~ **water recovery** WATER SUPP récupération d'eau de mer *f*

seal *vt* GEN sceller, *plug* colmater, *make hermetic* étancher, *waterproof* imperméabiliser, *block off* boucher

sealed[1] *adj* GEN scellé, *plugged* colmaté, *hermetically* étanche, hermétique, *waterproof* imperméabilisé, *blocked off* bouché

sealed[2]: ~ **ceiling** *n* NOISE plafond étanche *m*, plafond hermétique *m*; ~ **wall** *n* NOISE paroi étanche *f*

sealing *n* GEN scellement *m*, *waterproofing* imperméabilisation *f*, *of harmful substances* confinement *m*, *blocking off* bouchage *m*, *plugging* colmatage *m*; ~ **layer** WATER SUPP couche d'étanchéité *f*; ~ **liquid** ANALYSIS liquide d'étanchéité *m*; ~ **measure** WATER SUPP mesure d'étanchéité *f*; ~ **properties** *n pl* WATER SUPP propriétés étanches *f pl*; ~ **sheeting** SOIL revêtement d'imperméabilisation *m*

SEALOSAFE®: ~ **method** *n* SOIL *British solidification process* méthode SEALOSAFE® *f*; ~ **process** *n* WATER SUPP procédé SEALOSAFE® *m*

seasonal: ~ **heat storage** *n* RENEW ENERG *solar energy* stockage de chaleur saisonnier *m*

seasonalized: ~ **demand** *n* MANAGE demande saisonnalisée *f*

seawater *n* MARINE, RENEW ENERG, WATER SUPP eau de mer *f*

seaweed *n* MARINE algue *f*, algues *f pl*, algues marines *f pl*, goémon *m*, varech *m*, RENEW ENERG varech *m*, WATER SUPP algue *f*

seaworthiness *n* MARINE navigabilité en haute mer *f*

secondary[1] *adj* GEN secondaire

secondary[2]: ~ **activated-sludge tank** *n* WATER POLL cuve des boues activées secondaires *f*; ~ **adsorber** *n* AIR adsorbeur secondaire *m*; ~ **adsorption** *n* AIR, WATER POLL *of polymer in flocculation* adsorption secondaire *f*; ~ **air** *n* AIR, WATER POLL air secondaire *m*; ~**-air injection** *n* AIR *catalyst* injection d'air secondaire *f*; ~ **aluminium** *n BrE*, ~ **aluminum** *n AmE* WATER SUPP aluminium de récupération *m*; ~ **catalyst** *n* AIR catalyseur secondaire *m*; ~ **combustion chamber** *n* (*SCC*) WATER SUPP chambre de combustion secondaire *f*; ~ **cosmic radiation** *n* AIR rayonnement cosmique secondaire *m*; ~ **digestion tank** *n* WATER POLL cuve de digestion secondaire *f*; ~ **downpipe** *n* AIR *in packed-bed filter* descente secondaire *f*; ~ **energy source** *n* RENEW ENERG *of electricity* source d'énergie secondaire *f*; ~ **fiber** *AmE*, ~ **fibre** *BrE n* WATER SUPP fibre de récupération *f*; ~ **ion mass spectroscopy** *n* (*SIMS*) ANALYSIS spectroscopie de masse par émission ionique secondaire *f*; ~ **loading** *n* AIR *of waste gas* charge secondaire *f*; ~ **material** *n* WASTE matériau de récupération *m*, matériau secondaire *m*; ~ **packaging** *n* WASTE emballage de récupération *m*, emballage secondaire *m*; ~ **particle separator** *n* AIR séparateur de particules secondaires *m*; ~ **polishing pond** *n* WATER POLL bassin de finition *m*; ~ **pond** *n* WATER POLL bassin secondaire *m*; ~ **precipitator** *n* AIR dépoussiéreur secondaire *m*; ~ **raw material** *n* RENEW ENERG, WASTE matière première de récupération *f*; ~ **reaction time** *n* WATER POLL temps de réaction secondaire *m*; ~ **recycling** *n* WASTE recyclage secondaire *m*; ~**-resource container** *n* WASTE conteneur de matériaux de récupération *m*; ~ **resource content** *n* WATER SUPP teneur en matériaux de récupération *f*; ~ **resource fraction** *n* WATER SUPP fraction de matériaux de récupération *f*; ~ **resource quality** *n* WATER SUPP qualité de matériaux de récupération *f*; ~ **resources bale** *n* WATER SUPP balle de matériaux de récupération *f*; ~ **settlement** *n* SOIL tassement secondaire *m*, WATER SUPP décantation secondaire *f*; ~ **settler** *n* WATER POLL décanteur secondaire *m*; bassin de décantation finale *m*; ~ **settling tank** *n* WATER POLL décanteur secondaire *m*; ~ **sieving** *n* WATER SUPP tamisage secondaire *m*; ~ **sludge** *n* WATER POLL boues secondaires *f pl*; ~ **sound** *n* NOISE son secondaire *m*; ~ **treatment** *n* AIR traitement secondaire *m*; ~ **trickling filter** *n* WATER POLL lit bactérien secondaire *m*; ~ **waste avoidance** *n* WATER SUPP lutte contre la production de déchets secondaires *f*; ~**-waste avoidance** *n* WASTE mesure préventive contre la production de déchets secondaires *f*; ~ **waste-gas treatment** *n* AIR traitement secondaire des effluents gazeux *m*; ~ **wastewater treatment** *n* WATER POLL *biological* traitement secondaire des eaux résiduaires *m*

secretariat *n* POLICY *administrative unit* secrétariat *m*

sector: ~ **sampler** *n* ANALYSIS échantillonneur sectoriel *m*

sectoral: ~ **approach** *n* POLICY approche sectorielle *f*; ~ **authority** *n* POLICY organe compétent pour un secteur *m*; ~ **division** *n* POLICY répartition sectorielle *f*

security *n* SAFETY sécurité *f*, sûreté *f*; ~ **measure** SAFETY, mesure de sécurité *f*; ~ **system** SAFETY système de sécurité *m*

sediment[1] *n* MARINE, WATER POLL, WATER SUPP boues décantées *f pl*, dépôt *m*, sédiment *m*

sediment[2] *vt* AIR décanter, déposer, sédimenter, MARINE, WATER POLL, WATER SUPP sédimenter

sediment[3] *vi* AIR, WATER POLL se déposer

sedimentable[1] *adj* AIR décantable, sédimentable, MARINE, WATER POLL, WATER SUPP sédimentable

sedimentable[2]: ~ **material** *n* AIR matériau sédimentable *m*

sedimentation *n* GEN sédimentation *f*, décantation *f*, décantation statique *f*; ~ **curve** WATER POLL courbe de sédimentation *f*; ~ **glass** ANALYSIS vase de décantation *m*; ~ **pond** WATER SUPP bassin de décantation *m*; ~ **tank** WATER POLL bassin de sédimentation *m*

sedimented[1] *adj* AIR, WATER POLL, WATER SUPP décanté, déposé

sedimented[2]: ~ **parameter** *n* WATER POLL paramètres de décantation *m pl*; ~ **sludge** *n* WATER POLL boues décantées *f pl*

seeding *n* ANALYSIS ensemencement *m*; ~ **sludge** WATER POLL boues d'ensemencement *f pl*

seep *vi* GEN s'infiltrer, suinter

seepage *n* GEN fuite *f*, infiltration *f*; ~ **water** WASTE, WATER SUPP eau d'infiltration *f*

seeping *n* WATER POLL suintement *m*

Seger: ~ **cone** *n* ANALYSIS cône Seger *m*, montre Seger *f*

segment: ~ **baffle** *n* AIR chicane transversale *f*

segregate *vt* WATER SUPP séparer

segregation *n* WATER SUPP ségrégation *f*, *alloys* séparation *f*; ~ **berm** WATER SUPP berme de séparation *f*

select *vt* ANALYSIS sélectionner

selective[1] *adj* WATER POLL sélectif; ~ **catalytic** AIR catalytique sélectif

selective[2]: ~ **catalytic reduction** *n* (*SCR*) AIR, WATER SUPP réduction catalytique sélective *f*; ~ **catalytic reduction reactor** *n* AIR réacteur de réduction catalytique sélective *m*; ~ **collection** *n* WASTE collecte sélective *f*; ~ **filter** *n* ANALYSIS *of photometer* filtre sélectif *m*; ~ **ion-exchange resin** *n* WATER SUPP résine échangeuse d'ions sélective *f*;

~ **non-catalytic reduction** *n* (*SNCR*) WATER SUPP réduction sélective non catalytique *f*; ~ **refuse collection** *n* WASTE collecte sélective des déchets *f*

self[1]: ~**-cleaning** *adj* WATER POLL autonettoyant; ~**-draining** *adj* SOIL à drainage naturel

self[2]: ~**-adjusting floating weir** *n* MARINE barrage flottant autoréglable *m*, déversoir flottant autoréglable *m*; ~**-cleaning** *n* WATER POLL autonettoyage *m*; ~**-cleaning centrifugal screen** *n* WATER POLL grille centrifuge autonettoyante *f*; ~**-cleaning rotary drum screen** *n* WATER POLL dégrilleur rotatif autonettoyant *m*; ~**-compaction** *n* WATER SUPP compactage automatique *m*; ~**-composting** *n* WATER SUPP compostage automatique *m*; ~**-monitoring** *n* POLICY autosurveillance *f*, surveillance automatique *f*; ~**-priming centrifugal pump** *n* WATER POLL, WATER SUPP pompe centrifuge à amorçage automatique *f*; ~**-propelled skimmer** *n* MARINE écrémeur autopropulsé *m*, récupérateur autopropulsé *m*; ~**-purification** *n* SOIL, WATER POLL, WATER SUPP auto-épuration *f*; ~**-purification force** *n* SOIL, WATER SUPP force d'auto-épuration *f*; ~**-purifying capacity** *n* WATER POLL, WATER SUPP auto-épuration naturelle *f*; ~**-restoring capacity** *n* WATER POLL capacité de régénération *f*

SEM *abbr* ANALYSIS (*scanning electron microscopy*) microscopie à balayage électronique *f*, (*scanning electron microscope*) microscope à balayage électronique *m*

semi-arid: ~ **zone** *n* SOIL région semi-aride *f*

semiconductor *n* RENEW ENERG *of solar cell* semiconducteur *m*; ~ **component** RENEW ENERG *of solar cell* élément semiconducteur *m*, composant semiconducteur *m*; ~ **detector** ANALYSIS *measuring solid state ionization* détecteur à semiconducteur *m*

semidry: ~ **scrubber** *n* AIR épurateur semi-sec *m*

semi-finish: ~ **grinding** *n* WASTE broyage semi-fin *m*

semi-impermeable: ~ **layer** *n* WATER SUPP couche semi-perméable *f*

semipermeable[1] *adj* WATER POLL, WATER SUPP semi-perméable

semipermeable[2]: ~ **membrane** *n* WATER POLL, WATER SUPP membrane semi-perméable *f*

semiportable *adj* ANALYSIS semi-mobile

semirotary: ~ **pump** *n* WATER SUPP pompe demi-rotative *f*, pompe semi-rotative *f*

semisolid[1] *adj* WATER SUPP semi-solide

semisolid[2]: ~ **combustible waste** *n* WATER SUPP déchets combustibles semi-solides *m pl*; ~ **hazardous waste** *n* WASTE, WATER POLL, WATER SUPP déchets dangereux semi-solides *m pl*

semitrailer *n* WATER SUPP *waste transport* semi-remorque *f*

semivolatile: ~ **chlorinated hydrocarbon** *n* AIR, SOIL hydrocarbure chloré semi-volatil *m*; ~ **organic compounds** *n* (*SVOC*) SOIL composés organiques semi-volatils *m pl*

semi-wet: ~ **sorting** *n* WATER SUPP tri semi-humide *m*

sense *vt* ANALYSIS palper, sonder

sensing: ~ **device** *n* ANALYSIS, MARINE capteur *m*

sensitive *adj* AIR, ANALYSIS sensible

sensitivity *n* ANALYSIS *of detector* sensibilité *f*; ~ **to noise** NOISE sensibilité au bruit *f*

sensor *n* ANALYSIS, MARINE capteur *m*; ~ **module** ANALYSIS capteur modulaire *m*

sensory: ~ **adaptation** *n* ANALYSIS adaptation sensorielle *f*; ~ **analysis** *n* ANALYSIS analyse sensorielle *f*; ~ **fatigue** *n* ANALYSIS fatigue sensorielle *f*

separability: ~ **test** *n* ANALYSIS essai de séparabilité *m*

separable[1] *adj* WATER POLL, WATER SUPP séparable

separable[2]: ~ **roughage** *n* WATER SUPP matière indigestible séparable *f*

separate[1] *adj* WATER POLL à part, indépendant, séparé

separate[2]: ~ **canalization** *n* WATER POLL conduite distincte *f*; ~ **collection** *n* WATER SUPP collecte distincte *f*, collecte sélective *f*; ~ **collection by type** *n* WASTE collecte par type *f*; ~ **composting of dewatered sludges** *n* WATER POLL compostage distinct des boues déshydratées *m*; ~ **dewatering** *n* WATER POLL dénoyage à part *m*, déshydratation séparée *f*; ~ **disposal** *n* WATER SUPP élimination à part *f*, élimination distincte *f*; ~ **refuse collection** *n* WASTE collecte sélective des déchets *f*; ~ **sewerage system** *n* WATER POLL réseau séparatif *m*, WATER SUPP réseau d'assainissement séparatif *m*; ~ **treatment** *n* WATER POLL traitement à part *m*, traitement différencié *m*

separate[3] *vt* AIR, SOIL, WATER POLL, WATER SUPP *fluids* séparer, *solids* trier; ~ **by color** *AmE*, ~ **by colour** *BrE* WATER SUPP *glass* trier par couleur; ~ **electrolytically** WATER SUPP *dissolved metal in battery* séparer par voie électrolytique

separating: ~ **capacity** *n* ANALYSIS capacité de séparation *f*; ~ **cut** *n* SOIL excavation de séparation *f*; ~ **efficiency** *n* AIR rendement de séparation *m*; ~ **funnel** *n* ANALYSIS entonnoir de décantation *m*, entonnoir de séparation *m*; ~ **liquid** *n* ANALYSIS liquide de départ *m*

separation *n* WATER SUPP élimination des contaminants *f*; ~ **by color** *AmE*, ~ **by colour** *BrE* WATER SUPP *of recycled glass* tri par couleur *m*; ~ **capacity** AIR capacité de séparation *f*; ~ **effect** AIR effet de séparation *m*; ~ **layer** WATER SUPP couche de séparation *f*; ~ **process** AIR procédé de séparation *m*; ~ **properties** *n pl* AIR caractéristiques de séparation *f pl*; ~ **rate** WATER SUPP vitesse de séparation *f*; ~ **solvent** SOIL solvant de séparation *m*; ~ **stages** *n pl* WATER SUPP étages de séparation *m pl*; ~ **technology** WATER SUPP technique de séparation *f*

separator *n* AIR, MARINE, WASTE, WATER POLL séparateur *m*, WATER SUPP séparateur *m*, piège *m*; ~ **of oily waters** MARINE séparateur hydrocarbures/eau *m*

septic[1] *adj* WATER POLL, WATER SUPP septique

septic[2]: ~ **tank** *n* WATER POLL, WATER SUPP fosse septique *f*

sequencing: ~ **batch reactor** *n* (*SBR*) SOIL réacteur discontinu séquentiel *m*, WATER POLL réacteur séquentiel à alimentation discontinue *m*, réacteur séquentiel par cuvées *m*

sequential[1] *adj* WATER POLL séquentiel

sequential[2]: ~ **analyser** *BrE*, ~ **analyzer** *AmE n* ANALYSIS analyseur séquentiel *m*; ~ **dosage** *n* WATER POLL dosage séquentiel *m*

service: ~ **load** *n* WATER POLL, WATER SUPP charge utile *f*; ~ **mains** *n pl* WATER SUPP secteur *m*; ~ **water** *n* RENEW ENERG *of heat pump* eau motrice *f*, eau utile *f*

serviceable: ~ **life** *n* SAFETY *of machinery* durée de vie utile *f*

set[1] *n* ANALYSIS, WATER POLL, WATER SUPP batterie *f*, jeu *m*

set[2] *vt* ANALYSIS *net* caler

setting *n* ANALYSIS prise *f*, RENEW ENERG, SOIL implantation *f*, WATER SUPP prise *f*; ~ **time** WATER SUPP temps de prise *m*

settle *vi* WATER POLL, WATER SUPP se décanter, se déposer; ~ **out** WATER POLL, WATER SUPP se déposer

settleability *n* WATER POLL, WATER SUPP décantabilité *f*

settleable[1] *adj* WATER POLL, WATER SUPP décantable, sédimentable

settleable[2]: ~ **macrofloc** *n* WATER POLL macrofloc décantable *m*; ~ **sludge** *n* WATER POLL boues décantables *f pl*; ~ **solid** *n* WATER POLL matière solide décantable *f*

settled: ~ **apparent density** *n* ANALYSIS masse volumique apparente *f*; ~ **sludge** *n* WATER POLL boues décantées *f pl*; ~ **solids** *n* WATER POLL solides décantés *m pl*; ~ **wastewater** *n* WATER POLL eaux résiduaires décantées *f pl*; ~ **water** *n* WASTE, WATER POLL eaux usées décantées *f pl*

settlement *n* WASTE tassement *m*, WATER SUPP décantation *f*; ~ **speed** WATER SUPP rythme d'implantation *m*, vitesse d'établissement *f*; ~ **stoppage** SOIL arrêt du tassement *m*, WATER SUPP arrêt d'implantation *m*, cessation d'implantation *f*

settling *n* WATER POLL décantation *f*, WATER SUPP décantation primaire *f*, décantation *f*; ~ **basin** WATER POLL, WATER SUPP *for sewage* bassin de curage *m*, bassin de décantation *m*, bassin de tranquillisation *m*, fosse de sédimentation *f*; ~ **chamber** AIR chambre de décantation *f*, chambre de sédimentation *f*, WATER POLL, WATER SUPP chambre de décantation *f*, chambre de tranquillisation *f*; ~ **characteristics** *n pl* WATER SUPP caractéristiques de décantation *f pl*, propriétés de décantation *f pl*; ~ **conditions** *n pl* WATER POLL conditions de décantation *f pl*; ~ **lagoon** WATER POLL lagune de décantation *f*; ~ **material** WATER POLL *water-borne sediments* matière décantée *f*, produit de décantation *m*; ~ **out** WATER POLL séparation par sédimentation *f*; ~ **properties** *n pl* WATER SUPP caractéristiques de décantation *f pl*, propriétés de décantation *f pl*; ~ **speed** AIR vitesse de décantation *f*; ~ **tank** AIR *biological gas scrubbing* bassin de décantation *m*, WATER POLL bassin de décantation *m*, WATER SUPP bac de décantation *m*; ~ **test** ANALYSIS essai de décantation *m*; ~ **time** WATER POLL digestion des boues *f*, temps de sédimentation *m*, WATER SUPP temps de décantation *m*, temps de dépôt *m*; ~ **velocity** WATER POLL, WATER SUPP vitesse de décantation *f*, vitesse de sédimentation *f*; ~ **vessel** WATER POLL cuve de décantation *f*; ~ **zone** WATER POLL zone de décantation *f*

sewage *n* WATER POLL eaux usées *f pl*, eaux d'égout *f pl*; ~ **fungus** WATER POLL champignon des eaux résiduaires *m*; ~ **lagoon** WATER POLL, WATER SUPP étang de stabilisation *m*; ~ **plant** WATER POLL station d'épuration *f*; ~ **purification** WATER POLL, WATER SUPP épuration des eaux résiduaires *f*; ~ **scheme** WATER SUPP système d'épuration des eaux d'égout *m*; ~ **sludge** WATER POLL boues d'égout *f pl*, WATER SUPP boues d'épuration *f pl*; ~ **treatment** WATER SUPP épuration *f*, traitement des eaux d'égout *m*; ~ **treatment plant** WATER POLL station de traitement des eaux usées *f*; ~ **treatment plant residue** WATER SUPP résidu d'installation de traitement des eaux d'égout *m*

sewer *n* WASTE, WATER POLL, WATER SUPP égout *m*; ~ **cleaning** WATER SUPP curage des égouts *m*; ~ **system** WATER SUPP réseau d'égouts *m*; ~ **trench** WATER SUPP tranchée d'égout *f*

sewerage *n* WATER POLL, WATER SUPP *disposal* évacuation des eaux usées *f*; ~ **connection duty** POLICY droit de raccordement au tout-à-l'égout *m*; ~ **system** WATER SUPP réseau d'assainissement *m*

SFC *abbr* (*supercritical-fluid chromatography*) ANALYSIS CPS (*chromatographie en phase supercritique*)

SFE *abbr* (*supercritical-fluid extraction*) ANALYSIS extraction de fluide supercritique *f*

shackle *n* MARINE manille *f*

shadow: ~ **price** *n* POLICY prix de référence *m*

shaft: ~ **furnace** *n* AIR fourneau à cuve *m*, WATER POLL four à cuve *m*, WATER SUPP fourneau à cuve *m*

shake *vt* ANALYSIS *sampling bottle* agiter; ~ **in** SOIL mélanger par secousses

shaker: ~ **baghouse** *n* AIR filtre à poche à secousses *m*; ~ **mechanism** *n* AIR mécanisme de secouage *m*

shaking *n* AIR secousse *f*; ~ **screen** AIR *of packed-bed filter* tamis vibrant *m*

shallow[1] *adj* WATER SUPP peu profond

shallow[2]: ~ **water** *n* MARINE eau maigre *f*, eau peu profonde *f*, WATER POLL bas-fond *m*, eau maigre *f*, eau peu profonde *f*

shantytown *n* MANAGE, POLICY bidonville *m*

shape: ~ **recognition method** *n pl* WATER SUPP *sorting of plastics* méthode de reconnaissance des formes *f*

share: ~ **of retained water** *n* WATER SUPP pourcentage d'eau de rétention *m*; ~ **of surface waters** *n* WATER SUPP proportion des eaux de surface *f*

sharp: ~**-edge orifice** *n* ANALYSIS orifice en paroi mince *m*; ~**-edged particle** *n* AIR particule à bords aigus *f*

shavings *n pl* RENEW ENERG copeaux de rabotage *m pl*

shear[1] *vt* AIR, WASTE *liquid into droplets* cisailler; ~ **off** AIR, WASTE cisailler, découper

shear[2] *vi* AIR *liquid* se nébuliser

shearing *n* AIR, WASTE *of liquid* cisaillement *m*; ~ **action** WATER SUPP opération de cisaillage *f*; ~ **efficiency** AIR rendement de cisaillement *m*; ~ **force** AIR, WATER POLL force de cisaillage *f*, force de cisaillement *f*; ~ **off** AIR, WASTE cisaillement *m*; ~ **resistance** WATER POLL *of wastewater sludge* résistance au cisaillement *f*

sheet: ~ **absorber** *n* NOISE absorbeur stratifié *m*; ~ **erosion** *n* SOIL érosion en nappe *f*; ~ **glass** *n* WATER SUPP verre à vitres *m*; ~ **metal** *n* AIR, NOISE plaque de métal *f*; ~ **piling** *n* WATER SUPP rideau de palplanches *m*

sheeting *n* SOIL revêtement *m*

shell *n* AIR, NOISE enveloppe *f*; ~**-and-tube condenser** AIR condenseur tubulaire *m*; ~**-and-tube heat exchanger** AIR échangeur de chaleur tubulaire *m*

shielding: ~ **wall** *n* NOISE paroi de protection *f*

shift *vt* POLICY *tax* transférer

shifting: ~ **cultivation** *n* POLICY, SOIL culture nomade *f*

shock: ~ **loading** *n* WATER POLL chargement de choc *m*

shoot *n* RENEW ENERG pousse *f*; ~ **harvesting** RENEW ENERG récolte de troncs entiers *f*

shore: ~ **pollution** *n* MARINE pollution des côtes *f*; ~ **protection** *n* WATER SUPP protection du littoral *f*; ~ **reception facility** *n* MARINE hangar de débarquement *m*, installation de débarquement *f*, installation de réception *f*

shoreline *n* MARINE littoral *m*; ~ **clean-up** MARINE nettoyage du littoral *m*

short[1]: ~**-term** *adj* GEN à court terme

short[2]: ~**-glass-fiber-reinforced thermoplast** *n AmE*, ~**-glass-fibre-reinforced thermoplast** *n BrE* WATER SUPP thermoplastique renforcé à la fibre de verre courte *m*; ~**-term exposure limit** *n* (*STEL*) ANALYSIS limite d'exposition à court terme *f*; ~**-term measure** *n* MANAGE, POLICY mesure à court terme *f*; ~**-term monitoring** *n* AIR, ANALYSIS, MARINE, NOISE surveillance à court terme *f*; ~**-term storage** *n* RENEW ENERG stockage à court terme *m*, stockage de courte durée *m*; ~**-wave incoming radiation** *n* AIR rayonnement incident à onde courte *m*; ~**-wave sunlight** *n* AIR rayonnement solaire à onde courte *m*

shot: ~ **cleaning system** *n* AIR lavage par grenaillage *m*

shovel *n* MARINE cuiller *f*, pelle *f*, WATER SUPP pelle *f*

shred *vt* WASTE, WATER SUPP déchiqueter

shredded[1] *adj* WASTE, WATER SUPP déchiqueté

shredded[2]: ~ **plastics** *n pl* WATER SUPP plastique déchiqueté *m*; ~**-refuse landfill** *n* WASTE décharge pour déchets déchiquetés *f*; ~ **solid waste** *n* WASTE déchets solides déchiquetés *m pl*

shredder *n* WASTE déchiqueteuse *f*, dilacérateur *m*; ~ **waste** WASTE déchets de déchiquetage *m pl*

shredding *n* WASTE déchiquetage *m*, dilacération *f*; ~ **machine** WASTE déchiqueteuse *f*; ~ **tower** WASTE tour de déchiquetage *f*

shrinking: ~ **behavior** *AmE*, ~ **behaviour** *BrE n* WATER SUPP comportement au rétrécissement *m*, comportement de retrait *m*

shroud *n* SOIL enveloppe de protection *f*

SHS *abbr* (*static headspace analysis*) ANALYSIS analyse volumétrique statique *f*

shutoff: ~ **device** *n* WATER SUPP dispositif de coupure *m*, organe d'arrêt *m*; ~ **valve** *n* AIR vanne d'arrêt *f*

side: ~ **area** *n* WATER SUPP zone latérale *f*; ~**-channel compressor** *n* SOIL *vapour extraction* compresseur à canal latéral *m*; ~ **reaction** *n* AIR réaction secondaire *f*

sieve[1] *n* GEN crible *m*, tamis *m*; ~ **analysis** ANALYSIS analyse au tamis *f*, analyse par tamisage *f*, contrôle de granulométrie par tamisage *m*; ~ **fraction** WATER SUPP fraction de tamisage *f*; ~ **grate** AIR grille à barreaux *f*, grille de crible *f*; ~ **installation** WATER SUPP installation de criblage *f*; ~ **plate** AIR fond perforé *m*; ~ **residue** WATER SUPP résidu de tamisage *m*, refus de criblage *m pl*; ~ **size** WATER SUPP

taille de tamis *f*; **~-through fraction** WATER SUPP *composting* fraction tamisable *f*
sieve² *vt* GEN cribler, tamiser; **~ out** WATER SUPP éliminer par criblage, éliminer par tamisage
sieving *n* GEN criblage *m*, tamisage *m*
sift *vt* AIR, ANALYSIS, MARINE, WATER SUPP cribler, tamiser, SOIL tamiser
sifting *n* MARINE criblage *m*; **~ belt** MARINE bande de tamisage *f*, crible *m*
silage *n* SOIL ensilage *m*
silence *vt* NOISE amortir
silencer *n* *BrE* (*cf noise muffler AmE*) NOISE silencieux *m*
silencing *n* NOISE amortissement *m*
silica: **~ dust** *n* SAFETY poussière de silice *f*; **~ gel** *n* AIR, WATER POLL *adsorbent* gel de silice *m*
sill *n* WATER SUPP seuil *m*, base *f*, semelle *f*, sole *f*
silo *n* AIR, WATER SUPP silo *m*; **~ storage** WATER SUPP *plastics* stockage en silos *m*
silt *n* WATER POLL lais *m*, limon *m*, vase *f*, WATER SUPP limon *m*; **~ deposition** SOIL dépôt de limon *m*, WATER SUPP dépôt de limon *m*, dépôt de vase *m*; **~ trap** WATER POLL bac à boues *m*
siltage *n* WATER POLL *of lakes* colluvionnement *m*
silting: **~ up** *n* WATER POLL envasement *m*
simplified: **~ procedure** *n* POLICY procédure simplifiée *f*
SIMS *abbr* (*secondary ion mass spectroscopy*) ANALYSIS spectroscopie de masse par émission ionique secondaire *f*
simulation: **~ river systems** *n pl* MANAGE simulation bassins hydrographiques *f pl*
simultaneous: **~ denitrification** *n* WATER POLL dénitrification simultanée *f*; **~ measurement** *n* ANALYSIS mesure simultanée *f*; **~ operation** *n* ANALYSIS exploitation simultanée *f*; **~ read-out** *n* ANALYSIS affichage simultané *m*, lecture simultanée *f*
single¹: **~-crystalline** *adj* RENEW ENERG monocristallin; **~-cycle** *adj* AIR à cycle unique; **~-leaf** *adj* NOISE à lame unique; **~-sheet** *adj* NOISE *encapsulation* à couche unique; **~-stage** *adj* AIR mono-étage, AIR, WATER POLL, WATER SUPP à un étage, *axial fan* à étage unique
single²: **~-blade machine** *n* RENEW ENERG *wind turbine* éolienne monopale *f*; **~-crystalline silicon cell** *n* RENEW ENERG *photovoltaic* cellule au silicium monocristallin *f*; **~-cycle scrubber** *n* AIR laveur à cycle unique *m*; **~-event noise level** *n* NOISE niveau de bruit d'un événement isolé *m*; **~-event sound level** *n* NOISE niveau de bruit d'un événement isolé *m*; **~-material system** *n* WATER SUPP système à matériau unique *m*; **~ measurement** *n* ANALYSIS mesure unique *f*; **~-phase jet** *n* AIR *of jet scrubber* jet à phase unique *m*; **~-screw extruder** *n* WASTE, WATER SUPP extrudeuse à vis unique *f*; **~-stage activated sludge process** *n* WATER POLL traitement par boues activées en une seule étape *m*; **~-stage digester** *n* WATER POLL digesteur à un étage *m*, séparateur *m*; **~-stage fluidized-bed adsorber** *n* AIR adsorbeur à lit fluidisé à étage unique *m*; **~-stage peripheral pump** *n* WATER POLL pompe tourbillonnaire monocellulaire *f*, WATER SUPP pompe périphérique à un étage *f*; **~-unit cyclone** *n* AIR cyclone monocorps *m*; **~-wall corrugated fiberboard** *AmE*, **~-wall corrugated fibreboard** *BrE* *n* WATER SUPP carton ondulé simple rannelura *m*
sink¹ *n* WATER POLL récepteur *m*, WATER SUPP piège *m*, puits *m*, *domestic* évier *m*; **~-float flotation** WATER SUPP flottation gravimétrique *f*
sink² *vi* WATER SUPP s'enfonçer
sinkhole *n* WATER POLL aven *m*
sinking: **~ agent** *n* MARINE agent de surfactage *m*, agent tensioactif *m*; **~ material** *n* WATER POLL matière décantée *f*, produit de décantation *m*; **~ matter** *n* WATER POLL produit de décantation *m*
sinter¹: **~ oven** *n* SOIL *for pellet incineration* four de frittage *m*
sinter² *vt* SOIL fritter
sintering *n* SOIL frittage *m*
siphon¹ *n* WATER POLL, WATER SUPP siphon *m*
siphon² *vt* WATER POLL, WATER SUPP siphonner; **~ off** WATER POLL, WATER SUPP siphonner
site¹: **~-dependent** *adj* RENEW ENERG en fonction du site; **~-related** *adj* AIR, MARINE, NOISE, RENEW ENERG du site, local, SOIL lié à la décharge; **~-specific** *adj* AIR, NOISE, RENEW ENERG, WATER SUPP du site, local
site² *n* AIR site *m*, NOISE, SOIL décharge *f*, WASTE, WATER SUPP site *m*; **~ assessment** ANALYSIS évaluation de site *f*, évaluation sur site *f*; **~ characteristics** *n pl* SOIL caractéristiques de la décharge *f pl*; **~ characterization** WATER SUPP caractérisation du site *f*; **~ closure** WATER SUPP enceinte de site *f*; **~ criteria** *n pl* AIR caractéristiques du site *f pl*, SOIL critères de la décharge *m pl*, WATER SUPP critères d'implantation *m pl*; **~ operation** WATER SUPP aménagement d'un site *m*, exploitation d'un site *f*; **~ parameters** *n pl* ANALYSIS caractéristiques du site *f pl*; **~ plan** MANAGE plan de situation *m*; **~-related criterion** *n pl* RENEW ENERG paramètres spécifique du site *m pl*; **~-related noise** MARINE, NOISE bruit inhérent au site *m*, NOISE bruit du site *m*; **~-specific criteria** *n pl* RENEW ENERG caractéristiques du site *f pl*; **~-specific installation** AIR installation du site considéré *f*, NOISE installation spécifique du site *f*, RENEW ENERG installation adaptée au site *f*, SOIL installation propre à une décharge particulière *f*; **~-specific properties** *n pl* RENEW ENERG caractéristiques du site *f pl*; **~ targeted for decontamination** SOIL décharge destinée à être décontaminée *f*; **~ type** SOIL type de décharge *m*
situ: **in ~ vitrification** *n* (*ISV*) SOIL vitrification in situ *f*
size *vt* WASTE calibrer
sized *adj* WASTE calibré
sizing *n* WASTE calibrage *m*
skim *vt* MARINE, WATER POLL, WATER SUPP écrémer, écumer
skimmer *n* MARINE, WATER POLL, WATER SUPP écrémeur *m*, écumeuse *f*, écumoire *f*; **~ arm** WATER SUPP bras de râclage *m*
skimming *n* MARINE, WATER POLL, WATER SUPP écrémage *m*, écumage *m*; **~ barge** MARINE barge d'écrémage *f*; **~ barrier** MARINE barrage d'écrémage *m*, barrage récupérateur *m*; **~ head** MARINE tête d'écrémage *f*; **~ tank** MARINE cuve d'écrémage *f*, WATER POLL bac à écumes *m*
skin: **~ irritant** *n* SAFETY irritant cutané *m*
skip *n* WASTE benne *f*; **~ lorry** *BrE* WASTE camion benne *m*
skirt *n* MARINE lisière *f*
slag *n* WASTE mâchefer *m*, scories *f pl*, WATER SUPP laitier *m*, scories *f pl*; **~ discharge** WATER SUPP rejet de scories *m*; **~ from waste incineration plant** WASTE scories d'une usine d'incinération *f pl*; **~ generation** WATER SUPP production de mâchefer *f*, production de scories *f*; **~ recovery** WATER SUPP récupération des scories *f*, récupération du mâchefer *f*; **~ removal** AIR décrassage *m*, extraction des scories *f*, WATER SUPP extraction des scories *f*; **~ removal equipment** WATER SUPP matériel d'extraction des scories *m*; **~-removing plant** WATER POLL poste de décrassage *m*; **~ tank** SOIL cuve à laitier *f*
slagged: **~ dust** *n* AIR crasses *f pl*
slake *vt* AIR, WATER SUPP *lime* éteindre
slaked: **~ lime** *n* GEN chaux éteinte *f*; **~-lime mixing tank** *n* AIR cuve de mélange pour chaux éteinte *f*; **~ lime suspension** *n* AIR suspension de chaux éteinte *f*
slash: **~-and-burn agriculture** *n* AIR, POLICY agriculture sur brûlis *f*
slide: **~ gate** *n* WATER SUPP vanne *f*
sliding: **~ floor** *n* WATER SUPP plancher mobile *m*
sling *n* MARINE, WATER SUPP élingue *f*
slip: **~ erosion** *n* SOIL érosion provenant des mouvements de masse *f*
SLM *abbr* (*sound-level meter*) ANALYSIS, MARINE, NOISE, SAFETY sonomètre *m*
slop: **~ tank** *n* WASTE citerne des huiles usées *f*

slope *n* SOIL talus *m*, WATER SUPP pente *f*; **~ landfill** WATER SUPP décharge en pente *f*, décharge inclinée *f*; **~ leachate** WATER SUPP lixiviat de pente *m*; **~ leachate emission** WATER SUPP émission de lixiviats sur une pente *f*; **~ recess** WATER SUPP aménagement d'une pente *m*, enfoncement d'une pente *m*; **~ sealing** SOIL imperméabilisation des talus *f*; **~ water** WATER SUPP eau de pente *f*

slotted: **~ panel** *n* NOISE panneau rainuré *m*

sloughed: **~ biomass** *n* WATER POLL biomasse de séparation *f*

sloughing *n* WATER POLL desquamation *f*

slow: **~ composting** *n* WATER SUPP compostage lent *m*; **~ fermentation** *n* WASTE fermentation lente *f*; **~ filter process** *n* WATER SUPP procédé de filtration lent *m*; **~ filtration** *n* WATER POLL, WATER SUPP filtration lente *f*; **~ process filtrate** *n* WATER SUPP filtrat lent *m*; **~ process sand filter bed** *n* WATER SUPP lit de filtration lente sur sable *m*; **~ process sand filtration** *n* WATER POLL, WATER SUPP filtration lente sur sable *f*; **~ sand filter** *n* WATER POLL, WATER SUPP filtre lent à sable *m*; **~ sand filtration** *n* WATER POLL, WATER SUPP filtration lente sur sable *f*

sludge *n* AIR cambouis *m*, WATER POLL boue *f*, cambouis *m*, limon *m*, WATER SUPP boue *f*, vase *f*, limon *m*; **~ accumulation** WATER POLL accumulation de boues *f*; **~ age** WATER POLL âge des boues *m*; **~ barging** WATER POLL chargement des boues sur bateau-citerne *m*; **~ blanket** WATER POLL lit de boues *m*; **~-blanket filter** WATER POLL filtre à lit de boues *m*; **~ cake** WATER POLL gâteau de boue *m*; **~ centrifugation** WATER POLL centrifugation des boues *f*; **~ centrifuge** WATER POLL centrifugeuse à boues *f*; **~ characteristics** *n pl* WATER POLL caractéristiques des boues *f pl*; **~ collection** WASTE reprise des boues *f*; **~ composting** WATER POLL compostage des boues *m*; **~ conditioning** WATER POLL conditionnement des boues *m*; **~ contact clarifier** WATER POLL bassin de contact *m*; **~ dehydration** WATER POLL déshydratation de la boue *f*; **~ density index** (*SDI*) WATER POLL, WATER SUPP indice de volume des boues *m* (*IVB*); **~ dewatering equipment** WATER POLL matériel de déshydratation des boues *m*; **~ dewatering tank** WATER POLL vase à clarifier *m*; **~ digester** WATER POLL digesteur *m*; **~ digestion** WATER POLL digestion des boues *f*, temps de sédimentation *m*; **~-digestion tank** WATER POLL chambre de digestion des boues *f*; **~ disposal** WATER POLL évacuation des boues *f*; **~ draw-off valve** WATER SUPP vanne d'extraction des boues *f*; **~ drying** WATER POLL séchage des boues *m*; **~ drying area** WATER POLL aire de séchage des boues *f*; **~ drying bed** WATER POLL *in dewatering* bassin de décantation *m*, lit de séchage des boues *m*, WATER SUPP lit de séchage des boues *m*; **~ filtration** WATER SUPP filtration des boues *f*; **~ formation** WATER SUPP formation de boue *f*; **~-forming residue** WATER SUPP résidus formant des boues *m pl*; **~ freezing** WATER POLL congélation des boues *f*; **~ gas** WATER POLL biogaz *m*, gaz de curage *m*; **~ hopper** WATER POLL fosse à boues *f*, puits de boues *m*; **~ index** WATER POLL indice de boue *m*; **~-laden liquid** WATER SUPP liquide boueux *m*; **~ lagoon** WATER POLL lagune *f*; **~ liquor** WATER POLL liqueur de boues *f*, WATER SUPP liqueur de boues *f*, *sewage* liqueur surnageante *f*, liquide surnageant *m*; **~ outlet** AIR *of foam scrubber* sortie de boue *f*; **~ pond** WATER POLL bassin de boues *m*; **~ processing** WATER POLL traitement des boues *m*; **~ pump works** WATER POLL ouvrage de pompage des boues *m*; **~ recovery** WATER POLL récupération des boues *f*, valorisation des boues *f*; **~ recycling** WATER POLL recyclage des boues *m*; **~ removal** WATER POLL extraction des boues *f*; **~ residence time** (*SRT*) WATER POLL temps de séjour des boues *m*; **~ return** WATER POLL retour des boues *m*; **~ return flow** WATER POLL retour des boues *m*; **~ siphon** WATER POLL siphon à boues *m*; **~ stabilization** WATER POLL stabilisation des boues *f*; **~ sump** WATER POLL puisard à boues *m*; **~ thickening** WATER POLL épaississement des boues *m*; **~-to-oil reactor system** (*STORS*) WATER POLL réacteur boues-huile *m*; **~ treatment** SOIL, WATER POLL traitement des boues *m*; **~ volume** WATER POLL volume de boues *m*; **~ worm** WATER POLL ver des boues *m*; **~ yield** WATER POLL rendement de boues *m*

sluice *n* WATER SUPP vanne *f*, écluse *f*, pertuis *m*

sluiceway *n* RENEW ENERG *of tidal power plant* barrage mobile à vannes *m*, vannage *m*

slum *n* MANAGE, POLICY bidonville *m*

slurry *n* WATER POLL lisier *m*, WATER SUPP lisier *m*, purin *m*, suspension épaisse *f*; **~ pit** RENEW ENERG fosse à lisier *f*

small: **~ bag** *n* WATER SUPP sachet *m*; **~ landfill** *n* WATER SUPP décharge de faible capacité *f*

small-scale[1] *adj* RENEW ENERG à petite échelle

small-scale[2]: **~ installation** *n* RENEW ENERG installation à petite échelle *f*

smell *n* WATER POLL odeur *f*

smog *n* AIR brouillard photochimique *m*, brumée *f*, smog *m*, brouillard industriel *m*; **~ alarm** AIR alerte au brouillard photochimique *f*, alerte au smog *f*; **~ area** AIR zone de brouillard photochimique *f*; **~ period** AIR période de brouillard photochimique *f*; **~ risk** AIR risque de brouillard photochimique *m*; **~ warning** AIR alerte au brouillard photochimique *f*, alerte au smog *f*; **~ weather conditions** *n pl* AIR conditions météorologiques favorisant le brouillard photochimique *f pl*

smoke[1]: **~-preventing** *adj* AIR capnofuge

smoke[2] *n* AIR fumée *f*; **~ abatement** AIR réduction des fumées *f*; **~ build-up** AIR formation de fumée *f*; **~ concentration** AIR concentration de fumée *f*; **~-control system** AIR système anti-fumées *m*; **~ density meter** ANALYSIS fumimètre *m*, opacimètre *m*; **~ emissions** *n pl* AIR émissions de fumées *f pl*; **~ filter** AIR filtre à fumées *m*; **~ and heat exhaust installation** SAFETY installation d'évacuation des rejets de fumées et de chaleur *f*; **~ plume** AIR panache de fumée *m*; **~ scrubbing** AIR lavage des fumées *m*; **~ shade method** ANALYSIS indice de noircissement *f*; **~ spot number** AIR indice de noircissement par filtration sur papier *m*; **~ streamer** AIR collecteur de fumées *m*

smokeless *adj* SAFETY sans fumée

smokestack *n* AIR cheminée d'évacuation des fumées *f*

smooth *vt* SOIL lisser

smut *n* AIR fumeron *m*

snap: **~ sample** *n* ANALYSIS échantillon ponctuel *m*

SNCR *abbr* (*selective non-catalytic reduction*) WATER SUPP réduction sélective non catalytique *f*

SNG *abbr* (*synthetic natural gas*) RENEW ENERG gaz naturel de synthèse *m*

snifter: **~ valve** *n* WATER SUPP reniflard *m*

SO *abbr* (*sulfur oxide AmE, sulphur oxide BrE*) SO (*oxyde de souffre*), (*sulfur monoxide AmE, sulphur monoxide BrE*) AIR SO (*monoxyde de soufre*)

SO_2 *abbr* AIR (*sulfur dioxide AmE, sulphur dioxide BrE*) SO_2 (*dioxyde de soufre*), WATER SUPP (*sulfur anhydride AmE, sulphur anhydride BrE*) SO_2 (*anhydride sulfureux*)

SO_2/NO_x removal *n* (*desonox*) AIR élimination de SO_2/NO_x *f* (*desonox*)

SO_3 *abbr* AIR (*sulfur trioxide AmE, sulphur trioxide BrE*) SO_3 (*trioxyde de soufre*), WATER SUPP (*sulfuric anhydride AmE, sulphuric anhydride BrE*) SO_3 (*anhydride sulphurique*)

SO_4 *abbr* (*sulfur tetroxide AmE, sulphur tetroxide BrE*) AIR SO_4 (*tétroxyde de soufre*)

SO_x *abbr* (*sulfur oxides AmE, sulphur oxides BrE*) AIR, WATER SUPP SO_x (*oxydes de soufre*)

soak *vt* SOIL tremper

socket *n* WATER POLL, WATER SUPP *pipe end extension* cosse *f*, douille *f*, manchon *m*

soda *n* WATER POLL carbonate de sodium *m*, soude *f*, WATER SUPP *in glass recycling* soude *f*; **~ ash** WATER SUPP carbonate de soude *m*, cendres de soude *f pl*, cristaux de soude *m pl*, soude calcinée *f*

sodden *adj* SOIL détrempé

sodium *n* (*Na*) AIR sodium *m* (*Na*); ~ **absorption ratio** WATER SUPP rapport d'absorption du sodium *m* (*RAS*); ~ **chloride** MARINE, WATER POLL chlorure de sodium *m*; ~ **hydroxide** WATER POLL, WATER SUPP soude *f*; ~ **ion** WATER SUPP ion sodium *m*; ~ **removal** WATER POLL désodation *f*

SOFC *abbr* (*solid oxide fuel cell*) AIR pile à combustible à oxyde solide *f*

soft: ~ **detergent** *n* WATER POLL détergent doux *m*; ~ **instrument** *n* POLICY *information, labelling* instrument peu contraignant *m*; ~ **rot** *n* WASTE pourriture molle *f*; ~ **tourism** *n* MANAGE tourisme écologique *m*; ~ **water** *n* WATER SUPP eau douce *f*

soften *vt* WATER SUPP adoucir

softened *adj* WATER SUPP *water* adouci

softener *n* WATER SUPP adoucisseur *m*

softening *n* WATER SUPP adoucissement *m*; ~ **plant** WATER SUPP usine d'adoucissement *f*

software[1]: ~**-controlled** *adj* ANALYSIS à commande informatique, géré informatiquement

software[2] *n* ANALYSIS logiciel *m*

soil[1]: ~**-physical** *adj* SOIL physico-pédologique

soil[2] *n* SOIL sol *m*, terreau *m*, WASTE, WATER SUPP sol *m*; ~ **ameliorant** SOIL *conditioner* conditionneur de sol *m*; ~ **analysis** ANALYSIS, SOIL analyse du sol *f*; ~ **area** SOIL zone de sol *f*; ~**-bearing capacity** SOIL résistance à l'écrasement *f*; ~ **break-up** SOIL retournement du sol *m*; ~ **characteristics** *n pl* SOIL caractéristiques du sol *f pl*; ~ **column** SOIL coupe stratigraphique *f*; ~ **component** SOIL matériau du sol *m*; ~ **condition** SOIL nature du sol *f*; ~ **conditioner** SOIL conditionneur de sol *m*; ~ **conditioning** SOIL amendement *m*; ~ **conservation** SOIL conservation du sol *f*; ~ **constituent** SOIL constituant du sol *m*; ~ **contamination** SOIL contamination du sol *f*; ~ **covering** SOIL couverture du sol *f*; ~ **decontamination** SOIL *of radioactive soil* décontamination du sol *f*; ~ **degradation** SOIL dégradation du sol *f*; ~ **degradation control** MANAGE, SOIL lutte contre la dégradation des sols *f*; ~ **depletion** SOIL épuisement du sol *m*; ~ **deposit sampler** ANALYSIS échantillonneur de dépôt de sol *m*; ~ **dust** SOIL poussière de sol *f*; ~ **environment** SOIL milieu pédologique *m*; ~ **erosion** SOIL érosion des sols *f*; ~ **exchange** SOIL échange de sol *m*; ~ **exhaustion** SOIL épuisement du sol *m*, appauvrissement du sol *m*; ~ **filter** WATER SUPP filtre pour sol *m*; ~ **filter-layer** WATER SUPP couche filtrante du sol *f*; ~ **flushing** SOIL rinçage du sol *m*; ~ **formation** SOIL pédogenèse *f*; ~ **freezing** SOIL congélation du sol *f*; ~ **function** SOIL fonction du sol *f*; ~ **gas** SOIL gaz du sol *m*; ~ **horizon** SOIL horizon de sol *m*; ~ **improvement** SOIL amendement *m*; ~ **layer** SOIL couche de sol *f*; ~**-like material** SOIL simili-sol *m*; ~ **map** MANAGE, POLICY, SOIL carte des sols *f*; ~ **matrix** SOIL, WATER SUPP matrice du sol *f*; ~ **mechanics** SOIL mécanique du sol *f*; ~ **monitoring** MANAGE surveillance des sols *f*; ~ **particle** SOIL particule de sol *f*; ~ **passage** WATER SUPP passage dans le sol *m*; ~ **permeability** SOIL, WATER SUPP perméabilité du sol *f*; ~**-physical parameter** SOIL paramètre physico-pédologique *m*; ~ **pipe** WATER POLL descente *f*, WATER SUPP tuyau d'assainissement *m*; ~ **pollution** SOIL pollution du sol *f*; ~ **processing** SOIL traitement du sol *m*; ~ **properties** *n pl* SOIL propriétés du sol *f pl*; ~ **quality** SOIL qualité du sol *f*; ~ **reclamation** SOIL mise en valeur des sols *f*, mise en valeur des terres *f*; ~ **sampling** SOIL prélèvement d'échantillons de sol *m*; ~ **saturation** SOIL saturation du sol *f*; ~ **science** SOIL pédologie *f*; ~ **and sediment washing** SOIL lavage du sol et des sédiments *m*; ~ **slurry** SOIL sol boueux *m*; ~ **survey** SOIL enquête pédologique *f*; ~ **texture** SOIL texture du sol *f*; ~ **use** SOIL utilisation du sol *f*; ~ **vapor** *AmE*, ~ **vapour** *BrE* SOIL vapeur du sol *f*; ~ **vapour concentration** *BrE* SOIL concentration des vapeurs du sol *f*; ~ **vapour extraction** *BrE* SOIL *pneumatic measure* extraction des vapeurs du sol *f*; ~ **washer** SOIL laveur de sol *m*; ~ **water** SOIL eau de sol *f*, WATER SUPP eau du sol *f*; ~**-water balance** WATER SUPP bilan d'eau du sol *m*; ~ **water sampling** ANALYSIS prélèvement d'eau du sol *m*

soil[3] *vt* ANALYSIS, MARINE, WASTE souiller

soiled: ~ **plastics** *n pl* WASTE, WATER SUPP plastique souillé *m*

soiling *n* ANALYSIS, MARINE, WASTE souillure *f*

solar[1] *adj* RENEW ENERG solaire; ~**-driven** RENEW ENERG alimenté par l'énergie solaire; ~**-heated** RENEW ENERG chauffé par l'énergie solaire; ~**-operated** RENEW ENERG solaire; ~**-powered** RENEW ENERG alimenté à l'énergie solaire; ~**-supported** RENEW ENERG d'origine solaire; ~**-thermal** RENEW ENERG thermique solaire

solar[2]: ~ **architecture** *n* RENEW ENERG architecture solaire *f*; ~**-based renewable energy source** *n* RENEW ENERG source d'énergie renouvelable d'origine solaire *f*; ~ **cell** *n* RENEW ENERG cellule photovoltaïque *f*, cellule solaire *f*; ~ **cell module** *n* RENEW ENERG module de cellules photovoltaïques *m*; ~ **collector** *n* RENEW ENERG capteur solaire *m*, héliostat *m*; ~ **dish installation** *n* RENEW ENERG installation de réflecteurs solaires *f*; ~ **disposal of chemical wastes** *n* WATER SUPP élimination solaire des déchets chimiques *f*; ~ **district heating** *n* RENEW ENERG chauffage urbain alimenté par l'énergie solaire *m*; ~ **district heating supply** *n* RENEW ENERG chauffage urbain d'origine solaire *m*; ~**-driven well installation** *n* RENEW ENERG installation de puits solaire *f*; ~ **energy** *n* RENEW ENERG énergie solaire *f*; ~ **energy enterprise** *n* RENEW ENERG entreprise de fourniture d'énergie solaire *f*; ~ **energy resources** *n pl* RENEW ENERG ressources en énergie solaire *f pl*; ~ **energy use** *n* RENEW ENERG application de l'énergie solaire *f*; ~ **farm** *n* RENEW ENERG centrale solaire *f*; ~ **farm capacity** *n* RENEW ENERG puissance d'une centrale solaire *f*; ~**-heat power plant** *n* RENEW ENERG centrale héliothermique *f*, centrale solaire *f*; ~**-heated steam generator** *n* RENEW ENERG générateur de vapeur chauffé à l'énergie solaire *m*; ~ **heating** *n* RENEW ENERG chauffage solaire *m*; ~ **heating plant** *n* RENEW ENERG installation de chauffage solaire *f*; ~ **high-temperature destruction** *n* WATER SUPP *detoxification of wastes* destruction solaire à haute température *f*; ~**-hydrogen technology** *n* RENEW ENERG technologie couplant l'énergie solaire à l'hydrogène *f*; ~ **module** *n* RENEW ENERG *solar cell* module solaire *m*; ~ **pond** *n* RENEW ENERG bassin solaire *m*; ~ **power** *n* RENEW ENERG énergie hélio-électrique *f*, énergie solaire *f*; ~ **power plant** *n* RENEW ENERG centrale solaire *f*; ~ **radiation** *n* AIR, RENEW ENERG rayonnement solaire *m*; ~ **reactor** *n* WATER SUPP réacteur solaire *m*; ~**-supported district heating system** *n* RENEW ENERG système de chauffage urbain alimenté par l'énergie solaire *m*; ~ **technology** *n* RENEW ENERG technologie solaire *f*; ~ **thermal energy** *n* RENEW ENERG énergie thermique d'origine solaire *f*; ~ **tower** *n* RENEW ENERG récepteur solaire central *m*, tour solaire *f*; ~ **tower installation** *n* RENEW ENERG installation d'une tour solaire *f*; ~ **tower top** *n* RENEW ENERG sommet d'une tour solaire *m*

solarimeter *n* ANALYSIS pyranomètre *m*, solarimètre *m*

solenoid: ~ **valve** *n* AIR électrovanne *f*

solid: ~**-borne noise** *n* NOISE bruit porté dans les solides *m*; ~ **brick** *n* NOISE brique pleine *f*; ~**-containing waste air** *n* AIR effluent gazeux chargé de particules solides *m*; ~ **fiberboard** *AmE*, ~ **fibreboard** *BrE* *n* WATER SUPP carton rigide *m*; ~ **fuel** *n* AIR combustible solide *m*; ~**-liquid extraction** *n* ANALYSIS extraction solide-liquide *f*; ~ **material** *n* WASTE matériau solide *m*; ~**-matter loading** *n* AIR charge de matières solides *f*; ~**-matter separation** *n* AIR séparation de matière solide *f*; ~ **oxide fuel cell** *n* (*SOFC*) AIR pile à combustible à oxyde solide *f*; ~ **particle** *n* AIR,

WATER POLL, WATER SUPP particule solide *f*; ~ **particulate** *n* AIR, WATER POLL, WATER SUPP particule solide *f*; **~-phase extraction** *n* (*SPE*) ANALYSIS extraction en phase solide *f*; ~ **reacting agent** *n* ANALYSIS réactif solide *m*; ~ **residue** *n* WATER SUPP résidus solides *m pl*; ~ **state of aggregation** *n* AIR état solide *m*; ~ **waste** *n* WASTE, WATER POLL, WATER SUPP déchets solides *m pl*

solidification *n* AIR, SOIL, WASTE, WATER SUPP solidification *f*; ~ **with cement** WATER SUPP cimentation *f*; ~ **with organic polymers** WATER SUPP *hazardous waste treatment* solidification avec des polymères organiques *f*; ~ **process** WATER SUPP procédé de solidification *m*; ~ **technique** WATER SUPP technique de solidification *f*

solidified[1] *adj* AIR, SOIL, WASTE, WATER SUPP solidifié

solidified[2]: ~ **end product** *n* AIR produit final solidifié *m*; ~ **material** *n* WASTE matériau solidifié *m*; ~ **product** *n* WATER SUPP produit solidifié *m*; ~ **soil** *n* SOIL sol solidifié *m*; ~ **waste** *n* WASTE déchets solidifiés *m pl*

solidifier *n* MARINE solidificateur *m*

solidify *vt* AIR, SOIL, WASTE, WATER SUPP solidifier

solidifying: ~ **agent** *n* SOIL, WATER SUPP agent solidifiant *m*; ~ **waste** *n* WATER SUPP, WASTE déchets de solidification *m pl*

solidity *n* RENEW ENERG *of wind turbine* coefficient de plénitude *m*

solids: ~ **contents** *n* WATER POLL, WATER SUPP teneur en matières solides *f*; ~ **separator** *n* WATER POLL, WATER SUPP séparateur de matières solides *m*

solubility *n* AIR, SOIL, WATER POLL, WATER SUPP solubilité *f*

solubilization *n* AIR, SOIL, WATER POLL, WATER SUPP solubilisation *f*

solubilize *vt* AIR, SOIL, WATER POLL, WATER SUPP solubiliser

solubilized[1] *adj* AIR, SOIL, WATER POLL, WATER SUPP solubilisé

solubilized[2]: ~ **organics** *n pl* WATER POLL matière organique dissoute *f*

soluble[1] *adj* AIR, SOIL, WATER POLL, WATER SUPP soluble

soluble[2]: ~ **gas** *n* AIR gaz soluble *m*

solute *n* AIR soluté *m*

solution *n* ANALYSIS, WATER POLL, WATER SUPP solution *f*; ~ **tank** WATER POLL bac de dissolution *m*

solvability *n* AIR, SOIL, WATER POLL, WATER SUPP *of contaminant* solubilité *f*

solvate *vt* WATER POLL solvater

solvated *adj* WATER POLL solvaté

solvation *n* WATER POLL solvatation *f*

solvent[1]: **~-free** *adj* AIR exempt de solvent, sans solvant, WATER SUPP sans solvant; **~-laden** *adj* AIR chargé de solvants

solvent[2] *n* GEN solvant *m*; ~ **accumulation** AIR accumulation de solvants *f*; **~-based coating** WATER SUPP revêtement à base de solvants *m*; ~ **concentration** AIR concentration de solvants *f*; **~-containing waste air** AIR effluent gazeux chargé de solvants *m*; ~ **emissions** *n pl* AIR émissions de solvants *f pl*; **~-free paint** WATER SUPP peinture sans solvant *f*; **~-free waste air** AIR effluent gazeux exempt de solvants *m*; **~-laden air** AIR, SAFETY air chargé de solvants *m*; **~-laden waste air** AIR effluent gazeux chargé de solvants *m*; ~ **loading** AIR charge de solvants *f*; ~ **mix from benzene, toluol and xylene** WATER POLL mélange solvant à base de benzène, de toluol et de xylène *m*; ~ **recovery** AIR récupération de solvants *f*; ~ **recovery plant** WATER SUPP usine de récupération des solvants *f*; ~ **recovery unit** WATER SUPP usine de récupération des solvants *f*; ~ **regeneration** SOIL régénération d'un solvant *f*; ~ **residue** SOIL, WATER SUPP résidu de solvant *m*; ~ **tank** AIR cuve de solvants *f*; ~ **vaporization** AIR vaporisation de solvants *f*

solvents: ~ **reactor** *n* WATER SUPP réacteur à solvants *m*

solvolysis *n* WATER SUPP *plastics* dissolution chimique *f*, lyolyse *f*, solvolyse *f*

sone *n* ANALYSIS *phonometric unit* sone *m*

sonic *adj* NOISE, WATER SUPP sonique

soot[1]: **~-free** *adj* AIR sans suie

soot[2] *n* AIR suie *f*; ~ **catcher** AIR capte-suies *m*; ~ **coal** AIR houille fuligineuse *f*; **~-free incineration** AIR incinération sans suie *f*

sorb *vt* AIR, SOIL adsorber

sorbate *n* AIR, MARINE sorbat *m*

sorbent *n* AIR, MARINE sorbant *m*

sorption *n* AIR, WATER POLL, WATER SUPP sorption *f*; ~ **analysis** ANALYSIS analyse par sorption *f*; ~ **capacity** AIR, WATER SUPP capacité de sorption *f*; ~ **isotherm** AIR isotherme de sorption *f*

sort *vt* WASTE trier; ~ **according to size** WASTE trier par taille

sorted[1] *adj* WASTE trié

sorted[2]: ~ **fraction** *n* WASTE fraction triée *f*

sorting *n* WASTE tri *m*, triage *m*; ~ **analysis** ANALYSIS analyse par triage *f*; ~ **based on density** WASTE tri par densité *m*; ~ **belt** WASTE bande trieuse *f*; ~ **capacity** WASTE capacité de tri *f*; ~ **machine** WASTE trieuse *f*; ~ **personnel** WASTE personnel de tri *m*; ~ **plant** WASTE usine de tri *f*; ~ **of refuse** WASTE tri de déchets *m*; ~ **of rubbish** WASTE tri de déchets *m*

sound[1]: **~-conducting** *adj* NOISE phono-conducteur; **~-damped** *adj* NOISE insonorisé; **~-protecting** *adj* MARINE, NOISE, SAFETY insonorisant; **~-reflecting** *adj* NOISE réfléchissant le son

sound[2] *n* ANALYSIS, NOISE, WATER SUPP son *m*; **~-absorbent damper** NOISE amortisseur absorbant le son *m*, amortisseur de bruit *m*; **~-absorbent zone** NOISE zone absorbant le son *f*; ~ **absorber** MARINE, NOISE absorbeur acoustique *m*; **~-absorbing apron** NOISE chape insonorisante *f*; **~-absorbing barrier** NOISE *on roadside* écran acoustique *m*; **~-absorbing brick** NOISE brique insonorisante *f*; **~-absorbing casing** NOISE boîtier insonorisant *m*; **~-absorbing ceiling** NOISE plafond insonorisant *m*; **~-absorbing door** MARINE, NOISE porte insonorisante *f*, SAFETY porte insonorisante *f*, porte phono-absorbante *f*; **~-absorbing foam panel** SAFETY panneau insonorisant en mousse *m*; **~-absorbing glass** MARINE, NOISE, SAFETY verre insonorisant *m*; **~-absorbing hood** NOISE capot insonorisant *m*; **~-absorbing lining** NOISE revêtement insonorisant *m*; **~-absorbing material** NOISE matériau absorbant *m*, matériau insonorisant *m*; **~-absorbing panel** NOISE panneau d'insonorisation *m*, panneau phono-absorbant *m*; **~-absorbing properties** NOISE propriétés insonorisantes *f pl*; **~-absorbing wall** NOISE paroi phono-absorbante *f*; **~-absorbing wall lining** NOISE revêtement insonorisant *m*; **~-absorbing wallpaper** NOISE papier peint insonorisant *m*; ~ **absorption** NOISE absorption du son *f*; ~ **absorption coefficient** NOISE coefficient d'absorption acoustique *m*; ~ **attenuation** NOISE atténuation du son *f*; ~ **barrier** NOISE écran acoustique *m*; ~ **boring** ANALYSIS sondage *m*; **~-control system** NOISE silencieux *m*; ~ **damper** NOISE amortisseur acoustique *m*; ~ **damping** NOISE amortissement du son *m*, amortissement sonore *m*; ~ **deadener** NOISE amortisseur de bruit *m*; ~ **deadening** NOISE amortissement du son *m*, amortissement sonore *m*; **~-deadening material** NOISE matériau amortisseur du son *m*; ~ **emitter** NOISE émetteur sonore *m*; ~ **energy** NOISE énergie sonore *f*; **~-exposure meter** ANALYSIS sonomètre *m*; ~ **field** NOISE champ sonore *m*; ~ **generation** NOISE production de son *f*; ~ **immission** NOISE immission sonore *f*; ~ **incidence direction** NOISE direction de l'incidence du son *f*; **~-insulating wall** MARINE, NOISE, SAFETY paroi insonorisante *f*; ~ **insulation** NOISE isolation acoustique *f*; ~ **insulation material** NOISE matériau insonorisant *m*; ~ **intensity** MARINE, NOISE intensité acoustique *f*; ~ **intensity analyser** *BrE*, ~ **intensity analyzer** *AmE* ANALYSIS analyseur d'intensité acoustique *m*; ~ **intensity meter** ANALYSIS, NOISE intensimètre *m*; ~ **isolation** NOISE isolation acoustique *f*; ~ **level** NOISE niveau de bruit *m*, niveau sonore *m*; **~-level meter** (*SLM*) ANALYSIS, MARINE, NOISE, SAFETY sonomètre *m*; ~ **occurrence** NOISE phénomène acoustique *m*; ~ **permeability** NOISE perméabilité sonore *f*; ~ **pollution** NOISE pollution acoustique *f*; ~ **power** NOISE énergie acoustique *f*, puissance acoustique *f*; ~ **power level** NOISE niveau de puissance sonore *m*; ~ **pressure** NOISE pression acoustique *f*; ~ **pressure level** (*SPL*) NOISE niveau de pression acoustique *m*; ~ **pressure measuring** ANALYSIS mesure de pression acoustique *f*; ~-**pressure reference value** ANALYSIS valeur de référence de pression acoustique *f*; ~-**pressure spectrum** NOISE spectre de pression acoustique *m*; ~ **probe** ANALYSIS *sonic* sonde acoustique *f*; ~-**protecting window** NOISE fenêtre insonorisante *f*; ~ **radiation** NOISE rayonnement acoustique *m*; ~ **radiation pressure** NOISE pression de rayonnement acoustique *f*; ~ **radiometer** ANALYSIS radiomètre acoustique *m*; ~ **reduction index** NOISE indice de réduction du bruit *m*; ~ **reflection** NOISE réflexion du son *f*; ~ **silencer** *BrE* (*cf noise muffler AmE*) NOISE silencieux *m*; ~ **spectrum** NOISE spectre acoustique *m*; ~ **transmission** NOISE transmission sonore *f*; ~ **transmission class** (*STC*) NOISE classe de transmission acoustique *f*; ~ **transmission loss** NOISE perte sonore par transmission *f*; ~ **trap** NOISE piège acoustique *m*; ~ **vibration** NOISE vibration acoustique *f*; ~ **wave** NOISE onde sonore *f*

sound[3] *vt* ANALYSIS sonder

soundbox *n* NOISE phonocapteur *m*

sounding: ~ **boring** *n* ANALYSIS forage *m*, sondage *m*

soundless *adj* MARINE, NOISE, SAFETY insonore

soundness: ~ **certificate** *n* MANAGE certificat de solvabilité *m*

soundproof[1] *adj* MARINE, NOISE, SAFETY insonorisé

soundproof[2]: ~ **tile** *n* NOISE carreau insonorisant *m*

soundproof[3] *vt* MARINE, NOISE, SAFETY insonoriser

soundproofing[1] *adj* MARINE, NOISE, SAFETY insonorisant

soundproofing[2] *n* MARINE, NOISE, SAFETY insonorisation *f*

sour *adj* AIR *gas* corrosif

source[1]: ~-**oriented** *adj* SOIL axé sur les sources; ~-**related** *adj* AIR, MARINE, NOISE de la source, source, SOIL lié aux sources, WATER SUPP de la source, source

source[2] *n* AIR, WATER SUPP source *f*; ~ **area** AIR région émettrice *f*, région source *f*, zone source *f*; ~ **height** AIR *of chimney stack* hauteur de source *f*; ~ **of odor** *AmE*, ~ **of odour** *BrE* AIR source olfactive *f*; ~-**oriented measure** SOIL mesure axée sur les sources *f*

souring *n* RENEW ENERG pourrissage *m*

Soxhlet: ~ **extraction** *n* ANALYSIS extraction par tube de Soxhlet *f*; ~ **extractor** *n* ANALYSIS tube de Soxhlet *m*

space: ~ **utilization** *n* POLICY utilisation de l'espace *f*; ~ **velocity** *n* AIR *of catalyst* vitesse spatiale *f*

span *n* NOISE intervalle de mesurage *m*

sparger *n* WATER POLL arroseur *m*

spark: ~ **arrester** *n* AIR pare-étincelles *m*; ~ **ignition engine** *n BrE*, ~ **ignition motor** *n AmE* AIR moteur à allumage par étincelles *m*, moteur à essence *m*; ~-**over** *n* AIR décharge disruptive *f*

spatial: ~ **distribution** *n* AIR, WATER SUPP distribution géographique *f*, distribution spatiale *f*, répartition spatiale *f*; ~ **evaluation** *n* ANALYSIS évaluation spatiale *f*; ~ **pattern** *n* AIR modèle spatial *m*, WATER SUPP configuration spatiale *f*, profil spatial *m*; ~ **resolution** *n* AIR *nuclear technology*, WATER SUPP résolution spatiale *f*; ~ **trend** *n* AIR tendance spatiale *f*; ~ **variability** *n* AIR variabilité spatiale *f*

SPE *abbr* (*solid-phase extraction*) ANALYSIS extraction en phase solide *f*

special: ~ **industrial refuse** *n* WASTE déchets industriels spéciaux *m pl* (*DIS*); ~ **waste** *n* WASTE, WATER POLL déchets spéciaux *m pl*

specialist: ~ **laboratory** *n* ANALYSIS laboratoire spécialisé *m*

specific: ~ **coagulation** *n* WATER POLL coagulation spécifique *f*; ~ **conductance** *n* ANALYSIS conductance spécifique *f*; ~ **gravity** *n* WATER SUPP densité *f*; ~ **oxygen transfer capacity in clean water** *n* WATER POLL capacité de transfert d'oxygène spécifique de l'eau propre *f*; ~ **oxygen transfer capacity in wastewater** *n* WATER POLL capacité de transfert d'oxygène spécifique des eaux usées *f*; ~ **surface area** *n* WATER POLL surface spécifique *f*

specimen *n* ANALYSIS spécimen *m*

spectral: ~ **absorption coefficient** *n* ANALYSIS coefficient d'absorption spectrale *m*; ~ **analysis** *n* ANALYSIS analyse spectrale *f*; ~ **extinction coefficient** *n* ANALYSIS coefficient d'extinction spectrale *m*; ~ **half-width value** *n* ANALYSIS largeur de spectre à mi-hauteur *f*; ~ **line** *n* ANALYSIS raie spectrale *f*; ~ **region** *n* ANALYSIS zone spectrale *f*

spectrometer *n* ANALYSIS spectromètre *m*

spectrometric *adj* ANALYSIS spectrométrique

spectrometry *n* ANALYSIS spectrométrie *f*

spectrophotometer *n* ANALYSIS spectrophotomètre *m*

spectroscopic[1] *adj* ANALYSIS spectroscopique

spectroscopic[2]: ~ **method** *n* ANALYSIS méthode spectroscopique *f*; ~ **spectral region** *n* ANALYSIS zone spectroscopique *f*

spectroscopy *n* ANALYSIS spectroscopie *f*

speed: ~ **of escape** *n* ANALYSIS *gas density determination* vitesse d'échappement *f*; ~ **of sound** *n* NOISE vitesse du son *f*

speedometer *n* ANALYSIS vélocimètre *m*

spent: ~ **lye** *n* WASTE, WATER POLL, WATER SUPP lessive résiduaire *f*; ~ **oil** *n* WATER POLL huile épuisée *f*

spill *n* MARINE déversement *m*, déversoir *m*, rejet *m*, WATER POLL, WATER SUPP déversement *m*

spillage *n* MARINE, WATER POLL, WATER SUPP *of hazardous waste* déversement *m*

spilled: ~ **oil** *n* WATER POLL dispersion des hydrocarbures *f*

spillway *n* WATER POLL, WATER SUPP canal évacuateur de crues *m*

spin[1]: ~ **dryer** *n* WATER POLL centrifugeuse *f*

spin[2] *vi* WATER SUPP tournoyer

spiral: ~ **aerator** *n* WATER POLL aérateur spiral *m*; ~ **classifier** *n* SOIL classificateur à vis hélicoïdale *m*; ~ **module** *n* WATER POLL module à enroulement spirale *m*; ~ **plate** *n* AIR plaque spirale *f*; ~-**plate condenser** *n* AIR condenseur à plaques spirales *m*; ~-**plate heat exchanger** *n* AIR échangeur de chaleur à plaques spirales *m*

spiraling *AmE see spiralling BrE*

spiralling *adj BrE* AIR en spirale

SPL *abbr* (*sound pressure level*) NOISE niveau de pression acoustique *m*

splay[1] *n* AIR, WASTE, WATER POLL évasement *m*

splay[2] *vt* AIR, WASTE, WATER POLL évaser

splaying *n* AIR, WASTE, WATER POLL évasement *m*

spokesperson *n* POLICY porte-parole *m*

spontaneous: ~ **combustion** *n* SAFETY combustion spontanée *f*

spot: ~ **analysis** *n* ANALYSIS analyse par goutte *f*; ~ **measurement** *n* ANALYSIS mesure ponctuelle *f*; ~ **sample** *n* ANALYSIS échantillon localisé *m*, prélèvement instantané *m*

spotter: ~ **plane** *n* MARINE avion d'observation *m*

spray[1] *n* GEN atomiseur *m*; ~ **absorber** AIR absorbeur de particules *m*; ~ **absorption process** AIR *dry-wet* procédé d'absorption par atomisation *m*; ~ **aeration** AIR, WATER SUPP aération par pulvérisation *f*; ~ **aperture** MARINE buse de vaporisation *f*; ~ **boom** MARINE rampe de pulvérisation *f*; ~ **column** AIR *absorption device* colonne de pulvérisation *f*; ~-**column scrubber** AIR dépoussiéreur à pulvérisation *m*; ~-**dry absorber** (*SDA*) AIR absorbeur à pulvérisation par voie sèche *m*, absorbeur à séchage par pulvérisation *m*; ~-**dry scrubber** AIR dépoussiéreur à pulvérisation par voie sèche *m*; ~ **dryer** WATER POLL sécheur-pulvérisateur *m*; ~ **frame** AIR cadre de pulvérisation *m*; ~ **gun** MARINE pistolet pulvérisateur *m*; ~ **nozzle** AIR buse de vaporisation *f*, tuyère de pulvérisation *f*; ~ **reactor** AIR réacteur à pulvérisation *m*; ~ **tower** AIR tour à pulvérisation *f*;

~**-tower scrubber** AIR dépoussiéreur à pulvérisation *m*; ~ **washing** WATER SUPP nettoyage au jet *m*
spray[2] *vt* GEN atomiser, pulvériser, vaporiser
spraying *n* GEN atomisation *f*, arrosage *m*; ~ **cone** AIR cône d'aspersion *m*; ~ **zone** AIR zone d'aspersion *f*, zone de pulvérisation *f*
spread[1]: ~ **of noise** *n* NOISE propagation sonore *f*
spread[2] *vt* MARINE, SOIL, WATER POLL épandre
spread[3] *vi* MARINE, WATER POLL, WATER SUPP se répandre
spreading *n* AIR, MARINE, NOISE propagation *f*, SOIL épandage *m*, WATER POLL, WATER SUPP propagation *f*; ~ **ground** WATER POLL terrains d'épandage *m pl*
spring: ~ **acid shock** *n* AIR choc acide printanier *m*; ~ **maximum of fallout** *n* AIR maximum printanier des retombées *m*; ~ **tide** *n* MARINE, RENEW ENERG marée de vives eaux *f*
sprinkler *n* WATER POLL asperseur *m*, sprinkler *m*, WATER SUPP sprinkler *m*
sprinkling *n* WATER POLL aspersion *f*; ~ **filter** WATER SUPP lit percolateur *m*
squirrel: ~**-cage induction motor** *n* RENEW ENERG *of wind turbine*, WATER POLL *pump driving gear* moteur à induction à cage d'écureuil *m*
S-RIM *abbr* (*structural reaction injection moulding BrE*, *structural reaction injection molding AmE*) WASTE moulage par injection-réaction en moule ouvert *m*
SRT *abbr* (*sludge residence time*) WATER POLL temps de séjour des boues *m*
stability *n* GEN stabilité *f*
stabilization *n* GEN stabilisation *f*; ~ **pond** WATER POLL bassin de stabilisation *m*, étang d'oxydation *m*, WATER SUPP étang d'oxydation *m*
stabilize *vt* GEN stabiliser
stabilized[1] *adj* GEN stabilisé
stabilized[2]: ~ **sludge** *n* WATER POLL boues stabilisées *f pl*
stabilizer *n* GEN stabilisant *m*
stable[1] *adj* GEN stable
stable[2]: ~ **temperature layering** *n* AIR stratification stable des températures *f*
stack[1] *n* AIR, RENEW ENERG cheminée *f*; ~ **diameter** AIR diamètre de cheminée *m*; ~ **gas** AIR gaz de cheminée *m*; ~ **height** AIR hauteur de cheminée *f*; ~ **lining** AIR revêtement de cheminée *m*; ~ **orifice** AIR *opening* sortie de cheminée *f*; ~ **wall** AIR paroi de cheminée *f*
stack[2] *vt* AIR empiler
stage: ~ **of decomposition** *n* WASTE stade de décomposition *m*
staged: ~ **fuel feeding** *n* AIR alimentation étagée en combustible *f*
stages: ~ **incineration** *n* AIR incinération par étages *f*
stagnation *n* WATER POLL stagnation *f*
stand: ~**-by** *n* MARINE réserve *f*
standard[1]: ~ **conditions** *n pl* (*SC*) ANALYSIS conditions normales *f pl*; ~ **curve time-temperature** *n* ANALYSIS courbe normale temps-température *f*; ~**-rate trickling filter** *n* WATER POLL lit bactérien à débit moyen *m*; ~ **turbidity** *n* ANALYSIS turbidité normale *f*
standard[2]: ~ **temperature and pressure** *phr* (*STP*) ANALYSIS conditions normales de température et pression *f pl*, température et pression normales
standardized[1] *adj* SAFETY normalisé
standardized[2]: ~ **equipment** *n* SAFETY équipement normalisé *m*, matériel normalisé *m*
standards: ~ **implementation monitoring** *n* POLICY suivi des normes de mise en vigueur *m*
standing: ~**-wave flume** *n* WATER POLL canal jaugeur à ressaut *m*; ~**-wave tube** *n* NOISE tube à ondes stationnaires *m*
standpipe *n* AIR descente *f*
staple[1] *n* WASTE agrafe *f*, broche *f*
staple[2] *vt* WASTE agrafer
stapling *n* WASTE *of bales* agrafage *m*; ~ **equipment** WASTE matériel d'agrafage *m*; ~ **machine** WASTE brocheuse *f*; ~ **pliers** *n pl* WASTE pince d'agrafage *f*; ~ **wire** WASTE fil d'agrafage *m*
stapylococci *n* WATER POLL staphylocoques *m pl*
star: ~**-shaped procedure** *n* POLICY *five administrative units* structure en étoile *f*
starchy[1] *adj* RENEW ENERG amylacé
starchy[2]: ~ **plants** *n pl* RENEW ENERG *biomass conversion* plantes amylacées *f pl*; ~ **raw material** *n* RENEW ENERG *biomass conversion* matière première amylacée *f*
start: ~ **burner** *n* AIR brûleur de démarrage *m*
starting: ~ **operation** *n* AIR opération de démarrage *f*; ~ **point** *n* ANALYSIS point de départ *m*
state: ~ **of aggregation** *n* AIR état solide *m*
static[1] *adj* GEN statique
static[2]: ~ **air layer** *n* AIR couche d'air statique *f*; ~ **deflection** *n* NOISE déflexion statique *f*; ~ **head** *n* WATER POLL hauteur manométrique *f*, hauteur piézométrique *f*, pression d'eau *f*; ~ **headspace analysis** *n* (*SHS*) ANALYSIS analyse volumétrique statique *f*; ~ **headspace method** *n* ANALYSIS méthode volumétrique statique *f*; ~ **leachability test** *n* ANALYSIS, SOIL, WASTE essai de lixiviation statique *m*; ~ **leaching test** *n* ANALYSIS, SOIL, WASTE essai de lixiviation statique *m*; ~ **mixer** *n* WATER POLL mélangeur statique *m*; ~ **pressure** *n* AIR pression statique *f*; ~ **suction head** *n* WATER POLL hauteur d'aspiration statique *f*; ~ **suction lift** *n* WATER POLL hauteur d'aspiration statique *f*
stationary[1] *adj* AIR fixe, stationnaire, ANALYSIS stationnaire, WASTE *crushing unit* fixe, immobile, stationnaire
stationary[2]: ~**-bed adsorber** *n* AIR *adsorption process*, WATER POLL adsorbeur à lit fixe *m*; **container system** *n* (*SCS*) WASTE, WATER SUPP conteneur fixe *m*; ~ **emission source** *n* AIR source d'émissions fixe *f*; ~ **fluidized bed** *n* AIR, SOIL lit fluidisé fixe *m*; ~ **fluidized-bed oven** *n* AIR, SOIL, WASTE, WATER POLL four fixe à lit fluidisé *m*; ~ **sampler** *n* ANALYSIS échantillonneur stationnaire *m*; ~ **tank** *n* AIR réservoir fixe *m*
statistical[1] *adj* ANALYSIS statistique
statistical[2]: ~ **evaluation** *n* ANALYSIS évaluation statistique *f*; ~ **possibility of error** *n* ANALYSIS risque statistique d'erreur *m*
stator: ~**-rotor system** *n* AIR système stator-rotor *m*
statutory[1] *adj* POLICY, SAFETY statutaire
statutory[2]: ~ **power** *n* POLICY pouvoir statutaire *m*; ~ **powers** *n pl* POLICY *of administration* pouvoirs statutaires *m pl*; ~ **regulation** *n* POLICY, SAFETY réglementation statutaire *f*
STC *abbr* (*sound transmission class*) NOISE classe de transmission acoustique *f*
steam[1] *n* GEN vapeur *f*, vapeur d'eau *f*; ~ **air preheater** AIR réchauffeur de vapeur *m*; ~ **circulation** RENEW ENERG circulation de vapeur *f*; ~ **cleaning** MARINE nettoyage à la vapeur *m*; ~ **distillation** ANALYSIS évapodistillation *f*, distillation à la vapeur d'eau *f*; ~ **generator** WATER SUPP générateur de vapeur *m* (*GV*); ~ **power plant** POLICY, RENEW ENERG centrale thermique à vapeur *f*; ~ **pressure** SOIL *soil vapour extraction*, WATER POLL, WATER SUPP pression de vapeur *f*; ~ **pressure thermometer** ANALYSIS thermomètre à pression de vapeur *m*; ~ **process** RENEW ENERG procédé mettant en œuvre la vapeur *m*; ~ **stripping** SOIL extraction à la vapeur *f*, WATER POLL stripage à la vapeur *m*; ~ **turbine** RENEW ENERG turbine à vapeur *f*; ~**-vacuum stripping** SOIL extraction à la vapeur sous vide *f*
steam[2]: ~ **out** *vt* AIR, ANALYSIS, WATER SUPP évaporer
steel: ~ **band strapping** *n* WASTE ruban d'acier *m*; ~ **body** *n* AIR *of catalyst* corps en acier *m*; ~ **fiber** *AmE*, ~ **fibre** *BrE* *n* AIR *filter medium* fibre d'acier *f*; ~ **housing** *n* AIR caisson en acier *m*; ~ **reusable CKD container** *n* WASTE pièces détachées de conteneur en acier recyclables *f pl*
STEL *abbr* (*short-term exposure limit*) ANALYSIS limite d'exposition à court terme *f*

stemflow *n* AIR écoulement supercortical *m*, eau de ruissellement sur les troncs *f*
step: **~-by-step** *adj* ANALYSIS par étapes, par paliers, pas à pas
steppe: **~ region** *n* RENEW ENERG région de steppe *f*
stepped: **~ aeration** *n* WATER POLL, WATER SUPP aération étagée *f*
stereospecific: **~ catalyst** *n* AIR catalyseur stéréospécifique *m*
sterile *adj* ANALYSIS, WATER POLL stérile
sterilization *n* ANALYSIS, WATER POLL stérilisation *f*
Steuler: **~ process** *n* AIR procédé Steuler *m*
stickiness *n* AIR adhésivité *f*
sticky[1] *adj* AIR, WASTE adhésif, collant, gluant
sticky[2]: **~ deposit** *n* AIR dépôt adhésif *m*; **~ dust** *n* AIR poussière adhésive *f*, poussière collante *f*; **~ emissions** *n pl* AIR émissions collantes *f pl*
still *adj* WATER POLL mort
stilling: **~ chamber** *n* WATER SUPP bassin de tranquillisation *m*, chambre de tranquillisation *f*; **~ picket fence** *n* WATER SUPP herse de tranquillisation *f*
Stirling: **~ engine** *n BrE* (*cf Stirling motor AmE*) AIR, RENEW ENERG moteur à air chaud *m*, moteur à air dilaté *m*, moteur Stirling *m*; **~ motor** *n AmE* (*cf Stirling engine BrE*) AIR, RENEW ENERG moteur à air chaud *m*, moteur à air dilaté *m*, moteur Stirling *m*
stirrer *n* SOIL agitateur *m*
STN: **~-modem** *n* SAFETY modem RTC *m*
stock *n* POLICY gisement *m*
stoichiometric[1] *adj* AIR, ANALYSIS, WATER POLL stœchiométrique
stoichiometric[2]: **~ combustion** *n* AIR combustion stœchiométrique *f*; **~ oxygen supply** *n* AIR approvisionnement en oxygène stœchiométrique *m*
stoichiometry *n* AIR, ANALYSIS, WATER POLL stœchiométrie *f*
stoker *n* AIR *device* distributeur mécanique *m*, *person* chargeur *m*, chauffeur *m*
Stokes: **~'s law** *n* AIR, ANALYSIS loi de Stokes *f*
stone: **~-filled** *adj* WASTE empierré
stoneware: **~ pipe** *n* WATER POLL tuyau en grès *m*
stop *vt* WATER POLL, WATER SUPP engorger
stoppage *n* WATER POLL, WATER SUPP *in sewer* engorgement *m*, obstruction *f*
storable *adj* MARINE, RENEW ENERG, SAFETY, WATER POLL stockable
storage[1]: **~-free** *adj* RENEW ENERG sans stockage
storage[2] *n* MARINE, RENEW ENERG, SAFETY, WATER POLL stockage *m*; **~ area** SOIL *for polluted and remediated soils* zone de stockage *f*, WASTE aire de stockage *f*, surface de stockage *f*; **~ basin** WATER POLL réservoir de stockage *m*; **~ behavior** *AmE*, **~ behaviour** *BrE* WASTE, WATER SUPP comportement au stockage *m*; **~ capacity** AIR *of filter*, RENEW ENERG, WASTE capacité de stockage *f*; **~ chamber** WASTE salle de stockage *f*; **~ durability** WASTE durée d'emmagasinage *f*; **~ efficiency** RENEW ENERG rendement de stockage *m*; **~ facility** AIR installation de stockage *f*, WASTE capacité de stockage *f*, installation de stockage *f*, usine de stockage *f*; **~ filter** AIR filtre de stockage *m*; **~ function** RENEW ENERG fonction de stockage *f*; **~ gas** WASTE gaz de stockage *m*; **~ hydroelectric plant** RENEW ENERG centrale hydroélectrique avec retenue *f*, retenue hydroélectrique *f*; **~ medium** RENEW ENERG *of heat pump* fluide caloporteur *m*; **~ reservoir** RENEW ENERG *hydropower* retenue *f*, WATER SUPP bassin de retenue *m*, barrage réservoir *m*; **~ site** WASTE site de stockage *m*; **~ space** WASTE volume de stockage *m*; **~ tank** AIR cuve de stockage *f*, WASTE citerne de stockage *f*, cuve de stockage *f*; **~ technology** RENEW ENERG technique de stockage *f*; **~ volume** RENEW ENERG volume de stockage *m*
store[1] *vt* ANALYSIS *sample*, RENEW ENERG *solar energy*, SOIL, WASTE stocker; **~ intermediately** SOIL *excavated mass*, WASTE stocker provisoirement, stocker temporairement
store[2]: **~ energy** *vi* RENEW ENERG stocker l'énergie
stored[1] *adj* ANALYSIS, RENEW ENERG, SOIL, WASTE stocké
stored[2]: **~ energy** *n* RENEW ENERG énergie stockée *f*; **~ geothermal energy** *n* AIR, RENEW ENERG énergie géothermique stockée *f*; **~ heat** *n* RENEW ENERG chaleur stockée *f*; **~ solar energy** *n* RENEW ENERG énergie solaire stockée *f*; **~ water** *n* RENEW ENERG, WATER SUPP eau de réserve *f*, eau stockée *f*
storeroom *n* WASTE magasin de stockage *m*
storm: **~ overflow** *n* WATER SUPP trop-plein des eaux pluviales *m*; **~ water** *n* WATER POLL eau pluviale *f*, eau d'orage *f*, WATER SUPP eau pluviale *f*; **~-water main** *n* WASTE, WATER POLL, WATER SUPP égout pluvial *m*; **~-water overflow** *n* WATER POLL déversoir d'eaux pluviales *m*; **~-water overflow tank** *n* WATER POLL déversoir d'eaux pluviales *m*; **~-water retention tank** *n* WATER POLL bassin de rétention des eaux pluviales *m*; **~-water sewer** *n* WASTE, WATER POLL, WATER SUPP égout pluvial *m*; **~-water treatment** *n* WATER POLL traitement des eaux pluviales *m*
STORS *abbr* (*sludge-to-oil reactor system*) WATER POLL réacteur boues-huile *m*
stove: **~ heater** *n* AIR poêle *m*
STP *abbr* (*standard temperature and pressure*) ANALYSIS conditions normales de température et pression *f pl*, température et pression normales
strainer *n* MARINE filtre *m*, tamis *m*
strapping: **~ equipment** *n* WASTE matériel d'enrubannage *m*, matériel de banderolage *m*; **~ machine** *n* WASTE cercleuse *f*; **~ seal** *n* WASTE fermeture de cerclage *f*; **~ steel** *n* WASTE acier de cerclage *m*
stratosphere *n* AIR stratosphère *f*
stratospheric[1] *adj* AIR stratosphérique
stratospheric[2]: **~ ozone** *n* AIR ozone stratosphérique *m*; **~ ozone concentration** *n* AIR concentration d'ozone stratosphérique *f*; **~ ozone-layer depletion** *n* AIR appauvrissement de la couche d'ozone stratosphérique *m*; **~ ozone pollution** *n* AIR pollution de l'ozone stratosphérique *f*
straw: **~ bale** *n* RENEW ENERG *biomass conversion* balle de paille *f*; **~ burner** *n* RENEW ENERG brûleur à paille *m*; **~-burning plant** *n* RENEW ENERG *biomass conversion* usine brûlant de la paille *f*; **~-fired boiler** *n* RENEW ENERG chaudière à paille *f*; **~ pellet** *n* RENEW ENERG granulé de paille *m*
strawboard *n* WASTE carton paille *m*
stream[1] *n* AIR courant *m*; **~ aeration** WATER POLL aération des cours d'eau *f*
stream[2]: **~ in** *vi* AIR affluer
streamline *n* AIR ligne aérodynamique *f*, profil aérodynamique *m*; **~ flow** AIR écoulement non perturbé *m*
street: **~ cleaner** *n* WASTE *machine* balayeuse *f*, engin de nettoiement de la voie publique *m*; **~-cleaning vehicle** *n* WASTE véhicule de lavage des rues *m*, véhicule de nettoyage des chaussées *m*; **~ emissions level** *n* AIR, NOISE niveau d'émissions dans la rue *m*; **~ furniture** *n* POLICY mobilier urbain *m*; **~ refuse** *n* WASTE déchets de la voie publique *m pl*; **~ sweeper** *n* WASTE *machine* balayeuse *f*
streptococci *n pl* WATER POLL streptocoques *m pl*
stress *n* AIR, MARINE, SOIL contrainte *f*, perturbation *f*; **~ capacity of soil** SOIL résistance du sol aux agressions *f*
stretch: **~ film** *n* WASTE film étirable *m*, pellicule extensible *f*; **~ wrapping** *n* WASTE emballage extensible *m*
strict: **~ liability** *n* MANAGE, POLICY responsabilité objective *f*
string: **~ rot** *n* WASTE pourriture fibreuse *f*
stripping *n* WASTE *rolling mill* démontage *m*, démoulage *m*, WATER POLL stripage *m*; **~ pump** MARINE pompe d'assèchement *f*; **~-purging technique** ANALYSIS *for extraction of VOC from water sample* méthode d'extraction par

stripping *f*; ~ **test** ANALYSIS essai de décoloration *m*, essai de lavage *m*; ~ **tower** SOIL colonne de rectification *f*
stroboscope *n* ANALYSIS stroboscope *m*
strong: ~ **sample** *n* ANALYSIS échantillon fortement concentré *m*, échantillon fortement dosé *m*; ~-**sun period** *n* RENEW ENERG *solar energy* période de fort ensoleillement *f*
strontium: ~ **sulfate** *AmE*, ~ **sulphate** *BrE n* AIR *adsorbent* sulfate de strontium *m*
structural: ~ **engineering** *n* AIR, RENEW ENERG, WASTE génie civil *m*; ~-**engineering technology** *n* AIR technique de construction *f*, technologie de génie civil *f*, RENEW ENERG technique de génie civil *f*; ~ **insulation board** *n* NOISE carton renforcé isolant *m*; ~ **reaction injection molding** *AmE*, ~ **reaction injection moulding** *BrE n* (*S-RIM*) WASTE moulage par injection-réaction en moule ouvert *m*
structurally[1]: ~-**engineered** *adj* AIR, RENEW ENERG, WASTE de génie civil
structurally[2]: ~-**engineered barrier** *n* WASTE barrière intégrée *f*
structure *n* GEN structure *f*; ~-**borne absorber** NOISE absorbeur intégré *m*; ~-**borne noise** NOISE bruit de structure *m*; ~-**borne noise reduction** NOISE réduction intégrée du bruit *f*; ~-**borne noise transmission** NOISE transmission intégrée du bruit *f*; ~-**borne noise-proofing** NOISE isolation acoustique intégrée *f*
strut *n* WASTE mât *m*, pilier *m*
stub: ~ **cable** *n* SOIL câble de dérivation *m*
subcell *n* WASTE sous-cellule *f*
subcool *vt* AIR sous-refroidir
subdivision *n* POLICY *administrative unit* subdivision *f*
sub-index *n* POLICY sous-indice *m*
subject: ~ **to calibration** *adj* ANALYSIS soumis à étalonnage
sub-lethal[1] *adj* SAFETY sublétal
sub-lethal[2]: ~ **dose** *n* SAFETY dose sublétale *f*; ~ **effect** *n* SAFETY effet sublétal *m*
submains *n pl* WATER SUPP dérivation *f*
submerged: ~-**plant tank** *n* WATER POLL cuve d'usine immergée *f*; ~ **weir** *n* WATER SUPP déversoir noyé *m*
submersible: ~ **burner** *n* WATER POLL, WATER SUPP brûleur submersible *m*; ~ **pump** *n* ANALYSIS pompe submersible *f*
submicron *n* AIR *particle* sous-micron *m*; ~ **particulate filter** AIR filtre à particules submicroniques *m*; ~ **particulate matter** AIR matière sous forme de particules submicroniques *f*
submicropore *n* AIR pore submicronique *m*
subpopulation: ~ **collection** *n* SAFETY collecte sur un échantillon de population *f*; ~ **collective dose** *n* AIR dose collective reçue par un groupe de population *f*
subsequent: ~ **dewatering** *n* SOIL dénoyage ultérieur *m*
subsidence *n* SOIL subsidence *f*
subsidiarity *n* POLICY subsidiarité *f*; ~ **principle** POLICY principe de subsidiarité *m*
subsidize *vt* POLICY subventionner
subsidy *n* POLICY subvention *f*
subsoil: ~ **resources** *n pl* RENEW ENERG ressources du sous-sol *f pl*
substance *n* GEN substance *f*
substrate *n* WATER SUPP substrat *m*; ~ **utilization** WATER SUPP capture *f*
subsurface: ~ **diffusion** *n* WATER POLL *aeration system* diffusion souterraine *f*; ~ **repository** *n* WASTE dépôt souterrain *m*; ~ **water** *n* WATER SUPP eaux souterraines *f pl*
subterranean: ~ **water** *n* WATER SUPP eaux souterraines *f pl*
subterraneous: ~ **curtain** *n* SOIL, WASTE écran souterrain *m*; ~ **curtain plate** *n* SOIL, WASTE écran souterrain *m*
subtidal: ~ **zone** *n* MARINE zone subtidale *f*
suburbs *n pl* POLICY banlieue *f*
suck: ~ **in** *vt* AIR, WATER POLL, WATER SUPP aspirer
suction *n* AIR, WATER POLL, WATER SUPP aspiration *f*, succion *f*; ~ **blower** WASTE balayeuse aspiratrice *f*; ~ **box** AIR caisse aspirante *f*; ~ **collector** WASTE ramasseuse aspiratrice *f*; ~ **device** AIR installation d'aspiration *f*; ~ **filter** AIR filtre aspirant *m*, WASTE filtre à vide *m*, filtre d'aspiration *m*; ~ **and filter installation for dust and chippings** SAFETY aspirateur et filtre pour poussières et copeaux *m*; ~ **head** WATER POLL hauteur d'aspiration *f*; ~ **level** SOIL niveau d'aspiration *m*; ~ **lift** WATER POLL hauteur d'aspiration *f*; ~ **nozzle** WATER SUPP buse d'aspiration *f*, tuyère d'aspiration *f*; ~ **pipe** AIR conduite d'aspiration *f*; ~ **piping** WATER POLL tuyauterie d'aspiration *f*; ~ **port** WASTE orifice d'aspiration *m*; ~ **pump** ANALYSIS machine aspirante *f*, pompe aspirante *f*, pompe suceuse *f*, WATER POLL pompe aspirante *f*, WATER SUPP appareil déprimogène *m*, pompe aspirante *f*; ~ **side** WATER POLL *of pump* côté admission *m*; ~ **strainer** WATER SUPP filtre à succion *m*; ~-**type governor** AIR régulateur à dépression *m*
sulfate *AmE see sulphate BrE*
sulfide *AmE see sulphide BrE*
sulfite *AmE see sulphite BrE*
sulfur *AmE see sulphur BrE*
sulfuric *AmE see sulphuric BrE*
sulfurous *AmE see sulphurous BrE*
sulphate *n BrE* AIR, SOIL sulfate *m*; ~ **cellulose** *BrE* WATER SUPP cellulose au sulfate *f*; ~ **respiration** *BrE* WATER SUPP respiration du sulfate *f*
sulphide *n BrE* AIR sulfure *m*; ~ **precipitation** *BrE* WATER POLL, WATER SUPP précipitation de sulfure *f*
sulphite *n BrE* WATER SUPP bisulfite *m*; ~ **cellulose** *BrE* WATER SUPP pâte au bisulfite *f*
sulphur[1]: ~-**free** *adj BrE* AIR désulfuré, sans soufre; ~-**rich** *adj BrE* AIR riche en soufre
sulphur[2] *n* (*S*) WATER POLL, SOIL soufre (*S*); ~ **anhydride** *BrE* (SO_2) WATER SUPP anhydride sulfureux *m* (SO_2); ~ **budget** *BrE* AIR bilan de soufre *m*; ~ **content** *BrE* AIR teneur en soufre *f*; ~ **cycle** *BrE* AIR cycle du soufre *m*; ~ **dioxide** *BrE* (SO_2) AIR dioxyde de soufre *m* (SO_2); ~ **dioxide emissions** *n pl BrE* AIR émissions de dioxyde de soufre *f pl*; ~ **dioxide poisoning** *BrE* AIR empoisonnement au dioxyde de soufre *m*; ~ **dioxide reduction** *BrE* AIR réduction des émissions de dioxyde de soufre *f*; ~ **heptoxide** *BrE* AIR heptaoxyde de soufre *m*; ~ **monoxide** *BrE* AIR monoxyde de soufre *m* (*SO*); ~ **oxide** *BrE* (*SO*) AIR, WATER SUPP oxyde de soufre *m* (*SO*); ~ **oxides** *n pl* (SO_x) AIR, WATER SUPP oxydes de soufre *m pl* (SO_x); ~-**oxides scrubber** *BrE* AIR épurateur d'oxydes de soufre *m*, désulfureur *m*; ~ **recovery plant** *BrE* AIR, WATER SUPP installation de désulfuration *f*; ~ **tetroxide** *BrE* (SO_4) AIR tétraoxyde de soufre *m* (SO_4); ~ **trioxide** *BrE* (SO_3) AIR trioxyde de soufre *m* (SO_3)
sulphuric: ~ **acid** *n* (H_2SO_4) *BrE* AIR, SOIL, WATER POLL acide sulfurique *m* (H_2SO_4); ~ **anhydride** *n BrE* (SO_3) WATER SUPP anhydride sulfurique *m* (SO_3)
sulphurous[1] *adj BrE* AIR sulfureux
sulphurous[2]: ~ **acid** *n* (H_2SO_3) *BrE* AIR acide sulfureux *m* (H_2SO_3); ~ **combustible** *n BrE* AIR combustible sulfureux *m*; ~ **gas** *n BrE* AIR gaz sulfureux *m*; ~ **waste** *n BrE* WATER SUPP déchets sulfureux *m pl*
Sulzer-Ariqua: ~ **process** *n* AIR procédé Sulzer-Ariqua *m*
summertime: ~ **ozone value** *n* AIR indice estival d'ozone *m*
sump *n* WATER POLL puisard *m*
sunlight *n* RENEW ENERG ensoleillement *m*, insolation *f*; ~ **radiation** AIR rayonnement solaire *m*
sunny: ~ **region** *n* RENEW ENERG région ensoleillée *f*
superchlorination *n* WATER POLL surchloration *f*
supercritical: ~-**fluid chromatography** *n* (*SFC*) ANALYSIS chromatographie en phase supercritique *f* (*CPS*); ~-**fluid extraction** *n* (*SFE*) ANALYSIS extraction de fluide supercritique *f*; ~ **oxidation** *n* SOIL oxydation supercritique *f*; ~ **water** *n* SOIL eau supercritique *f*
superheated: ~ **steam** *n* AIR, RENEW ENERG, WATER SUPP vapeur surchauffée *f*

superheater *n* AIR, RENEW ENERG, WATER POLL, WATER SUPP surchauffeur *m*
supernatant: ~ **liquor** *n* WATER POLL liqueur surnageante *f*, liquide surnageant *m*, surnageant *m*
supersaturated *adj* AIR, WATER SUPP sursaturé
supersede *vt* POLICY détrôner
supersonic: ~ **boom** *n* NOISE bang supersonique *m*; ~ **sound** *n* NOISE son supersonique *m*
supervision *n* GEN surveillance *f*
supervisor *n* MANAGE, SAFETY inspecteur *m*
supervisory: ~ **authority** *n* POLICY autorité de tutelle *f*
supplementary: ~ **fuel** *n* AIR *thermal and catalytic combustion*, WATER POLL combustible d'appoint *m*; ~ **purification** *n* AIR épuration supplémentaire *f*
supplied: ~**-air breathing apparatus** *n* SAFETY appareil respiratoire fourni *m*
supplier *n* AIR, RENEW ENERG, WATER POLL, WATER SUPP fournisseur *m*
supply[1] *n* AIR, RENEW ENERG, WATER POLL, WATER SUPP alimentation *f*, approvisionnement *m*, fourniture *f*; ~ **pipe** WATER SUPP tuyau d'amenée *m*; ~ **potential** RENEW ENERG *of energy source* potentiel d'approvisionnement *m*; ~ **pump** WATER SUPP pompe alimentaire *f*
supply[2] *vt* AIR, RENEW ENERG, WATER POLL, WATER SUPP alimenter,fournir, approvisionner
support: ~ **body** *n* AIR corps support *m*, support *m*; ~ **element** *n* AIR *of bag filter* élément support *m*; ~ **ring** *n* AIR *of bag filter* anneau support *m*; ~ **wire cage** *n* AIR *of bag filter* cage d'écureuil support *f*
supporting: ~ **grate** *n* AIR grille support *f*; ~ **suspension** *n* SOIL suspension d'appui *f*
suppress *vt* MARINE, NOISE supprimer
surface[1]: ~**-active** *adj* AIR, SOIL, WATER POLL tensioactif
surface[2] *n* GEN surface *f*; ~**-active agent** AIR, MARINE, SOIL surfactant *m*, agent tensioactif *m*, WATER POLL surfactant *m*; ~**-active substance** SOIL *additive* substance tensioactive *f*; ~ **aeration** WATER POLL aération de surface *f*; ~ **application** WATER POLL application en surface *f*; ~ **area** AIR, SOIL, MARINE, WASTE, WATER POLL, WATER SUPP superficie *f*, surface *f*; ~ **catalysis** AIR catalyse de surface *f*; ~ **collector** RENEW ENERG collecteur de surface *m*; ~ **condenser** AIR condenseur de surface *m*; ~ **conductivity** AIR conductivité superficielle *f*; ~ **cover** SOIL couverture de surface *f*; ~ **dumping** WASTE décharge de surface *f*, rejet en surface *m*; ~ **film** WATER POLL chapeau *m*, pellicule de surface *f*; ~ **filter** AIR, WASTE, WATER POLL filtre de surface *m*; ~ **filtration** AIR, WASTE, WATER POLL filtration en surface *f*; ~ **filtration with membranes** WATER POLL, WATER SUPP filtration en surface sur membranes *f*; ~ **installation** WASTE installation de surface *f*; ~ **irradiation** SOIL irradiation de surface *f*; ~ **irrigation** SOIL irrigation de surface *f*; ~ **landfill** WASTE décharge à ciel ouvert *f*; ~ **layer** MARINE couche de surface *f*; ~ **lining** NOISE *sound-absorbing* revêtement de surface *m*; ~ **load** AIR charge de surface *f*; ~ **method** WASTE méthode de surface *f*; ~ **pollutant** AIR, WATER POLL polluant de surface *m*; ~ **protection film** WASTE pellicule de protection de surface *f*; ~ **protection tape** WASTE bande de protection de surface *f*; ~ **requirements** *n pl* RENEW ENERG besoins de surface *m pl*; ~ **reservoir** WATER POLL *indirect reuse of wastewater* réservoir de surface *m*; ~ **run-off** WATER SUPP ruissellement de surface *m*; ~ **seawater** RENEW ENERG eau de mer de surface *f*, eau de mer superficielle *f*; ~ **sensor** ANALYSIS capteur de surface *m*; ~ **skimmer** WATER SUPP racleur de surface *m*; ~ **smoothing** SOIL lissage de surface *m*; ~**-tension modifier** MARINE modificateur de tension superficielle *m*; ~ **water** WATER POLL, WATER SUPP eau de jour *f*; ~ **water management** MANAGE, WATER SUPP gestion des eaux superficielles *f*; ~ **water temperature** RENEW ENERG température de l'eau de mer superficielle *f*; ~**water treatment** WATER SUPP traitement des eaux de surface *m*
surfactant *n* AIR, MARINE surfactant *m*, WATER POLL, WATER SUPP agent tensioactif *m*, surfactant *m*, tensioactif *m*
surge: ~ **hopper** *n* SOIL trémie-tampon *f*
surrounding: ~**-field analysis** *n* ANALYSIS analyse du champ périphérique *f*
surveillance *n* GEN surveillance *f*
survey[1] *n* POLICY enquête *f*; ~ **map** MANAGE, POLICY, SOIL carte topographique *f*; ~ **method** POLICY enquête par sondage *f*; ~ **of resources** POLICY évaluation des ressources *f*
survey[2] *vt* ANALYSIS, MANAGE, SAFETY étudier
suspended[1] *adj* GEN en suspension, suspendu
suspended[2]: ~ **bacteria** *n pl* WATER POLL bactéries en suspension *f pl*; ~ **ceiling** *n* NOISE *resilient layer* plafond suspendu *m*; ~ **frame** *n* AIR *of bag filter* cadre suspendu *m*; ~ **liquid droplet** *n* AIR gouttelette de liquide en suspension *f*; ~ **liquid phase** *n* AIR phase liquide en suspension *f*; ~ **matter** *n* WATER POLL matière en suspension *f*; ~ **organic solid** *n* WATER POLL matière organique solide en suspension *f*; ~ **particle** *n* AIR particule en suspension *f*; ~ **particulate matter** *n* AIR particule en suspension *f*; ~ **and settleable solid** *n* WATER POLL matière solide en suspension et décantable *f*; ~ **solid** *n* WATER POLL matière solide en suspension *f*, matière en suspension *f*; ~ **solid particle** *n* AIR, ANALYSIS particule solide en suspension *f*; ~ **solid particle meter** *n* ANALYSIS compteur de particules solides en suspension *m*
suspension *n* WASTE suspension *f*; ~ **dryer** AIR séchoir à tissu flottant *m*, WATER POLL séchoir à bande *m*, séchoir à tapis *m*
sustain *vt* SAFETY *injury* supporter
sustainability *n* POLICY caractère durable *m*, viabilité *f*
sustainable[1] *adj* POLICY durable, viable, SOIL durable
sustainable[2]: ~ **business** *n* MANAGE affaire viable *f*, entreprise viable *f*; ~ **development** *n* POLICY développement durable *m*; ~ **economy** *n* POLICY économie viable *f*; ~ **forestry** *n* POLICY sylviculture viable *f*; ~ **recycling economy** *n* POLICY, RENEW ENERG économie de recyclage viable *f*; ~ **soil use** *n* SOIL utilisation durable des sols *f*
SVOC *abbr* (*semivolatile organic compounds*) SOIL composés organiques semi-volatils *m pl*
swallow: ~ **hole** *n* WATER POLL aven *m*
sweep[1]: ~ **coagulation** *n* WATER POLL *flocculation* coagulation par piégeage *f*
sweep[2] *vt* WASTE balayer
sweeping *n* WASTE balayage *m*
swell *n* MARINE houle *f*; ~ **test** MARINE essai de houle *m*
swelling *n* WATER POLL gonflement *m*
swinging: ~ **scraper** *n* WATER POLL bras racleur oscillant *m*, racleur à balancier *m*
symbiosis *n* WATER SUPP symbiose *f*
synchronous[1] *adj* ANALYSIS synchrone
synchronous[2]: ~ **generator** *n* RENEW ENERG *of wind turbine* alternateur synchrone *m*
synergetic: ~ **effect** *n* AIR, WATER SUPP effet synergique *m*
synergy *n* AIR synergie *f*
synthesis: ~ **gas generation** *n* WASTE *plastics* production de gaz de synthèse *f*
synthetic[1] *adj* WATER POLL de synthèse, synthétique
synthetic[2]: ~ **fertilizer** *n* SOIL, WATER POLL engrais chimique *m*; ~ **fiber** *AmE*, ~ **fibre** *BrE* *n* AIR *filter medium* fibre de synthèse *f*, fibre synthétique *f*, WASTE, WATER SUPP fibre synthétique *f*; ~ **lining** *n* WASTE revêtement synthétique *m*; ~ **material** *n* WASTE *landfill sealing* matériau de synthèse *m*, matériau synthétique *m*; ~ **natural gas** *n* (*SNG*) RENEW ENERG gaz naturel de synthèse *m*; ~ **oil** *n* RENEW ENERG huile de synthèse *f*; ~ **polymer** *n* WATER POLL polymère de synthèse *m*
system *n* ANALYSIS, MANAGE, POLICY système *m*, WATER SUPP

réseau *m*; ~ **boundary** MANAGE limite du système *f*; ~ **parameter** ANALYSIS paramètre du système *m*

systematic: ~ **possibility of error** *n* ANALYSIS risque systématique d'erreur *m*

T

Ta *abbr* (*tantalum*) WATER SUPP Ta (*tantale*)
tablet: ~ **press** *n* ANALYSIS *pellet press* pastilleuse *f*, presse à comprimés *f*; ~ **sorting and inspection machine** *n* WASTE machine de tri et d'inspection des produits pharmaceutiques *f*
tack: ~ **level** *n* WASTE degré d'adhésivité *m*
tacky: ~ **particulate components** *n pl* AIR éléments particulaires adhésifs *m pl*
tail: ~ **gas** *n* AIR gaz de queue *m*
tailings *n pl* WASTE déchets *m pl*, stériles *m pl*; ~ **container** *n* WASTE conteneur d'otons *m*, conteneur de déchets *m*; ~ **pond** *n* WATER POLL bassin à schistes *m*, bassin de refus *m*
tailpipe *n* AIR tubulure d'évacuation *f*, tuyau de sortie *m*; ~ **emissions** *n pl* AIR *carbon monoxide, nitrogen oxide* émissions en sortie de tuyau *f pl*
tamped: ~ **density** *n* ANALYSIS densité après damage *f*
tamper: ~-**proof closure** *n* SAFETY fermeture inviolable *f*; ~-**proofing** *n* SAFETY protection contre les manipulations *f*
tamping: ~ **volume meter** *n* ANALYSIS débitmètre *m*, voluménomètre de pilonnage *m*
tandem: ~ **detector** *n* ANALYSIS détecteur tandem *m*; ~ **mounting** *n* AIR *of cyclones* montage tandem *m*
tangential: ~-**entry cyclone** *n* AIR cyclone à admission tangentielle *m*
tank *n* AIR, WASTE cuve *f*, réservoir *m*, citerne *f*, WATER POLL bassin *m*, citerne *f*, WATER SUPP bâche *f*; ~ **adsorber** AIR adsorbeur de réservoir *m*; ~ **bottom** WATER POLL fond de cuve *m*; ~ **room** WASTE hall de stockage *m*
tanker *n* MARINE pétrolier *m*, wagon citerne
tantalum *n* (*Ta*) WATER SUPP tantale *m* (*Ta*)
tap: ~ **water** *n* WATER POLL, WATER SUPP eau de consommation *f*, eau de distribution *f*
tapered: ~ **aeration** *n* WATER POLL aération avec réduction progressive *f*, aération décroissante *f*
tapping *n* WATER SUPP captage *m*, prélèvements *m pl*
tar *n* MARINE goudron *m*; ~ **ball** MARINE aggloméré de goudron *m*; ~ **paper** WASTE papier goudron *m*
tare *n* WASTE tare *f*; ~ **weight** WASTE poids à vide *m*
target: ~ **droplet** *n* AIR gouttelette cible *f*; ~ **group** *n* POLICY groupe cible *m*; ~ **sample** *n* ANALYSIS échantillon cible *m*
taste: ~ **control** *n* WATER SUPP neutralisation du goût *f*
tasteless *adj* AIR *gas* insipide
tax *n* POLICY impôt *m*, taxe *f*; ~ **base** POLICY base d'imposition *f*, base fiscale *f*; ~ **burden** POLICY pression fiscale *f*; ~ **exemption** POLICY exemption fiscale *f*; ~ **incentive** MANAGE, POLICY incitation fiscale *f*; ~ **law** POLICY droit fiscal *m*; ~ **provision** POLICY provision pour impôt *f*; ~ **quota** POLICY quota fiscal *m*; ~ **rate** POLICY taux d'imposition *m*; ~ **rebate** POLICY déduction fiscale *f*, dégrèvement fiscal *m*; ~ **reduction** POLICY réduction d'impôt *f*; ~ **relief** POLICY abattement *m*, dégrèvement fiscal *m*, exonération fiscale *f*; ~ **return** MANAGE déclaration de revenus *f*; ~ **shift** POLICY report d'imposition *m*; ~ **tariff** POLICY barème d'imposition *m*
TC *abbr* (*total carbon*) ANALYSIS carbone total *m*
TCD *abbr* (*thermal conductivity detector*) ANALYSIS détecteur de conductivité thermique *m*
TCDD *abbr* (*tetrachlorodibenzoparadioxin*) SOIL TCDD (*tétrachlorodibenzoparadioxine*)
TCL_0 *abbr* (*total lethal concentration zero*) SAFETY concentration létale zéro totale *f*
TDL_0 *abbr* (*total lethal dose zero*) SAFETY dose létale zéro totale *f*
TDS *abbr* (*total dissolved solids*) WATER SUPP matière solide total dissoute *f*
tear: ~ **gas** *n* AIR gaz lacrymogène *m*; ~ **strength** *n* AIR *of filter* résistance à la déchirure *f*; ~-**tab lid** *n* WASTE couvercle à languette de déchirage *m*
tearing *n* AIR *of filter* déchirure *f*
technical[1] *adj* GEN technique
technical[2]: ~ **directive** *n* POLICY directive technique *f*; ~ **instruction on waste management** *n* WASTE instructions techniques sur la gestion des déchets *f pl*; ~ **potential** *n* RENEW ENERG *of energy source* potentiel technique *m*; ~ **regulation on noise protection** *n* SOIL réglementation sur la protection contre le bruit *f*; ~ **safety** *n* SAFETY sûreté technique *f*; ~ **safety demand** *n* SAFETY exigence de conformité aux critères techniques de sécurité *f*; ~ **safety measure** *n* SAFETY mesure technique de sécurité *f*
technically: ~ **adjustable** *adj* WASTE facile à adapter d'un point de vue technique
technique *n* GEN technique *f*
technology *n* GEN technique *f*, technologie *f*; ~ **application** GEN application technologique *f*; ~ **assessment** POLICY évaluation technologique *f*
telecontrol: ~ **system** *n* WATER POLL système de télécommande *m*
telemetering *n* ANALYSIS télémesure *f*
telemetry *n* ANALYSIS télémétrie *f*
telescope *n* ANALYSIS télescope *m*
TEM *abbr* (*transmission electron microscopy*) ANALYSIS microscopie électronique à transmission *f*
temper *vt* ANALYSIS détremper
temperature[1]: ~-**controlled** *adj* ANALYSIS à réglage de température; ~-**resistant** *adj* AIR *filter material* thermorésistant, SOIL thermostable; ~-**stable** *adj* ANALYSIS stable en température
temperature[2] *n* GEN température *f*; ~ **gradient** AIR, NOISE, WATER POLL gradient de température *m*, WATER SUPP gradient thermométrique *m*; ~ **inversion** AIR inversion de température *f*; ~ **measurement** ANALYSIS mesure de température *f*; ~ **probe** ANALYSIS sonde de température *f*; ~ **range** AIR plage de température *f*; ~ **resistance** AIR *of filter material* résistance aux écarts de température *f*, thermorésistance *f*; ~ **rise** AIR élévation de température *f*; ~ **sensor** ANALYSIS capteur de température *m*, palpeur pyrométrique *m*; ~ **stability** ANALYSIS stabilité de température *f*
temporal: ~ **dose distribution** *n* AIR échelonnement des doses *m*; ~ **evaluation** *n* ANALYSIS évaluation temporelle *f*; ~ **fluctuation** *n* AIR fluctuation temporelle *f*; ~ **resolution** *n* AIR, WATER SUPP résolution temporelle *f*; ~ **variation** *n* AIR variation temporelle *f*
temporary: ~ **deposit for hazardous waste** *n* WASTE dépôt temporaire de déchets dangereux *m*; ~ **hardness** *n* WATER POLL, WATER SUPP dureté temporaire *f*; ~ **remediation** *n* SOIL mesure correctrice provisoire *f*; ~ **storage** *n* WASTE stockage temporaire *m*; ~ **storage silo** *n* AIR silo de stockage temporaire *m*; ~ **tank** *n* AIR cuve temporaire *f*; ~ **threshold shift** *n* (*TTS*) NOISE déplacement de seuil temporaire *m*
tensile: ~ **strength** *n* AIR charge de rupture *f*, résistance à la traction *f*
tensiometer *n* ANALYSIS tensiomètre *m*

tension: ~ **member** *n* MARINE élément de tension *m*
teratogen *adj* SAFETY tératogène
tertiary: ~ **pond** *n* WATER POLL bassin tertiaire *m*; ~ **recycling** *n* WASTE recyclage tertiaire *m*; ~ **wastewater treatment** *n* WATER POLL traitement tertiaire des eaux usées *m*
Terylene® *n BrE* (*cf Dacron AmE*) AIR Dacron® *m*
test[1] *n* ANALYSIS essai *m*; ~ **conditions** *n pl* WASTE, WATER SUPP conditions d'essai *f pl*; ~ **crucible** ANALYSIS coupelle *f*; ~ **environment** ANALYSIS milieu d'essai *m*; ~ **filtration plant** WATER SUPP banc de filtration expérimental *m*; ~ **gas** ANALYSIS gaz d'essai *m*; ~ **gas calibrator** ANALYSIS étalonneur de gaz d'essai *m*; ~ **liner board** WASTE carton de revêtement expérimental *m*; ~ **paper** ANALYSIS papier indicateur *m*, papier réactif *m*; ~ **phase** RENEW ENERG *of renewable energy installation* stade expérimental *m*; ~ **portion** ANALYSIS prise d'essai *f*; ~ **probe** ANALYSIS échantillon d'essai *m*; ~ **sieving** ANALYSIS *for grain size analysis of dust* tamisage de contrôle *m*; ~ **solution** ANALYSIS solution d'essai *f*; ~ **tube** ANALYSIS tube à essai *m*; ~ **value** ANALYSIS valeur d'essai *f*; ~ **water** ANALYSIS eau d'essai *f*
test[2] *vt* ANALYSIS essayer, expérimenter
testing: ~ **aerator** *n* WATER POLL aérateur d'essai *m*; ~ **laboratory** *n* ANALYSIS laboratoire d'essai *m*; ~ **limit** *n* ANALYSIS limite d'essai *f*
tetrachlorodibenzoparadioxin *n* (*TCDD*) SOIL tétrachlorodibenzoparadioxine *f* (*TCDD*)
textile: ~ **waste** *n* WASTE déchets textiles *m pl*
TFT-LCD *abbr* (*thin film transistor liquid-crystal display*) ANALYSIS écran à cristaux liquides à transistor en couches minces *m*
theoretic: ~ **potential** *n* RENEW ENERG *of energy source* potentiel théorique *m*
theoretical: ~ **oxygen demand** *n* WATER POLL demande d'oxygène théorique *f*
therm: ~-**ionic detector** *n* ANALYSIS détecteur thermo-ionique *m*
thermal[1] *adj* RENEW ENERG thermique
thermal[2]: ~ **air cleaner** *n* AIR épurateur thermique *m*; ~ **analysis** *n* ANALYSIS analyse thermique *f*; ~ **balance** *n* AIR, WATER POLL bilan thermique *m*; ~ **capacity** *n* AIR capacité calorifique *f*, capacité thermique *f*; ~ **combustion** *n* AIR, WATER POLL combustion thermique *f*; ~ **combustion plant** *n* AIR centrale thermique *f*; ~ **conduction** *n* AIR, ANALYSIS conduction thermique *f*; ~ **conductivity detector** *n* (*TCD*) ANALYSIS détecteur de conductivité thermique *m*; ~ **decomposition** *n* WASTE décomposition thermique *f*; ~ **desorption** *n* ANALYSIS *pretreatment for pollutant analysis* désorption thermique *f*; ~ **desorption chamber** *n* ANALYSIS chambre de désorption thermique *f*; ~ **destruction** *n* SOIL *direct heating by radiation* destruction thermique *f*; ~ **discharge** *n* AIR, WASTE rejet thermique *m*; ~ **drying** *n* WATER POLL séchage thermique *m*; ~ **effluent** *n* AIR effluent thermique *m*; ~ **energy** *n* AIR, RENEW ENERG énergie thermique *f*; ~ **heat expansion** *n* AIR *of catalyst* dilatation thermique *f*; ~ **incinerator** *n* AIR *afterburner* incinérateur thermique *m*; ~ **load** *n* RENEW ENERG, WATER SUPP charge thermique *f*; ~ **oil** *n* RENEW ENERG huile thermique *f*; ~ **oil heater** *n* AIR poêle à mazout *m*; ~ **output** *n* AIR rendement thermique *m*; ~ **oxidation** *n* AIR *thermal combustion* oxydation thermique *f*; ~ **pollution** *n* AIR, WATER SUPP pollution thermique *f*; ~ **power plant** *n* AIR centrale thermique *f*; ~ **process** *n* WATER POLL traitement thermique *m*; ~ **reactor** *n* AIR réacteur thermique *m*; ~ **recovery** *n* WASTE récupération thermique *f*; ~ **regeneration** *n* WATER POLL régénération thermique *f*; ~ **spring** *n* RENEW ENERG source thermique *f*; ~ **treatment** *n* WASTE traitement thermique *m*; ~ **use** *n* WATER POLL *from stabilized sludge, exhaust gas* valorisation thermique *f*; ~ **waste** *n* WASTE déchets thermiques *m pl*; ~ **waste treatment** *n* WASTE traitement thermique des déchets *m*
thermally: ~-**aerated** *adj* ANALYSIS aéré thermiquement; ~-**isolated** *adj* ANALYSIS à isolation thermique, isolé thermiquement; ~-**stabilized** *adj* WATER POLL thermiquement stabilisé
thermistor *n* ANALYSIS thermistor *m*
thermo: ~-**softening contaminant** *n* WASTE contaminant thermoadoucissant *m*, polluant thermoadoucissant *m*
thermochemical: ~ **conversion** *n* RENEW ENERG *of biomass* conversion thermochimique *f*
thermocline *n* WATER POLL thermocline *f*
thermocouple *n* AIR *catalytic incineration* thermocouple *m*
thermoforming: ~ **packaging system** *n* WASTE système d'emballage par thermoformage *m*
thermogravimetry *n* ANALYSIS thermogravimétrie *f*
thermohygrograph *n* ANALYSIS thermohygrographe *m*
thermoluminescence: ~ **dosimeter** *n* (*TLD*) ANALYSIS dosimètre à thermoluminescence *m*
thermolysis *n* WASTE *feedstock recycling* thermolyse *f*
thermomagnetic *adj* ANALYSIS thermomagnétique
thermometer *n* ANALYSIS thermomètre *m*
thermophilic[1] *adj* WATER POLL thermophile; ~-**mesophilic** WATER POLL thermophile-mésophile
thermophilic[2]: ~ **bacteria** *n pl* WATER POLL bactéries thermophiles *f pl*; ~ **digestion** *n* WATER POLL digestion thermophile *f*; ~-**mesophilic digestion** *n* WATER POLL digestion thermophile-mésophile *f*; ~ **microorganism** *n* MARINE, WATER POLL microorganisme thermophile *m*
thermoplastic[1] *adj* WASTE thermoplastique
thermoplastic[2]: ~ **material** *n* AIR matériau thermoplastique *m*; ~ **solidification** *n* WASTE plastification à chaud *f*, thermoplastification *f*
thermoset *adj* WASTE thermodurci
thermosetting[1] *adj* WASTE thermodurcissable
thermosetting[2] *n* WASTE, WATER SUPP thermodurcissage *m*
thermoshrinkable *adj* WASTE thermorétractable
thermosphere *n* AIR thermosphère *f*
thermospray: ~ **interface** *n* ANALYSIS interface de thermo-injection *f*
thermostability *n* AIR thermostabilité *f*
thermostable *adj* AIR thermostable
thermostat *n* ANALYSIS thermostat *m*
thermostatic *adj* ANALYSIS thermostatique
thick: ~ **matter flow** *n* WATER POLL écoulement de matières épaisses *m*
thicken[1] *vt* WATER POLL, WATER SUPP épaissir, coaguler
thicken[2] *vi* WATER POLL, WATER SUPP s'épaissir, se coaguler
thickened[1] *adj* AIR, WATER POLL, WATER SUPP épaissi
thickened[2]: ~ **sludge** *n* WATER POLL boues épaissies *f pl*
thickener *n* WATER POLL, WATER SUPP épaississeur *m*, matière épaississante *f*, épaississant *m*; ~ **overflow tank** AIR cuve à débordement d'épaississeur *f*; ~ **pre-storage tank** AIR cuve de stockage préliminaire d'épaississeur *f*
thickening *n* AIR, WATER POLL, WATER SUPP épaississement *m*
thickness *n* WASTE *of landfill layers* épaisseur *f*
thin[1]: ~-**film** *adj* WASTE, WATER POLL à film mince, à pellicule fine; ~-**layered** *adj* ANALYSIS, WASTE, WATER POLL, WATER SUPP à couche mince, en couche mince; ~-**walled** *adj* ANALYSIS *sampling* à paroi mince
thin[2]: ~-**film evaporator** *n* WATER POLL évaporateur pelliculaire *m*; ~-**film module** *n* RENEW ENERG *of solar cell* module de cellules photovoltaïques en couche mince *m*; ~-**film photovoltaic cell** *n* RENEW ENERG cellule photovoltaïque en couche mince *f*; ~-**film transistor liquid-crystal display** *n* (*TFT-LCD*) ANALYSIS écran à cristaux liquides à transistor en couches minces *m*; ~ **layer** *n* ANALYSIS, WASTE, WATER POLL, WATER SUPP couche mince *f*; ~-**layer chromatography** *n* (*TLC*) ANALYSIS chromatographie en couche mince *f*; ~-**layer drying** *n* WATER POLL séchage sur couches minces *m*; ~-**layer electrophoresis** *n* ANALYSIS électrophorèse sur couches

minces *f*; **~-layer installation** *n* WASTE *surface installation* installation en couches minces *f*
thinning *n* AIR *of ozone layer* diminution *f*, RENEW ENERG éclaircie *f*
thiol *n* AIR, WASTE mercaptan *m*, thiol *m*
third: **~-party certification system** *n* MANAGE système de certification par tierce partie *m*; **~-party liability** *n* MANAGE responsabilité civile *f*
thixotrope: **~ grouting material** *n* SOIL matériau d'injection thixotrope *m*
thorough: **~ mixture** *n* AIR mélange intégral *m*
three: **~-way catalyst** *n* (*TWC*) AIR catalyseur à trois voies *m*, convertisseur catalytique à trois voies *m*; **~-way catalytic converter** *n* (*TWC*) AIR pot catalytique à trois voies *m*
threshold *n* POLICY, SAFETY seuil *m*; **~ of audibility** NOISE seuil d'audition *m*; **~ audiometry** SAFETY audiométrie de seuil *f*; **~ of harmfulness** SAFETY nocivité liminale *f*, seuil de nocivité *m*; **~ limit value** (*TLV*) AIR, NOISE, POLICY, SAFETY valeur seuil *f*, valeur limite *f*; **~ limit value at the work place** SAFETY valeur seuil sur le lieu de travail *f*; **~ limit value in the free environment** AIR valeur limite admissible à l'air libre *f*, valeur limite d'exposition à l'air libre *f*, NOISE valeur de seuil à l'air libre *f*, POLICY valeur de seuil dans l'environnement *f*; **~ odor** *AmE*, **~ odour** *BrE* WATER POLL, WATER SUPP odeur limite *f*; **~ odour concentration** *BrE* WATER POLL seuil de concentration olfactive *m*; **~ of toxicity** SAFETY seuil de toxicité *m*, toxicité liminale *f*; **~ value** POLICY, SAFETY *for dangerous substances* valeur seuil *f*
throughfall *n* AIR eau de pénétration dans les frondaisons *f*, pluie au sol *f*
throughput *n* AIR débit *m*, SAFETY production *f*, SOIL *precipitation* précipitation directe *f*, WASTE débit *m*, tonnage *m*, WATER POLL débit *m*, quantité passée *f*, WATER SUPP rendement *m*; **~ fluctuation** AIR fluctuation de débit *f*
throwaway: **~ pack** *n* WASTE emballage jetable *m*, emballage perdu *m*; **~ product** *n* WASTE produit jetable *m*, produit à jeter *m*, produit perdu *m*
thruput *AmE see throughput*
thrust *n* MARINE butée *f*, poussée *f*
TIC *abbr* (*total inorganic carbon*) ANALYSIS carbone inorganique total *m*
tidal[1] *adj* RENEW ENERG marémoteur
tidal[2]: **~ action zone** *n* MARINE zone de balancement des marées *f*; **~ barrage** *n* RENEW ENERG *of tidal power plant* usine marémotrice *f*; **~ basin** *n* MARINE bassin de marée *m*, RENEW ENERG *of tidal power plant* bassin à flot *m*, bassin de stockage *m*; **~ energy** *n* RENEW ENERG *energy source* énergie marémotrice *f*; **~ energy resources** *n pl* RENEW ENERG ressources en énergie marémotrice *f pl*; **~ power** *n* RENEW ENERG énergie marémotrice *f*; **~ power plant** *n* POLICY centrale marémotrice *f*, RENEW ENERG usine marémotrice *f*, centrale marémotrice *f*; **~ range** *n* RENEW ENERG amplitude de la marée *f*, hauteur des marées *f*; **~ wave** *n* MARINE raz-de-marée *m*
tide *n* MARINE, RENEW ENERG marée *f*
tiered: **~ biofilter** *n* AIR *biological sorption process* biofiltre à étages *m*; **~ drier** *n* WATER POLL four à soles étagées *m*, sécheur à étages *m*
tight[1] *adj* GEN *water-impermeable* étanche, étanche à l'eau
tight[2]: **~ layer** *n* WASTE couche étanche *f*
tightly: **~-compacted landfill** *n* WASTE décharge fortement compactée *f*
tillage *n* RENEW ENERG, SOIL préparation du sol *f*
tilling: **~ process** *n* RENEW ENERG, SOIL préparation du sol *f*
tilted: **~ mirror** *n* ANALYSIS miroir renversé *m*
time[1]: **~-dependent** *adj* ANALYSIS *sample*, RENEW ENERG en fonction du temps; **~-intensive** *adj* MANAGE long à mettre en œuvre, de longue durée
time[2]: **~-delayed vacuum** *n* AIR vide à retardement *m*; **~-intensive method** *n* AIR, MARINE, NOISE, SOIL méthode forte consommatrice de temps *f*, méthode longue à mettre en œuvre *f*, méthode de longue durée *f*; **~ interval** *n* ANALYSIS intervalle de temps *m*, moyenne à pondération temporelle *f*; **~-weighed average** *n* (*TWA*) ANALYSIS moyenne pondérée dans le temps *f*
tin[1] *n BrE* (*cf can AmE*) WASTE *container* boîte de conserve *f*, boîte en fer blanc *f*, boîte *f*, *material* étain *m*, fer blanc *m*; **~-plate waste** WASTE *from sheet metal* déchets de tôle *m pl*; **~-plated can** WASTE boîte de conserve en fer blanc *f*, boîte en fer blanc *f*
tin[2]: **~-plate** *vt* WASTE étamer
tip[1] *n* WASTE *action* déversement *m*, *machine* culbuteur *m*, *site* dépôt de déblais *m*; **~ filling method** WASTE méthode à l'avancement *f*
tip[2] *vt* WASTE déverser, déposer
tipped: **~-edge installation** *n* WASTE installation en tranchées *f*
tipping: **~ with compaction** *n* WASTE déversement avec compactage *m*; **~ ramp** *n* WASTE rampe à déversement *f*; **~ site** *n* WASTE site de décharge *m*; **~ table** *n* WASTE plate-forme *f*; **~ trough** *n* WATER POLL goulotte basculante *f*
titanium: **~ dioxide waste** *n* WASTE déchets de dioxyde de titane *m pl*
titer *AmE see titre BrE*
titrate *vt* ANALYSIS titrer
titration *n* ANALYSIS *volumetric analysis* titrage *m*; **~ calorimeter** ANALYSIS calorimètre de titrage *m*; **~ method** ANALYSIS méthode de titration *f*; **~ vessel** ANALYSIS bac de titrage *m*
titre *n BrE* ANALYSIS titre *m*
titrimeter *n* ANALYSIS titrimètre de phase gazeuse *m*
titrimetric[1] *adj* ANALYSIS titrimétrique
titrimetric[2]: **~ substance** *n* ANALYSIS substance titrimétrique *f*; **~ test** *n* ANALYSIS *for water* essai titrimétrique *m*
TL *abbr* (*tolerance level*) NOISE niveau seuil *m*
TLC *abbr* (*thin-layer chromatography*) ANALYSIS chromatographie en couche mince *f*
TLD *abbr* (*thermoluminescence dosimeter*) ANALYSIS dosimètre à thermoluminescence *m*
TLV[1] *abbr* (*threshold limit value*) AIR, NOISE, POLICY, SAFETY valeur seuil *f*, valeur limite *f*
TLV[2]: **~ at the workplace** *n* SAFETY valeur seuil sur le lieu de travail *f*; **~ in the free environment** *n* AIR valeur limite admissible à l'air libre *f*, valeur limite d'exposition à l'air libre *f*, NOISE valeur seuil à l'air libre *f*, POLICY valeur seuil dans l'environnement *f*
TOC[1] *abbr* (*total organic carbon*) ANALYSIS, WATER POLL carbone organique total *m*
TOC[2]: **~ analyser** *BrE*, **~ analyzer** *AmE n* ANALYSIS analyseur de carbone organique total *m*
tolerable *adj* SOIL *harmful substance concentration* tolérable, WASTE *residue emission* admissible
tolerance *n* SOIL *of harmful substance concentration* tolérance *f*; **~ level** (*TL*) NOISE niveau seuil *m*
toluene *n* WASTE toluène *m*
TOMS *abbr* (*total ozone mapping spectrometer*) AIR, ANALYSIS TOMS (*spectromètre imageur de l'ozone total*)
top *n* AIR haut *m*, tête *f*; **~-gas recycling** AIR recyclage des gaz de tête *m*; **~ height** AIR hauteur de tête *f*; **~ layer** WASTE couche de couverture *f*; **~ level of sludge** WATER POLL voile de boue *m*; **~ soil** WASTE sol de couverture *m*
torque *n* RENEW ENERG couple moteur *m*; **~ adjuster** AIR dispositif de réglage du couple moteur *m*
tort: **~ liability** *n* MANAGE responsabilité délictuelle *f*
tortfeasor *n* POLICY contrevenant *m*
total *n* AIR total *m*; **~ air quality** AIR qualité globale de l'air *f*, qualité totale de l'air *f*; **~ ammonia** WATER POLL ammoniac total *m*, ammoniaque totale *m*; **~ analysis** ANALYSIS analyse totale *f*; **~ capacity** RENEW ENERG puissance totale *f*, WATER POLL *of ion exchanger* capacité totale *f*; **~ carbon** (*TC*)

ANALYSIS carbone total *m*; ~ **carbon dioxide** AIR dioxyde de carbone total *m*; ~ **collection** AIR collecte totale *f*; ~ **deposition** AIR, SOIL, WATER SUPP dépôt total *m*; ~ **dissolved solids** (*TDS*) WATER SUPP matière solide total dissoute *f*; ~ **dust** AIR, ANALYSIS poussière totale *f*, volume total de poussière *m*; ~ **dust content** AIR teneur totale en poussières *f*; ~ **efficiency** RENEW ENERG rendement total *m*; ~ **flow** ANALYSIS *of nappe* débit global *m*; ~ **gas quantity** AIR, ANALYSIS quantité totale de gaz *f*, volume total de gaz *m*; ~ **hardness** WATER POLL, WATER SUPP dureté totale *f*; ~ **head** WATER POLL hauteur de chute totale *f*, pression manométrique totale *f*; ~ **inorganic carbon** (*TIC*) ANALYSIS carbone inorganique total *m*; ~ **lethal concentration zero** (TCL_0) SAFETY concentration létale zéro totale *f*; ~ **lethal dose zero** (TDL_0) SAFETY dose létale zéro totale *f*; ~ **noise** NOISE bruit total *m*; ~ **organic carbon** (*TOC*) ANALYSIS, WATER POLL carbone organique total *m*; ~ **organic carbon analyser** *BrE*, ~ **organic carbon analyzer** *AmE* (*TOC analyser*) ANALYSIS analyseur de carbone organique total *m*; ~ **organic compounds** AIR composés organiques totaux *m pl*; ~ **ozone mapping spectrometer** (*TOMS*) AIR, ANALYSIS spectromètre imageur de l'ozone total *m* (*TOMS*); ~ **pollutant emissions** *n pl* ANALYSIS ensemble des émissions polluantes *m*; ~ **removal** AIR dépoussiérage intégral *m*, extraction totale *f*; ~ **residue** ANALYSIS résidus totaux *m pl*; ~ **risk** SAFETY risque total *m*; ~ **sludge yield** WATER POLL rendement total de boue *m*; ~ **suspended solids** *n pl* (*TSS*) WATER POLL total des matières solides en suspension *m*; ~ **undissolved solids** *n pl* WATER POLL total des matières solides en suspension *m*; ~ **volume** ANALYSIS volume total *m*

tote: ~ **box** *n* WASTE bac de manutention *m*

tower *n* RENEW ENERG *of wind turbine* tour *f*; ~ **biofilter** AIR *biological sorption process* réacteur de biofiltration *m*; ~ **height** RENEW ENERG *of wind turbine* hauteur de tour *f*; ~ **system** WASTE système à tours *m*; ~**-type trickling filter** WATER POLL lit bactérien traditionnel *m*

towing: ~ **gear** *n* MARINE engin de remorquage *m*

towline *n* MARINE remorque *f*

town: ~ **and country planning** *n* POLICY aménagement du territoire *m*; ~ **gas** *n* AIR gaz de ville *m*; ~ **refuse** *n* WASTE résidu urbain *m*

toxic[1] *adj* GEN toxique

toxic[2]: ~ **agent** *n* GEN agent toxique *m*; ~ **cloud** *n* AIR nuage toxique *m*; ~ **degradation product** *n* WASTE produit de dégradation toxique *m*; ~ **dust** *n* AIR poussière toxique *f*; ~ **effect** *n* AIR, ANALYSIS, SAFETY, WATER SUPP effet toxique *m*; ~ **substance** *n* AIR, SAFETY, SOIL, WASTE substance toxique *f*; ~ **waste** *n* SAFETY, WASTE, WATER SUPP déchets toxiques *m pl*; ~ **waste gas** *n* AIR effluent gazeux toxique *m*

toxicant *n* AIR, SOIL produit toxique *m*

toxicity *n* GEN toxicité *f*; ~ **equivalent** ANALYSIS équivalent de toxicité *m*; ~ **potential** SOIL pouvoir toxique *m*; ~ **test** ANALYSIS essai de toxicité *m*

toxicological: ~ **assessment** *n* MANAGE, POLICY évaluation toxicologique *f*

toxicology *n* SAFETY toxicologie *f*

toxin *n* AIR, MARINE toxine *f*

trace: ~ **analysis** *n* ANALYSIS *of element* analyse en trace *f*; ~ **concentration** *n* AIR concentration de traces *f*, SOIL *of contaminants* traces *f pl*; ~ **element** *n* AIR élément à l'état de trace *m*, élément-trace *m*, ANALYSIS élément à l'état de trace *m*, WATER POLL, WATER SUPP oligo-élément *m*; ~ **element determination** *n* ANALYSIS détermination des oligo-éléments *f*, dosage des oligo-éléments *m*; ~ **gas** *n* AIR gaz à l'état de trace *m*; ~**-measuring device** *n* ANALYSIS appareil de mesure des traces *m*; ~ **metal** *n* SOIL métal en traces *m*

traceability *n* ANALYSIS traçabilité *f*

tracer *n* AIR indicateur *m*, traceur *m*

traces *n pl* ANALYSIS traces *f pl*

tracker *n* RENEW ENERG orienteur *m*, pointeur *m*

tracking: ~ **mechanism** *n* RENEW ENERG *of solar collector* mécanisme d'orientation *m*, mécanisme de pointage *m*, mécanisme de poursuite *m*; ~ **reflector** *n* RENEW ENERG *of solar tower* héliostat orientable *m*, réflecteur de poursuite *m*

trade: ~ **waste** *n* WASTE déchets du commerce *m pl*

tradeable[1] *adj* POLICY négociable

tradeable[2]: ~ **emissions rights** *n pl* POLICY droits d'émissions négociables *m pl*; ~ **environmental use rights** *n pl* POLICY droits d'usage de l'environnement négociables *m pl*

traditional: ~ **energy source** *n* RENEW ENERG source traditionnelle d'énergie *f*

traffic: ~ **noise** *n* NOISE bruit de la circulation routière *m*; ~ **plan** *n* POLICY plan de circulation *m*

transaction: ~ **costs** *n pl* POLICY coûts de transaction *m pl*

transboundary[1] *adj* AIR transfrontière

transboundary[2]: ~ **air pollution** *n* AIR pollution atmosphérique transfrontière *f*; ~ **waste movement** *n* WASTE mouvement transfrontière des déchets *m*

transfer[1]: ~ **frequency** *n* WASTE fréquence de transfert *f*; ~ **line** *n* AIR, ANALYSIS ligne de transfert *f*; ~ **mold technique** *AmE*, ~ **mould technique** *BrE* *n* WASTE technique de moulage par transfert *f*; ~ **pump** *n* MARINE pompe de transfert *f*; ~ **rate** *n* WATER POLL taux de transfert *m*; ~ **room** *n* WASTE salle de transit *f*; ~ **station** *n* WASTE poste de transit *m*, station de transit *f*

transfer[2] *vt* AIR, SOIL *harmful substance* transférer

transformation: ~ **rate** *n* AIR taux de conversion *m*, taux de transformation *m*

transformer *n* RENEW ENERG transformateur *m*

transient: ~ **flow** *n* WATER SUPP écoulement *m*

transition *n* POLICY transition *f*; ~ **lake** WATER SUPP lac en transition *m*; ~ **phase** POLICY phase de transition *f*; ~ **point** ANALYSIS point de transition *m*, température de transition *f*; ~ **regulation** POLICY réglementation temporaire *f*, réglementation transitoire *f*; ~ **temperature** ANALYSIS point de transition *m*, température de transition *f*; ~ **zone** MARINE, WATER POLL zone de transition *f*

transmission *n* AIR perméance *f*, ANALYSIS, RENEW ENERG transmission *f*; ~ **electron microscopy** (*TEM*) ANALYSIS microscopie électronique à transmission *f*; ~ **loss** NOISE perte par transmission *f*; ~ **performance rating** AIR indice de qualité de transmission *m*; ~ **spectrum** ANALYSIS spectre de transmission *m*

transmit *vt* ANALYSIS, RENEW ENERG transmettre; ~ **power** *vi* RENEW ENERG transmettre la puissance

transmittance: ~ **meter** *n* ANALYSIS compteur de transmittance *m*

transpiration *n* AIR transpiration *f*

transport[1] *n* SOIL, WASTE, WATER SUPP transport *m*; ~ **capacity** RENEW ENERG, WATER SUPP *of distribution network* capacité de transport *f*; ~ **container** WASTE conteneur de transport *m*; ~ **conveyor belt** WATER POLL convoyeur à bande *m*; ~ **model** WATER SUPP *of pollutant* modèle de transport *m*; ~ **packaging** WASTE emballage de transport *m*; ~ **safety** SAFETY sûreté de transport *f*; ~ **volume** WASTE, WATER SUPP volume de fret *m*, volume transporté *m*

transport[2] *vt* GEN transporter; ~ **away** SOIL emporter

transportable *adj* GEN transportable

transportation: ~ **source** *n* WATER SUPP source mobile *f*

transshipment *n* WATER SUPP transbordement *m*

transverse: ~ **entry** *n* AIR admission latérale *f*, admission tangentielle *f*; ~**-entry traveling-bed reactor** *AmE*, ~**-entry travelling-bed reactor** *BrE* *n* AIR réacteur à lit mobile à admission tangentielle *m*; ~ **strain** *n* SOIL contrainte tangentielle *f*

trap *n* AIR piège *m*, ANALYSIS piège *m*, séparateur *m*, NOISE piège *m*, WATER POLL piège *m*, siphon *m*, WATER SUPP siphon *m*

trapezoid: ~ **compost heap** *n* WATER SUPP tas de compost trapézoïdal *m*

trash: ~ **can** *n AmE* (*cf dustbin BrE, rubbish bin BrE*) WASTE boîte à ordures *f*, poubelle *f*
traveling *AmE see travelling BrE*
travelling: ~-**bed reactor** *n BrE* AIR réacteur à lit mobile *m*; ~ **distributor** *n BrE* WATER SUPP distributeur à va-et-vient *m*; ~ **grate** *n BrE* SOIL *for pellet incineration* grille mécanique *f*, grille sans fin *f*, WASTE grille sans fin *f*, grille mécanique *f*, WATER SUPP grille mécanique *f*, grille mobile *f*, grille sans fin *f*; ~-**grate installation** *n BrE* WATER SUPP installation à grille mobile *f*
trawl: ~ **net** *n* MARINE chalut *m*
trawling *n* MARINE chalutage *m*
tray *n* AIR, NOISE, WASTE plateau *m*; ~ **absorber** AIR absorbeur à plateaux *m*; ~ **packing** WASTE emballage sur plateau *m*; ~-**plate column** AIR colonne à plateaux *f*; ~ **tower** AIR *of wet scrubber* colonne à plateaux *f*
treat *vt* AIR traiter, MANAGE, WASTE épurer, traiter, WATER POLL traiter, WATER SUPP *water* épurer, traiter; ~ **finally** WATER SUPP procéder au traitement final
treated[1] *adj* GEN traité
treated[2]: ~ **effluent** *n* WATER POLL, WATER SUPP effluent traité *m*; ~ **sludge** *n* WATER POLL boues traitées *f pl*
treatment *n* GEN épuration *f*, traitement *m*; ~ **cell** SOIL compartiment d'assainissement *m*; ~ **duration** SOIL durée du traitement *f*; ~ **intensity** SOIL intensité du traitement *f*; ~ **plant** GEN usine de traitement *f*, station d'épuration *f*; ~ **process** GEN procédé de traitement *m*, procédé d'épuration *m*; ~ **time** SOIL durée du traitement *f*; ~ **unit** WATER SUPP station d'épuration *f*
treble: ~ **absorber** *n* NOISE absorbeur d'aigus *m*
tree: ~ **cover** *n* SOIL couverture végétale *f*
trench *n* MARINE fossé *m*; ~ **landfill** WASTE décharge en tranchée *f*; ~ **method** WASTE méthode des tranchées *f*
trenching *n* SOIL excavation de tranchées *f*
TRI *abbr* (*trichloroethylene*) AIR trichloréthylène *m*
trial: ~ **operation** *n* MANAGE, POLICY *of plant*, WATER SUPP exploitation expérimentale *f*
tributary *n* WATER POLL affluent *m*
trichloroethylene *n* (*TRI*) AIR trichloréthylène *m*
trichlorofluoromethane *n* AIR *halogenated hydrocarbon* trichlorofluorométhane *m*
trickle[1] *n* WATER SUPP ruissellement *m*
trickle[2] *vi* WATER SUPP couler goutte à goutte, ruisseler
trickling *n* AIR oxydation *f*, WATER SUPP ruissellement *m*; ~ **film** AIR *of wet precipitator* film ruisselant *m*; ~-**film flow** WATER POLL *of packed column* écoulement en film ruisselant *m*; ~ **film surface** WATER POLL surface de lit bactérien *f*; ~ **filter** WATER POLL *biofilter* lit bactérien *m*; ~ **filter bottom** WATER POLL fond de lit bactérien *m*; ~ **filter process** AIR *in biowasher* procédé à filtre ruisselant *m*, WATER POLL *biofiltration* procédé mettant en œuvre un lit bactérien *m*, procédé par lit bactérien *m*; ~ **water** WASTE, WATER SUPP *from plastics reprocessing plant* eau de ruissellement *f*
triglyceride *n* WATER POLL triglycéride *m*
trommel: ~ **screen** *n* WATER SUPP crible à tambour *m*
tropopause *n* AIR tropopause *f*
troposphere *n* AIR troposphère *f*
tropospheric[1] *adj* AIR troposphérique
tropospheric[2]: ~ **ozone** *n* AIR ozone troposphérique *m*; ~ **ozone concentration** *n* AIR concentration d'ozone troposphérique *f*; ~ **ozone pollution** *n* AIR pollution par l'ozone troposphérique *f*
trough *n* WATER SUPP goulotte *f*
troughed: ~ **chain conveyor** *n* WASTE, WATER SUPP convoyeur en auge *m*
truck: ~ **farming** *AmE* (*cf market gardening BrE*) SOIL maraîchage *m*; ~ **trailer** *n* WASTE remorque de camion *f*; ~ **transport** *n* WASTE transport en camion *m*, transport routier *m*
trunk: ~ **main** *n* WATER SUPP feeder principal *m*
trust: ~ **area** *n* POLICY zone protégée *f*
try: ~ **cock** *n* WATER SUPP robinet d'essai *m*, robinet de hauteur d'eau *m*, robinet de jauge *m*, robinet de regard *m*
TSS *abbr* (*total suspended solids*) WATER POLL total des matières solides en suspension *m*
TTS *abbr* (*temporary threshold shift*) NOISE déplacement de seuil temporaire *m*
tube *n* WATER SUPP tuyau *m*; ~ **bundle** AIR *of heat exchanger* faisceau de tubes *m*; ~ **cage** AIR cage tubulaire *f*; ~-**cage recuperator** AIR récupérateur à cage tubulaire *m*; ~ **support plate** AIR plaque support de tube *f*; ~ **wall** WATER SUPP paroi tubulaire *f*
tuber *n* RENEW ENERG tubercule *m*
tubular[1] *adj* AIR tubulaire
tubular[2]: ~ **drainage** *n* WATER SUPP évacuation tubulaire *f*; ~ **electrostatic precipitator** *n* AIR dépoussiéreur à tubes *m*, dépoussiéreur tubulaire *m*; ~ **ESP** *n* AIR *dry electrofilter* électrofiltre tubulaire *m*, dépoussiéreur à tubes *m*, dépoussiéreur tubulaire *m*; ~ **filter bag** *n* AIR filtre tubulaire à manche *m*, sac de filtre tubulaire *m*; ~ **module** *n* WATER POLL module tubulaire *m*; ~ **precipitator** *n* AIR *original electrostatic precipitator* dépoussiéreur tubulaire *m*
tumbling: ~ **station** *n* WASTE poste de déversement *m*, WATER SUPP poste de tonnelage *m*, *to remove sand* poste de dessablage *m*
turbid[1] *adj* WATER POLL, WATER SUPP trouble, turbide
turbid[2]: ~ **matter** *n* WATER POLL matière turbide *f*; ~ **wastewater** *n* WATER POLL eaux usées troubles *f pl*, eaux usées turbides *f pl*; ~ **wastewater sample** *n* ANALYSIS prélèvement d'eaux usées troubles *m*, prélèvement d'eaux usées turbides *m*; ~ **water** *n* WATER POLL eau trouble *f*, eau turbide *f*
turbidimeter *n* AIR, ANALYSIS, WATER POLL turbidimètre *m*, capteur turbidimétrique *m*
turbidimetric[1] *adj* AIR, ANALYSIS, WATER POLL turbidimétrique
turbidimetric[2]: ~ **analysis** *n* ANALYSIS *of water* analyse turbidimétrique *f*; ~ **titration** *n* ANALYSIS titrage turbidimétrique *m*
turbidimetry *n* AIR, ANALYSIS, WATER POLL turbidimétrie *f*
turbidity *n* AIR, ANALYSIS turbidité *f*, WATER POLL turbidité *f*, turbidité des eaux *f*; ~ **value** ANALYSIS indice de turbidité *m*
turbine *n* RENEW ENERG, WASTE turbine *f*; ~ **blade** RENEW ENERG *of wind turbine* aube de turbine *f*
turbo-aerator *n* WATER POLL turbo-aérateur *m*
turbocompressor *n* WATER POLL turbocompresseur *m*
turbogenerator *n* RENEW ENERG *in tidal power plant* turbo-alternateur *m*
turbomachine *n* WATER POLL, WATER SUPP turbomachine *f*
turbopump *n* WATER SUPP turbopompe *f*
turbowasher *n* WASTE turbolaveur *m*
turbulence *n* AIR, WASTE, WATER SUPP turbulence *f*
turbulent[1] *adj* WATER POLL turbulent
turbulent[2]: ~ **flow** *n* AIR, WATER SUPP écoulement turbulent *m*
TWA *abbr* (*time-weighed average*) ANALYSIS moyenne pondérée dans le temps *f*
TWC *abbr* (*three-way catalytic converter*) AIR catalyseur à trois voies *m*, pot catalytique à trois voies *m*
twin: ~ **cock** *n* WATER SUPP robinet à deux faces *m*; ~-**cutting shredder** *n* WASTE déchiqueteur à deux arbres *m*; ~-**screw extruder** *n* WASTE, WATER SUPP boudineuse à deux vis *f*, boudineuse à vis jumelée *f*
two: ~-**cycle engine** *n AmE*, ~-**stroke engine** *n BrE* AIR, WATER POLL moteur à deux temps *m*; ~-**phase jet** *n* AIR *of scrubber* jet diphasique *m*; ~-**stage activated sludge process** *n* WATER POLL traitement par boues activées en deux étapes *m*; ~-**stage activated sludge process with pre-denitrification** *n* WATER POLL traitement par boues activées en deux étapes avec prédénitrification *m*; ~-**stage aerobic thermophilic stabilization** *n* WATER POLL stabilisation thermophile aérobie en deux étapes *f*;

~-stage digester *n* WATER POLL digesteur à deux étages *m*; **~-stage fired heater** *n* AIR foyer à étage *m*; **~-stage gas scrubber** *n* AIR dépoussiéreur hydraulique à étage *m*, laveur à étage *m*; **~-stage recarbonation** *n* WATER POLL recarbonatation en deux étapes *f*; **~-stage thermophilic-mesophilic digestion** *n* WATER POLL digestion thermophile-mésophile en deux étapes *f*; **~-way cock** *n* WATER SUPP robinet à deux eaux *m*, robinet à deux voies *m*

Tyndall: **~ effect** *n* ANALYSIS effet Tyndall *m*

type[1]: **~-specific** *adj* GEN propre à un type, spécifique d'un type

type[2]: **~ specification** *n* WASTE spécification du type *f*

U

UCT: ~ **method** *n* (*University of Cape Town method*) WATER POLL méthode UCT *f* (*méthode de l'Université du Cape Town*)
UCV *abbr* (*upper calorific value*) AIR, WATER SUPP PCS (*pouvoir calorifique supérieur*)
ultimate: ~ **storage** *n* WASTE stockage définitif *m*
ultrafiltration *n* GEN ultrafiltration *f*
ultrapure: ~ **water** *n* SOIL *used in extraction* WATER SUPP eau ultrapure *f*
ultrared[1] *adj* (*UR*) ANALYSIS infrarouge (*IR*)
ultrared[2] *n* (*UR*) ANALYSIS infrarouge *m* (*IR*)
ultrasonic[1] *adj* ANALYSIS, SAFETY ultrasonique
ultrasonic[2]: ~ **extraction** *n* ANALYSIS extraction par ultrasons *f*; ~ **hazard** *n* SAFETY risque ultrasonique *m*; ~ **measurement** *n* ANALYSIS mesure par ultrasons *f*, mesure ultrasonique *f*
ultrasound *n* ANALYSIS, NOISE, SOIL *used in extraction* ultrason *m*; ~ **irradiation** SOIL irradiation ultrasonique *f*; ~ **pulverizer** ANALYSIS *for analysis of smallest concentration* pulvérisateur ultrasonique *m*; ~ **sensor** ANALYSIS capteur ultrasonique *m*; ~ **telemetering** ANALYSIS télémétrie par ultrasons *f*
ultratrace: ~ **analysis** *n* ANALYSIS ultramicroanalyse *f*
ultraviolet[1]: ~ **and visible** *adj* (*UV/VIS*) ANALYSIS ultraviolet et visible (*UV/VIS*)
ultraviolet[2] *n* (*UV*) GEN ultraviolet (*UV*); ~ **absorption** ANALYSIS, WATER SUPP absorption ultraviolette *f*; ~ **disinfection** WATER SUPP désinfection aux rayons ultraviolets *f*; ~ **filter** AIR, SAFETY filtre ultraviolet *m*; ~ **irradiation** WATER POLL, WATER SUPP irradiation aux ultraviolets *f*, rayonnement ultraviolet *m*; ~ **lamp** ANALYSIS lampe à lumière ultraviolette *f*; ~ **light** ANALYSIS lumière ultraviolette *f*; ~ **optical absorption** ANALYSIS absorption optique des ultraviolets *f*; ~ **oxidation** WATER SUPP oxydation aux ultraviolets *f*; ~ **photometry** ANALYSIS photométrie ultraviolette *f*; ~ **radiation** AIR rayonnement ultraviolet *m*; ~ **radiation intensity** AIR intensité du rayonnement ultraviolet *f*; ~ **radiator** WATER POLL émetteur de rayons ultraviolets *m*; ~ **rays** *n pl* SAFETY *risk for skin cancer* rayons ultraviolets *m pl*; ~ **resistance** WATER SUPP résistance aux ultraviolets *f*; ~ **spectroscopy** ANALYSIS spectroscopie ultraviolette *f*; ~ **and visible detector** ANALYSIS détecteur de rayonnement ultraviolet et visible *m*; ~ **and visible spectrometer** ANALYSIS détecteur de rayonnement ultraviolet et visible *m*, spectromètre à rayonnement ultraviolet et visible *m*; ~ **and visible spectrophotometry** ANALYSIS spectrophotométrie ultraviolette et visible *f*; ~ **and visible spectroscopy** ANALYSIS spectroscopie ultraviolette et visible *f*
ultraviolet-C: ~ **radiation** *n* (*UV-C radiation*) WATER SUPP rayonnement ultraviolet-C *m* (*rayonnement UV-C*)
UMI *abbr* (*universal measurement instrument*) ANALYSIS instrument de mesure universel *m*
unaerated: ~-**bed composting** *n* SOIL compostage sur lit non aéré *m*
unbalance: ~-**type vibrating screen** *n* WASTE tamis vibrant sans balancier *m*; ~-**type vibratory conveyor** *n* WASTE transporteur à secousses non équilibré *m*
unburnt: ~ **fuel** *n* AIR, WASTE carburant non brûlé *m*
uncertainty: ~ **of measurement** *n* ANALYSIS incertitude de mesure *f*
unchanged *adj* ANALYSIS *sample* inchangé
uncharged[1] *adj* AIR, WATER POLL non-chargé
uncharged[2]: ~ **particle** *n* AIR particule non chargée *f*
uncleaned[1] *adj* AIR, WASTE non-épuré, non-traité
uncleaned[2]: ~ **waste air** *n* AIR effluent non épuré *m*
uncoated[1] *adj* AIR, WATER SUPP non-revêtu
uncoated[2]: ~ **surface** *n* AIR, WATER SUPP surface non revêtue *f*
uncompacted[1] *adj* WASTE, WATER SUPP non-compacté
uncompacted[2]: ~ **tip** *n* WASTE, WATER SUPP décharge non compactée *f*
unconfined: ~ **groundwater** *n* WATER SUPP nappe libre *f*, nappe superficielle *f*; ~ **water** *n* WATER SUPP eau libre *f*
uncontrolled[1] *adj* GEN non-contrôlé, non-maîtrisé, non-jugulé
uncontrolled[2]: ~ **denitrification** *n* WATER POLL dénitrification non-contrôlée *f*; ~ **disposal** *n* WASTE, WATER SUPP décharge non-contrôlée *f*, décharge sauvage *f*, rejet sauvage *m*; ~ **dumping** *n* WASTE, WATER SUPP décharge non-contrôlée *f*, décharge sauvage *f*, rejet sauvage *m*; ~ **growth** *n* WATER POLL croissance désordonnée *f*; ~ **tipping** *n BrE* POLICY dépôt sauvage *m*, WASTE, WATER SUPP décharge sauvage *f*, dépôt sauvage *m*, rejet sauvage *m*; ~ **urbanization** *n* MANAGE urbanisation débridée *f*
uncultivated: ~ **area** *n* SOIL région inculte *f*, région non-cultivée *f*
undamped *adj* NOISE non-amorti
underaerated *adj* WATER POLL non-désaéré
underblast: ~ **chain grate stoker** *n* WATER SUPP foyer avec grille à chaîne à soufflage sous grille *m*; ~ **fan** *n* WATER SUPP soufflante sous grille *f*; ~ **zone** *n* WATER SUPP zone sous grille *f*
underdrain *n* WATER POLL collecteur noyé *m*
underdrainage *n* WATER POLL drainage souterrain *m*
underfloor: ~ **catalyst** *n* AIR catalyseur souterrain *m*
underflow *n* WATER SUPP courant de fond *m*, courant profond *m*, *in sluice* courant dérivé *m*, sous-verse *m*
underground[1] *adj* WASTE souterrain
underground[2]: ~ **deformation** *n* WASTE, WATER SUPP déformation souterraine *f*; ~ **depot** *n* WASTE, WATER SUPP dépôt souterrain *m*; ~ **disposal** *n* WASTE évacuation souterraine *f*; ~ **drainage** *n* WATER SUPP drainage profond *m*; ~ **hydrant** *n* WATER SUPP prise d'eau souterraine *f*; ~ **infiltration facility** *n* WATER SUPP station d'infiltration souterraine *f*; ~ **landfill for hazardous waste** *n* WASTE décharge souterraine de déchets dangereux *f*, enfouissement de déchets dangereux *m*; ~ **landfill for special waste** *n* WASTE décharge souterraine de déchets spéciaux *f*, enfouissement de déchets spéciaux *m*; ~ **landfilling** *n* WASTE mise en décharge souterraine *f*; ~ **sealing** *n* WASTE, WATER SUPP étanchéité souterraine *f*; ~ **storage** *n* WATER SUPP stockage souterrain *m*; ~ **wastewater disposal** *n* WATER POLL évacuation souterraine des eaux usées *f*; ~ **water** *n pl* WATER SUPP eaux souterraines *f pl*, nappe phréatique *f*
undergrowth *n* SOIL sous-bois *m*
undermine *vt* WATER SUPP affouiller
under-pressure: ~ **evaporator well** *n* SOIL puits d'évaporation sous pression *m*
underwashing *n* WATER SUPP affouillement *m*
underwater[1] *adj* MARINE, WATER POLL immergé, WATER SUPP immergé, noyé, sous-marin
underwater[2]: ~ **habitat** *n* MARINE habitat sous-marin *m*; ~ **suction pipe** *n* WATER SUPP conduite d'aspiration noyée *f*

underwriting: ~ **fee** *n* MANAGE commission de garantie *f*, commission de placement *f*
undigested: ~ **sludge** *n* WATER POLL boues brutes *f pl*
undiluted[1] *adj* MARINE non-délayé, non-dilué
undiluted[2]: ~ **leachate** *n* WASTE, WATER SUPP lessivat non dilué *m*, lixiviat non dilué *m*
undissolved[1] *adj* WATER POLL, WATER SUPP non-dissous
undissolved[2]: ~ **substance** *n* WATER POLL, WATER SUPP substance non-dissoute *f*
undisturbed[1] *adj* ANALYSIS non-perturbé, non-remanié
undisturbed[2]: ~ **sample** *n* ANALYSIS, SOIL échantillon non remanié *m*, échantillon non remué *m*
uneconomical *adj* MANAGE, POLICY non-rentable, non-économique
UNEP *abbr* (*United Nations Environmental Programme*) POLICY PNUE (*Programme des Nations Unies pour l'Environnement*)
unfiltered[1] *adj* AIR, ANALYSIS, WATER POLL non-filtré
unfiltered[2]: ~ **sample** *n* ANALYSIS échantillon non filtré *m*
unfinished: ~ **support** *n* WASTE *of landfill* support non-façonné *m*, WATER SUPP soubassement non-fini *m*
unground *adj* WASTE non-broyé, imbroyé
unit: ~ **dose container** *n* WASTE récipient doseur *m*; ~ **pack** *n* WASTE emballage individuel *m*; ~ **processes** *n pl* WATER POLL traitement unitaire *m*, WATER SUPP *contaminant removal by biological or chemical reaction* procédés chimiques de base *m pl*
United: ~ **Nations Environmental Programme** *n* (*UNEP*) POLICY Programme des Nations Unies pour l'Environnement *m* (*PNUE*)
universal: ~ **measurement instrument** *n* (*UMI*) ANALYSIS instrument de mesure universel *m*; ~ **shredder** *n* WASTE ciseau universel *m*; ~ **space** *n* AIR espace universel *m*
University: ~ **of Cape Town method** *n* (*UCT method*) WATER POLL méthode de l'Université du Cape Town *f* (*méthode UCT*)
unladen *adj* AIR *particle* déchargé
unlawful *adj* POLICY illicite
unleaded: ~ **gasoline** *n AmE* (*cf unleaded petrol BrE*) WATER SUPP essence sans plomb *f*; ~ **petrol** *n BrE* (*cf unleaded gasoline AmE*) WATER SUPP essence sans plomb *f*
unload *vt* MARINE *catch* débarquer, WASTE, WATER SUPP décharger
unloading *n* MARINE débarquement *m*, WASTE, WATER SUPP déchargement *m*; ~ **hall** WATER SUPP hangar de déchargement *m*; ~ **hopper** WASTE, WATER SUPP trémie de déchargement *f*
unmixed: ~ **recovery of harmful substances** *n* WATER SUPP récupération des substances dangereuses sans mélange *f*, récupération sélective des substances dangereuses *f*; ~ **waste recovery** *n* WASTE récupération de déchets non mélangés *f*
unpolluted *adj* AIR, MARINE, WATER POLL *air* non-pollué
unpressurized[1] *adj* WATER SUPP non-pressurisé, sans pression
unpressurized[2]: ~ **supply** *n* WATER SUPP fourniture sans pression *f*
unprocessed[1] *adj* WASTE, WATER SUPP non-traité
unprocessed[2]: ~ **wastes** *n pl* WATER SUPP déchets non traités *m pl*
unprofitable *adj* MANAGE, POLICY non-rentable
unsafe[1] *adj* SAFETY dangereux, hasardeux, peu sûr
unsafe[2]: ~ **environmental condition** *n* SAFETY menace pour l'environnement *f*, environnement peu sûr *m*, milieu peu sûr *m*
unsafeness *n* SAFETY insécurité *f*
unsaturated[1] *adj* AIR, SOIL, WASTE, WATER SUPP insaturé, non-saturé
unsaturated[2]: ~ **hydrocarbon** *n* AIR hydrocarbure insaturé *m*; ~ **polyester** *n* WASTE polyester non saturé *m*; ~ **polyester resin** *n* WASTE résine de polyester non saturée *f*; ~ **soil zone** *n* SOIL zone de sol non saturée *f*; ~ **zone** *n* WATER SUPP zone non saturée *f*
unslaked *adj* WATER POLL non-éteint, vif
unsorted[1] *adj* WASTE non-trié
unsorted[2]: ~ **waste** *n* WATER SUPP déchets non triés *m pl*
unspecific: ~ **coagulation** *n* WATER POLL coagulation non spécifique *f*
unstable: ~ **methane fermentation** *n* WASTE *landfill gas phase* fermentation méthanique instable *f*
unsupported: ~ **shrink wrapping** *n* WASTE emballage sous film thermorétractable sans support *m*
unthickened[1] *adj* WATER POLL non-épaissi
unthickened[2]: ~ **sludge** *n* WATER POLL boues non épaissies *f pl*
untreated[1] *adj* WASTE non-traité, WATER POLL, WATER SUPP *water* brut, non-traité
untreated[2]: ~ **refuse** *n* WASTE résidus non-traités *m pl*, déchets non-traités *m pl*; ~ **slag** *n* WASTE scories non-traitées *f pl*; ~ **sludge** *n* WATER POLL boues non-traitées *f pl*; ~ **wastewater** *n* WATER POLL eaux usées non-traitées *f pl*, eaux non-épurées *f pl*; ~ **water** *n* WATER POLL, WATER SUPP eau brute *f*, eau non-traitée *f*; ~ **water quality** *n* WATER POLL, WATER SUPP titre de l'eau brute *m*, titre de l'eau non-traitée *m*; ~-**water tank** *n* WATER POLL, WATER SUPP bâche d'eau brute *f*, bâche d'eau non-traitée *f*, réservoir d'eau brute *m*, réservoir d'eau non-traitée *m*
unusable[1] *adj* GEN inutilisable
unusable[2]: ~ **residue** *n* WATER SUPP *from recycling process* résidus inutilisables *m pl*
unwoven *adj* AIR non-tissé
up: ~-**current classifier** *n* WATER SUPP classificateur à courant ascendant *m*
upbrow: ~ **conveyor belt** *n* WASTE bande transporteuse ascendante *f*
upflow: ~ **filter** *n* WATER SUPP filtre à courant ascendant *m*, filtre à flux ascendant *m*; ~ **sludge blanket process** *n* WATER POLL refoulement de boue en nappe *m*
upgrading *n* AIR amélioration *f*, valorisation *f*, POLICY, WASTE, WATER POLL, WATER SUPP modernisation *f*
upper: ~ **bed** *n* AIR *catalyst* lit supérieur *m*; ~ **calorific value** *n* (*UCV*) AIR, WATER SUPP pouvoir calorifique supérieur *m* (*PCS*); ~ **course** *n* WATER SUPP cours supérieur *m*; ~ **surface** *n* WASTE face supérieure *f*, WATER SUPP surface supérieure *f*
upstream[1] *adj* AIR, WATER SUPP d'amont
upstream[2] *adv* AIR, WATER SUPP en amont
upward: ~ **surface installation** *n* WASTE installation de surface amont *f*, installation de surface en pente ascendante *f*
upwind: ~ **power plant** *n* RENEW ENERG centrale au vent *f*
UR *abbr* (*ultrared*) ANALYSIS IR (*infrarouge*)
urban: ~ **canalization** *n* WATER POLL canalisation urbaine *f*; ~ **capital stocks** *n pl* POLICY équipements urbains *m pl*; ~ **catchment** *n* WATER SUPP bassin urbain *m*; ~ **change** *n* POLICY mutations urbaines *f pl*; ~ **congestion** *n* POLICY manque d'espace *m*, pénurie d'espace *f*; ~ **core** *n* POLICY centre urbain *m*; ~ **ecosystem** *n* AIR, SOIL écosystème urbain *m*; ~ **fabric** *n* POLICY tissu urbain *m*; ~ **fringe** *n* POLICY banlieue *f*, couronne *f*; ~ **improvement** *n* POLICY rénovation urbaine *f*; ~ **mesh** *n* POLICY maillage urbain *m*; ~ **planning** *n* POLICY urbanisme *m*; ~ **solid waste** *n* WASTE déchets urbains solides *m pl*; ~ **sprawl** *n* POLICY urbanisation anarchique *f*; ~ **waste** *n* WASTE déchets urbains *m pl*; ~ **water management** *n* MANAGE, WATER SUPP gestion des eaux urbaines *f*
urbanization *n* MANAGE, POLICY urbanisation *f*
urbanized: ~ **area** *n* MANAGE, POLICY agglomération urbaine *f*
urea *n* WATER POLL urée *f*
usability *n* GEN possibilité d'utilisation *f*, aptitude à l'usage *f*, aptitude à l'emploi *f*, abilité *f*, utilisabilité *f*
usable[1] *adj* GEN utilisable

usable[2]: ~ **by-product** *n* AIR, MANAGE, WASTE, WATER SUPP sous-produit valorisable *m*, produit dérivé utilisable *m*, sous-produit utilisable *m*; ~ **energy form** *n* RENEW ENERG forme d'énergie consommable *f*; ~ **fraction** *n* WATER SUPP fraction utilisable *f*; ~ **portion** *n* WATER SUPP *of the hydrosphere for human use* part utilisable *f*; ~ **potential** *n* RENEW ENERG potentiel valorisable *m*; ~ **waste by-product** *n* WASTE sous-produit de déchet *m*; ~ **waste products** *n pl* WASTE déchets valorisables *m pl*, produits résiduels utilisables *m pl*

use[1]: ~**-related** *adj* GEN en fonction d'un usage, en fonction de la consommation

use[2] *n* GEN consommation *f*, jouissance *f*, usage *m*, utilisation *f*; ~ **of bioenergy** RENEW ENERG valorisation de la bioénergie *f*; ~ **of geothermal energy** RENEW ENERG valorisation de l'énergie géothermique *f*; ~ **of ocean currents** RENEW ENERG utilisation de l'énergie des courants marins *f*; ~ **possibility** AIR, RENEW ENERG possibilité d'utilisation *f*; ~ **prohibition** POLICY *of site*, WATER SUPP interdiction d'utilisation *f*; ~ **recommendation** POLICY *of site*, WATER SUPP mode d'emploi *m*; ~ **reduction** POLICY *of site*, WATER SUPP baisse d'utilisation *f*, réduction de la consommation *f*, réduction d'utilisation *f*; ~ **regulation** POLICY réglementation d'utilisation *f*; ~ **safety** SAFETY sécurité d'utilisation *f*, sûreté d'utilisation *f*; ~ **of tidal energy** RENEW ENERG valorisation de l'énergie marémotrice *f*; ~ **of treated sludge as fuel** WATER POLL valorisation des boues traitées comme combustible *f*; ~**-value analysis** MANAGE analyse de la valeur d'usage *f*; ~ **of wave energy** RENEW ENERG valorisation de l'énergie des vagues *f*; ~ **of wind energy** RENEW ENERG valorisation de l'énergie éolienne *f*

use[3] *vt* GEN consommer, valoriser; ~ **stored energy** RENEW ENERG consommer l'énergie stockée

used[1] *adj* GEN usagé

used[2]: ~ **aluminium** *BrE*, ~ **aluminum** *AmE n* WATER SUPP aluminium usagé *m*; ~ **battery** *n* WASTE batterie usagée *f*; ~**-car dump** *n* WATER SUPP dépôt d'épaves *m*; ~ **commercial packaging** *n* WASTE emballage industriel usagé *m*; ~ **glass** *n* WATER SUPP verre usagé *m*; ~ **material** *n* WASTE matériau usagé *m*; ~ **oil** *n* WATER SUPP huile résiduaire *f*, huile usée *f*; ~ **product** *n* WATER SUPP produit usagé *m*; ~ **storage battery** *n* WASTE accumulateur usagé *m*; ~ **tire** *AmE*, ~ **tyre** *BrE n* WATER SUPP pneu usé *m*; ~**-tyre carcass** *n BrE* WATER SUPP carcasse de pneus usés *f*; ~**-tyre disposal** *n BrE* WATER SUPP décharge de pneus *f*; ~ **tyre pyrolysis** *n BrE* WATER SUPP pyrolyse de pneus usés *f*

useful: ~ **area** *n* WASTE surface utile *f*, WATER SUPP zone utile *f*

user: ~ **charge** *n* POLICY droit à payer par l'usager *m*; ~**'s contract** *n* POLICY *of environmental label* contrat d'utilisation *m*; ~**-related function** *n* MANAGE fonction de service *f*

utility: ~ **scale** *n* RENEW ENERG échelle industrielle *f*

utilization *n* GEN valorisation *f*; ~ **cycle** WASTE cycle d'utilisation *m*

utilize *vt* GEN valoriser

UV *abbr* (*ultraviolet*) GEN UV (*ultraviolet*)

UV-C: ~ **radiation** *n* (*ultraviolet-C radiation*) WATER SUPP rayonnement UV-C *m* (*rayonnement ultraviolet-C*)

UV/VIS[1] *abbr* (*ultraviolet and visible*) ANALYSIS UV/VIS (*ultraviolet et visible*)

UV/VIS[2]: ~ **detector** *n* ANALYSIS détecteur de rayonnement ultraviolet et visible *m*; ~ **spectrometer** *n* ANALYSIS détecteur de rayonnement UV et visible *m*, spectromètre à rayonnement UV et visible *m*; ~ **spectrophotometry** *n* ANALYSIS spectrophotométrie UV et visible *f*; ~ **spectroscopy** *n* ANALYSIS spectroscopie UV et visible *f*

V

V *abbr* (*vanadium*) AIR V (*vanadium*)

vacuum *n* ANALYSIS, MARINE, SOIL vide *m*; **~ belt filter** AIR filtre à vide *m*; **~ breaker** WATER SUPP casse-vide *m*; **~ closing machine** WASTE machine de fermeture sous vide *f*; **~ concentration** WATER SUPP concentration à vide *f*, concentration sous vide *f*; **~ de-aerator** WATER POLL désaérateur sous vide *m*; **~ evaporator** WATER SUPP évaporateur à vide *m*, évaporateur sous vide *m*; **~ filling machine** WASTE machine de remplissage sous vide *f*; **~ film transport system** WASTE système de transport de films sous vide *m*; **~ filter** WASTE, WATER POLL, WATER SUPP filtre à vide *m*, filtre sous vide *m*; **~ filter dewatering** WATER POLL déshydratation par filtre sous vide *f*; **~ filtration** WASTE, WATER POLL, WATER SUPP filtration sous vide *f*; **~ flotation** WATER POLL flottation sous vide *f*; **~-formed package** WASTE emballage formé sous vide *m*; **~ freeze-dryer** WASTE lyophilisateur sous vide *m*; **~ gage** *AmE*, **~ gauge** *BrE* ANALYSIS déprimomètre *m*, jauge de vide *f*; **~ heat-sealer** WASTE thermocolleuse sous vide *f*, thermoscelleuse sous vide *f*; **~ pack** WASTE emballage sous vide *m*; **~ packaging machine** WASTE machine d'emballage sous vide *f*; **~ principle** ANALYSIS *of sampling* principe du vide *m*; **~ pump** AIR, ANALYSIS, MARINE, RENEW ENERG pompe à vide *f*; **~ sealing machine** WASTE thermoscelleuse sous vide *f*; **~ street-sweeper** WASTE balayeuse à vide *f*; **~ thermoforming machine** WASTE machine de thermoformage sous vide *f*; **~ truck** MARINE camion à vide *m*; **~ well** SOIL puits sous vide *m*

vadose: **~ water** *n* WATER SUPP eau suintée *f*, eaux vadoses *f pl*

valence *n* AIR, ANALYSIS, WATER POLL valence *f*; **~ force** AIR, ANALYSIS, WATER POLL force de valence *f*

valley: **~ landfill** *n* WASTE décharge en vallée *f*, décharge inclinée *f*

valorization *n* WASTE valorisation *f*

valuable: **~ substance** *n* WASTE matière précieuse *f*

value: **~ management** *n* MANAGE maîtrise de la valeur *f*

valve *n* WATER SUPP vanne *f*

valves: **~ and fittings** *n pl* WATER SUPP robinetterie *f*

vanadium *n* (*V*) AIR vanadium *m* (*V*)

van der Waals: **~ force** *n* AIR, WATER SUPP force de van der Waals *f*

vane *n* AIR ailette *f*; **~ pump** WATER POLL pompe multicellulaire *f*, pompe rotative à ailettes *f*

vapor *AmE see vapour BrE*

vaporization *n* AIR, RENEW ENERG vaporisation *f*

vaporize[1] *vt* AIR, RENEW ENERG vaporiser

vaporize[2] *vi* AIR, RENEW ENERG s'évaporer

vaporizer *n* AIR, RENEW ENERG vaporisateur *m*

vaporous[1] *adj* AIR à l'état de vapeur, en phase vapeur

vaporous[2]: **~ fuel** *n* AIR combustible en phase vapeur *m*; **~ pollutant** *n* AIR polluant en phase vapeur *m*; **~ solvent** *n* AIR solvant en phase vapeur *m*

vapour *n BrE* AIR, RENEW ENERG, SOIL vapeur *f*; **~ compressor** *BrE* AIR compresseur de vapeur *m*; **~ emissions** *n pl BrE* AIR émissions de vapeur *f pl*, émissions en phase vapeur *f pl*; **~ extraction** *BrE* SOIL extraction des vapeurs *f*; **~-laden emissions** *n pl BrE* AIR émissions chargées de vapeur *f pl*; **~ plume** *BrE* AIR, WASTE, WATER SUPP panache de vapeur *m*; **~ pressure** *BrE* SOIL, WATER POLL, WATER SUPP *of pump* pression de vapeur *f*; **~ pressure analysis** *BrE* ANALYSIS analyse de pression de vapeur *f*; **~ pressure determination** *BrE* ANALYSIS détermination de pression de vapeur *f*; **~ pressure measurement** *BrE* ANALYSIS mesure de pression de vapeur *f*; **~ pressure scale** *BrE* ANALYSIS échelle de pression de vapeur *f*; **~ riser** *BrE* AIR *of bubble-cap tray tower* colonne de vapeur montante *f*; **~ treatment** *BrE* WATER SUPP traitement à la vapeur *m*

variable[1] *adj* AIR variable

variable[2]: **~-gas throughput**, **~-gas thruput** *AmE n* AIR débit de gaz variable *m*; **~-gas throughput amount**, **~-gas thruput amount** *AmE n* AIR débit de gaz variable *m*; **~ gas waste-gas flow** *n* AIR écoulement variable d'effluents gazeux dans un courant gazeux *m*; **~-pitch propeller** *n* MARINE hélice à pas variable *f*; **~-pitch rotor blade** *n* RENEW ENERG *wind turbine* pale de rotor à pas réglable *f*; **~-speed motor** *n* WATER POLL moteur à vitesse variable *m*; **~-speed pump** *n* WATER POLL pompe à vitesse variable *f*

varnish: **~ waste** *n* WASTE déchets de peinture *m pl*; **~ waste sludge** *n* WASTE boues de déchets vernis *f pl*

vat *n* WATER SUPP *gas melt* cuve *f*

VAWT *abbr* (*vertical axis wind turbine*) RENEW ENERG éolienne à axe vertical *f*

VC *abbr* (*vinyl chloride*) AIR chlorure de vinyle *m*

vegetable: **~ crop** *n* SOIL culture maraîchère *f*; **~ waste** *n* WASTE épluchures de légumes *f pl*

vegetation: **~ cover** *n* SOIL couvert végétal *m*, tapis végétal *m*; **~ loss** *n* SOIL dépérissement de la végétation *m*

vegetational: **~ survey** *n* MANAGE, SOIL inventaire de la végétation *m*

vehicle: **~ capacity** *n* WASTE *waste transport* capacité d'un véhicule *f*; **~ shredder** *n* WASTE déchiqueteuse *f*, cisailles pour automobiles *f pl*

veil: **~ of pollution** *n* AIR voile de pollution *m*

velocity: **~ of flow** *n* ANALYSIS vitesse d'écoulement *f*; **~ head** *n* WATER POLL, WATER SUPP hauteur dynamique *f*, énergie cinétique *f*; **~ of migration** *n* AIR, ANALYSIS vitesse de migration *f*; **~ pressure** *n* AIR pression du vent *f*

vent *n* AIR, WATER SUPP évent *m*, aération *f*, bouche d'aération *f*, mise à l'air *f*, orifice de ventilation *m*, purge de l'air *f*; **~ condenser** WATER POLL condenseur de buées *m*; **~ muffler** *AmE* (*cf vent silencer BrE*) NOISE silencieux d'évent *m*; **~ silencer** *BrE* (*cf vent muffler AmE*) NOISE silencieux d'évent *m*

ventilate *vt* AIR, SAFETY, WASTE, WATER SUPP aérer, ventiler

ventilated: **~ nappe** *n* WATER SUPP nappe aérée *f*

ventilation *n* AIR, SAFETY, WASTE, WATER SUPP aérage *m*, aération *f*, ventilation *f*

ventilator *n* AIR, SAFETY, WASTE, WATER SUPP ventilateur *m*

venting *n* SOIL *of soil atmosphere* évacuation des gaz *f*

Venturi[1]: **~-shaped** *adj* AIR en forme de Venturi

Venturi[2]: **~ burner** *n* AIR brûleur à Venturi *m*; **~ flume** *n* WATER SUPP canal Venturi *m*; **~ meter** *n* AIR Venturimètre *m*; **~ nozzle** *n* AIR tuyère de Venturi *f*; **~ principle** *n* AIR principe du Venturi *m*; **~ scrubber** *n* AIR laveur à Venturi *m*, WASTE séparateur laveur *m*; **~ throat** *n* AIR col de Venturi *m*; **~ tube** *n* AIR, ANALYSIS tube Venturi *m*, Venturi *m*

verifiable *adj* ANALYSIS vérifiable

vermin *n* WASTE, WATER SUPP vermine *f*, parasite *m*

vertical: **~ axis wind turbine** *n* (*VAWT*) RENEW ENERG éolienne à axe vertical *f*; **~ cartoner** *n* WASTE encartonneuse verticale *f*; **~ cartoning machine** *n* WASTE encartonneuse verticale *f*; **~ degassing** *n* WASTE, WATER SUPP dégazage vertical *m*; **~ dispersion** *n* AIR dispersion verticale *f*; ~ **distribution** *n* AIR, SOIL distribution verticale *f*, répartition verticale *f*; ~ **filtering well** *n* WATER SUPP puits de filtration vertical *m*; ~ **fixed-bed adsorber** *n* AIR, WATER POLL adsorbeur à lit fixe vertical *m*; ~ **gas collection** *n* AIR, WASTE, WATER SUPP collecte verticale des gaz *f*; ~ **gas well** *n* WASTE, WATER SUPP puits de gaz vertical *m*; ~ **moving-bed adsorber** *n* AIR, WATER POLL adsorbeur à lit mobile vertical *m*; ~ **percolation well** *n* WATER SUPP puits de percolation vertical *m*; ~ **well** *n* WATER SUPP puits vertical *m*

vessel *n* AIR cuve *f*, MARINE bâtiment *m*; ~ **wall** AIR *of cyclone* paroi de cuve *f*

viable: ~ **bacteria** *n pl* WATER POLL bactéries revivifiables *f pl*, WATER SUPP bactéries viables *f pl*; ~ **microorganism** *n* MARINE, WATER POLL micro-organisme revivifiable *m*

vibrate *vi* GEN vibrer

vibrating[1] *adj* GEN vibrant

vibrating[2]: ~ **conveyor** *n* WATER SUPP transporteur à secousses *m*, transporteur vibrant *m*; ~ **screen** *n* SOIL, WASTE tamis à secousse *m*, tamis à vibrations *m*, tamis vibrant *m*; ~ **sieve** *n* WATER POLL tamis vibrant *m*; ~ **worm** *n* SOIL vis vibrante *f*

vibration *n* GEN vibration *f*; ~ **attenuation** NOISE atténuation des vibrations *f*; ~ **behavior** *AmE*, ~ **behaviour** *BrE* RENEW ENERG *of cable-braced solar model construction* comportement vibratoire *m*; ~ **damping** NOISE amortissement des vibrations *m*; ~ **isolation** NOISE isolation contre les vibrations *f*; ~ **isolator** NOISE isolateur de vibrations *m*; ~ **method** ANALYSIS méthode de vibration *f*; ~ **viscosimeter** ANALYSIS viscosimètre de vibration *m*

vibrator: ~ **screen** *n* SOIL, WASTE tamis à vibrations *m*, tamis vibrant *m*

vibrion *n* WATER POLL vibrion *m*

vibrometer *n* ANALYSIS vibromètre *m*

VIC *abbr* (*volatile inorganic compounds*) AIR, SOIL composés inorganiques volatils *m pl*

vine: ~ **shoot** *n* RENEW ENERG sarment de vigne *m*

vinyl: ~ **chloride** *n* (*VC*) AIR chlorure de vinyle *m*

virgin: ~ **material** *n* WASTE matériau vierge *m*, matière vierge *f*

viricidal *adj* WATER SUPP virulicide

virus *n* WATER POLL virus *m*

viscid *adj* AIR gluant, visqueux, WATER POLL visqueux

viscosimeter *n* ANALYSIS viscosimètre *m*

viscosity *n* AIR, WATER POLL, WATER SUPP viscosité *f*; ~ **meter** ANALYSIS viscosimètre *m*

viscous[1] *adj* AIR, WATER POLL, WATER SUPP visqueux

viscous[2]: ~ **liquid** *n* AIR, WATER POLL, WATER SUPP liquide visqueux *m*

visibility: ~ **depth** *n* ANALYSIS profondeur de champ *f*

visible[1] *adj* AIR, ANALYSIS visible

visible[2]: ~ **optical absorption** *n* ANALYSIS absorption optique visible *f*; ~ **plume** *n* AIR panache visible *m*

visual: ~ **analysis** *n* ANALYSIS analyse visuelle *f*; ~ **pack** *n* WASTE emballage transparent *m*; ~ **test** *n* ANALYSIS essai visuel *m*

vitrification *n* SOIL, WASTE vitrification *f*; ~ **process** WASTE procédé de vitrification *m*

vitrified[1] *adj* SOIL, WASTE vitrifié

vitrified[2]: ~ **clay pipe** *n* WATER POLL tuyau en grès *m*, WATER SUPP conduite en grès cérame vitrifié *f*; ~ **slag** *n* SOIL laitier vitrifié *m*

vitrify *vt* SOIL, WASTE vitrifier

VOC *abbr* (*volatile organic compounds*) AIR, SOIL COV (*composés organiques volatils*)

void *n* WATER POLL *in pump* vide *m*, WATER SUPP cavité *f*; ~ **space** WATER POLL *of trickling filter* espace interstitiel *m*; ~ **volume** AIR volume de vide *m*

voids: ~ **fraction** *n* AIR taux de vides *m*

volatile[1] *adj* AIR, SOIL, WATER POLL volatil

volatile[2]: ~ **chlorinated hydrocarbon** *n* AIR, SOIL hydrocarbure chloré volatile *m*; ~ **emissions** *n pl* AIR émissions volatiles *f pl*; ~ **fraction** *n* AIR particule volatile *f*; ~ **halogenated hydrocarbon** *n* AIR, WATER SUPP hydrocarbure halogéné volatil *m*; ~ **heavy metal** *n* AIR, WASTE, WATER POLL métal lourd volatil *m*; ~ **inorganic compounds** *n pl* (*VIC*) AIR, SOIL composés inorganiques volatils *m pl*; ~ **organic compounds** *n* (*VOC*) AIR, SOIL composés organiques volatils *m pl* (*COV*)

volatility: ~ **analysis** *n* ANALYSIS analyse de volatilité *f*

volatilization *n* SOIL, WASTE volatilisation *f*

volatilize *vt* SOIL, WASTE volatiliser

voltammetric *adj* ANALYSIS *detector* voltampèremétrique

voltammetry *n* ANALYSIS voltampèremétrie *f*

volume *n* GEN volume *m*; ~ **charge** AIR *electrostatic precipitator* charge volumique *f*; ~ **concentration** AIR concentration volumique *f*; ~ **filter** WATER POLL filtre volumique *m*; ~ **filtration** WATER POLL filtration volumique *f*; ~ **flow** AIR, ANALYSIS, WATER SUPP débit volumique *m*, débit-volume *m*; ~ **flow rate** AIR, ANALYSIS, WATER SUPP débit volumique *m*, débit-volume *m*; ~ **of hazardous waste** WASTE volume de déchets dangereux *m*; ~ **reduction** WASTE réduction de volume *f*; ~ **unit meter** (*VU meter*) ANALYSIS volumètre *m*, vumètre *m*; ~ **of water** WATER SUPP volume d'eau *m*; ~ **weight** WASTE *of refuse* poids en volume *m*

volumetric[1] *adj* ANALYSIS volumétrique

volumetric[2]: ~ **analysis** *n* ANALYSIS analyse volumétrique *f*; ~ **analysis by titration** *n* ANALYSIS analyse volumétrique par titrage *f*; ~ **determination** *n* ANALYSIS détermination volumétrique *f*; ~ **factor** *n* ANALYSIS facteur volumétrique *m*; ~ **flask** *n* ANALYSIS ampoule graduée *f*, flacon mélangeur *m*; ~ **gas analysis** *n* ANALYSIS analyse de gaz volumétrique *f*; ~ **loss** *n* WATER POLL, WATER SUPP perte volumétrique *f*

voluntary: ~ **agreement** *n* POLICY accord librement souscrit *m*; ~ **commitment** *n* POLICY engagement délibéré *m*

volute: ~ **chamber** *n* WATER SUPP *of centrifugal pump* canal collecteur *m*; ~ **insert** *n* WATER SUPP *of centrifugal pump* anneau d'écoulement *m*

vortex *n* AIR tourbillon *m*, vortex *m*; ~ **finder** AIR *of cyclone* détecteur de tourbillon *m*; ~ **flow** AIR *of cyclone* écoulement tourbillonnaire *m*; ~ **skimmer** MARINE récupérateur à tourbillon *m*

voucher *n* MANAGE pièce justificative *f*

VU: ~ **meter** *n* (*volume unit meter*) ANALYSIS volumètre *m*, vumètre *m*

vulcanize *vt* WASTE vulcaniser

vulcanized *adj* WASTE vulcanisé

vulnerability *n* POLICY vulnérabilité *f*

W

wadding *n* WASTE bourre *f*, capitonnage *m*, ouate *m*, rembourrage *m*
wagon: ~ **drill** *n* ANALYSIS jumbo *m*, sondeuse portée *f*
waiting: ~ **time** *n* ANALYSIS temps d'attente *m*
wake *n* MARINE sillage *m*
wall: ~ **depth** *n* SOIL profondeur de mur *f*, WASTE, WATER SUPP épaisseur de paroi *f*; ~ **hydrant** *n* WATER SUPP bouche murale *f*
Walther: ~ **process** *n* AIR procédé Walther *m*
WAO *abbr* (*wet air oxidation*) SOIL, WASTE, WATER POLL oxydation par voie humide *f*
warm: ~ **seawater** *n* RENEW ENERG eau de mer chaude *f*; ~**-water generation** *n* AIR production d'eau chaude *f*; ~**-water heater** *n* AIR chauffe-eau *m*
warning: ~ **level** *n* SAFETY niveau d'alerte *m*; ~ **signal** *n* SAFETY signal d'alarme *m*; ~ **system** *n* SAFETY système d'alerte *m*
wash[1]: ~ **bottle** *n* ANALYSIS flacon de lavage *m*, pissette *f*; ~ **water** *n* AIR, WATER SUPP eau de lavage *f*; ~**-water circulation** *n* AIR circulation d'eau de lavage *f*; ~**-water tank** *n* AIR réservoir d'eau de lavage *m*
wash[2] *vt* GEN laver; ~ **away** ANALYSIS charrier, emporter, entraîner, entraîner au lavage, GEN laver, lixivier, éliminer par lavage; ~ **out** GEN laver, évacuer par lavage, lessiver, WATER SUPP éliminer par lavage
washed: ~ **and squashed consumer waste cartons** *n pl* WASTE cartons d'emballage de produits de consommation lavés et compactés *m pl*
washing[1]: ~**-active** *adj* SOIL détergent-actif
washing[2] *n* AIR, SOIL, WASTE, WATER SUPP lavage *m*; ~ **agent** AIR agent de lavage *m*; ~ **agent circulation** AIR circulation d'agent de lavage *f*; ~**-agent processing** AIR traitement par un agent de lavage *m*; ~ **agent regeneration** AIR régénération par un agent de lavage *f*; ~ **away** AIR lavage, ANALYSIS entraînement *m*, SOIL, WATER SUPP lavage *m*; ~ **bottle** ANALYSIS flacon de lavage *m*, pissette *f*; ~ **cycle** AIR cycle de lavage *m*; ~ **operation** AIR opération de lavage *f*; ~ **out** AIR *of filter fibres*, SOIL, WASTE, WATER SUPP lavage *m*; ~ **process** AIR procédé de lavage *m*; ~ **soda** WATER SUPP cristaux de soude *m pl*; ~ **solution** AIR solution de lavage *f*; ~ **system** AIR système de lavage *m*
washings *n pl* SOIL résidus de lavage *m pl*
washout *n* AIR lavage *m*, ANALYSIS *water determination in gases* affouillement *m*, SOIL affouillement *m*, lavage *m*, lavage par la pluie, WASTE *of landfill* lavage *m*, WATER POLL, WATER SUPP sortie d'eau de lavage *f*, lessivage *m*, lavage *m*, lavage par la pluie; ~ **rate** AIR taux de lavage *m*
waste[1]: ~**-economical** *adj* WASTE produisant peu de déchets; ~**-fired** *adj* WASTE brûlant du biocombustible, chauffé par les déchets; ~**-free** *adj* WASTE sans déchet; ~**-generating** *adj* WASTE source de déchets
waste[2] *n* WASTE ordures *f pl*, déchets *m pl*; ~ **acid** WASTE acidé récupéré *m*, acide perdu *m*, acide résiduare *m*; ~ **air concentration** AIR concentration d'air résiduel *f*; ~ **air constituent** AIR composant d'air résiduel *m*; ~ **air fan** AIR ventilateur d'air résiduel *m*; ~ **air flow** AIR débit d'air résiduel *m*; ~ **air temperature** AIR température de l'air résiduaire *f*; ~ **air throughput**, ~ **air thruput** *AmE* AIR débit d'air résiduel *m*; ~ **avoidance** WASTE mesure de prévention contre les déchets *f*, lutte contre les déchets *f*; ~ **avoidance and management act** MANAGE, WASTE loi sur la prévention et la gestion de déchets *f*; ~ **bag** WASTE sac poubelle *m*; ~ **bale** WASTE déchets mis en balle *m pl*, balle de déchets *f*; ~ **body** WASTE carcasse *f*, masse de déchets *f*; ~ **bottle** WASTE bouteille perdue *f*; ~ **bunker** WASTE décharge aménagée *f*; ~ **charge** POLICY redevance sur les déchets *f*; ~**-coal mixture** WASTE *refuse-derived fuel* mélange déchets-charbon *m*; ~ **collection** WASTE enlèvement *m*, collecte des déchets *f*, ramassage des ordures *m*; ~ **collector** WASTE collecteur de déchets *m*; ~ **composition** WASTE composition des déchets *f*; ~ **container** WASTE bac à déchets *m*, conteneur de déchets *m*; ~**-containing pollutants** *n pl* WASTE déchets pollués *m pl*; ~ **crop** WASTE déchets agricoles *m pl*; ~ **crusher** WASTE broyeur de déchets *m*; ~ **dampness** WASTE humidité des déchets *f*; ~ **denitrification** WASTE dénitrification des déchets *f*; ~ **description sheet** POLICY, WASTE fiche déchet *f*; ~ **detector** WASTE détecteur de déchets *m*; ~ **discharge** WASTE rejet de déchets *m*; ~ **disinfection** WASTE désinfection des déchets *f*; ~ **disintegrator** WASTE désintégrateur de déchets *m*; ~ **disposal** MANAGE, POLICY, WASTE élimination des déchets *f*, élimination des ordures *f*, évacuation des déchets *f*; ~ **disposal act** MANAGE, POLICY, WASTE loi sur l'élimination de déchets *f*; ~ **disposal company** MANAGE, WASTE compagnie d'élimination des déchets *f*, entreprise d'élimination des déchets *f*, société de traitement de déchets *f*; ~ **disposal firm** MANAGE, WASTE compagnie d'élimination des déchets *f*, entreprise d'élimination des déchets *f*, société de traitement de déchets *f*; ~ **disposal plan** MANAGE, POLICY, WASTE plan d'élimination des déchets *m*, programme d'élimination des déchets *m*; ~ **disposal plant** WASTE centre d'enfouissement des déchets *m*; ~ **disposal regulation** POLICY réglementation applicable à l'élimination des déchets *f*; ~ **disposal tax** POLICY taxe sur l'élimination des déchets *f*; ~ **dump** WASTE décharge *f*, déchetterie *f*, décharge de déchets *f*, dépôt de déchets *m*; ~ **economical planning** MANAGE, WASTE gestion visant à limiter la production de déchets *f*; ~ **of energy** AIR, RENEW ENERG, WATER SUPP perte d'énergie *f*, gaspillage d'énergie *m*; ~ **exchange market** WASTE bourse des déchets *f*; ~**-fired plant** WASTE usine alimentée en biocombustible *f*; ~ **formation** WASTE formation de déchets *f*; ~ **fuel** WASTE biocarburant *m*; ~ **gas** AIR gaz résiduaire *m*, gaz rejeté *m*, effluent gazeux *m*, gaz d'échappement *m*; ~ **gas back pressure** AIR refoulement des gaz de combustion *m*; ~ **gas blower** AIR soufflante des gaz de combustion *f*, ventilateur d'effluents gazeux *m*; ~ **gas burner** AIR torchère *f*; ~ **gas cleaning** AIR épuration des gaz brûlés *f*, épuration des gaz perdus *f*, WASTE épuration des gaz brûlés *f*; WATER SUPP lavage des gaz brûlés *m*, lavage des gaz perdus *m*; ~ **gas component** AIR élément de gaz pollués *m*; ~ **gas composition** AIR composition des effluents gazeux *f*; ~ **gas dilution** AIR dilution des effluents gazeux *f*; ~ **gas emissions** *n pl* AIR émissions gazeuses *f pl*; ~ **gas extraction** AIR extraction des gaz pollués *f*; ~ **gas flow** AIR flux de gaz pollué *m*; ~ **gas incineration** AIR incinération de gaz pollués *f*; ~ **gas loading** AIR charge de gaz pollué *f*; ~ **gas plume** AIR panache d'effluents gazeux *m*; ~ **gas preheating** AIR préchauffage des gaz *m*; ~ **gas pressure** AIR *in stack* pression des rejets gazeux *f*; ~ **gas probe** AIR prélèvement de gaz pollués *m*; ~ **gas problem** AIR problème dû aux effluents gazeux *m*; ~ **gas purifier** AIR épurateur de gaz *m*; ~ **gas temperature** AIR température de gaz pollués *f*, température des effluents gazeux *f*, température des rejets gazeux *f*; ~ **gas throughput**, ~ **gas thruput** *AmE* AIR débit

de rejets gazeux *m*; ~ **gas turbidity** ANALYSIS *of dust, smoke* turbidité des effluents gazeux *f*; ~ **gas velocity** AIR *in stack* vitesse de gaz résiduels *f*; ~ **gas volume** AIR volume de rejets gazeux *m*; **~-generating factory** WASTE usine source de déchets *f*; ~ **generation** WASTE production de déchets *f*; ~ **generator** WASTE producteur de déchets *m*; ~ **glass** WASTE verre perdu *m*, WATER SUPP verre usagé *m*; ~ **glass container** WASTE conteneur de verre usagé *m*, conteneur de verre perdu *m*; ~ **glass cullet** WASTE calcin *m*; ~ **harmful to water** WATER POLL déchets nocifs pour l'eau *m pl*; ~ **heat** AIR, WASTE, WATER SUPP chaleur dissipée *f*, chaleur perdue *f*, chaleur résiduelle *f*; ~ **heat boiler** (*WHB*) AIR, WASTE, WATER POLL, WATER SUPP chaudière alimentée par des rejets thermiques *f*, chaudière de récupération de chaleur *f*; ~ **heat boiler system** AIR système de chaudière à récupération de chaleur *m*, WATER SUPP système de chaudière à chaleur de récupération *m*; ~ **heat recovery** AIR, WASTE, WATER POLL, WATER SUPP récupération de chaleur dissipée *f*, récupération de la chaleur dissipée par les déchets *f*; ~ **heat recovery boiler** AIR, WASTE, WATER POLL, WATER SUPP chaudière alimentée par des rejets thermiques *f*, chaudière de récupération de chaleur *f*; ~ **heat system** AIR, WASTE, WATER POLL, WATER SUPP système à récupération de chaleur *m*; ~ **heat use** AIR, WASTE, WATER POLL, WATER SUPP valorisation de chaleur de récupération *f*, valorisation des rejets thermiques *f*, consommation de chaleur de récupération *f*, consommation de chaleur résiduelle *f*, valorisation de chaleur résiduelle *f*; ~ **heat utilization** AIR, WASTE, WATER POLL, WATER SUPP valorisation de chaleur de récupération *f*, valorisation des rejets thermiques *f*, consommation de chaleur de récupération *f*, consommation de chaleur résiduelle *f*, valorisation de chaleur résiduelle *f*; ~ **incineration** WASTE, WATER SUPP incinération de déchets *f*, incinération d'ordures *f*; ~ **incineration at sea** WASTE incinération de déchets en mer *f*; ~ **incineration plant** WASTE usine d'incinération de déchets *f*; ~ **incineration slag** WASTE laitier d'incinération de déchets *m*; ~ **incinerator** WASTE incinérateur d'ordures *m*, incinérateur de déchets *m*; ~ **injection** WASTE injection de déchets *f*; ~ **inventory** WASTE inventaire des déchets *m*; ~ **land** POLICY terrain vague *m*, SOIL terre inculte *f*; ~ **landfill planning** WASTE aménagement de décharge *m*, planification de la mise en décharge *f*; ~ **landfilling** WASTE mise en décharge de déchets *f*; ~ **layer** WASTE couche de déchets *f*, strate de déchets *f*; ~ **lubricant** WASTE lubrifiant perdu *m*, lubrifiant pour déchets *m*, lubrifiant résiduel *m*; ~ **lye** WASTE, WATER POLL, lessive résiduaire *f*; ~ **management** MANAGE, POLICY, WASTE gestion de déchets *f*; ~ **management act** MANAGE, POLICY, WASTE loi sur la gestion de déchets *f*; ~ **management company** MANAGE, WASTE entreprise de gestion de déchets *f*; ~ **management concept** MANAGE type de gestion de déchets *m*; ~ **management firm** MANAGE, WASTE entreprise de gestion de déchets *f*; ~ **management market** MANAGE marché de la gestion de déchets *m*; ~ **management officer** MANAGE, POLICY, WASTE responsable de la gestion de déchets *m*; ~ **management planning** MANAGE planification de la gestion de déchets *f*; ~ **mass** WASTE masse de déchets *f*; ~ **material** WASTE matériau de rebut *m*; ~ **material return rate** WASTE taux de retour des matériaux de rebut *m*; ~ **material use rate** WASTE taux d'utilisation des matériaux de rebut *m*; ~ **materials** *n pl* WASTE déchets *m pl*; ~ **mixture** WASTE mélange de déchets *m*; ~ **neutralization** WASTE neutralisation de déchets *f*; ~ **oil** SOIL, WASTE huile usagée *f* huile usée *f*; ~ **oil act** POLICY, SOIL, WASTE loi sur les huiles usagées et les corps gras de récupération *f*, loi sur les rejets d'hydrocarbures *f*, loi sur les huiles usées *f*; ~ **oil collection** WASTE collecte des huiles usées *f*; ~ **oil preparation** WASTE préparation des résidus de bacs à graisse *f*; ~ **oil recovery** WASTE récupération des huiles usagées *f*, récupération des huiles usées *f*; ~ **paint** WASTE déchets de peinture *m pl*; ~ **paint sludge** WASTE boues de peinture *f pl*; ~ **paper** WASTE vieux papiers *m pl*; ~ **paper collection** WASTE, collecte de vieux papiers *f*; ~ **paper compressing press** WASTE presse pour vieux papiers *f*; ~ **paper preparation** WASTE préparation de vieux papiers *f*; ~ **paper recycling** WASTE, recyclage de vieux papiers *m*; ~ **plant** WASTE déchetterie *f*; ~ **plastics** *n pl* WASTE déchets plastiques *m pl*; ~ **pretreatment** WASTE prétraitement de déchets *m*; ~ **processing** WASTE traitement de déchets *m*, transformation de déchets *f*; ~ **processing plant** WASTE installation de traitement de déchets *f*; ~ **producer** WASTE producteur de déchets *m*; ~ **product** WASTE, WATER POLL effluent *m*, produit des déchets *m*; ~ **production** WASTE production de déchets *f*; ~ **products** *n pl* WASTE déchets *m pl*; ~ **pulp** WASTE pâte de déchets *f*; ~ **pyrolysis** WATER POLL pyrolyse des ordures *f*; ~ **quality** WASTE qualité des déchets *f*; ~ **raw material** RENEW ENERG *biomass conversion base material*, WASTE matière première résiduelle *f*; ~ **recovery** WASTE récupération de déchets *f*; ~ **reduction** WASTE réduction de déchets *f*; ~ **removal** WASTE extraction de déchets *f*; ~ **reprocessing** WASTE retraitement de déchets *m*; ~ **rubber** WASTE caoutchouc perdu *m*, caoutchouc usagé *m*, déchets de caoutchouc *m pl*; ~ **salts** *n pl* WATER SUPP sel résiduaire *m*; ~ **segregation** WASTE tri de déchets *m*; ~ **sludge mixture** WASTE mélange déchets-boues *m*; ~ **sorting** WASTE tri de déchets *m*; ~ **sorting plant** WASTE, WATER SUPP poste de tri de déchets *m*, usine de tri de déchets *f*; ~ **steam** WATER SUPP vapeur résiduelle *f*; ~ **storage** WASTE stockage de déchets *m*; ~ **stream** WASTE flux de déchets *m*; ~ **stuff** WASTE matériau de déchets *m*; ~ **substance** WASTE substance perdue *f*, substance rejetée *f*; ~ **tax** POLICY impôt sur les déchets *m*, taxe sur les déchets *f*; ~ **tip** *BrE* WASTE *place* dépotoir *m*, décharge *f*, *action* déversement de déchets *m*; ~ **tourism** POLICY, WASTE tourisme des déchets *m*; ~ **treatment** WASTE traitement de déchets *m*; ~ **treatment plant** WASTE station de traitement de déchets *f*, usine de traitement de déchets *f*; ~ **type** WASTE type de déchets *m*; ~ **utilization** WASTE valorisation de déchets *f*; ~ **utilization plant** WASTE usine de valorisation de déchets *f*; ~ **volume** WASTE volume de déchets *m*; ~ **washings** *n pl* WASTE stériles *m pl*

waste[3] *vt* GEN gaspiller

wastewater[1]: **~-free** *adj* AIR *spray absorption process* sans eaux usées

wastewater[2] *n* WATER POLL eaux usées *f pl*, *industrial* eaux d'égout *f pl*, eaux résiduaires *f pl*; ~ **analyser** *BrE*, ~ **analyzer** *AmE* ANALYSIS analyseur des eaux usées *m*; ~ **bed** WATER POLL nappe d'eaux usées *f*; ~ **clarification** WATER POLL clarification des eaux usées *f*; ~ **clarification plant** WATER POLL installation de clarification des eaux usées *f*; ~ **collection** WATER POLL collecte des eaux usées *f*; ~ **collection system** WATER POLL réseau de collecte des eaux usées *m*; ~ **collection tank** WATER POLL, WATER SUPP réservoir de collecte des eaux usées *m*; ~ **collector** WATER POLL collecteur d'eaux usées *m*; ~ **coloration** ANALYSIS coloration des eaux usées *f*; ~ **composition** WATER POLL composition des eaux usées *f*; ~ **control** WATER POLL épuration des eaux usées *f*; ~ **decontamination** WATER POLL décontamination des eaux usées *f*; ~ **discharge** WATER POLL rejet d'eaux usées *m*; ~ **disposal** WATER POLL, WATER SUPP évacuation des eaux usées *f*; ~ **disposal in the ocean** WATER POLL rejet des eaux usées en mer *m*; ~ **farm** WATER POLL champ d'épandage *m*; ~ **fish pond** WATER POLL bassin à poissons alimenté avec des eaux recyclées *m*; ~ **flow** WATER POLL écoulement d'eaux usées *m*; **~-free flue-gas cleaning** AIR, WASTE épuration des effluents gazeux sans eaux usées *f*; ~ **from flue-gas scrubbing** WATER POLL effluent liquide du lavage des gaz *m*; ~ **from waste incineration plant** WATER POLL eaux usées d'usine d'incinération *f pl*; ~ **fungus** WATER POLL champignon des eaux usées *m*; ~ **in ash removal**

WATER POLL eaux usées résultant de l'extraction des cendres *f pl*; ~ **incineration** WATER POLL incinération des boues résiduaires *f*; ~ **incineration with heat use** WATER POLL incinération des eaux usées avec consommation de chaleur *f*; ~ **incineration without heat use** WATER POLL incinération des eaux usées sans consommation de chaleur *f*; ~ **irrigation** WATER POLL irrigation avec des eaux recyclées *f*; ~ **load** WATER POLL charge en eaux usées *f*; ~ **monitoring station** ANALYSIS, WATER POLL station de traitement des eaux usées *f*; ~ **organism** WATER POLL organisme se développant dans les eaux usées *m*; ~ **outfall** WATER POLL déversoir des eaux résiduaires *m*; ~ **outflow** WATER POLL sortie des eaux usées *f*; ~ **production** WATER POLL production d'eaux résiduaires *f*; ~ **pump** WATER POLL pompe de station d'épuration *f*; ~ **purification** WATER POLL épuration des eaux usées *f*; ~ **purification plant** WATER POLL station d'épuration *f*; ~ **purification process** WATER POLL procédé d'épuration des eaux usées *m*; ~ **quality** WATER POLL qualité des eaux usées *f*; ~ **reuse** WATER POLL réutilisation des eaux après épuration *f*, recyclage des eaux usées *m*; ~ **sample** ANALYSIS, WATER POLL échantillon d'eaux usées *m*; ~ **sludge** WATER POLL boues d'eaux résiduaires *f pl*; ~ **sludge amount** WATER POLL volume des boues d'eaux résiduaires *m*; ~ **sludge catalysis** WATER POLL catalyse des boues d'eaux résiduaires *f*; ~ **sludge composting** WATER POLL compostage des boues d'eaux résiduaires *m*; ~ **sludge conditioning plant** WATER POLL usine de conditionnement des boues d'eaux résiduaires *f*; ~ **sludge delivery** WATER POLL écoulement de boues résiduaires *m*, débit d'eaux résiduaires *m*; ~ **sludge deposition ban** WATER POLL interdiction de mise en décharge des boues d'eaux résiduaires *f*; ~ **sludge dewatering equipment** WATER POLL poste de déshydratation des boues d'eaux résiduaires *m*; ~ **sludge incineration** WATER POLL incinération des boues d'eaux résiduaires *f*; ~ **sludge incineration plant** WATER POLL usine d'incinération des boues d'eaux résiduaires *f*; ~ **sludge treatment** WATER POLL traitement des boues d'eaux résiduaires *m*; ~ **sprinkling** WATER POLL aspersion d'eaux usées *f*; ~ **stream** WATER POLL écoulement d'eaux résiduaires *m*, écoulement d'eaux usées *m*; ~ **technology** WATER POLL technique de traitement des eaux usées *f*; ~ **temperature** WATER POLL température des eaux résiduaires *f*; ~ **toxicity** WATER POLL toxicité des eaux usées *f*; ~ **trail** WATER POLL *plume* traînée d'eaux usées *f*; ~ **treatment** WATER POLL traitement des eaux usées *m*; ~ **treatment fee** WATER POLL redevance d'épuration des eaux usées *f*; ~ **treatment plant** WATER POLL station d'épuration des eaux usées *f*; ~ **treatment plant installation** WATER POLL aménagement d'une station d'épuration d'eaux usées *m*; ~ **utilization** WATER POLL valorisation des eaux usées *f*; ~ **volume** WATER POLL volume d'eaux usées *m*; ~ **yield** AIR, WATER POLL production d'eaux usées *f*

water[1]: ~**-free** *adj* AIR, WATER SUPP anhydre, déshydraté, sans eau; ~**-insoluble** *adj* AIR, WASTE, WATER POLL, WATER SUPP insoluble dans l'eau; ~**-repellent** *adj* WASTE, WATER POLL, WATER SUPP hydrofuge; ~**-resistant** *adj* WASTE, WATER POLL, WATER SUPP étanche, resistant à l'eau , hydrofuge; ~**-saturated** *adj* WASTE, WATER POLL, WATER SUPP saturé en eau; ~**-soluble** *adj* AIR, WASTE, WATER POLL, WATER SUPP soluble dans l'eau

water[2] *n* GEN eau *f*; ~ **analysis** ANALYSIS analyse de l'eau *f*; ~ **authority** POLICY, WATER SUPP agence de l'eau *f*, compagnie des eaux *f*, service des eaux *m*; ~ **balance** POLICY bilan hydrique *m*, bilan hydrologique *m*, WATER SUPP *nappe* bilan d'eau *m*; ~**-based backing adhesive** WASTE renfort adhésif à l'eau *m*; ~**-bearing soil** SOIL, WATER SUPP terrain aquifère *m*; ~ **catchment** WATER SUPP captage d'eau *m*; ~ **circulation** WATER POLL, WATER SUPP circulation d'eau *f*; ~ **clearness** WATER POLL, WATER SUPP pureté de l'eau *f*, limpidité de l'eau *f*; ~ **column** ANALYSIS, WATER SUPP colonne d'eau *f*; ~ **consumption** WATER SUPP consommation d'eau *f*; ~ **content** ANALYSIS, WASTE teneur en eau *f*, WATER POLL *of wastewater sludge* taux d'humidité *m*, taux d'hygrométrie *m*, teneur en eau *f*, WATER SUPP degré d'hygrométrie *m*, teneur en eau *f*; ~ **control charge** POLICY taxe de gestion de l'eau *f*; ~ **cycle** WATER POLL cycle de l'eau *m*; ~ **damage** WATER SUPP dégât des eaux *m*; ~ **deficiency** WATER SUPP déficit hydrique *m*; ~ **delivery** WATER SUPP adduction d'eau *f*; ~ **demand** WATER SUPP demande d'eau *f*; ~ **distribution** WATER SUPP distribution d'eau *f*; ~ **distribution system** WATER SUPP réseau de distribution d'eau *m*; ~**-drawing engine** WATER SUPP moteur de pompage *m*, moteur de soutirage d'eau *m*; ~ **equivalent** WATER SUPP équivalent en eau *m*; ~ **evaporation** AIR, WATER SUPP évaporation d'eau *f*; ~ **examination** ANALYSIS examen de l'eau *m*; ~ **film** AIR pellicule d'eau *f*; ~ **filter** WATER SUPP filtre à eau *m*; ~ **flush** WATER SUPP chasse d'eau *f*; ~ **gage** *AmE*, ~ **gauge** *BrE* WATER SUPP échelle d'étiage *f*, indicateur de niveau d'eau *m*; ~ **hammer** WATER SUPP coup de bélier *m*; ~ **hardness** WATER POLL, WATER SUPP dureté de l'eau *f*; ~ **hardness degree** WATER POLL, WATER SUPP degré hydrotimétrique de l'eau *m*; ~ **head** WATER SUPP eau d'amont *f*; ~ **hose** WATER POLL, WATER SUPP lance d'arrosage *f*; ~ **hydrant** WATER SUPP bouche d'eau *f*; ~ **indicator** WATER SUPP indicateur de niveau d'eau *m*; ~ **inflow** WATER SUPP apport d'eau *m*, venue d'eau *f*; ~ **inlet** WATER SUPP orifice d'admission d'eau *m*; ~**-insoluble hazardous waste** WASTE, WATER POLL déchets dangereux insolubles dans l'eau *m pl*; ~ **intake** WATER POLL, WATER SUPP prise d'eau *f*; ~ **intake structure** WATER POLL, WATER SUPP ouvrage de prise d'eau *m*; ~ **jet** WATER POLL, WATER SUPP jet d'eau *m*; ~ **jet pump** WATER POLL, WATER SUPP trompe soufflante *f*; ~ **law** POLICY droit de l'eau *m*; ~ **level** MARINE, RENEW ENERG *tidal power plant*, WATER SUPP niveau d'eau *m*; ~ **level indicator** WATER SUPP indicateur de niveau d'eau *m*; ~ **level recession** MARINE, WATER SUPP baisse du niveau d'eau *f*; ~ **line** MARINE ligne de charge *f*; ~ **loss** WATER SUPP perte d'eau *f*; ~ **loss within distribution system** WATER SUPP perte interne au réseau de distribution d'eau *f*; ~ **main** WATER SUPP conduite d'adduction *f*, conduite de distribution d'eau *f*; ~ **management** MANAGE, WATER POLL, WATER SUPP gestion de l'eau *f*; ~ **management policy** MANAGE, POLICY politique de l'eau *f*; ~ **mark** WATER SUPP marque de niveau d'eau *f*; ~ **mass** RENEW ENERG *tidal energy* masse d'eau *f*; ~ **meter** WATER SUPP compteur d'eau *m*; ~ **outlet** WATER SUPP sortie d'eau *f*; ~ **penny** POLICY redevance sur l'eau *f*; ~ **pipe** WATER SUPP conduite d'eau *f*, tuyau d'eau *m*; ~ **poisoning** WATER POLL empoisonnement de l'eau *m*; ~ **pollutant** WATER POLL polluant de l'eau *m*; ~ **pollution** WATER POLL pollution des eaux *f*, pollution de l'eau *f*, contamination des eaux *f*; ~ **pollution control officer** MANAGE, POLICY, WATER POLL responsable de la qualité de l'eau *m*; ~ **pollution tax** POLICY taxe sur la pollution de l'eau *f*; ~ **power** RENEW ENERG énergie hydroélectrique *f*, houille blanche *f*, force hydraulique *f*; ~ **power plant** RENEW ENERG centrale hydroélectrique *f*; ~ **pressure** WATER SUPP charge d'eau *f*, pression de l'eau *f*; ~ **properties** *n pl* WATER SUPP propriétés d'une eau *f pl*; ~ **pump** RENEW ENERG, WATER SUPP pompe à eau *f*; ~ **purification** WATER POLL, WATER SUPP épuration de l'eau *f*, épuration des eaux *f*, purification de l'eau *f*, clarification de l'eau *f*; ~ **quality** WATER POLL, WATER SUPP qualité de l'eau *f*, titre de l'eau *m*; ~ **quality class** WATER POLL, WATER SUPP classe de qualité de l'eau *f*; ~ **quality grade** WATER POLL, WATER SUPP classe de qualité de l'eau *f*; ~**quality monitoring** WATER POLL, WATER SUPP surveillance de la qualité de l'eau *f*; ~ **recovery** WATER SUPP récupération d'eau *f*; ~ **regime** WATER SUPP régime des eaux *m*; ~ **reserve** WATER SUPP réserve d'eau *f*; ~ **resources** *n pl* RENEW ENERG, WATER SUPP ressources hydrologiques *f pl*, ressources en eau *f pl*, ressources hydriques *f pl*; ~ **resources management** MANAGE, WATER SUPP gestion des ressources en eau *f*; ~ **resources protection** POLICY sauvegarde des ressources en eau *f*; ~ **sample** ANALYSIS prélèvement d'eau *m*; ~ **sampler** ANALYSIS échantillonneur d'eau *m*; ~ **scrubbing** AIR dépoussiérage hydraulique *m*, lavage à l'eau *m*; ~**-soluble hazardous waste** WASTE, WATER POLL, déchets dangereux solubles dans l'eau *m pl*; ~ **spraying** AIR pulvérisation d'eau *f*, vaporisation d'eau *f*; ~ **supply** WATER SUPP alimentation en eau *f*, fourniture d'eau *f*; ~ **surface** RENEW ENERG *wave energy* surface de l'eau *f*; ~ **table** WATER SUPP nappe phréatique *f*; ~ **table pollution** WATER POLL pollution de la nappe phréatique *f*; ~ **treatment** WATER POLL, WATER SUPP traitement des eaux *m*; ~ **treatment engineering** WATER POLL, WATER SUPP ingénierie du traitement de l'eau *f*; ~ **turbine** RENEW ENERG turbine hydraulique *f*; ~ **vapor** *AmE*, ~ **vapour** *BrE* AIR, RENEW ENERG, SOIL vapeur d'eau *f*; ~ **vapour barrier** *BrE* AIR écran de vapeur d'eau *m*; ~ **vapour plume** *BrE* AIR panache de vapeur d'eau *m*

water[3] *vt* WATER SUPP alimenter en eau

water[4] *vi* WATER SUPP faire de l'eau

Water: ~, **Energy, Land, Materials, Manpower** *phr* (*WELMM*) MANAGE Water, Energy, Land, Materials, Manpower (*WELMM*)

waterborne: ~ **disease** *n* SAFETY, WATER POLL maladie d'origine hydrique *f*, maladie transmise par l'eau *f*

watercourse *n* WATER SUPP cours d'eau *m*

waterfall *n* WATER SUPP chute d'eau *f*

watering *n* AIR *of filter* humidification *f*, mise en eau *f*, WATER POLL, WATER SUPP arrosage

waterlogged: ~ **belt** *n* SOIL terre gorgée d'eau *f*; ~ **soil** *n* SOIL sol gorgé d'eau *m*

waterlogging *n* WATER POLL engorgement par l'eau *m*

waterproof *vt* WATER SUPP étancher

watershed *n* WATER SUPP bassin hydrographique *m*

watertight *adj* GEN étanche, étanche à l'eau, hermétique

waterwheel *n* RENEW ENERG *formerly for tidal power plant* roue à eau *f*

wave *n* MARINE, RENEW ENERG vague *f*; ~ **crest** RENEW ENERG crête de houle *f*, crête de vague *f*; ~ **energy** RENEW ENERG énergie des vagues *f*; ~ **energy conversion** RENEW ENERG conversion de l'énergie des vagues *f*; ~ **energy resources** *n pl* RENEW ENERG ressources de l'énergie des vagues *f pl*; ~ **energy use** RENEW ENERG valorisation de l'énergie des vagues *f*; ~ **erosion** MARINE, SOIL érosion par les vagues *f*; ~ **frequency** RENEW ENERG fréquence des vagues *f*; ~ **hollow** RENEW ENERG creux de vague *m*; ~ **movement** RENEW ENERG mouvement des vagues *m*; ~ **power plant** RENEW ENERG centrale utilisant l'énergie des vagues *f*

wavefront: ~ **length** *n* RENEW ENERG longueur du front d'onde *f*

wavelength *n* ANALYSIS longueur d'onde *f*; ~ **scale** ANALYSIS échelle de longueurs d'onde *f*

wax: ~ **paper** *n* WASTE papier ciré *m*

weak: ~ **sample** *n* ANALYSIS échantillon faiblement concentré *m*, échantillon faiblement dosé *m*; ~**-sun period** *n* RENEW ENERG *solar energy* période de faible ensoleillement *f*

weaken *vt* ANALYSIS affaiblir, atténuer

weakened *adj* ANALYSIS affaibli

weakening *n* ANALYSIS *optics* affaiblissement *m*, atténuation *f*

wear[1]: ~**-resistant** *adj* AIR, WASTE résistant à l'usure

wear[2]: ~**-resistant lining** *n* AIR revêtement résistant à l'usure *m*

weather *n* AIR, RENEW ENERG temps *m*; ~ **conditions** AIR, RENEW ENERG conditions météorologiques *f pl*; ~ **map** AIR, MARINE carte météorologique *f*

weathered: ~ **oil** *n* MARINE huile désagrégée *f*

weathering *n* MARINE, SOIL, WATER SUPP désagrégation *f*, altération *f*, dégradation *f*, détérioration due aux intempéries *f*; ~ **resistance** SOIL *of liner* résistance aux intempéries *f*

webbing *n* MARINE sangle *f*

WEC *abbr* (*wind energy conversion*) RENEW ENERG conversion de l'énergie éolienne *f*

WECS *abbr* (*wind energy conversion system*) RENEW ENERG système de conversion de l'énergie éolienne *m*

weed: ~ **control** *n* SOIL lutte contre les adventices *f*, lutte contre les mauvaises herbes *f*

weedkiller *n* SOIL désherbant *m*

weigh[1]: ~ **office** *n* WASTE poste de pesage *m*

weigh[2] *vt* ANALYSIS, WASTE peser; ~ **out** ANALYSIS, WASTE peser; ~ **out analytically** ANALYSIS peser par voie analytique

weighable: ~ **lysimeter** *n* ANALYSIS lysimètre flottant *m*

weighing *n* ANALYSIS, WASTE pesage *m*; ~ **and punnet filling** WASTE pesage et mise en flein

weight: ~**-filling machine** *n* WASTE machine de remplissage pondéral *f*

weighted: ~ **noise level** *n* NOISE niveau de bruit pondéré *m*; ~ **noise level indicator** *n* ANALYSIS, NOISE indicateur de niveau acoustique pondéré *m*; ~ **sound level** *n* NOISE niveau de bruit pondéré *m*

weir *n* MARINE, RENEW ENERG, WATER POLL trémail *m*, déversoir *m*; ~ **boom** MARINE récupérateur à déversoir *m*, récupérateur à seuil *m*; ~ **skimmer** MARINE récupérateur à déversoir *m*, récupérateur à seuil *m*

well *n* SOIL, WATER SUPP puits *m*; ~ **pipe** WATER SUPP conduite de puits *f*

Wellmann-Lord: ~ **process** *n* AIR *wet flue-gas desulphurization* procédé Wellmann-Lord *m*

WELMM *abbr* (*Water, Energy, Land, Materials, Manpower*) MANAGE WELMM (*Water, Energy, Land, Materials, Manpower*)

WEP *abbr* (*wet electrostatic precipitator*) AIR dépoussiéreur électrostatique par voie humide *m*

wet[1] *adj* AIR, SOIL humide, mouillé

wet[2]: ~ **acidic fallout** *n* AIR retombées acides humides *f pl*; ~ **air oxidation** *n* (*WAO*) SOIL, WASTE, WATER POLL oxydation par voie humide *f*; ~**-bulb temperature** *n* ANALYSIS température humide *f*; ~ **chemical method** *n* ANALYSIS méthode chimique par voie humide *f*; ~ **classifier** *n* WATER POLL séparateur par voie humide *m*; ~ **cleaning** *n* WATER POLL nettoyage par voie humide *m*; ~ **cleaning process** *n* AIR procédé d'épuration par voie humide *m*, procédé de dépoussiérage par voie humide *m*; ~ **collector** *n* AIR collecteur humide *m*; ~ **deposition** *n* AIR, WATER SUPP dépôt humide *m*; ~ **desulfurization process** *AmE*, ~ **desulphurization process** *BrE* *n* AIR procédé de désulfuration par voie humide *m*; ~**-dry scrubber** *n* AIR dépoussiéreur mixte *m*, dépoussiéreur semi-sec *m*; ~ **dust removal** *n* AIR dépoussiérage par voie humide *m*; ~ **electrofilter** *n* WASTE électrofiltre humide *m*; ~ **electrostatic filter** *n* AIR électrofiltre par voie humide *m*, filtre électrostatique de dépoussiérage par voie humide *m*; ~ **electrostatic precipitator** *n* (*WEP*) AIR dépoussiéreur électrostatique par voie humide *m*; ~ **flue-gas desulfurization plant** *AmE*, ~ **flue-gas desulphurization plant** *BrE* *n* AIR installation de désulfuration des gaz par voie humide *f*, usine de désulfuration des gaz par voie humide *f*; ~ **fraction of waste** *n* WASTE fraction humide des déchets *f*; ~ **grinding** *n* AIR, WASTE broyage par voie humide *m*; ~ **mixer** *n* WASTE malaxeur humide *m*; ~ **oxidation** *n* SOIL, WASTE, WATER POLL oxydation humide *f*, oxydation par voie humide *f*; ~ **period** *n* AIR saison humide *f*; ~ **precipitation** *n* AIR séparation par voie humide *f*; ~ **screening** *n* WATER POLL tamisage par voie humide *m*; ~**-scrubbed waste top gas** *n* AIR gaz de tête d'un dépoussiéreur par voie humide *m*; ~ **scrubber** *n* AIR dépoussiéreur hydraulique *m*, dépoussiéreur par voie humide *m*, laveur hydraulique *m*; ~ **scrubbing** *n* AIR dépoussiérage hydraulique *m*, lavage par voie humide *m*; ~**-scrubbing device** *n* AIR appareil de lavage par voie humide *m*, dispositif de lavage par voie humide *m*, dispositif de séparation par voie humide *m*, WASTE dispositif de séparation par voie humide *m*; ~ **sludge utilization** *n* WATER POLL valorisation des boues humides *f*; ~ **sorting** *n* WASTE tri par voie humide *m*; ~ **waste gas treatment** *n* AIR traitement des gaz par voie humide *m*; ~ **well** *n* WATER SUPP puits humide *m*

wet³ *vt* AIR humidifier

wetland *n* SOIL terrain marécageux *m*

wettability *n* AIR, SOIL mouillabilité *f*

wettable¹ *adj* AIR, SOIL humidifiable, mouillable

wettable²: ~ **dust** *n* AIR poussière humidifiable *f*, poussière mouillable *f*

wetted¹ *adj* AIR, SOIL humidifié, mouillé

wetted²: ~**-filter-type scrubber** *n* AIR dépoussiéreur à filtre par voie humide *m*

wetting¹ *adj* AIR, SOIL mouillé, mouillant

wetting² *n* AIR, SOIL mouillabilité *f*; ~ **agent** AIR, MARINE agent mouillant *m*

WHB *abbr* (*waste heat boiler*) AIR, WASTE, WATER POLL, WATER SUPP chaudière alimentée par des rejets thermiques *f*, chaudière de récupération *f*, chaudière de récupération de chaleur *f*

Wheatstone: ~ **bridge** *n* ANALYSIS pont de Wheatstone *m*

wheel: ~ **loader** *n* SOIL chargeur à roues *m*

whirling *n* AIR tourbillonnement *m*

white: ~ **noise** *n* NOISE bruit blanc *m*; ~ **product** *n* MARINE produit blanc *m*

wide¹: ~**-meshed** *adj* WATER POLL à grandes mailles

wide²: ~**-mouthed laboratory bottle** *n* ANALYSIS flacon de laboratoire à col large *m*

wild: ~ **land** *n* POLICY région non aménagée *f*; ~ **species** *n* POLICY espèce sauvage *f*

wilderness: ~ **area** *n* SOIL région désertique *f*

wildlife: ~ **conservation** *n* POLICY conservation de la faune et de la flore sauvage *f*, préservation de la faune et de la flore sauvage *f*

winch *n* MARINE treuil *m*

wind¹: ~**-favorable** *AmE*, ~**-favourable** *BrE adj* RENEW ENERG favorable au vent, venté

wind² *n* RENEW ENERG vent *m*; ~ **chart** AIR, MANAGE, MARINE, POLICY carte des vents *f*; ~ **conditions** *n pl* RENEW ENERG régime des vents *m*; ~**-diesel system** RENEW ENERG système couplant éolienne et groupe électrogène *m*; ~ **direction tracking mechanism** RENEW ENERG mécanisme de suivi des vents *m*; ~ **duration** RENEW ENERG durée d'une rafale *f*; ~ **energy** RENEW ENERG énergie éolienne *f*; ~ **energy conversion** (*WEC*) RENEW ENERG conversion de l'énergie éolienne *f*; ~ **energy conversion system** (*WECS*) RENEW ENERG système de conversion de l'énergie éolienne *m*; ~ **erosion** SOIL érosion éolienne *f*; ~ **exposure** AIR, WASTE exposition au vent *f*; ~ **farm** RENEW ENERG exploitation éolienne *f*, parc d'éoliennes *m*; ~ **farm capacity** RENEW ENERG puissance d'une exploitation éolienne *f*; ~**-favorable region** *AmE*, ~**-favourable region** *BrE* RENEW ENERG région ventée *f*; ~ **park** RENEW ENERG parc d'éoliennes *m*; ~ **park capacity** RENEW ENERG puissance d'un parc d'éoliennes *f*; ~ **power** RENEW ENERG énergie éolienne *f*; ~ **power plant** RENEW ENERG centrale éolienne *f*; ~ **power resources** *n pl* RENEW ENERG ressources de l'énergie éolienne *f pl*; ~ **power use** RENEW ENERG utilisation de l'énergie éolienne *f*; ~ **rose** AIR rose des vents *f*; ~ **sensor** ANALYSIS anémomètre *m*; ~ **turbine** RENEW ENERG éolienne *f*, turbine éolienne *f*; ~ **turbine blade** RENEW ENERG pale d'éolienne *f*; ~ **turbine generator** (*WTG*) RENEW ENERG groupe turbo-alternateur d'éolienne *m*, turbo-alternateur d'éolienne *m*; ~ **velocity** AIR *in stack*, RENEW ENERG vitesse du vent *f*

windmill *n* RENEW ENERG moulin à vent *m*

window: ~ **packaging** *n* WASTE emballage à fenêtre *m*

windy¹ *adj* RENEW ENERG venté

windy²: ~ **region** *n* RENEW ENERG *suitable for wind power* région ventée *f*

winter: ~ **grain** *n* RENEW ENERG céréale d'hiver *f*

wire *n* AIR fil torsadé *m*; ~ **cage** WASTE cage d'écurevil *f*

wood¹: ~**-free** *adj* WASTE sans bois

wood² *n* RENEW ENERG *biomass conversion* bois *m*; ~ **chip** RENEW ENERG plaquette de bois *f*, copeau de bois *m*; ~ **chip boiler** RENEW ENERG chaudière à copeaux de bois *f*; ~ **firing** RENEW ENERG chauffage au bois *m*; ~ **firing plant** RENEW ENERG installation chauffée au bois *f*; ~ **pellet** RENEW ENERG plaquette de bois *f*; ~ **pulp** WASTE pâte de bois *f*; ~ **waste** WASTE chutes de bois *f pl*, déchets de bois *m pl*, déchets ligneux *m pl*; ~ **wool** WASTE fibre d'emballage *f*, laine de bois *f*

woodland *n* RENEW ENERG boisement *m*

woody: ~ **waste** *n* WASTE déchets ligneux *m pl*

wool *n* AIR *filter fibre material* laine *f*

work: ~ **safety** *n* SAFETY sécurité du travail *f*; ~ **station** *n* SAFETY poste de travail *m*

working: ~ **capacity** *n* WATER POLL *of clarification tank* capacité de traitement *f*; ~ **face** *n* WASTE paroi avant *f*, front de taille *m*; ~ **medium** *n* RENEW ENERG *of heat pump* caloporteur *m*, vecteur *m*

workplace *n* SAFETY *bench, desk* poste de travail *m*, *factory, office* lieu de travail *m*, WASTE poste de travail *m*; ~ **atmosphere** AIR, SAFETY air des lieux de travail *m*

world: ~ **food problem** *n* POLICY problème alimentaire mondial *m*; ~ **register of genetic resources** *n* MANAGE registre mondial des ressources génétiques *m*

worm *n* SOIL, WATER POLL ver *m*; ~ **composting** SOIL lombricompostage *m*

wound: ~**-rotor induction motor** *n* WATER POLL *of pump driving gear* moteur à induction à rotor bobiné *m*

woven: ~**-wire cloth** *n* WATER POLL *of screen* toile métallique *f*

wraparound *n* WASTE bouclage *m*, serrage *m*

wrapping: ~ **machine** *n* WASTE machine à gainer les fils *f*, rubaneuse *f*; ~ **paper** *n* WASTE papier d'emballage *m*

wreck *n* MARINE *ship*, WASTE *car* épave *f*

wring *vt* WASTE essorer, tordre

wringer *n* WASTE essoreuse *f*

wringing *n* WASTE essorage *m*

WTG *abbr* (*wind turbine generator*) RENEW ENERG groupe turbo-alternateur d'éolienne *m*, turbo-alternateur d'éolienne *m*

Wye: ~ **branch** *n* WATER POLL connexion en étoile *f*; ~ **connection** *n* WATER POLL connexion en étoile *f*

X Y Z

xanthate: ~ **polymer complex** *n* WATER POLL complexe polymères-xanthate *m*
Xe *abbr* (*xenon*) AIR Xe (*xénon*)
xenobiotic *adj* SOIL xénobiotique
xenon *n* (*Xe*) AIR xénon *m* (*Xe*)
xeromorphic *adj* SOIL xéromorphe
xerophil *adj* SOIL xérophyle
xerophilous *adj* SOIL xérophyle
xerothermal *adj* AIR xérotherme
xerothermique *adj* AIR xérotherme
X-ray: ~ **protective glass** *n pl* SAFETY verre protecteur contre les rayons X *m*; ~ **protective glasses** *n pl* SAFETY lunettes de protection contre les rayons X *f pl*
X-rays *n pl* SAFETY rayons X *m pl*

yardang *n* SOIL yardang *m*
yarn *n* AIR *fabric filtration* fil *m*, filasse *f*
Yb *abbr* (*ytterbium*) SOIL Yb *m* (*ytterbium*)
ylem *n* SOIL ylem *m*
ylide *n* ANALYSIS ylure *m*
ytterbium *n* (*Yb*) SOIL ytterbium *m* (*Yb*)

zeolite *n* AIR zéolite *f*; ~ **catalyst** AIR catalyseur à zéolite *m*
zeolitic[1] *adj* AIR zéolitique
zeolitic[2]: ~ **absorbent** *n* AIR absorbant zéolitique *m*
zero: ~ **CO_2 emissions** *n pl* AIR absence totale d'émissions de CO_2 *f*, zéro émission de CO_2 *m*
zeta: ~ **potential** *n* (*ZP*) SOIL débit *m*
zinc *n* (*Zn*) AIR, SOIL, WATER POLL zinc *m* (*Zn*)
zip: ~**-lock bag** *n* WASTE sac à fermeture éclair *m*
Zn *abbr* (*zinc*) AIR, SOIL, WATER POLL Zn (*zinc*)
zone: ~ **of aeration** *n* SOIL zone d'aération *f*; ~ **of saturation** *n* SOIL zone de saturation *f*
zoning *n* SOIL, WATER SUPP zonage *m*; ~ **map** SOIL, WATER SUPP plan d'occupation des sols *m*; ~ **plan** POLICY plan d'affectation des sols *m*
zoogleal: ~ **layer** *n* WATER POLL floc zoogléal *m*
zooplankton *n* WATER POLL zooplancton *m*
ZP *abbr* (*zeta potential*) SOIL débit *m*